Biomineralization Mechanism of the Pearl Oyster, *Pinctada fucata*

Rongqing Zhang • Liping Xie • Zhenguang Yan

Biomineralization Mechanism of the Pearl Oyster, *Pinctada fucata*

Springer

Rongqing Zhang
School of Life Sciences
Tsinghua University
Beijing, China

Liping Xie
School of Life Sciences
Tsinghua University
Beijing, China

Zhenguang Yan
Chinese Research Academy
of Environmental Sciences
Beijing, China

ISBN 978-981-13-1458-2 ISBN 978-981-13-1459-9 (eBook)
https://doi.org/10.1007/978-981-13-1459-9

Library of Congress Control Number: 2018950551

This Springer imprint is published by the registered company Springer Nature Singapore Pte Ltd.
The registered company address is: 152 Beach Road, #21-01/04 Gateway East, Singapore 189721, Singapore

Contents

Chapter 1
Molecular Basis of Biomineralization in *Pinctada fucata*

Abstract Biomineralization is the accumulation and formation of minerals regulated by living organisms transforming into biological structures and tissues. This is an extremely widespread phenomenon since we've found many creatures in all six taxonomic kingdoms which could form biominerals, and more than 60 different types of them have been identified, such as silicates in diatoms and algae, carbonates in invertebrates, and carbonates and calcium phosphates in vertebrates. These minerals often form structures like sea shells and the bone in mammals and birds.

The study of biomineralization started at the first half of the twentieth century by the extensive use of optical microscopy, which helped a lot to identify various types and patterns of structures. It was not until the 1960s and 1970s, with the application of transmission electron microscopy (TEM), the scanning electron microscopy (SEM), radioisotope, the biochemical research on membrane transport, the analysis of bone-related organic materials, and the understanding of the ultrastructure of mineralized materials, was the initial idea of the biomineralization formed. Though some progress has been made in revealing the principles of these processes during the past few decades, there are still a lot of unknowns waiting for more exploration. The main content in this book primarily focused on the following aspects: the identification and function characterization of biomineralization-related genes, including those that encode shell matrix proteins and enzymes and participate in calcium metabolism; the cellular regulation such as cell signaling pathways during biomineralization process; and the ecological studies on biomineralization in *Pinctada fucata*.

Keywords *Pinctada fucata* · *Biomineralization* · *Molecular basis* · *Introduction* · *Calcium metabolism*

R. Zhang et al., *Biomineralization Mechanism of the Pearl Oyster, Pinctada fucata*,
https://doi.org/10.1007/978-981-13-1459-9_1

1.1 Introduction

Biomineralization is an extremely widespread phenomenon, the process of which is conducted by reaction and precipitation converting ions into solid minerals [1, 2]. Specifically, biomineralization refers to organisms using their synthetic organic macromolecules manipulating the process of the inorganic crystal nucleation, growth, and molecular arrangements, eventually producing minerals with special optical, magnetic, and/or mechanical properties [2, 3]. Biomineralization involves biologists, chemists, and geologists in interdisciplinary studies at the interfaces between Earth and life.

Almost all living creatures, from lower unicellular organisms or the plants to higher animals, are closely associated with the products of biomineralization, biominerals. But the biomineralization mechanisms involved in different biological systems are quite complicated and exhibit a large diversity, and more than 60 different biominerals have been identified, including bones responsible for supporting and movements in vertebrates; shells functioning on protecting soft tissues in bivalves, brachiopods, and foraminifers; otoliths in fishes and mammals for balancing movements; magnetosomes found in magnetotactic bacteria sensing the changes of geomagnetic field; etc. [4]. The main chemical compositions of these biominerals mentioned above contain calcium phosphate, calcium carbonate, and oxidation silicon. Among them, mollusk shells are composed of calcium carbonate or calcium magnesium carbonate, while the bones and teeth in mammals are mainly calcium and magnesium phosphate [4–6].

Biomineralization processes could divided into two fundamentally different categories based upon their degree of biological control, namely, biologically induced biomineralization and biologically controlled biomineralization [4, 6]. Bio-induced mineralization refers to the deposition of inorganic minerals formed by the reaction of the metabolites at the cell wall when they are expelled or into the cell wall, common in protists, alga, and lower prokaryotes. Characterized by the low-degree control of this process, the crystals or biominerals formed during bio-induced mineralization usually lack unique morphologies, most of which are irregular and disorganized. In contrast, in "biologically controlled" mineralization, the organism uses cellular activities to direct the nucleation, growth, morphology, and final location of the minerals which are deposited. While the degree of control varies across species, almost all controlled mineralization processes occur in an isolated environment. The results can be remarkably sophisticated, species-specific products that give the organism specialized biological functions [1, 7, 8]. In conclusion, elucidating the underlying mechanisms of biomineralization will lead to innovative understanding of the field of materials and medicine sciences and also provide significant technical supports for pearl shellfish farming and pearl production.

1.2 General Features of Biomineralization in *Pinctada fucata*

1.2.1 The $CaCO_3$ Crystalline Forms in Pinctada fucata

There are six different types of natural $CaCO_3$ biominerals, calcite, aragonite, vaterite, $CaCO_3 \cdot H_2O$, $CaCO_3 \cdot 6H_2O$, and amorphous calcium carbonate (ACC), the chemical compositions of which are exactly the same; nevertheless, their crystal structures are totally different. Among these different crystalline forms, calcite and aragonite are thermodynamically stable and most common in mollusk shells [9]. Aragonites could be induced in calcite crystallization system with the addition of 50 mM Mg^{2+}. Vaterite is the metastable status of $CaCO_3$ and the most unstable crystalline. And it could be transformed to calcite or aragonite spontaneously in aqueous solution. Vaterite has not been found in our animal model *P. fucata*, while some exceptions have been reported in other close-related species. A new morphology of vaterite was confirmed in lackluster pearls of *Hyriopsis cumingii* [10]. Moreover, it also formed in the abnormal shells of *Corbicula fluminea* [11]. ACC has been found in plant larvae and the initial stage of shell biomineralization. ACC is usually considered as the precursor in shell formation [12]. In spite of the rapid transformation from ACC to other stable crystals in aqueous solution, it is still relatively stable due to the adhesion of the macromolecules like polysaccharide to its surface. Organisms control the biomineralization process by the regulation of the formation, stabilization, destabilization, and transformation of ACC [13].

1.2.2 The Structure of the Shell

Pearl oysters, *Pinctada fucata*, famous as important economical pearl production species, are one of the best studied biomineralization models [14]. The shell can be divided into two distinct layers, the periostracum and the calcified layer (Figs. 1.1 and 1.2).

The periostracum is the organic layer of the hardened organic products on the outer surface of the shell [16]. The hardened part is mainly composed of insoluble Quinone tannin containing dopa [17]. Besides protecting the shell from the destruction of acidic substances and parasites, periostracum also plays important roles during biomineralization. It could block the extrapallial space on the ventral side, isolating the extrapallial fluid from the external environment, delimitating, and sealing the space where the biomineralization takes place thus provides a supersaturation niche for crystal deposition. What's more, the periostracum may also function as a substrate for Ca^{2+} deposition and even affect the formation of prismatic layer [18, 19].

The calcified layer could be subdivided into prismatic layer and nacreous layer, where $CaCO_3$ crystals are deposited as calcite and aragonite, respectively [20–

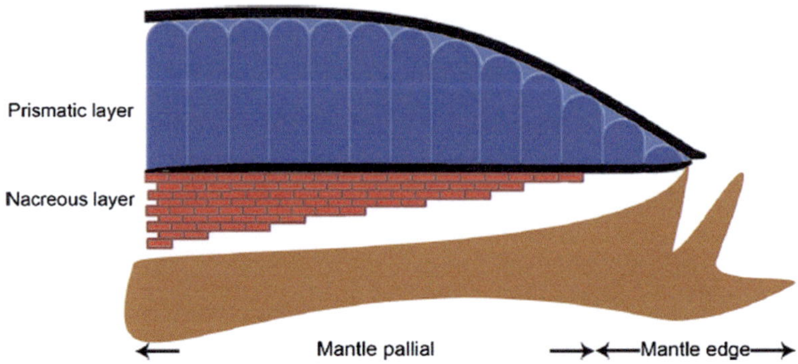

Fig. 1.1 Diagram of the cross section of the shell and the mantle of *P. fucata* [15]

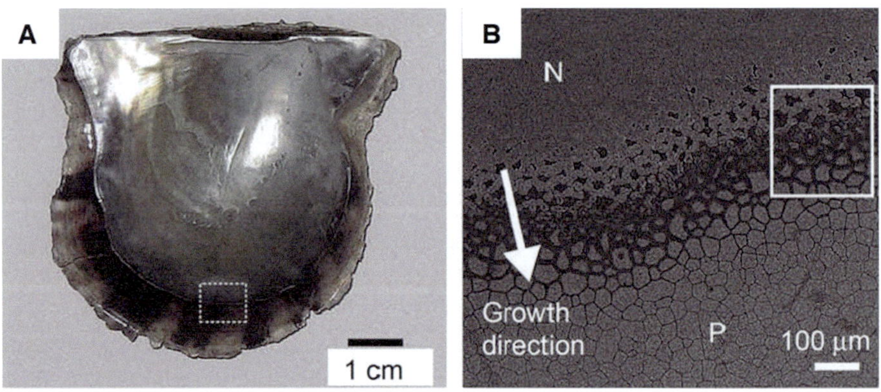

Fig. 1.2 The shells of *P. fucata* [23]
(**a**) The inner shell of *P. fucata* and (**b**) SEM pictures of the inner surfaces. *N* nacreous layer, *P* prismatic layer. The square frame indicated the growth zone

22]. Our concerns on biomineralization studies primarily focused on these two layers (Fig. 1.2). The shell calcification is mainly composed of calcium carbonate crystals with a total weight of more than 95% and much more smaller amount of organic substrates (mainly proteins and polysaccharides) [4].

The initiation of prisms occurs in the proximal region of the outer surface of the outer mantle fold in the pallial space bounded externally by the periostracum [24]. SEM pictures showed that prismatic layer is composed of a great amount of parallel column calcites perpendicular to the shell surface, the cross section of which are regular polygons, and the edges of these "polygons" are surrounded with organic matrix [24, 25] (Fig. 1.3). A lot of studies have proved that the organic matrix could regulate the growth of these calcites in prismatic layer and its main components are proteins and chitin [26, 27]. The cells form mantle edge are commonly believed to be responsible for the formation of prisms [17] (Fig. 1.1).

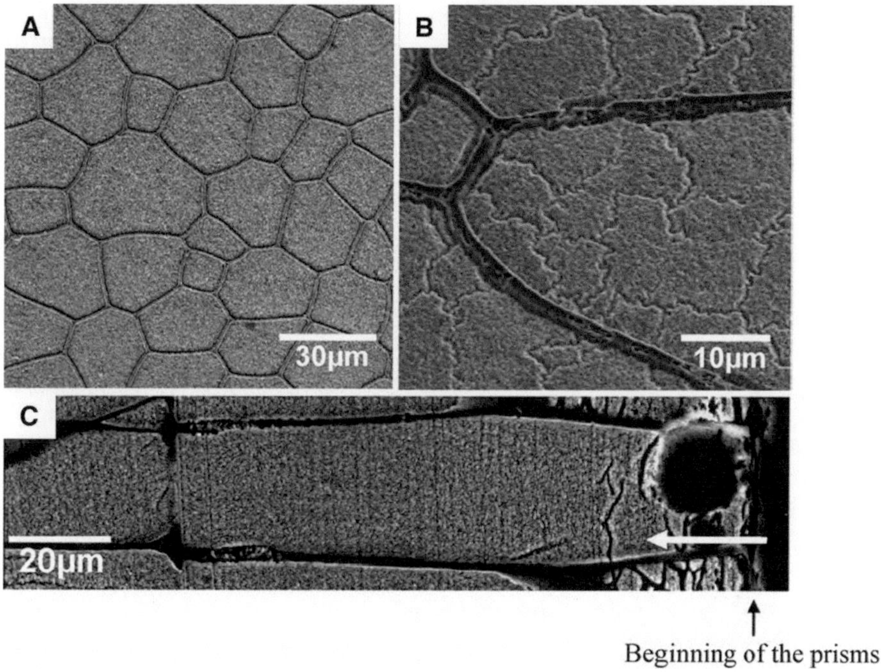

Fig. 1.3 SEM images of the prismatic layer of *P. fucata* etched by EDTA [28]
(**a**) Cross section of the prismatic layer in *P. fucata*, (**b**) enlarged view of the grooves, and (**c**) longitudinal cross section of the prismatic layer. The arrow indicated the growth direction.

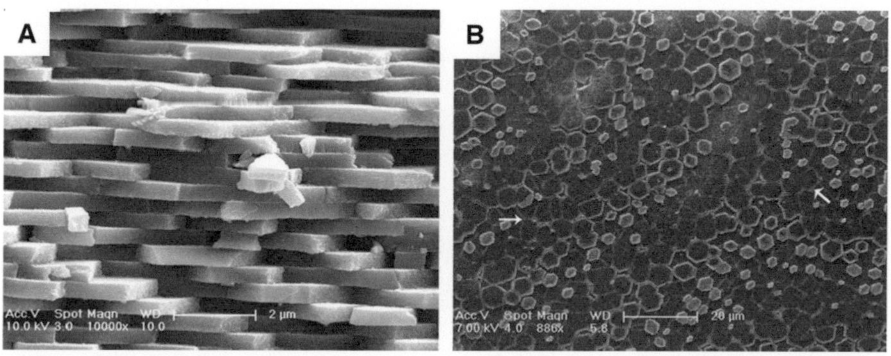

Fig. 1.4 Nacre structure of *P. fucata* [31]
(**a**) The cross section of nacreous layer, composed of continuous parallel lamellae; (**b**) each lamellae is composed of polygonal aragonitic tablets, sealing each other by the intertabular matrix.

The nacreous layer is composed of aragonite tablets, and the aragonite sheet is 0.4 to 0.5 μm in width and 5–10 μm in thickness [29] (Fig. 1.4). These are typical nanostructures. Besides regulating crystal formation and growth, the organic matrix filled in aragonites could spread the force, significantly improving the mechanical strength of the shell especially when subjected to external forces. The mechanical properties of the nacreous layer, also known as "dense slit brick structure," are approximately 3000 times as strong as the mineral aragonites, thus providing adequate protection for the soft tissue inside the shell [30].

1.2.3 The Process of Shell Formation

The main component of the shell is $CaCO_3$ of course; however, those macromolecules playing crucial roles in the formation of the delicate microstructures during biomineralization only account for less than 5% [32]. Studies on the regulation of matrix proteins during shell formation are one of the current research hot spots. Several different shell formation models were brought up recently [2, 33], but the specific mechanisms have not been clearly elaborated.

The life circle of *P. fucata* could be divided into six stages, including the fertilized egg, trochophore stage, D-shaped stage, umbonal stage, juvenile stage, and adult [34]. The polymorphism of $CaCO_3$ crystals and the shell microstructure changes a lot throughout these stages [35]. Prodissoconch I, forming during early D-shaped stage, is the original form of the shell and is considered to be composed of ACC [34, 35]. Prodissoconch II with a homogeneous structure of aragonite and a thinner layer of calcite appears at the lateral D-shaped stage or umbonal stage. The formation of prismatic and nacreous layer completes while entering the juvenile stage (Fig. 1.5).

Bar, 20 μm in (**a**), (**b**), and (**c**); bar, 200 μm in d. *P1* prodissoconch I shell, *P2* prodissoconch II shell, *D* dissoconch shell, *PL* prismatic layer, *NL* nacreous layer

Classically, the physiology of molluscan shell calcification can be described as a succession of compartments, the most important of which are the inner shell surface, extrapallial space, and outer mantle epithelium. The mantle tissue, coating the inner surface of the shell, with the other compartments being the extrapallial space and the shell, is the critical organ during mollusk shell formation and comprises internal and external epithelium, connective tissue, mantle muscle, and nerve fibers. The ridge between the outer and median folds is the periostracal groove, responsible for the formation of periostracum (Fig. 1.6).

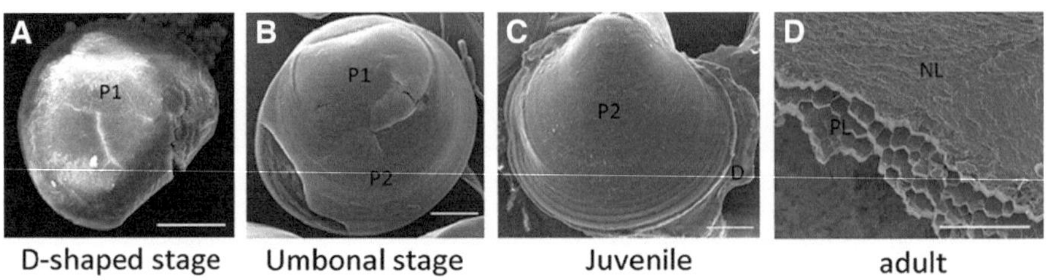

Fig. 1.5 The SEM pictures of three larval development stages in *P. fucata* [36]

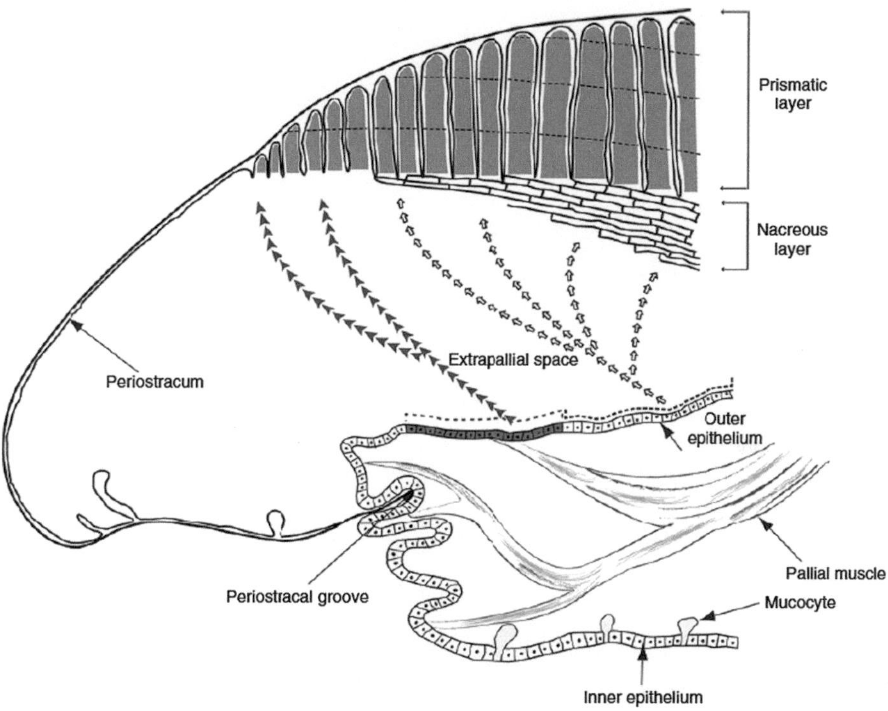

Fig. 1.6 Physiology of the shell calcification in a nacro-prismatic bivalve [17]

Precursor ions, calcium and bicarbonate, are transported via the hemolymph and then directed toward the outer mantle epithelium. After extrusion from the cytosols to extralpallial space, the liquid precursor ions, Ca^{2+} and CO_3^-, are gradually transited to solid $CaCO_3$ on the inner surfaces of the shell under the regulation of matrix macromolecules, especially matrix proteins [17] (Fig. 1.6). But how the organic matrix interacts with the mineral ions to produce this well-designed and fine-structured biomineral is still under debated and more investigation.

Extrapallial fluid (EPF) is a colorless, bloodlike fluid locating in the extrapallial space, where the blood, EPF, and ambient medium (e.g., seawater) carry out the frequent exchanges of substances [37]. The special roles EPF playing during shell formation hasn't gained enough attention initially, fortunately, with the developing of the research, scientists realized its importance. Some evidence suggested that matrix proteins may accomplish self-assembling in extrapallial space, which may explain the formation of shell-organic scaffolds [17, 38]. Whereas, this concept is still debated, and more proofs should be provided for further validation.

Moreover, we shouldn't ignore the functions of hemocytes, those free circulating cells of hemolymph, during shell formation and construction. The studies on hemocytes originally focused on its role in immune defense system [39] and tissue reparation process [40]. These functions have been well underlined during the past few decades, but their contribution to biomineralization may be greatly

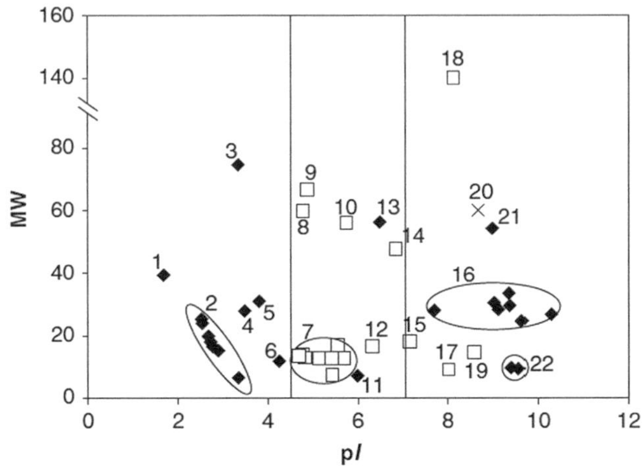

Fig. 1.7 The molecular weight (MW) and isoelectric point (pI) distribution of partial identified shell matrix proteins [17]
The theoretical MW and pI were calculated after identification and removal of the signal peptide. (1, aspein; 2, Asp-rich proteins; 3, ASP-1; 4, ASP-2; 5, MSI31; 6, prismalin-14; 7, N14/N16/pearlin/perline proteins masking AP7 and AP24; 8, MSI60; 9, mucoperlin; 10, nacrein from *P. fucata*; 11, MSI7; 12, dermatopontin; 13, tyrosinase-like 1; 14, nacrein from *T. marmoratus*; 15, perlucin; 16, shematrin proteins; 17, perlustrin; 18, lustrin A; 19, perlwapin; 20, N-66; 21, tyrosine-like2; 22, KRMPs)

underestimated. According to Mount et al., hemocytes could directly release the calcite crystals for successive remodeling at the biomineralization site [41]. But we still need more evidence to confirm whether this is a general metabolic pathway in bivalves or just a particular phenomenon in eastern oyster.

1.2.4 Shell Matrix Proteins

As mentioned in the introduction and detailed later, the organic matrix is a mixture of proteins, glycoproteins, chitin, and lipids. The matrix, especially matrix proteins, interacts with the mineral ions and controls the shape, deposition, and crystalline forms of the produced crystals.

The earliest systematic studies on shell matrix were conducted by Crenshaw et al. in 1972. They used EDTA, a calcium-chelating agent, effective at weak dilute acids to decalcify the shells of *Mercenaria mercenaria*. However, due to the limited conditions, their studies predominantly concentrated on the integral components, amino acid composition of the mixture, or its general functions during shell formation and reparation [42–45]. According to those successfully isolated and identified matrix proteins, we summarized three fundamental features of them (Fig. 1.7):

1. Certain amino acid residues tend to take up extremely high proportions, more commonly are asp, Gly, and Ser, sometimes including pro and Cys. The carboxyl group in side chain of asp may bind with Ca^{2+}, facilitating the formation of $CaCO_3$ crystal lattices. Gly and pro are more likely relevant to the flexibilities of protein structures, for example, proline could form a cycle to the polypeptide backbone. As for cysteine, it contributes to the formation of covalent disulfide bonds to other cysteine residues [46, 47].
2. The sequences of shell matrix proteins (SMPs) are usually composed of different domains. And these domains generally contain modular structures with distinct functions [48, 49]. The first identified matrix protein, nacrein, possesses a carbonic anhydrase domain [38]. Lustrin a contains ten highly conserved cysteine-rich domains and a whey domain [50]. Analyses of these genes of various modular proteins implied that gene assembly may be promoted by intronic recombination, i.E., the proteins were formed by exon shuffling [51, 52].
3. A large amount of matrix proteins have been posttranslationally modified, such as phosphorylated and/or glycosylated [50, 53]. And these posttranslational modifications often associate with the binding properties with Ca^{2+} [54, 55].

With the wide use of new technologies, over 40 different kinds of matrix proteins have been identified from different mollusks. But we still cannot illustrate the molecular mechanisms of shell formation; plenty of questions remain unsolved. The isolation of SMPs lays the foundation for its further function characterization. We went through three stages on this process:

1. As mentioned above, EDTA, ethylic acid, and water were frequently used to dissolve the shell, among which EDTA is the most widely used agent. This decalcification procedure yields two organic fractions, EDTA-soluble matrix (ESM) and EDTA-insoluble matrix (EISM). ESM and EISM have been proved to play different roles during shell formation. EISM often functions as the organic scaffold, providing crystal nucleation sites for $CaCO_3$ deposition, while ESM usually takes control of the crystal form and their morphologies [56–58]. Matrix protein Narein [59] MIS60 [60] lustrin A [50] MSI7 [50] N16 [61] N14 [47] p10 [62] etc. were isolated in this way. The main disadvantage of this method is that only the proteins with high abundancy could be identified, while those exhibiting a relative lower expression level but still crucial in biomineralization are excluded.
2. Then the search for new members benefited a lot from the major technical advances in molecular biology based on the cDNA sequences and primary structures of identified ones. PCR and rapid amplification of complementary DNA ends (RACE) were broadly applied, obtaining matrix protein KRMP [63] and Prisilkin-39 [64]et al. The principal factor limiting this method was that we could only get matrix proteins similar to previously identified ones in both sequences and functions.
3. Another strategy successfully developed was cDNA expression library. In most cases, genes encoding matrix proteins exhibit high expression levels in mantle tissue. With the help of the mantle cDNA library, the construction of the

suppression subtractive hybridization (SSH) library paves the way for the bulk cloning and characterization of new genes involved in biomineralization [35, 65]. Our group constructed three SSH libraries that represented genes expressing at three key points during shell formation in *P. fucata* in 2011 (Fig. 1.8) [35]. Finally, we obtained 227 unigenes in U–D library, 89 unigenes in J–U libraries, and novel matrix protein PfN23 [15], PfN44 [66], and DT252 (unpublished data). Since the fragments identified from SSH library were not long (shorter than 330 bp on average), we couldn't get much information for the establishment of the biomineralization model in mollusks. Much more efforts should be put to screen new matrix protein candidates.

In addition to the above methods, our group have identified 72 unique SMPs using liquid chromatography-tandem mass spectrometry (LC-MS/MS) analysis of proteins directly extracted from the shells of *P. fucata* combined with a draft genome (Fig. 1.9). These proteomic data not only increase the repertoires of the SMPs but also give us new inspiration on shell formation which may involve tight regulation of cellular activities and the extracellular microenvironment [21]. Moreover, based on mantle transcriptome and microarray, the gene expression profiles during larval development, 5 novel potential matrix protein candidates were identified for further function analysis, brought global perspectives to the relationship between gene expression profiles and larval shell development (Fig. 1.10) [36].

1.3 Calcium Metabolism During Shell Formation

The studies of shell and pearl formation can be divided into two aspects: the isolation, identification, and functional analysis of SMPs and calcium metabolism process, including Ca^{2+} absorption, transportation, storage, deposition, etc [67] However, more focus has been dragged to the purification and characterization of SMPs previously, while the mechanism of oyster calcium metabolism has been comparatively "neglected." Acting as a universal second messenger in various cells, calcium is an indispensable regulator during the act of fertilization and development of the whole life in all kinds of creatures [68]. Numerous functions of all types of cells are regulated by Ca^{2+} to a greater or lesser degree [69]. In a word, biomineralization implies the transport and control of calcium ions [67].

1.3.1 Calcium Absorption

The mechanism of Ca^{2+} metabolism could be depicted in Fig. 1.11. Precursor ions, Ca^{2+} and HCO_3^-, are taken up from the ambient medium with the help of the body epithelium, or gill, which can also originate from the food in *P. fucata*. Then these ions are transited by hemolymph, soon afterward directed toward the outer mantle

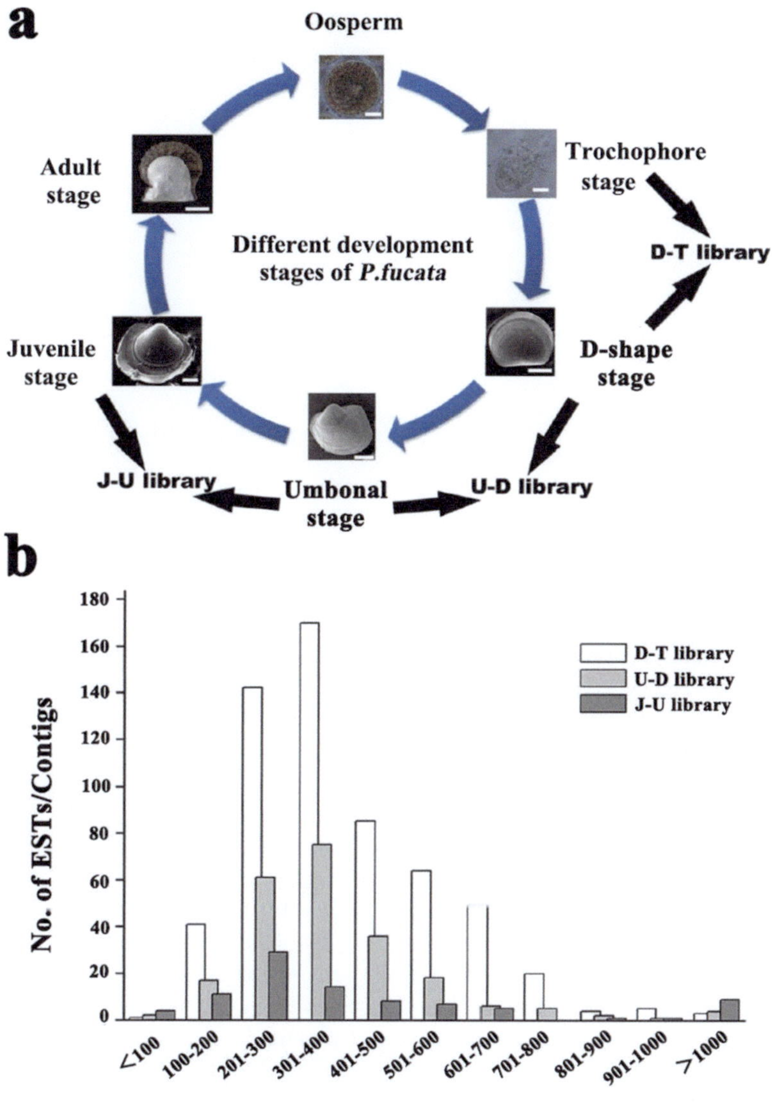

Fig. 1.8 Overview of the three SSH libraries [35]
(**a**) Schematic of SSH library construction. The D–T library was constructed using D-shaped stage larval cDNA as the tester and trochophore stage larval cDNA as the driver. The U–D library was constructed using umbonal stage larval cDNA as the tester and D-shaped stage larval cDNA as the driver. The J–U library was constructed using juvenile stage larval cDNA as the tester and umbonal stage larval cDNA as the driver. Bar = 15 mm in oosperm, 15 mm in trochophore and D-shaped stages, 40 mm in umbonal and juvenile stages, and 4 cm in the adult stage. (**b**) Length distribution of unigenes in the three SSH libraries. The white columns represent unigenes in the D–T library, gray columns represent unigenes in the U–D library, and the dark gray columns represent unigenes in the J–U library. Unigene-length frequencies for each library are plotted in 100-bp bins.

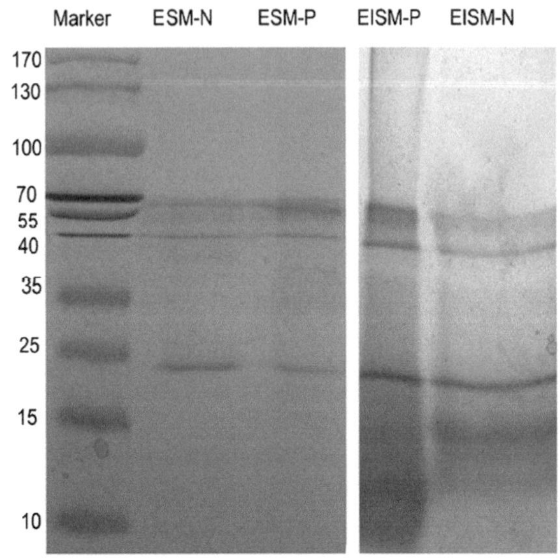

Fig. 1.9 SDS-PAGE of SMPs directly extracted from the shell for further LC-MS/MS analysis (*P* prismatic layer, *N* nacreous layer) [21]

epithelium. In other words, in addition to secreting SMPs for regulating shell formation, mantle is also indispensable to calcium absorption. Finally, large amounts of calcium ions are continuously deposited onto the framework under the precise regulation of matrix proteins.

After the injection of calcium isotope Ca^{45} into *Anodonta woodiana pacifica* (Heude), Ca^{45} was rapidly accumulated in gill, the radioactivity of which was the highest compared with other tissues or organs and lasted the longest. These results indicated that the gill exhibits the strong affinity and high metabolic rate to Ca^{2+}, which is quite crucial for the uptake, accumulation even storage of calcium. More to the point, the absorption of Ca^{2+} in mantle tissue is only second to gill. Furthermore, electrophysiologic [71] and isotopic studies [72] also showed similar effects that mantle epithelium did have high permeability to calcium. Calcium-positive reaction indicated numerous acidic mucous glands assembled in the inner mantle epithelium, suggesting the significant roles acid mucus is playing in Ca^{2+} interception, capture, and assurance. Due to the low concentration of Ca^{2+} in the surroundings, directly intake from environments is very critical for freshwater mollusks to accumulate calcium. It's difficult for Ca^{2+} to cross the mantle epithelium cell membrane in its original form, and this process needs ATP to provide energy. Therefore, there is a strong possibility that the carrier protein and the binding of Ca^{2+} induce a conformational (shape) change driving the ions to transport against the chemical gradient. But the calcium intake method may be different in mollusks living in seawater and hypersaline conditions, considering the similar ion concentration gradient between hemocytes and their living environments. Actually, the $CaCO_3$ solution is hypersaturated in both hemocytes and mantle cavity.

The digestive gland is the metabolic regulatory center in mollusks, besides participating in immune defense, detoxification, and elimination of xenobiotics;

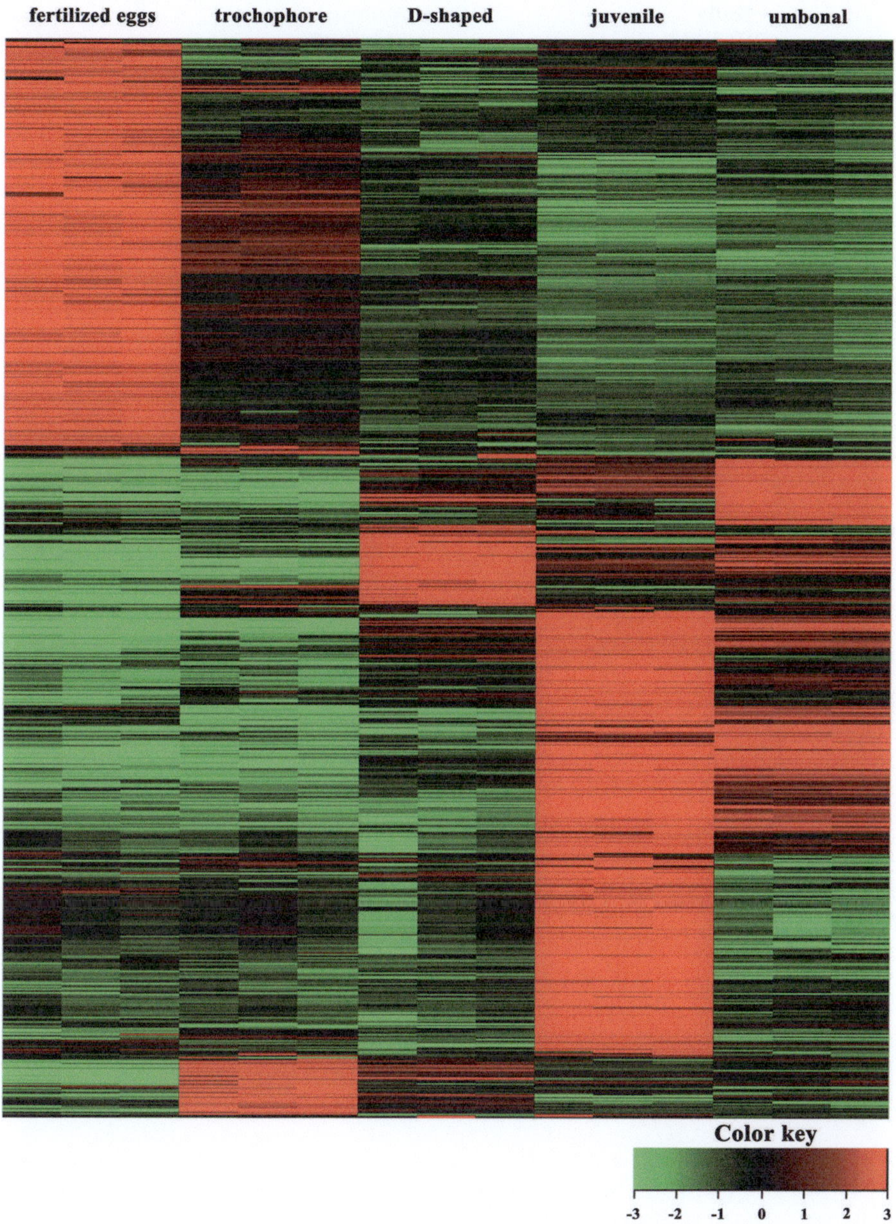

Fig. 1.10 Gene microarray analysis during the different larval development stages in *P. fucata* [36]

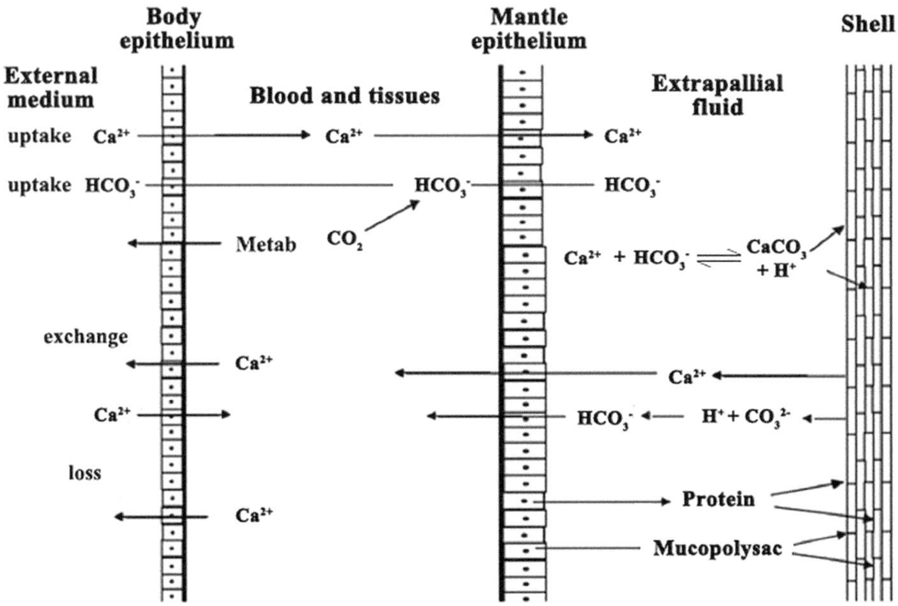

Fig. 1.11 The schematic diagram of calcium metabolism in mollusks [70]

it's also significant to homeostatic regulation of the internal medium (calcium balance, hemolymphatic pH, cell volume, etc.) [73]

Moreover, foot is involved in the calcium uptake process, too. Using the tracers like enzyme peroxidase and ionic lanthanum, paracellular uptake of Ca^{2+} was detected in the foot of the *Agriolimax reticulatus* [74]. On the other hand, electron probe microanalysis (EPMA) also detected the absorption of calcium and trace elements in byssus from mussels.

Thought with some progress, there are still lots of doubts and suspicion on the specific mechanism of calcium absorption waiting for exploration.

1.3.2 Calcium Transportation

There are two pathways in the transportation of Ca^{2+} between different epithelium cells: paracellular pathway [75–78] and transcellular pathway. The depolarization of epithelium cells indicated the apical barrier was permeable to Ca^{2+} when the concentration of calcium was raised from 1 mM to 6 mM the other side of the blood lymphocytes outside the mantle epithelium [79]. While with an exceeding of the equilibrium stage of 1–6 mM Ca^{2+}, calcium was then induced to transport to the other side shell surface [80]. These results together indicated at least part of the transepithelial calcium movements are transcellular. Bleher used lanthanum as an electron density marker in extracellular fluids of mantle epithelium cells in *Anodonta cygnea* and proved Ca^{2+} could transport to shell surface under the regulation of paracellular pathway [81]. Based on these published literatures, transcellular pathway includes simple diffusion, ion channels [79], intracellular transport with calcium-binding protein [82], calcium vesicles [83], secretory granules [84], etc. In Neff's opinions, after the specific binding with calcium-binding proteins, Ca^{2+} was transported into cell with the help of pinocytosis vesicles. Then the fusion of these vacuoles with lysosomes was catalyzed by acid hydrolases, leading to the degeneration of calcium-binding protein and release of free calcium through exocytosis [75].

Studies showed that different mantle epithelium cells from different areas would take different transporting approaches to release Ca^{2+} to extrapallial fluid, forming saturated $CaCO_3$ solution. Calcium ions diffuse mainly from the hemolymph through intercellular spaces between the epithelium cells, within the pallial line (outer mantle cells, Fig. 1.6), into the extrapallial cavity [78, 85].

Mount and his group reported a class of granulocytic hemocytes may be directly involved in shell crystal production for oysters. They observed some hemocytes could release $CaCO_3$ crystals which are subsequently remodeled [41]. Nair's research on quahog (*Mercenaria mercenaria*) also proved that calcium engaged in shell formation must be obtained from external environments and transported by blood lymphocytes or hemocytes to extrapallial fluid for further biomineralization. That is to say, free Ca^{2+}, calcium binding with its special chaperones, or $CaCO_3$ crystals could be involved in any part of this process. Equilibrium dialysis experiments in *M. mercenaria* also reveal that Ca^{2+} constitutes only 2% of the total Ca in plasma, while most of the Ca (85%) bound to macromolecules and the remaining 13% present as dialyzable low-molecular-weight moieties. They conclude that Ca^{2+} have to be transported with the combination of calcium-binding protein or some smaller organic molecules [37]. All in all, scientists haven't reached an agreement on specific transportation forms of calcium (free Ca^{2+}, the complex of Ca^{2+} and its specific binding proteins, or $CaCO_3$ crystals) or its transportation mechanisms, and much more research should be done to solve these problems.

1.3.3 Calcium Storage

Many tissues and organs are capable of storing calcium, among which the shell is the largest calcium storage one. EPMA detected the existence of calcium in various tissues, including ctenidial cells, gill epithelial cells, mantle epithelium cells, pore cells, hemolymph, digestive cells, basophilic cells, nephrocytes, etc [86–89] Furthermore, extracellular calcium granules are observed in mantle tissue and gonads as well [89]. Apparently, the function of these tissues mentioned above has not been limited to calcium metabolism; they may also be relevant to calcium storage.

Besides temporarily chelating with other metal ions in lysosomes from epitheliums, calcification particles could also function as repositories for metal ion storage in the gill of freshwater snails and marine gastropods [73]. Histochemical and X-ray microscopic analysis were performed to observe the calcification process in mantle from *M. margaritifera*, a famous pearl production mussel, and Ca, Mn, and P were found in forming crystals [90]. Based on those results, Pekkarine proposed that these crystals may participate in calcium storage, but not exclude the possibility that they may just assemble and immobilize the metal ions onto shell surface. While Jones and his group supported the opinion that the lysosomes in mantle cells were responsible for the storage and transcellular transport of Ca^{2+} [91] .

The calcium storage mechanism varies in different oysters, depending on their living circumstance. It seems that mollusks in seawater don't need to restore extra Ca^{2+}, while those living in freshwater have to transform the free calcium into calcospherites just in case. Calcospherites are universal in invertebrates like protozoa, platyhelminthes, arthropods, and mollusks, playing significant roles in calcium transportation, storage, and homeostasis. Generally, changes of Ca^{2+} concentration may cause different effects on calcium metabolism in mantle and pearl sac, and the uptake and entrapment of calcium didn't increase but decline when the concentration of Ca^{2+} was beyond a certain level, which may be closely related to the activity of calcium channel and Ca^{2+}-ATPase. Recently, more and more experiments have proved the calcium storage capacity of endoplasmic reticulum due to the identification of multiple calcium-binding proteins [92–94]. Every single calcium-binding protein is estimated to bind with 30 Ca^{2+}, since the concentration of Ca-binding protein could reach 30–100 mg/ml; the final concentration of calcium may be up to 10–15 mM. Generally speaking, as the calcium storage repository, endoplasmic reticulum is very critical to maintain the Ca^{2+} homeostasis between cytoplasm and extracellular matrix.

1.3.4 The Maintain of Calcium Balance

The fluxion of ions between extrapallial fluid and blood may be bilateral: On one hand, Ca^{2+}, mainly from the calcium cells in foot and mantle or the dissolution of $CaCO_3$ granules, could be directed toward shell surface during shell formation and reparation, accelerating shell growth. On the other hand, Ca^{2+} would migrate to tissues or return to external environments during intermolt stage [95]. The bidirectional transportation greatly maintains the dynamical equilibrium of calcium circulation, guaranteeing the homeostasis. The concentration of extracellular calcium is about 10 mM, 10,000 times higher than that inside the cell, precisely controlled by a series of regulatory proteins [37]. The regulation of calcium balance may comprise both extracellular and intracellular pathways. Extracellular circulation controls the intake and discharge of calcium between cytosol and extracellular matrix; the latter regulates the calcium fluxion between endoplasmic reticulum and cytosol.

Previous studies reveal that calcium metabolism is regulated by several different protein families (Fig. 1.12): the entrance of Ca^{2+} into cell is controlled by calcium channels locating at cytoplasm and membrane, while Ca^{2+}-ATPase and Na^{+}/Ca^{2+} exchanger is in charge of the efflux of calcium out of cell [96]; calcium-binding

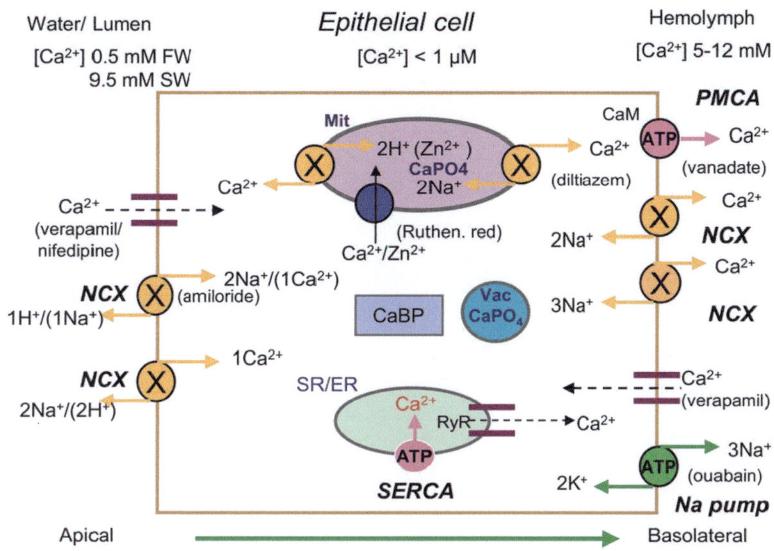

Fig. 1.12 A working model for the apical-to-basolateral transcellular calcium transport and proposed subcellular storage in crustacean epithelial cells (gills, antennal gland, and hepatopancreas) [97]

protein and EF-hand protein are responsible for Ca^{2+} recognition. Wheatly drafted a model of transcellular calcium transport in crustaceans where voltage-dependent calcium channels (VDCCs), plasma membrane Ca^{2+}-ATPase (PMCA), sarco/endoplasmic reticulum Ca^{2+}-ATPase, calmodulin, calreticulin, calsequestin, and other macromolecules all play pivotal roles.

All in all, biomineralization is a ubiquitous and exquisite process, and its production biominerals exhibit various functions in every aspect of individual development. More remarkably, the formation of these biominerals is conducted at room temperature, normal atmospheric pressure, and almost neutral environment, but the characteristics of them are unapproachable among other artificial inorganic materials synthesized under extremely difficult conditions, especially their extraordinary stiffness, great resilience, and mechanical properties. Shell and pearl are typical biominerals in mollusks. The main component of shell and pearl is nacre. The unique luster of pearl significantly improves its decorative and economical value. Moreover, its application as bone-repairing material has also been reported. Present studies suggest that nacre would have a bright prosperity for biomedical potential since it could promote bone regeneration in mammals without causing inflammatory reactions [98, 99]. As a result, studies on the forming mechanisms of these two typical biominerals—shell and pearl—would firstly help us intensify the knowledge of biomineralization process, provide better ideas, and create the foundation for the synthesis of biomaterials. Secondly, their excellent performances are under the regulation of shell matrix proteins, the knowledge of which we've just acquired a small portion. The study of SMPs will be one of the most enduring research fields.

Last but not least, the regulation of shell matrix protein themselves has been seldom reported, and this would be an advanced research direction with the assistance of in-depth investigation of signaling pathways.

References

1. S. Weiner, P.M. Dove, An overview of biomineralization processes and the problem of the vital effect. Rev. Mineral. Geochem. **54**, 1–29 (2003). https://doi.org/10.2113/0540001
2. L. Addadi, D. Joester, F. Nudelman, S. Weiner, Mollusk shell formation: a source of new concepts for understanding biomineralization processes. Chemistry (Weinheim an der Bergstrasse, Germany) **12**, 980–987 (2006). https://doi.org/10.1002/chem.200500980
3. A. Alivisatos, P. Biomineralization, Naturally aligned nanocrystals. Science **289**, 736–737 (2000)
4. H.A. Lowenstam, Minerals formed by organisms. Science **211**, 1126–1131 (1981)
5. M. Cusack, A. Freer, Biomineralization: Elemental and organic influence in carbonate systems. Chem. Rev. **108**, 4433–4454 (2008). https://doi.org/10.1021/cr078270o
6. S. Mann, On the nature of boundary-organized biomineralization (BOB). J. Inorg. Biochem. **28**, 363–371 (1986)
7. S. Mann, S. Weiner, Biomineralization: Structural questions at all length scales. J. Struct. Biol. **126**, 179–181 (1999). https://doi.org/10.1006/jsbi.1999.4131
8. A. Veis, Biomineralization – Cell biology and mineral deposition – Simkiss,K, Wilbur,Km. Science **247**, 1129–1130 (1990). https://doi.org/10.1126/science.247.4946.1129
9. N.H. de Leeuw, S.C. Parker, Surface structure and morphology of calcium carbonate polymorphs calcite, aragonite, and vaterite: An atomistic approach. J. Phys. Chem. B **102**, 2914–2922 (1998). https://doi.org/10.1021/jp973210f
10. L. Qiao, Q.L. Feng, Z. Li, Special vaterite found in freshwater lackluster pearls. Cryst. Growth Des. **7**, 275–279 (2007). https://doi.org/10.1021/cg060309f
11. N. Spann, E.M. Harper, D.C. Aldridge, The unusual mineral vaterite in shells of the freshwater bivalve Corbicula fluminea from the UK. Naturwissenschaften **97**, 743–751 (2010). https://doi.org/10.1007/s00114-010-0692-9
12. J. Aizenberg, G. Lambert, L. Addadi, S. Weiner, Stabilization of amorphous calcium carbonate by specialized macromolecules in biological and synthetic precipitates. Adv. Mater **8**, 222–225 (1996). https://doi.org/10.1002/adma.19960080307
13. J.T. Su et al., Structural characterization of amorphous calcium carbonate-binding protein: An insight into the mechanism of amorphous calcium carbonate formation. Biochem. J. **453**, 179–186 (2013). https://doi.org/10.1042/Bj20130285
14. Y. Yan et al., A novel matrix protein, PfY2, functions as a crucial macromolecule during Shell formation. Sci. Rep. **7**, 6021 (2017). https://doi.org/10.1038/s41598-017-06375-w
15. D. Fang et al., Novel basic protein, PfN23, functions as key macromolecule during nacre formation. J. Biol. Chem. **287**, 15776–15785 (2012). https://doi.org/10.1074/jbc.M112.341594
16. J.D. Taylor, W.J. Kennedy, The influence of the periostracum on the shell structure of bivalve molluscs. Calcif. Tissue Res. **3**, 274–283 (1969)
17. F. Marin, G. Luquet, B. Marie, D. Medakovic, Molluscan shell proteins: primary structure, origin, and evolution. Curr. Top. Dev. Biol. **80**, 209–276 (2008). https://doi.org/10.1016/S0070-2153(07)80006-8
18. A. Checa, A new model for periostracum and shell formation in Unionidae (Bivalvia, Mollusca). Tissue Cell **32**, 405–416 (2000). https://doi.org/10.1054/tice.2000.0129
19. H. Petit, W.L. Davis, R.G. Jones, H.K. Hagler, Morphological studies on the calcification process in the fresh-water mussel Amblema. Tissue Cell **12**, 13–28 (1980)
20. L. Addadi, S. Weiner, Biomineralization: a pavement of pearl. Nature **389**, 912–915 (1997)

21. C. Liu et al., In-depth proteomic analysis of shell matrix proteins of Pinctada fucata. Sci. Rep. **5**, 17269 (2015). https://doi.org/10.1038/srep17269
22. J. Xie et al., Influence of the Extrapallial fluid of Pinctada fucata on the crystallization of calcium carbonate and shell biomineralization. Cryst. Growth Des. **16**, 672–680 (2016). https://doi.org/10.1021/acs.cgd.5b01203
23. K. Saruwatari, T. Matsui, H. Mukai, H. Nagasawa, T. Kogure, Nucleation and growth of aragonite crystals at the growth front of nacres in pearl oyster, Pinctada fucata. *Biomaterials* **30**, 3028–3034 (2009). https://doi.org/10.1016/j.biomaterials.2009.03.011
24. H. Nakahara, G. Bevelander, The formation and growth of the prismatic layer of Pinctada radiata. Calcif. Tissue Res. **7**, 31–45 (1971)
25. H. Tong et al., In situ analysis of the organic framework in the prismatic layer of mollusc shell. Biomaterials **23**, 2593–2598 (2002)
26. F. Marin, B. Pokroy, G. Luquet, P. Layrolle, K. De Groot, Protein mapping of calcium carbonate biominerals by immunogold. Biomaterials **28**, 2368–2377 (2007). https://doi.org/10.1016/j.biomaterials.2007.01.029
27. M. Suzuki, H. Nagasawa, The structure-function relationship analysis of Prismalin-14 from the prismatic layer of the Japanese pearl oyster, Pinctada fucata. *FEBS J.* **274**, 5158–5166 (2007). https://doi.org/10.1111/j.1742-4658.2007.06036.x
28. T. Okumura, M. Suzuki, H. Nagasawa, T. Kogure, Characteristics of biogenic calcite in the prismatic layer of a pearl oyster, Pinctada fucata. Micron **41**, 821–826 (2010). https://doi.org/10.1016/j.micron.2010.05.004
29. C. Gregoire, Topography of the Organic Components in Mother-of-Pearl. J. Biophys. Biochem. Cytol. **3**, 797–808 (1957). https://doi.org/10.1083/jcb.3.5.797
30. K. Oaki, H. Imai, The hierarchical architecture of nacre and its mimetic material. Angew. Chem. Int. Ed. **44**, 6571–6575 (2005). https://doi.org/10.1002/anie.200500338
31. N.P. Gong et al., Immunolocalization of matrix proteins in nacre lamellae and their in vivo effects on aragonitic tablet growth. J. Struct. Biol. **164**, 33–40 (2008). https://doi.org/10.1016/j.jsb.2008.05.009
32. F. Nudelman, B.A. Gotliv, L. Addadi, S. Weiner, Mollusk shell formation: mapping the distribution of organic matrix components underlying a single aragonitic tablet in nacre. J. Struct. Biol. **153**, 176–187 (2006). https://doi.org/10.1016/j.jsb.2005.09.009
33. Y. Levi-Kalisman, G. Falini, L. Addadi, S. Weiner, Structure of the nacreous organic matrix of a bivalve mollusk shell examined in the hydrated state using cryo-TEM. J. Struct. Biol. **135**, 8–17 (2001). https://doi.org/10.1006/jsbi.2001.4372
34. R.A. Rose, S.B. Baker, Larval and spat culture of the Western-Australian Silverlip or Goldlip pearl oyster, Pinctada-Maxima Jameson (Mollusca, Pteriidae). Aquaculture **126**, 35–50 (1994). https://doi.org/10.1016/0044-8486(94)90246-1
35. D. Fang et al., Identification of Genes Directly Involved in Shell Formation and Their Functions in Pearl Oyster, Pinctada fucata. Plos One **6**., ARTN e218601371/journal.pone.0021860 (2011)
36. J. Liu et al., Microarray: a global analysis of biomineralization-related gene expression profiles during larval development in the pearl oyster, Pinctada fucata. BMC Genomics **16**, 325 (2015). https://doi.org/10.1186/s12864-015-1524-2
37. P.S. Nair, W.E. Robinson, Calcium speciation and exchange between blood and extrapallial fluid of the quahog Mercenaria mercenaria (L.). Biol. Bull. **195**, 43–51 (1998). https://doi.org/10.2307/1542774
38. H. Miyamoto, F. Miyoshi, J. Kohno, The carbonic anhydrase domain protein nacrein is expressed in the epithelial cells of the mantle and acts as a negative regulator in calcification in the mollusc Pinctada fucata. Zool. Sci. **22**, 311–315 (2005). https://doi.org/10.2108/zsj.22.311
39. I.M. Sokolova, Apoptosis in molluscan immune defense. Isj-Invertebr. Surv. J. **6**, 49–58 (2009)
40. A. Serpentini et al., Collagen study and regulation of the de novo synthesis by IGF-I in hemocytes from the gastropod mollusc, Haliotis tuberculata. J. Exp. Zool. **287**, 275–284 (2000). https://doi.org/10.1002/1097-010x(20000901)287:4<275::Aid-Jez2>3.0.Co;2-8

41. A.S. Mount, A.P. Wheeler, R.P. Paradkar, D. Snider, Hemocyte-mediated shell mineralization in the eastern oyster. Science **304**, 297–300 (2004). https://doi.org/10.1126/science.1090506
42. A.P. Wheeler, C.S. Sikes, K.W. Rusenko, J.E. Borbas, The role of structural features of oyster Shell organic matrix in biomineralization. Am. Zool. **27**, A52–A52 (1987)
43. A.P. Wheeler, C.S. Sikes, Regulation of carbonate calcification by organic matrix. Am. Zool. **24**, 933–944 (1984)
44. C.S. Sikes, A.P. Wheeler, The organic matrix from oyster Shell as a regulator of calcification Invivo. Biol. Bull. **170**, 494–505 (1986). https://doi.org/10.2307/1541857
45. M.A. Cariolou, D.E. Morse, Purification and Characterization of Calcium-Binding Conchiolin Shell Peptides from the Mollusk, Haliotis-Rufescens, as a Function of Development. J. Comp. Physiol. B. **157**, 717–729 (1988). https://doi.org/10.1007/Bf00691002
46. X.Y. Shen, A.M. Belcher, P.K. Hansma, G.D. Stucky, D.E. Morse, Molecular cloning and characterization of lustrin a, a matrix protein from shell and pearl nacre of Haliotis rufescens. J. Biol. Chem. **272**, 32472–32481 (1997). https://doi.org/10.1074/jbc.272.51.32472
47. M. Kono, N. Hayashi, T. Samata, Molecular mechanism of the nacreous layer formation in Pinctada maxima. Biochem. Biophys. Res. Commun. **269**, 213–218 (2000). https://doi.org/10.1006/bbrc.2000.2274
48. P. Bork, Shuffled Domains in Extracellular Proteins. Febs. Lett. **286**, 47–54 (1991). https://doi.org/10.1016/0014-5793(91)80937-X
49. R. Har-el, M. Tanzer, L. Extracellular matrix. 3: evolution of the extracellular matrix in invertebrates. FASEB J. **7**, 1115–1123 (1993)
50. X. Shen, A.M. Belcher, P.K. Hansma, G.D. Stucky, D.E. Morse, Molecular cloning and characterization of lustrin A, a matrix protein from shell and pearl nacre of Haliotis rufescens. J. Biol. Chem. **272**, 32472–32481 (1997)
51. L. Patthy, Exon shuffling and other ways of module exchange. Matrix Biol. **15**, 301–310; discussion 311–302 (1996)
52. L. Patthy, Genome evolution and the evolution of exon-shuffling–a review. Gene **238**, 103–114 (1999)
53. Y. Dauphin et al., In situ chemical speciation of sulfur in calcitic biominerals and the simple prism concept. J. Struct. Biol. **142**, 272–280 (2003)
54. J.C. Marxen, M. Hammer, T. Gehrke, W. Becker, Carbohydrates of the organic shell matrix and the shell-forming tissue of the snail Biomphalaria glabrata (say). Biol. Bull. **194**, 231–240 (1998). https://doi.org/10.2307/1543052
55. J.C. Marxen, W. Becker, Calcium binding constituents of the organic shell matrix from the freshwater snail Biomphalaria glabrata. Comp. Biochem. Phys. B. **127**, 235–242 (2000). https://doi.org/10.1016/S0305-0491(00)00261-3
56. Q.L. Feng et al., Crystallographic alignment of calcite prisms in the oblique prismatic layer of Mytilus edulis shell. J. Mater. Sci. **35**, 3337–3340 (2000). https://doi.org/10.1023/A:1004843900161
57. G. Falini, S. Albeck, S. Weiner, L. Addadi, Control of aragonite or calcite polymorphism by mollusk shell macromolecules. Science **271**, 67–69 (1996). https://doi.org/10.1126/science.271.5245.67
58. T. Samata et al., A new matrix protein family related to the nacreous layer formation of Pinctada fucata. Febs Letters **462**, 225–229 (1999). https://doi.org/10.1016/S0014-5793(99)01387-3
59. H. Miyamoto et al., A carbonic anhydrase from the nacreous layer in oyster pearls. Proc. Natl. Acad. Sci. U. S. A. **93**, 9657–9660 (1996). https://doi.org/10.1073/pnas.93.18.9657
60. S. Sudo et al., Structures of mollusc shell framework proteins. Nature **387**, 563–564 (1997). https://doi.org/10.1038/42391
61. M. Sarikaya, H. Fong, D.W. Frech, R. Humbert, Biomimetic assembly of nanostructured materials. Mater. Sci. Forum **293**, 83–97 (1999)
62. C. Zhang, S. Li, Z.J. Ma, L.P. Xie, R.Q. Zhang, A novel matrix protein p10 from the nacre of pearl oyster (Pinctada fucata) and its effects on both CaCO3 crystal formation and

mineralogenic cells. Mar. Biotechnol. **8**, 624–633 (2006). https://doi.org/10.1007/s10126-006-6037-1
63. C. Zhang, L.P. Xie, J. Huang, X.L. Liu, R.Q. Zhang, A novel matrix protein family participating in the prismatic layer framework formation of pearl oyster, Pinctada fucata. Biochem. Biophys. Res. Commun. **344**, 735–740 (2006). https://doi.org/10.1016/j.bbrc.2006.03.179
64. Y.W. Kong et al., Cloning and characterization of Prisilkin-39, a novel matrix protein serving a dual role in the prismatic layer formation from the oyster Pinctada fucata. J. Biol. Chem. **284**, 10841–10854 (2009). https://doi.org/10.1074/jbc.M808357200
65. H.L. Liu et al., Identification and characterization of a biomineralization related gene PFMG1 highly expressed in the mantle of Pinctada fucata. Biochemistry **46**, 844–851 (2007). https://doi.org/10.1021/bi061881a
66. C. Pan et al., A novel acidic matrix protein, PfN44, stabilizes magnesium calcite to inhibit the crystallization of aragonite. J. Biol. Chem. **289**, 2776–2787 (2014). https://doi.org/10.1074/jbc.M113.504027
67. M. Rousseau et al., Biomineralisation markers during a phase of active growth in Pinctada margaritifera. Comp. Biochem. Physiol. A Mol. Integr. Physiol. **135**, 271–278 (2003)
68. R. Tisi et al., Evidence for inositol triphosphate as a second messenger for glucose-induced calcium signalling in budding yeast. Curr. Genet. **45**, 83–89 (2004). https://doi.org/10.1007/s00294-003-0465-5
69. A. Takahashi, P. Camacho, J.D. Lechleiter, B. Herman, Measurement of intracellular calcium. Physiol. Rev. **79**, 1089–1125 (1999)
70. K.M. Wilbur, A.S.M. Saleuddin, *The Mollusca, Shell Formation* (Academic Press, New York, 1983)
71. L.B. Kirschner, A.L. Sorenson, M. Kriebel, Calcium and electric potential across the clam mantle. Science **131**, 735–735 (1960). https://doi.org/10.1126/science.131.3402.735-a
72. H. Petit, W.L. Davis, R.G. Jones, H.K. Hagler, Morphological-Studies on the Calcification Process in the Fresh-Water Mussel Amblema. Tissue Cell **12**, 13–28 (1980). https://doi.org/10.1016/0040-8166(80)90049-X
73. I. Marigomez, M. Soto, M.P. Cajaraville, E. Angulo, L. Giamberini, Cellular and subcellular distribution of metals in molluscs. Microsc. Res. Tech. **56**, 358–392 (2002). https://doi.org/10.1002/jemt.10040
74. T.A. Ryder, I.D. Bowen, Studies on transmembrane and paracellular phenomena in the foot of the slug agriolimax reticulatus (Mu). Cell Tissue Res. **183**, 143–152 (1977)
75. J.M. Neff, Ultrastructure of the outer epithelium of the mantle in the clam Mercenaria mercenaria in relation to calcification of the shell. Tissue Cell **4**, 591–600 (1972)
76. U. Karbach, Paracellular calcium-transport across the small-intestine. J. Nutr. **122**, 672–677 (1992)
77. U. Bielefeld, K.H. Kortje, H. Rahmann, W. Becker, The Shell-forming mantle epithelium of Biomphalaria-Glabrata (Pulmonata) – ultrastructure, permeability and cytochemistry. J. Mollus Stud. **59**, 323–338 (1993). https://doi.org/10.1093/mollus/59.3.323
78. R. Bleher, J. Machado, Paracellular pathway in the shell epithelium of Anodonta cygnea. J. Exp. Zool. A Comp. Exp. Biol. **301**, 419–427 (2004). https://doi.org/10.1002/jez.a.20065
79. J. Coimbra, J. Machado, P.L. Fernandes, H.G. Ferreira, K.G. Ferreira, Electrophysiology of the mantle of Anodonta-cygnea. J. Exp. Biol. **140**, 65–88 (1988)
80. A.M. Coimbra, P.L. Fernandes, K.G. Ferreira, H.G. Ferreira, Intracellular calcium in the cells of the outer mantle epithelium of Anodonta-cygnea. Comp. Biochem. Phys. A. **103**, 497–500 (1992). https://doi.org/10.1016/0300-9629(92)90278-X
81. R. Bleher, J. Machado, Paracellular pathway in the shell epithelium of Anodonta cygnea. J. Exp. Zool. Part A **301a**, 419–427 (2004). https://doi.org/10.1002/jez.a.20065
82. C. Lucu, Calcium-transport across isolated gill epithelium of Carcinus. J. Exp. Zool. **268**, 339–346 (1994). https://doi.org/10.1002/jez.1402680502
83. R.E. Akins, R.S. Tuan, Transepithelial Calcium-Transport in the Chick Chorioallantoic Membrane .2. Compartmentalization of Calcium during Uptake. J. Cell Sci. **105**, 381–388 (1993)

84. A. Ziegler, Ultrastructural evidence for transepithelial calcium transport in the anterior sternal epithelium of the terrestrial isopod Porcellio scaber (Crustacea) during the formation and resorption of CaCO3 deposits. Cell Tissue Res. **284**, 459–466 (1996). https://doi.org/10.1007/s004410050606
85. M. Carre et al., Calcification rate influence on trace element concentrations in aragonitic bivalve shells: evidences and mechanisms. Geochim. Cosmochim. Acta **70**, 4906–4920 (2006). https://doi.org/10.1016/j.gca.2006.07.019
86. P.E. Gibbs, J.A. Nott, A. Nicolaidou, M.J. Bebianno, The composition of phosphate granules in the digestive glands of marine prosobranch gastropods: variation in relation to taxonomy. J. Mollus Stud. **64**, 423–433 (1998). https://doi.org/10.1093/mollus/64.4.423
87. J.D. Thomson, B.J.S. Pirie, S.G. George, Cellular metal distribution in the pacific oyster, Crassostrea-Gigas (THUN) determined by quantitative x-ray microprobe analysis. J. Exp. Mar. Biol. Ecol. **85**, 37–45 (1985). https://doi.org/10.1016/0022-0981(85)90012-7
88. C.K. Yap, S.G. Tan, A study on the potential of the Periostracum of Perna viridis as a biomonitoring material for Pb in tropical coastal waters based on correlation analysis. Sains Malays. **40**, 809–819 (2011)
89. S.M. Adams, C.D. Shorey, M. Byrne, An ultrastructural and microanalytical study of metal-ion content in granular concretions of the freshwater mussel Hyridella depressa. Micron **28**, 1–11 (1997). https://doi.org/10.1016/S0968-4328(97)00003-6
90. M. Pekkarinen, I. Valovirta, Histochemical and X-ray studies on tissue concretions and shells of Margaritifera margaritifera (Linnaeus). J. Shellfish Res. **16**, 169–177 (1997)
91. R.G. Jones, W.L. Davis, Calcium-containing lysosomes in the outer mantle epithelial-cells of Amblema, a fresh-water mollusk. Anat. Rec. **203**, 337–343 (1982). https://doi.org/10.1002/ar.1092030304
92. H.A. Lucero, D. Lebeche, B. Kaminer, ERcalcistorin/protein-disulfide isomerase acts as a calcium storage protein in the endoplasmic reticulum of a living cell – comparison with calreticulin and calsequestrin. J. Biol. Chem. **273**, 9857–9863 (1998). https://doi.org/10.1074/jbc.273.16.9857
93. J. Meldolesi, T. Pozzan, The endoplasmic reticulum Ca2+ store: a view from the lumen. Trends Biochem Sci **23**, 10–14 (1998). https://doi.org/10.1016/S0968-0004(97)01143-2
94. P.J.E. Rowling, S.H. Mclaughlin, G.S. Pollock, R.B. Freedman, A single purification procedure for the major resident proteins of the Er lumen – Endoplasmin, bip, calreticulin and protein disulfide-isomerase. Protein Expres. Purif. **5**, 331–336 (1994). https://doi.org/10.1006/prep.1994.1049
95. P. Greenaway, Calcium regulation in freshwater crayfish Austropotamobius-Pallipes (Lereboullet) .1. Calcium balance in intermoult animal. J. Exp. Biol. **57**, 471 (1972)
96. A.M. Katz, Molecular biology of calcium channels in the cardiovascular system. Am. J. Cardiol. **80**, I17–I22 (1997)
97. G.A. Ahearn, P.K. Mandal, A. Mandal, Calcium regulation in crustaceans during the molt cycle: A review and update. Comp. Biochem. Physiol. A Mol. Integr. Physiol. **137**, 247–257 (2004). https://doi.org/10.1016/j.cbpb.2003.10.012
98. P. Andrade, R. Willoughby, S.A. Pomponi, R.G. Kerr, Biosynthetic studies of the alkaloid, stevensine, in a cell culture of the marine sponge Teichaxinella morchella. Tetrahedron Lett. **40**, 4775–4778 (1999). https://doi.org/10.1016/S0040-4039(99)00881-3
99. P. Westbroek, F. Marin, A marriage of bone and nacre. Nature **392**, 861–862 (1998). https://doi.org/10.1038/31798

Chapter 2
Identification and Characterization of Biomineralization-Related Genes

Abstract *Pinctada fucata*, famous as one of the most important economically bivalves all over the world, is also an excellent animal model for biomineralization study, the process of which is conducted by reaction and precipitation via converting ions into solid minerals where matrix protein functions as crucial regulators (Pertea G et al, Bioinformatics 19:651–652, 2003; Conesa A et al, 21:3674–3676, 2005). In general, the studies of pearl formation could be separated into two aspects: the isolation, identification, and functional analysis of shell matrix proteins and calcium metabolism process, including Ca^{2+} absorption, transportation, storage and deposition, etc (Ye et al. 34:W293–W297, 2006). We successfully constructed three suppression subtractive hybridization (SSH) libraries and established transcriptome database to obtain and analyze novel matrix proteins and their functions. In addition, we first isolated the full-length of CaM cDNA and characterized its functions. Other calcium metabolism-related proteins, such as calmodulin-like proteins and calcineurin, were also identified, the working mechanisms of which were studied in more detail at the same time. The primary techniques adopted in this chapter comprised RNA-seq, rapid amplification of cDNA ends (RACE), cell co-localization, in situ hybridization, RNA interference, $CaCO_3$ crystallization, etc. In summary, elucidating the underlying mechanisms of biomineralization will lead to the innovative understanding of medical sciences and materials and also provide significant technical supports for pearl shellfish farming and pearl production.

Keywords Biomineralization · Gene · Shell matrix proteins · Calcium metabolism

2.1 Materials and Methods

2.1.1 Ethics Statement

All of the studies were approved by the Animal Ethics Committee of Tsinghua University, Beijing, China.

R. Zhang et al., *Biomineralization Mechanism of the Pearl Oyster, Pinctada fucata*,
https://doi.org/10.1007/978-981-13-1459-9_2

2.1.2 Animals

The adult pearl oysters, live and healthy *P. fucata*, with a shell length 5–6 cm and 30–40 g of wet weight, were purchased from the Zhanjiang Pearl Farm (Guangdong, China). They were maintained in glass aquaria filled with aerated artificial seawater (Sude Instant Sea Salt, 3%) for a week under controlled temperature (20 ± 2 °C) before experiments. They were fed with spirulina powder every day.

Larval samples of *P. fucata* in Sects. 2.2.1 and 2.2.2 were gathered from the Guofa Pearl Farm in Beihai, Guangxi Province, China and Daya Bay Marine Comprehensive Experimental Station, Shenzhen, Guangdong Province, China, respectively. Fertilized eggs were harvested immediately after insemination. The trochophore, D-shaped, umbonal stage larvae and the juvenile samples were collected 17 h, 48 h, 14 days, and 35 days after a microscopic count guaranteed that >75% of the larvae had reached a particular growth stage. Fertilized eggs were incubated at 25 °C.

Rabbits were raised under standardized pathogen-free conditions in the Animal Care Facility at Beijing. The study protocol for experimental use of the animals was authorized by the Ethics Committee of National Center for Clinical Laboratories. The 3R (reduction, replacement, and refinement) principles of animal experiments were strictly followed.

2.1.3 RNA Extraction and cDNA Synthesis

Total RNA was extracted from the larval stages by SV total RNA isolation kit (Promega, USA), according to manufacturer's instruction. RNA quantity was assessed by measuring $OD_{260/280}$ with a NanoDrop Lite spectrophotometer (Thermo Scientific, USA) or an Ultrospec 3000 UV-visible spectrophotometer (Amersham Biosciences, USA). The integrity of RNA was detected by fractionation on 1.2% formaldehyde denatured agarose gel and staining with ethidium bromide. In the subtractive hybridization procedure, 0.5 μg of total RNA was applied to produce double-stranded cDNA using the SMARTTMPCR cDNA synthesis kit (Clontech, USA). We used 1 μg of total RNA as the template for reverse transcription to first-strand cDNA synthetization by MMLV-RT reverse transcriptase (Promega, USA) and oligo (dT) (Promega, USA) as the primer.

cDNA sequences were sequenced via an automated DNA sequencer (Applied Biosystems 3730XL, USA) and analyzed by the BLAST program available from the National Center for Biotechnology Information (http://www.ncbi.nlm.nih.gov/). Primer Premier 5.0 (PREMIER Biosoft International, USA) was used to identify the encoding proteins. The ClustalX (version 1.81, ftp://ftp-igbmc.u-strasbg.fr/pub/clustalx) and the Vector NTI (InforMax, Frederick, USA) were applied to perform multiple alignments and produce the phylogenetic tree.

2.1.4 Suppression Subtractive Hybridization and Library Construction

Suppression subtractive hybridization procedures were performed with different development stages. The D-T library was constructed using cDNA prepared from early D-shaped stage (20-hour-old larva) as the tester and the trochophore stage (14-hour-old larva) as the driver. The U-D library used cDNA prepared from the umbonal stage (7-day-old larva) as the tester and that from D-shaped stage as the driver. And the J-U library was obtained using cDNA from the juvenile stage (35-day-old larva) and umbonal stage as tester and driver.

We applied SSH using a PCR-Select^{TM} cDNA Subtraction kit (Clontech, USA), according to the manufacturer's instructions. We used primer *gapdh*F and *gapdh*R (see Supplementary Table 2.1 for primer details) to examine the transcription abundance of the housekeeping gene *gapdh* for evaluating the subtraction efficiency. The PCR-amplified subtracted cDNA was then inserted into the T/A cloning vector pGEM-T Easy Vector (Promega, USA) and transformed into Trans2-Blue Chemically Competent Cell (TransGen, China), following the manufacturer's instructions.

3515 clones from the 3 libraries were amplified by PCR using paired primers (Nested Primer1 and Nested Primer 2R) with the PCR-Select^{TM} cDNA Subtraction kit (Clontech, USA) to evaluate whether inserts were the enriched cDNA of the tester and then analyzed by electrophoresis on a 1.2% agarose gel to estimate the insertion size. The PCR products were combined with an equal volume of 0.6 N NaOH and spotted manually onto a nylon membrane. The membrane was hybridized with DIG-labeled cDNA probes synthesized for the driver and tester. The hybridized membrane was revealed to X-ray film to show the positive clones, which was conducted according to the manufacturer's instructions with the DIG high prime DNA labeling and detecting starter kit II (Roche, Switzerland). Finally, 2975 clones with positive signals from tester cDNA hybridization and negative signals from driver cDNA hybridization were selected from the 3 libraries. Furthermore, we submitted the raw sequences to the NCBI dbEST database (GenBank: GW602961-GW603856).

2.1.5 Sequence Analysis and Bioinformatics

We selected 2975 positive clones for sequencing, and 2923 available sequences of them were returned. These sequences were edited and then assembled by the programs PHRED, CROSS_MATCH, and CONSED (http://bozeman.mbt.washington.edu/). After removing the vector sequences, the remaining sequences with a minimum size of 100 bases were selected for further analysis. Remaining sequences were clustered using the TGICL [1] and assembled into 990 unigenes. We used the BLASTX search with the GenBank nr database to annotate those sequences; meanwhile, an e-value of 10e-05 was used to cut off annotations with low similarities.

Since most of the sequences returned no hits, so we used Blast2GO [2] for gene ontology (GO) assignment, the results of which were classified using WEGO [3]. Unigenes were searched against the transcriptome of mantle tissue with a cutoff e-value of 10e-06 by BLASTN annotation. Coding regions were identified by aligning the unigenes by BLASTN (cutoff e-value 10e-05) to protein databases in the priority order of GenBank nr, KEGG, Swiss-Prot, and COG. Unigenes aligned to databases with higher priority did not proceed to the next stage. It was not until all the circles were finished when the alignments got ended. Proteins with highest ranks in BLAST results were selected to determine the coding region sequences and these sequences would be translated into amino acid sequences. Thus, both the nucleotide sequences (5′-3′) and amino sequences of the unigene coding regions were acquired. Unigenes that showed no similarities to known genes were then investigated by ESTScan [4] of nucleotide sequences (5′-3′) and amino sequences of the coding regions. Unigenes with a coding region shorter than 150 base pairs and with no stop codon were discarded during the secreted protein research. Signal peptides were searched in the different remaining potential full-length coding regions using SignalP v3.0 [5]. Transmembrane proteins were removed using the TMHMM Server v. 2.0 (http://www.cbs.dtu.dk/services/TMHMM/) to predict transmembrane domains. Glycosylphosphatidylinositol-anchored proteins were discarded using the GPI modification predictor [6]. Signal peptide-positive unigenes were searched by TargetP [7] to remove proteins targeted to organelles. The remaining unigenes were considered to code for secreted proteins [8]. We also searched for known matrix proteins in *P. fucata* as a control.

XSTREAM [9] was used to isolate genes coding for proteins with tandem-arranged repeated units. Unigenes with a coding region shorter than 150 base pairs were discarded in this research. The settings used in this analysis were as the default, i.e., degeneracy, 0; TR significance, high; and min consensus match, 0.8. The known matrix proteins of *P. fucata* were included as the control.

2.1.6 **P. fucata** *Oligonucleotide Microarray*

Based on the adult mantle tissue transcriptome data obtained from our previous study, 58,940 probes (60 nt), representing 58,623 transcripts, were synthesized in situ. In Sect. 2.2.2, total RNA from different stages was separately reverse-transcribed and purified. The labeled cDNA samples were mixed and hybridized to the microarray, and the processed slides were scanned with an Agilent G2565CA Microarray Scanner (Agilent Technologies, USA).

The scanned microarray images were analyzed via the Agilent Software Feature Extraction. GeneSpring GX software (CapitalBio, USA) was used for quantile normalization and the statistical analysis. Cluster 3.0 software (CapitalBio, USA) was used to perform the cluster analyses on the entire data set. A principal component analysis (PCA) was used to determine the significantly differentially expressed genes between developmental periods. The significance analysis of microarrays

(SAM) software [10] was applied to perform the expression profile comparisons. Only genes whose changes were consistent (p-value < 0.05, fold change [FC] $\geq$ 2) were selected as differentially expressed genes. We also prepared a scatter plot. We used the non-parametric Spearman's rank correlation test to assess the correlation between the expression values measured by real-time RT-PCR and microarray for the set of candidate genes by SPSS 12.0 software (SPSS, USA). The functional enrichment of up- or downregulated genes during two consecutive developmental stages was assessed based on the GO [2] and KEGG [11] pathways annotation terms. Only the annotations with a q-value < 0.05 were considered significant.

2.1.7 Candidate Biomineralization-Related Gene Screening and Bioinformatics

In this study, among all unigenes upregulated from the umbonal to the juvenile stage, 1113 unigenes, with an expression increasement over 20-fold, were further analyzed to screen for candidate biomineralization-related genes contributing to larval shell formation. A total of 831 of these unigenes mentioned above were not similar to any known genes/proteins, or they had some similarities between hypothetical and uncharacterized proteins after a BLASTX search using the GenBank nr database. All of these unigenes were considered uncharacterized, and the nucleotide (5′-3′) and ESTScan were used to investigate and analyze the amino acid sequences of those coding regions [4]. The following bioinformatics tools were employed to search for secreted potential matrix proteins [8]. Unigenes with a coding region <240 base pairs were removed, and the remaining coding regions were searched for signal peptides using SignalP v4.1 [5]. Peptide-positive proteins targeted to organelles were removed after a TargetP [7] search, and transmembrane proteins were removed using the TMHMM Server v. 2.0 (http://www.cbs.dtu.dk/services/TMHMM/). XSTREAM [9] was used to isolate proteins with tandem-arranged repeated units using the default settings. The known *P. fucata* matrix proteins were checked as a control.

2.1.8 Gene Expression Analysis by RT-PCR Assay

Tissue expression pattern in the pearl oyster was investigated and analyzed by RT-PCR. Total RNA was isolated using the method described above for mantle edge, mantle pallial, gonad, adductor muscle, gill, hemocytes, foot, and viscus of adult individual *P. fucata*. Equal quantities (1 µg) of total RNA from different tissues were reverse-transcribed to first-strand cDNA using quant reverse transcriptase (Takara, Japan), following the manufacturer's instructions. The synthesized cDNA was used as a template for the following PCR reactions. Gene *GAPDH/Actin* was used as the positive control for cDNA preparations (see Supplementary Table 2.1 for primers details). Negative controls were conducted in the absence of the cDNA template to test for cross-contamination of the samples. All PCR products were subcloned and confirmed by sequencing.

2.1.9 Tissue Specificity Analysis by Semiquantitative PCR

Most matrix proteins were specifically expressed in mantle tissue, and the different regions of mantle pallial or edge referred to the formation of either nacreous or prismatic layer of the shell according to previous studies. As a result, a tissue specificity analysis by semiquantitative PCR or RT-PCR would help a lot to identify the candidate genes involved in shell or pearl formation. Primers (please refer to Supplementary Table 2.1 for more details) corresponding to the candidate genes were designed to investigate the tissue expression patterns in the pearl oyster, and housekeeping genes like *β-actin* or *GAPDH* were used as the positive control. Synthesized cDNA from gill, gonad, viscera, adductor muscle, foot, mantle edge, and the mantle pallial of adult oysters was used as the template for the PCR in every single reaction. All PCR products were verified and confirmed by sequencing.

2.1.10 In Situ Hybridization

Mantle tissue from adult *P. fucata* was removed and immediately fixed overnight in 4% paraformaldehyde containing 0.1% diethyl pyrocarbonate (Sigma, USA). In situ hybridization assay was performed on frozen mantle sections. The fragments were amplified with the designed primers of the certain genes (see Supplementary Table 2.1 for more details) and inserted in multiple cloning sites of the pGEM-T Easy Vector (Promega, USA) and digoxigenin-labeled RNA probes were generated using a DIG RNA Labeling Kit (Roche, Switzerland). The concrete protocol was executed as described previously [12]. To avoid false-positive signals, the hybridization temperature was increased to 50 °C.

2.1.11 RNAi Experiments

RNAi assay was performed as described in Suzuki et al. [13], with some modifications. Different designed primers (see Supplementary Table 2.1 for details) were used to amplify specific sequences from the first-strand cDNA for GFP, while pEGFP-C1 (Clontech, USA) was used as the template. The PCR products were purified using PCR product purification kit (Boster, China). A RiboMAX™ Large Scale RNA Production Systems (T7) kit (Promega, USA) was used to synthesize and purify the double-stranded RNA. RNase-free DNase I (Takara, Japan) was used to digest the remaining template DNA. The integrity and quantity of the dsRNA were detected as previously described. The dsRNA was then diluted to 40, 80, or 160 μg 100 μl^{-1} with PBS or water, and this 100 μl solution was injected into the adductor muscle of five 2-year-old individuals with a shell length of 5–6 cm. 100 μl PBS (water) and 40 μg GFP dsRNA in 100 μl PBS (water) were used as controls. Five individuals were used in each treatment.

Total RNA from the mantle tissue of each oyster 6 days after injection was extracted and synthesized first-strand cDNA, as mentioned above. Real-time quantitative PCR (RT-PCR) was used to quantity the expression levels of each gene, and *β-actin* was used as an internal reference. RT-PCR was conducted with the Mx3000PTM (Stratagene, USA) using an SYBRH Premix Ex TaqTMII kit (Takara, Japan), according to the manufacturer's instructions (see Supplementary Table 2.1 for primer details). Cycling parameters were 95 °C for 30 s (1 cycle) and 95 °C for 5 s, 55 °C for 30 s, and 72 °C for 30 s (40 cycles). Dissociation curves were analyzed to determine the purity of the product and specificity of amplification.

The shells of the injection groups were thoroughly washed with Milli-Q water 6 days after injection and air-dried. Shells were carefully cut into pieces that mounted on the scanner with the inner nacreous surface face-up, sputter-coated with 10-nm-thick gold, and analyzed using an FEI Quanta 200 scanning electron microscope. Areas at the center and edge of the shell (both prismatic and nacreous layer) were thoroughly examined and observed. Cross-sectional views were obtained by viewing from the edge of the shell.

2.1.12 Recombinant Protein Expression and Purification

The target gene was inserted into the recombinant plasmid vector like pET21b/pET28a/pGEX-4T-1/pmal-c5x, which was then transformed into *E. coli* cells (Transsetta, DE3) (Transgene, China). The recombinant cells were incubated at 37 °C for 3 h when the cell density (OD_{600}) reached 0.6–0.8 and then induced by 0.45 mM isopropyl 1-thio-β-D-galactopyranoside (IPTG; Sigma-Aldrich) at 15 °C for 12 h. After 24 h of cultivation, cells were harvested and resuspended with lysis buffer (20 mM Tris, 500 mM NaCl, 10% glycerol, and 20 mM imidazole; pH 7.5). Then cells were disrupted via ultrasonic dismembrator (Sonics & Materials Inc., USA) at 28% power with 4 s pulse on and 6 s pulse off repetition cycle on ice. After centrifugation 12000 g for 40 min at 4 °C, the insoluble fraction was removed, and the supernatant was collected for further purification using affinity chromatography or antibody preparation. The purified protein was stored in a pH 7.5 solution (10 mM Tris, 500 mM NaCl) at -80 °C before use. The protein concentration and quality was determined by Nanodrop or BCA assay kit (Pierce, USA). Slight modifications were applied in actual concrete operation during the purification of different recombinant proteins.

2.1.13 Preparation of Recombinant Protein Antibody

The insoluble precipitation mentioned above was redissolved and resuspended in denaturing lysis buffer (20 mM Tris, 500 mM NaCl, 6 M urea; pH 7.5). After the same centrifugation condition (12000 g for 40 min at 4 °C), the urea-soluble supernatant was collected and mixed with Ni–nitrilotriacetic acid (Ni–NTA) resin (CWBIO, China). The mixture was incubated at 4 °C for 1 h and then washed by washing buffer (20 mM Tris, 500 mM NaCl, 6 M urea, 40 mM imidazole; pH 7.5) to remove the heteropolymeric protein. Target protein was then eluted in elution buffer (20 mM Tris, 500 mM NaCl, 6 M urea, 200 mM imidazole; pH 7.5). Then the protein gel was injected into New Zealand rabbits. Anti-PfY2 polyclonal antibodies were raised in these rabbits following standard immunization procedures according to the manufacturer's instructions.

2.1.14 Ca^{2+}-Dependent Electrophoretic Migration and ^{45}Ca Overlay Analyses

Ca^{2+}-dependent electrophoretic shift assay was carried out according to the method of Burgess et al. [14]. The sample buffer, running buffer, and gels all contained 0.1 mM $CaCl_2$ or EDTA in the presence of SDS. ^{45}Ca overlay analysis followed the procedure of Maruyama et al. [15]. Purified recombinant oyster protein was transferred onto nitrocellulose membrane after electrophoresis as described above and then labeled with ^{45}Ca (Amersham, USA). Autoradiography of the ^{45}Ca-labeled proteins on the nitrocellulose membrane was obtained by a Storm 860 scanner (Amersham, USA).

2.1.15 Circular Dichroism Spectropolarimetry and Fluorescence Spectra

Circular dichroism (CD) spectroscopy was performed at 25 °C with constant N_2 flushing using a CD instrument (Jasco J-715, UK) calibrated with d_{10}-camphorsulfonic acid. The far-UV CD spectra of target proteins were measured from 190 to 250 nm in 100 mM KCl, 20 mM Hepes buffer, pH 7.5 in the presence of 2 mM $CaCl_2$ or EGTA and corrected using a blank buffer containing 100 mM KCl, 20 mM Hepes buffer, pH 7.5. All measurements were performed 10 min after sample preparation with the following instrument settings: response time, 0.5 s scan speed, 200 nm $\cdot$ min^{-1}; sensitivity, 100 millidegrees; 1 mm spectral band width; and an average of four scans.

The fluorescence emission spectra were collected at 25 °C using a Hitachi F-2500 (Japan) spectrofluorimeter according to the method described by VanScyoc et al.

[16] with a slight modification. All samples were diluted in 100 mM KCl, 20 mM Hepes buffer, pH 7.5 in the presence of 5 mM $CaCl_2$ or EGTA, and the final concentration of each protein is 10 μM. The fluorescence of the phenylalanine and tyrosine residues in oyster CLP and CaM was measured using an excitation and emission wavelength pairs (250/280 nm and 277/320 nm, respectively). Testing methods and conditions may be modified slightly based on the characteristics of target proteins.

2.1.16 Preparation of Recombinant Oyster CaLP and CaM Affinity Chromatography Columns and Affinity Chromatography

Recombinant oyster protein CaLP (rCaLP; 5 mg) (sample preparation as mentioned above but not heated at 90 °C) and recombinant protein CaM (rCaM) were coupled to 0.6 g of CNBr-activated Sepharose 4B (Amersham Biosciences, USA) on the basis of the manufacturer's instructions. The coupling efficiency was about 1 mg proteins per ml gel. Affinity chromatography was performed as described by Me'hul et al. [17] with some slight modifications. Five grams of oyster mantle and gill tissues was homogenized in 20 ml extraction buffer [10 mM Hepes, 150 mM NaCl, 0.1% (w/v) Triton X-100, 5 mM dithiothreitol, 0.5 mM phenylmethanesulfonyl fluoride, 10 μg/ml leupeptin, 10 μg/ml pepstatin, pH 7.5] at 4 °C, respectively, and centrifuged at 22000 g for 35 min at 4 °C. The supernatants were passed over 0.22 μm filters and adjusted to 2.5 mM $CaCl_2$ before chromatographic separation. Then, the supernatants were loaded onto the rCaLP and rCaM affinity columns pre-equilibrated with the extraction buffer but with an addition of $CaCl_2$ to a concentration of 2.5 mM at room temperature. Thereafter, the columns were washed with 40 column volumes of extraction buffer containing 2.5 mM $CaCl_2$. Elution was carried out with the help of the extraction buffer containing 5 mM EGTA. The eluted proteins were analyzed by 12.5% SDS/PAGE and silver stained.

2.1.17 Localization of CaM and CaLP in Primary Cultured Mantle Cell

The primary culture of *P. fucata* was performed as Sect. 2.1.2. The culture medium used in this experiment was 1 × L-15 dissolved in fresh seawater, and the medium was changed every 2 days. On the 7th day after initiation of mantle cell culture, epithelial-like cells formed a monolayer on the poly-L-lysine-coated dishes and were lysed for subcellular fraction analysis using a nuclear/cytosol extraction kit (Tianlai, China). Samples were then processed for Western blot analysis with antibodies of CaM and CaLP.

2.1.18 Subcellular Localization of CaM and CaLP in Mantle Epithelium

Small pieces (1 mm × 1 mm) of mantle located under the periostracum groove where biomineralization initiates [18] were fixed at 20 °C for 3 h with 1.5% glutaraldehyde and 3.4% paraformaldehyde in 0.1 M phosphate buffer, pH 7.4. The following steps were performed according to Caceres et al. described before [19] with some modifications. Anti-rabbit serum conjugated with 12-nm-diameter gold (Jackson ImmunoResearch Lab., USA), diluted in PBG-Tween (1:20), was used as the secondary antibody in this study, and the samples were observed under the Philips CM120 Biotwin transmission electron microscope.

2.1.19 Cell Culture and Transfection

HEK 293T cells were maintained in DMEM medium supplemented with 10% FBS, 1.5 g/l $NaHCO_3$, and 100 U penicillin/streptomycin solution at 37 °C in 5% CO_2 and 95% air on dishes or plates. Mammalian expression vectors were transfected into 293T cells using Vigorous transfection reagent (Vigorous Biotechnology, China) following the manufacturer's instructions. For transient transfection, 10 mg of DNA with 4 ml transfection reagent were added to 293T cells (>40% confluent for fluorescence confocal microscopy and >90% confluent for co-immunoprecipitation, respectively) in dishes or plates.

2.1.20 Fluorescence Confocal Microscopy

For intracellular localization analysis of EGFP and other fusion proteins, cells were grown on coverslips and fixed in a mixture of cold acetone and methanol (1:1) for 5 min at −20 °C 24 h after transfection. Images were visualized and captured with an Olympus inverted microscope equipped with a charge-coupled camera (Olympus FV500).

2.1.21 Preparation of Nuclear and Cytoplasmic Fractions

After 24 h transfection, subcellular fractionation of 293T cells was separated by a nuclear/cytosol extraction kit (Tianlai, China) in accordance with the instructions of the manufacturer. Briefly speaking, after transfection of different expression plasmids for 48 h, the cells were collected into a 1.5 ml centrifuge tube and centrifuged for 5 min at 800 g. The pellets were washed one more time, and after that the cells were collected by centrifugation. The pelleted cells were resuspended in 500 ml Cytosol Extraction Buffer A (CEB-A) and incubated on ice for 20 min. Then the cells were vortexed for 10 s every 5 min and centrifuged at 1000 g for 5 min at 4 °C. The pellets contained crude nuclei. The supernatant was transferred to a brand-new tube and further centrifuged at 12,000 g for 5 min at 4 °C. The supernatant was cytosol extracted and was transferred to another new tube. Glycerol (20%) was added to the tube and stored at -80 °C until use.

The crude nuclei pellets prepared previously were used to extract the nuclei protein. After removing the membrane components contaminated in the cytosol extraction, 100 ml of cold nuclear extraction buffer (NEB, USA) was added into the crude nuclei pellet, further vortexed and incubated on ice for 30 min, followed by centrifugation at 12,000 g for 5 min at 4 °C. The supernatant fraction containing the proteins extracted from the nucleus was transferred to a clean centrifuge tube and stored at −80 °C.

2.1.22 Co-immunoprecipitation

Cells were lysed in lysis buffer (Beyotime, China) containing 1 mM PMSF 48 hours after transfection. Furthermore, cell extracts were spun down at 14,000 g for 10 min at 4 °C 1 min after lysis. The extracts from each transfection were pooled into a 1.5 ml tube. For co-immunoprecipitation, 2 μg antibody of EGFP was introduced into the extracts and incubated with end-over-end mixing for 2 h at 4 °C. Protein A Sepharose was then added and mixed for 1 h at 4 °C. After the immunoprecipitates had been washed five times with 1 ml lysis buffer, the protein A Sepharose beads were collected and suspended in 2 × SDS sample buffer, boiled for 5 min, and subjected to Western blot analysis with relative antibody.

2.1.23 Domain–Domain Interaction Assay

The effects of the extra C-terminal tail on the interactions between the C- and N-domains of CaLP were assayed by CD spectra. CD spectra were collected on a Jasco J-715 spectropolarimeter. C-CaLP and N-CaLP and C2-CaLP and N-CaLP were mixed with the molecule ratio of 1:1, respectively. The concentration of each protein component was 10 μM. CD spectra of single protein and different protein mixtures were obtained by an average of eight scans collected between 250 and 190 nm. All spectra were recorded in buffer E (Pierce, USA) containing 2 mM $CaCl_2$ or 2 mM EDTA. All these experiments were repeated four times with a fresh batch of protein samples, and the data obtained were remarkably similar.

2.1.24 ANS Fluorescence Assay

The conformational changes in all proteins were further investigated by 8-anilino-1-naphthalenesulfonic acid (ANS, Sigma) binding fluorescence assay on a Hitachi F-2500 spectrofluorimeter. ANS fluorescence emission spectra were controlled with excitation wavelength set at 370 nm. All procedures were performed at 25 °C, using a Multi-Temp III water bath circulator (Amersham Biosciences, USA). All proteins were incubated with ANS dissolved in buffer E for 30 min before measurement. All measurements were conducted in buffer E in the presence of 5 mM $MgCl_2$, 2 mM EDTA, or 2 mM $CaCl_2$, with final concentrations of 20 μM for M-CaLP and W-CaLP; 30 μM for C-CaLP, N-CaLP, and C2-CaLP; 50 μM for ANS; excitation slit at 5 nm and emission slit at 10 nm; and an average of four scans.

2.1.25 Structural Stability Studies

Structural stability studies were performed under the instructions of the methods described by Ababou et al. [20, 21] with some modifications. Chemical unfolding at 25 °C with GuHCl was carried out by titrating an initial 2 mL of the folded protein placed in a 1 × 1cm cuvette containing a stir bar. The titration was executed by removing an aliquot of the folded protein solution and replacing the same volume from an unfolded protein stock prepared in buffer E containing 6M GuHCl. The unfolded protein stock was loaded into the titrator. After each automated increasing in the denaturant concentration, the samples were equilibrated overnight to ensure that all the proteins were unfolded completely. The native and denatured protein solutions both contained equal concentrations of proteins (for M-CaLP and W-CaLP, 10 μM; for C2-CaLP, N-CaLP, and C-CaLP, 30 μM). The GuHCl stock solution was prepared fresh daily in buffer E. The decrease in helical signal with the addition of GuHCl to the proteins was monitored using far-UV CD spectra from 250 to 190 nm recorded on a Jasco J-715 spectropolarimeter as described in Sect. 2.1.15. All measurements were performed in buffer E. Absorbance signals at 222 nm were selected to calculate the folding fractions of the proteins.

Observed ellipticities were transformed to apparent fractions of native-state protein (N protein), $f_{N,app}$, by using the equation:

$$f_{N,\mathrm{app}} = (\theta - \theta_D)/(\theta_N - \theta_D). \tag{2.1}$$

where θ is the sample ellipticity at a particular GuHCl concentration and θ_D and θ_N represent the native (treated with 0 M GuHCl) and irreversible complete denatured states of the proteins (denatured by 6 M GuHCl), respectively.

The fraction of both the D protein ($f_{D,app}$) and the N protein ($f_{N,app}$) are such that

$$f_{N,\mathrm{app}} + f_{D,\mathrm{app}} = 1. \quad (2.2)$$

2.1.26 Isolation of EDTA-Soluble Matrix Protein from Nacre in P. fucata

The outer prismatic layer of the shell from *P. fucata* was removed, and the inner nacreous layer was crushed to a fine powder. The powdered nacre was decalcified with Tris-buffered 0.5 M EDTA (pH 8.0) at 4 °C for 3 days with constant stirring. The ESM was obtained by centrifugation at 16,000 rpm for 30 min, followed by exhaustive dialysis against Tris-buffered Milli-Q water. The dialyzed solution was concentrated and analyzed by reduced and unreduced SDS-PAGE.

2.1.27 Preparation of Biotinylated CaLP and Protein Blotting

Biotinylation of CaLP was performed as described by Billingsley et al. [22]. The efficiency of biotinylated CaLP was proven by dot blots [23] using CaLP-binding proteins extracted from the mantle of the pearl oyster using affinity chromatography [24]. BSA, in equivalent amounts, was used as a negative control.

2.1.28 Protein Blotting

Proteins in the gel (reduced PAGE) were electroblotted to nitrocellulose (NC) membrane using the Multiphor II Western blotting system (Amersham, USA). After incubation for 1 h in blocking solution (5% nonfat dry milk/20 mM Tris–HCl, pH 7.4/150 mM NaCl/2 mM $CaCl_2$), the blot was incubated with biotinylated CaLP (20 μg/ml) for 2 h in blocking solution containing either 2 mM $CaCl_2$ or, for control studies, 5 mM EGTA. Thereafter, the blot reacted with alkaline phosphatase (AP)-avidin (Invitrogen, USA) dissolved in blocking solution for 1 h, followed by color development with NBT/BCIP (Invitrogen, USA).

2.1.29 $CaCO_3$ Crystallization Experiments In Vitro

The calcitic crystallization oversaturated solution was prepared according to Guofeng Xu et al. [25] Crystallization experiments were performed by adding samples to the freshly prepared crystallization solution on a slide at room temperature. Twenty-four hours after the crystallization, the reaction solution was removed, and the crystals were analyzed and characterized.

Raman spectrum, scanning electron microscopy (SEM), and optical microscopy were used to support the morphological observation and identification of the induced crystals. The Raman spectrum of the crystals was recorded with a Renishaw RM2000 spectrometer. The samples were positioned on a slide under a microscope (×50) focusing the laser beam on the sample while collecting the backscattered light. The Raman spectra showed a line with a wavelength of 514 nm provided by an argon laser limited to 4.6 mW power. Spectrum was recorded from 100 to 1500 cm^{-1}. The SEM photos were obtained using a Sirion200 scanning electron microscope (PEI). The cover glasses were coated with gold before imaging. For some larger crystals, an optical microscope (Leica DMIRB) could be used.

2.1.30 Immunohistological Staining

Immunohistological staining was carried out following Nudelman et al. [26] with minor modifications. A transversal section of the shell was incubated in 4% formaldehyde containing 0.5 M EDTA, pH 8, at 20 °C with gentle shaking for 7 days for demineralization. After demineralization, the samples were washed with Milli-Q water and incubated in rabbit serum of anti-recombinant protein. The serum was diluted to 1:20 in TBS and incubated for 1 h with 5% nonfat dry milk. After washing with TBS, the secondary antibody, AP-conjugated goat-anti-rabbit antibody (1:600, Sigma), was applied to the samples for 1 h. Then, the samples were washed with TBS, and the color was developed with NBT/BCIP (Invitrogen, USA). The negative control group was prepared with preimmune serum.

2.1.31 Cell Cycle Synchronization in G_0 Phase by Serum Starvation and Cell Cycle Analysis

COS-7 cells were grown in DMEM supplemented cell culture medium with 10% FBS, 1.5 g/l $NaHCO_3$, and 100 U penicillin/streptomycin solution at 37 °C in 5% CO_2. Cells were synchronized in G_0 by serum starvation (0.1% FBS) on confluent cultures for 48 h. Reentry into the cell cycle of G_0-arrested cells could be achieved by cultivation in 10% FBS DMEM [27].

Analysis of the cell cycle was performed by flow cytometry. In three independent experiments, COS-7 cells were cultured on 60 mm dishes. After a certain time required for protein synthesis and expression, 1.5×10^5 cells were harvested (in collecting co-transfected cells, cells expressing both of green and red fluorescence were gathered by the sorter), pelleted, and washed with phosphate-buffered saline three times. After being fixed in 1 ml of ice-cold 70% ethanol overnight at 4 °C, the cells were pelleted by centrifugation and incubated with RNAse (100 μg/ml) at 37 °C for 30 min. Propidium iodide (20 μl of 1 mg/ml) was added to each suspension, and the cells were passed through a fluorescence-activated cell sorter (FACS Calibur, BD Biosciences, USA); data were acquired using CellQuest (BD Biosciences, USA) software. Percentages of cells in different phases were calculated and analyzed directly by the software.

2.1.32 Calcineurin Enzymatic Activity Assay

Calcineurin (PP2B) enzymatic activity in different tissues including the digestive gland, mantle, gill, gonad, viscera, gill, adductor muscle, and gonad of the adult *P. fucata* oyster was measured using a nonradioactive serine/threonine phosphatase assay system (Promega, USA), which could determine the amount of free phosphate generated in a reaction by measuring and analyzing the absorbance of a molybdate–malachite green–phosphate complex. Tissue preparation was performed following the procedure recommended by the manufacturers.

Hemolymph withdrawn from the adductor muscle of six individuals with a disposable syringe was pooled to obtain 8 ml samples. The whole hemolymph was centrifuged at 200 g for 10 min, and the pellet was held on ice for further experiments. Other tissues sampled from six randomly selected adult oysters were pooled, respectively, and frozen in liquid nitrogen immediately after dissecting the animals. 0.5–1.0 g tissue from pool was powdered in liquid nitrogen and thawed on ice in three volumes of chilled phosphatase storage buffer (25 mM Tris–HCl, pH 7.5, 2 mM EDTA, 5 mg/L leupeptin, 10 mM β-mercaptoethanol, 1 mM benzamidine, 5 mg/L aprotinin, 1 mmol/L PMSF, 2 mg/L pepstatin). The mixtures were then homogenized for 30 s at 4 °C with a glass Teflon potter three times. The resulting homogenates were centrifuged for 1 h at 100,000 g to remove the insoluble particles. To reduce 90–96% of the endogenous phosphate from sample, 250 μl of the supernatants was added to the Sephadex® G-25 spin column provided by the manufacturer, which should be equilibrated with 10 ml of cold phosphatase storage buffer ahead of time. After centrifuging at 600 g for 5 min at 4 °C, the eluted samples were collected for further assay. Appropriate phosphate standards were set by diluting 1 mM phosphate standard with the supplied phosphate-free water to generate a standard curve for free phosphate.

In a 50 μL reaction system, the premix solution including 10 μL of PPase 5 × reaction buffer (250 mM imidazole, pH 7.2, 50 mM $MgCl_2$, 1 mM EGTA, 250 μg/mL CaM, 5 mM $NiCl_2$, 0.1% β-mercaptoethanol) and 5 μl of 1 mM phosphopeptide, RRA(pT)VA, was added into the provided 96-well plate, which was then placed at 37 °C for 3 min. 35 μl enzyme sample containing 3 μg protein was directly added into each well and incubated at 37 °C. Half an hour later, 50 μl molybdate dye/additive mixture was added to all wells to stop the reactions. The absorbance of the samples was measured using a plate reader with a 630 nm filter when color development is completed after another incubation at 37 °C for 30 min. The data were presented as a concentration (pM) of phosphate released per min per μg of tissue-specific protein.

2.1.33 Quantitative Measurement of Interleukin-2 (IL-2) in Serum by ELISA

The IL-2 assay was carried out with Human IL-2 ELISA Kit (Boster, China) following the manual instructions. Briefly, this assay employed and developed the quantitative sandwich enzyme immunoassay technique. A monoclonal antibody specific for human IL-2 was pre-coated onto a 96-well plate. Standards and samples diluted with sample solution (100 ml) were pipetted into the wells and incubated at 37 °C for 1.5 h. All standards and samples were run and repeated in triplicate. After washing with PBS buffer, biotin-labeled antihuman IL-2 antibody was added and incubated at 37 °C for 1 h. Then, 100 ml of prepared streptavidin solution was pipetted into the wells and incubated for another 30 min at 37 °C with gentle shaking. After thoroughly washing the wells, 90 ml TMB substrate solution was added to the wells. The color was changed from blue to yellow with the addition of 100 ml Stop Solution, and the intensity of the color was measured and analyzed at 450 nm immediately.

2.1.34 Quantitative Measurement of iNOS in Cultured Hemocytes

Protein extracts from lysates of cultured hemocytes were prepared, and iNOS enzyme activity was measured with the help of a commercially available iNOS detection kit (JIANCHEN, China). iNOS activity was detected in the hemocytes by the conversion of ʟ-arginine to ʟ-citrulline. The iNOS is Ca^{2+}-independent, and the activity of which was calculated from the difference between samples containing 1 mM EGTA and samples containing 1 mM EGTA with 1 mM NG-monomethyl-L-arginine (L-NMMA, a NOS inhibitor). The principle of this assay was based on the measurement of NO produced in the sample during a timed reaction. Then NO could produce a chromophore which could be further measured at 530 nm. One unit was defined as the amount of enzyme that would produce 1.0 nmol of NO per minute at 37 °C in 1.0 ml lysate of cultured hemocytes. Assays were repeated in triplicate.

2.1.35 NO Detection Assay

NO detection kit (JIANCHEN, China) was used for the measurement of NO accumulation in the culture medium system. At the indicated time points, the culture medium was collected and nitrite was measured. The assay was performed according to the manufacturer's instructions as follows: the culture medium from hemolymph was collected and centrifuged at 600 g for 10 min at 20 °C. The supernatants and standards (85 μl) were transferred onto a 96-well plate in triplicate and mixed with 5 ml (0.01 unit) of nitrate reductase and 10 μl of 2 mM NADH for 20 min at room temperature. Then 50 μl each of color reagents A [sulfanilamide (r-aminobenzenesulfonamide)] and B [N-(1-naphthyl) ethylenediamine dihydrochloride] were added to the above mixture. After shaking for 5 min, the absorbance of each well was measured and analyzed at 540 nm. The concentrations of samples were calculated referring to the standard curve.

2.1.36 Effect of Cyclosporin A (CsA) on Bacterial Clearance In Vivo

E. coli MG1655, a wild-type strain carrying type 1 fimbriae which was cleared effectively by circulating hemolymph in vivo [28], was applied in this study. After MG1655 was grown in LB culture medium at 37 °C for 12 h, bacterial cells were harvested by centrifugation at 5000 rpm for 10 min at 4 °C, then washed three times with cold PBS (137 mM NaCl, 2.7 mM KCl, 2 mM KH_2PO_4, 10 mM Na_2HPO_4, pH 7.4) buffer, and resuspended in PBS at an OD_{600} of 1 (1×10^8–2×10^8 CFU ml^{-1}). Bacterial clearance was studied with 50 ml of each bacterial suspension injection into the anterior adductor muscle of 15 individuals. Then the oysters were maintained within an aquarium at 20 °C. At 30, 90, and 150 min after the injection of bacterial suspension, 0.1 ml hemolymph samples were withdrawn from the anterior adductor muscles of five oysters separately. Each sample was placed in a tube containing 9.5 ml of sterile distilled water and thoroughly mixed to osmotically lyse the hemocytes. Tenfold serial dilutions in PBS of this lysate were placed onto LB agar. To detect the ability of bacterial clearance in the hemolymph when CN was inhibited, CsA was injected (20 mg/g soft tissue of pearl oyster) 6 h before the injection of the bacterial suspension. The hemolymph samples from CsA-treated group of 15 individuals were plated and incubated under the conditions mentioned above. The individual colonies grown on LB agar were counted after incubation at 37 °C for 10 hours. Three aliquots of hemolymph were taken from five oysters without any treatment as the basal number control of the endogenous bacteria.

2.1.37 *Southern Blot Analysis*

The Southern blot analysis was carried out to determine whether these isoforms originated from one or multiple genes. Total genomic DNA was extracted from the oyster adductor muscle using the universal genomic DNA extraction kit (Takara, Japan) according to the manufacturer's introductions. The quantity and content of DNA were determined by measuring A_{260nm} with a NanoDrop Lite spectrophotometer (Thermo Scientific, USA). Genomic DNA (20 μg) was digested with restriction endonuclease EcoRV and XhoI, respectively. After digestion for 2 h, the digested DNA was separated on a 0.7% agarose gel, then transferred to a nitrocellulose membrane, and hybridized with a DNA probe. Negative control was carried out without adding genomic DNA. The DNA probe labeling and hybridization were fulfilled by using DIG high prime DNA labeling and detection starter kit I (Roche, Sweden) in accordance with the manufacturer's protocol.

2.1.38 *Northern Blotting*

Total RNA (12 μg) from the mantle, digestive gland, muscle, gill, and gonad were transferred onto the S & S NYTRAN® SuPer-Charge nylon membranes (Schleicher & Schuell, USA). The membranes were cross-linked by UV cross-linker (CL-1000 Ultraviolet Crosslinker, USA). The *P. fucata* GNA probe was generated by random-primed labeling of a fragment containing the initially cloned 335 bp fragment, labeled with [α-^{32}P] dCTP using the Prime-a-Gene® Labeling System (Promega, USA) and purified by MicroSpink G-50 columns (Amersham Pharmacia, USA). Northern hybridization was performed at 68 °C for 2 h in ExpressHyb™ Hybridization Solution (Clontech, USA).

2.2 The Acquisition of Biomineralization Genes

2.2.1 *Identification of Genes Directly Involved in Shell Formation and Their Functions in Pearl Oyster,* **Pinctada fucata**

Identification and characterization of biomineralization-related genes during larval development could greatly enhance our understanding of shell formation. Genes involved in shell biomineralization were isolated via constructing three suppression subtractive hybridization (SSH) libraries representing genes expressed at key points during larval shell formation. A total of 2923 ESTs from these libraries were sequenced and gave us 990 unigenes. Unigenes coding for secreted proteins and proteins with tandem-arranged repeated units were screened and analyzed in the three SSH libraries. A set of sequences coding for genes involved in shell formation was attained. RT-PCR and in situ hybridization assays were performed on five genes to investigate their specific spatial expression in several tissues, especially the mantle tissue. They all showed a distinct expression pattern from known biomineralization-related genes. Specific inhibition of the five genes by RNA interference resulted in diverse defects of the nacreous layer, indicating that they all were involved in aragonite crystallization. Intriguingly, one gene (UD_Cluster94.seq.Singlet1) was restricted to the "aragonitic line." The current data has yielded for the first time, to our knowledge, a suite of biomineralization-related genes active during the different developmental stages of *P. fucata*, five of which were all responsible for nacreous layer formation. This provides an available starting point for isolating new genes involved in shell formation. The effects of genes on the formation of the "aragonitic line," and other areas of the nacreous layer, suggest a different control and regulation mechanism for aragonite crystallization initiation from that of mature aragonite growth.

2.2.1.1 SSH Library Composition and Characterization

Transcripts relevant to shell formation of *P. fucata* were isolated using an SSH procedure for the different development stages. That was the reason why three SSH libraries were constructed. The D-T library was constructed employing cDNA prepared from the trochophore stage (17-hour-old larvae) as the driver and that from early D-shaped stage (22-hour-old larvae) as the tester. The U-D library was obtained with cDNA prepared from the D-shaped stage (22-hour-old larvae) as the driver and that from umbonal stage (7-day-old larvae) as the tester, respectively. The J-U library was obtained by cDNA from the juvenile stage (35-day-old larvae) and the umbonal stage (7-day-old larvae) as tester and driver, respectively (Fig. 1.8a).

To evaluate the effectiveness of subtraction, we tested the reduction of GAPDH mRNA expression level in each library. These results indicated that GAPDH mRNA abundance appeared to decrease by a factor of at least 1:1024 in the subtracted library compared with the unsubtracted samples. Thus, we could conclude that we have successfully removed housekeeping genes in our three SSH libraries.

Differential cDNA fragments from the 3 libraries were screened and analyzed using cDNA dot blots of the 3515 clones (data not shown), i.e., we sequenced 990 positive sequences in the D-T library, 1180 in the U-D library, and 753 in the J-U library. Quality trimming and clustering were carried out, and these sequences produced 2376 contigs, which were assembled into 900 unigenes (see Table 2.1). In the D-T library, 484 sequences were singletons and library redundancy was 31%. In the U-D library, 177 sequences were singletons and library redundancy was 80%. In the J-U library, 70 sequences were singletons and library redundancy was 79%. The average cluster length was 396 bp in the D-T library, 376 bp in the U-D library, and 415 bp in the J-U library (Fig. 1.8b).

Table 2.1 Summary of the three libraries

	Tested clones	Positive clones	ESTs	Contigs	Unique clusters
D–T library	1227	1012	990	843	584
U–D library	1257	1196	1180	1130	227
J–U library	1031	767	753	403	89
Total	3515	2975	2923	2376	900

Gene Ontology of SSH Library Unigenes

Unigenes were annotated by BLASTX searches against the GenBank nr database to identify with an e-value >10e-05, with the result that 29% were found in the D-T library, 38% existed in the U-D library, and 47% in the J-U library. Blast2GO [2] and WEGO [3] were employed to annotate unigenes from the three libraries. Eighty-four (14%) D-T library unigenes, 52 (23%) U-D library unigenes, and 19 (21%) J-U library unigenes had hits in the gene ontology (GO) annotation and UniProt databases. What's more, we found that a large portion of subtracted genes were similar among the three libraries (data not shown). In the three libraries, the most highly represented classes for GO "molecular function" were binding, transporting, and catalytic activity. The majority of genes with GO "biological process" annotations were involved in cellular and physiological processes. The "cellular location" of the majority of genes was "cell."

Annotation of Unknown Unigenes

SSH procedures were carried out in the developmental stages, so the sequences found in libraries were related to metamorphosis, growth, and shell formation. We isolated genes according to their different functions from these libraries.

We mainly focused on genes involved in the biomineralization process of *P. fucata*; as a result, we retrieved all unigenes with shared similarity using the GenBank nr database. Fifteen unigenes in the D-T library were similar to genes involved in biomineralization, i.e., chitin synthase [29–31], chitin-binding protein [32–35], lectin [36–40], collagen [41–44], cysteine-rich protein [45], and fibrinogen [46–48]. Ten unigenes in U-T library returned a hit. They shared similarity with collagen, calmodulin, lectin, calcium-/calmodulin-dependent protein kinase kinase [49–53], BMP [54], ferritin-like protein [54–57], and matrix protein MSI7. In the J-U library, we found six unigenes by same approach. They shared high similarity with ferritin, lectin, matrix protein MSI60, and matrix protein MSI7. Much more interestingly, we found that DT_Cluster236 was similar to the nicotinic acetylcholine receptor in *Takifugu rubripes*. Furthermore, an ACC binding protein (ACCBP) was also found to possess a nicotinic acetylcholine receptor [58] in previous work. Given the similar functional domain of DT_Cluster236, it might also be involved in the control of the transition process of ACC to other stable crystallization forms (like calcite and aragonite) in shell formation. Therefore, we performed further in situ hybridization and RNAi analysis assay.

The mantle is the most crucial tissue that controls shell formation, and proteins expressed by mantle tissue would be more likely to be involved in shell biomineralization. Thus, we tried to screen and search some unigenes in the three SSH libraries against the transcriptome of the mantle tissue in *P. fucata* (data not published) by BLASTN annotation. 446 unigenes returned a hit with a cutoff e-value of 10e-06.

Secreted Proteins

Matrix proteins involved in the formation of the shell are mainly secreted proteins. We selected conceptually derived full-length gene products possessing a signal sequence, from the three substrate libraries. Based on the presence of conserved signal peptides and similarities (tandem-repeated units) to secreted proteins in GenBank (Fig. 2.1), we found that 52 (8.9%) unigenes encoding secreted proteins in the D-T library, 16 (7.0%) in the U-D library, and 7 (5.2%) in the J-U library.

Based on the BLASTX analysis with a cutoff e-value of 10e-05, we found that most of the unigenes coding for secreted proteins in three libraries, i.e., 88% in all, shared no similarities with sequences in the EST databases or GenBank nr. Some of them were similar to chitin-binding protein, sialic acid-binding lectin, and collagen- and glycine-rich shell matrix protein (MSI7), which were known to be involved in biomineralization. As a control, we also searched known matrix proteins in *P. fucata*, and 80% of them did encode secreted proteins.

Proteins with Tandem-Arranged Repeated Units

Another important characteristic of *P. fucata* biomineralization-related proteins is their modularity, which means that they are well organized in different functional domains. And many of the domains are composed of tandem-arranged repeated units. Thus, we retrieved and selected certain known matrix proteins in *P. fucata* as controls. XSTREAM [9] analysis showed that 74.6% of the matrix proteins contained tandem-arranged repeat units. Data from the three libraries were then searched, and we found 14 unigenes in the D-T library, 4 unigenes from the U-D library, and 6 unigenes in the D-T library, encoded for proteins with tandem-arranged repeated units (Fig. 2.2a–c).

Shematrin family was chosen as query sequences for control (Fig. 2.2d). XSTREAM suggested that all of them had tandem-arranged repeated units and most of them contained more than one. BLASTX search of the GenBank nr database was conducted to analyze the tandem-arranged repeated units of 24 unigenes in our 3 SSH libraries. Only one unigene (MU_Cluster65.seq.Singlet2) was proved to be a previously known *P. fucata* matrix protein MSI7, with 7 unigenes encoding for proteins similar to those found in other species, whereas the others showed no

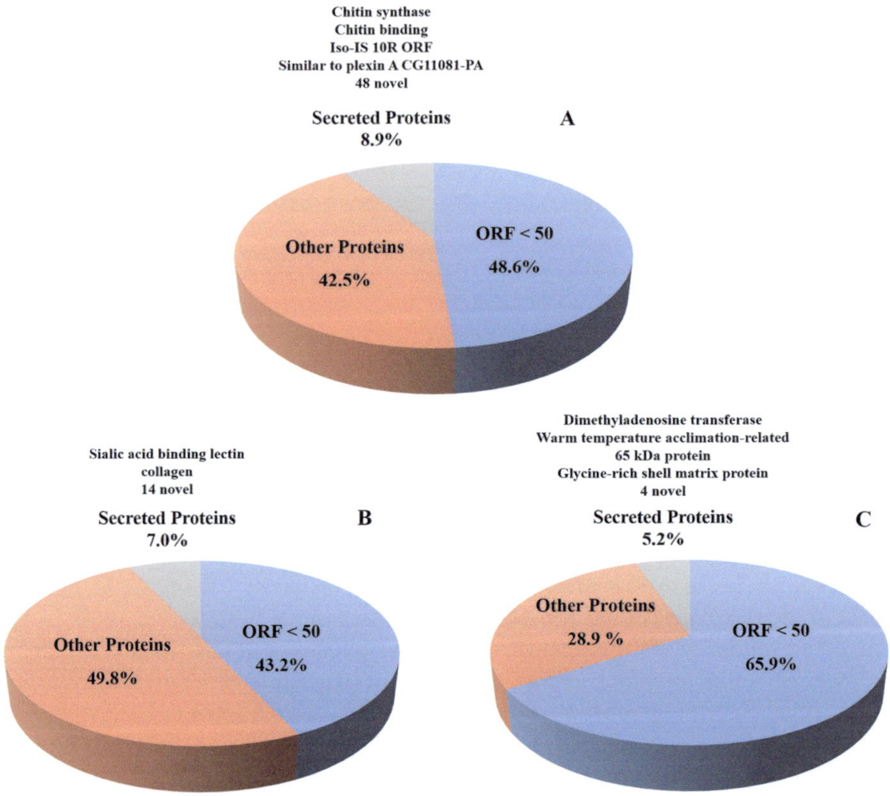

Fig. 2.1 Predicted secreted proteins in the three SSH libraries [59]
Unigenes with a coding region. 150 base pairs, processing a signal sequence, predicted to be secreted proteins by TargetP [7] on TMHMM Server v. 2.0 (http://www.cbs.dtu.dk/services/TMHMM/) and GPI modification predictor [6] together [8]. The unigenes coding for secreted matrix proteins were blasted and searched by BLASTX by the GenBank nr database with a cutoff e-value of 10e-05. (**a**) Unigenes from the D-T library. (**b**) Unigenes in the U-D library. (**c**) Unigenes from the J-U library

similarities to any known genes in any known species. Altogether, considering the analysis of secreted proteins found in the three libraries, we presumed that there were four secreted proteins with tandem-arranged repeated units, i.e., namely DT_Cluster252, DT_Cluster524.seq.Contig1, UD_Cluster94.seq.Singlet1, and IU_Cluster32.

Three SSH libraries were constructed to isolate unigenes expressing at different developmental stages during new shell formation. Changes of crystals and microstructures of the shell were relevant to biomineralization-related sequences in the D-T library involved in ACC formation, sequences in the U-D library correlated with aragonite formation, and sequences in the J-U library were involved in the formation of the microstructure of both nacreous and prismatic layers.

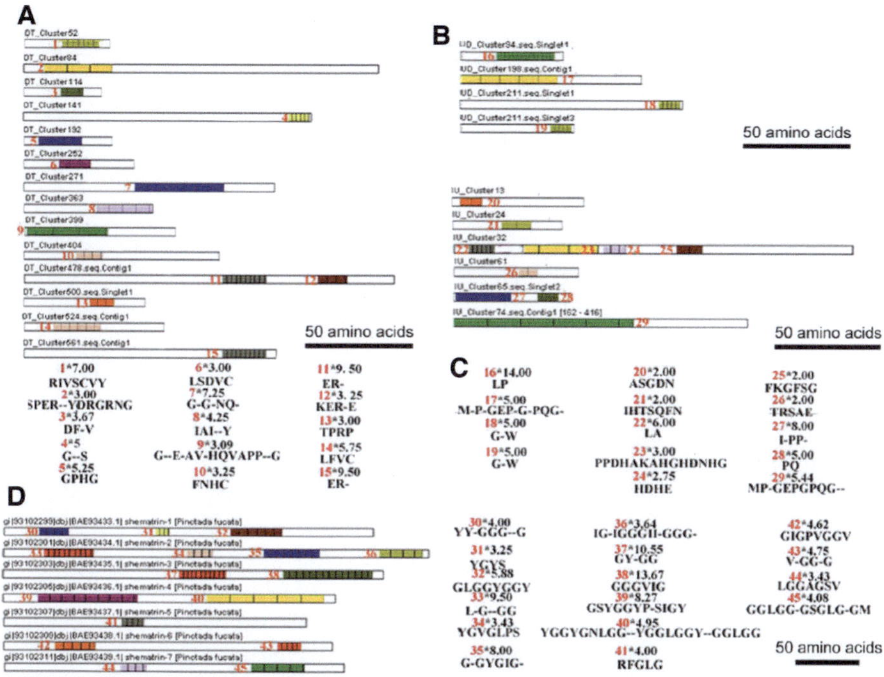

Fig. 2.2 Schematic of unigenes coding for proteins with tandem-arranged repeated units XSTREAM was conducted to isolate unigenes coding for proteins with tandem-arranged repeated units. Unigenes in the D-T library (**a**), U-D library (**b**), and J-U library (**c**), with a coding region longer than 150 base pairs, were consistent with this analysis. As a control, *shematrin* family genes found in *P. fucata* were analyzed by XSTREAM (**d**). The identities of tandem-arranged repeated units were marked by numbers on the schematic at the top left of each motif (red font). The copy number of repeated units was shown on the top right of the motif (black font). There was a label of identity*copy number on the top of each motif. Bars= 50 amino acids (in **a**, **b**, **c**, **d**)

Bioinformatics analysis indicated that only a small number of sequences in our three libraries were associated with GO terms. These results were similar to other non-model invertebrate EST data sets [60–63]. The main classes of annotated genes were the same in our three libraries and closely matched the mantle transcriptome analysis of the silver-lipped pearl oyster *Pinctada maxima* and the tropical abalone *Haliotis asinina*, where the public shared ESTs (removed genes encoding rRNAs, riboproteins, and mitochondrial proteins) were annotated with GO terms [64]. The cutoff e-value in our annotation pipeline was 10e-05 implying sequences with low similarities would be excluded in the GO hits. Sequences with low similarities and novel sequences were eliminated in the GO annotation, but the GO annotation showed that the main classes of sequences in the three libraries were putative genes expressed by the mantle which could control shell biomineralization directly.

After careful and thorough analysis of BLASTX results, we found that some sequences were similar to certain known biomineralization-related genes. In the D-T library, sequences were found with similarities to chitin-binding protein and chitin synthase. Chitin is an insoluble polysaccharide forming the highly structured framework of the shell. The chitinous material is already present in larval shell, presumably as a chitin–protein complex during the shell formation process of the marine bivalve mollusk *Mytilus galloprovincialis* [33]. Chitin synthase does play an important enzymatic role in the coordinated formation of larval bivalve shells [65]; therefore, a similar process may also exist during larval shell formation in *P. fucata*, where chitinous material was proved to be synthesized in the D-shaped stage.

ACC is considered to be the precursor of other stable crystalline aragonite and calcite, also an important component of the D-shaped larval shell. According to current studies, ACC binding protein (ACCBP) shares significant sequence similarities with a group of acetylcholine-binding proteins and controls the formation process of ACC in *P. fucata*. And our group has proved it could inhibit the growth of calcite and induce the formation of ACC. This capacity of ACCBP relies heavily on its acetylcholine-binding site. The sequence of DT_Cluster236 shared some similarities with nicotinic acetylcholine receptor in *Takifugu rubripes*, indicating that it had an acetylcholine-binding site which might participate in the transition of ACC to other crystallines in the D-shaped larval shell formation. As a result, we investigated its function in later experiments. What's more, we found some sequences in the U-D library similar to calcium metabolism-associated proteins (CaM, BMP, and CaM-KK). This could be explained by large amounts of calcium required during umbo shell formation. MSI60 participated in the nacreous layer formation during adult shell biomineralization and expressed in the J-U library. This result was consistent with previous work [66] that showed MSI60 expression was initiated in 31-day-old larvae. On the one hand, lectin, cysteine-rich protein, MSI7, fibrinogen, ferritin, and collagen were found to be involved in biomineralization. On the other hand, we found unigenes in our libraries did share some similarities with these biomineralization genes, suggesting us these unigenes might also participate in shell formation or reparation, and investigation to their structure–function relationships would greatly enhance our understanding of nacre formation and biomineralization. Further studies of these sequences are still ongoing.

Genes containing tandem-arranged repeated units and encoding for secreted proteins could be selected as potential matrix protein candidates, and this strategy was successfully used for screening genes involved in shell formation in *P. maxima* and *H. asinina* [64], and experimental results showed that these isolated genes were indeed critical to the formation of shell. In our libraries, the four unigenes coding for secreted proteins with tandem-arranged repeat units were crucial matrix protein candidates for direct participation during biomineralization. BLASTX searching found no significant similarities with any known sequences in the GenBank nr database. Since mantle tissue has been considered to directly control the formation and reparation of shell, we also performed in situ hybridization to find out their specific distribution in mantle tissue in order to understand their functions. It has been verified that the shell matrix protein genes have their regional expression patterns along the mantle, and the distributions of these genes suggested their functions in shell biomineralization. Together with the ACC control-related sequence DT_Cluster236 mentioned above, we analyzed the spatial distribution of the other selected genes in different target tissues, especially the mantle tissue. The results indicated that although they all expressed in the mantle but not typical nor distinctive, since some of them could also be detected in other atypical or biomineralization irrelevant tissues. DT_Cluster236, IU_Cluster32, and DT_Cluster252 were expressed not only by the mantle but also other tissues. This distribution pattern was different from known biomineralization-related genes. We suspected that proteins encoding by these genes may be secreted into the body fluids after synthesis in mantle tissue or undergo some fluid circulation so they could protect the oysters from abnormal crystallization in inappropriate locations, as ACCBP does. DT_Cluster236, IU_Cluster32, and DT_Cluster252 mainly expressed in both mantle pallial and mantle edge, whereas prismatic layer genes mostly expressed in the periostracal groove of the mantle, suggesting these genes were not involved in the prismatic layer formation. And this hypothesis was further supported by subsequent RNAi assay.

2.2.1.2 Spatial Distribution of Unigenes in Mantle Tissue

Bioinformatics analysis isolated a series of unigenes that might be involved in shell formation. However, if we didn't perform any functional test or assay, their roles during shell biomineralization would be doubtful and unclear. Consequently, we further analyzed DT_Cluster236, the one predicted participating in ACC transition, and the other four unigenes encoding secreted proteins with tandem-arranged repeated units at the same time. These five unigenes had a hit in BLASTN search against the transcriptome of mantle tissue (cutoff e-value 10e-06). We analyzed the expressions of these five unigenes using RT-PCR technique. The concrete expression patterns of the five genes were different, i.e., DT_Cluster32 and IU_Cluster236 could be detected in all eight tissues; DT_Cluster252 expressed in all tissues except the viscus, whereas UD_Cluster94.seq.Singlet1 and DT Cluster524.seq.Contig1 were only expressed in the mantle tissue, including mantle pallial and edge (Fig. 2.3a). We have already known that different shell types are formed by the ordered secretion of matrix proteins and other macromolecules along the mantle [67]; thus the analysis of the distribution of these genes by in situ hybridization via frozen sections of the mantle admitted of no delay. Both DT_Cluster524.seq.Contig1 and DT_Cluster236 were detected in all the mantle epithelial cells, except cells at the bottom of the periostracal groove (Fig. 2.3c, e). In situ expression of the DT_Cluster252 was detected locating to the mantle epithelial cells, with weaker signals in the outer epithelial cells of the middle fold at the periostracal groove and the inner epithelial cells of the outer fold (Fig. 2.3d). UD_Cluster94.seq.Singlet1 mainly expressed in the epithelial cells at the top of the outer fold, the inner epithelial cells of the middle fold, and the epithelial cells of the inner fold (Fig. 2.3f). As with DT_Cluster524.seq.Contig1 and DT_Cluster236, the expression of IU_Cluster32 was localized to most of the mantle epithelial cells. However, stronger signals appeared in epithelial cells at the bottom of the periostracal groove (Fig. 2.3g).

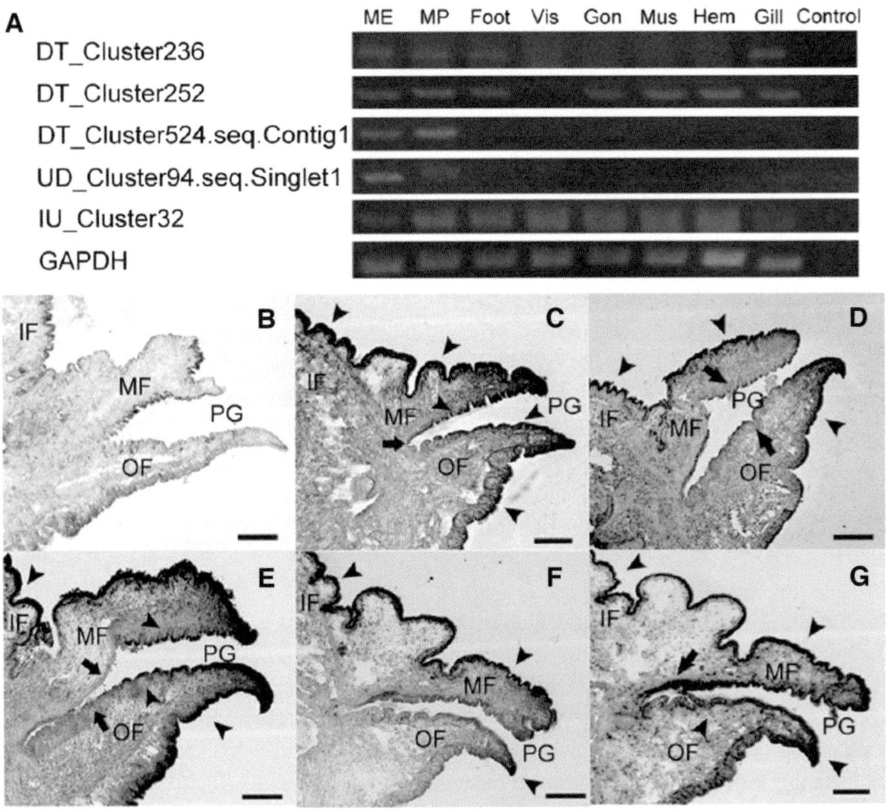

Fig. 2.3 Gene distribution patterns of selected genes in tissues of *P. fucata*
(**a**) Tissue-specific gene expression pattern of selected unigenes by RT-PCR analysis. Total RNA was extracted from mantle edge (ME), mantle pallial (MP), foot, viscus (Vis), gonad (Gon), adductor muscle (Mus), hemocytes (Hem), and gill. RT-PCR was performed without a template as a negative control (Control). The housekeeping gene GAPDH was used as a positive control. (**b**) A section of oyster mantle was hybridized with GFP antisense RNA probes as a negative control, and no hybridization signal (dark purple) was detected in this section. DT_Cluster236 (**c**) and DT_Cluster524.seq.Contig1 (**e**) were detected in all mantle epithelial cells (indicated by arrowheads), except cells at the bottom of the periostracal groove (indicated by arrows). (**d**) Expression of DT_Cluster252 was detected in mantle epithelial cells (indicated by arrowheads) with weaker signals in the outer epithelial cells of the middle fold at the periostracal groove and the inner epithelial cells of the outer fold (indicated by arrows). (**f**) Expression of UD_Cluster94.seq.Singlet1 was localized to the epithelial cells of the epithelial cells at the top of the outer fold, the inner epithelial cells of the middle fold, and the inner fold (indicated by arrowheads). (**g**) IU_Cluster32 was expressed nearly all the mantle epithelial cells (indicated by arrowheads), but stronger signals was detected in epithelial cells at the bottom of the periostracal groove (indicated by arrows). *OF* outer fold, *MF* middle fold, *IF* inner fold, *PG* periostracal groove. Scale bar, 0.5 mm

Hybridization with a control GFP antisense RNA probe didn't show any significant signals (Fig. 2.3b). Furthermore, it was notable that the expression profiles were entirely different to all known matrix proteins.

2.2.1.3 RNAi Knockdown of Unigenes

The analysis of the spatial distribution profiles of those unigenes in mantle tissue suggested that the five genes (DT_Cluster236, DT_Cluster252, DT_Cluster524.seq. Contig1, UD_Cluster94.seq.Singlet1, and IU_Cluster32) expressed by mantle tissue were directly engaged in shell formation. Then we investigated the functions of them in vivo by RNAi. Matrix protein KRMP, responsible for the formation of the outer calcitic prismatic layer, and nacrein, involved in the inner aragonitic layer, were selected as positive controls, and GFP, which was not expressed in *P. fucata*, was set as independent control. Double-stranded RNA (dsRNA) was specially designed based on the sequences of the eight genes and injected into the adductor muscle of *P. fucata* (2 years old, shell length = 5–6 cm). Furthermore, we performed real-time quantitative PCR (RT-qPCR) to measure the mRNA expression levels of these eight genes 6 days after dsRNA injection.

Apparently, compared with the GFP- and PBS dsRNA-injected groups, RT-PCR results showed that the gene expression levels of the 40 μg-injected groups were inhibited by approximately 50–60% and decreased by approximately 40% in the 80 μg-injected groups. We could see from Fig. 2.4 that the expression levels of the

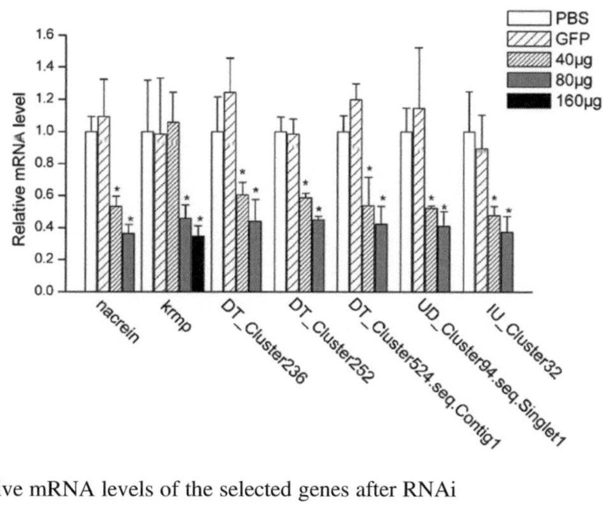

Fig. 2.4 Relative mRNA levels of the selected genes after RNAi
The expression levels of the selected genes were detected using real-time quantitative PCR and analyzed 6 days after injection. Five individuals were executed in every group. The relative mRNA level of PBS (water)-injected group was detected to a relative value of 1.0 as control. The asterisk represented a significant ($p < 0.005$) difference compared with the control groups. Because the relative mRNA expression level of KRMP couldn't be depressed in the 40 μg dosage-injected group, we must use 160 μg dsRNA in this group to ensure efficiency

eight genes did decrease with the increasing injection doses of dsRNA (from 40 μg to 80 μg), with the exception of the expression level of the 40 μg KRMP group remaining at 95% after 6 days, but the expression level of KRMP was reduced by about 35% with the addition of 160 μg dsRNA when compared with the negative control group (Fig. 2.4). Moreover, we detected and analyzed the expression of other six known biomineralization-related genes (Prismalin-14, aspein, and Prisilkin-39 participating in the formation of prismatic layer; Pearlin, Pif97, and N19 engaged in nacreous layer formation) as controls for the higher-dosage injection groups (160 μg KRMP dsRNA-injected group and 80 μg from the other five groups). The gene expression levels of these six genes did not increase or decrease significantly, confirming that our RNAi assay did not interrupt other general gene expressions in the mantle tissue (data not shown).

To further dig the role of these genes during shell biomineralization in depth, the inner surface structures of the shells were observed 6 days after injection, i.e., internal prismatic layer, nacreous layer, and the transition section "aragonitic line" also known as the boundary of the prismatic and nacreous layer. The surfaces of shells in the PBS- and GFP dsRNA-injected groups shared the same, normal, and well-defined microstructures compared with the non-treated groups. Small rectangular or hexagonal flat tablets of aragonites were packed together in nacreous layer to exhibit a stair-like growth pattern (Fig. 2.5a1, a2). The boundary "aragonitic line" consisted of calcitic prisms on one side and flat tablets of aragonites on the other side (Fig. 2.5l1, l2). Prisms were built up as a well-compacted smooth structure in the prismatic layer (Fig. 2.5o). Surfaces of the internal nacreous layers were changed dramatically to disordered forms in the five dsRNA-injected groups, i.e., nacrein, DT_Cluster252, DT_Cluster236, IU_Cluster32, and DT_Cluster524.seq.Contig1. The normal structures were seriously disturbed either by random accumulation of crystals (DT_Cluster236, DT_Cluster252, and DT_Cluster524.seq.Contig1, Fig. 2.5d1–i2) or the weird changed shapes of tablets (nacrein and IU_Cluster32, Fig. 2.5b, c, j1–k2). These abnormal phenotypes were much more obvious in the higher-dosage (80 μg)-injected groups compared with the 40 μg dsRNA-injected groups (Fig. 2.5b–k2). However, shell surfaces of the "aragonitic line" and prismatic layer were normal and almost like the non-treated control. These results indicated that the four newly identified genes were involved in the formation of the nacre and internal nacreous layer.

According to Fig. 2.5m1, m2, some cube-shaped tablets were observed at the nacreous side of the "aragonitic line," disrupting shell structures in the lower-dosage (40 μg) UD_Cluster94.seq.Singlet1 dsRNA-injected group (Fig. 2.5m1, m2). But the edges of the cube-shaped tablets were obscured by the changed shape of the tablets and some different-shaped, with the fusion of crystals causing the formation of some bigger tablets (Fig. 2.5n1, n2) in the high-dosage (80 μg)-injected group. Otherwise, the surfaces of the prismatic and nacreous layer were just like normal shells. These results implied that this gene could modify the shape of the tablets on the "aragonitic line" but had no effects on other sections of the shell. The shell surface in the 40 μg KRMP dsRNA-injected group remained normal as expected, which was consistent with the RT-PCR result that KRMP expression level hadn't been decreased significantly. In the 80 μg KRMP dsRNA-injected groups, calcitic prisms that formed the prismatic layer were abnormal. The surface of the calcitic prisms became lacunose, and their borders were broken with a little fusion (Fig. 2.5p). With the increasing amount of injection (160 μg group), similar changes appeared in calcites from prismatic layer (Fig. 2.5q). Structures of the "aragonitic line" and internal nacreous layer were normal in all KRMP dsRNA-injected groups compared with the negative control group. There were no significant disorganized structures in prismatic layers in the remaining groups. SEM pictures suggested that there was a strong possibility that the disturbance or disorder of growth in both layers could lead to the secretion of some regulatory proteins in mantle tissue to overcome inordinate growth in one shell layer. As a result, we injected 80 μg nacrein dsRNA and 80 μg KRMP dsRNA together to confirm our hypothesis. We found that the structures changed in both prismatic and nacreous layer, and the abnormal growth with disordered microstructures observed in each layer just matched the phenotypes in the separately injected groups.

To investigate the roles of these five genes in shell formation, dsRNAs were injected into the oyster to knock down their specific expression levels. The relative mRNA expression levels of these genes decreased to 50–60% in lower-dosage (40 μg or 80 μg only for KRMP)-injected groups and to 40% in higher-dosage (80 μg or 160 μg only for KRMP)-injected groups (Fig. 2.4). Together with the SEM images of the inner surface of the shells (Fig. 2.5), we found that with more dsRNA injected, the relative expression levels of these genes were decreased with concentration gradient effects and the phenotype of shell defects became increasingly noticeable. The injection of KRMP needed more dsRNA compared with the other genes because the KRMP family has more than one homolog in *P. fucata*. So the dsRNA was designed to affect all of the members. And the expression level we tested and analyzed by RT-PCR was that of the four known genes. Therefore it is not amazing that more dsRNA was required. The prismatic layer showed lots of defects on the border of calcite crystals after the knock down of KRMP genes. Furthermore, the calcitic prisms formed a loosely organized structure. As a result, we supposed that the abnormal shell structures might be caused by the poor connection between the framework and the prisms [45, 68].

Nacrein has been considered to play an important role as a main synthesizer and carrier of $CaCO_3$ in aragonite crystallization [69]. However, further studies implied that the specific involvement of nacrein in the nucleation of aragonite crystals was skeptical [70–72]. According to previous research, shell defects only appeared in the nacreous layer, and this situation was consistent with the earlier injection of nacrein antibody experiments [73]. The results of our nacrein RNAi experiments supported the former view, but not the latter—the hypothesis that nacrein was engaged in the formation of nacreous layer. On the one hand, all of the five gene candidates could influence the nacreous layer formation as expected. On the other hand, the specific abnormal phenotypes of nacreous layer were not exactly identical in the injected groups, which could be generally classified into two categories: random accumulation of aragonite crystals, as seen in the DT_Cluster524.seq.Contig1-, DT_Cluster252-, and DT_Cluster236-injected groups, and changes of crystal

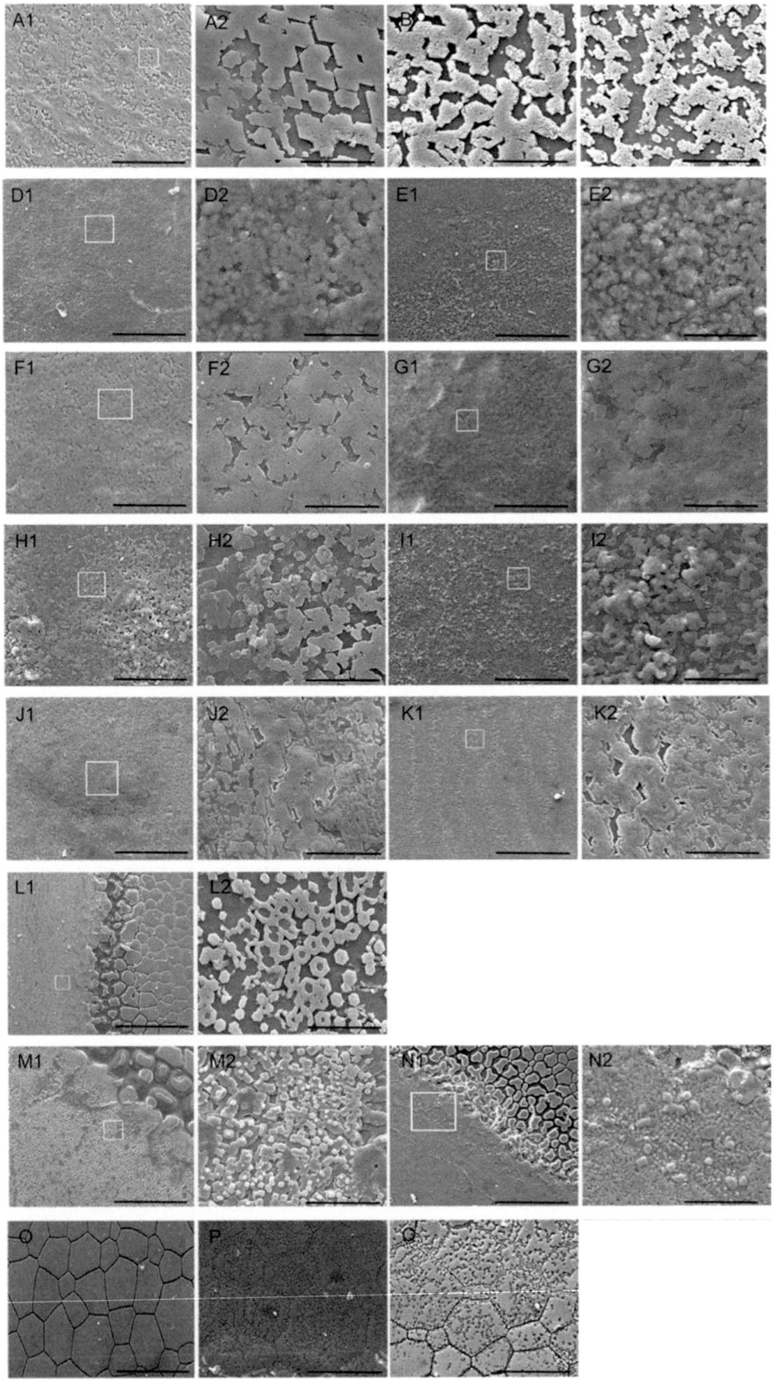

Fig. 2.5 Effects of selected genes on shell growth [59]
The mRNA relative expression levels of the selected genes were significantly decreased by RNAi. (**a**1) SEM image of the internal nacreous layer surface of the normal shell, with stair-like

tablets, as seen in the other two unigene dsRNA-injected groups. We've already known that there might be two different antagonistic mechanisms that could explain the control process of shell formation: growth inhibition and crystal nucleation [74]. Nacrein, Pif, ACCBP, and Prisilkin-39 have been well studied in the shell formation of *P. fucata*. Not surprisingly, by observing the shell defects induced by repression of those genes encoding shell matrix proteins and their functions in $CaCO_3$ crystallization [13, 58, 73, 75], we noted that repression of genes inducing the nucleation of aragonite crystals, i.e., Pif and nacrein, would lead to changes in aragonitic tablet shapes; however, repression of genes that could induce or facilitate aragonite crystal growth, like ACCBP and Prisilkin-39, would cause random accumulation of crystals. Based on these outcomes, we made a hypothesis that the changes of tablet shape were induced by the imperfection of crystal nucleation and the random accumulation was caused by crystal overgrowth.

The "aragonitic line" is the sharp division between calcite and aragonite, where the longitudinal growth of calcitic prisms is inhibited and new aragonitic crystals are emerged. It is very inspiring that the knockdown of UD_Cluster94.seq.Singlet1 influenced the formation of "aragonitic line." The mineralogy, growth modalities, size, and shape of biocrystals are completely changed. However, little is known about this boundary, because it is a very narrow section in the inner surface of the shell, which is very difficult to split from the shell and very hard to directly extract proteins from this place. Our U-D library was then come out to isolate genes engaged in the initiation of nacre formation, which also occurred at this "aragonitic line." Therefore it was not surprising that UD_Cluster94.seq.Singlet1 participated in "aragonitic line" formation. UD_Cluster94.seq.Singlet1 affected the inner surface of shell on the aragonitic side of the line, where the aragonitic tablets were barely formed and inhibited the progress of crystal growth. The tablets we reviewed were hexagonal and some even with a hollow structure. Nevertheless, the normal mature aragonitic tablets from the inner surface of the internal nacreous layer were much bigger and well organized with no hollow structure. In short, unigene UD_Cluster94.seq.Singlet1 couldn't affect the internal nacreous layer, but led to

Fig. 2.5 (continued) microstructures, and this was the normal growth pattern. (**b**–k2) SEM images of the internal nacreous layer surface of the shells in dsRNA-injected groups, showing that the growth of the tablets was disrupted. (**l1**) SEM image of the "aragonitic line" indicated the growth of nacreous tablets on the left side and calcites on the other right side. (**m1**–**n2**) SEM images of the "aragonitic line" in the UD_Cluster94.seq.Singlet1-injected group. (**o**) SEM image of the normal prismatic layer of the *P. fucata* shell. (**p**–**q**) SEM images of the KRMP-injected groups. Low-dosage (80 μg for KRMP, 40 μg for the other genes) injection of nacrein (**b**), DT_Cluster236 (**d1**), DT_Cluster252 (**f1**), DT_Cluster524.seq.Contig1 (**h1**), IU_Cluster32 (**j1**) UD_Cluster94.seq. Singlet1 (**m1**), and KRMP (**p**). High-dosage (160 mg for KRMP, 80 μg for the other genes) injection of nacrein (**c**), DT_Cluster236 (**e1**), DT_Cluster252 (**g1**), DT_Cluster524.seq.Contig1 (**i1**), IU_Cluster32 (**k1**), UD_Cluster94.seq.Singlet1 (**n1**), and KRMP (**o**) (**a2**, **d2**, **e2**, **f2**, **g2**, **h2**, **i2**, **j2**, **k2**, **l2**, **m2**, and **n2**) enlargement of the box in (**a1**, **d1**, **e1**, **f1**, **g1**, **h1**, **i1**, **j1**, **k1**, **l1**, **m1**, and **n1**). Bar= 10 mm in **a2**, **b**, **c**, **f2**, **g2**, **h2**, **j2**, **k2**, **l2**, **d2**, **e2**, **i2**, **m2**; bar = 50 mm in **a1**, **f1**, **g1**, **h1**, **j1**, **n2**, **p**, **q**; bar = 100 mm in **d1**, **e1**, **i1**, **k1**, **m1**, **o**, **l1**, **n1**

the changes of the nacreous side of the "aragonitic line." Thus, the controlling mechanism of aragonitic crystal initiation is different from mature aragonite crystal growth. The aberrant textures observed in this dsRNA-injected groups were very similar to those reviewed in some diseased specimens [76], suggesting that the pearl oyster may share the same defective mechanism under natural conditions. Further structure–function studies of this unigene will be continuing, and we truly hope that this will bring us closer and deeper understandings of biomineralization and mollusk shell formation. What's more, we would like to use morpholino oligos to specifically knock down the expression level of this unigene in larvae, which would definitely help us analyze the function of the certain type of genes during shell formation [59].

Altogether, there are two different methods of investigation to profile the mechanism of shell formation: one is to investigate its main components, such as the proteins directly involved in formation; the other one is to learn the formation pathways. This study mainly focused on the proteins participating in biomineralization, but the pathways are also very crucial in understanding the shell formation. And we will identify the concrete regulation pathways in our subsequent work.

2.2.2 Microarray: A Global Analysis of Biomineralization-Related Gene Expression Profiles During Larval Development in the Pearl Oyster, **Pinctada fucata**

Biomineralization-related genes have been proved to play very important roles regulating shell formation. But most previous work just focused on their functions in adult oysters, only small counts of studies investigated the biomineralization process during larval development when the shell initially starts to construct and finally forms until the juvenile stage in dissoconch shells. Here, for the first time, we intend to report a comprehensive and global gene analysis during larval development in *P. fucata* based on the microarray data and try to reveal the close relationship between biomineralization-related genes and shell formation. Based on the mantle transcriptome database of *P. fucata*, 58,940 probes (60 nt), representing 58,623 transcripts, were synthesized. The global gene expression profiles of the fertilized egg; trochophore, D-shaped, umbonal stage larvae; and juveniles were analyzed by microarray performance. The expression patterns of the biomineralization-related genes were changing in accordance with their regulatory function during shell formation. Matrix proteins chitin PFMG2 and synthase were highly expressed at the D-shaped stage, and yet PFMG6, PfN23, and PFMG8 were significantly upregulated from the umbonal stage, indicating their different roles in regulating the formation of periostracum, Prodissoconch I, or Prodissoconch II shells. However, the majority of matrix proteins were expressed at high levels at the juvenile stage when the shells have formed both a calcitic prismatic layer and an aragonitic nacreous layer as adults. Furthermore, we also identified and characterized five new genes which were significantly upregulated in juveniles. These genes were specifically expressed in the mantle tissue and coded for secreted proteins with tandem-arranged repeated units as most matrix proteins do. The RNAi knockdown experiments resulted in disrupted nacreous and prismatic shell layers, suggesting their potential crucial roles in shell formation. Our results provide a global perspective on larval gene expression patterns of *P. fucata* and propose a novel mechanism of how biomineralization-related genes regulate the formation of larval shell. These results could increase our knowledge on biomineralization-related genes and highlight some new aspects of shell formation mechanisms.

2.2.2.1 Global Gene Expression Analysis

The raw and normalized fluorescence microarray data have been deposited in the GEO database under accession number GSE63824. A principal component analysis (PCA) on the entire probe set divided all 15 sample pools into 5 different groups (Fig. 2.6), which were corresponded to the 5 larval developmental stages, respectively. The first and second components explained 70.93% of the variation in the entire data set. Stages of larval development were explained by the first principal components (PC1), explaining 48.99% of the variation, with low component loading for earlier stages and high component loading for later stages. A curved "horseshoe effect" along the second axis was obvious, which separated the D-shaped and trochophore stage larvae on the opposite side from fertilized eggs and juvenile

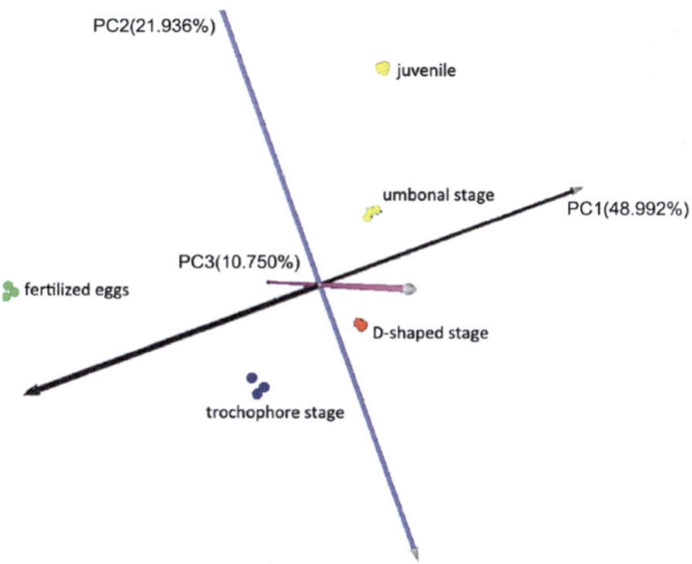

Fig. 2.6 3D score plot using the first 3PCs identified by principal component analysis on the entire larval gene probe set
Fifteen pools of samples were divided into five groups, associated with the five larval developmental stages. A triplicate was included at each stage point

stage. After a hierarchical clustering analysis, comparable results were obtained associated with similarities in the expression patterns (Fig. 1.10). These results demonstrated the different gene expression patterns in accordance with different developmental stages. Meanwhile, the three biological replicate pools at each developmental stage shared a similar pattern, increasing the reliabilities and repeat abilities of our work.

A. Transcriptional Changes Across Larval Stage Transitions

Comparative analysis was achieved by SAM software to clarify the changes in expression profiles between two contiguous developmental stages [77]. Among all 58,749 unigenes, the greatest amount of differentially expressed genes was identified between fertilized eggs and trochophore stage larvae, with 23,300 significant genes (11,021 downregulated and 12,279 upregulated). However, the stage when the fewest genes changed was between the D-shaped stage and trochophore stage, with 14,197 significant genes (6312 downregulated and 7885 upregulated). Overall, the expression profiles of most genes have changed dramatically during larval development (Fig. 2.7).

B. Comparison of Fertilized Eggs and Trochophore Stage Larvae

Seventeen different KEGG pathways (q-value <0.05) were significantly enriched between fertilized eggs and trochophore stage larvae. The majority was correlated to cellular progression (e.g., ko03030://DNA replication, ko04110://cell cycle, ko03420://nucleotide excision repair, ko03430://mismatch repair, ko03440://homologous recombination). The genes which changed dramatically between these two stages were related to GO: 0005634//nuclear and GO: 0005622//intracellular processes in a cellular component GO term analysis.

C. Comparison of Trochophore Stage and D-Shaped Stage Larvae

As many as 31 different KEGG pathways (q-value <0.05) were significantly enriched between the trochophore and D-shaped stages. Three pathways related to shell formation and biomineralization were enriched, such as ko04020://calcium signaling pathway, ko04310://Wnt signaling pathway, and ko00350://tyrosine metabolism. Chitin synthase [31] was greatly upregulated with fold change (FC) values of CUST_24765, 75.81-fold increase; CUST_58227, 84.49-fold increase; CUST_56685, 34.93-fold increase; and CUST_39082, 60.81-fold increase, suggesting their potential roles controlling shell formation, because chitinous materials are synthesized during the D-shaped stage and contribute a lot to the shell framework. Matrix protein PFMG2 [10] (CUST_40154, 5.25-fold increase) was also upregulated at this stage and of great possibilities of engagement in the formation of Prodissoconch I.

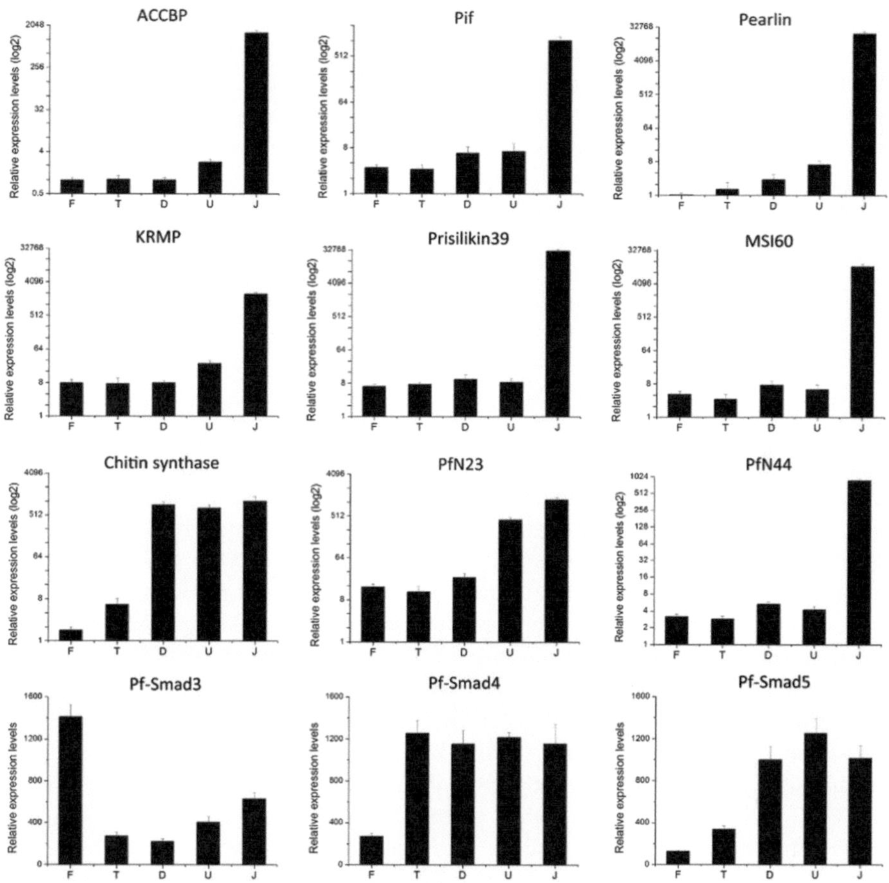

Fig. 2.7 Analysis on the particular expression of 11 genes by RT-qPCR
Relative gene expression levels (log2) of matrix proteins ACCBP, Pif, Pearlin, KRMP, Prisilkin-39, PfN23, chitin synthase, MSI60, and PfN44 were described in detail before, while relative gene expression levels (linear) of transcription factors Pf-smad3, Pf-smad4, and Pf-smad5 showed different patterns. All of these results were positively associated with the microarray data. *F* fertilized eggs, *T* trochophore stage, *D* D-shaped stage, *U* umbonal stage, *J* juvenile

D. Comparison of D-Shaped Stage and Umbonal Stage Larvae

Only ten KEGG pathways (q-value <0.05) were significantly enriched compared with D-shaped and umbonal stage larvae. Most of the terms were referred to metabolic processes (e.g., ko00480://glutathione metabolism ko00260://glycine, serine, and threonine metabolism; ko00982://drug metabolism cytochrome P450, ko00591://linoleic acid metabolism). One of the most interesting things was that the tyrosine metabolic pathway was also involved in these stages. PFMG6 and PFMG8 [10] were upregulated at the beginning of the D-shaped stage and then upregulated to a greater degree at the umbonal stage.

E. Comparison of Umbonal Stage Larvae and Juveniles

A total of 20 KEGG pathways (q-value <0.05) were remarkably enriched between umbonal stage larvae and juveniles, and proteins engaged in tyrosine metabolic pathway increased the most. A cellular component GO analysis identified 16 terms for genes with most significantly changed. The terms "extracellular matrix" and "extracellular region" were referred to matrix proteins. As we have expected before, most known matrix proteins were upregulated to a greater level at the juvenile stage, indicating their particular and important roles in regulating the formation of either the outer calcite prismatic layer or the inner aragonite nacreous layer.

F. Expression of the Tyrosine Metabolic Pathway from the Trochophore Stage to Juveniles

Of all of the KEGG pathways predicted to participate in the comparison between two consecutive developmental stages, the tyrosine metabolic pathway (ko00350) was persistently active from the trochophore stage to juveniles with a q-value <0.05. Among all 202 contigs from the transcriptome engaged in this pathway, 70 were up- or downregulated from the trochophore stage to the D-shaped stage, 79 were up- or downregulated from the D-shaped stage to the umbonal stage, and 96 were up- or downregulated from the umbonal stage to juveniles. Twenty-six genes were engaged in all three comparisons (Table 2.2). The most significantly expressed gene was tyrosinase (CUST_54654), 716.82-fold upregulated from the umbonal stage to juveniles, as well as two other tyrosinase-like proteins (tyrosinase-like protein 1, CUST_11152, and tyrosinase-like protein 2, CUST_41072) (Table 2.3). A previous research reported that tyrosinase was located in the prismatic layer of the shell [78] and expressed specifically in the outer epithelial cells of the middle fold in mantle tissue, where periostracum formation occurred [79]. In addition to tyrosinase, melanin and the final metabolic product were also considered to play certain roles during cuticle sclerotization in insects. Altogether, these results implied the multiple and important functions of the tyrosine metabolic pathway during *P. fucata* larval development, shell and periostracum formation also included.

G. Expression of Biomineralization-Related Genes During Larval Development

Although matrix proteins are crucial for the regulation of the biomineralization processes, such as pearl and shell formation in mollusks, the number of identified and characterized matrix protein genes is still limited [59]. Normal GO or KEGG term analysis cannot perfectly match all the matrix proteins. Therefore, we identified all genes engaged in biomineralization in the oyster transcriptome, including the majority of known *P. fucata* matrix proteins like PFMG family, tyrosinase family, shematrin family, and other proteins mentioned above, and analyzed their expression

Table 2.2 Consecutively up- or downregulated unigenes of the tyrosine metabolic pathway between the trochophore larval stage and juveniles

Probe name	Unigene name	Homologous gene	Species
CUST_54694	Unigene54694	Tyrosinase	*Pinctada fucata*
CUST_58138	Unigene58178	astacin-like protein	*Pinctada fucata*
CUST_6716	Unigene6721	Gastric triacylglycerol lipase	*Pinctada fucata*
CUST_51407	Unigene51446	Caffeoyl-CoA O-methyltransferase	*Pinctada fucata*
CUST_11562	Unigene11570	cytosolic glutathione S-transferase 1	*Oesophagostomum dentatum*
CUST_18671	Unigene18681	mu class glutathione S-transferase	*Crassostrea gigas*
CUST_25915	Unigene25932	glutathione S-transferase sigma class protein	*Crassostrea gigas*
CUST_28047	Unigene28066	glutathione S-transferase class mu	*Cyphoma gibbosum*
CUST_39791	Unigene39819	glutathione S-transferase	*Chironomus tentans*
CUST_54455	Unigene54495	glutathione S-transferase	*Branchiostoma belcheri tsingtauense*
CUST_56433	Unigene56473	microsomal glutathione S-transferase	*Venerupis philippinarum*
CUST_53080	Unigene53119	glutathione S-transferase pi	*Mytilus edulis*
CUST_53245	Unigene53284	microsomal glutathione S-transferase 1, isoform CRA_c	*Homo sapiens*
CUST_29733	Unigene29754	Temptin	*Haliotis discus discus*
CUST_12898	Unigene12906	Temptin	*Haliotis discus discus*
CUST_22403	Unigene22418	Temptin	*Haliotis discus discus*
CUST_18929	Unigene18939	amiloride binding protein 1	*Danio rerio*
CUST_23905	Unigene23920	Semicarbazide-sensitive amine oxidase	*Equus caballus*
CUST_27372	Unigene27390	C. briggsae CBR-GST-11 protein	*Caenorhabditis briggsae*
CUST_36101	Unigene36123	ShTK domain containing protein	*Brugia malayi*
CUST_44222	Unigene44255	amiloride binding protein 1	*Bos Taurus*
CUST_47036	Unigene47071	ovo peroxidase	*Lytechinus variegatus*
CUST_56279	Unigene56319	oxidase/peroxidase	*Aedes aegypti*
CUST_54925	Unigene54965	xanthine dehydrogenase/oxidase	*Tribolium castaneum*
CUST_54967	Unigene55007	melanogenic peroxidase	*Sepia officinalis*
CUST_56212	Unigene56252	Aspartate aminotransferase	*Bacillus stearorhermophilus*

profiles between two consecutive developmental stages [10, 24, 30, 58, 78–81]. Moreover, the expression levels of other biomineralization-related proteins like molluskan prismatic and nacreous layer 88 kDa protein [81], calmodulin-like protein [24], ferritin-like protein [82], and N151 [83] were also detected in Table 2.3. It was very interesting to find that most of the identified and characterized *P. fucata* matrix proteins were significantly upregulated between the umbonal stage and juveniles, such as the shell matrix protein shematrin-4 (CUST_24804) at 2977.7 FC. And most of these genes were expressed at low levels in fertilized eggs to the

Table 2.3 Expression profiles of biomineralization-related genes between two consecutive developmental stages identified in *P. fucata*

Probe name	Unigene name	Gene name	T/O FC	D/T FC	U/ DFC	J/U FC
CUST_24804	Unigene24820	Shematrin-4	–	–	–	2977.7↑
CUST_58195	Unigene58235	Pif	–	–	–	2116.9↑
CUST_38575	Unigene38603	KRMP 1	–	–	–	949.73↑
CUST_31	Unigene31	Shematrin-5	–	–	–	943.32↑
CUST_27219	Unigene27237	Pearlin	–	–	1.90	823.02↑
CUST_54654	Unigene54694	Tyrosinase		0.40↓	0.44↓	716.82↑
CUST_53622	Unigene53661	Prismalin-14	0.95	0.89	0.93	324.48↑
CUST_50742	Unigene50781	Shematrin-6	–	–	–	256.39↑
CUST_41072	Unigene41100	Tyrosinase-like protein 2	0.99	0.95	0.43↓	196.87↑
CUST_53862	Unigene53901	PFMG3	6.48↑	0.09↓	1.30	187.5↑
CUST_44067	Unigene44100	shell matrix protein	1.06	1.21	0.79	176.49↑
CUST_56052	Unigene56092	ACCBP 1	–	–	–	119.53↑
CUST_36021	Unigene36043	Prisilkin-39	0.77	1.21	0.84	119.48↑
CUST_31135	Unigene31156	MSI60-related protein	–	–	–	102.96↑
CUST_27662	Unigene27680	PFMG9	–	–	–	76.88↑
CUST_11152	Unigene11160	Tyrosinase-like protein 1	–	–	3.48↑	52.10↑
CUST_55983	Unigene56023	PFMG4	–	2.24↑	1.99	46.13↑
CUST_47372	Unigene47407	Shematrin-7	3.85↑	0.42↓	0.58	42.42↑
CUST_9246	Unigene9253	Calmodulin-like protein			0.79	40.16↑
CUST_57434	Unigene57474	Nacrein	1.54	1.24	0.63	8.38↑
CUST_367	Unigene367	molluscan prismatic and nacreous layer 88 kDa protein	–	–	–	5.06↑
CUST_50058	Unigene50096	PFMG6	–	2.03↑	5.82↑	4.53↑
CUST_1708	Unigene1709	PFMG11	–	–	–	2.94↑
CUST_45587	Unigene45621	PFMG8	0.40↓	2.1↑	13.42↑	2.66↑
CUST_24765	Unigene24781	Chitin synthase	3.39↑	75.81↑	0.83	1.34
CUST_45506	Unigene45540	PFMG1/7	0.85	0.84	0.92	1.63
CUST_40154	Unigene40182	PFMG2	0.41↓	5.25↑	1.15	1.56
CUST_53588	Unigene53627	N151	0.49↓	0.71	0.85	0.74
CUST_52216	Unigene52255	PFMG12	0.83	0.48↓	3.33↑	1.81
CUST_56226	Unigene56266	Ferritin-like protein	0.87	0.99	1.59	0.51

T/O FC fold-change (FC) in gene expression comparing trochophore stage larvae with fertilized eggs by Significance Analysis of Microarrays analysis; D/T ratio, U/D ratio, J/U ratio are similar ↑, up-regulated with FC$\geq$ 2, and ↓, down-regulated with FC $\leq$05

umbonal stage but at high levels in juveniles, with the exception of PFMG1/7 (CUST_45506), ferritin-like protein (CUST_56226), and PFMG12 (CUST_52216) [10, 80]. We also detected and analyzed several genes homologous to those associated with biomineralization in other molluskan species. Biomineralization-related genes, such as Clp1 protein (CUST_56502, *Crassostrea gigas*), Clp3 protein (CUST_57568, *C. gigas*) [84], linkine (CUST_53921, *P. margaritifera*) [85], and KRMP-7 (CUST_47674, *Pinctada margaritifera*), were upregulated in a similar mode between the umbonal stage and juvenile stage (data not shown).

PFMG2 (CUST_40154) and chitin synthase (CUST_58227) were greatly expressed during the D-shaped stage. PFMG6 and PFMG8 were initially upregulated at the D-shaped stage but then significantly upregulated at the umbonal stage, indicating their potential roles in regulating the formation of Prodissoconch II.

2.2.2.2 Gene Expression Profiles and Shell Formation During Larval Development

The $CaCO_3$ crystal polymorphism and the shell layer structure are changing throughout the whole developmental stages [86]. Prodissoconch I is believed to be comprised of ACC at the early D-shaped stage, but Prodissoconch II, composed of both calcite and aragonite, just appears during the umbonal stage. Finally, the dissoconch shell, with an outer calcitic prismatic layer and an inner aragonitic nacreous layer, forms at the juvenile stage [86]. Anyway, changes in matrix protein expression patterns reflect their roles regulating shell formation to a great extent.

Chitin is an insoluble polysaccharide comprising the body structure framework of insects and mollusks [87]. Chitinous materials appear in the D-shaped stage and widely distributed in larval shells of the oyster *Mytilus galloprovincialis*, where it is predicted to form a chitin–protein complex [33]. Chitin synthase could stimulate mammalian chondrocyte proliferation via regulating the synthesis of extracellular matrix components [84], which is also a critical matrix protein responsible for the formation of structural framework in shells [30, 31, 88]. In our study, the expression of chitin synthase was significantly upregulated at the D-shaped stage and remained high expression level since then (Table 2.3), which might be relevant to the synthesis of chitinous materials and other biological processes in larval shell. Results of the RT-qPCR analysis showed a similar expression mode (Fig. 2.7). In addition, the calcium signaling pathway was also predicted to be engaged in this stage, due to the large calcium requirement to form Prodissoconch I. Furthermore, matrix protein PFMG2 was significantly upregulated at this stage, somehow suggesting its potential role in the formation of Prodissoconch I, which still needs more investigations.

Most of biomineralization-related genes were expressed at relative low levels in umbonal stage, except for a few genes, such as PFMG6 and PFMG8. They were originally upregulated since the D-shaped stage and then upregulated to a greater extent from the umbonal stage. Based on previous research, the knockdown of PfN23 would interrupt the larval shell formation process since the early stage [59, 89]. The microscopic structures of Prodissoconch I and Prodissoconch II vary from those of dissoconch shells [66], and expression of PfN23, PFMG6, and

PFMG8 may significantly affect early formation of larval shells. Furthermore, a KEGG pathway analysis predicted that some amino acid metabolic pathways were very active between the D-shaped and umbonal stage reflecting the rapid larval growth at this time.

Microarray results showed that the majority of biomineralization-related genes were significantly upregulated at the juvenile stage with large FCs. These genes play critical roles regulating the formation of both prismatic and nacreous layers (Table 2.3). In addition to the contigs listed above, other contigs from the same biomineralization-related genes in transcriptome showed similar expression patterns across stages, in spite of some differences with fold change values. What's more, gene expression profiles of chitin synthase, ACCBP, Pearlin, KRMP1, Prisilkin-39, Pif, PfN23, MSI60, PfN44, and transcription factors, such as Pf-smad3, Pf-smad4, and Pf-smad5 (GenBank accession nos. EU137731, KF307635, and KC462554), were also analyzed by RT-qPCR (Fig. 2.7). This result confirmed the positive correlation with microarray performance. Our data indicated that the highest expression level changes in most biomineralization-related genes were in accordance with the formation of the prismatic and nacreous layers instead of the construction of Prodissoconch I and II.

Furthermore, the tyrosine metabolic pathway truly has multiple roles in a wide arrange of biological processes, including innate immunity, pigmentation, sclerotization, and wound healing [90–93]. Above all, the final product is melanin instead of tyrosinase, or other products from this metabolic pathway play crucial roles during cuticle sclerotization in insects [93, 94]. Shell is the last barrier for mollusks to external environment, and it possesses a similar function to that of the sclerotized cuticle in insects [10]. Tyrosinase, especially expressed in outer epithelial cells of the middle fold of the mantle, is responsible for pigmentation, construction, and the periostracum shell covering [78, 79]. In our research, the tyrosine metabolic pathway was continuously high active from the D-shaped stage larvae to juveniles, until the shell was initially constructed and gradually transformed into a typical bilayer structure. Among them, tyrosinase (CUST_54654, 716.82-fold increase), tyrosinase-like protein 2, (CUST_41072, 196.87-fold increase), and tyrosinase-like protein 1 (CUST_11152, 52.10-fold increase) were all upregulated at first time in the larval umbonal through the juvenile stage (Table 2.3) and sharing a similar role as the majority of other known matrix proteins. Two homologous oxidase genes (CUST_56279 and CUST_47036) shared a similar and interesting expression pattern: both of them were dramatically upregulated at the D-shaped stage (46.35- and 44.82-fold increases) and downregulated at the umbonal stage (204.60- and 223.99-fold decreases) and further upregulated again (112.70- and 219.42-fold increases) from the juvenile stage. These results indicated that multiple functions of this tyrosine metabolic pathway during larval development in *P. fucata*, including the formation of the periostracum in shell and other complex biological processes. These results accurately reflected the relationship between the formation of prodissoconch or dissoconch shells and biomineralization-related genes. As a

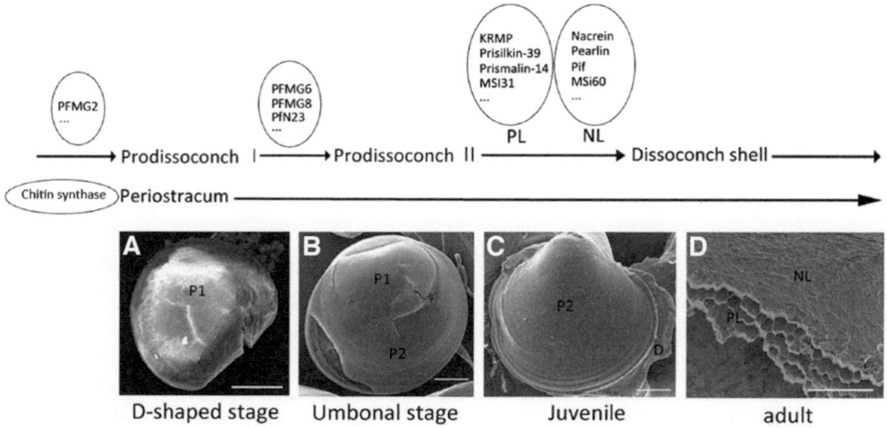

Fig. 2.8 Gene expression patterns on how biomineralization-related genes regulating the formation of shell in *P. fucata* [95]
Bar, 20 μm in *a*, *b*, *c*; bar = 200 μm in *d*. *P1* Prodissoconch I shell, *P2* Prodissoconch II shell, *D* Dissoconch shell, *PL* prismatic layer, *NL* nacreous layer

result, elaborating the working mechanism and significance of these genes regulating either the framework or the $CaCO_3$ crystals during shell formation is of great urgency. Based on the global expression profiles and potential functions we predicted, we propose a model on how different biomineralization-related genes regulate the larval shell formation process (Fig. 2.8).

2.2.2.3 Identifying the Candidate Genes Involved in Shell Formation

Most known proteins relevant to shell formation were upregulated with at least 20-fold changes from the umbonal stage to juveniles. But the functions of biomineralization-related genes with similar expression patterns were still unclear. We speculated that some of them may get involved in pearl or shell formation process, which deserved to be further studied. As a result, we investigated those upregulated genes using some bioinformatics tools and screening criterions to select some potential matrix proteins. Among all 8786 upregulated genes between these 2 stages, the expression of 1113 unigenes increased with a FC $\geq$ 20. Besides the known matrix proteins and biomineralization-related genes as mentioned above, 747 of these dramatically upregulated genes were not similar to any other known proteins after annotation, yet another 84 of them shared few similarities with uncharacterized or hypothetical proteins without defined functions.

As we have stated previously, 80% of the known matrix proteins encode secreted proteins with signal peptides, and 74.6% of them possess tandem-arranged repeated units [59], indicating some localized and modular features in accordance with their functions. These 831 genes were thoroughly analyzed and separated into 3 different classes. The remaining 138 genes were further analyzed by a SignalP search after excluding the genes without an open reading frame (ORF) containing 80 amino acid, and 35 secreted proteins with signal peptides were remained. A XSTREAM screening analysis was then performed, and five genes encoding secreted proteins also containing tandem-arranged repeated units were selected as potential matrix protein candidate. The FCs in these genes were unigene35118_TP, 206.96-fold increase; unigene56675_TP, 152.68-fold increase; unigene18749_TP, 100.016-fold increase; unigene51738_TP, 21.94-fold increase; and unigene34354_TP, 22.37-fold increase. After a BLASTp search, we found that unigene56675_TP shared 45% identities with nacre uncharacterized shell protein 5 (NUSP5, GenBank accession no. P86967) in *P. margaritifera* [96], but the other four genes were not similar to any defined genes or known proteins. Since the functions of NUSP5 are still unclear, the roles of all these five candidates need further analysis and investigations.

Identifying and characterizing secreted proteins containing tandem-arranged repeated units has become recognized as a useful strategy and criterion for screening genes engaged in shell formation of *Haliotis asinina*, *P. fucata*, and *Pinctada maxima* [59, 64]; PfN23 and PfN44 have been identified to be involved in shell formation by these methods [89, 97]. In this research, we want to investigate the genes which contain tandem-arranged repeated units and encode secreted proteins from the overregulated gene pool (FC $\geq$ 20) between the umbonal and juvenile stage, when the majority of biomineralization-related genes were significantly upregulated and the dissoconch shell has been formed. Finally, we identified five new potential matrix protein candidates with great possibility to participate in biomineralization. Based on the tissue distribution analysis and RNAi knockdown experiment, these genes showed potential biomineralization-related characters not only in their gene expression patterns but also in functions.

2.2.2.4 Expression Patterns of the Candidate Genes in Different Tissues

The relative mRNA expression level of the five candidate genes probably involved in shell formation was analyzed in different tissues by semiquantitative PCR. As we expected before, each gene exhibited tissue-specific distribution pattern (Fig. 2.9). Unigene56675 was expressed specifically in the mantle pallial where matrix proteins responsible for the formation of nacreous layer were secreted [13, 74, 98]. Unigene51738 and unigene34354 were specifically expressed in the prismatic layer-related tissue of mantle edge [74, 75, 99]. Furthermore, unigene18749 was detected in both the mantle edge and pallial, indicating its dual roles involved in both prismatic and nacreous layers [97, 100]. On the one hand, unigene35118 exhibited the highest expression in gill but also showed a relative high expression in the mantle edge based on RT-qPCR analysis (data not shown), probably due to the chitin-binding domain predicted in its coding protein. On the other hand, chitin is a critical organic molecule relevant to biomineralization [101], especially shell formation in

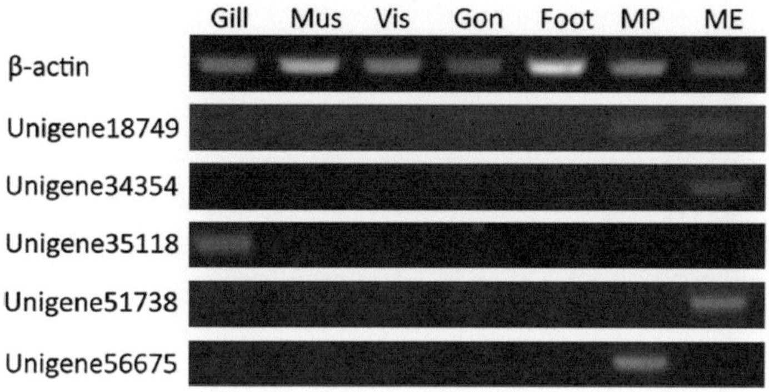

Fig. 2.9 Tissue distribution patterns of the five potential matrix protein candidates in *Pinctada fucata* by semiquantitative PCR analysis
Total RNA in adult oyster gill, adductor muscle (Mus), viscera (Vis), gonad (Gon), foot, mantle pallial (MP), and mantle edge (ME) was extracted and reverse-transcribed as templates separately. The expression level of gene *β-actin* gene was used as a positive control

mollusks [31, 33, 87, 88], and there are many matrix proteins participating in shell framework formation, such as Prisilkin-39 [75], the chitin-binding ability of which has been proved strong. Therefore, the role of unigene35118 still needs further investigation.

2.2.2.5 RNAi Knockdown of the Candidate Genes

The RNAi knockdown assay has been widely and maturely used to investigate the regulatory roles of matrix proteins in $CaCO_3$ crystallization during shell formation [32]. The microstructures of the prismatic and nacreous layer could be disrupted due to the downregulated matrix proteins [13]. That's how the functions of candidate genes were investigated in RNAi experiments in vivo. An 80 μg aliquot of double-stranded RNA (dsRNA) was specifically designed according to each gene and injected into the adductor muscle of adult *P. fucata* with similar size. The efficiency of RNAi was tested 6 days later by RT-qPCR, leading to a 40–60% decrease in expression level of the corresponding genes, with no significant effects on other matrix proteins like pif, nacrein, and KRMP (Fig. 2.10).

The inner surface of the shells was scanned and observed by SEM 6 days after injection, and both the prismatic and nacreous layers were observed separately (Fig. 2.11). The prismatic and nacreous layers of the shells in GFP-injected group were normal, same as negative control PBS-injected group, as seen in untreated oysters, (Fig. 2.11a, e). The knockdown of unigene51738 and unigene34354 would result in disrupted phenotypes in prismatic layer. And the calcite in prisms became lacunose after the downregulation of unigene51738 (Fig. 2.11d), while the downregulation of unigene34354 resulted in an improper overgrowth of organic

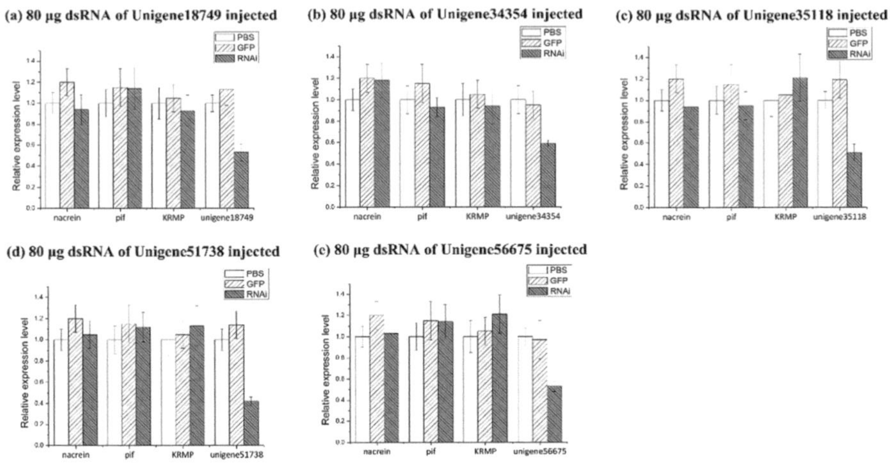

Fig. 2.10 mRNA relative expression levels of the selected candidates and matrix proteins were knocked down by RNAi

framework in prismatic layer (Fig. 2.11b). The inhibition of the chitin-binding domain contained in unigene35118 induced the absence of shell framework and lacunose prismatic surface (Fig. 2.11c). These results revealed their potential functions on controlling the prismatic layer formation.

On the other hand, the decreased expression of unigene56675 and unigene18749 would induce the abnormal and different $CaCO_3$ depositions in nacreous layer (Fig. 2.11f, g) at the same time, indicating their potential roles in controlling the nacreous layer formation. All of these results revealed the strong relationship between these potential matrix protein candidates and shell formation, but their primary structures and functions during the regulation process involved in biomineralization still need further investigation and analysis.

Unigene51738 and unigene34354 were specifically expressed in mantle edge which was considered relevant to the formation of shell prismatic layer [45, 74, 75]. The downregulation of these two genes caused the defects in shell prismatic layer. SEM images showed that the calcitic surface was tending lacunose, whereas the organic framework was still clear and visible after inhibiting unigene51738 (Fig. 2.11d). But the downregulation of unigene34354 induced an abnormal inner surface of the prismatic layer, which was probably due to the activating overgrowth of the organic framework or the inhibition deposition of calcites (Fig. 2.11b). Unigene56675 was found expressing mainly in mantle pallial responsible for regulating the formation of shell nacreous layer [13, 74, 98]. The inhibition of this unigene56675 led to disrupted crystal deposition in nacreous layer (Fig. 2.11g). Unigene18749 was detected in both the mantle edge and pallial of the oyster, suggesting its dual roles in both the prismatic and nacreous layer formation. The injection of 80 μg dsRNA for RNAi resulted in disrupted morphologies in nacreous layer (Fig. 2.11f), with no significant effects in prismatic layer (data not shown).

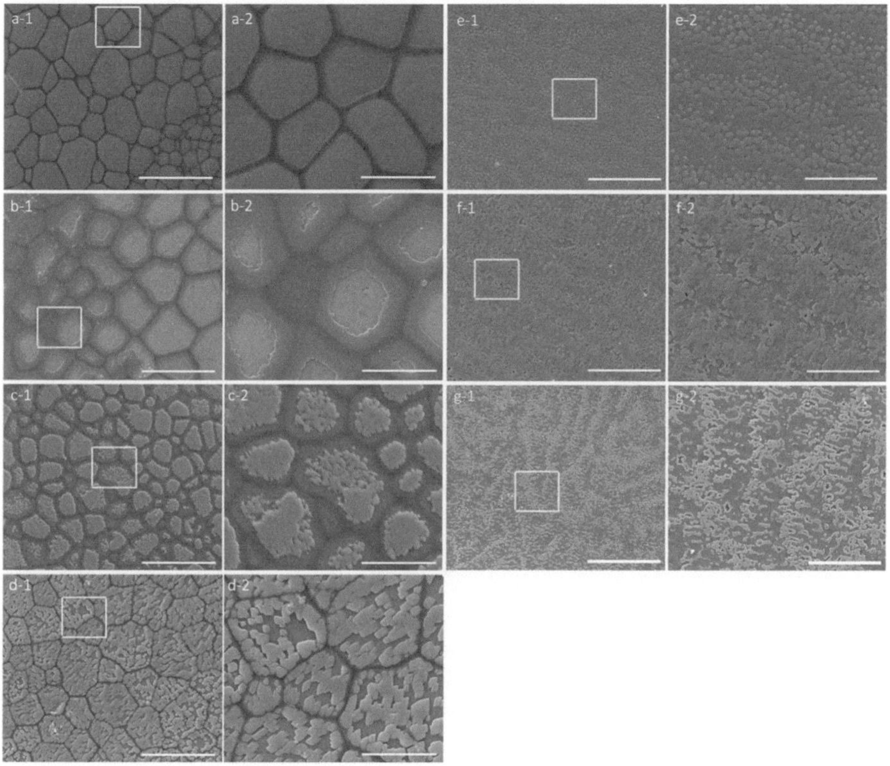

Fig. 2.11 Influences of the candidates on regulating $CaCO_3$ crystals on shell surface [95]
The relative mRNA expression levels of the candidate genes were significantly decreased after the injection of dsRNA in vivo. (**a**) Scanning electron microscopic (SEM) images of normal shell prismatic layer. (**b–d**) SEM images of the prismatic layers after RNAi treatment: unigene34354 (**b**), unigene35118 (**c**), and unigene51738 (**d**). (**e**) SEM images of the nacreous layer in normal shells. (**f–g**) SEM images of the nacreous layers after RNAi treatment: unigene18749 (**f**) and unigene56675 (**g**). **a**-2 showed an enlargement of the white box in **a**-1, as do **b**-2, **g**-2. Bar, 50 μm in **a**-1, **b**-1, **c**-1, **d**-1, **e**-1, **f**-1, and **g**-1; bar = 20 μm in **a**-2, **b**-2, **c**-2, **d**-2, **e**-2, **f**-2, and **g**-2

Unigene35118 the one containing chitin-binding domain was expressed at its highest level in gill in which place chitin was at filament shape [102]. Nevertheless, its expression level in mantle edge was still higher than other tissues. We've already known chitin is an important organic molecule in biomineralization, the formation of shell in mollusks included. The abnormal prismatic layer caused by the inhibition of unigene35118 was probably associated with the absence of shell framework (Fig. 2.11c). Further investigation of this gene would probably give us some new inspiration on the mechanisms of chitin mineralization. Taken all these results together, our data suggest the potential of these candidates on regulating shell formation [95].

2.3 The Significant Roles Calmodulin and Calmodulin-Like Protein (CLP) Play During Calcium Metabolism

*2.3.1 Cloning and Expression of a Pivotal Calcium Metabolism Regulator: Calmodulin Involved in Shell Formation from Pearl Oyster (*Pinctada fucata*)*

The main component of bivalve shell is $CaCO_3$, product of calcium metabolism. The process of calcium uptake, transportation, and recruitment is highly regulated and mediated by many factors. Among these factors, calmodulin (CaM) is a pivotal multifunction regulator of calcium metabolism existing in nearly all organisms, often considered to play an essential role in shell formation. In our research, we isolated the full-length CaM cDNA from *Pinctada fucata*. The protein encoded by oyster calmodulin shares high similarity with that in vertebrates. What's more, the oyster CaM mRNA exhibited the highest expression level in gill, a crucial organ participating in calcium uptake during oyster calcium metabolism. In situ hybridization results indicated that oyster CaM located specifically distributed at the outer epithelium cells of the dorsal region and the folds of mantle tissue, revealing that CaM truly engaged in regulation of calcium secretion and transportation. Furthermore, oyster CaM also exhibited a strong calcium-binding ability and typical characterization of Ca^{2+}-dependent electrophoretic shift. In summary, we identified and characterized a pivotal calcium metabolism regulator of the oyster playing a crucial role in calcium uptake, transportation, and secretion during shell formation.

2.3.1.1 Cloning and Sequence Analyses of *P. fucata* CaM cDNA

We obtained a 377 bp cDNA fragment CaM1 by specially designed two degenerate oligonucleotides (F1 and R1) based on the conserved sequence of CaM in other species. Based on this sequence, another specific primer F2 was synthesized to get the unknown remaining 3′ terminal sequence of CaM. Same method was applied to get the 5′ terminal sequence. As we could see from Fig. 2.12, the full-length cDNA sequence of CaM was 739 bp, with a 69 bp 5′ untranslated region and a 203 bp 3′ untranslated region, an open reading frame of 447 bp. Two putative polyadenylation signals (AATAAA) were followed by a stop codon (TGA) and 15 bp upstream of poly (A) was situated at the 3′ noncoding region. A Kozak sequence (PuNNATGPu) was also found surrounding the methionine initiation codon (AAAATGG) of CaM cDNA. This cDNA sequence has successfully been submitted to GenBank with the accession number AY341376.

The deduced CaM protein comprises of 149 amino acids with an isoelectric point of 4.14 and a calculated molecular mass of approximately 16.8 kDa. CaM found in *P. fucata* did not contain cysteine and tryptophan residues just like vertebrate and other mollusk CaM proteins. Generally speaking, CaM is a fairly conserved protein from the evolutionary standpoint. Multiple protein sequence alignment analysis

```
GGCTTTTGATACTGACAACTTAAACCCTACACCCAGTCGTACTTCATCCCCTGATATTAA   60
TAGGATAAAATGGCCGATCAGCTGACAGAGGAGCAGATCGCAGAATTCAAGGAAGCGTTC  120
          M  A  D  Q  L  T  E  E  Q  I  A  E  F  K  E  A  F
AGTTTGTTTGACAAGGACGGCGATGGTACAATAACAACCAAGGAGCTGGGTACAGTAATG  180
 S  L  F  D  K  D  G  D  G  T  I  T  T  K  E  L  G  T  V  M

AGGTCACTAGGTCAGAATCCAACCGAGGCTGAACTTCAGGATATGATCAACGAAGTAGAC  240
 R  S  L  G  Q  N  P  T  E  A  E  L  Q  D  M  I  N  E  V  D

GCTGATGGAAACGGAACCATCGACTTCCCAGAATTCCTTACAATGATGGCGCGAAAGATG  300
 A  D  G  N  G  T  I  D  F  P  E  F  L  T  M  M  A  R  K  M
AAAGACACCGACTCGGAAGAGGAAATTAGAGAAGCTTTCAGAGTGTTCGATAAAGATGGC  360
 K  D  T  D  S  E  E  E  I  R  E  A  F  R  V  F  D  K  D  G

AATGGTTTCATTAGTGCAGCTGAATTACGTCACGTGATGACAAATCTCGGAGAAAAACTT  420
 N  G  F  I  S  A  A  E  L  R  H  V  M  T  N  L  G  E  K  L

ACGGATGAGGAAGTAGACGAGATGATTCGTGAAGCTGATATTGACGGTGACGGACAGGTT  480
 T  D  E  E  V  D  E  M  I  R  E  A  D  I  D  G  D  G  Q  V

AATTATGAAGAATTTGTGAAAATGATGATGTCGAAATGAAATAAACATTTCATCGGAACG  540
 N  Y  E  E  F  V  K  M  M  M  S  K  *
GCTTTGCTACGCACGTCATAAACGCAATTTTGTCAAACTTTTTACAGCGTACTTTACATT  600
GTAATACATTGAATATTATACAACTTATCAGGAATCCAATGCATTTGTGTCTATTGAATT  660
ATATTTTACGAAAGGTGTTAACTTTATTTATATCATGAAACAATAAACTAGAAACGACTT  720
GGAAAAAAAAAAAAAAAAA                                         739
```

Fig. 2.12 Nucleotide and its deduced amino acid sequence of CaM cDNA in *P. fucata*
The stop codon was labeled with an asterisk, and the possible polyadenylation signal sequence in 3′ untranslated region was underlined. This cDNA sequence has successfully been submitted to GenBank (accession number: AY341376)

(Fig. 2.13) revealed that the CaM sequence of *P. fucata* shared high similarities with other homologs isolated no matter from vertebrate or invertebrate animals. Only four residues (Phe^{100}, Lys^{144}, Met^{147}, and Ser^{148}) concentrating mainly at the carboxyl terminus varied from vertebrate CaM, although the identity of the nucleotide sequence of the open reading frame was not significant. Phe^{100} and Ser^{148} were treated as the typical residue of invertebrate CaM [103] and the characteristic residue in other mollusks [104]. In other words, there were only two amino acid residues altered in CaM during the evolution process from pearl oyster to vertebrate. Furthermore, only two residues (Lys^{144} and Met^{147}) were different from *Aplysia* CaM

```
Apc   (1)   -ADQLTEEQIAEFKEAFSLFDKDGDGTITTKELGTVMRSLGQNPTEAELQ
Mes   (1)   -ADQLTEEQIAEFKEAFSLFDKDGDGTITTKELGTVMRSLGQNPTEAELQ
Pat   (1)   -ADQLTEEQIAEFKEAFSLFDKDGDGTITTKELGTVMRSLGQNPTEAELQ
Pfa   (1)   MADQLTEEQIAEFKEAFSLFDKDGDGTITTKELGTVMRSLGQNPTEAELQ
Hsa   (1)   MADQLTEEQIAEFKEAFSLFDKDGDGTITTKELGTVMRSLGQNPTEAELQ
Bot   (1)   -ADQLTEEQIAEFKEAFSLFDKDGDGTITTKELGTVMRSLGQNPTEAELQ
Rno   (1)   MADQLTEEQIAEFKEAFSLFDKDGDGTITTKELGTVMRSLGQNPTEAELQ
            51                                               100
Apc   (50)  DMINEVDADGNGTIDFPEFLTMMARKMKDTDSEEEIREAFRVFDKDGNGF
Mes   (50)  DMINEVDADGDGTIDFPEFLTMMARKMKDTDSEEEIREAFRVFDKDGDGF
Pat   (50)  DMINEVDADGDGTIDFPEFLTMMARKMKDTDSEEEIREAFRVFDKDGDGF
Pfa   (51)  DMINEVDADGNGTIDFPEFLTMMARKMKDTDSEEEIREAFRVFDKDGNGF
Hsa   (51)  DMINEVDADGNGTIDFPEFLTMMARKMKDTDSEEEIREAFRVFDKDGNGY
Bot   (50)  DMINEVDADGNGTIDFPEFLTMMARKMKDTDSEEEIREAFRVFDKDGNGY
Rno   (51)  DMINEVDADGNGTIDFPEFLTMMARKMKDTDSEEEIREAFRVFDKDGNGY
            101                                            149
Apc   (100) ISAAELRHVMTNLGEKLTDEEVDEMIREADIDGDGQVNYEEFVTMMTSK
Mes   (100) ISAAELRHVMTNLGEKLTDEEVDEMIREADIDGDGQVNYEEFVKMMTSK
Pat   (100) ISAAELRHVMTNLGEKLTDEEVDEMIREADIDGDGQVNYEEFVTMMTSK
Pfa   (101) ISAAELRHVMTNLGEKLTDEEVDEMIREADIDGDGQVNYEEFVKMMMSK
Hsa   (101) ISAAELRHVMTNLGEKLTDEEVDEMIREADIDGDGQVNYEEFVQMMTAK
Bot   (100) ISAAELRHVMTNLGEKLTDEEVDEMIREADIDGDGQVNYEEFVQMMTAK
Rno   (101) ISAAELRHVMTNLGEKLTDEEVDEMIREADIDGDGQVNYEEFVQMMTAK
```

Fig. 2.13 Multiple sequence alignment of oyster calmodulin and CaMs from other typical vertebrates and invertebrates. Apc, (*Aplysia californica*) CaM, CAA40207; Mes, sea anemone (*Metridium senile*) CaM, MCXAM; Pat, scallop (*Patinopecten* sp.) CaM, MCSW; Pfa, pearl oyster (*P. fucata*) CaM, AY34137; Hsa, human (*Homo sapiens*) CaM, P02593; Bot, bovine (*Bos taurus*) CaM, MCBO; Rno, rat (*Rattus norvegicus*) CaM, CAA32119

compared with other mollusk species [105]. Nevertheless, it was still interesting that four residues (Asn^{61}, Asn^{97}, Lys^{144}, and Thr^{147}) in *P. fucata* were converted to Asp ^{61}, Asp^{97}, Thr^{144}, and Thr^{147}, respectively, in *Patinopecten* [106], a closer species to our pearl oyster *P. fucata* in an evolutionary standpoint. In addition, multiple sequence alignment analysis also indicated that Lys^{144} was fairly conserved in sea anemone and pearl oyster same time. As Met^{147} was only existed in *P. fucata*, it could be viewed as a unique residue for oyster.

2.3.1.2 Expression Pattern Analysis

Tissue-specific distribution of oyster CaM in different tissues was detected via semiquantitative RT-PCR to investigate the function of CaM in pearl oyster. RT-PCR reactions were carried out based on RNA samples from mantle, gonad, adductor muscle, and gill. Oyster CaM was expressed in all tested tissues as shown

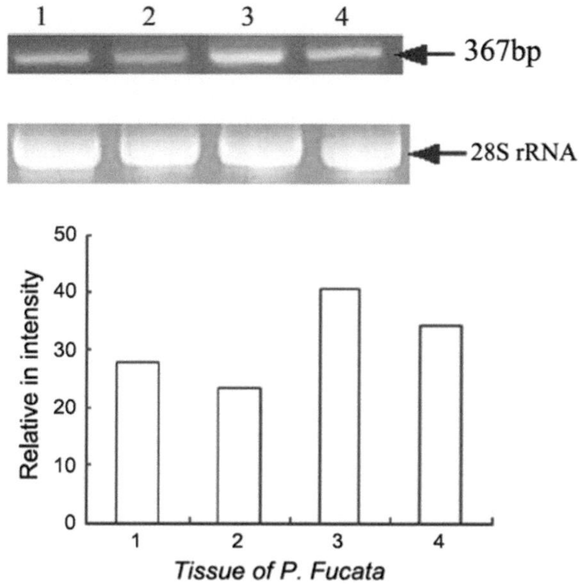

Fig. 2.14 Tissue distribution of CaM in *P. fucata* Semiquantitative RT-PCR reactions were carried out with RNA samples from adult tissues in *P. fucata*. The PCR product (367 bp) was indicated by an arrow, and the band relative intensity was evaluated by densitometry. The 28S rRNA was used as an internal reference control for equal quantities of every sample. *1*, mantle; *2*, gonad; *3*, gill; *4*, muscle

in Fig. 2.14. However, the relative expression level of CaM was higher in gill, adductor muscle and mantle other than gonad with identical expression of 28S ribosomal RNA in all samples. Data from three independent experiments were similar.

Shell formation requires large amounts of calcium [107], and biomineralization also needs the effective transportation and control of calcium ions. According to current findings, gill is the major organ functioning on calcium uptake from the external environment [107, 108], and CaM has been considered to be a critical regulatory factor in membrane Ca^{2+}-ATPase system or "calcium sink" in the gill of scallop [109]. What's more, CaM has been reported engaged in ciliary arrest of calcium [110]. The results of in situ hybridization showed that the mRNA of oyster CaM was widely expressed in epithelial cells of the branchial filaments (data not shown). Since L-type Ca^{2+} channels modulated by CaM were suggested taking part in calcium transportation for calcification in some known marine invertebrates [111, 112], the highest expression level of oyster CaM in gill supported the hypothesis that CaM was most likely to function as an essential regulatory factor during calcium uptake and accumulation in gill.

Mantle is another crucial organ regulating metal ion metabolism, and actively participating in the secretion of calcium and other ions for $CaCO_3$ crystal growth during shell formation [113]. In situ hybridization analysis was further performed to detect the precise distribution pattern of CaM in mantle tissue. Strong hybridization signals were detected mainly at the outer epithelial cells of the inner fold, the outer epithelial cells of the middle fold, and the inner epithelial cells of the outer fold of mantle (Fig. 2.15a). In addition, they were also found in the outer epithelial cells of

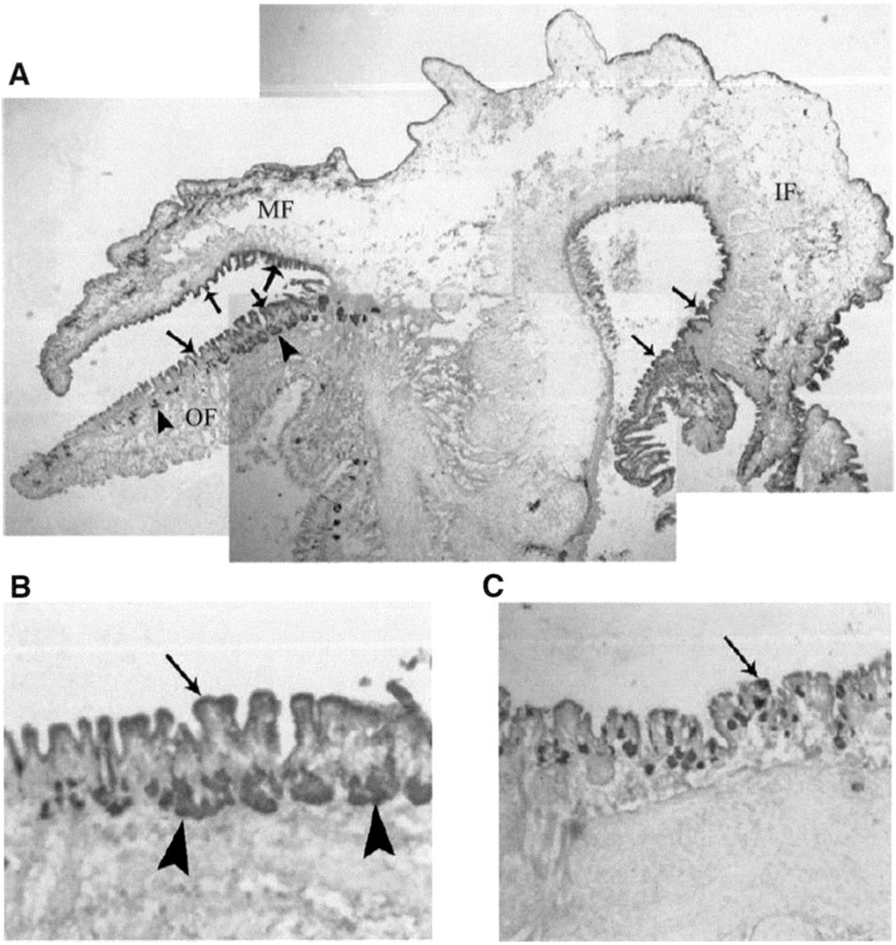

Fig. 2.15 In situ hybridization of oyster CaM mRNA in the mantle of *P. fucata* [52]
Hybridization signals (arrows and arrow heads) showed in all three folds of the mantle (**a**). Three overlapping pictures of the same section were taken to observe the distribution of hybridization signal over the whole tissue. Strong hybridization signals were indicated in the inner epithelial cells (arrow heads) and the basement cells of outer fold of the mantle (arrow) in (**b**). Hybridization signals were also detected in outer epithelial cells of the dorsal region by arrows in (**c**). *OF* outer fold, *MF* middle fold, *IF* inner fold. Magnifications are 25× (**a**) and 100× (**b** and **c**)

the dorsal region (Fig. 2.15c) and the basement cells of the outer fold (Fig. 2.15b), but yet the control sense probe during hybridization showed no signals (data not shown). Surprisingly, no hybridization signal was tested in the outer epithelial cells of the outer fold. Based on these data we got, we proposed a hypothesis that CaM might be involved in the CaM-dependent calcium transportation driven by Ca^{2+}-ATPase and participate in the calcium secretion regulation during shell formation. What's more, the oyster CaM also showed high expression in muscle, in accordance

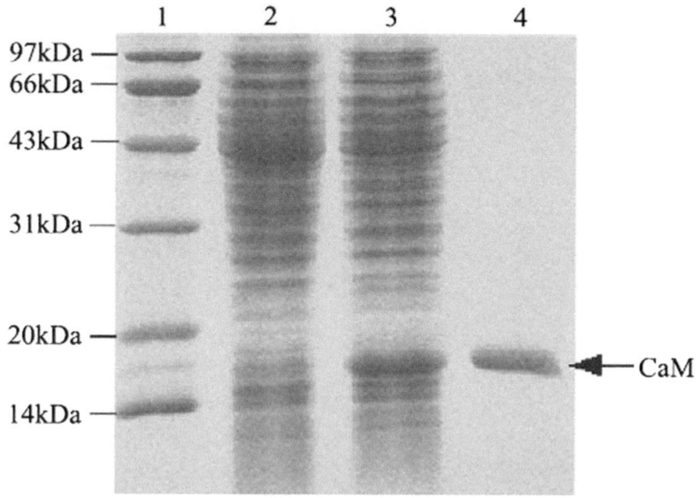

Fig. 2.16 Expression of recombinant *P. fucata* CaM in the culture supernatant detected by 15% SDS-PAGE with sample buffer containing 5 mM EGTA and stained by Coomassie Brilliant Blue R-250
Lane 1, protein molecular mass marker; lane 2, proteins of BL21 (DE3); lane 3, proteins of pET-15b/CaM induced by 0.5 mM IPTG for 2.5 h; lane 4, purified recombinant CaM by phenyl-Sepharose column. The molecular mass in kDa was indicated on the left of the gel, and the purified recombinant CaM was shown by the arrow on the right

with the fact that CaM was a dispensable modulator for muscle contraction controlling shell closing and opening. Besides, oyster CaM was also detected in gonad, suggesting its dual roles in oyster procreation.

In summary, although we have analyzed the expression level of oyster CaM in different tissues and investigated its precise expression site in the gill and mantle, the detailed regulatory mechanisms of CaM in calcium uptake, accumulation, and release during shell formation remain further study.

2.3.1.3 Expression and Purification of *P. fucata* Recombinant CaM

The coding region of *P. fucata* CaM was cloned into pET-15b expression vector and further expressed in *E. coli.* BL21 (DE3). The recombinant expressed protein CaM was detected by SDS-PAGE and accounted approximately 21% of the total bacteria soluble proteins as shown in Fig. 2.16. We must assure that there was only one single band could be observed on 15% SDS-PAGE, the sample buffer of which should contain 5 mM EGTA after purification by phenyl-Sepharose hydrophobic chromatography. The apparent molecular mass of this single band was 17 kDa, consistent with the expected molecular mass of oyster CaM. After multiple attempts, we found that direct isolation and purification of CaM from tissues were very complicated and the yield was extremely low. But luckily, the expression and purification of CaM in

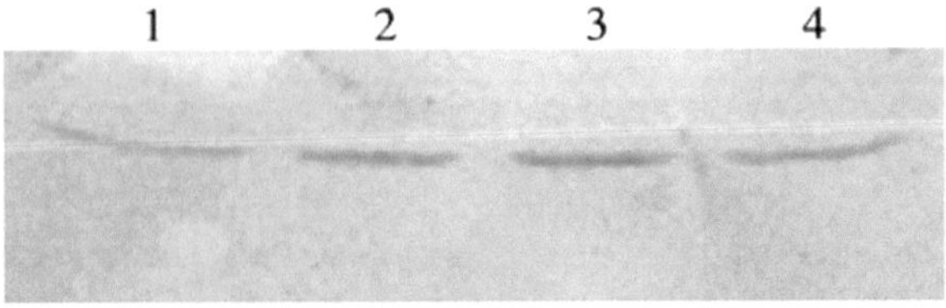

Fig. 2.17 Western blot analysis of the expression of CaM in *P. fucata*
Lane 1, purified recombinant oyster CaM; lanes 2, 3, and 4, oyster tissues, mantle, muscle, and gill, respectively

E. coli cells were much easier, and the yield was even higher than traditional methods. The expression and purification of chicken, human, and rat CaM in *E. coli* have been reported [114–117], and here, we successfully expressed and purified a mollusk CaM in *E. coli* for the first time, and its antibody was prepared for further analysis. The purified recombinant CaM and CaMs from oyster tissues including mantle, muscle, and gill were further detected by Western blot analysis with rabbit polyclonal antibody of the recombinant protein CaM. Only one visualized immunoreactive band was detected at 17 kDa (Fig. 2.17), consistent with the results of semiquantitative RT-PCR.

2.3.1.4 Ca^{2+}-Dependent Electrophoretic Shift and ^{45}Ca Overlay Assay

Ca^{2+}-dependent electrophoretic migration and ^{45}Ca overlay analyses were carried out to further confirm the identity of the purified protein and its calcium-binding ability, respectively. When the purified recombinant proteins were loaded with the sample buffer for further electrophoresis, both running buffer and gels containing 0.1 mM $CaCl_2$, a single band with an approximate molecular mass of 14 kDa was detected as Fig. 2.18a showed, whereas, the protein band migrated with another protein of 17 kDa when $CaCl_2$ was replaced with EDTA (Fig. 2.18b). The mobility of calcium-induced electrophoresis was a useful tool to identify CaM [14], and this assay successfully confirmed the purified protein was CaM indeed. Besides, the ^{45} Ca overlay analysis implied its calcium-binding property later (Fig. 2.19). The CaM isolated and purified from vertebrates migrated with an apparent molecular mass of 21 and 15 kDa in the absence or presence of calcium, respectively, in SDS-PAGE [14]. But we should note that substitutions of amino acid in the C-terminal residues of CaM in *P. fucata* might influence the identity conformation slightly and cause a smaller electrophoretic shift than vertebrate CaM. Additionally, the recognition and interaction of oyster CaM with other target proteins like binding calcium, such as calcium pump, CaM-dependent protein kinase, and calcium channels, may be different from the vertebrate CaM, on account of the presence of calcium-induced conformational change, the detailed mechanism of which remains further study [52].

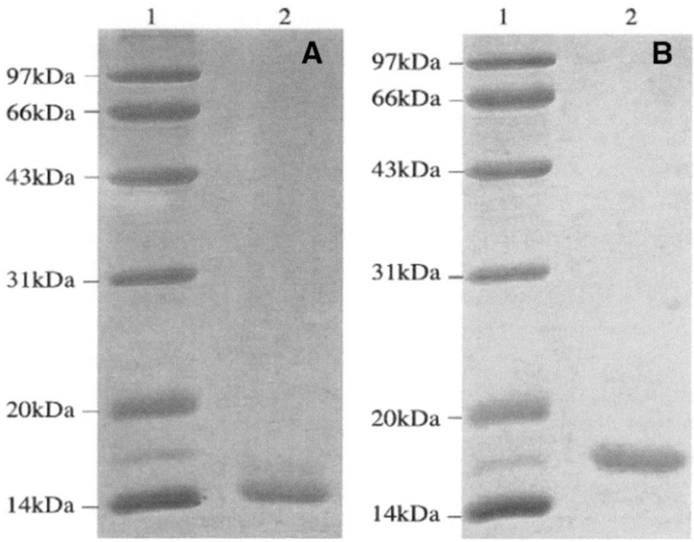

Fig. 2.18 Ca^{2+}-dependent electrophoretic migration of purified recombinant *P. fucata* CaM Purified recombinant oyster CaM (lane 2) was run on a 15% SDS-PAGE in the presence of Ca^{2+} (**a**) or EDTA (**b**). Sample buffer, running buffer, and gels all contained 0.1 mM $CaCl_2$ or 0.1 mM EDTA. Lane 1, protein molecular mass marker; molecular mass in kDa was shown on the left of the gels

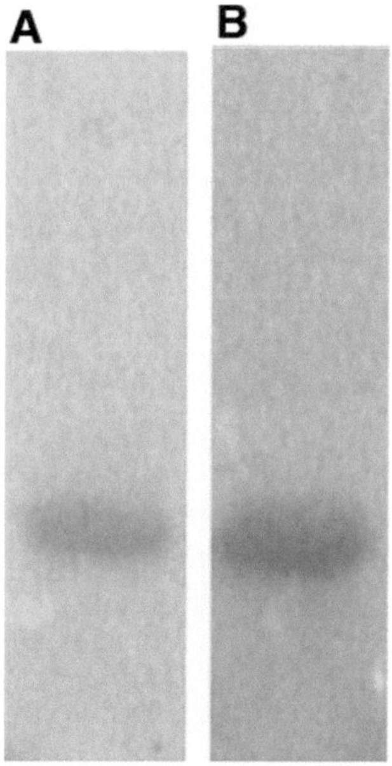

Fig. 2.19 Identification of calcium-binding characterization of oyster CaM on nitrocellulose membrane after SDS electrophoresis [52]
A indicated the protein band stained by Amido black after autoradiography. *B* was an autoradiograph of the transferred nitrocellulose membrane

2.3.2 *cDNA Cloning and Characterization of a Novel Calmodulin-Like Protein from Pearl Oyster* Pinctada fucata

Calcium metabolism in oysters is a very complex and sophisticated physiological and biochemical process. However, we know little about the regulation mechanism of this pathway in oyster. Our previous research indicated that calmodulin (CaM) might play a regulatory role during oyster calcium metabolism. In this study, a full-length cDNA encoding a novel calmodulin-like protein (CaLP) with a long C-terminal sequence was obtained and identified from pearl oyster *Pinctada fucata*, further expressed and purified in *Escherichia coli* and characterized in vitro. The distribution of CaLP mRNA was ubiquitous in all tested tissues, with its highest expression level in mantle, a critical organ controlling calcium secretion. To find out the specific expression site, we performed in situ hybridization, the strong signals of which were detected in the outer epithelial cells of the middle fold, both inner and outer epithelial cells of inner fold, and the dorsal region of mantle. The oyster CaLP protein, with four putative Ca^{2+}-binding domains, is highly heat-stable and has a potential high affinity for calcium binding. CaLP also exhibited strong Ca^{2+}-binding activity, significant Ca^{2+}-induced conformational changes, and typical Ca^{2+}-dependent electrophoretic shift. In addition, Ca^{2+}-dependent affinity chromatography analysis indicated that CaLP isolated from *P. fucata* possessed the important interactions with some different target proteins from those of oyster CaM no matter in the gill or mantle. In conclusion, our results indicated that the oyster CaLP was a novel member of the CaM superfamily and suggested that the function of oyster CaLP protein may be different from CaM during the regulation of oyster calcium metabolism.

2.3.2.1 Cloning of a Full-Length cDNA Encoding a Calmodulin-Like Protein from *P. fucata*

A 377 bp PCR product called CaLP1 sharing high similarities with oyster CaM was cloned from the gill in *P. fucata* using degenerate oligonucleotides based on the conserved regions of CaM sequence. 5′- and 3′-RACE were performed separately to obtain the full length of CaLP. Finally, the splicing sequence was confirmed by sequencing, the result of which matched perfectly as we predicted before. The complete full-length CaLP cDNA consisting of a complete poly (A) tail was 757 bp, with a 130 bp 5′ untranslated sequence, a 483 bp open reading frame, a stop codon named TGA, a 146 bp 3′ untranslated region, and a poly (A) tail of 18 nucleotides as shown in Fig. 2.20. Then the cDNA sequence was successfully submitted to GenBank with the accession no. AY663847.

```
  1  CGGGGGATCATCGTCCATACGTCCGTACTAGGTTCCAAAGTTAGAATAATCATCAGGAATGAT
 64  ATTTGTTTACATTTATATGTAGAATTCATTCATCAGGAAGAATACAGACACGGACAGAAACTT
127  CTAAATGGCGGAAGATCTCACAGAAGAACAAATCGCAGAGTTCCGAGAGGCGTTCCACCTCTTT
             M  A  E  D  L  T  E  E  Q  I  A  E  F  R  E  A  F  H  L  F
191  GACAAAGATGGCAGTGGGAGTATATCAGCCGAAGAGCTCGGAACAGTCATGAGATCTCTAGGT
      D  K  D  G  S  G  S  I  S  A  E  E  L  G  T  V  M  R  S  L  G
254  CAAAACCCCAATGAACAAGAATTACAAGATCTCGTGGAAGAAATCGACACAGACGGCAATGGA
      Q  N  P  N  E  Q  E  L  Q  D  L  V  E  E  I  D  T  D  G  N  G
317  GAGGTTGATTTCAACGAGTTCTTGGCCATGATGGCTAAGAAAATGAAAGACACCGACTCGGAA
      E  V  D  F  N  E  F  L  A  M  M  A  K  K  M  K  D  T  D  S  E
380  GAGGAAATTCGGGAAGCCTTCAGAGTGTTCGATAGGGATGACAAAGGTTTTATTACTGCTTCA
      E  E  I  R  E  A  F  R  V  F  D  R  D  D  K  G  F  I  T  A  S
443  GAATTAAAACATATAATGACGACACTGGGTGAGAAAATGGACGACGAAGAAGCAGAAGAGATG
      E  L  K  H  I  M  T  T  L  G  E  K  M  D  D  E  E  A  E  E  M
506  GTAGCGGCTGCAGATATTGATGGAGATGGAGAAATAAATTATGAGGAGTTTGTCAAAATGATT
      V  A  A  A  D  I  D  G  D  G  E  I  N  Y  E  E  F  V  K  M  I
567  TCAATGAAAGACACCGATCAAGAAGAACAGCAACAAGAAAATAAATGATAATGAATTAGACAA
      S  M  K  D  T  D  Q  E  E  Q  Q  Q  E  N  K  *
632  ACAAAAATTGGACTTCAAAATTTGGCATCAATGAAGCGATGTATAAACTGTTGTTATATCTTT
695  TTGTTATCGATAACGTTGTATGTAAATAAAAGCTTCATATTTCTTAAAAAAAAAAAAAAAAAA
```

Fig. 2.20 Nucleotide and its deduced amino acid sequence of CaLP cDNA in *P. fucata*
The possible polyadenylation signal sequence in the 3' untranslated region was underlined and the stop codon was marked with an asterisk and this cDNA sequence has been successfully submitted to GenBank with accession no. AY663847

2.3.2.2 Sequence Analysis of Oyster CaLP Protein

The deduced CaLP protein in *P. fucata* consisted of 161 amino acids with a calculated molecular mass of approximately 18.3 kDa and a predicted isoelectric point of 4.04. Multiple sequence analysis showed that this protein shared 67% identity and 87% similarity with the *P. fucata* CaM protein. CaLP had 93.9% similarity with oyster CaM when excluded the extra C-terminal end segment of 12 amino acids. Both CaM and CaLP possessed only one Tyr residue, and neither of them had Trp nor Cys residue. In addition, the predicted secondary structures of both proteins (Fig. 2.21a) were also pretty close (helix $\approx$ 57%, beta-sheet $\approx$ 3.7%, and coil $\approx$ 38%). All these results suggested that the oyster CaLP protein was closely associated with oyster CaM. Interestingly, a classic structural feature of this novel CaLP was noticed: there were 12 extra hydrophilic amino acid residues locating at the end of C-terminal (Fig. 2.21b), implying that CaLP may play a special role in Ca^{2+}-mediated cellular process. What's more, CaLP contained four putative Ca^{2+}-binding EF-hand domains (Fig. 2.21b) predicted from the protein family database in Sanger Institute (http://www.sanger.ac.uk/Software/Pfam). Among them, the Ca^{2+}-binding residues (X, Y, Z, -Y, -X, -Z) in the 2nd and 4th EF-hand domains seemed much more conserved than those in the 1st and 3rd domains if compared with oyster CaM. The physiological function and specific identification of different substrates may benefit from the structural variations in EF-hand domains. Comparisons of the amino acid composition between the EF-hand domain [118] and calcium-binding domain in CaLP exhibited a great correlation of Ca^{2+}-binding ligand positions. There was one exception on Lys residue in domain 3 at ligand position Z, which may lead to a weaker potential of calcium binding in this loop

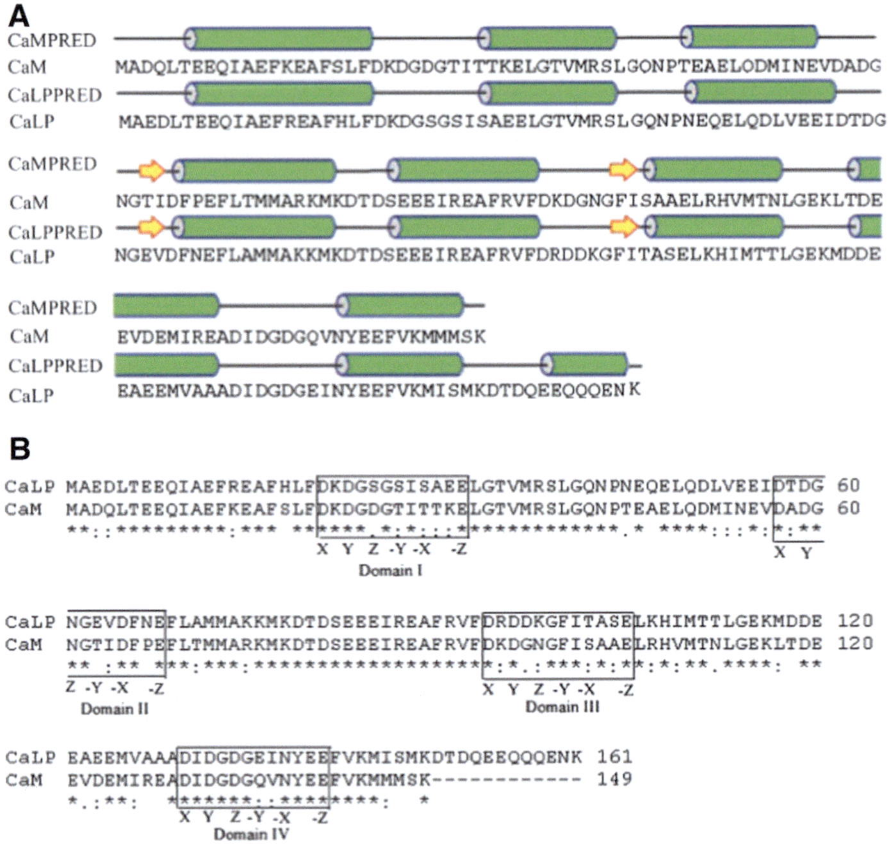

Fig. 2.21 The predicted secondary structures (A) and multiple alignment of the amino acid sequence of CaLP and CaM (B) in *P. fucata* [24]
CaMPRED and CaLPPRED represented the predicted secondary structures of CaM and CaLP, respectively. Green barrels showed the predicted α-helices, yellow arrows indicated the predicted β-strands, and black lines marked the predicted random coils. The four Ca^{2+}-binding domains were boxed; homologous and identical amino acids were showed by dots and stars, respectively. X, Y, Z, -Y, -X, and -Z indicated the Ca^{2+}-binding ligand residues

compared that in oyster CaM. The good news was the four acidic residues (Asp and Glu) found in domains 2 and 4, an interesting cue for higher calcium-binding potential. The region between the 2nd and 3rd EF-hand domains was the most conserved region of the flexible central helix in CaLP, contributing to the binding properties of CaM to other target proteins [119–121]. Conversely, the least homology region was between the 3rd and 4th calcium-binding sites. In summary, CaLP in *P. fucata* showed several putative phosphorylation sites with higher scores predicted by NetPhos 2.0 Server [122], including five serine, three threonine locating on the 1st Ca^{2+}-binding domain, and one tyrosine residues, which might be phosphorylated by myosin light-chain kinase [123] at the same place of oyster CaM. We could

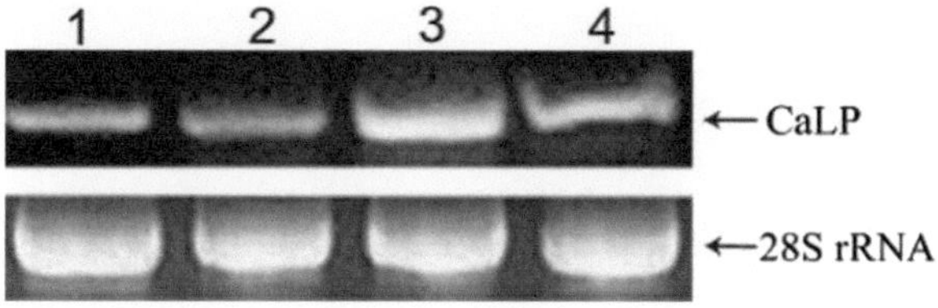

Fig. 2.22 Distribution pattern of CaLP mRNA in tissues from *P. fucata*
Agarose gel electrophoresis was performed to analyze the RT-PCR products from the muscle (*1*), gonad (*2*), mantle (*3*), and gill (*4*). The PCR product of CaLP (486 bp) was showed by an arrow. The 28S rRNA was used as an inner control of equal quantities of total RNA used in this experiment

conclude that all these residues may affect the interaction activity between CaLP and its target proteins. We should also notice that Thr80 and Ser82 in the central α-helix of CaM, a crucial region for its interaction with target CaM-dependent proteins, were fairly conserved in CaLP. Moreover, Tyr139 standing in the 4th Ca^{2+}-binding domain that could be phosphorylated by epidermal growth factor, insulin receptor, and Src family kinases [124] was also very conserved in oyster CaLP.

2.3.2.3 Gene Expression Analysis and In Situ Hybridization

RT-PCR reactions were performed to detect the mRNA distribution of CaLP in oyster tissues such as mantle and gonad. We obtained a 486 bp RT-PCR product with the specifically designed primers (G1 and G2), total RNA of tested tissues as templates. We could see from Fig. 2.22 that the mRNA of CaLP was detected in almost every tested tissues, with its highest expression level in mantle, an important organ for metal ion metabolism, secretion, and shell formation [113]. Similar data were obtained from three independent repeated experiments.

In situ hybridization was performed to get the precise expression site of CaLP in mantle tissue. Strong hybridization signals showed in both middle fold and inner fold, especially the outer epithelial cells from the two folds mentioned above (Fig. 2.23a). Additionally, the outer epithelial cells were responsible for periostracum secretion [125]. Nevertheless, only weaker signals could be detected in inner epithelial cells of the outer fold where CaM was highly expressed [52]. We also detected strong signals in outer epithelial cells of the dorsal region of the mantle (Fig. 2.23b), a place controlling the secretion of nacre [125], and the negative control sense probe showed no signal (data not shown). We've already known that calcium is the main component of shell, also a critical intracellular second messenger. Ninety percent of oyster shells is $CaCO_3$, the product of calcium metabolism with a small amount of macromolecules. The analysis and observations above also proved that CaLP was a modular just like CaM, both of which could regulate the complicated and delicate process of calcium metabolism in oyster, including Ca^{2+} absorption, accumulation, transportation, secretion, and deposition.

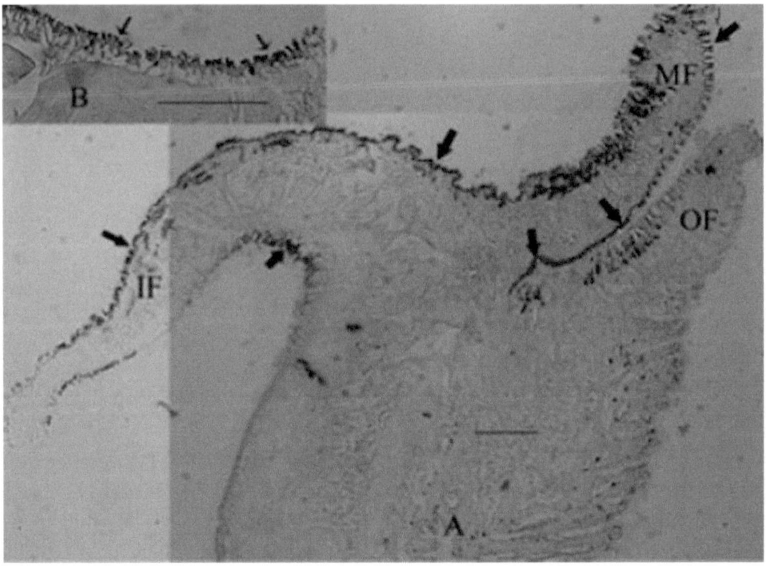

Fig. 2.23 In situ hybridization of oyster CaLP in mantle of *P. fucata*
Three overlapping pictures of the same section were taken to better observe the specific distribution of the hybridization signals. Strong hybridization signals were presented by arrow heads in (**a**). Other signals were also shown in the outer epithelial cells of the dorsal region of the mantle (arrow heads) in (**b**). *OF* outer fold, *MF* middle fold, *IF* inner fold. Scale bar, 0.2 mm

2.3.2.4 Expression and Purification of *P. fucata* Recombinant CaLP

For better understanding and characterizing the functions of CaLP during calcium metabolism, we expressed CaLP with a His-tag in *E.coli* and applied nickel metal affinity chromatography for the further purification of this recombinant protein. The recombinant CaLP protein showed high heat stability as shown in Fig. 2.24. After the further purification steps, only a single band with > 95% purity appeared on 15% SDS/PAGE stained by Coomassie Brilliant Blue R-250. The relative molecular mass of the band was approximately 18 kDa in accordance with the previous predicted molecular mass. This recombinant CaLP was still homogeneous on polyacrylamide gel electrophoresis with the addition of either Ca^{2+} or EGTA as Fig. 2.25 showed. Though we have attempted to exclude the His-tag, the strong hydrophilicity of 12 extra amino acids at C-terminal, which was absent in CaM, counteracted us (data not shown).

Purified recombinant oyster CaLP and CaM were run on a 15% SDS-PAGE in the addition of Ca^{2+} or EGTA. The sample buffer was added with 2.5 mM $CaCl_2$ or EGTA. M, protein molecular mass markers. The molecular mass in kDa was indicated on the left of the gel

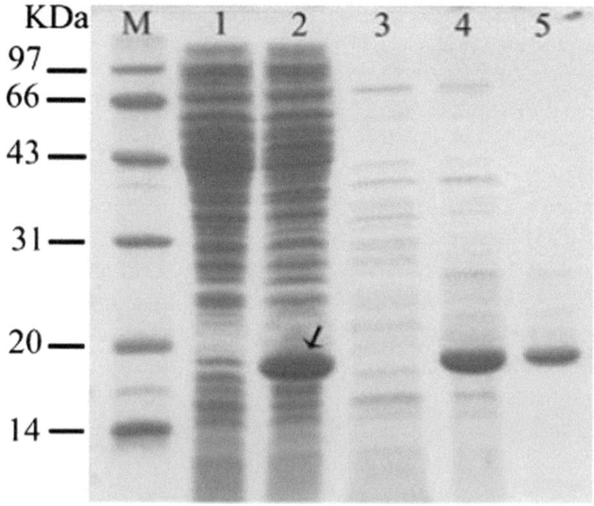

Fig. 2.24 Expression and purification of recombinant *P. fucata* CaLP and heat stability profile of CaLP detected by 15% SDS-PAGE and stained by Coomassie Brilliant Blue R-250
Arrows represented the induced proteins after addition of IPTG. M, protein molecular mass markers; lane 1, un-induced whole-cell lysate; lane 2, whole-cell lysate induced by 0.5 mM IPTG for 2.5 h; lane 3, un-induced whole-cell lysate heated at 90 °C for 10 min; line 4, whole-cell lysate induced by IPTG after heat treatment 90 °C for 10 min; line 5, purified recombinant CaLP by nickel affinity chromatography column. The molecular mass in kDa was indicated on the left of the gel

Fig. 2.25 Ca^{2+}-dependent electrophoretic migration of the purified recombinant CaM and CaLP.

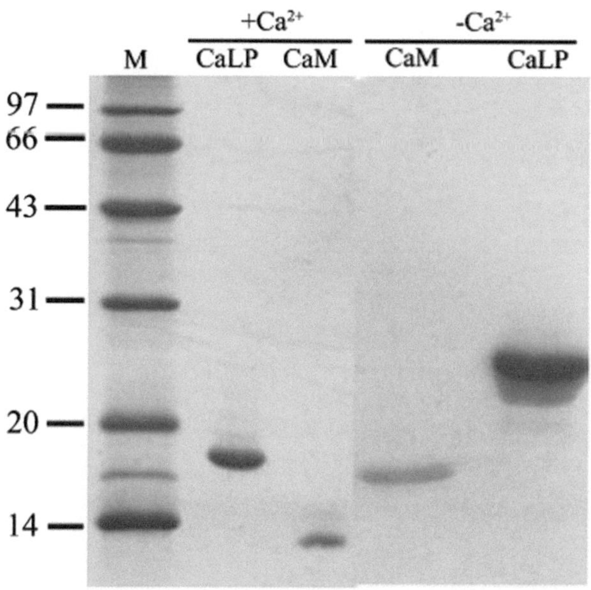

2.3.2.5 Ca^{2+}-Dependent Electrophoretic Shift and Calcium-Binding Properties of Oyster CaLP

Calcium-induced electrophoretic mobility is a helpful method in the characterization of CaM functions, so it was carried out to confirm whether the oyster CaLP protein was a CaM-like protein or not. As we could see from Fig. 2.25, both CaLP and CaM showed an apparent calcium-dependent mobility, suggesting their intimate relationships. Finally, CaM and CaLP showed up as a single band with the apparent molecular weight of 14 kDa and 18 kDa, respectively, in the addition of calcium. But the absence of Ca^{2+} led to a shift stopped at 17 kDa and 25 kDa, respectively (Fig. 2.25). The shift in the band with the addition of calcium was not only caused by the conformational changes within CaLP but also from the extra positive charges during the affinity binding of calcium. Then this binding property was investigated by ^{45}Ca overlay assay. As we could see from Fig. 2.26, both CaM and CaLP showed a strong ability of calcium binding in vitro, indicating that CaLP may be a novel Ca^{2+} sensor or a regulator for calcium arrest and temporary just like CaM [24].

2.3.2.6 CD Spectroscopy and Fluorescence Assay

An important method for determining protein secondary structure in solution was introduced to our research. CD spectra in the far-UV region (190–250 nm) were

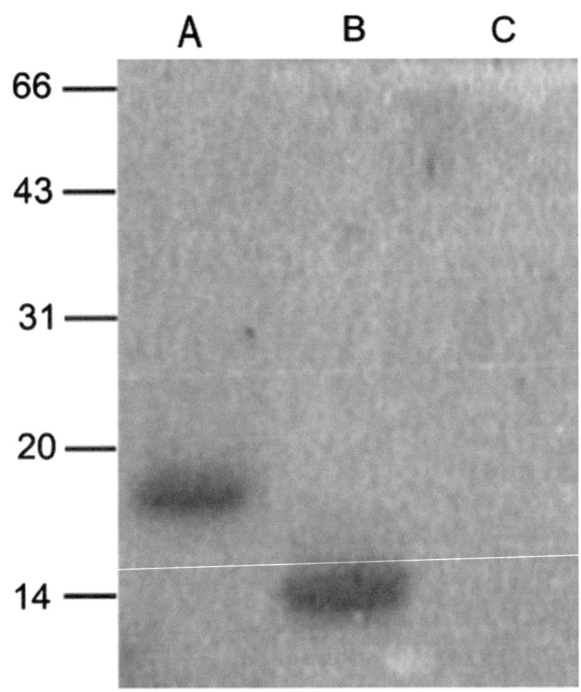

Fig. 2.26 The binding activity of calcium to oyster CaM and CaLP on nitrocellulose membrane after SDS electrophoresis *A*, *B*, and *C* were autoradiographs of the transferred nitrocellulose membrane of oyster CaLP, CaM, and BSA (as a negative control), respectively. The molecular mass in kDa was showed on the left of the membrane

calculated to analyze the secondary structures of CaLP. The change in secondary structures of CaLP and CaM was similar with the addition of Ca^{2+} or EGTA as Fig. 2.27 showed. Double-negative peaks at 209 nm and 220 nm came out in both proteins, along with an increase in α-helix on calcium binding (the value of $\Delta\varepsilon_{220}$ increased 79% and 112% in CaLP and CaM, respectively) after adding 2 mm $CaCl_2$. These results indicated that Ca^{2+} could induce the reorganization and great changes in composition of the secondary structure elements within CaLP. Whereas, when $CaCl_2$ was replaced by EGTA, both proteins seemed to exhibit partial unfolding, and the percentage of α-helix of both proteins was decreased at the same time. Figure 2.27 also implied that CaLP was more unfolded in the Ca^{2+}-free status compared with oyster CaM as high lightened by a slight blue shift of the peak at 209 nm.

The conformational changes and calcium binding of CaM and CaLP were further analyzed by monitoring intrinsic phenylalanine and tyrosine fluorescence. Intrinsic phenylalanine fluorescence spectra (with excitation at 250 nm and emission at 280 nm) were exhibited in Fig. 2.28 A. The phenylalanine fluorescence emission of oyster CaLP and CaM upon calcium binding decreased about 51% and 32%, respectively. This fluorescence quenching could be partially originated from the energy transfer to the tyrosine residues around. When Ca^{2+} was excluded in solution, the fluorescence of Phe went down, which might be caused by the unfolded state in Ca^{2+}-free status. The tyrosine fluorescence emission (with excitation at 277 nm and emission at 320 nm) showed an increase to 0.28- and 1.6-fold after the calcium binding of CaLP and CaM, respectively, in Fig. 2.28b. The different changes of intrinsic phenylalanine and tyrosine fluorescence in CaLP may be caused by the

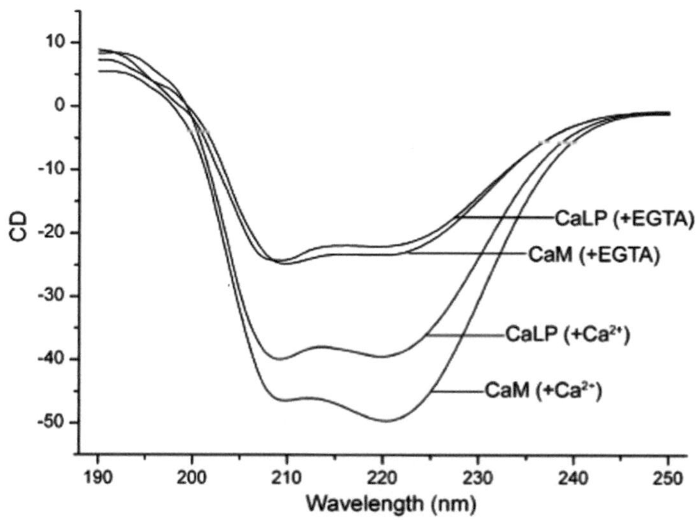

Fig. 2.27 CD spectra of oyster CaLP and CaM with the addition of Ca^{2+} or EGTA. The spectra of oyster CaLP and CaM were recorded in 20 mM Hepes, 100 mM KCl buffer, pH 7.5 with 2 mM $CaCl_2$ or EGTA and corrected by a blank buffer just as mentioned above. The concentration of both proteins was 10 μM

extra C-terminal of CaLP. Unfortunately, we couldn't just measure Ca^{2+} binding to only the C- or N-terminal domain of these two proteins via monitoring intrinsic phenylalanine or tyrosine fluorescence by the proposal described by VanScyoc et al. [16] on account that the Tyr residue in the 3rd EF-hand domain of rat CaM was replaced by Phe at the same place of oyster CaM and CaLP. The Tyr fluorescence

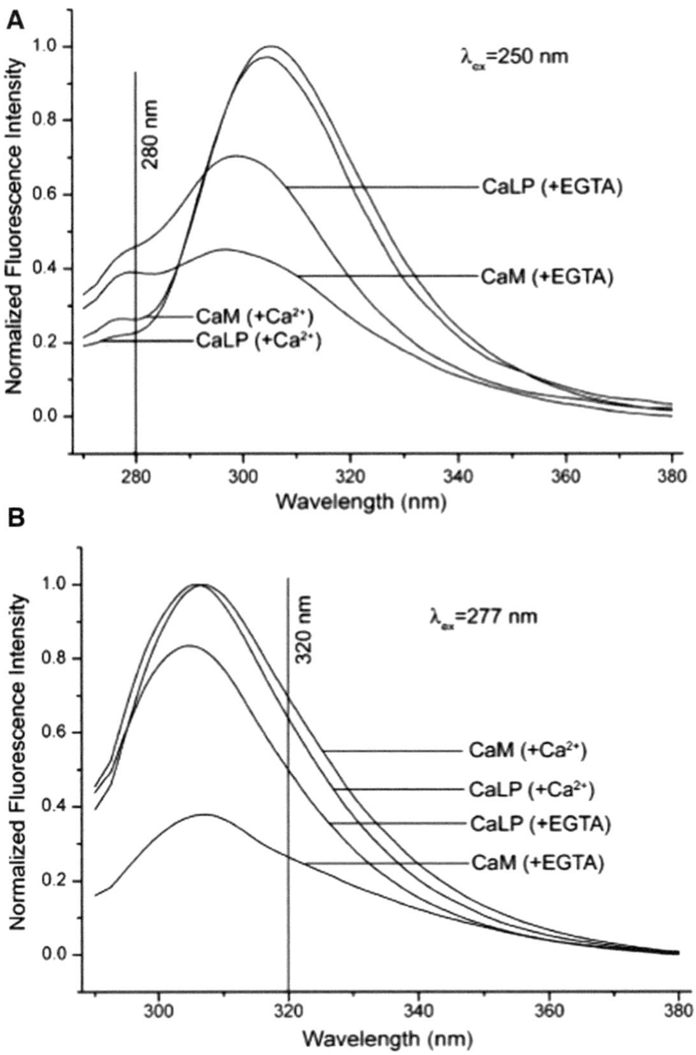

Fig. 2.28 Normalized [(f-f_{min}) / (f_{max}–f_{min})] emission fluorescence spectra of oyster CaLP and CaM [24]
The fluorescence of the phenylalanine (**a**) and tyrosine residues (**b**) in oyster CaLP and CaM was analyzed by an excitation and emission wavelength pair (250/280 nm and 277/320 nm, respectively). CaM and CaLP were diluted in 20 mM Hepes, 100 mM KCl buffer, pH 7.5 with the addition of 5 mM $CaCl_2$ or EGTA, and the final concentration of both proteins was 10 μM

reflected Ca^{2+} could only bind to the 4th EF-hand domain in oyster CaM and CaLP, while this happened in 3rd and 4th EF-hand domains in VanScyoc's case. Due to the substitution of Phe, the Phe fluorescence reflected only Ca^{2+} binding occurred at the 1st, 2nd, and 3rd EF-hand domains in oyster CaM and CaLP, while the 1st and 2nd EF-hand domains in VanScyoc et al.

2.3.2.7 CaLP and CaM Chromatography of Extracts from Oyster Mantle and Gills

Considering the high similarities of CaLP and CaM shared in predicted amino acid sequence, it was with great possibility that these proteins would have identical potential binding sites or target proteins. Potential CaM- and CaLP-binding proteins from the extracts of gill and mantle tissues, two organs directly participating in oyster calcium metabolism, were further compared by Ca^{2+}-dependent affinity chromatography. Figure 2.29 indicated that more proteins, no matter from the mantle

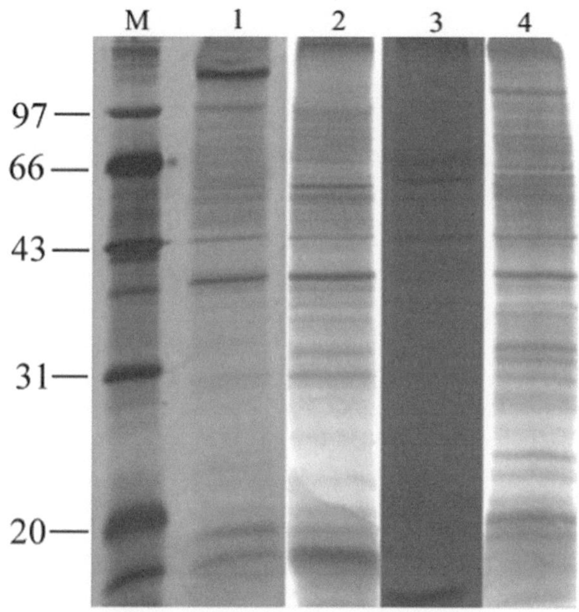

Fig. 2.29 Ca^{2+}-dependent affinity chromatography of extracts from mantle and gill by CaM and CaLP affinity columns
The extraction from oyster tissues of the mantle and gill was loaded on CaM and CaLP affinity columns. After washing with extraction buffer containing 2.5 mM $CaCl_2$, bound proteins were eluted with extraction buffer containing 5 mM EGTA. The eluted proteins were then separated by 12.5% SDS-PAGE and silver stained. M, protein molecular mass markers; lane 1, mantle proteins eluted from CaLP affinity column; line 2, mantle proteins eluted from CaM affinity column; line 3, gill proteins eluted from CaLP affinity column; line 4, gill proteins eluted from CaM affinity column. The molecular mass in kDa was indicated on the left of the gel

or gill, were retained by CaM affinity column than CaLP. Besides, oyster CaM affinity column could react with more target proteins in gill than mantle, which was also consistent with previous findings that oyster CaM gene exhibited a higher expression level in gill [52]. On the contrary, oyster CaLP could bind more proteins in mantle than gill, indicating that CaLP would be quite active in calcium metabolism-related processes in the mantle. Few differences were recognized in the overall patterns of proteins retained by CaLP column or CaM. In future work, we would investigate whether oyster CaLP was able to active CaM-dependent enzymes, carry out two-dimensional electrophoresis, and use proteomic strategy to identify and characterize the different affinity purified proteins which could interact with oyster CaLP and CaM protein in oyster mantle and gill, and this would definitely help a lot to clarify the details of oyster CaLP on regulating calcium metabolism.

Generally speaking, we successfully identified and characterized a full-length cDNA coding for a novel CaM-like protein from *P. fucata*. The oyster CaLP dose has a lot in common with CaM, including strong calcium-binding capacity, high heat stability, Ca^{2+}-dependent electrophoretic shift properties, and Ca^{2+}-induced conformational changes. But the expression pattern of CaLP and CaM was different in major tissues, and CaLP could interact with proteins various from those with CaM, suggesting its different but still important roles in calcium metabolism.

2.3.3 Significance of the Extra C-Terminal Tail of CaLP, a Novel Calmodulin-Like Protein Involved in Oyster Calcium Metabolism

Oyster (*Pinctada fucata*) calmodulin-like protein (CaLP), containing an extra hydrophilic tail (150D–161K) at the C-terminal domain, is a new protein engaged in the regulation of calcium metabolism in oyster. A truncated mutant, M-CaLP, without the extra C-terminus was cloned and expressed in *Escherichia coli* to study the significance of the extra tail to the interactions with targets of CaLP and the Mg^{2+}-/Ca^{2+}-related conformational changes in CaLP protein. The conformational properties of M-CaLP were investigated with fluorescence spectroscopy and CD, the results of which were compared with the intact CaLP and CaM. Results from the far-UV CD indicated that the extended fragment strongly affected the conformational transformations induced by Ca^{2+}, but effects on the Mg^{2+}-induced changes were fairly weaker. However, only mild changes in intrinsic fluorescence spectra of tyrosine and phenylalanine between CaLP and M-CaLP were observed when binding Ca^{2+} or Mg^{2+}. On the other hand, these results suggested that the exposure of the hydrophobic patches of CaLP could be decreased obviously by the extra tail. Furthermore, affinity chromatography indicated that the ability for targets binding of CaLP was strongly affected by its additional tail. In conclusion, from all these results, it was deduced that the extended tail might act as an important role for the interactions of CaLP to its targets in vivo.

```
                                      EF-hand loops
  1 MAEDLTEEQIAEFREAFHLF DKDGSGSISAEE LGTVMRSL  40
 41     GQNPNEQELQDLVEEI DTDGNGEVDFNE FLAMMAKK  76
 77    MKDTDSEEEIREAFRVF DRDDKGFITASE LKHIMTTL 113
114     GEKMDDEEAEEMVAAA DIDGDGEINYEE FVKMISMK 149
150     DTDQEEQQQENK                           161
```

Fig. 2.30 Amino acid sequence of pearl oyster CaLP protein in *P. fucata*
The four EF-hand loops were boxed, and the extra hydrophilic tail (150D–161K) compared with typical CaM sequence was underlined

2.3.3.1 Expression and Purification of the Truncated Oyster CaLP Mutant

Different from typical CaM proteins, there was an extra 12-amino-acid-long (150D–161K) hydrophilic fragment at the C-terminal of the oyster CaLP (Fig. 2.30), which demonstrated that some physiological functions of CaLP might be different from those we already knew. In order to figure out the functions of the extended C-terminal tail in the Mg^{2+}- and Ca^{2+}-promoted structure conformation changes of CaLP and in the relationship between CaLP and the target proteins, a truncated CaLP mutant lacking the extra residues 150D–161K ($CaLP_{(1-149)}$, named M-CaLP) was cloned by PCR. The M-CaLP sequence was then cloned into the pET-15b vector followed by expression in *E. coli.* BL21 (DE3). The recombinant M-CaLP protein had a proportion about nearly 25% of the total soluble extract separated by SDS-PAGE in Fig. 2.31. Only one single band with >95% purity appeared on the 15% SDS-PAGE in which sample buffer contained 2.5 mM EGTA after staining by Coomassie Blue R-250. After purification by phenyl-Sepharose hydrophobic chromatography, the apparent molecular mass was about 20 kDa with no Ca^{2+}, corresponding to a target protein expression level of 20 mg/L LB culture. We've tried to express the oyster CaLP without the His_6-tag in our previous researches, but the protein just failed to be purified by hydrophobic chromatography, even if we increased the salt concentration to promote the ligand-binding capability of CaLP. Luckily, in this paper, the truncated CaLP mutant was successfully expressed and purified by phenyl-Sepharose hydrophobic chromatography, suggesting that the great hydrophilicity of the extended 12 residues at the C-terminal in CaLP may dramatically reduce the exposure of the hydrophobic patches of CaLP induced by Ca^{2+}.

2.3.3.2 Ca^{2+}-Dependent Electrophoretic Shift

Ca^{2+}-induced electrophoretic shift analysis, a useful technique in characterizing CaM, was applied to explore the change of shift pattern induced by Ca^{2+} in the truncated CaLP mutant. The results of the Ca^{2+}-dependent electrophoretic shift of the recombinant CaM, CaLP, and M-CaLP with or with no Ca^{2+} were shown in

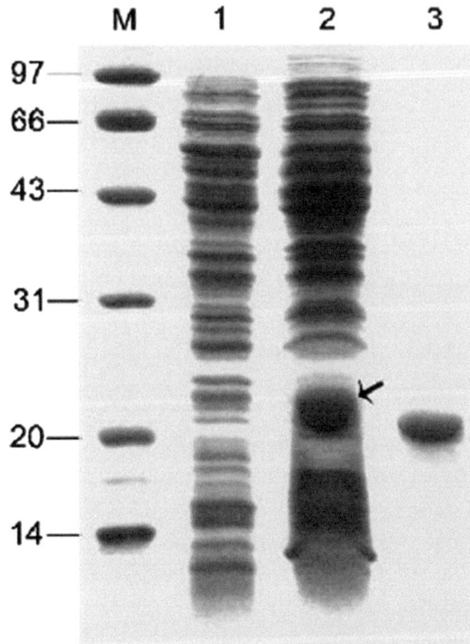

Fig. 2.31 Expression of recombinant *P. fucata* truncated mutant of CaLP (M-CaLP) in culture supernatants
The protein was detected by 15% SDS-PAGE with sample buffer containing 5 mM EGTA and stained by Coomassie Brilliant Blue R-250. Arrow represented the induced proteins after the addition of IPTG. M, protein molecular mass markers; lane 1, un-induced whole-cell lysate; lane 2, whole-cell lysate induced by 0.5 mM IPTG for 3.5 h; lane 3, purified recombinant M-CaLP by phenyl-Sepharose column. The molecular mass in kDa was shown on the left of the gel

Fig. 2.32. The predicted molecular weights of M-CaLP, CaLP (with His_6-tag) and CaM were 16.8 kDa, 19.4 kDa, and 16.8 kDa, respectively. However, after binding with Ca^{2+}, the oyster CaM, M-CaLP, and CaLP displayed as a single band with an apparent molecular weight of about 14 kDa, 18 kDa, and 14 kDa, respectively, while if there was no Ca^{2+}, the apparent molecular weight turned to be 17 kDa, 25 kDa, and 20 kDa, suggesting that the extended segment of CaLP had an important influence on the Ca^{2+}-related structure change of CaLP protein. Using Mg^{2+} to replace Ca^{2+}, all proteins would exhibit little mobility change (data not shown). Then it could be deduced that the mobility change of the bands were regulated by the Ca^{2+}-induced conformation shifts in these three proteins, as well as the extra positive charges provided by Ca^{2+} binding.

2.3.3.3 Circular Dichroism Spectroscopy

It has been reported that CaM often gets significant conformational shifts after binding Ca^{2+} or Mg^{2+}, followed by binding to its target proteins and regulating their activities [127–130]. As a result, far-ultraviolet circular dichroism (far-UV CD) spectroscopy was conducted to test the Ca^{2+}-/Mg^{2+}-induced M-CaLP conformational shifts, which was then compared with those of CaM and CaLP. In the far-UV CD spectra (Fig. 2.33a–c), there was a similar change in the secondary structure of CaM (Fig. 2.33a), CaLP (Fig. 2.33b), and M-CaLP (Fig. 2.33c) with the existence of

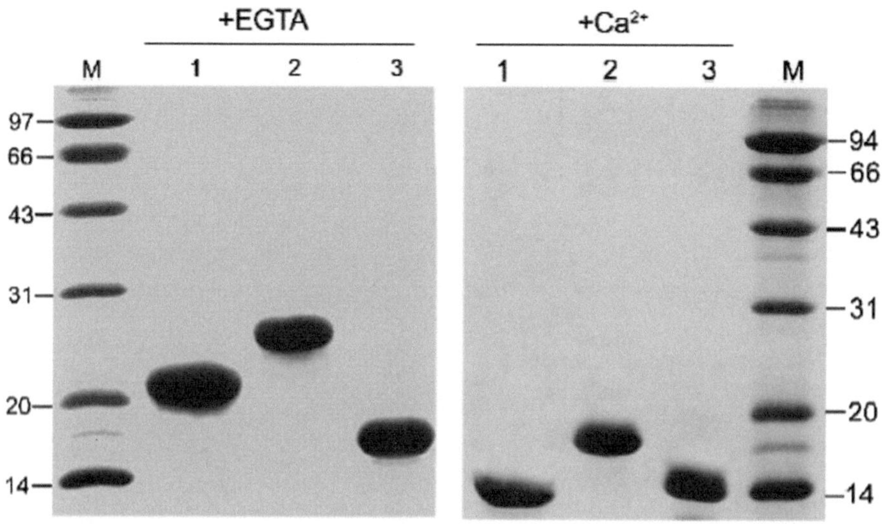

Fig. 2.32 Ca^{2+}-dependent electrophoretic migration of the purified recombinant CaM, CaLP, and M-CaLP in *P. fucata* [126]
The purified recombinant oyster M-CaLP, CaLP, and CaM were run on a 15% SDS-PAGE with the addition of Ca^{2+} or EGTA. The sample buffer was added with 2.5 mM $CaCl_2$ or EGTA. M, protein molecular mass markers; line 1, M-CaLP; line 2, CaLP; line 3, CaM. The molecular mass in kDa was indicated on the left of the gel

EDTA, Ca^{2+}, or Mg^{2+}. When 5 mM $MgCl_2$ or 2 mM $CaCl_2$ was supplied, double-negative peaks at 220 nm and 209 nm appeared in all proteins, while the α-helical contents increase as the value of $\Delta\varepsilon_{220}$ raised about 53%, 79%, and 112% in M-CaLP, CaLP, and CaM, respectively. While when binding Mg^{2+}, the $\Delta\varepsilon_{220}$ increases, respectively, about 24%, 28%, and 97% in M-CaLP, CaLP, and CaM. Results above indicated that both Mg^{2+} and Ca^{2+} would promote obviously conformational shifts in which the content of the secondary structures would change within these proteins, with the effect induced by Ca^{2+} being stronger than that of Mg^{2+}. Our data also indicated that the effect of the extended segment of CaLP on the Ca^{2+}-induced conformational shifts in the secondary structure was fairly stronger than that of Mg^{2+}. However, all proteins appeared to be partial unfolded because the α-helical compositions decreased when $CaCl_2$ was replaced by EDTA. As shown in Fig. 2.33d, there were only inconspicuous differences in the conformational changes of CaM between Mg^{2+} and Ca^{2+} binding. However, a large difference appeared in the magnitudes of the Ca^{2+}-/Mg^{2+}-induced conformational changes either in M-CaLP or CaLP, or as for M-CaLP, the differences in Ca^{2+}-/Mg^{2+}-induced conformational changes were weaker than that of CaLP. It could be demonstrated that the existence of the extended tail of CaLP might cause this phenomenon.

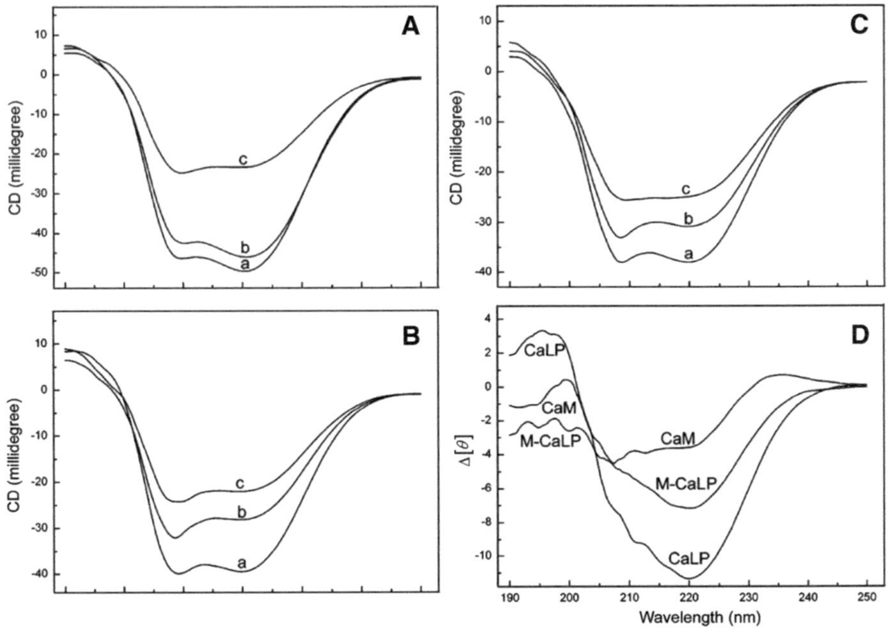

Fig. 2.33 Ca^{2+}-/Mg^{2+}-induced conformational changes of the oyster CaM (**a**), CaLP (**b**), and M-CaLP (**c**) with the addition of Ca^{2+}, Mg^{2+}, or EDTA measured by far-UV CD spectroscopy. **d** represented the difference far-UV CD spectra of the oyster CaM, CaLP, and M-CaLP after binding Ca^{2+} and Mg^{2+}. The difference CD curves of the oyster CaM, CaLP, and M-CaLP were generated by subtracting the spectroscopy in Mg^{2+}-binding state from that of Ca^{2+}-binding state, respectively. The spectra were recorded in 20 mM Hepes, 100 mM KCl buffer, pH 7.5, with the addition of 2 mM $CaCl_2$ (line a), 5 mM $MgCl_2$ (line b), or 2 mM EDTA (line c), and corrected by a blank buffer containing 20 mM Hepes buffer, 100 mM KCl, pH 7.5. The final concentration of each protein was 10 μM

2.3.3.4 Intrinsic Phenylalanine and Tyrosine Fluorescence Assay

Intrinsic tyrosine and phenylalanine fluorescence was performed to monitor the Ca^{2+} and Mg^{2+} binding and further investigate their effects on the conformational shifts of the CaM, CaLP, and M-CaLP. Intrinsic phenylalanine fluorescence spectra contained an excitation at 250 nm and an emission at 280 nm, as displayed in Fig. 2.34a–c. After binding Ca^{2+}, the phenylalanine fluorescence emission reduced nearly to 32%, 51%, and 52% for CaM (Fig. 2.34a), CaLP (Fig. 2.34b), and M-CaLP (Fig. 2.34c), respectively. Replacing Ca^{2+} with Mg^{2+} would lead to a decrease of the emission of phenylalanine fluorescence of M-CaLP, CaLP, and CaM to about 9.6%, 6.6%, and 8.2%, respectively. It was suggested that both Mg^{2+} and Ca^{2+} could facilitate the phenylalanine fluorescence quenching in all these proteins. Simultaneously, it could also be concluded that the extra fragment in CaLP could mildly affect the phenylalanine fluorescence quenching induced by Mg^{2+} in CaLP protein in which the fluorescence quenching might be caused by energy shift to the tyrosine

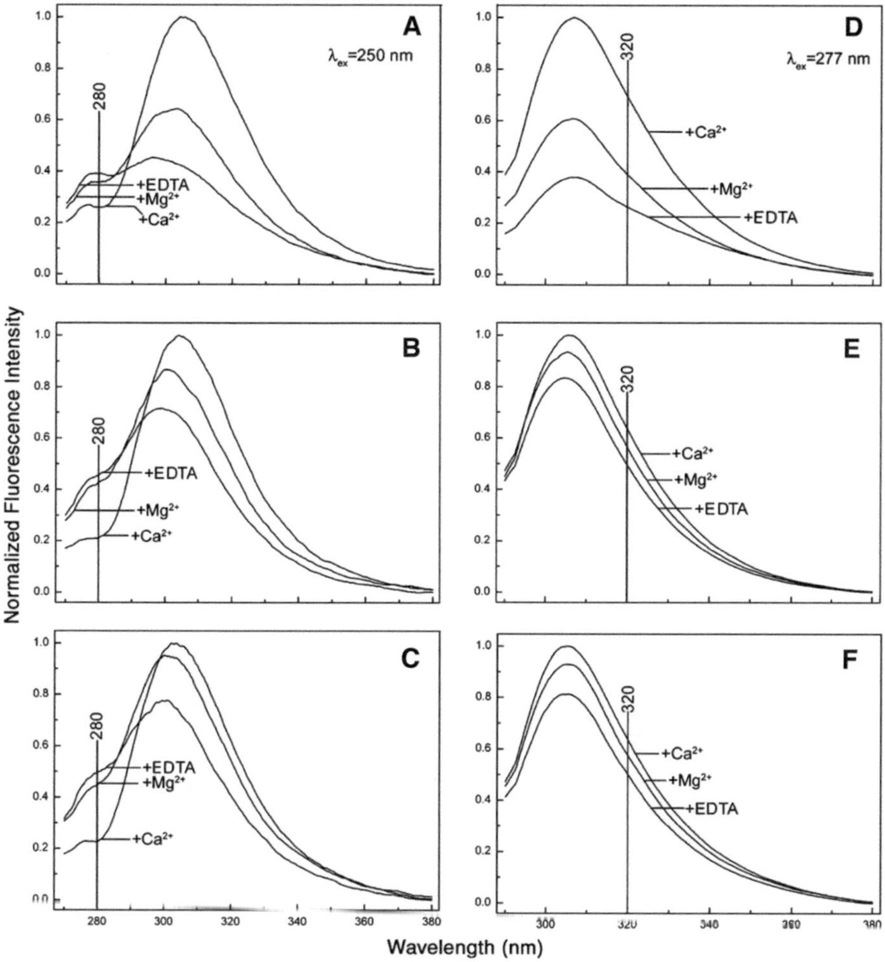

Fig. 2.34 Normalized [(f-f_{min})/(f_{max}-f_{min})] emission fluorescence spectra of the oyster M-CaLP, CaLP, and CaM

The fluorescence of the phenylalanine (**a**–**c**) and tyrosine residues (**d**–**f**) in the oyster CaM (**a** and **d**), CaLP (**b** and **e**), and M-CaLP (**c** and **f**) was analyzed by an excitation and emission wavelength pair (250/280 nm and 277/320 nm, respectively). M-CaLP, CaM, and CaLP were diluted in 20 mM Hepes, 100 mM KCl buffer, pH 7.5, with the addition of 2 mM $CaCl_2$, 5 mM $MgCl_2$, or 2 mM EDTA, and the final concentration of each protein was 10 μM

residue around [16]. The fact that M-CaLP and CaLP were relatively more unfolded than CaM without Ca^{2+} may induce the high Phe fluorescence of CaLP and M-CaLP in the absence of Ca^{2+}. Figure 2.34d–f showed the tyrosine fluorescence emission (excitation at 277 nm and emission at 320 nm) for CaM (Fig. 2.34d), CaLP (Fig. 2.34e), and M-CaLP (Fig. 2.34f). The tyrosine fluorescence emission for M-CaLP, CaLP, and CaM increases about 0.26-fold, 0.28-fold, and 1.6-fold, respectively, after binding Ca^{2+}, while as for M-CaLP, CaLP, and CaM, it increases about

0.14-fold, 0.13-fold, and 0.48-fold, respectively. The results above demonstrated that the extended tail had little effects on the Ca^{2+}-/Mg^{2+}-promoted change of the intrinsic fluorescence of CaLP. However, it couldn't be denied that the extra segment might affect the interactions of the C/CN domains and conformational changes in CaLP. Several previous researches have explored the Mg^{2+}-/Ca^{2+}-induced conformational shifts in CaM [131–137]. However, there was still some controversy about the Ca^{2+}-/Mg^{2+}-binding sites of CaM [138]. Several studies suggested that Ca^{2+} and Mg^{2+} occupied and competed at the same binding site in CaM [132, 133, 135, 136, 139, 140] and there might exist auxiliary sites for binding Ca^{2+} besides the four Ca^{2+} critical sites important for binding enzyme [134, 141, 142]. On the contrary, it had been demonstrated from the electrospray ionization mass spectrometry analysis that Mg^{2+} did not occupy the sites for Ca^{2+} in CaM [143]. Therefore it was quite necessary to figure out whether the extra tail of CaLP functioned as a possible auxiliary binding site for Mg^{2+}/Ca^{2+} or not and explore whether Mg^{2+} did have some effects on the Ca^{2+}-dependent activity of CaLP.

2.3.3.5 8-Anilino-1-Naphthalenesulfonate (ANS) Fluorescence Measurements

ANS, widely used as an external fluorescent detector with a rather low quantum yield in water, however, turns to be quite fluorescent in nonpolar solvent, which makes it a general method to test the surface hydrophobicity of proteins [144]. Figure 2.35 showed the fluorescence emission of ANS for M-CaLP, CaLP, and CaM with the existence of EGTA or Ca^{2+}. In these proteins, the hydrophobic patches exposure to the polar solvent environment could be upregulated by Ca^{2+}, yet there was little effect for Mg^{2+} on the exposure of hydrophobic patches (data not shown). Table 2.4 concluded the recorded changes in maximum wavelength of emission (λmax) and fluorescence intensity. As for CaM, there is an obvious increase in intensity of the ANS fluorescence (about 16.5-fold) but a little reduction of λmax (12 nm) with the existence of Ca^{2+}. The M-CaLP and CaLP, however, appeared to possess a little increased intensity (8.7-fold, 4.8-fold, respectively) but a great reduction of λmax ($\sim$30 nm, $\sim$29 nm respectively) when binding Ca^{2+}. CaM usually underwent a structure change to a separated "open" conformation in Ca^{2+}-saturated condition, from a four-helix bundle-like "closed" structure in Ca^{2+}-free condition [145]. The conformational shift resulted from the exposure of the hydrophobic patches induced by Ca^{2+} in CaM was the important original event in the interaction between CaM and its targets [146, 147]. Our data above indicated that the extra fragment led to the reduction of the energetically unfavorable exposure of a huge nonpolar surface of CaLP; on the contrary, the deletion of the extended fragment could obviously rise the exposure degree of the hydrophobic patches in CaLP. As a result, the interactions between CaLP and its target proteins and the stability of CaLP might be under regulation by this extra fragment. The functions of the extra fragment in the stability of CaLP still remain under investigation at present. Since a high similarity (93.9%) existed between the oyster CaM and M-CaLP, the

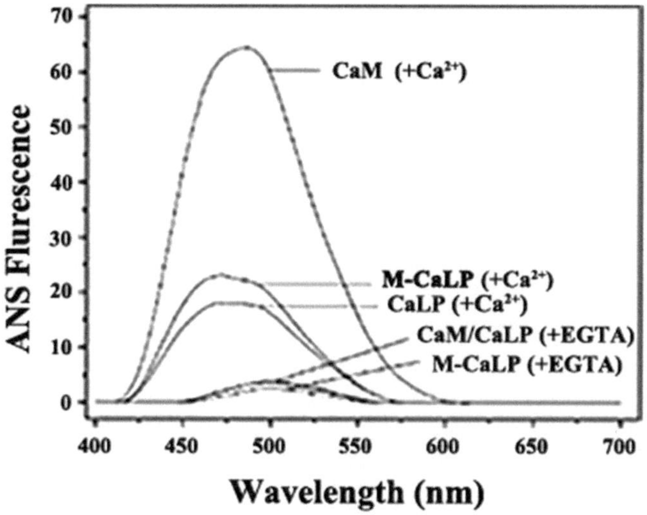

Fig. 2.35 Conformational changes of the oyster M-CaLP, CaLP, and CaM measured by ANS binding with the addition of Ca^{2+} or EGTA
Fluorescence emission spectra were monitored with excitation wavelength set at 370 nm for ANS, and all measurements were performed using 10 μM each protein, ANS at a final concentration of 50 μM, and 20 mM Hepes, 100 mM KCl buffer, pH 7.5, in the presence of 2 mM $CaCl_2$ or EGTA at 25 °C

Table 2.4 Fluorescence spectral changes of ANS in the presence or absence of Ca^{2+}

Protein	$\Delta\lambda_{max}$	$I_{max}(+Ca^{2+})/I_{max}(-Ca^{2+})$
CaM	12	16.5
CaLP	29	4.8
M-CaLP	30	8.7

Δλ max indicated the difference of maximum fluorescence emission of ANS in the addition and absence of Ca^{2+}. I_{max} $(+Ca^{2+})/I_{max}$ $(-Ca^{2+})$ implied the ratio of the maximum fluorescence intensity of ANS in the presence of Ca^{2+} to that of in the absence of Ca^{2+}

ANS fluorescence with Ca^{2+} became stronger in CaM than that in M-CaLP, suggesting the formation of a relatively more exposed hydrophobic surface of CaM. The mechanism of these changes were still not clear enough, which, however, could be deduced to be the result of the sequence changes of EF-hand domains between M-CaLP and CaM [126].

2.3.3.6 CaLP and M-CaLP Chromatography of Extracts from Oyster Mantle and Gills

CaM has been found as a calcium-binding protein with high conservation in all eukaryotic cells. The conservation of CaM sequence seems to result from the sum of requirements dictated by a wide spectrum of CaM-related proteins rather than the strict structural requirements essential for activating any one protein [148]. The existence of the extra tail in CaLP might have important influences on the variety of proteins/enzymes that function as targets of CaLP. M-CaLP-binding or CaLP-binding candidate proteins in the extracts of the oyster mantle or gill issues, two tissues that are thought to directly participate in the calcium metabolism, were investigated by Ca^{2+}-dependent affinity to further study the effects of the extra tail on target interactions of CaLP. It was demonstrated in Fig. 2.36 that proteins from the mantle were retained by CaLP affinity chromatography more easily than by M-CaLP. On the contrary, in the gills the binding activity of M-CaLP seemed more obvious than that of CaLP. There were few main difference bands in overall patterns of proteins held by CaLP affinity chromatography column compared with those

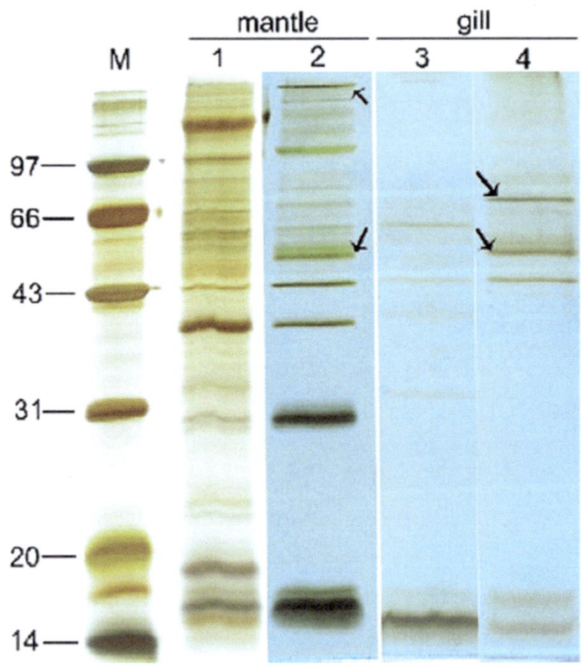

Fig. 2.36 Ca^{2+}-dependent affinity chromatography of extracts from the oyster tissues of mantle and gill by M-CaLP and CaLP affinity columns
The extracts from the oyster tissues of mantle and gill were loaded on CaLP and M-CaLP affinity columns. After washing with abundant extraction buffer containing 2.5 mM $CaCl_2$, the bound proteins were eluted with extraction buffer containing 5 mM EGTA. The eluted proteins were separated by 12.5% SDS-PAGE and silver stained. M, protein molecular mass markers; lane 1, mantle proteins eluted from CaLP affinity column; line 2, mantle proteins eluted from M-CaLP affinity column; line 3, gill proteins eluted from CaLP affinity column; line 4, gill proteins eluted from M-CaLP affinity column. The molecular mass in kDa was shown on the left of the gel. Arrows represented the major difference proteins retained by M-CaLP affinity column compared with the proteins eluted from CaLP affinity column in gills and mantle

obtained from M-CaLP affinity chromatography column from the gill or mantle. However, there were still some slight difference bands retained by CaLP affinity chromatography column compared with those from M-CaLP affinity chromatography column in mantle. The data suggested that the extended part at C-terminal would greatly activate the target selectivity of the wild CaLP protein. It had a 50% identity with the 12 amino acid residues from 77 M to 90 K, locating within the central helical area of CaM, even though the extended segment of CaLP had no homolog within any other database. Since the flexible central helix (AA residues 65–92) connecting the two domains of CaM plays crucial roles in the characteristics of CaM for binding different target proteins [149], it could be assumed that the existence of the extra segment may act as a predicted repeat sequence derived partially from the central helical part of CaM inducing the increased target selectivity of CaLP in vivo.

Considering that the high level of amino acid sequence similarity truly existed between M-CaM and CaLP, it could be speculated that these two proteins might have a lot of similar target proteins. However, it seemed that more proteins might have interactions with CaM in gill compared with that of M-CaLP (data not shown), which suggested that the exact structure or conformation of CaM might be different from that of M-CaLP. Furthermore, since a lot of major channel proteins and enzymes like sarcoplasmic reticulum Ca^{2+} release channel [150], adenylyl cyclase [151] and inducible nitric oxide [152], were all under regulation by CaM in Ca^{2+}-free condition, it would be difficult to explore these targets that could interact with M-CaLP or CaLP by Ca^{2+}-dependent affinity chromatography.

As a rather complicated and highly regulated biochemical and physiological process, the calcium metabolism in oyster relied mainly on existence of various regulatory proteins in same or different tissues. CaLP and CaM in mantle and gill might contribute a great and efficient regulatory function to the complicated process. Furthermore, the extended tail of CaLP might have great effects on target protein interactions and protein stability of CaLP. Finally, the role of CaLP in regulating the oyster calcium metabolism-related proteins/enzymes, for example, CaM-dependent protein kinases, calcium channel, or pump, could also be regulated by the extra region. Further investigation of functions of the extended area in stabilizing and regulating roles of CaLP, and additional identification of the exact target proteins which have interactions with M-CaLP or CaLP in the mantle or gill, would be performed to figure out the specific mechanism of the function–structure connections and the regulation functions of this new protein during the calcium metabolism process and Ca^{2+} signaling pathway in oyster.

In conclusion, these results demonstrated that the extended C-terminal tail (150D–161K) of CaLP could significantly regulate the conformational shifts induced by Mg^{2+}/Ca^{2+} in the composition of the secondary structures of CaLP and greatly reduce the exposure degree of hydrophobic patches in intact CaLP protein. Therefore, the extra tail would obviously affect the interactions between CaLP and the proteins targets. All results demonstrated that the extra tail act as a critical role in the interactions between CaLP and its target proteins in vivo.

2.3.4 *Localization of Calmodulin and Calmodulin-Like Protein and Their Functions in Biomineralization in* P. fucata *[49]*

Calmodulin (CaM) and calmodulin-like protein (CaLP) are two important proteins engaged in biomineralization. Their localizations in *Pinctada fucata* mantle epithelia were thoroughly studied by immunogold electron microscopy and Western blot (WB) analysis of the nuclear/cytosol fraction of primary cultured *P. fucata* mantle cells in detail. The results revealed a completely distinct distribution of these two proteins from the subcellular level. CaM was distributed throughout both the cytoplasm and nucleus of the mantle epithelium cells, but CaLP was only detected in the cytoplasm. The functions of these two proteins in biomineralization were further analyzed by shell regeneration assay. The mRNA relative expression level of CaM and CaLP were highly increased in different organelles of the mantle epithelium during this process. What's more, overexpression of these two proteins and a mutant of calmodulin-like protein (M-CaLP), the one lacked an extra C-terminal tail in MC3T3-E1 cells, facilitated the expression of osteopontin, which was a typical biomineralization marker for osteoblasts. All of the results implied that the distribution pattern of CaM and CaLP are totally different in mantle epithelium and they could affect the biomineralization process from different levels. The extra C-terminal tail of CaLP is of great importance for its functions during biomineralization in *P. fucata*.

2.3.4.1 Distribution of CaM and CaLP in Primary Cultured Mantle Cell

Forty-eight hours after initiation of the mantle cell cultures, typical fibroblast-like cells and epithelial-like cells were observed to migrate from the explant (Fig. 2.37a). On the 7th day after initiation of mantle cell culture, the epithelial-like cells formed a monolayer on the poly-L-lysine-coated dishes (Fig. 2.37b).

To avoid the cross-reaction between CaM and CaLP antibodies and the proteins, the purified antibodies of CaM and CaLP were tested by Western blot analysis. The negative cross-reaction results indicated that the two antibodies were specific for CaM and CaLP, respectively (data not shown). Western blot on the extracts of cultured cells showed that CaM was distributed throughout the nucleus and cytoplasm of the mantle epithelium but CaLP was distributed only in the cytoplasm (Fig. 2.38). Immunogold electron microscopy confirmed this distribution pattern (Fig. 2.39).

2.3.4.2 Expression of CaM and CaLP During Shell Regeneration

Western blot analysis indicated that, compared with the mantle before the notching, the expressions of CaM and CaLP were activated and reached the highest level on the 3rd and 4th days (Fig. 2.40a). Moreover, results of the expression analysis by

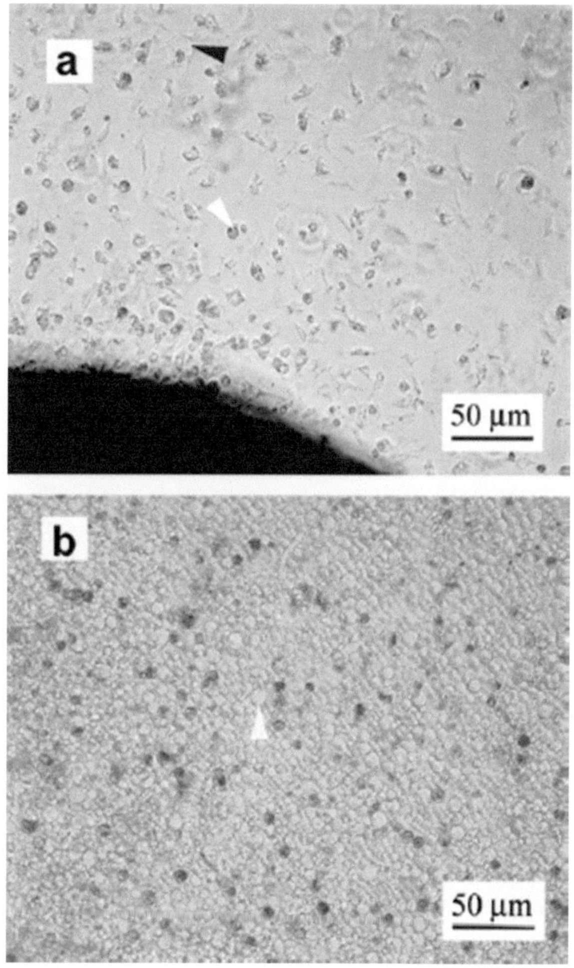

Fig. 2.37 Morphology of the cultured mantle cells (**a**) Cells migrated from mantle tissue after 48 h culture; (**b**) a monolayer of epithelia formed after 7 days' culture. Black arrow shows fibroblast-like cells and white arrow shows epithelial-like cells

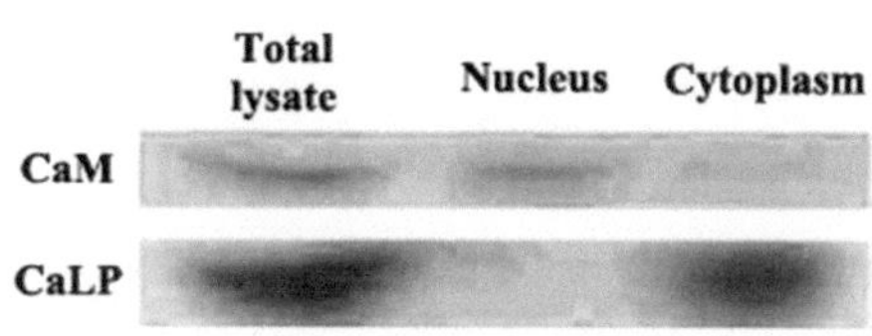

Fig. 2.38 Western blot analysis of CaM and CaLP in nuclear/cytosol extraction from primary cultured mantle epithelia
Total lysates of mantle cell and fractions of nucleus/cytoplasm were separated by SDS/PAGE, immunoblotted, and revealed with anti-CaM or anti-CaLP antibody

RT-PCR at mRNA level were almost the same as what we obtained by Western blot, only that the highest level was reached on the 2nd and 3rd days (Fig. 2.40b). Localization of CaM and CaLP by immunogold electron microscopy demonstrated

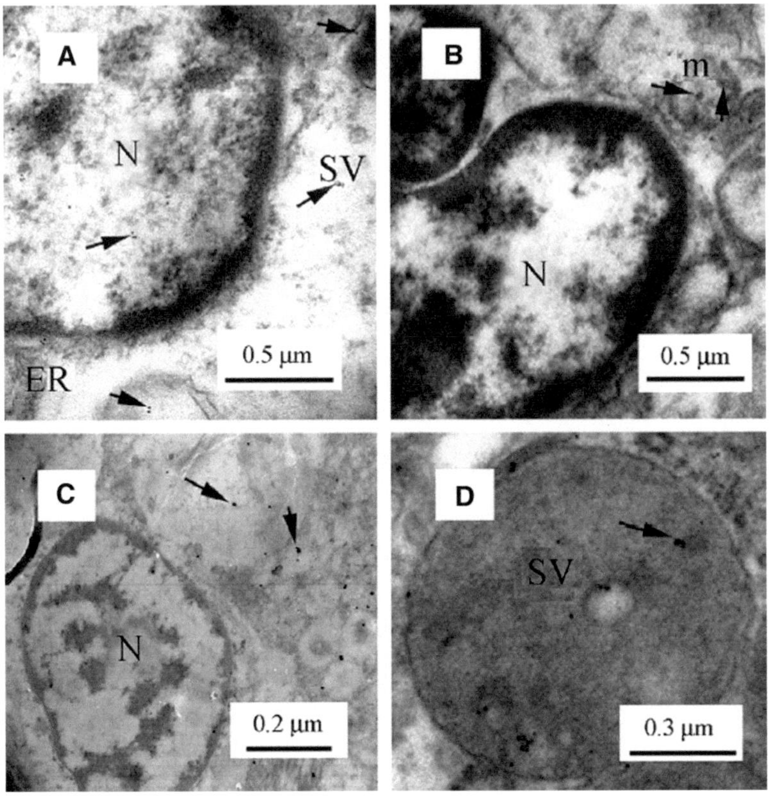

Fig. 2.39 Subcellular localization of CaM (a and b) and CaLP (c and d) in epithelium of mantle by immunogold [49]
(**a**) CaM in nucleus (N), secretory vesicles (SV), and endoplasmic reticulum (ER) indicated by arrows; (**b**) CaM in mitochondria (m); (**c**) no CaLP in nucleus and the existence of CaLP in cytoplasm (arrows); (**d**) CaLP in secretory vesicles (SV) indicated by an arrow

that during the shell regeneration, the CaM existed mainly in the nucleus, mitochondria, and ER, while CaLP existed mainly in the secretory vesicles (Fig. 2.41).

2.3.4.3 Biological Functions of CaM, CaLP, and M-CaLP in Transfected MC3T3-E1 Cells

CaM, CaLP, and M-CaLP Promote Proliferation of MC3T3-E1 Cells

We examined the cell proliferation-promoting effects of CaM, CaLP, and M-CaLP in transfected MC3T3-E1 cells. Cell proliferation was analyzed by MTT assay. As shown in Fig. 2.42, transfections of CaM, CaLP, or M-CaLP, particularly CaM, significantly promoted cell proliferation when compared with cells transfected with the empty vector. Furthermore, the proliferation-promoting effect of M-CaLP was

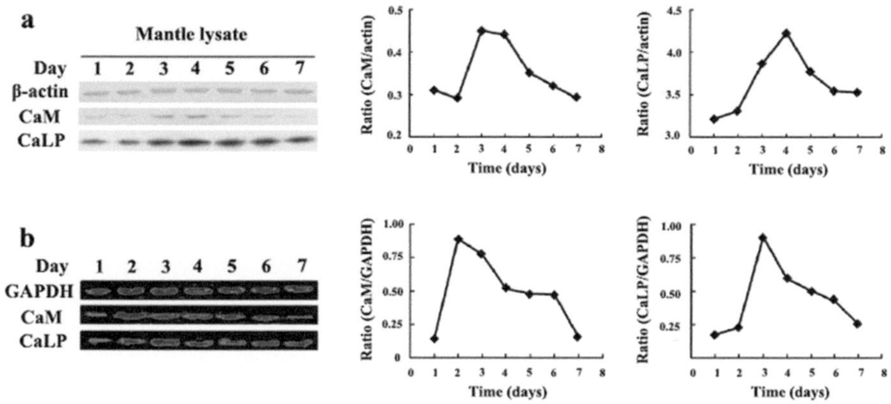

Fig. 2.40 CaM and CaLP expressions during shell regeneration
(**a**) Total mantle lysates of mantle tissue were separated by SDS/PAGE, immunoblotted, and revealed with anti-CaM/CaLP antibody. The curves on the right demonstrate that the protein expression level of CaM and CaLP peaked on the 3rd and 4th days, respectively. (**b**) Relative mRNA expression levels of CaM and CaLP detected by RT-PCR for 7 days. The curves demonstrate that the mRNA expression level of CaM and CaLP peaked on the 2nd and 3rd days, respectively

more significant than that of wild-type CaLP. These results indicated that CaM plays a more significant role than CaLP in MC3T3-E1 proliferation.

ALP Activity in Transfected MC3T3-E1 Cells

Alkaline phosphatase is thought to be important to biomineralization [153, 154] and often acts as a phenotypic marker for the osteoblast. Previous studies suggested that many cytokines induced MC3T3-E1 cells to develop into osteoblast-like cells in vitro and expressed alkaline phosphatase [155]. Our results showed that MC3T3-E1 cells transiently transfected by CaLP experienced a significant increase in ALP activity, but the MC3T3-E1 cells transfected by CaM and M-CaLP did not exhibit that significant increase when compared with those transfected by CaLP (Fig. 2.43). This difference suggested that CaLP plays a significant role in cell differentiation.

Expression of Osteopontin (*OP*) in Transfected MC3T3-E1 Cells

As shown by RT-PCR analysis (Fig. 2.44), CaM and CaLP transfected cells (especially CaLP) showed higher mRNA levels of *OP* than empty vector transfected cells. But the M-CaLP, which lacks the C-terminal, obviously exhibited much lower

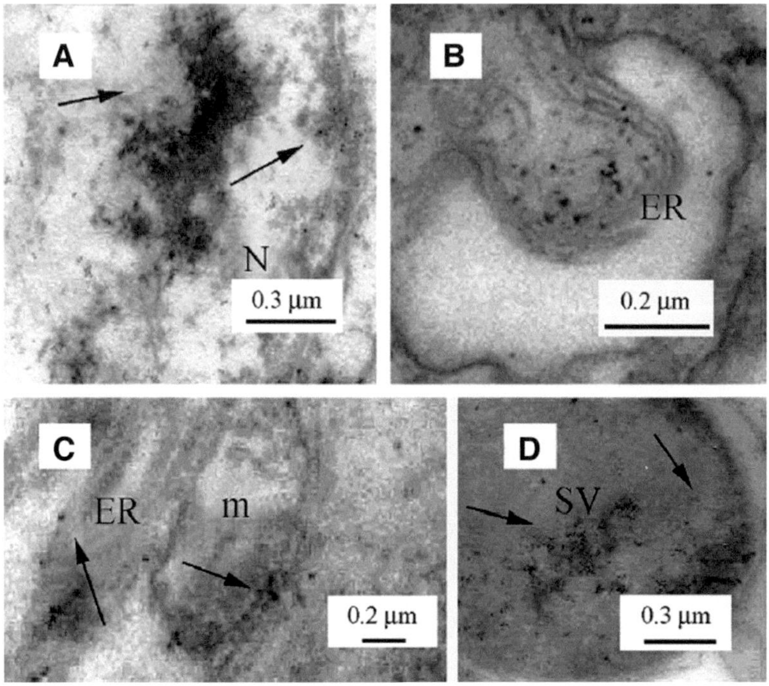

Fig. 2.41 Subcellular localization of CaM (a, b, and c) and CaLP (d) in epithelium of mantle during shell regeneration induction.
(**a**) Expression of CaM (arrows) increased obviously in nucleus (N); (**b** and **c**) expression of CaM (arrows) increased obviously in endoplasmic reticulum (ER) and mitochondria (m); (**d**) expression of CaLP (arrow) increased obviously in secretory vesicles (SV)

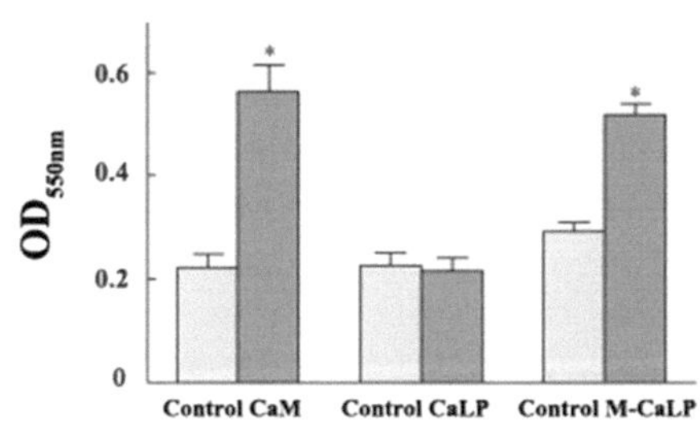

Fig. 2.42 Expression of CaM, CaLP and M-CaLP in MC3T3-E1 cells
The effect of CaM was the most significant and CaLP did not significantly promote cell proliferation, $*p < 0.01$

OP mRNA level compared with wild-type CaLP. These data indicated that expression of CaLP stimulated the expression of the bone matrix protein *OP* in MC3T3-E1 cells.

Previous research on calmodulin indicated that it directly affects osteoblast differentiation by mediating the Ca^{2+} signal to a multitude of different enzyme systems [156–161]. Localization of calmodulin has been investigated in a broad range of different tissues and cultured cells, and it is closely related with tissue types,

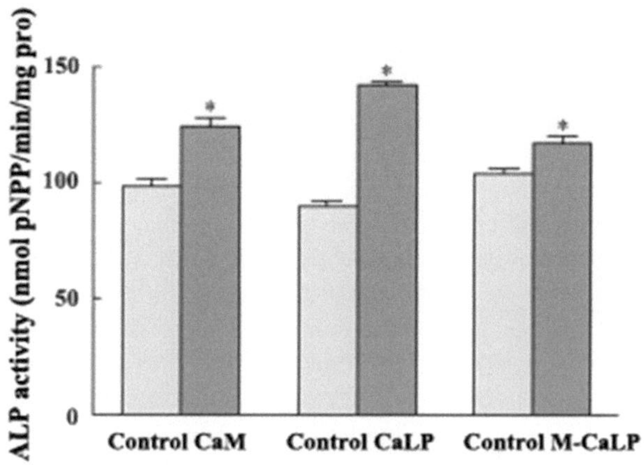

Fig. 2.43 ALP activity in MC3T3-E1 cells transfected with CaM, CaLP, and M-CaLP
ALP activity was highest in the CaLP experimental group, $*p < 0.01$

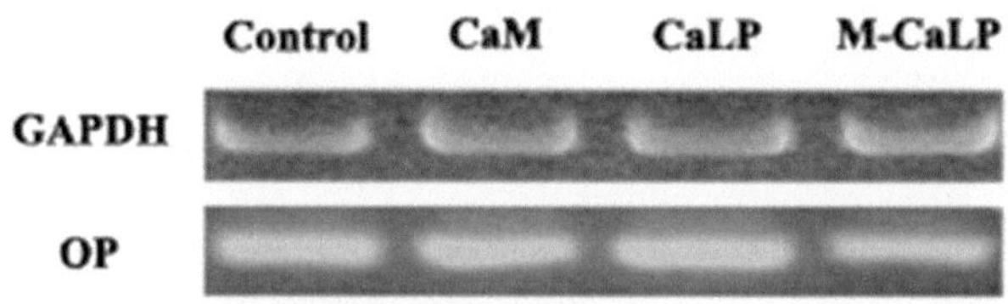

Fig. 2.44 RT-PCR for OP which is the biomineralization-related gene after the transient transfection of CaM, CaLP, and M-CaLP in MC3T3-E1 cells
Compared with the control, which was transfected with vectors, the CaM and CaLP transfected cells showed higher levels of OP mRNA, especially the CaLP transfected cells. But the effect of M-CaLP obviously lost the stimulating functions compared with CaLP

physiological processes, and developmental stages. For example, in the epidermis of the *Salmo trutta*, the calmodulin is localized in the superficial epithelial cells, indicating that it might have physiological significance in the activation of the microvillar skeleton and play an important role in controlling the permeability of the skin epithelium [162]. In mammalian tissues such as rat's liver and cortex, calmodulin is localized in the cytoplasm and nucleus where it plays an important role in the nuclear process and glycogen metabolism [163]. In mitotic cells, it is in the mitotic spindle and is excluded from the nucleus in interphase cells, but there is no specific enrichment in the cytoplasm [164]. The localization of calmodulin in spindle pole body of *Schizosaccharomyces pombe* indicated that it plays an essential role in chromosome segregation [165].

Preliminary studies of physiological process of shell regeneration showed that 6 days were required for a thin sheet of mineralized tissue to cover the shell notch [166] and initial mineral deposition occurred about 24–48 h after the creation of the notch in the marine snail, *Tegula*. In the present study, RT-PCR analysis of CaM and CaLP confirmed the time course of the shell regeneration in *P. fucata*. Localization of CaM in the intact mantle and its obvious increase in the nucleus, mitochondria, and endoplasmic reticulum (ER) after the shell regeneration indicated that CaM actively takes part in the shell regeneration. The transient transfection of CaM in MC3T3-E1 cells clearly confirmed its functions in cell proliferation.

Biomineralization systems such as coral, tooth, eggshell, and shell of mollusk revealed that a number of small, round secretory bodies were responsible for the biomineralization process [167]. Coral calicoblast cells contain membrane bound vesicles with homogeneously fine granular contents, and these secretory vesicles may be the sites of organic matrix synthesis [168]. There are many secretory vesicles present in the newly deposited predentin matrix in teeth and inside the eggshell mantle [169, 170]. We found an obviously discrepant localization of CaLP from the CaM, especially the enrichment of CaLP in secretory vesicles after the shell regeneration, which indicated the important effects of CaLP in biomineralization, though the precise mechanism of it needs to be unveiled [49].

Several biomineralization markers for bone cell differentiation and mineralization have been identified such as osteopontin (OP), osteonectin (ON), and octeocalcin (OC). OP, a highly acidic phosphoprotein, is synthesized and secreted by osteoblasts. OP has been demonstrated as a multifunctional protein that can trigger signal transduction processes in osteoclasts [171] and regulate fibroblast cell attachment and the calcification of extracellular matrix [172]. OP production is frequently augmented when cell signaling pathways are activated by a variety of Ca^{2+}-mediated or Ca^{2+}-dependent stimuli [173]. In this study, mRNA expression of OP gene was significantly enhanced in the MC3T3-E1 cells transfected with wild-type CaLP, and it was very interesting to note that M-CaLP with its C-terminal truncated lost the differentiation- and mineralization-promoting functions by ALP and RT-PCR analysis, suggesting the C-terminal of M-CaLP is a crucial functional domain of the protein.

Our previous research showed that Ca^{2+}/Mg^{2+} could induce conformational changes of CaM, CaLP, and M-CaLP, which are more obvious in CaLP and M-CaLP than CaM, and the changes in M-CaLP are smaller than those in CaLP. Although the extra tail in CaLP shares no homolog with any other domains, it shows 50% identity with a 12 amino acid sequence located in the central helical region of CaM. Since the flexible central helix contributes greatly to the functional characteristics of CaM binding to various target proteins [149], we assume that the existence of this extra tail may represent a repeated domain derived from a part of central helical region of CaM, and this extra region may increase target selectivity of CaLP in vivo by binding and activating completely different target enzymes/proteins from CaM [174].

Taken together, our data indicate that CaM and CaLP in *P. fucata* are both involved in cell proliferation and differentiation. But the different subcellular localization of these two proteins and function of CaLP in differentiation and biomineralization seem to be closely related with the extra 12 amino acid C-terminal tail of CaLP through interactions with target proteins.

2.3.5 Significance of the C-Terminal Globular Domain and the Extra Tail of the Calmodulin-Like Protein (Pinctada fucata) in Subcellular Localization and Protein–Protein Interaction

Calmodulin (CaM) was reported to play very significant roles in many physiological processes and shared high conservative properties in different species. In an early research, we successfully obtained the full cDNA length of CaM gene and a new calmodulin-like protein (CaLP) with an extra C-terminal domain from *Pinctada fucata*, and then it was expressed in *Escherichia coli*. In this paper, fluorescence confocal microscopy was applied to investigate the protein–protein interaction between CaM/CaLP and $p21^{Cip1}$ derived from mammalian cells, to reveal the different properties of the two proteins in vivo. It was showed that both the C-terminal globular domain and the extra tail of CaLP were quite critical in its cytoplasmic sequestration. The most attractive phenomenon was that the translocating of CaLP from the cytoplasm to nucleus could be promoted by transfection of $p21^{Cip1}$, which was not the case for CaM. However, in co-immunoprecipitation on different mutants of CaLP with $p21^{Cip1}$ and fluorescence confocal microscopy experiments, it was showed that the C-terminal globular domain of CaLP was responsible for CaLP's transportation to nucleus from cytoplasm.

2.3.5.1 C-Terminal of CaLP Ensures the Sequestration of Protein in the Cytoplasm

To obtain insight into the biological difference between CaLP and CaM, it was quite necessary to explore their subcellular localization. Fluorescence imaging of GFP-CaLP and GFP-CaM in HEK293T cells showed that CaM distributed homogeneously within cells, while CaLP primarily located in cytoplasm (Fig. 2.45), suggesting that CaM and CaLP might play different roles in cells. In our previous research of CaM and CaLP cloned from *P. fucata*, CaLP had 87% similarity and 67% identity with CaM. However, the gene expression assays by ISH, Ca^{2+}-dependent affinity chromatography by affinity columns and CD spectroscopy in Ca^{2+}-/Mg^{2+}-related conformational changes told the differences between the two proteins in vitro [174]. It was not surprise to find the ubiquitous distribution of CaM-WT due to the high conservation and extensive appearance of CaM localization in many other species. What attracted us most was the cytoplasmic distribution of CaLP-WT. To confirm the domain which was responsible for the localization of CaLP, different mutants were established to investigate the domain accounting for the cytoplasm sequestration. The CaLP-Δ 150-161 deletion mutant, truncated in the tail, failed to rest in the cytoplasm. To ensure the influences of the molecular weight of the extra tail in CaLP-WT on transportation of CaLP-WT into nucleus through the nuclear pore complexes (NPCs), a CaM + tail mutation, with an extra tail of CaLP added to

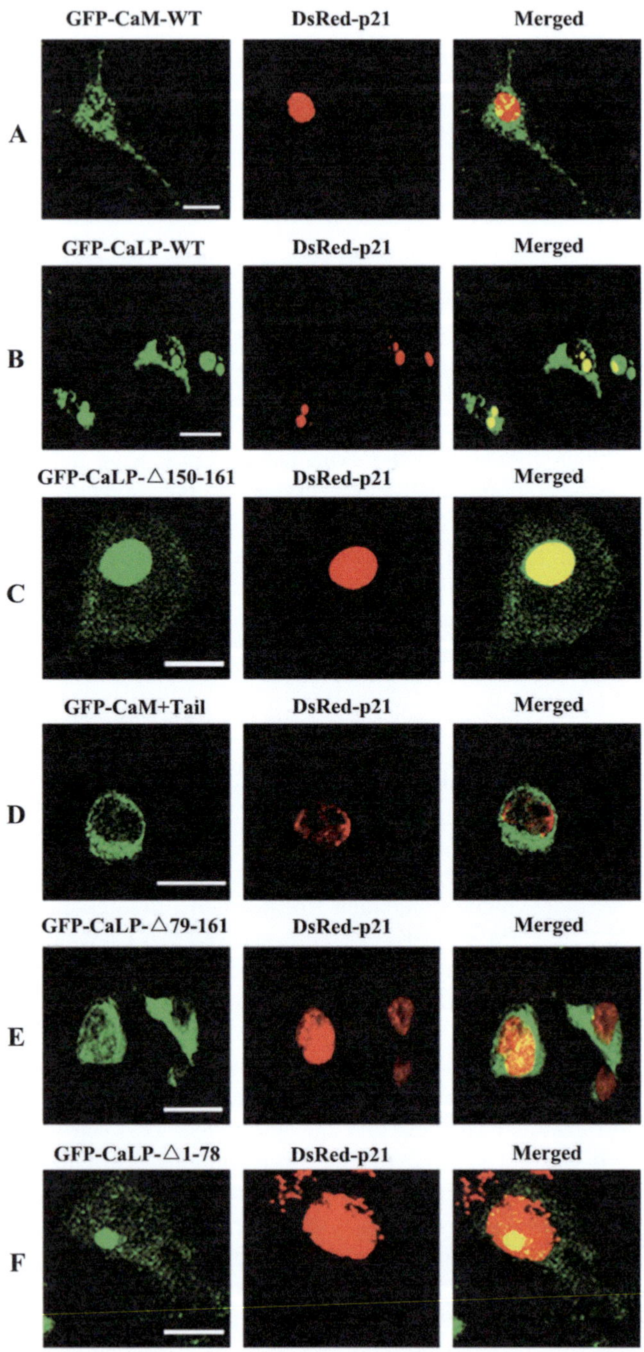

Fig. 2.45 Subcellular localization of $p21^{Cip1}$ with CaM, CaLP, and their mutants [126]. (**a**, **d**, **e**) Subcellular localization of CaM, CaM + tail, and CaLP-Δ 79-161 was not influenced by co-transfection with $p21^{Cip1}$. (**b**, **c**, **f**) CaLP-WT, CaLP-Δ 150-161, and CaLP-Δ 1-78 were stimulated by $p21^{Cip1}$ to translocate from the cytoplasm to nucleus. Scale bar: 10 mm

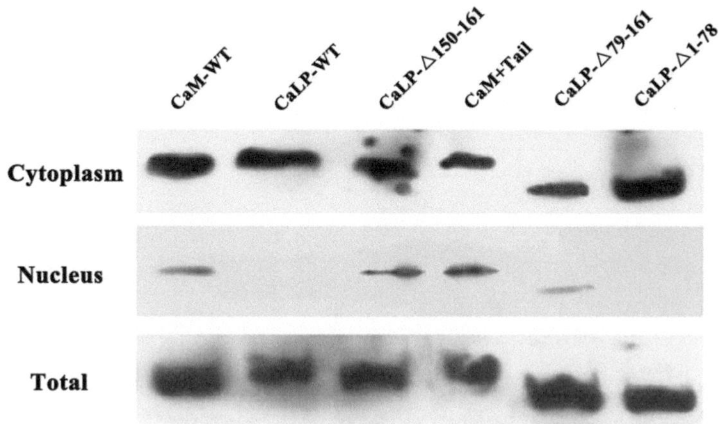

Fig. 2.46 Localization of CaM, CaLP and the mutants in cytoplasm and nuclear cell fractions in 293T cells
Proteins within various fractions of 293T cells were investigated by SDS-PAGE and immunoblotting. Upper lane: cytoplasm fractions for fusion proteins. Middle lane: nucleus fractions for fusion proteins. Lower lane: total expression of exogenous GFP protein. The antibody for GFP was used for the Western immunoblots, and what the figure shown was representative for three experiments

the C-terminal of CaM, was built and the distribution of CaM + tail resembled that of CaM. Early NMR researches showed that a flexible central nonhelical domain 182 linked the two globular domains of CaM. So in this study, the CaM (*P. fucata*) was separated into two parts from 79th amino acid which has been reported in research of *Ikura* et al. [175]. The CaLP-Δ 1-78 mutant, which still had the C-terminal globular domain and the extra tail, had a truncated N-terminal globular domain of CaLP-WT and located in the cytoplasm; however, the CaLP-Δ 79-161 mutant which had a truncated C-terminal extra tail of CaLP-WT and globular domain was found to locate throughout nucleus and cytoplasm. From the experiments of immunoprecipitation of cytoplasmic and nuclear fractions from the wild type and these mutants, it was further proved the C-terminal globular domain in CaLP plus the extra tail did have significance in protein localization (Fig. 2.46). The results of the fluorescence confocal microscopy suggested that the extra tail and the C-terminal globular domain determined the sequestration of CaLP in cytoplasm.

2.3.5.2 Overexpression of p21^{Cip1} Stimulates CaLP Translocation from Cytoplasm to Nucleus

It has been reported that p21^{Cip1} could directly bind to CaM in mammalian cells [176–178]. To verify whether the CaM-WT in *P. fucata* also had the ability to bind to p21^{Cip1}, subcellular co-localization after overexpression of p21^{Cip1} with CaM in HEK 293T was tested by co-immunoprecipitation and fluorescence confocal microscopy. It was showed that the wild-type CaM in *P. fucata* bound to $p21^{Cip1}$ in vivo. After being co-transfected with $p21^{Cip1}$, it was amazing to observe that CaLP could translocate to nucleus from the cytoplasm and co-localize with $p21^{Cip1}$ (Fig. 2.47). Different pEGFP link-coupled mutants stated above were co-transfected

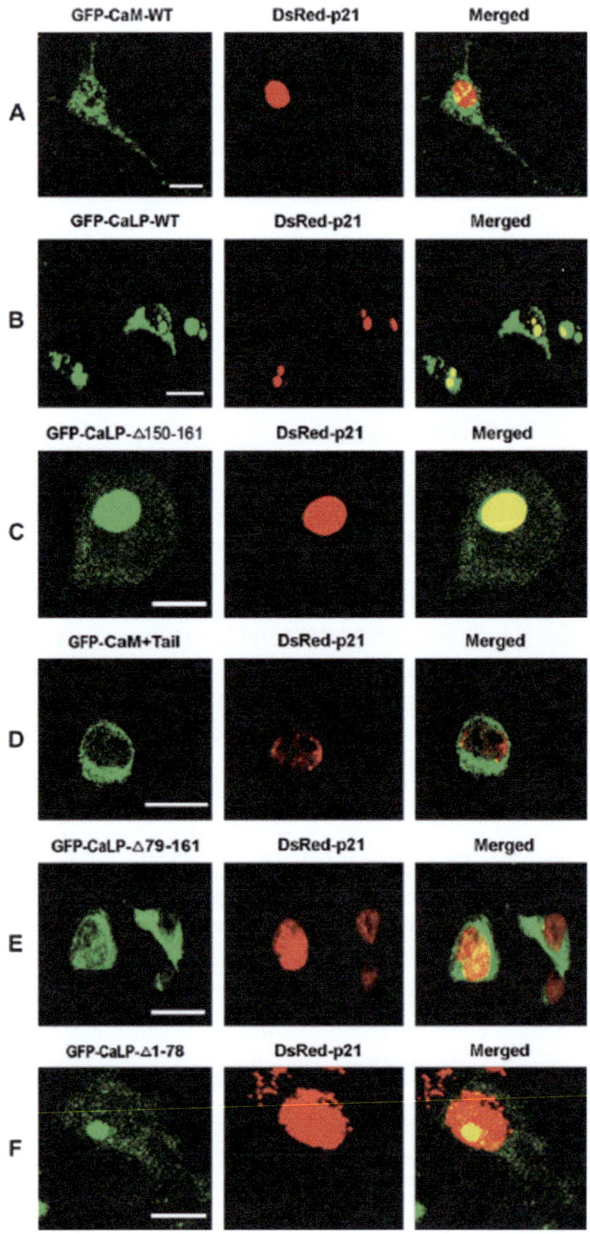

Fig. 2.47 Subcellular localization of $p21^{Cip1}$ with CaM, CaLP, and their mutants
(**a**, **d**, **e**) Subcellular localization of CaM, CaM + tail, and CaLP-Δ 79-161 was not affected by co-transfection with $p21^{Cip1}$. (**b**, **c**, **f**) CaLP-WT, CaLP-Δ 150-161, and CaLP-Δ 1-78 were stimulated by $p21^{Cip1}$ to translocate from the cytoplasm to the nucleus. Scale bar: 10 mm

with $p21^{Cip1}$ to further investigate which domains in CaLP were essential for translocation. As shown in Fig. 2.47, when co-transfected with $p21^{Cip1}$, (GFP)-CaLP-Δ 150-161 translocated into the nucleus and co-localized with $p21^{Cip1}$, which was identical to that of CaLP-Δ 1-78. However, when co-transfected with CaLP-Δ 79-161, which lacked the C-terminal globular domain, the recombinant protein lost its property to locate in the nucleus with disability to co-localize with $p21^{Cip1}$. CaM + tail built by adding a tail of CaLP (aa 150-161) to the C-terminal of CaM stayed throughout all the nucleus and cytoplasm, without any effects that the transfection of $p21^{Cip1}$ had on its translocation.

Co-immunoprecipitation was applied to test the interaction between the domains of CaLP and $p21^{Cip1}$. CaLP-WT, CaM-WT, CaLP-Δ 1-78, and CaLP-Δ 150-161 were all found to be co-immunoprecipitated with $p21^{Cip1}$, while CaLP-Δ 79-161 couldn't in vivo in Fig. 2.48. It was fairly interesting to observe that though CaM-WT could be co-immunoprecipitated with $p21^{Cip1}$, the interaction with p21 ^{Cip1} still failed after adding a tail of CaLP to CaM. Therefore, we proposed that the tail of CaLP might have some interaction with CaM which affected its interaction with p21, while more evidences were required.

It has been proved that calmodulin (CaM) could upregulate the Ras/Raf/MEK/ERK pathway [179]. As a result, we analyzed the phosphorylation degree of ERK1/2 in 293T cells after co-transfection with (GFP)-CaM-WT, CaLP-WT, CaLP-Δ 150-161, CaLP-Δ 79-161, CaLP-Δ 1-78, CaM + tail, and (Red)-$p21^{Cip1}$ to explore the functional relationships between different domains of CaLP, CaM, and $p21^{Cip1}$. When transfected with $p21^{Cip1}$, the expression level of p-ERK1/2 was reduced, but when co-transfected with CaM-WT, CaLP-WT, CaLP-Δ 150-161, or CaLP-Δ 1-78, the p-ERK1/2 expression level was further rescued; however, the transfection with CaLP-Δ 79-161 and CaM + tail had no effects on that of p-ERK1/2 (Fig. 2.49). In conclusion, these experiments obviously implied that the C-terminal globular domain of CaLP played significant roles in its cellular translocation and protein interaction with $p21^{Cip1}$.

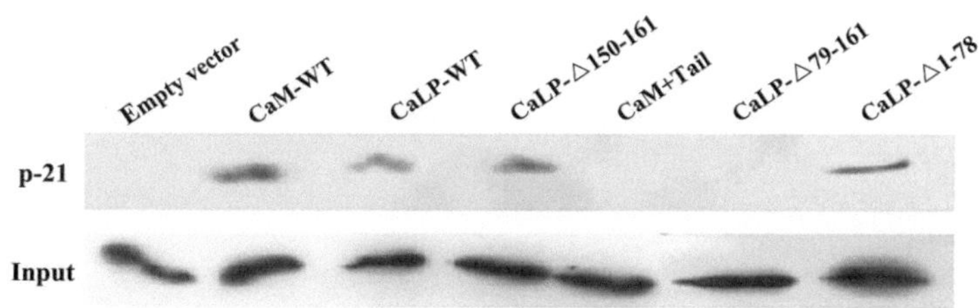

Fig. 2.48 Analysis of the p21-binding domain of CaLP
293T cells were lysed and immunoprecipitated with anti-GFP antibody. The presence of co-immunoprecipitated $p21^{Cip1}$ was tested by Western blotting. Immunoprecipitation of empty vector of EGFP co-transfected with p21 was used as negative control, and the same amount of EGFP proteins was loaded in co-immunoprecipitation which were showed in the lower panel. The figure shown was representative for three experiments

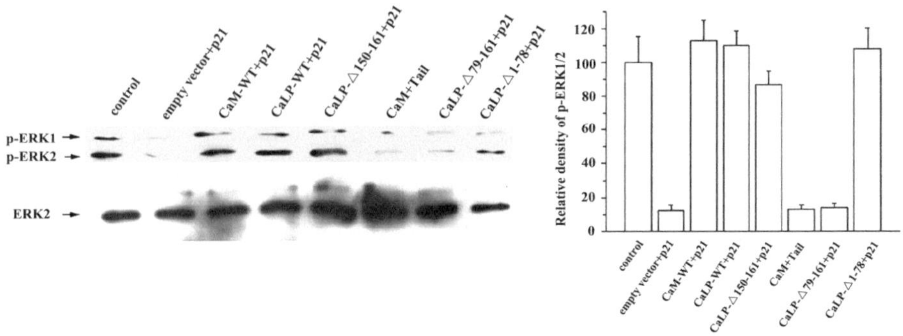

Fig. 2.49 Phosphorylation of ERKs after co-transfection with $p21^{Cip}$ and different mutants to investigate the functional interactions between p21 and its different domains
ERK2 was acted as the inner control and the relative density of ERK was quantified by ImageJ semiquantitative image analysis software. The figure shown was representative for three independent experiments

Nuclear envelope locating nuclear pore complexes (NPCs) could regulate the transaction of a large account of nucleocytoplasmic shuttling proteins in eukaryotic cells. The NPC is a large, multiprotein complex with a molecular weight of approximately 125 MDa. There is a 10-nm-diameter diffusion channel in the center of the NPC, allowing free transit of molecules with molecules smaller than 60 kDa in size [180, 181]. Nucleocytoplasmic trafficking of molecules larger than this limit requires extra energy and trafficking signal sequences embedded inside the protein primary structure. As an example, the nuclear localization signal (NLS) and the M9 signal [182–185] were two of the best characterized import signals. M9 binding proteins, functioning as cytosolic factors, could help with the signal-mediated import together with Ran GTPase [186, 187]. There were no predicted import signals in CaM, while it was small enough to go directly through the diffusion channel. In previous studies, we detected that CaM would diffuse freely through NPCs with nuclear-located CaM-binding proteins functioning as a sink for Ca^{2+}–CaM using fluorescence imaging and fluorescence recovery followed by photobleaching studies of fluoresceinated CaM microinjection into intact cells. The relative concentration and affinity of CaM-binding proteins between nucleus and cytoplasm and the intracellular concentration of free Ca^{2+} have important influences on the equilibrium state of CaM in cytoplasm and nucleus.

According to fluorescence confocal microscopy results, CaM + tail failed to stay in the cytoplasm, since the extra 12aa tail of CaLP was not large enough to be limited in the cytoplasm. DNA sequence analysis indicated that there were no obvious import signals in CaLP; on the contrary CaLP-Δ 1-78 with a deleted N-terminal globular domain exhibited intact ability of staying in the cytoplasm. Considering the ubiquitous distribution of CaLP-Δ 79-161, these results suggested that the 12aa tail together with the C-terminal guaranteed the CaLP sequestration in cytoplasm. The pathway calmodulin functions on and the specific mechanism of Ca^{2+}–CaM binding to its target proteins have been revealed in some previous biophysical researches [188, 189]. There were four Ca^{2+}-binding sites in CaM, two of which were in a globular N-terminal domain, separated by a yielding a-helix linker from a C-terminal globular domain, which contained the other two sites. It was believed that each domain would adapt an open conformation and a hydrophobic pocket would be exposed to help CaM bind to target proteins with the addition of Ca^{2+}. During the interaction between CaM and its target proteins, the C-terminal hydrophobic pocket of CaM would initially bind to the hydrophobic parts in target N-terminal sequence after which the C-terminal of the CaM-binding domain would interact with the N-terminal globular domain. As a result, CaM then collapsed and wrapped around the peptide, inducing a high affinity complex to form. Co-immunoprecipitation of different mutants of CaLP with $p21^{Cip1}$, together with fluorescence confocal microscopy assay after co-transfection, indicated that the high level of $p21^{Cip1}$ could promote the translocating of CaLP-WT, CaLP-Δ 150-161, and CaLP-Δ 1-78. And we inferred that the C-terminal globular domain of CaLP contributed to its translocation [126]. The nucleocytoplasmic transportation of various proteins like CaM was triggered by protein phosphorylation of insulin receptor tyrosine kinase [190], casein kinase II [191], and epidermal-growth-factor-receptor tyrosine kinase [192]. Thus, further analysis must be performed to investigate whether CaLP was directly phosphorylated or just released from the interacting protein in cytoplasm and then diffused into the nucleus due to the transfection of $p21^{Cip1}$. In this paper, the interacting sites with target proteins and phosphorylation sites most probably locate in the C-terminal globular domain of CaLP.

2.3.6 The Extra C-Terminal Tail Is Involved in the Conformation, Stability Changes, and the N/C-Domain Interactions of the Calmodulin-Like Protein from Pearl Oyster **Pinctada fucata**

Calmodulin-like protein (PfCaLP) in pearl oyster *Pinctada fucata*, with an extra D150-K161 tail at its C-terminal, is a new protein functioned as a regulator during calcium metabolism process. The aim of this research is gaining insight upon the conformation properties of the PfCaLP N/C-domain, particularly the relative stability of its C-domain and the intact PfCaLP protein, effects on the domain–domain communications within PfCaLP, and the exact effects of the extra tail on the conformational changes induced by Ca^{2+}/Mg^{2+}. Our results indicated that there was an obviously strong interaction between hydrophilic tail and the C-domain of PfCaLP. The extra tail further affected the rate of migration and exposure of the hydrophobic domains in the intact PfCaLP molecule and N/C-domain interactions by affecting the C-domain conformational transformation. Besides, the stability of C-domain and PfCaLP could be actively regulated by the tail. Results from this research could help us elucidate our previous conclusions that the tail of PfCaLP did play critical roles in interaction between PfCaLP and its target proteins during calcium metabolism of oyster.

2.3.6.1 The Isolation of Recombinant Proteins

M-CaLP, W-CaLP, and strings of CaLP domains, like C-CaLP, N-CaLP, and C2-CaLP, were expressed in *E. coli* successfully. It was showed in Fig. 2.50 that the high heat stability of all the recombinant proteins separated and detected by SDS-PAGE. After heating the lysates for 10 min at 90 °C, the main bacterial host-soluble proteins could be denatured to precipitation and then be removed out with all recombinant proteins still keeping in the supernatants.

Since H-CaLP exhibited a rather low affinity to phenyl-Sepharose because of the strong hydrophilicity of the C-terminally extra tail [24, 174], a modified protocol partially in terms of Newton et al. [193] was applied to the purification of C-CaLP and W-CaLP. After heat treating, C-CaLP and W-CaLP proteins were separated by DEAE-Sepharose anion exchange chromatography followed by being eluded with 0.3–0.35M NaCl, after which fractions with W-CaLP were gathered and then isolated using the phenyl-Sepharose chromatography. When this two-step purification was finished, W-CaLP would be purified successfully with high yields and purity (Fig. 2.50a). The phenyl-Sepharose binding affinities of M-CaLP and PfCaM are stronger than that of W-CaLP, indicating that the conformational changes induced by Ca^{2+} in W-CaLP generated less hydrophobic surface than M-CaLP and PfCaM for the existence of a hydrophilic tail [174, 194]. These two proteins, C2-CaLP and N-CaLP, could be obtained with phenyl-Sepharose at low salt concentration.

2.3.6.2 Calcium-Induced Electrophoretic Mobility Shift

CaM has a typical electrophoretic mobility change induced by calcium, in which the protein appears to have different apparent molecular weights with or without Ca^{2+} on SDS/PAGE [14]. The Ca^{2+}-dependent electrophoretic change profiles of M-CaLP, H-CaLP, and W-CaLP were shown in Fig. 2.51a. When compared to their predicted molecular weight, the apparent molecular mass of M-CaLP, H-CaLP, and W-CaLP reduces by nearly 17%, 7%, and 7%, respectively, after binding Ca^{2+}; however, the apparent molecular weight increased about 25%, 30%, and 31%, respectively, without Ca^{2+}. Furthermore, the Ca^{2+}-induced electrophoretic shift pattern of CaLP didn't change after connecting to a His_6-tag. Figure 2.51b showed the Ca^{2+}-dependent electrophoretic shift profiles of C2-CaLP, N-CaLP, and C-CaLP. Compared with their predicted molecular weight, the apparent molecular mass of C2-CaLP and N-CaLP decreases about 20% and 42% in Ca^{2+}-saturated form; however, C-CaLP

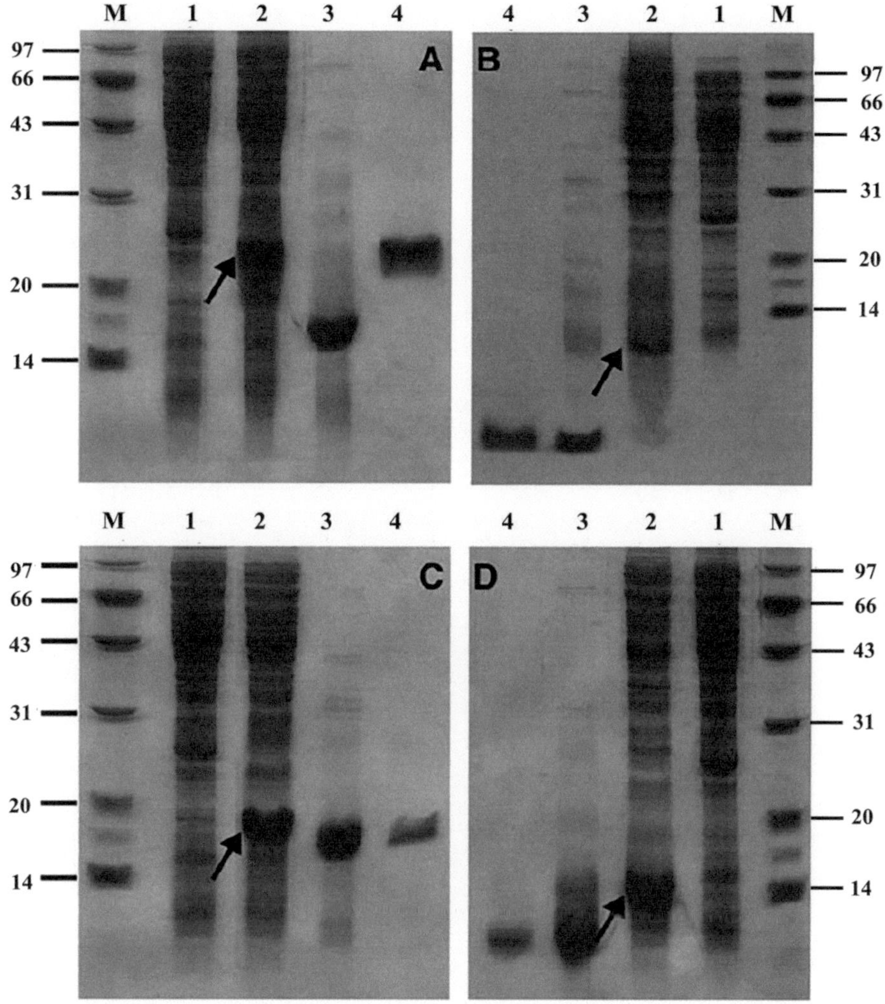

Fig. 2.50 Expression and purification of recombinant *P. fucata* CaLP and its domains in *E. coli* Arrows represented the induced proteins after IPTG inducing. (**a**) W-CaLP; (**b**) N-CaLP; (**c**) C-CaLP; (**d**) C2-CaLP. The molecular mass in kDa was shown on both sides of the gel. M, protein molecular mass markers; lane 1, un-induced whole-cell lysate; lane 2, whole-cell lysate induced by 0.5 mM IPTG for 3.5 h; lane 3, whole-cell lysate induced by IPTG after heat treatment for 10 min at 90 °C with the addition of 5 mM Ca^{2+}; lane 4, purified recombinant W-CaLP and its domains eluted from phenyl-Sepharose chromatography column by 1 mM EGTA, 50 mM Tris–HCl, and pH 7.5 (A) or 1 mM $CaCl_2$, 50 mM Tris–HCl, and pH 7.5 (B, C, and D)

mass increases nearly 82%, while the values increase approximately 73%, 42%, and 89% for C2-CaLP, N-CaLP, and C-CaLP in Ca^{2+}-free status, respectively. The different electrophoretic shift patterns induced by Ca^{2+} between C2-CaLP and

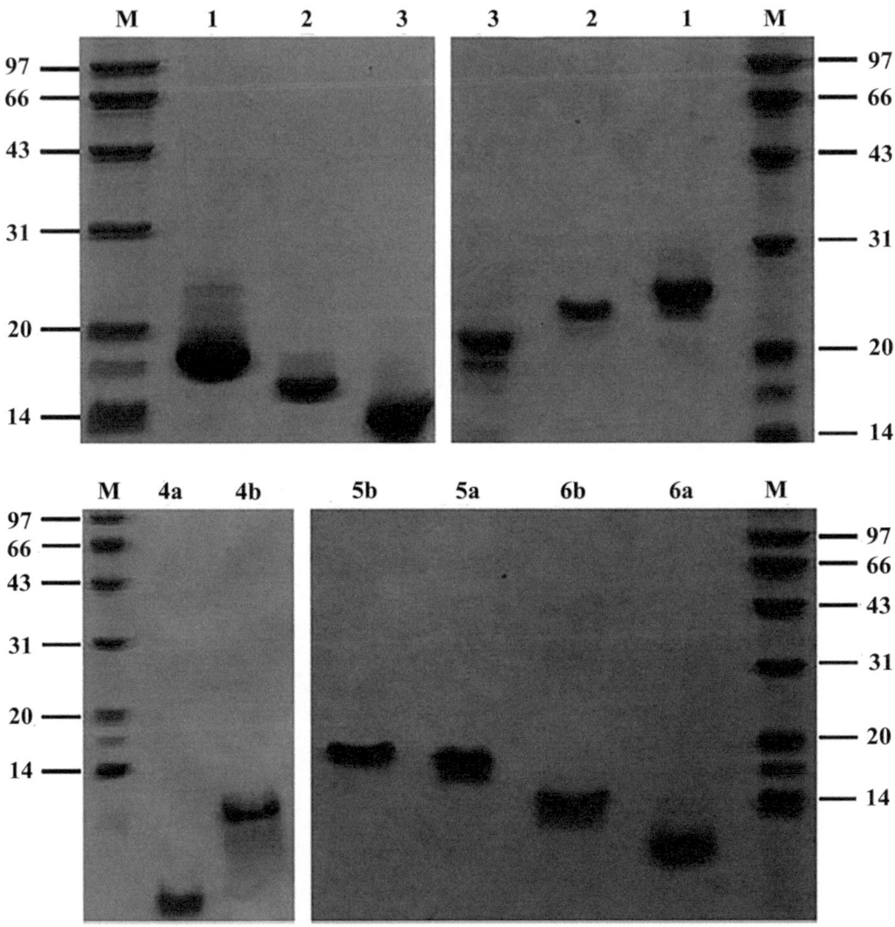

Fig. 2.51 Ca^{2+}-dependent electrophoretic migration profiles of the recombinant *P. fucata* CaLP, M-CaLP and CaLP domains

All the purified recombinant proteins were run on a 15% SDS/PAGE in the presence of Ca^{2+} or EGTA. (A) Both the gel and the sample buffer were added with 2.5 mM $CaCl_2$ or EGTA; (B) only the sample buffer was added with 2.5 mM $CaCl_2$ or EGTA; a, in the presence of Ca^{2+}; b, in the presence of EGTA. M, protein molecular mass markers; lines 1–6, H-CaLP, W-CaLP, M-CaLP, N-CaLP, C-CaLP, and C2-CaLP, respectively. The molecular mass in kDa is indicated on both sides of the gel

C-CaLP suggest the existence of different Ca^{2+}-induced conformational transformations due to the extra C-terminal tail. Another phenomenon was that as Mg^{2+} replaced Ca^{2+}; however, all of these proteins seemed to lose the mobility shift (data not shown).

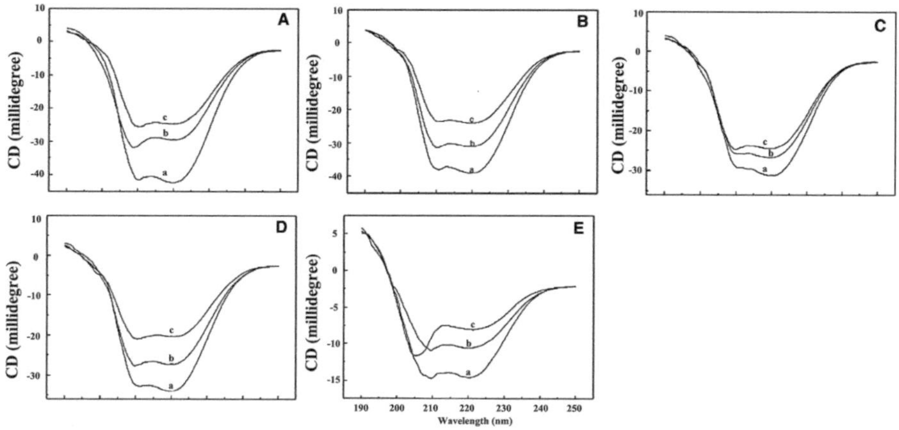

Fig. 2.52 Far-UV CD spectroscopy of Ca^{2+}-/Mg^{2+}-induced conformational changes in W-CaLP (**a**), M-CaLP(**b**), N-CaLP(**c**), C-CaLP(**d**), and C2-CaLP (**e**) with the addition of Ca^{2+}, Mg^{2+}, or EDTA [196]
The spectra were recorded in 20 mM Hepes, 100 mM KCl buffer, pH 7.5, with the addition of 2 mM $CaCl_2$ (line a), 5 mM $MgCl_2$ (line b), or 2 mM EDTA (line c), and corrected by a blank buffer containing 20 mM Hepes, 100 mM KCl, pH 7.5. The final protein concentration was 10 μM for M-CaLP and W-CaLP and 30 μM for C-CaLP, N-CaLP, and C2-CaLP

Table 2.5 Circular dichroism changes at 222 nm of M-CaLP, W-CaLP, C-CaLP, N-CaLP, and C2-CaLP

	θ_{222} (millidegrees) measured in				
Protein	Ca^{2+}	Mg^{2+}	EDTA	$\theta_{222}^{Ca}/\theta_{222}^{EDTA}$	$\theta_{222}^{Mg}/\theta_{222}^{EDTA}$
W-CaLP	41.66	29.18	23.81	1.75	1.23
M-CaLP	38.59	30.21	23.60	1.64	1.28
N-CaLP	30.77	26.29	24.02	1.28	1.09
C-CaLP	33.42	27.06	20.24	1.65	1.34
C2-CaLP	14.40	10.05	8.02	1.80	1.25

CD measurements were carried at 25 °C in 20 mM Hepes, 100 mM KCl buffer, and pH 7.5, containing 2 mM CaC12, 5 mM MgC12, or 2 mM EDTA

2.3.6.3 Far-Ultraviolet Circular Dichroism Measurements

The conformational changes induced by Ca^{2+}/Mg^{2+} of W-CaLP, domains, and mutants were further investigated with far-UV CD. There was an obvious increase in negative ellipticity at 208 and 222 nm in all the proteins (Fig. 2.52), suggesting that both Mg^{2+} and Ca^{2+} could promote conformation changes (in α-helical content) in all of the protein samples tested. The ellipticity changed at 222 nm of all proteins, which were summarized in Table 2.5. Significant conformation changes also appeared in all proteins when binding to Ca^{2+} and Mg^{2+}, in which the effects of Mg^{2+} were weaker than that of Ca^{2+} in all cases. Furthermore, greater conformational changes in C-CaLP could be induced by both Ca^{2+} and Mg^{2+} than in N-CaLP in accordance with the finding that TR_2C showed a weaker intensity upregulation than of TR_1C with Ca^{2+} binding [195]. The most prominent and attracting property of the CD spectra consisted in that there were also two EF-hand Ca^{2+}-binding motifs in C2-CaLP, while only an almost complete unfolded structure appeared with the presence of EDTA, indicating that there might exist a strong interaction between the C-domain and C-terminally extra tail of CaLP.

2.3.6.4 The Extra Tail Influences the CaLP N/C-Domain Interactions

Previous research have proved that far-UV CD was widely used to investigate the domain–domain interactions of CaM by comparing the sum of the ellipticities of individual domain with that of a stoichiometric mixture of the C- and N-domains of CaM [197–199]. In this paper, far-UV CD was also applied to analyze how the tail could affect the domain–domain interactions in CaLP. As shown in Fig. 2.53a, the sum of the individual CD spectra of C-CaLP and N-CaLP was the same as that of the equimolar mixture of C-CaLP and N-CaLP, suggesting that no significant interactions existed between the C/N-domain in CaLP. When Ca^{2+} was thoroughly mixed with C2-CaLP and N-CaLP, the ellipticity became higher than that of the sum of these individual spectra (Fig. 2.53c), indicating that interactions truly existed between the C- and N-domains of CaLP without its extra tail. However, no significant secondary structure changes appeared during the combination of C-CaLP and N-CaLP nor in C2-CaLP and N-CaLP after the addition of EDTA (Fig. 2.53b and d). In conclusion, this process may follow a Ca^{2+}-binding manner through which the tail could affect the interactions between C-domain and N-domain in CaLP.

2.3.6.5 Intrinsic Phenylalanine and Tyrosine Fluorescence Assay

The Mg^{2+}- and Ca^{2+}-binding influences on the conformational transformations of W-CaLP, N-CaLP, M-CaLP, C2-CaLP, and C-CaLP were further studied by testing their intrinsic fluorescence of phenylalanine and tyrosine. The N-domain of CaLP possessed five phenylalanine residues, with no tryptophan or tyrosine. The CaLP C-domain contained one tyrosine and four phenylalanine residues instead of tryptophan. What's more, there was no aromatic residue locating in the tail of CaLP either. Intrinsic fluorescence of phenylalanine spectra was measured and exhibited in Fig. 2.54a–e. The reduction in phenylalanine fluorescence emission in each protein on Mg^{2+}/Ca^{2+} binding was shown in Table 2.6. Both Mg^{2+} and Ca^{2+} could promote the quenching of phenylalanine fluorescence in all proteins, which might be partially induced by the energy transferred to the tyrosine around [16]. Comparing W-CaLP with M-CaLP, the additional tail would just mildly influence the phenylalanine fluorescence quenching induced by Ca^{2+}/Mg^{2+} in the CaLP protein. By contrast, the additional C-terminal tail could influence the quenching of phenylalanine

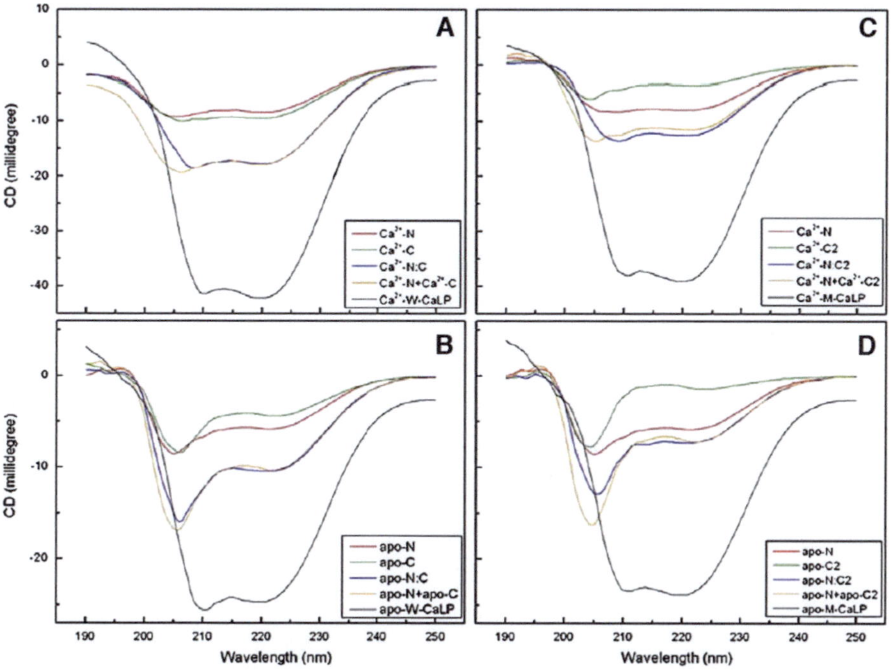

Fig. 2.53 Far-UV CD spectra of un-complexed and 1:1 mixture of C-CaLP, N-CaLP, and C2-CaLP
(**a** and **b**) Unmixed N-CaLP and C-CaLP and the sum of the two spectra, as well as spectra of intact W-CaLP and a 1:1 mixture of N-CaLP and C-CaLP; (**c** and **d**) 1:1 mixture of C2-CaLP and N-CaLP with the sum of the spectra recorded individually and the spectra of M-CaLP. The concentration of protein was 10 μM in all groups; all spectra were performed in 20 mM Hepes, buffer 100 mM KCl, pH 7.5, with the addition of 2 mM$CaCl_2$ (**a** and **c**) or 2 mM EDTA (**b** and **d**)

fluorescence induced by the Ca^{2+}/Mg^{2+} in the C-domain of CaLP (Fig. 2.54d, e), among which the effect of Ca^{2+} was much more obvious than that of Mg^{2+}. Furthermore, there was no evident red or blue shift with any maximal emission wavelength observed in neither C2-CaLP nor C-CaLP.

The absorption spectra of PfCaM containing a single tyrosine residue at position 139 have been well studied [24, 52, 174]. Figure 2.54f–i showed the emission of tyrosine for W-CaLP, its domains, and mutants. After the binding with Ca^{2+}, tyrosine fluorescence emission in W- and M-CaLP (Fig. 2.54f, g), C2-CaLP (Fig. 2.54i), and C-CaLP (Fig. 2.54h) increased to 0.28-fold, 0.26-fold, 0.36-fold, 0.23-fold, and 0.23-fold, respectively, while the binding of Mg^{2+} led to an increase about 0.14-fold, 0.13-fold, 0.16-fold, and 0.07-fold as the same succession mentioned above. The data above indicated that the extra tail contributed little to changes of the intrinsic fluorescence induced by Mg^{2+}/Ca^{2+} in CaLP yet showed significant effects on the conformational changes of the CaLP C-domain.

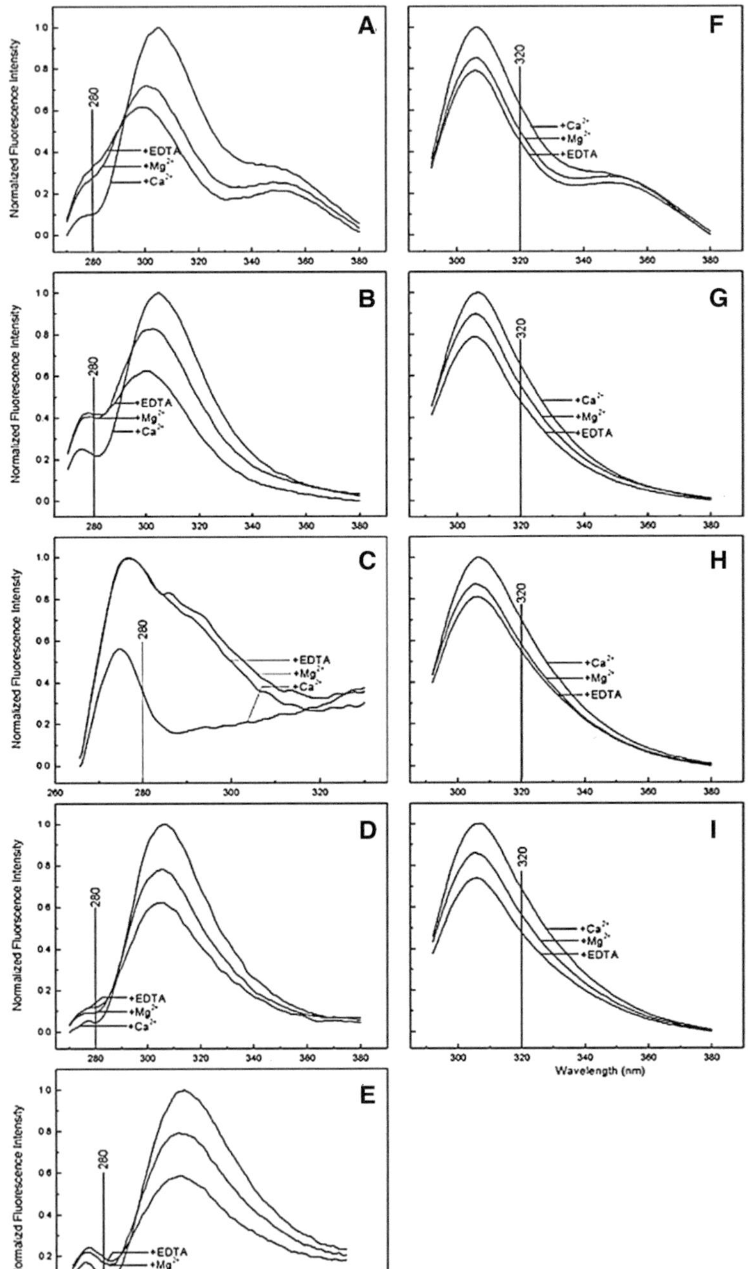

Fig. 2.54 Normalized [(f-f_{max})]/(f_{max}-f_{min})] emission fluorescence spectra of M-CaLP, W-CaLP, C-CaLP, N-CaLP, and C2-CaLP

The fluorescence of the phenylalanine (**a**–**e**) and tyrosine residues (**f**–**i**) in W-CaLP (**a** and **f**), M-CaLP (**b** and **g**), N-CaLP (**c**), C-CaLP (**d** and **h**), and C2-CaLP (**e** and **i**) was monitored using an

Table 2.6 The reduction of phenylalanine fluorescence intensity with the addition of Ca^{2+} or Mg^{2+} relative to that with EDTA

Protein	$+Ca^{2+}$ (%)	$+Mg^{2+}$ (%)
W-CaLP	51	6.6
M-CaLP	52	9.6
N-CaLP	44	0.5
C-CaLP	61	21
C2-CaLP	49	12

2.3.6.6 The Extended C-Terminal Region Influences the Hydrophobic Patches Exposure of CaLP and Its C-Domain

The formation of hydrophobic patches in CaM after binding with Ca^{2+} was the most important structural property of revealing the Ca^{2+}-induced activation mechanism [200–202]. The external fluorescent probe ANS, a widely used external fluorescent probe with a very low quantum yield in water, was extensively used to monitor the event, but it turned to be very fluorescent in nonpolar solvent. This characteristic made it a useful method to test the surface hydrophobicity in protein [144]. Figure 2.55 demonstrated the ANS fluorescence emission of W-CaLP, its domains and mutant with Mg^{2+}, Ca^{2+}, or EDTA. The hydrophobic patch exposure to the polar environment could be enhanced by Ca^{2+}, but Mg^{2+} almost had no effects in all these cases. We could see from Table 2.7 the changes in maximum wavelength of emission (λ_{max}) and maximum fluorescence intensity in all proteins after binding with ANS. After M-CaLP or W-CaLP binding with Ca^{2+}, a similar decrease of λ_{max} (about 30 nm) but distinct change in maximum fluorescence emission intensity (I_{max}) appeared. For C-CaLP and N-CaLP, their cases indicated different increase in I_{max} intensity while no significant λ_{max} change after Ca^{2+} binding. However, the reasons were still not clear for these differences. The sequence variations of EF-hand motifs between C-CaLP and N-CaLP might contribute to this phenomenon. As for C2-CaLP and C-CaLP, CaLP exhibited more enhancement of ANS intensity in C-CaLP induced by Ca^{2+} than that in C2-CaLP, which could be partially explained by the interaction disruption between the tail of CaLP and its C-terminal domain, leading to the complete unfolding of C2-CaLP with the addition of EDTA. In conclusion, the extra tail could obviously affect the Ca^{2+}-promoted hydrophobic patch exposure of the whole CaLP by influencing C-domain conformational shift.

Fig. 2.54 (continued) excitation and emission wavelength pair (250/280 nm and 277/320 nm, respectively). All the proteins were diluted in 20 mM Hepes, 100 mM KCl buffer, pH 7.5, with the addition of 2 mM $CaCl_2$, 5 mM $MgCl_2$, or 2 mM EDTA. The final protein concentration was 10 μM for M-CaLP and W-CaLP and 30 μM for C-CaLP, N-CaLP, and C2-CaLP

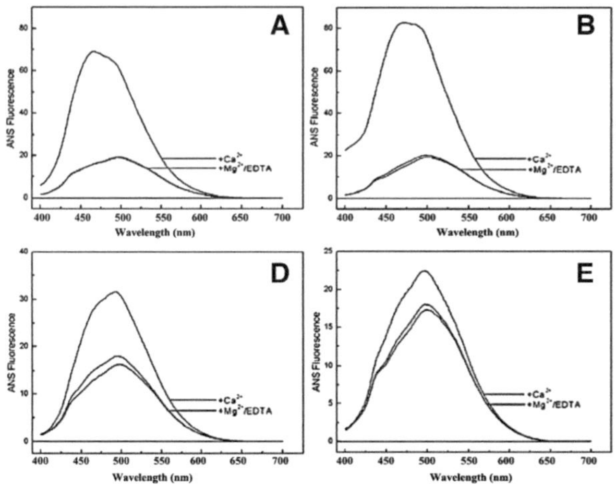

Fig. 2.55 Conformational changes of M-CaLP, W-CaLP, and CaLP domains tested by ANS binding with the addition of Ca^{2+}, Mg^{2+} or EDTA
(**a**)W-CaLP; (**b**) M-CaLP; (**c**) N-CaLP; (**d**) C-CaLP; (**e**) C2-CaLP. Fluorescence emission spectra were measured with excitation wavelength set at 370 nm for ANS and all measurements were carried out using 20 μM for M-CaLP and W-CaLP and 30 μM for C-CaLP, N-CaLP, and C2-CaLP. ANS was at a final concentration of 50 μM and 20 mM Hepes buffer, 100 mM KCl, pH 7.5, with 2 mM $CaCl_2$, 5 mM $MgCl_2$, or 2 mM EDTA at 25 °C

Table 2.7 Fluorescence spectral changes of ANS with the addition of Mg^{2+}, Ca^{2+}, or EDTA

Protein	$\Delta\lambda_{max}$	$I_{max}(+Ca^{2+})/I_{max}(+EDTA)$	$I_{max}(+Mg^{2+})/I_{max}(+EDTA)$
W-CaLP	29	3.66	1.02
M-CaLP	30	4.31	1.05
N-CaLP	0	1.40	1.02
C-CaLP	3	193	1.11
C2-CaLP	2	1.29	1.04

Δλmax indicated the difference between maximum fluorescence emission of ANS in the presence of $Ca^{2+}$and that in EDTA. $I_{max}(+Ca^{2+})/I_{max}(+EDTA)$ or $I_{max}(+Mg^{2+})/I_{max}(+EDTA)$ indicated the ratio of the maximum fluorescence intensity of ANS with the addition of Ca^{2+} or Mg^{2+} to that of EDTA

2.3.6.7 The Extra C-Terminal Tail Affects the Structural Stability of CaLP

Far-UV CD using chemical denaturation (GuHCl) assay were applied to monitor the structural stabilities of W-CaLP, its mutant and domains in the absence (apo-state) or presence (holo-state) of Ca^{2+}. The differences of the far-UV CD signals at 222 nm of the apo- and holo- state of all the recombinant proteins were demonstrated in Fig. 2.56. All the proteins in holo-state were substantially more stable than those in apo-state. The denaturation curves for holo-M-CaLP and holo-W-CaLP clearly indicated two resolvable transition states (Fig. 2.56a). If GuHCl <3.5M, holo-M-

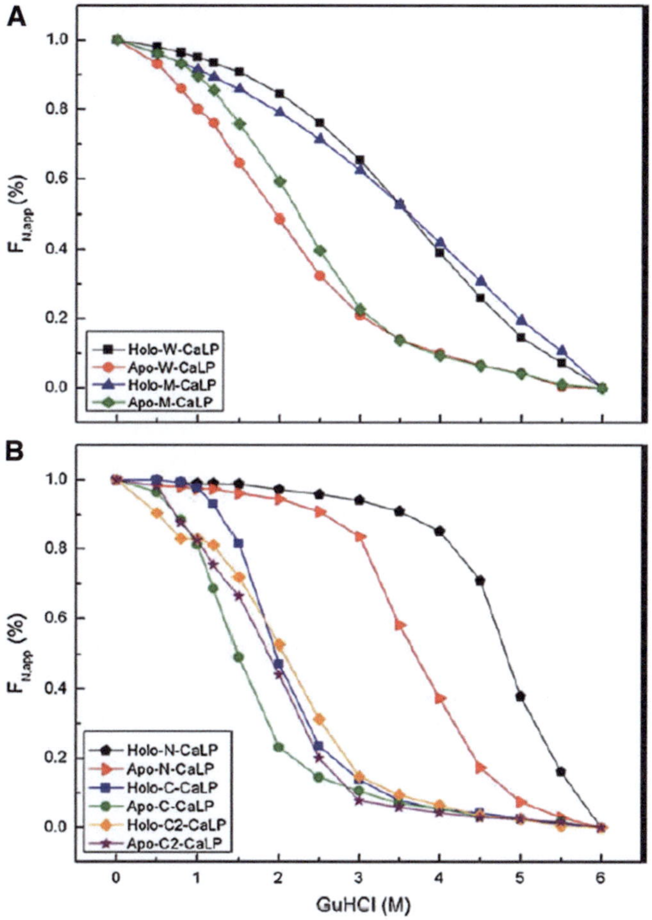

Fig. 2.56 Structural stability of recombinant *P. fucata* CaLP, its mutant, and various domains CD spectra signals at 222 nm were used to analyze the folding fractions of the proteins. Schematic representation of the GuHCl-induced unfolding of M-CaLP and W-CaLP (**a**) and C-CaLP, N-CaLP, and C2-CaLP (**b**) with 2 mM $CaCl_2$ (holo-state) or 2 mM EGTA (apo-state). Unfolding was tested at 25 °C in 20 mM Hepes, 100 mM KCl, pH 7.5. $f_{N,app}$, fraction of folded protein

CaLP was less stable than holo-W-CaLP, which suggested that the molecular stability of W-CaLP could be increased by the extra tail, whereas, when GuHCl concentration increased (GuHCl>3.5M), holo-M-CaLP was mildly more stable than holo-W-CaLP, suggesting that the molecular stability above this GuHCl concentration could be decreased by the tail. Apo-W-CaLP showed a less stable state than apo-M-CaLP, which implied that the tail might repressed the chemical stability of W-CaLP without Ca^{2+}. Therefore, we speculated that the flexible C-terminal tail contributed actively to control the stability of W-CaLP.

We could clearly see from Fig. 2.56b that holo-N-CaLP exhibited the highest stability. And it could still remain at least 91% of its native state even with the addition of 3.5 M GuHCl. There has also been a broad transition from 1.5 to 6 M GuHCl in amplitude corresponding to the complete unfolding of holo-N-CaLP protein. The denaturant Cm value (midpoint of the unfolding transition) for holo-C-CaLP and holo-N-CaLP was 1.95 M and 4.8 M, respectively. The Cm value for apo-C-CaLP and apo-N-CaLP was 1.5 M and 3.7 M, respectively. What's more, we should note that apo-N-CaLP was still more stable than holo-C-CaLP under these circumstances. C2-CaLP showed different stability patterns compared with its holo- and apo-state. As the concentration of GuHCl decreased to less than 1.8 M in holo-state, holo-C-CaLP was still much more stable than holo-C2-CaLP; however, the increasing concentration of GuHCl led to a slightly stable state of holo-C2-CaLP than that in holo-C-CaLP. As in apo-state, both apo-C2-CaLP and apo-C-CaLP exhibited similar stabilities as the concentration of GuHCl reduced to less than 1.0 M. However, the increasing concentration of GuHCl made apo-C-CaLP less stable than apo-C2-CaLP, suggesting that the tail could regulate the stability of the intact CaLP molecule by acting on the C-domain.

In summary, our results elaborated that the N-domain and C-domain of CaLP did have different conformational changes on Ca^{2+} and Mg^{2+} binding. And a strong interaction truly existed between the C-domain and the tail of CaLP. What's more, the extra C-terminal tail, regulating the stability of the C-domain of CaLP and the conformational changes induced by Ca^{2+}/Mg^{2+} as well as Ca^{2+}-dependent electrophoretic migration rate would affect the stability and conformational properties of the intact CaLP molecule [196]. Our present study is definitely facilitating to demonstrate our previous findings that the tail plays crucial roles in CaLP-target interaction.

2.3.7 *Biomineralization: Functions of Calmodulin-Like Protein in the Shell Formation of Pearl Oyster*

Calmodulin-like protein (CaLP) has been considered to be engaged in the shell formation of many oysters. However, no further studies of this protein have ever been performed. In our research, the in vitro crystallization experiment indicated that CaLP could modify the morphology of calcites. In addition, aragonites could be induced in the addition of CaLP and a nacre protein (at 16 kDa), which was detected, isolated, and purified from the EDTA-soluble matrix of nacreous layer. These results were consistent with that of immunohistological staining experiment where CaLP was detected not only in the organic layer sandwiched between calcitic prismatic layer and aragonitic nacreous layer but also around the calcitic prisms of the prismatic layer. Taken all these results together, we concluded that (1) CaLP, as a component of the organic layer, could induce the nucleation of aragonites through binding with the 16 kDa protein and (2) CaLP may regulate the growth of calcites in the outer prismatic layer of the shell.

2.3.7.1 Expression and Purification of Recombinant CaLP

As shown in Fig. 2.57, the expressed recombinant CaLP reached approximately 20% of the total bacterial soluble proteins. After heating at 90 °C for 10 min, most bacterial soluble proteins were denatured, while the recombinant CaLP demonstrated high heat stability and was purified by anion exchange chromatography. After purification, only a single band was observed on SDS-PAGE, and in the presence of Ca^{2+}, its relative molecular mass was 17 kDa which was consistent with the predicted molecular mass.

2.3.7.2 Isolation of EDTA-Soluble Matrix (ESM) from Nacre

The analysis of ESM by SDS-PAGE is shown in Fig. 2.58. Many protein bands can be seen under the reduced conditions, while only a few bands appeared on the unreduced gel and a majority of the proteins aggregated at about 60 kDa.

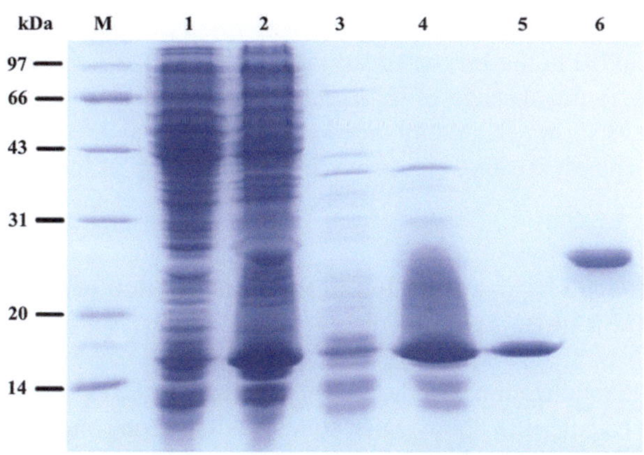

Fig. 2.57 Expression and purification of recombinant CaLP detected by 15% SDS-PAGE and stained by CBB R-250
M, molecular weight markers; lane 1, un-induced whole-cell lysate; lane 2, induced whole-cell lysate; lane 3, un-induced whole-cell lysate with heat treatment; lane 4, induced whole-cell lysate with heat treatment; lane 5, purified recombinant CaLP in the presence of 2.5 mM CaCl2; lane 6, purified recombinant CaLP in the presence of 2.5 mM EGTA

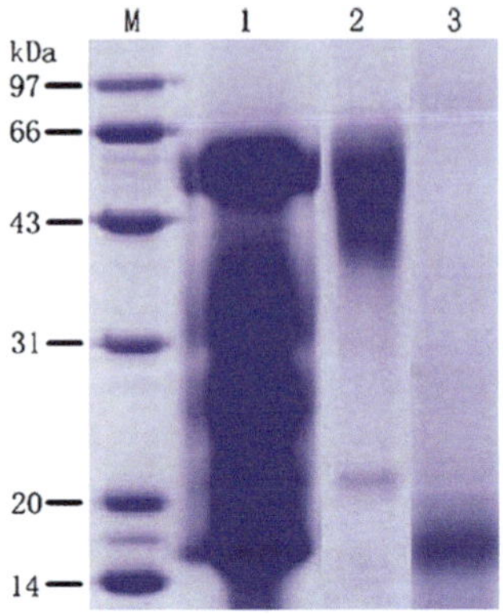

Fig. 2.58 SDS-PAGE of ESM isolated from *Pinctada fucata* nacre
M, molecular weight markers; lane 1, ESM under reduced conditions; lane 2, ESM under unreduced conditions; lane 3, the purified 16 kDa protein

2.3.7.3 Efficiency of Biotinylated CaLP and Detection of CaLP-Binding Proteins

The efficiency of biotinylated CaLP was proven by dot blots. In the presence of biotinylated CaLP, the CaLP-binding proteins presented a definite positive signal, while BSA, as a negative control, was not detected (data not shown). This indicates that the biotinylated CaLP can serve as a powerful tool to detect CaLP-binding proteins. To detect CaLP-binding protein in ESM, protein blotting was performed (Fig. 2.59). The molecular weight markers (rabbit phosphorylase b, 97 kDa; BSA, 66 kDa; rabbit actin, 43 kDa; bovine carbonic anhydrase, 31 kDa; trypsin inhibitor, 20 kDa; hen egg white lysozyme, 14 kDa) and *P. fucata* calmodulin [52], a classical acidic protein [to detect nonspecific absorptions between the avidin (alkaline) and acid proteins], were electroblotted together with ESM for control studies. Five mM EGTA was included in the incubated solutions to detect Ca^{2+}-independent binding. Protein blotting in the absence of biotinylated CaLP was also performed (data not shown). Fig. 2.59 showed that only one band at 16 kDa can be seen in the presence of 2 mM Ca^{2+} and no positive signal was observed in any of the controls.

2.3.7.4 Polymorph Transformation in the Combination of CaLP and the 16 kDa Protein

The results of crystallization, conducted in a mixture of CaLP (2.5 μg/ml) and the 16 kDa protein (2.5 μg/ml), showed that some needle-shaped aragonites (Fig. 2.60d, e) and plated vaterites (Fig. 2.60f) had been induced. All the control experiments performed in the presence of filtrate (Fig. 2.60a), mixture of BSA and CaLP (Fig. 2.60b), or mixture of BSA and the 16 kDa protein (Fig. 2.60c) at the same concentration produced typical rhombohedra of calcite. The aragonite and vaterite were validated by Raman spectra (Fig. 2.60g).

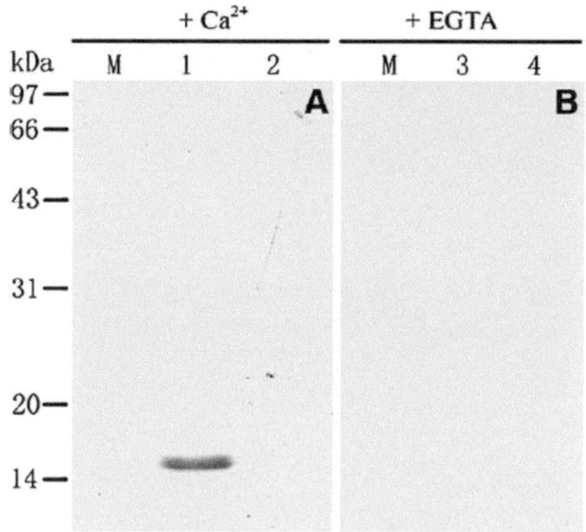

Fig. 2.59 Detection of CaLP-binding proteins in ESM (separated by reduced SDS-PAGE) by protein blotting
(**a**) blotting in the presence of 2 mM Ca^{2+}; (**b**), blotting in the presence of 5 mM EGTA; M, molecular weight markers; lane 1 and lane 3, ESM; lane 2 and lane 4, calmodulin

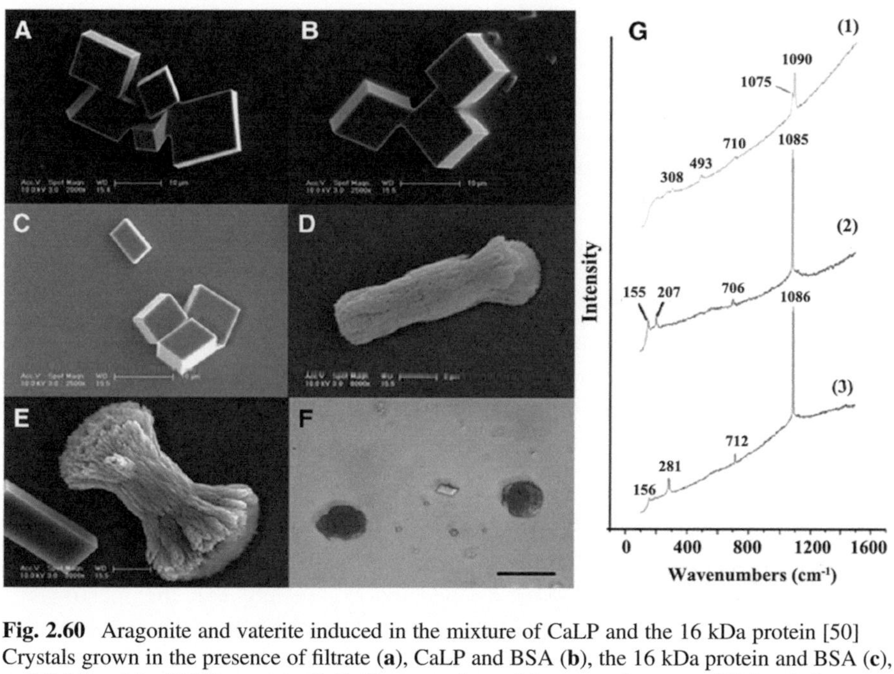

Fig. 2.60 Aragonite and vaterite induced in the mixture of CaLP and the 16 kDa protein [50]
Crystals grown in the presence of filtrate (**a**), CaLP and BSA (**b**), the 16 kDa protein and BSA (**c**), and CaLP and the 16 kDa protein (**d–f**). Concentrations of these proteins were all 2.5 μg/ml. **a–c** are controls; **d** and **e** show aragonite (SEM); F shows vaterite (optical microscope). Scale bars in **d** and **e** are 2 μm, and the other bars are 10 μm. G is the Raman spectra of induced crystals, (1) vaterite, (2) aragonite, and (3) calcite, with the characteristic peaks of calcite at 281, 712, and 1086 cm^{-1}; of aragonite at 207, 706, and 1085 cm^{-1}; and of vaterite at 308, 1075 (shoulder peak), and 1090 cm^{-1}

Fig. 2.61 SEM images of calcite modifications by CaLP
Calcite crystals grown in the presence of 20 μg/ml BSA (**a**, control), 5 μg/ml CaLP (**b**), 10 μg/ml CaLP (C), and 20 μg/ml CaLP (D). Scale bars, 10 μm

2.3.7.5 Calcite Modification by CaLP

To investigate the influences of CaLP on the morphology of calcite, the crystals precipitated in vitro were examined by SEM. In the control experiments performed with the filtrate (data not shown) or BSA (Fig. 2.61a), the crystals obtained were the typical rhombohedra of calcite with smooth crystal faces. In the presence of CaLP, the morphology of calcite was drastically changed. At low concentrations of CaLP (<2 μg/ml), however, no significant effect on the morphology of the calcite was observed. The effects of CaLP became apparent at a concentration of 5 μg/ml, when all the crystals were changed with rounded edges and visualized steps (Fig. 2.61b). The crystals reached their most modified state at the concentration of 20 μg/ml, when the edges and some crystal faces disappeared completely and the macro-steps formed on the surface of the calcite were distinct (Fig. 2.61d).

2.3.7.6 Immunohistological Staining

CaLP was detected in the organic layer (sandwiched between the nacre and the prismatic layer) (Fig. 2.62b, black arrows) and around the prisms (Fig. 2.62b, white arrows). No positive signal was observed in the control sample.

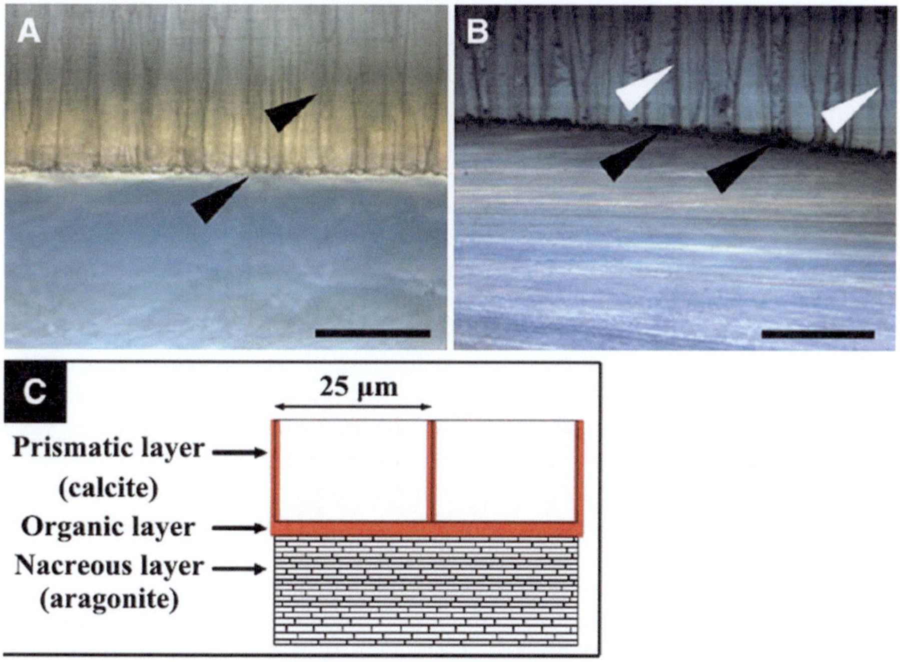

Fig. 2.62 In situ localization of CaLP in transversal section of the shell of *Pinctada fucata* [50] The upper compartment is the prismatic layer and the lower is the nacre. (**a**) control sample stained with preimmune serum. The black arrow indicates no positive signal was detected in the organic layer. (**b**) sample stained with anti-CaLP serum. The black arrow indicates the localization of CaLP in the organic layer, and the white arrow indicates CaLP around the prisms. Scale bars: 100 μm. (**c**) schematic drawing of CaLP (red) localization in the shell

Our previous study [24, 52, 174] showed that CaLP was involved in biominer alization, but no direct evidence was ever presented. To advance our understanding about the functions of CaLP in shell formation, this study was performed.

Calmodulin is an important constituent of cellular signal transduction pathways and some recent studies showed that calmodulin seemed to have a close relationship with biomineralization [203, 204]. Identification of the binding proteins is a crucial step to revealing the functions of calmodulin (or CaLP) [205], and for many years the CaLP-binding proteins were believed to be located intracellularly. But in recent years, extracellular calmodulin (or CaLP) and many calmodulin (or CaLP)-binding proteins have been identified, and they have also been believed to play important roles in extracellular process [206, 207]. Therefore, we extracted ESM from the nacre of the pearl oyster and detected the CaLP-binding proteins in it. The 16 kDa protein was found.

To study the influences of the complex of CaLP and the 16 kDa protein on crystallization, we have attempted to purify the 16 kDa protein. Unfortunately, the proteins of ESM were cross-linked and aggregated under unreduced conditions

(Fig. 2.58, lane 2), so it was difficult to purify the 16 kDa protein. Finally, preparative electrophoresis was chosen, and the results indicated that the 16 kDa protein was purified successfully from the ESM (Fig. 2.58, lane 3). After renaturation (extensive dialysis) [208], CaLP affinity column [24] was used to validate the CaLP-binding ability of the 16 kDa protein. The analysis showed that the renatured 16 kDa protein was able to bind with CaLP (data not shown), indicating that the 16 kDa protein was renatured at least in part.

Protein interaction is a common way that matrix protein achieves its regulatory role in biomineralization [209]. Pearlin can induce aragonite crystallization through interacting with pearl keratin in the nacreous layer [210], and the mixture of N14 and N66 can promote the nucleation of aragonite crystals [211]. In this study, when CaLP and the 16 kDa protein were added to the crystallization solution, respectively, no significant influence on the calcite growth was observed. However, when they were mixed in the crystallization solution, some aragonites and vaterites were induced (Fig. 2.60 D–F). The results strongly indicated an interaction between the two proteins. The interaction may induce an exposure of a hydrophobic surface in CaLP in such a way that this interaction participates in the nucleation of aragonite crystal [212].

Although aragonite and vaterite were induced in the crystallization solution, the solution does not guarantee the total absence of calcite nuclei. Many calcites were produced together with aragonite and vaterite. This is agreeable with an earlier study in which the organic matrix could induce nucleation of crystal, but calcite formed predominantly [213]. Thus, the characteristics of aragonite or vaterite could not be detected by X-ray diffraction in this work, due to interference of high-intensity calcite peaks. On the other hand, the preparative method of the 16 kDa protein was so harsh that it may influence the functions of the 16 kDa protein, resulting in the decrease in quantity of aragonite in the crystallization solution.

We have also performed crystallization experiments in an aragonitic crystallization solution [98] using CaLP and the 16 kDa protein with similar sets. However, all the experimental solutions presented spherical and needle-like aragonite, and no significant results were observed (data not shown). We infer that the complex of CaLP and the 16 kDa protein is associated with the framework proteins at the interface. When the growth of nacre is initiated, the protein complex can induce the nucleation of aragonite and maybe in such a way that the complex participates in the location of aragonitic nucleation sites on the organic surface. However, in the aragonitic crystallization solution, the roles of the protein complex were obscured by Mg^{2+}.

In vitro studies showed that the soluble protein fractions extracted from the shell can influence the nucleation, polymorph, orientation, and morphology of growing $CaCO_3$ crystals [214]. In this study, when CaLP was added singly to the crystallization solution, it demonstrated drastic effects on the growth of calcite (Fig. 2.61) in a concentration-dependent manner, while BSA, as a negative control, had no significant influence (Fig. 2.61a). This strongly suggests that CaLP was occluded in the shell. When we attempted to detect CaLP in the ESM of nacre by Western blot, no positive signal was detected (data not shown). This was consistent with the results of immunohistological staining in which CaLP was detected not in the nacre, but in the organic layer and around the prisms of the prismatic layer (Fig. 2.62).

Based on the results of crystallization experiments and immunohistological staining, firstly, we deduce that CaLP is a component of the organic layer and it can induce the nucleation of aragonite at the interface through binding with the 16 kDa protein. Secondly, CaLP is also localized in the prismatic layer and may regulate the growth of calcite. Multifunction is a familiar characteristic of many shell proteins [212], and we infer that CaLP shares it.

In addition, cell signaling is a putative function of shell proteins [212], but few studies have been performed on it. Besides CaLP in this study, a receptor for growth factors was also found in the shell [215]. The presence of these signal-related proteins in the shell matrix is interesting, and the reason for it remains to be elucidated.

2.3.8 *Investigation of Phosphorylation Site Responsible for CaLP (*P. fucata*) Nucleocytoplasmic Shuttling Triggered by Overexpression of p21^{Cip1}*

Calmodulin (CaM) is a highly conservative and unique Ca^{2+}-binding protein regulating intracellular Ca^{2+} concentration functioning as a sensor of Ca^{2+} in eukaryotic cells. Being such an important signal sensor, we have good reasons to believe CaM undergoes many posttranslational modifications, one of which is its phosphorylation. Our previous studies showed that calmodulin-like protein (CaLP) and CaM cloned from *Pinctada fucata* have many various characteristics even though they shared high similarities to each other. And we have narrowed down that the C-terminal domains of CaM and CaLP were in control of their discrepant subcellular localizations and shuttling of CaLP when it was co-transfected with p21^{Cip1}, which was usually viewed as an essential cell cycle-regulating protein. In this paper, for the first time, we predicted the potential phosphorylation site involved in shuttling and confirmed it by fluorescence confocal microscopy. Together with FACS results, we further studied the releasing ability between wild-type and point-mutated CaLP arrested from cell cycle induced by p21^{Cip1} overexpression. We proposed that the CaLP locating in cytoplasm was phosphorylated; however, the overexpression of p21^{Cip1} in mammalian cells led to the dephosphorylation and translocation of CaLP to nucleus triggered by certain signals by carrying out pull-down analysis and phosphorylation status of CaLP in cytoplasm of transfected COS-7 cells with CaLP alone as well as phosphorylation status of CaLP in nuclear of co-transfected COS-7 cells with CaLP and p21^{Cip}.

2.3.8.1 Mutation of Y139A Changed CaLP Localization and Its Translocation

CaM-WT distributed widely in cells, but CaLP-WT was converted in cytoplasm [49, 126] as Fig. 2.64a–h showed. Interestingly, CaLP, instead of CaM, would translocate from cytoplasm to nucleus and co-localize with $p21^{Cip1}$ after the co-transfection with $p21^{Cip1}$, for which we've confirmed before that the C-terminal globular domain was responsible [49, 126]. Moreover, we found one tyrosine site predicted to be potentially phosphorylated in CaLP (Fig. 2.63) by bioinformatics.

To identify the crucial roles P-(Y139) of CaLP played during its subcellular localization, and confirm the interesting nuclear–cytoplasm translocation mechanism, CaLPY139A alone, as well as CaLPY139A together with $p21^{Cip1}$, was transfected to 293T cells, separately for further comparison between CaM-WT and CaLP-WT. Apparently, the mutant was detected in both cytoplasm and nucleus, and the translocation phenotype also disappeared (Fig. 2.64i–l). These results indicated that the Y139 of CaLP was a critical site and the phosphorylation status might be very crucial for the different characteristics of CaM from CaLP.

2.3.8.2 CaLPY139A Failure in Releasing Cells from G_0/G_1 Arrest

The ability of CaLP, CaLPY139A, and CaM to release COS-7 cells from G_1 arrest induced by the overexpression of $p21^{Cip1}$ was investigated to confirm our previous observations on Y139's crucial role during its translocation from cytoplasm to nucleus. The results of the representative profiles of $p21^{Cip1}$ overexpression on cell cycle and synchronization of COS-7 cells by serum starvation were indicated in Fig. 2.65. Cells in G_2/M were indicated by the right tail peak, the intermediate area indicating cells in S phase, and the large initial peak in the left represented cells in G_0/G_1 phase. A typical cell cycle was released from serum starvation as negative control in Fig. 2.65. The cells would enter G_2 stage normally 24 hours later. The distribution

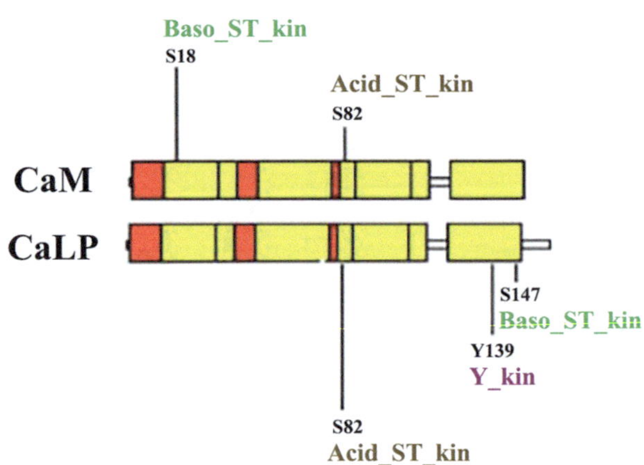

Fig. 2.63 Potential phosphorylation sites of CaLP and CaM predicted by Scansite (medium stringency)
The site was narrowed in the Y139 locating in the C-terminal globular domain

of cell cycle in COS-7 cells after transfection with p21^{Cip1} alone experiencing serum starvation showed the arrest in G_0/G_1 phase 24 hours after transfection and failed to enter to the next stage. When co-transfected with CaLP-WT and p21^{Cip1} after serum starvation, the FACS results suggested that the percentage of cells in S phase significantly increased compared with those transfected with p21^{Cip1} alone (50.11± 6.12% vs 21± 4.1%). However, in the co-transfection group of p21^{Cip1} and CaLPY139A after serum starvation, the rescuing of cell cycle arrested by p21^{Cip1} failed at the G_0 phase. These data demonstrated that Y139 regulated the releasing of cells from G_0/G_1 phase induced by the overexpression of p21^{Cip1}. Another interesting phenomenon we should notice was that when compared with the FACS figures of co-transfection of

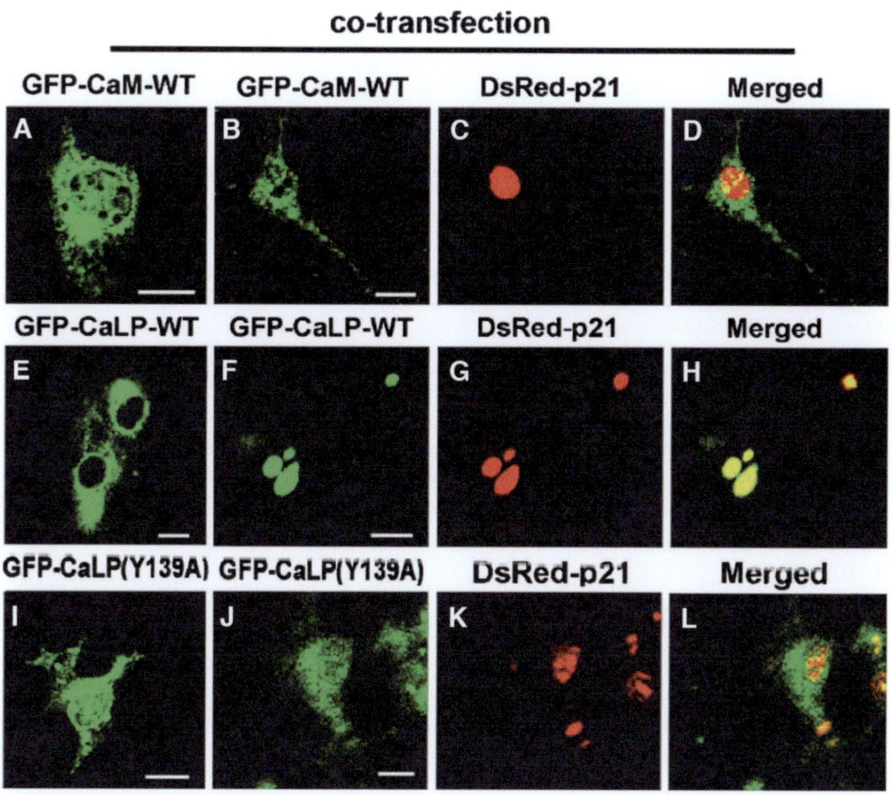

Fig. 2.64 Subcellular localization of wild-type CaLP, wild-type CaM, and wild-type CaLPY139A before and after co-transfection of p21^{Cip1} [216]
(**a**, **e**, **i**) Subcellular localization of wild-type CaM, CaLP, and CaLPY139A in 293T cells. 293T cells were transfected with different recombinant plasmids and visualized by fluorescence microscopy 24 h after the transfection. (**a**) Distribution of wild-type CaM were detected in both cytoplasm and nucleus. e Cytoplasmic localization of wild-type CaLP. (**i**) CaLP transferred into nucleus after Y139 was mutated to A139. (**f–h**) Subcellular co-localization of p21^{Cip1} with WT-CaLP. CaLP was stimulated by p21^{Cip1} to translocate from cytoplasm to nucleus. (**b–d**, **j–l**) Localization of p21^{Cip1} with wild-type CaM and CaLPY139A. The mutant's localization was not influenced by co-transfection with p21^{Cip1} just as CaM

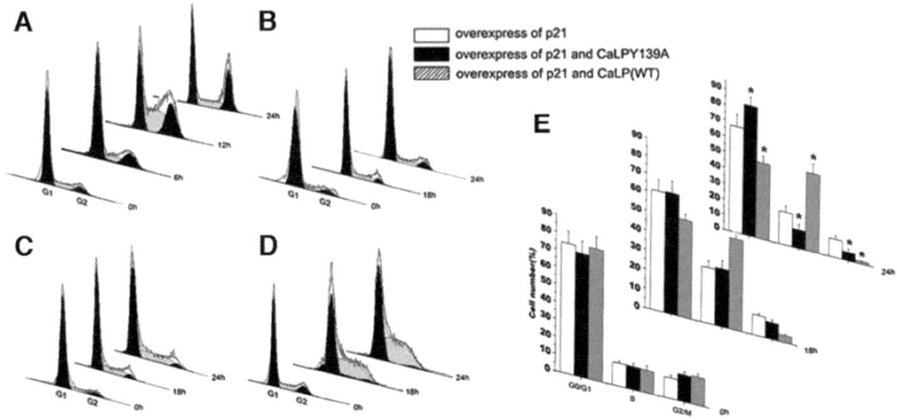

Fig. 2.65 Effect of Y139 mutation on cell cycle distribution
(**a**) The typical cell cycle released from serum starvation for 0, 6, 12, and 24 h. At 6 h, S phase cells increased and so did the G_2 phase; at 24 h, cells entered into G_2 normally. (**b**) Cell cycle distribution of $p21^{Cip1}$-transfected COS-7 cells after serum starvation. The cell cycle was blocked in G_0/G_1 phase and failed to enter into the next stage. (**c**) Cell cycle distribution of CaLPY139A and $p21^{Cip1}$ co-transfected COS-7 cells after serum starvation. CaLPY139A failed to rescue the cell cycle checked by $p21^{Cip1}$ transfection. (**d**) Cell cycle distribution of CaLP-WT and $p21^{Cip1}$ co-transfected COS-7 cells. The FACS results showed that the transfection of CaLP-WT made the amount of S phase cells much more than control. (**e**) The statistical analysis of different cell cycle phases of different time periods. Bars represented mean + SD of three independent experiments, each carried out three times. $^{*}P < 0.001$, statistically significant differences

CaLP/$p21^{Cip1}$ and CaM/$p21^{Cip1}$, the cells in G_0/G_1 arrest couldn't be released or normally impelled to the subsequent phases after the overexpression of CaM; however, the overexpression of CaLP overexpression made it possible to transit from G_1 to S, while the entry to the next phase failed: cells were all hindered in S phase (Fig. 2.66).

2.3.8.3 Protein–Protein Interaction Between $p21^{Cip1}$ and CaLP Nuclear/Cytoplasm Fractions

We were wondering whether the phosphorylation of CaLP was different in nucleus and cytoplasm and whether this different status was the reason during CaLP translocation and sequestration based on our previous investigation. In this paper, the cytoplasm and nuclear fractions were separated from the cultured COS-7 cells co-transfected with $p21^{Cip1}$ and CaLP together, as well as those transfected with CaLP alone to confirm our hypothesis. We performed GST pull-down assay of the GST-p21 complex, and the recombinant CaLP was isolated and purified by positive interaction detected by CaLP polyclonal antibody (Fig. 2.67a, upper panel). The middle panel of Fig. 2.67a showed that the cytoplasmic fractions of COS-7 cells only transfected with CaLP failed to interact with GST-p21 after pull-down analysis. In Fig. 2.67b, CaLP-GFP expressed in cytoplasm was firstly immunoprecipitated by

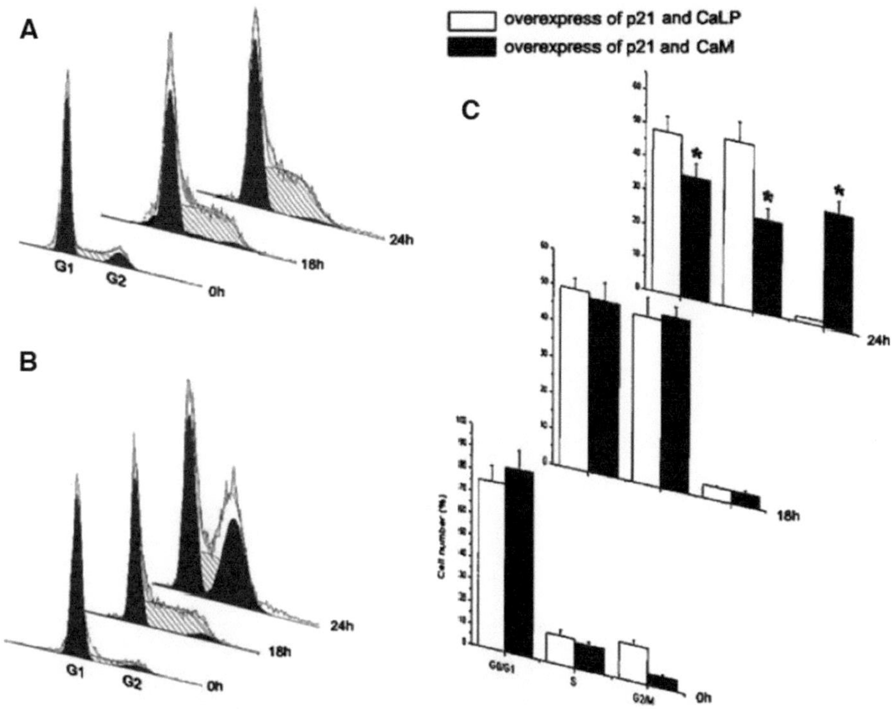

Fig. 2.66 Comparison between CaM and CaLP of their effects on cell cycle distribution
This was analyzed by flow cytometry of COS-7 cells synchronized by serum starvation and transfected by different recombinant plasmids for different time periods. Results indicated one representative experiment performed in triplicate. a Cell cycle released from serum starvation for 24 h. At 24 h, although the cells of S phase elevated greatly, G2 phase cells were still hardly seen. b Compared with the CaLP, after co-transfection of CaM and p21^{Cip1}, the distribution of cell cycle returned to normal after 24 h transfection. c The statistical analysis of different cell cycle phases of different time periods. Bars represented mean + SD of three independent experiments, each performed three times. *$P < 0.001$, statistically significant differences

GFP antibody and exhibited a phosphorylation state. The positive signal in the upper panel indicated that CaLP was phosphorylated in tyrosine. Moreover, we performed the same experiment of CaLP in nucleus. After immunoprecipitation with antibody of p21^{Cip1} followed by the immunoblot with CaLP or tyrosine antibody, we found that even though the CaLP in nucleus could be co-immunoprecipitated with p21^{Cip1}, the CaLP combining with p21^{Cip1} was not phosphorylated as least on its tyrosine site. These results were crucial evidences for our previous hypothesis that the CaLP phosphorylation status in nucleus and cytoplasm was totally different, which was also important for its subcellular localization and translocation when co-transfected with p21^{Cip1}.

For the first time, we successfully set the threshold of prediction as "low stringency" and obtained our results in Table 2.8 via Scansite 2.0 to predict the phosphorylation sites of CaM and CaLP cloned in *P. fucata*. Y139 was the only tyrosine

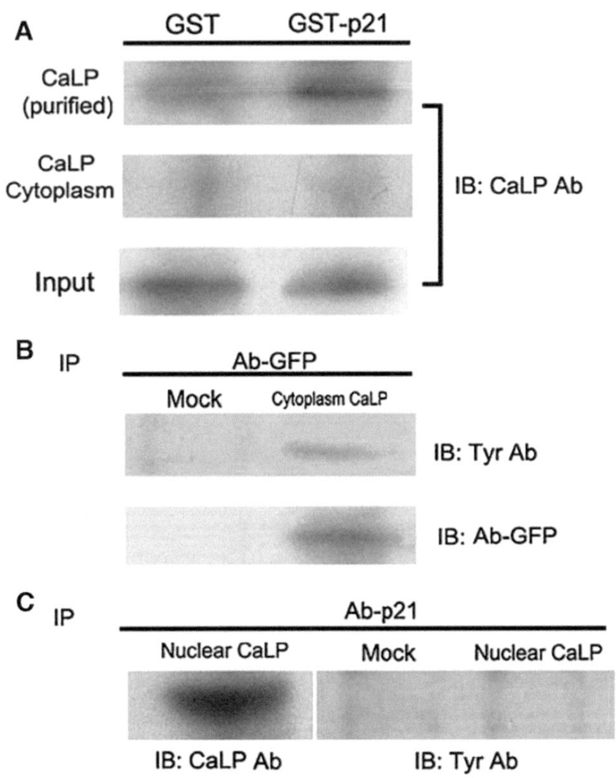

Fig. 2.67 Protein–protein interaction between cytoplasmic/nuclear fractions of CaLP and p21^{Cip} (**a**) Upper panel was GST pull-down assay for GST-p21 and recombined protein CaLP expressed in *E. coli* (purified as described before). GST-p21 was initially incubated with glutathione-Sepharose beads. After sufficient elution, the beads were then incubated with recombined CaLP expressed in *E. coli*. After several rounds of sufficient elution, the complexes of proteins were boiled, and the phosphorylation status of CaLP was detected by its polyclonal antibody. The middle panel was GST pull-down assay of cytoplasm GFP-CaLP and GST-p21 expressed in COS-7 cells. Glutathione-Sepharose beads without p21^{Cip} were acted as controls and incubated with the CaLP. Input represented CaLP amount used in the experiment by being blotted with antibody of CaLP. (**b**) Phosphorylation assay of cytoplasm CaLP. Total CaLP were immunoprecipitated by GFP-Ab and the state of phosphorylation was detected by the Tyr-Ab. The lower panel represented the band detected by Tyr-Ab the same as CaLP. Mock indicated that no CaLP plasmid was transfected. (**c**) Co-IP of p21^{Cip} with nuclear CaLP and the phosphorylation state of the co-immunoprecipitated CaLP. After the immunoprecipitation with Abp21^{Cip}, the nuclear CaLP was then blotted with antibodies of CaLP and Tyr. Mock indicated that neither p21^{Cip1} nor CaLP plasmid was successfully transfected

site for potential phosphorylation both in CaM and CaLP. CaM could only be phosphorylated by Src Kinase, while CaLP could be phosphorylated by many other kinases, including Lck kinase, IRK, and PDGFR kinase. Furthermore, we set the threshold of prediction as "medium stringency," and results indicated that Y139 and S147 in the C-terminal globular domain could be phosphorylated in CaLP, while CaM couldn't.

Table 2.8 Prediction of the phosphorylation sites of CaM and CaLP (*P. fucata*) with "low stringency" by Scansite 2.0

		Kinase group		Kinase		Score	
Site		CaM	CaLP	CaM	CaLP	CaM	CaLP
18	Ser	Acid-ST-kin	–	CK-1	–	0.5260	–
		Baso-ST-kin		CaMKII		0.6026	
				PKC-ζ		0.5166	
27	Ser	–	Baso-ST-kin		PKC ε	–	0.4993
35	Thr	Baso-ST-kin	–	PKC-δ	–	0.5072	–
58	Thr	–	Acid-ST-kin	–	CK-2	–	0.5657
71	Thr	Baso-ST-kin	–	PKC-ζ	–	0.6011	–
80	Thr	Baso-ST-kin	–	PKC-ζ	–	0.4764	–
		Acid-ST-kin		CK-2		0.5457	0.5289
*82	Ser	Acid-ST-kin		CK-2		0.4729	0.4561
111	Thr	Baso-ST-kin		PKC-ζ		0.5975	0.5975
118	Thr	Acid-ST-kin	–	CK-2	–	0.5097	–
*139	Tyr	Y-kin		Src Kinase		0.5838	0.4779
		–	Y-kin	–	PDGFR kin	–	0.5046
					Lck kinase		0.6005
					IRK		0.7029
*147	Ser	–	Baso-ST-kin	–	PKC ε	–	0.4245
					PKC-α/β/γ		0.5210
151	Thr	–	Acid-ST-kin	–	CK-2	–	0.5159

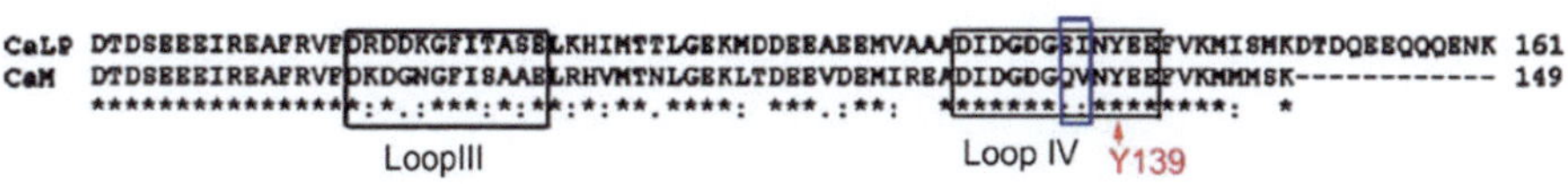

Fig. 2.68 Diagram of amino acid sequence alignment from 79 aa to 161 aa for CaLP and 79 aa to 149 aa for CaM

The 3rd and 4th Ca^{2+}-binding domains were boxed and indicated as loops III and IV; homologous and identical amino acids were indicated by dots and stars, respectively. Although the Y139 was highly conserved in sequence in both CaLP and CaM, the site of aa136 and aa137 was still variable in these two proteins, and these two amino acids may affect the phosphorylation ability of these two proteins via affecting protein's secondary structure

We've mentioned before that this domain took the control of the subcellular localization and translocation of CaLP at which Y139 and S147 were just located. As illustrated in Fig. 2.68, multiple sequence alignment was performed with some modification between CaM and CaLP. The alignment results indicated that despite that Y139 was conserved in CaLP and CaM, the site of aa136 and aa137 was greatly variable between the two proteins. What's more, we could also infer that these two amino acids could affect the protein's space configuration and then influence the phosphorylation ability of Y139 based on the results from fluorescence confocal microscopy and point mutation [216].

It is widely known that the transition of both active and inactive state could be obtained from various situation of nucleocytoplasmic shuttling, phosphorylation, and/or other processes involved in signal transduction [217]. The signal of tyrosine kinase has great effects on cell differentiation and growth. The studies of tyrosine phosphorylation in CaM are becoming more and more crucial, because CaM works as a significant factor that controls the two critical physiological progresses. It has been proven by reliable evidences that there are 2 Tyr phosphorylation sites as well as 12 Thr phosphorylation sites and 4 Ser sites in mammalian CaM [218]. It is notable that dephosphorylated and phosphorylated CaM maintains in balanced states in vivo and CaM is believed to quite probably be dephosphorylated [219]; however, there is little understanding about enzymes functioning on its dephosphorylation. Furthermore, the mechanism in CaM phosphorylation remains relatively unknown to some extent, although a lot of transportation systems and enzymes have been investigated in vitro [220–226]. The main differences of the phosphorylation reactions in various CaM lied in the affinity to other proteins and the kinetics of enzyme, but there is little report of the effect of phosphorylation state on the subcellular localization of CaM. The similar phosphorylation effects on the subcellular localization between CaLP and CaM in *P. fucata* were demonstrated in this article. It could be proposed from the results of phosphorylation prediction and the point mutation that the CaM truly existed in a balance and dynamic between dephosphorylation and phosphorylation states; however, the Y139 in CaLP would guarantee a high phosphorylation state of CaLP so that it could be maintained in the cytoplasm. The great tendency of CaLP to be phosphorylated could be destroyed by CaLP139A; therefore, the subcellular localization of CaLP could be similar to that of CaM localizing in both nucleus and cytoplasm.

It was proven that CaM's function in regulation had great effects on cell cycle for CaM must attend two critical check points during cell cycle: the G_2/M transition as well as the G_1/S boundary. A series of serine/threonine protein kinases, the activities of which were dependent on cyclin binding [227–229], named cyclin-dependent kinases (Cdks) were acting as crucial regulators during cell cycle. The activity of Cdks was highly associated with the progress of cell cycle, which could be regulated through many factors, and one of the most significant methods was binding with the cyclin-dependent kinase inhibitors (CKIs) combining with the cyclin/Cdk or Cdk complex so as to prevent the activation of the complex. In the CKIs, $p21^{Cip1}$ was broadly studied. $P21^{Cip1}$ could regulate the progression of G_1/S and G_2/M transition by combining with both Cdk2 and Cdk4. In this study, the transformation of cell cycle after co-transfecting with $p21^{Cip1}$ and/or CaLPY139A or CaLP might be caused by the direct connection of $p21^{Cip1}$ to Cdk complex that was regulated by the interactions between CaLP and $p21^{Cip1}$. There was a mutation of CaLP139A in the C-terminal region, which would influence the interaction between CaLP and p21 ^{Cip1} [49, 126]. According to the fact that CaLPY139A failed to recover the cell cycle, it could be deduced that it was the critical C-terminal globular region that led to the failure in rescuing the cell cycle.

Furthermore, it was proved from the FACS results that overexpression of CaM in *P. fucata* could indeed promote the arrested cell cycle to normal cell cycle, coinciding with the results of early reports [230] in which radioimmunoassay method and dot blot hybridization were applied. It was found that CaM critically influenced the CHOK1 cells' return to cell cycle from silent state arrested by an anti-calmodulin molecule W13. In Fig. 2.66, cell cycle could just get across G_0/G_1 block with the help of overexpressed CaLP; however, the cell cycle couldn't entirely recover. This phenomenon demonstrated that even though CaLP in *P. fucata* could bind with $p21^{Cip}$, the signal transduction of CaLP still differed in mammals considering that this overexpressed CaLP couldn't absolutely replace CaM in blocked cell cycles.

There have been many examples of phosphorylation affecting the association and dissociation of two proteins. In the PDGF stimulation of skeletal muscle, the phosphorylation of Y521 on Munc18c occurred along with the dissociation of the Munc18c protein from Syntaxin 4 in a time frame consistent with Glut4 translocation [231]. Another example is nucleophosmin (NPM)/B23. It is also a multifunctional protein in a wide amount of cellular functions phosphorylated by several different kinases such as nuclear kinase II and CK2. It was localized to centrosomes, but after the phosphorylation by Cdk2/cyclin E, it would dissociate from centrosomes [232]. Based on these evidences and our results on protein interaction between CaLP and $p21^{Cip1}$ in different fractions of cytoplasm and nuclei, we proposed a model to explain the interesting translocation of CaLP when co-transfected with $p21^{Cip1}$ (demonstrated in Fig. 2.69): After the transfection with CaLP alone, the phosphorylation in the Y139 by some kinase affected the subcellular localization of CaLP and sequestered CaLP in cytoplasm in COS-7 cells. But when co-transfected with $p21^{Cip1}$, the overexpression of $p21^{Cip1}$ would trigger the dephosphorylation of CaLP, and the high affinity of unphosphorylated CaLP to $p21^{Cip1}$ would make the CaLP strongly combine with $p21^{Cip1}$ and then enter into the nucleus.

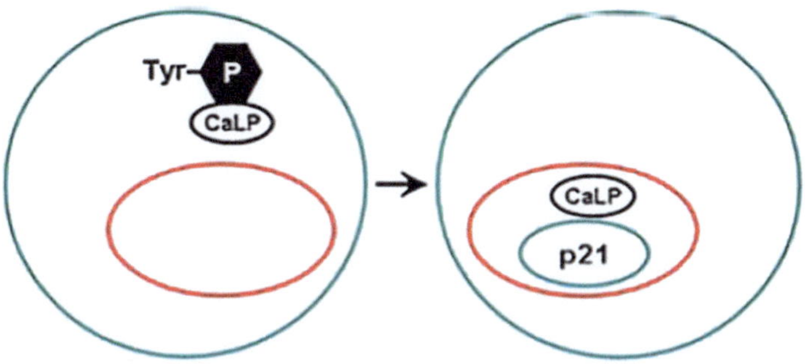

Fig. 2.69 A model explaining the translocation of CaLP when co-transfected with $p21^{Cip1}$
The left circle represented the COS-7 cells transfected with CaLP alone, and the phosphorylation in the Y139 induced the sequestering of CaLP in cytoplasm. The right circle illustrated the COS-7 cells co-transfected with $p21^{Cip1}$ and the CaLP. Overexpression of $p21^{Cip1}$ triggered the dephosphorylation of CaLP, and the high affinity of unphosphorylated CaLP with $p21^{Cip1}$ made the CaLP combine with $p21^{Cip1}$ in nucleus

2.3.9 *Cloning and Characterization of a Homologous Ca^{2+}/ Calmodulin-Dependent Protein Kinase PSKH1 from Pearl Oyster* Pinctada fucata

Many effects of Ca^{2+} signaling are regulated by the Ca^{2+}/calmodulin complex and its acceptors, the Ca^{2+}/calmodulin-dependent protein kinases, PSKH1 included. Studies of the proteins participating in the calcium metabolism in oysters will help us elucidate and demonstrate the mechanism of pearl formation. In this study, we try to describe the acquisition of a full-length PSKH1 cDNA isolated from pearl oyster *Pinctada fucata*. Oyster PSKH1 shares 48% similarity with rat CaM kinase I and 65% homology with human PSKH1 according to their amino acid sequences and contains a calmodulin-binding domain. The results of semiquantitative RT-PCR and in situ hybridization indicated that oyster PSKH1 mRNA was highly expressed in gill epithelial cells and outer epithelial cells of the mantle pallial. These studies have provided important information and new prospects on the complex Ca^{2+} signaling mechanism of calcium metabolism.

2.3.9.1 Cloning of the cDNA Encoding Oyster PSKH1

A 470 bp size product was obtained by degenerate PCR strategies using mantle total RNA as a template. A BLAST search revealed that the product shared high similarity with human PSHK1. The full-length cDNA encoding oyster PSKH1 was obtained using a pair of gene-specific primers based on the 470 bp size sequence. By using touchdown PCR, 5′-RACE yielded a 5′ untranslated region (5′ UTR) and an N-terminus open reading frame (ORF) weighing 477 bp, along with 3′-RACE, a fragment weighing 719 bp and 3′ UTR. The full-length cDNA of oyster PSKH1 weighs 1534 bp with the start site at nt 111 (ATG) and the stop codon at nt 1355 (TGA). The deduced oyster PSKH1 protein consists of 415 amino acid residues (Fig. 2.70) with a calculated molecular mass of 47.151 kDa and a pI of 9.69.

2.3.9.2 Motif Analysis of Oyster PSKH1 Protein

The phosphorylation of CaMKs is critical for transducing intracellular Ca^{2+} signals [233]. Four classes of phosphorylation sites were identified in oyster PSKH1 protein, including 8 sites for casein kinase I [S-X-X-S/T (364-367, 367-370, 370-373, 373-376, 376-379, 379-382, 382-385, 385-388, which occur near the C-terminus)] [234], 3 sites for protein kinase A [R/K-X-S/T (75-77, 357-359, 372-374)] [235], 15 sites for protein kinase C [S/T-X-R/K (61-63, 77-79, 171-173, 195-197, 224-226, 342-344, 359-361, 364-366, 367-369, 373-375, 376-378, 379-381, 382-384,

Fig. 2.70 *P. fucata* PSKH1 cDNA and the deduced amino acid sequence Coding cDNA is numbered on the left, and the deduced amino acid sequence is numbered on the right. A putative CBD is in boldface (amino acids 343-361). This cDNA sequence has been submitted to GenBank (accession number: AY818357)

```
   1                                            tggattcatataacatgact
  21 ttattaatttctcttgcactcggaacaaatgatcccctcaactttaattgaggggatgaa
  81 ggacttaattccctataatatcctttcctaatgggttgtatgagtagtaagttagtgccc
                                     M  G  C  M  S  S  K  L  V  P   10
 141 gaggggccatcgggaaaccaagcagttgtcgaagtttttaacaatgaaaggaataaggaa
      E  G  P  S  G  N  Q  A  V  V  E  V  F  N  N  E  R  N  K  E   30
 201 tacaacaggcagaatccacaccagaacaggggccctcgggatccgaccaaggaccctaaa
      Y  N  R  Q  N  P  H  Q  N  R  G  P  R  D  P  T  K  D  P  K   50
 261 aatgctggacccccacccgagggacagaggtccaacagaaaggtgaagaaatacagggat
      N  A  G  P  P  P  E  G  Q  R  S  N  R  K  V  K  K  Y  R  D   70
 321 aaattcgaccctcgagttacagctaagtatgacataaaagcattgattggaaggggaat
      K  F  D  P  R  V  T  A  K  Y  D  I  K  A  L  I  G  R  G  N   90
 381 ttcagcaaggtcgtgcgtgttgaacacagggtaactaaacaaccctatgcaatcaagatg
      F  S  K  V  V  R  V  E  H  R  V  T  K  Q  P  Y  A  I  K  M  110
 441 atagacagagtgcaagggaaagaagttttcgaatcggaggtggctgtcctccgccgagtt
      I  D  R  V  Q  G  K  E  V  F  E  S  E  V  A  V  L  R  R  V  130
 501 aaacactcctatatcattcagctcattgaagttttcgaaacaaaagataaagtgtacatg
      K  H  S  Y  I  I  Q  L  I  E  V  F  E  T  K  D  K  V  Y  M  150
 561 gtaatggagcttgctacaggtggtgaattatttgataggattatagctaaaggttccttc
      V  M  E  L  A  T  G  G  E  L  F  D  R  I  I  A  K  G  S  F  170
 621 acagagagggatgctacacgggtattgaacatggttctagatggtgtgaagtacttacac
      T  E  R  D  A  T  R  V  L  N  M  V  L  D  G  V  K  Y  L  H  190
 681 gggctaggcatcactcatcgggatttaaaacccgaaaatcttttatactatcatcctgga
      G  L  G  I  T  H  R  D  L  K  P  E  N  L  L  Y  Y  H  P  G  210
 741 catgattctaaaatcatgatcactgatttcggtctcagttctacgagaaaaggaccagaa
      H  D  S  K  I  M  I  T  D  F  G  L  S  S  T  R  K  G  P  E  230
 801 aatttcatgcggactacatgtgggaccccagaatacattgcaccggaaatcatcgcccgg
      N  F  M  R  T  T  C  G  T  P  E  Y  I  A  P  E  I  I  A  R  250
 861 aaaccttacatgtgtcaagtggacatgtgggctgtaggggtgataacctacattctactc
      K  P  Y  M  C  Q  V  D  M  W  A  V  G  V  I  T  Y  I  L  L  270
 921 agtggaacgatgccgtttgacgatgaaaataaaactaggttgtatagactcatcctcaag
      S  G  T  M  P  F  D  D  E  N  K  T  R  L  Y  R  L  I  L  K  290
 981 gccaagtatagctatgccggagagcattggaaggatgtgtctgcacaagccaaagacttt
      A  K  Y  S  Y  A  G  E  H  W  K  D  V  S  A  Q  A  K  D  F  310
1041 atagacaaactgttggtagtgagtcccggggatcgcctgtcggcagcggacgctctcaaa
      I  D  K  L  L  V  V  S  P  G  D  R  L  S  A  A  D  A  L  K  330
1101 catcaatggttgatttcgaatgctgcttcgtccagtaacaagaaccttcatcgtacaatc
      H  Q  W  L  I  S  N  A  A  S  S  S  N  K  N  L  H  R  T  I  350
1161 agccagaatcttatacaccgccaatctacgcgtgcaaacagcaccaagtcagccaagtcg
      S  Q  N  L  I  H  R  Q  S  T  R  A  N  S  T  K  S  A  K  S  370
1221 acgagatctacgaaatcaaacaaatccaaccgatctggacgttcgttgcgttccgagcac
      T  R  S  I  K  S  N  K  S  N  R  S  G  R  S  L  R  S  E  H  390
1281 cgccgggttatgccggacgaaatcgatgaactccatcgtgacccggacgtgcaagcagat
      R  R  V  M  P  D  E  I  D  E  L  H  R  D  P  D  V  Q  A  D  410
1341 ttggcatctctgggatgacacacttatttttgtgatgaaaaatgcaaatcagggtcattt
      L  A  S  L  G  *                                             415
1401 tgtatttatagcaactttatgagtaaccatgttatagtcaatgtggtgctttaagattca
1461 aattaaggtatatactgtatatactgaattccgacatggcatttaaaaaaaaaaaaaaa
1501 aaaaaaaaaaaaaa     1534
```

385-387)] [236], and 3 sites for GSK3 phosphorylation recognition, including 2 overlapping regions [X-X-S/T-X-X-X-S/T (358-364, 365-371, 368-374)]. There were four *N*-glycosylation sites (NFSK, NKTN, NSTK, and NKSN) [237], three *N*-myristoylation sites (GQRSNR, GLGITH, and GLSSTR) [238], and two glycosaminoglycan attachment sites (KSAK and RSGR).

2.3.9.3 Analysis of Domains and Structure of Oyster PSKH1 Protein

The protein–protein BLAST results through the NCBI website reveal that oyster PSKH1 shows high identity with human PSKH1 (65%) and human CaMKI (48%) in protein sequence. As shown in Fig. 2.71 a, sequence alignment reveals that oyster PSHK1 contains 15 invariant amino acid residues and 11 conserved subdomains within the catalytic core (80-334) of the CaMKs family [239, 240]. Subdomain I, at the N-terminus of the kinase domain, contains a consensus motif Gly-X-Gly-X-X-Ser-X-Val, a signature for the protein kinases ATP-binding region [241]. An invariant Lys88 lies in subdomain II which appears to be directly involved in the phosphotransfer reaction and possibly mediates proton transfer [240]. The highly conserved subdomain VI includes two invariant residues (Asp198 and Asn203) that lie within the consensus motif HRDLKXXXN (named a catalytic loop, Asp198 is the key amino acid accepting protons). In subdomains VII and VIII, two conserved motifs "Asp-Phe-Gly" (residues 219-221) and "Ala-Pro-Glu" (residues 244-246) are commonly considered as the key protein kinase catalytic domain indicators [242]. Subdomains IX and X are poorly conserved subdomains except for Arg321 in subdomain X.

The striking similarity between oyster PSKH1 and human PSKH1 provides an approach for analyzing the domains and structure of oyster PSKH1. A sketch of the structure of PSKH1 (Fig. 2.71b) was drawn according to reported results for human PSKH1. Outside the catalytic domain, an Src homogenous domain 4 (SH4) motif (M-G-C-X-X-X-K) [243] was identified in the NH2-terminal region containing two amino acid residues Gly2 and Cys3, as the myristorylation and palmitoylation sites which were reported to be required for Golgi targeting [244] . Amino acid residues 21-29 (NRGPRDPTK) harbor the core binding consensus for Src homology domain 3 (SH3), in which I Src domain (R/K-X-X-P-X-X-P) and II Src domain (P-X-X-P-XR/K) cross. The motif was found to participate in several biological processes as diverse as signal transduction pathways, cytoskeleton organization, membrane traffic, or organelle assembly [244]. Studies of a double-truncated mutant of human PSKH1 (Δ78-360, homologous to Δ62-339 of oyster PSKH1) suggested that a homodimerization domain is located on the conserved catalytic domain. A similar autophosphorylation process is determined by this domain on human PSKH1 [242]. The C-terminus of oyster PSKH1 has a special domain which contains eight continuous CK1 phosphorylation sites. The speciality of this domain indicates that oyster PSHK1 is involved in cell division, DNA repair, and glycogen metabolism.

2.3.9.4 Expression of Oyster PSKH1 mRNA in Tissues

The function of oyster PSKH1 in vitro was studied by analyzing the expression patterns of PSKH1 in different oyster tissues using semiquantitative RT-PCR. A pair of specific primers (F3 and R3) was used to perform RT-PCR using mantle, gill, muscle, and gonad tissues. The 512 bp product was then inserted into the pGEM-T

Easy Vector and confirmed by sequencing. As shown in Fig. 2.72, PSKH1 mRNA is expressed in all tested tissues, with higher expression in mantle and gill tissues.

The expression locations within different tissues were detected by in situ hybridization. Strong PSKH1 hybridization signals are detected in the outer epithelial cells of the middle fold of the mantle, the outer epithelial cells of the inner fold of the mantle (Fig. 2.73a), and the epithelial cells of the gill (Fig. 2.73b). Some interesting positive signals also occurred in the outer epithelial cells of the mantle pallial but was absent in the outer fold of the mantle. Although some positive signals were also observed in muscle and gonad, their intensities were too weak (data not shown). Hybridization with the control sense probe yielded no hybridization signals (data not shown).

Sequence alignment revealed a high homology between oyster PSKH1 and CaMKs, which indicates possible similarities in their domain structure and functions. The CBD, an important domain for CaMKs' activation by Ca^{2+}/CaM [245], was not obvious within the deduced amino acid sequence of oyster PSKH1. Computer predictions suggested that a stretch of 19 residues (NKNLHRTISQNLIHRQSTR, amino acids 343-361, Fig. 2.70, boldface) can form an amphiphilic α-helix, which is a typical CBD structure (Fig. 2.71b). Human PSKH1 studies [242] suggest that its autophosphorylation takes place between amino acids 372 and 381, corresponding to residues 351-359 of oyster PSKH1 which is inside the putative CBD, implying the possibility that Ca^{2+}/CaM binding is regulated by autophosphorylation [246], as has been proved in other CaMKs [247–249]. Actually, the autophosphorylated residue between residues 351 and 359 is limited to Ser351 of oyster PSKH1. The presence of the CBD in oyster PSKH1 indicates that PSKH1 plays a role in translating and coordinating the dynamic second messenger, Ca^{2+}, into the appropriate cellular responses via phosphorylation [250]. Many CaMKs have a primary amino acid sequence (the autoinhibitory domain, AID) partially overlapped by the CBD that interacts with and inhibits the conserved catalytic domain [245]. However, since PSKH1 does not attach to any of broad-range substrates in vitro, the determination of the AID in PSKH1 is complicated [242]. Further studies are needed to describe the domain structure of oyster PSKH1.

Pearl formation is a complex process, involving the uptake, transport, and recruitment of Ca^{2+}. The oyster mantle is the key organ responsible for calcium metabolism and mineral processing in pearls [113]. The strong hybridization signals of oyster PSKH1 in the outer epithelial cells of the middle fold and in the inner fold of mantle further validate the semiquantitative RT-PCR results. Previous studies proved that CaM, an activator of CaMKs, may participate in the process of shell and pearl formation [52]. Compared with the CaM mRNA signals in the mantle, an especially high expression of oyster PSKH1 mRNA occurs in the outer epithelial cells of the mantle pallial, a key tissue involved in nacreous secretions [251]. The oyster PSKH1 mRNA is also highly expressed in epithelial cells of the gill, a major organ for calcium uptake from external medium. Changes of the intracellular Ca^{2+}

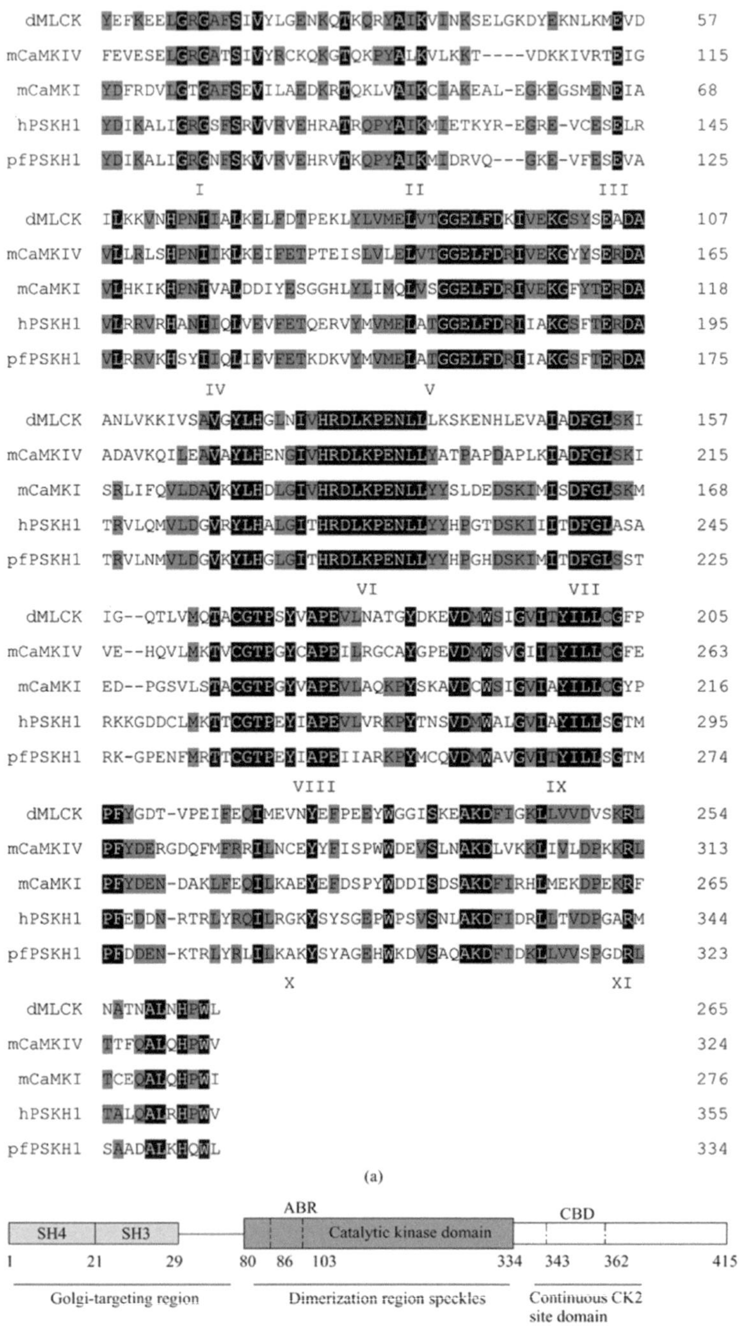

Fig. 2.71 Analysis of domains and structure of oyster PSKH1 protein
(**a**) Alignment of the catalytic core of members of the Ca2+-/CaM-dependent protein kinases family

concentration are decoded by its binding proteins and their following acceptors such as the CaMKs [250], which phosphorylate and regulate multiple cellular targets that contribute to neurotransmission, neuronal plasticity, cell excitability, gene expression, cell secretion, and cell shape [252, 253]. The distribution of the PSKH1 mRNA in oyster tissues indicates that PSKH1 may play an important role in the complex calcium signaling pathways in oyster calcium metabolism during pearl formation. However, the detailed mechanism needs further study.

2.4 The Identification and Function Analysis of Calcineurin

2.4.1 *Cloning, Characterization, and Immunolocalization of Two Subunits of Calcineurin from Pearl Oyster (*Pinctada fucata*)*

Calcineurin (CN) containing a catalytic subunit (CN A) and a regulatory subunit (CN B) is a multifunctional protein engaged in many important physiological processes. Here, we obtained the two subunits of CN (Pf-CN A and Pf-CN B) from pearl oyster *Pinctada fucata* and, for the first time, reported its expression and distribution patterns in different developmental stages and its enzymatic activity as well as the immunolocalization in various tissues of adult oyster. The Pf-CN A was widely located in all of the tested tissues including the mantle, digestive gland, gills, gonad, foot, and adductor muscle, also with strong signals detected in the gill, gonad, mantle, and foot. We must notice that Pf-CN A was mainly detected at both lateral surface of inner mantle fold and the inner epithelial cells of basal periostracal groove, two places responsible for shell construction and periostracum formation, respectively. What's more, strong signals were also detected in the base of gill filaments and epithelial cells of the branchial filaments. Taken all results together, we suggested that Pf-CN may participate in the whole development in *P. fucata* and play crucial roles in many different physiological activities, especially shell formation. Our research could provide important prospects to get better understandings of biomineralization.

Fig. 2.71 (continued) with the corresponding region of oyster PSKH1. The 11 conserved subdomains of CaMKs in the sequences including oyster PSKH1 are indicated by Roman numerals. dMLCK, *Dictyostelium discoideum* myosin light-chain kinase; mCaMKIV, rat calcium-/calmodulin-dependent protein kinase type IV; mCaMKI, rat calcium-/calmodulin-dependent protein kinase type I; hPSKHl, human PSKH1; pfPSKHl, *P. fucata* PSKH1. (**b**) Putative domain structure of oyster PSKH1. ABR, ATP-binding region; CBD, calmodulin-binding domain; SH3, Src homology domain 3; SH4, Src homology domain 4

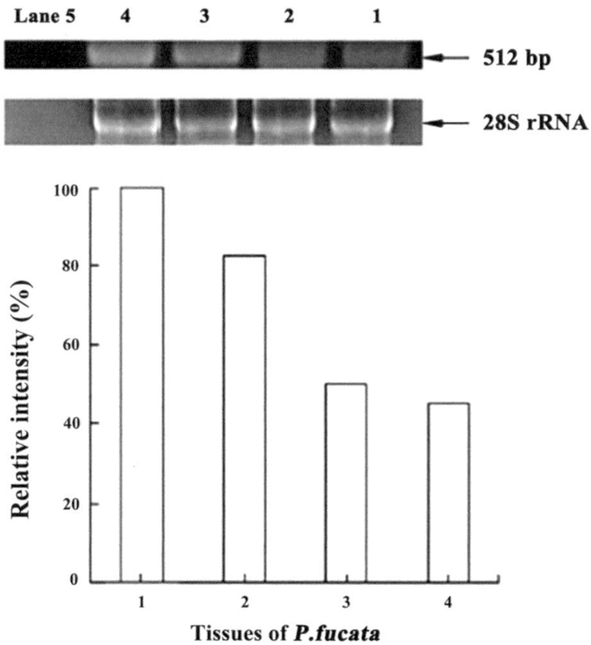

Fig. 2.72 Expression of PSKH1 mRNA in tissues of *Pinctada fucata* Semiquantitative RT-PCR was done with RNA samples from the adult tissues of pearl oyster. The PCR product (512 bp) is indicted by an arrow with the relative intensities of the bands evaluated by densitometry. The 28S rRNA was used as a control of equal quantities of total RNA used in RT-PCR. Lane 1, mantle; lane 2, gill; lane 3, muscle; lane 4, gonad; lane 5, negative control

2.4.1.1 Cloning and Sequence Analyses of the Oyster Pf-CN A

We obtained the 783 bp Pf-CN A cDNA fragment by RT-PCR from the mantle of *P. fucata* with two degenerate primers APF and APR. Based on this confirmed sequence, we designed and synthesized two gene-specific primers A3RSP and A5RSP for 3′-RACE and 5′-RACE, respectively, to amplify the unknown nucleotide sequence of 3′- and 5′-terminal cDNA. Then another two specific primers ACF4 and ACR2 were applied to confirm the sequence spliced by RACE. The PCR products were then cloned and sequenced, which matched well the sequence obtained earlier.

The complete 2043 bp cDNA sequence of Pf-CN A consisted a 1461 bp ORF, a 436 bp 3′ untranslated sequence, and a poly (A) tail of 14 nucleotides as indicated in Fig. 2.74a, which was successfully submitted to GenBank with the accession no. EU797510. The deduced CN A protein encoded a 487 amino acid protein with a calculated molecular mass of approximately 55.2 kDa and a predicted isoelectric point of 5.68. Pf-CN A showed 78%, 80%, and 92% identities to other CN A homologs from human, African clawed toad, and scallop, respectively (Fig. 2.75a). Multiple amino acid sequence analysis run on BLAST showed that Pf-CN A had a typical primary structure of CN A proteins in other species. The well-conserved amino acid domains in CN A protein comprised a catalytic domain (aa 40-314), a CN B-binding domain (aa 326-370), and a calmodulin (CaM)-binding domain (aa 389-415) as well as an autoinhibitory domain (AI) (aa 459-480). The COOH and NH_2 termini of CNA showed high variances between species even CN A

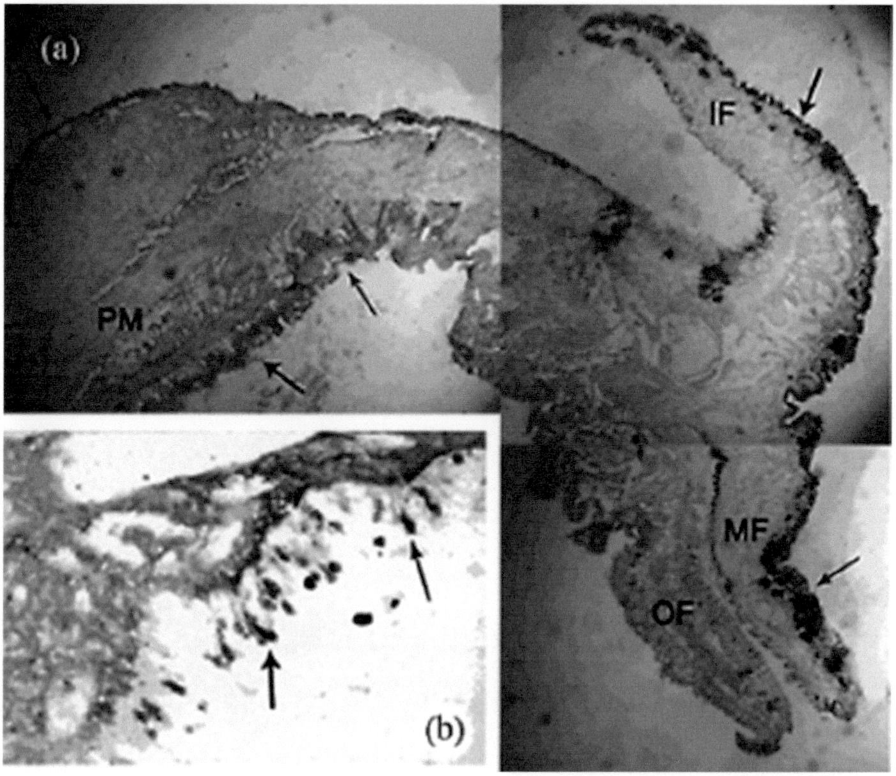

Fig. 2.73 In situ hybridization of oyster PSKH1 mRNA in the mantle and gill of *Pinctada fucata* Hybridization signals (arrow) were observed in epithelial cells of the middle fold and of the inner fold of the mantle (**a**) and in the gill epithelial cells (**b**). Signals also occurred in the outer epithelial cells of the mantle pallial (**a**). Magnifications are 25×. *IF* inner fold, *MF* middle fold, *OF* outer fold, P*M* mantle pallial

genes inside same organism [254]. CN A has been classified into three categories: α-, β-, and γ-type in mammals. The first one, CN Aα isoform, known as brain-type, is extensively distributed; a conserved character of an N-terminal polyproline motif is found in Aβ isoform, and Aγ is specifically expressed in the testis of mouse [255] and human [255]. Although CN A was isolated from the scallop testis, the high similarity with mammalian brain-type CN made it a brain-type instead of a testis-specific isoforms [256] in mollusk. Multiple CN A sequence alignment between *Mizuhopecten yessoensis* and *P. fucata* exhibited relative high identities of C-terminus, N-terminus, CN B-binding domain, catalytic domain, AI domain, and CaM-binding domain as 62.5%, 66.7%, 93.3%, 93%, 95.5%, and 88.9%, respectively (Fig. 2.79a). Another interesting thing was that the linker sequence between CN B-binding and catalytic domain, as well as the linker between AI domain and CaM-binding, was 100% and 97.9%, respectively, while the linker region between CaM-binding and CN B-binding domain exhibited lower sequence consistency.

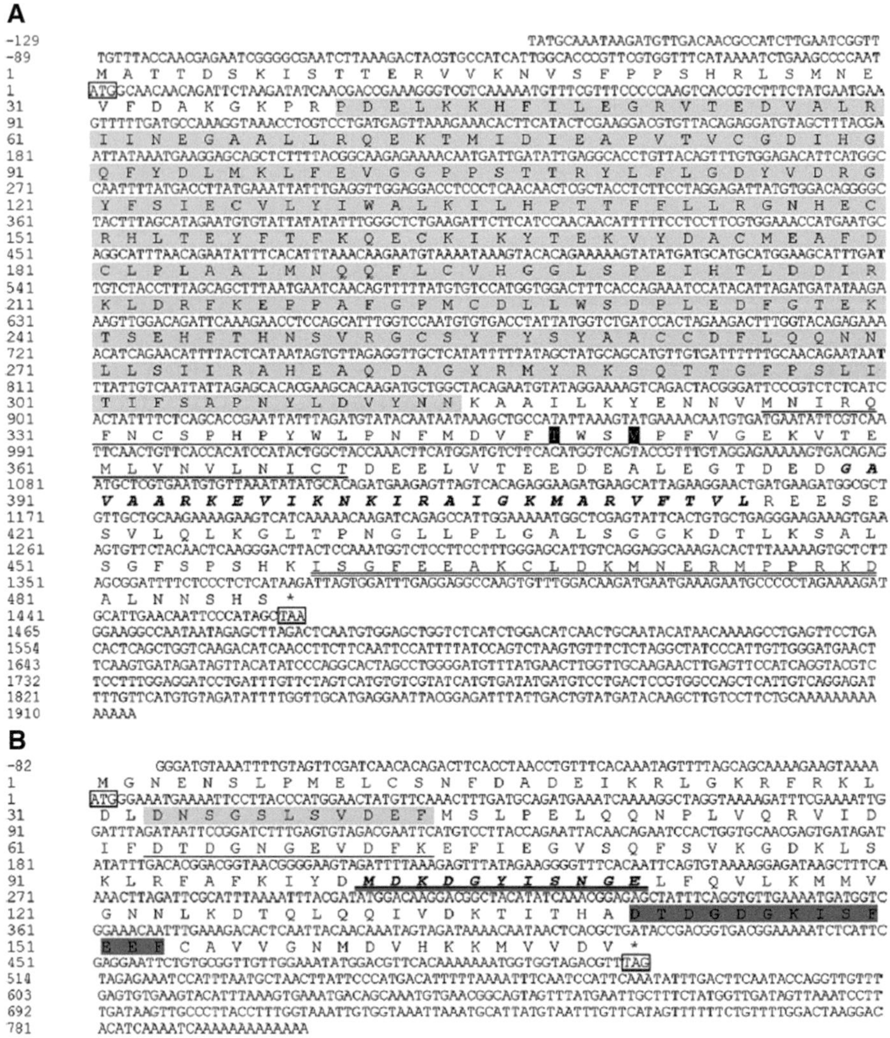

Fig. 2.74 cDNA sequences and deduced amino acid sequences of calcineurin two subunits: Pf-CN A (A) and Pf-CN B (B) in *P. fucata*

Pf-CN A possessed a putative catalytic domain (shaded), a CN B-binding domain (underlined), a CaM-binding domain (in bold and italicized), and an autoinhibitory domain (double underlined). In CN B-binding domain (underlined), two putative residues T and V (white on black shade) were essential for Pf-CN A interaction with FKRP and cyclophilin; the four EF-hand-type calcium-binding sites from 1 to 4 in Pf-CN B were light shaded, underlined, double underlined, and black shaded, respectively. GenBank accession numbers for calcineurin A (EU797510) and calcineurin B (EU797511)

A

	BmCA	DmCA	CeCA	DsCA	HsCA	XlCA	MyCA	PfCA	BtCA	RnCA
BmCA	100	74	76	45	76	81	83	83	80	80
DmCA		100	69	42	66	69	70	70	68	68
CeCA			100	45	71	73	73	74	73	73
DsCA				100	45	46	45	46	45	45
HsCA					100	79	78	78	77	77
XlCA						100	79	80	94	94
MyCA							100	92	78	79
PfCA								100	80	80
BtCA									100	100
RnCA										100

Fig. 2.75 The construction of the phylogenetic tree of *P. fucata* calcineurin A and B with those of other species

The tree was produced by the ClustalX program. GenBank accession numbers for these amino acid sequences of calcineurin A were as follows: PfCA, CN A [*Pinctada fucata*] (GenBank accession number EU797510); MyCA, CN A [*Mizuhopecten yessoensis*] (BAA94542); BmCA, CN A [*Bombyx mori*](NP_001037025); DmCA, CN A [*Drosophila melanogaster*](CAA54807); HsCA, calcineurin A β isoform [*Homo sapiens*] (EAW54496); BtCA, protein phosphatase 3, α isoform [*Bos taurus*] (NP_777212); RnCA, calcineurin A α [*Rattus norvegicus*] (BAA14083); XlCA, CN A [*Xenopus laevis*] (BAB19673); CeCA, abnormal CHEmotaxis family member (tax-6) [*Caenorhabditis elegans*] (NP_001021292); DsCA, CN A [*Dictyostelium discoideum*] (CAA65935). GenBank accession numbers for amino acid sequences of calcineurin B are as follows: PfCB, [*Pinctada fucata*] (EU797511); MyCB, CN B [*Mizuhopecten yessoensis*] (BAA94543); BmCB, CN B [*Bombyx mori*] (NP_001037026); DmCB, calcineurin subunit B 2 isoform [*Drosophila melanogaster*] (Q24214); HsCB, calcineurin 3 B β isoform [*Homo sapiens*] (NP_671709); BtCB, calcineurin 3 Bα1 isoform [*Bos taurus*] (NP_777008); RnCB, calcineurin 3 Bα isoform, type 1 [*Rattus norvegicus*] (NP_059005); XlCB, CN B [*Xenopus laevis*] (AAH82858); DsCB, CN B [*Dictyostelium discoideum*] (CAC20026); CeCB, calcineurin B-1 isoform [*Caenorhabditis elegans*] (AAO33925)

Furthermore, the NH_2 terminus of Pf-CNA was not as rich in proline as that CNA β-isoform [257–259] in mammalian. Additionally, asparagine and glutamine were also not abundant in Pf-CNA but in *Dictyostelium discoideum* CN A [260]. However, the Pf-CNA from mantle shared no typical or significant similarities with those from mammalian isoforms; it showed highly homologous to the brain-type scallop CN A, which could also be explained by the speculation that the isoforms from mammalians appeared after the divergence of mammalian from the mollusks and then diverged rapidly for specific functions in the testis, brain, and other tissues [256] from the evolution standpoint.

Cyclosporin A [261] and tacrolimus (FK506), as two most commonly used immunosuppressants, could potently inhibit the activity of phosphatase according to Klee et al. These two compounds could specifically bind with their cellular receptors, cyclophilin and FKBP, respectively, to form a complex to interact with CN B-binding domain of CN A and further induce dramatic bone loss through the block of CN [262]. Thr [345] and Val [348] were two typical putative residues important for interaction with cyclophilin and FKRP in Pf-CN A's CN B-binding domain.

2.4.1.2 Cloning and Sequence Analyses of the Oyster Pf-CN B

The 171 bp product of Pf-CN B was amplified by RT-PCR with two degenerate oligonucleotide primers (BFP2 and BRP2) with the template of total mantle RNA. Based on this sequence, another two gene-specific primers B3RF2 and B5RR2 were designed to amplify the 3′- and 5′-terminal nucleotide sequence of the putative oyster CN B cDNA. Then two specific confirming primers BCF2 and BCR2 in accordance with the 5′ and 3′ UTR sequence of putative CN B were specially synthesized to confirm the sequence obtained from RACE. The PCR products were subcloned to pMD19-T vector and then sequenced, which matched well with the sequence expected earlier.

The complete cDNA sequence of CN B containing the poly (A) tail was 886 bp in length, an 82 bp 5′ untranslated region and a 278 bp 3′ untranslated region also included, as Fig. 2.74b showed, which was further successfully submitted to GenBank with the accession no. EU797511. The deduced 170 amino acid Pf-CN B protein comprised 4 conserved EF-hand-type calcium-binding motifs, EF-1 (aa 33-43), EF-2 (aa 64-73), EF-3 (aa 101-111), and EF-4 (aa 141-153), with a calculated molecular mass of approximately 19.2 kDa and a predicted isoelectric point of 4.60. The sequence of oyster CN B shared intensive similarities with those of CN B from amphibian, mollusk, and mammalian (90% to African clawed toad, 98% identity to scallop, and 78% to human), as indicated in the phylogenetic tree analysis from Fig. 2.75b. The high conservation percentage throughout evolution from yeast to mammalians enabled it an important functional interchange of CN B subunits between *N. crassa* and mammalian [263].

2.4.1.3 Gene Expression Profile of Pf-CN A and Pf-CN B in Developmental Stages of *P. fucata*

CN has been proved to play critical roles during development and differentiation [254]. Semiquantitative PCR was performed to analyze the relative expression level of Pf-CN A and Pf-CN B during various developmental stages in pearl oyster (fertilized egg, cleavage, gastrula stage, D-shaped larval stage, attachment stage, and adult). Both two CN subunits expressed constitutively with only a small fluctuation in quantity during the different developmental stages of oyster (Fig. 2.76). The expression of the two Pf-CNA and Pf-CN B subunits increased in accordance with the developmental process from fertilized egg to gastrula but underwent a decrease from D-shaped stage to adult. Meantime, the change of Pf-CN B was almost synchronized with Pf-CN A in every detected stage. The intake of food and shell formation was the symbol of D-shaped larval stage, while the start of byssus secretion indicated the beginning of the attachment stage. These results suggested that the important roles CN played during the development of pearl oyster.

2.4.1.4 PP2B Activity

The CN enzymatic activity in the tissues of *P. fucata*, including the gonad, gill, hemocyte, foot, mantle, and adductor muscle, was shown in Fig. 2.77, which was identical to our immunolocalization analysis (see below). The highest level of CN activity appeared in the gonad. Moreover, the activity was also relatively higher in the hemocyte, gill, mantle, and foot than that in the viscera or adductor muscle. In scallop, CN A mRNA expression level in the testis was significantly higher than that in the mantle, digestive gland, and gill [264], which was consistent with our findings

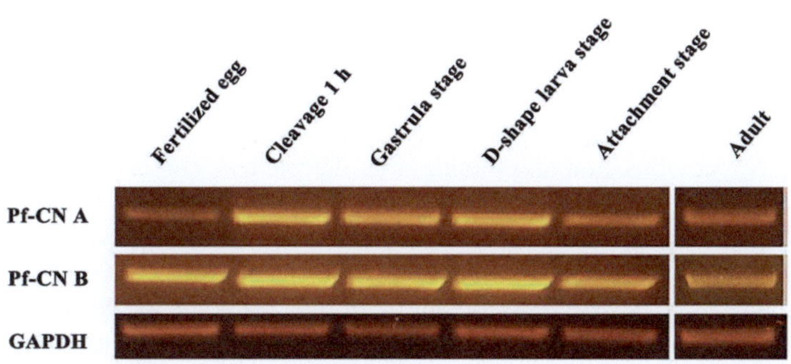

Fig. 2.76 Relative expression level of the Pf-CN A and Pf-CN B in six various developmental stages in *P. fucata*
RT-PCR was performed to analyze the expression of Pf-CN A and Pf-CN B in different developmental stages in *P. fucata*. Glyceraldehyde-3-phosphate dehydrogenase (GAPDH), as a widely known housekeeping gene, was used as a positive control. This was the representative of three repeated independent RT-PCR analysis results

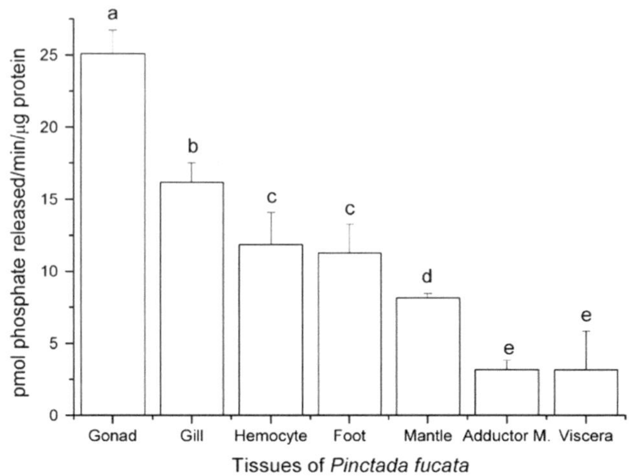

Fig. 2.77 Calcineurin enzymatic activity in different tissues from *P. fucata*
Each assay was repeated three times based on tissue preparations. Results expressed as means ± SE were analyzed using one-way ANOVA at the 5% level. Values with dissimilar alphabetic letters differed significantly ($p < 0.05$)

on CN enzymatic activity just showed. Our group has successfully isolated and identified a novel calmodulin-like protein with an extended tail at the C-terminal from the mantle of *P. fucata* [24, 196]. Since CaM was the pivotal regulator of CN, we intended to investigate the potential interactions between CN and CaLP in further research. Therefore, immunohistochemical analysis was performed to gain the precise distribution of Pf-CN in *P. fucata*.

2.4.1.5 Immunolocalization of Pf-CN A

Immunohistochemical analysis was carried out to better analyze the Pf-CN distribution pattern in different tissues from *P. fucata*. Figure 2.78 implied that CN A was located in tissues including the gonad, gill, digestive gland, adductor muscle, mantle, and foot. No obvious immunostaining was detected in control groups (data not shown).

Strong positive immunostaining signals were detected in all surfaces of inner, middle, and outer mantle folds, especially in outer epithelial cells of the outer and middle fold near the base of the periostracal groove, the epithelial cells of the middle fold edge, as well as the epithelial cells of lateral but not medial surface of the inner fold (Fig. 2.78a). The mantle has always been considered as the crucial tissue participating in shell regeneration and formation [74, 265]. Epithelial cells in this tissue are supposed to secret matrix proteins, regulate the transportation of various metal ions through ion channels, and help establish a supersaturated environment in extrapallial cavity for further biomineralization-related processes. What's more, epithelial cells from the base of periostracal groove also control the secretion of some organic macromolecules for the formation and thickening of the periostracum

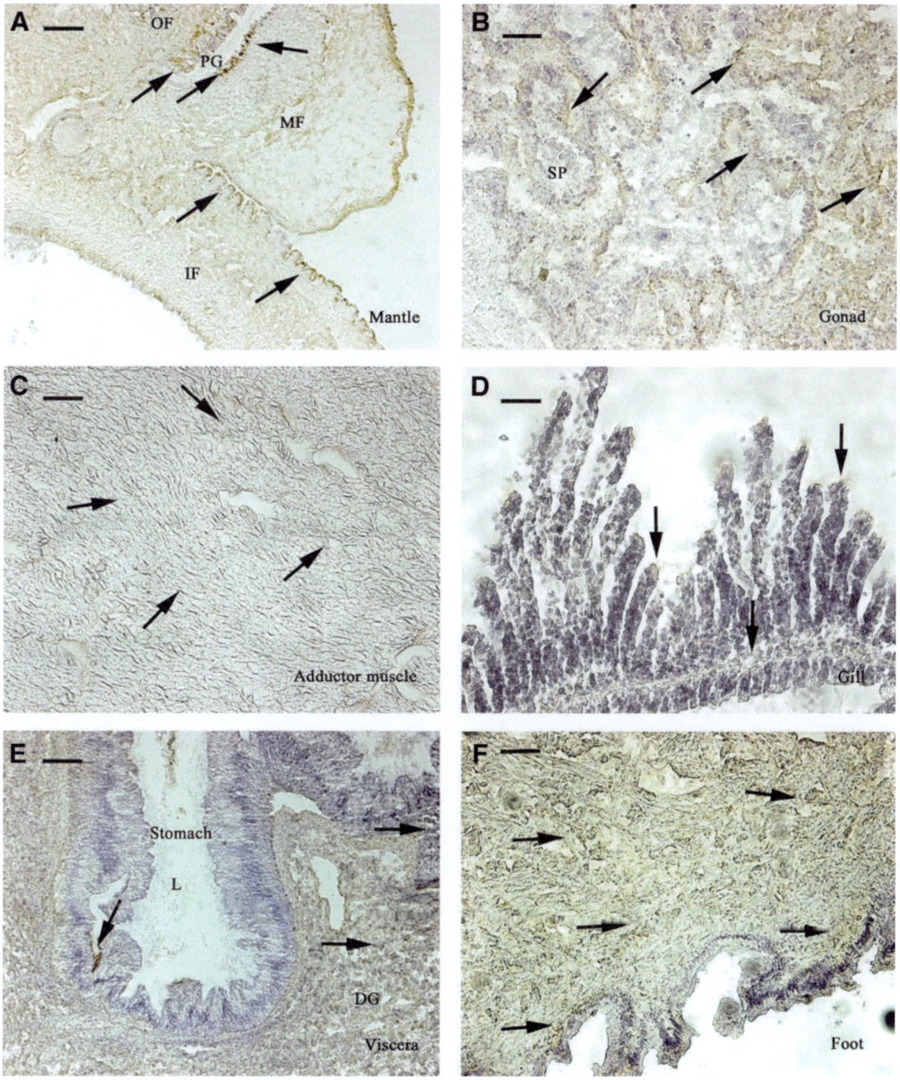

Fig. 2.78 Immunohistochemical localization of CN A in different tissues in *P. fucata* [268] (**a**) Mantle; (**b**) gonad; (**c**) adductor muscle; (**d**) gills; (**e**) viscera; (**f**) foot. Immunostaining signals were indicated by arrows. Bars 100 μm in (**b**), (**c**), and (**f**), 50 μm in (**d**), and 200 μm in (**a**) and (**e**). *PG* periostracal groove, *SP* spermatogenic duct, *L* lumen, *IF* inner fold, *MF* middle fold, *OF* outer fold

[266], covering the shell surface to protect oysters and forming a compact compartment of scaffold [18] for further shell formation. Moreover, there are a large number of intercellular ion channels and Ca^{2+} channels [267] in the mantle, responsible for the regulation of various ion transportations for shell formation. Since Pf-CN A was mainly detected in lateral surface of the inner fold and basal periostracal groove,

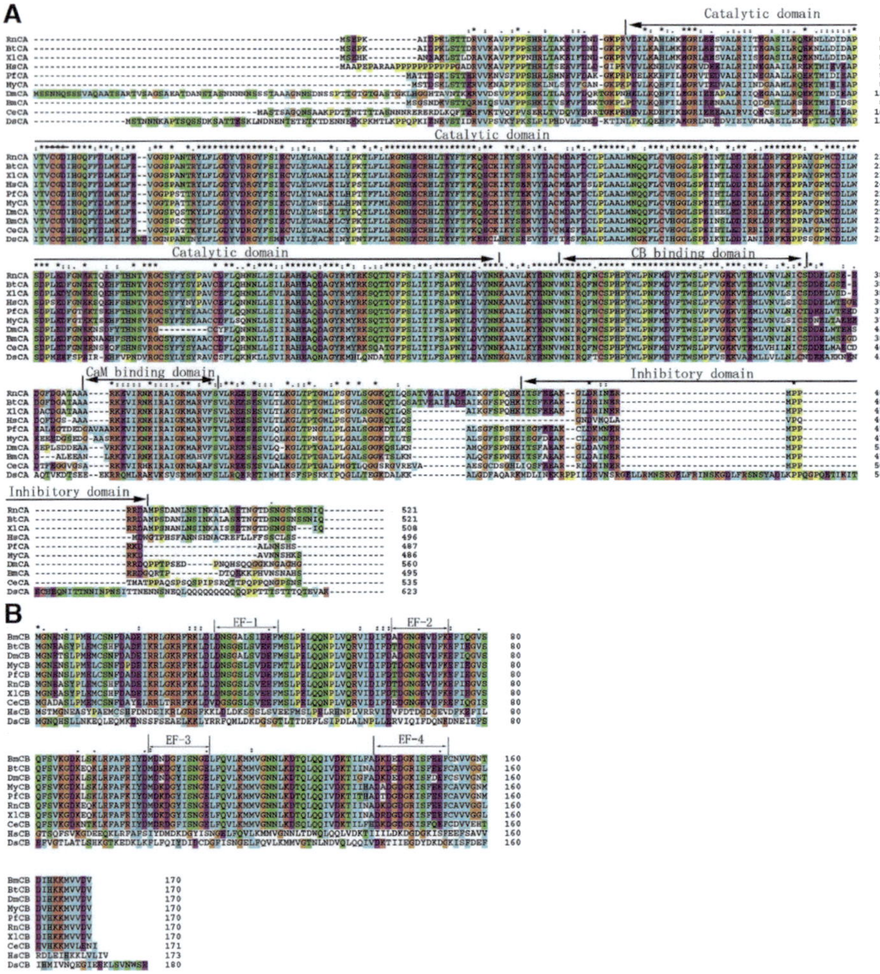

Fig. 2.79 Multiple sequence alignments of Pf-CN A (A) and Pf-CN B (B) with those in other species
For abbreviated protein names, see the legend of Fig. 2.75. □, identity; :, close similarity; ·, more distant similarity

there was strong possibility that it may be involved in shell formation. CN has been proved to be engaged in bone formation (another example for biomineralization) by regulating the expression of related genes in mammals. Thus, further investigation needs performing to demonstrate whether the CN in mantle from mollusk could also achieve the same physiological functions.

Strong positive immunostaining signals appeared in the epithelia of the spermatogenic duct in the gonad (Fig. 2.78b). CN Aγ, known as a testis-specific isoform, reached its highest expression level immediately after meiosis [269] in mammals, also detected in the nuclei of both elongated and round spermatids [270], as well as the postacrosomal region and the flagella of mammalian spermatozoa [271]. These results indicated that testis-specific calcineurin is associated with spermatogenesis. However, the particular functions of Pf-CN in gonad still remain further investigation.

As we could see from Fig. 2.78d, strong signals were observed in the base of gill filaments and epithelial cells of branchial filaments. We've already known that the gills in bivalve contained a series of demibranchs and each demibranch possessed two lamellae, which were formed by rows of ciliated filaments [272]. The gills of aquatic mollusks were important interface for the uptake of various ions from its surroundings [273]. After absorption, the dissolved ions would be quickly transported to other organs through blood or pericardial fluid [274]. Generally speaking, most of the Ca^{2+} absorbed from external environment [107, 108] were utilized and circulated with the help of gill during biomineralization. Then calcium should be stored and transported to certain biomineralization place for further shell formation. In gills of freshwater mussels (*Anodonta*, *Unio*, *Ligumia*), there were detected remarkably high levels of calcium concretions [275]. In this paper, the strong hybridization signals detected in gill inspired us to determine its crucial roles in uptake and accumulation process of calcium [268].

In summary, we isolated and identified two highly homologous cDNAs of CN from *P. fucata*, named as Pf-CN A and Pf-CN B, respectively. Expression analysis of these two genes indicated that their expression was sustained during the whole developmental stages in *P. fucata*. The immunolocalization and enzymatic activity assay indicated that CN was extensively distributed in all the tested tissues. Our findings would provide some novel thoughts on its unique functions during biomineralization as well as the comparisons between mollusk and other species, like mammalians.

*2.4.2 Calcineurin Plays an Important Role in the Shell Formation of Pearl Oyster (*Pinctada fucata*)*

Calcineurin (CN) is a multifunctional protein engaged in many important physiological processes in mammalians, but its function still remains largely unknown. In this paper, the changes of enzymatic activity of CN were determined during the shell regeneration in *Pinctada fucata* with the help of shell regeneration system. CN was activated immediately and continuously in this process. The microstructures of inner shell surface were closely observed by SEM after the inhibition of CN with intramuscular injection of the immunosuppressant inhibitor cyclosporine A (CsA), and the growth speed of shell regeneration was also measured. Results indicated that the regeneration speed was delayed at first and crystal (including both calcite and aragonite) morphologies also turned abnormal after CN's inhibition. Moreover, the expression of *P. fucata* BMP-2 in mantle exhibited the same decreasing trend after the injection of CsA. In vitro secretion level of proteoglycans (PGs) in primary mantle cells was also decreased when cells are exposed to CsA. All these results, for the first time, indicated that CN participated in shell formation mainly through regulating the expression of Pf-BMP-2, which controlled the secretion of PGs/GAGs in mantle.

2.4.2.1 The Quick Response of Enzymatic Activity of CN to the Shell Injury and to CsA

The quick enzymatic activity responses of CN to shell injury were detected (Fig. 2.80). At 12 h in incising group (Ic) or 6 h in incising with vehicle (used for dissolving the CsA) injection group (Ic + V) post-shell injury, the enzymatic activity of CN gained an increase and reached the maximum at 36 h (Ic and Ic + V), while this index in incising with CsA injection group (Ic + CsA) underwent a decrease at 6 h post-shell injury and declined to the minimum at 36 h. This indicated that CN could be immediately activated soon after shell injury, and this kind of mobilization could be significantly stopped ($p < 0.05$) by CsA. In the meantime, vehicle would not effect the enzymatic activity.

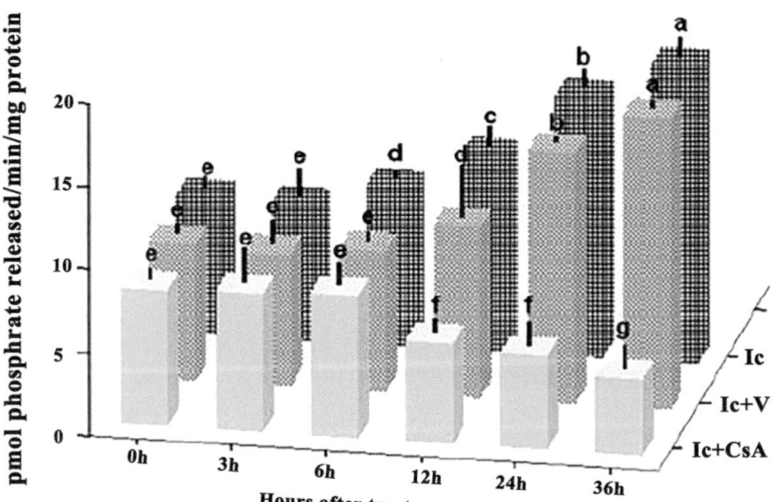

Fig. 2.80 The quick response of CN enzymatic activity to CsA and shell injury
Values (means ± SE, $n = 3$) from group Ic, Ic + CsA (20 μg/g of pearl oyster), and Ic + V were analyzed at the 5% level at particular time point. Different lowercase letters indicated significant difference ($p < 0.05$)

2.4.2.2 The Changes of CN Activity During the Shell Regeneration and Influence of CsA

The enzymatic activity of CN was measured after shell injury at particular time point to detect its responses in long-term shell regeneration process (Fig. 2.81). The results indicated that its enzymatic activity gradually increased and reached the top 6 days (Ic and Ic + V) after the removal of shell edge, then keeping a higher level during the 26-day regeneration process. However, the injection of CsA after the initiation of shell regeneration led to a decrease on the activity of CN, which was continuous within the whole detecting stage. This trend indicated that CN could be rapidly stimulated after shell injury and sustained a high expression level within the whole testing period, but this stimulation could be interrupted instantly by CsA administration. The vehicle used for dissolving CsA had no obvious effects on enzymatic activity.

2.4.2.3 The Effect of CsA on the Speed of Shell Regeneration

Fifteen individuals in every group were treated as only incising and incising only with one time injection of CsA to measure the normal speed and the effects of CsA on regeneration speed, respectively (Fig. 2.82). The length of the regenerated shell in two groups was measured at particular time point 6 days after injury, which was

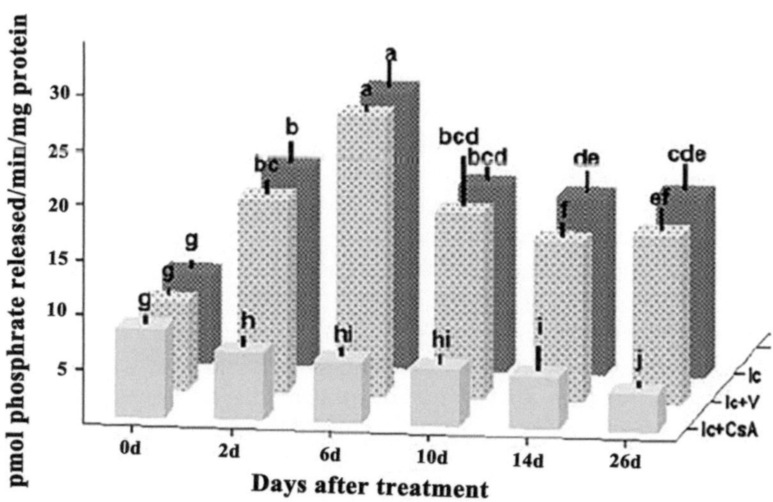

Fig. 2.81 Changes of enzymatic activity of CN during shell regeneration and under the influence of CsA

Values (means ± SE, $n = 3$) from group Ic, Ic + CsA (2.5 μg/g of pearl oyster), and Ic + V at each time point were measured at the 5% level. Different lowercase letters indicated significant difference ($p < 0.05$)

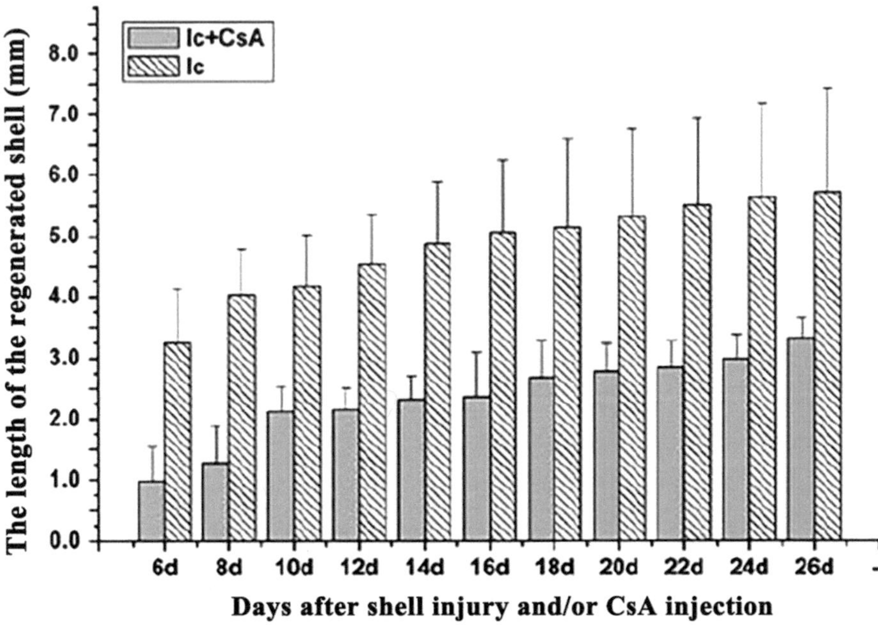

Fig. 2.82 The normal speed of shell regeneration and the effects of CsA on regeneration speed Two groups of animals were treated with and without CsA injection (20 μg/g of pearl oyster) after shell injury. Values indicate the means ± SE ($n = 10$)

increasing continuously over time during the whole detecting period (Fig. 2.82). On the contrary, the injection of CsA significantly delayed the regeneration speed ($p < 0.05$). The average increased length of the incising group was 0.22 mm/day, while the injection of CsA inhibited this speed to 0.13 mm/day. It was interesting to see the final increased length of the regenerated shell in two groups was equal. And the length difference between them was primarily because of the inhibition of CN caused by CsA injection, which was in consistent with the tendency of CN activity during 26 days after the initiation of shell regeneration. It seemed that CsA mainly function at the initial of shell regeneration.

Fourteen days after treatment, the lengths of regenerated shell in groups of incising with both two times and seven times CsA injection were shorter than those in other groups of incising with/without two times and seven times vehicles of injection ($p < 0.05$) in Fig. 2.83. And the lengths of five groups at 26 days after treatments exhibited almost the same trend as those at 14 days after treatments. The lesser amount of CsA injected, the faster the regenerated shell would grow.

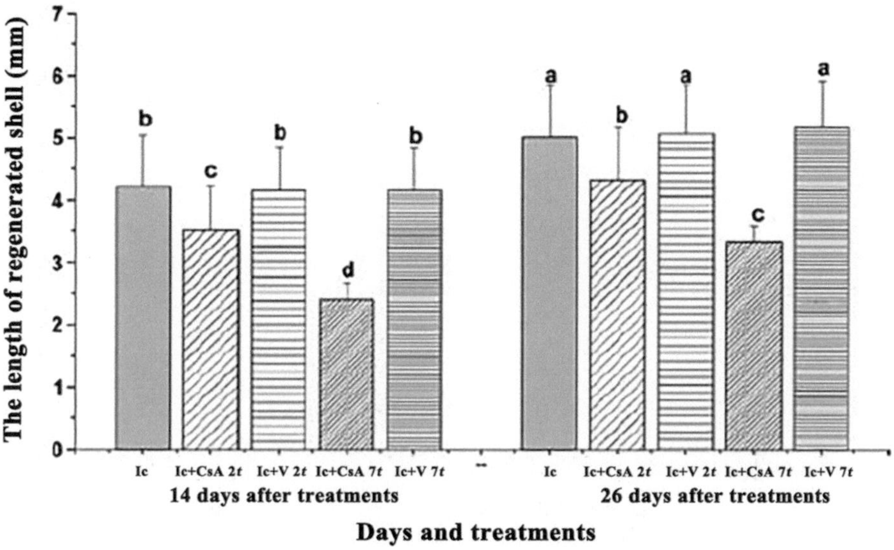

Fig. 2.83 The dose effect of CsA on the regeneration speed of shell
The quantity of CsA injected was 2.5 μg/g of pearl oyster, and the same volume of vehicle was injected as control. *Ic* incising group. *Ic + CsA 2t* incising with two times CsA injection every other day from day 0 to day 4. *Ic + V 2t* incising with twice vehicle injection every other day from day 0 to day 4. *Ic + CsA 7t* incising with seven times CsA injection every other day from day 0 to day 14. *Ic + V 7t* incising with seven times vehicle injection every other day from day 0 to day 14. Values (means ± SE, $n = 14$-16) from all groups at 14 and 26 days were analyzed at the 5% level. Different lowercase letters indicated significant difference ($p < 0.05$)

2.4.2.4 The Effect of CsA on the Ultrastructure of the Inner Shell Surface

Injury was made at the edge of shell as Fig. 2.84a1, a2 showed. Collected shells were prepared into several sections for SEM (Fig. 2.84a3, a4) at specific observation time point. The regenerated membranes and deposition of crystals were showed in Fig. 2.84b1, b2. Calcites were the first deposited crystalline on recently formed membranes, also the main component of the first major layer—prismatic layer (Fig. 2.84c1, c2). The newly formed calcites were round with smooth surface. Nacreous layer was overlaid as the shell thickened (Fig. 2.84d1). The aragonite layer in regenerated shell was also smooth and similar to normal phenotype (Fig. 2.84d2).

The comparison between the microstructures of inner surface in normal shell and the regenerated shell with CsA injection was observed to detect the effects of CsA. The images of group Ic + CsA 7t were displayed in Fig. 2.85. Since no significant changes occurred in Ic + CsA 2t group, the pictures of this group weren't shown.

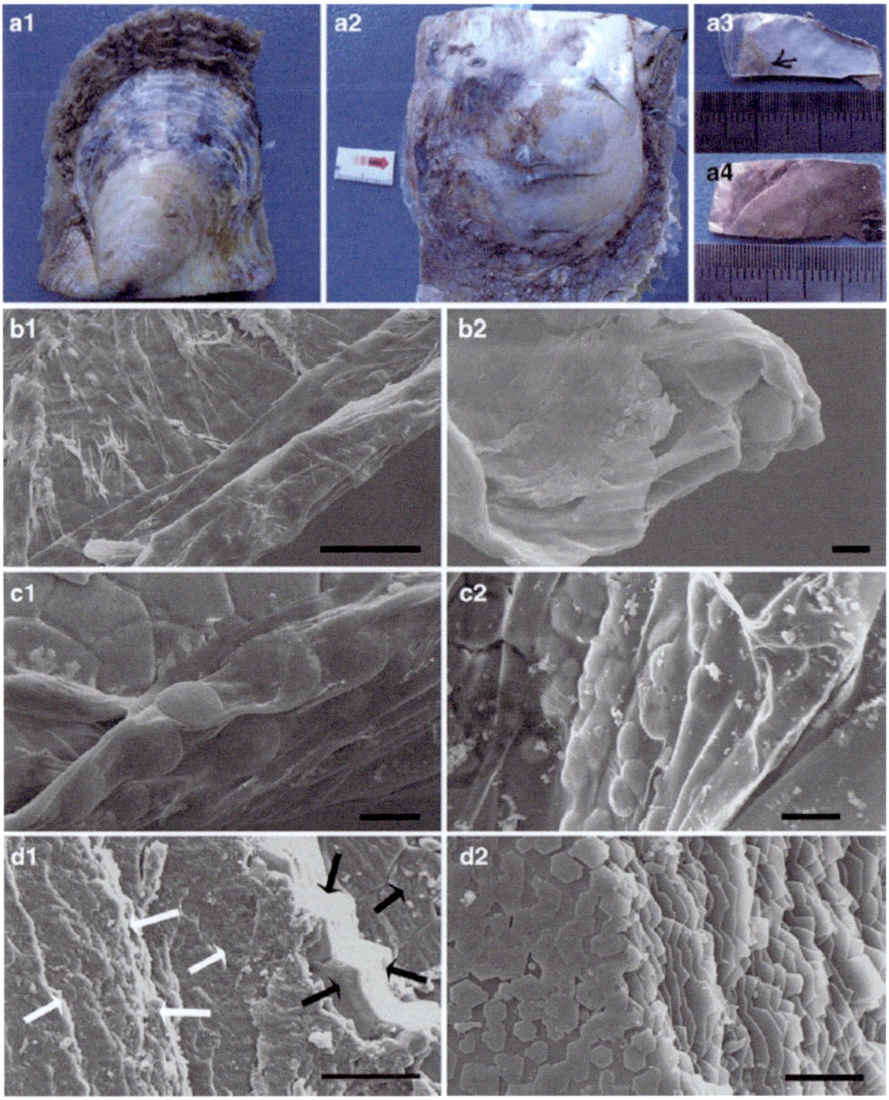

Fig. 2.84 The initiation of shell regeneration, the measurement of shell regeneration speed, and scanning electron micrographs of the inner shell surface [276]
(**a**1) The shell regeneration was initiated by the removal of a piece of shell on the edge opposite to the side with foot. (**a**2) A very thin, transparent organic membrane was observed through naked eyes. The length of growing membranes was measured through vernier calipers at the longest point to represent the speed of shell regeneration. (**a**3, **a**4) The collected shells were cut into sections for SEM observation. **b**1, **b**2 indicated the inner surface of organic membrane in regenerated shell near growing edge. 26 days. Bars = 50 μm in B1, 10 μm in **b**2. **c**1, **c**2 the prismatic layer of regenerated shell near growing edge and the morphology of newly deposited calcites. 26 days. Bars =10 μm in **c**1 and **c**2. **d**1, **d**2 the prismatic layer was indicated by black arrow and nacreous layer was indicated by white arrow in the inner shell surface from group Ic. 45 days. Bars =50 μm in **d**1, 10 μm in **d**2

Dishing in the central surface (Fig. 2.85 a1 and a2), instead of the normal, round, and plump calcites (Fig. 2.84 c1 and c2), appeared in the 7th CsA injection group. The abnormal crystals were mainly observed in aragonites, widely distributed in different areas of inner shell after the injection of CsA. Compared with the normal big and mature aragonites, newly formed crystals turned incomplete and hollow 26 days after CsA injection (Fig. 2.85 b1 and b2). What's more, a large number of melting and incomplete crystals appeared in calcite–aragonite crossed regions after CsA injection (Fig. 2.85 c1 and c2). All the surface of aragonites in these two regions was significantly different from normal ones with complete and smooth surface (Fig. 2.84d1 and d2).

2.4.2.5 *Pf-BMP-2* Gene of Mantle Tissue Response to the Depressed CN by CsA Injection

The relative expression level of Pf-BMP-2 was detected at 0, 3, 6, 12, 24, and 36 h after CsA and vehicle administration. As we could see from Fig. 2.86, the expression of Pf-BMP-2 underwent a decrease 12 hour after CsA injection and hit the bottom at day 36 ($p < 0.05$), but sustained stable only decreased a little in negative vehicle injection group (Fig. 2.86). Moreover, the relative expression values at

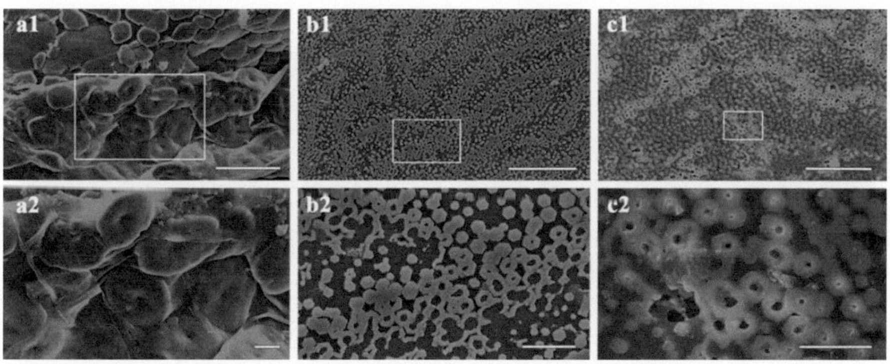

Fig. 2.85 Scanning electron micrographs of *P. fucata* shell [276]
All the SEM images were shown from group Ic + CsA 7t without the SEM images from group Ic + CsA 2t. a1: The prismatic layer of regenerated shell exhibited a large pitted surface. 14 days after shell injury and CsA injection. a2 was the enlargement of box in a1. Bars = 50 μm in a1, 10 μm in a2. b1, b2 Surface of nacreous layer in aragonite distributed regions exhibited hollow and incomplete crystals. 26 days after CsA injection. b2 was the enlargement of box in b1. Bars = 50 μm in b1, 10 μm in b2. c1, c2 Incomplete and central melting crystals were observed in the surface of nacreous layer in the aragonite–calcite crossed regions. 26 days after CsA injection. c2 was the enlargement of box in c1. Bars = 50 μm in c1, 10 μm in c2

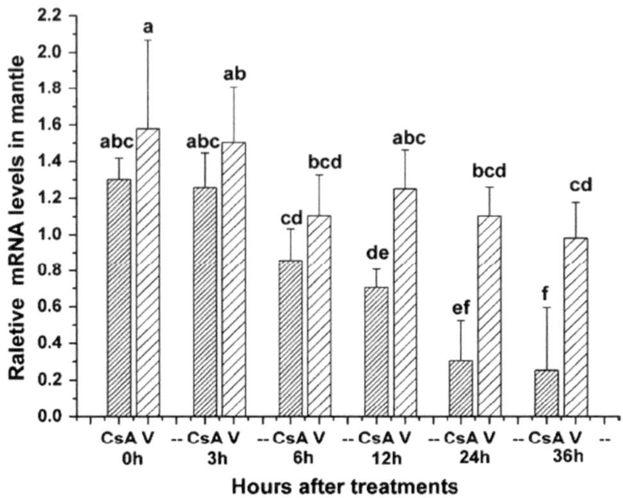

Fig. 2.86 mRNA relative expression level of Pf-BMP-2 in mantle tissue after CsA and vehicle administration
Gene expression analysis of Pf-BMP-2 at different detecting time points after CsA (20 μg/g of pearl oyster) and vehicle administration. The housekeeping gene GAPDH was applied as the control. The expression levels of Pf-BMP-2 were means of three assays ($n = 3$), which were calculated in accordance with that of the expression recorded for control (shown as 100%). Bars represented the means ± SD. Significance of the differences ($p < 0.05$) were shown as alphabetic letters

days 12, 24, and 36 in CsA injection group were smaller ($p < 0.05$). Apparently, the expression of Pf-BMP-2 could be significantly inhibited by the injection of CsA.

2.4.2.6 The Concentration of PGs/GAGs in Primary Cultures of the Mantle Cells

The concentration of PGs/GAGs in the medium of the cultured mantle cells incubated with CsA, vehicle, and MBSS was measured at 3, 6, 12, and 36 h, respectively. This value in three groups almost sustained stable at every detecting time point after treatments as Fig. 2.87 showed, whereas the values after incubation with CsA were significantly lower than those in other two groups ($p < 0.05$). There were no significant effects of vehicle on the secretion of PGs/GAGs. When CN was inhibited by CsA, the in vitro secretion of PGs/GAGs in mantle cells was reduced either. Trypan blue exclusion assay indicated that the number of mantle cells almost sustained stable within the whole experimental hours.

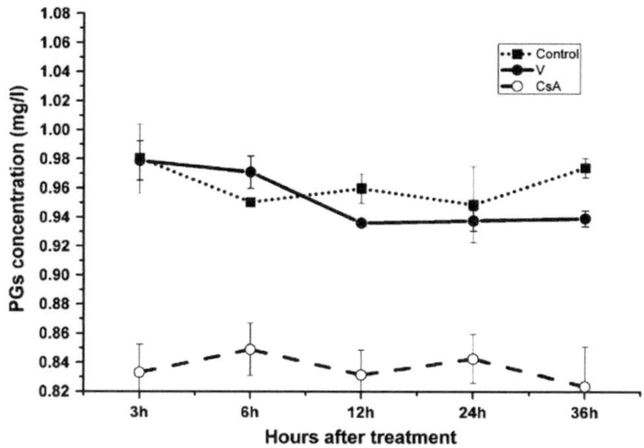

Fig. 2.87 The concentration of PGs/GAGs in primary mantle cells was detected at different time points after incubation with or without CsA and/or vehicle
Values (means $\pm$ SE, $n = 3$) from three groups at each time point were analyzed at the 5% level. The values at each time point after incubation with CsA was significantly lower than those in other two groups ($p < 0.05$)

CN is a multifunctional protein involved in the regulation of many physiological processes of various organisms from the lower animals to mammals [254]. CN has been proved to take part in chondrogenesis and bone metabolism via CN/NFAT signal pathway [277, 278] in mammals. Nevertheless, whether it participates in the regulation of shell biomineralization or not still needs more investigation. In this study, we found that shell injury could significantly induce the activation of CN, and this activating state maintained a relative high level during the shell repair process.

In summary, all the results have provided direct evidences to demonstrate that mollusk CN could regulate the formation of shell probably by homologous signaling pathway in mantle. To be more detailed, the possible cytoplasmic NFAT couldn't translocate to nuclei causing the decreased expression of Pf-BMP-2 after the injection of CsA. That was how the secretion ability of epithelial cells in mantle was inhibited. Therefore, those abnormal morphologies and phenotypes of crystals in shell surface and the slowly regeneration speed occurred. Moreover, it is also extremely important to elaborate how the decreased secretion of PGs induced the central vacuum of aragonites in inner shell surface. On the other hand, the abnormal morphologies of crystals may be not only relevant to the decrease expression of PGs/GAGs but also the wrong expression or secretion of some matrix proteins as well.

2.4.3 *Calcineurin Mediates the Immune Response of Hemocytes Through NF-κB Signaling Pathway in Pearl Oyster (*Pinctada fucata*)*

As a multifunctional protein, calcineurin (CN) mediates the immune response in mammals through diverse signaling pathways, while the function of CN in the immune response of molluskan hemocytes remains unknown. In the present study, we determined the distribution of CN in various tissues and the gene expression levels of *Pf-CNA* and *Pf-CNB* in hemocytes of the pearl oyster *Pinctada fucata*. After the preparation of hemocyte monolayers, we measured the degradation level of IκBα, the enzymatic activity of CN, the activity of iNOS, and the production of NO and IL-2 to reveal how hemocytes response to the challenges of lipopolysaccharide (LPS) and cyclosporin A (CsA). CN activity in hemocytes was quite sensitive to the stimulation of LPS and the inhibition of CsA. Most importantly, IκBα degradation in hemocytes was induced by LPS but weakened by CsA, thus elevating the activity of iNOS and increasing the production of NO. Additionally, we found that the synthesis of IL-2 was increased by LPS stimulation but was apparently attenuated by CsA. In vivo bacterial clearance experiments suggested that CsA significantly inhibited the ability of bacterial clearance in pearl oyster. For the first time, we showed that CN mediated the immune responses in molluskan hemocytes by activating NF-κB signaling pathway.

2.4.3.1 Distribution of CNA in Various Tissues and the Expression Levels of Pf-CNA and Pf-CNB Gene in Hemocytes of *P. fucata*

The rabbit polyclonal antibodies against CNA from *Pinctada fucata* were prepared, after cloning the two subunits of CN from mantle. To determine the distribution of CN in various tissues, the antibodies against Pf-CNA were employed for Western blot analysis. The result revealed that CNA was widely expressed in various tissues of pearl oyster (Fig. 2.88). In particular, the expression level of CNA in hemocytes was significantly ($p < 0.01$) higher than that in other tested tissues.

CN, also known as protein phosphatase 2B (PP2B), is the only serine-/threonine-specific protein phosphatase regulated by both CaM and calcium and acts as a heterodimeric enzyme composed of a CNA and a CNB [279–281]. Previous studies demonstrated that CN is widely distributed in various tissues and in many cell types of different organisms from lower species to higher species [282] and is a multifunctional protein participating in numerous important physiological processes. CN is widely expressed in mammal tissues, with the highest levels found in the brain [254]. The two CN subunits display in multiple isoforms [283], three isoforms (α, β, and γ) [280] in CNA and two isoforms (CNB1 and CNB2) in CNB. Among them, CNA α, β and CNB1 are relevant in T cells, while CNA γ and CNB2 are found almost exclusively in the testis [284]. Generally, the molar ratio of CNA/CNB is

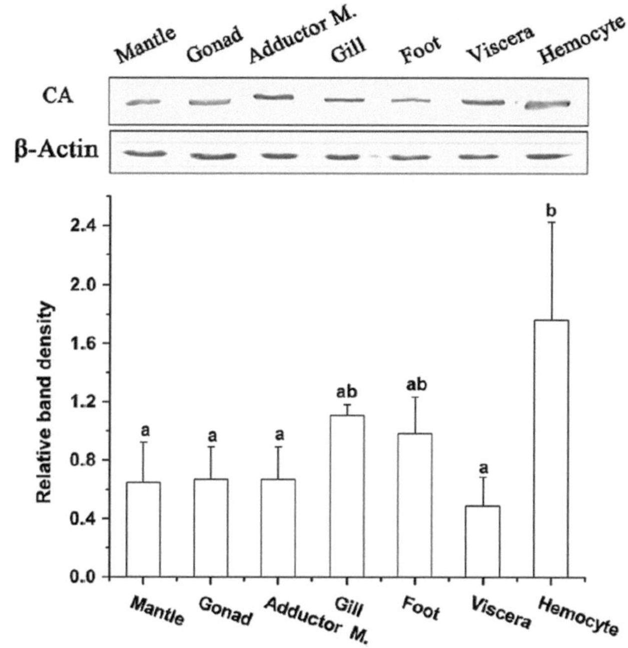

Fig. 2.88 Protein expression of Pf-CNA in the tissues of the mantle, adductor muscle, gonad, viscera, gill, hemocytes, and foot of *P. fucata*
Protein extract (15 mg) from tissues mentioned above was loaded. Protein extract corresponding to 3×10^6 cells was loaded in hemocyte lane. Western blot analyses were performed in triplicate. Values were expressed relative to control (β-actin) and were means ± SE of three analyses ($p < 0.01$)

equal to one, and the two units bind tightly in the cytoplasm. However, it was found that more CNB than CNA exists in tumor tissue and other tissues or even only CNB is present [285, 286]. So CNB seemed to play other important roles besides acting as a regulatory subunit of CNA [285]. Therefore, the isoforms have their particular tissue localization and function as crucial mediators in many pathological and physiological processes.

In mollusks, two subunits were identified from the gonad of scallop [256] and the mantle of pearl oyster [268]. Our current study showed that expression level of CN in hemocytes was the highest. According to RT-PCR analysis, the molar ratio of CNA/CNB in hemocytes was almost equal to one (Fig. 2.89). The highest expression level of CN in hemocytes indicated that CN plays important roles in many ways. In mollusks, the hemocyte participates in innate immunity and transportation of necessary metabolites for many physiological processes. Here, we explored the function of CN in mediating the immune response using both in vivo and in vitro experiments.

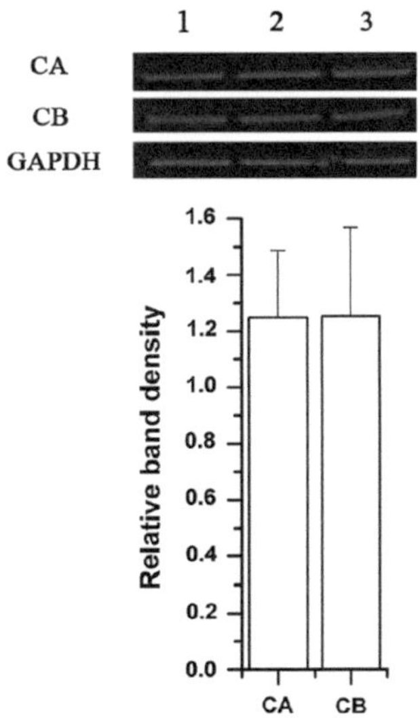

Fig. 2.89 Expression level of Pf-CNA and Pf-CNB gene in hemocytes from *P. fucata*
Relative band density analysis of RT-PCR ($n = 3$) showed no significant difference between Pf-CNA and Pf-CNB gene expression in hemocytes of *P. fucata* ($p > 0.05$). The pearl oyster housekeeping gene GAPDH was used as a positive control

2.4.3.2 Response of Hemocyte CA Activity and Protein Level to LPS and CsA In Vitro

Lipopolysaccharide (LPS) is a cell wall component of Gram-negative bacteria and regarded as a potent activator of innate immunity in mammals [287, 288] by activating Toll-like receptor-4 (TLR-4). LPS receptor recognizes pathogen-associated molecular patterns (PAMP) and then stimulates the downstream cascades through the cell signaling pathway. Two immunosuppressants, CsA and FK506 (tacrolimus), are commonly used in mammals for medical treatment, potently by inhibiting the enzyme activity of PP2B [281, 289]. The two drugs sterically hamper the access to the active site of CN by binding to their cellular receptors, cyclophilin and FKBP, respectively [290, 291], to form complexes which subsequently bind to the CNA/CNB composite surface [292, 293].

Figure 2.90 showed the comparison of CN enzyme activities of the lysate in cultured hemocytes with different treatments. Three groups of the prepared hemocytes monolayers correspond to preincubated with CsA for 1 h and then exposed to LPS (group CsA + LPS), only exposed to LPS (group LPS), and cultured without any treatment as control. The lysate of cultured hemocyte was collected for detection of PP2B enzyme activities at 0, 2, 5, 15, 30, and 90 min after the treatments. As a result, the activities of CN in cultured hemocytes immediately increased at 2 min ($p < 0.05$) and quickly recovered to basal level at 5 min. The activities of CN in the cultured hemocytes were significantly inhibited ($p < 0.05$), when the hemocytes were preincubated with CsA, even LPS was showed to fail to activate CN. The inhibition of CsA on CN activity was maintained throughout the tested period. Therefore, CN was very sensitive to both the stimulation of LPS and the inhibition of CsA.

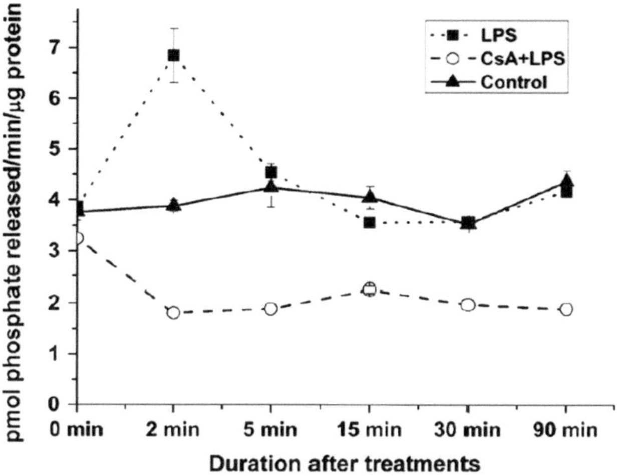

Fig. 2.90 CN enzymatic activity in the lysate of cultured hemocytes with different treatments At 0, 2, 5, 15, 30, and 90 min, the lysates of the prepared hemocyte monolayer from three groups (group CsA + LPS, 1 h after preincubation with CsA and then exposed to LPS; group LPS, only exposed to LPS, and control without any treatment) were collected and the enzymatic activity of CN was detected. Each assay was replicated three times based on sample preparations. Values were expressed as means ± SE were analyzed at the 5% level

FK506 and CsA have been proven to be valuable probes for studying signal transduction in T cells [294, 295]. Although the sequence information about cyclophilin of *P. fucata* was absent, our sequence analysis reveals that [268] Thr348 and Val351 in the CN B-binding domain of Pf-CNA are two putative residues that are critical for interaction with cyclophilin and FKRP. In addition, the result of the CN activity analysis was also showed to be sensitive to inhibition by CsA.

Meanwhile, the protein expression level of Pf-CNA was also determined by Western blot analysis. As shown in Fig. 2.91, the expression level of Pf-CNA in hemocytes kept steady after LPS treatment throughout the tested period ($p > 0.05$). Therefore, CN mediated the cellular physiological process mainly by regulating the CN activity rather than altering the expression level of CN.

2.4.3.3 IκBα Degradation in Hemocytes Was Induced by LPS and Attenuated by CsA

The degradation levels of IκBα were measured by Western blot with mouse monoclonal antibody against IκBα (human). As shown in Fig. 2.92, the amount of the 37 kDa IκBα significantly increased from 15 min to 90 min, after the hemocytes were exposed to LPS (5 μg/ml). When the hemocyte monolayer was preincubated with CsA (1 μg/ml) for 1 hour before challenged with LPS, the expression levels of

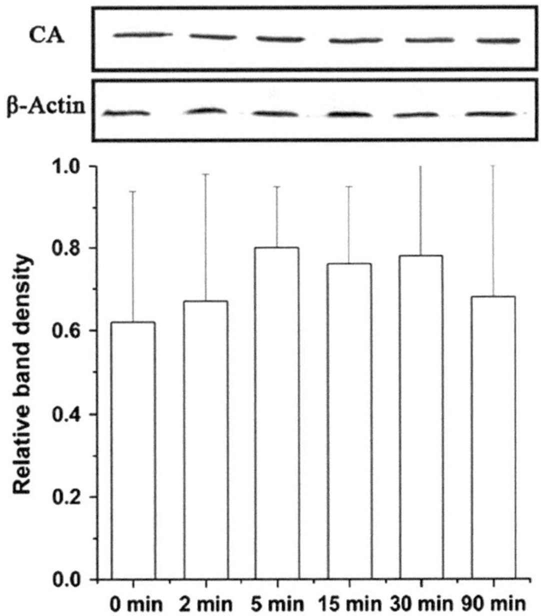

Fig. 2.91 Protein expression level of CNA in whole lysates of cultured hemocytes after exposure to LPS
The expression levels of CNA (n = 3, means ± SE) were calculated relative to that of the expression recorded for the control (shown as 100%). The comparison between relative band densities at each time point showed no significant difference ($p < 0.05$)

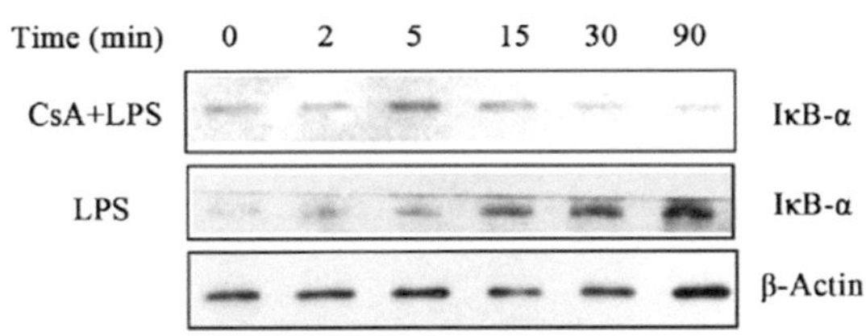

Fig. 2.92 IκBα degradation in hemocytes was induced by LPS and attenuated by CsA
The levels of IκBα degradation in the collected lysate of cultured hemocytes from group LPS and group CsA + LPS were detected by Western blot analysis with anti-IκBα antibody. Equal loading (10 μg protein per lane) of membranes was confirmed by re-probing with anti-β-actin antibody at each time point

IκBα changed little from 0 to 90 min. It shows that LPS can induce a significant upregulation of IκBα in hemocytes and such induction was inhibited when CN was blocked by CsA.

Besides mediating NFAT signaling pathway, CN is also considered responsible for the regulation of NF-κB signaling pathway, where CN cooperated with protein kinase C (PKC) to activate the transcription activity of NF-κB and IκB kinase (IKK) in T cells [296]. As for the remaining cells, NF-κB was retained in cytosol because it would become inactive after binding with other specific IκB proteins, IκBα. Once IκBα was activated by IKK, NF-κB was then released, activated, and translocated to the nucleus for further binding with κB promoter elements to initiate the expression of its target genes, such as iNOS and IL-2. This mechanism that also existed in non-T lymphocytes may be a common mechanism of NF-κB transcriptional activation (review in [297]). The degradation status of IκBα could be analyzed and evaluated by detecting the fluctuation of IκBα released to cytoplasm. And this process could be inferred that accordingly dissociated NF-κB translocated to the nucleus to reprogram the expression of its target genes.

PKC signaling pathway of molluskan hemocytes could regulate the production of NO according to previous research [298, 299]. A novel gene encoding an IκB member named Cg-IκB exhibited most known IκB protein characteristics [300]. We confirmed that the release of IκBα was stimulated by LBS, which could be inhibited by CsA. In LPS-treated cells, we didn't detect the transient decrease of IκBα from 0 to 5 min. But NF-κB signaling pathway was activated due to the continuous increase of IκBα from 15 to 90 min. However, we couldn't assure that the degradation of IκBα just occurred before the translocation of NF-κB to nucleus.

2.4.3.4 The Activity of iNOS and the Production of NO

NO is generated by the inducible isoform of nitric oxide synthase (iNOS), also characterized as a famous modulator in regulation of many important physiological processes. NO has been proved participating in the regulation of neurotransmission, vasodilatation and blood pressure, apoptosis, and nonspecific immune response.

As reviewed before [301], the expression of iNOS could be initiated by a lot of inducers through various signaling pathways, including cytokines and LPS, such as interleukin-1β (IL-1β), tumor necrosis factor-α (TNF-α), interleukin-6 (IL-6), and interferon-γ (IFN-γ). Other compounds could also attenuate or extend the influences of the expression of iNOS induced by LPS/cytokine. The conditions on the expression of iNOS and the effects of the stimuli showed are cell- and species-specific [301], which relies mainly on the protein expression level. An extensive amount of signaling pathways have been elaborated to activate and inhibit iNOS expression in different cells and organisms, but under any circumstances, the expression of iNOS is always regulated by the inhibition and activation of its transcription factor NF-κB [302]. Researchers have successfully established the linker between NFAT and iNOS in cardiomyocytes [303, 304]. CN was proved to regulate the expression of iNOS through IKK and NF-κB activity [305] in mouse peritoneal macrophages.

In this paper (Fig. 2.93), the activity of iNOS in the lysate of cultured hemocytes kept increasing, reaching the maximum at 12 h ($p < 0.05$) and returned to baseline 24 h later in cultured hemocytes treated with LPS but without the preincubation of CsA. The increasing activity of iNOS could be clearly weakened by LPS. If cells were preincubated with CsA and then challenged with LPS, the activity of iNOS would also increase and peaked at 12 h but the extent was much lower than that in only LPS-treated group.

The production of NO (μmol/L) in hemocyte culture medium at each detecting time point after treatments was shown in detail in Fig. 2.94. The concentration of NO in the culturing medium started to increase and reached the maximum 24 hours later

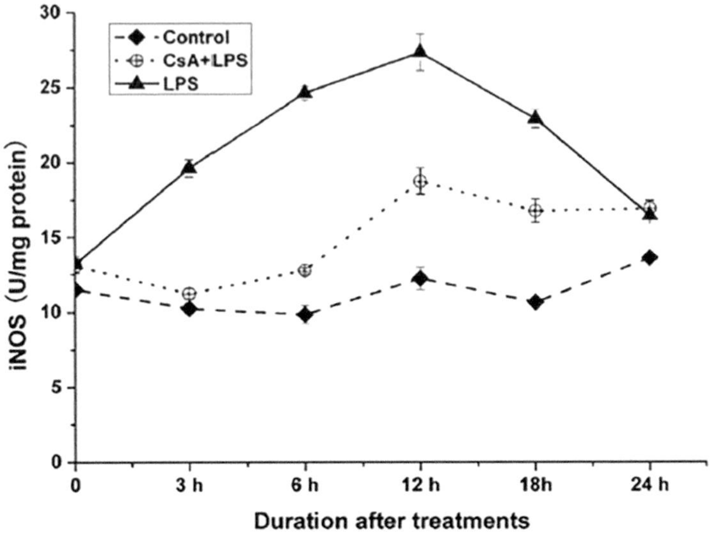

Fig. 2.93 Responses of the iNOS activity to LPS without or with CsA preincubation
The activity of iNOS of hemocyte lysates from group LPS, group CsA + LPS, and control was measured at each time point. Values ($n = 3$, means ± SE) in group LPS from 3 h to 18 h were significantly higher than those in the other two groups. Values in group CsA þ LPS from 12 h to 18 h were higher than those in control at same time points ($p < 0.05$)

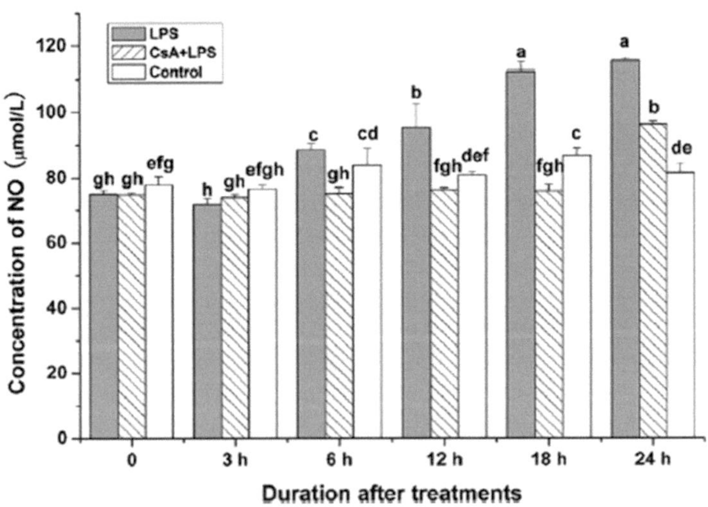

Fig. 2.94 The production of NO (μmol/L) in the culture media of hemocytes at each sampling time after treatments
The culture media were collected from three groups: group LPS, group CsA þ LPS, and control group at 0, 3, 6, 12, 18, and 24 h. Values ($n = 3$, means ± SE) with different small alphabetic letter differed significantly ($p < 0.05$)

($p < 0.05$) 6 hours after the LPS exposure. If the hemocytes were preincubated with CsA and then challenged with LPS, the concentration of NO kept a low status from 0 to 18 h, with a little increase than control 24 hours after treatment ($p < 0.05$), suggesting that the LPS-induced increased production of NO was suppressed when CN was inhibited by CsA.

NO was also confirmed to be an essential regulator in innate immune responses of mollusks. Two NOS isoforms binding anti-eNOS and anti-iNOS antibodies from vertebrates were detected in hemocytes of *Mytilus galloprovincialis*, and the NO production was also increased by IL-2 in winter [306], but the expression of iNOS exhibited no seasonal fluctuation. We've already known that PKC signaling pathway regulated the production of NO [298, 299]. Our results suggested that the degradation of IκBα was induced by the increase of LPS in molluskan hemocytes; then NO was synthesized during the initial expression of iNOS resulted from the translocating of the dissociated NF-κB into nucleus.

2.4.3.5 Response of Hemocyte IL-2 Level to LPS and CsA In Vitro

After the cultured hemocytes were challenged with LPS without CsA preincubation, Fig. 2.95 exhibited the increasing production level of IL-2 at a certain time, and this expression level was significantly decreased with the introduction of CsA. The effects of CsA and LPS on hemocytes started to appear 6 hours later, and the expression of IL-2 in LPS group started to increase and reach the peak at

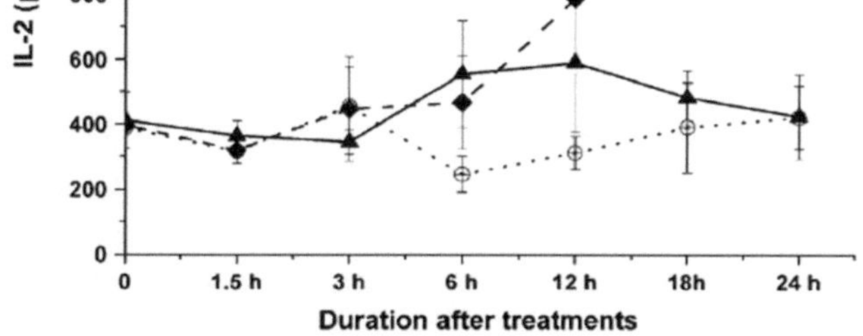

Fig. 2.95 Response of IL-2 in the culture media of hemocytes treated with LPS, with and without CsA preincubation
The expression level of IL-2 in the culture medium of hemocytes from three groups was detected by Human IL-2 ELISA Kit. Values ($n = 3$) were means ± SE ($p < 0.05$)

24 h. Moreover, the expression of IL-2 in CsA + LPS group was the least ($p < 0.05$), while no difference ($p < 0.05$) was detected in both control group and CsA + LPS group at 6 h and 12 h.

Transcription factors, such as NF-κB, AP-1, NFAT, and Oct-2, could be activated through their own signaling pathway and then bind with the enhancer and/or promoter to activate the expression of IL-2 in activated cells. IL2 in vertebrates could stimulate the proliferation and differentiation of B cells for increasing the cytolytic activities in NK cells, as well as the growth of T cells for cytokine secretion [307].

Despite that the function of IL-2 was not exactly known in mollusk hemocytes, the studies on antihuman IL-2 and other isoforms did provide some helpful hints. Among the cytokine-like molecules detected in the hemolymph of mollusks [308, 309], IL-2-like peptide found in the soluble fraction of hemolymph of the sea mussel *M. galloprovincialis* showed immune reactivity with antihuman IL-2. The utilization of antihuman IL-2R antibodies [310, 311] also helped us find three types of their receptor subunits IL-2Rα, IL-2Rβ, and IL-2Rγ. Furthermore, the recombinant human IL-2 could induce the expression of IL-2Rα in hemocytes [311], and LPS was used to induce the expression of IL-2Rα [310, 311]. Based on the results we got, it was indicated that after the rapid activation of CN induced by LPS, the IL-2 was secreted; CsA then suppressed the translocation of NF-κB from cytosol to nucleus through inhibiting the activity of CN; and finally, the initiation of IL-2 and other cytokines were hampered.

2.4.3.6 Effect of CsA on Bacterial Clearance in Hemolymph

Bacterial clearance experiments in vivo (Fig. 2.96) indicated that pearl oyster had strong abilities to clear exogenous *E. coli* from hemolymph circulation within 90 min. Negative blank control was set where hemolymph samples were excluded from any treatment with bacteria. The amount of bacteria still remaining in hemolymph circulation achieved the scale in blank control within 150 min. However, if CN was suppressed 6 hours after CsA injection, the bacterial clearance ability of oysters would decrease dramatically at 90 min ($p < 0.05$) and the number of hemolymph in CsA pre-treated oysters also reached that of control ($p > 0.05$). These results indicated that the longer after CsA injection, the lesser inhibition effects on clearance could be detected. After incubation for 30, 90, and 150 min, the percentage of bacteria in control group was 35%, 16%, and 4.3%, respectively, and 56%, 31%, 16% in CsA injection group, respectively. Another interesting thing was that the number of hemolymph in CsA preincubated oyster was much lower compared with the control group ($p > 0.05$). According to previous research, the suspension of bacteria 30 min after injection was applied as an equilibration period of hemolymph [28]; moreover, this time also indicated the baseline (the initial value) of the number of hemolymph. We speculated that the slower clearance of *E. coli* from hemolymph circulation at 30 min was partially due to the inhibition of CN.

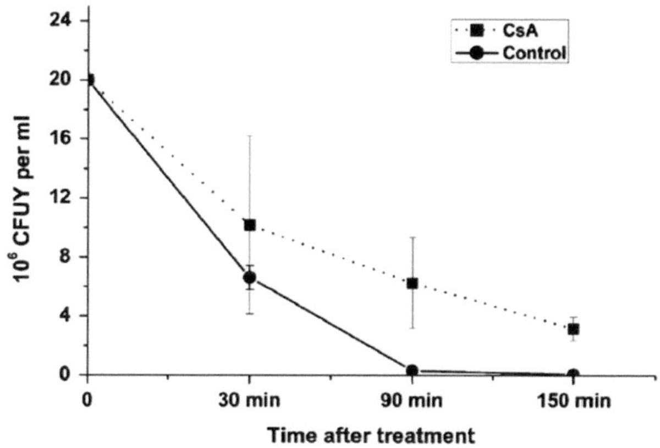

Fig. 2.96 Survival of fimbriated strain MG1655 of *E. coli* from pearl oyster hemolymph in vivo At 30, 90, and 150 min after the injection of bacterial suspension, the number of culturable bacteria per milliliter of hemolymph from pearl oysters, with or without pre-treatment with CsA, was compared. Values are means ± SE of five analyses ($p < 0.05$). At zero time, the value represents the number of *E. coli* in prepared bacterial suspension (2×10^8 CFU/ml)

Generally speaking, the decreased expression of cytokines, including IL-2, and the release of NO to serum resulted from the inhibition of CN followed by the reduced clearance of bacteria from hemolymph as Fig. 2.97 showed. In other words, initially, CN was immediately activated by LPS through enhancing its enzymic activity instead of its expression in cytosol. Then the activated CN, very likely synergized with PKC, stimulated the expression of IKK where IκBα was dissociated with NF-κB. After the translocation from cytosol to cell nucleus, NF-κB was programmed to bind with the promoter of iNOS and IL-2. As a result, both the synthesis of IL-2 and the production of NO made an increase. In this case, the immune response was initiated. Nevertheless, the stimulation of LPS was attenuated by CsA via the inhibition of CN, leading to the reduction of in vivo bacterial clearance in pearl oyster.

It has been widely known that CN could regulate a large amount of cellular responses through mediating the dephosphorylation of NFATc family members from yeast to mammalians. But it was suggested that the Ca^{2+}-/CN-dependent NFATc family, only found in genomes of vertebrates, arose following the recombination of an ancient precursor with a Rel domain in the initial of vertebrate life [313]. Actually, there is no genetic information about NFAT similar factors in mollusks up to now. In yeast, the function of Crz 1 p/Tcn 1 p is considerably similar to NFATc proteins in mammalians, which could directly translocate from cytoplasm to the nucleus with the activation of CN [314, 315]. Besides CN/NF-κB signaling pathway elaborated in this paper [312], it is still unclear whether CN-NFAT similar factors truly exist to regulate the innate immune responses of mollusks. Apparently, further investigation is quite necessary to demonstrate this mystery. In conclusion, our results demonstrated that CN could regulate immune responses of molluskan hemocytes by activating NF-κB signaling pathway.

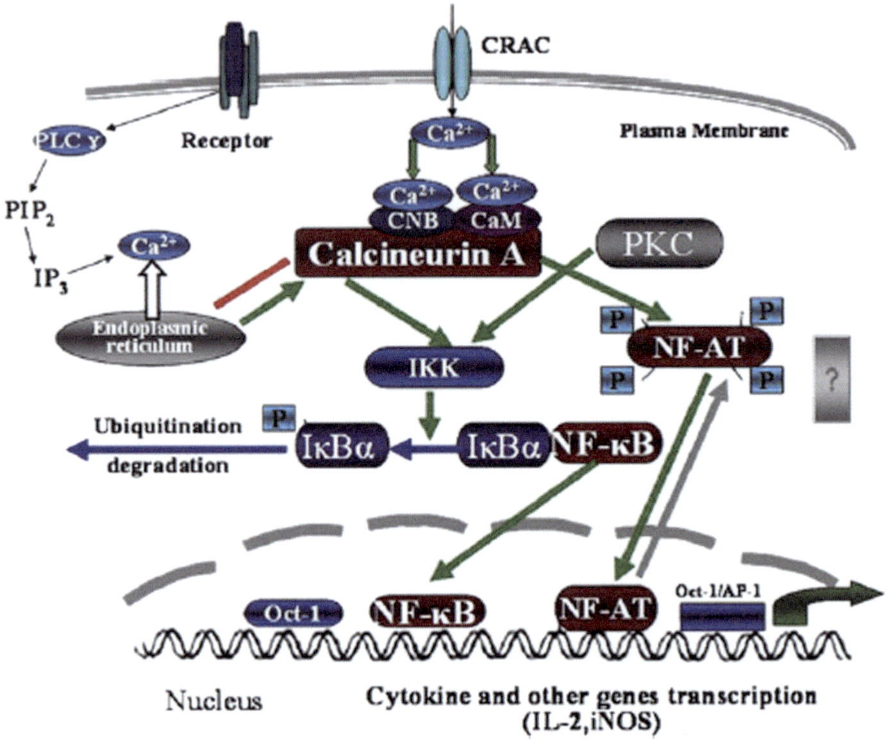

Fig. 2.97 CN regulated immune responses of molluskan hemocytes via activating NF-κB signaling pathway [312]
Originally, LPS immediately activated CN by elevating its enzymatic activity rather than its expression in cytosol. Then the activated CN, very likely synergized with PKC, stimulated the expression of IKK by which IκBα was phosphorylated. Later on IκBα underwent site-specific ubiquitination and degradation by proteosome complex. Finally, NF-κB was dissociated and translocated from cytosol into nucleus to bind with the promotor of iNOS and IL-2. Consequently, the synthesis of IL-2 and the production of NO underwent an increasment. However, CsA attenuated the stimulation of LPS via the inhibition of CN, which accordingly resulted in the decrease of *in vivo* bacterial clearance in pearl oyster. NFAT with a question mark meant that CN/NFAT pathway in molluskan hemocytes was not clearly demonstrated yet. (/): activation pathway

2.5 Preliminary Exploration of Calcium Channel and Calcium Transportation

*2.5.1 Cloning, Characterization, and Expression Analysis of Calcium Channel β-Subunit from Pearl Oyster (*Pinctada fucata*)*

The absorption, transportation, and localization of calcium underlie the foundation of biomineralization, and the entry of Ca^{2+} into epithelial cells is the principal step during shell formation. Nevertheless, the related mechanism of Ca^{2+} transportation is poorly demonstrated no matter at the gene or protein level. L-type voltage-dependent calcium channels might be engaged in calcium transportation for biomineralization in some marine invertebrates. In our research, the full-length cDNA of a voltage-dependent calcium channel β-subunit from *Pinctada fucata* (PCaβ) was cloned, and its amino acid sequence was deduced meantime. PCaβ shared 51%–67% apparently sequence identities with voltage-dependent calcium channel β subunits from other species. However, PCaβ was much shorter than other voltage-dependent calcium channel β subunits especially at the carboxyl terminus, suggesting that it is likely a truncated β-subunit isoform. Semiquantitative RT-PCR analysis revealed that PCaβ was expressed in all tissues we tested and it showed a relative higher expression level in hemolymph and gill than other tissues, indicating that L-type voltage-dependent calcium channels were responsible for Ca^{2+} entry into hemocytes and Ca^{2+} absorption in the gill. The relevant mRNA expression level of PCaβ was predominant in both inner and middle folds of mantle epithelium, indicating that L-type voltage-dependent calcium channels might participate in the absorption of Ca^{2+} from ambient medium in mantle. All these results implied that voltage-dependent calcium channels participate in the process of Ca^{2+} uptake and transportation during biomineralization.

2.5.1.1 The Cloning and Sequence Analysis of PCaβ

A 399 bp PCR product named Caβ1, which showed high similarity with voltage-dependent calcium channel β subunits from other species, was amplified from the mantle tissue of *P. fucata* using degenerate oligonucleotide primers derived from the conserved regions of calcium channel β-subunit nucleotide sequences. On the basis of this sequence, two specific gene primers (BGSP3 and BGSP4) were synthesized and used to amplify the 5′ and 3′ nucleotide sequences of PCaβ cDNA, respectively, by RACE. To confirm the sequence obtained by RACE, the two specific primers LCabQF1 and LCabQR1, respectively, corresponding to the 5′ UTR and 3′ UTR of PCaβ mRNA, were designed, and RT-PCR was performed. The PCR products were cloned and sequenced, the sequences of which clearly matched with the sequences expected from the results of 5′ RACE and 3′ RACE. As shown in Fig. 2.98, the complete PCaβ cDNA sequence including the poly (A) tail derived from the mRNA of pearl oyster was 1457 bp. It contained a 206 bp 5′ untranslated sequence, an open reading frame consisting of 1173 bp, a TGA stop code, a 75 bp 3′ untranslated sequence, and a poly(A) tail. A putative polyadenylation signal (AATAAA) was recognized at nucleotide 1418, which was 18 nucleotides upstream of the poly (A) tail. This cDNA sequence was submitted to GenBank under accession no. EF154452.

The deduced amino acid sequence of the oyster voltage-dependent calcium channel β comprised 391 amino acids with a calculated molecular mass of 43.1 kDa and an isoelectric point of 9.77. It contained several features common to all voltage-dependent calcium channel β subunits [316]: (i) a β-interaction

```
                    GAGAATCGCTGATACCTTGCTCCGAGGGGAAAATGAACGAAGCGAATGTTAACAGT   56
TTTCTAAAATATGAGGGCAGATCTGAAGAATTGGAGTAGATCAGAGACTTCAGAAACTGCGGGCAGTTTATCTGA  131
AAGCTCCTTATATTGGCAAACTCCAGGTGAGACGTTAGCTCTGTGAGGGTACATGTGGGTGACAAGCAGTACAGC  206
ATGGCTCTTGACGATAAACCAAGTAAACCAAAGAAATCTGCCTTCACCAAGACGGCAAAAACTCACATGATACAA  281
 M  A  L  D  D  K  P  S  K  P  K  K  S  A  F  T  K  T  A  K  T  H  M  I  Q
CAGCCAAGTGTGGAGTCACAAGATTCTACATACAGCCAACCTTCCTCCGATCTGTCAGTAGATGACGATAATCCA  356
 Q  P  S  V  E  S  Q  D  S  T  Y  S  Q  P  S  S  D  L  S  V  D  D  D  N  P
CTCCGTACAGAAACAAGTAGACAAGCTCAAGAACAACTGAACAAAGCTAGGACAAAACCTGTGGCGTTTGCTGTG  431
 L  R  T  E  T  S  R  Q  A  Q  E  Q  L  N  K  A  R  T  K  P  V  A  F  A  V
CGAACAAATGTGTCGTTTAATGGCGGCCCGGAAACAGATTGTCCTGTGCCGGGGCAGCAAGTTTCCTTTGAAATG  506
 R  T  N  V  S  F  N  G  G  P  E  T  D  C  P  V  P  G  Q  Q  V  S  F  E  M
AAAGACTTCCTGCATATAAAGGAGAAATTTAACAATGAATGGTGGATAGGACGACCAGTGAAGGAGGGGTGTGAT  581
 K  D  F  L  H  I  K  E  K  F  N  N  E  W  W  I  G  R  P  V  K  E  G  C  D
GTGGGATTTATACCTAGCCCTGCCAAACTAGGAAATATAAAGATCCAGGTCCCGAGTGGTTCAAAAGCAGAAGTA  656
 V  G  F  I  P  S  P  A  K  L  G  N  I  K  I  Q  V  P  S  G  S  K  A  E  V
CCAAACTCCGTGTCTGGTATAGAATCCATGTTACCAGGGAGTGGTACAGGAGTGGAAGCCGAGAATGGAGTGGAG  731
 P  N  S  V  S  G  I  E  S  M  L  P  G  S  G  T  G  V  E  A  E  N  G  V  E
AGTACCGTTGGAGAGGACAGTGACAGTCTGGGAAACCCCAAACCCAAAACCAGTATCAACCCCCCAACCAAAGAA  806
 S  T  V  G  E  D  S  D  S  L  G  N  P  K  P  K  T  S  I  N  P  P  T  K  E
AAGAAAAAGCCTTTCTTTAAAAAGAATGAGACAGTTCCACCTTATGAAGTCGTGCCTTCCATGCGGCCTGTAGTA  881
 K  K  K  P  F  F [K  K  N  E  T  V  P  P  Y  E  V  V  P  S  M  R  P  V  V]
TTGATAGGTCCCTCACTTAAGGGGTATGAAGTCACAGACATGATGCAGAAAGCTTTATTCGACTTCCTCAAACAT  956
[L  I  G  P  S  L  K  G  Y  E  V  T  D  M  M  Q  K  A  L  F  D  F] L  K  H
CGTTTTGAAGGAAGGATAATCATTACTCGTGTTACGGCAGACATTTCGTTAGCAAAACGTTCCTTACTCAATAAT 1031
 R  F  E  G  R  I  I  I  T  R  V  T  A  D  I  S  L  A  K  R  S  L  L  N  N
CCCAGTAAACGTGGATTGATTAACAGTCGAACAACTGGTATAGGTGAGGTACAGGTCGAAATAGAGAGGATATTT 1106
 P  S  K  R  G  L  I  N  S  R  T  T  G  I  G  E  V  Q  V  E  I  E  R  I  F
GACCTCGCTCGTTCCTTACAGTTGGTTGTTCTGGACTGTGACACAATCAACCACCCCTCCCAGTTAGCAAAGACT 1181
 D  L  A  R  S  L  Q  L  V  V  L  D  C  D  T  I  N  H  P  S  Q  L  A  K  T
TCCTTAGCTCCTATTATTGTATATGTCAAAATAGCATCACCAAAGGTATTACAGCGATTAATCAAATCTCGAGGA 1256
 S  L  A  P  I  I  V  Y  V  K  I  A  S  P  K  V  L  Q  R  L  I  K  S  R  G
AAATCCCAAAGTCGGAATATGAATGTACAGTTAGTGGCTGCTGATAAGCTCAATCAATGCAGTCAGCACCTACAA 1331
 K  S  Q  S  R  N  M  N  V  Q  L  V  A  A  D  K  L  N  Q  C  S  Q  H  L  Q
CGTCATTTCTTCATTCTATTACGGTCGATGGTAGCCATGGTATGTACATGACAGTATGCACCTTTTGTAAATTTG 1406
 R  H  F  F  I  L  L  R  S  M  V  A  M  V  C  T  *
CAATCGTTTATGAATAAATTGGTCTTCAGGTGACCGAAAAAAAAAAAAAAA                         1457
```

Fig. 2.98 Nucleotide and deduced amino acid sequence of *Pinctada fucata* voltage-dependent calcium channel β-subunit (PCaβ) cDNA

Numbers on the right indicate nucleotide positions. The stop codon is marked with an asterisk and the possible polyadenylation signal sequence in the 3′ untranslated region is underlined. BID (β-interaction domain) is boxed. This cDNA sequence has been submitted to GenBank (accession no. EF154452)

domain (BID) required for binding to the α1 subunit [317, 318], KKNETVPPYEVVPSMRPVVLIGPSLKGYEVTDMMQKALFDF (the underlined residues have been identified as particularly important for binding to α1 subunits, and the two serines in bold are potential protein kinase C phosphorylation sites) [319] and (ii) two highly conserved functional domains, SH3 domain (domain 94-131) in D2 and GK domain (domain 221-371) in D4 [316]. As shown in Fig. 2.99, similar to other voltage-dependent calcium channel β subunits, PCaβ contains five domains, of which D2 and D4 were conserved, whereas D1, D3, and D5 were variables. The presence of these features indicates that the isolated cDNA fragment encoded a voltage-dependent calcium channel β subunit. As shown in

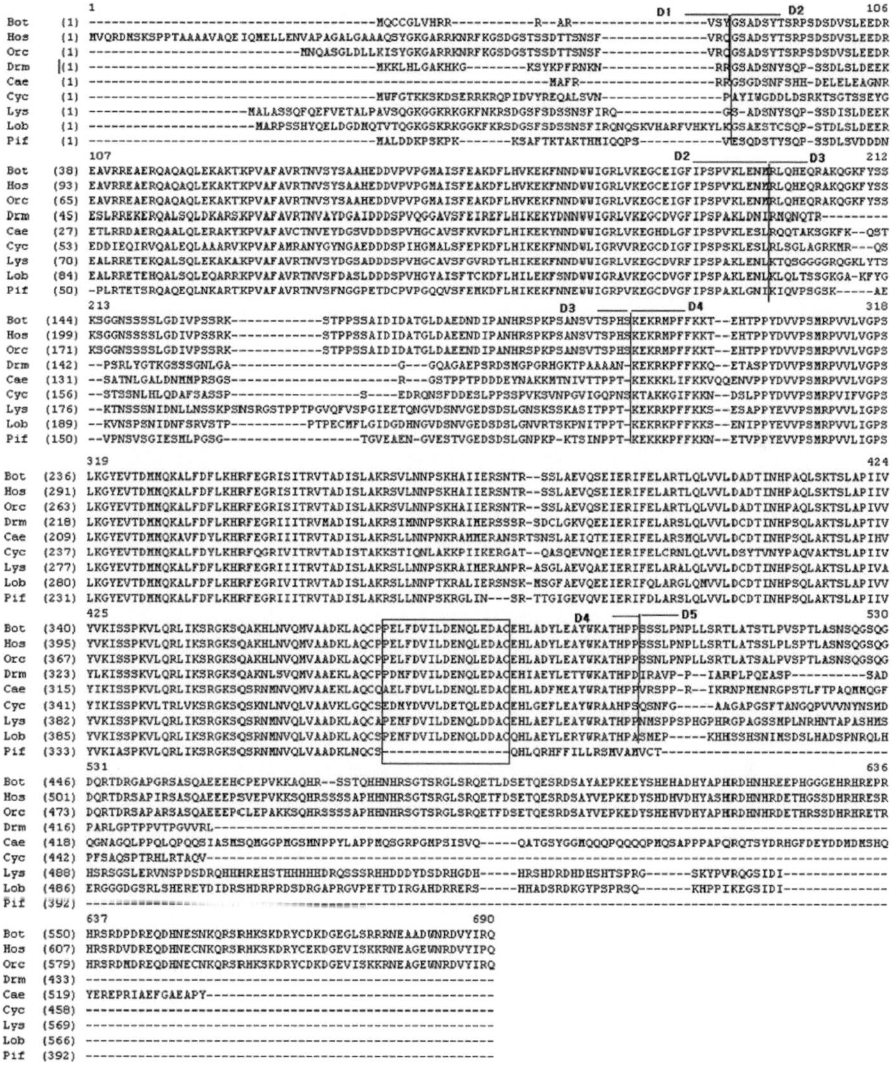

Fig. 2.99 Alignment of sequence of *Pinctada fucata* voltage-dependent calcium channel β-subunit (PCaβ) with other calcium channel β subunits [320]

Alignment of overall sequence of PCaβ with β subunits from principal invertebrates and vertebrates. D1–D5 indicate the five regions of β subunits, and their boundaries are delimitated by these vertical lines. The putative deleted amino acid sequences are boxed. Bot, *Bos taurus* (AAF26681); Hos, *Homo sapiens* (NP_963890); Orc, *Oryctolagus cuniculus* (P54288); Drm, *Drosophila melanogaster* (NP_523546); Cae, *Caenorhabditis elegans* (AAB53056); Cyc, *Cyanea capillata* (AAB87751); Lys, *Lymnaea stagnalis* (AAO83844); Lob, *Loligo bleekeri* (BAB88220); and Pif, *Pinctada fucata* (EF154452)

Fig. 2.100, the complete amino acid sequence alignment of PCaβ showed the highest identity with the mollusks *Loligo bleekeri* (67%) and *Lymnaea stagnalis* (67%) and the lowest identity with *Cyanea capillata* (54%) and *Schistosoma mansoni* (51%). The homology between PCaβ and mammalian voltage-dependent calcium channel β subunits was lower (60%) than those within mammalian β subunits (67%). If the GK domain was only taken into account, PCaβ showed 71%–89% identities with calcium channel β subunits from other species. If the SH3 domain was only taken into account, PCaβ showed 61%–88% identities with

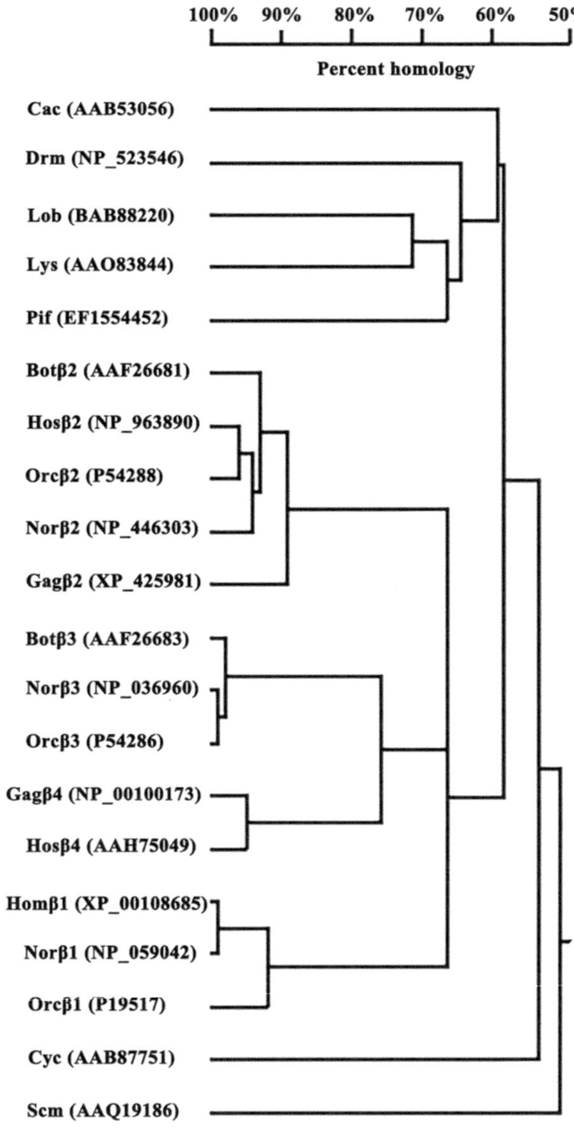

Fig. 2.100 Structural homology comparisons between voltage-dependent calcium channel β-subunit from *Pinctada fucata* (PCaβ) and other β subunits from principal invertebrates and vertebrates Percentages of sequence identity, as determined by the ClustalX program using a full amino acid sequence, are indicated. The accession numbers of these sequences found in the GenBank database (National Center for Biotechnology Information) are given in parentheses. Hos, *Homo sapiens*; Bot, *Bos taurus*; Hom, *Macaca mulatta*; Orc, *Oryctolagus cuniculus*; Nor, *Rattus norvegicus*; Gag, *Gallus gallus*; Scm, *Schistosoma mansoni*; Drm, *Drosophila melanogaster*; Lob, *Loligo bleekeri*; Lys, *Lymnaea stagnalis*; Pif, *Pinctada fucata*; Cae, *Caenorhabditis elegans*; and Cyc, *Cyanea capillata*

calcium channel β subunits from other species. These results suggest that PCaβ belong to voltage-dependent calcium channel β subunits. However, the full length of PCaβ was only 70%–80% that of other calcium channel β subunits. As shown in Fig. 2.99, compared with other calcium channel β subunits, PCaβ has a shorter carboxyl terminus. In D4, PCaβ appeared to have a 17 amino acid sequence skipping from the 371st amino acid onward; this sequence was conserved in other calcium channel β subunits. The sequence skipping formed an incomplete D4 domain and a truncated D5. Thus, PCaβ is likely to be a truncated voltage-dependent calcium channel β-subunit isoform.

2.5.1.2 Tissue-Specific Attribution of PCaβ

To investigate the function of voltage-dependent calcium channels in vivo, the tissue-specific expression of oyster voltage-dependent calcium channel β-subunit mRNA in different tissues was examined by semiquantitative RT-PCR analysis. RT-PCR was performed with RNA samples from the mantle, gill, digestive gland, and muscle tissues and hemolymph. A 1157 bp RT-PCR product was obtained with FBF4 and FBR4 and the total RNA of various tissues as template; however, the negative control exhibited no products (data not shown). The PCR products were then inserted into a pGEM-T Easy Vector and confirmed by sequence analyses. As shown in Fig. 2.101, PCaβ mRNA was expressed in all the tissues tested. The expression levels of PCaβ in the gill tissue and hemolymph were significantly higher than those in the gonad, mantle, and digestion gland tissues ($p < 0.05$).

The mantle of *P. fucata* is a key tissue responsible for oyster biomineralization, e.g., it participates actively in the secretion of calcium and other ions for mineral growth in shell formation [113]. To examine the precise expression site of PCaβ mRNA in the mantle tissue of *P. fucata*, in situ hybridization was performed on the mantle tissue. As shown in Fig. 2.102, strong hybridization signals were detected on both sides of the inner fold and on the outer side of the middle fold of the mantle. However, no hybridization signal was detected on either side of the outer fold of the mantle. Hybridization with the control sense probe yielded no hybridization signals (data not shown).

In this study, we cloned and characterized a voltage-dependent calcium channel β-subunit cDNA from *P. fucata* and analyzed its expression and distribution patterns to investigate possible roles of voltage-dependent calcium channels in mollusk shell formation. The amino acid sequence deduced from the cDNA sequence contains a putative BID, which interacts with the α1 subunit and forms a α1/β complex [321]. Although PCaβ has the key domains of the voltage-dependent calcium channel β subunit, it is generally 70–80% the length of voltage-dependent calcium channel β-subunits from other species in the NCBI database. For example, the voltage-dependent calcium channel β-subunit of *Loligo bleeker* has 514 amino acids (BAB88219), and that of *Lymnaea stagnalis* has 568 amino acids

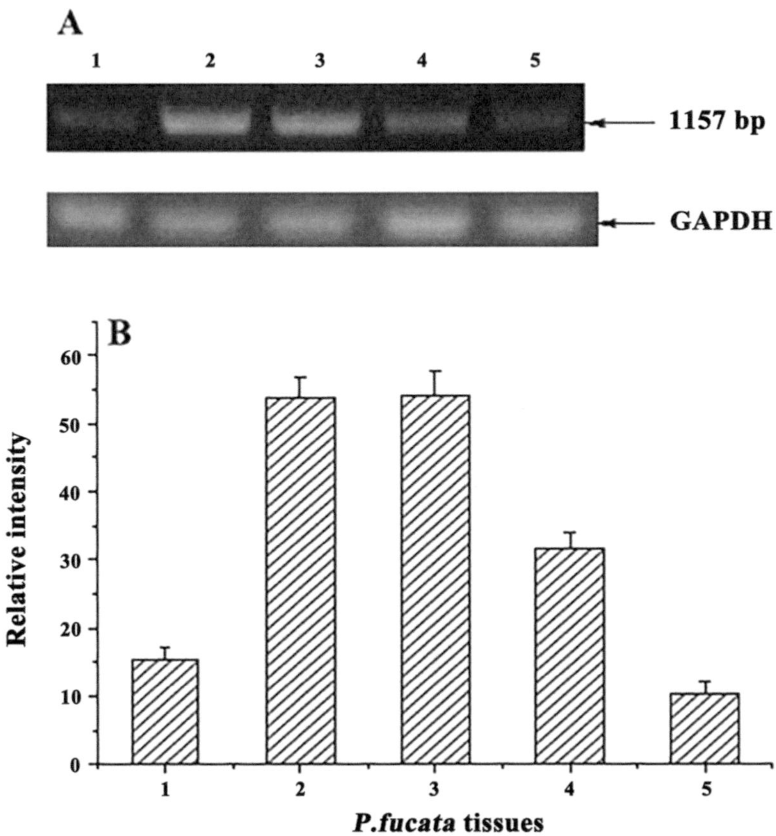

Fig. 2.101 Expression of voltage-dependent calcium channel β-subunit (PCaβ) mRNA in tissues of *Pinctada fucata*
(**a**) Semiquantitative RT-PCR was done with RNA samples from the adult tissues of *P. fucata*. The PCR products (1157 bp) were indicated by an arrow, and the GAPDH mRNA expression level was used as an internal control. (**b**) Densitometry analysis of PCaβ mRNA expression in tissues of *P. fucata*. The results were standardized using results of the internal GAPDH control. The results are expressed as mean ± SD of three independent experiments. Data were analyzed by one-way analysis of variance (ANOVA) using Tukey's test. Bars: *1*, gonad; *2*, gill; *3*, hemolymph; *4*, mantle; and *5*, digestion gland

(AAO83844). From the 371st amino acid onward, the carboxyl terminus of PCaβ showed low homology with those of other voltage-dependent calcium channel β subunits; however, this region was conserved among other voltage-dependent calcium channel β subunits. Recently, several truncated β-subunit isoforms have been identified. A truncated β3 subunit resulted from deletion of a 20 bp exon resulting in a reading frame shift and translation break [322]. A similar truncated β4 mRNA was identified in chicken, resulting from a 59 bp exon skipping [323]. A truncated β3 mRNA was previously reported in the mouse brain [324]. This results in a protein

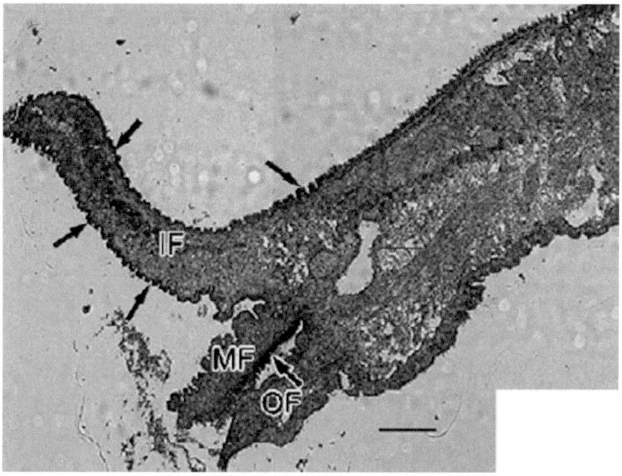

Fig. 2.102 In situ hybridization of *P. fucata* voltage-dependent calcium channel β-subunit (PCaβ) mRNA in mantle
Strong hybridization signals were detected in the inner and outer epithelial cells of the inner fold (arrows) and in the outer epithelial cells of the middle fold of the mantle (arrows). To view the distribution of hybridization signals in the whole tissue, two overlapping pictures of the same section were taken together. *OF* outer fold, *MF* middle fold, *IF* inner fold. Scale bars: 0.2 mm

lacking the β-interaction domain (BID) sequence. Although PCaβ has both BID and the GK domain, it is likely to have a truncated C-terminal and a premature stop codon. Consequently, PCaβ is likely to be a truncated voltage-dependent calcium channel β-subunit isoform.

The results of the semiquantitative RT-PCR analysis show that PCaβ mRNA is expressed in all the tissues tested; the wide tissue distribution of PCaβ suggests that PCaβ is involved in many subunit arrangements with the α1 subunit. The tissue distribution of PCaβ, to a certain extent, indicates the distribution pattern of L-type voltage-dependent calcium channels in *P. fucata* because the β-subunit is always connected to the α1 subunit [325]. Actually, voltage-dependent calcium channels are expressed in almost all mammalian tissues and organs.

Shell formation needs a large amount of calcium, which are absorbed from various body surfaces such as the mantle, gill, and byssus of mollusks and transported to the biomineralization site mainly through hemolymph [107, 326]. In marine and freshwater mollusks, the gill tissue has been shown to be important in Ca $^{2+}$ uptake [107, 108]. The semiquantitative RT-PCR analysis showed that PCaβ mRNA has a significantly higher expression level in the gill tissue than in the gonad, mantle, and digestion gland tissues, suggesting that L-type voltage-dependent calcium channels are proteins responsible for Ca^{2+} absorption from seawater.

Hemolymph is used as the transport vehicle, circulating high concentrations of calcium and carbonate to the biomineralization site [107]. However, to date, there are few reports on voltage-dependent calcium channels in invertebrate hemolymph.

In our study, the highest expression level of the voltage-dependent calcium channel β-subunit was detected in hemolymph, which suggests that L-type voltage-dependent calcium channels are involved in Ca^{2+} entry into the hemocytes. In addition, recent observations suggest that hemocytes contain amorphous calcium carbonate (ACC) and are directly involved in shell crystal formation in oyster [327]. Ca^{2+} and ACC transports via hemolymph both require Ca^{2+} entry into the hemocytes. Ca^{2+} and ACC are transported to the calcifying epithelium, namely, the mantle of *P. fucata* the key tissue responsible for the secretion of inorganic and organic matrixes to the extrapallial space in which $CaCO_3$ crystals are deposited. Calcium and bicarbonate from hemolymph are then brought into by the outer membrane epithelial cells of the mantle.

Compared with the PCaβ mRNA expression level in the gill tissue and hemocytes as determined by semiquantitative RT-PCR analysis, which in the mantle was low, we suspected that mantle may not be the most important tissue responsible for Ca^{2+} absorption from an external medium. There actually exist different mechanisms of Ca^{2+} transport from the mantle to the extrapallial space because the epithelial cells in different parts of the mantle vary anatomically, histochemically, and functionally. In bivalves, it has been proposed that the passive diffusion of calcium into the extrapallial cavity through transport channels formed between cells is performed by epithelial cells within the pallial line of the mantle [328], whereas the transcellular pathway (passive and diffusional) is performed by those outside the pallial line, in the marginal zone, to transport calcium through cells to the biomineralization site. Obviously, further study of the distribution of voltage-dependent calcium channels in different parts of the mantle is needed.

The mantle edge in bivalves usually consists of three folds: outer, middle, and inner folds. The outer epithelial cells of the outer fold of the mantle, facing the shell and extrapallial space, are not in direct contact with the ambient medium, but they participate in biomineralization [212]. In situ hybridization showed no visible hybridization signals in the outer epithelial cells of the outer fold, suggesting that voltage-dependent calcium channels are not involved in Ca^{2+} transport from the outer epithelial cells into the extrapallial space. The epithelial cells of the inner and middle folds are in direct contact with the ambient medium; strong signals suggest that voltage-dependent calcium channels mostly participate in Ca^{2+} absorption from seawater [320].

In conclusion, the cloned L-type voltage-dependent calcium channel β-subunit is likely a truncated β-subunit isoform. It is suggested that voltage-dependent calcium channels are responsible for Ca^{2+} absorption from seawater in the gill tissue and in the inner epithelial cells of the mantle but not for direct Ca^{2+} transport from the outer epithelial cells of the mantle into the extrapallial space. In addition, L-type voltage-dependent calcium channels play an important role in Ca^{2+} entry into hemolymph. The detailed mechanisms of L-type voltage-dependent calcium channel in Ca^{2+} absorption and transport need further study.

2.5.2 *Cloning, Characterization, and Expression Patterns of Three Sarco-/Endoplasmic Reticulum Ca^{2+}-ATPase Isoforms from Pearl Oyster (*Pinctada fucata*)*

A large quantity of calcium is badly needed for shell biomineralization. Sarco-/endoplasmic reticulum Ca^{2+}-ATPase (SERCA) is a widely known protein functioning on supporting the calcium homeostasis. What is its specific working mechanism during biomineralization in pearl oyster? Three SERCA isoforms, namely, PSERA, PSERB, and PSERC, were successively cloned from *Pinctada fucata*. The cDNAs of the three isoforms were isolated by RT-PCR and RACE. PSERA comprised 3568 bp encoding 1007 amino acids; PSERB was 3953 bp encoding 1024 amino acids, and PSERC consisted of 3450 bp encoding 1000 amino acids. The three isoforms exhibited high homology (65%–87%) with SERCAs from other species. In accordance with the results from other invertebrates, Southern blot analysis indicated that the three isoforms were derived from a single gene which was also associated with SERCA1, SERCA2, and SERCA3 of vertebrates. The splicing mechanism of the three isoforms was identical to these isoforms of vertebrate SERCA3. Semi-quantitative RT-PCR was then performed to investigate the distribution patterns of the three isoforms. The results revealed that PSERA was detected in foot and adductor muscle and was very likely to be a muscle-specific isoform, PSERB was extensively detected in all tested tissues and was a potential "housekeeping" SERCA isoform, and PSERC was expressed in all the other tissues except foot and adductor muscle, with its highest expression level in the gill and mantle, indicating that it was a non-muscle-specific isoform and might participate in calcium homeostasis during shell formation.

2.5.2.1 Isolation and Sequence Analysis of cDNA Sequence Encoding SERCA

We obtained a 1112 bp cDNA product F1 with two degenerate oligonucleotides (L1 and L2), the template of which was from total adductor muscle RNA. Two gene-specific primers were designed and synthesized based on the known fragment for further 5′- and 3′-RACE. The product of 5′-RACE was a fragment of 1510 bp, while the products of 3′-RACE comprised three fragments with different sizes: 1176 bp, 1298 bp, and 1680 bp. Then we designed and used another three different pairs of specific primers (L3 and L4, L3 and L5, L3 and L6) in accordance with the 5′ and 3′ untranslated sequences separately to confirm sequences of putative SERCA cDNA the sequence obtained from RACE. The PCR products were then sequenced to confirm whether it could match the expected sequence from 5′- and 3′-RACE.

The full-length cDNAs of the three isoforms, PSERA, PSERB, and PSERC, were 3568 bp, 3953 bp, and 3450 bp, respectively. PSERA had an open reading frame (ORF) of 3021 bp encoding 1007 amino acids with a calculated molecular mass of approximately 111.2 kDa and a 325 bp 3′ untranslated region; PSERB possessed an

ORF of 3072 bp encoding a 1024 amino acid protein with a calculated molecular mass of approximately 112.8 kDa and a 659 bp 3′ untranslated region; and PSERC had a 3000 bp ORF encoding a 1000 amino acid protein with a calculated molecular mass of approximately 110.3 kDa and a 228 bp 3′ untranslated region. The three isoforms share a common 219 bp at 5′ untranslated region. These cDNA sequences have been successfully submitted to GenBank with accession nos. EF488285, EF488286, and EF488287, respectively.

PSERA had a stop codon TAA and a poly (A) tail, but without poly (A) addition signal AATAAA; PSERB was similar to PSERA but had a poly (A) addition signal AATAAA ahead of the poly (A) tail; interestingly, both stop codon TAA and the poly (A) addition signal AATAAA were found in PSERC, but poly (A) tail was missing. The three isoforms were almost identical except their 3′ terminals where cDNA sequence just diverged at nucleotide 2980. What's more, it should be noted that there was another 156 bp common nucleotide fragment between PSERA and PSERC, locating downstream at the divergence site. The deduced amino acid sequences of PSERA, PSERB, and PSERC all shared common domains in other SERCAs from other species, including the fluorescein isothiocyanate site, phosphorylation region and the 5′-pfluorosulfonylbenzoyladenosine binding site, phospholamban-binding motif, tryptic sites, and thapsigargin sites (Fig. 2.103). Both PSERA and PSERC owned ten transmembrane domains like other known SERCAs in other species; however, PSERB just had a potential extra transmembrane domain (Fig. 2.104).

A homology search was performed with the BLASTP program against all protein sequences in GenBank. The deduced amino acid sequences of the three isoforms showed similar homologies to each other (97%–99%), and each of them was matched with a large amount of SERCAs from various species no matter in invertebrates or vertebrates. They exhibited the highest homology with SERCA in mollusk sea scallop (84%–87%) and the least homology with crayfish (75%–77%). Furthermore, they also exhibited 66%–76% identities with mammalian SERCA1, SERCA2, and SERCA3. The phylogenetic tree we built showed a separate branching between vertebrate SERCAs and invertebrate SERCAs. SERCA from *P. fucata* was much closer to invertebrate SERCAs than vertebrate SERCAs phylogenetically as we could see from Fig. 2.105.

2.5.2.2 Southern Blot Analysis of the SERCA Gene Copy Number

The similarity analysis of amino acid sequences among the three isoforms suggested that they may originate from a same gene. So we performed Southern blotting to confirm this hypothesis. The genomic DNA was extracted and digested by two different restriction endonuclease enzymes, XhoI and EcoRV, neither of which could cut the DNA sequence of the three isoforms. After digestion, they were then hybridized with a specifically designed cDNA probe from nucleotide 2530 to 2970, a highly conserved region among them. No significant band was detected in the negative control group (data not shown). These results showed one specific hybridization band in each DNA lane, indicating that these isoforms were truly encoded by one identical gene (Fig. 2.106).

2.5.2.3 Tissue Distribution Analysis of Isoforms by Semiquantitative RT-PCR

Semiquantitative RT-PCR was performed to determine the tissue expression of the three isoforms in *P. fucata*. RT-PCR was then carried out with RNA samples from the mantle, adductor muscle, digestive gland, gill, foot, and gonad. A common forward primer L7 and three reverse primers L8, L9, and L10 specifically designed for every isoform were used to amplify all the isoforms. As shown in Fig. 2.107, PSERA was detected in only the adductor muscle (higher) and foot; PSERB was detected in all the tested tissues and PSERC was found in all other tissues except the foot and adductor muscle. And the relative expression level of PSERB was similar in each tissue. PSERC exhibited higher expression level in gill other than in remaining tested tissues.

Our present studies described the identification, characterization, and tissue distribution of three different SERCA isoforms in pearl oyster *P. fucata*. The three isoforms were different from each other especially in their C-terminal amino acids, but all of them were encoded by one single gene. Although several SERCA genes have been successfully identified from the mollusk, such as *Yesso scallop* and sea scallop, SERCA was the first gene obtained from *P. fucata*.

It has been reported that SERCA pumps consist of three genes, SERCA1, SERCA2, and SERCA3, in mammals and birds [332]. SERCA1 produces two isoforms, SERCA1a and SERCA1b, by alternative splicing [330, 333, 334]; SERCA2 gives rise to SERCA2a, SERCA2b, and SERCA2c isoforms [335–337], while SERCA3 has six isoforms, including SERCA3a–f [338–340]. In addition, SERCAs was identified and characterized, and their isoforms have been proven encoded by one gene in some invertebrates, i.e., the crayfish heart muscle and axial muscle isoforms [341, 342]. *Artemia* 4.5 kb and 5.2 kb isoforms [343] also belong to this category. Our results confirmed the previous hypothesis that there was a unique ancestral SERCA gene that could produce the three genes in vertebrates and a single gene in invertebrates [343].

Artemia and crayfish SERCA may have a similar splicing mechanism to the vertebrate SERCA2 gene, where the donor splicing site of the penultimate exon can either be recognized and fused to the last exon, giving rise to the mRNA coding for the shorter protein, or remains unrecognized, in which case a poly (A) site is recognized before the last exon of the gene and the mRNA coding for the longer protein is originated [342]. *P. fucata* SERCA gives rise to three isoforms, and their splicing mechanism is likely to be similar to that of the vertebrate SERCA3 gene, where the penultimate exon can be alternatively excluded, partially included, or totally included, thus generating, respectively, three isoforms with different C-terminals [339]. Among the three isoforms of *P. fucata* SERCA, there is likely

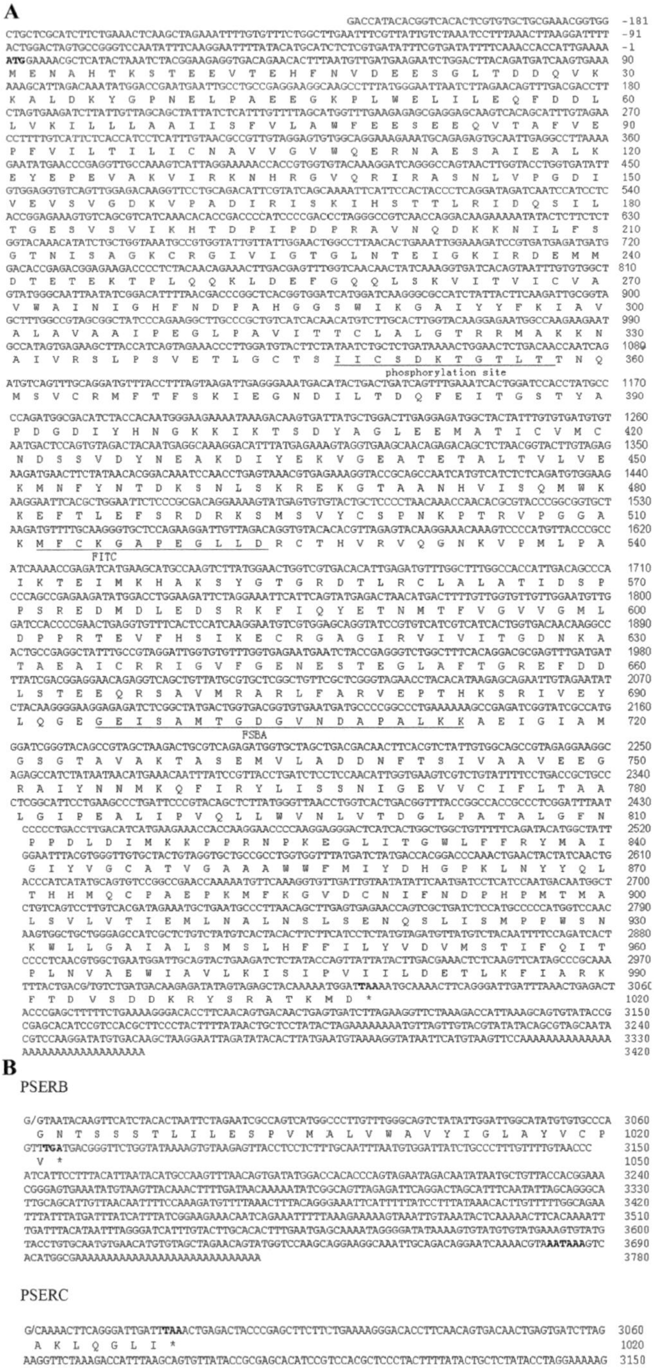

Fig. 2.103 Nucleotide and deduced amino acid sequences of three *Pinctada fucata* SERCA isoforms

Three isoforms have successfully been submitted to GenBank with accession nos. PSERA,

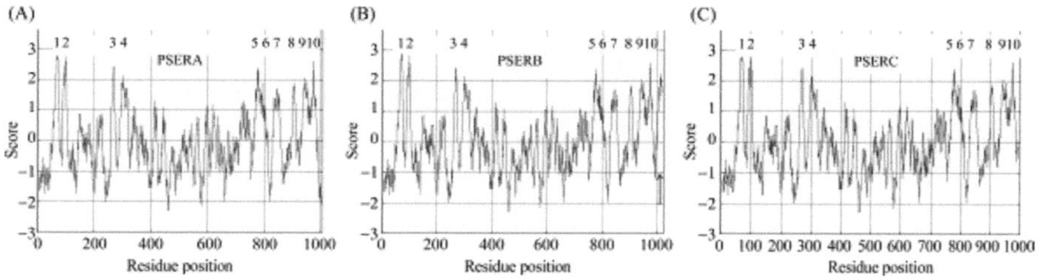

Fig. 2.104 Comparison of hydropathy plots of the SERCA isoform PSERA (**a**), PSERB (**b**), and PSERC (**c**) from *Pinctada fucata*
Hydrophobicity scores were determined by the method raised by Kyte and Doolittle [329] using a window of 13 residues (http://cn.expasy.org/tools/protscale.html). Putative transmembrane domains were numbered and marked on the basis of the model of Brandl et al. [330]. Arrows showed the extra potential transmembrane domains

to be an optional exon existing downstream of the conserved point of divergence. In PSERA, this exon is partially included; in PSERB this exon is totally included, while in PSERC this exon is excluded. Further research at the genomic level is necessary to disclose the exact splicing mechanism of the three isoforms.

However, little has been reported about the biological significance of the identification of multiple SERCA isoforms in invertebrate. Previous research didn't focus on function differences among different isoforms although their tissue distribution pattern was quite different [342, 344]. The divergent C-terminal of the three isoforms exhibited poor homology. The C-terminal extension attributed a lot to the prominent hydrophobic property of PSERB, and it was very likely that this extension could form the 11th transmembrane domain. Semiquantitative RT-PCR results indicated that PSERB showed a wide expression in various tissues; thus it was difficult to combine this extra domain with those particular tissues. On the contrary, PSERA and PSERC were highly expressed in certain particular tissues, respectively. The differential of 3′ terminal from SERCA might be engaged in the location of SERCA in different cell types or different sites of the same cell.

SERCA1 and SERCA3 exhibit a restricted expression pattern, while SERCA2 is extensively expressed in mammals. SERCA1a is the main isoform of fast-twitch skeletal muscles [345], while isoform SERCA1b is detected in neonatal muscles

Fig. 2.103 (continued) EF488285; PSERB, EF48828; and PSERC, EF488287. Nucleotides and amino acids were numbered to the right of the sequence. (A) Common sequences of the three isoforms. The phosphorylation site, the fluorescein isothiocyanate site (FITC), and the 5-′-pfluorosulfonylbenzoyladenosine binding site (FSBA) were underlined and labeled. Two typical tryptic sites R188 and R505 existing in sarco-/endoplasmic reticulum Ca^{2+}-ATPase (SERCA)1 and SERCA2 were marked in bold and underlined. Typical phospholamban-binding motif of SERCA2 was underlined italicized. Putative thapsigargin sites were indicated in bold. (B) Differential 3′ terminals of the three isoforms. Slash in the sequence at nucleotide 2980 was the site where the three isoforms diverged from each other. Stop codon was showed by asterisks. Common sequences of PSERA and PSERC were indicated in bold and italics

(A)

	CafATP1	CafATP2	HosATP2	MumATP2	GagATP2	GagATP3	HosATP3	MumATP3	MiyATP	PlmATP	PSERA	PSERB	PSERC	PosATP	PrcaATP	PrchATP	GagATP1	HosATP1	OrcATP1
CafATP1	100	85	81	81	81	76	72	73	75	75	75	74	76	74	74	72	90	96	96
CafATP2		100	94	94	90	76	74	74	74	74	74	73	75	73	73	71	84	84	84
HosATP2			100	99	94	76	74	74	70	69	71	71	71	69	69	69	80	80	80
MumATP2				100	94	76	73	74	70	70	71	71	71	69	69	69	80	80	80
GagATP2					100	76	74	75	70	70	72	72	72	69	70	70	81	81	80
GagATP3						100	78	78	68	68	68	68	68	67	67	68	76	76	75
HosATP3							100	93	65	65	66	66	66	66	66	67	73	73	72
MumATP3								100	66	66	66	66	66	66	66	66	73	73	72
MiyATP									100	96	85	84	86	76	76	75	75	74	74
PlmATP										100	86	84	87	77	77	75	74	74	74
PSERA											100	97	99	77	76	75	75	75	75
PSERB												100	97	75	75	75	74	73	73
PSERC													100	77	77	75	76	75	75
PosATP														100	89	87	73	73	74
PrcaATP															100	97	73	74	74
PrchATP																100	72	72	72
GagATP1																	100	91	90
HosATP1																		100	97
OrcATP1																			100

(B)

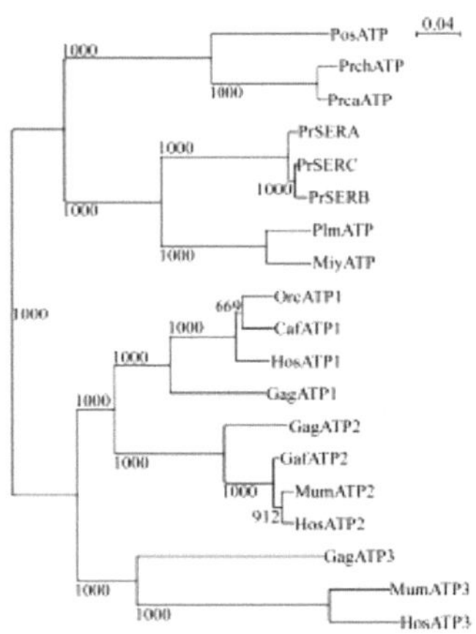

Fig. 2.105 Homology comparison and phylogenetic analysis of the three SERCA isoforms from *Pinctada fucata* with SERCAs from other species
(**a**) Matrix indicated the percentage of identities of the aligned SERCAs determined by ClustalX program based on the full amino acid sequence. (**b**) Phylogenetic tree showed the evolutionary relationship between SERCAs and other different species according to (**a**). Values at branch points were bootstrap percentages with 1000 replications. Accession numbers of these sequences could be found in GenBank database were the following: HosATP1, *Homo sapiens* SERCA1 (O14983); HosATP2, *Homo sapiens* SERCA2 (NP_733765); HosATP3, *Homo sapiens* SERCA3 (Q93084); OrcATP1, *Oryctolagus cuniculus* SERCA1 (P04191); MumATP2, *Mus musculus* SERCA2 (O55143); MumATP3, *Mus musculus* SERCA3 (Q64518); MiyATP, *Mizuhopecten yessoensis* SERCA (BAA37143); PrcaATP, *Procambarus clarkii* axial muscle SERCA (AAB82291); PrchATP, *Procambarus clarkii* heart muscle SERCA (AAB82290); PlmATP, *Placopecten magellanicus* SERCA (AAC63909); GagATP3, *Gallus gallus* SERCA3 (NP_990222); GagATP2, *Gallus gallus* SERCA2 (B40812); GagATP1, *Gallus gallus* SERCA1 (NP_990850); CafATP2, *Canis familiaris* SERCA2 (NP_001003214); CafATP1, *Canis familiaris* SERCA1 (XP_849777); PosATP, *Porcellio scaber* SERCA (AAN77377), PSERA (EF488285), PSERB (EF488286), and PSERC (EF488287)

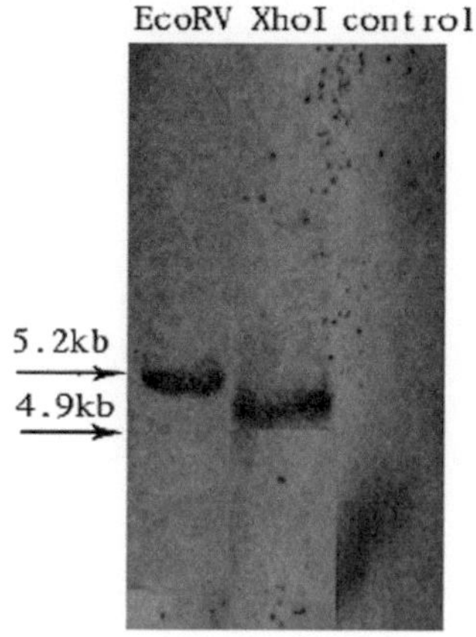

Fig. 2.106 Southern blot analysis of the copy number of sarco-/endoplasmic reticulum Ca^{2+}-ATPase gene in *Pinctada fucata* [331]
Genomic DNA (20 μg) was digested with XhoI or EcoRV, respectively, then separated on a 0.7 % agarose gel, and after transferred to a new nitrocellulose membrane; then it was hybridized with a specifically designed probe from nucleotide 2530 to 2970 of the isoform PSERA

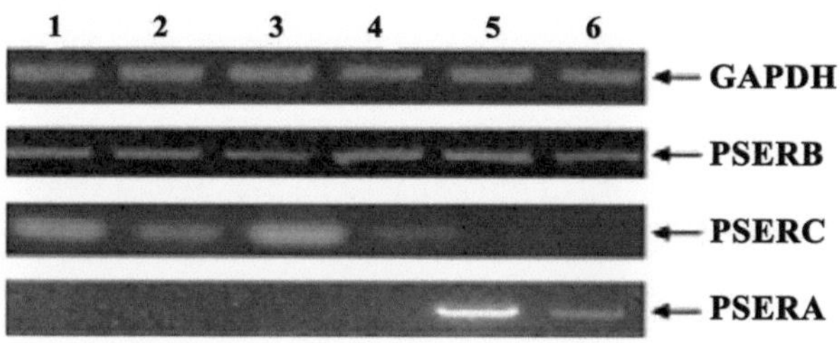

Fig. 2.107 Tissue distribution and mRNA expression analysis of the three sarco-/endoplasmic reticulum Ca^{2+}-ATPase isoforms in different tissues from *Pinctada fucata*
Semiquantitative RT-PCR was carried out with RNA samples from the adult tissues in *Pinctada fucata*. PCR products of the PSERA (346 bp), PSERB (734 bp), and PSERC (271 bp) were indicated by the arrows, and the expression of GAPDH (252 bp) was used as an internal control. *1*, mantle; *2*, gonad; *3*, gill; *4*, digestive gland; *5*, adductor muscle; *6*, foot

[330, 337]. SERCA2a mainly expresses in slow-twitch cardiac, skeletal, and smooth muscles [335, 346, 347]; SERCA2b is widely expressed and considered as the housekeeping calcium pump [348, 349]. SERCA3 is present in platelets and other blood-forming cell lines as well as many endothelial and epithelial cells [350, 351]. Furthermore, the isoforms of SERCA3 are rarely alone and always co-expressed with SERCA2 [350]. In this paper, PSERB is widely distributed in various tissues and functions as a potential "housekeeping gene"; its distribution pattern is similar to that of SERCA2b according to previous research. PSERA is detected in foot and adductor muscle, which also exhibits a relative higher expression level in the adductor muscle. PSERC is detected in other tissues except the foot or adductor muscle, and the highest expression in the gill and mantle, suggesting us it might be a non-muscle-specific isoform. The adductor muscle in *P. fucata* controls the opening and closing of the shell, and the foot regulates shell movement. In other words, muscle cells are indispensable for both foot and adductor muscle. Therefore,

PSERA is very likely to be a muscle-specific isoform, while SERCA mainly functions on helping relax contracted muscles [352–354]. The high expression levels of PSERA in adductor muscle suggested that the function of PSERA mainly focused on the muscle relaxation of *P. fucata*. A lot of studies have revealed that mantle is the key tissue controlling calcium secretion to the mineralization site after the uptake of calcium from seawater [354]. Gill is another pivotal tissue for calcium uptake from seawater [355]. They both participate in the active calcium metabolism, and calcium homeostasis is particularly vital to them. High expression of PSERC in the gill and mantle suggests that it may play essential roles in calcium homeostasis during biomineralization. Apparently, it is very essential to perform further investigations on the concrete functions of PSERC in calcium homeostasis of *P. fucata*, with which the mechanism of calcium uptake, deposition, and transportation may be clearly understood. Further work associated with the specific location of PSERC at the mantle and gill and in vitro expression of the three isoforms based on their cDNA sequences remains further conducted [331].

All in all, the three identified SERCA isoforms PSERA, PSERB, and PSERC from the pearl oyster *P. fucata* exhibited high homology with SERCAs from other species. The splicing mechanism of these three isoforms in *P. fucata* is similar to SERCA3 in mammals; the three isoforms exhibit different expression patterns in various tissues. Isoform PSERC probably plays a key role in calcium homeostasis during oyster biomineralization.

2.5.3 *Cloning, Characterization, and Expression Analysis of Calreticulin from Pearl Oyster* **Pinctada fucata**

Calreticulin is a unique calcium-binding protein with multiple functions mostly located in the sarcoplasmic/endoplasmic reticulum. A large amount of calcium is absorbed from the medium and transported to mineralization sites during biomineralization in pearl oyster. This paper describes the cloning of the full-length cDNA of calreticulin from *Pinctada fucata*, namely, PCRT. PCRT encodes a deduced 414 amino acid protein, which includes a predicted 17 amino acid signal peptide and an endoplasmic reticulum retrieval sequence HDEL. The protein shows 63%–76% sequence identity and shares some common characteristics with calreticulins from other species. Semiquantitative RT-PCR indicates that PCRT is ubiquitously expressed in all tissues tested with the highest expression in the hemolymph and the mantle. In situ hybridization analysis of PCRT in the mantle showed strong signals in the inner fold, the inner side of middle fold, and the inner side of outer fold of the mantle epithelium. All these results suggest PCRT might be involved in Ca^{2+} transport and storage during oyster biomineralization.

2.5.3.1 Cloning of PCRT

A 533 bp PCR product named CRT1, showing high similarity with calreticulin from other species, was amplified from the mantle of *Pinctada fucata* using degenerate oligonucleotide primers derived from the conserved regions of the calreticulin nucleotide sequences. Two specific gene primers (GCRTF1 and GCRTR1) were synthesized based on this sequence and used to amplify the 3′- and 5′-nucleotide sequences of the PCRT cDNA by RACE reactions. To confirm the sequence obtained by RACE, two specific primers (CRTQF1 and CRTQR1) were designed corresponding to the 5′ UTR and 3′ UTR of PCRT mRNA. The PCR products were cloned and sequenced and found to match well the sequence expected from the results of the 5′- and 3′-RACE.

As shown in Fig. 2.108, the complete PCRT cDNA sequence including the poly (A) tail derived from the mRNA of pearl oyster was 1935 bp. The sequence

```
   AAGCAGTAGTATCAACGCAGAGTCATCTGGGCTCAGAAGTCAACAATCATACAGCGGAGGGAGCAACGTTTACAGCGACCAGCAAGA 87
ATGAAAGCTGTTTTTCTGCTTGTGTTAGTCGGAGTTTTATACTCCGACGCTACTATTCATTTCGAAGAAAAGTTTGATAGTGATGGTTGG 177
[M  K  A  V  F  L  L  V  L  V  G  V  L  Y  S  D  A] T  I  H  F  E  E  K  F  D  S  D  G  W  30
AAAGACAAATGGGTAGATTCCACAAATAAAGGAGCCGAGCAAGGAAAGTTTGAATGGACAGCTGGAAAGTTTTATGGGGATGCTGATAAG 267
 K  D  K  W  V  D  S  T  N  K  G  A  E  Q  G  K  F  E  W  T  A  G  K  F  Y  G  D  A  D  K  60
GACAAAGGAATTCAGACTTCCCAAGATGCAAAGTTCTATGGAATTTCAGCAAAACTTGATAAACCCTTCAGCAATGAGGGAAAGAAGCTA 357
 D  K  G  I  Q  T  S  Q  D  A  K  F  Y  G  I  S  A  K  L  D  K  P  F  S  N  E  G  K  K  L  90
GTTGTCCAATTTACTGTTAAGCACGAGCAGAACATTGACTGTGGTGGAGGGTATTTGAAGGTATTCTCCAGCGATTTGGACCAGGCAGGC 447
 V  V  Q  F  T  V  K  H  E  Q  N  I  D  C  G  G  G  Y  L  K  V  F  S  S  D  L  D  Q  A  G  120
ATGCATGGAGACTCCCCTTATCTTATCATGTTTGGGCCTGATATTTGTGGACCAGGGACAAAGAAAGTCCATGTCATCTTCAACTACAAA 537
 M  H  G  D  S  P  Y  L  I  M  F  G  P  D  I  C  G  P  G  T  K  K  V  H  V  I  F  N  Y  K  150
GGAAAGAATTTACTTGTCAAGAAAGAAATCCGATGCAAGGATGATGTATTTACCCATTTGTACACTTTGATTGTAAATTCTGACAACACG 627
 G  K  N  L  L  V  K  K  E  I  R  C  K  D  D  V  F  T  H  L  Y  T  L  I  V  N  S  D  N  T  180
TACGAAGTCAGGATTGATAACGAGAAGGCTGAGAGTGGAAATTTAGAGGAAGACTGGGACTTTTTACCACCCAAGAAAATTAAGGATCCT 717
 Y  E  V  R  I  D  N  E  K  A  E  S  G  N  L  E  E  D  W  D  F  L  P| P  K  K  I  K  D  P  210
                                                                    N          A1
GAGGCTAAGAAGCCTGAAGATTGGGATGACCGTGAGAAAATTGATGATCCTGATGACACAAAACCCGAGGACTGGGACAAGCCAGAACAC 807
 E  A  K  K  P  E  D  W  D  D  R  E  K  I  D  D  P  D  D  T  K  P  E  D  W  D  K  P  E  H  240
           A1                              A2                                      A3
ATCCCTGACCCAGATGCCAAGAAACCCGACGACTGGGATGATGAAATGGACGGTGAATGGGAACCCCCACAGATTGATAACCCCGAGTAC 897
 I  P  D  P  D  A  K  K  P  D  D  W  D  D  E  M  D  G  E  W  E  P  P  Q  I  D  N  P  E  Y  270
         A3                                                   B1
AAGGGAGAATGGAAACCAAAACAGATTGATAATCCAGCATACAAAGGAAAATGGATTCATCCAGAAATAGATAACCCTGAATATGTTGCT 987
 K  G  E  W  K  P  K  Q  I  D  N  P  A  Y  K  G  K  W  I  H  P  E  I  D  N  P  E  Y  V| A  300
                  B2                                         B3                   P    C
GATGATAAATTATATAAATATGACGATATTGGTGCCATAGGATTTGATTTATGGCAGGTAAAATCAGGAACAATTTTTGATAACATTATA 1077
 D  D  K  L  Y  K  Y  D  D  I  G  A  I  G  F  D  L  W  Q  V  K  S  G  T  I  F  D  N  I  I  330
ATAACAGATGATATAAAAGAAGCAGAGGAAGATGCCAAATTATGGAGTGCGACCAAAGATGCTGAAAAGAAAATGAAAGATCAACAGGAT 1167
 I  T  D  D  I  K  E  A  E  E  D  A  K  L  W  S  A  T  K  D  A  E  K  K  M  K  D  Q  Q  D  360
GAGGAAGAAAGGAAACAGAGAGAAGAGGAAGAAAAGAAGAGGAAAGAAGAAGAGAAGGATAGTAAAAAAGATGACGATGATGATGATGAC 1257
 E  E  E  R  K  Q  R  E  E  E  E  K  K  R  K  E  E  E  K  D  S  K  K  D  D  D  D  D  D  D  390
GATGCTGATGATGATGAAGAAAAGGATGACGACAAGAAAGAGGATAGCGATGATCATGACCATGATGAATTATAAGAATAGTTTATAACT 1347
 D  A  D  D  D  E  E  K  D  D  D  K  K  E  D  S  D  D  H  D [H  D  E  L]  *                 420
AGGACTATTATCTAATATTATTTATGTTCCATTTCTTTTTTTACCTTAATAATAATGGGGGAAATTATTGTAGTAGAAAGACTTTTAAAT 1437
GGAATTTAACGTAAATTTGCATAAGTGGATTAAACTTGAACCAAAAATAACTACGCAGCTATTAAAGTGTCAGATCCGAAATAGTTGACA 1527
CTCTTTTCTATTTTGTCATTTTGTTTTCCCATTTTCAAACATGCCTCATTTGTGACTTGTTGATTATTCTCATTTTATTGTGTGAATACA 1617
TGTTAGTGCATGTGTATGCATATGATTTACATGTAACAGGTGAGTGGAGTTGTATGTGTACAAGAAGATGAATGAATGACTCGAGTGAAT 1707
GAGTTGGGATGTGAATACTGCCTTAAACCAGAGATCTTCATAATTTTTGTGATGAGTTTTATGCATGACATTTTGTAATTTTTTTTTTGA 1797
GGAATCAAACTTTGTTACTATTGTTATATTATCGAGAAAAGCATAAAATTGTCAAGAGCTCTGGTTTATATATAAGTGTTAAGCTAGAAT 1887
AATTTTCAGGAATAAACACTTGATTTTTAATATAAAAAAAAAAAAAAA                                           1977
```

Fig. 2.108 Nucleotide and deduced amino acid sequence of the *Pinctada fucata* calreticulin cDNA (PCRT)

Numbers on the right indicate nucleotide or amino acid positions. The stop codon is marked with an asterisk, and the possible polyadenylation signal sequence in the 3′ untranslated region is underlined. The ER retrieval signal and the putative signal peptide are boxed. The two calreticulin family signature sequences are in italics. The two triplicate repeats are underlined and marked with A1–A3 and B1–B3 contained an 87 bp 5′ untranslated sequence, an open reading frame comprising 1242 bp, a TAA stop codon, a 603 bp 3′ untranslated sequence, and a poly (A) tail. A putative polyadenylation signal (AATAAA) was recognized at the nucleotide position 1897. This sequence has been submitted to GenBank with the accession number EF5513344.

2.5.3.2 Sequence Analysis of PCRT

The deduced oyster calreticulin includes 414 amino acids with a calculated molecular mass of 48 kD and an isoelectric point of 4.24. The deduced amino acid sequence contained several features common to calreticulins. The fact is that a putative 17 amino acid signal peptide sequence was located at the N-terminal of this protein. Apart from the signal peptide, the amino acid sequence encoded by PCRT can be divided into three domains. Two potential calreticulin family signature motifs, KHEQNIDCGGGYL and IMFGPDICG, which were highly conserved among calreticulins as well as three conserved cysteines at residues 104, 136, and 162 was located in the N-domain. The P-domain was rich in proline and included two triplicate repeats (A1–A3, B1–B3). The C-domain was abundant in aspartic and glutamic acids and terminated with the ER retrieval sequence HDEL (Fig. 2.108).

The presence of these features indicated that the isolated cDNA PCRT encoded the calreticulin of *Pinctada fucata*. PCRT had an ER retrieval signal HDEL, which was different from the typical retrieval signal KDEL and identical to those of some higher plants, *Drosophila melanogaster*, *Schistosoma mansoni*, and so on.

As shown in Fig. 2.109, the calreticulin from *Pinctada fucata* showed high homology with calreticulins from other species. Complete amino acid sequence alignment of PCRT showed the best identity with mollusks *Aplysia California* (the homology is 76%) and less identity with *Drosophila melanogaster* (the homology is 63%). The homology between PCRT and mammalian calreticulin is 67%. These results suggest that calreticulin is a highly conserved protein and PCRT belongs to the calreticulin family.

2.5.3.3 Expression Pattern of PCRT in Different Tissues

To investigate the function of calreticulin in *Pinctada fucata*, the tissue-specific expression of oyster calreticulin mRNA in different tissues was examined by semiquantitative RT-PCR. RT-PCR was performed with RNA samples from the mantle, gill, digestion gland, gonad, and hemolymph. A 402 bp product was obtained with specific primers (CRTSF1 and CRTSR1), and the total RNA of various tissues as the template, while the negative control exhibited no product (data not shown). PCR products were then inserted into pMD18-T vector and confirmed by sequence analyses.

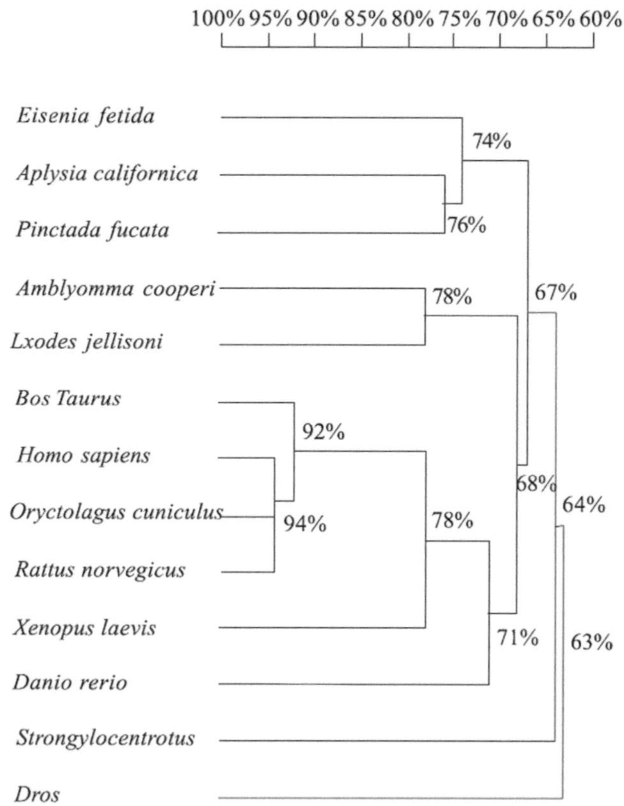

Fig. 2.109 Structural homology comparisons between PCRT and other calreticulins from principal invertebrates and vertebrates
Percentages of sequence identity, as determined by the ClustalX program using the full amino acid sequence, are indicated. Accession numbers of these sequences found in the GenBank database (National Center for Biotechnology Information): *Homo sapiens* (NP_004334), *Oryctolagus cuniculus* (NP_001075704), *Bos taurus* (NP_776425), *Rattus norvegicus* (NP_071794), *Danio rerio* (AAH46906), *Strongylocentrotus* (AAD55725), *Drosophila melanogaster* (BAA85379), *Amblyomma cooperi* (AY395248.1), *Eisenia fetida* (ABI74618), *Aplysia californica* (AAB24569), *Ixodes jellisoni* (AAR29949), *Xenopus laevis* (AAH46699), *Pinctada fucata* (EF5513344)

As shown in Fig. 2.110, PCRT mRNA was expressed in all the tissues tested. The expression levels of PCRT in mantle and hemolymph were highest among these tissues. Three independent experiments showed similar results.

2.5.3.4 Expression of PCRT mRNA in Mantle

The mantle of *Pinctada fucata* is a crucial tissue responsible for oyster biomineralization, since it actively participates in the secretion of calcium and other ions for

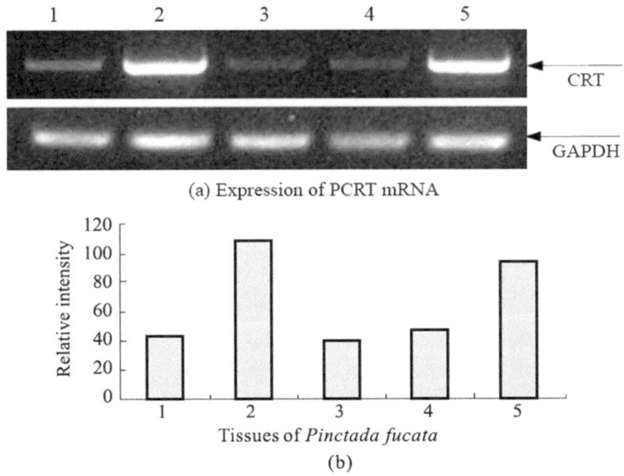

Fig. 2.110 Expression of PCRT mRNA in different *Pinctada fucata* tissues
(**a**) Semiquantitative RT-PCR was done with RNA samples from adult tissues of *Pinctada fucata*. The PCR products are indicated by an arrow and the GAPDH mRNA expression level was used as an internal control. (**b**) Densitometry analysis of the PCRT mRNA expression in tissues of *Pinctada fucata*. The results were standardized by the internal GAPDH control. The results are expressed as mean ± SD of three individual experiments. The data was analyzed by the one-way analysis of variance (ANOVA) using Tukey's test. Bars: 1, gonad; 2, mantle; 3, gill; 4, digestion gland; 5, hemolymph

mineral growth [113]. In situ hybridization of the mantle was performed to determine the precise location of the PCRT mRNA in the mantle tissue of *Pinctada fucata*. As shown in Fig. 2.111, strong hybridization signals were detected on both sides of the inner fold, the inner side of the middle fold, and the inner side of the outer fold of the mantle. However, no hybridization signal was detected in the outer side of the outer fold of the mantle. Hybridization with the control sense probe did not yield any hybridization signals (data not shown).

In this study, we cloned and characterized the full-length calreticulin cDNA from *Pinctada fucata* and analyzed its expression and distribution patterns to investigate its possible roles during mollusk biomineralization. The deduced amino acid sequence from PCRT contains several features, and it shares significant identity with a number of calreticulins from other organisms in the database of NCBI.

It is notable that the ER retrieval sequence of calreticulin of *Pinctada fucata* is HDEL, which differs from the classical ER retrieval signal, KDEL, found in many calreticulins from mammalians and other vertebrates [356]. *Euglena*, a kind of protozoan, has the ER retrieval signal KDEL. Additionally, the retrieval sequences of calreticulins from several higher plants (including barley and maize), *Caenorhabditis elegans*, *Drosophila melanogaster*, and *Schistosoma mansoni* are all HDEL [357–361]. Interestingly, the ER retrieval signal of calreticulin from another mollusk, *Aplysia california*, is KDEL [362]. What is more, calreticulins from several ticks have an ER retrieval signal HEEL [363], while two calreticulins

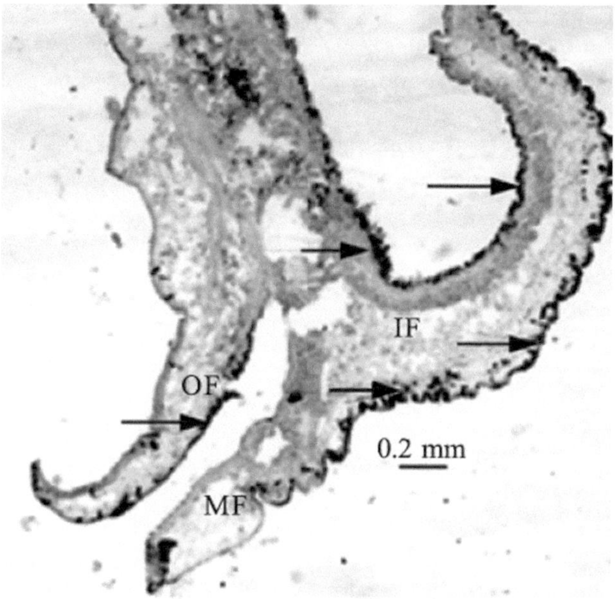

Fig. 2.111 In situ hybridization of PCRT mRNA in mantle
Strong hybridization signals were detected in the inner and outer epithelial cells of the inner fold, the inner epithelial cells of the middle fold, and the inner epithelial cells of outer fold of the mantle (arrows). *OF* outer fold, *MF* middle fold, *IF* inner fold

from two elegans have no ER retrieval signal [364]. These investigations indicate that although most calreticulins have the retrieval signal KDEL, and calreticulins from closely related species likely have the same ER retrieval signal, the ER retrieval signal of calreticulin is species-specific.

In some ticks and hookworms, the classical calreticulin ER retrieval signal is replaced by HEEL. This phenomenon suggests that calreticulins from these parasites might be secreted into the hosts rather than being retained in the ER [363]. The calreticulins from filarial parasites *Dirofilaria immitis*, *Onchocerca volvulus*, and *Litomosoides sigmodontis* all lack the ER retrieval signal and may exist outside the ER. However, the calreticulins found at the cell surface, in the bloodstream, and in the extracellular matrix all have the retrieval signals [365–367]. Hence, the classical ER retrieval signal is not necessary for calreticulin relocating from the ER to other sites. The species differentiation of the ER retrieval signal of calreticulin and its role in locating calreticulin need further study.

The results of the semiquantitative RT-PCR showed that the PCRT mRNA was expressed in all tested tissues, suggesting that calreticulin plays versatile roles in *Pinctada fucata*. Three individual experiments all showed the expression levels of PCRT in the mantle and hemolymph are higher than those in the other tissues.

Hemolymph is an important transport vehicle for biomineralization, moving the high concentrations of calcium and carbonate to the mineralization site. The form of calcium transported through the hemolymph is not yet completely known. Previous investigations have suggested that the transport involves calcium ions, binding to calcium-binding proteins, and amorphous calcium carbonate (ACC) [327, 368]. Calreticulin is a unique calcium-binding protein with a high-affinity calcium-binding domain and a high-capacity calcium-binding domain. In this study, the highest expression of calreticulin was detected in the hemolymph, which suggests that calreticulin might be involved in calcium transport in the hemolymph. A large amount of calcium may bind to calreticulin for transport to the mineralization site.

The mantle is a key tissue involved in biomineralization through calcium translocation, storage, and secretion [355]. The high expression of calreticulin in the mantle indicates calreticulin might participate in these processes. The mantle edge in the bivalve includes the outer, middle, and inner folds. The outer epithelial cells of the outer fold of the mantle, facing the shell and the extrapallial space, participate in biomineralization directly [212]. In situ hybridization showed no visible hybridization signals in the outer epithelial cells of the outer fold, suggesting that calreticulin may not be involved in Ca^{2+} secretion into the extrapallial space from the outer epithelial cells of the outer fold. The epithelial cells of the inner and middle folds communicate directly with the ambient medium and are responsible for absorbing Ca^{2+} from the seawater. The mantle is also considered to be an important tissue for calcium storage [113]. A large amount of calcium absorbed by the gill, mantle, and other tissues is transported by the hemolymph. A bulk of calcium may be temporarily stored in the mantle. The strong signals detected on both sides of the inner fold, the inner side of the middle fold, and the inner side of the outer fold of the mantle indicate that calreticulin is involved in calcium storage in the mantle tissue of *Pinctada fucata*.

We isolated the full-length cDNA of calreticulin from *Pinctada fucata*. The results suggest that calreticulin might be responsible for Ca^{2+} transport in the hemolymph and calcium storage in the mantle. The detailed mechanisms by which calreticulin participates in calcium storage and transport need further study.

2.5.4 Molecular Cloning and Distribution of a Plasma Membrane Calcium ATPase Homolog from the Pearl Oyster **Pinctada fucata**

Plasma membrane calcium ATPase (PMCA) plays a critical role in transporting Ca^{2+} out of the cytosol across the plasma membrane which is essential both in keeping intracellular Ca^{2+} homeostasis and in biomineralization. In this paper we cloned and localized a gene encoding PMCA from the pearl oyster *Pinctada fucata*. This PMCA shares similarity with other published PMCAs within the functional domains. Reverse transcription-polymerase chain reaction analysis shows that it is expressed ubiquitously. Furthermore, in situ hybridization reveals that it is expressed in the inner epithelial cells of the outer fold and in the outer epithelial cells of the middle fold, as well as the edge near the shell, which suggests that PMCA may be involved in calcified layer formation. The identification and characterization of oyster PMCA can help to further understand the structural and functional properties of molluskan PMCA, as well as the mechanism of maintaining Ca^{2+} homeostasis and the mechanism of mineralization in pearl oyster.

2.5.4.1 Isolation and Sequence Analysis of cDNA Sequence Encoding Pearl Oyster PMCA

The complete cDNA sequence including the poly (A) tail is 4517 bp. It contains a 226 base 5′ untranslated sequence, an open reading frame (ORF) consisting of 3567 bp (Fig. 2.112), a TAA stop, a 704 base 3′ untranslated sequence, and a poly (A) tail of 17 nucleotides. A putative polyadenylation signal (AATAA) is recognized at position 4198, which is located in the upstream of the poly (A) tail separated by 297 nucleotides. This cDNA sequence has been submitted to NCBI with the accession no. EF121960. The ORF of this cDNA clone encodes a protein consisting of 1189 amino acid residues. A search of the NCBI GenBank database using the BLAST algorithm indicated that this pearl oyster PMCA sequence is homologous with many PMCAs, including PMCA1, PMCA2, PMCA3, and PMCA4, and shares the highest homology (63% identity at the amino acid level) with red swamp crayfish PMCA3 isoforms (NCBI GenBank accession no. AAR28532).

The pearl oyster PMCA has very high identity with the human PMCA3 isoform 3 (NCBI GenBank accession no. NP_001001344), the crayfish PMCA3 isoform (AAR28532), and the *C. elegans* PMCA (AAR00672) in some regions as shown in Fig. 2.113, such as phosphorylation site, the fluorescein isothiocyanate (FITC) site (assumed to be part of the ATP-binding site), the FSBA site (5-′-pfluorosulphonylbenzoyladenosine/r-(N-2-chloroethyl-N-methylamino) benzylamine ATP-binding site), TM1, TM2, TM4, TM5, and TM6. Other regions exhibit less identity, such as TM3, TM7, TM8, TM9, TM10, and the CaM (calmodulin) binding site. Two regions where major differences between sequences occur are the region between TM2 and TM3, which is the phospholipid interaction site [369], and the C-terminus immediately next to the 3′-sequence of the CaM-binding site, which includes a number of regulatory sites for protein kinase [370].

2.5.4.2 Phylogenetic Relationship

Based on the alignment of the full-length amino acid sequences of pearl oyster PMCA, other known PMCAs, and SERCA (sarco-/sendo-plasma reticulum calcium ATPase), the phylogenetic tree was constructed and showed that the pearl oyster PMCA is more closely related to other eukaryotic PMCAs than to mollusk SERCA (as shown in Fig. 2.114). Interestingly, it is more closely related to vertebrate and arthropod PMCAs than to the PMCA of *C. elegans*.

TCTGATACGACTGTCATAGGGCAGTGTTTTGTTGATGTTTTGAGCTCACATAACGTAAAACTCTTTGATTTGATTGATGTAAAGTGATATTG
CCCCCACCACCACCTCTTTGATCCACCCCTGACGACTTGTTACCGCAGATAATTTTCTGGATCGTCTAATCACACCGTCTAAGGTGATTGAT
GGACAGATTTCAGGTGTGTATTGTGGATAGATAACCTCTGCA
1 **ATG**GCTGGAGATGAAATGGACGGTATCGGCCTGTCACGGCCGGCCGACGGCCAGTATGGGGTTACGCTCATGGAACTCCGTTCCCTC
88 ATGGAGAAAAGGAAAAAAGAGGGATACGACACCATACAGGAAAAATATGGTGGTGTTATAGAACTCTGCAAAAAATTATTCACCTCG
175 CCAAACGAAGGCCTCTCAGGGGCGCCGGCTGATCTAGAAGACAGACGGACTGTCTTTGGTTCCAATGTTATACCTCCTAAACCTCCT
262 AAAACATTTCTACAATTAGTGTGGGAAGCCCTACAGGACGTTACATTAATTATTCTTCTAGTAGCTGCATTAATTTCCTTAGGTTTA
349 TCATTTTACAAGCCTCCAAAGGTTGAAGGATCAGAGGACGAGAGTTCGGAGTCGGAGGCTGGATGGATAGAAGGTGTAGCAATTTTA
436 GGTGCCGTAATAATTGTAGTGTTAGTGACAGCATTCAATGACTATCAGAAAGAAAAACAATTTAGAGGACTTCAAAGTAAAATTGAA
523 CATGAACATCAGTTTTCTGTGATTCGGGGAGGACAGGATAAAAATATTCCAGTAGGGGAGATTGTTGTAGGGGATATATGCCAAGTC
610 AAATATGGTGATCTGCTACCAGCTGATGGTGTTATTATACAAAGTAATGATTTAAAAATAGATGAAAGTTCGCTCACAGGAGAGTCT
697 GACCATGTCAAAAAGGGGGAGGAAATTGATCCCATGCTCCTATCAGGTACTCATGTTATGGAGGGGAGTGGTAAGATGTTGGTCACG
784 GCCGTTGGTGTTAACTCACAGACCGGTATTATCTTTGCTCTGCTCGGTGCCTCTGCCGAGGAGGAGAAAAAAGCCAAGAAAGAGAAA
871 AATAAAAACAAGGGAGATGATCCAAATGCTGGTGCACGGTCTCAGGGCGGAGTCAATGTACCATTGCTTACTAGCGATGCAGATGCA
958 AAAGTGGATATAGACACCATTGACCCCCAAATCCCTCATGGAAACTCTCATAATAACACCAACGCCAATGTAACCAAGAACGGAGAA
1045 ACGGGAGACAATGAGGAGACAAAGACTAAGATAGAACCTCAACGCAAGGAAAAATCTGTTCTCCAGGCTAAATTGACCAAACTCGCC
1132 ATTCAAATCGGTTATGGAGGTACTGCCATAGCTGTTCTCACTGTGGTACTGTTAATCGTCAAATACTGCATCATAGAATTTGTTGTA
1219 AATAAAAGGGAGTGGAAGTCTGATGAGACGACAGAGCATATTGAGTATTTTGTCAGCTACTTTATCATTGGTGTGACTGTGTTAGTG
1306 GTAGCTGTACCTGAAGGGCTTCCATTAGCTGTTACTCTGTCTTTGGCTTACTCAGTCAGGAAAATGATGGATGACAACAACTTGGTA
1393 CGACATTTAGACGCCTGTGAGACAATGGGTAATGCTACGGCTATTTGTTCAGATAAGACCGGTACTCTAACAACAAACAGAATGACA
1480 GTGGTGCAAAGCTATATAGGAGGGGCTCACTACAAGTCAACTCCAAACTTTAGTACACTTGCTAAAACATTCCAAGATCTGATTGCT
1567 CCCAGTGTGGCCATTAATAGTGGATATACATCACGCATAATGCATGATCCAGATGGAGGTCTACCCAAGCAGATAGGTAATAAGACA
1654 GAGTGTGCCCTGCTCGGCTTCGTAATGGATCTCAAACAGGATTACGAGGCTATCCGACACGAGACCCCCGAGGAAGCCCTGTTTAAG
1741 GTCTACACCTTTAATTCTGTACGGAAGTCCATGAGCACAGTTATCGAAATCAAAAATGGCTACAGACTCTTCACGAAAGGAGCTTCA
1828 GAGATTGTCCTGAAAAAGTGTTCATACATCCTGGACTGTCATGGTAACCCAAACAAGTTTTCTGTAGAAGATCAGGAGTCAATGGTA
1915 TCCAAGGTGATCGAGCCCATGGCAAGTGACGGTCTCAGGACTATCTGTGTGGCGTACAAGGATTTTGTTAATGGTACCCCTGAATCT
2002 GAGAACCAGGAGCAGTTTAAGGGTGACATAGACTGGGAAGATGAGGATTTGGTCGTTTCTGGTCTGACGTGTCTCATGGTGGTTGGT
2089 ATAGAGGATCCTGTCAGAGACGAGGTACCAGCCAGTATTATGAAGTGTAAACATTCAGGTATCTGTGTGCGTATGGTGACAGGTGAC
2176 AATGTTAACACTGCACGTTCTATCGCCTCTAAATGTGGAATCCTGACACCTGGTGAAGACTTTCTGGTTTTAGACGGCAAAGAATTC
2263 AATCGTAGGATCAGAGACGGTAGCGGAGAGGTATCTCAGGAACTGTTTGACAAGGTCTGGCCCAAGCTTCGCGTTCTGGCACGATCA
2350 TCACCTCAGGATAAGTACACTTTGGTTAAAGGCATCATCGACAGTAAGCTGAGTTCAAACCGAGAAGTGGTGGCTGTAACAGGAGAT
2437 GGTACAAATGATGGACCGGCACTGAAGAAAGCTGATGTTGGATTTGCAATGGGTATATCAGGAACTGATGTAGCTAAGGAAGCCTCT
2524 GACATTATTCTGACTGACGACAATTTTACCAGTATTGTGAAGGCCGTAATGTGGGGTAGAAACGTCTACGACAGTATCGCCAAATTT
2611 CTCCAGTTTCAGTTGACTGTAAACATTGTGGCTGTGTTAGCAGCTTTCCTCGGAGCATGCATCATCAGTGACAGTCCTCTGAAGGCC
2698 ATACAGATGTTGTGGGTTAACCTGATCATGGACACGCTCGCTTCACTAGCCCTCGCCACCGAGCTGCCAGGAGAAGAATTACTGAAC
2785 CGTAAACCCTACGGCAGGACCAAGCCACTTATATCGCGCAACATGATGAAAAATATCCTCGGCCATGCCGTATACCAGCTCACTGTT
2872 ATATTTGTTGTTCTATTTGCAGGGTCTAGTTTATTTGATATAGACGATGGAATAGAAAAACAGAAATTACATGGCCCTGCTACACAA
2959 CACTTTACCATTATATTTAATGTTTTCGTAATGATGACTCTGTTTAATGAAGTTAATTCACGTAAAATCCATGGGCAGCGAAATGTT
3046 TTTGAAGGTCTGAGGAGAAATCCTGTGTTCATTGGAATTTGGATCGGTACATTCGTAGCTCAGATTATACTTGTTCAATTTGGGGGC
3133 ATAGCCTTCAAGACAGCACCACTTTCAATAGATCAATGGATGTGGTGCTTTTTCTTGGGTGTGGGTGTTCTACTCTGGGGACAGCTA
3220 CTGACATGTATTCCAACTCATAAATTACCCAAGAAGATTTTCTCGTGGGGTCGACACCACAAGGATGAACTGGACCATGTGGACAGC
3307 CCTGACCTACATCTGGATGATGACGTGGGTCGCCGCGGCCAGATTCTGTGGCTCCGCGGCTTGACGCGGCTCCAGACACAGATTCGT
3394 GTTATTAATGCTTTCCGCATGGGCATAGATGCACGATATGACAGTCATAGTTTATCCTCAATCCACAGTTACCATAATTTGCAAACC
3481 CATAGAAAACCACACACGTCTACAAGTCCTCGTGTTGTCACCAGCCCTTTGGTCACAGGGGAGGGCGATCACGCAGAGTCCAACTTC
3568 **TAA**
TCTACCAAATTGTTTGTGTTTATCTTGTACCTGACTTTTTAAATAACATGCCAGACAAAGATAAACCTGAAATCTACAAAGCCTCTTTGGAA
ACCTGTGCTAGATTTTCTGTTTCATTACTCAGAATGGTGAACTAAATTTTTTTCTCATTTTGTTGTTTCACCCCCACATTATATTATGTTCT
CTCTTTATTTGTCTTTTACTATTCATGTTAAAAGTGGTAAAGTTATAATGTTCAGATTTCTGTATTTACTACACAATTGTGTCAAGATTATA
CACTTGAGCAGTTTTGCATACCATTTTCAGTAATGAGATGATACTTAGATTTGTCATTTTTAGATAGAATTTGGTGATTTTATACCACTAGG
GCATTGAAGGGATGACCATTTGGTTTCTATGTTTAATAATGATACCAAGTGTTGTGATTGTGCTCCCTTTATTTTTAATTTGATATTCAGAG
AAAAGTGTCTGTGATTTCTTCAAAGAGTTTCTTTGACCTTTAATATAGCCAAACAAGACTAGCAGTTTATTATTAGATCAGTCTTTCTGAGT
GAAGGTCAAGGCAACTTTGTTATCATTATTATTCCTTATTTTCATGTAGAACAAATATTTTGACTTTAATATTAGCTTATAATGATAATGCA
TAATACCGATAAAGGTTATTTTAGACATGGTATTTATTATTAAACACATTCCATAGGATGAAAAAAAAAAAAAAAAA

Fig. 2.112 Nucleotide of pearl oyster gonad plasma membrane Ca^{2+}-ATPase. The nucleotides are numbered on the left
The putative start codon ATG and the stop codon TAA are in bold. The primers used during sequencing and expression analysis are underlined

2.5.4.3 Gene Expression Analyses of Pearl Oyster PMCA

To investigate the functions of pearl oyster PMCA in vivo, tissue-specific expressions of PMCA mRNA in different tissues were examined by RT-PCR with specific primers (PBF and PBR). Samples of total RNA from the adductor muscle, digestive gland, gill, gonad, mantle, and foot tissues of *P. fucata* were prepared separately.

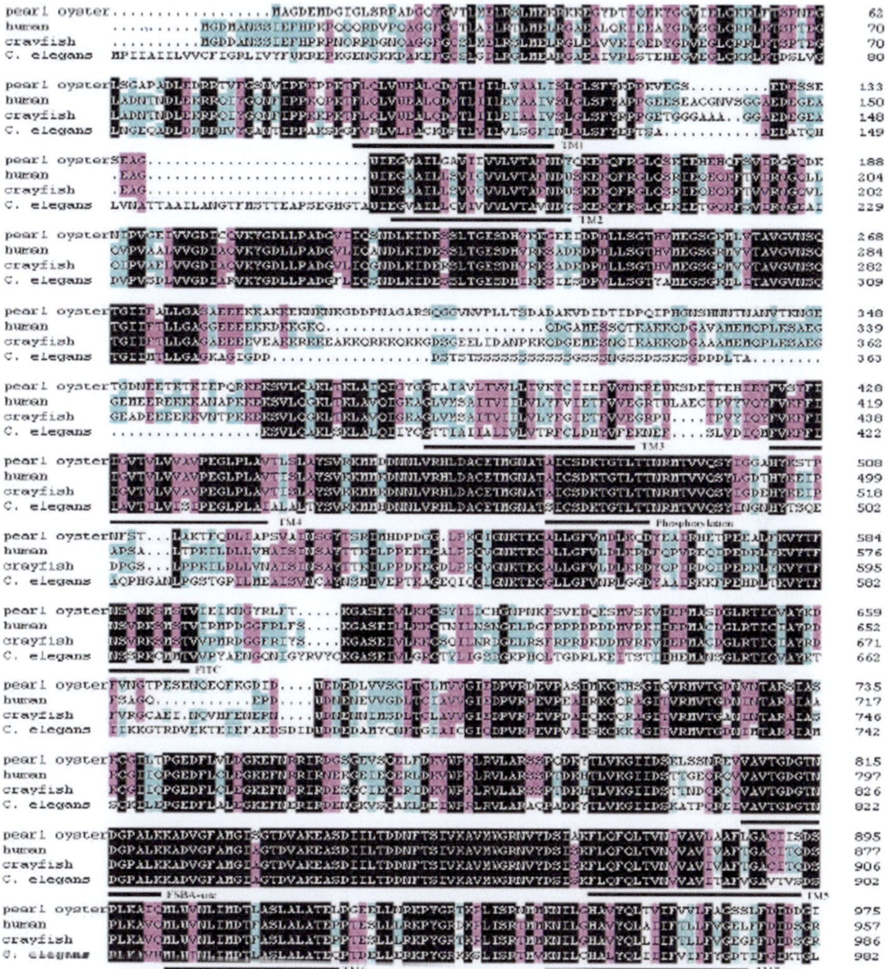

Fig. 2.113 Comparison of the deduced amino acid sequence of pearl oyster gonad plasma membrane Ca^{2+}-ATPase with other species

Human PMCA3b (NCBI GenBank accession no. NP_001001344), crayfish PMCA3 (AAR28532), and *C. elegans* PMCA (AAR00672). The identical sequences are shaded. The amino acids are numbered on the right. The transmembrane regions, as well as the phosphorylation site, FITC site (assumed to be part of the ATP-binding site), the FSBA site (5′-pfluorosulphonylbenzoyladenosine/r-(N-2-chloroethyl-N-methylamino) benzylamine ATP-binding site), and the calmodulin-binding site, are underlined and labeled on the right

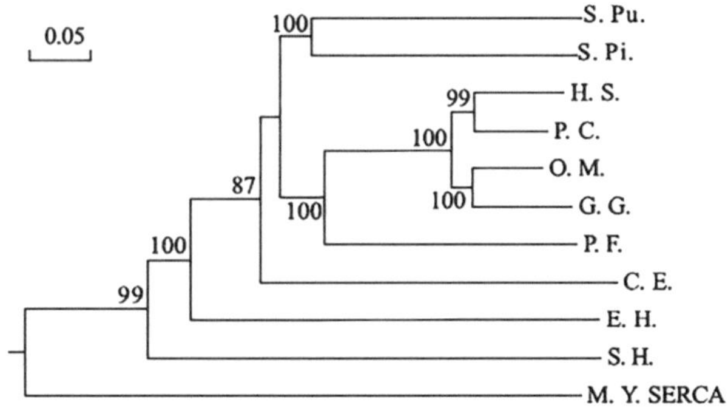

Fig. 2.114 Phylogenetic tree showing protein primary sequence relationships between the pearl oyster PMCA predicted protein and other selected calcium ATPases
Bootstrap values (1000 replicates) are indicated at the nodes. C.E., *Caenorhabditis elegans* PMCA (NCBI GenBank accession no. AAR00672); E. H., *Entamoeba histolytica* PMCA (AAC28745); G. G., *Gallus gallus* PMCA 2 (XP 414301); H.S., *Homo sapiens* PMCA 3b (NP 001001344); M. Y. SERCA, *Mizuhopecten yessoensis* sarcoplasmic reticulum Ca^{2+}-ATPase (AB002112); O. M., *Oreochromis mossambicus* PMCA (AAK15034); P. C., *Procambarus clarkii* PMCA3 (AAR28532); S. H., *Sterkiella histriomuscorum* PMCA (AAR85356); S. Pi., *Stylophora pistillata* PMCA (AAR13013); S. Pu., *Strongylocentrotus purpuratus* PMCA (NP 001028822)

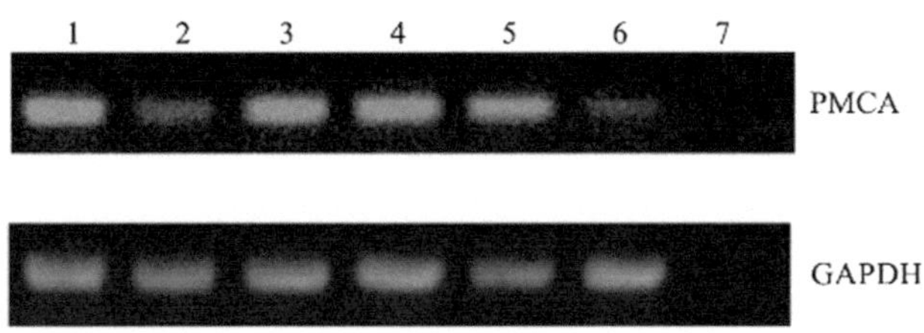

Fig. 2.115 RT-PCR for analysis of PMCA gene expressions in *P. fucata*
Lane 1, adductor muscle; lane 2, digestive gland; lane 3, gill; lane 4, gonad; lane 5, mantle; lane 6, foot; and lane 7, negative control. The housekeeping gene GAPDH was included as a positive control

The results show that the PMCA transcript is broadly expressed, with lower expression in the digestive gland and foot tissues than in other tissues (Fig. 2.115).

The precise expression site of PMCA mRNA in the mantle of *P. fucata* was determined using in situ hybridization that was performed with frozen sections of the mantle (Fig. 2.116). Strong signals PMCA mRNA were detected in the inner epithelial cells of the outer fold and the outer epithelial cells of the middle fold, as well as the edge near the shell, but hybridization signal was absent in the inner fold.

Intracellular Ca^{2+} homeostasis is essential to all cells because Ca^{2+} is an important secondary messenger in various processes. Ca^{2+} is also intimately involved in biomineralization. Pearl oysters need to closely regulate the Ca^{2+} concentration in their blood and tissues, as well as in the biomineralization process

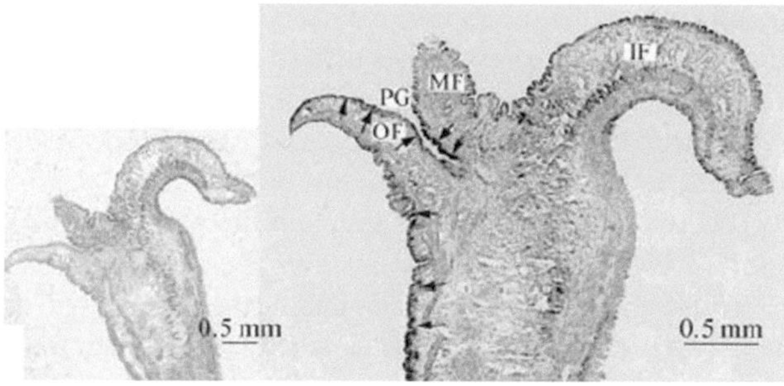

Fig. 2.116 Detection of PMCA mRNA in the mantle of *P. fucata* by in situ hybridization Hybridization signals are indicated by the arrows. The left picture shows a control section stained with no probe. *PG* periostracal groove, *OF* outer fold, *MF* middle fold, *IF*, inner fold

in shell and pearl formation. The mechanism that regulates the intra- and extracellular Ca^{2+} concentration in pearl oyster is not yet completely known.

PMCAs were believed to play a critical role in transferring Ca^{2+} across the cells' plasma membrane both in vertebrates and in invertebrates to maintain intracellular Ca^{2+} homeostasis. Furthermore, it is believed that PMCA is involved in the biomineralization process. This study cloned and characterized a gene encoding the plasma membrane calcium ATPase from pearl oyster *P. fucata*.

The amino acid sequence of pearl oyster PMCA is similar to that of previously published calcium ATPase. PMCA has ten transfer-membrane regions and four main units protruding into the cytoplasm [371]. The first unit encompasses the first 80–90 N-terminal amino acids, the second unit includes the phospholipid-interacting site [369], the third unit includes the catalytic site, and the fourth unit includes a number of regulatory sites, including the calmodulin-binding site and the substrate domains for protein kinase [370, 372]. All of them are found in the pearl oyster PMCA, which suggests that the pearl oyster PMCA belongs to the PMCA superfamily. Furthermore, the phylogenetic tree shows that the pearl oyster PMCA is more closely related to other eukaryote PMCAs than to scallop SERCA, confirming its relation to the PMCA superfamily.

The important function of PMCA in keeping the intracellular Ca^{2+} homeostasis was indicated by the abroad expression of pearl oyster PMCA in all tissues examined in the experiments. In situ hybridization of PMCA shows that it is expressed in the mantle sub-tissue. Thus, the oyster PMCA may be involved in the formation of the calcified layer because of its expression in the inner epithelial cells of the outer fold and the edge near the shell. The fact that there was no PMCA expression in the inner fold is unexpected. Even if the cells of the inner fold are not involved in the biomineralization, they still need to maintain Ca^{2+} homeostasis. We propose the following hypotheses to explain this paradox: PMCA has four isoforms each having various spliced modes. These isoforms have a tissue-specific distribution in mammals [373, 374]. Thus, there may be more than one isoform of PMCA in pearl oyster with other isoforms expressed abundantly in the inner fold, perhaps with another Na^{+} $/Ca^{2+}$ exchange mechanism expressed there.

The phylogenetic tree of the protein primary sequence shows that the pearl oyster PMCA is more closely related to vertebrates than to *C. elegans*. This is similar to the result for the PMCA of *Stylophora pistillata* and *Procambarus clarkii* [375, 376]. Analysis of the deduced amino acid sequence showed that this gene encodes highly conserved functional domains, which may function in calcium transport. The pearl oyster PMCA is conserved, which makes it a good model to study the function of PMCA in biomineralization. Thus, research on plasma membrane calcium ATPase cloned from *P. fucata* can further understand the mechanism for maintaining intracellular Ca^{2+} homeostasis and the biomineralization mechanism.

2.5.5 Cloning and Characterization of a Novel G Protein β-Subunit of Pearl Oyster (Pinctada fucata) and Its Interaction Sites with Calmodulin

The cDNA clone coding for a novel G protein β-subunit of β_1 subclass, pfGβ_1, was isolated from the pearl oyster (*Pinctada fucata*). The deduced amino acid sequence of this pfGβ_1 (341 amino acids) shares high homology to great pond snail (*Lymnaea stagnalis*) Gβ_1 and Northern European squid (*Loligo pealeii*) indeed, in spite that it derives from human and bovine (*Bos taurus*). The well-conserved amino acid domains in G protein β subunit, seven WD repeats, were detected in the deduced amino acid sequence. Alignment analysis indicated that the initial amino acid residues in variable fragments of the seventh WD motif were different from any other known Gβ. The prediction of 3D structure of pfGβ revealed that pfGβ belonged to β-propeller family proteins whose members usually comprised 4-8 antiparallel β-sheets resembling the blades of a propeller. Furthermore, Northern blotting analysis and in situ hybridization showed that the hybridization signals of pfGβ mRNA were extensively distributed in various tissues except muscle, especially with abundance in epithelia of outer fold of the mantle, gonad, and gill. What's more, we investigated and analyzed the interactions between calmodulin (CaM) and Gβγ, and specific amino acid residues which might be crucial for the binding of Gβγ to CaM were also identified and characterized. Furthermore, the functional analysis of protein–protein interaction suggested that the binding of CaM and Gβγ did increase the activity of alkaline phosphatase (ALP), an important indicator for biomineralization in MC3T3-E1 cells. The results showed that the ALP activity of the mutants of pfGβγ the one impairing the interactions of Gβγ with CaM was much higher than negative control group but lower than the WTC group. All in all these results indicated that the Gβγ might interact with CaM and perform significant physiological functions in regulating cellular functions.

2.5.5.1 Isolation of the *P. fucata* Gβ cDNA

A cDNA product of 335 bp was obtained by RT-PCR with a pair of degenerate primers (GBF and GBR) using *P. fucata* mantle cDNA as template. Sequence analysis of the cDNA fragment revealed that it most likely encodes a Gβ homolog. To obtain the full-length cDNA of the pfGβ protein, gene-specific primers (GSP1 and GSP2) were synthesized to screen the cDNA library by PCR. After four rounds of screening, only one positive tube yielded the 208 bp product. Then we seeded the phages and performed the in situ hybridization by random-primed labeling probe with [α-^{32}P] dCTP. The positive clone 11 was isolated, and sequence analysis showed that the longest cDNA insert (1056 bp) contained a 224 bp 3′ UTR fragment which includes a canonical polyadenylate signal and a poly (ad) tail (Fig. 2.117). The deduced amino acid sequence contained a perfect match peptide sequence of G-β-subunit (Fig. 2.117). Although this result verified that the clone 11 encodes Gβ subunit, it apparently does not contain the full-length cDNA as it lacks the start codon. Therefore, we designed the GSP3 and used 5′ RACE to obtain the start codon and 5′ UTR (Fig. 2.117).

2.5.5.2 Sequence, Phylogenetic, and Predicted Structure Analysis

The nucleotide sequence was submitted to GenBank with the accession no. AY524667. The sequence of oyster Gβ-subunit shared high similarity with those of Gβ subunits from mollusk, amphibian, and mammal (94% identity to squid, 90% to tiger salamander, and 86% to human) (Fig. 2.118a). The well-conserved amino acid domains in G protein β subunit, seven WD repeats, were founded in the deduced amino acid sequence. Each WD repeat has a region of variable length, flowed by a core of more or less constant length bracketed by GH and WD [377]. Alignment analysis showed that the beginning amino acid residues in variable fragment of the seventh WD repeat 300-303 (VLRQ) is different from any other Gβ (Fig. 2.119).

To investigate evolutionary relationships between pfGβ and known Gβ isoforms, phylogenetic tree was constructed with ClustalX program [379] with the neighbor-joining (NJ) algorithm of [380] (Fig. 2.118b). The phylogenetic tree revealed that it is closely related to Gβ and belongs to a member of $G\beta_1$ class. Within the Gβ class, pfGβ was fairly closed to great pond snail (*Lymnaea stagnalis*) and Northern European squid (*Loligo pealeii*). However, pfGβ was diverged from bovine (*Bos taurus*) and human.

The 3D structure of pfGβ was predicted by the following Internet address: http://www.expasy.org/swissmod/SWISSMODEL.html. The results showed that pfGβ also belongs to h-propeller family of proteins whose members contain 4-8 antiparallel h-sheets resembling the blades of a propeller [381]. The h-propeller of Gβ consists of seven β-sheets, each containing four antiparallel strands radiating outward from a central core with approximate sevenfold symmetry (Fig. 2.120). The

A RT-PCR (335 bp) □□□□□ (444-778 bp)
Clone 11 (1056 bp) ○□□□□□□○○○○○(341-1396 bp)
5'RACE (1-694 bp) ......○□□□□□□ (1-694 bp)
Full-length cDNA (1396 bp) ......○□□□□□□○○○○○(1-1396 bp)

B

ACGCCGGGACAGGAAGAGAGAAAATTGTTTATGAATGCCGGGGAAATCTAGCGTCGATTTTGTGTCTATAT -146
GACCATTTGACGGAGGTTTCGATCTCCATTTTATATTTTTCATATTTATTTTGTCATCTCACCTAAGTAGCCACC
ATGAGTTCAGAATTAGAAGCATTGCGCCAAGAGACGGAACAACTCAAAAACCAAATTAGGGAAGCACGGAAGGCA 75
M S S E L E A L R Q P T E Q L K N Q T R E A R K A 25
GCGGCAGACACCACACTCAATCAAGCTGCAGTCAATGTAGAACAAGTAGGCAGAATTCAAATGCGCACGCGTCGC 150
A A D T T L N Q A A V N V E Q V G R I Q M R T **_R_** **_R_** 50
ACCCTGAGAGGTCATTTGGCCAAAATATATGCCATGCATTGGGCCTCCGACTCCAGAAATCTAGTCTCTGCTTCA 225
_T_ **_L_** R G H L A K I Y A M H W A S D S R N L V S A S 75
CAAGATGGTAAATTAATTGTATGGGATGGTTACACAACAAACAAGGTACATGCCATCCCGTTGAGGTCCAGTTGG 300
Q D G K L I V W D G Y T T N K V H A I P L R S S W 100
GTGATGACCTGTGCGTACGCGCCATCAGGGAGTTACGTAGCATGTGGAGGTCTGGACAATATATGTTCTATCTAT 375
V M T C A Y A P S G S Y V A C G G L D N I C S I Y 125
AGTCTCAAGACAAGGGAAGGAAACGTACGAGTCAGTCGAGAATTACCGGGTCACACAGGCTATCTATCATGTTGT 450
S L K T R E G N V R V S R E L P G H T G Y L S C C 150
AGATTTTTGGATGATAACCAGATTGTGACCAGCTCGGGAGACATGTCATGTGCATTATGGGACATAGAGACAGGT 525
R F L D D N Q T V T S S G D M S C A L W D I E T G 175
CAGCAGACGACAGCATTTACAGGACACACAGGAGATGTGATGAGTCTGTCTTTGTCACCAGACATGAGAACGTTT 600
Q Q T T A F T G H T G D V M S L S L S P D M R T F 200
GTGTCTGGAGCGTGTGATGCCTCCGCCAAGCTTTGGGACATCAGAGACGGCATGTGTAAACAAACTTTCTCTGGA 675
V S G A C D A S A K L W D I R D G M C K Q T F S G 225
CACGAATCTGATATAAATGCAATTACATACTTTCCAAATGGCTACGCCTTTGCCACTGGATCTGACGATGCCACT 750
H E S D I N A I T Y P P N G Y A F A T G S D D A T 250
TGCCGTCTGTTTGACATCAGAGCAGACCAGGAGATCGGCATGTATTCTCACGATAACATTATCTGCGGCATCACA 825
C R L F D I R A D Q E I G M Y S H D N I I C G I T 275
TCTGTGGCCTTCTCTAAGTCCGGACGTCTCTTACTCGGAGGGTATGACGACTTCAACTGTAACGTTTGGGATGTG 900
S V A F S K S G R L L L G G Y D D F N C N V W D V 300
CTCAGACAGGAAAGAGCTGGTGTGTTGGCTGGCCATGACAACCGTGTTAGTTGTCTGGGAGTGACCGAGGACGGT 975
L R Q E R A G V L A G H D N R V S C L G V T E D G 325
ATGGCCGTAGCAACGGGATCATGGGATAGCTTCCTCCGGATTTGGAACTGACGACAATAAGAAACTAGGGGAATG 1050
M A V A T G S W D S F L R I W N * 341
GGCAGGACTATGCAAAATAGACAAAGAAGGCAAAGCCAGGAAAATTGTTGGATGCAGCTGTGTTGTTGTTAAGGA 1125
CTATGGCAAAGAGTCCCAAGCCCTGGATGTACTCAACGACTATTTCAAGACCAAGAAGTAAACTCTAACATTCAG 1200
ACAATAAAGTACATCAACTATAAAAAAAAAAAAAAAAAAAAAAAAAAAAAAA 1250

Fig. 2.117 The oyster G protein β-subunit cDNA and deduced amino sequence
The conserved amino sequence residues, based on which degenerate primers were designed, are bold, and the corresponding to nucleotides is shaded. The gene-specific primers used for screening cDNA library are underlined. The 5′ RACE primer is in the box. The stop codon is indicated by an asterisk. The mutant amino acid residues are bold and underlined

inner strands of each sheet run roughly parallel to central axis and form the walls of a water-fill tunnel, whereas the out strands tilt as they radiate away from the center. Each WD repeat forms one β-propeller. However, a single WD repeat does not directly correspond to a single stable fold domain; rather, the most variable segment of each WD repeat, corresponding to its N-terminal portion, forms the outermost, C-terminal strand of one blade, whereas the conserved core of the same repeat forms the innermost strand of the nest blade. The overall propeller structure of β is closed

A

	Sequence identities (%)												
	P. fu	*A. ti*	*C. el*	*H. am*	*D. me*	*S. av*	*L. pe*	*L. st*	*X. la*	*D. re*	*Rat*	*B. ta*	*H. sa*
P. fu	100	85	90	88	85	88	94	93	80	86	86	86	86
A. ti		100	85	96	83	84	96	84	84	98	96	96	96
C. El			100	91	87	89	87	87	82	86	85	86	86
H. Am				100	86	91	86	86	79	86	86	86	86
D. me					100	91	84	82	77	84	83	83	84
S. av						100	86	86	80	86	85	86	86
L. pe							100	91	79	86	85	85	86
L. St								100	80	84	83	83	84
X. La									100	84	84	84	85
D. Re										100	98	98	99
Rat											100	98	99
B. ta												100	100
H. sa													100

B

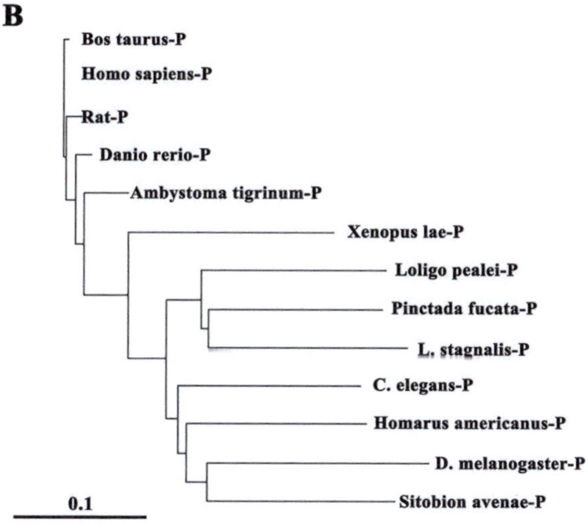

Fig. 2.118 The phylogenetic analysis of Gβ_1 subunit in *Pinctada fucata* and their homologies from other spices

(**a**) Matrix indicating the percentage identities of the aligned Gβ_1; (**b**) phylogenetic tree showing the evolutionary relationship of oyster Gβ_1 with proteins. (The N-J tree was generated by using the method N-J of program ClustalX1.81.) P.fu, oyster (*Pinctada fucata*) Gβ_1, AY524667; A.ti, tiger salamander (*Ambystoma tigrinum*) Gβ_1, AAG31060; C.el, *Caenorhabditis elegans* Gβ_1, NP_496508; H.am, American lobster (*Homarus americanus*) Gβ_1, 045040; D.me, (*Drosophila melanogaster*) Gβ_1, fruit fly, NP_525090; S.av, English grain aphid (*Sitobion avenae*) Gβ_1, AA046882; L.pe, Northern European squid (*Loligo pealeii*) Gβ_1, P23232; L.st, great pond snail (*Lymnaea stagnalis*) Gβ_1, Q08706; X.la, African clawed frog (*Xenopus laevis*) Gβ_3, AAH56002; D. re, zebra fish (*Danio rerio*) Gβ_1, NP_997774; Rat, Norway rat (*Rattus norvegicus*) Gβ_1, NP_112249; B.ta, cow (*Bos taurus*) Gβ_1, NP_786971; H.sa, human (*Homo sapiens*) Gβ, AAA35922

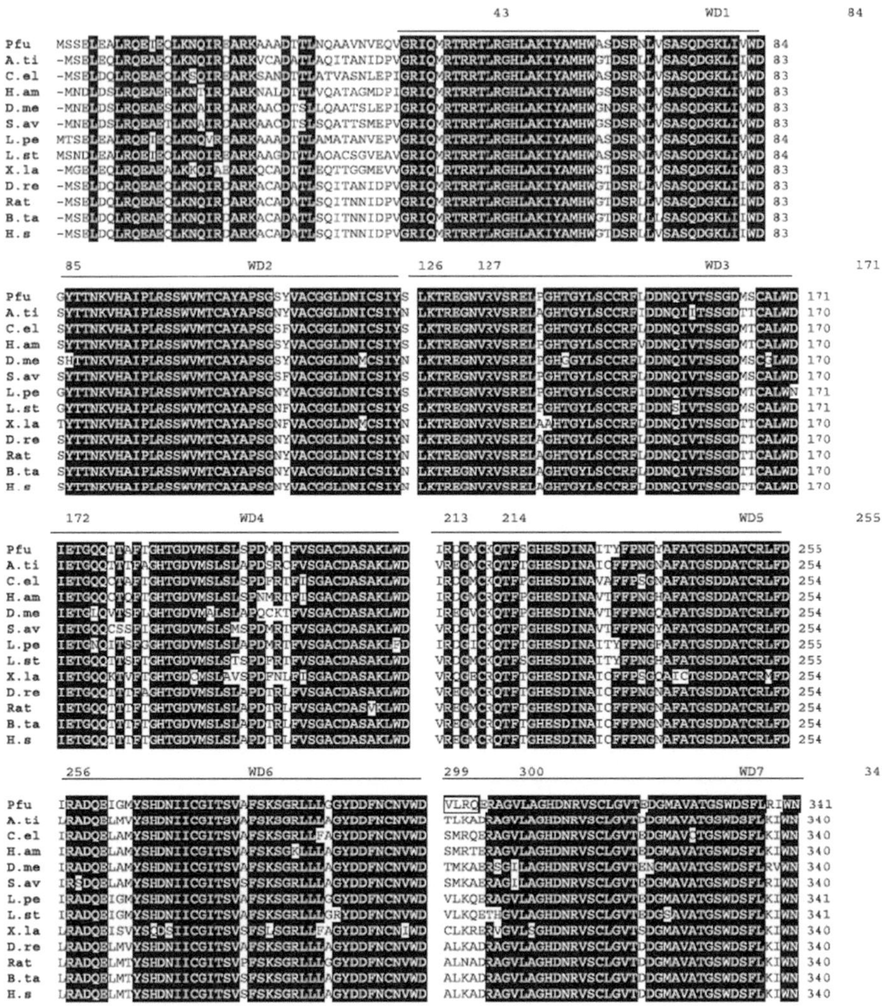

Fig. 2.119 Alignment of the sequence of oyster Gβ_1 with other Gβ from mollusk to mammalian [378]
The seven WD repeats are indicated by overlining. Abbreviations and sources are according to Fig. 2.118

by juxtaposition of strands from the first to the seventh WD repeat. This structure may contribute to the resistance to cleavage by trypsin despite many of tryptic cleavage sites [381]. As shown in Fig. 2.120, yellow highlights the different structure, the variable region of the seventh WD repeat which is different from any other Gβ subunits that may be involved in the closing β-propeller and formation.

However, pfGβ was unique among those which belong to Gβ family in the following criteria. First, in bivalve mollusk, deduced amino acid sequence of pfGβ_1 is first reported. Second, the variable region of the seventh WD repeat

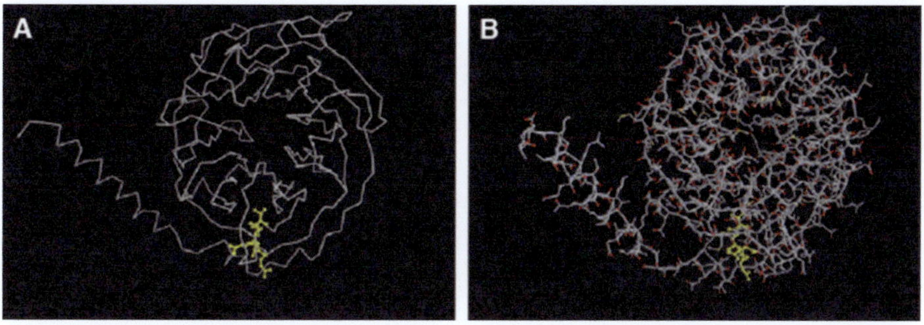

Fig. 2.120 The 3D structure prediction from the deduced amino acids
(**a**) Stick-and-stick model of the seven WD repeats in pfGβ. Each WD repeat forms one β-propeller, (**b**) ball-and-stick model of the seven WD repeats in pfGβ. The β-propeller domain (yellow) highlights the different structures, the variable region of the seventh WD repeat which is different from any other Gβ subunits and may involve in the closing β-propeller and formation. (For interpretation of the references to color in this figure legend, the reader is referred to the web version of this article)

which may involve in the closing β-propeller and formation is different from any other Gβ subunits. Third, the analysis of the phylogenetic tree showed that pfGβ_1 is different from any other members of Gβ_1. Based on these results we concluded that pfGh1 probably defines a novel subclass in G protein β subunit.

2.5.5.3 The Distribution and Expression of pfGβ mRNA in Tissues

Distribution of pfGβ was detected by in situ hybridization in five different tissues specifically the gill, mantle, gonad, digestion gland, and muscle. The pfGβ mRNA hybridization signals were dark brown, and the background was not stained. As shown in Fig. 2.121a–d, the positive signals only distributed in the cytoplasm with hybridization-negative nuclei, and they were detected in all five tissues. Strong pfGβ mRNA hybridization signals were observed in gill epithelial cells and gonad and mantle epithelial cells of both sides in out fold, while weak signals were observed in digestive gland, and positive signals were not presented in muscle. On the negative control, *P. fucata* pfGβ mRNA hybridized signal was not detected.

Northern blotting analysis of pfGβ transcripts in five different tissues was carried out with ^{32}P-labeled pfGβ as a probe. As shown in Fig. 2.122, the signal of pfGβ mRNA was observed at many tissues except muscle. The pfGβ mRNA was expressed abundantly in the gill, gonad and mantle and at low levels in the digestive gland. This is total agreement with the in situ hybridization results indicating that pfGβ is present in a limited number of tissues.

The gill is important to metabolism, which assimilates nutrients that particularly dissolve calcium from the water [382]. found that abalone gill membranes have highest binding with calcitonin gene-related peptide (CGRP) and likely

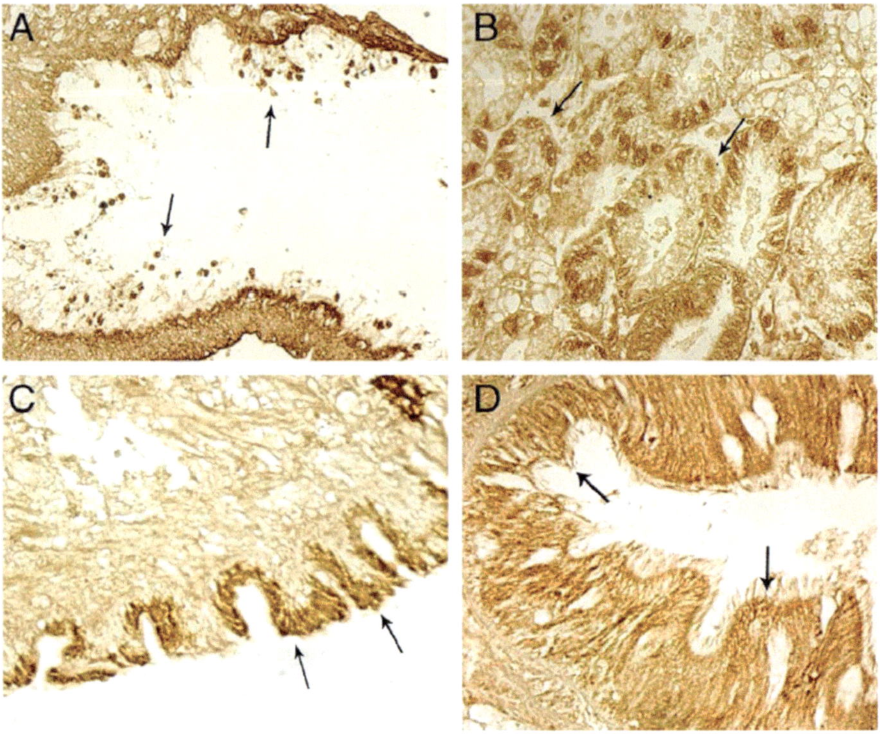

Fig. 2.121 In situ hybridization of oyster Gβ mRNA in different tissues of *P. fucata*
Among these tissues, hybridization signals (arrows) were observed in epithelial cells of gill (**a**), spermatogenic duct (**b**), the outer fold in mantle (**c**), and digestive gland (**d**). Original magnifications are ×400 (**a**) and ×200 (**b**, **c**, and **d**)

participate in the control of hydromineral metabolism in aquatic species [383]. suggested that the common beta-subunit (G beta) was present at all ages during development and was more prominent around the periphery of the tubules in younger animals but then became more evident in the cytoplasm of germ cells with increasing age. The high expression of pfGβ mRNA in gill and gonad are consistent with these important functions. Additionally, pfGβ mRNA hybridization signal is also expressed in other tissues, implying that the pfGβ signal transduction pathway existed in *P. fucata* widely.

2.5.5.4 Mutation Analysis of the Putative pfGβ-Binding Domain of CaM

G protein β subunits must depend upon γ subunits to fold correctly and its signals transduction in vitro also bound to G subunits [384]. We also cloned mouse γ_2 subunit and ligated expression vector. Figure 2.123a shows the in vitro translated wild and mutant Gβ proteins separated by Western blot, using anti-His-tag

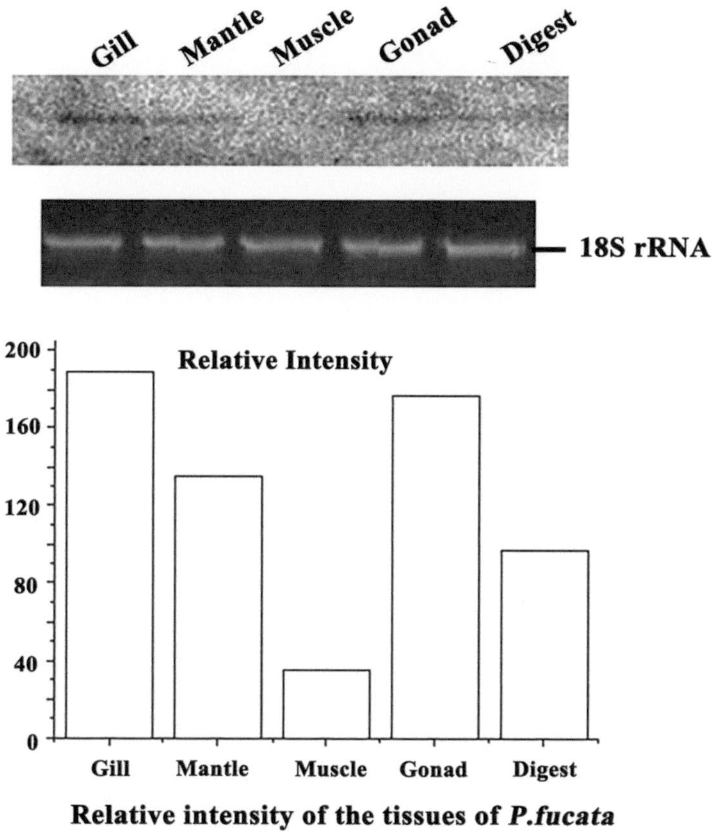

Fig. 2.122 Northern blot of β-subunit of G protein in various tissues in *Pinctada fucata* and the relative intensity of tissues of *P. fucata*

monoclonal antibody. The in vitro translated protein mixtures were incubated with anti-HA-tag monoclonal antibody, and the agarose beads were also added. The precipitated proteins were detected by Western blot, using Gβ rabbit anti-β-subunit antibody. As shown in Fig. 2.123b, unlike the wild-type Gβ protein, the binding affinities of DT and MT for CaM were dramatically reduced. The data suggested that the mutant residues are critical to the interaction with CaM.

Previous studies have revealed that G protein subunits can act as potent inhibitors of the Ca^{2+}–CaM-stimulated phosphodiesterase activity [385], probably through interactions of Gβγ with CaM. In the present report, we show the direct binding of Gβγ with CaM. By modifying specific amino acids in the region, we identified some key residues (R49A, R50A; deletion 49-53) that are crucial to the binding of Gβγ to CaM and also found the primary structure similarities of CaM-binding domain of Gβγ. For example, the identified β CaM-binding domain contains a high percentage of hydrophobic residues and an excess of positively charged residues, a property found in all CaM-binding peptides [386]. The basic residues contribute to the CaM

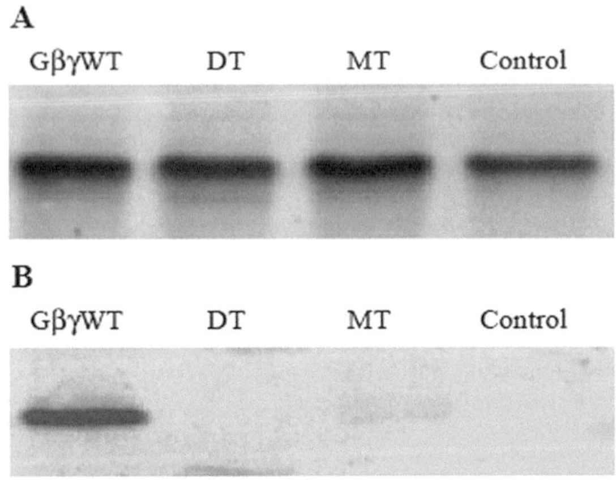

Fig. 2.123 Mutation analysis of putative CaM-binding domain of pfGβ subunit
(**a**) Similar amounts of wild-type (WT), mutant (deletion 49–53 amino acids, DT; R49, 50A, MT), and positive control proteins were obtained by in vitro translation in 293-T cell line. (**b**) CaM binding of in vitro translated pfGβ protein, its mutants, and negative control. Compared with the wild-type pfGβ protein, binding ability to CaM of the mutant proteins was dramatically reduced

binding via electrostatic interaction with acidic residues in CaM, while the hydrophobic amino acids seem to play a more important role in CaM binding through interactions with the hydrophobic patches of globular domains of CaM [387]. The mutants, DT (lacking N-terminal deletion 49-53 amino acids) and MT (R49A, R50A), showed that they both failed to associate with CaM (Fig. 2.123). The result suggests that the N-terminal 49-53 amino acid sequence is necessary to associate with CaM [378].

2.5.5.5 Effects of ALP Activity in Transfected MC3T3-E1

Alkaline phosphatases (ALPs) are a family of cell surface glycoproteins that catalyze the hydrolysis of phosphomonoesters with release of inorganic phosphate. ALP is an important enzyme in biomineralization [388], and its activity is also an important physiological functional index involving in biomineralization in MC3T3-E1 in vitro, even though, in vivo, 1,25$(OH)_2D_3$ may provide a mechanism for rapid control of the secretion of ALP to alter calcium movement [389]. We also assessed the effects of ALP activity in WT, WTC, DTC, MTC, and CaM. The data showed that the ALP activity in WTC is significantly higher than that of the other groups ($P < 0.05$). No prominent difference among the groups of DTC, MTC, and CaM was observed ($P > 0.05$), while they were all higher than that of the control group (only transfected vector without cDNA fragment) (Fig. 2.124).

In WT group, overexpression of Gβγ may cause general changes in the metabolic state of the cell and activate various signal transduction pathways, including

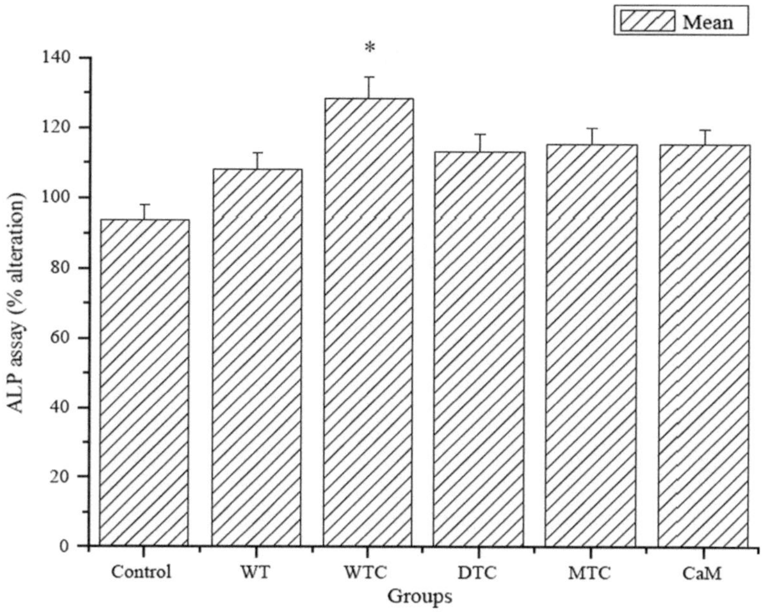

Fig. 2.124 The interaction of Ghβ subunits with CaM by measuring the ALP activity in co-transfected MC3T3-E1
ALP activity in WTC is significantly higher than that of the other groups ($P < 0.05$). There are no differences among the groups of WT, DTC, MTC, and CaM ($P > 0.05$), while they are all higher than that of the control group (only transfected vector without cDNA fragment). (*) $P < 0.05$, compared with any other group

stimulating MAPK kinase pathway through Ras and Raf [390] and causing permanent changes in Ca^{2+} channels state [391], which all improved the proliferation and further improved the ALP activity. In WTC group, the affinity of Gβγ subunits and CaM is sufficiently high so that much of the free βγ in the cells should be in the CaM-bound form. Therefore, an increase in intracellular Ca^{2+} concentration could selectively regulate the interactions of Gβγ with serials of other proteins through binding with CaM. CaM not only increases ALP activity but also participates in the biomineralization [153]. In DTC and MTC groups, mutants Gβγ failed to associate with CaM and, as a result, did not activate the downstream signal molecules. So there are no differences among the DTC, MTC, and CaM groups, but all of their ALP activities are higher than the control because of action with CaM.

In conclusion, we have cloned oyster Gβ-subunit and found it widely expressed in many tissues. By comparing the deduced amino acid sequence among Gβ family members, we considered that pfGβ should be classified in a novel subunit within $G\beta_1$ class. Further research indicates that pfGβ might interact with CaM and point to the important physiological function in modulating cellular functions.

2.5.6 *Cloning, Characterization, and Distribution of an mRNA Encoding a* H^+*-ATPase α-Subunit in the Mantle of Pearl Oyster,* **Pinctada fucata**

Mitochondrial ATP synthase, containing two major units (F_0 and F_1), controls the production of most cellular ATP. The structure and function of the active complex (five subunits (αβγδε)) are widely known, but the character of its α-subunit playing in catalytic mechanism still needs further investigation, especially in bivalve animals. We are the first group that successfully cloned and identified the full cDNA length of the mitochondrial H^+-ATP synthase α-subunit gene in *Pinctada fucata* RT-PCR. This mitochondrial H^+-ATP synthase α-subunit is composed of 1991 nucleotides in *Pinctada fucata*, the translation initiation site of which starts at nt 48 (ATG) and the stop codon at nt 1660 (TAA), encoding a polypeptide of 553 amino acids sharing high similarities (81% identity to fruit fly, 82% to carp, and 83% to human) to that of other animals. What's more, the α/β signal transduction domain, as the well-conserved amino acid domain, the sequence alignment analysis of which indicated that two residues (Asp358 and Asn359) were different from any other known ATP synthase α-subunits. In situ hybridization analysis was then carried out to detect the distribution pattern of mitochondrial H^+-ATP synthase in various tissues in *P. fucata*. Our work will definitely provide great properties for further improvement on pearl energy metabolism to increase the production and quality of pearls.

2.5.6.1 Cloning and Sequence Analyses of *P. fucata* ATPase α-Subunit cDNA

The sequence of ATPase α-subunit was obtained by mantle cDNA as template and two primers (ATPF1 and ATPR1), which was derived from the conserved amino acid segments, using PCR technique. A 164 bp product of appropriate size was then subcloned to T-vector and sequenced. Sequence alignments indicated that it most likely encoded an ATPase α-subunit homolog. Gene-specific primers based on the known sequence obtained from the original fragment were designed, respectively, to achieve the full-length cDNA of the pfATPase α-subunit by 5′- and 3′-RACE using touchdown PCR (Fig. 2.125). After sequence assembly, we found that 5′ untranslated regions (UTR) contained 47 bp and yielded a canonical polyadenylation signal and poly (dA) tail in 3′ UTR. In conclusion, the full-length cDNA was 1991 bp in size with the translation initiation site at nt 48 (ATG) and a stop codon at nt 1660 (TAA), encoding a 553 amino acid polypeptide (Fig. 2.125). The deduced amino acid sequence matched perfectly to the peptide sequence of ATPase α-subunit (Fig. 2.125).

```
A: RT-PCR   (164 bp)                      ▲▲▲▲               (1344 - 1507 bp )
5′-RACE   (1458 bp)               ○○○○▲▲▲                 (1-1458 bp )
3′-RACE   (558 bp)                     ▲▲▲------            (434 -1991 bp )
Full-length cDNA (1991 bp)  ○○○○▲▲▲▲------            (1-1991 bp )
B:
                                   ACGCGGGGATATTGGTCGAAACCGCACACATACGACAAGTAGACAAA     -47
ATGCTTTCGGCCAGATTAGCAGCAAGTCTCGTAAGACAGCTTCCCAGGGCTGCACCAAAGGTTTGCCAACATGCC   75
M  L  S  A  R  L  A  A  S  L  V  R  Q  L  P  R  A  A  P  K  V  C  Q  H  A   25
CTGGGAGCTGGACTTATTTCTGCCAAAAAGTTTTCCACATCTACACATCACCATACTGCAGGAGCTAGTGCGGAA   150
L  G  A  G  L  I  S  A  K  K  F  S  T  S  T  H  H  H  T  A  G  A  S  A  E   50
GTGTCATCCATCTTGGAGGAGAGGATTTTGGGACAGACAACACAAGCTGGATTAGAGGAAACCGGTCGTGTACTA   225
V  S  S  I  L  E  E  R  I  L  G  Q  T  T  Q  A  G  L  E  E  T  G  R  V  L   75
TCCATCGGAGACGGTATTGCTCGTGTTTATGGTCTAAAGAATATTCAGGCAGAAGAAATGGTAGAATTCTCCAGT   300
S  I  G  D  G  I  A  R  V  Y  G  L  K  N  I  Q  A  E  E  M  V  E  F  S  S   100
GGTTTAAAGGGTATGGCCTTGAACTTGGAGAGAGATAACGTAGGAGTTGTCGTATTTGGTAACGATAAATTAATT   375
G  L  K  G  M  A  L  N  L  E  R  D  N  V  G  V  V  V  F  G  N  D  K  L  I   125
AAAGAAGGAGACATCGTAAAAAGAACAGGAGCTATTGTGGATGTCCCCGTAGGAAAGGAATTATTAGGAAGAGTT   450
K  E  G  D  I  V  K  R  T  G  A  I  V  D  V  P  V  G  K  E  L  L  G  R  V   150
GTTGATGCTTTAGGAAATCCTATTGATGGAAAGGGACCAATCACATCACCTGACAGACAGAGGGTAGGAGTTAAG   525
V  D  A  L  G  N  P  I  D  G  K  G  P  I  T  S  P  D  R  Q  R  V  G  V  K   175
GCCCCTGGTATCATCCCCCGTATCTCAGTAAAGGAACCTATGCAGACCGGTATTAAGACTGTAGATAGCTTAGTA   600
A  P  G  I  I  P  R  I  S  V  K  E  P  M  Q  T  G  I  K  T  V  D  S  L  V   200
CCTATTGGTAGAGGTCAGAGAGAACTCATCATCGGTGACAGGCAGACTGGCAAAACCGCTATTGCAATTGACACA   675
P  I  G  R  G  Q  R  E  L  I  I  G  D  R  Q  T  G  K  T  A  I  A  I  D  T   225
ATTATTAACCAGAAGAGGTTCAACGATGGAACAAATGAGAAGGCCAAGCTTTATTGTATCTATGTTGCTATTGGT   750
I  I  N  Q  K  R  F  N  D  G  T  N  E  K  A  K  L  Y  C  I  Y  V  A  I  G   250
CAGAAGAGATCCACAGTGGCTCAGATTGTAAAGAGGCTTACTGATGCAGATGCCATGAAGTACACTGTGATTGTC   825
Q  K  R  S  T  V  A  Q  I  V  K  R  L  T  D  A  D  A  M  K  Y  T  V  I  V   275
AGTGCTACAGCTTCTGATGCTGCTCCCTTACAGTATCTGGCTCCTTACTCGGGATGTGCCATGGGAGAATTCTTC   900
S  A  T  A  S  D  A  A  P  L  Q  Y  L  A  P  Y  S  G  C  A  M  G  E  F  F   300
AGAGATAACGGCATGCACGCCCTCATCATCTACGACGATCTGTCAAAACAGGCTGTAGCTTACCGTCAAATGTCA   975
R  D  N  G  M  H  A  L  I  I  Y  D  D  L  S  K  Q  A  V  A  Y  R  Q  M  S   325
CTACTTTTGAGACGTCCCCCCGGTCGTGAAGCCTACCCTGGTGATGTGTTCTACCTCCATTCACGTTTGCTGGAG   1050
L  L  L  R  R  P  P  G  R  E  A  Y  P  G  D  V  F  Y  L  H  S  R  L  L  E   350
CGTGCTGCCAAGATGAATGACGATAACGGTGGTGGATCCCTCACTGCTCTCCCTGTCATCGAGACCCAGGCTGGT   1125
R  A  A  K  M  N  D  D  N  G  G  G  S  L  T  A  L  P  V  I  E  T  Q  A  G   375
GATGTGTCAGCCTACATCCCCACTAACGTCATCTCTATCACAGACGGACAGATCTTCTTGGAGACTGAGTTGTTC   1200
D  V  S  A  Y  I  P  T  N  V  I  S  I  T  D  G  Q  I  F  L  E  T  E  L  F   400
TACAAGGGTGTACGTCCAGCAATTAATGTAGGATTATCTGTCAGTAGGGTAGGATCGGCAGCCCAAACAAAGGCC   1275
Y  K  G  V  R  P  A  I  N  V  G  L  S  V  S  R  V  G  S  A  A  Q  T  K  A   425
ATGAAACAGGTAGCAGGCTCCATGAAGTTGGAGTTGGCCCAGTACAGAGAAGTGGCAGCTTTCGCTCAGTTTGGA   1350
M  K  Q  V  A  G  S  M  K  L  E  L  A  Q  Y  R  E  V  A  A  F  A  Q  F  G   450
TCTGATTTGGATCAGGCTACACAGAATTTACTGAACAGGGGAGTTCGTCTGACCGAGTTACTCAAACAGGCTCAG   1425
S  D  L  D  Q  A  T  Q  N  L  L  N  R  G  V  R  L  T  E  L  L  K  Q  A  Q   475
TATGTACCAATGGCCATTGAAGAACAAGTAGCAGTTATCTATGCTGGTGTCAAGGGACGTTTAGACAAAGTTGAT   1500
Y  V  P  M  A  I  E  E  Q  V  A  V  I  Y  A  G  V  K  G  R  L  D  K  V  D   500
CCATCTAGAATCACAGAATTTGAAGCAGCTTTTGTATCACACATCAGAGGAAGTCAGCAAGAATTATTAGGTCAG   1575
P  S  R  I  T  E  F  E  A  A  F  V  S  H  I  R  G  S  Q  Q  E  L  L  G  Q   525
ATCAGGAAAGACGGTCAAATCACAGAAGCCTCCGACGCTAAATTAAAGGAGGTTGTCACACAGTTCTTGTCCACC   1650
I  R  K  D  G  Q  I  T  E  A  S  D  A  K  L  K  E  V  V  T  Q  F  L  S  T   550
TTCGAAGCATAAATTCATGTGACATTAAACATTTTCTCAAACGAATGGTGATATAGACAGTGCGGTTTTCCTTGT   1725
F  E  A  *
GACACTTATTGACTTAGAGTCTATTTATATCAGTGCTAACTTTGAAAGTATGTTGAATTGATTTGTACATGTAAT   1800
GTCCCTTGTTTAAAAGATCTTTATAAACATTACATGGCAGTGACTGTTTTACATTGTTTTTGATGAATTCAGACA   1875
ACTGTTGATGTATAGATGACATTATTAAAGGGATATTATTAAAAAAAAAAAAAAAAAAAAAAAAAAAAA         1944
```

Fig. 2.125 Full-length cDNA of PfATPase α-subunit and its deduced amino acid sequence in *P. fucata*

The conserved amino acid sequence residues were applied to design the degenerate oligonucleotides. Primers were in bold and the corresponding nucleotides were shaded. The 5′-RACE primer was in the box, while 3′-RACE primer was underlined. The stop codon was indicated with an asterisk

2.5.6.2 Sequence, Phylogenetic, and Predicted Structure Analysis

The full-length nucleotide sequence was successfully submitted to the GenBank with accession no. DQ986328. ATPase α-subunit is a considerably well-conserved protein from the aspect of evolutionary. Multiple protein sequence alignment analysis indicated that the α-subunit of pfATPase exhibited high similarities with those of ATPase α-subunits from *Xenopus laevis* (*X. la*), carp, mammalian, and fruit fly (Fig. 2.126a). The α/β signal transduction domain, an indispensable residue for cooperative catalysis between α- and β subunits, is the most conserved domain [392–394], alignment analysis of which showed that two different residues (Asp358 and Asn359) were different from any other known ATPase α-subunits (Fig. 2.127). These could be referred as characteristic residues in pearl oysters.

To find out the evolutionary relationships between the ATPase α-subunit in *P. fucata* and other known ATPase α-subunit isoforms, we constructed a phylogenetic tree using the ClustalX program [379] with the neighbor-joining (NJ) algorithm of Saitou and Nei [380] (Fig. 2.126b). The phylogenetic tree indicated that the *P. fucata* ATPase α-subunit was closely associated with its homolog gene in fruit fly. In addition to the ATPase α-subunit, pfATPase was closely related to carp and fruit fly. However, pfATPase α deviated from rat and mouse.

a

Groups	Sequence identities (%)								
	P. fucata	**Carp**	**Fruit fly**	**Gallus**	***X. la***	**Cow**	**Homo**	**Mus**	**Rattus**
P. fucata	100	82	81	82	83	83	83	83	83
Carp		100	80	88	90	90	90	91	90
Fruit fly			100	79	81	81	80	80	80
Gallus				100	93	92	92	92	92
X. la					100	92	92	92	93
Cow						100	97	98	97
Homo							100	97	96
Mus								100	98
Rattus									100

b

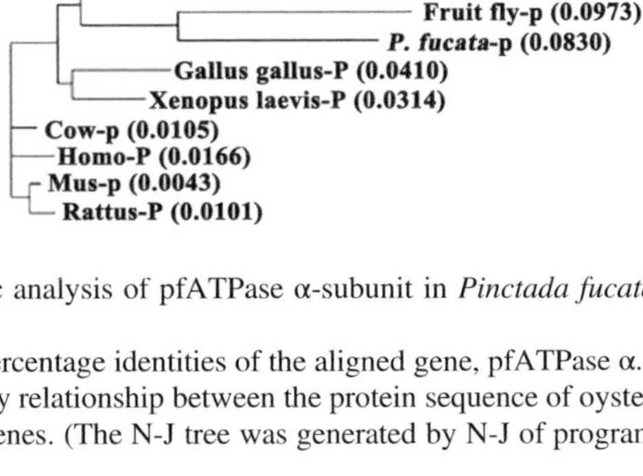

Fig. 2.126 Phylogenetic analysis of pfATPase α-subunit in *Pinctada fucata* and their homologs from other species
(**a**) Matrix showed the percentage identities of the aligned gene, pfATPase α. (**b**) Phylogenetic tree indicated the evolutionary relationship between the protein sequence of oyster pfATPase α-subunit and other homologous genes. (The N-J tree was generated by N-J of program ClustalX 1.81)

Fig. 2.127 Alignment between the oyster pfATPase α-subunit sequence and other ATP synthase α-subunits from invertebrates to mammalian The amino acid domains were marked by overlining. The 358th and 359th amino acids, Asp (D) and Asn (N), respectively, were different from other species. (**a**) N-terminal coupled domain (1-22 aa), (**b**) nucleotide-binding domain (257-303 aa), (**c**) α/β signal transduction domain (322-388 aa)

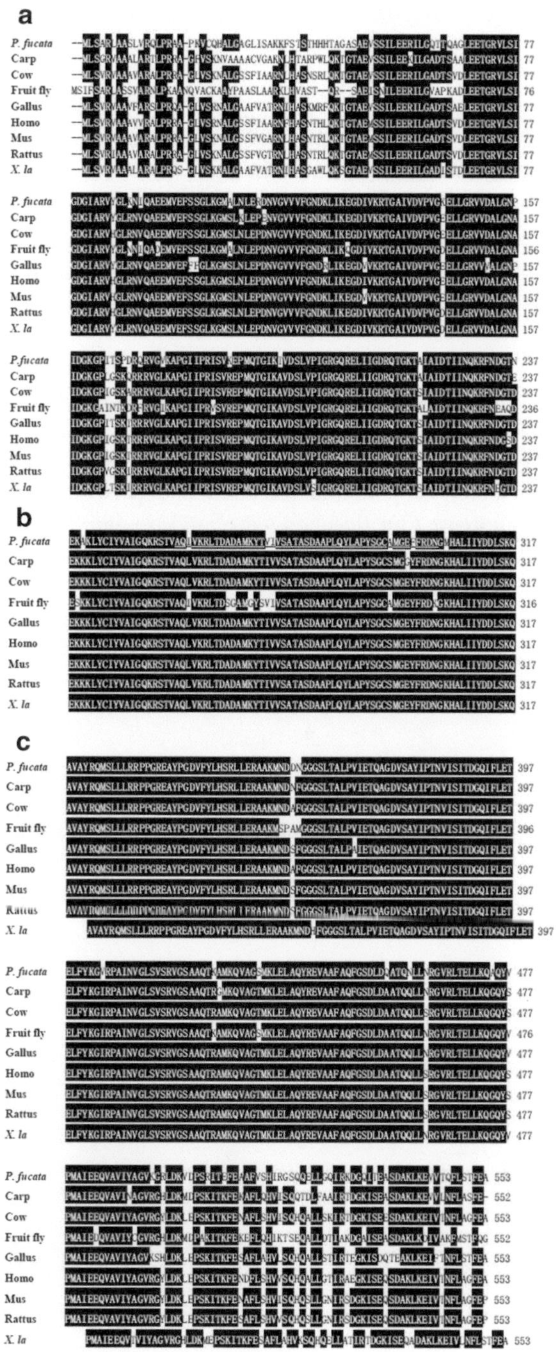

2.5.6.3 Distribution and Expression of pfATPase α-Subunit mRNA in Tissues

The specific distribution pattern of pfATPase α-subunit was investigated by in situ hybridization. PfATPase mRNA was widely expressed in almost every tissue in *P. fucata*, including the gill, digestion gland, mantle, gill, adductor muscle, spermatogenic epithelia, and epithelia. The positive hybridization signal of pfATPase α-subunit should be dark brown, and the background was not stained. As we can see from Fig. 2.128a–e, positive signals were only detected in cytoplasm, while hybridization-negative nuclei appeared in all five tissues. Interestingly, positive signals were not distributed in muscle. We detected very strong signals in gill epithelial cells, mantle epithelial cells, and mantle gland; however, weaker signals were detected in the gonad and digestive gland. And the pfATPase mRNA hybridization signal didn't appear in negative control group. In summary, the strong expression signals mainly appeared in gill epithelial and mantle cells which were considered responsible for shell and periostracum formation, as well as calcium metabolism. Our findings were in close agreement with previous studies [24, 378].

In conclusion, pfATPase α-subunit is very unique among those ATPase α-subunit family members in terms of the following criteria. First of all, the deduced amino acid sequence of pfATPase α hasn't been identified in other bivalve mollusks. Second, the

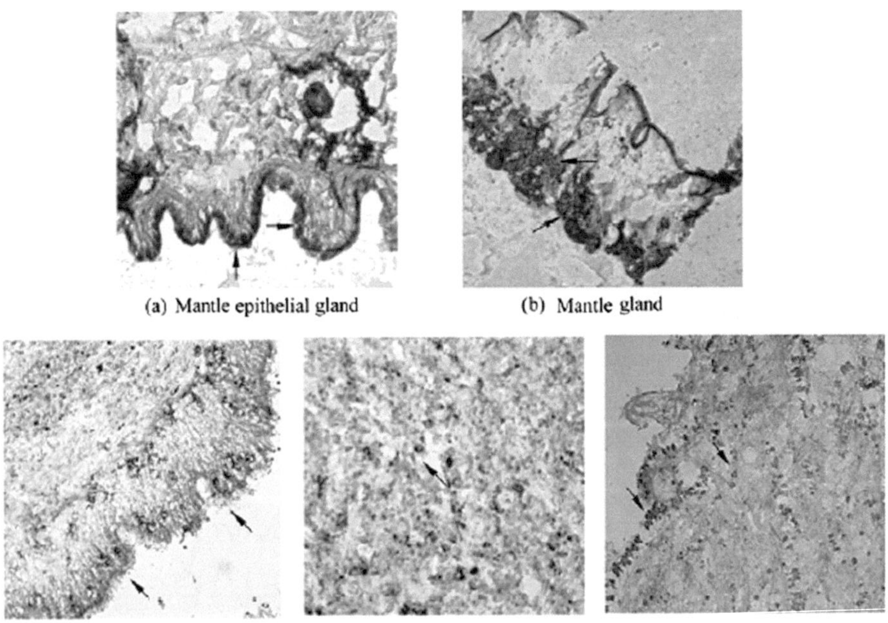

Fig. 2.128 In situ hybridization of oyster pfATPase α-subunit in various tissues of *P. fucata* Among these tissues, hybridization signals (arrows) were detected in the mantle epithelial cells (**a**), mantle gland (**b**), digestive gland (**c**), spermatogenic duct (**d**), and gill (**e**)

α/β signal transduction domain the one believed to be engaged in cooperative catalysis was quite different from any other known ATPase α-subunits. Last but not least, the construction of phylogenetic tree also proved that the pfATPase α-subunit was a very different kind of any other known members in its family. Thus, the pfATPase α-subunit probably defines a brand-new ATPase α-subunit that has never been reported before. The physiological function of this pfATPase α-subunit, especially its α/β signal transduction domain, remains further studies.

2.5.7 *Cloning and Characterization of an mRNA Encoding F1-ATPase Beta-Subunit Abundant in Epithelial Cells of Mantle and Gill of Pearl Oyster,* Pinctada fucata

In oyster biomineralization, large amounts of calcium are absorbed from external media, transported to the mineralization site, and finally deposited via a matrix-mediated process. All these activities are very energy intensive; therefore, investigations of the energy metabolism pathways of different oyster tissues will facilitate understanding of oyster biomineralization physiology. A full-length cDNA encoding the F_1-ATPase beta-subunit (the F_1-β-subunit, a major catalytic subunit of F-ATPase) from the pearl oyster (*Pinctada fucata*) was cloned using the homology strategy with a pair of degenerated primers based on the conserved regions of other animals' F_1-β-subunit genes. Sequencing and structural analyses showed that the obtained sequence shared high identity with other animals' F_1-β-subunits and had a unique phosphorylation site of PKC and CK II on the external surface of the putative protein. Results from semiquantitative reverse transcription-polymerase chain reaction and in situ hybridization demonstrated this oyster F1-β-subunit mRNA is abundant in the gill and mantle and distributed widely in the periostracal groove, the outer folder, and the dorsal region of the mantle and in the gill epithelial cells. These tissues were the main regions that participate in biomineralization processes such as calcium uptake, transport, and matrix secretion. The results indicate that tissues involved in biomineralization have stronger energy metabolic processes and that F_1-ATPase might play an important role in oyster biomineralization by providing energy transport.

2.5.7.1 Cloning of the cDNA Encoding Oyster F_1-β-Subunit

A 660 bp PCR product was amplified from the mantle of *P. fucata*, using degenerate oligonucleotide primers. Sequence analyses using the BLAST algorithm revealed that the product fragment encoded the F_1-β-subunit. Based on this sequence, two specific gene primers (FBF2 and FBR2) were used to amplify the 3′- and 5′-nucleotide sequences of pfATPB cDNA. Touchdown PCR with 5′-RACE yielded a 1154 bp fragment including a short 5′ untranslated region (5′ UTR), while 3′-RACE yielded a 744 bp fragment with a 247 bp 3′ UTR.

The full-length cDNA shown in Fig. 2.129 is 1838 bp in size, with a translation start site at nt +20 (ATG) and the stop codon at nt +1589 (TAA). This cDNA sequence has been submitted to GenBank with the accession number DQ358698. The deduced pfATPB amino acid sequence consisted of 523 amino acid residues (Fig. 2.129) with an estimated molecular mass of 56.6 kDa and pI of 5.38.

2.5.7.2 Amino Acid Sequencing and Motif Analyses

The deduced amino acid sequence of oyster F_1-β-subunit shares high similarity with F_1-β-subunits from sources ranging from bacteria to mammalian (63% identity to *Escherichia coli*, 80% to *Drosophila melanogaster*, and 78% to *Homo sapiens*). Three putative conserved domains, including the ATP-synt_ab, the ATP-synt_ab_C, and the ATP-synt_ab_N, were identified by searching the NCBI Conserved Domain Database website, indicating the deduced protein is an oyster F_1-β-subunit. Two functionally significant domains for ATP synthesis were also detected in pfATPB (Fig. 2.130). The DELSEED domain responsible for binding the γ-subunit, which couples the mechanical rotation with conformational change of the F_1 portion in ATP synthesis [395], was sited on the C-terminal region of the F_1-β-subunit. The nucleotide-binding domain GXXXXGKT named Walker motif A was also conserved [396, 397]. Additionally, the signature sequence [P-S/A/P-L/I/V-D/N/H-x-[314]- [314]-S-x–S] of the ATP synthase alpha- and beta-subunits was also sited at the end of the ATP-synt_ab domain in pfATPB.

Several biologically significant sites that are not in the human F_1-β-subunit protein were detected in the deduced pfATPB protein including a casein kinase II (CK II) phosphorylation site [S/T-X-D/E] located on β252-255 and two protein kinase C (PKC) phosphorylation sites [S/T-X-R/K] located on β155-157 and β247-249. The predicted 3D structure of the oyster F_1-β-subunit showed that these additional phosphorylation sites were all sited on external surfaces of the F_1-ATPase (Data not shown). Since both CK II and PKC are ubiquitous kinases that widely participate in physiological processes by means of phosphorylation modification [398, 399], these particular phosphorylation sites on the external surface of oyster F_1-β-subunit might act as new regulatory sites for ATP synthesis.

To investigate the evolutionary relationship between pfATPB and other known F_1-β-subunit isoforms, a phylogenetic tree was constructed using the ClustalX program. The phylogenetic tree in Fig. 2.131 shows that pfATPB indeed belongs to the F_1-ATPase beta-subunit family and is most closely related to that from *Drosophila melanogaster*.

AGTCTCCTCTACTGTCGCCATGATGCATGCAGTGAGACGAGCGTGTGCCGGGGTCTTTAAGACCTCGAATAGCTTTATT 79

M M H A V R R A C A G V F K T S N S F I 20

GGAACAACTTCAAACACATGCTCGCACGCAAAAGTCATACCTTCATATTTATCTACCCGAAGGCACTATGCAGCGGAACCCAAGGCAGCG 169

G T T S N T C S H A K V I P S Y L S T R R H Y A A E P K A A 50

GCAGCTACCACGCCCGGAAAAGTGGTCTCTGTCATCGGAGCTGTCGTAGACGTACAATTTGAGGACAGTCTACCTCCCATTCTCAATGCC 259

A A T T P G K V V S V I G A V V D V Q F E D S L P P I L N A 80

TTAGAAGTTAAAAATAGAAAGCCCAGGCTCATCCTGGAAGTAGCGCAGCATTTAGGTGAGAACACAGTAAGAACCATTGCTATGGACGGT 349

L E V K N R K P R L I L E V A Q H L G E N T V R T I A M D G 110

ACACAGCGAGTCGTTCGTGCTATGGAGTGTATACACTCTGGTTTCCCCATCAGGATCCCTGTAGGACCAGCTACCTTAGGAAGAATTATA 439

T E G V V R G M E C I D S G F P I R I P V G P A T L G R I I 140

AATGTCATTGGTGAACCCATTGATGAAACGGGCCCAGTACAAACTGACAAATTACTGCCAATTCACGCTGAAGCCCCAGAATTTGTAGAA 529

N V I G E P I D E R G P V Q T D K L L P I H A E A P E F V E 170

ATGAGTGTAGAACAAGAGATCTTAGAAACCGGAATCAAAGTTGTAGACTTGCTTGCTCCTTATGCCAAGGGAGGTAAAATTGGTCTGTTT 619

M S V E Q E I L E T G I K V V D L L A P Y A K G G K I G L F 200

GGAGGAGCAGGAGTGGGTAAGACTGTATTGATTATGGAACTGATTAACAATGTA**GCCAAAGCTCACGGAGGTTACTC**TGTGTTTGCCGGT 709

G G A G V G K T V L I M E L I N N V A K A H G G Y S V F A G 230

GTCGGAGAGAGAACACGTGAAGGAAACGATCTCTACCACGAAATGATCACGTCTAAAGTCATCAGCTTGACAGATGATACCTCAAAGGTA 799

V G E R T R E G N D L Y H E M I T S K V I S L T D D T S K V 260

TCGCTGGTATACGGACAGATGAACGAGCCCCCAGGCGCCCGTGCTCGTGTCGCTCTCACTGGACTTACTGTAGCCGAATATTTCCGTGAT 889

S L V Y G Q M N E P P G A R A R V A L T G L T V A E Y F R D 290

CAAGAAGGACAGGACGTGCTGCTTTTCATTGACAACATTTTCCGTTTCACCCAGGCCGGTTCAGAGGTGTCCGCTTTGTTAGGACGTATC 979

Q E G Q D V L L F I D N I F R F T Q A G S E V S A L L G R I 320

CCCTCTGCCGTAGGTTACCAACCCACTCTGGCCACTGACATGGGTACCATGCAGGAAAGAATTACCACAACCAAGAAAGGATCCATCACA 1069

P S A V G Y Q P T L A T D M G T M Q E R I T T T K K G S I T 350

TCTGTACAGGCTATCTACGTGCCCGCTGACGATTTGACAGATCCTGCCCCAGCCACAACATTCGCTCACTTGGACGCCACCACTGTATTG 1159

S V Q A I Y V P A D D L T D P A P A T T F A H L D A T T V L 380

TCCCGTGGTATCTCAGAGTTGGGTATCTACCCAGCTGTCGATCCCCTTGATTCTACATCACGTATCTTAGATCCCAATGTTGTAGGACAG 1249

S R G I S E L G I Y P A V D P L D S T S R I L D P N V V G Q 410

GAACATTACAGAGTAGCCCGTGGTGTACAGAAAATCTTACAGGACTTCAGATCACTTCAGGATATCATT**GCTATTTTGGGTATGGATGA**A 1339

E H Y R V A R C V Q K I L Q D F R S L Q D I I A I L G M D E 440

TTGTCCGAGGAAGACAAATTAACCGTCACCAGAGCCAGAAAAATTCAGAGATTCTTGTCACAACCTTTCCAGGTCGCTGAAGTTTTCACA 1429

L S E E D K L T V T R A R K I Q R F L S Q P F Q V A E V F T 470

GGATCAGCAGGAAAATATGTGCCACTCAAGGAAACCATCTTAGGGTTCCAACAGATTCTTCAAGGTGACTACGATCATTTGCCTGAGGTA 1519

G S A G K Y V P L K E T I L G F Q Q I L Q G D Y D H L P E V 500

GCTTTCTACATGGTGGGTAACATCGAAGAAGTCGTACAAAAGGCAGAAAGACTGGCCGAGGAGCAGTCATAATGACAGAAATGAAAATAA 1609

A F Y M V G N I E E V V Q K A E R L A E E Q S * 523

ACTTCTAATGTGTAGCGTTACTACTGACATTAAAGTGTGGTGTCGTAAGGCATTCTCTATGGATCAATATTTATATGACTATTCATAGGG 1699

TTCCATTTATCATAGTGCTGGATTGACTGATATCTGACAGAAAATGAGTTGTTCTTTTGTGATGAAATATGAGCTGTTTGTAAACTAAAT 1789

GTTTATTGAACAGTTTATTAAATATAAGAAATAAAAAAAAAAAAAAAAA 1838

Fig. 2.129 The pfATPB cDNA and its deduced amino acid sequence

The conserved amino sequence residues used to design the degenerate primers for RT-PCR are bold. The gene-specific primers used for RACE are underlined. The asterisk indicates the stop codon

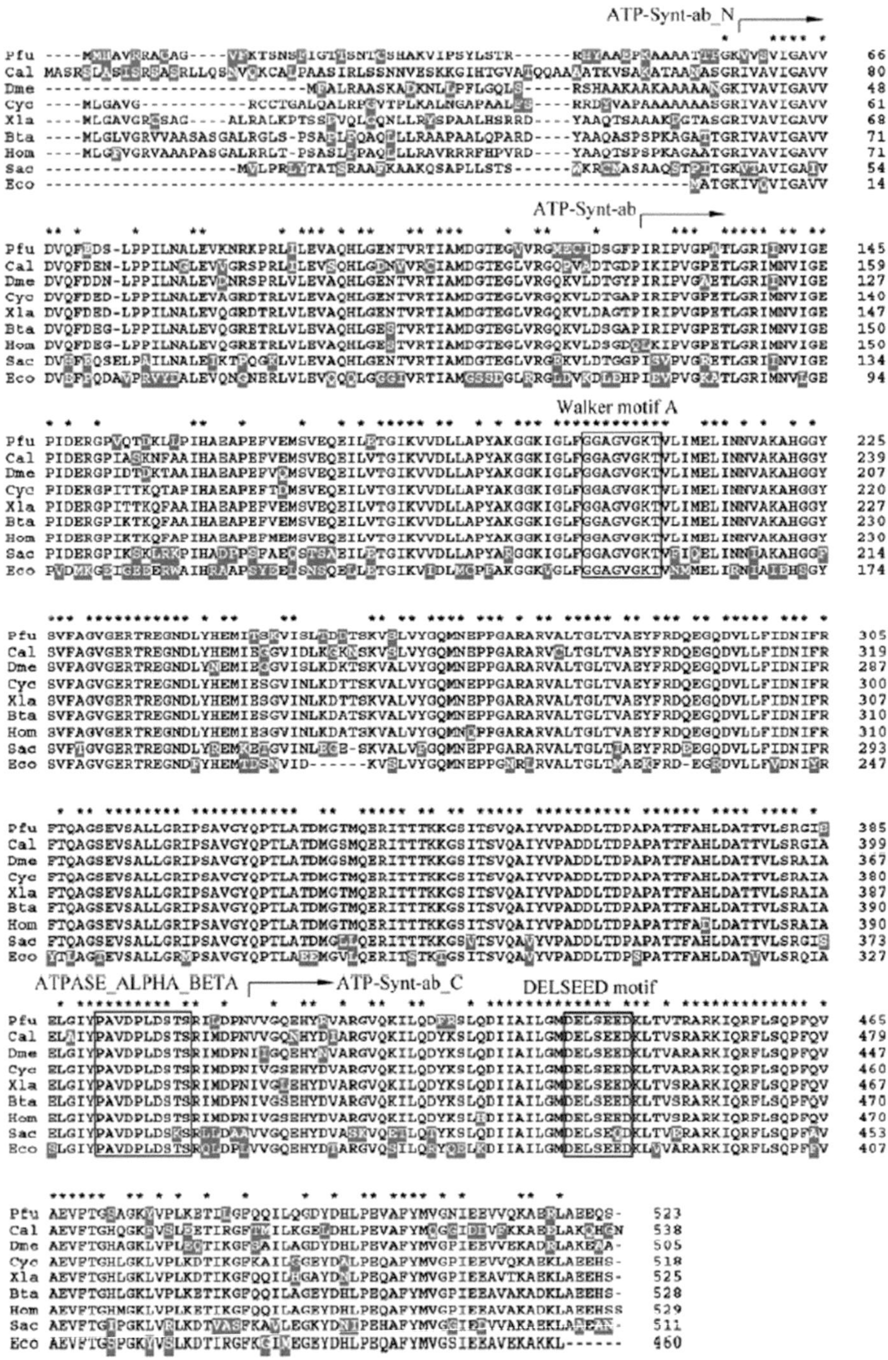

Fig. 2.130 Alignment of the protein sequence of pfATPB with other F_1-β-subunits from bacteria to mammalian

2.5.7.3 Spatial Distribution and Expression of Oyster F_1-β-Subunit mRNA in Tissues

The expression of pfATPB mRNA in various tissues was studied using semiquantitative RT-PCR. The data in Fig. 2.132 show that the pfATPB mRNA was expressed in all tissues, with the highest expression levers in the gill and mantle. Previous results have indicated that oyster gill is the major organ for calcium uptake from external media with the membrane Ca^{2+}-ATPase system playing a key role in the process [52, 108–110]. High expression of pfATPB mRNA in the gill might be related to the heavy energy demand of the calcium uptake process. Furthermore, in situ hybridization demonstrated that the strongest pfATPB mRNA hybridization signals were found in the epithelium of the gill (Fig. 2.133b), indicating that the calcium uptake might be mainly in the epithelium cells.

In situ hybridization was also carried out in the oyster mantle. Strong hybridization signals occurred in the periostracal groove, the outer folder, and the dorsal region of the mantle (Fig. 2.133a). The periostracal groove, located at the distal edge of the mantle between the outer and middle folds, was the main structure responsible for periostracum formation, with the epithelial cells of these two folds perhaps having different active statuses [400, 401]. In this work, relatively higher positive signals were detected on the inner face of the outer folder, suggesting this layer had more active energy metabolism and may provide the majority of the matrix secretion in the periostracal groove of *P. fucata*. Additionally, the distribution patterns of pfATPB mRNA in the outer folder and in the dorsal region of the mantle were similar to that of several key matrix proteins, indicating that F_1-ATPase might participate in matrix protein secretion by supplying energy for the oyster biomineralization.

In conclusion, the mRNA encoding the F_1-β-subunit in a bivalve pearl oyster, *P. fucata*, was cloned and characterized. Sequence analyses identified distinct phosphorylation sites, which might be possible regulatory sites for the F_1-β-subunit of pearl oyster. Studies of the spatial distribution and expression of pfATPB demonstrated that functionally different tissues had different expression levels of pfATPB mRNA. The results provide an informative energy metabolic perspective of tissues involved in oyster biomineralization and indicate that F_1-ATPase might participate in this process by providing the energy for the process. This investigation contributes more details of the physiological properties of oyster biomineralization.

Fig. 2.130 (continued) The sequence alignment was performed using the ClustalX program. Asterisks indicate positions that have a single, fully conserved residue. Three well-conserved regions of F_1-β-subunit, including the Walker motif A (GXXXXGKT/S), the DELSEED motif, and the ATPASE_ALPHA_BETA signature domain, are in the box. Pfu, oyster (*Pinctada fucata*) F_1-β-subunit, 781482; Cal, (*Caenorhabditis elegans*) F_1-β, AAA19068; Dme, fruit fly (*Drosophila melanogaster*) F_1-β, AAF59391; Xla, African clawed frog (*Xenopus laevis*) F_1-β, AAH46741; Cyc, cyprinoid fish (*Cyprinus carpio*) F_1-β, Q9PTY0; Bta, cow (*Bos taurus*) F_1-β, NP_786990; Hom, human (*Homo sapiens*) F_1-β, AAA51808; Sac, yeast (*Saccharomyces cerevisiae*) F_1-β, NP_012655; Eco, bacteria (*Escherichia coli*) F_1-β, AAA24737

	Sequence identities (%)								
	Bta	Hom	Cye	Xla	Dme	Cal	Eco	Sac	Pfu
Bta	100	95	89	91	83	78	64	71	80
Hom		100	86	89	82	78	63	71	78
Cye			100	92	85	78	65	74	82
Xla				100	83	79	64	72	81
Dme					100	78	67	75	80
Cal						100	60	67	74
Eco							100	64	63
Sac								100	71
Pfu									100

(a)

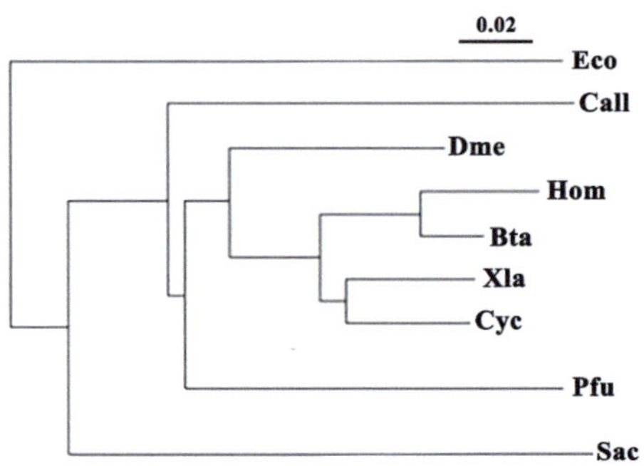

(b)

Fig. 2.131 Phylogenetic analysis of pfATPB and its homology with other species: (**a**) Matrix indicating the percentage identities of the aligned F_1-β-subunit; (**b**) phylogenetic tree showing the evolutionary relationship of F_1-β-subunit with proteins. (The N-J tree was generated using the N-J method in ClustalX1.81.) Abbreviations and sources are described in Fig. 2.130

This work was modified from the papers published by our group (Chen et al., 2005, 142, 142–152; Fan et al., 2007a, 104, 47-54; Fan et al., 2007b, 39, 722-730; Fang et al., 2011; Fang et al., 2008a, 32, 920-927; Fang et al., 2009, 11, 270-279; Fang et al., 2008b, 18, 405-412; Li et al., 2010a, 12, 100–110; Li et al., 2009, 153, 43-53; Li et al., 2010b, 28, 253-260; Li et al., 2005, 272, 4899-4910; Li et al., 2006, 144, 463-471; Li et al., 2004, 138, 235-243; Liu et al., 2015, 16, 325; Wang et al., 2008, 1784, 1514–1523; Yan et al., 2007, 1770, 1338–1344). The related contents are reused with the permission.

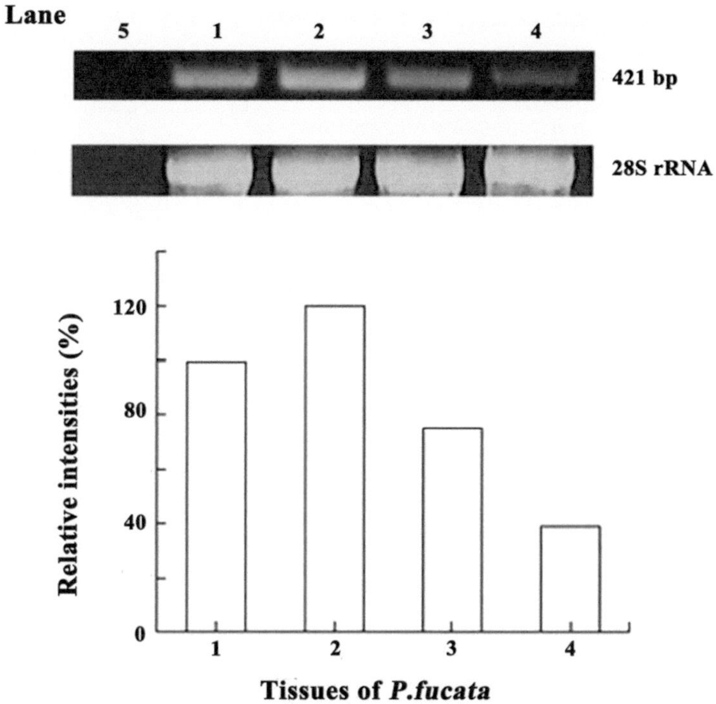

Fig. 2.132 Expression of pfATPB mRNA in *P. fucata* tissues
Semiquantitative RT-PCR was done with RNA samples from adult tissues of *Pinctada fucata*. The expression of the PCR product (421 bp) is shown with the relative intensities of the bands evaluated by densitometry. The 28S rRNA was used as a control with equal quantities of total RNA used in RT-PCR. Lane 1, mantle; lane 2, gill; lane 3, digestive gland; lane 4, gonad; lane 5, negative control

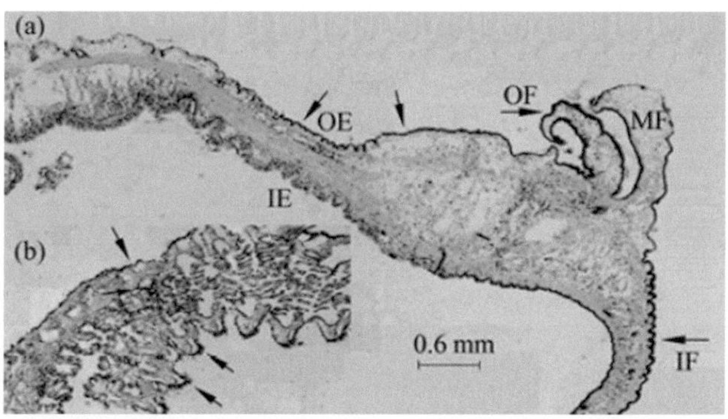

Fig. 2.133 In situ hybridization of oyster ATP5B in the mantle and gill of *P. fucata*
(**a**) Hybridization signals (arrow) were observed in epithelium cells along the periostracal groove, the outer folder, and the dorsal region of the mantle. (**b**) Intensive signals occurred in the gill epithelium cells. IF, inner fold; MF, middle fold; OF, outer fold; OE, outer epithelium; IE, inner epithelium

Supplementary Table 2.1

Chapter	Primer name	Primer sequence
2.2.1	gapdhF:	TCTGCTGATGCTCCTATGTTTG
	gapdhR	TCTGGTATAACTTTGCCTACGG
	ingapdhF	TCTGCTGATGCTCCTATGTTTG
	ingapdhR:	TCTGGTATAACTTTGCCTACGG
	inD1F:	CGTCCCAGCCCTACTTTTTA
	inD2R:	TGACCCTCATTGGATGTTCT
	inD3F:	TGTCCAGTATAAGTCTGTCGA
	inD3R:	TTAAAAGGTCAAAGGTCAAG
	inUF:	TTGTGAGTTGAACTGGTCGGGATC
	inUR:	GAAGCAGTGGCAGCGGAAGC
	inIF:	TGACCCTCGCACTGGTATTG
	inIR:	GATTGGCTCTGATTCGTCCT
	dsD1F:	GCGTAATACGACTCACTATAGGGAGA GATTCAGGGGGAAGCAC
	dsD1R:	GCGTAATACGACTCACTATAGGGAGA ATACGACGCAAAGCCAC
	dsD2F:	GCGTAATACGACTCACTATAGGGAGA CGTCCCAGCCCTACTTTTTA
	dsD2R:	GCGTAATACGACTCACTATAGGGAGA TGACCCTCATTGGATGTTCT
	dsD3F:	GCGTAATACGACTCACTATAGGGAGA TGTCCAGTATAAGTCTGTCGA
	dsD3R:	GCGTAATACGACTCACTATAGGGAGA TTAAAAGGTCAAAGGTCAAG
	dsUF:	GCGTAATACGACTCACTATAGGGAGA TTGTGAGTTGAACTGGTCGGGATC
	dsUR:	GCGTAATACGACTCACTATAGGGAGA GAAGCAGTGGCAGCGGAAGC
	dsIF:	GCGTAATACGACTCACTATAGGGAGA TGACCCTCGCACTGGTATTG
	dsIR:	GCGTAATACGACTCACTATAGGGAGA GATTGGCTCTGATTCGTCCT
	dsKF:	GCGTAATACGACTCACTATAGGGAGA AAGTTCGCCGCTGTTTTGGC
	dsKR:	GCGTAATACGACTCACTATAGGGAGA CATATCCACCACTGGATCCTGG
	dsNF:	GCGTAATACGACTCACTATAGGGAGA GGCTTTGGCGACGAACCGGA
	dsNR:	GCGTAATACGACTCACTATAGGGAGA ACACGGGGGAGTGGTCAGGG
	dsEF:	GCGTAATACGACTCACTATAGGGAGA ATGGTGAGCAAGGGCGAGGAG

(continued)

Chapter	Primer name	Primer sequence
	dsER	GCGTAATACGACTCACTATAGGGAGA TTACTTGTACAGCTCGTCCATG
	qD1F:	GAAGCACGAATCCTAGAG
	qD1R:	TTCACATACGACGCAAAG
	qD2F:	TCCGCAATATAATGAACAAGT
	qD2R:	TAGTAGACGCCACAATGAA
	qD3F:	GTTTGTCTGTTTGTCTGTCT
	qD3R:	CCCTATCACTTCCCATTCA
	qUF:	AGGATGATGGATCAAGCAT
	qUR:	GTGGAGGAAGTAGCAGTAA
	qIF	TAAAGGTCATAGGTCAAAGGT
	qIR	GCTCTGATTCGTCCTCTTA
	qKF:	AAGAAATGTCACCCTTGGGATTGG
	qKR	AATCATCGCCACCATATCCATCG
	qNF:	GGCTTTGGCGACGAACCGGA
	qNR	ACACGGGGGAGTGGTCAGGG
	qPifF	TGCTGCCATCACGTGAGTATG
	qPifR	GACTTCCCTTTCTCACACTTCCA
	qPeF	TACTCATACTGCTGGATA
	qPeR	TATCATCATCGGTGTAAC
	qN19F	CATCAACAAGTTCTCATTACAG
	qN19R	ATCTCCTACGGCTTACAG
	qP39F:	CTGGAATGAGAGGATATG
	qP39R:	TGCTGCTGTAATAACTATA
	qAF	TGATAGTGAAGACGATGA
	qAR	RTGTCATCATCATCATCATC
	qP14F:	TAAGACCATTCTACGGCTAT
	qP14R:	ACCATATCCTCCATATAATCCT
	qβ-actin-f:	CTCCTCACTGAAGCCCCCCTCA
	qβ-actin-R:	ATGGCTGGAATAGGGATTCTGG
2.2.2	actin-F	CTCCTCACTGAAGCCCCCCTCA
	actin-R	ATGGCTGGAATAGGGATTCTGG
	18749-F	GCGAAATGCAGAAACTAGCAG
	18749-R	TAGCTGATGTTGGAGGAGACC
	34354-F	GTGTTCTCTGCACCATCTGTTA
	34354-R	CTGACTCATCTGTTGCCGAAT
	35118-F	CCTTGATGAACACCTGCTGAA
	35118-R	AATCGTAGTTCCACCTCTATTGC
	51738-F	TCACGATCATGTGAATGCAGGA
	51738-R	GTTCCAGTTATTCCACCAGTTGTTC
	56675-F	AAGACGACACTACCAATAGCAA
	56675-R	TACCTTCCGTACATTCTGTTGT
	GFP-dsF	GCGTAATACGACTCACTATAGGGAGAATGGTGAGCAAGGGCGAGGAG
	GFP-dsR	GCGTAATACGACTCACTATAGGGAGATTACTTGTACAGCTCGTCCATG

(continued)

Chapter	Primer name	Primer sequence
	18749-dsF	GCGTAATACGACTCACTATAGGGAGAGCGAAATGCAGAAACTAGCAG
	18749-dsR	GCGTAATACGACTCACTATAGGGAGATACGTGATGTTGGAGGAGACC
	34354-dsF	GCGTAATACGACTCACTATAGGGAGAACGAATGGTCAATGGAGACAG
	34354-dsR	GCGTAATACGACTCACTATAGGGAGACTGACTCATCTGTTGCCGAAT
	35118-dsF	GCGTAATACGACTCACTATAGGGAGAACAGGATGCTACGAATTGGC
	35118-dsR	GCGTAATACGACTCACTATAGGGAGACCCGTATTCCATTTCCATTG
	51738-dsF	GCGTAATACGACTCACTATAGGGAGAGCAGTCCTATTTGCCGTGTTA
	51738-dsR	GCGTAATACGACTCACTATAGGGAGATGTTCCTGTTCCATCTGTTCC
	56675-dsF	GCGTAATACGACTCACTATAGGGAGAAGAAATTCATCGAACGCACC
	56675-dsR	GCGTAATACGACTCACTATAGGGAGATTTTGTGGGCCTTCTGTACC
	RT-18S -F	GGGAATCAGGGTTCGATTCC
	RT-18S -R	CCTGCTGCCTTCCTTGGA
	RT-Pif-F	TGCTGCCATCACGTGAGTATG
	RT-Pif-R	GACTTCCCTTTCTCACACTTCCA
	RT-ACCBP-F	GACATGGAACAAAGATGGTGGA
	RT-ACCBP-R	CTGTGGCTGGAATGGTTGG
	RT-P39-F	CTGGAATGAGAGGATATG
	RT-P39-R	TGCTGCTGTAATAACTATA
	RT-Pearlin-F	TGCGGACGTTACTCATACTGCT
	RT-Pearlin-R	CCACATCTAAGCCACTCATACCTCT
	RT-MSI60-F	AGGTGGACGAAGAGGTAGAGGTAG
	RT-MSI60-R	AGCGGCAGCGGCAACATC
	RT-KRMP1-F	TTTCCTCCTCCTTGGAGCATTC
	RT-KRMP1-R	CCAAAGTTGTAATCATCGCCACC
	RT-Tyrosinase-F	AGGAGGGATGCCATTTGATA
	RT-Tyrosinase-R	TGGAATGAATTATGGTGTTTGG
	RT-Chitin syn-F	CCAAATTGCCAACCATCTG
	RT-Chitin syn-R	GAAAGACTCTCCCTCAGACTATGG
	RT Smad5-F	ATGGGCGGAGAAAGCCGTAGA
	RT Smad5-R	GCCCTTACGATGGGACACTTG
	RT Smad3-F	TATTGAAGCACCAGTGTTGCCTCC
	RT Smad3-R	TCTAAAGGTGGGCTTGGTGAAACG
	RT Smad4-F	CCTGGATAGGTTCTGTTTGGG
	RT Smad4-r	TCTGACGGTGACACTGACGAA
2.3.1	forward primer F1	ATYGCWGARTTYAARGARGC
	reverse primer R1	CCRTCWCCATCAATRTCHGC
	forward primer F2	GAATTCCTTACAATGATGGC
	reverse primer R2	TTTCGACATCATCATTTTCAC

(continued)

Chapter	Primer name	Primer sequence
	oligo-dT adaptor	$TCGAATTCGGATCCGAGCTCT_{17}V$
	primer G1	TCAAGGAAGCGTTCAGTTT
	primer G2	CCGTCACCGTCAATATCAG
	F3	GGATCCATGGCCGATCAGCTGACAGAG
	R3	CGGGATCCTCATTTCGACATCATCATTTTCAC
2.3.2	oligo-dT adaptor primer	TCGAATTCGGATCCGAGCTCVT17
	primer F1	ATYGCWGARTTYAARGARGC
	primer R1	CCRTCWCCATCAATRTCHGC
	primer LS11	TCTCGTGGAAGAAATCGACA
	primer R2	TCGAATTCGGATCCGAGCTC
	primer LS12	CACAGACGGCAATGGAGAGG
	LSG1	CTACCATCTCTTCTGCTTCTTCGTCGTCGTCC
	LSG2	CCAAGAACTCGTTGAAATCAACC
	primer UPM	CTAATACGACTCACTATAGGGCAAGCAGTGGTAACAACGCAGAGT; CTAATACGACTCACTATAGGGC
	primers NUP	AAGCAGTGGTAACAACGCAGAGT
	P3	GGAAGAATACAGACACGGACAG
	P4	ATAACAACAGTTTATACATCGCTTC
	G1	ATGGCGGAAGATCTCACAGAAGAACAAA
	G2	TCATTTATTTTCTTGTTGCTGTTC
	P5	GGATCCATGGCGGAAGATCTCACA
	P8	CAGCTCGAGTTTATTTTCTTGTTGCTGTTC
2.3.3	P5	GGA TCC ATGGCG GAA GAT CTC ACA
	P10	CGG GAT CCT CAT TTC ATT GAAATC ATT TTG AC
2.3.4	RTCaM-f	ATGGCCGATCAGCTGACAGA
	RTCaM-r	TCATTTCGACATCATCATTT
	RTCaLP-f	ATGGCGGAAGATCTCACAGA
	RTCaLP-r	TCATTTATTTTCTTGTTGCT
	RTGAPDH-f	TATTTCTGCACCGTCTGCTG
	RTGAPDH-r	ATCTTGGCGAGTGGAGCTAA
	RTOP-f	ACACTTTCACTCCAATCGTCC
	RTOP-r	TGCCCTTTGGCTTGTTGTCC
	RTGAPDH-f	TATTTCTGCACCGTCTGCTG
	RTGAPDH-r	ATCTTGGCGAGTGGAGCTAA
	RTCaM-f	CGCGGATCCATGGCCGATCAGCTGACAGA
	RTCaM-r	CCGCTCGAGTCATTTCGACATCATCATTT
	RTCaLP-f	CGCGGATCCATGGCGGAAGATCTCACAGA
	RTCaLP-r	CCGCTCGAGTCATTTCATTGAAATCATTTTGAC
	RTM-CaLP-f	CGCGGATCCATGGCGGAAGATCTCACAGA
	RTM-CaLP-r	CCGCTCGAGTCATTTCATTGAAATCATTTTGAC
2.3.5	(GFP)-CaMWT-f	CCGCTCGAGCTATGGCCGATCAGCTGACAGAG

(continued)

Chapter	Primer name	Primer sequence
	(GFP)-CaMWT-r	CGGGGTACCTCATTTCGACATCATCATTTT
	(GFP)-CaLP-WT-f	CCGCTCGAGCTATGGCGGAAGATCTCACAGAA
	(GFP)-CaLP-WT-r	CGGGGTACCTCATTTATTTTCTTGTTGCTG
	terminal reverse oligonucleotide	CGGGGTACCTCATTTCATTGAAATCATTTTGAC
	CaLP-D-f	CCGCTCGAGCTATGGCGGAAGATCTCACAGAA
	CaLP-D-r	CGGGGTACCTCATTTCATTTTCTTAGCCATCAT
	CaLP-D2-f	CCGCTCGAGCTATGGACACCGACTCGGAAGAGGAA
	reverse primers1	TTCTTCTTGATCGGTGTCTTTCGACATCATCATTTTT
	reverse primers2	GGGGTACCTCATTATTTTCTTGTTGCTGTTCTTCTTGATCGGT
	Red-f	CCGCTCGAGATGTCAGAACCGGCTGGGGAT
	Red-r	CGGGGTACCGTGGGCTTCCTCTTGGAGAAGAT
2.3.6	P5	GGATCCATGGCGGAAGATCTCACA
	P6	CGGGATCCTCATTTCATTGAAATCATTTTGAC
	M2	GGATCCATGGACACCGACTCGGAAGAGGAA
	P10	CGGATCCTCATTTCATTGAAATCATTTTGAC
	M1	CGGGATCCTCATTTCATTTTCTTAGCCATC
2.3.8	forward oligonucleotide	GGAGAAATAAATGCTGAGGAGTTTGTC
	reverse oligonucleotides	GACAAACTCCTCAGCATTTATTTCTCC
2.3.9	forward primer F1	GGVAARGGVAGYTTYGGVAAG
	reverse primer R 1	ATGTAGTCHGGRGTBCCRCA
	R2	TTCACCACCTGTAGCAAGCTCCATTACC
	F2	CACGGGTATTGA ACATGGTTCTAG
	F3	CCTATGCAATCA AGATGATAG
	R3	CCCCGGGACTCACTACCAAC
2.4.1	APF	CCNGTNACNGTNTGYGGNGA
	APR	GGNARCCARTANGGRTGNGG
	A3RSP	AGAGCACACGAAGCACAAGATG
	A5RSP	CCCCTGTCCACATAATCTCC
	ACF4	TGGTTTCATAAAATCTGAAGCCCCA
	ACR2	ATGAGCTGGCCACGGAGTC
	BFP2	GGNGARGTNGAYTTYAARGARTT
	BRP2	ARRTTRTTNCCNACCATCAT
	B3RF2	GTGTTGAAAATGATGGTCGG
	B5RR2	AGCCGTCCTTGTCCATATCG
	BCF2	CTGTTTCACAAATAGTTTTAGCAGC

(continued)

Chapter	Primer name	Primer sequence
	BCR2	GGTAAGGGCAACTTATCAAAGG
	AE5	TGCCGCCCATATGGCAACAACAGATTCTAAG
	AE33	CCGGAATTCTCAGTGATGATGATGATGATGGCTATGGGAATTGTTCAATG
	ART55-2	ATTATTAGAGCACACGAAGCAC
	ART33-2	AGGAGACCATTTGGAGTAAGTC
	BRTF1	TAGCAGCAAAAGAAGTAAAAATG
	BRTR1	GTTCACATTTGCTGTCATTTCAC
	GAPDH5	GCCGAGTATGTGGTAGAATC
	GAPDH3	CACTGTTTTCTGGGTAGCTG
2.4.2	Pf-BMP-25	TTCAAACGAAAACTGTGCCC
	Pf-BMP-23	ATGCCACTCGTTTTCGGATA
	GAPDH5	GCCGAGTATGTGGTAGAATC
	GAPDH3	CACTGTTTTCTGGGTAGCTG
2.4.3	ART55-2	ATTATTAGAGCACACGAAGCAC
	ART33-2	AGGAGACCATTTGGAGTAAGTC
	BRTF1	TAGCAGCAAAAGAAGTAAAAATG
	BRTR1	GTTCACATTTGCTGTCATTTCAC
	GAPDH5	GCCGAGTATGTGGTAGAATC
	GAPDH3	CACTGTTTTCTGGGTAGCTG
2.5.1	BF8	ARYAATGAYTGGTGGAT
	BR8	GCYTTYTGCATCATRTCT
	BGSP3	TCCCAGACTGTCACTGTCCTCTCCAACG
	BGSP4	ATGAAGTCGTGCCTTCCATGCG
	GAPF	GCCGAGTATGTGGTAGAATC
	GAPR	CACTGTTTTCTGGGTAGCTG
	LCabQF1	GAGAATCGCTGATACCTTGC
	LCabQR1	GTGCATACTGTCATGTACATAC
	FBF4	ATGGCTCTTGACGATAAACCAAG
	FBR4	TCATGTACATACCATGGCTAC
2.5.2	primer L1	CTSACMACMAACCAGATGTC
	primer L2	ACVGCVGTDCCDGADCCCAT
	RACE	TGGTGACGGTGTGAATGATGCCCCG
	3′-RACE	TAGCCATCTCCTCAAGTCCAGCATAATC
	L3	GCTGCG AAACGGTGGCTGCTCG
	L4	ATTATACCTTTTACATTCATAAGTG
	L5	ATTTACGTTTTGATTCCTGTCTG
	L6	GCGGAATTCGATTTAATCGACTC
	L7	CATAGCCCGCAAATTTACTGAC
	L8	ATTATACCT TTTACATTCATAAGTG
	L9	ATTTACGTTTTGATTCCTGTCTG
	L10	GCGGAATTCGATTTAATCGACTC
	L11	GCCGAGTATGTGGTAGAATC
	L12	CACTGTTTTCTGGGTAGCTG

(continued)

Chapter	Primer name	Primer sequence
2.5.3	CRTF1	CAYGAGCAGAACATCGACTG
	CRTR1	TTCCAYTCVCCCTTGTACTC
	GCRTF1	CTAAGAAGCCTGAAGATTGGGATGACCG
	GCRTR1	TGGTGACGGTGTGAATGATGCCCCG
	CRTQF1	CAGAAGTCAACAATCATACAGCG
	CRTQR1	GAAAAGAGTGTCAACTATTTCGG
	CRTSF1	GACAACACGTACGAAGTCAG
	CRTSR1	ATCGTCATATTTATATAATTTATCATC
	GAPF	GCGGAATTCGATTTAATCGACTC
	GAPR	CACTGTTTTCTGGGTAGCTG
2.5.4	PAJF1	GATCCHATGCTVCTCTCWG
	PAJR1	CGACAYTCBGTYTTATTGCC
	PAF1	GGATATACATCACGCATAATG
	PAR1	CATCATDACGAANGTGTTGAAGAT
	PA5R1	TACATTGACTCCGCCCTGAGACCGTGC
	PA5R2	GCACCGAGCAGAGCGAAGATAATACCAG
	PA3F1	ATCATGGACACGCTCGCTTCA
	PA3F2	ACATGATGAAAAATATCCTCGGC
	PF1	TTCAGGTGTGTATTGTGGA
	PR1	TAGAAGTTGGACTCTGCGT
	PBF1	GGCTGTAACAGGAGATGGT
	PBR1	TAGTGAAGCGAGCGTGTCC
	GAPDHF	GATGGTGCCGAGGAGTATGTGGTA
	GAPDHR	CGTTGATTATCTTGGCGAGTG
2.5.5	GBF	TGGGTI ATG(A/G)C(C/A)TGTGCITA
	GBR	A(A/G/C)(T/C)TTG(G/A)(C/T/A)I(G/C)(A/T)(G/A)GCATCACA
	GSP1	CTCAAGACAAGGGAAGGAAA
	GSP2	CTGGTGACAAAGACAGACTC
	GSP3	CCTGTAAATGCTGTCGTCTGCTGACCT
	BeF	CGCGGATCCATGAGTTCAGAATTAGAAGCATTG
	BeR	CCGCTCGAGCTATCAGTTCCA AATCCGGAGGAAGC
	G2F	CCGGAATTCGGATGGCCAGCAACAACAACGCCAGC
	G2R	CCGCTCGAGTCATTA AAGGATGGCGCAGAAGAAC
	CaMF	CGCGGATCCATGGCCGATCAGCTGACAGAG
	CaMR	CCGCTCGAGTCAAGCATAATCTGGCACATCATACCCTTTCGACATCATCAT TTTC
2.5.6	ATPF1	ATGAAG(T/C)T(G/C)GA(G/A)TT(G/A)GCICAGT
	ATPR1	AC(A/T)(G/C)(C/T)(A/T/G/C)ACTTGTTCTTCAAT
	5¢ATP	AGTAACTCGGTCAGACGAACTCCCCT
	3ATP	ACAGGGGAGTTCGTCTGACCGAGTTAC
2.5.7	FBF1	GCC AA(G/A)GC(T/C)CATGG(T/A)GG(A/T)TACTC
	FBR1	TCATCCATACCCA(A/G)(A/G)ATGGCA
	FBF2	CTGACGATTTGACAGATCC
	FBR2	CAGTGGTGGCGTCCA AGTGAGC

(continued)

Chapter	Primer name	Primer sequence
	FBF3	AAGCAGTGGTATCAACGCAGA G
	FBR3	TGATCCATAGAGAATGCCTTA CG
	FBF4	AGAGGTGTCCGCTTTGT
	FBR4	TCTGGTGACGGTTCCTTTGT

References

1. G. Pertea et al., TIGR Gene Indices clustering tools (TGICL): a software system for fast clustering of large EST datasets. Bioinformatics **19**, 651–652 (2003). https://doi.org/10.1093/bioinformatics/btg034
2. A. Conesa et al., Blast2GO: a universal tool for annotation, visualization and analysis in functional genomics research. Bioinformatics **21**, 3674–3676 (2005). https://doi.org/10.1093/bioinformatics/bti610
3. J. Ye et al., WEGO: a web tool for plotting GO annotations. Nucleic Acids Res. **34**, W293–W297 (2006). https://doi.org/10.1093/nar/gkl1031
4. C. Iseli, C.V. Jongeneel, P. Bucher, ESTScan: a program for detecting, evaluating, and reconstructing potential coding regions in EST sequences. Proc Int. Conf. Intell. Syst. Mol. Biol., 138–148 (1999)
5. H. Nielsen, J. Engelbrecht, S. Brunak, G. vonHeijne, Identification of prokaryotic and eukaryotic signal peptides and prediction of their cleavage sites. Protein Eng. **10**, 1–6 (1997). https://doi.org/10.1093/protein/10.1.1
6. B. Eisenhaber, P. Bork, F. Eisenhaber, Prediction of potential GPI-modification sites in proprotein sequences. J. Mol. Biol. **292**, 741–758 (1999). https://doi.org/10.1006/jmbi.1999.3069
7. O. Emanuelsson, H. Nielsen, S. Brunak, G. von Heijne, Predicting subcellular localization of proteins based on their N-terminal amino acid sequence. J. Mol. Biol. **300**, 1005–1016 (2000). https://doi.org/10.1006/jmbi.2000.3903
8. O. Emanuelsson, S. Brunak, G. von Heijne, H. Nielsen, Locating proteins in the cell using TargetP, SignalP and related tools. Nat. Protoc. **2**, 953–971 (2007). https://doi.org/10.1038/nprot.2007.131
9. A.M. Newman, J.B. Cooper, XSTREAM: a practical algorithm for identification and architecture modeling of tandem repeats in protein sequences. BMC Bioinformat. **8**, Artn 382 (2007). https://doi.org/10.1186/1471-2105-8-382
10. H. Miyamoto et al., The diversity of shell matrix proteins: genome-wide investigation of the pearl oyster, Pinctada fucata. Zool. Sci. **30**, 801–816 (2013). https://doi.org/10.2108/zsj.30.801
11. M. Kanehisa, S. Goto, S. Kawashima, A. Nakaya, The KEGG databases at GenomeNet. Nucleic Acids Res. **30**, 42–46 (2002). https://doi.org/10.1093/nar/30.1.42
12. W.Q. Huang et al., Immunohistochemical and in situ hybridization studies of gonadotropin releasing hormone (GnRH) and its receptor in rat digestive tract. Life Sci. **68**, 1727–1734 (2001)
13. M. Suzuki et al., An acidic matrix protein, Pif, is a key macromolecule for nacre formation. Science **325**, 1388–1390 (2009)
14. W.H. Burgess, D.K. Jemiolo, R.H. Kretsinger, Interaction of calcium and calmodulin in the presence of sodium dodecyl-sulfate. Biochim. Biophys. Acta **623**, 257–270 (1980). https://doi.org/10.1016/0005-2795(80)90254-8

15. K. Maruyama, T. Mikawa, S. Ebashi, Detection of calcium-binding proteins by Ca-45 autoradiography on nitrocellulose membrane after sodium dodecyl-sulfate gel-electrophoresis. J Biochem-Tokyo **95**, 511–519 (1984). https://doi.org/10.1093/oxfordjournals.jbchem.a134633
16. W.S. VanScyoc et al., Calcium binding to calmodulin mutants monitored by domain-specific intrinsic phenylalanine and tyrosine fluorescence. Biophys. J. **83**, 2767–2780 (2002). https://doi.org/10.1016/S0006-3495(02)75286-7
17. B. Mehul, D. Bernard, L. Simonetti, M.A. Bernard, R. Schmidt, Identification and cloning of a new calmodulin-like protein from human epidermis. J. Biol. Chem. **275**, 12841–12847 (2000)
18. A. Checa, A new model for periostracum and shell formation in Unionidae (Bivalvia, Mollusca). Tissue Cell **32**, 405–416 (2000). https://doi.org/10.1054/tice.2000.0129
19. A. Caceres, P. Bender, L. Snavely, L.I. Rebhun, O. Steward, Distribution and subcellular localization of calmodulin in adult and developing brain tissue. Neuroscience **10**, 449–461 (1983)
20. A. Ababou, R.A. Shenvi, J.R. Desjarlais, Long-range effects on calcium binding and conformational change in the N-domain of calmodulin. Biochemistry **40**, 12719–12726 (2001). https://doi.org/10.1021/bi010405b
21. A. Ababou, J.R. Desjarlais, Solvation energetics and conformational change in EF-hand proteins. Protein Sci. **10**, 301–312 (2001). https://doi.org/10.1110/ps.33601
22. M.L. Billingsley, K.R. Pennypacker, C.G. Hoover, D.J. Brigati, R.L. Kincaid, A rapid and sensitive method for detection and quantification of calcineurin and calmodulin-binding proteins using biotinylated calmodulin. Proc. Natl. Acad. Sci. U. S. A. **82**, 7585–7589 (1985)
23. R. Jahn, W. Schiebler, P. Greengard, A quantitative dot-immunobinding assay for proteins using nitrocellulose membrane filters. P. Natl. Acad. Sci-Biol. **81**, 1684–1687 (1984). https://doi.org/10.1073/pnas.81.6.1684
24. S. Li, L.P. Xie, Z.J. Ma, R.Q. Zhang, cDNA cloning and characterization of a novel calmodulin-like protein from pearl oyster Pinctada fucata. FEBS J. **272**, 4899–4910 (2005). https://doi.org/10.1111/j.1742-4658.2005.04899.x
25. G.F. Xu, N. Yao, I.A. Aksay, J.T. Groves, Biomimetic synthesis of macroscopic-scale calcium carbonate thin films. Evidence for a multistep assembly process. J. Am. Chem. Soc. **120**, 11977–11985 (1998). https://doi.org/10.1021/ja9819108
26. F. Nudelman, B.A. Gotliv, L. Addadi, S. Weiner, Mollusk shell formation: mapping the distribution of organic matrix components underlying a single aragonitic tablet in nacre. J. Struct. Biol. **153**, 176–187 (2006). https://doi.org/10.1016/j.jsb.2005.09.009
27. D. Tedesco, J. Lukas, S.I. Reed, The pRb-related protein p130 is regulated by phosphorylation-dependent proteolysis via the protein-ubiquitin ligase SCFSkp2. Genes Dev. **16**, 2946–2957 (2002). https://doi.org/10.1101/gad.1011202
28. L. Canesi, C. Pruzzo, R. Tarsi, G. Gallo, Surface interactions between Escherichia coli and hemocytes of the Mediterranean mussel Mytilus galloprovincialis Lam. leading to efficient bacterial clearance. Appl. Environ. Microbiol. **67**, 464–468 (2001). https://doi.org/10.1128/Aem.67.1.464-468.2001
29. G.E. Hofmann, M.J. O'Donnell, A.E. Todgham, Using functional genomics to explore the effects of ocean acidification on calcifying marine organisms. Mar. Ecol. Prog. Ser. **373**, 219–225 (2008). https://doi.org/10.3354/meps07775
30. M. Suzuki, S. Sakuda, H. Nagasawa, Identification of chitin in the prismatic layer of the shell and a chitin synthase gene from the Japanese pearl oyster, Pinctada fucata. Biosci. Biotechnol. Biochem. **71**, 1735–1744 (2007). https://doi.org/10.1271/bbb.70140
31. I.M. Weiss, V. Schonitzer, N. Eichner, M. Sumper, The chitin synthase involved in marine bivalve mollusk shell formation contains a myosin domain. FEBS Lett. **580**, 1846–1852 (2006). https://doi.org/10.1016/j.febslet.2006.02.044
32. A. Shechter et al., A gastrolith protein serving a dual role in the formation of an amorphous mineral containing extracellular matrix. P. Natl. Acad. Sci. U.S.A. **105**, 7129–7134 (2008). https://doi.org/10.1073/pnas.0800193105

33. I.M. Weiss, V. Schonitzer, The distribution of chitin in larval shells of the bivalve mollusk Mytilus galloprovincialis. J. Struct. Biol. **153**, 264–277 (2006). https://doi.org/10.1016/j.jsb.2005.11.006
34. A. Wynn, T.H. Shafer, Four differentially expressed cDNAs in Callinectes sapidus containing the Rebers-Riddiford consensus sequence. Comp. Biochem. Physiol. B **141**, 294–306 (2005). https://doi.org/10.1016/j.cbpc.2005.04.002
35. H. Inoue, T. Ohira, N. Ozaki, H. Nagasawa, A novel calcium-binding peptide from the cuticle of the crayfish, Procambarus clarkii. Biochem. Biophys. Res. Commun. **318**, 649–654 (2004). https://doi.org/10.1016/j.bbrc.2004.04.075
36. K. Mann, F. Siedler, Amino acid sequences and phosphorylation sites of emu and rhea eggshell C-type lectin-like proteins. Comp. Biochem. Physiol. B **143**, 160–170 (2006). https://doi.org/10.1016/j.cbpb.2005.11.003
37. K. Mann, F. Siedler, Ostrich (Struthio camelus) eggshell matrix contains two different C-type lectin-like proteins. Isolation, amino acid sequence, and posttranslational modifications. Bba-Proteins Proteom. **1696**, 41–50 (2004). https://doi.org/10.1016/j.bbapap.2003.09.006
38. M.R. Illies, M.T. Peeler, A.M. Dechtiaruk, C.A. Ettensohn, Identification and developmental expression of new biomineralization proteins in the sea urchin Strongylocentrotus purpuratus. Dev. Genes Evol. **212**, 419–431 (2002). https://doi.org/10.1007/s00427-002-0261-0
39. I.M. Weiss, S. Kaufmann, K. Mann, M. Fritz, Purification and characterization of perlucin and perlustrin, two new proteins from the shell of the mollusc Haliotis laevigata. Biochem. Biophys. Res. Commun. **267**, 17–21 (2000). https://doi.org/10.1006/bbrc.1999.1907
40. K. Muramoto, H. Yako, K. Murakami, S. Odo, H. Kamiya, Inhibition of the growth of calcium-carbonate crystals by multiple lectins in the celomic fluid of the Acorn Barnacle Megabalanus-Rosa. Comp. Biochem. Physiol. B **107**, 401–409 (1994). https://doi.org/10.1016/0305-0491(94)90204-6
41. Y.K. Kim et al., Mineralisation of reconstituted collagen using polyvinylphosphonic acid/polyacrylic acid templating matrix protein analogues in the presence of calcium, phosphate and hydroxyl ions. Biomaterials **31**, 6618–6627 (2010). https://doi.org/10.1016/j.biomaterials.2010.04.060
42. J.D. Hartgerink, E. Beniash, S.I. Stupp, Self-assembly and mineralization of peptide-amphiphile nanofibers. Science **294**, 1684–1688 (2001). https://doi.org/10.1126/science.1063187
43. J.L. Arias, M.S. Fernandez, J.E. Dennis, A.I. Caplan, Collagens of the chicken eggshell membranes. Connect. Tissue Res. **26**, 37–45 (1991). https://doi.org/10.3109/03008209109152162
44. S.C. Benson, N.C. Benson, F. Wilt, The organic matrix of the skeletal spicule of sea-urchin embryos. J. Cell Biol. **102**, 1878–1886 (1986). https://doi.org/10.1083/jcb.102.5.1878
45. C. Zhang, L.P. Xie, J. Huang, X.L. Liu, R.Q. Zhang, A novel matrix protein family participating in the prismatic layer framework formation of pearl oyster, Pinctada fucata. Biochem. Biophys. Res. Commun. **344**, 735–740 (2006). https://doi.org/10.1016/j.bbrc.2006.03.179
46. J. Watanabe, M. Akashi, Formation of hydroxyapatite provides a tunable protein reservoir within porous polyester membranes by an improved soaking process. Biomacromolecules **8**, 2288–2293 (2007). https://doi.org/10.1021/bm0702915
47. S. Areva et al., Effect of albumin and fibrinogen on calcium phosphate formation on sol-gel-derived titania coatings in vitro. Chem. Mater. **14**, 1614–1621 (2002). https://doi.org/10.1021/cm0107272
48. B. Huang et al., Facilitated receptor-recognition and enhanced bioactivity of bone morphogenetic protein-2 on magnesium-substituted hydroxyapatite surface. Sci. Rep. **6**, 24323 (2016). https://doi.org/10.1038/srep24323
49. Z. Fang et al., Localization of calmodulin and calmodulin-like protein and their functions in biomineralization in P. fucata. Prog. Nat. Sci. **18**, 405–412 (2008). https://doi.org/10.1016/j.pnsc.2007.11.011

50. Z.G. Yan et al., Biomineralization: functions of calmodulin-like protein in the shell formation of pearl oyster. Bba-Gen Subjects **1770**, 1338–1344 (2007). https://doi.org/10.1016/j.bbagen.2007.06.018
51. T.Q. Le, M. Gochin, J.D.B. Featherstone, W. Li, P.K. DenBesten, Comparative calcium binding of leucine-rich amelogenin peptide and full-length amelogenin. Eur. J. Oral Sci. **114**, 320–326 (2006). https://doi.org/10.1111/j.1600-0722.2006.00313.x
52. S. Li et al., Cloning and expression of a pivotal calcium metabolism regulator: calmodulin involved in shell formation from pearl oyster (Pinctada fucata). Comp. Biochem. Physiol. B **138**, 235–243 (2004). https://doi.org/10.1016/j.cbpc.2004.03.012
53. J. Keith et al., Comparative-analysis of macromolecules in mollusk shells. Comp. Biochem. Physiol. B **105**, 487–496 (1993). https://doi.org/10.1016/0305-0491(93)90078-J
54. K.L. Taft, G.C. Papaefthymiou, S.J. Lippard, Synthesis, structure, and electronic-properties of a mixed-valent dodecairon oxo complex, a model for the biomineralization of ferritin. Inorg. Chem. **33**, 1510–1520 (1994). https://doi.org/10.1021/ic00085a049
55. I. Yamashita, Fabrication of a two-dimensional array of nano-particles using ferritin molecule. Thin Solid Films **393**, 12–18 (2001). https://doi.org/10.1016/S0040-6090(01)01083-5
56. K.K.W. Wong, S. Mann, Biomimetic synthesis of cadmium sulfide-ferritin nanocomposites. Adv. Mater. **8**, 928 (1996). https://doi.org/10.1002/adma.19960081114
57. P. Santambrogio, S. Levi, A. Cozzi, B. Corsi, P. Arosio, Evidence that the specificity of iron incorporation into homopolymers of human ferritin L- and H-chains is conferred by the nucleation and ferroxidase centres. Biochem. J. **314**, 139–144 (1996)
58. Z.J. Ma et al., A novel extrapallial fluid protein controls the morphology of nacre lamellae in the pearl oyster, Pinctada fucata. J. Biol. Chem. **282**, 23253–23263 (2007). https://doi.org/10.1074/jbc.M700001200
59. D. Fang, et al., Identification of genes directly involved in shell formation and their functions in pearl oyster, Pinctada fucata. PLoS One **6**, ARTN e21860 (2011). https://doi.org/10.1371/journal.pone.0021860
60. M.S. Clark, et al., Insights into shell deposition in the Antarctic bivalve Laternula elliptica: gene discovery in the mantle transcriptome using 454 pyrosequencing. BMC Genomics **11**, Artn 362 (2010). https://doi.org/10.1186/1471-2164-11-362
61. T. Takahashi, et al., An EST screen from the annelid Pomatoceros lamarckii reveals patterns of gene loss and gain in animals. BMC Evol. Biol. **9**, Artn 240 (2009). https://doi.org/10.1186/1471-2148-9-240
62. J.C. Vera et al., Rapid transcriptome characterization for a nonmodel organism using 454 pyrosequencing. Mol. Ecol. **17**, 1636–1647 (2008). https://doi.org/10.1111/j.1365-294X.2008.03666.x
63. J.H. Leu, et al., Comparative analysis of differentially expressed genes in normal and white spot syndrome virus infected Penaeus monodon. BMC Genomics **8**, Artn 120 (2007). https://doi.org/10.1186/1471-2164-8-120
64. D.J. Jackson et al., Parallel evolution of nacre building gene sets in molluscs. Mol. Biol. Evol. **27**, 591–608 (2010). https://doi.org/10.1093/molbev/msp278
65. V. Schonitzer, I.M. Weiss, The structure of mollusc larval shells formed in the presence of the chitin synthase inhibitor Nikkomycin Z. BMC Struct. Biol. **7**, Artn 71 (2007). https://doi.org/10.1186/1472-6807-7-71
66. Y. Miyazaki, T. Nishida, H. Aoki, T. Samata, Expression of genes responsible for biomineralization of Pinctada fucata during development. Comp. Biochem. Physiol. B **155**, 241–248 (2010). https://doi.org/10.1016/j.cbpb.2009.11.009
67. D.J. Jackson, et al., A rapidly evolving secretome builds and patterns a sea shell. BMC Biol. **4**, Artn 40 (2006). https://doi.org/10.1186/1741-7007-4-40
68. J. Liang et al., Identification and characterization of the lysine-rich matrix protein family in Pinctada fucata: indicative of roles in shell formation. Mar. Biotechnol. **18**, 645–658 (2016). https://doi.org/10.1007/s10126-016-9724-6

69. A. Matsushiro, T. Miyashita, Evolution of hard-tissue mineralization: comparison of the inner skeletal system and the outer shell system. J. Bone Miner. Metab. **22**, 163–169 (2004). https://doi.org/10.1007/s00774-003-0484-9
70. Z.Y. Yu, L.P. Xie, S.W. Lee, R.Q. Zhang, A novel carbonic anhydrase from the mantle of the pearl oyster (Pinctada fucata). Comp. Biochem. Physiol. B-Biochem. Mol. Biol. **143**, 190–194 (2006). https://doi.org/10.1016/j.cbpb.2005.11.006
71. M. Suzuki et al., Characterization of Prismalin-14, a novel matrix protein from the prismatic layer of the Japanese pearl oyster (Pinctada fucata). Biochem. J. **382**, 205–213 (2004). https://doi.org/10.1042/bj20040319
72. T. Miyashita, R. Takagi, H. Miyamoto, A. Matsushiro, Identical carbonic anhydrase contributes to nacreous or prismatic layer formation in Pinctada fucata (Mollusca : Bivalvia). Veliger **45**, 250–255 (2002)
73. N.P. Gong et al., Immunolocalization of matrix proteins in nacre lamellae and their in vivo effects on aragonitic tablet growth. J. Struct. Biol. **164**, 33–40 (2008). https://doi.org/10.1016/j.jsb.2008.05.009
74. F. Marin, G. Luquet, B. Marie, D. Medakovic, Molluscan shell proteins: primary structure, origin, and evolution. Curr. Top. Dev. Biol. **80**, 209–276 (2008). https://doi.org/10.1016/S0070-2153(07)80006-8
75. Y.W. Kong et al., Cloning and characterization of Prisilkin-39, a novel matrix protein serving a dual role in the prismatic layer formation from the oyster Pinctada fucata. J. Biol. Chem. **284**, 10841–10854 (2009). https://doi.org/10.1074/jbc.M808357200
76. J.P. Cuif, Y. Dauphin, Occurrence of mineralization disturbances in nacreous layers of cultivated pearls produced by Pinctada margaritifera var cumingi from French Polynesia. Comparison with reported shell alterations. Aquat. Living Resour. **9**, 187–193 (1996). https://doi.org/10.1051/alr:1996022
77. V.G. Tusher, R. Tibshirani, G. Chu, Significance analysis of microarrays applied to the ionizing radiation response. P. Natl. Acad. Sci. U.S.A. **98**, 5116–5121 (2001). https://doi.org/10.1073/pnas.091062498
78. K. Nagai, M. Yano, K. Morimoto, H. Miyamoto, Tyrosinase localization in mollusc shells. Comp. Biochem. Physiol. B-Biochem. Mol. Biol. **146**, 207–214 (2007). https://doi.org/10.1016/j.cbpb.2006.10.105
79. C. Zhang, L.P. Xie, J. Huang, L. Chen, R.Q. Zhang, A novel putative tyrosinase involved in periostracum formation from the pearl oyster (Pinctada fucata). Biochem. Biophys. Res. Commun. **342**, 632–639 (2006). https://doi.org/10.1016/j.bbrc.2006.01.182
80. H.L. Liu et al., Identification and characterization of a biomineralization related gene PFMG1 highly expressed in the mantle of Pinctada fucata. Biochemistry **46**, 844–851 (2007). https://doi.org/10.1021/bi061881a
81. S. Berland et al., Coupling proteomics and transcriptomics for the identification of novel and variant forms of mollusk shell proteins: a study with P. margaritifera. ChemBioChem **12**, 950–961 (2011). https://doi.org/10.1002/cbic.201000667
82. Y. Zhang et al., A novel ferritin subunit involved in shell formation from the pearl oyster (Pinctada fucata). Comp. Biochem. Physiol. B-Biochem. Mol. Biol. **135**, 43–54 (2003). https://doi.org/10.1016/s1096-4959(03)00050-2
83. S. Kinoshita, et al., Deep sequencing of ESTs from nacreous and prismatic layer producing tissues and a screen for novel shell formation-related genes in the pearl oyster. PLoS One **6**, ARTN e21238 (2011). https://doi.org/10.1371/journal.pone.0021238
84. F. Badariotti et al., The phylogenetically conserved molluscan chitinase-like protein 1 (Cg-Clp1), homologue of human HC-gp39, stimulates proliferation and regulates synthesis of extracellular matrix components of mammalian chondrocytes. J. Biol. Chem. **281**, 29583–29596 (2006). https://doi.org/10.1074/jbc.M605687200
85. C. Joubert, et al., Temperature and food influence shell growth and mantle gene expression of shell matrix Proteins in the pearl oyster Pinctada margaritifera. PLoS One **9**, ARTN e103944 (2014). https://doi.org/10.1371/journal.pone.0103944

86. R.A. Rose, S.B. Baker, Larval and spat culture of the Western-Australian silverlip or goldlip Pearl Oyster, Pinctada-Maxima Jameson (Mollusca, Pteriidae). Aquaculture **126**, 35–50 (1994). https://doi.org/10.1016/0044-8486(94)90246-1
87. G. Falini, S. Fermani, Chitin mineralization. Tissue Eng. **10**, 1–6 (2004). https://doi.org/10.1089/107632704322791646
88. I.M. Weiss, F. Luke, N. Eichner, C. Guth, H. Clausen-Schaumann, On the function of chitin synthase extracellular domains in biomineralization. J. Struct. Biol. **183**, 216–225 (2013). https://doi.org/10.1016/j.jsb.2013.04.011
89. D. Fang et al., Novel basic protein, PfN23, functions as key macromolecule during nacre formation. J. Biol. Chem. **287**, 15776–15785 (2012). https://doi.org/10.1074/jbc.M112.341594
90. F. Aguilera, C. McDougall, B.M. Degnan, Evolution of the tyrosinase gene family in bivalve molluscs: Independent expansion of the mantle gene repertoire. Acta Biomater. **10**, 3855–3865 (2014). https://doi.org/10.1016/j.actbio.2014.03.031
91. M. Cieslak, M. Reissmann, M. Hofreiter, A. Ludwig, Colours of domestication. Biol. Rev. **86**, 885–899 (2011). https://doi.org/10.1111/j.1469-185X.2011.00177.x
92. L. Cerenius, B.L. Lee, K. Soderhall, The proPO-system: pros and cons for its role in invertebrate immunity. Trends Immunol. **29**, 263–271 (2008). https://doi.org/10.1016/j.it.2008.02.009
93. S.O. Andersen, Insect cuticular sclerotization: a review. Insect Biochem. Mol. **40**, 166–178 (2010). https://doi.org/10.1016/j.ibmb.2009.10.007
94. A. Rabatel, et al., Tyrosine pathway regulation is host-mediated in the pea aphid symbiosis during late embryonic and early larval development. BMC Genomics **14**, Artn 235 (2013). doi: https://doi.org/10.1186/1471-2164-14-235
95. J. Liu et al., Microarray: a global analysis of biomineralization-related gene expression profiles during larval development in the pearl oyster, Pinctada fucata. BMC Genomics **16**, 325 (2015). https://doi.org/10.1186/s12864-015-1524-2
96. B. Marie et al., Different secretory repertoires control the biomineralization processes of prism and nacre deposition of the pearl oyster shell. P. Natl. Acad. Sci. U.S.A. **109**, 20986–20991 (2012). https://doi.org/10.1073/pnas.1210552109
97. C. Pan et al., A novel acidic matrix protein, PfN44, stabilizes magnesium calcite to inhibit the crystallization of aragonite. J. Biol. Chem. **289**, 2776–2787 (2014). https://doi.org/10.1074/jbc.M113.504027
98. T. Samata et al., A new matrix protein family related to the nacreous layer formation of Pinctada fucata. FEBS Lett. **462**, 225–229 (1999). https://doi.org/10.1016/S0014-5793(99)01387-3
99. M. Yano, K. Nagai, K. Morimoto, H. Miyamoto, Shematrin: a family of glycine-rich structural proteins in the shell of the pearl oyster Pinctada fucata. Comp. Biochem. Physiol. B-Biochem. Mol. Biol. **144**, 254–262 (2006). https://doi.org/10.1016/j.cpbp.2006.03.004
100. M. Yano, K. Nagai, K. Morimoto, H. Miyamoto, A novel nacre protein N19 in the pearl oyster Pinctada fucata. Biochem. Biophys. Res. Commun. **362**, 158–163 (2007). https://doi.org/10.1016/j.bbrc.2007.07.172
101. X.F. Lai, H. Gao, J. Kong, Q.Y. Wang, Identification and molecular characterization of a chitin binding-like protein from Fenneropenaeus Chinensis (Osbeck, 1765). Crustaceana **85**, 1133–1143 (2012). https://doi.org/10.1163/156854012x650250
102. M.A. Gregory, R.C. George, T.P. McClurg, The architecture and fine structure of gill filaments in the brown mussel, Perna perna. S. Afr. J. Zool. **31**, 193–207 (1996)
103. H.J. Yuasa, T. Suzuki, M. Yazawa, Structural organization of lower marine nonvertebrate calmodulin genes. Gene **279**, 205–212 (2001). https://doi.org/10.1016/S0378-1119(01)00755-7
104. F. Friedberg, A.R. Rhoads, Evolutionary aspects of calmodulin. IUBMB Life **51**, 215–221 (2001)

105. M.E. Swanson, S.F. Sturner, J.H. Schwartz, Structure and expression of the Aplysia-Californica Calmodulin gene. J. Mol. Biol. **216**, 545–553 (1990). https://doi.org/10.1016/0022-2836(90)90382-V
106. H. Toda et al., Amino-acid-sequence of Calmodulin from Scallop (Patinopecten) adductor muscle. J Biochem-Tokyo **90**, 1493–1505 (1981)
107. M. Rousseau et al., Biomineralisation markers during a phase of active growth in Pinctada margaritifera. Comp Biochem Physiol A Mol Integr Physiol **135**, 271–278 (2003)
108. V.K. Dimitriadis, G.P. Domouhtsidou, E. Raftopoulou, Localization of Hg and Pb in the palps, the digestive gland and the gills in Mytilus galloprovincialis (L.) using autometallography and X-ray microanalysis. Environ. Pollut. **125**, 345–353 (2003). https://doi.org/10.1016/S0269-7491(03)00122-2
109. E.W. Stommel, R.E. Stephens, H.R. Masure, J.F. Head, Specific localization of scallop gill epithelial calmodulin in cilia. J. Cell Biol. **92**, 622–628 (1982). https://doi.org/10.1083/jcb.92.3.622
110. E.W. Stommel, R.E. Stephens, Cyclic-Amp and calcium in the differential control of Mytilus gill cilia. J. Comp. Physiol. A. **157**, 451–459 (1985). https://doi.org/10.1007/Bf00615145
111. A.T. Marshall, Calcification in hermatypic and ahermatypic corals. Science **271**, 637–639 (1996). https://doi.org/10.1126/science.271.5249.637
112. D. Zoccola et al., Cloning of a calcium channel alpha 1 subunit from the reef-building coral, Stylophora pistillata. Gene **227**, 157–167 (1999). https://doi.org/10.1016/S0378-1119(98)00602-7
113. M. Pekkarinen, I. Valovirta, Histochemical and X-ray studies on tissue concretions and shells of Margaritifera margaritifera (Linnaeus). J. Shellfish Res. **16**, 169–177 (1997)
114. J.A. Rhyner, M. Koller, I. Durusselgerber, J.A. Cox, E.E. Strehler, Characterization of the human calmodulin-like protein expressed in Escherichia-Coli. Biochemistry **31**, 12826–12832 (1992). https://doi.org/10.1021/bi00166a017
115. J.A. Putkey, G.R. Slaughter, A.R. Means, Bacterial expression and characterization of proteins derived from the chicken Calmodulin Cdna and a Calmodulin processed gene. J. Biol. Chem. **260**, 4704–4712 (1985)
116. S. Matsuki et al., High-level expression of complementary-DNA encoding rat calmodulin in Escherichia-Coli. Biotechnol. Appl. Biochem. **12**, 284–291 (1990)
117. N. Hayashi, M. Matsubara, A. Takasaki, K. Titani, H. Taniguchi, An expression system of rat calmodulin using T7 phage promoter in Escherichia coli. Protein Expr. Purif. **12**, 25–28 (1998). https://doi.org/10.1006/prep.1997.0807
118. B.J. Marsden, G.S. Shaw, B.D. Sykes, Calcium-binding proteins elucidating the contributions to calcium affinity from an analysis of species variants and peptide-fragments. Biochem. Cell Biol. **68**, 587–601 (1990)
119. J.A. Putkey, T. Ono, M.F.A. Vanberkum, A.R. Means, Functional-significance of the central helix in calmodulin. J. Biol. Chem. **263**, 11242–11249 (1988)
120. J. Gulati, A. Babu, H. Su, Y.F. Zhang, Identification of the regions conferring calmodulin-like properties to Troponin-C. J. Biol. Chem. **268**, 11685–11690 (1993)
121. T. Yuan, H.J. Vogel, Substitution of the methionine residues of calmodulin with the unnatural amino acid analogs ethionine and norleucine: Biochemical and spectroscopic studies. Protein Sci. **8**, 113–121 (1999)
122. N. Blom, S. Gammeltoft, S. Brunak, Sequence and structure-based prediction of eukaryotic protein phosphorylation sites. J. Mol. Biol. **294**, 1351–1362 (1999). https://doi.org/10.1006/jmbi.1999.3310
123. H.W. Davis, D.L. Crimmins, R.S. Thoma, J.G.N. Garcia, Phosphorylation of calmodulin in the first calcium-binding pocket by myosin light chain kinase. Arch. Biochem. Biophys. **332**, 101–109 (1996). https://doi.org/10.1006/abbi.1996.0321
124. G. Benaim, A. Villalobo, Phosphorylation of calmodulin – functional implications. Eur. J. Biochem. **269**, 3619–3631 (2002). https://doi.org/10.1046/j.1432-1033.2002.03038.x

125. A. Garciagasca, R.I. Ochoabaez, M. Betancourt, Microscopic anatomy of the mantle of the pearl oyster Pinctada-Mazatlanica (Hanley, 1856). J. Shellfish Res. **13**, 85–91 (1994)
126. Z. Fang et al., Significance of the C-terminal globular domain and the extra tail of the calmodulin-like protein (Pinctada fucata) in subcellular localization and protein-protein interaction. Cell Biol. Int. **32**, 920–927 (2008). https://doi.org/10.1016/j.cellbi.2008.04.007
127. I. Niki, H. Yokokura, T. Sudo, M. Kato, H. Hidaka, Ca2+ signaling and intracellular Ca2+ binding proteins. J Biochem-Tokyo **120**, 685–698 (1996)
128. A. Crivici, M. Ikura, Molecular and structural basis of target recognition by calmodulin. Annu. Rev. Bioph. Biom. **24**, 85–116 (1995). https://doi.org/10.1146/annurev.bb.24.060195.000505
129. P. James, T. Vorherr, E. Carafoli, Calmodulin-binding domains – just 2-faced or multifaceted. Trends Biochem. Sci. **20**, 38–42 (1995). https://doi.org/10.1016/S0968-0004(00)88949-5
130. M. Ikura, Calcium binding and conformational response in EF-hand proteins. Trends Biochem. Sci. **21**, 14–17 (1996). https://doi.org/10.1016/0968-0004(96)80879-6
131. M.C. Kilhoffer, J.G. Demaille, D. Gerard, Tyrosine fluorescence of ram testis and octopus calmodulins. Effects of calcium, magnesium, and ionic strength. Biochemistry **20**, 4407–4414 (1981)
132. J. Haiech, Appendix – theoretical approach to study of multiple ligand-binding to a macromolecule. Biochemistry **20**, 3894–3897 (1981)
133. S. Iida, J.D. Potter, Calcium-binding to calmodulin – cooperativity of the calcium-binding sites. J Biochem-Tokyo **99**, 1765–1772 (1986)
134. M. Milos, J.J. Schaer, M. Comte, J.A. Cox, Calcium Proton and calcium magnesium antagonisms in calmodulin – microcalorimetric and potentiometric analyses. Biochemistry **25**, 6279–6287 (1986). https://doi.org/10.1021/bi00368a067
135. M.D. Tsai, T. Drakenberg, E. Thulin, S. Forsen, Is the binding of magnesium(Ii) to calmodulin significant – an Investigation by Magnesium-25 nuclear-magnetic-resonance. Biochemistry **26**, 3635–3643 (1987). https://doi.org/10.1021/bi00386a057
136. S.Y. Ohki, M. Ikura, M.J. Zhang, Identification of Mg2+-binding sites and the role of Mg2+ on target recognition by calmodulin. Biochemistry **36**, 4309–4316 (1997). https://doi.org/10.1021/bi962759m
137. T. Wolf, B. Solomon, D. Ivnitski, J. Rishpon, G. Fleminger, Interactions of calmodulin with metal ions and with its target proteins revealed by conformation-sensitive monoclonal antibodies. J. Mol. Recognit. **11**, 14–19 (1998). https://doi.org/10.1002/(Sici)1099-1352(199812)11:1/6<14::Aid-Jmr382>3.0.Co;2-2
138. A. Malmendal et al., When size is important – accommodation of magnesium in a calcium binding regulatory domain. J. Biol. Chem. **273**, 28994–29001 (1998). https://doi.org/10.1074/jbc.273.44.28994
139. K.B. Seamon, Calcium-dependent and magnesium-dependent conformational states of calmodulin as determined by nuclear magnetic-resonance. Biochemistry **19**, 207–215 (1980). https://doi.org/10.1021/bi00542a031
140. Y. Ogawa, M. Tanokura, Calcium-binding to calmodulin – effects of ionic-strength, Mg-2+, Ph and temperature. J Biochem-Tokyo **95**, 19–28 (1984)
141. J.S. Mills, J.D. Johnson, Metal-ions as allosteric regulators of calmodulin. J. Biol. Chem. **260**, 5100–5105 (1985)
142. M. Milos, M. Comte, J.J. Schaer, J.A. Cox, Evidence for 4 capital and 6 auxiliary cation-binding sites on calmodulin – divalent-cation interactions monitored by direct binding and microcalorimetry. J. Inorg. Biochem. **36**, 11–25 (1989). https://doi.org/10.1016/0162-0134(89)80009-1
143. D. Lafitte, J.P. Capony, G. Grassy, J. Haiech, B. Calas, Analysis of the ion-binding sites of calmodulin by electrospray-ionization mass-spectrometry. Biochemistry **34**, 13825–13832 (1995). https://doi.org/10.1021/bi00042a014
144. M. Cardamone, N.K. Puri, Spectrofluorimetric assessment of the surface hydrophobicity of proteins. Biochem. J. **282**(Pt 2), 589–593 (1992)

145. J. Engel, W. Taylor, M. Paulsson, H. Sage, B. Hogan, Calcium binding domains and calcium-induced conformational transition of SPARC/BM-40/osteonectin, an extracellular glycoprotein expressed in mineralized and nonmineralized tissues. Biochemistry **26**, 6958–6965 (1987)
146. T. Tanaka, H. Hidaka, Hydrophobic regions function in calmodulin-enzyme(S) interactions. J. Biol. Chem. **255**, 1078–1080 (1980)
147. M. Ikura, M. Osawa, J.B. Ames, The role of calcium-binding proteins in the control of transcription: structure to function. Bioessays **24**, 625–636 (2002). https://doi.org/10.1002/bies.10105
148. J.A. Putkey et al., Genetically engineered calmodulins differentially activate target enzymes. J. Biol. Chem. **261**, 9896–9903 (1986)
149. L. Tabernero et al., The structure of a calmodulin mutant with a deletion in the central helix: implications for molecular recognition and protein binding. Structure **5**, 613–622 (1997). https://doi.org/10.1016/S0969-2126(97)00217-7
150. H.C. Yang, M.M. Reedy, C.L. Burke, G.M. Strasburg, Calmodulin interaction with the skeletal-muscle sarcoplasmic-reticulum calcium-channel protein. Biochemistry **33**, 518–525 (1994). https://doi.org/10.1021/bi00168a017
151. R.C. Venema, H.S. Sayegh, J.D. Kent, D.G. Harrison, Identification, characterization, and comparison of the calmodulin-binding domains of the endothelial and inducible nitric oxide synthases. J. Biol. Chem. **271**, 6435–6440 (1996)
152. Y. Salomon, Citation classic – a highly sensitive adenylate-cyclase assay. Cc/Life Sci, 20–20 (1982)
153. M. Kondo, In vitro effect of chlorpromazine on the mineralization of tooth germ in mice--comparison with that of retinoic acid and HEBP. Nihon Yakurigaku Zasshi **97**, 85–95 (1991)
154. L.D. Quarles, D.A. Yohay, L.W. Lever, R. Caton, R.J. Wenstrup, distinct proliferative and differentiated stages of murine MC3T3-E1 cells in culture – an invitro model of osteoblast development. J. Bone Miner. Res. **7**, 683–692 (1992)
155. T. Bellido, V.Z.C. Borba, P. Roberson, S.C. Manolagas, Activation of the Janus kinase/STAT (signal transducer and activator of transcription) signal transduction pathway by interleukin-6-type cytokines promotes osteoblast differentiation. Endocrinology **138**, 3666–3676 (1997). https://doi.org/10.1210/en.138.9.3666
156. Y. Sasaki, H. Hidaka, Calmodulin and cell-proliferation. Biochem. Biophys. Res. Commun. **104**, 451–456 (1982). https://doi.org/10.1016/0006-291x(82)90658-1
157. C.D. Rasmussen, A.R. Means, Calmodulin is involved in regulation of cell-proliferation. EMBO J. **6**, 3961–3968 (1987)
158. M. Zayzafoon, K. Fulzele, J.M. McDonald, Calmodulin and calmodulin dependent kinase II alpha regulate osteoblast differentiation by controlling c-fos expression. J. Biol. Chem. **280**, 7049–7059 (2005). https://doi.org/10.1074/jbc.M412680200
159. J.J. Bergh, Y.H. Xu, M.C. Farach-Carson, Osteoprotegerin expression and secretion are regulated by calcium influx through the L-type voltage-sensitive calcium channel. Endocrinology **145**, 426–436 (2004). https://doi.org/10.1210/en.2003-0319
160. Y. Wang, M.S. Simonson, Voltage-insensitive Ca2+ channels and Ca2+/calmodulin-dependent protein kinases propagate signals from endothelin-1 receptors to the c-fos promoter. Mol. Cell. Biol. **16**, 5915–5923 (1996)
161. W.B. Bowler et al., Signaling in human osteoblasts by extracellular nucleotides – their weak induction of the c-fos proto-oncogene via Ca2+ mobilization is strongly potentiated by a parathyroid hormone/cAMP-dependent protein kinase pathway independently of mitogen-activated protein kinase. J. Biol. Chem. **274**, 14315–14324 (1999). https://doi.org/10.1074/jbc.274.20.14315
162. G. Zaccone, S. Fasulo, L. Ainis, A. Contini, Localization of calmodulin positive immunoreactivity in the surface epidermis of the Brown Trout, Salmo-Trutta. Histochemistry **91**, 13–16 (1989). https://doi.org/10.1007/Bf00501904
163. J.F. Harper et al., Localization of calmodulin in rat-tissues. P. Natl. Acad. Sci-Biol. **77**, 366–370 (1980). https://doi.org/10.1073/pnas.77.1.366

164. M. Zavortink, M.J. Welsh, J.R. Mcintosh, The distribution of calmodulin in living mitotic cells. Exp. Cell Res. **149**, 375–385 (1983). https://doi.org/10.1016/0014-4827(83)90350-6
165. M.J. Moser, M.R. Flory, T.N. Davis, Calmodulin localizes to the spindle pole body of Schizosaccharomyces pombe and performs an essential function in chromosome segregation. J. Cell Sci. **110**, 1805–1812 (1997)
166. C. Reedmiller, The initial calcification process in shell-regenerating Tegula (Archaeogastropoda). Biol. Bull. **165**, 265–275 (1983). https://doi.org/10.2307/1541368
167. S. Weiner, P.M. Dove, An overview of biomineralization processes and the problem of the vital effect. Rev. Mineral. Geochem. **54**, 1–29 (2003). https://doi.org/10.2113/0540001
168. J.H. Vandermeulen, Studies on reef corals .3. Fine-structural changes of calicoblast cells in Pocillopora-Damicornis during settling and calcification. Mar. Biol. **31**, 69–77 (1975). https://doi.org/10.1007/Bf00390649
169. Y. Kogaya, K. Furuhashi, Ultrastructural distribution of acidic glycosaminoglycans associated with matrix vesicle-mediated calcification in mouse progenitor predentine. Calcif. Tissue Int. **37**, 36–41 (1985). https://doi.org/10.1007/Bf02557676
170. J.E. Dennis, D.A. Carrino, K. Yamashita, A.I. Caplan, Monoclonal antibodies to mineralized matrix molecules of the avian eggshell. Matrix Biol. **19**, 683–692 (2000). https://doi.org/10.1016/S0945-053x(00)00118-9
171. G.K. Hunter, H.A. Goldberg, Modulation of crystal-formation by bone phosphoproteins – role of glutamic acid-rich sequences in the nucleation of hydroxyapatite by bone sialoprotein. Biochem. J. **302**, 175–179 (1994)
172. A. Miyauchi et al., Recognition of osteopontin and related peptides by an alpha-V-beta-3 integrin stimulates immediate cell signals in osteoclasts. J. Biol. Chem. **266**, 20369–20374 (1991)
173. D.T. Denhardt, X.J. Guo, Osteopontin – a protein with diverse functions. FASEB J. **7**, 1475–1482 (1993)
174. S. Li, L.P. Xie, Q.X. Meng, R.Q. Zhang, Significance of the extra C-terminal tail of CaLP, a novel calmodulin-like protein involved in oyster calcium metabolism. Comp. Biochem. Physiol. B **144**, 463–471 (2006). https://doi.org/10.1016/j.cbpb.2006.04.010
175. M. Ikura et al., Secondary structure and side-chain H-1 and C-13 resonance assignments of calmodulin in solution by heteronuclear multidimensional nmr-spectroscopy. Biochemistry **30**, 9216–9228 (1991). https://doi.org/10.1021/bi00102a013
176. M. Taules et al., Calmodulin is essential for cyclin-dependent kinase 4 (Cdk4) activity and nuclear accumulation of cyclin D1-Cdk4 during G(1). J. Biol. Chem. **273**, 33279–33286 (1998). https://doi.org/10.1074/jbc.273.50.33279
177. A. Rodriguez-Vilarrupla et al., Binding of calmodulin to the carboxy-terminal region of p21 induces nuclear accumulation via inhibition of protein kinase C-mediated phosphorylation of Ser153. Mol. Cell. Biol. **25**, 7364–7374 (2005). https://doi.org/10.1128/Mcb.25.16.7364-7374.2005
178. A. Rodriguez-Vilarrupla, et al., Identification of the nuclear localization signal of p21(cip1) and consequences of its mutation on cell proliferation. FEBS Lett. **531**, 319–323 (2002). Pii S0014-5793(02)03549-4. https://doi.org/10.1016/S0014-5793(02)03549-4
179. N. Agell, O. Bachs, N. Rocamora, P. Villalonga, Modulation of the Ras/Raf/MEK/ERK pathway by Ca2+, and calmodulin. Cell. Signal. **14**, 649–654 (2002). Pii S0898-6568(02)00007-4. https://doi.org/10.1016/S0898-6568(02)00007-4
180. U.F. Greber, L. Gerace, Depletion of calcium from the lumen of endoplasmic-reticulum reversibly inhibits passive diffusion and signal-mediated transport into the nucleus. J. Cell Biol. **128**, 5–14 (1995). https://doi.org/10.1083/jcb.128.1.5
181. V. Doye, E. Hurt, From nucleoporins to nuclear pore complexes. Curr. Opin. Cell Biol. **9**, 401–411 (1997). https://doi.org/10.1016/S0955-0674(97)80014-2
182. W.M. Michael, M.Y. Choi, G. Dreyfuss, A nuclear export signal in Hnrnp A1 – a signal-mediated, temperature-dependent nuclear-protein export pathway. Cell **83**, 415–422 (1995)

183. H. Siomi, G. Dreyfuss, A nuclear-localization domain in the Hnrnp A1 protein. J. Cell Biol. **129**, 551–560 (1995). https://doi.org/10.1083/jcb.129.3.551
184. F. Weighardt, G. Biamonti, S. Riva, Nucleo-cytoplasmic distribution of human hnRNP proteins: a search for the targeting domains in hnRNP A1. J. Cell Sci. **108(Pt 2)**, 545–555 (1995)
185. I.G. Macara, Transport into and out of the nucleus. Microbiol. Mol. Biol. R. **65**, 570 (2001). https://doi.org/10.1128/Mmbr.65.4.570-594.2001
186. D. Gorlich, Nuclear protein import. Curr. Opin. Cell Biol. **9**, 412–419 (1997). https://doi.org/10.1016/S0955-0674(97)80015-4
187. G. Dreyfuss, Nuclear export and import of proteins and RNAs. Mol. Biol. Cell **8**, 1357–1357 (1997)
188. M. Brini, E. Carafoli, Calcium signalling: a historical account, recent developments and future perspectives. Cell. Mol. Life Sci. **57**, 354–370 (2000). https://doi.org/10.1007/Pl00000698
189. D. Chin, A.R. Means, Calmodulin: a prototypical calcium sensor. Trends Cell Biol. **10**, 322–328 (2000). https://doi.org/10.1016/S0962-8924(00)01800-6
190. J.L. Joyal, D.L. Crimmins, R.S. Thoma, D.B. Sacks, Identification of insulin-stimulated phosphorylation sites on calmodulin. Biochemistry **35**, 6267–6275 (1996). https://doi.org/10.1021/bi9600198
191. D.B. Sacks, H.W. Davis, J.P. Williams, J.M. Mcdonald, Phosphorylation by casein kinase-II alters the biological-activity of calmodulin. FASEB J. **6**, A335–A335 (1992)
192. A. Benguria, O. Hernandezperera, M.T. Martinezpastor, D.B. Sacks, A. Villalobo, Phosphorylation of calmodulin by the epidermal-growth-factor-receptor tyrosine kinase. Eur. J. Biochem. **224**, 909–916 (1994). https://doi.org/10.1111/j.1432-1033.1994.00909.x
193. D.L. Newton, M.H. Krinks, J.B. Kaufman, J. Shiloach, C.B. Klee, Large-scale preparation of calmodulin. Prep. Biochem. **18**, 247–259 (1988). https://doi.org/10.1080/00327488808062527
194. S.R. Martin et al., Conformational and metal-binding properties of androcam, a testis-specific, calmodulin-related protein from Drosophila. Protein Sci. **8**, 2444–2454 (1999)
195. W. Gonzalez, J. Miksovska, Ca2+ and Mg2+ induced conformational changes in downstream regulatory antagonist modulator (DREAM) measured by extrinsic hydrophobic probes. Biophys. J. **102**, 507a–507a (2012)
196. Q. Wang et al., The extra C-terminal tail is involved in the conformation, stability changes and the N/C-domain interactions of the calmodulin-like protein from pearl oyster Pinctada fucata. Bba-Proteins Proteom. **1784**, 1514–1523 (2008). https://doi.org/10.1016/j.bbapap.2008.06.021
197. C.F. Shuman, R. Jiji, K.S. Akerfeldt, S. Linse, Reconstitution of calmodulin from domains and subdomains: influence of target peptide. J. Mol. Biol. **358**, 870–881 (2006). https://doi.org/10.1016/j.jmb.2006.02.017
198. W. Drabikowski, H. Brzeska, S.Y. Venyaminov, Tryptic fragments of calmodulin Ca-2+-induced and Mg-2+-induced conformational-changes. J. Biol. Chem. **257**, 1584–1590 (1982)
199. R.D. Brokx, H.J. Vogel, Peptide and metal ion-dependent association of isolated helix-loop-helix calcium binding domains: studies of thrombic fragments of calmodulin. Protein Sci. **9**, 964–975 (2000)
200. M. Zhang, T. Tanaka, M. Ikura, Calcium-induced conformational transition revealed by the solution structure of apo calmodulin. Nat. Struct. Biol. **2**, 758–767 (1995). https://doi.org/10.1038/nsb0995-758
201. H. Kuboniwa et al., Solution structure of calcium-free calmodulin. Nat. Struct. Biol. **2**, 768–776 (1995). https://doi.org/10.1038/nsb0995-768
202. B.E. Finn et al., Calcium-induced structural-changes and domain autonomy in calmodulin. Nat. Struct. Biol. **2**, 777–783 (1995). https://doi.org/10.1038/nsb0995-777
203. M. Zayzafoon, K. Fulzele, J.M. McDonald, Calmodulin and calmodulin-dependent kinase IIalpha regulate osteoblast differentiation by controlling c-fos expression. J. Biol. Chem. **280**, 7049–7059 (2005). https://doi.org/10.1074/jbc.M412680200

204. L. Sun et al., Calcineurin regulates bone formation by the osteoblast. P. Natl. Acad. Sci. U.S.A. **102**, 17130–17135 (2005). https://doi.org/10.1073/pnas.0508480102
205. R.E. Zielinski, Calmodulin and calmodulin-binding proteins in plants. Annu. Rev. Plant Physiol. **49**, 697–725 (1998). https://doi.org/10.1146/annurev.arplant.49.1.697
206. G.H. Mao, L.X. Hou, C.B. Ding, S.J. Cui, D.Y. Sun, Characterization of a cDNA coding for an extracellular calmodulin-binding protein from suspension-cultured cells of Angelica dahurica. Planta **222**, 428–437 (2005). https://doi.org/10.1007/s00425-005-1558-9
207. G. Crocker, R.A. Dawson, C.H. Barton, S. Macneil, An extracellular role for calmodulin-like activity in cell-proliferation. Biochem. J. **253**, 877–884 (1988)
208. H.R. Masure, D.J. Oldenburg, M.G. Donovan, R.L. Shattuck, D.R. Storm, The Interaction of Ca-2+ with the calmodulin-sensitive adenylate-cyclase from Bordetella-Pertussis. J. Biol. Chem. **263**, 6933–6940 (1988)
209. M. Michenfelder et al., Characterization of two molluscan crystal-modulating biomineralization proteins and identification of putative mineral binding domains. Biopolymers **70**, 522–533 (2003). https://doi.org/10.1002/bip.10536
210. A. Matsushiro et al., Presence of protein complex is prerequisite for aragonite crystallization in the nacreous layer. Mar. Biotechnol. **5**, 37–44 (2003). https://doi.org/10.1007/s10126-002-0048-3
211. M. Kono, N. Hayashi, T. Samata, Molecular mechanism of the nacreous layer formation in Pinctada maxima. Biochem. Biophys. Res. Commun. **269**, 213–218 (2000)
212. F. Marin, G. Luquet, Molluscan shell proteins. Comptes. Rendus. Palevol. **3**, 469–492 (2004). https://doi.org/10.1016/j.crpv.2004.07.009
213. J.W. Nelson et al., Mucinophilic and chemotactic properties of pseudomonas-aeruginosa in relation to pulmonary colonization in cystic-fibrosis. Infect. Immun. **58**, 1489–1495 (1990)
214. A.M. Belcher et al., Control of crystal phase switching and orientation by soluble mollusc-shell proteins. Nature **381**, 56–58 (1996). https://doi.org/10.1038/381056a0
215. I.M. Weiss, W. Gohring, M. Fritz, K. Mann, Perlustrin, a Haliotis laevigata (abalone) nacre protein, is homologous to the insulin-like growth factor binding protein N-terminal module of vertebrates. Biochem. Biophys. Res. Commun. **285**, 244–249 (2001). https://doi.org/10.1006/bbrc.2001.5170
216. Z. Fang et al., Investigation of phosphorylation site responsible for CaLP (P. fucata) nucleo-cytoplasmic shuttling triggered by overexpression of p21Cip1. Mar. Biotechnol. (NY) **11**, 270–279 (2009). https://doi.org/10.1007/s10126-008-9142-5
217. S. Beirer, T. Hofer, Control of signal transduction cycles: general results and application to the Jak-Stat pathway. Genome Inform. **17**, 152–162 (2006)
218. G. Benaim, V. Cervino, A. Villalobo, Comparative phosphorylation of calmodulin from trypanosomatids and bovine brain by calmodulin-binding protein kinases. Comp. Biochem. Physiol. C **120**, 57–65 (1998). https://doi.org/10.1016/S0742-8413(98)00006-1
219. Y.D. Plancke, E. Lazarides, Evidence for a phosphorylated form of calmodulin in chicken brain and muscle. Mol. Cell. Biol. **3**, 1412–1420 (1983)
220. J.P. Williams et al., Tyrosine-phosphorylated calmodulin has reduced biological-activity. Arch. Biochem. Biophys. **315**, 119–126 (1994). https://doi.org/10.1006/abbi.1994.1479
221. M.K. Saville, M.D. Houslay, Phosphorylation of calmodulin on Tyr(99) selectively attenuates the action of calmodulin antagonists on Type-I cyclic-nucleotide phosphodiesterase activity. Biochem. J. **299**, 863–868 (1994)
222. D.B. Sacks, H.W. Davis, D.L. Crimmins, J.M. Mcdonald, Insulin-stimulated phosphorylation of calmodulin. Biochem. J. **286**, 211–216 (1992)
223. M. Quadroni, P. James, E. Carafoli, Isolation of phosphorylated calmodulin from rat-liver and identification of the in-vivo phosphorylation sites. J. Biol. Chem. **269**, 16116–16122 (1994)
224. P.I. Palomo-Jimenez, S. Hernandez-Hernando, R.M. Garcia-Nieto, A. Villalobo, A method for the purification of phospho(Tyr)calmodulin free of nonphosphorylated calmodulin. Protein Expr. Purif. **16**, 388–395 (1999). https://doi.org/10.1006/prep.1999.1092

225. E. Leclerc et al., Serine/threonine phosphorylation of calmodulin modulates its interaction with the binding domains of target enzymes. Biochem. J. **344**, 403–411 (1999). https://doi.org/10.1042/0264-6021:3440403
226. C. Corti et al., Tyrosine phosphorylation modulates the interaction of calmodulin with its target proteins. Eur. J. Biochem. **262**, 790–802 (1999). https://doi.org/10.1046/j.1432-1327.1999.00441.x
227. D.O. Morgan et al., Control of eukaryotic cell cycle progression by phosphorylation of cyclin-dependent kinases. Cancer J. Sci. Am. **4**, S77–S83 (1998)
228. D.O. Morgan, Cyclin-dependent kinases: Engines, clocks, and microprocessors. Annu Rev Cell Dev Bi **13**, 261–291 (1997). https://doi.org/10.1146/annurev.cellbio.13.1.261
229. D.O. Morgan, Principles of Cdk regulation. Nature **374**, 131–134 (1995). https://doi.org/10.1038/374131a0
230. J.G. Chafouleas, L. Lagace, W.E. Bolton, A.E. Boyd 3rd, A.R. Means, Changes in calmodulin and its mRNA accompany reentry of quiescent (G0) cells into the cell cycle. Cell **36**, 73–81 (1984)
231. M. Umahara et al., Tyrosine phosphorylation of Munc18c regulates platelet-derived growth factor-stimulated glucose transporter 4 translocation in 3T3L1 adipocytes. Endocrinology **149**, 40–49 (2008). https://doi.org/10.1210/en.2006-1549
232. K. Shinmura, P. Tarapore, Y. Tokuyama, K.R. George, K. Fukasawa, Characterization of centrosomal association of nucleophosmin/B23 linked to Crm1 activity. FEBS Lett. **579**, 6621–6634 (2005). https://doi.org/10.1016/j.febslet.2005.10.057
233. S. Verploegen, J.W.J. Lammers, L. Koenderman, P.J. Coffer, Identification and characterization of CKLiK, a novel granulocyte Ca++/calmodulin-dependent kinase. Blood **96**, 3215–3223 (2000)
234. F. Meggio, L.A. Pinna, Phosphorylation of phosvitin by casein kinase-2 provides the evidence that phosphoserines can replace carboxylic amino-acids as specificity determinants. Biochim. Biophys. Acta **971**, 227–231 (1988). https://doi.org/10.1016/0167-4889(88)90196-6
235. D.B. Glass, M.R. Elmaghrabi, S.J. Pilkis, Synthetic peptides corresponding to the site phosphorylated in 6-Phosphofructo-2-Kinase Fructose-2,6-Bisphosphatase as substrates of cyclic nucleotide-dependent protein-kinases. J. Biol. Chem. **261**, 2987–2993 (1986)
236. J.R. Woodgett, K.L. Gould, T. Hunter, Substrate-specificity of protein-kinase-C – use of synthetic peptides corresponding to physiological sites as probes for substrate recognition requirements. Eur. J. Biochem. **161**, 177–184 (1986). https://doi.org/10.1111/j.1432-1033.1986.tb10139.x
237. R. Marshall, D. Glycoproteins, Annu. Rev. Biochem. **41**, 673–& (1972). https://doi.org/10.1146/annurev.bi.41.070172.003325
238. D.A. Towler, J.I. Gordon, S.P. Adams, L. Glaser, The biology and enzymology of eukaryotic protein acylation. Annu. Rev. Biochem. **57**, 69–99 (1988)
239. S.K. Hanks, A.M. Quinn, T. Hunter, The protein-kinase family – conserved features and deduced phylogeny of the catalytic domains. Science **241**, 42–52 (1988). https://doi.org/10.1126/science.3291115
240. S.K. Hanks, T. Hunter, Protein kinases .6. The eukaryotic protein-kinase superfamily – kinase (Catalytic) domain-structure and classification. FASEB J. **9**, 576–596 (1995)
241. R.K. Wierenga, W.G.J. Hol, Predicted nucleotide-binding properties of P21 protein and its cancer-associated variant. Nature **302**, 842–844 (1983). https://doi.org/10.1038/302842a0
242. G. Brede, J. Solheim, G. Troen, H. Prydz, Characterization of PSKH1, a novel human protein serine kinase with centrosomal, Golgi, and nuclear localization. Genomics **70**, 82–92 (2000). https://doi.org/10.1006/geno.2000.6365
243. G. Brede, J. Solheim, E. Stang, H. Prydz, Mutants of the protein serine kinase PSKH1 disassemble the Golgi apparatus. Exp. Cell Res. **291**, 299–312 (2003). https://doi.org/10.1016/j.yexcr.2003.07.009
244. S.M. Thomas, J.S. Brugge, Cellular functions regulated by Src family kinases. Annu Rev Cell Dev Bi **13**, 513–609 (1997). https://doi.org/10.1146/annurev.cellbio.13.1.513

245. T.R. Soderling, J.T. Stull, Structure and regulation of calcium/calmodulin-dependent protein kinases. Chem. Rev. **101**, 2341–2351 (2001). https://doi.org/10.1021/cr0002386
246. A. Hudmon, H. Schulman, Structure-function of the multifunctional Ca2+/calmodulin-dependent protein kinase II. Biochem. J. **364**, 593–611 (2002). https://doi.org/10.1042/Bj20020228
247. J.H.P. Skene, Gap-43 as a calmodulin sponge and some implications for calcium signaling in axon terminals. Neurosci. Res. **13**, S112–S125 (1990)
248. T. Meyer, P.I. Hanson, L. Stryer, H. Schulman, Calmodulin trapping by calcium-calmodulin dependent protein-kinase. Science **256**, 1199–1202 (1992). https://doi.org/10.1126/science.256.5060.1199
249. Y.C. Liu, D.R. Storm, Regulation of free calmodulin levels by neuromodulin – neuron growth and regeneration. Trends Pharmacol. Sci. **11**, 107–111 (1990)
250. P. De Koninck, H. Schulman, Sensitivity of CaM kinase II to the frequency of Ca2+ oscillations. Science **279**, 227–230 (1998). https://doi.org/10.1126/science.279.5348.227
251. S. Sudo et al., Structures of mollusc shell framework proteins. Nature **387**, 563–564 (1997). https://doi.org/10.1038/42391
252. H. Schulman, P.I. Hanson, T. Meyer, Decoding calcium signals by multifunctional caM kinase. Cell Calcium **13**, 401–411 (1992). https://doi.org/10.1016/0143-4160(92)90053-U
253. M. Isshiki, R.G.W. Anderson, Function of caveolae in Ca2+ entry and Ca2+-dependent signal transduction. Traffic **4**, 717–723 (2003). https://doi.org/10.1034/j.1600-0854.2003.00130.x
254. F. Rusnak, P. Mertz, Calcineurin: form and function. Physiol. Rev. **80**, 1483–1521 (2000)
255. T. Muramatsu, P.R. Giri, S. Higuchi, R.L. Kincaid, Molecular-cloning of a calmodulin-dependent phosphatase from murine testis – identification of a developmentally expressed nonneural isoenzyme. P. Natl. Acad. Sci. U.S.A. **89**, 529–533 (1992). https://doi.org/10.1073/pnas.89.2.529
256. M. Uryu, A. Nakatomi, M. Watanabe, R. Hatsuse, M. Yazawa, Molecular cloning of cDNA encoding two subunits of calcineurin from scallop testis: Demonstration of stage-specific expression during maturation of the testis. J Biochem-Tokyo **127**, 739–746 (2000)
257. D. Guerini, C.B. Klee, Cloning of human calcineurin A: evidence for two isozymes and identification of a polyproline structural domain. Proc. Natl. Acad. Sci. U. S. A. **86**, 9183–9187 (1989)
258. T. Kuno et al., Evidence for a second isoform of the catalytic subunit of calmodulin-dependent protein phosphatase (calcineurin A). Biochem. Biophys. Res. Commun. **165**, 1352–1358 (1989)
259. P.R. Giri, S. Higuchi, R.L. Kincaid, Chromosomal mapping of the human genes for the calmodulin-dependent protein phosphatase (calcineurin) catalytic subunit. Biochem. Biophys. Res. Commun. **181**, 252–258 (1991)
260. H. Dammann, S. Hellstern, Q.Y. Husain, R. Mutzel, Primary structure, expression and developmental regulation of a Dictyostelium calcineurin A homologue. Eur. J. Biochem. **238**, 391–399 (1996). https://doi.org/10.1111/j.1432-1033.1996.0391z.x
261. E. Csanky et al., Successful renal transplantation following lung transplantation: a survey of the first Hungarian case. Orv. Hetil. **148**, 2147–2151 (2007). https://doi.org/10.1556/OH.2007.28129
262. D.A. Sass et al., Alendronate prevents cyclosporin A-induced osteopenia in the rat. Bone **21**, 65–70 (1997). https://doi.org/10.1016/S8756-3282(97)00071-9
263. K. Ueki, R.L. Kincaid, Interchangeable associations of calcineurin regulatory subunit isoforms with mammalian and fungal catalytic subunits. J. Biol. Chem. **268**, 6554–6559 (1993)
264. I. Boutet, D. Moraga, L. Marinovic, J. Obreque, P. Chavez-Crooker, Characterization of reproduction-specific genes in a marine bivalve mollusc: influence of maturation stage and sex on mRNA expression. Gene **407**, 130–138 (2008). https://doi.org/10.1016/j.gene.2007.10.005
265. C. Zhang, R.Q. Zhang, Matrix proteins in the outer shells of molluscs. Mar. Biotechnol. **8**, 572–586 (2006). https://doi.org/10.1007/s10126-005-6029-6

266. J.D. Taylor, W.J. Kennedy, The influence of the periostracum on the shell structure of bivalve molluscs. Calcif. Tissue Res. **3**, 274–283 (1969)
267. M. Carre et al., Calcification rate influence on trace element concentrations in aragonitic bivalve shells: evidences and mechanisms. Geochim. Cosmochim. Ac **70**, 4906–4920 (2006). https://doi.org/10.1016/j.gca.2006.07.019
268. C.Z. Li et al., Cloning, characterization and immunolocalization of two subunits of calcineurin from pearl oyster (Pinctada fucata). Comp. Biochem. Physiol. B **153**, 43–53 (2009). https://doi.org/10.1016/j.cbpb.2009.01.008
269. T. Muramatsu, R.L. Kincaid, Inhibition of NF-AT signal transduction events by a dominant-negative form of calcineurin. Biochem. Biophys. Res. Commun. **218**, 466–472 (1996). https://doi.org/10.1006/bbrc.1996.0083
270. M. Moriya, K. Fujinaga, M. Yazawa, C. Katagiri, Immunohistochemical localization of the calcium calmodulin-dependent protein phosphatase, calcineurin, in the mouse testis – its unique accumulation in spermatid nuclei. Cell Tissue Res. **281**, 273–281 (1995). https://doi.org/10.1007/s004410050424
271. J.S. Tash et al., Identification, characterization, and functional correlation of calmodulin-dependent protein phosphatase in sperm. J. Cell Biol. **106**, 1625–1633 (1988). https://doi.org/10.1083/jcb.106.5.1625
272. I. Marigomez, M. Soto, M.P. Cajaraville, E. Angulo, L. Giamberini, Cellular and subcellular distribution of metals in molluscs. Microsc. Res. Tech. **56**, 358–392 (2002). https://doi.org/10.1002/jemt.10040
273. S. Rajalakshmi, A. Mohandas, Copper-induced changes in tissue enzyme activity in a fresh-water mussel. Ecotox Environ. Safe **62**, 140–143 (2005). https://doi.org/10.1016/j.ecoenv.2005.01.003
274. W.E. Robinson, D.K. Ryan, metal interactions within the kidney, gill, and digestive gland of the hard clam, Mercenaria-mercenaria, following laboratory exposure to cadmium. Arch. Environ. Contam. Toxicol. **15**, 23–30 (1986)
275. K. Pynnonen, D.A. Holwerda, D.I. Zandee, Occurrence of calcium concretions in various tissues of fresh-water mussels, and their capacity for cadmium sequestration. Aquat. Toxicol. **10**, 101–114 (1987). https://doi.org/10.1016/0166-445x(87)90017-8
276. C.Z. Li et al., Calcineurin plays an important role in the shell formation of pearl oyster (Pinctada fucata). Mar. Biotechnol. **12**, 100–110 (2010). https://doi.org/10.1007/s10126-009-9204-3
277. J. Aramburu, A. Rao, C.B. Klee, Calcineurin: from structure to function. Curr. Top. Cell. Regul. **36**, 237–295 (2000)
278. J. Aramburu, J. Heitman, G.R. Crabtree, Calcineurin: a central controller of signalling in eukaryotes – workshop on the calcium/calcineurin/NFAT pathway: Regulation and function. EMBO Rep. **5**, 343–348 (2004). https://doi.org/10.1038/sj.embor.7400133
279. A.A. Stewart, T.S. Ingebritsen, A. Manalan, C.B. Klee, P. Cohen, Discovery of a Ca2+- and calmodulin-dependent protein phosphatase: probable identity with calcineurin (CaM-BP80). FEBS Lett. **137**, 80–84 (1982)
280. C.B. Klee, H. Ren, X. Wang, Regulation of the calmodulin-stimulated protein phosphatase, calcineurin. J. Biol. Chem. **273**, 13367–13370 (1998)
281. C.B. Klee, T.H. Crouch, M.H. Krinks, Calcineurin: a calcium- and calmodulin-binding protein of the nervous system. Proc. Natl. Acad. Sci. U. S. A. **76**, 6270–6273 (1979)
282. G.R. Crabtree, Generic signals and specific outcomes: signaling through Ca2+, calcineurin, and NF-AT. Cell **96**, 611–614 (1999). https://doi.org/10.1016/S0092-8674(00)80571-1
283. C.B. Klee, G.F. Draetta, M.J.C. Hubbard, Adv. Enzymol. Ramb. **61**, 149–200 (1988)
284. J.O. Liu, Calmodulin-dependent phosphatase, kinases, and transcriptional corepressors involved in T-cell activation. Immunol. Rev. **228**, 184–198 (2009). https://doi.org/10.1111/j.1600-065X.2008.00756.x

285. Y.L. Wang, Y. Wang, L. Tong, Q. Wei, Overexpression of calcineurin B subunit (CnB) enhances the oncogenic potential of HEK293 cells. Cancer Sci. **99**, 1100–1108 (2008). https://doi.org/10.1111/j.1349-7006.2008.00799.x
286. W. Li, R.E. Handschumacher, Identification of two calcineurin B-binding Proteins: tubulin and heat Shock Protein 60. Bba-Proteins Proteom. **1599**, 72–81 (2002). Pii S1570-9639(02)00402-8. https://doi.org/10.1016/S1570-9639(02)00402-8
287. B. Beutler, Innate immunity: an overview. Mol. Immunol. **40**, 845–859 (2004). https://doi.org/10.1016/j.molimm.2003.10.005
288. C. Alexander, E.T. Rietschel, Bacterial lipopolysaccharides and innate immunity. J. Endotoxin Res. **7**, 167–202 (2001). https://doi.org/10.1179/096805101101532675
289. D.E. Butkus, G.A. Herrera, S.S. Raju, Successful renal-transplantation after cyclosporine-associated hemolytic-uremic syndrome following bilateral lung transplantation. Transplantation **54**, 159–162 (1992)
290. J.J. Siekierka, S.H.Y. Hung, M. Poe, C.S. Lin, N.H. Sigal, A cytosolic binding-protein for the immunosuppressant Fk506 has peptidyl-prolyl isomerase activity but is distinct from cyclophilin. Nature **341**, 755–757 (1989). https://doi.org/10.1038/341755a0
291. R.E. Handschumacher, M.W. Harding, J. Rice, R.J. Drugge, Cyclophilin – a specific cytosolic binding-protein for cyclosporin-A. Science **226**, 544–547 (1984). https://doi.org/10.1126/science.6238408
292. Q. Huai et al., Crystal structure of calcineurin-cyclophilin-cyclosporin shows common but distinct recognition of immunophilin-drug complexes. P. Natl. Acad. Sci. U.S.A. **99**, 12037–12042 (2002). https://doi.org/10.1073/pnas.192206699
293. J.P. Griffith et al., X-ray structure of calcineurin inhibited by the immunophilin immunosuppressant Fkbp12-Fk506 complex. Cell **82**, 507–522 (1995). https://doi.org/10.1016/0092-8674(95)90439-5
294. D.A. Fruman, C.B. Klee, B.E. Bierer, S.J. Burakoff, Calcineurin phosphatase-activity in lymphocytes-T is inhibited by Fk-506 and Cyclosporine-A. P. Natl. Acad. Sci. U.S.A. **89**, 3686–3690 (1992). https://doi.org/10.1073/pnas.89.9.3686
295. B.E. Bierer, Y.J. Jin, D.A. Fruman, V. Calvo, S.J. Burakoff, FK 506 and rapamycin: molecular probes of T-lymphocyte activation. Transplant. Proc. **23**, 2850–2855 (1991)
296. S.A. Trushin, K.N. Pennington, A. Algeciras-Schimnich, C.V. Paya, Protein kinase C and calcineurin synergize to activate IkappaB kinase and NF-kappaB in T lymphocytes. J. Biol. Chem. **274**, 22923–22931 (1999)
297. S. Baksh, S.J. Burakoff, The role of calcineurin in lymphocyte activation. Semin. Immunol. **12**, 405–415 (2000). https://doi.org/10.1006/smim.2000.0221
298. B. Wright, A.H. Lacchini, A.J. Davies, A.J. Walker, Regulation of nitric oxide production in snail (Lymnaea stagnalis) defence cells: a role for PKC and ERK signalling pathways. Biol. Cell. **98**, 265–278 (2006). https://doi.org/10.1042/Bc20050066
299. A. Novas, R. Barcia, J.I. Ramos-Martinez, Implication of PKC in the seasonal variation of the immune response of the Hemocytes of Mytilus galloprovincialis lmk. and its role in interleukin-2-induced nitric oxide synthesis. IUBMB Life **59**, 659–663 (2007). https://doi.org/10.1080/15216540701553235
300. C. Montagnani, Y. Labreuche, J.M. Escoubas, Cg-I kappa B, a new member of the I kappa B protein family characterized in the pacific oyster Crassostrea gigas. Dev. Comp. Immunol. **32**, 182–190 (2008). https://doi.org/10.1016/j.dci.2007.06.001
301. H. Kleinert, P.M. Schwarz, U. Forstermann, Regulation of the expression of inducible nitric oxide synthase. Biol. Chem. **384**, 1343–1364 (2003). https://doi.org/10.1515/Bc.2003.152
302. S. Ghosh, M.J. May, E.B. Kopp, NF-kappa B and rel proteins: Evolutionarily conserved mediators of immune responses. Annu. Rev. Immunol. **16**, 225–260 (1998). https://doi.org/10.1146/annurev.immunol.16.1.225
303. K. Obasanjo-Blackshire et al., Calcineurin regulates NFAT-dependent iNOS expression and protection of cardiomyocytes: Co-operation with Src tyrosine kinase. Cardiovasc. Res. **71**, 672–683 (2006). https://doi.org/10.1016/j.cardiores.2006.05.026

304. J. Heger, G. Euler, iNOS – another cardiac target of calcineurin. Cardiovasc. Res. **71**, 612–614 (2006). https://doi.org/10.1016/j.cardiores.2006.07.007
305. Y. Kim et al., Ca2+/calmodulin-dependent protein phosphatase calcineurin mediates the expression of iNOS through IKK and NF-kappa B activity in LPS-stimulated mouse peritoneal macrophages and RAW 264.7 cells. Biochem. Biophys. Res. Commun. **314**, 695–703 (2004). https://doi.org/10.1016/j.bbrc.2003.12.153
306. A. Novas, R. Barcia, J.I. Ramos-Martinez, Nitric oxide production by haemocytes from Mytilus galloprovincialis shows seasonal variations. Fish Shellfish Immunol. **23**, 886–891 (2007). https://doi.org/10.1016/j.fsi.2007.04.007
307. T.A. Waldmann, The multi-subunit interleukin-2 receptor. Annu. Rev. Biochem. **58**, 875–911 (1989). https://doi.org/10.1146/annurev.bi.58.070189.004303
308. E. Ottaviani, A. Franchini, C. Franceschi, Presence of several cytokine-like molecules in molluscan hemocytes. Biochem. Biophys. Res. Commun. **195**, 984–988 (1993). https://doi.org/10.1006/bbrc.1993.2141
309. T.K. Hughes Jr., E.M. Smith, J.A. Barnett, R. Charles, G.B. Stefano, LPS stimulated invertebrate hemocytes: a role for immunoreactive TNF and IL-1. Dev. Comp. Immunol. **15**, 117–122 (1991)
310. A. Cao, J.I. Ramos-Martinez, R. Barcia, In vitro effects of LPS, IL-2, PDGF and CRF on haemocytes of Mytilus galloprovincialis lmk. Fish Shellfish Immunol. **16**, 215–225 (2004). https://doi.org/10.1016/S1050-4648(03)00080-9
311. R. Barcia, A. Cao, J. Arbeteta, J.I. Ramos-Martinez, The IL-2 receptor in hemocytes of the sea mussel Mytilus galloprovincialis lmk. IUBMB Life **48**, 419–423 (1999). https://doi.org/10.1080/713803540
312. C.Z. Li et al., Calcineurin mediates the immune response of hemocytes through NF-kappa B signaling pathway in pearl oyster (Pinctada fucata). Fish Shellfish Immunol. **28**, 253–260 (2010). https://doi.org/10.1016/j.fsi.2009.11.001
313. H. Wu, A. Peisley, I.A. Graef, G.R. Crabtree, NFAT signaling and the invention of vertebrates. Trends Cell Biol. **17**, 251–260 (2007). https://doi.org/10.1016/j.tcb.2007.04.006
314. A. Stathopoulos-Gerontides, J.J. Guo, M.S. Cyert, Yeast calcineurin regulates nuclear localization of the Crz1p transcription factor through dephosphorylation. Genes Dev. **13**, 798–803 (1999). https://doi.org/10.1101/gad.13.7.798
315. A.M. Stathopoulos, M.S. Cyert, Calcineurin acts through the CRZ1/TCN1-encoded transcription factor to regulate gene expression in yeast. Genes Dev. **11**, 3432–3444 (1997). https://doi.org/10.1101/gad.11.24.3432
316. M.R. Hanlon, N.S. Berrow, A.C. Dolphin, B.A. Wallace, Modelling of a voltage-dependent Ca2+ channel B subunit as a basis for understanding its functional properties. FEBS Lett. **445**, 366–370 (1999). https://doi.org/10.1016/S0014-5793(99)00156-8
317. M. DeWaard, V.E.S. Scott, M. Pragnell, K.P. Campbell, Identification of critical amino acids involved in alpha(1)-beta interaction in voltage-dependent Ca2+ channels. FEBS Lett. **380**, 272–276 (1996). https://doi.org/10.1016/0014-5793(96)00007-5
318. M. Dewaard, M. Pragnell, K.P. Campbell, Ca2+ channel regulation by a conserved beta-subunit domain. Neuron **13**, 495–503 (1994). https://doi.org/10.1016/0896-6273(94)90363-8
319. D. Walker, M. De Waard, Subunit interaction sites in voltage-dependent Ca2+ channels: role in channel function. Trends Neurosci. **21**, 148–154 (1998). https://doi.org/10.1016/S0166-2236(97)01200-9
320. W. Fan et al., Cloning, characterization and expression analysis of calcium channel beta subunit from pearl oyster (Pinctada fucata). J. Biosci. Bioeng. **104**, 47–54 (2007). https://doi.org/10.1263/jbb.104.47
321. N.L. Brice et al., Importance of the different beta subunits in the membrane expression of the alpha 1A and alpha 2 calcium channel subunits: studies using a depolarization-sensitive alpha 1A antibody. Eur. J. Neurosci. **9**, 749–759 (1997). https://doi.org/10.1111/j.1460-9568.1997.tb01423.x

322. R. Hullin et al., Cardiac L-type calcium channel beta-subunits expressed in human heart have differential effects on single channel characteristics. J. Biol. Chem. **278**, 21623–21630 (2003). https://doi.org/10.1074/jbc.M211164200
323. H. Hibino et al., Direct interaction with a nuclear protein and regulation of gene silencing by a variant of the Ca2+-channel beta(4) subunit. P. Natl. Acad. Sci. U.S.A. **100**, 307–312 (2003). https://doi.org/10.1073/pnas.0136791100
324. M. Murakami, U. Wissenbach, V. Flockerzi, Gene structure of the murine calcium channel beta 3 subunit, cDNA and characterization of alternative splicing and transcription products. Eur. J. Biochem. **236**, 138–143 (1996). https://doi.org/10.1111/j.1432-1033.1996.t01-1-00138.x
325. L. Birnbaumer et al., Structures and functions of calcium channel beta subunits. J. Bioenerg. Biomembr. **30**, 357–375 (1998). https://doi.org/10.1023/A:1021989622656
326. P.S. Nair, W.E. Robinson, Calcium speciation and exchange between blood and extrapallial fluid of the quahog Mercenaria mercenaria (L.). Biol. Bull. **195**, 43–51 (1998). https://doi.org/10.2307/1542774
327. A.S. Mount, A.P. Wheeler, R.P. Paradkar, D. Snider, Hemocyte-mediated shell mineralization in the eastern oyster. Science **304**, 297–300 (2004). https://doi.org/10.1126/science.1090506
328. J.M. Neff, Ultrastructure of the outer epithelium of the mantle in the clam Mercenaria mercenaria in relation to calcification of the shell. Tissue Cell **4**, 591–600 (1972)
329. J. Kyte, R.F.A. Doolittle, Simple method for displaying the hydropathic character of a protein. J. Mol. Biol. **157**, 105–132 (1982). https://doi.org/10.1016/0022-2836(82)90515-0
330. C.J. Brandl, N.M. Green, B. Korczak, D.H. MacLennan, Two Ca2+ ATPase genes: homologies and mechanistic implications of deduced amino acid sequences. Cell **44**, 597–607 (1986)
331. W.M. Fan et al., Cloning, characterization, and expression patterns of three sarco/endoplasmic reticulum Ca2+-ATPase isoforms from pearl oyster (Pinctada fucata). Acta Biochim. Biophys. Sin. **39**, 722–730 (2007). https://doi.org/10.1111/j.1745-7270.2007.00330.x
332. A.G. Lee, J.M. East, What the structure of a calcium pump tells us about its mechanism. Biochem. J. **356**, 665–683 (2001). https://doi.org/10.1042/0264-6021:3560665
333. Y.L. Zhang et al., Characterization of cDNA and genomic DNA encoding SERCA1, the Ca2+-ATPase of human fast-twitch skeletal muscle sarcoplasmic reticulum, and its elimination as a candidate gene for Brody disease. Genomics **30**, 415–424 (1995). https://doi.org/10.1006/geno.1995.1259
334. M. Schleef et al., The gene encoding sarcoplasmic reticulum calcium ATPase-1 (Atp2a1) maps to distal mouse chromosome 7. Mamm. Genome **7**, 788–788 (1996)
335. M.V. Heyen et al., Structure and organization of the mouse Atp2a2 gene encoding the sarco (endo)plasmic reticulum Ca2+-ATPase 2 (SERCA2) isoforms. Mamm. Genome **11**, 159–163 (2000)
336. P. Gelebart, V. Martin, J. Enouf, B. Papp, Identification of a new SERCA2 splice variant regulated during monocytic differentiation. Biochem. Biophys. Res. Commun. **303**, 676–684 (2003). https://doi.org/10.1016/S0006-291x(03)00405-4
337. C.J. Brandl, S. Deleon, D.R. Martin, D.H. Maclennan, Adult forms of the Ca-2+ Atpase of sarcoplasmic-reticulum – expression in developing skeletal-muscle. J. Biol. Chem. **262**, 3768–3774 (1987)
338. E. Poch et al., Functional characterization of alternatively spliced human SERCA3 transcripts. Am. J. Physiol. Cell Physiol. **275**, C1449–C1458 (1998)
339. L. Dode et al., Structure of the human sarco/endoplasmic reticulum Ca(2+)-ATPase 3 gene – promoter analysis and alternative splicing of the SERCA3 pre-mRNA. J. Biol. Chem. **273**, 13982–13994 (1998). https://doi.org/10.1074/jbc.273.22.13982
340. C. Chaabane et al., Sarco/endoplasmic reticulum Ca2+ ATPase type 3 isoforms (SERCA3b and SERCA3f): Distinct roles in cell adhesion and ER stress. Biochem. Biophys. Res. Commun. **345**, 1377–1385 (2006). https://doi.org/10.1016/j.bbrc.2006.05.054

341. Z.P. Zhang, D.G. Chen, M.G. Wheatly, Cloning and characterization of sarco/endoplasmic reticulum Ca2+-atpase (SERCA) from crayfish axial, muscle. J. Exp. Biol. **203**, 3411–3423 (2000)
342. D.D. Chen, Z.P. Zhang, M.G. Wheatly, Y.P. Gao, Cloning and characterization of the heart muscle isoform. of sarco/endoplasmic reticulum Ca2+ ATPase (SERCA) from crayfish. J. Exp. Biol. **205**, 2677–2686 (2002)
343. R. Escalante, L. Sastre, Similar alternative splicing events generate 2 sarcoplasmic or endoplasmic-reticulum Ca-Atpase isoforms in the crustacean Artemia-franciscana and in vertebrates. J. Biol. Chem. **268**, 14090–14095 (1993)
344. A.M. Campbell, P.D. Kessler, D.M. Fambrough, The alternative carboxyl termini of avian cardiac and brain sarcoplasmic-reticulum endoplasmic-reticulum Ca2+-Atpases are on opposite sides of the membrane. J. Biol. Chem. **267**, 9321–9325 (1992)
345. B. Korczak et al., Structure of the rabbit fast-twitch skeletal muscle Ca2+-ATPase gene. J. Biol. Chem. **263**, 4813–4819 (1988)
346. J. Lytton, M. Westlin, S.E. Burk, G.E. Shull, D.H. Maclennan, Functional comparisons between isoforms of the sarcoplasmic or endoplasmic-reticulum family of calcium pumps. J. Biol. Chem. **267**, 14483–14489 (1992)
347. A.M. Gunteskihamblin, J. Greeb, G.E. Shull, A novel Ca-2+ pump expressed in brain, kidney, and stomach is encoded by an alternative transcript of the slow-twitch muscle sarcoplasmic-reticulum Ca-Atpase gene – identification of Cdnas encoding Ca-2+ and other cation-transporting Atpases using an oligonucleotide probe derived from the Atp-binding site. J. Biol. Chem. **263**, 15032–15040 (1988)
348. J. Lytton, A. Zarainherzberg, M. Periasamy, D.H. Maclennan, Molecular-cloning of the mammalian smooth-muscle sarco(endo)plasmic reticulum Ca-2+-Atpase. J. Biol. Chem. **264**, 7059–7065 (1989)
349. J. Lytton, D.H. MacLennan, Molecular cloning of cDNAs from human kidney coding for two alternatively spliced products of the cardiac Ca2+-ATPase gene. J. Biol. Chem. **263**, 15024–15031 (1988)
350. F. Wuytack, L. Dode, F. BabaAissa, L. Raeymaekers, The SERCA3-type of organellar Ca2+ pumps. Biosci. Rep. **15**, 299–306 (1995). https://doi.org/10.1007/Bf01788362
351. K.D. Wu, W.S. Lee, J. Wey, D. Bungard, J. Lytton, Localization and quantification of endoplasmic-reticulum Ca2+-Atpase isoform transcripts. Am. J. Physiol. Cell Physiol. **269**, C775–C784 (1995)
352. Y. Nagata et al., cDNA cloning and predicted primary structure of scallop sarcoplasmic reticulum Ca2+-ATPase (vol 119, pg 777, 1998). Comp Biochem. Phys. B **121**, 361–362 (1998)
353. G. Inesi, C. Sumbilla, M.E. Kirtley, Relationships of molecular-structure and function in Ca-2 +-transport Atpase. Physiol. Rev. **70**, 749–760 (1990)
354. J.H. Cho, J. Bandyopadhyay, J. Lee, C.S. Park, J. Ahnn, Two isoforms of sarco/endoplasmic reticulum calcium ATPase (SERCA) are essential in Caenorhabditis elegans. Gene **261**, 211–219 (2000)
355. P.S. Beirao, J.H.M. Nascimento, Sodium-dependent and calcium-dependent mechanisms in the action-potential of the secretory epithelium of a clam mantle. J. Exp. Biol. **145**, 395–402 (1989)
356. M. Opas, E. Dziak, L. Fliegel, M. Michalak, Regulation of expression and intracellular-distribution of calreticulin, a major calcium-binding protein of nonmuscle cells. J. Cell. Physiol. **149**, 160–171 (1991). https://doi.org/10.1002/jcp.1041490120
357. M.J. Smith, Nucleotide sequence of a Drosophila melanogaster gene encoding a calreticulin homologue. DNA Seq. **3**, 247–250 (1992)
358. M.J. Smith, A C. elegans gene encodes a protein homologous to mammalian calreticulin. DNA Seq. **2**, 235–240 (1992)
359. R.M. Napier et al., Purification, sequencing and functions of calreticulin from maize. J. Exp. Bot. **46**, 1603–1613 (1995). https://doi.org/10.1093/jxb/46.10.1603

360. J. Khalife et al., Cloning of the gene encoding a schistosoma-mansoni antigen homologous to human Ro Ss-a autoantigen. Mol. Biochem. Parasitol. **57**, 193–202 (1993). https://doi.org/10.1016/0166-6851(93)90195-4
361. F.Q. Chen, P.M. Hayes, D.M. Mulrooney, A.H. Pan, Identification and characterization of Cdna clones encoding plant calreticulin in barley. Plant Cell **6**, 835–843 (1994)
362. T.E. Kennedy, D. Kuhl, A. Barzilai, J.D. Sweatt, E.R. Kandel, Long-term sensitization training in aplysia leads to an increase in calreticulin, a major presynaptic calcium-binding protein. Neuron **9**, 1013–1024 (1992). https://doi.org/10.1016/0896-6273(92)90062-I
363. D.C. Jaworski et al., A Secreted calreticulin protein in ixodid tick (Amblyomma-americanum) saliva. J. Insect Physiol. **41**, 369–375 (1995). https://doi.org/10.1016/0022-1910(94)00107-R
364. N. Tsuji, T.H. Morales, V.V. Ozols, A.B. Carmody, R. Chandrashekar, Molecular characterization of a calcium-binding protein from the filarial parasite Dirofilaria immitis. Mol. Biochem. Parasitol. **97**, 69–79 (1998). https://doi.org/10.1016/S0166-6851(98)00131-5
365. T. Sueyoshi, B.A. Mcmullen, L.L. Marnell, T.W. Duclos, W. Kisiel, A new procedure for the separation of protein-Z, prothrombin fragment-1.2 and calreticulin from human plasma. Thromb. Res. **63**, 569–575 (1991). https://doi.org/10.1016/0049-3848(91)90184-X
366. J.M. Holaska et al., Calreticulin is a receptor for nuclear export. J. Cell Biol. **152**, 127–140 (2001). https://doi.org/10.1083/jcb.152.1.127
367. F.A. Arosa, O. de Jesus, G. Porto, A.M. Carmo, M. de Sousa, Calreticulin is expressed on the cell surface of activated human peripheral blood T lymphocytes in association with major histocompatibility complex class I molecules. J. Biol. Chem. **274**, 16917–16922 (1999). https://doi.org/10.1074/jbc.274.24.16917
368. C. Lucu, Calcium-transport across isolated gill epithelium of carcinus. J. Exp. Zool. **268**, 339–346 (1994). https://doi.org/10.1002/jez.1402680502
369. E. Zvaritch et al., Mapping of functional domains in the plasma membrane Ca2+ pump using trypsin proteolysis. Biochemistry **29**, 8070–8076 (1990)
370. A. Enyedi, A.K. Verma, A.G. Filoteo, J.T. Penniston, Protein kinase C activates the plasma membrane Ca2+ pump isoform 4b by phosphorylation of an inhibitory region downstream of the calmodulin-binding domain. J. Biol. Chem. **271**, 32461–32467 (1996)
371. D. Guerini, The significance of the isoforms of plasma membrane calcium ATPase. Cell Tissue Res. **292**, 191–197 (1998). https://doi.org/10.1007/s004410051050
372. P. James et al., Identification and primary structure of a calmodulin binding domain of the Ca-2 + pump of human-erythrocytes. J. Biol. Chem. **263**, 2905–2910 (1988)
373. T.P. Stauffer, D. Guerini, E. Carafoli, Tissue distribution of the four gene products of the plasma membrane Ca2+ pump. A study using specific antibodies. J. Biol. Chem. **270**, 12184–12190 (1995)
374. E. Carafoli, T. Stauffer, The plasma-membrane calcium-pump – functional domains, regulation of the activity, and tissue-specificity of isoform expression. J. Neurobiol. **25**, 312–324 (1994). https://doi.org/10.1002/neu.480250311
375. D. Zoccola et al., Molecular cloning and localization of a PMCA P-type calcium ATPase from the coral Stylophora pistillata. Bba-Biomembranes **1663**, 117–126 (2004). https://doi.org/10.1016/j.bbamem.2004.02.010
376. Y.P. Gao, M.G. Wheatly, Characterization and expression of plasma membrane Ca2+ ATPase (PMCA3) in the crayfish Procambarus clarkii antennal gland during molting. J. Exp. Biol. **207**, 2991–3002 (2004). https://doi.org/10.1242/jcb.01101
377. I. GarciaHiguera et al., Folding of proteins with WD-repeats: comparison of six members of the WD-repeat superfamily to the G protein beta subunit. Biochemistry **35**, 13985–13994 (1996). https://doi.org/10.1021/bi9612879
378. L. Chen et al., Cloning and characterization of a novel G protein beta-subunit of pearl oyster (Pinctada fucata), and its interaction sites with calmodulin. Comp. Biochem. Physiol. B **142**, 142–152 (2005). https://doi.org/10.1016/j.cbpc.2005.02.013
379. J.D. Thompson, T.J. Gibson, F. Plewniak, F. Jeanmougin, D.G. Higgins, The CLUSTAL_X windows interface: flexible strategies for multiple sequence alignment aided by quality

analysis tools. Nucleic Acids Res. **25**, 4876–4882 (1997). https://doi.org/10.1093/nar/25.24.4876

380. N. Saitou, M. Nei, The neighbor-joining method – a new method for reconstructing phylogenetic trees. Mol. Biol. Evol. **4**, 406–425 (1987)
381. J. Sondek, A. Bohm, D.G. Lambright, H.E. Hamm, P.B. Sigler, Crystal structure of a G-protein beta gamma dimer at 2.1 angstrom resolution (vol 379, pg 369, 1996). Nature **379**, 847–847 (1996)
382. L. Duvail, E. Lopez, M. Fouchereau-Peron, Characterization of binding sites for calcitonin gene related peptide in abalone gill. Peptides **20**, 361–366 (1999). https://doi.org/10.1016/S0196-9781(99)00043-1
383. S. Lamsamcasalotti, M. Onoda, V. Papadopoulos, M. Dym, Developmental expression of Gtp-binding proteins in rat testes. J. Reprod. Fertil. **99**, 487–495 (1993)
384. S.R. Sprang, G protein mechanisms: Insights from structural analysis. Annu. Rev. Biochem. **66**, 639–678 (1997). https://doi.org/10.1146/annurev.biochem.66.1.639
385. T. Katada, K. Kusakabe, M. Oinuma, M. Ui, A novel mechanism for the inhibition of adenylate-cyclase via inhibitory Gtp-binding proteins – calmodulin-dependent inhibition of the cyclase catalyst by the beta-gamma-subunits of Gtp-binding proteins. J. Biol. Chem. **262**, 11897–11900 (1987)
386. B. Ulbricht, T. Soldati, Production of reagents and optimization of methods for studying calmodulin-binding proteins. Protein Expr. Purif. **15**, 24–33 (1999). https://doi.org/10.1006/prep.1998.0983
387. P.M. Bayley, W.A. Findlay, S.R. Martin, Target recognition by calmodulin: Dissecting the kinetics and affinity of interaction using short peptide sequences. Protein Sci. **5**, 1215–1228 (1996)
388. T. Szuwart, H. Kierdorf, U. Kierdorf, J. Althoff, G. Clemen, Tissue differentiation and correlated changes in enzymatic-activities during primary antler development in fallow deer (Dama-dama). Anat. Rec. **243**, 413–420 (1995). https://doi.org/10.1002/ar.1092430403
389. A. Nagata, T. Komoda, Y. Sakagishi, Relationship between the uptake of calcium or phosphorus and alkaline-phosphatase activity induced by certain modulators in rat organs. Calcif. Tissue Int. **45**, 173–181 (1989). https://doi.org/10.1007/Bf02556061
390. A. Ito, T. Satoh, Y. Kaziro, H. Itoh, G-protein beta-gamma subunit activates Ras, Raf, and Map kinase in Hek-293 cells. FEBS Lett. **368**, 183–187 (1995). https://doi.org/10.1016/0014-5793(95)00643-N
391. T. Ivanina, Y. Blumenstein, E. Shistik, R. Barzilai, N. Dascal, Modulation of L-type Ca2+ channels by G beta gamma and calmodulin via interactions with N and C termini of alpha(1C). J. Biol. Chem. **275**, 39846–39854 (2000). https://doi.org/10.1074/jbc.M005881200
392. W.V. Yotov, R. St-Arnaud, Cloning and functional expression analysis of the alpha subunit of mouse ATP synthase. Biochem. Biophys. Res. Commun. **191**, 142–148 (1993)
393. H. Omote, N.P. Le, M.Y. Park, M. Maeda, M. Futai, Beta-Subunit Glu-185 of Escherichia-coli H+-Atpase (Atp synthase) is an essential residue for cooperative catalysis. J. Biol. Chem. **270**, 25656–25660 (1995)
394. M. Jounouchi, M. Maeda, M. Futai, The alpha subunit of ATP synthase (F0F1): the Lys-175 and Thr-176 residues in the conserved sequence (Gly-X-X-X-X-Gly-Lys-Thr/Ser) are located in the domain required for stable subunit-subunit interaction. J. Biochem. **114**, 171–176 (1993)
395. H. Noji, R. Yasuda, M. Yoshida, K. Kinosita Jr., Direct observation of the rotation of F1-ATPase. Nature **386**, 299–302 (1997). https://doi.org/10.1038/386299a0
396. J.E. Walker, M. Saraste, M.J. Runswick, N.J. Gay, Distantly related sequences in the alpha-subunits and beta-subunits of Atp synthase, myosin, kinases and other Atp-requiring enzymes and a common nucleotide binding fold. EMBO J. **1**, 945–951 (1982)
397. C. Ramakrishnan, V.S. Dani, T. Ramasarma, A conformational analysis of Walker motif A [GXXXXGKT (S)] in nucleotide-binding and other proteins. *Protein Eng.* **15**, 783–798 (2002). https://doi.org/10.1093/protein/15.10.783

398. D.B. Parekh, W. Ziegler, P.J. Parker, Multiple pathways control protein kinase C phosphorylation. EMBO J. **19**, 496–503 (2000). https://doi.org/10.1093/emboj/19.4.496
399. D.W. Litchfield, B. Luscher, Casein kinase-II in signal-transduction and cell-cycle regulation. Mol. Cell. Biochem. **128**, 187–199 (1993)
400. M. Yoshida, E. Muneyuki, T. Hisabori, ATP synthase – a marvellous rotary engine of the cell. Nat. Rev. Mol. Cell Biol. **2**, 669–677 (2001). https://doi.org/10.1038/35089509
401. H. Miyamoto, F. Miyoshi, J. Kohno, The carbonic anhydrase domain protein nacrein is expressed in the epithelial cells of the mantle and acts as a negative regulator in calcification in the mollusc Pinctada fucata. Zool. Sci. **22**, 311–315 (2005). https://doi.org/10.2108/zsj.22.311

Chapter 3
Identification (Characterization) and Function Studies of Matrix Protein from the Oyster *Pinctada fucata*

Abstract Shell matrix protein (SMP) which is extracted from shell and extrapallial fluid matrix protein (EPFMP) which is extracted from extrapallial fluid have important functions in biomineralization during shell formation of *Pinctada fucata*. In the past decades, the functions of SMPs and EPFMPs were gradually revealed. In 2015, our group identified 72 unique SMPs in *Pinctada fucata* by liquid chromatography-tandem mass spectrometry (LC-MS/MS) analysis of proteins extracted from the shells of *P. fucata* aligned with a draft genome. In this chapter, we introduced SMPs from nacreous layer (P14, N40, and PfN23), from prismatic layer (KRMP, KRMP-3, and Prisilkin-39), from both prismatic and nacreous layer (MSI7, PfN44, and PfY2), EPFMP and extracellular matrix protein expressed by mantle (EFCBP and Ferritin). For example, P14 plays a crucial role during nacre biomineralization. N40 could stimulate the nucleation of aragonite drastically by serving as a nucleation site. The basic protein PfN23 might be a key accelerator during the regulation of crystal growth in nacre. KRMP protein family plays important roles in the framework formation of prism.

The assays of functional study include RNA interference, antibody injection, notching experiment, gene expression analysis by quantitative real-time PCR, immunolocalization with gold particles, immunofluorescence *in vivo* and recombinant protein purification, chitin and calcium carbonate crystal binding assay, calcium carbonate precipitation assay, *in vitro* calcium carbonate crystallization assay, transition of ACC to stable crystals, CD spectroscopy and fluorescence quenching *in vitro*.

Keywords SMPs · EPFMPs · Biomineralization · Functional analyses

3.1 Introduction of Matrix Protein of *Pinctada fucata*

Extracellular matrix protein (ECMP), for short, matrix protein of *Pinctada fucata*, includes shell matrix protein (SMPs) which is extracted from shell and extrapallial fluid matrix protein (EPFMP) which is extracted from extrapallial fluid. ECMPs

R. Zhang et al., *Biomineralization Mechanism of the Pearl Oyster, Pinctada fucata*,
https://doi.org/10.1007/978-981-13-1459-9_3

almost have tandem repeat units which are consist of two to four amino acids; what's more, the primary structures of ECMPs are almost modular structure.

Pinctada fucata (pearl oysters), one of the most important economical pearl production species in China and Japan, is also one of the best studied biomineralization models [1]. The shells of *Pinctada fucata* are composed of calcite as the outer prismatic layer, aragonite as the inner nacreous layer, and cuticle in the outside. The biomineralized products possess admirable mechanical [2] and biological properties [3] compared to pure calcium carbonate. The shells include 95% $CaCO_3$ and almost 5% organic macromolecules like proteins, polysaccharides, and lipids [1]. Shell matrix proteins (SMPs) were thought to play important roles in crystal nucleation, polymorphism, morphology, and organization of calcium carbonate crystallites during shell formation [4].

Since the cloning of the first SMP, nacrein from *P. fucata* in 1996 [5], MSI60 [6], N16 [7], Prismalin-14 [8], Shematrin [9], lysine (K)-rich matrix protein (KRMP) [9], Aspein [10], tyrosinase [11], N40 [12], Prisilkin-39 [13], PfN23 [14], and PfN44 [15] have been cloned and characterized. SMPs have been found to possess several functionalities: (1) they facilitate calcite (Aspein [16]) or aragonite crystallization (N40 [12]), (2) they act as framework proteins (Shematrin [17] and Prisilkin-39 [13]), and (3) they guide calcium carbonate assembly (N16 [18]). SMPs are very important for biomineralization of shell which could regulate unordered calcium carbonate to shell which is regular and mechanically strong (Table 3.1).

3.2 Materials and Methods

3.2.1 Extraction of SMPs from Shell

Collect shells from *P. fucata*, clean and immerse them in 5% sodium hydroxide for 24 h, and rinse in the deionized water to avoid possible contamination of soft tissue remnant in the inner surface of shells. The two layers of shells, the inner nacreous layer and the outer prismatic layer, were separated mechanically by polish and then were washed and air-dried. The fragments were pulverized and were then decalcified by 0.8 M ethylenediaminetetraacetic acid (EDTA, pH 8.0) at 4 °C for 60 h with continuous agitation. To extract the soluble matrix, the supernatant was collected by centrifugation at 13,000 rpm for 30 min at 4°C and was then desalted by ultrafiltration (3 K). For extraction of the insoluble matrix, rinse the above precipitation with water and treat it with denaturing solution (30 mM Tris–HCl, 1% sodium dodecyl sulfate (SDS), 10 mM dithiothreitol, pH 8.0,) for 30 min at 100 °C. After short centrifugation, the denatured samples were ready to be applied on 12% SDS-polyacrylamide gels. Proteins were stained with Coomassie Brilliant Blue and were quantified by Nanodrop.

Table 3.1 Shell matrix proteins found in *P. fucata*

Specie	Protein	PI	Molecular mass (KDa)	Functional prediction
P. fucata (N)	PfN23	11.17	23.4	Promote aragonite formation
P. fucata (N)	Nacrein	6.85	48.2	Carbonic anhydrase; Ca^{2+} binding
P. fucata (N)	MSI60	4.8	6.17	Framework protein
P. fucata (N)	Pif97	4.65	97	Promote aragonite formation
P. fucata (N)	Pif80	4.99	80	Promote aragonite formation
P. fucata (N)	MSI7	5.98	7.3	Framework protein
P. fucata (N)	N16	5.14	16	Promote aragonite formation
P. fucata (N)	P10		6.8	Induce aragonite formation
P. fucata (N)	N40		38	Induce aragonite nucleate
P. fucata (N)	N19	8.04	23	Inhibit crystals formation
P. fucata (N)	P14		14.5	Induce aragonite formation
P. fucata (P)	MSI31	3.8	32.85	Framework protein
P. fucata (P)	Aspein	1.67	39.3	Promote calcite formation
P. fucata (P)	Prismalin-14	4.24	11.9	Framework protein; inhibit crystals formation
P. fucata (P)	Prisilkin-39	8.83	39.3	Framework protein
P. fucata (P)	KRMP family	9.4–9.6	9.5–9.8	
P. fucata (P)	Shematrin family	7.7–10	27–30	Framework protein
P. fucata (P)	Tyrosine1,2	6.5–9	54–56	
P. fucata (N/P)	MSI7	5.98	7.3	Framework protein
P. fucata (N/P)	PfN44	4.25	44.1	Inhibit aragonite deposition
P. fucata (EF)	ACCBP	4.62	270	Ca^{2+} binding
P. fucata (N/P)	Alv[a]	11.34	26.5	Framework protein
P. fucata (N/P)	PNU9[a]			

(continued)

Table 3.1 (continued)

Specie	Protein	PI	Molecular mass (KDa)	Functional prediction
P. fucata	PNU7[a]			
P. fucata	PfY2[a]			

P prism, *N* nacre, *EF* extrapallial fluid
[a]Unpublished

3.2.2 Extraction of the EPFMP from Extrapallial Fluid

The EPF proteins were extracted from the extrapallial space which is located between the mantle center and the shell by inserting a 0.8 mm × 50 mm syringe needle through the byssal foramen. The tips were inserted slowly and kept away from the mantle. Samples collected were transferred to a 50 ml centrifuge tube that was placed on ice. EPF was filtered with a 70 μm nylon cell strainer (BD Falcon, Bedford, MA, USA) and centrifuged at 800 g and 4 °C for 5 min to remove the cells and impurity after extraction. Then the supernatant was filtered by a 0.22 μm MILLEXGP filter unit (SLGP033RB) and stored at −80 °C until use.

3.2.3 Identification of the Complete Gene Sequence

Total RNA was extracted from the mantle tissue of *P. fucata* using TRIZOL (Life Technologies, USA), according to the manufacturer's instructions. Rapid amplification of cDNA ends (RACE) was operated by using a SMARTer™ RACE cDNA amplification kit (Clontech, Japan). RACE3 and RACE5 primers were designed with the primers according to the 3′- and 5′-RACE kits, respectively. Confirm the full-length sequence using the primers Full1 and Full2. The amino acid sequence deduced from the gene sequence was determined using the ExPASy website. The signal peptide was predicted by the SignalP 3.0 website.

3.2.4 Recombinant Protein Expression and Purification

A recombinant plasmid was constructed by inserting the target gene into the vector-like pET21b/pET28a/PGEX-4T-1/Pmal-c5x, which was then transformed into E. coli cells. The expression host was incubated at 37 °C in LB medium containing ampicillin. When a D600 of 0.6 was reached, the culture temperature may be different from 37 °C or 16 °C, and protein expression was induced by the addition of IPTG. After being induced for 4 h or 10 h, the cells were collected by centrifugation at 8000 g at 4 °C for 6 min. The harvested cells were resuspended by Tris–NaCl buffer (10 mM Tris, 500 mM NaCl, pH 7.5) and subjected to ultrasonication, and the protein was purified using affinity chromatography and was desalted using AKATA with desalted column. The purified protein was stored in a pH 7.5 solution (10 mM Tris, 500 mM NaCl) at −20 °C before use. The protein concentration and quality was determined by Nanodrop.

3.2.5 Preparation of Polyclonal Antibodies Against the Recombinant Protein

Recombinant protein with His/mbp/GST tag at the N terminus was overexpressed in *E. coli*, and elution fractions containing recombinant protein were tested by SDS-PAGE and were concentrated and desalted by AKATA against Tris–NaCl buffer, pH 7.4. Polyclonal antibodies against recombinant protein were injected in New Zealand rabbits following standard immunization procedures and then were affinity purified. The specificity of the antibodies was tested by Western blots against the purified protein, the EDTA-insoluble matrix extracted from the entire shell of *P. fucata*, and the polyhistidine.

3.2.6 Gene Expression Analysis by RT-PCR

Using the methods described before, total RNA was extracted from the mantle edge, mantle pallial, gonad, foot, gill, viscus, hemocytes, and adductor muscles of adult *P. fucata*. Equal quantities of total RNA of different tissues were reverse-transcribed into cDNA by using Quant Reverse Transcriptase (Tiangen, China) following the manufacturer's instructions. RT-PCR was conducted with the primer pairs to amplify target gene and actin (used as a positive control for cDNA preparations), respectively. To detect cross-contamination of the samples and avoid false-positive results, a negative control was performed without the cDNA template. The qPCR was performed on an ABI PCR amplifier (StepOnePlus™, Life Technologies, USA) following the SYBR® Premix Ex Taq™ (TaKaRa, Japan) protocol. The experimental process is as follows: 95 °C, 30 s; 95 °C, 5 s; and 60 °C, 30 s for 35 cycles.

3.2.7 In Situ Hybridization

In situ hybridization of mRNA of target gene was conducted on frozen sections of the mantle tissue that had been fixed in 4% paraformaldehyde with 0.1% diethyl pyrocarbonate (Sigma, USA) overnight before. The digoxigenin-labeled probe was amplified from the around 300 bp DNA fragment with the primer pair following High Prime DIG random labeling (Roche Applied Science, Switzerland) protocol. The procedures of hybridization were mainly conducted following protocol with some modifications [19].

3.2.8 Detection of Native Protein in Shell Extracts

The presence of native protein in different extracts of the prism and the nacre of the shell of *P. fucata* was tested by ELISA and Western blotting and immunofluorescence. During the ELISA, known amounts (0.1–5 ng) of target protein were used as a reference. The experiments were repeated three times to get reliable results. Protein samples from different shell extracts were loaded into SDS-PAGE and transferred to a PVDF membrane (Millipore, USA). Anti-recombinant protein antibody was used at dilutions of 1:1000–1:2000, and HRP-conjugated goat anti-rabbit antibody (Calbiochem, Germany) was used as secondary antibody at 1:10000–1:5000. The dot was incubated with Luminata Crescendo Western HRP substrate (Millipore, USA) and then was exposed to X-ray film for about 1 min. Control experiments were performed without the first antibody step.

3.2.9 Immunolocalization with Gold Particles

The immunogold-labeling assay was performed as described by Marin et al. [20] with some modifications. The anti-protein antibodies were used at dilutions of 1:200. Goat anti-rabbit antibodies were diluted (1:400) and were coupled to 5 nm gold particles (Sigma, USA) used as secondary antibodies. Negative controls were performed avoiding the first antibody step. Samples were spattered with carbon and observed by an FEI Sirion2000 scanning electron microscope (SEM; Netherlands) in the backscattering mode.

3.2.10 In Vivo Inhibition of Native Protein

The RNAi assay was conducted as described in Suzuki et al. [21] with little modifications. Primers were designed to amplify specific DNA fragment from the first strand cDNA. The PCR products were purified using an EasyPure Quick Gel Extraction Kit (TransGen Biotech, China). The dsRNA was synthesized and purified by a RiboMAX^{TM} large-scale RNA production system (T7) kit (Promega, USA).

RNase-free DNase I (Takara, Japan) was used to remove the template DNA. The dsRNA was diluted to low dosage and high dosage which is different in different papers, with PBS, and then injected into the adductor muscle. Three individuals were used for each treatment to avoid individual variation. Total RNA of the mantle tissue from each oyster was collected 6 days after injection and used to synthesize the first strand cDNA, as described before. Real-time quantitative PCR was used to quantify the expression levels of each gene, where actin was used as an internal reference. Quantitative PCR was conducted with an SYBRH Premix Ex Taq^{TM} II kit (Takara, Japan), according to the protocol. The cycling process were 95 °C for 30 s (1 cycle), 95 °C for 5 s, 55 °C for 30 s, and 72 °C for 30 s (40 cycles). As the control, the expression levels of the PBS injected group were set as a relative value of 1.0. The shells of the injection groups were thoroughly washed with Milli-Q water and then air-dried 6 days after injection. Shells were cut into pieces and then sputter-coated with 10 nm thick gold and observed by an FEI Quanta 200 scanning electron microscope.

3.2.11 Chitin and Calcium Carbonate Crystal Binding Assay

Chitin-binding assays were conducted as described by Inoue et al. [22], and the calcium carbonate crystal binding assay was performed as described by Suzuki et al. (1) with some modifications. BSA or target protein (200 μg each) was incubated with 20 mg of chitin and calcite or aragonite (Sigma, USA). The solution was removed after centrifugation, and the insoluble mixture was then washed by distilled water and Tris–NaCl buffer for three times. Each washing was mixed with denaturation buffer and then was boiled for 10 min. The final insoluble residue was also boiled in 100 μl denaturation buffer for 10 min, after that the supernatant was separated by centrifugation for 5 min at 12,000 rpm. All washing or supernatant agents were subjected to SDS-PAGE on a 12% gel. The gel was then stained with Coomassie Brilliant Blue after electrophoresis.

3.2.12 Calcium Carbonate Precipitation Assay

The effect of recombinant protein on the rate of calcium carbonate precipitation was tested according to the method of Suzuki et al. [8], with some modifications. The sample solution (10 μl) was mixed with 100 μl of 100 mM sodium bicarbonate (pH 8.5). Then 100 μl of 100 mM calcium chloride was added to the mixed solution, and the formation of calcium carbonate precipitate was detected every half minute for 5 min based on the measured absorbance at 570 nm by a spectrophotometer (Bio-Rad 680, USA).

3.2.13 *In Vitro* Calcium Carbonate Crystallization Assay

Saturated calcium bicarbonate solution was prepared to imitate native crystallization environment following the method of Xu et al. [23] with little modifications. CO_2 gas was bubbled into the mixture of calcium carbonate solution for 6 h. The excess $CaCO_3$ solid was removed by filtration. An aragonitic crystallizing solution was also produced like calcite crystallizing solution but adding 50 mM magnesium chloride to the first solution. Recombinant protein and negative control protein were added to the freshly prepared crystallization solution onto a siliconized glass slide. The crystallization solution was removed after 24 h, and air-dried crystals were observed by an FEI Quanta 200 scanning electron microscope and Raman spectroscopy which was recorded at an excitation wavelength of 514 nm. The spectra were collected in the range of 100–1400 cm^{-1} using a Renishaw RM2000 spectrometer three times for 20 s.

3.2.14 Transition of ACC to Stable Crystals

50 mM CaC12 together with 100 mM $MgCl_2$ and 100 μg/ml PfN44 was prepared (solution A). 50 mM Na_2CO_3 was prepared (solution B). Solutions A and B were cooled on ice for 1 h. An equal volume of solution B was quickly added to solution A and mixed well at 4 °C. After 0, 24, or 48 h, deposited crystals were washed by acetone, dried, and analyzed by X-ray diffraction and FTIR. This analysis was done by three times to get a consistent result.

3.2.15 CD Spectroscopy and Fluorescence Quenching

CD measurements were carried with a JASCO-600 spectropolarimeter using cuvettes with a 5 mm path length. PfN44 concentration was 200 μg/ml. Fluorescence spectra were collected using a Hitachi F-2500 spectrofluorometer, with a 1 cm path length cell and excitation wavelength of 280 nm. PfN44 concentration was 5 μM. The CD spectroscopy and fluorescence quenching analyses were done three times to get a consistent result.

3.2.16 Primers

Refer to Supplementary Table 3.1 for details.

3.3 In-depth Proteomic Analysis of Shell Matrix Proteins of *Pinctada fucata*

The shells of pearl oysters, *Pinctada fucata,* possess calcite and aragonite and have excellent mechanical properties. These shells are formed with the regulation of macromolecules, particularly shell matrix proteins (SMPs). Identification and characteristic of diverse SMPs will shed light on understanding biomineralization process. Herein, we identified 72 unique SMPs by liquid chromatography-tandem mass spectrometry (LC-MS/MS) analysis of proteins which were extracted from the shells of *P. fucata* along with a draft genome. Among 72 SMPs, 17 SMPs existed in both the prismatic and nacreous layers. What's more, according to the predicted domains found in the SMPs, we hypothesize that besides controlling crystal organization and $CaCO_3$ crystallization, these proteins might regulate the extracellular microenvironment and communicate between mantle cells and the extracellular matrix (ECM) [24]. The SMPs were identified by immunohistological localization techniques in the mantle, shells, and synthetic calcite. Taken together, these proteomic data could increase the repertoires of the shell matrix proteins in *P. fucata* and imply that shell formation in *P. fucata* may be relevant tightly to the regulation of the extracellular microenvironment and cellular activities.

3.3.1 Protein Composition in the Prismatic and Nacreous Matrix

The shell of *P. fucata* is composed of prism and the nacre (Fig. 3.1a). The prismatic layer consist of prisms with length of 10–40 μm wrapped by the organic sheath (Fig. 3.1b), and the nacreous layer is composed by stacked hexagonal nanotablets of which side lengths of 0.5–3 μm (Fig. 3.1c). In order to extract SMPs, shells were cleaned with 5% NaOH to exclude possible contamination from outside organic matter. Then the separated shells with prism and nacre were dissolved with EDTA, isolating soluble and insoluble extracts. EDTA could chelate Ca^{2+} to dissolve the shell and release the organic matrices finally. In this study, we observed the shell after 60 h and found that it can be fully dissolved. The harvest of organic matrices from the shell was around 1.5–3.5 mg/g (depending on the concentration of proteins obtained from certain amount of mashed shell powders).

The soluble and insoluble extractive were subjected to SDS-PAGE and then was stained by coomassie brilliant blue (Fig. 3.1d). Protein bands of the gel were cut off and then digested with trypsin. By LC-MS/MS, sequences of peptide fragments were blasted against the proteome translated from *P. fucata* draft genome [25]. Ultimately, via bioinformatics analysis including BLAST, SMART, and InterProScan, we identified 72 different SMPs from 144 whole proteomes, among which 36 and 19 SMPs are solely found in the prism and the nacre, respectively, while 17 SMPs are found in both layers [24]. It should be paid attention that these identified proteins

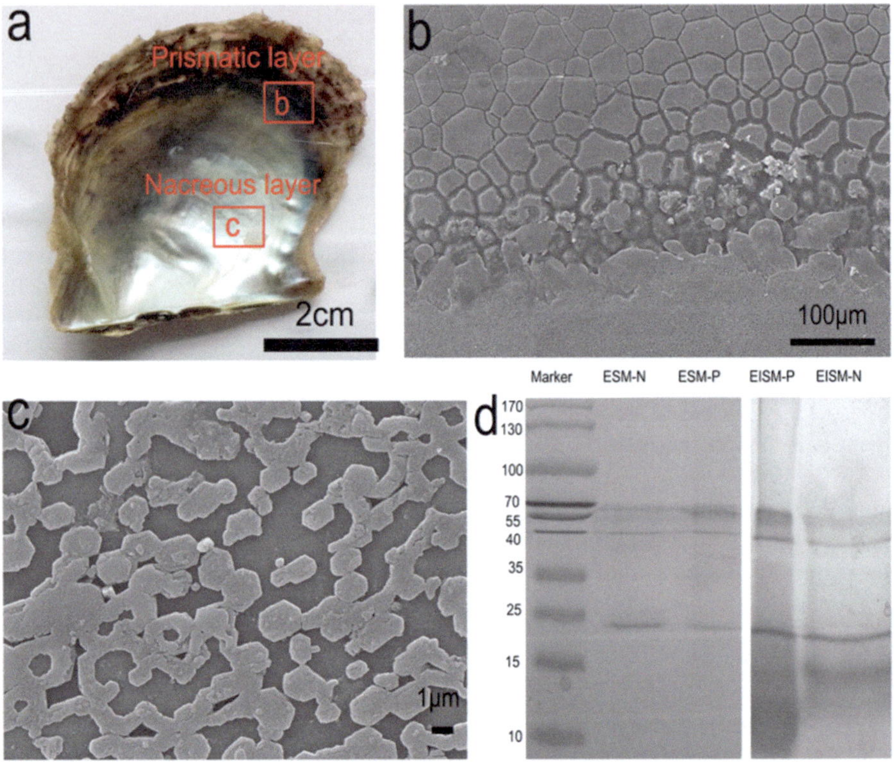

Fig. 3.1 (**a**) Optical image shows the prismatic and nacreous layer from a typical shell (red box is detected by SEM); (**b**) SEM indicates the shell surfaces of *Pinctada fucata*: the prismatic (**b**) and nacreous layer (**c**); (**d**) SDS-PAGE of the four groups of extracted proteins from *P. fucata* (*ESM* EDTA-soluble extracts, *EISM* EDTA-insoluble extracts, *P*, prismatic layer, *N*, nacreous layer)

are relevant to the existence of lysine and arginine residues which are available for trypsin cleavage in the protein sequences [26]. For instance, Aspein and Prisilkin-39 lack trypsin cleavage sites, making them undiscovered under standard proteomic detection [26]. Drake et al. discovered 36 skeletal organic matrix proteins from the coral, *Stylophora pistillata*. Thirty-one were found with tryptic digestion, whereas the remaining five were obtained after proteinase K digestion [27]. That is to say, the use of different digestion reagents only slightly influence the number of detected SMPs, and proteins found by trypsin as a digestion reagent discover most proteins in the shell.

Intrinsic disorder (ID) refers to segments or the whole proteins that have no predicted 3D structures, while such disorder sometimes existing in the native state proteins. ID domains are crucial molecular features that are related to the formation of mollusk nacre; John Evans observed 39 mollusk aragonite-associated protein sequences, containing more than one region of intrinsic disorder or unfolding region

[28]. This researcher suggested that the intrinsically disordered domains are significant for matrix assembly [28]. Therefore, we used IUPRED to check the 35 unique SMPs obtained by proteomics. The consequences showed that 22 out of 35 sequences were predicted to have more than one intrinsic disorder domain (data not shown). Through XSTREAM, 7 out of 35 sequences were suggested to have tandem repeats (data not shown). It is noteworthy that repetitive low complexity domains (RLCDs) are crucial but not the only character for intrinsically disordered proteins.

In terms of the blastp results from the National Center for Biotechnology Information (NCBI) database, SMPs were distinguished into two groups: proteins with homology (e-value$\leq 10^{-5}$) (Table 3.2) and proteins lacking homology (e-value$>10^{-5}$) (Table 3.3). Compared with the shell proteomics of *Pinctada margaritifera* [29], a coevolutionary species with *P. fucata*, the numbers of proteins found in both layers were notably improved. Shematrin-1, Shematrin-2, Shematrin-7, Mpn88, nacrein, nacrein-like, PTyr, PTyr1, and PNU1-9 existed in both layers, implying their potential functions in the formation of both layers in *P. fucata*. However, the proteins in both layers of *P. margaritifera* were nacrein, Shematrin-8, and nacre uncharacterized shell protein (NUSP18), suggesting that the molecular toolkits contribute to formation of the prismatic and nacreous layers were totally different. Sequence alignments from different mollusks and metazoans (data not shown) showed that peroxiredoxin, copper amine oxidase, and chitinase were highly conserved among the Metazoa. FN3 domain-containing proteins and EGF domain-containing proteins were highly conserved in the *Pinctada* family.

3.3.2 Immunolocalization of Proteins

To further confirm the SMPs in vivo and in vitro, immunolocalization experiments were executed. Western blotting using polyclonal antibodies as first antibody raised against the mixed shell matrix proteins of *P. fucata* suggested ETDA-soluble matrices (ESMs) and ETDA-insoluble matrices (EISMs) all detected by the antibodies (data not shown). Immunohistochemical results clearly showed that the SMPs existed in the mantle edge and mantle pallial (Fig. 3.2 a1) while showed no signal in the control group (Fig. 3.2a2). Immunogold detections of the prism suggested that the matrix proteins exhibited in both the prismatic tablets (Fig. 3.2b1) and the chitin layer (Fig. 3.2b3). In the nacre of *P. fucata*, antibodies exhibited a specific signal on nacreous tablets and localized in the interlamellar matrix which separated nacre tablets (Fig. 3.2b4) and in the nacre tablets (Fig. 3.2b2), while the control had no gold nanoparticle signal (data not shown). Immunolabeling synthetic calcite was executed to verify the function of extracted proteins on the growth of $CaCO_3$. In the control experimental group without extracted proteins, no fluorescence signal was detected under the same microscopy settings (data not shown). In contrast, all four

Table 3.2 Shell matrix proteins extracted from the shells of *P. fucata* with blast homology

Shell layer	Protein ID	Protein name	Protein homology	Domain
P,N	pfu_aug1.0_7401.1_31382.t1	Mpn88	*pfu* Mpn88-lack7	–
P,N	pfu_aug1.0_33972.1_19513.t1	Nacrein	*pfu* nacrein	CA
P,N	pfu_aug1.0_8238.1_17260.t1	Nacrein-like	*pfu* nacrein-like	CA
P,N	pfu_aug1.0_24266.1_33421.t1	Shematrin-1	*pfu* Shematrin-1	GY-rich
P,N	pfu_aug1.0_114185.1_13709.t1	Shematrin-2	*pfu* Shematrin-2	GYV-rich
P,N	pfu_aug1.0_411.1_00293.t1	Shematrin-7	*pfu* Shematrin-7	GLS-rich
P,N	pfu_aug1.0_109634.1_42492.t1	PTyr	*pfu* PfTy (tyrosinase-like protein)	Tyrosinase
P,N	pfu_aug1.0_10251.1_39018.t1	PTyr1	*pfu* tyrosinase-like protein 1	Tyrosinase
P	pfu_aug1.0_14219.1_03462.t1; pfu_aug1.0_3035.1_59110.t1	PFMG1	*pfu* mantle protein 10 (PFMG1)	Complement component C1q
P	pfu_aug1.0_15935.1_54245.t1	Prismalin-14	*pfu* Prismalin-14	GY-rich, D-rich
P	pfu_aug1.0_89559.1_56770.t1	PGRP2	*pfu* glycine-rich protein 2-like	G-rich
P	pfu_aug1.0_275.1_14939.t1	PAmylase	*pfu* amylase	A-rich
P	pfu_aug1.0_5122.1_59706.t1	KRMP4	*pfu* KRMP4	K-rich
P	pfu_aug1.0_3950.1_30572.t1	Shematrin-5	*pfu* Shematrin-5	GLY-rich
P	pfu_aug1.0_3212.1_37533.t1	PTyr1.1	*pfu* tyrosinase-like protein 1	Tyrosinase
P	pfu_aug1.0_12145.1_17832.t1	PTyrB1.1	*pmax* tyrosinase B1.1	Tyrosinase
P	pfu_aug1.0_21093.1_62062.t1	PTyr2	*pmar* tyrosinase-like protein 2	Tyrosinase
P	pfu_aug1.0_16905.1_25558.t1	PTyr2.1	*pmar* tyrosinase 2	Tyrosinase
P	pfu_aug1.0_13989.1_32380.t1	PClp1	*pmar* Clp1	Glyco_hydro_18
P	pfu_aug1.0_10761.1_31980.t1	PClp3	*pmar* Clp3	Glyco_hydro_18
P	pfu_aug1.0_20170.1_11291.t1	PPP-10	*pfu* PPP-10	S,Y-rich
P	pfu_aug1.0_89.1_57967.t1	Alveoline-like protein (Alv)	*pmar* Alveoline-like protein	V-rich
P	pfu_aug1.0_5205.1_38088.t1	PCopper	*pmar* Copper amine oxidase	Cu_amine_oxid
P	pfu_aug1.0_1824.1_37129.t1	PLiprin-α	*Cgi* liprin-alpha-1	SAM domain
P	pfu_aug1.0_3120.1_51891.t1	Beta-actin	*pfu* beta-actin	actin
P	pfu_aug1.0_17386.1_69084.t1	SGMP1	*pfu* SGMP1 (cement-like protein)	S,G-rich
P	pfu_aug1.0_12300.1_32179.t1	PTIMP3	*pmar* TIMP3	TIMP
P	pfu_aug1.0_8833.1_02656.t1	PTIMP	–	TIMP

(continued)

Table 3.2 (continued)

Shell layer	Protein ID	Protein name	Protein homology	Domain
N	pfu_aug1.0_13237.1_61173.t1;	Pif177	*pfu* Pif177	vWA+ChtBD2
	pfu_aug1.0_220503.1_50183.t1			
N	pfu_aug1.0_277.1_07463.t1	N16	*pfu* N16 (Pearlin)	–
N	pfu_aug1.0_16924.1_69026.t1;	N19	*pfu* N19	–
	pfu_aug1.0_2495.1_22991.t1;			
	pfu_aug1.0_2495.1_22993.t1;			
	pfu_aug1.0_36504.1_12255.t1			
N	pfu_aug1.0_170.1_36333.t1	MSI80	*Pfu*MSI80	A-rich
N	pfu_aug1.0_5926.1_52657.t1	MSI60	*pfu* MSI60	A,D-rich
N	pfu_aug1.0_13405.1_25160.t1	MRNP	*pfu*MRNP	M-rich
N	pfu_aug1.0_3120.1_51891.t1	Actin	*pfu* actin	Actin
N	pfu_aug1.0_6671.1_24057.t1	Peroxiredoxin	*pfu* peroxiredoxin	-
N	pfu_aug1.0_579791.1_57830.t1	Polyubiquitin	*pfu* polyubiquitin	Ubiquitin

groups with the presence of extracted proteins showed fluorescence signals, suggesting they could be attached in/on the $CaCO_3$. Especially, SMPs extracted from prismatic layers at approximately 1 $\mu g \cdot ml^{-1}$ had no significant effect on the morphology of $CaCO_3$ crystals and were uniformly distributed (Fig. 3.2c1 and c2). EDTA-soluble matrix from nacre was focused in the corn of crystals (Fig. 3.2c3). EDTA-insoluble matrix from nacre changed the rhombohedral crystals into 5–10 μm spherical particles. What's more, the fluorescence distributed at the edge of particles (Fig. 3.2c4). These results suggest that SMPs secreted by the mantle cells and are finally transported onto the shells. Besides, SMPs can influence the $CaCO_3$ crystallization process. Extensive studies have indicated that SMPs, with a single protein or proteins group, take participate in the nucleation, morphology, and polymorphism of $CaCO_3$ [18, 30]. The immune assay using antibodies against the SMPs as first antibody suggests that SMPs from different parts of shells play their distinct roles during $CaCO_3$ crystallization, leading to being distributed in/on $CaCO_3$ with different patterns. The in vitro crystallization experiments were executed under a basic condition (pH 8.0), which is similar to the pH of seawater (pH 8.2); hence, the experiment may provide clue about the mechanism behind in vivo mineralization.

Table 3.3 Shell matrix proteins extracted from the shells of *P. fucata* without blast homology [24]

Shell layer	Protein ID	Protein name	Domain	pI	MW kDa
P, N	pfu_aug1.0_10759.1_31979.t1	PNU1	ChtBD2	5.69	98.2
P, N	pfu_aug1.0_14699.1_32469.t1	PNU2	D-rich	4.85	43.6
P, N	pfu_aug1.0_14887.1_32490.t1	PNU3	Glyco_hydro_18	6.07	33.8
P, N	pfu_aug1.0_17316.1_18451.t1	PNU4	vWA	4.84	7.12
P, N	pfu_aug1.0_287428.1_50307.t1	PNU5	vWA	8.96	17.1
P, N	pfu_aug1.0_4881.1_67066.t1	PNU6	D-rich	3.51	21.4
P, N	pfu_aug1.0_75094.1_56521.t1	PNU7	–	5.40	13.8
P, N	pfu_aug1.0_7598.1_38581.t1	PNU8	–	9.86	11.6
P, N	pfu_aug1.0_86755.1_42180.t1	PNU9	–	8.20	9.6
P	pfu_aug1.0_1065.1_36810.t1	PU1	Cu2_monooxygen	9.49	53.6
P	pfu_aug1.0_10761.1_31980.t1	PU2	Glyco_hydro_18	8.85	58.5
P	pfu_aug1.0_11437.1_03076.t1	PU3	Fibronectin type 3	5.82	43.2
P	pfu_aug1.0_12760.1_53855.t1	PU4	vWA+ CCP	7.74	45.0
P	pfu_aug1.0_13143.1_39453.t1	PU5	Fibronectin type 3	4.90	33.2
P	pfu_aug1.0_155246.1_57288.t1	PU6	Fibronectin type 3	9.81	16.0
P	pfu_aug1.0_164724.1_21334.t1	PU7	–	11.6	29.0
P	pfu_aug1.0_1843.1_37145.t1	PU8	ChtBD2+CCP+ laminin_G_3	5.13	21.8
P	pfu_aug1.0_20998.1_69384.t1	PU9	–	5.14	9.3
P	pfu_aug1.0_24840.1_62334.t1	PU10	Zn^{2+} binding + CCP	4.71	75.3
P	pfu_aug1.0_289511.1_28801.t1	PU11	–	9.43	21.5
P	pfu_aug1.0_853.1_22356.t1	PU12	EGF	8.30	38.4
P	pfu_aug1.0_9037.1_60542.t1	PU13	–	7.89	9.1
P	pfu_aug1.0_9053.1_67966.t1	PU14	Q-rich	10.2	77.3
P	pfu_aug1.0_9430.1_38910.t1	PU15	Fibronectin type 3	5.12	18.1
P	pfu_aug1.0_154829.1_06667.t1	PU16	Fibronectin type 3	9.63	13.6

(continued)

Table 3.3 (continued)

Shell layer	Protein ID	Protein name	Domain	pI	MW kDa
N	pfu_aug1.0_12674.1_61081.t1	NU1	–	4.82	5.28
N	pfu_aug1.0_18113.1_25698.t1	NU2	Zf-Tim10_DDP	8.25	70.1
N	pfu_aug1.0_2111.1_22872.t1	NU3	–	7.98	29.2
N	pfu_aug1.0_21286.1_69411.t1	NU4	–	8.15	17.1
N	pfu_aug1.0_2323.1_15782.t1	NU5	ChtBD2, GS-rich	4.69	184.3
N	pfu_aug1.0_23607.1_62238.t1	NU6	–	8.48	7.93
N	pfu_aug1.0_24634.1_54990.t1	NU7	-GSM-rich	11.79	35.4
N	pfu_aug1.0_28.1_57930.t1	NU8	–	8.76	54.8
N	pfu_aug1.0_4561.1_44973.t1	NU9	–	8.76	46.5
N	pfu_aug1.0_954.1_58317.t1	NU10	Laminin_G_2	9.62	17.8

Aberration: *P* prismatic layer, *N* nacreous layer, "*P, N*" means that the SMPs exist in both of two layers, *Pfu Pinctada fucata, Pmax Pinctada maxima*, *Pmar Pinctada margaritifera*, *Cgi Crassostrea gigas*, *CA* carbonic anhydrase, *EGF* epidermal growth factor-like, *VWA* von Willebrand factor (vWF) type A domain, *ChtBD2* chitin-binding domain 2, *CCP/SCR/SUSHI*, the complement control protein, *TIMP* tissue inhibitor of metalloproteinase, *SAM* sterile alpha motif, Zf-Tim10_DDP, zinc finger (Znf) domain. "–" represents undetected
Notes: The protein ID is the gene ID in the draft genome of *Pinctada fucata,* which is accessed at http://marinegenomics.oist.jp/genomes/viewer?project_id=20&current_assembly_version=ver1.0

3.3.3 Identification and Quantification of Matrix Genes by Real-Time PCR

To identify and quantify the proteins found according to our proteomic analysis, real-time PCR was conducted. As we know, the mantle edge contributes to the formation of the periostracum and the prismatic layer; nevertheless, the mantle pallial is responsible for the formation of nacreous layer [29]. Therefore, we detected the relative gene expression of 21 selected genes found in the proteomic analysis, in the mantle edge, and mantle pallial and other tissue of *P. fucata* (Fig. 3.3a). Referred to previous research, six developmental stages have been distinguished across the entire *P. fucata* life cycle, containing descriptions of the fertilized egg, trochophore stage, D-shaped stage, umbonal stage, juvenile stage, and adult stage [31]. The shell layer structure, calcium carbonate crystal polymorphisms, and the expression of SMPs were different in different stages [31]. Almost all SMPs represented a dramatic increase until the adult stage. For instance, the relative expression level of Prisilkin-39 and Pif in the adult stage is 119.48 and 2116.9 times that in the juvenile stage, respectively [31]. Therefore, we extract all RNA from the mantle of adult oysters in this study. SMPs encoded by the 21 genes include one valine (V)-

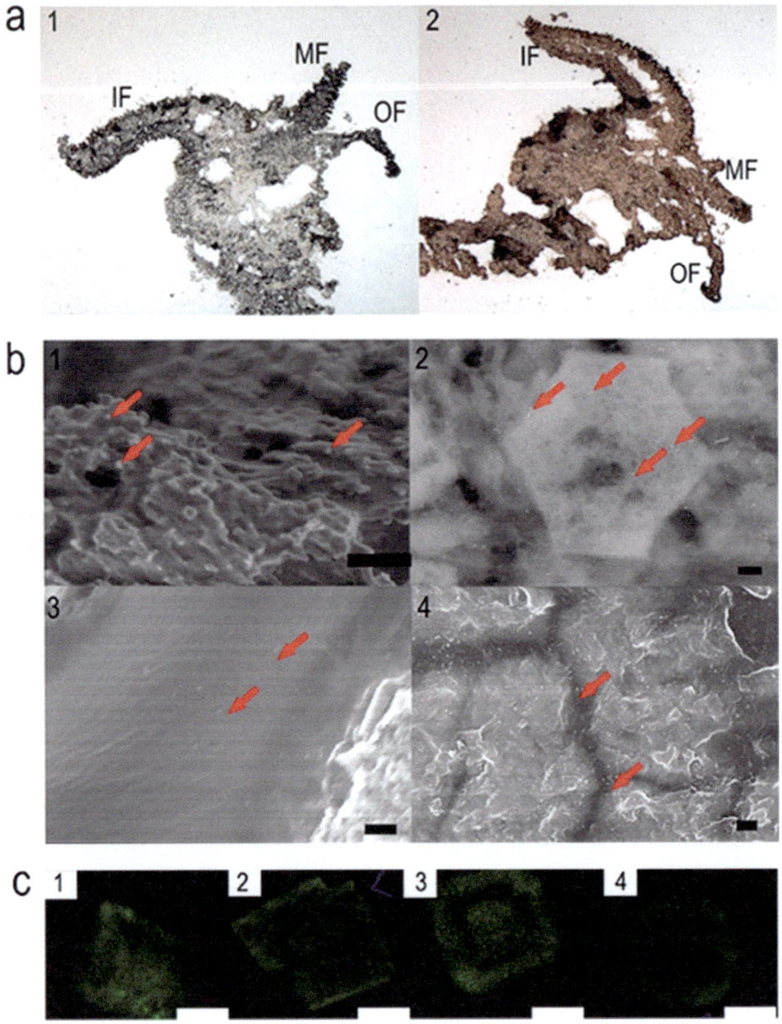

Fig. 3.2 Immunolocalization of the shell matrix proteins (SMPs) in *P. fucata*. Polyclonal antibody raised against whole extracted proteins is used to identify EDTA-soluble matrices (ESMs) and EDTA-insoluble matrices (EISMs) as first antibody. (**a**) Immunohistochemical localization of SMPs in the mantle epithelia: control group without the first antibody (a1); mantle sections addition of the first antibody (a2) (MF, middle fold; OF, outer fold; IF, inner fold); (**b**) immunogold labeling of SMPs on the EDTA-processed prismatic and nacreous layers. Layers with the first antibody (b1–c4). b1 and b3 are prism and b2 and b4 are nacre. The red arrowheads point gold nanoparticles. (Scale bars, 200 nm) (**c**) Confocal fluorescence laser scanning microscopy images of synthetic calcite in vitro after adding immunolabeled SMPs. Calcite with 1 μg·mL^{-1} ESM-P (c1), EISM-P (c2), ESM-N (c3), EISM-N (c4) (Microscopy parameters are the same. Scale bars, 10 μm)

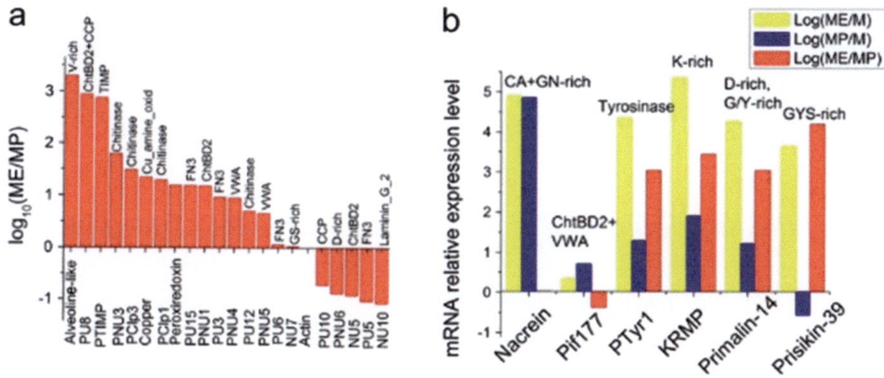

Fig. 3.3 Real-time PCR of chosen SMPs with functional domains or repeats indicates relative gene expression in the mantle pallial and mantle edge of *P. fucata*: (**a**) Relative gene expression of these 21 selected SMPs in the mantle edge referenced with mantle pallial. Copper=copper amine oxidase (**b**) Relative gene expression of six well-known SMPs (The longitudinal coordinates show the values of $\log_{10}$(ME/MP), *ME* mantle edge, *MP* mantle pallial)

rich protein (alveoline-like protein), one aspartic acid (D)-rich protein (PNU6), one glycine and serine (GS)-rich protein (NU7), one peroxiredoxin, one copper amine oxidase, one tissue inhibitor of metalloproteinase (PTIMP), two complement control protein (CCP) proteins (PU8 and PU10), two von Willebrand factor type A (vWA) proteins (PNU4 and PNU5), one laminin G protein (NU10), three chitin-binding proteins (PU8, PNU1, and NU5), four fibronectin type III (FN3) proteins (PU3, PU5, PU6, and PU15), and four chitinase (Clp1, Clp3, PNU3, and PU12). Among all 21 tested genes [24], 18 genes (*alveoline-like*, *PU8*, *PTIMP*, *PNU3*, *Clp1*, *Clp3*, *copper amine oxidase*, *peroxiredoxin*, *PU15*, *PNU1*, *PU3*, *PU12*, *PU6*, *PU10*, *PNU6*, *NU5*, *PU5*, and *NU10*) were highly expressed in the mantle pallial or/both mantle edge, compared to the muscle (data not shown), suggesting they were likely to participate in biomineralization process. However, *PNU4*, *PNU5*, and *NU7* were lowly expressed in the mantle edge and mantle pallial, so they may originate from other cells like hemocytes [32] and finally be involved in biomineralization [33]. Log (ME/MP) is the relative expression in the mantle edge compared to the mantle pallial, implying roles of genes in the formation of the prismatic or/both nacreous layers. The results suggested that most proteins existed in the prismatic layer were highly expressed in the mantle edge (Fig. 3.3a). Unexpectedly, PU10 and PU5 were found existing in the prismatic layer, while the corresponding genes were more highly expressed in the mantle pallial than mantle edge (Fig. 3.3a), indicating their additional roles in nacre formation. In the same way, previous studies found that Shematrin-5, a prism-related SMP, had much higher expression in the mantle pallial than in the mantle edge [34]. What's more, six well-studied genes were also detected by real-time PCR (Fig. 3.3b). *Nacrein* and *Pif177* genes are related to the formation of the nacre and presented high expression in the mantle pallial (~10^5 and ~tenfold compared to the muscle, respectively). *Tyrosinase-1*(*Tyr-1*), *Primalin-14*, *Prisikin-39*, and *KRMP* are genes related to the formation of the prism and showed high expression in the mantle edge (~10^4, ~10^5, ~10^4, and ~10^3 time compared to the muscle, respectively). Hence, these data verify that these genes are highly expressed in the calcifying tissues, and the corresponding proteins are involved in the shells of *P. fucata*.

3.3.4 Putative Functions of the SMPs According to Domains

To get insight into the functions of SMPs in biomineralization, analysis of their sequences and domains is executed. SMPs involved in biomineralization have several distinct features: (1) they are abundant in some specific amino acids including aspartic acid, glycine, glutamic acid, and serine; (2) they have RLCDs (repeated low complexity domains) and flexible secondary structures; and (3) they have multiple modularity. From well-studied SMPs, biomineralization-related proteins included carbonic anhydrases and aragonite-binding, chitin-binding, vWA, and D-rich domains. According to the blast results, some well-known SMPs are found in the shells, confirming the effectiveness of our method. Generally, the Shematrin family [17] and tyrosinase family [11] were involved in the most abundant proteins in the prismatic layer, and others like nacrein [5], chitinase-like protein 1 (Clp1) [29], Clp3 [29], KRMP [9], amylase (GenBank, AGN55420.1), alveoline-like protein [29], glycine-rich protein 2-like (PGRP2) [35], Prismalin-14 [8], liprin-α protein, PPP-10 [36], mantle protein 10 (PFMG1) [37], Mpn88 [26], cement-like protein (SGMP1) [29], tissue inhibitor of metalloproteinases (TIMPs) [38], actin (GenBank, ACD99707.1), and copper amine oxidase [29]. Nacrein [5], Pif177 [21], N16 [7], and N19 [39] were abundant in the nacre, and others were MSI80 (GenBank, BAL45933.1), methionine-rich nacre protein (MRNP) [40], MSI60 [6], Mpn88 [26], Actin (GenBank, ACD99707.1), peroxiredoxin [29], and polyubiquitin [41] (Table. 3.2). The SMPs that have been thoroughly studied previously will not be discussed in this work [24].

What's more, we recognized some domains that are thought to play significant roles in the biomineralization process. These domains contain Cu2_monooxygen, Glyco_hydro_18, chitin-binding domain 2 (ChtBD2), epidermal growth factor-like (EGF), complement control protein (CCP/SUSHI), TIMP, and laminin_G_2/3. Referred with domains existed in *P. margaritifera*, Glyco_hydro_20, Kunitz-like domains, and EF-hand were not found. Kunitz domains (InterPro accession number, IPR002223) are the functional domains of proteins that depress the function of protein-degrading enzymes. It is noteworthy that the proteins extracted from prismatic layers possess more diverse domains than those from nacre. According to the domains, the SMPs were divided into the following two groups:

1. Proteins that putatively regulate the extracellular microenvironment: chitinase, carbonic anhydrase, chitin-binding proteins, and TIMP

The microenvironment, the temporary small-scale environment around mantle cells, and organs are involved in the formation of shells, like pH, framework (mostly chitin), and proteinase in the seawater. Hence, the extracellular microenvironment is important for shell formation. Three types of proteins are involved in the microenvironment: carbonic anhydrase, chitin relevant proteins, and proteinase.

Carbonic anhydrase take participate in controlling pH by converting CO_2 to HCO_3^- and existed in nacrein. Chitin is the major framework on which $CaCO_3$ grows [42]. Four proteins (PClp1, PClp3, PNU3, and PU2) have Glyco_hydro_18 domains (IPR001223), which is one of the families of glycoside hydrolases that hydrolyze the glycosidic bond between carbohydrates or between a carbohydrate and a noncarbohydrate moiety. Especially, chitin oligosaccharides could be hydrolyzed by chitinase. Real-time PCR results showed that the mantle edge and mantle pallial had high expression levels of mRNAs of *PClp1*, *PClp3*, and *PNU3*, suggesting their important roles in shell formation (Fig. 3.3a). A very recent microarray study suggested that the D-shaped stage of *P. fucata* had highly expressed of chitinases when the shell were first formed [31]. Besides chitinase, another domain relevant to chitin is chitin-binding domain_2 (IPR002557). Four proteins (PNU1, PU8, Pif177, and NU5) have this domain, indicating that chitin-binding ability is important in both layers. The mantle edge had higher mRNA expression levels of *PNU1* and *PU8* than in the mantle pallial, while *N-U5* was the opposite. *Pif177*, an important gene in nacre formation, had a higher mRNA expression levels in the mantle pallial than in the mantle edge (Fig. 3.3b). In complicated seawater, all secreted SMPs will partly degraded by microorganism proteinases. Action must be taken to solve this problem. Tissue inhibitors of metalloproteinase (PTIMP and PTIMP3) were found, which may mix with extracellular matrix metalloproteinases including collagenases and irreversibly devitalize them (IPR001820). The extracellular regions of vertebrate species had members of this family. It has been suggested that the effect of TIMP in *P. martensii* on nacre formation is to depress matrix metalloproteinases (MMP) activity by degrading most components of the extracellular matrix [38]. Therefore, the TIMP could ensure a "safe" microenvironment for other SMPs to execute their activities via creating suitable ratios of MMP to TIMP. Inhibition of MMP activity in humans occurs in a 1:1 stoichiometric relationship, and an imbalanced MMP to TIMP ratio may lead to various diseases [43]. In *P. martensii*, knocking down of TIMP by double-strand RNA interference leads to abnormal nacre formation. However, the "suitable" MMP-TIMP ratio in the formation of nacre has yet unknown. In fact, water-soluble extracts from the nacre of *P. margaritifera* had proteinase inhibitory effect against proteinase K [44]. What's more, TIMP also existed in the shells of *C. gigas* [45] and *M. coruscus* [46]. Whereas, there are not found of other protease inhibitors, including Kunitz-like protease inhibitor and WAP (whey acidic protein) domains proteins [47] in the shells of *P. fucata*. There is also a guess that the content of proteinase inhibitors in the shell is too low to detect. The problem may be solved by RNA-seq of mantle cells in the future.

2. Extracellular matrix (ECM) relevant proteins: laminin proteins, fibronectin-related proteins, EGF proteins, and liprin-α proteins

The shells keep growing and can be repaired after injury of cut, showing there might be interaction between cells of different systems from *P. fucata* and the extracellular matrix (ECM). Proteomic data of *C. gigas* showed that oyster shell matrix formation is not simply by self-assembling silk-like proteins but via diverse proteins by complex combination and modification processes that may include hemocyte and exosomes [45]. In fact, we found that the RNA transport and ECM receptor proteins were the most abundant in shell matrix. In accord with previous results, four ECM-related domains (fibronectin-related, laminin, EGF proteins, and liprin-α) in the SMPs were identified that may regulate communication between cells and the extracellular matrix.

Fibronectin type III (FN3) repeat regions were found in five uncharacterized proteins from the prismatic layer, PU3, PU5, PU6, PU15, and PU16, (Fig. 3.3a). Interestingly, PU3, PU5, and PU6 are one of the most expressed proteins in the mantle pallial (data not shown), even though they exist in the prismatic layer. PU3, PU5, PU6, and PU15 have high expression in both mantle edge and mantle pallial, indicating their crucial roles in the formation of both layers. FN3, an approximately 100 amino acid region, contains different tandem repeats of binding sites to DNA, heparin, or the cell surface. Most of proteins with FN3 participated in cell surface binding or are cytokine receptors (IPR003961) or receptor protein of tyrosine kinases. The beta-sandwich structure of FN3 closely likes that of immunoglobulin domains [48]. Remarkably, a gene coding for a fibronectin-like protein in *C. gigas* had high expression at the early developmental stage when larval shells are formed in accord with chitin synthase [45]. The shells of various biomineralization species contained fibronectin, including *P. margaritifera* [29], *C. gigas* [45], *M. coruscus* [46], and *A. millepora* [49], suggesting that fibronectin is vital but yet unknown roles.

PU8 and NU10 contain a laminin_G (LG) domain, which is reported to mediate attachment, migration, and organization of cells into tissues when embryonic development due to its function of interacting with other extracellular matrix components (IPR001791). This domain has about 180–200 residues and exists in many extracellular and receptor proteins. LG modules have been predicted to interact with cellular receptors including $\alpha_6\beta_1$ integrins, sulfated carbohydrates, and other extracellular ligands [50]. Proteins containing laminin_G domain are found in the shells of *C. gigas* [45], *M. coruscus* [46], and *L. gigantea* [51].

PU12 has two epidermal growth factor-like domains (EGF), which commonly exist in extracellular proteins (IPR000742). EGF could exhibit six conserved cysteine residues linked by three sulfide bonds. These domains are relevant to the apoptosis, immune system, and Ca^{2+} binding. EGF proteins are found in the shells of *P. margaritifera, C. gigas* [45], *M. coruscus* [46], and *L. gigantea* [51].

Another protein in this category is PLiprin-α, a kind of the leukocyte common antigen-related (LAR) protein tyrosine phosphatase-interacting protein. This protein could bind to the tyrosine phosphatase LAR and the LAR locates at cell focal adhesions. This binding may adjust the disassembly of focal adhesion and then help orchestrate cell-matrix coactions (IPR029515) [52]. However, this protein only exists in the shells of *C. gigas* [45].

Notably, FN3, laminin_G, EGF, and liprin-α proteins are capable of interacting with integrins which are transmembrane receptors which mediate cell-cell and cell-ECM interactions [53]. Integrin was found in the matrix proteins of *S. pistillata* which is a biomineralization model of coral species [27]. What's more, $\alpha_v\beta_6$ integrin was reported to be expressed by ameloblasts and participated in regulating amelogenin deposition and enamel biomineralization [54]. These studies indicate that the integrin has a strong participation in biomineralization. In fact, gene expression of *integrin* in the mantle of *P. fucata* is higher than that in the gonad, foot, muscle, and gill through real-time PCR (data not shown).

3. Other Proteins of Interest

Except for the above two families of proteins, some genes are also important according to their high expression in the mantle. There are an acidic protein, four VwA proteins, three complement control protein (CCP), a V-rich protein, and two enzymes (tyrosinase and copper amine oxidase) discussed as below.

PNU6 is an extremely acidic protein with predicted pI of 3.51 and has poly(D). The unique primary sequence suggests its role during the $CaCO_3$ formation, potentially anchoring Ca^{2+} via the polyD domain, concentrating local Ca^{2+} and assisting calcite precipitation. The effect of PNU6 may be resembled to the polyD-containing protein Aspein, an unusually acidic matrix protein existing in *P. fucata* [16].

Interactions between protein-protein are important for shell formation as framework proteins, acidic proteins, and other proteins must cooperate to fulfill the complicated process of biomineralization. VwA domains (IPR002035) of extracellular eukaryotic proteins could mediate adhesion through metal ion-dependent adhesion sites (MIDAS), which are also found in Pif177, PNU4, PU4, and PNU5. Three out of the four proteins exist in both layers, implying the importance of protein-protein interaction in formation of both layers. VWA proteins are also found in the shells of *P. margaritifera* [29], *C. gigas* [45], *M. coruscus* [46], and *L. gigantea* [51]

The complement control protein (CCP) modules (also titled as short consensus repeats (SCRs) or SUSHI repeats) are composed of approximately 60 amino acid residues. They are found in many kind of complement and adhesion protein [55] and are also found in matrix protein PU4, PU8, and PU10. Some of these proteins in this group play roles in the molecular basis of the blood group antigens which lie on the outside of the red blood cell membrane as surface markers (IPR000436). The CCP proteins suggest a putative relationship between the immune system and shells formation. CCP proteins exist in the shells of *P. margaritifera* [29], *C. gigas* [45], *M. coruscus* [46], and *S. pistillata* [27].

Alveoline-like protein (Alv), a protein with characteristic of V-rich, has only been found in the other two related species *P. maxima* and *P. margaritifera* [29]. Real-time PCR suggests that it is highly expressed (almost 10^3–10^4 times compared to the muscle) in both the mantle edge and mantle pallial (Fig. 3.3a), suggesting its crucial roles in the formation of both prism and nacre. Whereas, the function of Alv during $CaCO_3$ crystallization is poorly understood.

Tyrosinase is an oxidase that could control the production of melanin through hydroxylation of a monophenol to o-quinone (IPR002227). It has been reported that there are 21 tyrosinase genes in the genome of *P. fucata* [56]. As expected, there are four tyrosinase proteins identified in the prismatic layer and two tyrosinase proteins exist in both layers. Tyrosinase is also identified in the shells of *P. margaritifera* [29], *C. gigas* [45], and *M. coruscus* [46]. Although the specific roles of tyrosinase are still under study, it is speculated that this protein has distinctive functions in melanogenesis of pigmented shells [11]. What's more, a tyrosinase gene is possibly involved in the biogenesis of larval shell in *C. gigas* [57].

Copper amine oxidase could catalyze the oxidation of many kinds of biogenic amines like neurotransmitters, histamine, and xenobiotic amines (IPR000269). They have multifunctions including cell differentiation and growth, detoxification, wound healing, and cell signaling in eukaryotes. Copper amine oxidase in *P. fucata* has expressed highly in the mantle edge and has lower level in the mantle pallial. Whereas, its role has not thoroughly been investigated in mollusks. A previous report of eastern oysters indicated that the amine metabolic was enriched in SMPs via gene ontology enrichment analysis [33]. This protein has not been studied in other mollusks shells.

3.3.5 The Hint of Proteomics on the Shell Mineralization Mechanism

Some unexpected proteins have been discovered from our proteomic analysis, implying the complication of biomineralization in pearl oyster. The increased SMPs offer help to refine the previously proposed model of "chitin-silk fibroin gel proteins-acidic macromolecules" [42]. It has been previously proposed that the biomineralization mechanism of nacre includes the following four stages: (1) matrix assembly, (2) the first-formed mineral phase, (3) nucleation of sporadic aragonite tablets, and (4) tablets growth and formation of the mature tissue [42]. The matrix is formed by β-chitin layers at the first stage, with a gel composing of silk-like protein. Through proteomics analysis, we distinguished four chitinases in both shells layers which is an enzyme that could catalyze the hydrolysis of linkages of β-1,4-N-acetyl-d-glucosamine in chitin polymers or oligomers. Surprisingly, we did not identify any chitin synthase from our proteomic analysis, even though chitin synthase genes can be identified in *P. fucata* genome [56]. These results suggest that chitin synthases might locate inside the cell or on the cell membrane, whereas chitinases are secreted to the outside of cell to distribute on the chitin network. Chitin-binding proteins could interact with both chitin and minerals. For instance, Pif protein with chitin domains have been reported to participate in the combination of polysaccharide template and the inorganic phase and in the regulated nucleation of the initial mineral phase [30]. Pif proteins in vivo are indicated to be able to interact with other proteins like N16, participating in the formation of the lamellar sheet of nacre [21]. Silk-like proteins contain abundant Gly and Ala or just in Gly. Similar proteins cover MSI60, MSI80, Shematrins, PGRP2, PAmylase, SGMP1, NU5, KRMP4, and NU7. Silk-like proteins are discovered in both the prismatic and nacreous layers, suggesting their importance during shell formation. One effect of silk-like proteins is to work as a mild inhibitor of mineralization [42]. An in vitro crystallization experiment suggests that recombinant KRMP3 depresses the $CaCO_3$ precipitation, affects the crystal morphology of calcite, and inhibits the aragonite growth in vitro, and these effects are almost totally due to the lysine-rich region. The Gly/Tyr-rich region in KRMP3 could bind chitin [58]. After that, the second stage of the first-form mineral phase, which usually consists of amorphous calcium carbonate (ACC), is formed. ACC has been considered as the precursor of biominerals, which is found in many living organisms like nacre [59]. Mollusks need high dosages of Ca^{2+} and CO_3^{2-} from the seawater to ACC. CO_3^{2-} is concentrated by carbonic anhydrase including nacrein, and Ca^{2+} is concentrated by acidic proteins like Pif [21] and Aspein [16]. ACC is unstable referenced to calcite and thus requires to be stabilized by specialized macromolecules including Pif [30]. We found several acidic proteins in the shell, including PNU2 and PNU6. Nucleation of individual tablets begins at the third stage, requiring some nucleators. Asp-rich proteins are indicated to play important role in this stage. The final stage is the growth of the tablets to form mature shell, and the morphology of the final biominerals is considered to be thermodynamically driven [60]. At the final stage, SMPs are combined into the final biominerals. Notably, some SMPs like Pif have multiple effects and can function during more than one stage. What's more, peroxiredoxin, tyrosinase, and copper amine oxidase could contribute to modification of SMPs, being related to the massive structure and mechanical properties of shells.

Although this model could explain biomineralization largely from crystal growth perspective, many proteins like FN3 proteins, CCP proteins, and EGF proteins found in our proteomic analysis are not included to crystal growth until now. Recently, researches show that eastern oyster, *Crassostrea virginica*, forms its shell via a series of coordinated events including hemocyte cells and ECM [61]. Indeed, primary mantle cell cultures of *P. fucata* could also precipitate amorphous calcium carbonate in vitro, indicating the ability of mantle cells to participate in biomineralization and shell formation processes [62]. Thus, it is convictive to hypothesize that the shell formation process is relevant to ECM-related proteins which are secreted by the mantle cells. These ECM-related proteins are a part of SMPs and may have multiple functions. In fact, a highly expressed ECM-related protein, osteopontin, in the bone that is glycosylated and rich in acidic residues takes part in a number of cellular processes like immune response and apoptosis in addition to its main role in biomineralization [63].

Although we notice that domains could not necessarily represent the exact function of whole proteins, the results in the present study could provide hints on the further study of the diverse shell matrix proteins and improve our understanding of biomineralization.

3.4 Shell Matrix Protein from Nacre Layer

3.4.1 Extraction and Purification of Matrix Protein P14 from the Nacre of Pearl Oyster **Pinctada fucata**

P14, a soluble matrix protein with a calculated molecular mass of 14.5 kDa, was separated from fragmented nacre of pearl oysters (*Pinctada fucata*) which were treated with 10% NaOH solution to extract the nacre matrix proteins and study their effect on the $CaC0_3$ crystal further. The protein was identified by gel exclusion chromatography and reversed-phase high-performance liquid chromatography after demineralized by 10% acetic acid. The X-ray diffraction pattern of P14 crystals suggests that P14 plays a crucial role during nacre biomineralization. P14 could induce aragonite formation by stimulating $CaCO_3$ crystal formation and accelerating aragonite precipitation. Calefaction of the acid-insoluble nacre residue, named as conchiolin, in 10% sodium dodecyl sulfate solution with addition of 10% β-mercaptoethanol solution for 10–20 min at approximately 100 °C produces two other soluble proteins having molecular masses of 19.4 kDa and 25.0 kDa, respectively. The present study implies that these two proteins are linked to the insoluble organic matrix via disulfide bridges as the extraction yield increases when adding β-mercaptoethanol to the medium.

3.4.1.1 Isolation and Purification of P14

P14 was isolated from the acid soluble matrix (ASM) using reversed-phase high-performance liquid chromatography (HPLC) and gel exclusion chromatography. The absorbance at 280 nm (data not shown) indicated that the ASM was a compound which contains more than one protein in one peak. These proteins were then purified using gradient elution with acetonitrile through the HR CI8 of HPLC column, which could separate one main peak to several smaller and single peaks (Fig. 3.4). The SDS-PAGE results showed that the main peak was a purified protein which has apparent molecular mass of 14.5 kDa, the so-called P14 (Fig. 3.5). At the same time, P14 subjected into native PAGE showed two different bands (Fig. 3.6), which indicates that P14 has two different components which were not shown in the denaturalized condition.

3.4.1.2 Effect of P14 on $CaCO_3$ Crystallization

The function of P14 on the CaC03 crystallization process was examined by the X-ray diffraction pattern of crystals grown in crystallization solution with and without ASM and P14 (Fig. 3.7). The experiments results suggest that P14 and

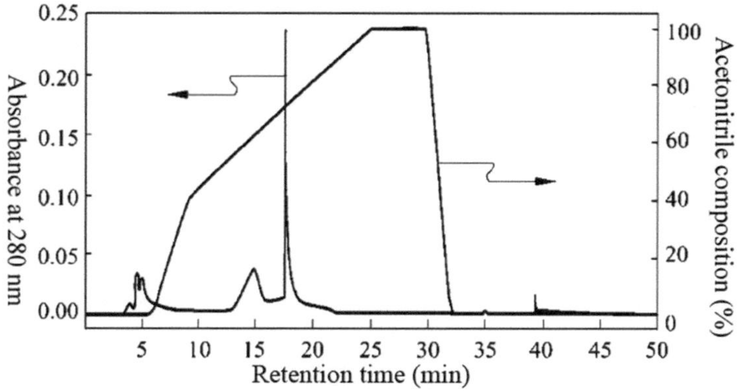

Fig. 3.4 Reversed-phase HPLC analyses of ASM which were first purified by gel exclusion chromatography in present study. 150 μl samples were eluted with a gradient of 0–100% acetonitrile with flow rate of 1.5 ml/min. The main peak was collected and concentrated to obtain purified P14

Fig. 3.5 Analysis by 15% SDS-PAGE of proteins (ASM) and P14 extracted from nacre of *P. fucata*. Lane 1, ASM from nacre of *P. fucata*; lane 2, protein band after first separation by gel exclusion chromatography; lane 3, purified P14 with an apparent molecular weight of 14.5 kDa; and lane 4, protein marker of molecular mass standards

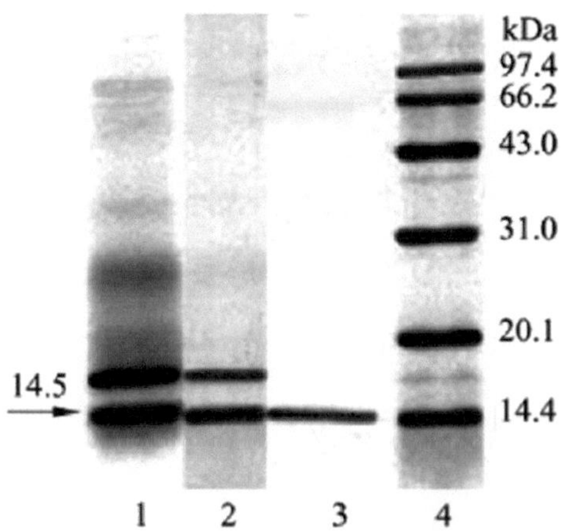

Fig. 3.6 Indication of purified P14 in 12.5% native PAGE. Lane 1, BSA solution with a dosage of 1 mg/ml; lane 2, 0.1 mg/ml purified P14

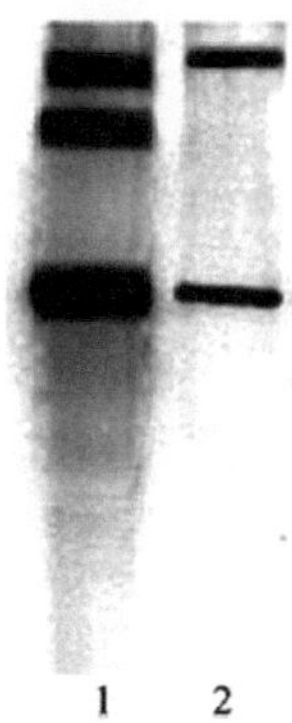

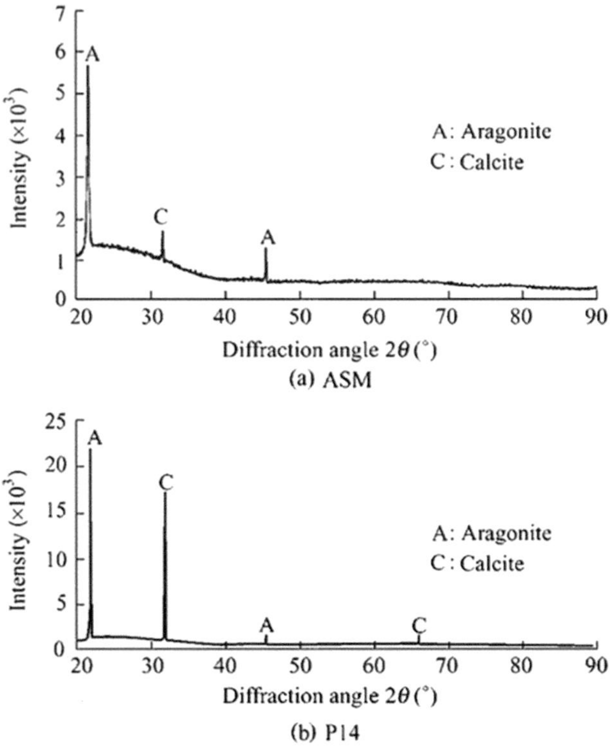

Fig. 3.7 X-ray diffraction pattern of $CaCO_3$ crystals mineralized on the glass surface with (**a**) ASM and (**b**) P14

ASM from the nacre of *Pinctada fucata* induced the formation of aragonite. What's more, P14 stimulated CaC03 formation and accelerated the precipitation rate of aragonite.

3.4.2 *A Novel Matrix Protein p10 from the Nacre of Pearl Oyster (*Pinctada fucata*) and Its Effects on Both $CaCO_3$ Crystal Formation and Mineralogenic Cells*

A novel matrix protein, named as p10 according to its apparent molecular mass of 10 kDa, was separated from the nacreous layer of pearl oyster (*Pinctada fucata*) via reverse-phase high-performance liquid chromatography [64]. In vitro crystallization assays indicated that p10 could stimulate the calcium carbonate crystals nucleation and induce aragonite formation, suggesting that it might play a crucial role during nacre biomineralization. As nacreous layer is known to possess osteogenic factors, two mineralogenic cell lines MRC-5 fibroblasts and MC3T3-E1 preosteoblasts were

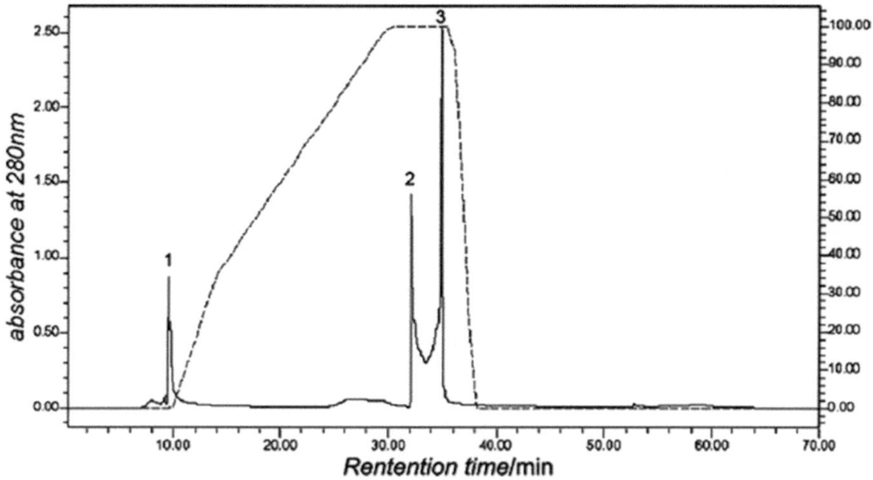

Fig. 3.8 Reverse-phase HPLC profile of the total extracted proteins. 100 μl samples were eluted with an elution gradient from 0% to 100% acetonitrile (dashed line indicates) with a flow rate of 0.8 ml/min. Peak 3 (p10) was collected and concentrated for further analysis

used to study the biological activity of p10. The results implied that p10 could increase alkaline phosphatase vitality, an early differentiation marker of osteoblast, while the activity of MRC-5 and MC3T3-E1 remained stabilized after treatment of p10. All in all, the findings resulted in identification of a novel matrix protein from the nacreous layer of *P. fucata* that plays an important role in both the biomineralization and in the differentiation of the cells involved in biomineralization.

3.4.2.1 Characterization and Purification of p10

Total nacre matrix proteins were extracted from the nacreous layer of *P. fucata* with phosphate buffer (pH 7.0). A major band of the proteins with apparent molecular mass of 10 kDa was observed in these extracted proteins. The total nacre matrix protein was then subjected to reverse-phase HPLC and this 10 kDa protein was purified with three main peaks appeared in the result (Fig. 3.8). SDS-PAGE analysis (Fig. 3.9a, lanes 2 and 5) showed that peak 3 was a purified and single protein which has a molecular mass of 10 kDa so-called p10. p10 showed a single mass spectrum protonated ion peak at m/z 6813.82 (Fig. 3.9b). The yield of p10 detected by the BCA method was approximately 50 ng/g from nacre powder.

3.4.2.2 Effect of p10 on Calcium Carbonate Precipitation

A series of precipitation experiments with or without p10 showed that the procedure could be divided into the following stages (Fig. 3.10). Firstly, $CaCl_2$ was added to

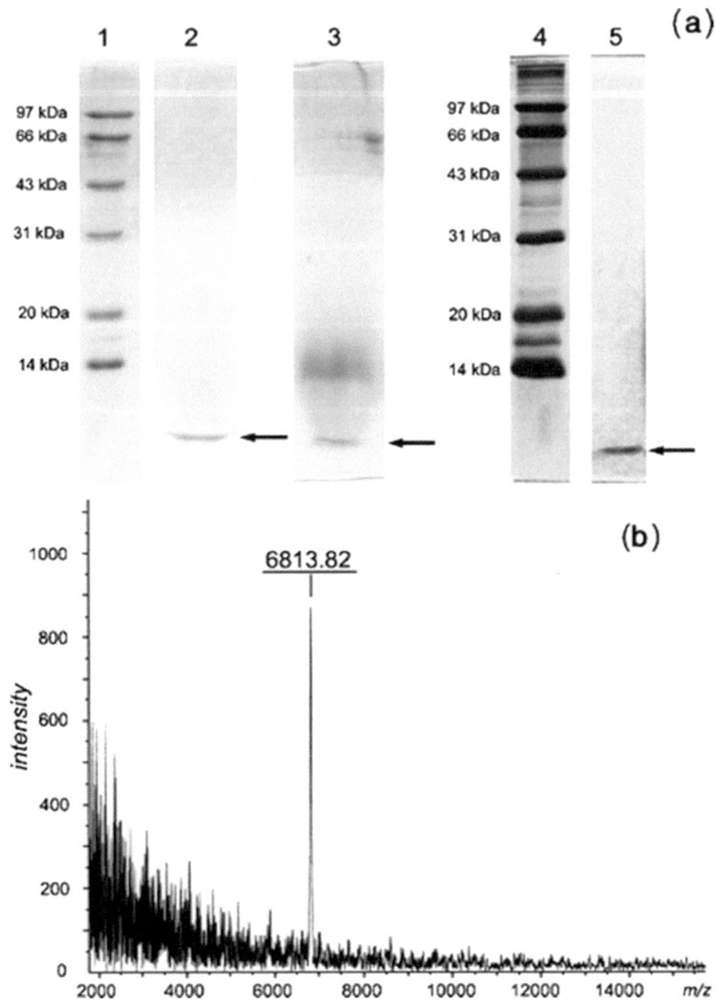

Fig. 3.9 (**a**) SDS-PAGE and (**b**) MALDI-TOF mass spectrometry of the protein p10 purified by reverse-phase HPLC. Lane 1 and lane 4, molecular weight markers with masses directed on the left. Lane 2 and lane 5, purified p10 with an apparent molecular mass of about 10 kDa (black arrow points). Lane 3, protein band pattern of whole extract protein, including p10 (black arrow indicates). The gels containing lanes 1–3 were dyed with Coomassie Brilliant Blue, while the gels containing lane 4 and lane 5 were dyed with silver salt [64]

$NaHCO_3$ to form saturated solution along with pH dropped instantly. This sudden change might be related to the formation of a compound from the ionic species, other than nucleation. Secondly, there was a slight rise of the pH, resulting to a relatively stable stage for a few seconds. Finally, precipitation and nucleation occurred

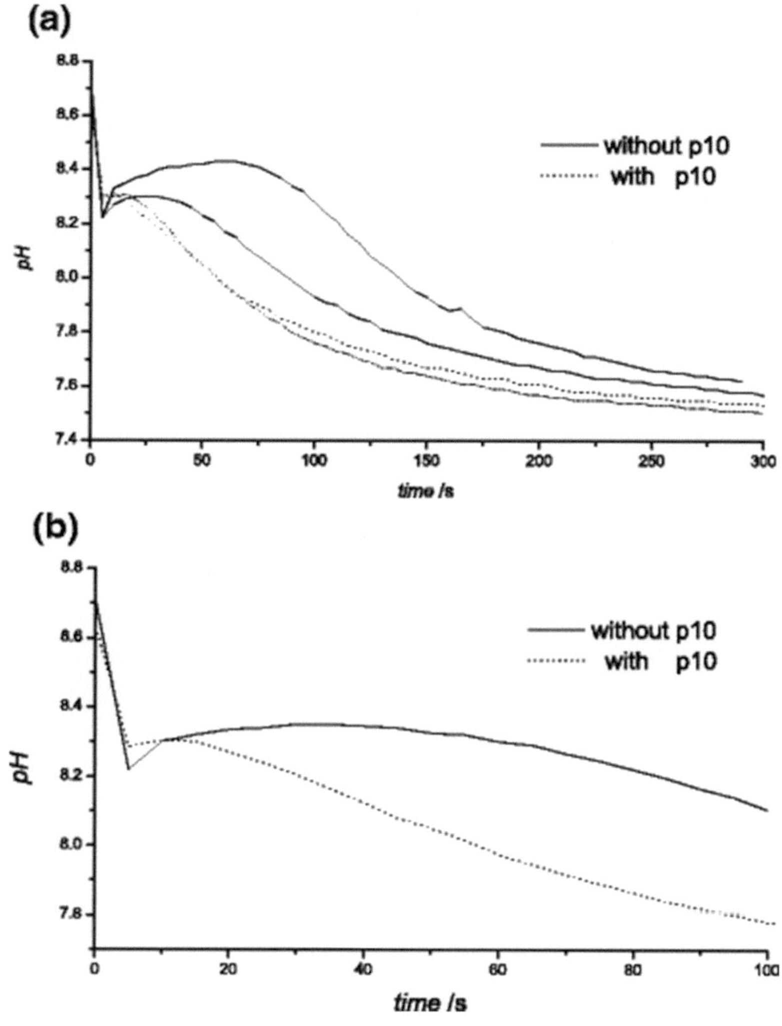

Fig. 3.10 Detections of calcium carbonate precipitation of a saturated solution. (**a**) Effect of p10 on the precipitation rate of calcium carbonate. The pH of the solution was recorded as a distinction of precipitation. (**b**) Averages of the data in (**a**) in the first 100 s of the pH calculations

[65]. The duration of the stable period was decreased with p10 in solution, compared with control group. What's more, the slope of the precipitation curve of calcium carbonate from experimental group was also slightly steeper than that of control group during the first 100 s (Fig. 3.10b). All these appearances suggested that p10 could stimulate the precipitation rate of calcium carbonate; in other words, there is no inhibition on crystal growth from p10.

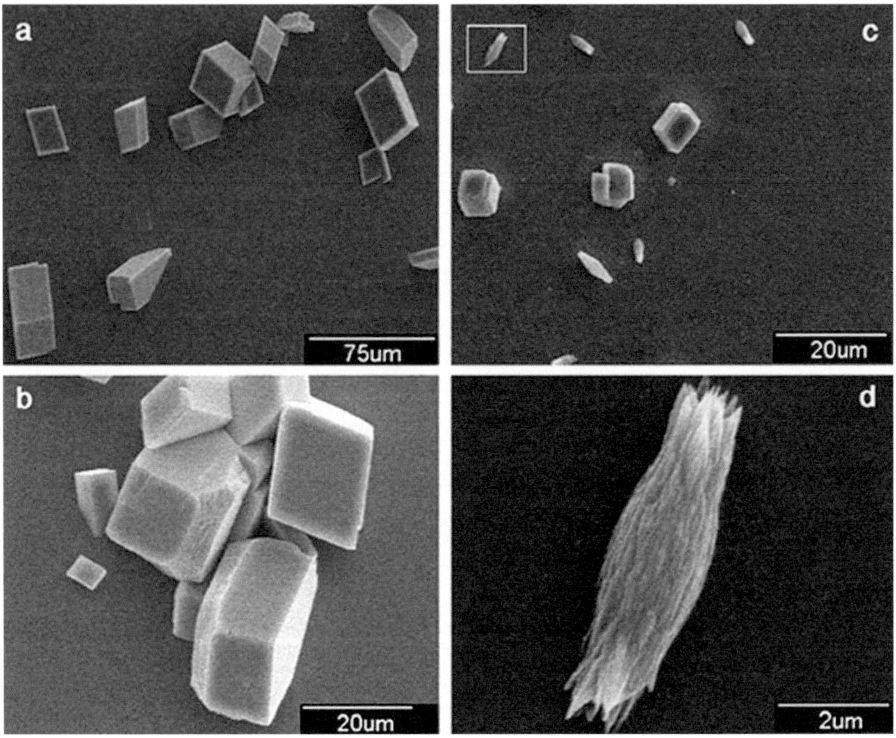

Fig. 3.11 SEM images of crystals mineralized in supersaturated calcium carbonate solution with addition of (**a**) Milli-Q water (control), (**b**) BSA protein (control), or (**c**) p10. (**d**) Enlarged image of the boxed part in (**c**)

3.4.2.3 Effect of p10 on Calcium Carbonate Crystallization

To investigate the influence of p10 on calcium carbonate crystallization, the morphology of crystals grown in the presence of p10, or no protein was observed using SEM (Fig. 3.11). The crystals grown without p10 (both with BSA and with no protein) all exhibited forming rhombohedral morphology of calcite (Fig. 3.11a, b). Nevertheless, crystals grown with p10 showed two different kinds of morphologies (Fig. 3.11c). One kind is the exhibited rhombohedral morphology of the calcite, while the other crystals grown in the presence of p10 displayed a needlelike, cluster morphology (Fig. 3.11d), which is the classical morphology of aragonite [64].

The small needlelike crystals grown with p10 detected to have the characteristic Raman peaks of aragonite at 206, 703, and 1085 cm^{-1} (Fig. 3.12a), whereas the rhombohedral crystals displayed characteristic Raman peaks of calcite at 282, 712, and 1086 cm^{-1} (Fig. 3.12b) [66]. The X-ray diffraction spectrum which could analyze the crystals is shown in Fig. 3.13 The strongest diffraction intensity was

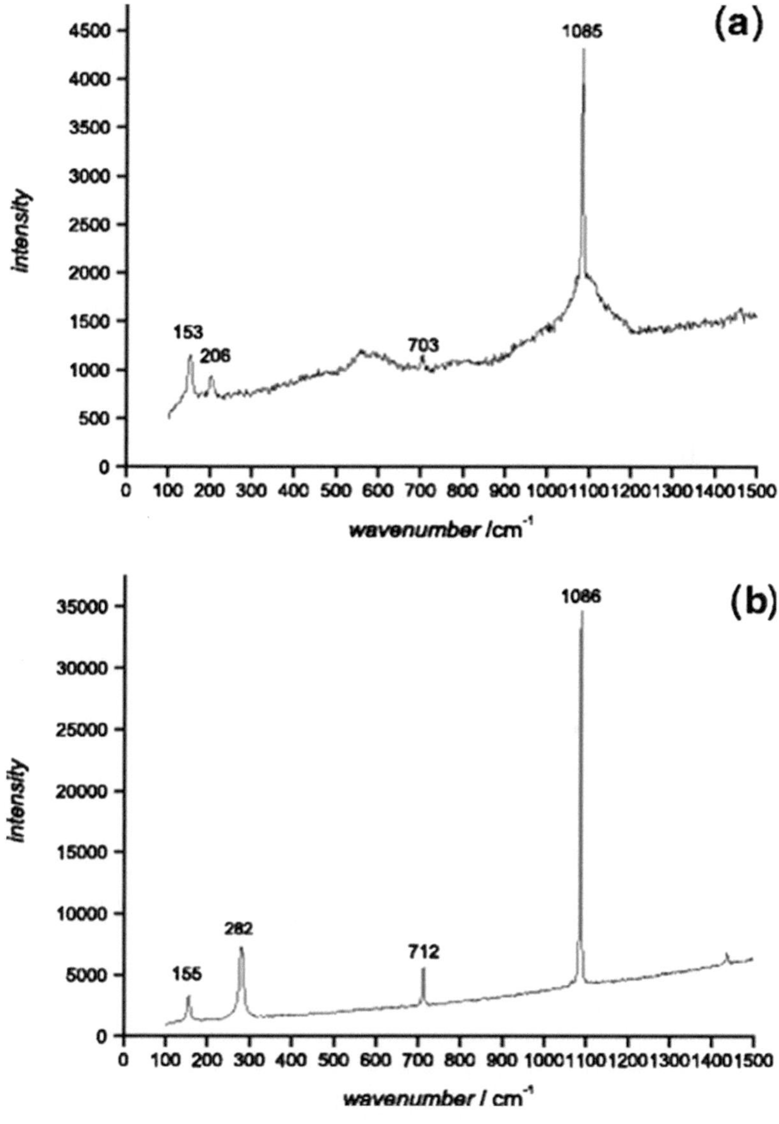

Fig. 3.12 Raman spectra of crystals of different morphology. (**a**) Raman spectrum of a needlelike aragonite crystal formed with p10 which determine polymorph. Characteristic Raman spectrum of aragonite is 206, 703, and 1085 cm^{-1}. (**b**) Raman spectrum of a rhombohedral calcite crystal. Characteristic Raman peaks for calcite are at 282, 712, and 1086 cm^{-1}

the calcite (104) and the calcite (208) was the next strongest for all the crystal samples. The diffraction intensity of aragonite (022) could only be detected in the crystals grown with p10 (Fig. 3.13c) which suggest that p10 could induce the aragonite formation.

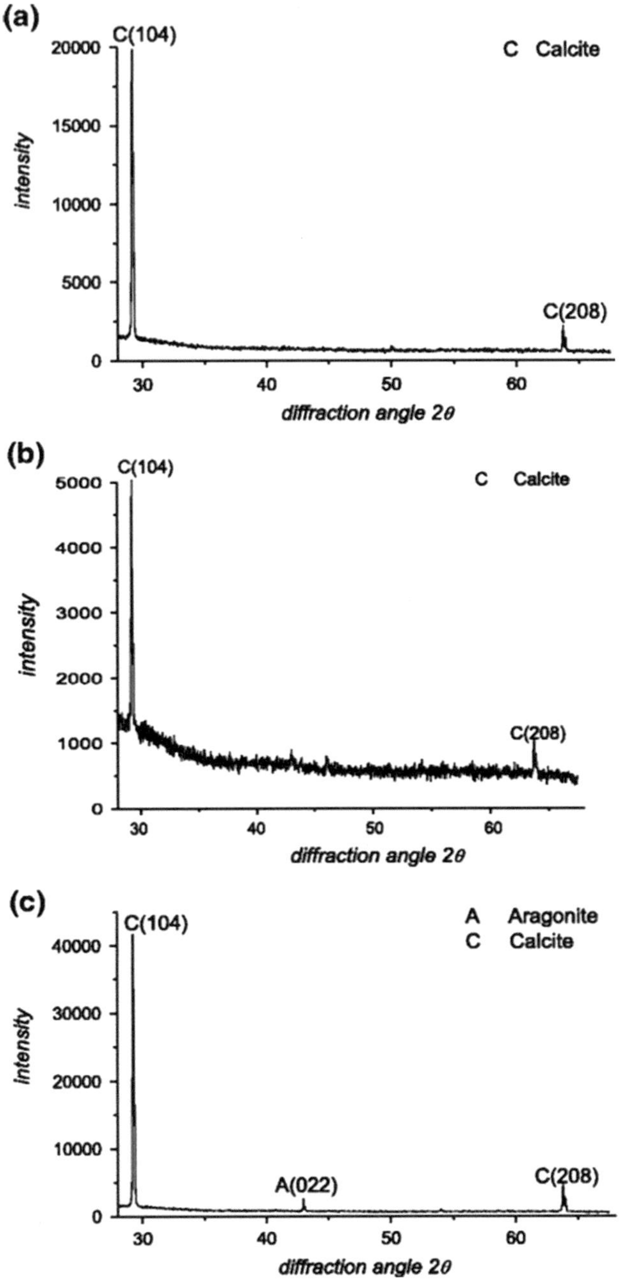

Fig. 3.13 X-ray diffraction pattern of the calcium carbonate crystals formed on the glass surface (**a**) without any protein, with (**b**) BSA, or (**c**) with p10

3.4.2.4 Effects of p10 on Cell Proliferation and Differentiation [64]

Nacre matrix was thought to contain osteogenic signal molecules, so two cell types including MC3T3-E1 (a preosteoblast cell line) and MRC-5 (a fibroblast cell line) which has the ability to differentiate into osteoblasts were used to investigate the biological activity of the p10. ALP activity was measured to estimate differentiation ability of the cells. ALP was expressed in the early stage of osteoblast development process and is regarded as an accepted indicator of osteoblast differentiation. High ALP activity is generally deemed as the first marker of osteoblast maturation [67] [68]. The effects of p10 on ALP activity in both cell types were concentration-dependent (Fig. 3.14a). The ALP activities of MRC-5 cells increase intensively ($p < 0.01$) at dosage of 5 and 10 μg/ml of p10, but higher p10 concentrations (30 and 50 μg/ml) did not increase the ALP activity of MRC-5. The effects of p10 on ALP activity in MC3T3-E1 cells were slightly different from those in MRC-5 cells. A concentration of 5 μg/ml of p10 had no significant effect on ALP activity ($p = 0.44$), whereas higher dosage of p10 led to ALP activity increasing significantly ($p < 0.01$), and the maximum effects are achieved at a dosage of 50 μg/ml. The increase of ALP activity of these two cell types indicated that p10 could affect the differentiation course of fibroblasts and preosteoblasts during bone formation.

MTT assay, an enzymatic activity test based on quantification of activity of the mitochondrial dehydrogenase [69], was used to display the effects of p10 on cell viability (Fig. 3.14b). In both cell types, MTT response increased signally ($p < 0.01$) with the dosages of p10 changed and achieved maximum influence at a dosage of 50 μg/ml of p10. It appeared that p10 could heighten the viability of the both types of cells. The MTT mitochondrial test could provide universal measure of the number of viable cells and of their mitochondrial activity in vitro; cellular metabolic viability can even be detected in the absence of cell proliferation [69]. To know whether the increasement in the MTT response was the effect of the activation of cell metabolism or cell proliferation, we compared the results data of MTT assay with the data of total protein content in ALP activity assay (Fig. 3.14c). The total concentrations of p10 and total protein contents of both cell types were not significantly different from control group ($p > 0.05$) except the concentration at 50 μg/ml of p10 in MRC-5 cells ($p = 0.008$), which shown that the total protein content increased slightly with p10. Thus, this response of MTT would be on account of the metabolism activity increase of viable cells but not to an increase of cell number. Therefore, we assumed that p10 does not affect the proliferation process but only the differentiation process.

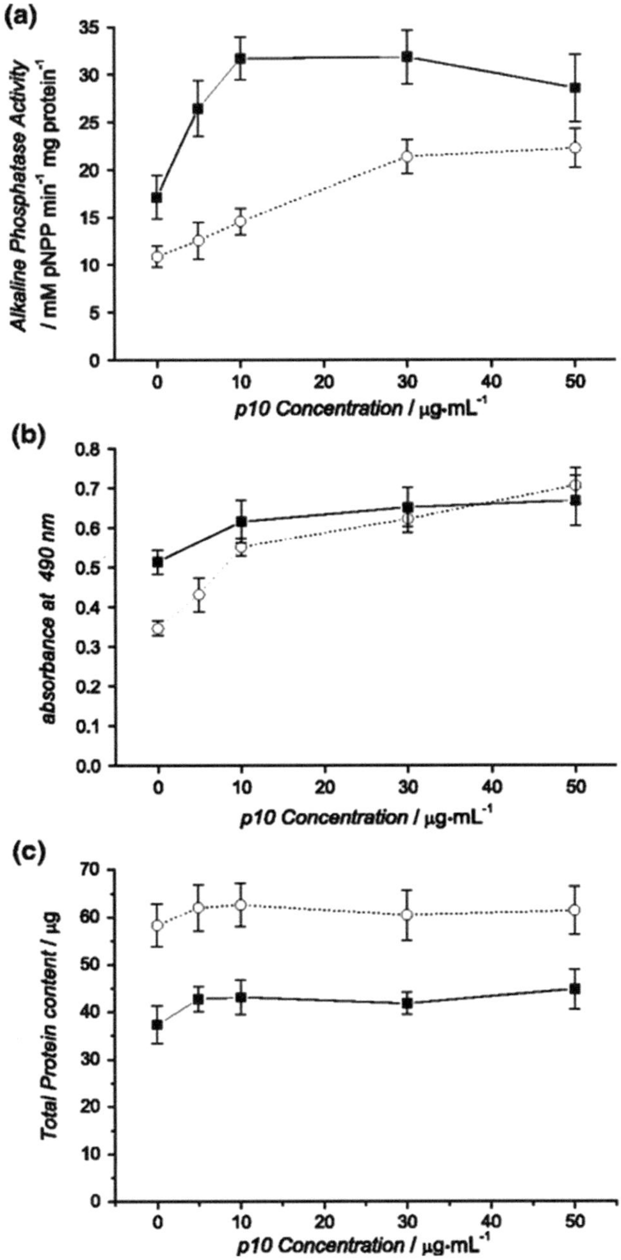

Fig. 3.14 Influence of p10 on (**a**) ALP activity, (**b**) MTT response, and (**c**) all protein content in MRC-5 cells (—■—) and MC3T3-E1 cells (····○····) which were treated with different dosages of p10 for 7 days. Results are recorded as the mean T SD for six wells

3.4.3 N40, a Novel Nonacidic Matrix Protein from Pearl Oyster Nacre, Facilitates Nucleation of Aragonite In Vitro

A novel nonacidic matrix protein extracted from pearl oyster nacreous layer has been purified through cation-exchange chromatography. It was named as N40 for the nacre protein of about 40 kDa [12]. According to the extraction method (solution with Tris-buffered Milli-Q water) and amino acid component (Gly- and Ala-rich), N40 was thought to be a traditional "insoluble matrix protein." Crystallization assays showed that N40 could stimulate the nucleation of aragonite drastically. Until now, among the macromolecules that have been identified from the shells of *Pinctada fucata*, N40 is a novel protein that can induce aragonite nucleation by itself, with no need of adsorption to a substrate. Therefore, the present study suggests that the nonacidic shell protein (possible a traditional "insoluble framework protein") can also directly take part in aragonite nucleation and even serve as a nucleation site. It is an important supplement to the classic biomineralization theory before, in which only the soluble acidic proteins of the shell are generally thought to function as a nucleation site.

3.4.3.1 N40 Protein Purification and Partial Characterization

The total nacre proteins were loaded into a CM-Sepharose fast flow column, and the N40-containing fractions were selected (Fig. 3.15). The purified N40 was subjected by SDS-PAGE with CBB R-250 and silver staining. As shown in Fig. 3.16 (lanes 2 and 3), only one band with molecular mass of 40 kDa was visualized on the gels. The identification by MALDI-TOF MS showed that the accurate molecule mass of the protein (N40) was 38.3 kDa (Fig. 3.17) subsequently. The analysis result of

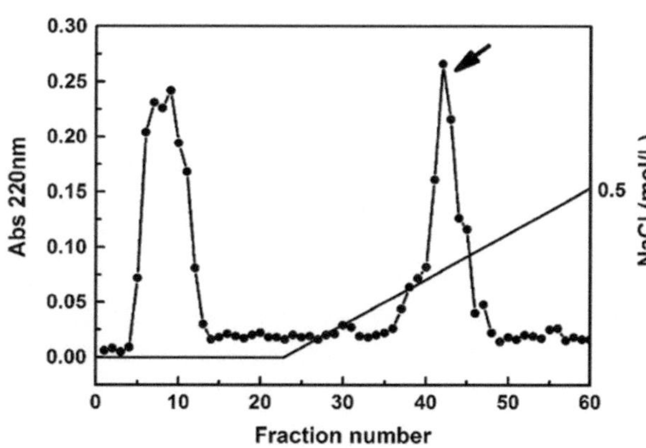

Fig. 3.15 Chromatography profile of the N40 protein in a CM-Sepharose fast flow column. The basic elution buffer was 25 mM sodium phosphate buffer (pH = 7.0). A linear gradient elution buffer of 0~0.5 M NaCl was added to the starting buffer. The arrow points the N40 fraction

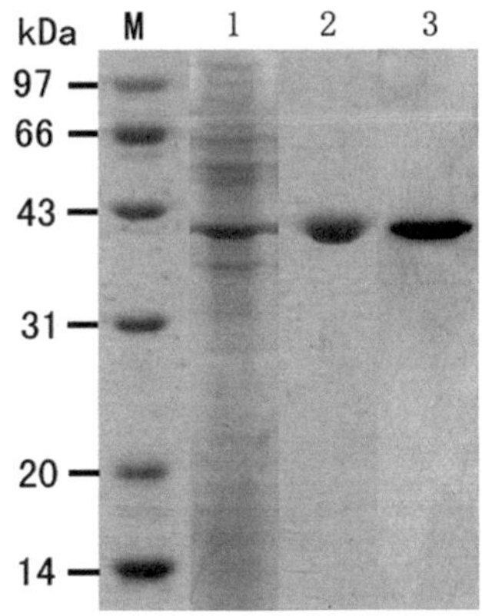

Fig. 3.16 SDS-PAGE image of N40 purification: M, markers indicating molecular mass; lane 1, total nacre proteins dyed by CBB; lane 2, purified N40 stained with CBB; lane 3, purified N40 stained with silver. The molecular mass (kDa) is indicated on the left of the gel

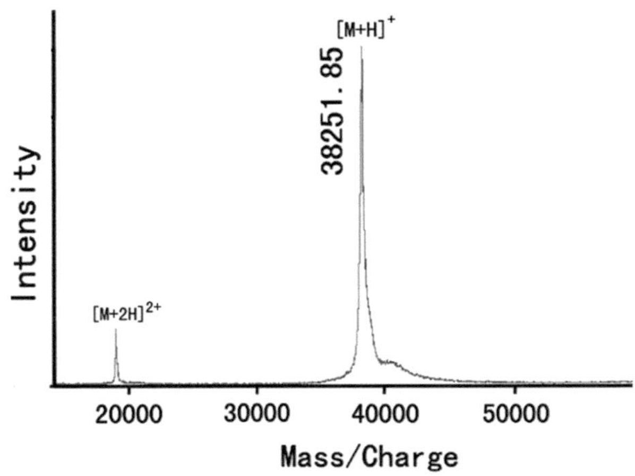

Fig. 3.17 MALDI-TOF mass spectrum profile of purified N40

Table 3.4 Amino acid analysis of N40 composition

Amino acid	Concn, mol %	Amino acid	Concn, mol %	Amino acid	Concn, mol %
Asx	6.45	Ser	8.33	Leu	6.27
Glx	3.17	Cys	6.87	Tyr	1.84
Gly	10.90	Val	9.88	Phe	6.28
Ala	15.73	Met	1.76	Lys	6.60
Thr	9.50	Ile	1.87	Arg	4.55

amino acid composition is presented in Table 3.4. The results show that the percentage of basic residues (Lys and Arg) and acidic residues (Asx and Glx) were 11.2 and 9.6 mol %, respectively. In addition, we also noted that N40 is rich in Ala and Gly (26.6 mol %) [12].

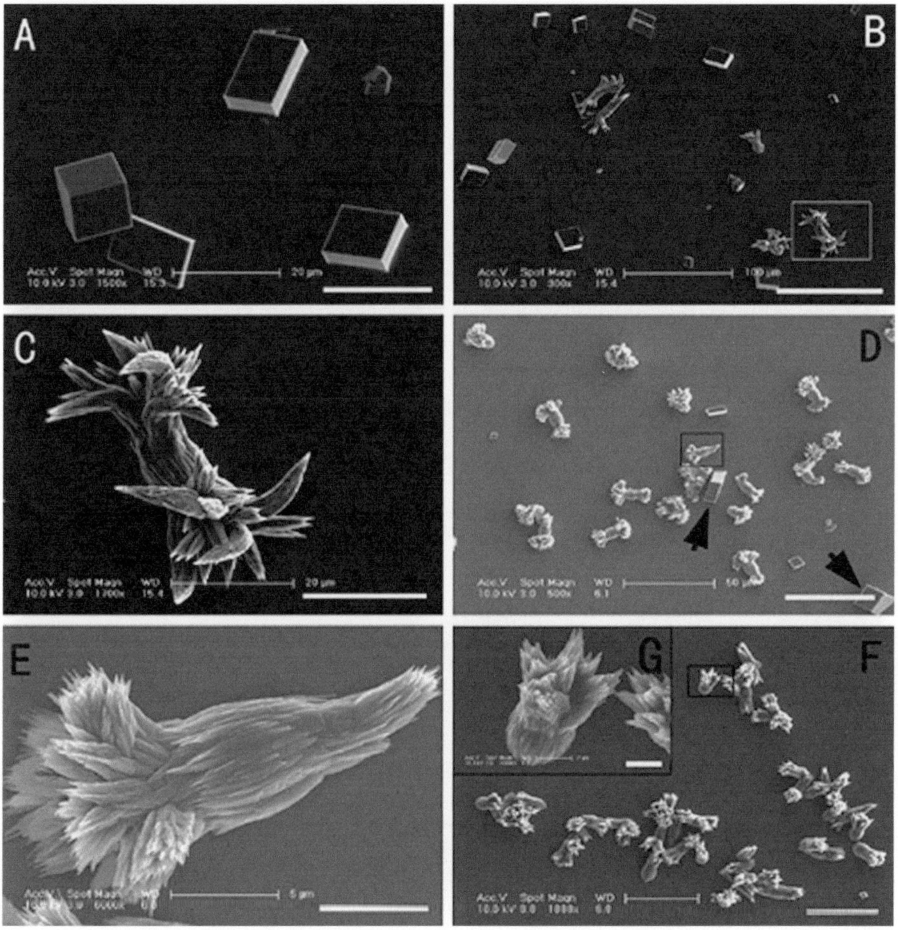

Fig. 3.18 SEM images of in vitro crystallization assays with addition of N40. Crystals were formed with 30 μg/ml BSA (**a**), 10 μg/ml N40 (**b**), 20 μg/ml N40 (**d**), and 30 μg/ml N40 (**f**). The black arrows in (**d**) show that the calcite crystals retained in the solution. (**c**), (**e**), and (**g**) are amplified images of the boxed part in images (**b**), (**d**), and (**f**), respectively. Scale bars: **a/c/f**, 20 μm; **b**, 100 μm; **d**, 50 μm; **e**, 5 μm; **g**, 2 μm

3.4.3.2 The Effect of N40 on Aragonite Growth

To investigate the effect of N40 on the crystal growth, in vitro crystallization experiments were conducted, and the precipitated crystals were examined using SEM and Raman spectrum. As shown in Fig. 3.18, N40 could induce aragonite crystals, while the crystals of the control group in the absence of protein (data not shown) or in the presence of BSA (Fig. 3.18a) were all the typical calcite of rhombohedra. In the presence of 10 μg/ml N40, some aragonites (about 30%) appeared together with calcite (Fig. 3.18b). When a higher dosage (20 μg/ml) of

N40 was present, the aragonite crystals turned into dominant (about 80%) in the whole crystallization solution (Fig. 3.18d). The N40 protein displayed its maximal influence on the crystal polymorph at a dosage of 30 μg/mL, where there was total aragonite observed in the crystallization solution, and the induced crystals were shown to have no calcite (Fig. 3.18f) [12].

3.4.3.3 The Effect of N40 on Polymorph Determination

To investigate the polymorph of the induced crystals in the crystallization solution, the Raman spectrum was used to characterize the induced crystals. As shown in Fig. 3.19, the aragonites induced in the presence of N40 were validated (upper panel), while the crystals grown in the control experiments were shown to be calcite (lower panel). Because Raman spectra can only identify the single crystals on the cover glass, it is difficult to reveal the characteristic of whole crystals in the

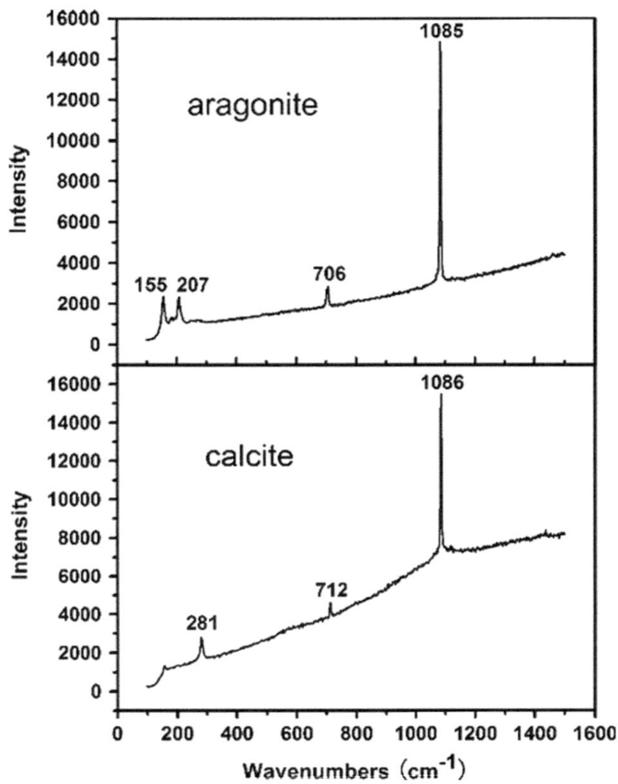

Fig. 3.19 Raman spectra of the crystals formed in the crystallization solution. The aragonite (induced with addition of N40) and calcite (control experiments with no protein) were confirmed with the characteristic peaks of aragonite at 207, 706, and 1085 cm^{-1} and of calcite at 281, 712, and 1086 cm^{-1}

crystallization solution, so we collected the crystals grown on the cover glass and characterized them by FTIR. The crystals induced by N40 at a concentration of 30 μg/ml demonstrated the characteristic peaks of aragonite (Fig. 3.20a) in the FTIR spectra which indicates the crystals induced with 30 μg/ml N40 were all aragonite, while the crystals in the control group displayed the typical peaks of calcite in the spectra (Fig. 3.20c). In the presence of N40 at a dosage of 10 μg/ml, there shown characteristic peaks of calcite (Fig. 3.20b, indicated by circles) and aragonite (Fig. 3.20b, indicated by stars) simultaneously.

3.4.4 Novel Basic Protein, PfN23, Functions as Key Macromolecule During Nacre Formation

The excellent microstructure of nacreous layer (mother of pearl) indicates the beauty of nature. Proteins exist in nacre were thought to be "natural hands" that could regulate nacre formation. In the classical theory of nacre formation, nucleation of the main minerals of nacre, calcium carbonate, is induced by the acidic proteins in nacreous layer. Whereas, the basic proteins were not involved to be components of nacre. Herein, we found that a novel basic protein, named as PfN23, was a crucial accelerator in the regulation over crystal growth in nacre [14]. The expression location, in situ immunostaining, and in vitro immunodetection assays suggested that PfN23 distributed within calcium carbonate crystals of the nacre. The expression of PfN23 was knocked down in adults via double-stranded RNA injection, resulting in a disordered nacreous layer surface in adults. Inhibiting the translation of PfN23 in embryos by morpholino oligomers generated the arrest of larval development. The in vitro crystallization experiment showed that PfN23 stimulates the rate of calcium carbonate precipitation and induced the aragonite crystals formation with characteristics similar to nacre. What's more, we constructed the expression vectors of peptides and truncations of different regions of this protein and discovered that the positively charged C-terminal region was a crucial region for the function of PfN23. All in all, the basic protein PfN23 might be a key accelerator during the regulation of crystal growth in nacre. This present study provides a valuable balance to the classic theory that only acidic proteins could control calcium carbonate deposition in nacre.

3.4.4.1 Purification and Bioinformatics Analyses of PfN23

According to the expression sequence tag in the suppression subtractive hybridization library from the D-shaped stage [70] in *Pinctada. fucata*, we acquired a 1220 bp transcript with a 5′-untranslated region of 77 bp, an open reading frame of 702 bp which encodes a deduced 233 amino acids protein, and a 518 bp 3′-untranslated region (GenBank^{TM} accession number JN995665). The deduced mature protein had a predicted molecular mass of 23.4 kDa, and it was called as PfN23. A BLASTx

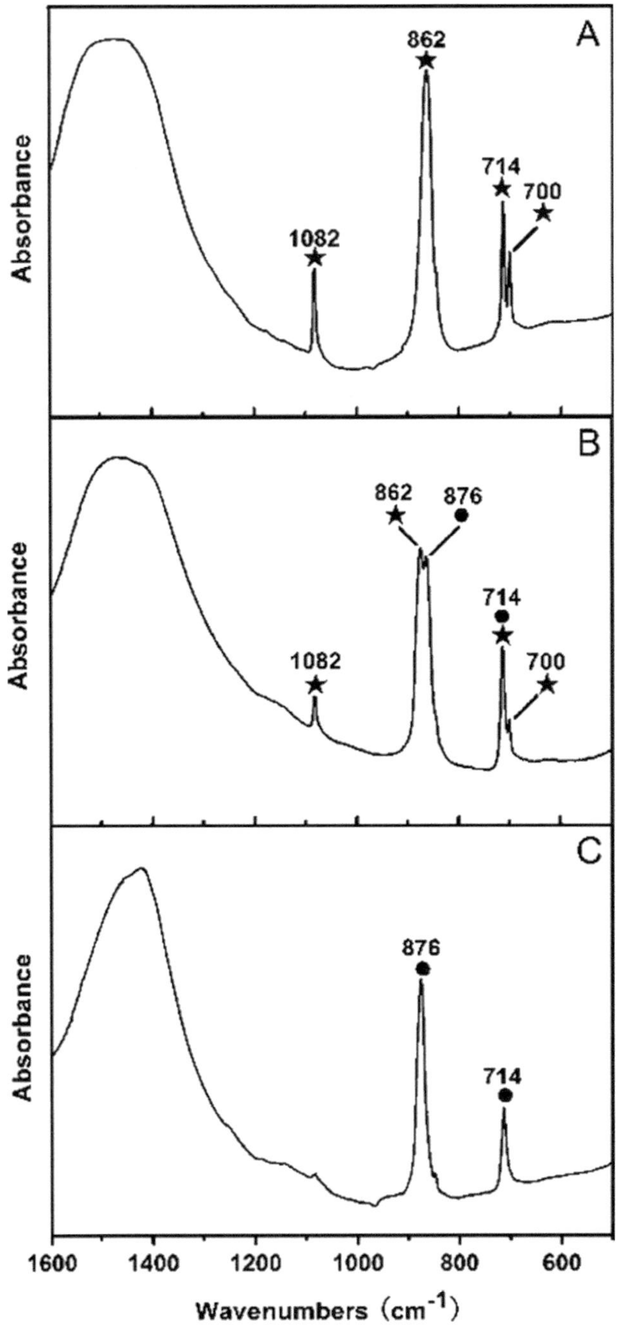

Fig. 3.20 FTIR spectra of the crystals grown on the silicified cover glass in crystallization assays. The crystals were induced with addition of N40 at dosage of 30 μg/ml (**a**, dominated as aragonite), 10 μg/ml BSA (**b**, compound of calcite and aragonite), and 30 μg/ml BSA (**c**, control experiment, confirmed as calcite). The characteristic peaks of calcite (pointed by circles) are at 876 and 714 cm $^{-1}$ and that of aragonite (shown by stars) are at 1082, 862, 714, and 700 cm^{-1}

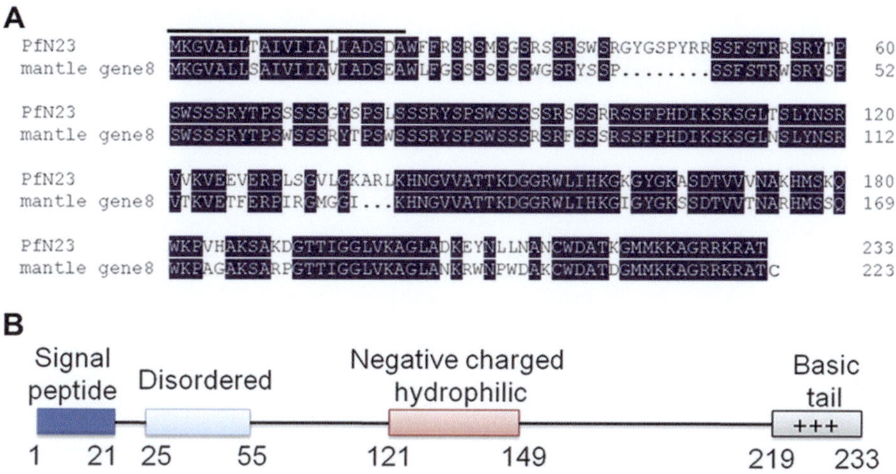

Fig. 3.21 Translated amino acid sequences and bioinformatics analysis of PfN23. (**a**) The amino acid sequence of PfN23 was aligned with another protein named as mantle gene 8. The predicted signal peptide in N-terminal is marked by overbar. (**b**) Functional regions of PfN23 showing the putative signal sequence, a disordered region, a negatively charged and hydrophilic domain, and a basic tail in C-terminal

search against the GenBank™ database shown that PfN23 was similar with mantle gene 8 of *Pinctada fucata*, and there were no homologous genes in other species (Fig. 3.21a). However, mantle gene 8 had not been previously identified before, so the function of PfN23 was still unclear. The sequence analyses showed that PfN23 is a basic protein with a predicted isoelectric point of 11.17. PfN23 was rich in positively charged residues (Arg + Lys, 19.8%) and Ser (22.2%), and it was predicted to have a signal peptide, a disordered region, a hydrophilic and negatively charged region, and a C-terminal basic region (Fig. 3.21b) [14].

3.4.4.2 Detection of PfN23 in the Shell

The specific expression location of PfN23 was tested in different *P. fucata* target tissues using qRT-PCR (Fig. 3.22a). PfN23 expression location was found to be specific to the mantle pallial which is related to the nacreous aragonite layer formation, while it was not detected in the mantle edge which corresponded to the prismatic calcite layer. Biochemical analyses were performed to explore whether native PfN23 exists in either the prismatic layer and/or the nacreous layer. Polyclonal antibodies raised against recombinant PfN23 were used to detect native PfN23 in ESM and EISM of shell using an ELISA. PfN23 was found in the ESM of the shell that is known to be crucial for calcium carbonate formation (data not shown). To further investigate the presence of PfN23 in different shell layers, EDTA extracts of the prismatic and nacreous layers were subjected into SDS-PAGE followed by Western blotting. PfN23 was detected exclusively in the ESM fractions

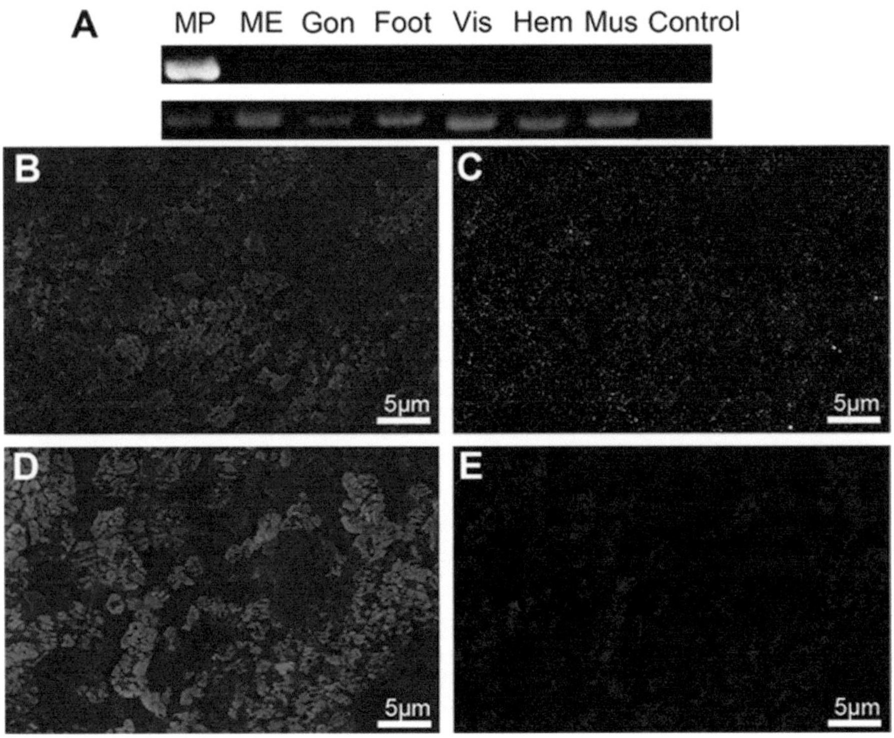

Fig. 3.22 Detection of native PfN23 in *Pinctada fucata*. (**a**) Gene expression of PfN23 in different tissues by RT-PCR analysis. Total RNA was extracted from different tissues including mantle edge (ME), mantle pallial (MP), gonad (Gon), foot, viscus (Vis), hemocytes (Hem), and adductor muscle (Mus) and was prepared for RT-PCR. RT-PCR with no template was used as a negative control group (control). The housekeeping gene GAPDH was detected as a positive control. (**b–e**), Observation of native PfN23 in shells by immunogold staining on the nacre surface. (**b**), SEM image of immunogold staining on the nacreous layer surface. (**d**), SEM image of the nacreous layer surface in control group. (**c** and **e**), Images of backscattered electron mode scanning electron microscope (SE-BSE) of the same region shown in (**b** and **d**), respectively [14]

of the nacre (data not shown). In the control experiment, no signal was observed when anti-PfN23 antibody was not used. The immunodetection assay results confirmed the gene expression analyses that PfN23 expression is located in nacreous layer. To further analyze the microstructural distribution of native PfN23 in the shell, the nacreous and prismatic layers were immunogold-labeled using anti-PfN23 antibody as the primary antibody and 15 nm diameter gold-labeled antibody as the secondary antibody. High atomic number gold elements could be scanned in the backscattered electron mode using scanning electron microscope. PfN23 was observed on the whole surface of the nacreous layer equally (Fig. 3.22b, c). In the control group, no staining was found when anti-PfN23 antibodies were not used (Fig. 3.22d, e).

3.4.4.3 In Vivo Function of PfN23 During Nacre Formation

To further investigate the role of PfN23 in nacre formation, morpholino soaking assays and a combination of RNAi were conducted. Designed dsRNA of PfN23 was injected into the adductor muscle of *P. fucata* to explore the effects of PfN23 on the nacre formation in adults. Compared with the PBS-injected group, the PfN23 RNA expression levels decreased ~50% in the 40 μg-injected group and 70% in the 80 μg-injected group. At the same time, the expression level of PfN23 was invariable in the GFP dsRNA-injected groups as a control (Fig. 3.23a). We also tested the expression levels of other matrix proteins like Aspein, N19, Prismalin-14, Pearlin, and Prisilkin-39. The expression of these matrix proteins was also not affected by the injection of PfN23 dsRNA (data not shown). The surface structure of the nacreous layer from each injected group was explored using SEM. In the GFP dsRNA-injected group, the surface of the nacreous layer showed a normal stair-like growth pattern (Fig. 3.23b). Small flat tablets could be seen clearly on the surface (Fig. 3.23c, arrowhead). In the PfN23 dsRNA low dosage-injected group (40 g), the surface also had a natural stair-like growth pattern (Fig. 3.23d). However, the aragonite tablets were covered with excessive accumulated crystals (Fig. 3.23e, arrowhead). The deposited crystals did not form the characteristic rectangular or hexagonal flat tablets. In the high-concentration (80 μg) dsRNA-injected group, this abnormal phenomenon was more remarkable (data not shown). However, the morphology of the prismatic layers had no change in the GFP and PfN23 dsRNA-injected groups (data not shown). The gene was expressed in the D-shaped stage of development particularly, so the function of PfN23 was identified using morpholinos. We designed two translational-blocking morpholinos to block the translation process of PfN23 (Fig. 3.24a), called morpholino A (Mo A) and morpholino B (Mo B). A random morpholino (Mo R) was designed as a negative control. The embryos of two-cell to four-cell stage were immersed with 5 or 10 μM morpholinos to inhibit the translation of PfN23 [14].We executed a Western blotting 32 h after treatment to detect PfN23 expression. As Fig. 3.24b had shown, the expression of PfN23 had no change when the Mo R or Mo A was used compared with untreated groups. In contrast, Mo B could decrease the expression level of PfN23, and few proteins were observed in the 10 μM-treated group. Larvae from all the groups could grow to the trochophore stage within 10 h. The larvae in the untreated, Mo A-treated, and Mo R-treated groups could grow to the D-shaped stage, resulting in a capital D-like morphology after 32 h (Fig. 3.24c). However, in the Mo B-treated groups, the development of major larvae was inhibited. In addition, there is no shell on the arrested larvae (Fig. 3.24c). In the Mo R or Mo A groups, major larvae (~80%) could develop to the D-shaped stage. However, when treated with 5 μM Mo B, only 29.8% (187/628) of larvae grew to the D-shaped stage, whereas 53.8% (338/628) of larvae were stopped before they were covered with shells. Compared with the 5μM treatment, fewer larvae (2.4%, 12/509) grew to the D-shaped stage, while more larvae (74.9%, 381/509) were inhibited in the 10 μM Mo B-treated groups (Fig. 3.24d).

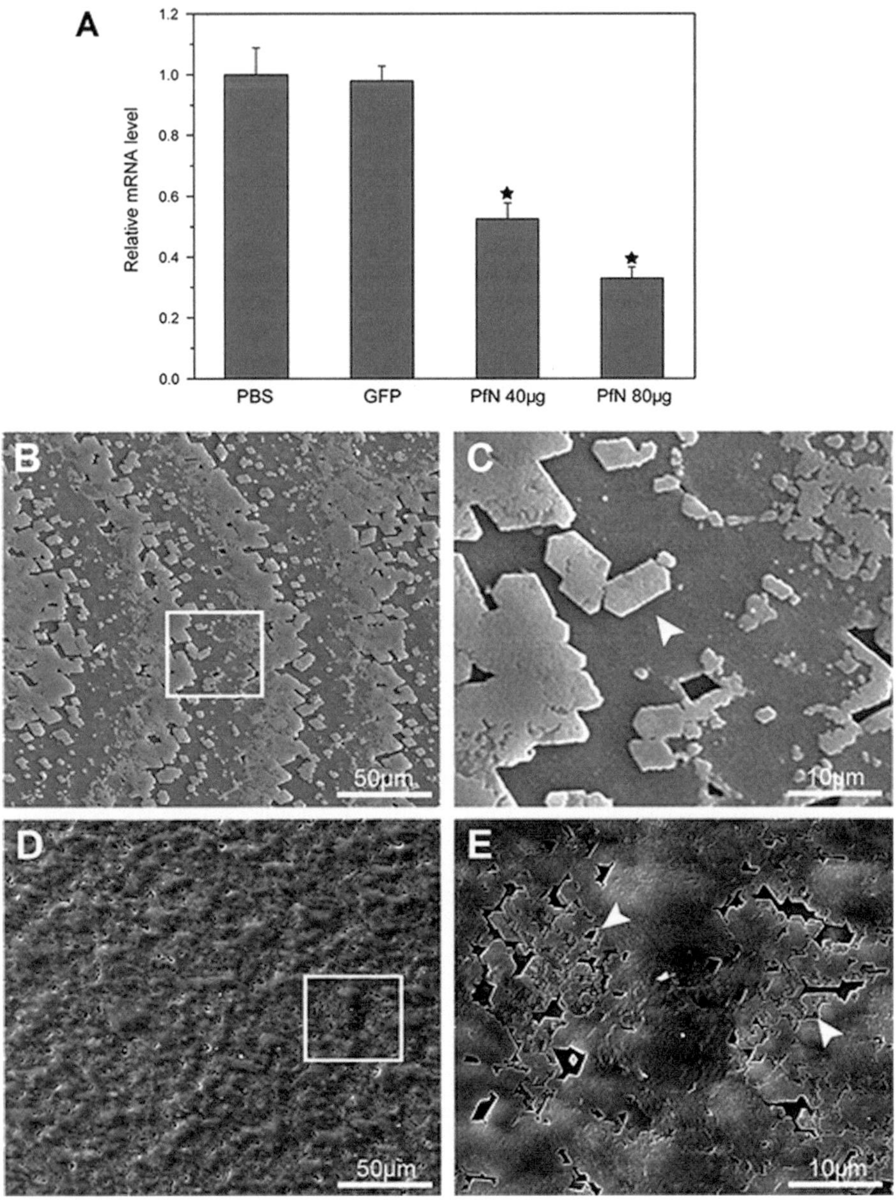

Fig. 3.23 Functions of PfN23 on the nacreous layer growth. (**a**), Expression levels of PfN23 were depressed by RNAi. In the control group, the relative expression levels of the PBS-injected groups were recorded to a relative value of 1.0. The star indicates a significant ($p < 0.001$) difference compared between RNAi groups the PBS-injected groups. (**b**), SEM image of shell surface of GFP dsRNA-injected group representing the stair-like growth pattern. (**d**), SEM image of the 40 μg of PfN23 dsRNA-injected group. The construct of nacre tablet was disturbed. (**c** and **e**), Enlargement of the box indicated in (**b** and **d**), respectively. Arrowheads point the flat tablets on the surface of nacreous layer

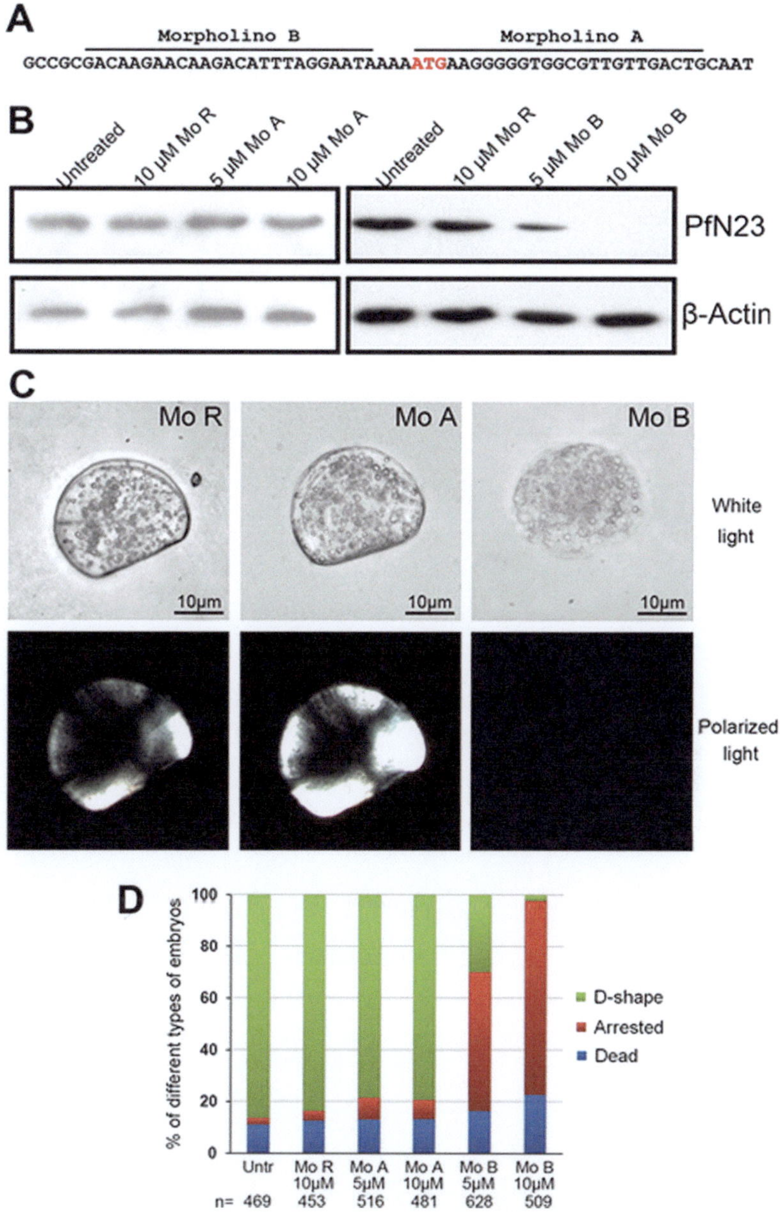

Fig. 3.24 Degradation of PfN23 using antisense morpholino oligomers. (**a**), Positions of the designed antisense morpholino oligomers and the ATG initiation codon are labeled as red. (**b**), Western blot analysis with anti-PfN23 antibody representing that Mo B blocked the translation of PfN23 mRNA. β-Actin was detected as a loading control. (**c**), Phenotypes of morpholino oligomer-treated larvae 32 h after fertilization. The samples were observed under white polarized light to identify the formation of shells. (**d**), Frequency of the phenotypes obtained with different dosages of Mo. The number shows the number of all analyzed embryos. Mo R, random morpholino oligomers; Mo A, morpholino A; Mo B, morpholino B

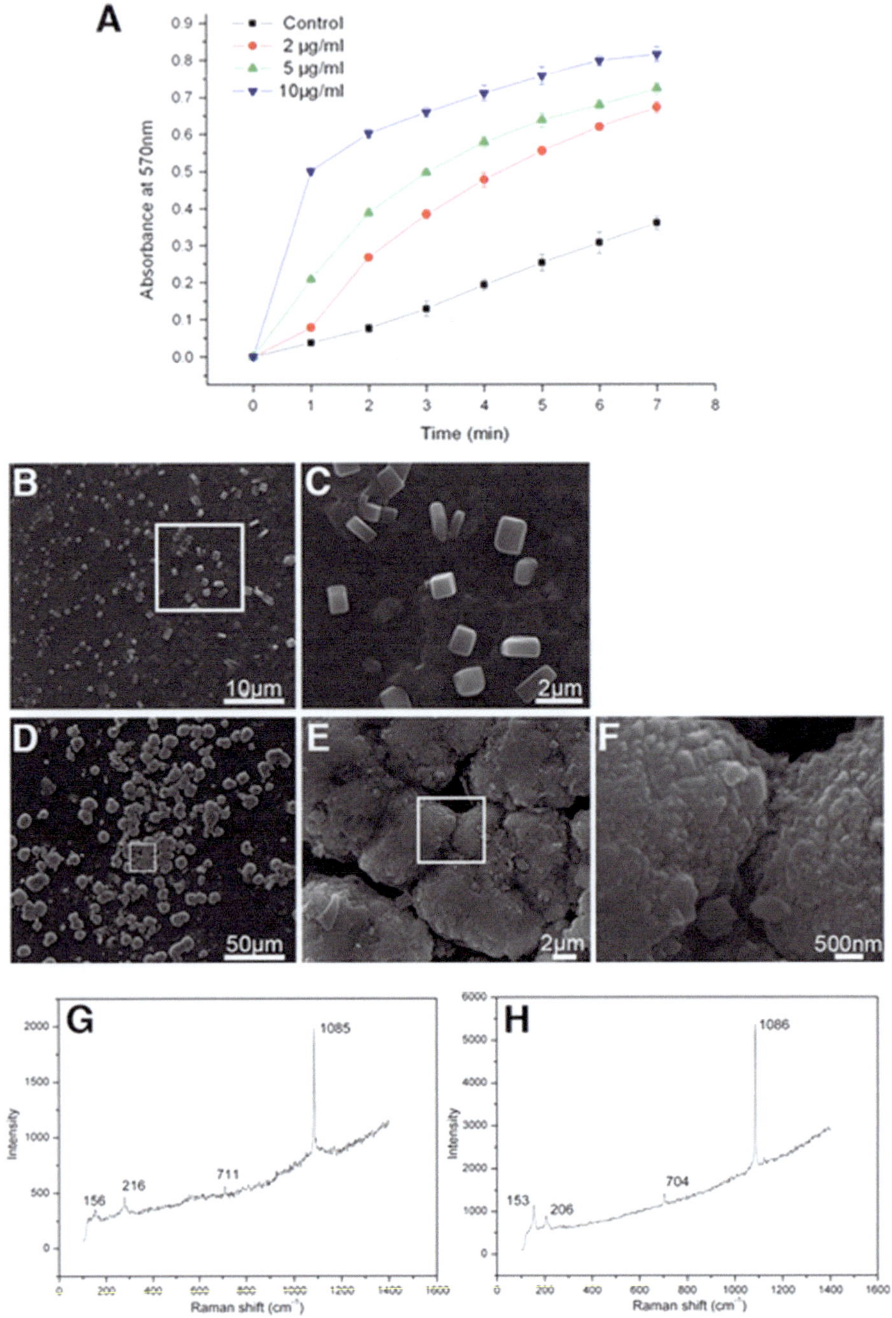

Fig. 3.25 Functions of PfN23 during calcium carbonate crystallization. (**a**), Accelerative activity of PfN23 on calcium carbonate precipitation. ■, BSA (10 μg/ml) was used as a negative control; ●, 2 μg/ml PfN23;▲, 5 μg/ml PfN23;▼, 10 μg/ml PfN23 was shown. (**b–e**) Functions of PfN23 on the

3.4.4.4 In Vitro Effect of PfN23 in Calcium Carbonate Crystallization

To further investigate the function of PfN23 during calcium carbonate crystallization, we detected its binding ability with the main component of shell including calcite, aragonite, and chitin and found that PfN23 could bind to aragonite and calcite (data not shown). We then tested its effect in calcium carbonate crystallization. A series of precipitation experiments was also conducted to identify the effect of PfN23 on the calcium carbonate precipitation rate. Compared with the control, PfN23 could increase the precipitation rate, and this effect occurred in a dosage-dependent manner (Fig. 3.25a). To explore the effect of PfN23 on the calcium carbonate crystals growth, crystallization experiments in vitro with or without magnesium were performed. The effect of PfN23 on the growth of calcite was analyzed in the magnesium-free system. The precipitated crystals were all representative rhombohedra of calcite in the control group with BSA, while PfN23 could induce the formation of vaterite in the magnesium-free system. The percentages of the vaterite crystals in the deposited crystals were about 25% with 10 μg/ml PfN23, 44% with 30 μg/ml PfN23, and 65% with 50 μg/ml PfN23, respectively (data not shown). Magnesium was introduced to the crystallization system to induce the formation of aragonite which is the composition of the nacreous layer. Compared with calcite crystallization system, different dosage of PfN23 had little effect on aragonite formation. Even with a high dosage of 50 μg/ml, the typical aragonite crystals of a normal needle-shaped morphology were deposited (data not shown). According to the crystallization models, aragonite is formed in a silk gel phase in the nacre. As PfN23 is a protein in the ESM of the shell, these findings suggested that the functionality of PfN23 requires a silk gel. We tested this hypothesis by conducting the crystallization in vitro on a silk film. In the presence of 50 μg/ml BSA, there is typical calcite crystals deposited on the film (Fig. 3.25b, c) which were confirmed to be calcite using Raman spectra (Fig. 3.25g). However, these crystals were smaller (~2 μm), and they had rounder edges comparing with the crystals (~10 μm) formed without the silk film (data not shown). With 50 μg/ml of PfN23, the deposited crystals amalgamated to form tablets that were similar to the microstructure of nacre instead of separate particles (Fig. 3.25D). What's more, these tablets were formed by small sheets of crystals (Fig. 3.25e, f). The Raman spectra found that these crystals were all aragonite (Fig. 3.25h). These results showed that PfN23 could induce the formation of aragonite tablets in the silk gel.

Fig. 3.25 (continued) crystals growth were conducted on the silk film. (**b**) 50 μg/ml BSA was used as a negative control. (**d**), The system with 50 μg/ml PfN23. (**c**, **e**, and **f**), SEM images of the amplification of the box in (**b**), (**d**), and (**e**), respectively. (**g**) and (**h**), Raman spectra analysis of the crystals collected in (**b**) and (**d**), respectively. Scale bars, 50 μm in **d**, 10 μm in **b,** 2 μm in **c** and **e**, and 500 nm in F [14]

3.4.4.5 Functions of Different Regions of PfN23

We constructed peptides and truncations of the predicted disordered domain (Peptide, Pep-N; mutation, ΔN), hydrophilic and negatively charged domain (Peptide, Pep-M; mutation, ΔM), and C-terminal basic region (peptide, Pep-C; mutation, ΔC) to explore the function of these regions (Fig. 3.26a). The Pep-M and Pep-N had little effect on the morphology of the formed crystals in the in vitro crystallization experiments (data not shown). What is notable is that the Pep-N could inhibit calcium carbonate crystals growth, and Pep-M had little effect of calcium carbonate crystals growth rate (Fig. 3.26b). The peptide of C-terminal basic region (Pep-C) could induce the formation of vaterite and promote the growth of calcite comparing with Pep-M and Pep-N (Fig. 3.26b, d, e). The proportion of the round-shaped vaterite in the formed crystals were around 27% with 10 μg/ml Pep-C, 35% with 30 μg/ml Pep-C, and 44% with 50 μg/ml Pep-C. This peptide could induce the formation of small hexagonal vaterite crystals without connection in silk gel (Fig. 3.26f, g). What's more, these crystals were quite similar to those crystals that deposited with the predicted disordered region or negatively charged or hydrophilic region truncations (data not shown). The protein could not induce vaterite formation or promote the crystallization without basic region (Fig. 3.26c, h, i). The crystals formed with ΔC (Fig. 3.26h, i) were similar to those formed with ΔN without silk film (data not shown). The basic region truncation had little effect on the crystal deposition even with silk gel (Fig. 3.26j, k). All in all, this positively charged region is suggested to be a crucial region for the function of PfN23, and the other regions are also obbligato to achieve the full function of PfN23.

3.5 Shell Matrix Protein from Prismatic Layer

3.5.1 A Novel Matrix Protein Family Participating in the Prismatic Layer Framework Formation of Pearl Oyster, Pinctada fucata

Figuring out the molecular component and the formation mechanism of shells matrix framework is of great use for biomineralization in mollusk shell. The cDNAs sequence encoding a novel matrix protein family, KRMP, was cloned from the mantle tissue of pearl oyster, *Pinctada fucata*. Analysis of the putative amino acid sequences suggested that KRMP is rich in lysine, glycine, and tyrosine, and the predicted isoelectric points of KRMPs are larger than any other identified shell matrix protein. The translated amino acid sequences of KRMP can be separated into three regions, including an N-terminal signal peptide, a lysine-rich basic domain which could interact with acidic proteins or CO_3^{2-}, and a Gly/Tyr-rich region which is thought to be involved in the protein cross-link by quinone-tanning process [9]. RT-PCR and in situ hybridization indicated that KRMP mRNA was highly

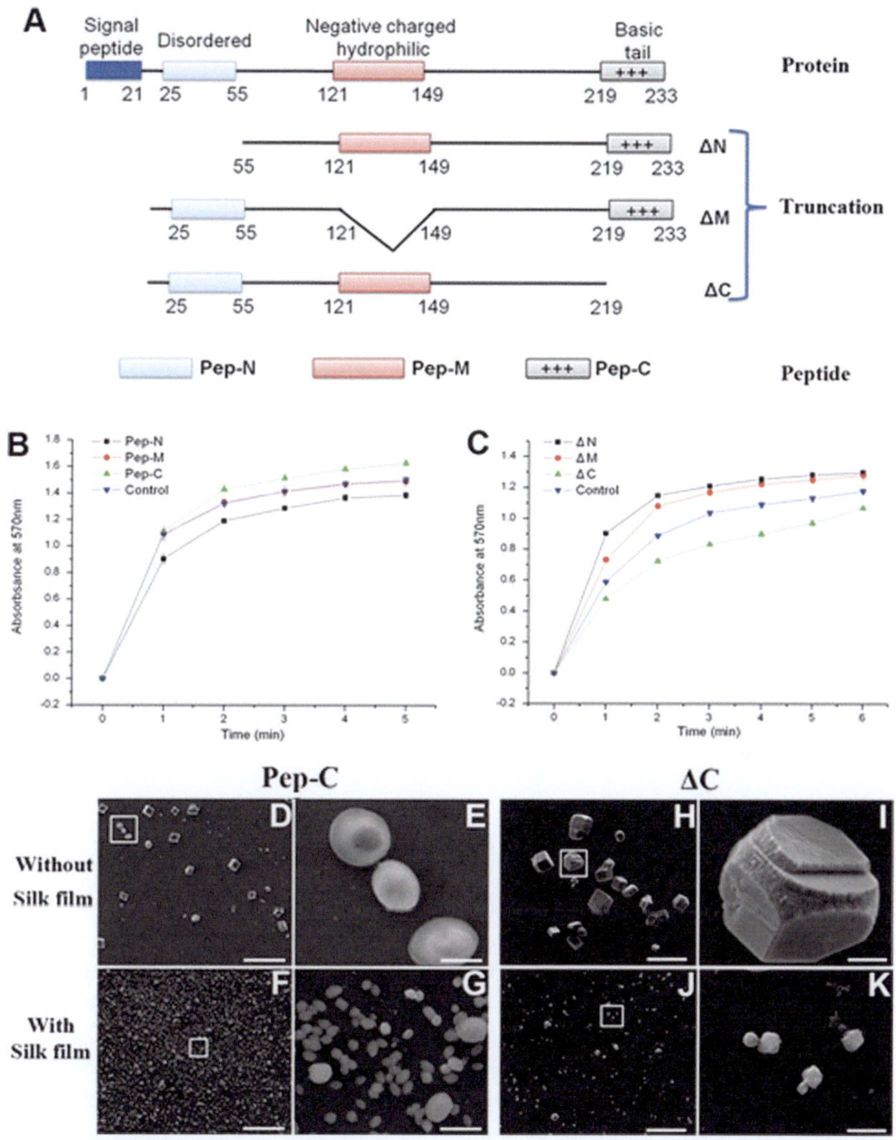

Fig. 3.26 Functions study of different domains of PfN23. (**a**), Schematic representation of peptides and truncations. Numbers mean the position of amino acid. (**b** and **c**), Accelerative activities of truncations and peptides on calcium carbonate precipitation. Records of the turbidity during the assayed solutions are shown. (**b**), Effects of peptides. ■, 10 μg/ml Pep-N; ●, 10 μg/ml Pep-M; ▲, 10 μg/ml Pep-C was present; ⍰, 10 μg/ml BSA was added as negative control. (**c**), Functions of truncations. ■, 10 μg/ml ΔN; ●, 10 μg/ml ΔM; ▲, 10 μg/ml ΔC was present; ⍰, 10 μg/ml BSA was added as negative control. (**d–k**), In vitro crystallization assays in the presence of peptides and truncations. 10 μg/ml Pep-C was studied with (**f**) or without (**d**) the silk film. 10 μg/ml ΔC was studied with (**j**) or without (**h**) the silk film. **e**, **g**, **i**, and **k**, Amplification of boxes in **d**, **f**, **h**, and **j**, respectively

expressed in the mantle edge which was thought to participate in the prismatic layer formation. Taken together, it is reasonable that KRMP is a matrix protein family playing roles in the framework formation of prism.

3.5.1.1 Identification of the Full-Length cDNA of KRMP

A 263 bp cDNA product was obtained by the 5′-RACE with degenerate primer YD5P. Sequencing analyses indicated that the primer YD5P anneal with a kinder different sequence of cDNA through the 5′-RACE procedure subsequently. The primer YD5P encoding GGYGGF/L did not match in four nucleotide positions, besides the corresponding sequence of cDNA turned out to be translated to be GGYGPG. Based on the sequence, a gene-specific primer FSP was also synthesized and the 3′ cDNA was obtained by 3′-RACE procedure. Three different cDNA clones were detected from the PCR product. To confirm the sequence obtained from 3′-RACE, full amplification of the full-length cDNA was performed with the *P. fucata* mantle cDNA library as a template. All three cDNA clones were acquired again, which matched well the sequence obtained from the results of 3′-RACE.

As shown in Fig. 3.27, full-length cDNA sequences of these three clones include the 500 bp poly(A) tail. Each of the cDNA sequences includes an open reading frame encoding 98 or 101 amino acids. The open reading frames start with the ATG as start codon at the nucleotide position 40. These cDNA sequences have been submitted to GenBank and have the accession nos. DQ114788, DQ114789, and DQ114790, respectively [9].

3.5.1.2 Amino Acid Sequences of KRMP

The location of signal peptide cleavage site was analyzed using hidden Markov models and neural networks trained on eukaryotes supported by CBS prediction servers. Both the methods predicted that the first 20 amino acids of the predicted amino acid sequences of all the three cDNA clones include a signal peptide. These predicted signal peptides share resemblance with the signal peptides of MSI7 [71] and MSI31 [6]; two matrix proteins are both involved in the prism framework formation of *P. fucata*, besides Aspein [10] and Asprich family [72] which are unusually acidic matrix proteins of prism of *P. fucata* and *Atrina rigida*. Thus, these three cDNA clones are deemed to encode three different matrix proteins, which compose a protein family. As three matrix proteins are all rich in lysine (Table 3.5), they are named as KRMP (lysine (K)-rich matrix protein). KRMP are also rich in glycine and tyrosine residues besides lysine residues (Table 3.5), which were also found to have high property in the crude matrix proteins from the prism of *P. fucata*. The isoelectric point of mature KRMP which excludes the putative signal peptides is predicted between 9.5 and 9.8; it is the most basic matrix protein among all the entirely sequenced molluscan shell proteins known before [5–8, 10, 71–83]. The most basic shell matrix protein studied before is N66 that had an isoelectric point of

```
KRMP-1:
1    ACTTGACTCCACCGATATTTTTGAAATATACGGTTCAGA
40   ATGAAGTTCGCCGCTGTTTTGGCCGTTTTCCTCCTACTTGGAGCATTCGGTGCCGATGGATACTGGCATAAACCTAATCTGAATATCTGC
1    [M  K  F  A  A  V  L  A  V  F  L  L  L  G  A  F  G  A  D  G] Y  W  H  K  P  N  L  N  I  C
130  TGGTGGAAGCTGAAGTGGTGCCTGAAGAAATGTCATCCCTGGGATTGGAAATGCAAAAAGAAATGTTATTGGAAGTATAAATGGTGCCTT
31    W  W  K  L  K  W  C  L  K  K  C  H  P  W  D  W  K  C  K  K  K  C  Y  W  K  Y  K  W  C  L
220  CATAAATTTGGTGGACATTATCCGTATGGTGGTTATGGCGCCGGATCCAGTGGTGGATACGGATATGGTGACGACGATTACACGTCTGGT
61    H  K  F  G  G  H  Y  P  Y  G  G  Y  G  A  G  S  S  G  G  Y  G  Y  G  D  D  D  Y  T  S  G
310  GGATACGGATATGGTCATCGTAAATACAAATACTAAACACAATGTTACAGATCCAGGAACCTTCAGCAACTAGAAGCATTCAATAGACAA
91    G  Y  G  Y  G  H  R  K  Y  K  Y  *
400  AATAATTATGATAGTATTCCATTATCAGCAATCCTTTGTAAAATCATTTTTGTTGAACATGATTAAATATACTCAGTGACAAAAAAAAAA
490  AAAAAA

KRMP-2:
1    ACTTGACTCCACCGATATTTTTGAAATATACGGTTTAGA
40   ATGAAGTTCGCCGCTGTTTTGGCCGTTTTCCTCCTACTTGGAGCATTCGGTGCCGATGGATACTGGCATAAACCTAATCTGAATATCTGC
1    [M  K  F  A  A  V  L  A  V  F  L  L  L  G  A  F  G  A  D  G] Y  W  H  K  P  N  L  N  I  C
130  TGGTGGAAGCTGAAGTGGTGCCTGAAGAAATGTCATCCCTGGGATTGGAAATGCAAAAAGAAATGTTATTGGAAGTATAAATGGTGCCTT
31    W  W  K  L  K  W  C  L  K  K  C  H  P  W  D  W  K  C  K  K  K  C  Y  W  K  Y  K  W  C  L
220  CATAAATTTGGTGGGCATTATCCGTATGGTGGTTATGGCCCCGGATCCAGTGGTGGATACGGATATGGTGACGACGATTACACGTCTGGT
61    H  K  F  G  G  H  Y  P  Y  G  G  Y  G  P  G  S  S  G  G  Y  G  Y  G  D  D  D  Y  T  S  G
310  GGATACGGATATGGTCATCGTAAATACAAATACTAAACACAGTGTTACAGATCCAGGAACCTTCATCAACTAGAAGCATTCAGTGGACAA
91    G  Y  G  Y  G  H  R  K  Y  K  Y  *
400  AATTATTTTGATAGTATTCCATTATAAGCAATTCTTTGTAAAATCACTTTTGTTGAACATGATTAAATATACTCAGTGACAAAAAAAAAA
490  AAAA

KRMP-3:
1    ACTTGACTCCACCGATATTTTTGAAATCTACGGTTTAGA
40   ATGAAGTTCGCCGCTGTTTTGGCCGTTTTCCTCCTCCTTGGAGCATTCAGTGCAGATGGAATCTGGCATAAGCCTAATCTGAATATCTGC
1    [M  K  F  A  A  V  L  A  V  F  L  L  L  G  A  F  S  A  D  G] I  W  H  K  P  N  L  N  I  C
130  TGGTGGAAGCTGAAGTGGTGCCTCAAGAAATGTCACCCTTGGGATTGGAAATGCAAAAAGAAATGTTTTTGGAAGTATAAATGGTGCCTT
31    W  W  K  L  K  W  C  L  K  K  C  H  P  W  D  W  K  C  K  K  K  C  F  W  K  Y  K  W  C  L
220  AAGAAATTTGGTGGACATTTTCCGTATGGTCCAGGATCCAGTGGTGGATATGGATTCGATGGATATGGTGGCGATGATTACAACTTTGGC
61    K  K  F  G  G  H  F  P  Y  G  P  G  S  S  G  G  Y  G  F  D  G  Y  G  G  D  D  Y  N  F  G
310  TATGGTCATCGTAAATACAAATATTAAACACATTGTTACAGATCCAGAAACCGTCATCAACTAGAAACATTCAATGGACAAAATTAATTT
91    Y  G  H  R  K  Y  K  Y  *
400  GGATAGTATTCTTATATCAGCAATCCATCCTTTGTAAAATCATTTTTGTTGAACATGATTAAATTCATACTCAAAAAAAAAAAAAAAAA
```

Fig. 3.27 The cDNA sequence and putative amino acid sequence of the KRMP-1, KRMP-2, and KRMP-3 of the pearl oyster, *P. fucata*. Brackets indicate the predicted signal peptide. The stop codon is marked with an asterisk. The nucleotides shaded are the sequence that degenerate primer YD5P annealed to. These cDNA sequences have been uploaded to GenBank (Accession Nos. DQ114788, DQ114789, and DQ114790)

8.68 [77]. Further motif analyses indicated that no potential O- or N-glycosylation sites were found in the deduced sequence. Sequence alignment showed that the deduced amino acid sequence of KRMP-2 and KRMP-1 had no difference except for residue 74, and the KRMP-3 showed more than 80% amino acid sequence identical with KRMP-1 (or KRMP-2) (Fig. 3.28a). These sequences can be divided into three regions, including the C-terminal Gly/ Tyr-rich region, the lysine-rich basic region, and the N-terminal signal peptide (Fig. 3.28). Homology search was carried out against all the protein sequences in GenBank with the BLAST program. The sequences of Gly/Tyr-rich region showed some resemblance with the GGY cuticle protein 1 of *B. mori* (accession no. BAB62815) and the cement protein Pc-1 from *P. californica* (accession no. AAY29115). These proteins are all quinone-tanned proteins involved in the formation of the insoluble extracellular structure like the cuticle [84] and the mineralized tube [85].

Table 3.5 Amino acid component of KRMP [9]

Amino acid	KRMP-1		KRMP-2		KRMP-2	
	aa number	mol%	aa number	mol%	aa number	mol%
Ala	1	1.23	0	0	0	0
Cys	6	7.41	6	7.41	6	7.69
Asp	4	4.94	4	4.94	4	5.13
Glu	0	0	0	0	0	0
Phe	1	1.23	1	1.23	5	6.41
Gly	14	17.28	14	17.28	12	15.38
His	5	6.17	5	6.17	4	5.13
Ile	1	1.23	1	1.23	2	2.56
Lys	14	17.28	14	17.28	15	19.23
Leu	4	4.94	4	4.94	4	5.13
MET	0	0	0	0	0	0
Asn	2	2.47	2	2.47	3	3.85
Pro	3	3.7	4	4.94	4	5.13
Gln	0	0	0	0	0	0
Arg	1	1.23	1	1.23	1	1.28
Ser	3	3.7	3	3.7	2	2.56
Thr	1	1.23	1	1.23	0	0
Val	0	0	0	0	0	0
Trp	8	9.88	8	9.88	8	10.26
Tyr	13	16.05	13	16.05	8	10.26
Calculated M.W.	9798		9824		9513	
Calculated p*I*	9.58		9.58		9.82	

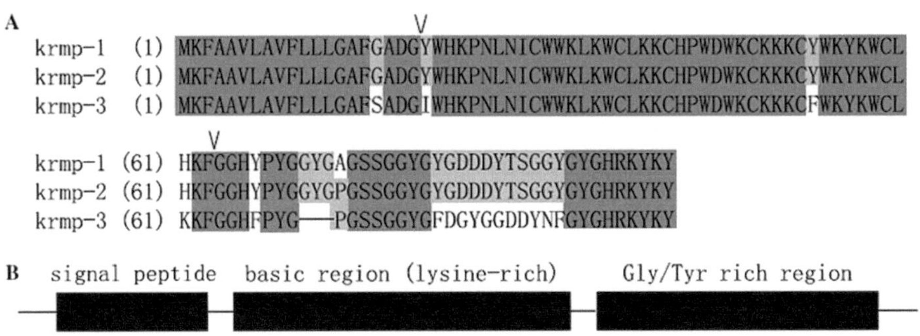

Fig. 3.28 (**a**) Alignment of the putative amino acid sequences of KRMP family. Arrowheads point the borders between the different regions in (**b**). (**b**) Schematic representation of the three regions of KRMP

3.5.1.3 Gene Expression Analyses of KRMP [9]

To explore the functions of KRMP in vivo, gene expressions of KRMP and other three known matrix proteins of *P. fucata* as positive control (MSI31 [6], MSI60 [6],

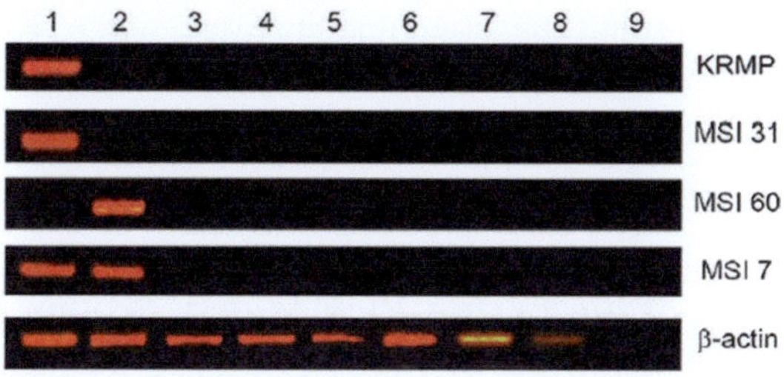

Fig. 3.29 RT-PCR analysis of KRMP and some other gene expressions of shell matrix proteins in *P. fucata*. Lane 1, the mantle edge; lane 2, the mantle pallial; lane 3, foot; lane 4, gill; lane 5, gonad; lane 6, digestive gland; lane 7, adductor muscle; lane 8, hemocytes; and lane 9, negative control. The housekeeping gene β-actin was detected as a positive control

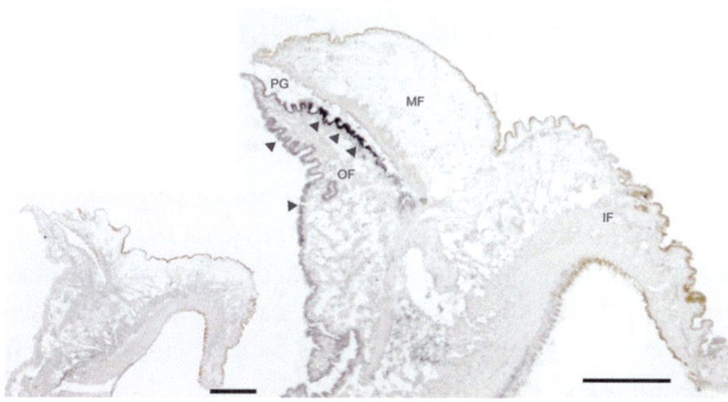

Fig. 3.30 Expression of KRMP in the mantle of *P. fucata* via in situ hybridization. Hybridization signals are pointed by arrows. The left inset represents a negative control section incubated with sense probe. *PG* periostracal groove, *OF* outer fold, *MF* middle fold, *IF* inner fold. Scale bars, 0.5 mm

and MSI7 [71]) were examined by qRT-PCR in different tissues with specific primers (kmpF and kmpR). Samples of total RNA from different tissues were prepared separately from the mantle pallial, the mantle edge, foot, gonad, gill, digestive gland, hemocytes, and adductor muscle tissues from *P. fucata*. The results showed that KRMP transcript was exclusively expressed in the mantle edge, which is related to the formation of the periostracum and the prismatic layer of the shell (Fig. 3.29). Similarly, mRNA of MSI31which is rich in Gly of the prismatic layer was expressed in the mantle edge exclusively, while mRNA of MSI60 which is a framework protein of the nacreous layer was expressed exclusively in the mantle pallial, and the mRNA of MSI7 was expressed in both the pallial and the edge of the mantle tissue. Experiments were repeated twice with similar results. It suggested that KRMP only participate in the formation of periostracum and/or the prismatic layer matrix.

To investigate the more precise expression location of KRMP in the mantle edge of *P. fucata*, hybridization in situ was performed using frozen sections of the mantle (Fig. 3.30). Strong hybridization signals of KRMP mRNA were observed in the inner epithelial cells from the outer fold and weaker signals were found in the outer epithelial cells from the outer fold. No hybridization signal was observed in the middle and inner fold. Hybridization with a negative control sense probe showed no significant signals. This result suggested that the expression of KRMP mRNA was similar to that of Prismalin-14 partially [8], indicating that KRMP is a prismatic layer matrix protein family.

3.5.2 Dual Roles of the Lysine-Rich Matrix Protein (KRMP)-3 in Shell Formation of Pearl Oyster, Pinctada fucata

Matrix proteins play crucial roles in shell formation. Our group firstly cloned three cDNAs encoding lysine-rich matrix protein from pearl oyster, *Pinctada fucata* in 2006. Whereas, the functions of KRMPs are still not fully understood. What's more, KRMPs possess two functional domains, the basic domain and the Gly/Tyr domain. According to the modular organization, the functions of their two domains were poorly characterized. Moreover, KRMPs were then found in other two species, *P. maxima* and *P. margaritifera*, which suggested that KRMPs might be crucial for shell formation. In this research, the function and characterization of KRMP-3 [58] and the two functional domains were studied in vitro by purification of recombinant glutathione S-transferase (GST) tagged KRMP-3 and another two KRMP-3 deletion mutants. Western blot and immunofluorescence indicated that the native KRMP-3 existed in the EDTA-insoluble matrix composition of the prismatic layer and distributed in both of the organic sheet and the prismatic sheath. Recombinant KRMP-3 (rKRMP-3) could bind tightly to chitin and this binding capacity was relevant with the Gly/Tyr-rich region. rKRMP-3 could repress the $CaCO_3$ precipitation, influence the crystal morphology of calcite, and inhibit the aragonite growth in vitro, which was almost entirely due to the lysine-rich region. The results show direct evidence of the roles of KRMP-3 in shell biomineralization. The functional rBR domain was found to take part in the growth control of crystals, and the rGYR region was found to contribute to binding chitin.

3.5.2.1 Expression and Purification of rKRMP-3, rBR, and rGYR

Sequence-verified pGEX-4T-1 vectors which were inserted by either the KRMP-3 or GYR or BR were transformed for expression in *E. coli* (Fig. 3.31a). After addition of IPTG, the three GST-tag proteins were effectively expressed (Fig. 3.31b). The major bands of rKRMP3, rGYR, and rBR on the SDS-PAGE gel migrated have molecule mass of 30 ~ 43 kDa, which correspond to the molecular masses of the recombinant proteins: 35 kDa (rKRMP-3), 30.2 kDa (rGYR), and 30.8 kDa (rBR) (26 kDa GST plus either 9, 4.2, or 4.8 kDa, respectively). The molecular mass of rBR is slightly greater than rGYR. In the meantime, we purified GST using the same method as a negative control. Traditionally, most matrix proteins have been separated directly from the soluble shell powder in various agents, including EDTA [5], PBS [64], ddH2O [12], or weak dilute acids [66]. Among the separated proteins, most were

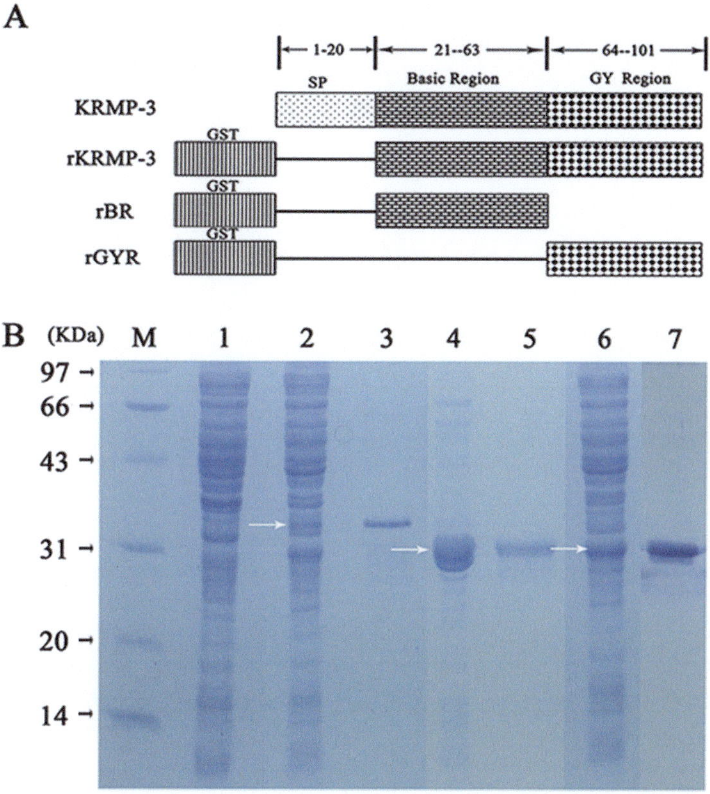

Fig. 3.31 Recombinant proteins of KRMP-3 derivatives. (**a**) Schematic representation of KRMP-3, rKRMP-3, rBR, and rGYR. KRMP-3, full length of wild-type KRMP-3 possessing the signal peptide (SP) sequence and total 101 amino acid residues; rKRMP-3, KRMP-3 lacks the SP sequence and has 81 amino acid residues and tagged with an affinity GST in N-terminal; rBR, the basic region of KRMP-3 with 43 amino acid residues (21–63) and tagged with an affinity GST N-terminally; rGYR, the Gly/Tyr-rich region of KRMP-3 with the C-terminal 38 amino acid residues (64–101) and tagged with GST in N-terminal as well. (**b**) Production and purification of recombinant KRMP-3, GYR, and BR in *E. coli*. Arrows mean the induced proteins by IPTG. M, markers indicating protein molecular mass; lane 1, whole cell lysate uninduced by IPTG; lanes 2, 4, and 6, whole cell lysate of rKRMP-3, rGYR, and rBR induced by 0.8 mM IPTG, respectively; lanes 3, 5, and 7, purified rKRMP-3, rGYR, and rBR eluted from GSTrap FFcolumn, respectively [58]

acidic matrix proteins with few basic matrix proteins that usually constitute the insoluble organic framework part and were barely solubilized by weak dilute acids or EDTA. When treated with denaturing agent at high temperature, most basic matrix proteins separated from the insoluble components were detected to be denatured and inactive. Then the recombinant proteins seemed to be a viable alternative to study matrix proteins. *E. coli* and *P. pastoris* expression systems were selected in our study. A GST-tagged KRMP-3 was successfully expressed in *E. coli* after the attempt to express a His-tagged KRMP-3 in *E. coli* and *P. pastoris*

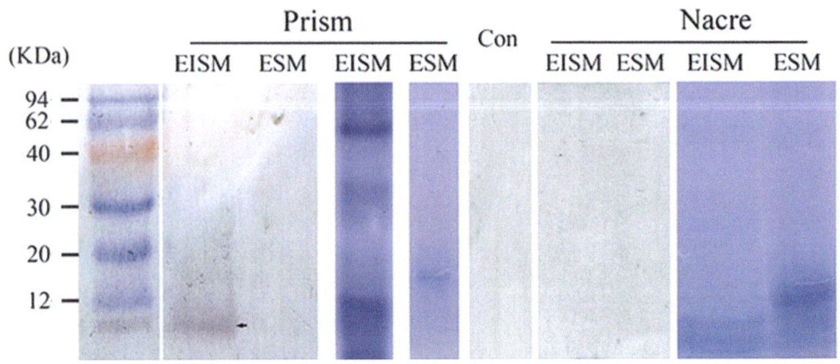

Fig. 3.32 Western blot analysis using anti-rKRMP-3 as primary antibody. The EDTA extracts from the insoluble matrices (EISM) and soluble matrices (ESM) of the prismatic layer from shells. The EISM and ESM of the nacreous layer were conducted to SDS-PAGE and subjected to Western blot analysis with anti-rKRMP-3 antibody or preimmune serum as primary antibody. The arrow shows that KRMP3 was found in the EISM composition of the prismatic layer. The con lane is the control group incubated with preimmune serum as primary antibody

failed. The two function regions of KRMP-3 were also expressed and purified to investigate their functions. After the expression and purification of the three GST-tagged proteins, we tried to excise the GST tag using proteolytic; nevertheless, low recovery of the target proteins held back to research rKRMP-3, rBR, and rGR excluding the GST tag. Similar challenge has been reported to express and purify recombinant matrix proteins PFMG1 [37] and MSI7 [9]. All things considered, purified rKRMP-3, rBR, and rGR with the GST-tag were used to explore the function of these three proteins in vitro.

3.5.2.2 Immunolocation of Native KRMP-3 in the Shell

We detected the distribution of native KRMP-3 in the shell with the anti-rKRMP-3 antibody or preimmune serum as the primary antibody. Western blot analysis indicated that the single intense band with apparent molecular mass of about 10 KDa corresponds to the native KRMP-3. What's more, KRMP-3 was detected in the EDTA-insoluble matrix (EISM) extracts from the calcite layer, but not in the EISM extracts from the nacreous layer or the EDTA-soluble matrix (ESM) from the two layers (Fig. 3.32). The shell of *P. fucata* displays a representative prismatic and nacreous microstructure. The outer structure is the prismatic layer and an organic sheet next, and the inner structure is the nacreous layer (Fig. 3.33a). The immuno-fluorescence localization of KRMP-3 on thin adult shell cross-sections displayed that native KRMP-3 existed in the prismatic sheath and the organic sheet (Fig. 3.33c, d). A small intensity of background staining was viewed in the control sample (Fig. 3.33b) but much weaker than that of experimental group in Fig. 3.33c, d. From dorsal to ventral the content of KRMP-3 increased following with the sheet becoming thicker gradually (Fig. 3.33c, d). Same phenomenon also occurred in several different *P. fucata* shells (data not shown). The results of Western blot

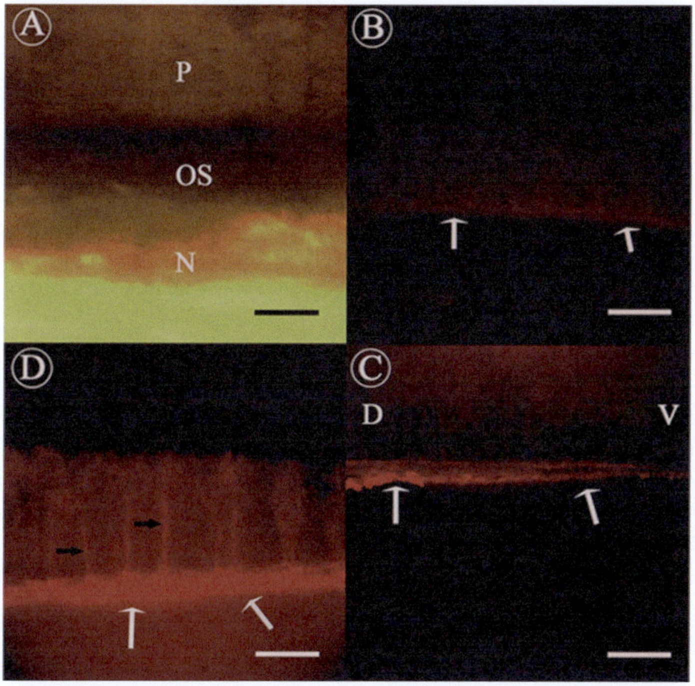

Fig. 3.33 Immunofluorescence distribution of native KRMP-3 in the adult shells of *P. fucata*, with anti-rKRMP-3 antibody. Shells were cleaned and modified and then decalcified and incubated with anti-rKRMP-3 antibody (**c**, **d**) or preimmune serum (**b**). (**a**), White light micrograph of the shell. The shell exhibits a representative nacroprismatic microstructure, with columnar calcitic prisms in the upper and nacreous layer beneath prism. These two calcified layers are generally isolated by an organic layer. *P* prismatic layer, *OS* organic sheet, *N* nacreous layer. (**b**), Negative control incubated with preimmune serum. The white arrows in (**b**) indicated only a small amount of background immunofluorescence was detected in the organic sheet. (**c**), Incubated with anti-rKRMP-3 antibody The white arrows pointed positive signal in the organic sheet. Moreover, the positive signal of the ventral to dorsal shows the distribution of organic sheet; *D* dorsal, *V* ventral. (**d**), Enlargement images of the sample. Positive signal found not only in the organic sheet (white arrows) but also in the prismatic sheath (black arrows indicate). Scale bars in (**a**, **d**), 12.5 μm and in (**b**, **c**), 50 μm

analyses and the immunofluorescence showed that KRMP-3 was localized at the prismatic sheath [86] and the organic sheet [12, 78, 87], which was suggested to be a component of the EISM. This result was supported by the data obtained from in situ hybridization and RT-PCR [9]. KRMP-3 is the second matrix protein to be found in the organic layer after the discovery of Prisilkin-39.

3.5.2.3 Chitin-Binding Ability

Chitin is the chief component of the insoluble matrix of prismatic layer [88, 89]. KRMP was proven to exist in the insoluble matrix. Therefore, we

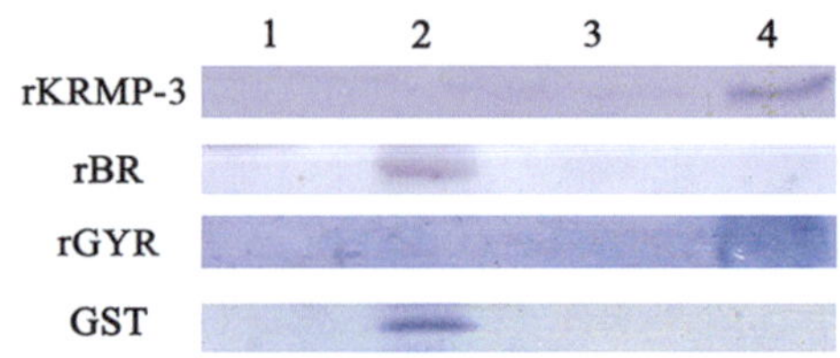

Fig. 3.34 Chitin-binding experiment. Lane 1, samples after water-washed; lane 2, samples of 0.2 M NaCl washings; lane 3, samples after 1 M acetic acid-washed; lane 4, extract with detergent-SDS/β-mercaptoethanol at 100 °C for ~10 min. GST was applied as a negative control

performed the chitin-binding assay of rKRMP-3, rGYR, and rBR, with GST as negative control protein. GST and rBR were washed through with saline and no binding ability was detected in the subsequent process. In contrast, rGYR and rKRMP-3 were not eluted by the acidic solution and saline washings but eluted by the detergent solution washing (Fig. 3.34). The results showed that KRMP-3 could combine with chitin and the ability was related to the GY region. Proteins containing Rebers–Riddiford (RR) [90] regions have been shown to have ability to bind chitin in vitro [22, 91]. The Rebers–Riddiford motif is the most widely spread in chitin-binding proteins in arthropod cuticles [92]. However, proteins without the RR sequences in mollusks also showed chitin-binding ability in vitro. For example, Prismalin-14 could bind to chitin because of its GY-rich region. The GYR of KRMP and other matrix protein regions are also named as RLCD (repetitive, low-complexity domain). RLCDs are found to have a key role to play in structural construction in eggshells and cuticles [93], and this may be the reason why these constructions show strong mechanical toughness [17, 94]. The GYR of KRMP-3 was shown to combine with chitin in our study, which provided supplement evidence of the function of RLCDs. The chitin-binding regions of chitinase are found to form β-sheet structure to bind chitin [88]. From known matrix proteins knowledge, the GY region of KRMP-3 and Prismalin-14 may possibly be cross-linked with quinone tanning to form β-sheets. It is possible that Prismalin-14 [8], Prisilkin-39, MSI31 [6], and the Shematrin family [17] which have Gly-rich sequences such as GGY are also involved in the formation of the prism (i.e., β-sheet structure) [58].

3.5.2.4 Morphology of the Calcite and Aragonite Crystals Induced by rKRMP-3, rBR, and rGYR In Vitro

To explore the functions of KRMP-3 in crystallization, the interactions of rKRMP-3, rBR, and rGYR with the growth of calcites were examined. The biominerals can be divided into subangular, subrounded, angular, and rounded according to the shapes generally. In our experiments, we found that the morphologies of crystals made a difference with the addition of rBR, rKRMP-3, and rGYR. 40 μg/ml GST was added as a negative control experiment with induced crystals as typically angular with smooth surface (Fig. 3.35a). The average width and length were 10.92 μm and

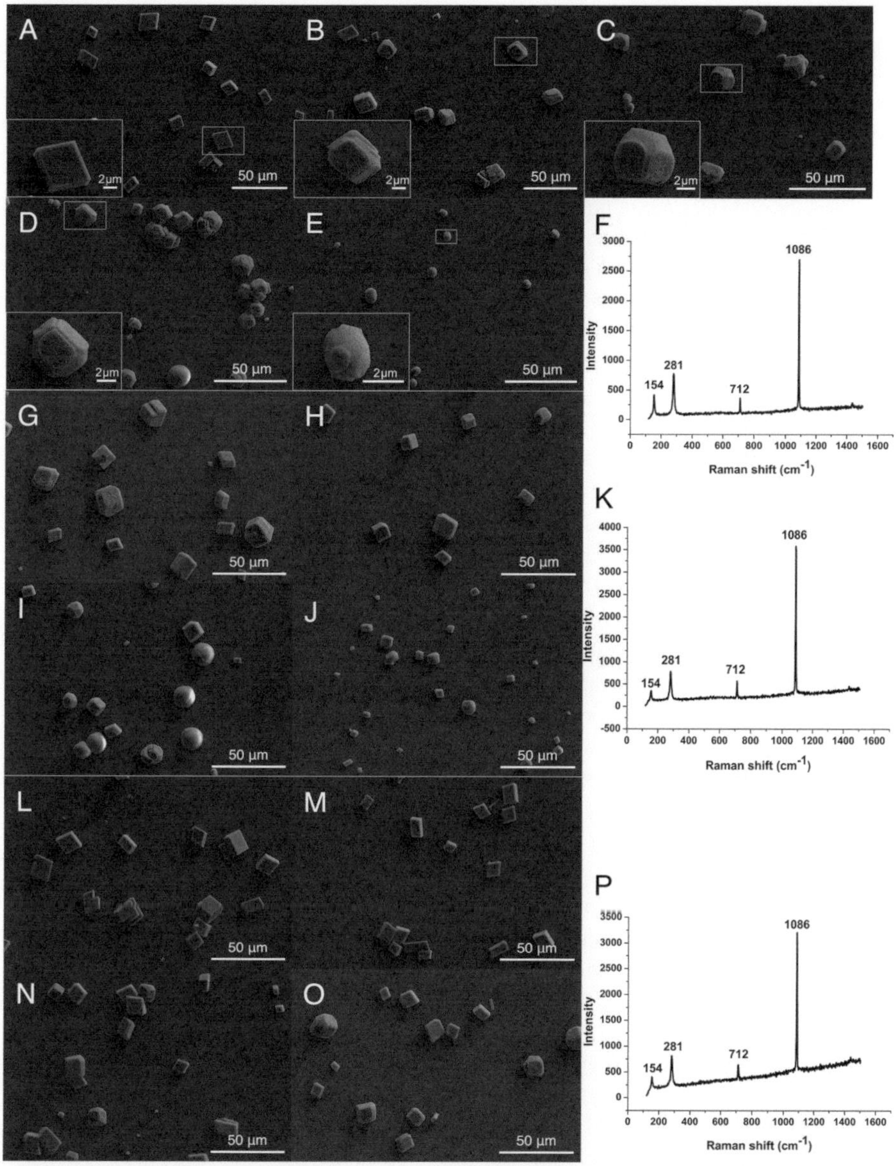

Fig. 3.35 SEM images and Raman spectrum analysis of in vitro crystallization of calcite system with addition of rKRMP-3, rBR, and rGYR. Crystals were formed with 40 μg/ml GST (**a**), 2 μg/ml rKRMP-3 (**b**), 10 μg/ml rKRMP-3 (**c**), 20 μg/ml rKRMP-3 (**d**), 40 μg/ml rKRMP-3 (**e**), 2 μg/ml rBR (**g**), 10 μg/ml rBR (**h**), 20 μg/ml rBR (**i**), 40 μg/ml rBR (**j**), 2 μg/ml rGYR (**l**), 10 μg/ml rGYR (**m**), 20 μg/ml rGYR (**n**), and 40 μg/ml rGYR (**o**). The crystals boxed up in left bottom of the image are enlargement images of the smaller boxed part, respectively. **f**, **k**, and **p** are the Raman spectrum of the crystals with 40 μg/ml rKRMP-3 (**e**), 40 μg/ml rBR (**j**), and 40 μg/ml rGYR (**o**), respectively

13.27 μm, respectively. With the concentration of 2 μg/ml of rKRMP-3, the shape of the calcite crystals changed. All crystals displayed rounded edges and hierarchy that could be visualized (Fig. 3.35b). With increasing concentration of rKRMP-3 (10–40 μg/ml), the shape of calcite changed from angular to round gradually (Fig. 3.35c–e). At dosage of 40 μg/ml of rKRMP-3, most calcites obtained changed to be round crystals and their size get smaller (Fig. 3.35e). Moreover, the morphologies of the crystals also changed in the system with rGYR (Fig. 3.35l–o) and rBR (Fig. 3.35g–j) gradually. In order to quantify the sizes and shapes of crystals appropriately, we observed crystals at low magnification (data not shown). With 40 μg/ml of rKRMP-3, 97.3% of the formed crystals were sphere with an average diameter of about 11.82 μm. With 40 μg/ml of rBR, subsphere and sphere crystals took up 70.3% of the crystals, with an average diameter of about 10.01 μm. With 40 μg/ml of rGYR, subsphere and sphere crystals occupied 33.8% of the crystals with an average diameter of about 12.36 μm. The Raman spectrum analyses of the formed crystals from the three groups (E, J, O) showed all of them were calcite with characteristic absorption peaks of 281, 712, and 1086 cm^{-1} (Fig. 3.35f, k, p).

In vitro crystallization studies have shown that natural and recombinant matrix proteins have effect on the crystal morphology [4, 13, 22, 95, 96]. Lysine residues, which are rich in the KRMP BR region, have been implicated to have ability of controlling crystal formation in biomimetic silicification crystallization before [97]. What's more, the addition of two lysine-rich peptides were found to perform circular crystals in $CaCO_3$ precipitation experiments [89]. In addition, the prismatic layer was shown to be disintegrated after KRMP genes were knocked down by dsRNA injection [14]. The dose effect of rKRMP-3 on calcite growth showed that the shape of calcite crystals changed with the addition of 2–20 μg/ml rKRMP-3 with the crystal angular gradually disappeared. The crystals became sphere and crystal sizes get smaller with higher dosage of KRMP-3 (40 μg/ml). Moreover, the system containing rBR showed consistent effects on crystal morphology as shown in the rKRMP-3 system, although rGYR showed minor influence on the crystals. All in all, our results are consistent with the reports before.

The presence of Mg^{2+} in the crystallization system tend to form aragonite [4], so 50 mM Mg^{2+} was added to the aragonite precipitation solution. GST was used as a negative control in the control group, with the formation of sphere-shaped crystals (Fig. 3.36a). With 2 μg/ml rKRMP-3, no obvious effect on the crystals was found (Fig. 3.36b). With 10 μg/ml rKRMP-3, no crystals but the material precipitated like a film was observed in the field of vision (Fig. 3.36d, e). We further examined the film-like material with Raman spectra and found the characteristic absorption peaks at 558.9 and 1092.6 cm^{-1}, not identical with any kinds of crystals (Fig. 3.36f). That is, there were no shaped crystals in the conduction, and we couldn't distinguish the material. No typical crystals were found with 2 μg/ml of rBR (Fig. 3.36g) and a few crystals formed in the presence of 10 μg/ml of rBR (Fig. 3.36h), while typical crystals formed in the presence of 2 μg/ml of rGYR (Fig. 3.36i). No typical crystals were observed only at 10 μg/ml of rGYR (Fig. 3.36j). Raman spectrum showed that the formed typical crystals in Fig. 3.36a and b/i were aragonite (Fig. 3.36c). The

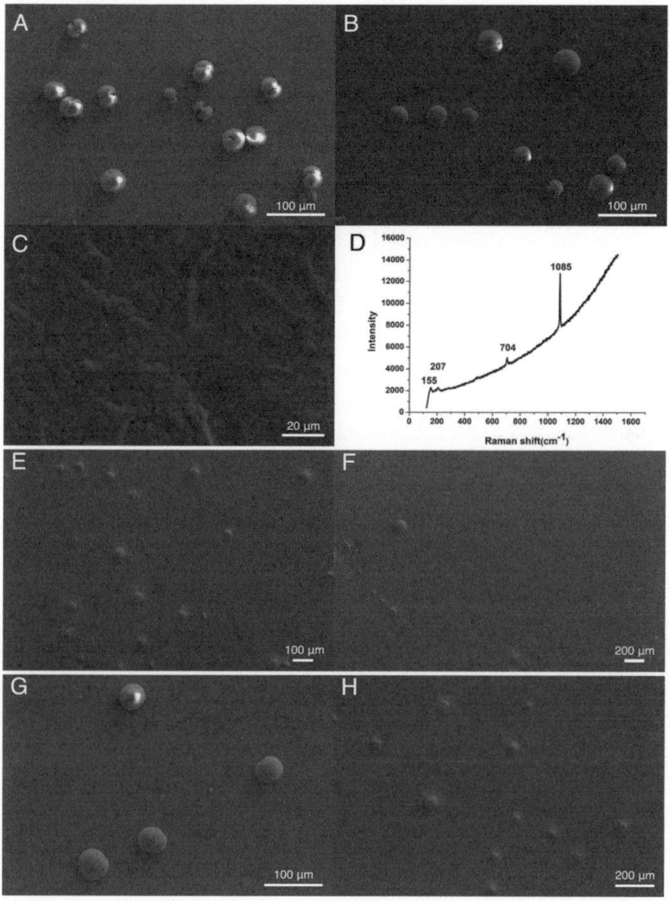

Fig. 3.36 SEM images and Raman spectrum analysis of in vitro crystallization of aragonite system with addition of rKRMP-3, rBR, and rGYR. Crystals were formed with 50 mM $MgCl_2$ and 10 μg/ml GST(**a**), 2 μg/ml rKRMP-3 (**b**), 10 μg/ml rKRMP-3 (**d**, **e**), 2 μg/ml rBR (**g**), rGYR (**i**), 10 μg/ml rBR (**h**), and rGYR (J). Raman spectrum indicated that the typical crystals formed in A/ B/ I are aragonite, with the special peaks at 207, 704, and 1085 cm^{-1}. (**c**). Raman spectra of the crystals precipitated with 10 μg/ml rKRMP-3 (**c/d**) have the absorption peaks at 558.9 and 1092.6 cm^{-1} (**f**)

concentration effect of rKRMP-3 showed strong inhibition in the aragonite growth system in vitro. rBR showed stronger inhibition of the formation of aragonite compared with rKRMP-3. Even with 2 μg/ml of rBR, there is no typical crystal formed. rGYR was observed to show weaker inhibition ability of aragonite crystallization compared with the activity of rKRMP-3 and rBR.

3.5.2.5 Effect of rKRMP-3, rBR and rGYR on Calcium Carbonate Precipitation

The function of rKRMP-3 and its two regions on calcium carbonate precipitation was investigated in vitro. In the blank control, the absorbance value at 570 nm which is an indicator of the precipitation changed and reached about 0.1 in 5 min. The precipitation rate drastically changed with the addition of rKRMP-3. The highest absorbance value did not even exceed 0.045 in the presence of 2 μg/ml rKRMP-3. The absorbance at every minute reduced following with the dosage of rKRMP-3 present increased. The similar effect was found when rBR was present in the system. Rbr and rKRMP-3 inhibited calcium carbonate precipitation dose dependently and almost inhibited this precipitation completely with about 20 μg/ml protein (Fig. 3.37a, b). To examine the contribution of the two domains to the inhibitory ability, 10 μg/ml rKRMP-3, rGYR, and rBR were added to the precipitation system, respectively. The inhibitory effect of rKRMP-3 and rBR was similar, while the inhibitory effect of rGYR was feeble (Fig. 3.37c). At the same time, GST had no influence on the precipitation of calcium carbonate. The results showed that rKRMP-3 could inhibit $CaCO_3$ precipitation significantly, inhibit the growth of aragonite, and change the morphology of calcite in vitro.

The results of the inhibitory ability on calcium carbonate precipitation of rBR and rKRMP-3 could explain the effect of protein on the crystals growth. The finding that rKRMP-3 controls calcite growth and inhibits aragonite growth is accordant with the shell localization of protein. The matrix protein from the prismatic layer could inhibit aragonite growth and control calcite growth. Two antagonistic mechanisms are crystal nucleation and growth inhibition in molluscan shell formation. KRMP-3 has an effect on growth inhibition in theory. This result is similar to that of Prisilkin-39 and different from that of major acidic matrix proteins. The functions identified are relative to a lysine-rich basic domain of KRMP-3, which may interact with the CO_3^{2-} or HCO_3^- during formation of the shell. What's more, the lysine side chains can participate in intermolecular cross-linking by Schiff's base conjugates [75]. The lysine-rich region in KRMP-3 is also suggested to interact with the anionic groups of sugars in the matrix glycoproteins and polysaccharides or acid proteins, which would be a supplement of the interaction with CO_3^{2-} or HCO_3^-.

3.5.3 *Cloning and Characterization of Prisilkin-39, a Novel Matrix Protein Serving a Dual Role in the Prismatic Layer Formation from the Oyster* Pinctada fucata

Shells were formed by mollusks out of $CaCO_3$ and matrix is composed of biomacromolecules. Understanding the role of matrices may afford clues on the biomineralization mechanism. Herein, a cDNA sequence with 1401 bp in full length encoding a novel matrix protein was cloned from the mantle of the pearl oyster,

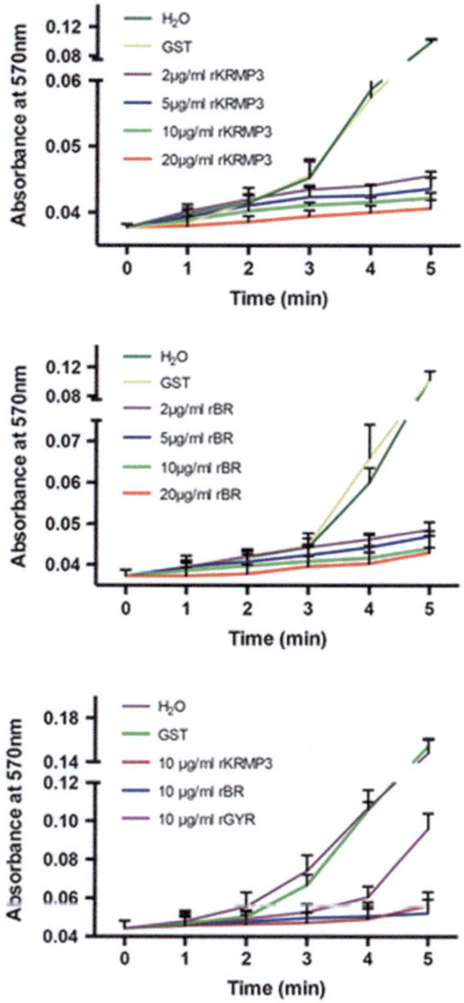

Fig. 3.37 Inhibition of $CaCO_3$ precipitation in vitro with rKRMP-3 (**a**), rBR (**b**), and rGYR. 2 μg/ml rKRMP-3 and rBR is effective, and the inhibitory vitality is dose-dependent. Inhibitory activity of 10 μg/ml rKRMP-3 is similar to 10 μg/ml rBR, whereas the activity of 10 μg/ml rGYR is lower (**c**). 20 μg/ml GST had no influence on calcium carbonate precipitation

Pinctada fucata. The putative protein (Prisilkin-39) [13], which has a calculated molecular mass of 39.3 kDa and a predicted isoelectric point of 8.83, was fully identified, and its function in biomineralization was studied by both in vivo and in vitro crystal growth experiments. Prisilkin-39 is a highly repeating protein with a novel composition of Gly, Tyr, and Ser residues. Expression of Prisilkin-39 was distributed in columnar epithelial cells of the mantle edge, contributing to the prismatic layer formation of calcite. Immunohistochemical in situ and immunodetection in vitro had shown the existence of a characteristic pattern of Prisilkin-39 in the sheaths around the prisms and organic sheet. Prisilkin-39 could bind tightly with chitin which is an insoluble polysaccharide that provides the highly structured framework of the shells. Antibody injection in vivo led to dramatic morphological damage in the inner shell surface structure, where a large amount of $CaCO_3$ was formed in a disordered manner. In addition, Prisilkin-39 strictly inhibited the aragonite precipitation in vitro. In conclusion, Prisilkin-39 is the first protein proven to have dual function, involved in both of the chitinous framework building and the crystal growth regulation during the prismatic layer formation. These findings may shed light on the rare group of basic matrices and their functions during regulation of the molluscan shell.

3.5.3.1 Isolation and Sequence Analysis of cDNA for Prisilkin-39 [13]

A DNA fragment of ~1000 bp was obtained by the 3′-RACE process with degenerate primer of YGS-F1. Then the sequences analysis indicated that the degenerate primer YGS-F1 annealed with the sequence of cDNA encoding GYGGYS peptide. According to the 3′ sequence, we designed and synthesized a gene-specific antisense primer YGS-R1, and then a cDNA fragment was obtained by the 5′-RACE process. To explore the sequence combination of the results of 3′-RACE and 5′-RACE, another PCR of the full-length cDNA was conducted with the *P. fucata* mantle cDNA as a template. We obtained a positive confirmed cDNA clone YGS7 including the poly (A) tail with 1401 bp in length (Fig. 3.38). It possesses a 5′-untranslated sequence of 59-base, an open reading frame consisting of 1218 bp nucleotides, an in-frame stop codon TAA, and a 3′-untranslated sequence of 121-base. The position of the first start codon is at nucleotide 60, with a purine (A) three nucleotides upstream which is consistent with the Kozak consensus sequence of initiation of transcription in eukaryotic species [98, 99]. A putative polyadenylation signal AATAAT is recognized within this cDNA sequence at the nucleotide position of 1305, which are 76 nucleotides upstream from the poly (A) tail.

3.5.3.2 Analysis of Putative Amino Acid Sequence of Prisilkin-39

The deduced protein Prisilkin-39 contains 406 amino acid residues and has a putative molecular mass of ~39.3 kDa before any posttranslational modification. Prisilkin-39 is identified by high proportions of Gly (29.7%), Tyr (23.3%), and Ser (18.6%), which accounted together for over 70% of the all amino acid residues (Table 3.6). Besides, the amino acid component of Prisilkin-39 is also similar to that of most matrix proteins of the prismatic layer from *P. fucata* [100]. Different from many other shell matrix proteins that are singularly rich in Asp and Glu residues [8, 10, 72, 101, 102], the content of these acidic amino acids in Prisilkin-39 is drastically low, with only one Asp residue. Therefore, Prisilkin-39 is a part of the rare group of basic matrix proteins with an apparent theoretical pI of 8.83.

Another characteristic feature of the Prisilkin-39 sequence is its modular structure, possessing a signal peptide, a short alanine-rich hydrophobic region in N-terminal, two highly repetitive regions, and a histidine-rich basic tail in C-terminal (Fig. 3.39). The first 19 amino acids are predicted as signal peptide, which is similar with the signal peptides of MSI31 [6] and MSI7 [103], two matrix

```
1       acgcgggggttttcactgcagtttcgaactactaaagacaggttacaaaattgtcaca
60   atg aaa gga ttc ctg acg ctc ttg ctg gtt tgt gct ata ctc agt act ggg tat tgt caa
1     M   K   G   F   L   T   L   L   L   V   C   A   I   L   S   T   G   Y   C   Q
120  agt cgt cga aga gct gca ctt act gga tta gta gct ggt gcc acc att ggc gct tta gct
21    S   R   R   R   A   A   L   T*  G   L   V   A   G   A   T   I   G   A   L   A
180  agt ggt ggt tta ggt gcc ggt gct ggt ggt ttc ggt gta gga ggt ttc cca gta gga gtt
41    S   G   G   L   G   A   G   A   G   G   F   G   V   G   G   F   P   V   G   V
240  ggt gcg gtc gga att cca gtt gct gtt ggc ggt ggt ata ccg tat ggt tac ggc gga tat
61    G   A   V   G   I   P   V   A   V   G   G   G   I   P   Y   G   Y   G   G   Y
300  agc ggg tat ggc tac ggc tat cca gct ggt gga tat ggt gga tac agt tat ggc tat cca
81    S   G   Y   G   Y   G   Y   P   A|1 G   G   Y   G   G   Y   S   Y   G   Y   P
360  act ggt gga tac ggt gga tac agc tac gga tat cca act ggt ggg tac gga gga tac agc
101   T|2 G   G   Y   G   G   Y   S   Y   G   Y   P   T|3 G   G   Y   G   G   Y   S
420  tac gga tat cca act ggt gga tac gga gga tac agc tac ggt tac cca acg gga gga tac
121   Y   G   Y   P   T|4 G   G   Y   G   G   Y   S   Y   G   Y   P   T|5 G   G   Y
480  agc gga tac agc tac gga tat cca acg gga gga tac agc gga tac agc tac gga tat cca
141   S   G   Y   S   Y   G   Y   P   T|6 G   G   Y   S   G   Y   S   Y   G   Y   P
540  acg gga gga tac agc gga tac agc tac gga tat cca acg gga gga tac agc gga tac agc
161   T|7 G   G   Y   S   G   Y   S   Y   G   Y   P   T|8 G   G   Y   S   G   Y   S
600  tac gga tat cca acg gga gga tac agc gga tac agc tac gga tat cca acg gga gga tac
181   Y   G   Y   P   T|9 G   G   Y   S   G   Y   S   Y   G   Y   P   T|10 G  G   Y
660  agt gga tac agc tat cca acg ggc gga tac agt gga tac agc tac agt tca acc cca gga
201   S*  G   Y   S   Y   P   T|11 G  G   Y   S   G   Y   S*  Y   S   S*  T   P|12 G
720  tat gga tat tac ggc tca ggg tcc gga atg ggt gga atg cgt tca ggg tat agt tat tac
221   Y   G   Y   Y   G   S   G   S   G   M   G   G   M   R   S   G   Y   S*  Y   Y
780  agc agc ccc gcc cca tca tat tac agt tct ggt agt atg aca cca gga tac gga tat tac
241   S   S   P   A   P   S*  Y   Y   S   S*  G   S*  M   T   P   G   Y   G   Y   Y
840  agc tca ggc tct gga ata gga gga gga atg ggt tca gga tat agc tat tat agc agc ccc
261   S   S   G   S*  G   I   G   G   G   M   G   S   G   Y   S*  Y   Y   S   S   P
900  gcc cct tca tat tac agt tct agt gtg agt cca gga tac gga tat tat ggc tca ggc tct
281   A   P   S*  Y   Y   S   S   S   V   S*  P   G   Y   G   Y   Y   G   S   G   S*
960  gga atg aga gga tat ggt tat tac agt agc agc aca cct atg tat tac ggt tct aga agt
301   G   M   R   G   Y   G   Y   Y   S   S   S   T   P   M   Y   Y   G   S   R   S*
1020 acg ggt tat ggt ccc ttt tct tct gga ttg gga gga atg ggt tca ggg tat agt tat tac
321   T*  G   Y   G   P   F   S   S   G   L   G   G   M   G   S   G   Y   S*  Y   Y
1080 agc agc agc acc cca tca tat tac agt tct ggt agt atg aca cca gga tac gga tat tac
341   S*  S   S   T   P   S*  Y   Y   S*  S*  G   S*  M   (T) P   G   Y   G   Y   Y
1140 ggc agt act agc tat cca gga cca ggg tat ggc tcc tac agt tat cgt aca act agc tat
361   G   S   T*  S   Y   P   G   P   G   Y   G   S   Y   S*  Y   R   (T) (T)* S* Y
1200 caa ccc tcc tct tat gga tat tcc tct tat gga acc acg tat cca gga cac ggg cat tgg
381   Q   P   S   S*  Y   G   Y   S   S*  Y   G   T*  T   Y   P   G   H   G   H   W
1260 cat ggt cac aag gac tgt taa aaggaccttaacaagatggacattaataatgtttcccgactctatccaa
401   H   G   H   K   D   C 406
1330 ttgtaattttatcgttgtcatatgagaaatgcagcattattaaagtgttattttcgaaaaaaaaaaaaaaa 1401
```

Fig. 3.38 Nucleotide sequence and putative amino acid sequence of Prisilkin-39
Numbers and boldface numbers on the left of sequence are positions of the nucleotides in the Prisilkin-39 cDNA sequence and the amino acid residues in the putative protein, respectively. The predicted signal peptide (residues 1–19) is marked in boldface italic type. The initiation codon (atg), the stop codon (taa), and the predicted polyadenylation signal (aataat) are boxed up. Primers designed for cloning are indicated in gray shading. Protein sequence of Prisilkin-39 possesses 12 tandem repeats of which the first repeat is underlined and the last amino acid residue of other repeat is shown by a vertical line and numbered 1–12. Four histidine residues and one lysine residue in the C-terminal are double underlined. The predicted phosphorylation sites are indicated by asterisks, and the possible O-glycosylation sites of three threonine residues are circled [13].

Table 3.6 Amino acid component of Prisilkin-39, KRMP, and Shematrin-1

	Prisilkin-39		KRMP-1		Shematrin-1	
Amino acid	Residues	mol %	Residues	mol %	Residues	mol %
Gly	115	29.72	14	17.28	129	38.05
Tyr	90	23.26	13	16.05	48	14.16
Ser	72	18.60	3	3.7	29	8.55
Pro	29	7.49	3	3.7	14	4.13
Thr	23	5.94	1	1.23	8	2.36
Ala	13	3.36	1	1.23	21	6.19
Met	8	2.07	0	0	1	0.29
Val	8	2.07	0	0	29	8.55
Arg	7	1.81	1	1.23	5	1.47
leu	5	1.29	4	4.94	16	4.72
His	4	1.03	5	6.17	1	0.29
Ile	4	1.03	1	1.23	13	3.83
Phe	3	0.78	1	1.23	11	3.24
Gln	2	0.52	0	0	4	1.18
Asp	1	0.26	4	4.94	1	0.29
Lys	1	0.26	14	17.28	3	0.88
Trp	1	0.26	8	9.88	0	0
Cys	1	0.26	6	7.41	0	0
Asn	0	0.00	2	2.47	6	1.77
Glu	0	0.00	0	0	0	0
Total	387		101		339	
M_r	39,323		9798		31,912	
pI	8.83		9.58		9.13	

proteins contributed in the prismatic layer formation of *P. fucata*, and the signal peptides of Aspein [10] and the Asprich family [5] which are extremely acidic matrix proteins found in the prismatic layer of *P. fucata* and *Atrina rigida*. In view of the high content of hydrophobic amino acids in the first 60 residues of the mature protein without signal peptide (Fig. 3.39b, arrowheads), the hydropathy of Prisilkin-39 was further verified with the method of Kyte and Doolittle accessed on line. The analysis suggested that the only hydrophobic region of the protein is located at the N terminal, consisting with residue number 20–77 in the alanine-rich region (Fig. 3.3b). After this region, almost 80% of the total Prisilkin-39 sequence is separated by two consecutive regions named as GYS region I and GYS region II (Fig. 3.39c, d). They are organized in tandem with highly conserved repeats, resulting in Prisilkin-39 as one of the most repetitive molluscan shell proteins known. GYS region I consists of 12 repetitive sequence regions, 10 of which are GGY (G/S) GY (G/S) YGYP (T/A) region. There are high degrees of sequence similarity in these elements, suggesting that they are possible to have similar folding model. Meanwhile, GYS region II is also characterized with random repetitive

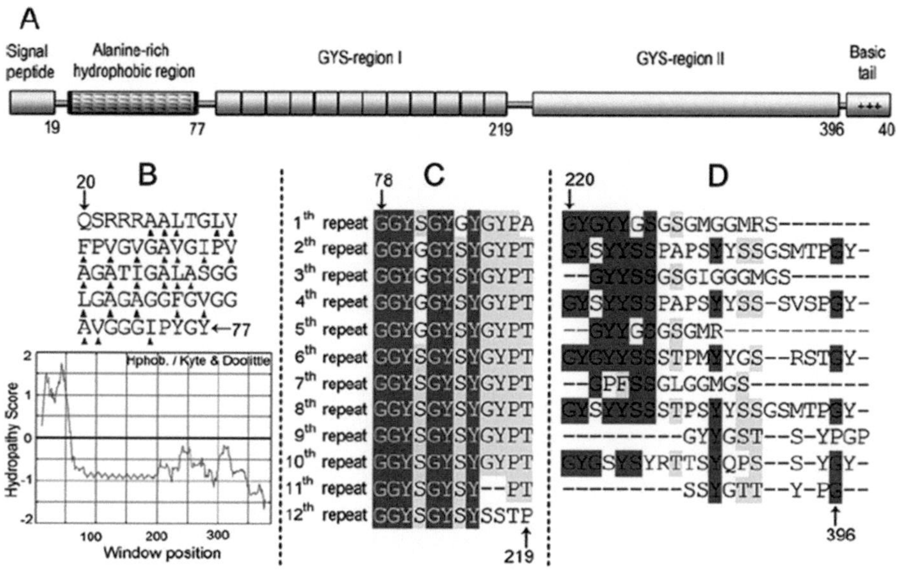

Fig. 3.39 Schematic representation of protein Prisilkin-39
(**a**) Schematic exhibition of the regional structure of Prisilkin-39. The protein possesses a signal peptide, a short alanine-rich hydrophobic region in N-terminal (residues 20-77), two highly repetitive regions, GYS region I (residues 78-219), GYS region II (residues 220-396), and a histidine-rich basic tail in C-terminal (residues 397-406). (**b**) Amino acid sequence of the alanine-rich hydrophobic region (upper region) and the corresponding hydrophobicity profile of Prisilkin-39 (lower part). Amino acids with high hydropathy scores are marked by arrowheads. Hydrophobicity analysis was conducted with the method of Kyte and Doolittle (window = 19). (**c**) Sequence alignments of these 12-fold tandem repeats in a consensus sequence of GYS region I. Consistent amino acid residues are shaded, and the same glycine and tyrosine residues are highlighted. (**d**) Sequence alignment in GYS region II. The arrowheads and number point the amino acid range of each region in (**b**–**d**) [13]

elements, which are composed by Gly-Tyr, Tyr-Ser, Tyr-Tyr, and Gly-Ser dipeptides, while there seems to be disorder in terms of the organization of these different dipeptides in this section, leading to GYS region II less repetitive than GYS region I. These two repeated regions are followed with a segment of ten amino acids enriched in polar residues. Four histidine residues and one lysine residue are compactly located in the C-terminal of Prisilkin-39. This basic tail has some similarity with the tail of the Shematrin protein family [17]. It is reasonable that in the extrapallial fluid with pH of 7.4, where matrix combines and shell mineralization occurs [75], these residues' side chains are preferred to be positively charged and thus able to interact with carbonate ions or negatively charged amino acid residues of other acidic proteins [101].

Further bioinformatics analysis of the Prisilkin-39 sequence indicates that some serine and threonine residues of Prisilkin-39 may either be O-glycosylated or phosphorylated (as shown in Fig. 3.38), a character commonly found in proteins

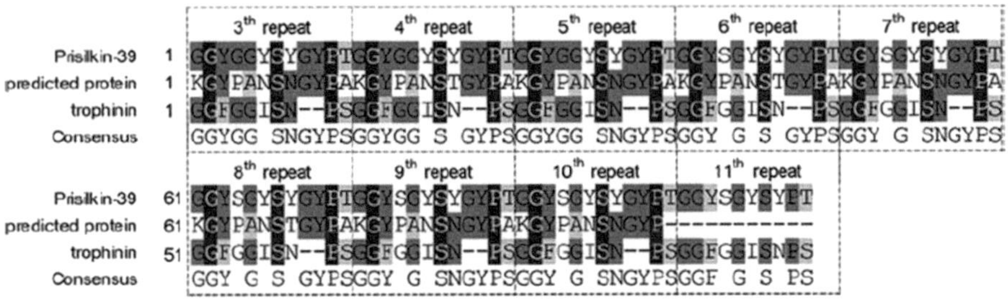

Fig. 3.40 Alignment of repeat numbers 3–11 of Prisilkin-39 with these two similar tandem repeated proteins

involved in mineralized tissues. Drastically, three potential O-glycosylation sites and 24 phosphorylation sites of 28, based on NetOGlyc3.1 [104] and NetPhos2.0 [105], respectively, are compactly distributed in GYS region II. This may suggest that GYS region II seems to have a more flexible and more complicated folding structure, as it is more inclined to be posttranslational modificated than the GYS region I.

The Blastp (protein with protein BLAST) search [106] against GenBank in NCBI website showed no remarkable homology of Prisilkin-39 with any other known protein. A low similarity to Prisilkin-39 existed in a predicted protein from *Ostreococcus lucimarinus* and a protein trophinin which is from mouse [107] (Fig. 3.40). Both proteins possess a conserved glycine-rich region which is repeated in tandem like in Prisilkin-39 10 and 13 times, respectively, whereas there have been no reports of any known functions.

The proteins shown in the figure are the translated protein from *O. lucimarinus* CCE9901 (residues 55–149) with NCBI accession number of ABO98608 and the trophinin protein of *Mus musculus* (residues 894–990) with NCBI accession number of NM001002272. Amino acids are numbered at the left, and gaps are marked by (-). Same residues are highlighted or shaded in black, and other conserved residues are shaded.

3.5.3.3 Gene Expression Analysis

We investigate the expression of Prisilkin-39 in several tissues of *P. fucata* via RT-PCR (Fig. 3.41a) and the expression being found only in the mantle edge, consistent with the prismatic calcite shell layer, while not in the mantle pallial relevant to the nacreous aragonite layer or in any other tissues. To explore the more precise expression sites of Prisilkin-39 mRNA in the mantle edge in *P. fucata*, we conducted in situ hybridization with frozen sections of the mantle tissue. Expression of Prisilkin-39 in situ was distributed in the inner epithelial cells of the outer fold and the outer epithelial cells in the middle fold at the bottom of the

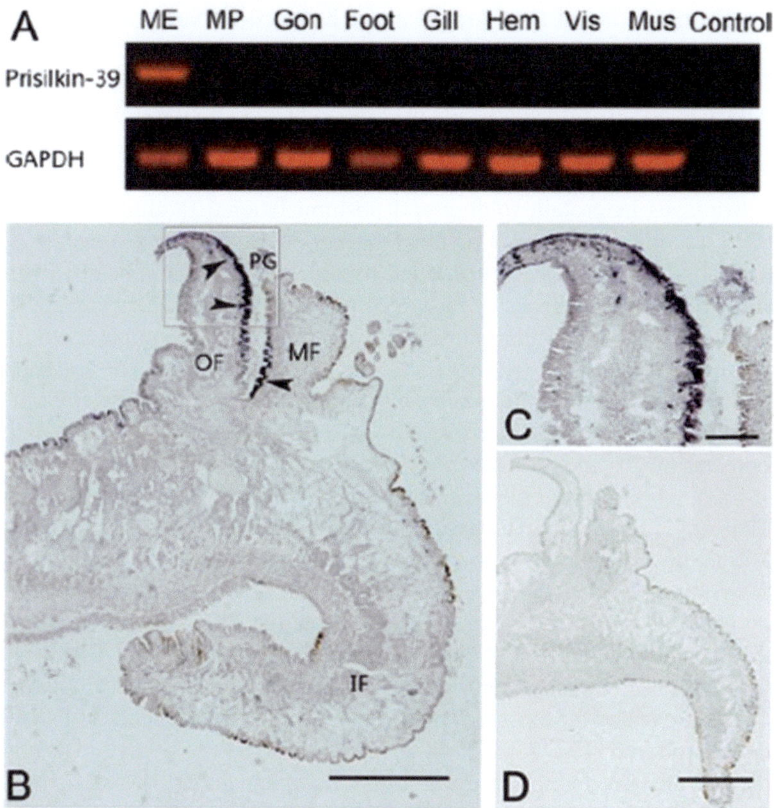

Fig. 3.41 Gene expression level of Prisilkin-39 and its location in the columnar epithelium of the mantle tissue
(**a**) Relative tissue specific gene expression of Prisilkin-39 with RT-PCR analysis. Total RNA was extracted from mantle edge (ME), mantle pallial (MP), gonad (Gon), foot, gill, hemocytes (Hem), viscus (Vis), and adductor muscle (Mus), respectively. The housekeeping gene GAPDH was detected as a positive control. After reverse transcription, PCR was conducted with primer sets for Prisilkin-39 and GAPDH presented under "experimental process." (**b**) Distribution of Prisilkin-39 mRNA in the mantle of *P. fucata* by hybridization in situ. Hybridization signals (dark purple) pointed by arrowheads existed in the inner epithelial cells of the outer fold and the outer epithelial cells in the middle fold. *OF* outer fold, *MF* middle fold, *IF* inner fold, *PG* periostracal groove. Scale bars, 500 μm. (**c**) Amplification of the box in (**b**). Scale bars, 100 μm. (**d**) Control section incubated with a sense probe. Bar, 500 μm

periostracum groove (Fig. 3.41b, c), while no hybridization signal was found at the dorsal mantle or the inner fold. Incubation with a control probe had no significant signals (Fig. 3.41d). What's more, it is worthy to note that this expression profile was similar to that of Prismalin-14 [8] and KRMP in part [9], which existed in the prismatic layer of *P. fucata*, indicating the probable involvement of Prisilkin-39 during the process of prismatic layer formation.

3.5.3.4 Identification of Prisilkin-39 in the Extracts of Nacre and Prisms from *P. fucata*

Biochemical analyses were conducted to explore whether native Prisilkin-39 exist in the prismatic and/or nacreous layer of *P. fucata*. Yeast which is a largely used eukaryotic expression system was picked to produce recombinant proteins as a result of its advantages in protein folding and posttranslational modification in present study. The recombinant model of the Prisilkin-39 protein was transformed into *P. pastoris* GS115 cells and then purified by nitrilotriacetic acid beads. Polyclonal antibodies which were raised against recombinant Prisilkin-39 were applied for immunodetection of native Prisilkin-39 in proteins of EDTA soluble matrix (ESM) and EDTA-insoluble matrix (EISM) of the shells from *P. fucata*. The reactions were executed with ELISA and immune-dot blot assay. As Fig. 3.42 shows, Prisilkin-39 was detected only in the EISM while not in the ESM of the shells. EISM proteins were conventionally thought to contribute to the construction of the structural organic framework and the physical properties of the shells during biomineralization. Thus, it is reasonable to assume that Prisilkin-39 works as a framework composition in the prismatic layer according to its presence pattern and amino acid component (Gly/Tyr-rich).

What's more, the existence of Prisilkin-39 in the prismatic layer was further explored and identified by SDS-PAGE and Western blotting with EDTA extracts from separated nacreous layer and prismatic layer. The gels stained by Coomassie Brilliant Blue revealed the existence of numerous different thin bands together with a smeared background (Fig. 3.43a), which shown the special behavior of total shell proteins. All extracts exhibited distinctively different patterns, and many matrix proteins were poorly subjected on acrylamide gels stained with Coomassie Blue or silver nitrate staining (data not shown). In addition, Stains-All stained gel images consistent with the samples Fig. 3.43 shown represented that both the recombinant Prisilkin-39 and the prism EISM extracts stained pink negatively, and in turn it provides evidence that native Prisilkin-39 is not a vast phosphoprotein in the EISM extracts of prism (Fig. 3.43b). Western blot results suggested that Prisilkin-39 was detected only in the EDTA-insoluble fractions of the prisms after boiling with denaturing solution for 15 min at 100 °C. Neither the parts of nacre nor the EDTA-soluble matrix of the prisms exhibited any recognition with anti-Prisilkin-39 (Fig. 3.43c). The biochemical distribution of Prisilkin-39 did, however, accordant with the results of previous gene expression analyses. What's more, additional bands with higher molecular mass (about 60 and 80 kDa) were also discovered, which could be consistent with multimers of the immunoreactive band or to Prisilkin-39 modified posttranslationally as tested previously of Caspartin [74]. Meanwhile, this banding pattern was also found in the results of Western blotting analyses of Prisilkin-39 of different tissues. As Fig. 3.43d had shown, Prisilkin-39 mainly lies in the mantle and extrapallial fluids and some is detected in the homolymph and

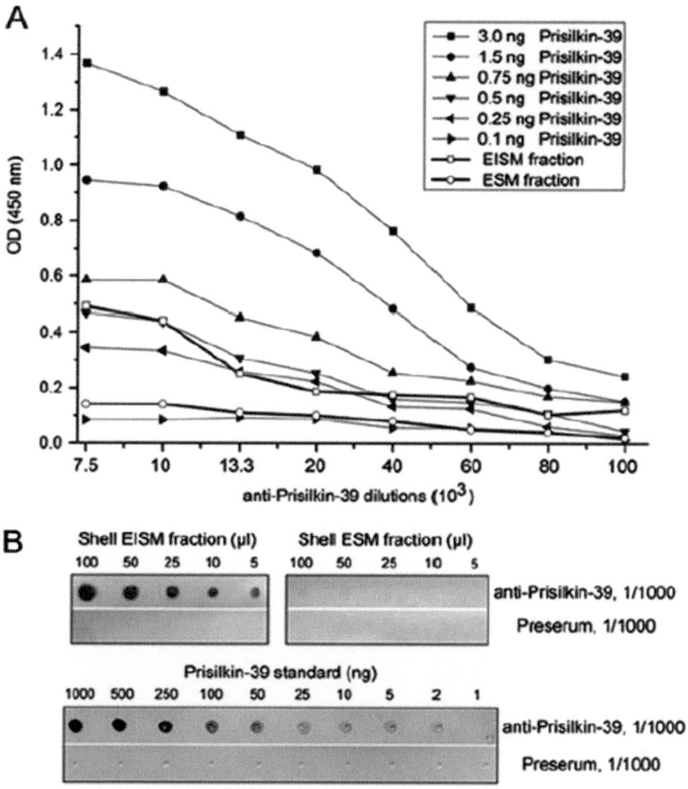

Fig. 3.42 Detection of native Prisilkin-39 in different extracts of shells
The purified anti-Prisilkin-39 antibody was applied against EDTA-soluble matrix (ESM) and EDTA-insoluble matrix (EISM) of the shells from *P. fucata*, using ELISA (**a**) and dot immune blot assay (**b**). For more procedure details, see "Experimental Procedures." The extracts were compared with known content of recombinant Prisilkin-39, which were diluted continuously by EDTA and treated as a positive control. Both the ELISA curve (**a**) and the blot (**b**) indicate that strong signals are found in the EISM but not in the ESM of the shells.

viscus. The exhibition of native Prisilkin-39 in the mantle, extrapallial fluid, and the homolymph suggested the possible route of protein transport (from the mantle tissue to the extrapallial fluid via the homolymph) before Prisilkin-39 had been involved in the mineralized prisms.

3.5.3.5 Chitin-Binding Vitality of Prisilkin-39

It is suggested that framework proteins could interact with chitin which is a major insoluble component of the organic framework on basis of the models proposed by Weiner and his co-workers [42, 108, 109]. In consistent with the notion that Prisilkin-39 works as a structural role in the formation of the framework during

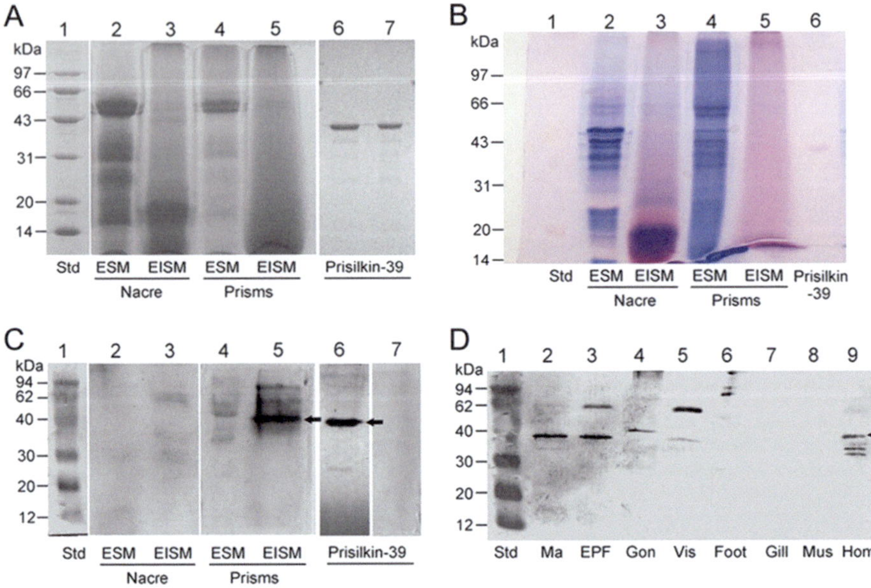

Fig. 3.43 Existence of Prisilkin-39 in specific shell layer and different tissues of *P. fucata* Prisilkin-39 was identified in EDTA extracts of isolated nacre and prism of the shell in 12% SDS-polyacrylamide gels (**a** and **b**) and Western blotting (**c**). The gels were dyed by Coomassie Blue (**a**) and the carbocyanine dye Stains-All (**b**). Lane 1, molecular mass standards (Std); lane 2, EDTA-soluble matrix of the nacreous layer; lane 3, the denatured segment of the EDTA-insoluble matrix of the nacreous layer; lane 4, EDTA-soluble matrix of the calcitic prisms; lane 5, the denatured fraction of the EDTA-insoluble matrix of the calcite prismatic layer; lanes 6 and 7, recombinant Prisilkin-39. (**c**), Western blots after electrotransfered to polyvinylidene difluoride membrane. Lanes 1-6 were incubated with the anti-Prisilkin-39 antibody (dilution 1:10,000), and lane 7 was reacted with the rabbit preserum as the negative control. The arrowheads (on the right side of the lanes) point the position of the Prisilkin-39 immunoreactive bands. Prisilkin-39 mainly exist in the EDTA-insoluble matrix of the prismatic layer. (**d**), Western blotting to explore the tissue distribution pattern of Prisilkin-39. Lane 1, molecular mass standards (Std); lane 2, mantle (Ma); lane 3, extrapallial fluid (EPF); lane 4, gonad (Gon); lane 5, viscus (Vis); lane 6, foot; lane 7, gill; lane 8, muscle (Mus); lane 9, homolymph (Hom)

shell grown. Chitin-binding assay was conducted during which a solution of Prisilkin-39 or BSA was mixed with powdered chitin, and the insoluble mixture was then washed with distilled water, saline, and a boiling denaturing solution. Each washing was subjected in SDS-PAGE (Fig. 3.44). BSA which was used as a negative control was completely washed out with water and saline, and no band could be found in the SDS washing. While Prisilkin-39 could not be washed off by water and saline but by the SDS denaturing solution, which is accordant with our previous finding that Prisilkin-39 could only be found in the EISM fractions after being boiled with similar denaturing solution in high temperature (Fig. 3.43). These findings revealed that Prisilkin-39 could bind tightly to chitin, and with this ability, it may thus work as a framework part participating in the shell formation.

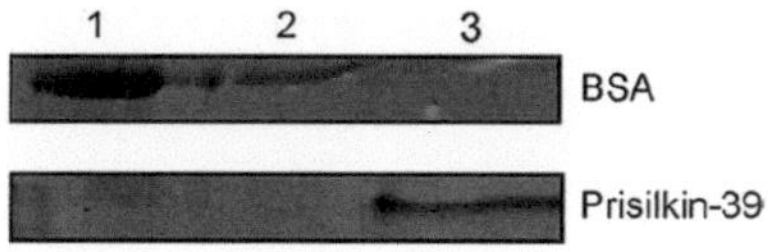

Fig. 3.44 Chitin-binding experiment for Prisilkin-39
The gel was subjected under reducing conditions and dyed with Coomassie Brilliant Blue. Lane 1, sample after water washed; lane 2, 0.2M NaCl washings; and lane 3, incubated with SDS/β-mercaptoethanol at 100 °C for 10 min. Prisilkin-39 could only be washed with the SDS-denaturing solution. BSA is studied as a negative control

3.5.3.6 Immunofluorescence of Prisilkin-39 in *P. fucata* [13]

To investigate the microstructural distribution of native Prisilkin-39 protein in decalcified shells, sections of the shells were incubated with the anti-Prisilkin-39 antibodies. The cross-section of the shells from *P. fucata* represents a classical bilayer structure, with a prismatic texture of long, columnar prisms of calcite (Fig. 3.45a and c, upper compartment) and a nacreous texture consists of flat aragonite tablets (Fig. 3.45a, lower part). In addition, the organic sheet (40 μm thickness) lies between the nacre and the prismatic layer. According to its unique interfacial location, this sheet is thought to play a crucial role during the transition process of two layers with different crystal polymorph. The in situ labeling results of the same sample, which was observed under white light (Fig. 3.45a) and under fluorescent light (Fig. 3.45b), respectively, made sure the prism-specific location of Prisilkin-39. The positive signal had shown drastic intense in the organic sheet, perpendicular to the axis of the calcite prism (Fig. 3.45d, arrows), and in the organic sheaths compacting the prisms, in which the main content is chitin [88, 110] (Fig. 3.45d, white arrowheads indicate). No signals were found in control groups with preimmune serum (Fig. 3.45c). As we know, Prisilkin-39 is the first matrix protein that was detected to be involved in the organic sheet. This also suggests another function of this protein that Prisilkin-39 is not just a structural matrix protein binding with chitin but also might regulate crystalline deposition.

3.5.3.7 In Vivo Exploration of the Function of Prisilkin-39 in Mineralization

To further investigate the role of Prisilkin-39 in shell biomineralization and study why a large amount of Prisilkin-39 is localized in the organic sheet, an in vivo antibody inhibition experiment was conducted. We injected affinity-purified antibodies into the extrapallial fluid of *P. fucata* to inhibit the physiological functions of Prisilkin-39. Compared with the control group (Fig. 3.46a, b), the surface of the nacre lamellae in the low-dose anti-Prisilkin-39-injected group changed to be ragged

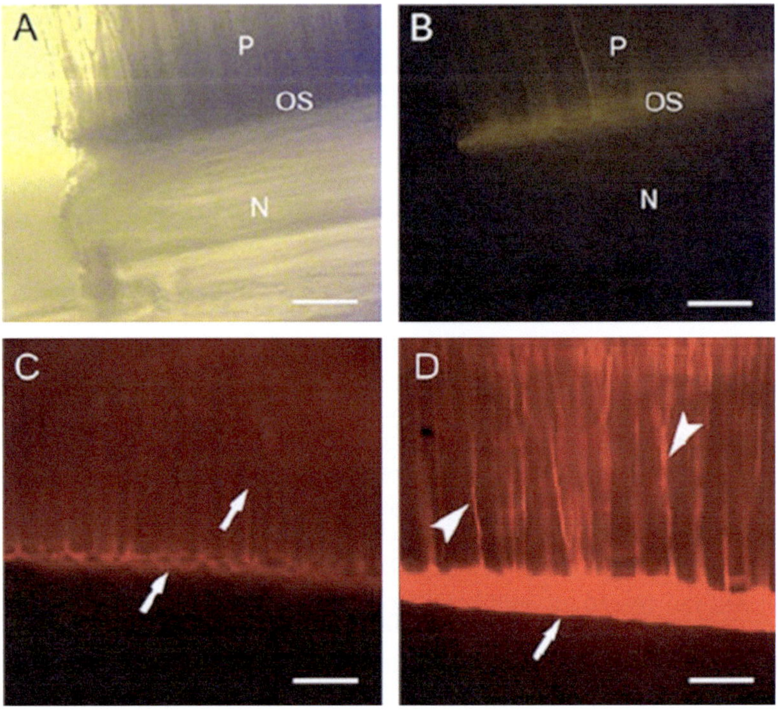

Fig. 3.45 Immunofluorescence distribution of native Prisilkin-39 in cross-section of the shell from *P. fucata*
Normal shells of *P. fucata* was fixed and decalcified and then incubated with anti-Prisilkin-39 antibody (**a**, **b**, and **d**) or normal rabbit preimmune serum (**c**). The upper segment is the prismatic layer, and the lower part is the nacre. For procedure details, see "Experimental Procedures." (**a**), Differential compartment contrast microscopy image of the cross-section of a sample incubated with anti-Prisilkin-39. (**b**), Accordant image of A visualized by fluorescence light. *P* prismatic layer, *OS* organic sheet, *N* nacreous layer. (**c**), Control group incubated with preimmune serum of rabbit. The white arrows show no positive signal was observed in the organic sheet or in the prismatic sheath. (**d**), Experimental group stained with anti-Prisilkin-39 antibody. The arrows point the location of Prisilkin-39 in the organic sheet and around the prisms. Scale bars, 50 μm

and disordered texture (Fig. 3.46c, d). The disordered texture was characterized by the random stacking of crystals on the original tablet (Fig. 3.46d, black arrow indicate), disturbing the stair-like growth model of the inner shell. This abnormal phenomenon turned out to be more drastic in the high-dose group (Fig. 3.46e, f), where thickened irregular tablets were combined and linked together, making the boundaries unclear and fused. What's more, the chemical composition analyses of energy-dispersive X-ray spectroscopy indicated that the irregular precipitations consisted of carbon, oxygen, and calcium (Fig. 3.46k), while the nacre lamellae of the preimmune-injected group kept a normal form and smooth surfaces (Fig. 3.46g–j). These findings revealed that large amounts of calcium carbonate could be abnormally deposited from the extrapallial fluid as a result of the inhibition of the normal physiological functions of native Prisilkin-39 in *P. fucata*, suggesting that Prisilkin-39 should work as an obbligato negative regulator of aragonite growth to stop its ectopic precipitation during shell formation.

3.5.3.8 In Vitro Crystallization with Addition of Prisilkin-39

To explore the effect of Prisilkin-39 on the calcium carbonate crystals growth, two individual in vitro crystallization experiments examining the formation of aragonite and calcite were established, respectively. In the crystallization system of aragonite precipitation, a high Mg^{2+} dosage (50 mM) was added to imitate the composition of the extrapallial fluids in *P. fucata*. First, in the control experiments conducted with the filtrate (data not shown) or BSA, many large needle-shaped crystals were preferentially formed, along with some small rodlike polyhedral crystals (Fig. 3.47a, b). Raman analyses confirmed the aragonitic character of the former crystal and its favorable stability, as shown by a special sharp peak at 1085 cm^{-1} (upper spectra), while the small crystals had a broad peak with low intensity of 1085 cm^{-1} (lower spectra), suggesting the presence of low crystallinity sediments (Fig. 3.47p). On the other hand, with addition of Prisilkin-39 at a dosage of 30 μg/ml, aragonite growth was completely inhibited, and no other typical calcium carbonate crystals were found (Fig. 3.47c, d). At higher amplification, these different-shaped structures are obvious (Fig. 3.47e). Especially, samples were still focus on the deposition of mostly 100–200 nm spherules, along with clusters of amorphous gel-like together. Raman spectra indicated that the amorphous deposits in this condition are not typical aragonite or calcite, as shown with a low intensity peak at 1085 cm^{-1} (Fig. 3.47q, lower spectra). In addition, the lower dosage of Prisilkin-39 (15 μg/ml) induced the formation of vaterite crystals, ~20 μm in size each (Figure 3.47f, g and q, upper compartment), instead of the total inhibition of aragonite growth.

The function of Prisilkin-39 on calcite, formed from the Mg^{2+}-free crystalline system, was tested to be less significant compared to that on aragonite, because it only led to a change in crystal morphology (Fig. 3.47 j–o and s). Not like the aragonitic system, typical calcite crystals could be formed normally from the saturated solution, even though the presence of Prisilkin-39 is at a quite high dosage of 50 μg/ml (Fig. 3.47n, o). Whereas, the morphology of calcite, especially the edges and corners of calcite, was in a dosage-dependent manner, while the crystals formed in control experiments executed with the filtrate (data not shown) or BSA (Fig. 3.47h, i, and r) were the typical calcite rhombohedra with smooth crystal faces. To summarize, Prisilkin-39, as a basic matrix protein from the prismatic layer and organic sheet, represents a much stronger inhibitory ability on the

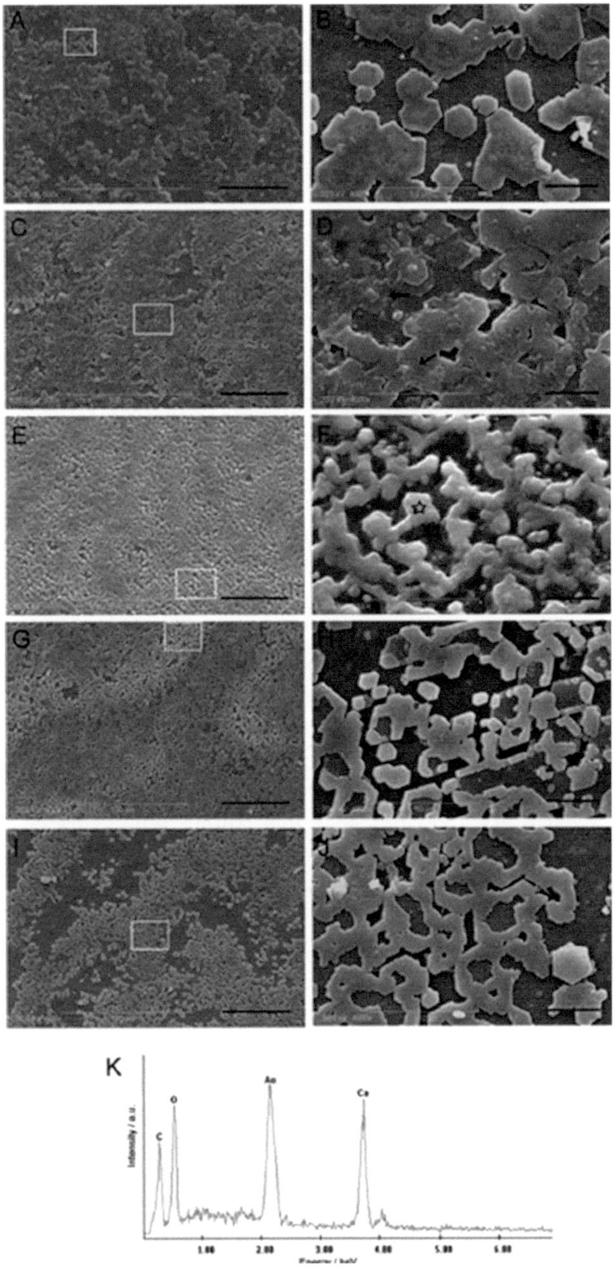

Fig. 3.46 The function of Prisilkin-39 on the growth of nacre lamellae
In the extrapallial fluid where shell biomineralization happens, the physiological functions of native Prisilkin-39 were depressed by its antibody. (**a**), SEM image of the inner surface of normal shell from the oyster *P. fucata*. The stair-like growth model of nacre can be found. (**b**), Amplification of the box in **a** indicating flat and regular tablets. (**c**), SEM image of the inner surface of the shell in

in vitro crystallization of aragonite system than that of calcite system. We speculated that Prisilkin-39, together with other acidic nucleating proteins, may play a crucial role in the process of calcite formation from the Mg^{2+}-containing extrapallial fluid during the prismatic layer mineralization and may also take part in the process of crystalline transition between the two distinct layers, possibly by inhibiting aragonite precipitation at the edge of growing shell of the prismatic layer.

3.5.3.9 In Vitro Depression Activity on Calcium Carbonate Precipitation and Crystal Binding Ability

A series of precipitation assays with or without Prisilkin-39 were conducted. The function of Prisilkin-39 on the precipitation rate of calcium carbonate was decided by recording the increase of absorbance at 570 nm in a saturated $CaCO_3$ solution every 30 s for 5 min. The presence of Prisilkin-39 in the saturated $CaCO_3$ solution led to a significantly decreased rate of precipitation in a dosage-dependent manner compared to that of the control (Fig. 3.47a). Considering the inhibitory ability of Prisilkin-39 on calcium carbonate precipitation, we examined its ability to bind calcium and its transference behavior in calcium-dependent electrophoresis. There is no positive results observed by neither the Stains-All method (Fig. 3.43B, lane 6) nor the electrophoretic shift assay (data not shown), suggesting that Prisilkin-39, as a basic protein, probably had no affinity to calcium. Whereas, crystal-protein binding assays by incubating rhodamine-labeled Prisilkin-39 with crystals indicated that fluorescence distributed only on the edges of calcite, while more intense fluorescence reactivity was found on the whole surfaces of aragonite (Fig. 3.48b, c). The different binding ability of Prisilkin-39 on aragonite and calcite is consistent with the results of in vitro crystallization experiments. At this point, Prisilkin-39 seems to be capable of affecting crystal growth by direct binding to a particular surface, depending on different crystal phases and different crystal faces. Taking together, Prisilkin-39 may thus effect as a crystal binding rather than calcium-binding protein during the process of calcium carbonate formation.

Fig. 3.46 (continued) low-dosage anti-Prisilkin-39-injected group. The stair-like growth edge became unclear and merged. (**d**), Amplification of the box in **c** indicating the coarse surface and disordered organization of the tablets. Random stacking of crystals on the original tablet can be found (arrows point). (**e**), SEM image of the inner surface of the shell of high dosage anti-Prisilkin-39-injected group. A large amount of low crystallinity of calcium carbonate were accumulated in an uncontrolled manner. (**f**), Amplification of the box in (**e**) showing the thickened and disorder tablets. (**g**), SEM image of the inner surface of the shell of low-dosage preimmune serum-injected group. (**h**), Amplification of the box in **g** suggesting the normal shape and smooth surface of these tablets. (**i**), SEM image of the inner surface of the shell of high-dosage preimmune serum-injected group. (**j**), Amplification of the box in I. (**k**), Energy-dispersive X-ray spectroscopy analysis of the tablet (asterisk) in (**f**). Scale bars, 50 μm in **a**, **c**, **e**, **g**, and **i**; 5 μm in **b**, **d**, **f**, **h**, and **j**

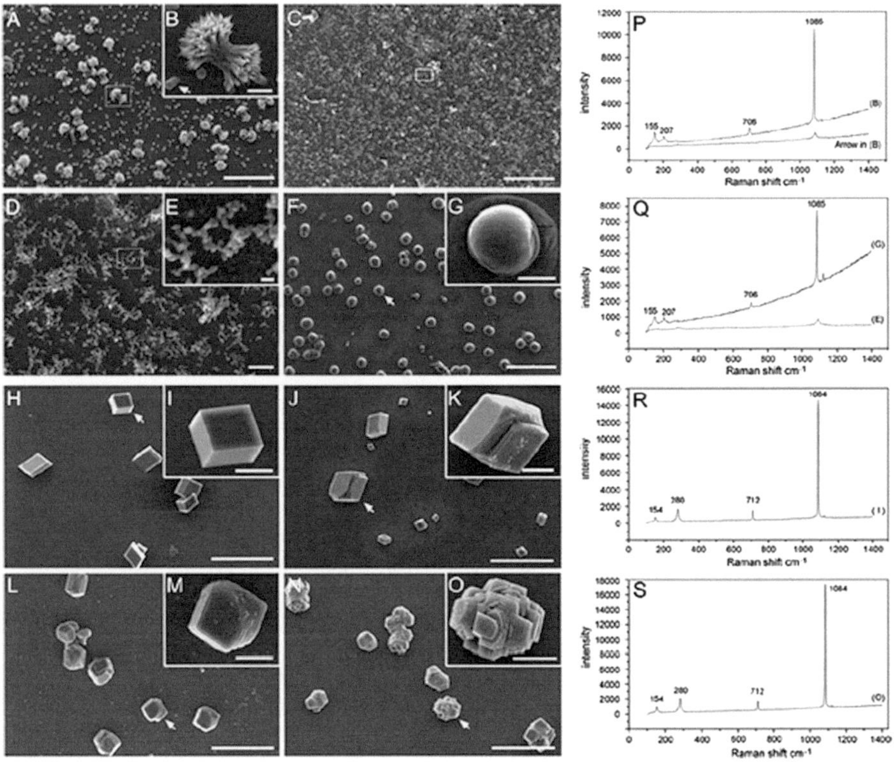

Fig. 3.47 SEM images of in vitro crystallization experiments with addition of Prisilkin-39 and their Raman spectra
The function of Prisilkin-39 on calcium carbonate crystals growth was examined in two crystallization systems. (**a**–**g**), Crystallization with 50 mM Mg^{2+} in aragonite growth; (**a**) and (**b**), Crystals are formed with 30 μg/ml BSA as negative control, needle-shaped crystals with large size (aragonites), and small rodlike crystals are boxed and amplified in (**b**). (**c**–**e**), Crystals in the presence of 30 μg/ml Prisilkin-39 (**d** and **e**) are amplified images of the boxed marked in (**c**) and (**d**), respectively. No aragonite crystals were found, while amorphous gel-like aggregates and nanoscale spherules can be found in (**e**). (**f** and **g**), Crystals formed with 15 μg/ml Prisilkin-39. A spherical crystal is pointed by an arrow in (**f**) and amplified in (**g**). (**h**–**o**), Crystallization without Mg^{2+} of calcite growth. Calcite crystals were formed in the presence of 30 μg/ml BSA in (**h**), 10 μg/ml Prisilkin-39 in (**j**), 30 μg/ml Prisilkin-39 in (**l**), and 50 μg/ml Prisilkin-39 in (**n**). **i**, **k**, **m**, and **o** are the amplified images of the crystal pointed by arrows in **h**, **j**, **l**, and **n**, respectively. (**p**–**s**), Raman spectra of the crystals grown in these assays. Scale bars, 100 μm in **a**, **c**, and **f**; 50 μm in **h**, **j**, **l**, and **n**; 10 μm in **b**, **d**, **g**, **i**, **k**, **m**, and **o**; 1 μm in **e**

(**a**), Variation in the turbidity of the experimental solutions are shown. ●, 2 μg/ml Prisilkin-39; ▲, 5 μg/ml Prisilkin-39; ▲ , 10 μg/ml Prisilkin-39. In the in vitro crystal-protein binding assay, Prisilkin-39 indicated different binding patterns to single calcite (**b**) and aragonite (**c**). No fluorescence was observed when equimolar BSA and rhodamine were involved (**d**). Scale bars in **b**–**d**, 20 μm

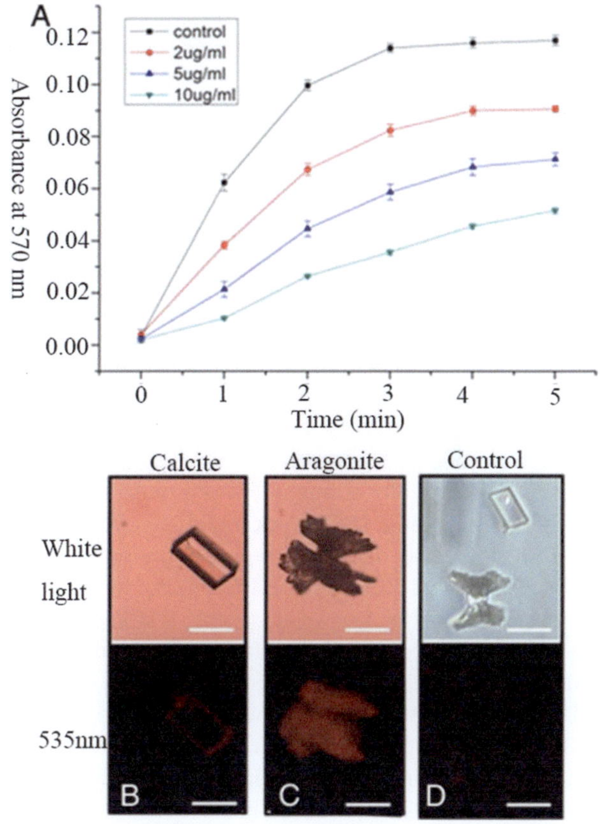

Fig. 3.48 Depression activity of Prisilkin-39 on calcium carbonate precipitation

Reports about the structure-function relationship of matrix proteins and their effect in calcification may afford clues to the puzzle of how the shell, an excellent mechanical material, grows [111], while only few shell matrix proteins have been cloned and identified their roles in mineralization evaluated. A basic difficulty has been the existence of a technical obstacle that traditional fractionations usually fail to separate and purify matrix proteins, possibly because of difficulty of behaviors in gels [112]. Until now, only three proteins, named Prismalin-14, Caspartin, and casprismin, have been identified and purified from the prismatic layer of mollusks by classical biochemical methods [8, 113]. More recently, the application of molecular biology techniques promoted the identification of putative transcripts associated with mineralization. Indeed, most of the fully known sequences of shell matrix were obtained through this strategy [112]. Several genes are suggested to be involved in prismatic layer formation in mollusks, but only few molecular evidence has been given [9, 17, 114]. Further explores of these genes have been difficult, as recombinant proteins are quite difficult to construct, presumably as their excessively uneven amino acid composition. In the present research, we have cloned a gene encoding a novel shell matrix protein from a mollusk with a putative molecular weight of

39 kDa, as suggested by SDS-PAGE. Because this protein was found in extracts of prisms of *P. fucata* and represents some silk fibroin-like features of the silk hydrogel composition in the mineralization microenvironment [42], we name this protein as Prisilkin-39.

Prisilkin-39 is the second macromolecule until now characterized in mollusks to contain capability to bind chitin. So far, the Rebers–Riddiford chitin-binding theory is the most widely spread motif in chitin-binding proteins existed in the arthropod cuticle [115–117]. Whereas, no such domain is found in any shell matrix protein of mollusks. Thus, a different kind of domain should be applicative for matrix proteins to bind to chitin directly. One potential region is the Gly/Tyr-rich consensus domain. In the N-terminal compartment of Prisilkin-39 (Fig. 3.38), GYS region I is identified by 10 tandem repeats of every 12-amino acid sequence GGY(G/S)GY(G/S)YGYP (T/A). As far as we know, there are few reports of this kind of motif organization in mollusks, just found so far solely in Lustrin A and mucoperlin [74, 75]. Even though the significance of the repeated sequence has not been characterized, it may be relevant to the regular arrangement of atoms to the repeated structure of polymers in the insoluble organic framework including chitin [8, 118]. The predicted secondary structure indicated that GYS region I prefers to form β-strand structure. What's more, the Rebers–Riddiford chitin-binding model and the chitin-binding region of chitinases are also known to form β-strand structure that could bind chitin [95]. Meanwhile, GYS region I showed some resemblance with the other chitin-binding sequence in mollusks including the Gly/Tyr-rich region of Prismalin-14, related to the Gly/Tyr arrangement. Thus, we presume that the GYS region I of Prisilkin-39, along with the hydrophobic sequence before, may possess a structural binding domain and participate in the direct interaction with chitin. The data shown here is also consistent with the previous work that suggested that the Gly/Tyr-rich region might possess a novel chitin-binding motif of molluscan matrix proteins [95]. Our further researches on synthesized peptides are currently in progress.

After the analysis of the amino acid sequence and hydropathy (Fig. 3.39), Prisilkin-39 apparently possesses an amphiphilic geometry consisting of an N-terminal hydrophobic and a C-terminal hydrophilic domain. On basis of a previous study of the matrix protein ansocalcin [111] during eggshell biomineralization, this amphiphilic feature facilitates self-assembly, which subsequently can give rise to a hydrophilic exterior surface including the basic side chains in the C-terminus tail. Such a character may then provide specific interactions with a specific acidic protein that was gathered to function as a nuclear factor or with crystal faces influencing crystal growth. In addition, the analysis of GYS region II that is more liable to posttranslational modifications suggests a quite exquisite structure of this peptide is required for functional matrix-matrix or matrix-mineral interactions. Whereas, the predicted secondary structure showed that GYS region II is mainly composed of random coil, which generally leads to the flexibility and elasticity of the molecular conformation. We hypothesize that this character is highly favorable to the regulatory activities of Prisilkin-39 during crystallization.

Distribution of Prisilkin-39 was found only in the mantle edge but not in the mantle pallium or other tissues (Fig. 3.41a), suggesting the involvement of Prisilkin-39 in calcite prismatic layer formation when shell grows. This expression pattern was further proven by the results of immunofluorescence location in situ and immunodetection in vitro. All the results shown above are direct genetic and biochemical proof that basic matrix proteins can also participate in the prismatic layer construction, even if they had never been successfully extracted and purified before. A striking distribution profile of native Prisilkin-39 in the shell was found in Fig. 3.45. Before this study, matrix protein in the organic sheet had never been cloned and characterized in mollusks. Therefore, observation of Prisilkin-39 in the organic sheet, with a drastic crystalline transition between nacreous layer and prismatic layer, suggested another possible role of Prisilkin-39 during shell formation. It is well known that multifunction is a common feature of many shell matrix proteins like Lustrin A [75], MSI60 [6], and nacrein [5], and we assume that Prisilkin-39 shares it. In support of this notion, we tested its biomineralization function by using both in vivo antibody inhibition experiment and in vitro precipitation of calcium carbonate formation. To explore the crystal precipitation rate, the ionic component of the extrapallial fluid of marine mollusks, with a high Mg^{2+} content (50 mM) compared to Ca^{2+} (10mM), must be taken into consideration.

Mg^{2+} is harmful in solution for calcite precipitation; this is why aragonite crystals become main component in Mg^{2+}-rich solutions [96]. Hence, to get an efficient process of calcitic prismatic layer calcification, some regulatory factors are needed to override this spontaneous condition to form calcite. Therefore, we presume that Prisilkin-39 is one of the molecules that effect as a regulator of calcite growth, as our results in vitro accordant well with the in vivo detection that Prisilkin-39 showed a powerful inhibitory ability on the aragonite precipitation but not of calcite. It is noteworthy that the two aspects that Prisilkin-39 takes part in, that is, intercolumnar framework formation and aragonite growth depression, are thoroughly consistent to the construction of the prismatic layer. Additionally, crystal binding assays (Fig. 3.48) further revealed a different binding pattern for Prisilkin-39 to aragonite and calcite. Whereas, the regulation of crystal growth should be more complex and subtle, and the minute manner by which Prisilkin-39 inhibits aragonite formation has yet to be explained.

How are the prisms with calcite in the growing shell edge of *P. fucata* formed? Even though the mechanism of growth of the prism, especially at the initiation stage beneath the inner periostracum, is still unclear, our study proposed here affords important clues to these complicated physiological processes. We do think that both acidic and basic matrix proteins are both required to function in a cooperative way to finish this work. Without a comprehensive model of the prismatic layer construction, any imply for the role of Prisilkin-39 must be hypothetical. First of all, basic matrix proteins could bind tightly to a chitinous framework to construct a regular platform for crystal precipitation. Second, some proteins including Prisilkin-39, according to their amphiphilic character, also play an interfacial role by connecting together the framework and acidic matrix proteins, which are achieved by basic residues of Prisilkin-39. Meanwhile, acidic proteins induce calcite nucleation on the condition that the spontaneous nucleation of aragonite and aragonite growth are prohibited by Prisilkin-39 at the same mineralization site. Another important aspect to mention is that the property and function of Prisilkin-39 we shown here are remarkably consistent with that of the silk-like proteins in the shells, which are widely believed to possess three basic functions in biomineralization as follows: (1) pre-filling the microenvironment formed with two layers of β-chitin; (2) some of these molecules are finally fixed into the mineralized framework when shell grow; and (3) they may work as an inhibitor and help prevent disordered crystallization at surplus nucleation sites. Although silk-like proteins have not been identified in prisms before present study, we temporarily infer that the hypothesized model for nacre mineralization by Addadi et al. [42] is also to some degree comparable with the prismatic layer biomineralization involved in some fundamental steps, i.e., the chitinous framework formation and crystal growth termination. Therefore, according to Marin et al. [119], "the classical paradigm in molluscan shell biomineralization was to consider that the control of shell synthesis was performed primarily by two antagonistic mechanisms, crystal nucleation, and growth inhibition." We think that Prisilkin-39 takes part in growth inhibition.

Taken together, based on the structural and functional features of Prisilkin-39, we hypothesize a molecular model for this protein. This molecule has four functional regions. GYS region I, together with the alanine-rich hydrophobic region, constitutes a structural domain. They form a β-strand structure and help Prisilkin-39 bind directly to chitin and so are mainly responsible for the toughness of the shell structure. GYS region II and the basic tail of the C-terminal are predicted to make their secondary structure more flexible, and they may bind to certain crystal surfaces or acidic matrix proteins neatly regulating mineral deposition by means of specific growth inhibition. Further researches on the structure-function relationship, as well as antisense RNA assays, are necessary to understand the physiological function of Prisilkin-39 during the formation of the prismatic layer of *P. fucata*.

3.6 Shell Matrix Protein from Both Prismatic and Nacre Layer

3.6.1 A Novel Matrix Protein Participating in the Nacre Framework Formation of Pearl Oyster, Pinctada fucata

To understand the molecular component is of great interest for both pearl formation mechanism and biomineralization in mollusk shell. A cloned cDNA encoding an MSI31 analog, termed MSI7 according to its estimated molecular mass of 7.3 kDa, was separated from the pearl oyster, *Pinctada fucata* [71]. This novel protein shares resemblance with MSI31, a prismatic framework protein in *P. fucata*. It is special that MSI7 is much shorter in mass, possessing only the Gly-rich sequence that has been proposed to be crucial for Ca2q binding. In situ hybridization result indicated that MSI7 mRNA was expressed specifically at the folds and outer epithelia of the mantle, suggesting that MSI7 play roles in the framework formation of both the nacre and prism. In vitro assay on the function of MSI7 indicated that it induces the nucleation and precipitation of $CaCO_3$. All in all, we have cloned and identified a novel matrix protein in the pearl oyster, which may play a key role in regulating the texture of nacre.

3.6.1.1 cDNA Cloning and Sequence Analyses of *MSI7*

We isolated seven positive cDNA clones from the cDNA library after screening of four rounds. Sequence analyses revealed that one of cDNA clones had a 285 bp open reading frame (Fig. 3.49), encoding a 95 amino acid residues protein. The first 19 amino acid residues of the deduced protein were predicted to be a typical signal peptide. The protein contained a hydrophobic core (h-) region of 13 amino acids, suggested to be a most essential part indicating for membrane targeting and insertion. The h-region was flanked by a basic lysine residue at the N terminus and a polar (c-) region in its C-terminal side. This structure is also resembled to the known signal peptides which were found in proteins processed in the ER and secreted out the cell subsequently. Sequence alignment showed the deduced signal sequence of MSI7 was consistent with the predicted signal peptide of MSI31 except for residues 17. The amino acid sequence of MSI7 also shares high homology with MSI31

```
 GCACGAGGCAGATAAAAGGTCGGTGCCCAACCACAGTCATGAAACCCTTTGTTACACTC   59
                                        [M  K  P  F  V  T  L    7
GCAAGCTTGATCGTCTTAATTGCCTCTGTTTCTGCCGGCGGAGATGGTGACTATGGCAAA   119
 A  S  L  I  V  L  I  A  S  V  S  A] G  G  D  G  D  Y  G  K    27
TACGGCGGTGTCAGTTACGGACCTGGCATCAACCTTGGAGGTGGTAGTCTAAGTGCCGGA   179
 Y  G  G  V  S  Y  G  P  G  I  N  L  G  G  G  S  L  S  A  G    47
GGAGGTGTTATCGGGCTTGGAGGCGTCGGAGGTGGTAGTCTAGGTGCCGGAGGCGTCGGC   239
 G  G  V  I  G  L  G  G  V  G  G  G  S  L  G  A  G  G  V  G    67
CCAGTTGGTGGAATTACCATATGGACTCACCTTTTCACCTGGATATACCTTTTTCATGGA   299
 P  V  G  G  I  T  I  W  T  H  L  F  T  W  I  Y  L  F  H  G    87
TGTTTGATATTCATTTTTTTGCAATAAATGGCCGGGCACATACAAAAAAAAAAAAAAAAA   359
 C  L  I  F  I  F  L  Q  *                                      95
AAA                                                            362
```

Fig. 3.49 The nucleotide sequence of the cloned cDNA and putative amino acid sequence of the glycine-rich MSI7 of the pearl oyster, *P. fucata*. The putative signal peptide was enclosed by brackets. The cDNA sequence of MSI7 has been uploaded to GenBank (accession number: AF516712) [71]

```
MSI7    (1)    MKPFVTLASLIVLIASVSAGGDGDYGKYGGVSYGPGINLGGGSLSAGGGV
MSI31   (1)    MKPFVTLASLIVLIASASADGYDDYKKYGSVGYGPGISLGGGGLGGGGGI

MSI7    (51)   IGLGGVG------GGSLGAGGVGPVGG-----------ITIWTHLFTWIY
MSI31   (51)   ISVGGGGGLGGGLGGGLGCGLVGVGGGGLIGGGFGPGRVSGTINAGGGVF

MSI7    (84)   LFHGCLIFIFLQ--------------------------------------
MSI31   (101)  ASGSGLGGLSPAGRGAAQGAATLSALQIASGRPGRVSGVSVGTGGGRAVV

MSI7    (96)   --------------------------------------------------
MSI31   (151)  SGSATPVGGFGVPYGGYGYNYGVPSYGVGLPSYGVSLPSYGVGLGGGYGG

MSI7    (96)   --------------------------------------------------
MSI31   (201)  YGYGLDLASFQGSTYGNLATGQINTAVVAFHMAVLLESMEAESDTEVDTE

MSI7    (96)   --------------------------------------------------
MSI31   (251)  DMDSEEDMESEEDTESEEDTESEEDTESEEDMESEEDMDSESSVVDQVSM

MSI7    (96)   ----------------------------------
MSI31   (301)  VYPNHFTGDVLFCQVRLQELEFPALEALVSVVLE
```

Fig. 3.50 Alignment of the amino acid sequences of MSI7 and MSI31. The identical amino acid residues were shaded. Alignment gaps of the sequences are shown by dotted lines. Sequence of MSI31 was downloaded from GenBank (accession number: D86073)

(Fig. 3.50). According to the presence of the signal peptide and its high similarity with MSI31, MSI7 is suggested to be an extracellular matrix protein.

The deduced mature MSI7 includes 76 amino acid residues. The protein sequence is rich in glycine (35.5%). The second and third abundant amino acids are leucine of 10.5% and isoleucine of 9.2%. MSI7 has a deduced molecular mass of 7.3 kDa. The protein could be divided into three regions that are the glycine-rich region (from Y14 to G52), the hydrophilic region (from G1 to S13), and the C-terminal part (from I53 to Q76). MSI7 has a hydrophilic region in the N-terminal composed by 13 amino acid residues. Behind the hydrophilic region, there was a glycine-rich domain of 39 amino acid residues, which contained three GG and three GGG repeats. The C-terminus of MSI7 is a hydrophobic region composing of 24 amino acid residues. There is a cysteine residue in this hydrophobic region that may form disulfide bonds with cysteine residue from other proteins. No glycosylation site or carbonic anhydrase active-site sequence was found in the amino acid sequence. All in all, this protein possesses a hydrophobic and acidic character, with a pI of 5.98.

Search FASTA algorithm with the Swiss-Prot protein database showed lower degrees of resemblance of MSI7 and mouse type II cytokeratin 1 (KRT1, accession number Swiss-Prot P04104), Petunia (*Petunia hybrida*) glycine-rich cell wall structural protein 1 precursor (GRP-1, accession number Swiss-Prot P09789), human

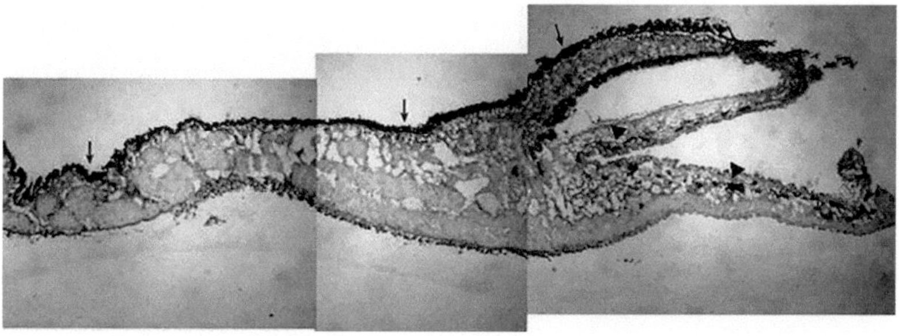

Fig. 3.51 Expression of MSI7 mRNA in the mantle of *P. fucata* by in situ hybridization. Strong hybridization signal is detected in the outer epithelia of the mantle (arrows indicate) and cells (arrowheads indicate) in both of the middle fold and inner fold in the mantle. Three overlapping pictures of the same section were jointed and provided a whole view of the location of hybridization signal in the whole tissue. Original amplification: 25×

type I cytokeratin 10 (KRT10, accession number Swiss-Prot P13645), and human loricrin (LOR, accession number Swiss-Prot P23490). Keratin plays a role in the formation of cytoskeleton, GRP-1 is a structural protein that composes cell walls, and LOR participates in epidermal cell envelope [120–122].

3.6.1.2 In Situ Hybridization

To determine the expression of MS17 mRNA in tissues, we used frozen sections of pearl oyster. Strong MSI7 mRNA hybridization signals were detected in cells along the outer edge of the mantle.

No hybridization signal was detected at the inner epithelia of the mantle or any other tissues. Hybridization with the control sense probe yielded no hybridization signal (data not shown). Figure 3.51 shows the MSI7 mRNA is expressed in both the dorsal region of the mantle and the mantle edge. Under the basilar membranes, the cells with positive signals were detected at the inner side of the middle fold and the outer side of the inner fold. This appears to be the first report of matrix proteins expressed in the middle fold and inner fold of the mantle.

3.6.1.3 Expression and Purification of MSI7 [71]

The coding sequence of oyster MSI7 was inserted into pGEX-4T-1 and incubated in *E. coli* BL21 (DE3). The purified recombinant protein was subjected to SDS-PAGE. As Fig. 3.52a had shown, the GST-MSI7 purified by affinity purification with only one band observed on the SDS-PAGE. The deduced molecular mass of the band is about 33.5 kDa, which is consistent with the calculated molecular mass of the GST-tagged MSI7. The purified protein and whole cell of untransformed BL21

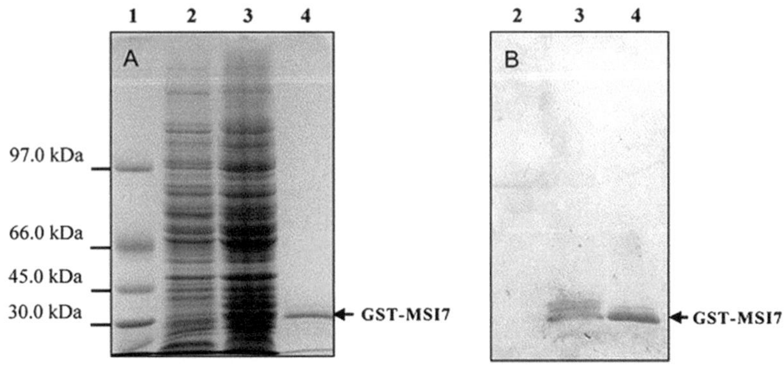

Fig. 3.52 SDS-PAGE and Western blotting analyses of the recombinant MSI7 protein. (**a**) SDS-PAGE gel image for analysis of recombinant MSI7. Lane 1: protein marker indicating molecular mass; lane 2: proteins of whole BL21 (DE3); lane 3: whole proteins of pGEX-4T-1y MSI7 induced by IPTG for 4 h; lane 4: purified GST-MSI7 protein. The molecular mass in kDa is indicated on the left of the gel and the purified GST-MSI7 is pointed by arrow on the right. (**b**) Western blotting analysis of the expressed MSI7 in different samples. Note the purified GST-MSI7 represents immunoreactive signal, which is pointed by arrow on the right

(DE3) were compared via Western blotting analyses with anti-GST rabbit polyclonal antibody. Only one visualized band about 33.5 kDa was found which is the GST-tagged MSI7 (Fig. 3.52b).

3.6.1.4 Effect of GST-MSI7 on $CaCO_3$ Crystallization

The pH drops suddenly after $CaCl_2$ is added to $NaHCO_3$ to form the saturated solution. The reason of this sudden change is not clear but was conjectured due to the formation of a complex by different ionic species other than nucleation [123]. As Fig. 3.53 had shown, the addition of GST-MSI7 in the saturated solution leads to sharp decrease of pH suddenly, which is measured by the rapid precipitation of $CaCO_3$. In contrast, the nucleation of crystals was occurred after few minutes in pure $CaCO_3$ solution or in the $CaCO_3$ solution with Tris. The addition of GST-MSI7 obviously led to a much sharper decrease of pH in the saturated solution than with GST, indicating that MSI7 has a promotion effect on the $CaCO_3$ precipitation. In crystallization experiment system, crystals formed with the addition of GST-MSI7 displayed morphological changes compared with crystals formed in the control group (Fig. 3.53a–b); at the same time, crystals formed with the treatment of Tris or GST showed no morphological change compared with the natural $CaCO_3$ crystals (data not shown). What's more, hexagonal crystals were observed with GST-MSI7 (Fig. 3.54d). Taken together, these results suggested that MSI7 affected the nucleation and $CaCO_3$ crystals growth.

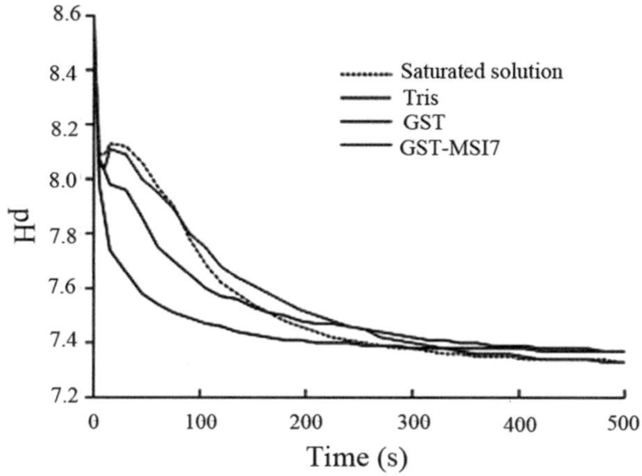

Fig. 3.53 Data of $CaCO_3$ precipitation in saturated solution. Note the addition of GST-MSI7 in the saturated solution leads to the rapid decrease of pH without any delay time, which suggests the accelerated precipitation of $CaCO_3$

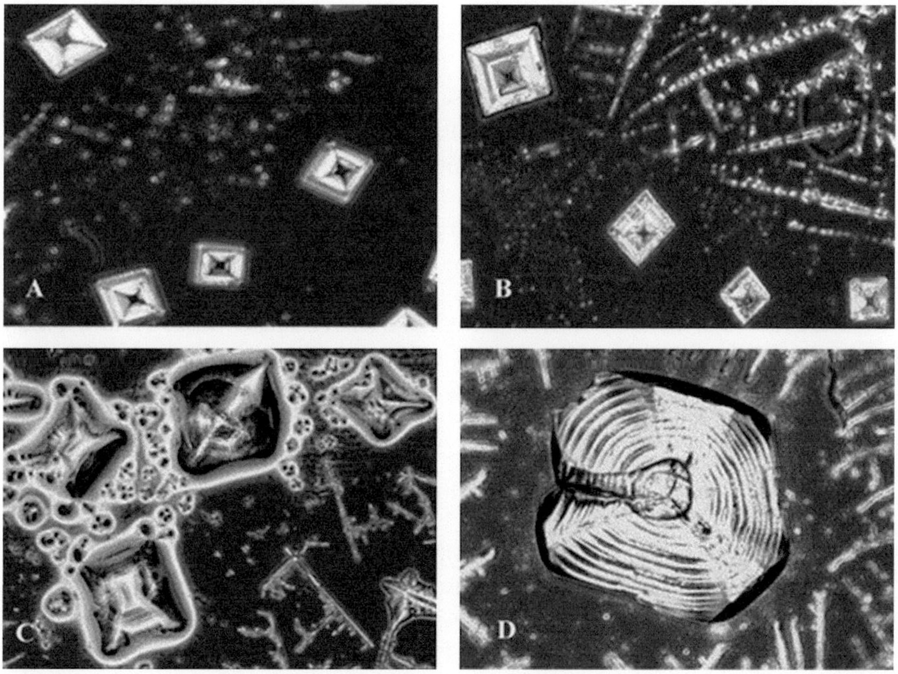

Fig. 3.54 Microscopic images of crystals formed in saturated $CaCO_3$ solution. (**a**)–(**c**) Crystals grown in the presence of Tris solution, GST, and GST-MSI7, respectively, (100×). Note the shape changes of the crystals with addition of GST-MSI7. (**d**) Hexagonal crystal grown with addition of GST-MSI7 (magnification, 200×)

3.6.2 *The Structure–Function Relationship of MSI7, a Matrix Protein from Pearl Oyster* Pinctada fucata

We previously identified a matrix protein, MSI7, from pearl oyster *Pinctada fucata*. According to the structural analysis, the DGD site in the N-terminal of MSI7 is crucial for its role in the shell formation. In this study, we expressed a series of recombinant MSI7 proteins, including the wild type and several mutants directed at the DGD site, using an *E. coli* expression system to reveal the structure–function relationship of MSI7. Furthermore, in vitro crystallization, crystallization speed assay, and circular dichroism spectrometry were carried out. Results indicated that wild-type MSI7 could induce the nucleation of aragonite and inhibit the crystallization of calcite. However, none of the mutants could induce the nucleation of aragonite, but all of them could inhibit the crystallization of calcite to some extent. And all the proteins accelerated the crystallization process. Taken together, the results indicated that MSI7 could contribute to aragonite crystallization by inducing the nucleation of aragonite and inhibiting the crystallization of calcite, which agrees with our prediction about its role in the nacreous layer formation of the shell. The DGD site was critical for the induction of the nucleation of aragonite.

3.6.2.1 Effect of Recombinant MSI7 Protein and Its Mutants on Calcium Carbonate Precipitation

As Fig. 3.55 had shown, GST-MSI7 increased calcium carbonate precipitation rate in a dose-dependent way. With high concentration of GST-MSI7, the rate of calcium carbonate precipitation sped up gradually, and with 40 μg/ml of GST-MSI7, the acceleration effect of GST-MSI7 was remarkable compared with control group especially. And all the mutants also could accelerate calcium carbonate precipitation rate in a dose-dependent way in a similar way as that of GST-MSI7 (data not shown).

3.6.2.2 Effect of Recombinant MSI7 Proteins on Crystallization

To explore the effect of recombinant MSI7 and mutants on the morphology of formed crystals in vitro were examined using SEM. Crystals in the control groups (with GST only or with no protein) were completely rhombohedral with a typical morphology of calcite (Fig. 3.56a). In the presence of 20 μg/ml GST-MSI7, there are a fraction of crystals (about 10%) displayed a characteristic cluster needlelike morphology of aragonite besides the typical morphology of calcite (Fig. 3.56b, c). With the increased concentration of GST-MSI7 (40 and 80 mg/ml), the shape of a part of calcite crystals (about 35%) changed gradually (Fig. 3.56d, e).

With addition of GST-DD, only calcite was found with morphology modified drastically. With 10 μg/ml GST-DD, no remarkable effect on the calcite morphology was observed (Fig. 3.57a). The function of GST-DD became prominent at a dosage

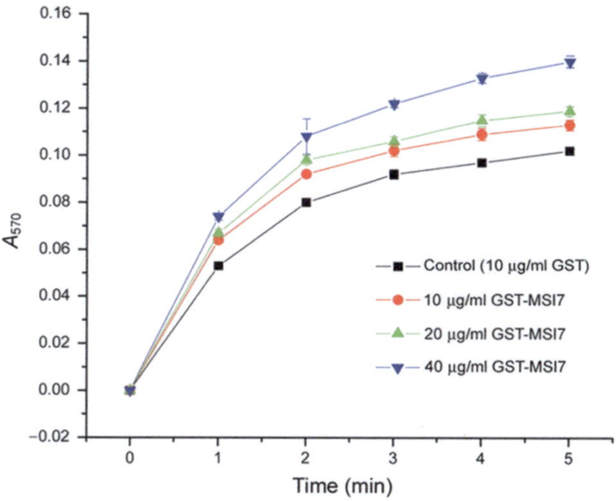

Fig. 3.55 Functions of GST-MSI7 on calcium carbonate precipitation changes in the turbidity of the assayed solutions are recorded

of 20 or 40 μg/ml where all the crystals displayed rounded edges and apparent stairs (Fig. 3.57b, c). With a concentration of 80 μg/ml, the morphology of crystals reached the most modified level, when the edges and the surfaces of some crystals were completely modified with the macro-steps dispreading on the calcite surface widely (Fig. 3.57d). The results suggested the effect of GST-DD on the morphology of all the calcite is a dose-dependent way.

With addition of GST-DDD, there are all calcite and the morphology of some of calcite (~35%) was modified similar to GST-MSI7 in a dose-independent way (data not shown). There is no aragonite induced but some of the calcite (~ 35%) was modified like GST-MSI7 in a dose-independent way with GST-DGYD (data not shown). The form of various calcium carbonate crystals above was detected by Raman microscopy. The needlelike crystals formed with 20 μg/ml GST-MSI7 displayed characteristic Raman bands of aragonite at 206, 705, and 1085 cm^{-1} (Fig. 3.58a), whereas the rhombohedral crystals in all groups shown a characteristic Raman bands of calcite at 282, 712, and 1086 cm^{-1} (Fig. 3.58b).

3.6.2.3 CD Spectroscopy

CD spectroscopy is an effective approach that could explore proteins' secondary structure character in solution. Once interacting with Ca^{2+}, some matrix proteins present apparent conformational changes to regulate the nucleation, orientation, polymorphism, and morphology of calcium carbonate crystals. CD spectra in the far-UV region (190–250 nm) were conducted with EDTA or $CaCl_2$ (final concentration of 2.5 mM) to explore the secondary structure features of GST-MSI7 and the

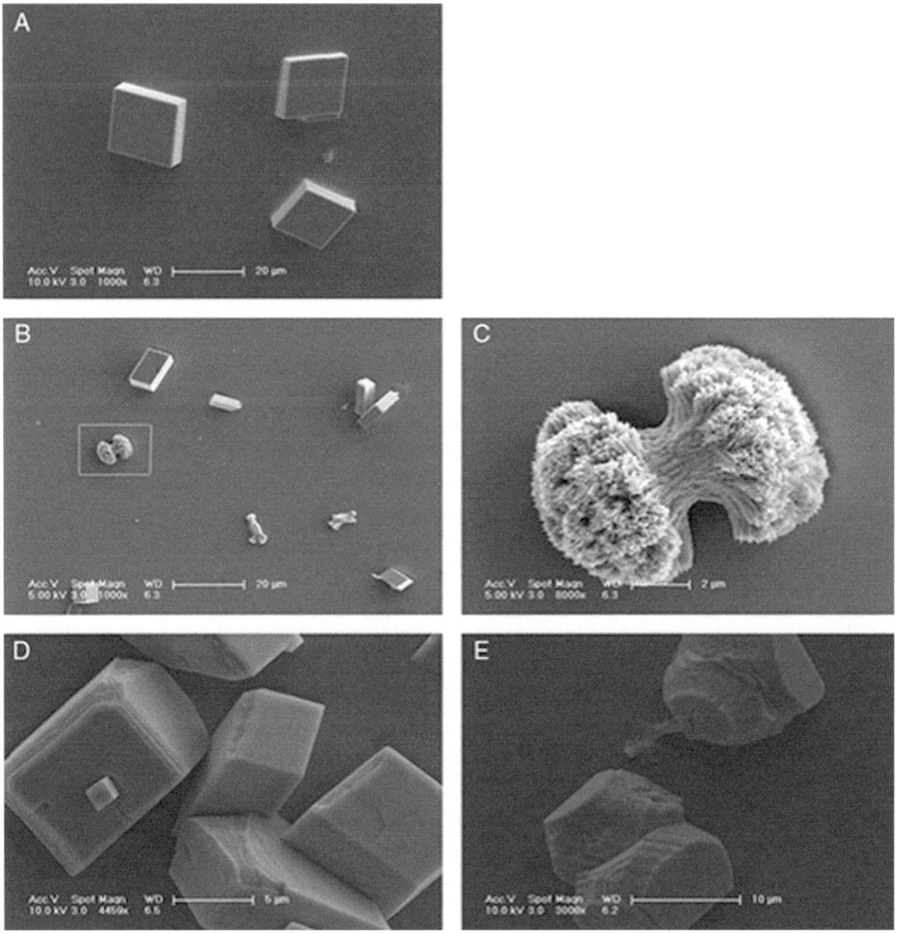

Fig. 3.56 SEM images of in vitro crystallization assays in the presence of GST-MSI7. Crystals were formed with (**a**) 20 μg/ml GST; (**b**) 20 μg/ml GST-MSI7; (**d**) 40 μg/ml GST-MSI7; (**e**) 80 μg/ml GST-MSI7; (**c**) is an amplified image of the boxed part in the (**b**)

mutants, with the final concentration of 0.3 μg/ml proteins. As Fig. 3.59a had shown, GST-MSI7 is composed of disordered structures and β-sheet; GST-DD is composed of mainly disordered structures; GST-DDD has both β-sheet and disordered structures; and GST-DGYD is composed mainly of α-helix with β-sheet structures. When there is $CaCl_2$ not EDTA, all the proteins present conformational changes of hypsochromic shifting. Whereas, the conformational transformations of three proteins (except GST-DGYD) were not significant compared with that of GST as a control (data not shown). When EDTA was replaced with $CaCl_2$, GST-DGYD present partial unfolding as the α-helix proportion was decreased (Fig. 3.59b).

Fig. 3.57 SEM images of calcite modified by GST-DD. Calcite crystals formed in the presence of different dosages of GST-DD. (**a**) 10 μg/ml. (**b**) 20 μg/ml. (**c**) 40 μg/ml. (**d**) 80 μg/ml

3.6.3 A Novel Acidic Matrix Protein, PfN44, Stabilizes Magnesium Calcite to Inhibit the Crystallization of Aragonite

Magnesium is widely believed to control calcium carbonate deposition in the shells of pearl oysters. Matrix proteins in the shell contribute to regulate nucleation and growth of calcium carbonate crystals. Whereas, there is no direct evidence proving a connection between matrix proteins and magnesium. Herein, we cloned and identified a novel acidic matrix protein called PfN44 that influenced aragonite formation of the shell in the pearl oyster *Pinctada fucata*. Executing immunogold labeling assays, we found the existence of PfN44 in both the nacreous and prismatic layers [15]. During shell repair, PfN44 was inhibited, while other matrix proteins were upregulated. Repressing the function of PfN44 by RNA interference led to the deposition of unsmooth nacreous tablets with crystals overgrowth in the nacreous layer. By in vitro circular dichroism spectra and fluorescence quenching assay, we found that PfN44 could bind to both calcium and magnesium with a stronger affinity to magnesium. During in vitro calcium carbonate crystallization and mineralization of amorphous calcium carbonate, PfN44 influenced the magnesium content of

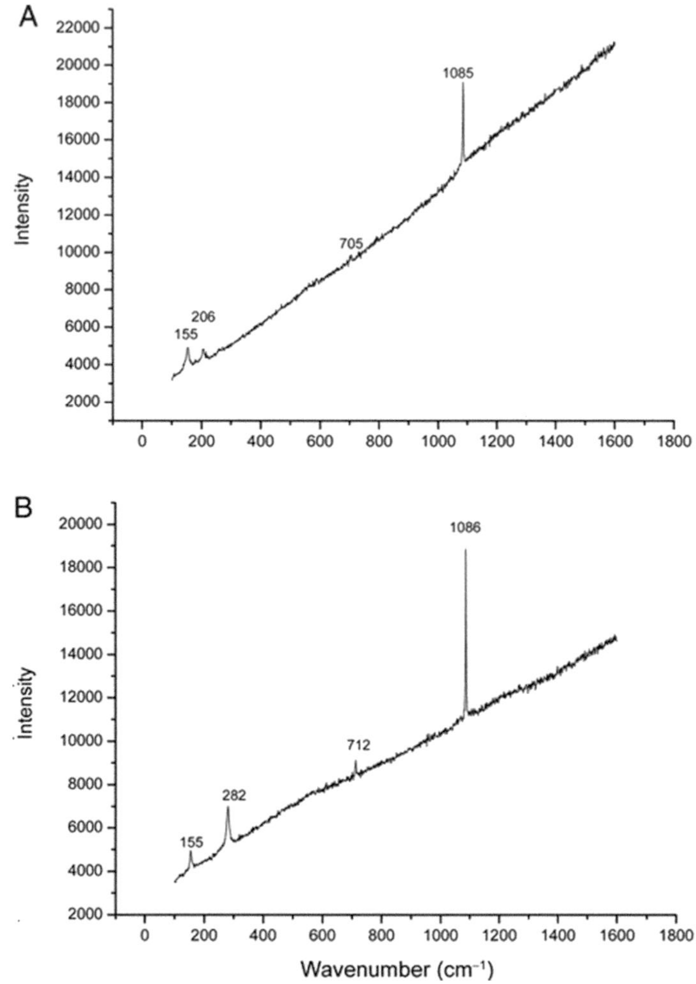

Fig. 3.58 Raman spectra profile of crystals with different shape. (**a**) Raman spectrum of an individual needlelike aragonite crystal formed with 20 μg/ml of GST-MSI7. (**b**) Raman spectrum of a typical hexahedral calcite crystal

crystalline carbonate polymorphs and stabilized calcite with magnesium to repress aragonite deposition. All in all, our results indicated that by stabilizing magnesium calcite to repress aragonite deposition, PfN44 participated in the shell formation of *P. fucata*. These results extend our understanding of the links between matrix proteins and magnesium.

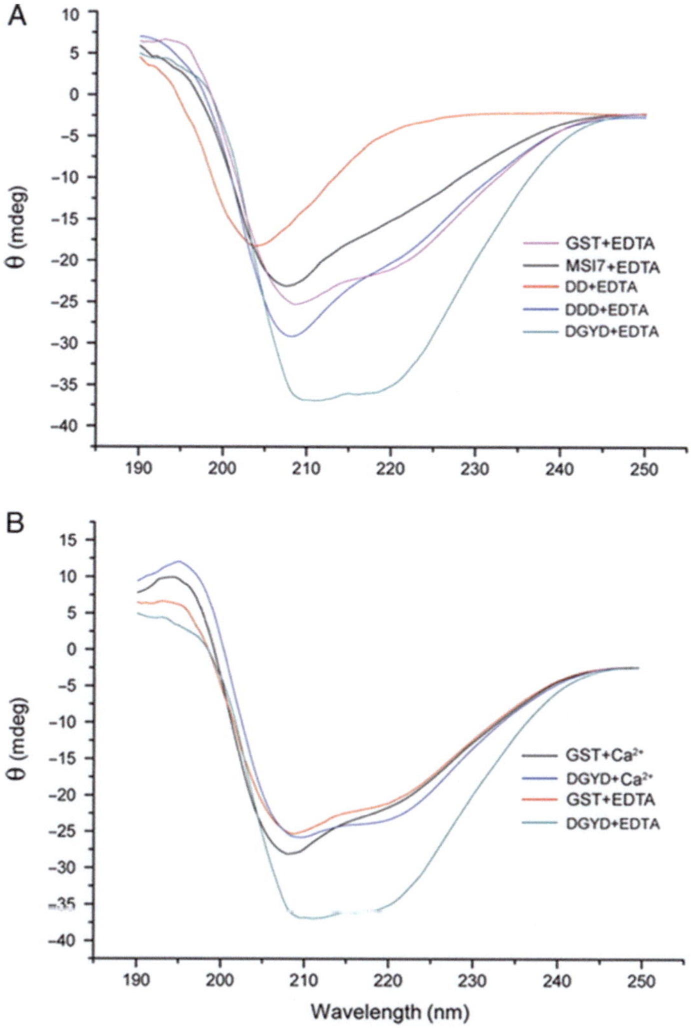

Fig. 3.59 Far-UV CD spectra profile of GST-MSI7 and mutated proteins. (**a**) Far-UV CD spectra of all proteins in the presence of EDTA. The final dosage of proteins is 0.3 mg/ml. The final dosage of EDTA is 2.5 mM. (**b**) Calcium ion resulted in conformational change of GST-DGYD with GST as a control. The final dosage of proteins is 0.3 mg/ml and the final dosage of calcium ion or EDTA is 2.5 mM

3.6.3.1 Identification and Bioinformatics Analyses of PfN44

We found a transcript exists in both the suppression transcriptome libraries and substrate hybridization by searching the mantle tissue transcriptome with suppression substrate hybridization library of D-shaped periods [70]. The transcript sequence had a tandem repeat domain which is deduced to be a character of matrix

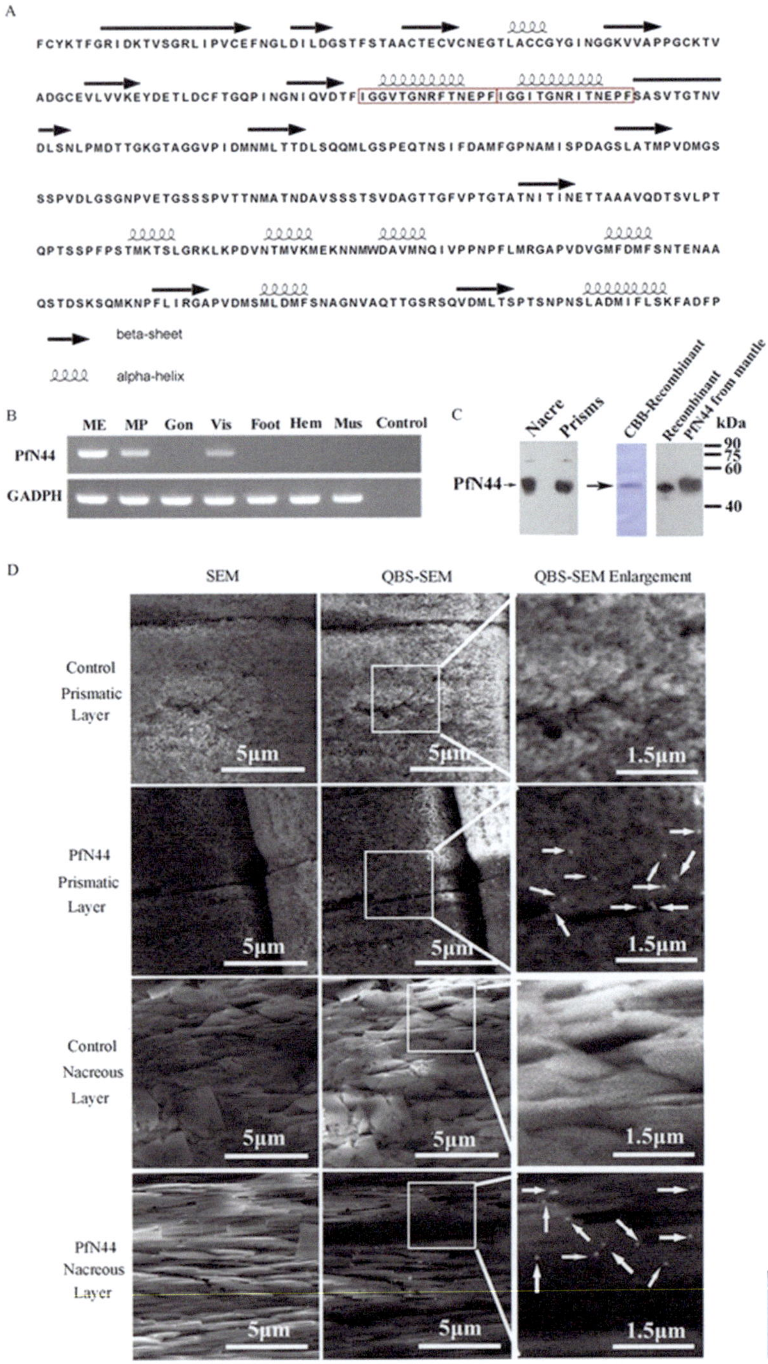

Fig. 3.60 Indication of PfN44 as a matrix protein. (**a**), Secondary structure predicted for PfN44. Red box, repeated region. (**b**), Relative tissue-specific gene expression of PfN44 via RT-PCR. Total RNA was extracted from mantle edge (ME), mantle pallial (MP), gonad (Gon), viscus (Vis), foot,

proteins [112]. We identified this 1617 bp transcript (GenBank™ accession number KC238310) from mantle tissue cDNA [15]. The sequence included a 5′-untranslated region of 75 bp and a 222 bp 3′-untranslated region and an open reading frame of 1320 bp which coded a protein of 439-amino acid. The deduced protein was predicted to contain a signal peptide of 17 amino acids according to SignalP 3.0 prediction. The mature protein was called as PfN44 on the basis of calculated molecular mass is 44.1 kDa. PfN44 was an acidic protein with a predicted theoretical isoelectric point of 4.25 analyzed by ProtScale. The amino acid sequence of PfN44 was only similar to predicted protein CGI_10014775 of the *Crassostrea gigas* after a BLASTp search against the GenBank™ nr database. Whereas, hypothetical protein CGI_10014775 has not been characterized before, so the function of PfN44 was unclear. PfN44 displayed both α-helix and β-sheet in its secondary structure with Jpred prediction (Fig. 3.60a).

3.6.3.2 Identification of PfN44 as a Matrix Protein in the Shell of *P. fucata*

The expression level of PfN44 in different *P. fucata* tissues was analyzed by qRT-PCR, and PfN44 expression was detected in the viscus, mantle pallial, and mantle edge (Fig. 3.60b). The mantle tissue is suggested to be related to shell formation that the mantle edge plays roles in prismatic layer formation, and the mantle pallial is important for formation of the nacre [124]. These results suggested that PfN44 take participation in shell formation of both the nacre and prism, although that the expression of PfN44 in the viscus tissue have not been analyzed thoroughly.

To investigate whether PfN44 exists in the nacreous and/or prismatic layers, polyclonal antibodies against recombinant PfN44 were established for immunodetection of native PfN44 in the shell. EDTA-soluble matrix (ESM) from separated nacreous layer and prismatic layer was analyzed by Western blotting. As Fig. 3.60c had shown, PfN44 was detected in the ESM of both prismatic and nacreous layers. On the contrary, no signal was detected without the anti-PfN44 as a control (data not shown). PfN44 protein was also found in the extract of viscus by Western blotting (data not shown). The purified recombinant PfN44 was analyzed by SDS-PAGE stained with Coomassie Brilliant Blue staining, and the purity of

Fig. 3.60 (continued) hemocytes (Hem), and adductor muscle (Mus). RT-PCR with no template was the negative control group (control). Housekeeping gene GAPDH was used as the positive control. (**c**), Confirmation of PfN44 in the shell. Native PfN44 exists in EDTA-soluble shell matrix by Western blots. Recombinant protein and mantle tissue proteins were also tested as contrast. CBB, Coomassie Brilliant Blue. (**d**), Location of native PfN44 in shells by immunogold staining. SEM images of prismatic layer prisms and nacreous layer tablets with PfN44 in the prism and nacre. Control, with no primary antibody. Arrows indicates PfN44 in the shell. QBS-SEM, backscattered electron mode scanning electron microscope images [15] recombinant protein was sufficient for studies in vitro (Fig. 3.60c). Proteins expressed by *P. fucata* could be modified posttranslationally by phosphorylation or glycosylation, so that we compared the molecular mass of recombinant PfN44 with native protein extracted from mantle tissue. As Fig. 3.60c had shown, recombinant and native PfN44 showed a similar molecular mass, suggesting that native PfN44 was not highly phosphorylated or glycosylated in *P. fucata*.

The microstructural location of PfN44 in shells was explored using immunogold labeling. Slightly notched prismatic layer and nacreous layer were incubated with rabbit antibody against PfN44 as the primary antibody and 5 nm gold-labeled anti-rabbit antibody as the secondary antibody. The white spots in Fig. 3.60d observed by scanning electron microscopy in backscattered mode are gold elements. PfN44 existed in the nacreous tablets of the nacre and prisms of the prismatic layer (Fig. 3.60d). Staining was not found without antibody against PfN44 (Fig. 3.60d) or with rabbit preimmune serum as the primary antibody (data not shown) in control groups. Taken together, PfN44 existed in shell and might participate in shell formation.

3.6.3.3 In Vivo Function of PfN44 During Shell Formation

To explore the functions of PfN44 in shell formation, we conducted experiments in which shells were notched until nacreous layer induces shell repair. As Fig. 3.61a had shown, gene expression of PfN44 increased at 6 h after shell notching conduction and then declined. PfN44 expression level decreased to approximate 10% at 48 h after shell notching. We also analyzed expression change of two other matrix proteins, lysine-rich matrix protein 3 (KRMP3) and nacrein as positive control. The expression nacrein had increased to 135% at 6 h after notching. On the contrary, nacrein expression level increased again at 48 h after notching. The expression tendency of KRMP3 was resemble to nacrein after notching.

The in vivo function of PfN44 during shell formation was further explored by RNAi assay. Synthesized dsRNAs of PfN44 were injected into adductor muscles of *P. fucata* to identify the functions of PfN44 in shell formation. GFP dsRNA or PBS was injected in control groups. Relative mRNA levels of PfN44 were analyzed 7 days after injection. PfN44 expression level in the GFP dsRNA-injected group was the same as the PBS-injected group, while PfN44 expression decreased to about 45% in the group with 40 μg of dsRNA and 30% in the group with 80 μg of dsRNA (Fig. 3.61b). Both of the prismatic and nacreous layers of the treated groups were observed using SEM. The structure of the nacreous layer in the GFP dsRNA-injected group exhibited a normal flat nacreous tablet and a stair-like growth pattern (Fig. 3.61c). The surface of the nacreous layer of the 40 μg PfN44 dsRNA-injection group had shown that deposited tablets were porous with the overgrown crystals. What's more, these abnormal nacreous tablets were smaller than normal tablets.

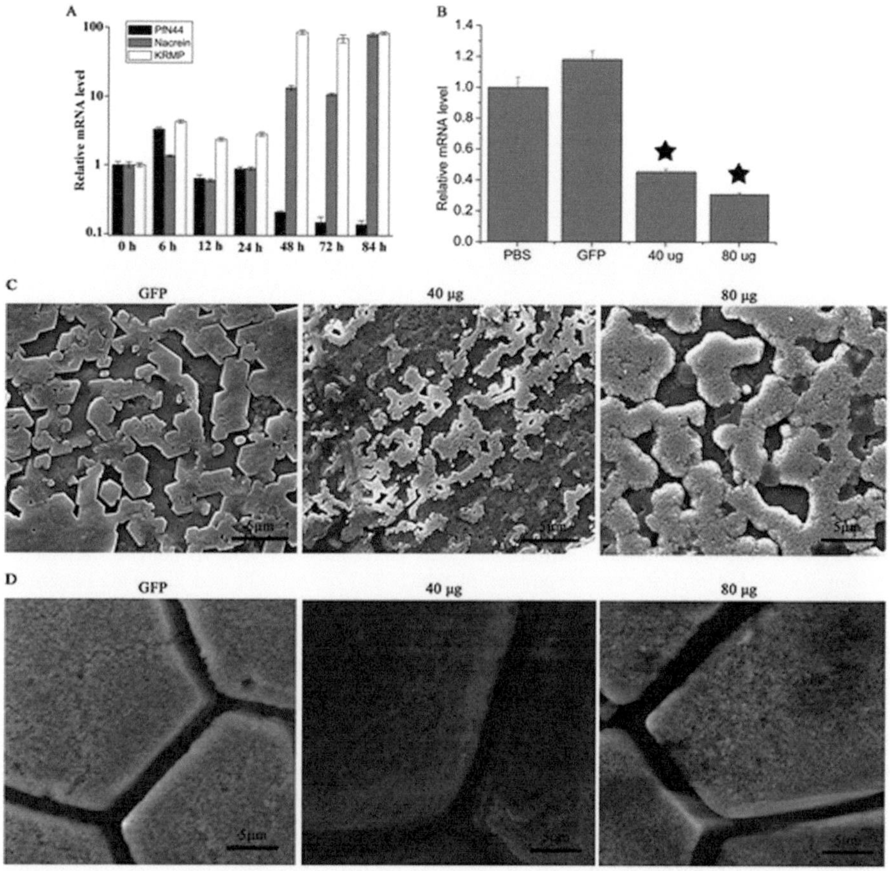

Fig. 3.61 In vivo effects of PfN44. (**a**), PfN44 expression during shell notching. Expression of the control groups (0 h) gave a relative value of 1.0. The y axis indicated a log standard. (**b**), PfN44 expression inhibited by double-strand RNA. Control, PBS-injected groups had a relative value of 1.0. The nontarget control was GFP dsRNA group. Star, significant ($p < 0.001$) difference compared to PBS-injected groups. (**c**), SEM images of inner surface in nacreous layer. GFP dsRNA-injected group indicated normal stair-like growth; 40 and 80 μg of PfN44 dsRNA-injected groups represented overgrowth. Scale bars, 5 μm. D, SEM images of inner surface in prismatic layer. Scale bars, 5 μm

These changes of the nacreous tablets were more prominent in the 80 μg injection group with deposited tablets larger than those normal tablets. The abnormal calcium carbonate deposition was out of control of the feedback system when the expression of PfN44 was extremely deficient, so that in the 80 μg of dsRNA-treated group, the deposited tablets were larger than the normal one. The surface of the prismatic layer had no obvious change with the interference of PfN44 compared with control group (Fig. 3.61d).

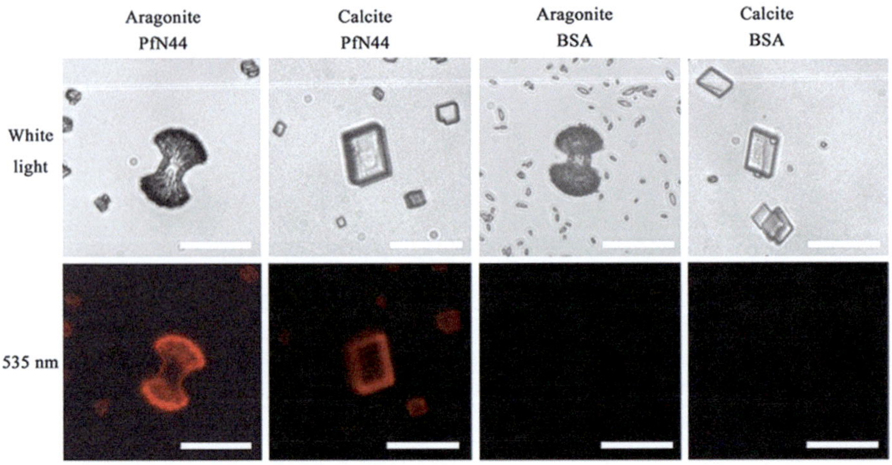

Fig. 3.62 Binding ability profiles of PfN44. PfN44 was marked with rhodamine and then reacted with calcite or aragonite. The rhodamine-labeled BSA was added as a control. Scale bars, 20 μm. 109 crystals were examined of which 105 crystals represented the same pattern in the calcite-binding assay. 124 crystals were examined of which 114 crystals showed the same pattern as represented in the aragonite-binding assay

3.6.3.4 Binding Ability of PfN44 with Main Shell Components

To investigate the mechanisms of how PfN44 affected the shell formation, the assay of binding ability of PfN44 with the main mineral components of the shell, i.e., calcite and aragonite, was conducted [15]. PfN44 labeled with rhodamine was incubated with calcite or aragonite, and surfaces of calcite and aragonite were labeled with rhodamine indicating there is binding between calcium carbonate crystals and PfN44 (Fig. 3.62).

3.6.3.5 Effect of PfN44 on Calcium Carbonate Crystallization

In vitro calcite and aragonite crystallization system was used to explore the function of PfN44 on calcium carbonate crystallization. In the crystallization in vitro assays, the effects of proteins were analyzed. The calcite crystallization system without magnesium was conducted to investigate the effects of PfN44 on calcite formation. Crystals deposited in saturated calcium bicarbonate solution were typical calcite rhombohedra with 2.5 μg/ml BSA as a control (Fig. 3.63a). The morphology of deposited crystal had no obvious change in presence of 1 or 2.5 μg/ml PfN44. Raman spectra analyses indicated that all deposited crystals are calcite (Fig. 3.63b).

To explore the function of PfN44 on aragonite deposition, in vitro crystallization assays were conducted with 10 mM magnesium to induce formation of aragonite.

Deposited crystals were slightly notched calcite-like crystals and needlelike spindles with 10 mM magnesium and BSA as a control (Fig. 3.63a). Raman spectra analyses suggested that the two kinds of crystals were calcite and aragonite, respectively (Fig. 3.63b). In the presence of 1 μg/ml PfN44, fewer aragonite crystals were deposited with the size decreased compared with the control group. Deposited crystals were all calcite with the addition of 2.5 μg/ml PfN44, as distinguished by Raman spectra analyses. This result indicated that PfN44 could inhibit aragonite deposition. To confirm whether PfN44 inhibited aragonite deposition, we conducted crystallization in vitro assays with 50 mM magnesium, which was sufficient to induce all deposited crystals to be aragonite. As Fig. 3.63a had shown, formed crystals were all aragonite with a round morphology and needlelike spindles when 2.5 μg/ml BSA was added as a control. Two kinds of crystals were formed with the addition of 1 μg/ml PfN44 in which one was similar to the aragonite deposited in the control group and the other was dumbbell-shaped calcite which were confirmed by Raman spectra. In the presence of 2.5 μg/ml PfN44, deposited crystals were all calcites of dumbbell shape. These results indicated that aragonite deposition was inhibited by PfN44.

We also explored the elements of deposited crystals via energy-dispersive X-ray spectroscopy. The ratio of Ca^{2+}/Mg^{2+} decreased in formed calcite with PfN44. With addition of 10 mM magnesium, the ratio of Ca^{2+}/Mg^{2+} was 8.4 in calcite and decreased to 5.2 in the presence of 1 μg/ml and 5.13 with addition of 2.5 μg/ml PfN44. In crystallization system of 50 mM magnesium, the ratio of Ca^{2+}/Mg^{2+} in formed calcite was 3.2 in the presence of 1 μg/ml PfN44. The ratio of Ca^{2+}/Mg^{2+} decreased to 0.92 with 2.5 μg/ml PfN44. The increased magnesium content in calcite suggested that PfN44 could stabilize magnesium calcite to affect calcium carbonate crystallization.

3.6.3.6 Interactions Between Magnesium and PfN44

To figure out whether PfN44 interacted with calcium thus control calcium carbonate crystallization, we analyzed near-UV CD spectra assay of PfN44 with or without Ca^{2+}. Without calcium (1 mM EDTA), PfN44 showed typical secondary structure of α-helix and β-sheet consistent with the prediction of secondary structure in Fig. 3.60a. With addition of 1 mM calcium, the peak of the α-helix had no change. With 2.5 mM calcium in PfN44 solution, the peak of the α-helix at 208 nm had a hypochromatic shift. Following with higher dosages of calcium, this CD spectrum shift became more obvious. This result indicated that calcium could increase the hydrophobicity of the PfN44 by forming a more flexible α-helix secondary structure of PfN44. Carbonate deposition in Mollusca is also related to magnesium [125], so that we analyzed the near-UV CD spectra of PfN44 along with magnesium. The α-helix peak shifted to a shorter wavelength in a dosage-dependent manner with 1 mM and higher dosages of magnesium (Fig. 3.64a). Ca^{2+} could not induce the shift until at 2.5 mM or more, while Mg^{2+} could make the blue shift at only 1 mM. The shift caused by Mg^{2+} could be observed in the presence of

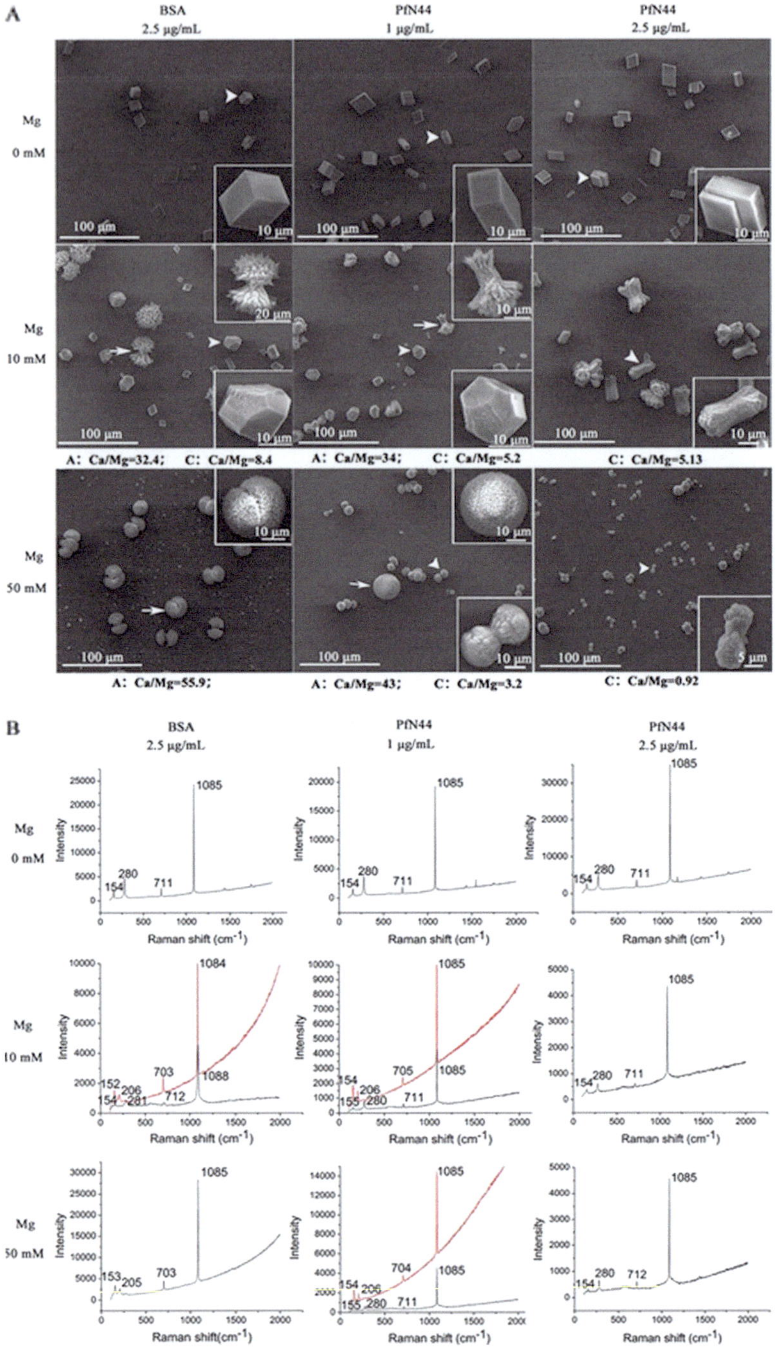

Fig. 3.63 Functions of PfN44 on crystal growth of calcite and aragonite. (**a**), SEM images of deposited crystals. Amplified images represent crystals indicated by arrows or arrowheads. The

lower concentrations Mg^{2+} than Ca^{2+}, indicating a stronger interaction between Mg^{2+} and PfN44. To confirm the interaction of PfN44 with Ca^{2+} and Mg^{2+}, we conducted the intrinsic fluorescence of PfN44 in the presence or absence of Ca^{2+} and Mg^{2+}. With the addition of 2.5, 5, or 10 μM calcium, the intrinsic tryptophan fluorescence of PfN44 quenched about 5.4%. At the same time, the quenching was 22.3% with 50 μM calcium and 29.5% with 500 μM calcium (Fig. 3.64b). As Fig. 3.64b had shown, quenching of 2.5, 5, or 10 μM magnesium groups was higher than quenching of calcium group. Quenching with 10 μM magnesium was 14.8% higher than that of calcium. However, intrinsic fluorescence was not further changed when PfN44 was incubated with 50 or 500 μM magnesium. The intrinsic fluorescence no further changed because quenching by magnesium was saturated at a definite dosage. The substantial different quenching levels indicated that PfN44 had a stronger interaction with magnesium than calcium.

3.6.3.7 Functions of PfN44 on the Transition of ACC to Stable Crystals

ACC is the precursor of calcium carbonate crystallization in mollusks [126]. To figure out how PfN44 inhibited aragonite deposition, we investigated the effect of PfN44 in the transition of ACC to stable calcium carbonate crystals. We first analyzed the effect of PfN44 on the transition process of ACC to the stable crystals using FTIR. ACC could transform to calcite and aragonite at 24 or 48 h of crystallization in the presence of 50 mM magnesium. Characteristic peaks at 700 and 854 nm indicated deposited aragonite, and peaks of 714 and 874 nm meant calcite. The FITR spectrum of deposited crystals showed only characteristic peaks at 717 and 874 nm in the presence of 50 μg/ml PfN44, suggesting that the deposited crystals were all calcite (Fig. 3.64c).

To further analyze the polymorphs of formed crystals, we analyzed the formed crystals by X-ray diffraction. ACC transformed to aragonite or Mg^{2+}-bearing calcite at 24 or 48 h crystallization with addition of 50 mM magnesium (Fig. 3.64d). There was no aragonite formed in deposited crystals when 50 μg/ml PfN44 was present in the crystallization system. On the contrary, ACC transformed to Mg^{2+}-bearing calcite of rhombohedral and monoclinic calcium carbonate without magnesium. The proportion of Mg^{2+}-bearing calcite crystals increased twofold greater than crystals amount deposited without PfN44. What's more, the proportion of Mg^{2+}-

Fig. 3.63 (continued) control group was added with 2.5 μg/ml BSA. With no magnesium, deposited crystals were totally calcite. PfN44 could not change crystals morphology. With addition of 10 or 50 mm magnesium, PfN44 induced calcite-like crystals formation and repressed aragonite deposition. Crystal compositions were analyzed by energy-dispersive X-ray spectroscopy. *A* aragonitem *C* calcite. At least 50 crystals images in each spot were observed. The images shown in (**a**) are typical of these crystals. (**b**), Raman spectra profile of deposited crystals in (**a**). Red lines, crystals indicated by arrows; black lines, crystals pointed by arrowheads

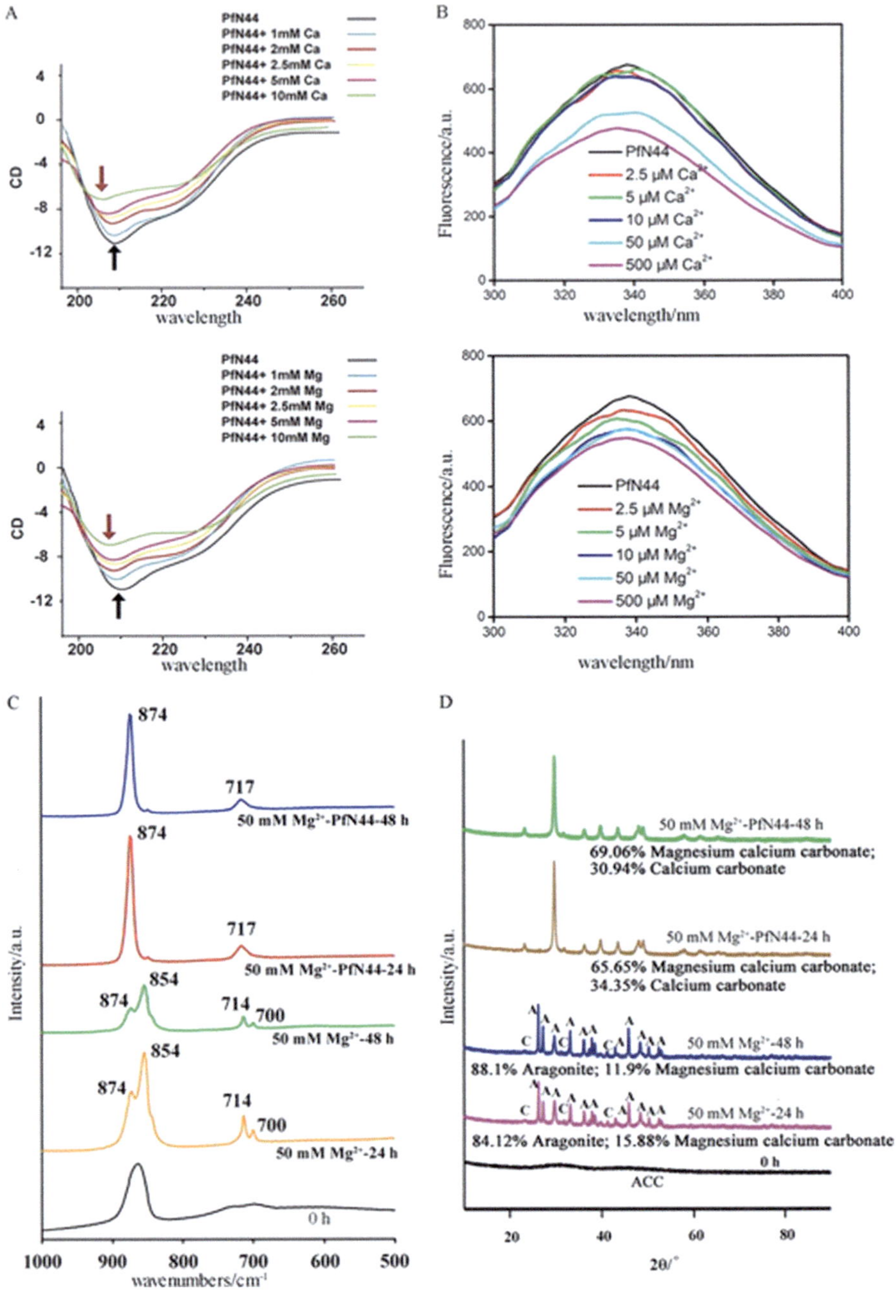

Fig. 3.64 Interaction between PfN44 and magnesium stabilizes magnesium calcite. (**a**), Near-UV CD spectra of PfN44 with (1 mm EDTA) or without calcium or magnesium. Arrows, α-helix peaks of 208 nm. Black arrows, native protein; red arrows, proteins with addition of calcium or magnesium. (**b**), Intrinsic fluorescence of PfN44 with (1 mm EDTA) or without calcium or magnesium. Representation of IR (**c**) and X-ray diffraction spectra (**d**) precipitates in solutions with the indicated dosages of magnesium and different crystallization times

bearing calcite changed from 15.88 ± 0.4% to 11.9 ± 0.2% in the absence of PfN44 at 48 h (p value = 0.001); no difference with the hypothesis that Mg^{2+}-bearing calcite is a thermodynamically unstable form. The ratio of deposited Mg^{2+}-bearing calcite changed from 65.7 ± 1.0% to 69.1 ± 0.7% in the presence of PfN44 from 24 to 48 h (p value = 0.0089), indicating that PfN44 stabilized the Mg^{2+}-bearing calcite and increased its content. These results suggested that, in the transition of ACC to the stable crystals, PfN44 stabilized the Mg^{2+}-bearing calcite to inhibit aragonite formation.

3.6.4 A Novel Matrix Protein, PfY2, Functions as a Crucial Macromolecule During Shell Formation

Biomineralization, including shell formation, is drastically regulated by matrix proteins. PfY2, a matrix protein existing in the ethylenediaminetetraacetic acid (EDTA)-soluble matrix of both prismatic layer and nacreous layer, was found by our group via microarray. It may play dual roles in biomineralization. Whereas, the molecular mechanism is still unknown. In present study, we explored the function of PfY2 on crystallization in vivo and in vitro, indicating that it might work as a negative regulator during shell formation. Notching experiment suggested that PfY2 was involved in shell repairing and regenerating procedure. Inhibition of PfY2 gene influenced the structure of prism and nacre layer together, confirming its dual roles during shell formation. Recombinant protein rPfY2 drastically suppressed $CaCO_3$ precipitation, contributed to the crystal nucleation process, resulted in the morphology changes of crystals, and inhibited the transition of amorphous calcium carbonate (ACC) to stable calcite and aragonite in vitro. Our results may provide new clue on the biomineralization inhibition process [127].

3.6.4.1 Identification and Bioinformatics Analysis of PfY2

Current researches suggest that most studied matrix proteins and proteins relevant to shell formation could be upregulated by over 20-fold from the umbonal stage to juveniles [31]. The full length of PfY2 cDNA sequence (GenBank™ accession number KY436033) is 493 nucleotide, possessing a 5′-untranslated region of 47 bp, an 336 bp ORF encoding a deduced 111-amino acid protein, and a 110 bp 3-′-untranslated region (Fig. 3.65a). The BLASTx alignment against the GenBank™ nr database indicated that PfY2 showed no resemblance to any known genes in any known species, and the function of PfY2 still remained unclear. The deduced mature protein had a putative molecular mass of 11.9 kDa.

3.6.4.2 The Expression Levels and Distribution Pattern of PfY2

On one hand, to understand and evaluate the potential effects of PfY2 protein, the relative expression of PfY2 in tissues was examined by RT-PCR, such as the foot, gonad, gill, mantle pallial, mantle edge, viscus, and adductor muscle. RT-PCR expressions showed that PfY2 represented the highest expression level in mantle edge, second in mantle pallial (Fig. 3.65b). It is generally thought that mantle tissue contributes to shell formation. And the mantle edge is closely relevant with prismatic layer, while the mantle pallial is reported to participate in the formation of nacreous layer [6, 128].

On the other hand, in situ hybridization was conducted as described above to detect the specific expression site of PfY2 mRNA in frozen sections of mantle tissue. The hybridization signals were found in the epithelial cells from both the outer fold and middle fold (Fig. 3.66b). The control group had no significant signals (Fig. 3.66a). This result was also accordant with the RT-PCR detection analyses.

In general, PfY2 expressed in the mantle tissue of *P. fucata* in particular and very likely participated in the formation of both nacreous layer and prismatic layer.

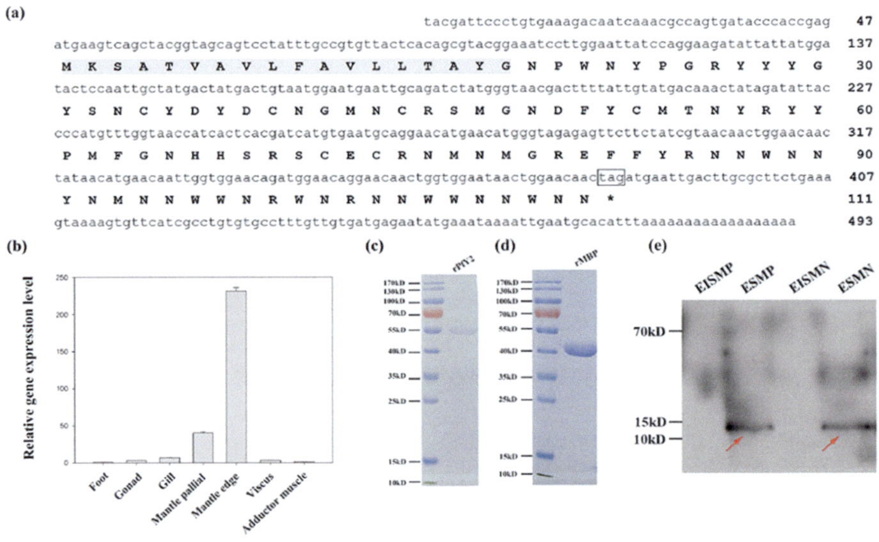

Fig. 3.65 Detection of PfY2 as a matrix protein
(**a**) The cDNA sequence and putative amino acid sequence of PfY2. The shaded region is the signal peptide. Black box indicates the transcription termination site. (**b**) Expressional analysis of PfY2 in different tissues by RT-PCR. Values of RT-PCR are means ± standard deviation of three repetition. (**c**) SDS-PAGE of recombinant protein PfY2 at 55 kDa with MBP tag in N terminal. (**d**) SDS-PAGE of MBP protein at 40 kD. (**e**) Identification of native PfY2 in the shells by Western blotting. The red arrow pointed the band of the native PfY2 in shell compositions. *EISMP* EDTA-insoluble segment from prismatic layer, *ESMP* EDTA-soluble section from prismatic layer, *EISMN* EDTA-insoluble segment from nacreous layer, *ESMN* EDTA-soluble matrix from nacreous layer [127]

3.6.4.3 Expression and Purification of rPfY2 and rMBP Proteins

In consideration of the relative low expression level of native PfY2, its recombinant expression in vitro looks like to be a viable alternative to study. The recombinant protein was sufficiently purified to study in vitro functions (Fig. 3.65c, d). The molecular weight of rPfY2 (55 kD) is almost the same to the calculated value (PfY2 with His-tag 12 kD plus MBP tag of 40 kD), suggesting that rPfY2 was not drastically glycosylated or phosphorylated. Additionally, mass spectrum was executed to reconfirm the protein sequence of rPfY2 (data not shown). We also produced and purified rMBP as negative control (Fig. 3.65d) for further function research. In this study, the recombinant protein MBP-PfY2 having a His_6-tag in C-terminal, rPfY2, was used for further function study.

3.6.4.4 Detection of PfY2 as a Matrix Protein in *P. fucata* Shells

Gene expression level analyses were not enough to ensure that native PfY2 protein truly existed in the nacreous layer and/or the prismatic layer. Therefore, we used the polyclonal antibody against recombinant protein PfY2 to immunodetect native PfY2. As shown in Fig. 3.65e, PfY2 could be found in EDTA-soluble matrix from both nacreous layer and prismatic layer by Western blotting. In control groups, no signal was identified in the absence of the anti-PfY2.

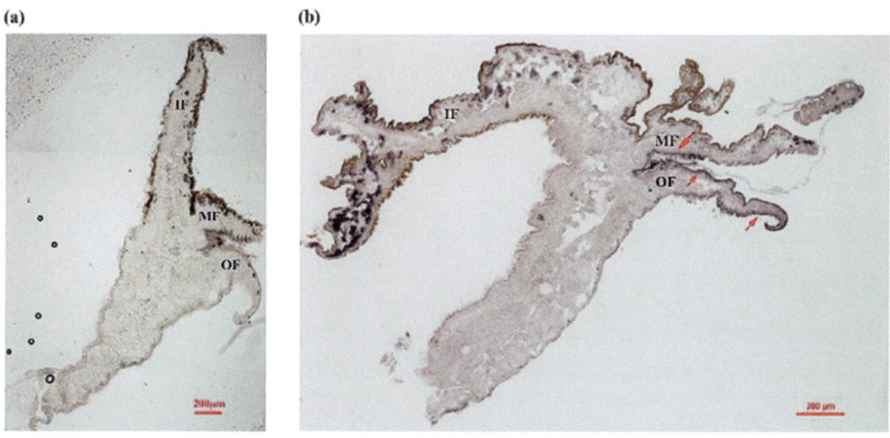

Fig. 3.66 Detection of PfY2 in mantle tissue by in situ hybridization
(**a**) Positive signals were pointed by arrows (**b**). *OF* outer fold, *MF* middle fold, *IF* inner fold. Scale bars, 200 μm. Negative control group represented no significant signals

3.6.4.5 In Vivo Function Study of PfY2 During Shell Formation and Repair

We executed shell notching experiments to explore the roles of PfY2 during shell reparation and formation, which also participate in better understanding the complicated natural shell repair process. The expression levels of PfY2 upregulated 12 h after notching, arriving the highest at 36 h. PfY2 expression upregulated to about 200% of the value before notching. Although slightly reducing occurred in the following hours, the expression level of PfY2 still increased compared with the negative control group. We also tested the relative gene expression levels of other matrix proteins, including KRMP and nacrein, after shell notching (Fig. 3.67a). Other than PfY2, the increasing level of the other two proteins shown was much more acute. KRMP was upregulated drastically only 6 h after notching, while the expression of nacrein did not increased until 24 h later. What's more, the expression level of nacrein kept upregulating over time. In contrast, the expression model of KRMP and PfY2 was similar. They both increased immediately after notching, underwent a slight decrease, and finally arrived a relative balance stage.

The effect of PfY2 in shell formation was further studied by RNA interference. The microstructures of the shells could be destroyed after the downregulation of particular matrix proteins. We synthesized the specific double-strand RNA to knock down the expression level of PfY2 by injecting them into the adductor muscles of oysters. Oysters in the control groups were injected with GFP double-strand RNA or water only. Seven days after injection, total RNA of mantle tissues was extracted and the expression level of PfY2 was detected by RT-PCR. The expression level of PfY2

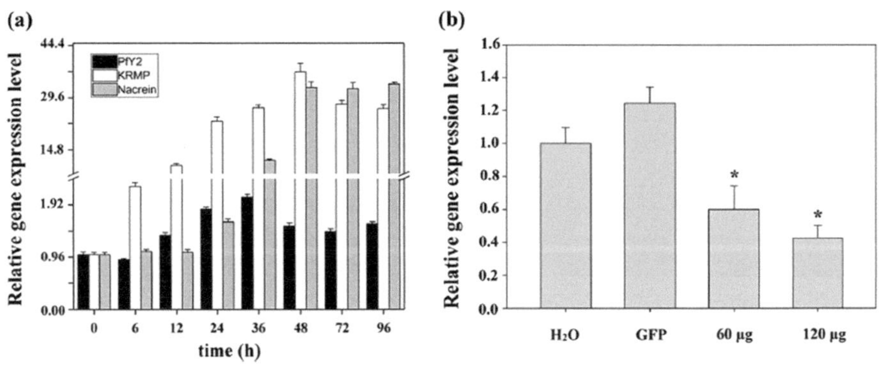

Fig. 3.67 In vivo functions study of PfY2 [127]
(**a**) The expression of PfY2 after shell notching. Expression of the negative control (0 h) set a relative value of 1.0. (**b**) PfY2 expression inhibited by RNA interference. The expression of negative control group of water-injected oyster was set as a relative value of 1.0. The oysters in nontarget control group were injected by GFP dsRNA. The expression of PfY2 was significantly ($p < 0.05$) repressed after the injection of specific PfY2 dsRNA with both 60 μg and 80 μg groups related with the oysters injected with water. Asterisk indicates significant difference ($p < 0.05$)

in the GFP dsRNA-injected group was resembled to the water-injected group. Whereas, the expression of PfY2 was reduced to almost 60% in the group injected by 60 μg PfY2 dsRNA and 40% in the group injected by 120 μg PfY2 dsRNA in relation to the groups injected with water (Fig. 3.67b). The appearance of both prismatic and nacreous layers in all treated groups was detected under SEM. In control groups (water-injected group and GFP dsRNA-injected group), prismatic layer suggested normal smooth surfaces (Fig. 3.68a–d), with the clear visible edges and the regular hexagonal structures of nacreous tablets in nacreous layer (Fig. 3.69a–d). Compared to those normal phenotypes, the injection of PfY2 dsRNAs resulted in the destruction of microstructures in both prismatic layer and nacreous layer. The appearance of prismatic layer became rougher and polyporous after injecting PfY2 dsRNA (Fig. 3.68e–h), whereas the shape of the nacreous layer was disordered with the overgrowth of the crystals above the nacreous tablets proposed to grow normally (Fig. 3.69e–h). These variations in both prismatic and nacreous layers were more obvious in the group injected with 120 μg PfY2 dsRNA, suggesting PfY2's dual roles in shell repair and formation.

3.6.4.6 Binding Abilities of rPfY2 with Mineralization Elements

Interactions with the main shell components including calcite and aragonite are one of the most important characters of calcification-related matrix proteins. Thus, we explored the binding abilities of rPfY2 with calcite and aragonite to analyze the mechanisms of how rPfY2 influenced the shell formation. rPfY2 was reacted with prepared calcite and aragonite powders for 12 h. The dosage of every kind of protein, including BSA, rMBP, and rPfY2 used in binding assays, was 40 μg/ml. No matter in which crystal system rPfY2 was added, it could be found in both calcite and aragonite crystals even all the crystals had been washed by water and protein storage buffer three times each time (Fig. 3.70a lane 3, Fig. 3.70b lane 3). Although rMBP (Fig. 3.70a lane 2, Fig. 3.70b lane 2) was also observed in calcites and aragonites, the contents of which had already bound to the crystals were much fewer compared to the washing fractions. Taken together, the content of binding protein rPfY2 (Fig. 3.70a, lane 3) in calcite system was much more than the rMBP (Fig. 3.70a, lane 2) and the BSA (Fig. 3.70a, lane 1) binding to the calcium carbonate crystals. Same binding effects also existed in aragonite crystalline system meanwhile (Fig. 3.70b). All in all, this binding ability is the foundation of matrix proteins' function on crystal nucleation, crystal polymorphism, crystal orientation, and crystal morphology [14, 15, 119, 129, 130].

3.6.4.7 Calcium Carbonate Crystallization with Addition of rPfY2 In Vitro

The shell of pearl oyster, *P. fucata*, is an organic mineral assembly controlled by calcium carbonate with organic matrix complex. In vitro calcium carbonate

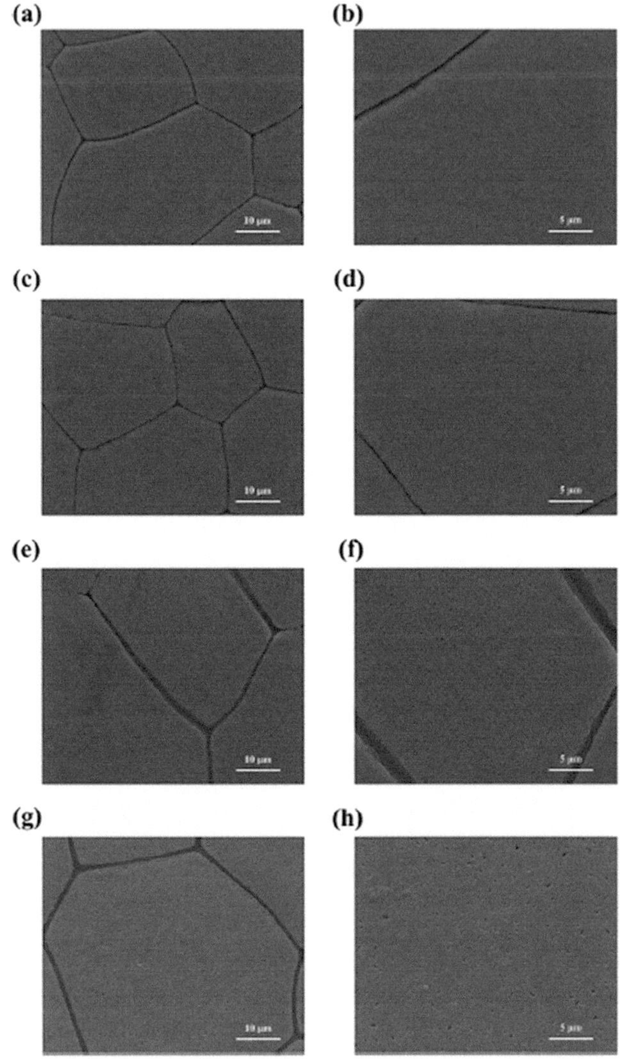

Fig. 3.68 Functions of the inhibition of PfY2 on prismatic layer of inner shells
(**a**, **b**) SEM images of the normal prismatic layers of oyster shells injected with water. (**c**, **d**) SEM images of the prismatic layers of shells in GFP-injected group. (**e**, **f**) SEM images of the prismatic layers of shells in 60 μg PfY2 dsRNA-injected group. (**g**, **h**) SEM images of the prismatic layers of shells in 120 μg PfY2 dsRNA-injected group. (**b**, **d**, **f**, **h**) represented an enlargement of (**a**, **c**, **e**, **g**), respectively

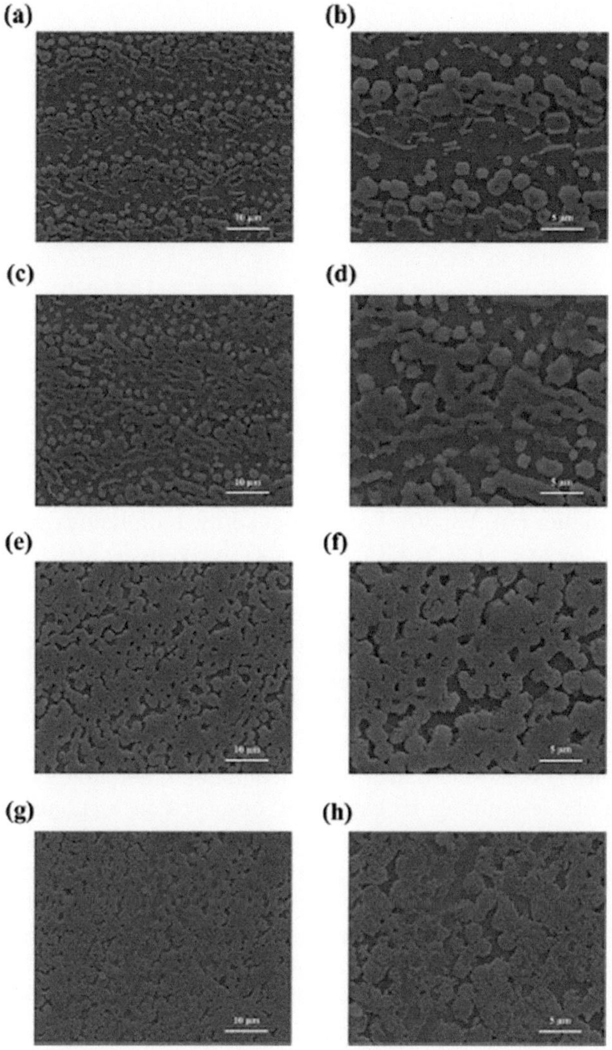

Fig. 3.69 Functions of the inhibition of PfY2 on the nacreous layer of inner shells
(**a**, **b**) SEM images of the natural nacreous layers of shells from oysters injected by water. (**c**, **d**) SEM images of the nacre surface in GFP-injected groups. (**e**, **f**) SEM images of the nacreous layers in group injected with 60 μg PfY2 dsRNA. (**g**, **h**) SEM images of the nacre surface in shells of 120 μg PfY2 dsRNA-injected group. (**b**, **d**, **f**, **h**) exhibited an enlargement of (**a**, **c**, **e**, **g**) respectively

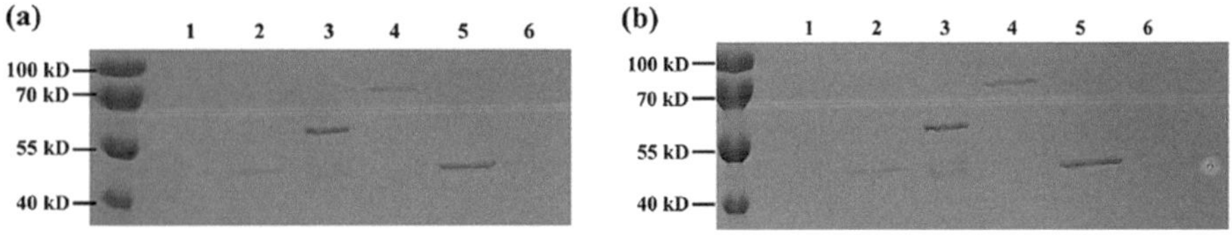

Fig. 3.70 Binding abilities of rPfY2 with shell components
Lanes 1–3, BSA, MBP, and rPfY2 binding with calcite (**a**) and aragonite (**b**) after washed by water and protein elution buffer for three times, respectively. Lanes 4–6, surplus unbound BSA, MBP, and rPfY2 in first wash segment after reaction with calcite (**a**) and aragonite (**b**), respectively.

crystallization experiment has been widely executed to mimic the biomineralization process during shell formation. First of all, we examined the effects of rPfY2 on calcite with no addition of Mg^{2+}. In negative control groups, the shape of crystals were typical rhombohedra with addition of 80 μg/ml BSA as predicted, Raman spectrum indicating they were typical calcite (Fig. 3.71a–c). These results agree with our previous studies [14, 15, 30]. We also found the morphology and identified the crystal morphology via SEM and Raman, respectively, in the MBP group to exclude the effects of the tag (80 μg/ml) and found that MBP had no influence on the crystallization of calcite (Fig. 3.71e–f). On the contrary, the addition of rPfY2 changed the shapes of calcite drastically (Fig. 3.71g–o). With the increasing dosage of rPfY2 from 0 to 80 μg/ml, the corners and edges of the crystals changed rounder. It seemed like that the calcite crystals assembled and rPfY2 might induce an aggregation of crystal particles to some extent. When the dosage of rPfY2 arrived to 80 μg/ml, the crystals were transited into a dumbbell morphology (Fig. 3.71m, n), which was quite resembling to aragonites observed in aragonite crystallization system in vitro. Whereas, the crystal formed had not been changed on the basis of Raman spectrum, with the special peaks of 155, 280, 711, and 1085 cm^{-1}, suggesting that they were still calcites (Fig. 3.71i, l, o). And those dumbbell-shaped calcites were also composed of small calcite crystals with higher magnification, just like the crystals grown with the addition of the lower dosage of rPfY2 (Fig. 3.71n). Those results suggested that rPfY2 could directly regulate the growth and morphologies of calcite, indicating its significance during the biomineralization procedure.

Then magnesium was added into the crystallization system to induce aragonite formation, the main composition of the nacreous layer. The typical aragonites were precipitated crystals mixed with needlelike spindles, whereas, the introduction of rPfY2 did not change the shape or crystal form (Fig. 3.72). The effects of rPfY2 mainly concentrated on affecting the crystal number and size. The scale of aragonite crystals became smaller and smaller with the increasing concentration of rPfY2, from 30 to 80 μg/ml (Fig. 3.72g–o). Meanwhile, the number of the crystals also increased when the concentration of rPfY2 increased. Taken together, we suspected that it was reasonable that rPfY2 could participate in the nucleation of aragonite crystallization. In conclusion, rPfY2 could both influence the crystallization procedure of calcite and aragonite in vitro, suggesting its importance and dual roles during biomineralization.

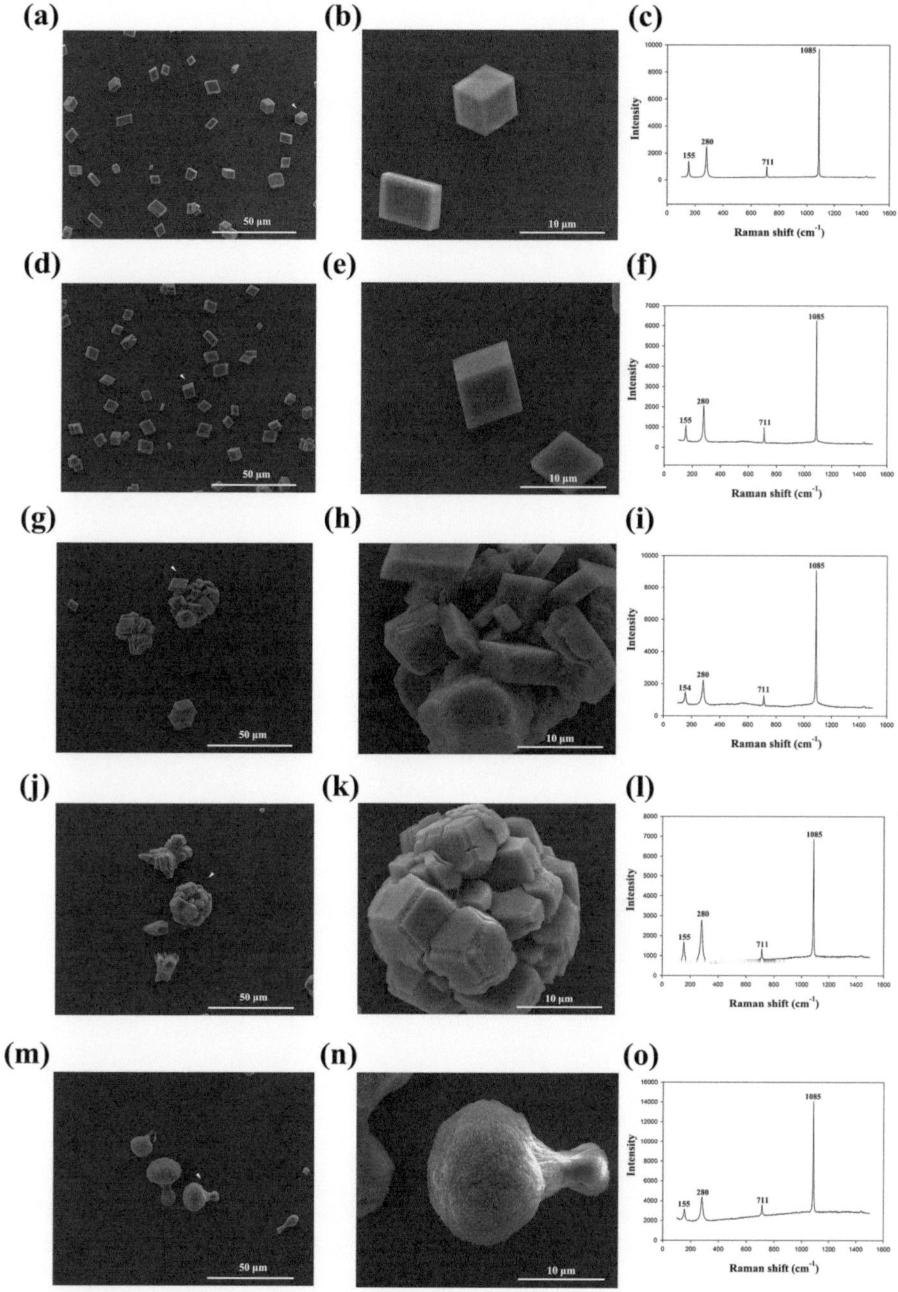

Fig. 3.71 SEM images of calcite crystallization in vitro with addition of rPfY2 and Raman spectrum analyses of crystals

Crystals formed with addition of 80 μg/ml BSA (**a**, **b**); 80 μg/ml MBP (**d**, **e**); 30 μg/ml rPfY2 (**g**, **h**); 50 μg/ml rPfY2 (**j**, **k**); 80 μg/ml rPfY2 (**m**, **n**). (**b**, **e**, **h**, **k**, **n**) were the magnifications of the crystals pointed by arrows in (**a**, **d**, **g**, **j**, **m**), respectively. (**c**, **f**, **i**, **l**, **o**) represented the Raman spectrum of the crystals grown with 80 μg/ml BSA (**b**), 80 μg/ml MBP (**e**), 30 μg/ml rPfY2 (**h**), 50 μg/ml rPfY2 (**k**), and 80 μg/ml rPfY2 (**n**), respectively

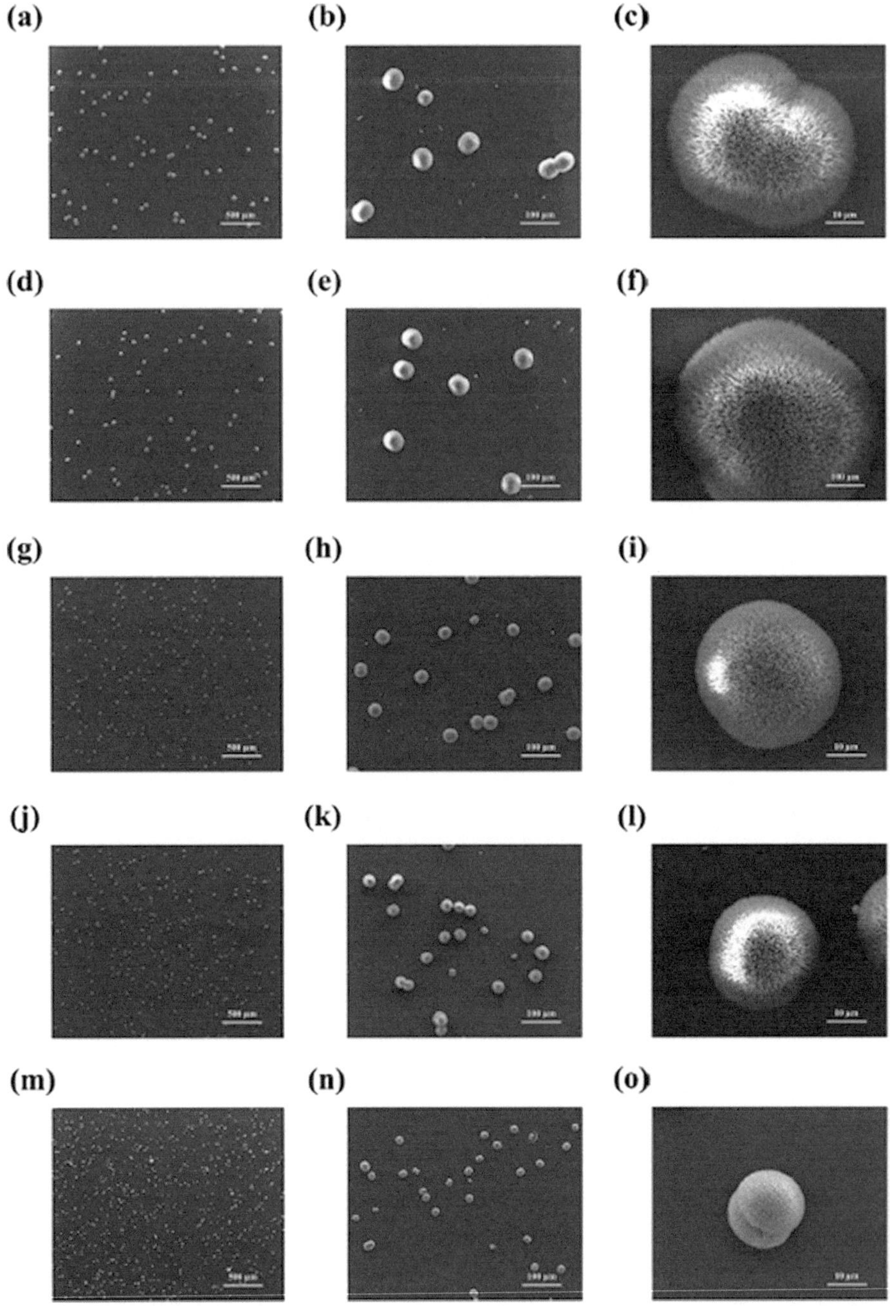

Fig. 3.72 SEM images of in vitro aragonite crystallization with addition of rPfY2. Crystals formed in the presence of 80 μg/ml BSA (**a–c**); 80 μg/ml MBP (**d–f**); 30 μg/ml rPfY2 (**g-i**); 50 μg/ml rPfY2 (**j–l**); 80 μg/ml rPfY2 (**m–o**). The images of medial and right columns were amplificated images of the crystals in the center

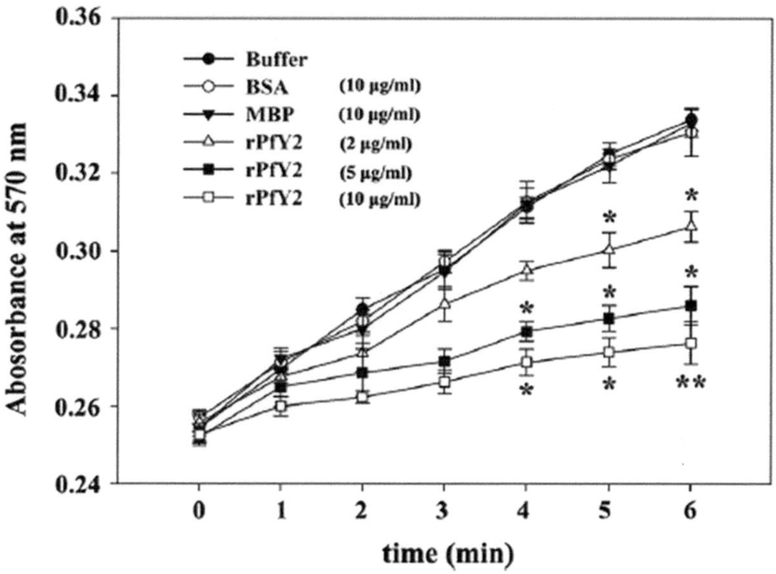

Fig. 3.73 In vitro repression of rPfY2 on $CaCO_3$ crystallization precipitation
$CaCO_3$ precipitation rate was recorded with the absorbance of crystals of 570 nm. 2 μg/ml rPfY2 was functional ($p < 0.05$) to inhibit the precipitation rate 5 min after addition compared with the buffer group (the negative control group), and this inhibitory capacity was dose-dependent *, $p < 0.05$; **, $p < 0.01$

3.6.4.8 In Vitro Repression Activity of rPfY2 on Calcium Carbonate Precipitation

Regulation in calcium carbonate precipitation is another important effect of matrix proteins. And this precipitation rate was recorded with the absorbance at 570 nm. This value of all control groups gradually increased and arrived approximately 0.33 in 6 min (Fig. 3.73). The absorbance values changed significantly after introducing rPfY2, yet the changing tendency of MBP group and BSA group was similar as the negative control (buffer). The highest absorbance value did not reach 0.31 in the presence of 2 μg/ml of rPfY2. In three control groups, the absorbance values increased gradually along with time, while it kept decreasing at every minute as the content of rPfY2 is increasing. What's more, it was noteworthy that there is a positive correlation between the reducing rate and the dosage of rPfY2. The precipitation rate was significantly inhibited 5 min after the introduction of 2 μg/ml rPfY2 ($p = 0.044$), while 10 μg/ml rPfY2 could effectively inhibit this rate only 4 min after the mixture ($p = 0.039$). What's more, the p value of higher dosage rPfY2 group was lower at the same moment. Thus, this inhibitory capacity was dosage-dependent [127].

After reaction for 24 h, we detected the crystals by SEM and observed that the crystal size in the presence of 10 μg/ml rPfY2 was much smaller than those in the control group (Fig. 3.74a–f), when the final dosage of CaCl2 and $NaHCO_3$ was 50 mM. In addition, the edges and corners of the rhombohedral calcite were repressed with rPfY2. The length of sides of the crystals was 2–3 μm (Figure 3.74e, f) on average while the length of sides of calcite in control groups was about 6–7 μm. What's more, the number of the crystals increased in the presence of rPfY2, suggesting that PfY2 might take part during the nucleation process of crystallization.

Fig. 3.74 SEM images of crystals grown in the crystal growth inhibition assays
SEM images of crystals grown with 10 μg/ml BSA (**a**, **b**); 10 μg/ml MBP (**c**, **d**); 10 μg/ml rPfY2 (**e**, **f**). (**b, d, f**) were the magnifications of the crystals indicated by arrows in (**a**, **c**, **e**), respectively. The final dosage of $CaCl_2$ and $NaHCO_3$ was 50 mM in (**a**–**f**) and 5 mM in (**g**–**i**). (**g**) 10 μg/ml BSA group; (**h**) 10 μg/ml MBP group; (**i**) 10 μg/ml rPfY2 group. Nearly no crystal had grown in the presence of 10 μg/ml rPfY2 when the dosage of Ca^{2+} reduced to 5 mM

Whereas, when the final dosage of $CaCl_2$ and $NaHCO_3$ reduced to 5 mM, nearly no obvious crystals had grown compared with the negative control group (Fig. 3.74g–i). Taken together, rPfY2 could effectively inhibit the calcium carbonate precipitation rate in vitro along with an obvious dosage gradient effect. We proposed that PfY2 might be a negative regulator participating in calcium carbonate crystallization.

3.6.4.9 Effects of rPfY2 on Transformation of ACC to Stable Crystals

ACC is the precursor of main anhydrous polymorphs, calcite, vaterite, and aragonite [131–133]. We executed the ACC transition assay to explore the roles of PfY2 played in the transition of ACC to stable crystals by X-ray diffractometer. All the ACC transited to calcite after 4 h in the negative control group and MBP group (Fig. 3.75a, b), while rPfY2 inhibited this process. Only 62% of the ACC turned into calcite, 38% of them remained to be vaterite in the presence of 50 μg/ml rPfY2 after 4 h (Fig. 3.75c). And there was no vaterite found in neither of the control groups. What's more, this ratio contained almost unchanged after 24 h (Fig. 3.75f), suggesting that rPfY2 could stabilize the formation of vaterite and thus inhibit the transition from ACC to calcite.

With 50 mM magnesium, almost 20% of the total ACC transited to aragonite in both MBP and negative control group after 4 h of reaction, and the remaining crystals forms were still calcium carbonate (Fig. 3.76a, b). When with the addition of 50 μg/ml rPfY2 in the crystallization system, aragonite disappeared in deposited crystals and was replaced by calcium carbonate (Fig. 3.76c). After 24 h of crystallization, almost all of the ACC transited to aragonite in MBP and negative control

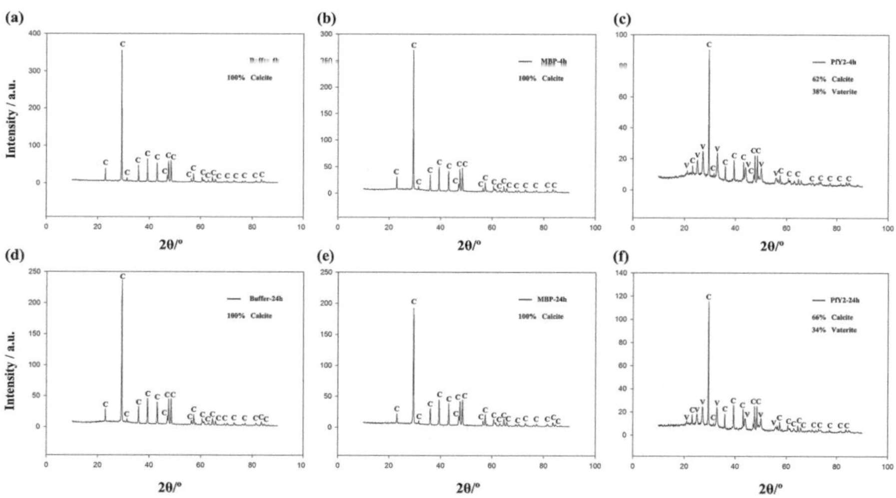

Fig. 3.75 The influences of rPfY2 on ACC transition to calcite
XRD was executed to detect the percentage of the crystals form ACC transformed to. The final dosage of rPfY2 was 50 μg/ml. Equal MBP was added to the reaction system as a control

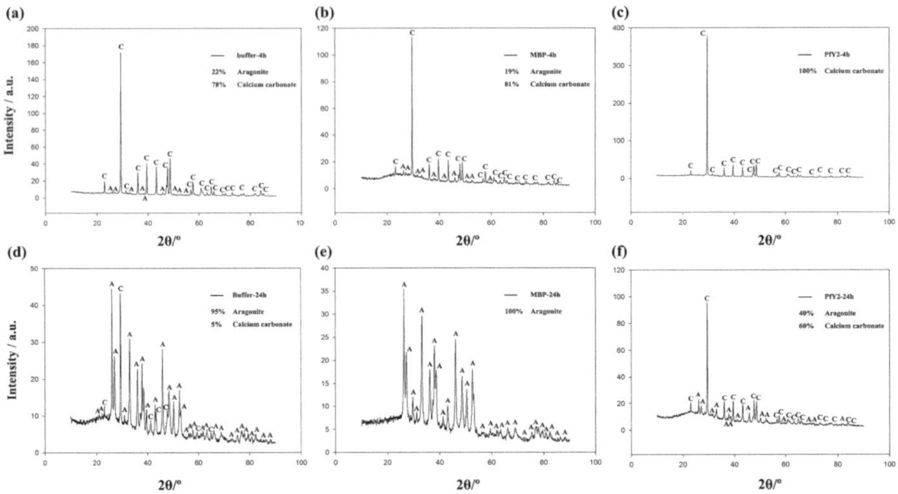

Fig. 3.76 The functions of rPfY2 during ACC transition to aragonite
XRD was executed to detect the percentage of the crystals ACC transited to. The final dosage of rPfY2 was 50 μg/ml. Equal MBP was introduced to the reaction system as control

group (Fig. 3.76d, e). Whereas, the addition of PfY2 repressed this process, and only 40% of the ACC transited to aragonite (Fig. 3.76f), suggesting that rPfY2 could stabilize calcium carbonate and repress the transition of ACC to aragonite.

PfY2 was detected as a potential matrix protein candidate by microarray of the global gene expression profiles in larval development of *P. fucata* in our previous study. Most known genes involved in biomineralization, including nacrein, ACCBP, Pearlin, KRMP, PfY2, etc., showed resemble expression model under which all of them increased significantly in juveniles [31]. Accordingly, the initial shell with typical structure began to shape and grow after entering this stage, which is closely relevant with the activation of the expression of these genes. A series of genomic and biochemical analyses suggested that PfY2 was highly expressed in both mantle edge and mantle pallial (Figs. 3.65b and 3.66b). On basis of previous studies, proteins of the EDTA-soluble fraction are thought to localize within calcium carbonate crystals and contribute to regulate their nucleation, phase, orientation, and morphology, while proteins of the EDTA-insoluble fraction are known to localize around calcium carbonate crystals and take part in the formation of the shell framework [4, 7, 77, 134–137]. PfY2 was found in EDTA-soluble segment from both prismatic layer and nacreous layer (Fig. 3.65e), indicating its dual roles during the shell formation procedure of *P. fucata* [58].

We analyzed the effects of PfY2 in *P. fucata* adults in vivo by shell notching assay and RNAi experiment. Notching brings destruction to shell, mimicking the shell formation stage in which biomineralization-related genes increased. The gene expression level of PfY2 was upregulated at 12 h after notching, arriving the peak at 36 h and then with a slight decline in the following hours. In general, the expression model of PfY2 suggested its participation and regulation on $CaCO_3$ precipitation. As shown in the rate determination experiment of $CaCO_3$ precipitation assay in vitro (Fig. 3.73) and the transformation of ACC to stable crystals (Figs. 3.75 and 3.76), PfY2 clearly influenced crystal deposition and repressed $CaCO_3$ crystallization. According to these performances, PfY2 presumably worked as a negative regulator in biomineralization. As shells cannot grow infinitely, there must be some "safety guard" that slows down shell growth, functioning as an inhibitor to repress crystal growth and control their morphologies or orientation. Our lab indicated extrapallial fluid proteins and some other matrix proteins, including PfN44, and played paradoxical roles to balance both $CaCO_3$ crystallization and shell biomineralization [15, 138]. Therefore, we speculated that the slight increase of the "negative regulator" PfY2 during shell notching mainly contributed to keep the balance of shell growth. RNAi experiment showed that after the inhibition of PfY2, the "negative regulator," many small nacreous tablets with the overgrowth of crystals appeared on the surface of nacreous layer (Fig. 3.69e–h). Since shell formation is under the regulation of many different matrix proteins, there is no doubt that they effect together to contain the balance of shell growth and reparation. Thus, the destroyed structures in prismatic layer might be the results of shell overgrowth or shell defects made by the disorder of the regulatory network of matrix protein.

On basis of ACC transition experiment, the content of vaterite in transition system was almost unchanged either 4 h or 24 h after the addition of rPfY2 (Fig. 3.75c, f), consistent with rPfY2's inhibition effects on $CaCO_3$ precipitation rate. Whereas, vaterite has not been detected in our animal model *P. fucata*, while there are some reports in other close-related species. A new morphology of vaterite was confirmed in lackluster pearls of *Hyriopsis cumingii* [139]. What's more, vaterites also formed in the unusual shells of *Corbicula fluminea* [140]. In these few reports, vaterite formation in mollusks is usually closely relevant with the unusual or abnormal biomineralization. Therefore, we infer that vaterite is a kind of transient phase in shell formation, and when its further transformation is stopped, it could be detected. This might be the reason why we could not detect vaterite in normal shell from *P. fucata*. And we could speculate that maybe PfY2's regulation on crystallization depends on its stabilization property of this intermediate vaterite phase.

Studies executed on $CaCO_3$ mineral interaction domains have found several crucial criteria which confirm a $CaCO_3$ mineral recognition sequence like the presence of hydrogen-bonding donor/acceptor residues (i.e., Ser, Tyr, Gln, Lys, Thr, Asn, Arg) suggesting putative sites for carbonate and/or mineral surfaces water interactions [83, 141–143]. rPfY2 is a protein rich in Asn, based on which we inferred that rPfY2 could interact with CO_3^{2-} at the first place to recruit ions for nucleation locally followed by binding with $CaCO_3$ microparticles, therefore inhibiting the growth and assemble of crystal particle. That might be the reason why A570 values of the rPfY2 group was lower compared with the control group (Fig. 3.73). This could also explain the microparticles assembled in calcite crystallization assay in Fig. 3.71g–n. In this hypothesis, rPfY2 might increase local dosage of ions to stimulate nucleation process when existed in a supersaturation system.

Once $CaCO_3$ nucleus formed and grown into a certain size, then rPfY2 bound to the certain surfaces of $CaCO_3$ particles, forming steric hindrance to repress the ions absorption process and stop the aggregation of $CaCO_3$ particles; therefore the crystal growth process was inhibited (Figs. 3.72 and 3.74a–f). What's more, it is necessary for us to assure that when the dosage of Ca^{2+} was low enough like 5 mM, rPfY2 could repress the growth of nearly all crystals (Fig. 3.74g–i). Whereas in increasing the dosage of rPfY2 in ACC transition experiments, the binding and inhibitory capabilities became stronger, which might restrict the transition from vaterite to calcite; thus the unstable vaterite phase could remain for a long time in Fig. 3.75.

Taken together, these results confirmed that the "safety guard" PfY2 that takes part in $CaCO_3$ crystallization worked as a negative regulator in this process and shell formation. This could also demonstrate that the expression of PfY2 was upregulated in juveniles rather than D-shaped stage according to its inhibition on ACC transition, which might result in the disruption of the formation of prodissoconch I. Thus, PfY2 plays a crucial role of sophisticated regulation in the shell formation and biomineralization.

In conclusion, PfY2, a novel matrix protein located in EDTA-soluble matrix from both prismatic and nacreous layers, plays crucial roles in shell biomineralization. Our results in this work indicate that PfY2 is an important matrix protein participating in the regulation of biomineralization. And it is meaningful to reveal the inhibition mechanism in shell formation and may give us new inspiration on the synthesis of artificial nacre.

3.7 Extrapallial Fluid Matrix Protein

3.7.1 A Novel Extrapallial Fluid Protein Controls the Morphology of Nacre Lamellae in the Pearl Oyster, Pinctada fucata

The nacre of mollusk shell is known of its superior mechanical abilities and precisely controlled biomineralization procedure. However, the question about how mollusks regulate the morphology of nacre lamellae remains unknown. Herein, a novel 38 kDa extrapallial fluid (EPF) protein, referred as amorphous calcium carbonate-binding protein (ACCBP), may partially solve this question [144]. Even though sequence analysis suggested ACCBP is a part of the acetylcholine-binding protein family, it actively contributed to the shell mineralization process. In vitro, ACCBP can repress the growth of calcite and induce the amorphous calcium carbonate formation. When ACCBP functions were inhibited in vivo, the nacre lamellae formed in a screw-dislocation pattern and few crystallinity $CaCO_3$ precipitated in the EPF. Crystal binding assays further suggested that ACCBP could identify different $CaCO_3$ crystal phases and crystal faces. With this ability, ACCBP could regulate the morphology of nacre lamellae by repressing the growth of unexpected aragonite crystal faces and meanwhile contain the stability of $CaCO_3$-supersaturated fluid by ending the nucleation and growth of calcite. What's more, the crystal growth inhibition ability of ACCBP was confirmed to be directly relevant to its acetylcholine-binding sites. Our results indicate that a "safeguard mechanism" of unexpected crystal growth is inevitable for shell microstructure formation.

3.7.1.1 Isolation and Characterization of ACCBP [144]

EPF has the ability to inhibit $CaCO_3$ crystalline, and it has been found that the organic components of the EPF can effectively inhibit $CaCO_3$ crystallization even at low dosage. According to our research on EPF, EPF can further delay the transition of amorphous calcium carbonate (ACC) which is the most unstable state of $CaCO_3$ to stable crystals [145, 146]. To explore the abnormal stability of ACC, we isolated ACC from the EPF by centrifugation and then decalcified the ACC using EDTA. SDS-PAGE analyses of the EDTA-soluble component showed a single 38 kDa protein band that didn't exist in the SDS-PAGE of the EPF (Fig. 3.77). This enriched protein was referred to as amorphous calcium carbonate-binding protein (ACCBP). A protein band with a molecular mass of about 76 kDa could be observed when the HPLC-purified ACCBP was subjected to a SDS-PAGE gel in unreduced conditions (Fig. 3.77). This result indicated that ACCBP might exist as a homodimer under native condition.

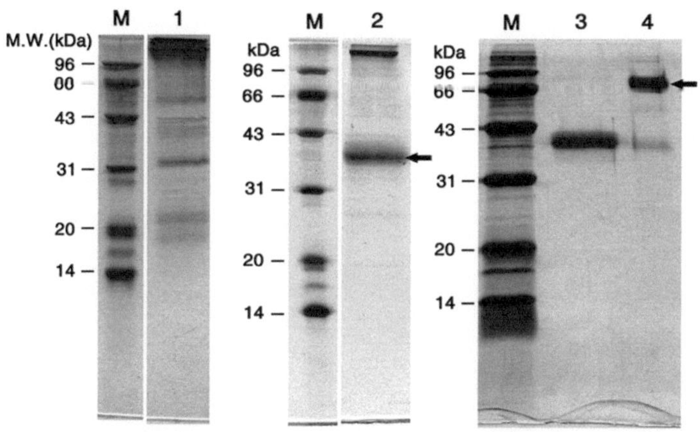

Fig. 3.77 SDS-PAGE of extrapallial fluid proteins and purified ACCBP. SDS-PAGE was subjected on 15% acrylamide gels. The gels were dyed with Coomassie Brilliant Blue. Lane 1, proteins of extrapallial fluid. Lane 2, proteins concentrated in ACC. ACCBP is highly gathered (arrow). Lanes 3 and 4, ACCBP purified by HPLC: lane 3, ACCBP under reduced conditions; lane 4, ACCBP under nonreduced conditions, and the ACCBP dimer can be found (arrow) [144]

Partial amino acid sequence of N-terminal of ACCBP (KCDYPEAKLL/KFLLDDYEKL/VRPVP) was obtained by Edman degradation and Western blotting. According to the N-terminal sequence, 5′-RACE and 3′-RACE were conducted to obtain the cDNA sequence of ACCBP (GenBank™ accession no. DQ473430). The 930 bp cDNA sequence encoded a 240-amino acid residues protein with a typical transcription termination sequence (AATAAA) which verified the integrity of the cDNA sequence (Fig. 3.78). The mature protein excluding signal peptide starts with Lys and has a calculated molecular mass of 24,389 Da.

The sequence analyses of ACCBP suggested that it is an acidic protein with a predicted pI of 4.56 and is rich in Asp and Glu (acidic residues) at nearly 16%. However, compared with many other matrix proteins separated from the shell matrix of bivalves like Caspartin, ACCBP is not rich in Asx and has high mol property of positively charged residues (Arg + His + Lys ~11%). In fact, the amino acid component of ACCBP is average in any residue except for Asp and Leu (~11%) which are the most abundant residues. The BLAST search against GenBank™ indicated that ACCBP has 30% sequence similarity with the extracellular ligand-binding region of the α7 nicotinic acetylcholine receptors (nAChR) of *Drosophila melanogaster* (Fig. 3.78). ACCBP has the residues sequence which is conserved in nAChRs, including those residues suggested to be involved in acetylcholine (ACh)-binding protein, and the disulfide bond is formed by adjacent cysteine residues Cys195 and Cys196, as well as the conserved cysteine loop region constituted between Cys133 and Cys145 [147]. There have been many reports about the soluble analog of the extracellular domain in the α7 nAChR [148, 149], like the ACh-binding protein (AChBP) separated from the glia cells of *Lymnaea stagnalis* [148]. AChBP is reported to modulate neuronal synaptic transmission actively. Other than the α7 nAChR and AChBP, ACCBP could form a homodimer by a disulfide bond between Cys2 residues of two proteins (data not shown) and has two predicted glycosylation sites at Asn29 and Asn184.

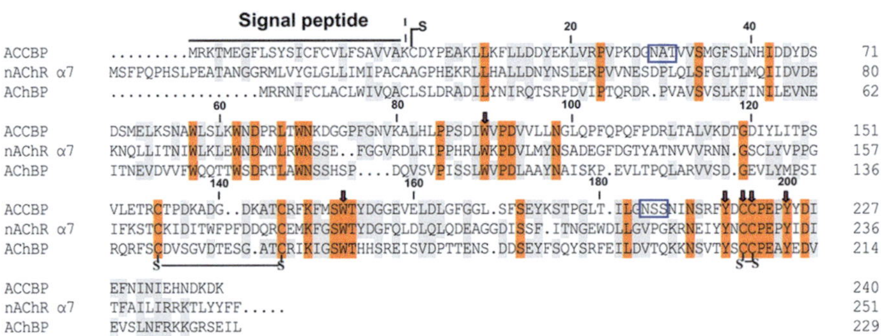

Fig. 3.78 Amino acid sequences of ACCBP and alignment with the acetylcholine-binding region of the α7 nAChR and AChBP. The amino acid sequences of ACCBP were compared with that of the acetylcholine-binding region of the α7 nAChR (*D. melanogaster*) and AChBP (*Lymnaea stagnalis*). Orange shaded, residues conserved in the nAChR family; gray, homologous residues of the three proteins; purple arrows, residues considered to participate in acetylcholine binding; blue box indicates potential N-glycosylation sites

Expression of ACCBP in different tissues reveals that ACCBP exists mainly in the extrapallial fluid and hemolymph with some detected in the adductor muscle (Fig. 3.79a). Although qRT-PCR analyses of ACCBP gene expression suggest that ACCBP is expressed in the mantle, the viscus, and the adductor muscle (Fig. 3.80), there is no positive signal observed in the viscus and the mantle by Western blotting (Fig. 3.79a). On the contrary, ACCBP was detected weakly by Western blotting of extracts from mantle and viscus which were enriched with ACC (data not shown). This result might be explained as follows: ACCBP is secreted out to the body fluid once ACCBP is synthesized in these tissues which cause low content of ACCBP to be detected on Western blotting with equal quantities of whole extract proteins.

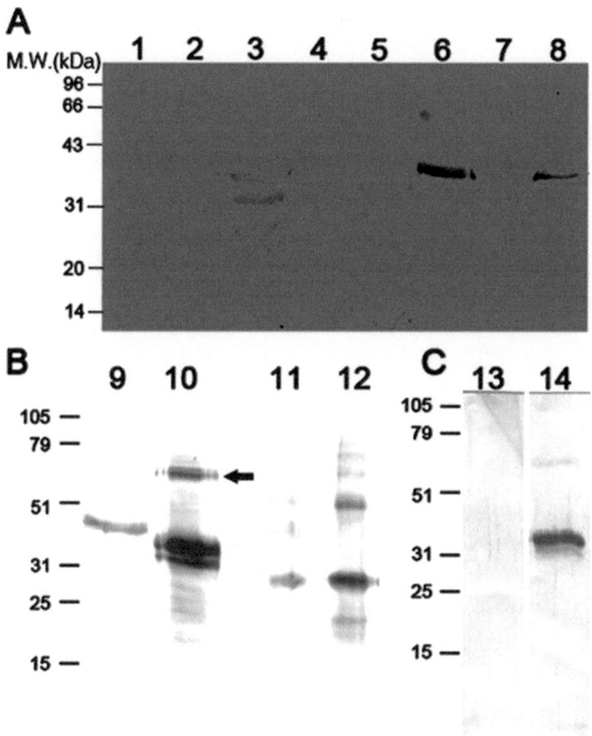

Fig. 3.79 Western blotting analyses of ACCBP and rACCBP. SDS-PAGE was subjected on 12% acrylamide gels. (**a**), Tissue location analysis of ACCBP. Lane 1, soluble extract matrix of the viscus; lane 2, soluble extract matrix of the mantle; lane 3, soluble extract matrix of the adductor muscle; lane 4, EDTA-soluble matrix of shells; lane 5, EDTA-insoluble matrix of shells; lane 6, total proteins extracted from the EPF; lane 7, soluble extract matrix of the gill; lane 8, total proteins extracted from the hemolymph. (**b**), Blast of ACCBP and rACCBP. Lane 9, ACCBP. Lane 10, rACCBP; arrow indicates unreduced rACCBP homodimer. Lane 11, TFMS-treated ACCBP. Lane 12, TFMS-treated rACCBP. (**c**), Analyses of ACCBP and rACCBP with biotin-labeled concanavalin A. Lane 13, ACCBP is not identified by biotin-concanavalin A; lane 14, rACCBP is identified by biotin-concanavalin A

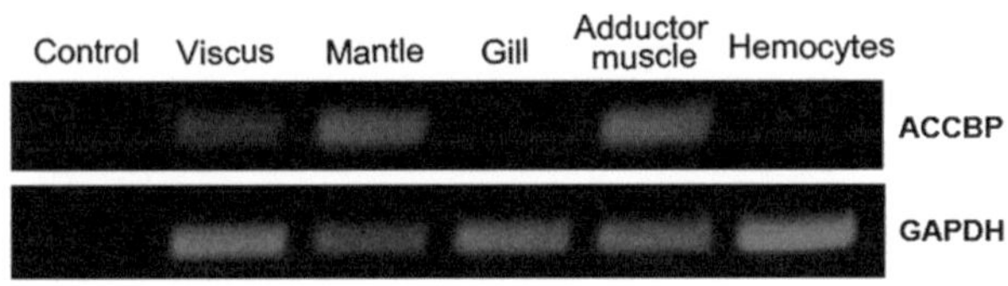

Fig. 3.80 RT-PCR analyses of ACCBP in different tissues. RT-PCR was executed on RNA extracted from viscus, mantle, gill, adductor muscle, and hemocytes. GAPDH was detected as the positive control. Specific primer designed for ACCBP and GAPDH are shown under "Experimental Procedures"

3.7.1.2 Effect of ACCBP on $CaCO_3$ Crystallization In Vitro and Interaction Between ACCBP and Calcite

To investigate whether ACCBP contributes to stable ACC in the EPF, we conducted crystal growth experiments in vitro. The formation of calcite can totally inhibit with just 0.2 μM ACCBP. The infrared spectra of all reagents show broad v1 and v2 absorption peaks of ACC at 864 and 1070 cm^{-1}, respectively. The v3 band split is around 1450 cm^{-1}, which suggests that only ACC is present (Fig. 3.81f). S.E. images displayed that the ACC precipitate was controlled by uniform spherical particles of 200–300 nm diameters, though there are particles as small as 50 nm in diameter (Fig. 3.81a, b). ACC will transform to a more stable crystal phase rapidly in water via dissolution of itself and formation of the stable phase under normal conditions [145, 146]. The stability of ACC can be maintained by delaying the dissolution of ACC or inhibiting the nucleation and growth of the stable crystals [150, 151]. The concentration of Ca^{2+} was adjusted to 8 mM in this experiment which meant one ACCBP molecule needs to stabilize 40,000 $CaCO_3$ molecules to stabilize ACC by inhibiting the dissolution of ACC like polycarboxylates and covering the whole surface of ACC particles [151]. The average molecular mass of polycarboxylates is ~70 kDa, resembling to the molecular mass of ACCBP dimer. Compared with the function ratio of polycarboxylates to Ca^{2+} (~1:1750), the function of ACCBP was far more effective. The stabilization mechanism of ACC by ACCBP was further studied with calcite growth inhibition experiment. As Fig. 3.82a, b had showed, ACCBP could rapidly inhibit the newly emerging calcite growth which suggests ACCBP could inhibit the growth of the more stable crystal of $CaCO_3$ and stabilize the intermediate state ACC consequently.

The interaction between ACCBP and calcite was further analyzed with a series of crystal binding assays. RA-α-Bgt (α-Bgt, a specific blocker of nAChRs [152]) was introduced to visualize ACh-binding site of ACCBP. Incubated RA-α-Bgt-labeled ACCBP with freshly synthesized calcite in a solution of saturated $CaCO_3$ has been identified by Raman spectra. BSA and RA-α-Bgt at the same dosages were used as a control. Fluorescence distributed on the total surface of the calcite in the ACCBP added groups at 1 h after incubation (Fig. 3.83a) but wasn't detected on those of the control groups (Fig. 3.83f). To affirm the above results, natural ACCBP was labeled with rhodamine directly and reacted with $CaCO_3$ crystals as above. As Fig. 3.83b

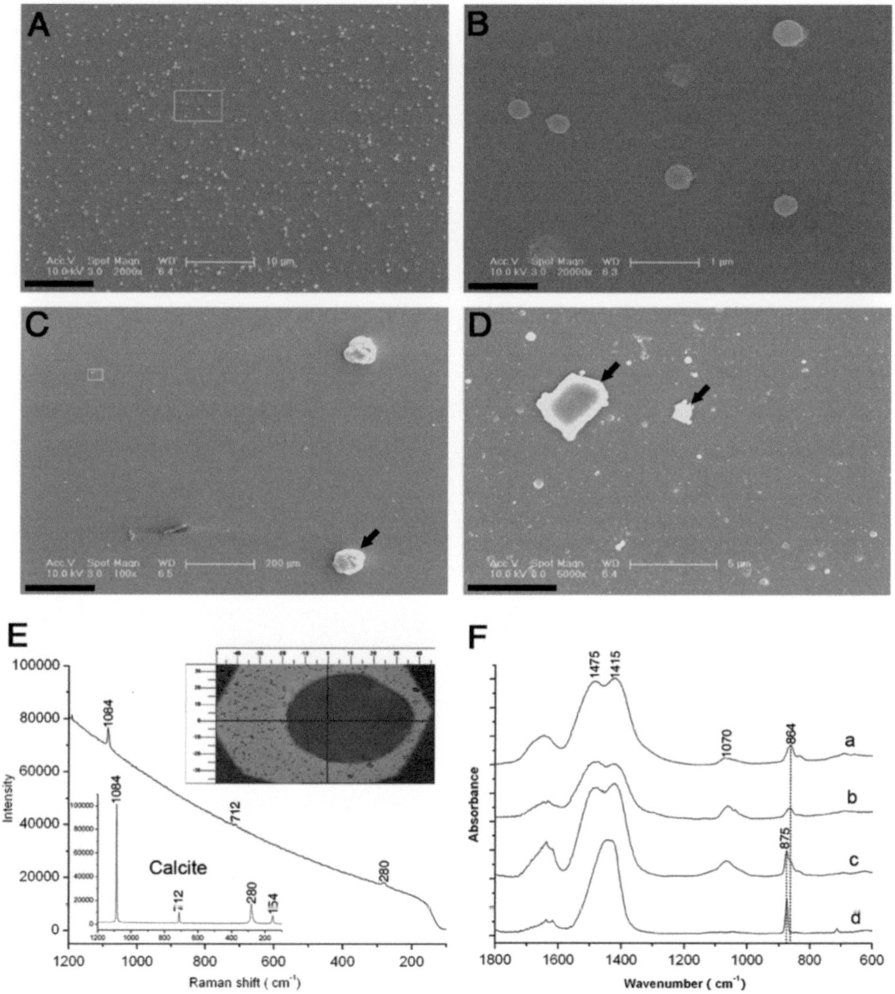

Fig. 3.81 In vitro crystal growth assays. (**a**), When 0.2 μM ACCBP was added, only spheres of amorphous calcium carbonate were found in S.E. images. Bar scale, 10 μm. (**b**), Amplification of the box in (**a**). Bar scale, 1 μm. **C**, When the 0.2 μm ACCBP and 0.4 μm α-Bgt were added, different morphologies were found. Some crystals, around 50 μm in diameter, indicated (arrow). Bar scales, 200 μm. (**d**), Amplification of the box in (**c**). The shape of some crystals was obviously different from typical ACC and much bigger in diameter (arrows). Bar scale, 5 μm. (**e**), Raman spectrum analyses of the crystal pointed with an arrow in (**c**) before S.E. images. The inset spectrum is the Raman spectrum of formed calcite. The inset figure is the crystal represented under white light before motivated with the laser. (**f**), Infrared spectrum of the segments of in vitro crystal growth experiments. *a*, 0.2 μm with ACCBP; *b*, with 0.2 μm ACCBP and 0.4 μm BSA; *c*, with 0.2 μm ACCBP and 0.4 μm α-Bgt; *d*, with 0.2 μm BSA

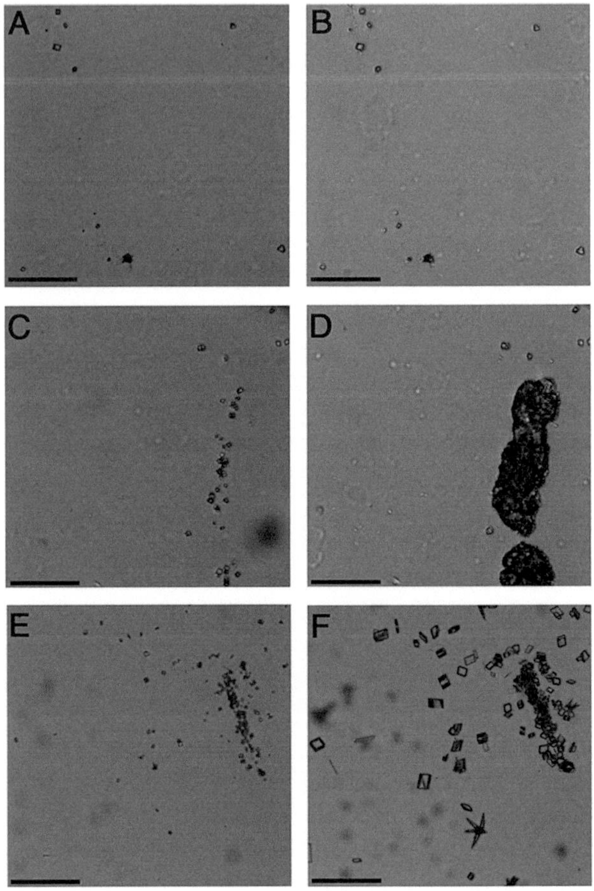

Fig. 3.82 Crystal growth inhibition assays. (**a**, **c**, and **e**), Calcite seeds had grown on a glass slide. Bar scales, 100 μm. (**b**), The same visual field with A 24 h after adding 0.2 μm ACCBP. Bar scales, 100 μm. (**d**), The same visual field with C 24 h after adding 0.2 μm ACCBP and 0.4 μm α-Bgt. Bar scales, 100 μm. (**f**), The same visual field with E 24 h after adding 0.2 μm BSA. Bar scales, 100 μm

had shown, same binding patterns as the RA-α-Bgt-labeled ACCBP occurred. Taken together, it appears that ACCBP can inhibit the growth of calcite by coating its whole surface.

3.7.1.3 Functions of ACCBP in Nacre Mineralization and Interaction Between ACCBP and Aragonite

ACCBP is expressed by the mantle and locates mainly in the EPF (data not shown). As nacre formation is a crystalline process that occurs in the EPF, the functions of ACCBP as a $CaCO_3$ crystalline inhibitor became an interesting question in this process. Purified polyclonal antibody against ACCBP (anti-ACCBP) was then injected into the extrapallial space of *P. fucata* at quantity of 0.5 g or 1.5 g of protein per gram weight of oyster every day. The same dosage of preimmune serum of rabbit was injected as a control. The oysters were collected and the inner surfaces of the

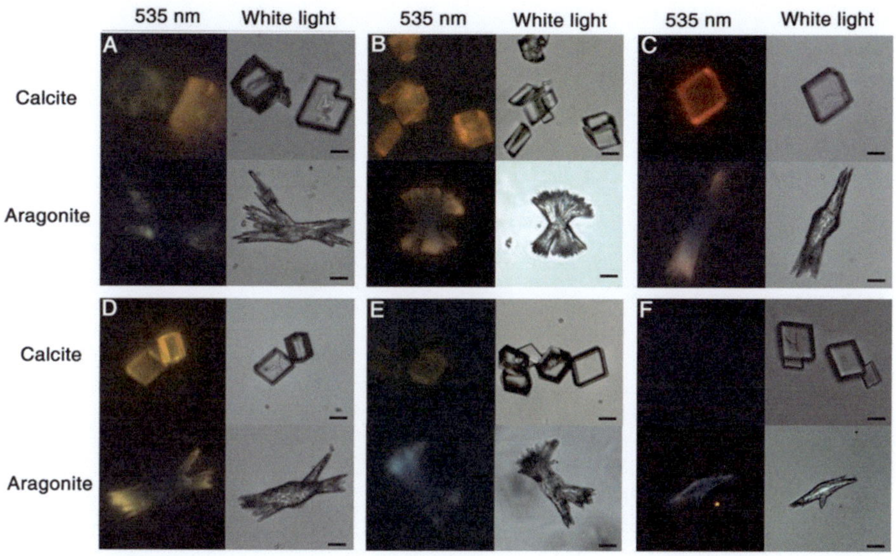

Fig. 3.83 Crystal binding assays of ACCBP and rACCBP. (**a**), Binding models of RA-α-Bgt-labeled ACCBP. Bar scales, 10 μm. (**b**), Binding models of RA-ACCBP. Bar scales, 10 μm. (**c**), Binding models of RA-rACCBP. Bar scales, 10 μm. (**d**), Binding models of RA-α-Bgt-labeled rACCBP. Bars, 10 μm. (**e**), Binding models of rACCBP labeled with both α-Bgt and RA-α-Bgt (molar ratio 9:1). Bar scales, 10 μm. (**f**), No fluorescence was found when adding equimolar BSA and RA-α-Bgt. Bar scales, 10 μm

nacreous layers were observed by S.E. 5 days later after injection. In low dosage group, the nacreous lamellae were obviously overgrown as a screw-dislocation pattern (Fig. 3.84e and f). In high dosage group, there is a large amount of low crystallinity precipitation on the nacreous lamellae (Fig. 3.84i, j). Elemental analyses showed that these precipitations are $CaCO_3$ (data not shown). The lamellae surface appeared rough (Fig. 3.84m–p) and jagged at the margins (in the high dosage group, Fig. 3.84o and p) compared with the untreated oyster in the control groups (Fig. 3.84a and b); these changes were unconspicuous compared with that in the antibody-treated groups (Fig. 3.84e, f, i, and j). These results indicate that lamellar aragonite growth was disturbed to some extent after the function of ACCBP was inhibited. What's more, the $CaCO_3$-supersaturated EPF changed into so unstable that a large amount of low crystallinity $CaCO_3$ occurred.

To identify the abnormal structures of nacre from the anti-ACCBP-injected groups which were directly related to the inhibition of ACCBP functions, the nacre from the untreated oysters (N-nacre), the low dosage oysters (L-nacre), and the high dosage oysters (H-nacre) were incubated with RA-ACCBP. As Fig. 3.84c had shown, fluorescence cannot be detected on the inner surface of N-nacre. On the contrary, strong fluorescence was observed on the nacre from L-nacre and H-nacre, but the location was restricted to the abnormal structures on the nacre (Fig. 3.84g, h, k, and l). These results suggest that the abnormal structures are bound by ACCBP

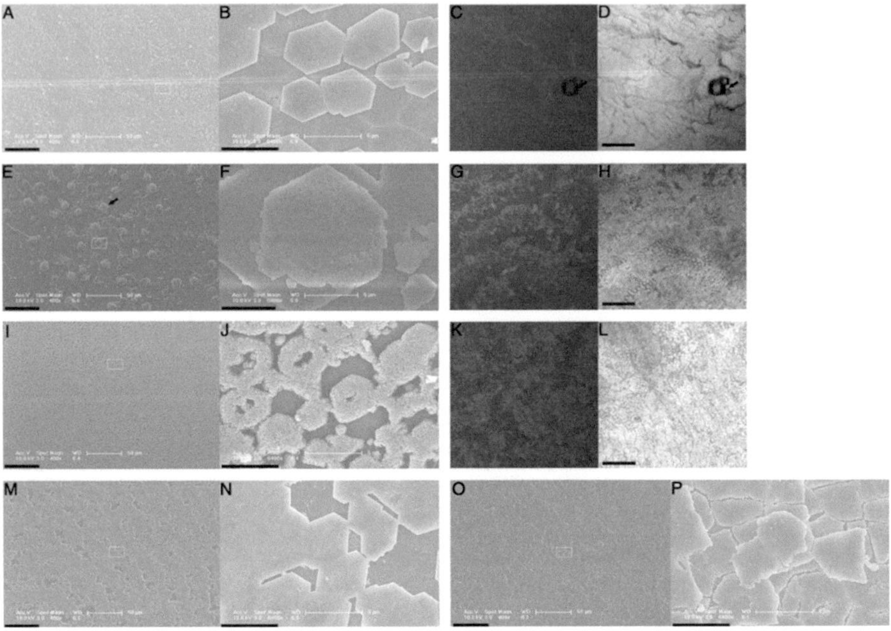

Fig. 3.84 The function of antibody against ACCBP on nacre lamellae growth. (**a**), S.E. images of the inner surface of normal shells nacre from the oyster, *P. fucata* (*N* nacreous layer). The stair-like growth model of nacre can be observed. Bar scales, 50 μm. (**b**), Amplification of the box in (**a**). Bar scale, 5 μm. (**c**), N-nacre and formed calcite were mixed with RA-ACCBP and detected under a fluorescence microscope. ACCBP could not bind to the nacre surface, while could bind to the synthesized calcite (arrow indicates). (**d**), The accordant image of (**c**) with white light. Bar scale, 50 μm. (**e**), S.E. image of the inner surface of nacre from the low dosage group (L-nacre). The lamellae are significantly larger than normal and screw-dislocation patterns can be seen (arrow indicates). Bar scale, 50 μm. (**f**), Amplification of the box in (**e**). Bar, 5 μm. (**g**), L-nacre was reacted with RA-ACCBP. When motivated at 535 nm, strong orange fluorescence shined on the disordered structures. (**h**), The corresponding image of G with white light. Bar scale, 50 μm. (**i**), S.E. image of the inner surface of nacre from the high dosage group (H-nacre). Many low crystallinity $CaCO_3$ formed on the nacre lamellae. Bar scale, 50 μm. (**j**), Amplification of the box in (**i**). Bar scale, 5 μm. (**k**), H-nacre was added with RA-ACCBP. When motivated at 535 nm, fluorescence arose on the low crystallinity $CaCO_3$ segment while not arising on the normal lamellae. (**l**), The accordant image of (**k**) with white light. Bar scale, 50 μm. (**m**), S.E. image of the inner shell surface from the control group of low dosage. (**n**), Amplification of the box in (**m**). Bar scale, 5 μm. (**o**), S.E. image of the inner shell surface from the high dosage control group. The stair-like growth model was inhibited. Bar scale, 50 μm. (**p**), Amplification of the box in (**o**). Margins of the lamellae became zigzag. Bar scale, 5 μm

and growth of nacre is inhibited at an early stage. However, ACCBP does not influence normal nacre lamellae growth. This result also explains why the quantity of ACCBP of the shell matrix is so few that no ACCBP detected via Western blotting of total matrix proteins (Fig. 3.84a). The nacre binding assays suggested that ACCBP could interact with aragonite (the crystal phase in nacre) rather than calcite. Hence, we labeled the ACCBP as above and interacted with synthesized aragonite.

Interestingly, the binding of ACCBP on aragonite was mainly limited to the needle-like region rather than on the surface of calcite (Fig. 3.84a–c). The phenomenon suggests that ACCBP could recognize different $CaCO_3$ crystal forms and crystal faces. ACCBP might be able to inhibit undesired crystals growth and thereby modify the surface of nacre lamellae with this capacity.

3.7.1.4 ACCBP Crystal Growth Inhibition Capacity Related to the Acetylcholine-Binding Site

ACCBP has an ACh-binding site same as that of α7 nAChRs and AChBP, but the functions of ACh-binding site were in a totally different way from them. It is of interest to explore the role of the ACh-binding site in function of ACCBP. In crystal binding assays, after ACCBP was incubated with calcite or aragonite and the RA-α-Bgt was added 1 h later, no fluorescence was detected (data not shown). This observation indicates that, when ACCBP adheres onto a crystal, the ACh-binding site is not accessible to the α-Bgt; therefore, the ACh-binding site was important for ACCBP function. In further studies, when crystal growth in vitro assays were performed with a compound of ACCBP and α-Bgt at 1:2 at molar ratio, S.E. images suggested that the $CaCO_3$ precipitates mainly remained amorphous without uniform size as the crystals formed with addition of ACCBP alone (Fig. 3.81c, d). Many nonspherical crystals much bigger than typical ACC exist. There are also a few huge spherical crystals with diameters of ~50 μm. Raman spectra of these huge crystals exhibited characteristic peaks at 280 cm^{-1}, 712 cm^{-1}, and 1084 cm^{-1} of calcite (Fig. 3.81e). The peaks are broad and weak and the baseline is high because of the low crystallinity. FTIR spectrum of the precipitate shows the characteristic v2 absorption peak of calcite at 875 cm^{-1}, but the rest of the spectrum shows typical bands of ACC (Fig. 3.81f). This spectrum suggests the presence of multiple crystal polymorphs. There was no peak shift observed when BSA replaced α-Bgt (Fig. 3.81f). In a crystal growth inhibition assay, α-Bgt inhibits the capacity of ACCBP of inhibition of crystal growth (Fig. 3.82c, d). Therefore, α-Bgt has a function on the ability of ACCBP to affect crystal growth. What's more, α-Bgt has no obvious influence on the binding of RA-rACCBP or RA-ACCBP to crystals in an experiment like that of Fig. 3.83 (data not shown). All in all, these results suggest that the ACh-binding site does not influence the binding of ACCBP with crystals but has functions in crystal growth inhibition.

Although rACCBP could interact with $CaCO_3$ crystals in the same way as ACCBP (Fig. 3.83c) and decreases the calcite growth rate obviously (Fig. 3.85d, e), it cannot inhibit the formation of calcite effectively (Fig. 3.85a, b). These results indicate that the effect of the ACh-binding site of rACCBP may be prevented. What's more, in the crystal binding experiments, we found that, even mixed with RA-α-Bgt for as long as 24 h, the RA-α-Bgt-labeled ACCBP experiment group (Fig. 3.83a) was heterogeneous and the fluorescence was much weaker than that of rACCBP group (Fig. 3.83d), which was incubated with RA-α-Bgt for only 1 h with the same concentration. As mentioned above, α-Bgt has no obvious effect on the

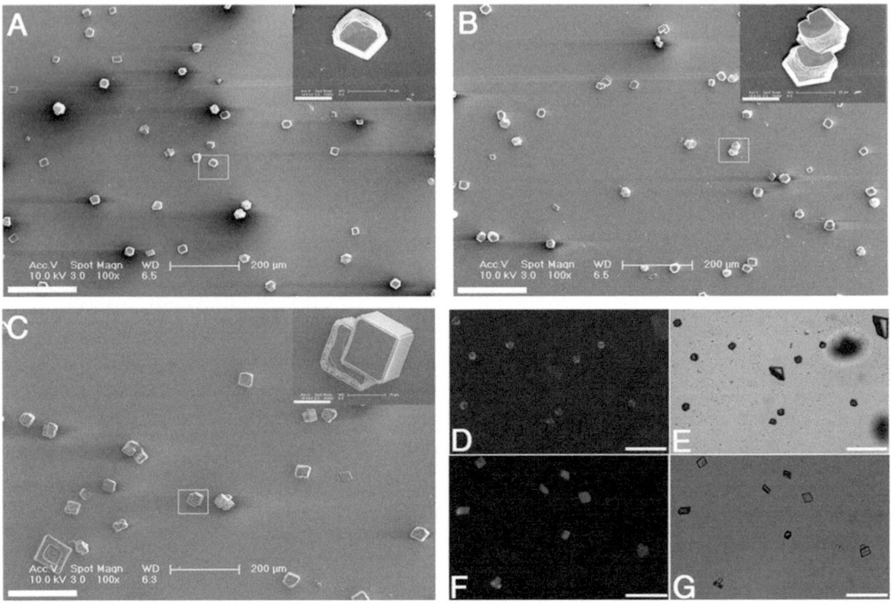

Fig. 3.85 In vitro crystal growth experiments with rACCBP. (**a**), S.E. micrograph of $CaCO_3$ crystal growth in the presence of 0.2 μm rACCBP. Bar, 200 μm. The inset is an enlargement of the box. Bar, 20 μm. (**b**), S.E. micrograph of $CaCO_3$ crystal growth in the presence of 0.2 μm ACCBP and 0.4 μm α-Bgt. Bar, 200 μm. The inset is an enlargement of the box. Bar, 20 μm. (**c**), S.E. micrograph of $CaCO_3$ crystal growth in the presence of 0.2 μm BSA. Bar, 200 μm. The inset is an enlargement of the box. Bar, 20 μm. (**d**), Fluorescence micrograph of $CaCO_3$ crystal growth in the presence of 0.1 μm rACCBP labeled with RA-α-Bgt. The crystals containing bound rACCBP (with orange fluorescence) are smaller than crystals lacking bound rACCBP (without fluorescence). Bar, 100 μm. (**e**), The corresponding image of D under white light. (**f**), Fluorescence micrograph of $CaCO_3$ crystal growth in the presence of 0.1 μm RA-α-Bgt and 0.1 μm BSA. The sizes of the crystals are nearly uniform, and no fluorescence can be seen. Bar, 100 μm. (**g**), The corresponding image of (**f**) under white light

crystal binding abilities of ACCBP, so the heterogeneous and weak fluorescence of RA-α-Bgt-labeled ACCBP might be caused of ineffective labeling of ACCBP with RA-α-Bgt compared with the labeling of rACCBP. To confirm this prediction, the crystal binding assays were conducted with rACCBP labeled with both RA-α-Bgt and α-Bgt (molar ratio ~9:1). The results suggested that the fluorescence of partially labeled rACCBP group (Fig. 3.83e) was much weaker and heterogeneous compared with the fully labeled group, while resembling to RA-α-Bgt-labeled ACCBP group. Taken together, the weak fluorescence of ACCBP group labeled with RA-α-Bgt might be related to lower binding ability of RA-α-Bgt to the natural protein.

What's more, a significant difference between rACCBP and ACCBP is their different migration patterns on SDS-PAGE. As Fig. 3.79b had shown, the molecular mass of rACCBP is about 34 kDa which is smaller than ACCBP. When natural rACCBP and ACCBP were treated by TFMS, the molecular masses of these molecules were reduced to the same size (Fig. 3.79b). Therefore, the major post-translational modifications of native ACCBP are glycosylation, while concanavalin A recognizes rACCBP only but not native ACCBP (Fig. 3.79c). What's more, the two bands of rACCBP were shown because of the different glycosylation on Asn184 (data not shown). The results here suggest that both the component of the posttranslational modifications and the size of rACCBP and ACCBP are different. Some modifications of the native protein might cause the structure of the ACh-binding site of ACCBP different from that of the rACCBP and less accessible to α-Bgt. The different structure of the ACh-binding site consequently leads to the different inhibition activity in the crystal growth of native ACCBP and rACCBP.

3.7.2 Structural Characterization of Amorphous Calcium Carbonate-Binding Protein: An Insight into the Mechanism of Amorphous Calcium Carbonate Formation

ACC (amorphous calcium carbonate) plays an important role in biomineralization process for its function as a precursor for calcium carbonate biominerals. However, it is unclear how biomacromolecules regulate the formation of ACC precursor in vivo. In the present study, we used biochemical experiments coupled with bioinformatics approaches to explore the mechanisms of ACC formation controlled by ACCBP (ACC binding protein) [153]. Size-exclusion chromatography, chemical cross-linking experiments, and negative staining electron microscopy reveal that ACCBP is a decamer composed of two adjacent pentamers. Sequence analyses and fluorescence quenching results indicate that ACCBP contains two Ca^{2+}-binding sites. The results of in vitro crystallization experiments suggest that one Ca^{2+}-binding site is critical for ACC formation and the other site affects the ACC induction efficiency. Homology modeling demonstrates that the Ca^{2+}-binding sites of pentameric ACCBP are arranged in a fivefold symmetry, which is the structural basis for ACC formation. To the best of our knowledge, this is the first report on the structural basis for protein-induced ACC formation, and it will significantly improve our understanding of the amorphous precursor pathway.

3.7.2.1 ACCBP Is a Decamer Consisting of Two Adjacent Pentamers

To investigate the mechanism of ACC formation with the addition of ACCBP, the recombinant ACCBP of oligomeric state, which is as active as the native protein (data not shown), was studied. As Fig. 3.86a had shown, size-exclusion chromatography indicates that the elution volume of ACCBP is 7.68 ml, meaning that the apparent molecular weight of ACCBP is about 270 kDa. The calculated molecular mass of monomeric ACCBP is about 27.1 kDa, and ACCBP should work as a decamer accordingly [153].

SDS/PAGE analyses showed that the apparent molecular weight of monomeric ACCBP is about 30 kDa (Fig. 3.86b). After cross-linking assays conducted with 0.02% glutaraldehyde for 10 and 30 min, there are polymers of ACCBP forms with apparent molecular weights of approximately 300, 150, 120, 90, and 60 KDa and consistent with a decamer, pentamer, tetramer, trimer, and a dimer (Fig. 3.86c). There are two types of dimers: one type is probably formed via cross-linking and the other type is formed by an intermolecular disulfide bond of the C2 residues according to our previous study. Cross-linking experiments performed with 0.1% glutaraldehyde of 10 min show only one higher molecular weight form of about 300 kDa, suggesting that decamer is a stable state. As the dosage of ACCBP in this assay is only 50 μg/ml which was detected by Western blots, it is reported that the

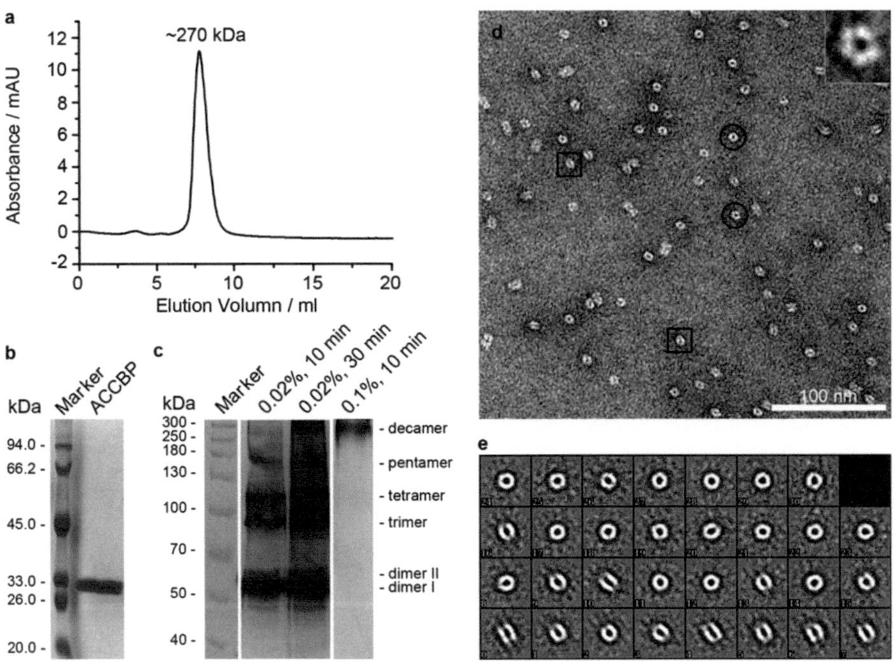

Fig. 3.86 The oligomeric condition of ACCBP. (**a**) Size-exclusion chromatography indicating that the putative molecular mass of ACCBP is about 270 kDa. *mAU* milli-absorbance units. (**b**) SDS-PAGE of recombinant ACCBP. Protein marker, protein ladder (Fermentas SM0441). (**c**) Chemical cross-linking of ACCBP analyzed by Western blotting. Marker, protein ladder (Thermo Scientific 26625). The higher molecular mass of ACCBP species and their speculated identities are shown. In (**b**) and (**c**), the molecular mass is kDa is indicated on the left-hand side. (**d**) Negative stain EM image of ACCBP protein. The inset image is a classic particle in a fivefold symmetry. The circles show the top view of ACCBP and the squares show the side view of ACCBP. (**e**) Classification of the selected particle images of ACCBP [153]

higher molecular weight species not the high concentration and purified on behalf of more stable protein–protein interactions. According to these combined data, it is summarized that ACCBP exists as a stable oligomer in solution, mainly probably as a decamer.

As ACCBP is a member of the acetylcholine binding proteins which are always existed as homopentamers [154], we predicted that ACCBP also assembles as a homopentamer in organisms, while the chemical cross-linking data and size-exclusion chromatography show that ACCBP exists as a decamer. The negative staining EM images helped to explain this difference. As raw negative staining EM images had shown, over 80% of the particles are rings with an inner diameter of about 2 nm and an outer diameter of about 8 nm, with about 10% of the particles composed of two layers, and few particles having lower image contrast indicate a pentagon with five subunits (Fig. 3.86d). By dividing these particle images into 30 groups, we found that there are two distinct categories (Fig. 3.86e). In considering the homogeneity of ACCBP indicated by the single sharp peak in SEC, the two distinct categories mainly occurring due to different orientations of the protein on the carbon film are conceivable. The ringlike classes and the dual layer classes correspond to a top and a side view of ACCBP. Therefore, these results suggest that ACCBP is a decamer composed of two adjacent pentamers.

3.7.2.2 ACCBP Contains Two Ca^{2+}-Binding Sites Which Are Important for Its Mineralization Activity

ACCBP should interact with Ca^{2+} ions strongly to induce ACC formation oligomeric. Thus, we then analyzed the Ca^{2+}-binding sites of ACCBP. An amino acid sequence with aspartic acid residues isolated by one neutral amino acid may compose part of Ca^{2+} -binding sites [108]. Sequence analyses indicate two potential Ca^{2+}-binding sites with the domain of D-X-D (X, neutral residue). The first predicted Ca^{2+}-binding site is composed of amino acid residues 44–48 (DYDSD) and the second site is composed of residues 139–141 (DGD) (Fig. 3.87). The first

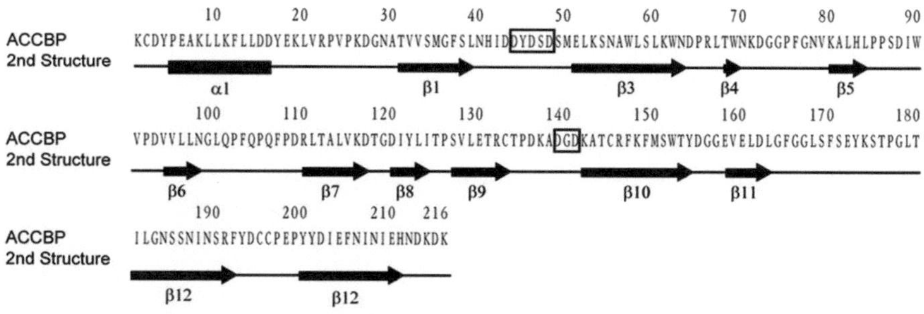

Fig. 3.87 The predicted Ca^{2+}-binding sites in the sheet–loop–sheet motifs (α, α-helix; β, β-strand) which is indicated under the sequence. The predicted Ca^{2+}-binding sites are indicated by square boxes [153]

predicted Ca^{2+}-binding site located at the N-terminal disordered sequence of ACCBP, which is suggested to be a site for mineralization activity [155].

The secondary structure of ACCBP was predicted by the PSIPRED program [156]. The predicted structure of ACCBP is composed of 12 β-sheets and an α-helix (Fig. 3.87) with the α-helix located in the N terminus. Both of the two Ca^{2+} -binding sites are located in abnormal sheet–loop–sheet motifs instead of classical helix–turn–helix motifs. The first site appears in the loop region between sheets β1 and β2, and the second site locates in the loop region intermediated in sheets β8 and β9.

We constructed two site-directed mutations, D46S and D139S, to investigate whether the predicted binding sites are exactly Ca^{2+}-binding sites. Mutations of an aspartic acid to a serine residue replace an uncharged hydroxyl by a negatively charged carboxyl. These two site-directed mutations then were expected to perturb the Ca^{2+}-binding capacity of the two sites. The intrinsic fluorescence measured by rACCBP, D139S, and D46S mutant indicates that D46S single mutation extremely alters the intrinsic fluorescence of the protein; however, the D139S mutation makes only slight change of the intrinsic fluorescence (Fig. 3.88a). The reason of large change in intrinsic fluorescence by the D46S mutant is that the amino acid next to Asp46 is Tyr45, which is fluorophore sensitive to its microenvironment [157]. Another supposed reason of this difference is that D46S mutation alters the oligomeric state of ACCBP, exposing tryptophan amino acid residues to the solution. What's more, aspartic acid consistently coats the surface of protein because of its negatively charged R group along with the mutation of a residue on the surface hardly affecting the tertiary structure obviously. It is suggested that neither of the two mutations cause a significant change in ACCBP tertiary structure. The presence of 1 mM Ca^{2+} could quench the intrinsic fluorescence of rACCBP by 26% (Fig. 3.88b), which is almost the same level of fluorescence quenching induced by 10 mM Ca^{2+}, indicating that ACCBP has a strong Ca^{2+}-binding ability. With 1 mM Ca^{2+}, the intrinsic tryptophan fluorescence of the D46S mutant was quenched by 4% (Fig. 3.88c), while the quenching level induced by 10 mM Ca^{2+} is about 7%. The intrinsic tryptophan fluorescence of the D139S mutant was quenched by the addition of 1 mM Ca^{2+} (Fig. 3.88d), whereas the level of quenching induced about 33% by 10 mM Ca^{2+} is 46%. These extremely different levels of quenching indicate that both of the identified sites are Ca^{2+}-binding sites.

In vitro crystallization assays with the site-directed ACCBP mutants indicate that both of the Ca^{2+}-binding sites are significant for the function of ACCBP. Micro-Raman spectra and SEM images show that nearly all the spherical precipitates formed in the presence of rACCBP are ACC with diameter of about 25 μm (Fig. 3.89). The spherical deposits formed with ACCBP look like vaterite morphologically under the SEM. Whereas, micro-Raman spectra suggest that these precipitates are exactly ACC on the basis of the Raman spectra of synthetic ACC and synthetic vaterite (data not shown). The results of XRD analyses indicate that the precipitate induced by ACCBP is a component of ACC and few calcites with no vaterite are detected (data not shown). In our previous study, we have proved that

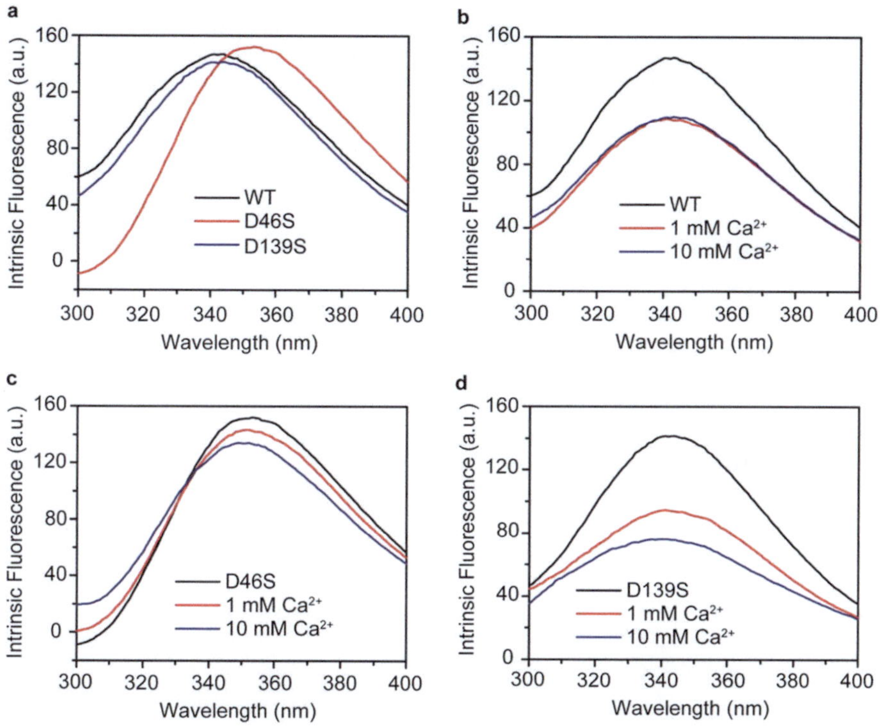

Fig. 3.88 Functions of the D46S and D139S mutations of the tertiary structure and the Ca^{2+}-binding ability of ACCBP. (**a**) Intrinsic fluorescence of wild-type (WT) ACCBP protein and its mutations. (**b**) Fluorescence quenching assays executed with 5 μM wild-type ACCBP. (**c**) Fluorescence quenching assays conducted with 5 μM D46S ACCBP protein. (**d**) Fluorescence quenching assays executed with 5 μM D139S ACCBP. *a.u.* arbitrary units

vaterite-like spherical precipitates formed with purified ACCBP isolated from extrapallial fluid are ACC [144]. Therefore, we inferred that rACCBP only stabilizes ACC and the activity is the same as the native protein. What is noteworthy is that when these ACC spherulites transform into vaterite, there will be morphological changes in the crystallization experiments in vitro if continued for another 24 h (data not shown). As reported before, ACC could transform from vaterite into crystalline calcite under the control of phospholipid and polymers [158, 159]. It is conceivable that ACC may also transform from vaterite into calcite in the presence of ACCBP. Micro-Raman spectra and SEM images also indicate that over 80% of the particles formed with D139S mutant are still ACC spherulites with a smaller diameter of about 20 μm, while almost all the particles formed with D46S mutant are calcite. Taken together, the first Ca^{2+}-binding site composing of residues 44–48 is key for ACC formation, while the second Ca^{2+}-binding site including residues of 139–141 is important for accelerating ACC formation.

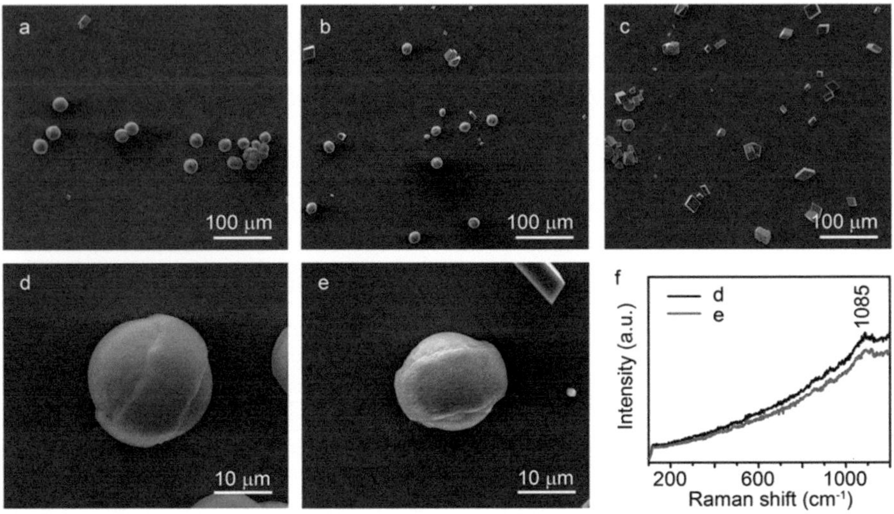

Fig. 3.89 SEM images of the in vitro crystallization assay. In vitro crystallization with the addition of wild-type ACCBP (**a**), D139S ACCBP (**b**), and D46S ACCBP (**c**). (**d**) Amplification of a classical ACC spherulite shown in (**a**). (**e**) Amplification of a typical ACC spherulite indicated in (**b**). (**f**) Raman spectrum of the precipitates. Spectrum profile indicates the Raman spectrum of an ACC spherulite shown in (**a**). Spectrum profile indicates the Raman spectrum of an ACC spherulite in (**b**). *a.u.* arbitrary units

3.7.2.3 The Ca^{2+}-Binding Sites of ACCBP Are Arranged with a Fivefold Symmetry, Which Is Essential for ACC Formation

In the conventional template hypothesis, acidic matrix proteins could control the biomineralization process by providing a nucleating template for the crystallization system [42]. The negatively charged residues, such as carboxylate of aspartic acids, are well arranged to form a highly structured surface to control the polymorph of calcium carbonate and interact with the Ca^{2+} [160, 161]. The atoms of the crystalline phase are organized in a periodic or repeating array over large atomic distances. On the contrary, the biogenic ACC phase is short for long-range atomic order and the short-range atomic order is not uniform in different ACC samples [162–164]. Therefore, ACC cannot be induced by a classical nucleating template.

Gower and colleagues hypothesized that PILP (polymer-induced liquid precursor) is an intermediate phase of the ACC precursor [165, 166]. In the PILP hypothesis, ovalbumin (pI = 4.7), which may serve as a significant stabilization agent for a temporary mineral precursor, stabilizes LACC (liquid ACC) by wastage stabilization and de-emulsification [167]. ACCBP is also an acidic protein with predicted pI of 4.62, and it is possible that ACCBP may also induce the ACC precursor by stabilizing LACC. Sequence analyses showed that ACCBP contains 24 aspartic acid and 10 glutamic acid residues. Both of the single mutations loss only one negatively charged residue and change the pI value from 4.62 to 4.68 slightly. Thus, both the mutants could still stabilize LACC by de-emulsification and depletion stabilization to induce ACC formation. Whereas, the crystallization experiments in vitro show that both the single mutations perturb the function of ACCBP obviously (Fig. 3.89). It is convincing that ACCBP could not induce the ACC precursor via stabilizing LACC with its negative surface. Whereas, Pipich et al. [168, 169] have reported that 2.5 mg/ml ovalbumin could induce ACC formation after about 2.5 h in the calcium carbonate crystallization process, and these amorphous particles could transform into crystalline phase after about 4 h. Wang et al. [169] have reported that in an experiment of in vitro crystallization, most of the deposits induced by 0.5 mg/ml ovalbumin are calcite particles with a coat of ACC protein/peptide aggregate. These results of present study suggest that most of the deposits induced by 0.5 μM ACCBP are all ACC, and these ACC particles could maintain for over 24 h (Fig. 3.89a). In short, ACCBP could induce ACC formation more efficiently. Politi et al. [170] also found that Asp-rich proteins isolated from the prism of mollusk are more efficient at inhibiting calcite nucleation and could produce ACC at lower dosages than poly-Asp of similar size. Almost 27% of amino acids of ovalbumin are acidic, while only 15.7% of amino acids of ACCBP are acidic. This result suggested that ACCBP could induce ACC formation by its negatively charged surface along with the Ca^{2+}-binding sites. Therefore, the tertiary structure of the protein, especially the distribution of Ca^{2+}-binding sites, is crucial for the ACC-induction function of ACCBP.

Homology model was used to support information on ACC formation with the addition of ACCBP. The homology model of monomeric ACCBP was first constructed. The B chain of a pentameric α-7 nicotinic receptor (PDB code 3SQ6) was used as a template because of its high-sequence identity of 32% compared with ACCBP. The alignment between these two proteins was then uploaded to the SWISS-MODEL Workspace, and the calculated structure of ACCBP was constructed by the alignment method. Although the QMEAN Z-score is only 3.212 because of the low sequence identity, this value means that there are no great errors with the structure. The homology modeling results suggest that β-sheets of ACCBP compose a barrel-like structure (Fig. 3.90a). These two Ca^{2+}-binding sites are close to one another locally, indicating that they might cooperate in controlling ACC formation. Both of the Ca^{2+}-binding sites are seated in a flexible loop of sheet–loop–sheet motif, suggesting that ACCBP could induce ACC precursors by a short-range atomic order. In short, monomeric ACCBP could not induce ACC only with the two Ca^{2+}-binding sites.

As ACCBP is a decamer constituted of two adjacent pentamers while its homologous proteins are all pentamers, the homology model of homopentameric ACCBP was performed with the ligand-binding domain from the pentameric α-7 nicotinic receptor (PDB code 3SQ6, sequence identity of 32%). The alignment between these two proteins was submitted to the SWISS-MODEL Workspace, and the tertiary structure of ACCBP was predicted by the project method. The QMEAN Z-score suggests that there are no great errors in the structure. The homology modeling indicates that homopentameric ACCBP should be a ringlike structure with an inner

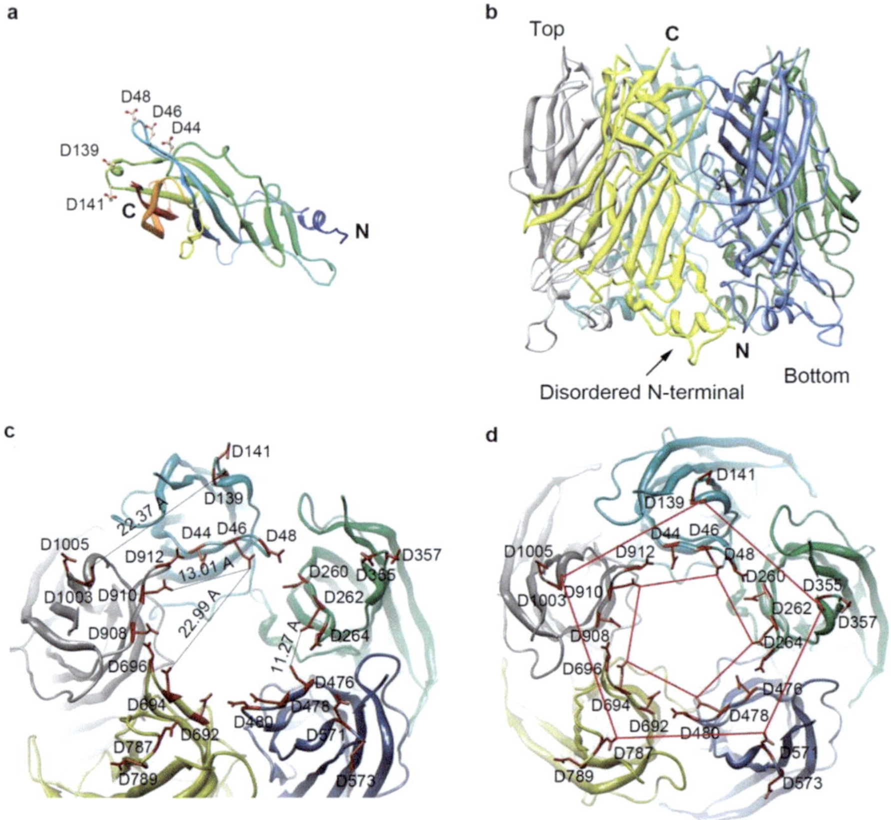

Fig. 3.90 Homology modeling of ACCBP monomeric and pentameric. (**a**) Homology modeling of ACCBP monomeric. (**b**) Homology modeling of ACCBP pentameric. (**c**) Tilted view of pentameric ACCBP. (**d**) Vertical view of pentameric ACCBP. *C* C-terminal, *N* N-terminal

diameter of ~2.1 nm and an outer diameter of ~7.5 nm, which is accordant with the negative staining EM images (Fig. 3.90b). The disordered N terminus is located at the bottom side of this protein and is exposed to the solution which could interact with other proteins to prompt two pentamers to form a decamer. This may be the cause why ACCBP is a decamer instead of a pentamer in solution. ACCBP has a predicted glycosylation site at Asn29, and this posttranslational modification could prevent the interaction between these disordered sequences. We therefore infer that ACCBP is a pentamer in vivo.

The Ca^{2+}-binding sites are found on the highest side of homopentameric ACCBP (Fig. 3.90c). These binding sites are organized like an outer equilateral pentagon with side length of about 22.37 Å and an inner ring in which the distance of adjacent Ca^{2+}-binding sites is about 11.27, 13.01, and 20.99 Å. The distance between the contiguous Ca^{2+}-binding sites is extremely different from the distance between contiguous calcium ions in ACC which is under 4 Å [162]; these Ca^{2+}-binding sites of pentameric ACCBP are not arranged in a repeating or periodic way. In short, these binding sites cannot form a classical template for ACC nucleating.

It is convincible to consider that ACC is an amorphous, not a stable, phase. According to the rule of symmetry, the crystalline phases display only two-, three-, four, or sixfold axes of symmetry. The template of crystalline phase should be organized in the same symmetry. ACC is short of such symmetry because of its disordered atomic structure, so that the symmetry request of the template for ACC would also be different. The homology modeling suggested that both the outer Ca^{2+}-binding sites and the inner Ca^{2+}-binding sites are surprisingly in a fivefold symmetry, which is extremely different of the symmetry of the template for the crystalline stage (Figure 3.90d). ACCBP could gather Ca^{2+} ions in a limited space because of its great Ca^{2+}-binding ability. Whereas, a template of fivefold symmetry cannot support nucleating template for any crystalline stage. Therefore, the gathered Ca^{2+} ions would be combined in a disordered structure. ACC is formed when the disordered calcium carbonate aggregation grows. Whether the ACC formation starts with a metastable liquid phase or prenucleation clusters [166, 171], ACCBP can remain the Ca^{2+} of the metastable liquid phase in a disordered structure or the prenucleation clusters, and then the PILP changes from the liquid to solid phase or prenucleation clusters aggregate and then ACC forms. The fivefold symmetry of Ca^{2+}-binding sites might be the structural basis of the ability of ACCBP to induce formation of ACC.

The classical template hypothesis suggested that a specific nucleating template could induce only one specific crystalline phase in a specific atomic order. On the contrary, ACCBP with a fivefold symmetry could only control the symmetry of the Ca^{2+} ions cluster and induce different ACC precursors of different atomic orders. During transformation and transportation, ACCBP could also stabilize ACC constantly with the atomic order of ACC precursor changes all the time. A fivefold symmetry controlling ACC formation is economical and efficient, and it is also constituted with the biogenic polyamorphism of ACC and its lasting change during the biomineralization process.

3.7.3 Transformation of Amorphous Calcium Carbonate Nanoparticles into Aragonite Controlled by ACCBP

Polymorph regulation during shells or pearl formation has been an intriguing question, especially for those animals of interest for human consumption. The polymorph changing of calcium carbonate regulated by amorphous calcium carbonate-binding protein (ACCBP) which is an extrapallial fluid (EPF) protein from the pearl oyster characterized by our group in 2007 is explored in this research. FTIR and TGA analyses indicate that ACCBP could decrease the bound water composition of amorphous calcium carbonate (ACC), indicating that ACCBP may participate in biogenic anhydrous ACC formation [136]. In vitro crystallization and

ACC transition experiments indicate that ACCBP could induce aragonite formation by an ACC precursor in Mg/Ca = 1 and Mg/Ca = 2 solutions in low temperature. Raman, ICP, FTIR, and XPS analyses of the ACC nanoparticles in the initial stage in the ACC transformation assay indicate that this polymorph changing may be regulated by promoting the surface Mg/Ca ratio of the ACC, rather than by changing the bulk Mg/Ca ratio or the short-range ordered structure. These results indicate that the polymorph chosen in nacre or pearl growth may be regulated not only by the nucleating template of the matrix but also by the physicochemical functions of EPF proteins.

3.7.3.1 ACCBP Induces Formation of Less Hydrated ACC

The FTIR spectra of ACC formed at 0 h in the ACC transformation experiment indicate that ACCBP could decrease the 3400 cm^{-1}/866 cm^{-1} intensity ratio (Fig. 3.91a), suggesting that ACCBP may reduce the bound water content. ACC formed with different dosages of ACCBP was further analyzed using TGA (Fig. 3.91b). On the basis of the thermal behavior of ACC measured by DSC (data not shown) and the data in previous research [150], the weight loss under 100 °C in the TGA is a result of the loss of free water, the weight loss above 100 °C is owing to the release of bound water, and the weight loss at about 690 °C is owing to the degeneration of calcium carbonate. The weight loss (ΔW) among 100 °C and 350 °C is thought as the total content of bound water in the ACC sample, and the weight (W) at about 350 °C is regarded as the content of calcium carbonate in the ACC sample. The molar masses of H_2O and calcium carbonate are 18 g mol^{-1} and 100 g mol^{-1}, respectively, and the bound water quantity per mole of calcium carbonate could be computed by $\Delta W \times 100/(W \times 18)$. The results show that the sediment prepared without protein contains 1.0 mol of water per mole calcium carbonate. With addition of 0.25 μM ACCBP, the water content of the ACC does not change, while in the presence of 0.5 μM and 1 μM ACCBP, the water content decreases to

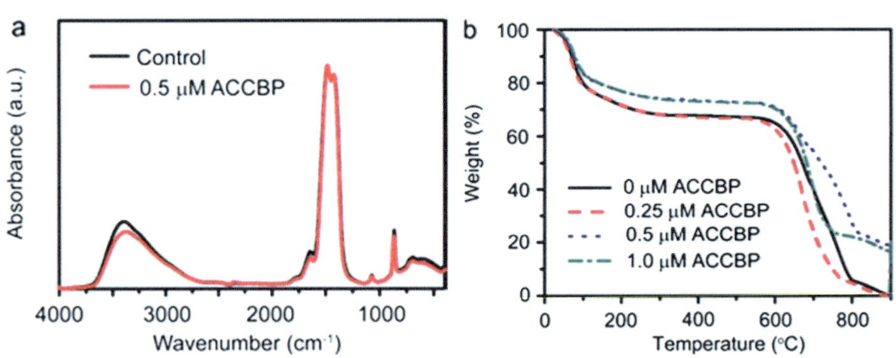

Fig. 3.91 FTIR spectrum (**a**) and TGA (**b**) of ACC stimulated in solution with or without ACCBP [136]

0.82 mol of water per mole calcium carbonate. The biogenic ACC is essentially anhydrous, while the biogenic stable ACC includes 1 mol of bound water per mole calcium carbonate. ACCBP reduces the bound water content of ACC and is involved in the formation process of biogenic anhydrous ACC in vivo.

3.7.3.2 ACCBP Induces Aragonite Formation via ACC Precursor in Low Mg/Ca Solutions at Low Temperature

The in vitro crystallization assay was conducted at 6 °C to avoid protein degradation or denaturation (Fig. 3.92). An SEM outfitted with an energy dispersive spectrometer (EDS) was utilized to observe the morphology of the sediments, and Raman spectroscopy was utilized to characterize the polymorph. ACC was identified with a broad peak at ~1085 cm^{-1}, vaterite with peaks at ~1074 and 1090 cm^{-1}, aragonite with peaks at ~155, ~204, and ~1085 cm^{-1}, and calcite with peaks at ~154, ~281, and ~1085 cm^{-1}. The Raman and SEM results suggest that ACC spherulites (about 9.5 μm in diameter) form with ACCBP, while only calcite particles are observed in

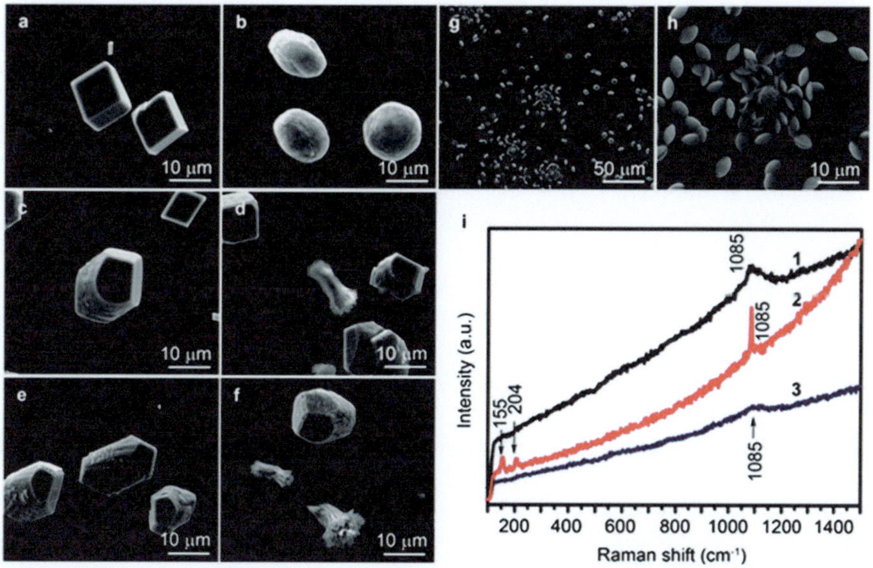

Fig. 3.92 The in vitro crystallization assay. SEM images of calcium carbonate segments had grown for 24 h in the solution conditions as follows: (**a**) 0.5 μM BSA, (**b**) 0.5 μM ACCBP, (**c**) 0.5 μM BSA and Mg/Ca = 1, (**d**) 0.5 μM ACCBP and Mg/Ca = 1, (**e**) 0.5 μM BSA and Mg/Ca = 2, and (**f**) 0.5 μM ACCBP and Mg/Ca = 2. (**g**) SEM image of calcium carbonate crystals had grown for 4 h in solution with 0.5 μM ACCBP and Mg/Ca = 1. (**h**) Amplification of the central region in (**g**). (**i**) Raman spectrum of the calcium carbonate segments. Spectrum 1 indicates an ACC spherulite in (**b**), spectrum 2 is an aragonite particle in (**d**), and spectrum 3 represents an ACC spherulite in (**h**) [136]

the control groups (Fig. 3.92a, b). The addition of Mg^{2+} alters the result dramatically. There are calcite particles, and some aragonite particles are formed instead of ACC spherulites in the solution having ACCBP and an Mg/Ca molar ratio of 1 (Mg/Ca = 1), while only calcite particles are formed in the control group (Fig. 3.92c, d). More aragonite particles are observed in the solution with Mg/Ca = 2 and ACCBP (Fig. 3.92e, f). Mg^{2+} could stabilize ACC, and calcite is preferred in solutions with a Mg/Ca molar ratio under 4 at a temperature of 6 °C. Thus, it is extraordinary that ACC disappears and aragonite was observed in the presence of both ACCBP and Mg.

EDS indicates that calcite had grown in the solution with Mg/Ca molar ratio at 1 and ACCBP has higher magnesian component with an average of 36.4 ± 7.4 mol % in randomly selected particles, compared with 27.9 ± 3.0 mol% of randomly selected calcite particles in the control group (data not shown). Considering that high-magnesian calcites are observed in vitro via ACC precursors, the aragonites may form Mg/Ca = 1 and Mg/Ca = 2 solutions by consuming the ACC precursor with the control of ACCBP.

Observations of early-stage samples from crystallization in vitro group support this hypothesis. Only dish-like calcium carbonate spherulites (~6 μm in diameter) are observed at 4 h in the solution with Mg/Ca ratio of 1 and ACCBP (Fig. 3.92g, h). A broad peak at ~1085 cm^{-1} of the Raman spectrum affirms that these spherulites are ACC (Fig. 3.92i). Obviously, these ACC spherulites are the precursors of aragonite or calcite particles at 24 h [136].

To affirm that ACCBP could induce the transformation from ACC precursor to aragonite, an ACC transformation assay was conducted (Fig. 3.93). The crystals were collected after 0, 12, 24, 48, and 72 h and analyzed by FTIR. ACC is characterized by a broad peak at ~866 cm^{-1} ($\nu2$), aragonite has absorption peaks at ~856 cm^{-1} ($\nu2$), 713 cm^{-1} ($\nu4$), and 700 cm^{-1} ($\nu4$) (a weak band), and calcite have absorption peaks at ~874 cm^{-1} ($\nu2$) and ~713 cm^{-1} ($\nu4$). The analyses show that all the samples at 0 h are ACC (Fig. 3.93). In the solution with Mg/Ca ratio of 1, a strong peak observed of calcite is ~874 cm^{-1}, and the peak at ~866 cm^{-1} of ACC decreases at 12 h; as time goes on, the 874 cm^{-1}/866 cm^{-1} intensity ratio increases persistently, and the peak of ~866 cm^{-1} disappears eventually (Fig. 3.93a). These results show that the entire ACC is transformed to form calcite in the control group. In the solution with Mg/Ca ratio of 1 and ACCBP, strong absorption of ~874 cm^{-1} and a weak shoulder peak at ~856 cm $^{-1}$ are detected at 12 h; the 874 cm^{-1}/866 cm^{-1} intensity ratio increases along time, and the absorption at ~856 cm^{-1} changes slightly (Fig. 3.93b). Most of the ACC is again transformed to form calcite, but a small amount of aragonite is also observed. In the solution with Mg/Ca ratio of 2, strong absorption of ~874 cm^{-1} is observed at 12 h, a characteristic peak of aragonite at 856 cm^{-1} appears weakly at 48 h, and the 856 cm^{-1}/874 cm^{-1} intensity ratio does not change during 48 h and 72 h (Fig. 3.93c). Most of the ACC transforms to calcite, and few of the final mixture is aragonite. In the solution with Mg/Ca ratio of 2 and ACCBP, the peak of ~874 cm^{-1} is broader at 12 and 24 h than control group, implying that ACC is more stable with ACCBP. The 856 cm^{-1}/874 cm^{-1} intensity ratio significantly increases at 48 h, indicating that a

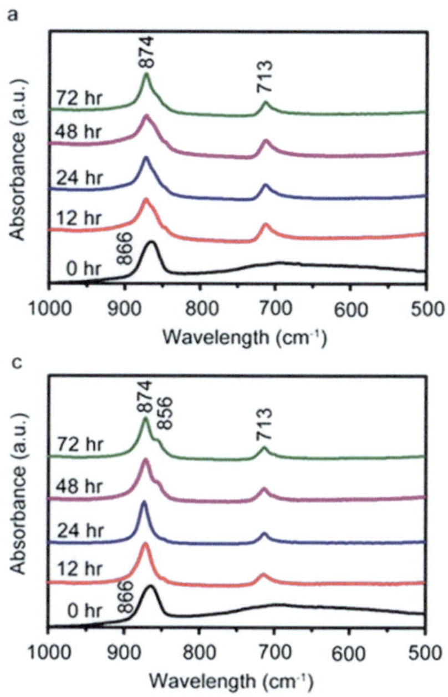

Fig. 3.93 FTIR analyses of the contributions in the ACC transition experiment. FTIR of the precipitates formed at different times from the solutions with the addition of (**a**) Mg/Ca = 1, (**b**) Mg/Ca = 1 and 0.5 μM ACCBP, (**c**) Mg/Ca = 2, and (**d**) Mg/Ca = 2 and 0.5 μM ACCBP

large amount of aragonite forms at this period. The 856 cm^{-1}/874 cm^{-1} intensity ratio increases persistently up to 72 h, indicating that aragonite forms continuously, and most of the final crystals is aragonite (Fig. 3.93d). ACC is transformed to aragonite in the presence of ACCBP and Mg/Ca ratio of 2.

The final crystals collected on the fourth day of the ACC transition experiment were analyzed by XRD. In the solution with Mg/Ca ratio of 1 and ACCBP, the calcite content of the final precipitates is greater than 95%, the same as the control group (Fig. 3.94). The results suggest that ACCBP induces only few aragonite formations in the Mg/ Ca ratio of 1 solution. In the solution with Mg/Ca ratio of 2 and ACCBP, more than 87% of the crystals is aragonite, while over 90% of the crystals is still calcite in the control experiment, which shows that ACCBP could induce aragonite formation in the ACC transformation assay. In the ACC transformation assay II, ACC was formed first without ACCBP or Mg^{2+}, isolated from the initial solution by centrifugation, and then resuspended in the solution of Mg/Ca at 1, 2, or 5, with or without ACCBP (data not shown). The XRD analyses showed that ACCBP could induce aragonite formation in solution with Mg/Ca ratio of 5, while no aragonite was observed in the control group. The result affirms again that ACCBP could induce aragonite formation with the help of Mg^{2+}.

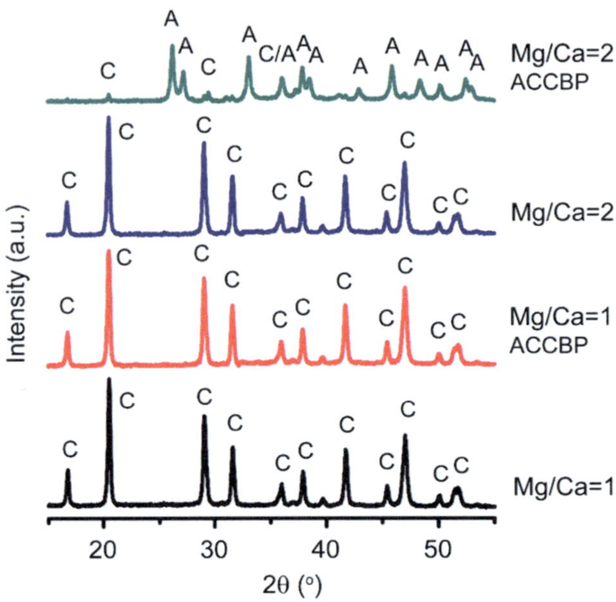

Fig. 3.94 XRD spectra of the final crystals in the ACC transformation assay. *A* aragonite, *C* calcite

3.7.3.3 Mg/Ca Ratio on the Surface Initial-Stage ACC Is Controlled by ACCBP

To investigate how ACCBP induces aragonite formation by the ACC precursor in solutions with low Mg^{2+} dosages, the initial ACC sediments collected at 0 h in the ACC transition experiment were analyzed. SEM images find that all the ACC samples constitute of spherulites with diameters of 20–100 nm (data not shown). ACCBP does not regulate morphology or the size of ACC at any Mg/Ca ratio. Raman spectrum with only a broad peak at ~1085 cm^{-1} affirms that all the precipitates are ACC (Fig. 3.95a). The peaks of ~1085 cm^{-1} in all spectra are similar, and no discrete peak is observed in the range of 100–300 cm^{-1} even if each spectrum peak represents the average of 10 separate scans. These virtually identical spectra suggest that no significant difference was detected by Raman spectroscopy. The FTIR spectra confirm that all the precipitates are ACC with the split peaks at ~1429 and ~1479 cm^{-1} ($\nu3$) (showing the asymmetric stretch of the carbonate ion), the weak and broad peak at ~1075 cm^{-1} ($\nu1$) (showing the symmetric stretch in non-centrosymmetric structures), the peak of ~866 cm^{-1} ($\nu2$) (indicating out-of-plane bending), and a weak and broad peak at ~713 cm^{-1} ($\nu4$) (representing in-plane bending) (Fig. 3.95b). The peak of ~866 cm^{-1}, the split peaks of ~1429 and ~1479 cm^{-1}, and the intensity ratio of 866 cm^{-1}/713 cm^{-1} are not affected by ACCBP. The virtually identical FTIR spectra show that the local environment of the

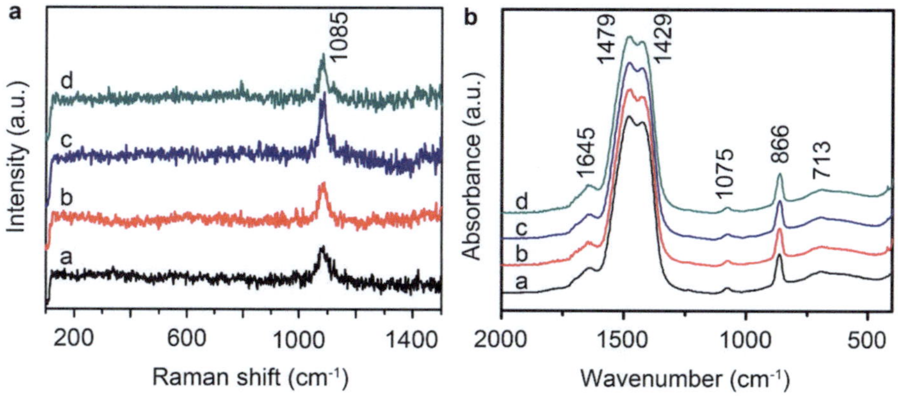

Fig. 3.95 Raman spectra profile (**a**) and FTIR spectra profile (**b**) of the initial-stage segments in the ACC transformation assay. The precipitates were collected in the following solution conditions: (a) Mg/Ca = 1, (b) 0.5 μM ACCBP and Mg/Ca = 1, (c) Mg/Ca = 2, and (d) 0.5 μM ACCBP and Mg/Ca = 2

carbonate group in ACC are identical, indicating that the local atomic orders of the ACC precipitates are identical. It is improbable that ACCBP induces aragonite formation via regulating the short-range order structure of the initial-stage ACC. This hypothesis is accordant with the recent report that the amorphous phase precursor of abalone nacre is not poorly crystalline aragonite or proto-aragonite.

ACCBP includes 11.11% Asp residues, and aspartic acid could increase the Mg^{2+} content of ACC. Thus, it is suggested that ACCBP could influence the magnesium content of ACC. To confirm the hypothesis, the bulk Mg/Ca ratio of ACC was analyzed via ICP. In the solution with ACCBP and Mg/Ca = 1, the ratio was 0.166 ± 0.007, resembling to the control value of 0.171 ± 0.002. In the solution with ACCBP and Mg/Ca = 2, the ratio was 0.251 ± 0.021, resembling to the control value of 0.257 ± 0.006 (Fig. 3.96a). ACCBP could not affect the bulk Mg/Ca ratio of ACC. The Mg/Ca ratio on the surface of ACC was further analyzed by XPS. In the solution with ACCBP Mg/Ca = 1, the ratio was 0.44 ± 0.10, compared with 0.28 ± 0.09 of the control. In the solution with ACCBP and Mg/Ca = 2, the ratio was 0.45 ± 0.22, compared with 0.27 ± 0.08 of the control (Fig. 3.96b). This tendency was always identical in the three independent experiment groups. XPS suggests that ACCBP might increase Mg/Ca ratio on the surface. Thinking of the effects of Mg^{2+} on crystallization, ACCBP might control the polymorph switching by affecting the local Mg^{2+} content of the surface of the ACC.

To affirm that ACCBP could increase the surface Mg/Ca ratio of ACC, a series of fluorescence quenching assays was conducted. With addition of 10 μM, Ca^{2+} quenched the intrinsic tryptophan fluorescence by about 6% (Fig. 3.97a), similar to the level of quenching induced by 50 or 500 μM Ca^{2+}. Presuming it is static quenching, then the binding constant should be same to the quenching constant. The

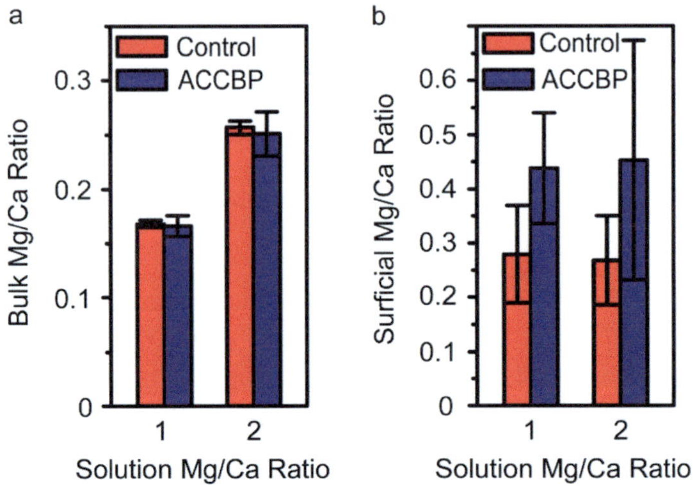

Fig. 3.96 Elemental analyses of the initial-stage segments in the ACC transition experiment. (**a**) Bulk Mg/Ca ratio of ACC explored by ICP. (**b**) Surface Mg/Ca ratio of ACC indicated by XPS

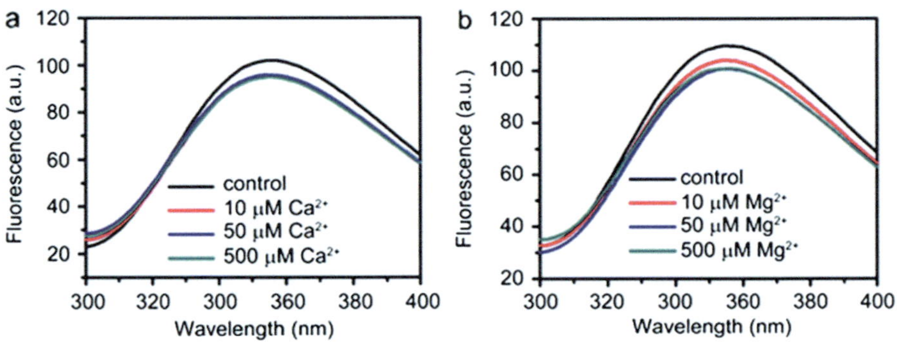

Fig. 3.97 Fluorescence quenching assay. (**a**) Intrinsic fluorescence of 5 μM ACCBP dropped with 10, 50, or 500 μM Ca^{2+}. (**b**) Intrinsic fluorescence of 5 μM ACCBP dropped with 10, 50, or 500 μM Mg^{2+}

binding ability for Ca^{2+} would be bigger than 6.4×103 L mol^{-1} according to the Stern–Volmer relationship, indicating that Ca^{2+} is able to bind with ACCBP. This is accordant with the ability of ACCBP to bind on the surface of ACC and calcite. The addition of 10, 50, and 500 μM Mg^{2+} quenches intrinsic fluorescence by 5%, 8%, and 8%, respectively (Fig. 3.97b), indicating that the binding ability of Mg^{2+} is greater than 1.7×103 L mol^{-1}. The fact that ACCBP could bind both Mg^{2+} and Ca^{2+} affirms that ACCBP might recruit Mg^{2+} on the surface of ACC.

3.7.4 *Influence of the Extrapallial Fluid of* Pinctada fucata *on the Crystallization of Calcium Carbonate and Shell Biomineralization*

Extrapallial fluid (EPF) is distributed between the mantle and the shell, and it is suggested to play crucial roles in shell biomineralization of *Pinctada fucata*. Whereas, few researches have been reported on the biomineralization effect of EPF [138]. In present work, $CaCO_3$ crystallization assays suggested that EPF proteins could not only regulate the morphology but also control the phase transition of calcium carbonate by the ability of different proteins to specifically bind to calcite or aragonite. In crystal growth inhibition assays, when the final dosage of $CaCl_2$ and $NaHCO_3$ decreased from 50 to 5 mM, the effect of EPF proteins (50 μg/ml) changed from significantly promoting over the number of $CaCO_3$ crystals (448 ± 28) over the control (128 ± 20) to repressing the precipitation of $CaCO_3$. The deposition rate of $CaCO_3$ was also stopped by EPF proteins. In vivo, when EPF was extracted every day for 20 days, the nacreous platelet in the nacre was destroyed, and calcite deposited disorderly due to the decrease of EPF proteins. What's more, EPF also worked as an external signal to regulate the expression variation of shell-related genes to participate in nacre and prism formation. Taken together, our findings suggest that EPF proteins not only participate in the nucleation, morphology, growth, and phase transition of $CaCO_3$ but also play a dual role in both prism–nacre transition and shell biomineralization, which is mediated by a combination of different proteins and changes of their secondary structure and conformation.

3.7.4.1 The Binding of EPF Proteins to Calcite and Aragonite

The average dosage of native EPF proteins separated was 0.2–0.4 mg/ml. The observations in SDS-PAGE of native EPF (Fig. 3.98, lane 5) found only one obvious band that was bigger than 170 kDa, which suggested that the dosage of native EPF proteins was too low to observe [138]. After the EPF was condensed (cEPF) with a 3 kDa centrifugal ultrafiltration tube, the protein bands were more obvious (Fig. 3.98, lanes 1 and 6). The proteins in EPF ranging from 10 to >170 kDa may play an important role in $CaCO_3$ crystallization and shell formation.

One of the direct methods to investigate whether a protein is crucial for $CaCO_3$ crystallization is to detect its binding ability to $CaCO_3$ crystals. To confirm that EPF proteins could bind to crystals, we wash with water for 3 times and with 500 mM NaCl to wash off the proteins after the binding assays. Major proteins were eluted away from the calcite and aragonite by water (Fig. 3.98, lanes 2 and 7). Only one band with molecular mass over 170 kDa was observed in the 500 mM NaCl elution buffer (Fig. 3.98, lanes 3 and 8). As expected, several protein bands were found in the 1 M acetic acid solutions, which indicated specific binding ability to calcite or aragonite (Fig. 3.98, lanes 4 and 9). These results suggest that EPF contains not only proteins binding to both calcite and aragonite (Fig. 3.98, bands A, C, D, and E in lane

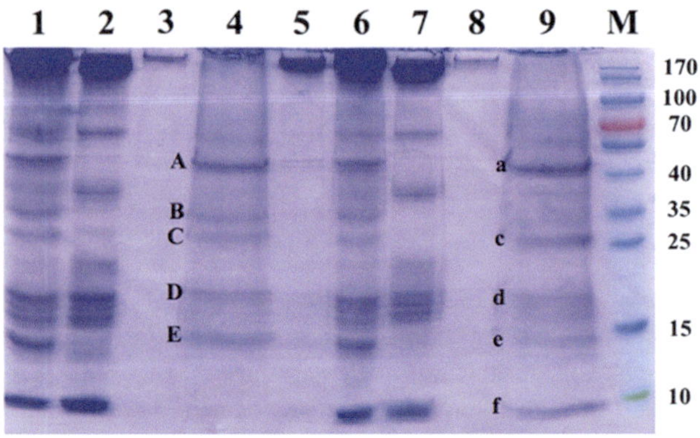

Fig. 3.98 SDS-PAGE image of EPF proteins extracted of *P. fucata* (lane 5) and proteins binding to calcite and aragonite among EPF. Lane 5 indicated the native EPF from *P. fucata*. Lanes 1 and 6 had shown EPF that was concentrated by a 3 kDa Millipore Amicon Ultra-15 centrifugal filter device as a positive control. Lanes 2–4 were water washings of EPF from calcite, 500 mM NaCl washings after incubation with calcite, and 1 M acetic acid dissolution after incubation with calcite, respectively. Lanes 7–9 indicated the corresponding washings after incubation with aragonite. Bands A–E and a–f had shown proteins that specifically bound to calcite and/or aragonite. The side numbers are the molecular masses (kDa) of the protein makers (M)

4 and bands a, c, d, and e in lane 9) but also binding solely to calcite (band B) or aragonite (band f). And the amino acids analyses results showed that proteins only binding to aragonite are rich in Leu, Glu, and Gly, while proteins binding solely to calcite are more enriched in Leu, Lys, and Val (data not shown). In view of that, the shell of *P. fucata* consists of aragonite and calcite; these proteins may play a role in the respective formation of each component. To investigate the effect of EPF proteins on $CaCO_3$ crystallization, we conducted the $CaCO_3$ crystallization assays in vitro.

3.7.4.2 Effect of EPF on $CaCO_3$ Crystallization In Vitro

Without EPF, only the precipitation of rhombohedral calcite crystals was found (Fig. 3.99a). The addition of EPF caused a morphological change in the crystals. The corners and edges of the rhombohedral calcite were weakened with 10 μg/ml EPF (Fig. 3.99b). What's more, the crystals changed into rods with 30 μg/ml EPF (Fig. 3.99c). When the dosage of EPF reached 50 μg/ml, the crystals changed larger and were transformed into a dumbbell morphology, an effect that was resemble to that phenomenon of aragonite in vitro (Fig. 3.99), even though Raman spectra showed distinctive peaks at 151, 281, 711, and 1085 cm^{-1}, which suggested that they were still calcite (data not shown). However, EPF was supposed to control the nacre growth in vivo. To cause the formation of aragonite, we also conducted crystallization assays with EPF on calcite (data not shown), organic, and silk membrane from the inner shell, as researched previously (data not shown), while

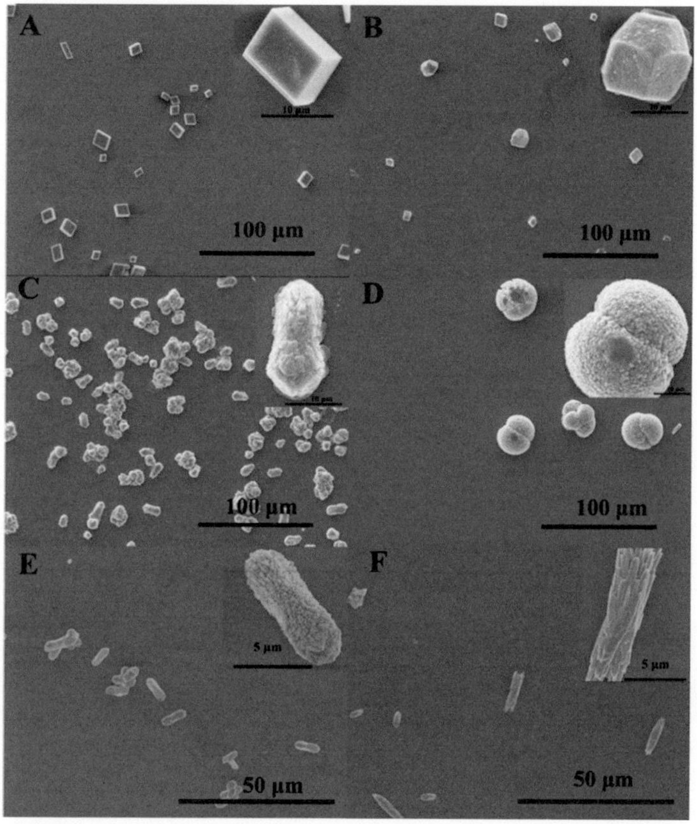

Fig. 3.99 SEM images of $CaCO_3$ crystals collected from the crystallization assays: (**a–d**), Crystals are formed in the presence of ASW or 10, 30, and 50 µg/mL EPF proteins, respectively; (**e** and **f**), Crystals had grown in the presence of 30 µg/mL EPF proteins binding to calcite and/or aragonite, respectively. The insets on the corner are high-amplification images of the associated particles [138]

none of these experiments work, unfortunately. Furthermore, we studied the influence of EPF proteins binding to aragonite or calcite on $CaCO_3$ crystallization, respectively. EPF proteins binding to calcite showed an effect resemble to the influence of native EPF on $CaCO_3$ crystals (Fig. 3.99e), whereas EPF proteins binding to aragonite caused the formation of aragonite crystals (Fig. 3.99f) which were confirmed by Raman spectra (data not shown). Then to affirm the exact contents of the crystals, the crystals were collected and analyzed using XRD. The results suggested that the crystals formed with EPF proteins binding to aragonite included 37.6% calcite and 62.4% aragonite, while only calcite was detected with EPF proteins binding to calcite (data not shown). These results suggested that the EPF could affect the formation of aragonite and the morphology of calcite through different proteins in vitro.

3.7.4.3 Inhibition of EPF in $CaCO_3$ Crystallization

The $CaCO_3$ crystallization rate was detected via the absorbance at 570 nm. EPF obviously decreased this rate, compared with the influence of the same volume of ASW (Fig. 3.100a). The effect in the crystallization rate by EPF was more lightening. To observe whether the inhibition was induced by the biomacromolecules in EPF, we compared the precipitation rates of micromolecules (mEPF) and biomacromolecules (cEPF) with that of ASW. The results suggested that biomacromolecules had a suppression effect on the precipitation rate of $CaCO_3$ resembling to that of EPF under the same dosages, while the micromolecules indicated effects resembling to those of ASW (data not shown). After crystallization of 24 h, the crystals formed with EPF were more uniform and had a smaller morphology and volume modification compared with crystals formed with ASW (Fig. 3.100c, d). At the same time, there were 448 $\pm$ 28 crystals in EPF, which was significantly more than the 128 $\pm$ 20 crystals in the ASW group (Fig. 3.100b). We inferred that biomacromolecules in EPF induced this result. As expected, the crystal number formed in the presence of biomacromolecules (cEPF) was always greater and the crystals were more structured and uniformly distributed than those induced with micromolecules (mEPF) (Fig. 3.100b), which suggested that the biomacromolecules in EPF could support nucleation sites in $CaCO_3$ crystallization when the dosage of $CaCl_2$ and $NaHCO_3$ was 50 mM. On the contrary, when the final dosage of $CaCl_2$ and $NaHCO_3$ was reduced to 5 mM, no distinct crystals formed in EPF after 24 h compared with the ASW group (Fig. 3.100e, f). Thus, we inferred that EPF proteins took part in the nucleation of $CaCO_3$ crystallization and inhibited $CaCO_3$ crystallization and shell formation.

EPF contains proteins that are both identical and distinct from those from the shell of *P. fucata* (data unpublished), which directly demonstrates that the EPF participates in shell formation or acts as a library of shell proteins. Proteins extracted from the shells of mollusks have been shown to extensively control the nucleation, morphology, and phase transformation of $CaCO_3$; however, the macromolecules of the EPF, which is located between the mantle and shell, have not received enough attention. The EPF proteins, in combination with Ca^{2+} or CO_3^{2-}, on the one hand function as nucleation sites in $CaCO_3$ crystallization (Fig. 3.100b) and control the morphology, volume, and phase of crystals (Fig. 3.99); on the other hand, the EPF proteins inhibit the precipitation rate and crystallization of $CaCO_3$ (Fig. 3.100a, f). These guarantee that the EPF not only participates and regulates the formation of shell but also inhibits abnormal shell growth when necessary. The shell protects mollusks from environmental changes and natural enemies, while the pearl also helps to avoid injury from foreign bodies. Shell and pearl biomineralization, which are similar processes, represent a complicated and highly controlled balance of biomacromolecules, which requires high energy consumption [172]. Mollusks must slow or inhibit the shell biomineralization as soon as the shell matures and then activate biomineralization immediately upon shell injury or invasion by foreign

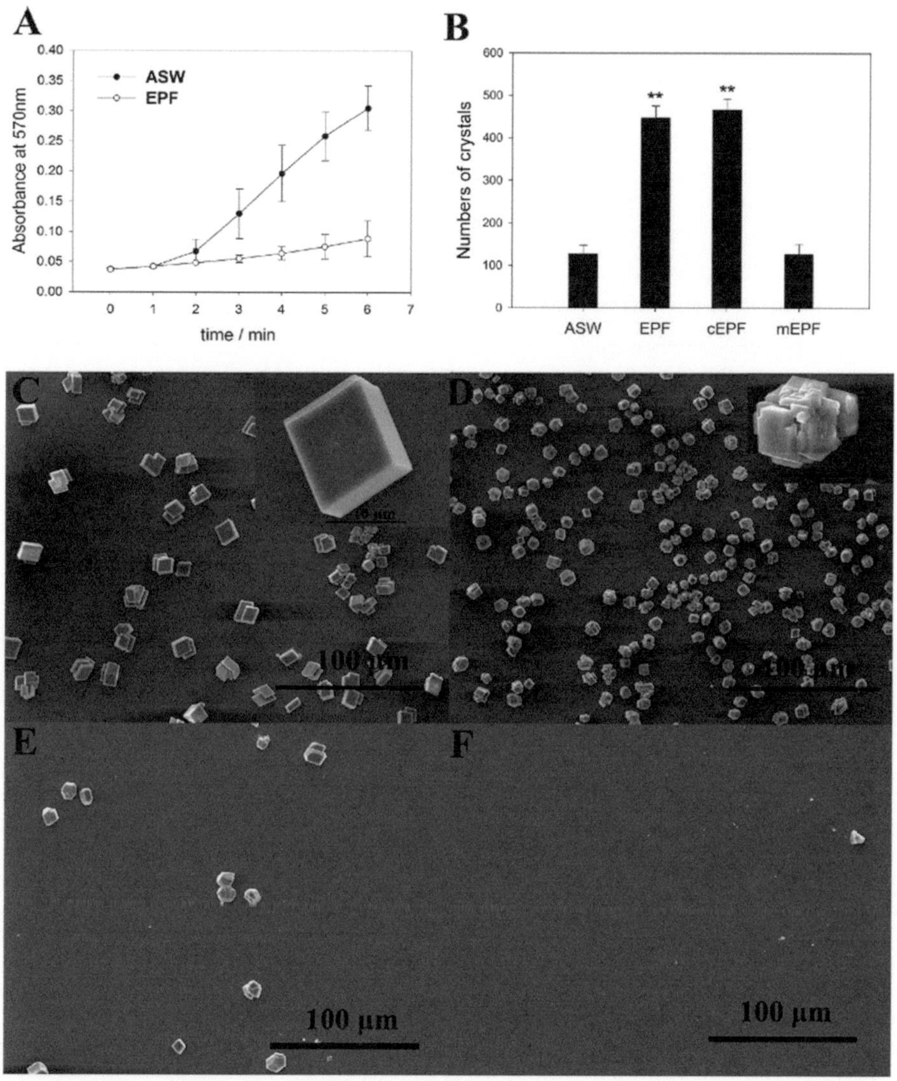

Fig. 3.100 Precipitation rate of $CaCO_3$ and SEM images of crystals formed in the crystal growth inhibition assays: (**a**), $CaCO_3$ precipitation rate shown by the crystals' absorbance at 570 nm (drastic differences occurred at 3 min ($p < 0.05$) and at 4, 5, and 6 min ($p < 0.01$)); (**b**), number of crystals grown in the crystal growth inhibition assays calculated in a $214 \times 162\ \mu m^2$ area (cEPF means EPF concentrated by a 3 kDa Millipore Amicon Ultra-15 centrifugal filter device; mEPF indicates the effluent after a 3 kDa Millipore Amicon Ultra-15 centrifugal filter device); (**c**–**f**), SEM pictures of crystals formed in the presence of ASW (**c** and **e**) or EPF (**d** and **f**). The final dosage of EPF was 50 µg/mL, and the same content of ASW was added as a control. The final dosage of $CaCl_2$ and $NaHCO_3$ was 50 mM in (**c**) and (**d**) and 5 mM in (**e**) and (**f**). The corners of (**c**) and (**d**) showed high-amplification images of the particles. The crystallization occurs at room temperature (approximately 20 °C)

bodies or bacteria. The EPF just plays a paradoxical role to achieve this. The results demonstrated that 50 μg/ml EPF inhibited or slowed the crystallization of $CaCO_3$ in 24 h, while the concentration of Ca^{2+} and HCO_3^- was reduced to 5 mM. ICP results showed that the concentration of Ca^{2+} in EPF of *P. fucata* was approximately 8.95 ± 0.30 mM, which was similar to that of ASW (approximately 8.75 ± 0.23 mM), and they were constant as previously reported [173]. The concentration of HCO_3^- in EPF was about 3.6 mM, while 2 mM in seawater. In contrast, the concentration of EPF proteins was 200–400 μg/ml, which was enough to inhibit and control the precipitation of $CaCO_3$ regularly. This guarantees the $CaCO_3$ will not precipitate only if it is transported to the right sites.

3.7.4.4 Interaction Between EPF Proteins, Ions, and Crystals

The effect of EPF macromolecules on $CaCO_3$ crystallization might be relative to the interaction of Ca^{2+} and CO_3^{2-}. Thus, we investigated the binding between EPF proteins and ions by intrinsic fluorescence quenching. The presence of $CaCl_2$ caused quenching of intrinsic fluorescence of EPF proteins when the dosage of Ca^{2+} reached 10 mM. The presence of 10 mM Na_2CO_3 led to the intrinsic fluorescence quenching of EPF proteins and caused a red shift, which suggested a conformational change of the EPF proteins (Fig. 3.101a, b). The presence of NaCl and pH change had no obvious influence on the quenching of intrinsic fluorescence of EPF proteins (data not shown). What's more, compared with the natural secondary structure of EPF proteins, both $CaCl_2$ and Na_2CO_3 could cause a red shift at the 208 nm, which suggested the secondary structure changes in EPF proteins (Fig. 3.101c, d), while the presence of NaCl and pH variation could not (data not shown). Further ITC experimental results affirmed the interaction between EPF proteins and Ca^{2+} showed that the binding constant (N) was 11.7 ± 2.43 (Fig. 3.101e). These results suggest that EPF proteins could significantly interact with Ca^{2+} and CO_3^{2-} and may influence the crystallization of $CaCO_3$ through conformational changes of their secondary structure.

The results of amino acids analyses and SDS-PAGE and $CaCO_3$ crystallization experiments suggested that different proteins of EPF take part in the formation of calcite and aragonite. What's more, both the aragonite and calcite crystals could cause a significant intrinsic fluorescence quenching of EPF proteins (Fig. 3.102a, b). These results showed a similar but not the identical quenching, indicating their interactions may be distinctive. The CD spectrum also indicated that the presence of calcite or aragonite could cause a red shift at the peak of 208 nm compared with the native EPF proteins (Fig. 3.102c, d). All of these results suggested that EPF proteins could interact with the different crystals and might influence through different proteins and their secondary structure change.

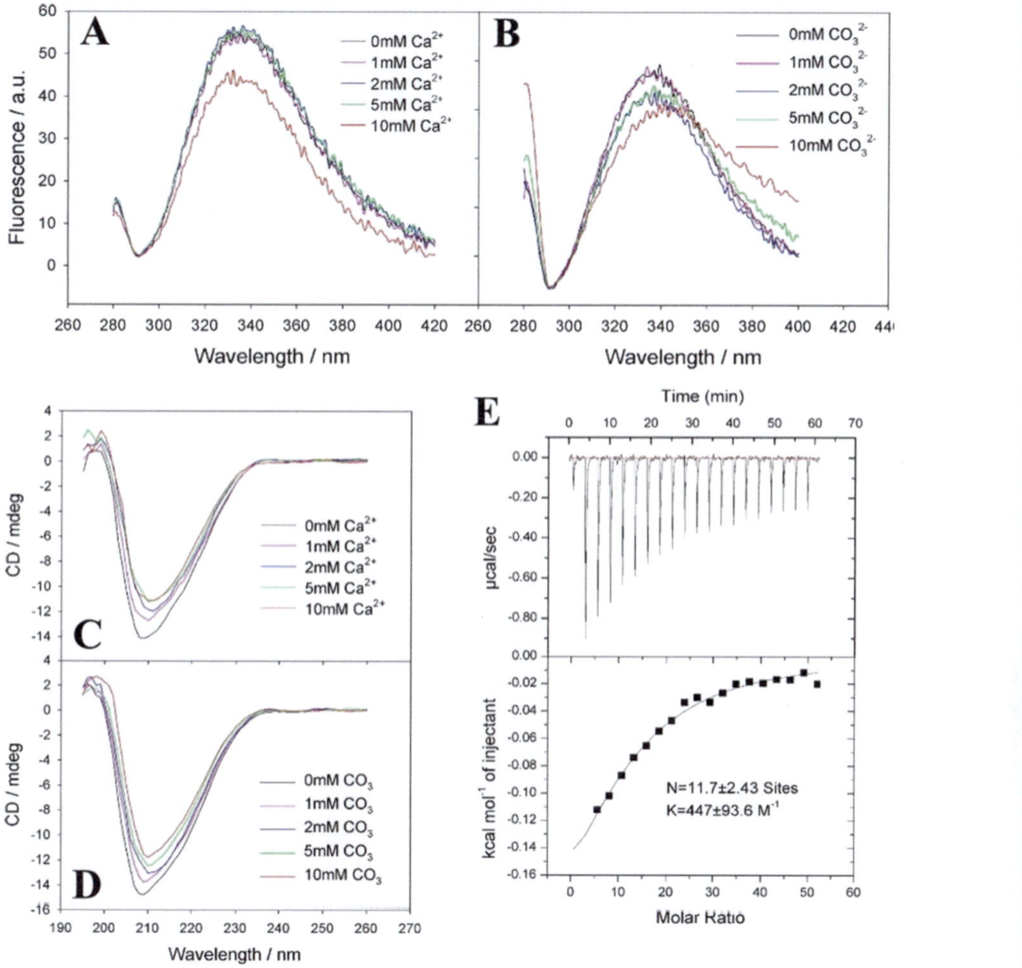

Fig. 3.101 Binding between EPF proteins, Ca^{2+}, and CO_3^{2-}. (**a**) and (**b**) indicated the change in the intrinsic fluorescence of EPF proteins when different dosages of Ca^{2+} or CO_3^{2-} were added. (**c**) and (**d**) indicated the variation of CD spectra of EPF proteins in the presence of ions. **e** indicated the ITC results between EPF proteins and Ca^{2+}. *N* number of binding sites, *K* indicates the binding constant

3.7.4.5 Effect of EPF on Shell Biomineralization In Vivo

The function of EPF on $CaCO_3$ precipitation suggested its important role in shell formation. The inner surface of nacre was also observed while EPF was separated daily. The native structure of nacre is regular aragonite lamellae (Fig. 3.103a, b, g). When the EPF was isolated daily for 10 days, random deposition occurred on the aragonite lamellae (Fig. 3.103c, d). What's more, a random depositional texture coated the aragonite slices when the EPF was isolated for 20 days (Fig. 3.103e, f). The crystal polymorph of the random deposition was calcite, on the basis of Raman

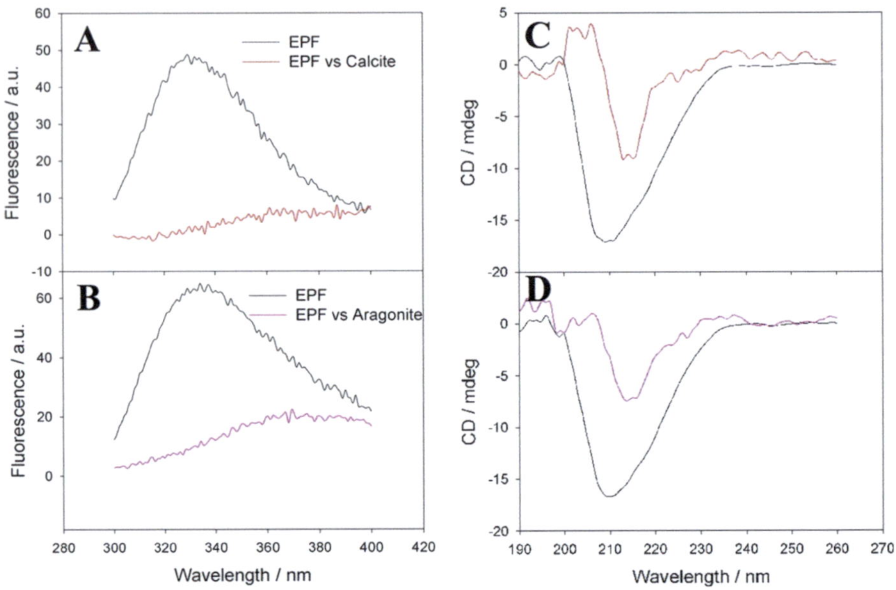

Fig. 3.102 Binding between EPF proteins and crystals. (**a**) and (**b**) indicated the intrinsic fluorescence variation of EPF proteins in the presence of 0.01 g of calcite or aragonite. (**c**) and (**d**) represented the change of CD spectra of EPF proteins in the presence of 0.01 g of calcite or aragonite

spectra (Fig. 3.103h), which directly indicated that EPF was crucial for the formation of nacre. To illuminate the mechanism, we also test the dosage of proteins and Ca^{2+} in EPF. The results indicated that the concentration of Ca^{2+} is 8.95 ± 0.30 mM during the extraction process, while the dosage of EPF proteins decreased to 0.10 ± 0.01 mg/ ml from 0.24 ± 0.01 the next day and stayed at 0.09 ± 0.01 mg/ml in the following days. The pH also maintained at 7.76 ± 0.01. Therefore, we referred that the random deposition was induced by the reduction of the EPF proteins. Thus, EPF inhibited the abnormal deposition of $CaCO_3$ and influenced the formation of the aragonite lamellae. The prismatic layer was also regulated, but the effect was much weaker (data not shown) [138].

3.7.4.6 Effect of EPF Extraction on the Expression of Shell-Related Genes

Shell-related genes are supposed to control and regulate shell formation; thus, we want to know the influence of EPF extraction on gene expression, because EPF already has a distinct effect on the shell formation. To detect the effect of EPF on gene expression, six genes, including EPF (ACCBP, PfN23), prism-related (P39, KRMP1-3), and nacre-related (Pif, nacrein) genes, were tested after EPF was isolated for 3 and 12 h. The change in gene expression was significant after EPF

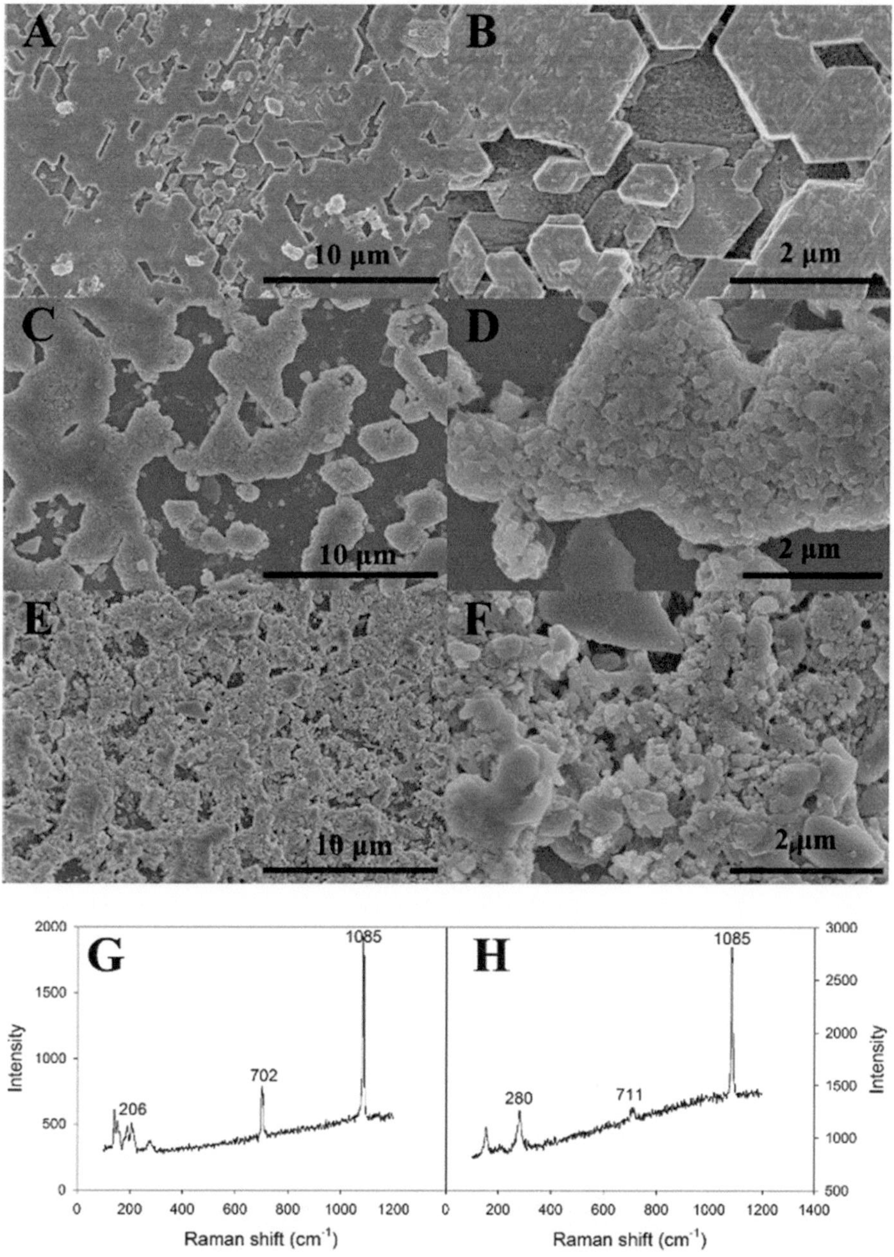

Fig. 3.103 SEM images indicating the inner face structure of nacre: inner face of natural nacre (**a**, **b**); surface of nacre after EPF extraction for 10 days (**c**, **d**) and 20 days (**e**, **f**); Raman spectrum of the natural surface of nacre (**g**) and after EPF extracted for 20 days (**h**)

extraction. The expression level of PfN23 and ACCBP obviously declined after the EPF is isolated for 3 h. The expression levels of P39 and Pif were obviously enhanced during EPF extraction, while KRMP1-3 indicated an up and down trend (Fig. 3.104). Therefore, we speculated that the changes in the EPF may occur as an external signal that influenced the expression of shell-related genes. However, the expression patterns were not fit strictly, and it might be caused by the complicated functions of these proteins.

3.8 Extracellular Matrix Protein Expressed by Mantle

3.8.1 A Novel Extracellular EF-Hand Protein Involved in the Shell Formation of Pearl Oyster

Mollusk shell formation is a complicated and highly controlled calcium metabolism process. Previous studies revealed that several EF-hand calcium-binding proteins actively participate in the regulation of shell mineralization. In this study, we cloned a full-length cDNA encoding a novel extracellular EF-hand calcium-binding protein (named EFCBP) from the pearl oyster, *Pinctada fucata*, according to the EF-hand motifs of calmodulin. Although it shares high similarity with the calmodulin family in its EF-hand signatures [174], EFCBP has just two EF-hand motifs and belongs to a new separate group from the other EF-hand proteins according to a phylogenetic analysis. EFCBP is specifically expressed in shell mineralization-related tissues, viz., the mantle, the gill, and the hemocytes. Moreover, its expression responds

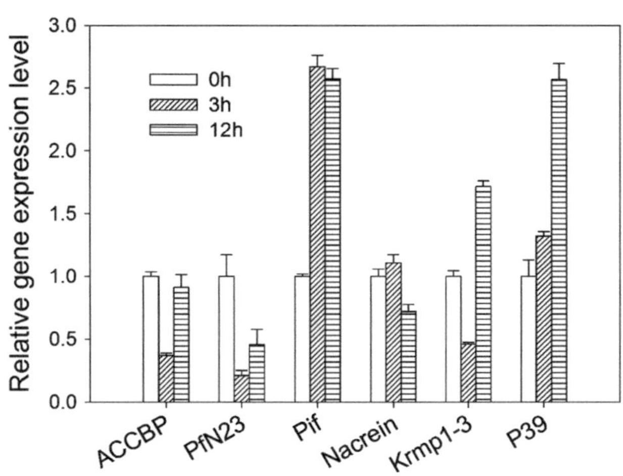

Fig. 3.104 Function of EPF extraction on the relative expression of biomineralization-related genes. The relative gene expression level was tested for 0, 3, and 12 h after extraction of the EPF. ACCBP and PfN23 were proteins existed in the EPF. Pif and nacrein existed in nacre. Lysine (K)-rich matrix protein (Krmp1-3) and Prisilkin-39 (P39) existed in the prism

quickly only to the shell damage, but not to the damage of other tissues and the infection of the lipopolysaccharides from *E. coli*. These results suggest that EFCBP might be an important regulator of shell formation. This finding may help to better understand the functions of EF-hand proteins on the regulation of mollusk shell formation.

3.8.1.1 Obtain of the Full-Length cDNA Sequence and Analyses of the Deduced Amino Acid Sequence

The 3′-RACE experiment with the EFP5 primer augmented a specific PCR product ~167 bp in length, whereas the 3′-RACE amplification with the EFP5′ primer gave no PCR product. A potential EF-hand motif is encoded by the 167 bp PCR product. According to this sequence, the residual of the full-length cDNA sequence was cloned via the 5′-RACE reaction. To affirm the accuracy of the cDNA sequence, an amplification of the full-length cDNA was conducted with the primer pair of EFgsp3 and EFgsp5 (Fig. 3.105), with the mantle cDNA library of *P. fucata* as a template. The result was accordant with that of the RACE. As Fig. 3.105 had shown, the full-length cDNA sequence excluding the poly(A) tail was 525 bp in length, with an open reading frame (ORF) encoding 129 amino acid residues. The predicted

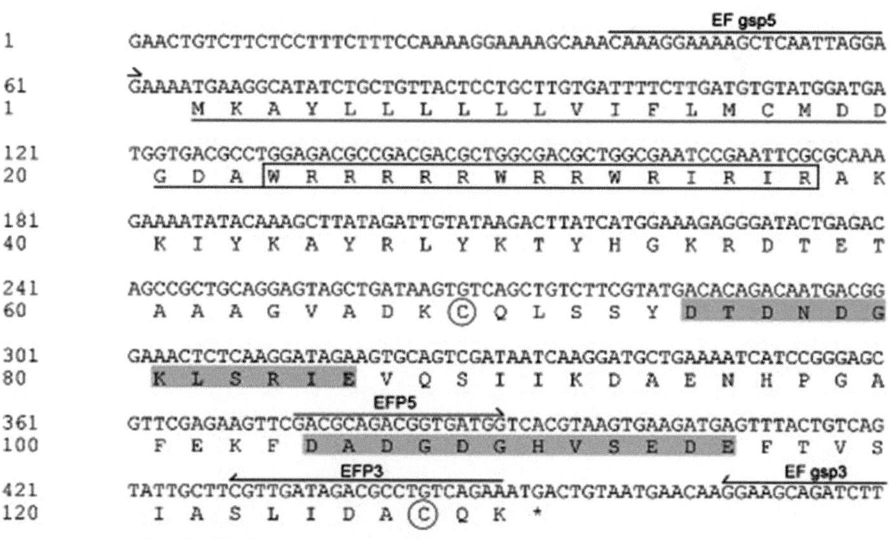

Fig. 3.105 The cDNA sequences and the putative amino acid sequences of EFCBP from the pearl oyster, *P. fucata*. The predicted signal peptide is underlined. The low complexity region rich in arginine is enclosed in the box. The two EF-hand calcium-binding domains are shaded. The two cysteine residues that might construct a disulfide bond are circled. The primers designed for cloning and RT-PCR are indicated with arrows [174]

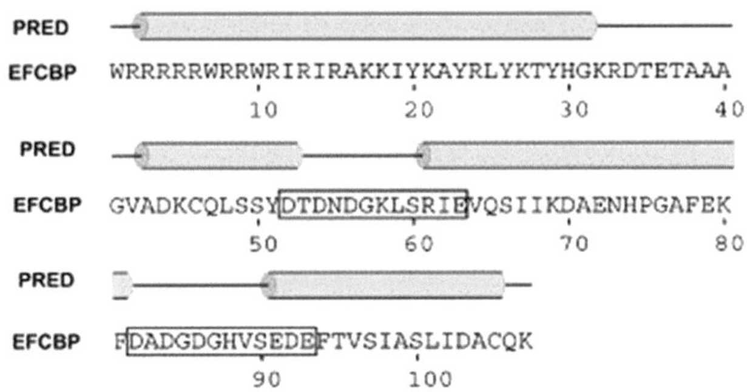

Fig. 3.106 Secondary structure prediction of EFCBP. PRED means the predicted secondary structure of EFCBP. The gray barrels mean the predicted α-helices, the black lines show the predicted random coils, and the two EF-hand regions are boxed

polyadenylation signal (AATAAA) started from the 53rd nucleotide downstream of the stop codon (TGA). The cDNA sequence has the accession number DQ494416 in GenBank [174].

The predicted signal peptide was the first 22 amino acids of the deduced protein. The mature protein excluding the signal peptide was constituted of 107 amino acid residues with a calculated molecular weight of 12.5 kDa and a predicted isoelectric point of 9.78. Analyses of its amino acid composition showed the two cysteine residues might form an intramolecular disulfide bond. Protein domain search against the PROSITE database found the existence of two EF-hand calcium-binding motifs from the 74th to the 85th amino acid residue and from the 104th to the 115th amino acid residue. What's more, a low complexity region with high percentage of arginine was observed from the 24th to the 37th amino acid, and the N terminus was also rich in lysine (Fig. 3.105). We called the protein as EFCBP (EF-hand calcium-binding protein). There are five coil regions (29.9%), four α-helix regions (70.1%), and no β-sheet in the predicted secondary structures of EFCBP. As Fig. 3.106 had shown, each EF-hand motif was besieged by two α-helixes, which was accordant with the characteristic conformation of helix–loop–helix in the canonical EF-hand domain [175].

A homology search using the BLAST program against the entire protein sequence database in GenBank showed that EFCBP had the highest homology with the calconectin (DQ352042.1) of *P. margaritifera* and was also analogous to the PFMG1 (DQ104255) and PFMG6 (DQ104260.1) of *P. fucata*. Even though calconectin has the highest identity, it has no signal peptide and seems to be intracellular. The PFMG1 and PFMG6 have a predicted signal peptide at the N terminus, with two EF-hand motifs located at the C terminus; thus, from the viewpoint of subcellular localization and protein structure, EFCBP might be more homological with PFMG1 and PFMG6. What's more, as Fig. 3.107 had shown, the EF-hand characters of EFCBP have high similarity with those of calmodulin and

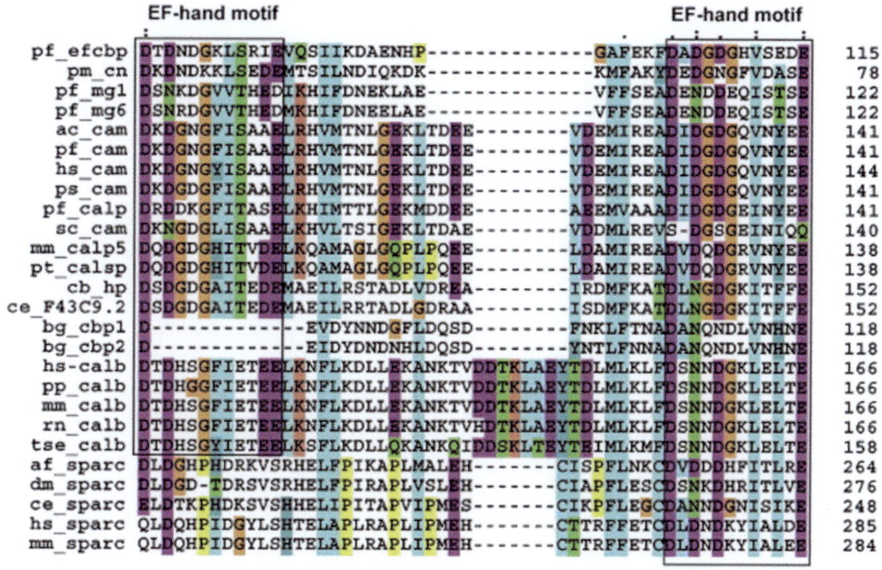

Fig. 3.107 Protein sequence alignment among the EF-hand signatures of EFCBP, calconectin, PFMG1, PFMG6, calmodulin and calmodulin-like protein family, calbindin, and SPARC. Numbers mean the positions of the amino acid residues in each sequence of genes. Gaps (–) are added to optimize the alignment. The two EF-hand calcium-binding models are enclosed in the box. Homologous and conservative amino acid sequences are indicated with dots. *pf_efcbp Pinctada fucata* EFCBP (DQ494416), *pm_cn Pinctada margaritifera* calconectin (DQ352042.1), *pf_mg1 Pinctada fucata* PFMG1 (DQ104255), *pf_mg6 Pinctada fucata* PFMG6 (DQ104260.1), *ac_cam Aplysia californica* calmodulin (P62145), *pf_cam Pinctada fucata* calmodulin (AAQ20043), *hs_cam Homo sapiens* calmodulin (CAA36839), *ps_cam Patinopecten sp.* calmodulin (P02595), *pf_calp Pinctada fucata* calmodulin-like protein (AAV73912), *sc_cam Saccharomyces cerevisiae* calmodulin (P06787), *mm_calp5 Macaca mulatta* calmodulin-like 5 (XP_001104766), *pt_calsp Pan troglodytes* calmodulin-like skin protein (XP 001144681), *cb hp Caenorhabditis briggsae* hypothetical protein CBG14543 (CAE68649), *ce_F43C9.2 Caenorhabditis elegans* F43C9.2 (NP_508818), *bg_cbp1 Biomphalaria glabrata* Calcium-binding protein 1 (AAV91525), *bg_cbp2 Biomphalaria glabrata* calcium-binding protein 2 (AAV91522), *hs_calb Homo sapiens* calbindin 1 (NP_004920), *pp_calb Pongo pygmaeus* calbindin (Q5R4V1), *mm_calb Mus musculus* calbindin-28K (NP_033918), *rn_calb Rattus norvegicus* calbindin-d28k (AAA40852), *tse_calb Trachemys scripta elegans* calbindin-D28K (AAO49508), *af_sparc Artemia franciscana* SPARC (BAB20042), *dm_sparc Drosophila melanogaster* BM40-SPARC CG6378-PA (NP_651509), *ce_sparc Caenorhabditis elegans* osteonectin (sparc)-related protein 1(AAB88325), *hs_sparc Homo sapiens* SPARC (CAG33080), *mm_sparc Mus musculus* SPARC (CAJ18514)

calmodulin-like protein family, like SPARC, calbindin, and some other calcium-binding proteins, while, a phylogenetic analysis of their EF-hand signatures showed that EFCBP did not belong to any of the above families. EFCBP, PFMG1, PFMG6, and calconectin might compose another new EF-hand protein family with two EF-hand motifs (Fig. 3.108).

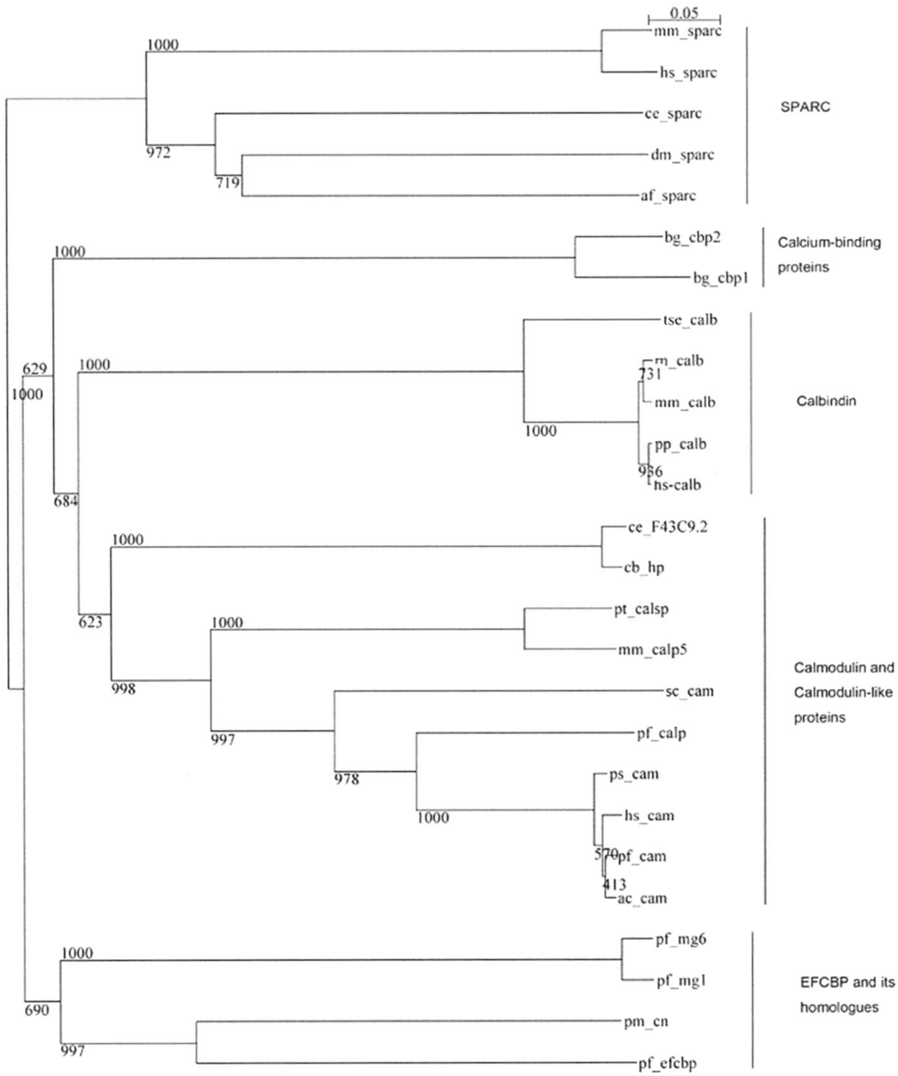

Fig. 3.108 A phylogenetic tree analyses of the EF-hand characters of EFCBP, calconectin, PFMG1, PFMG6, calmodulin and calmodulin-like protein family, calbindin, and SPARC. The tree is hypothesized with the ClustalX program via using the method of neighbor-joining analyses. The input alignment is the same with Fig. 3.107. Bootstrap values (1000 replicates) are suggested at the nodes. The abbreviations of protein names are same with Fig. 3.107

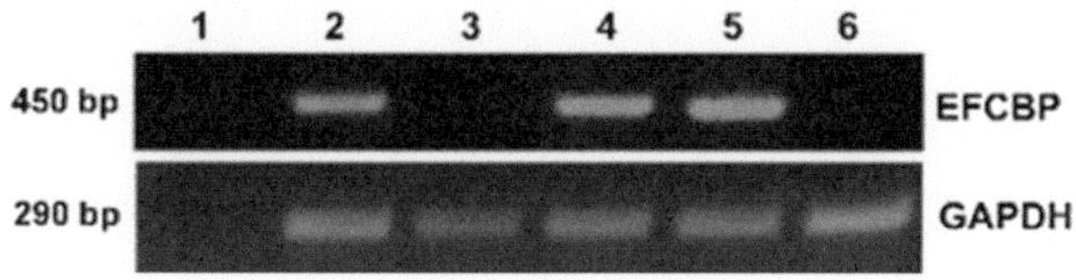

Fig. 3.109 RT-PCR analyses of relative EFCBP gene expression in different tissues of the oyster, *P. fucata*. Lane 1, negative control; lane 2, mantle; lane 3, viscus; lane 4, gill; lane 5, hemocytes; lane 6, adductor muscle. The housekeeping gene GAPDH is used as a positive control. The PCR synthesis of EFCBP is 450 bp in length, and the full-length of GAPDH is 290 bp

3.8.1.2 Gene Expression Analyses by RT-PCR and In Situ Hybridization Assay

As PFMG1, PFMG6, and calconectin are mantle-specific proteins and have been found to regulate the biomineralization process; we referred that EFCBP might play an analogous role. Thus, we analyzed the mRNA expression level of EFCBP in different tissues. RT-PCR was conducted with RNA samples from the mantle, gill, hemocytes, viscus, and adductor muscle and the gene-specific primers of EFCBP (EFgsp5 and EFgsp3). The housekeeping gene GAPDH was utilized as a positive control. As Fig. 3.109 had shown, the mRNA of EFCBP is expressed highly in the mantle, hemocytes, and gill of the oyster. These tissues, mantle in especial, actively take part in the calcium metabolism and shell formation.

The precise mRNA expression location of EFCBP in the mantle tissue was then analyzed by in situ hybridization. As Fig. 3.110b had shown, strong hybridization signals were observed in the outer epithelial cells of the middle fold, the inner epithelial cells of the outer fold, and the epithelial cells of the mantle pallial. These cells participate in different process of shell formation. Hybridization with a sense probe was conducted as a negative control with no signals detected (Fig. 3.110a).

These results of gene expression analyses showed that EFCBP was expressed by the cells and tissues participating in oyster biomineralization specifically. Thus, it might be a protein related to shell formation.

3.8.1.3 Gene Expression Analyses of EFCBP After Shell Notching, LPS Infection, and Tissue Damages

To further investigate the function of EFCBP in the biomineralization of *P. fucata*, a shell notching assay was performed. We cut a v-shaped notch on the shell margin close to the adductor muscle of the oyster to induce rapid shell repair, and then the expression levels of EFCBP was detected by RT-PCR at different times after shell notching. It was found that the expression level of EFCBP was significantly upregulated after shell repair began. After around the sixth hour, the mRNA

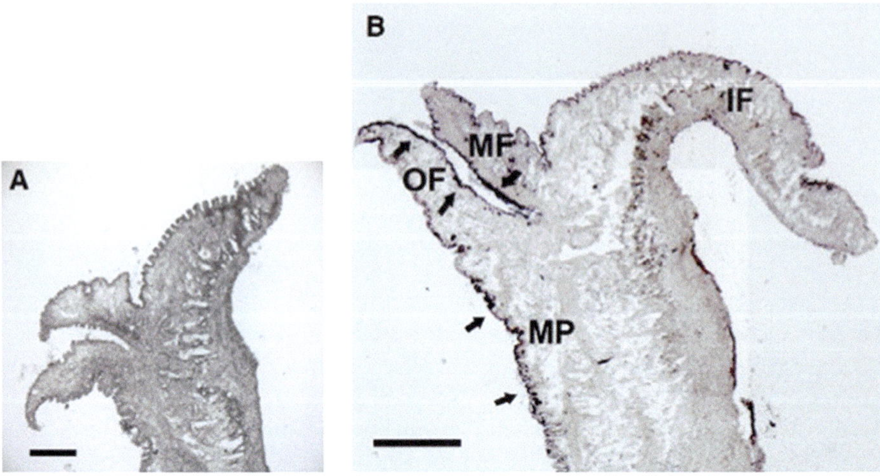

Fig. 3.110 In situ hybridization analysis of distribution of EFCBP gene expression of the mantle in *P. fucata*. (**a**) The sections of the oyster mantle are incubated with the sense RNA probes labeled with digoxigenin, which is used as a negative control. (**b**) The frozen sections of the oyster mantle are incubated with the antisense RNA probes labeled with digoxigenin. Three overlapping images of the same section are taken. Strong hybridization signals occurred in the inner epithelial cells of the outer fold, the epithelial cells of the mantle pallial, the outer epithelial cells of the middle fold (arrows indicate). *OF* outer fold, *MF* middle fold, *IF* inner fold, *MP* mantle pallial. Scale bars, 0.5 mm

expression level reached a peak and then gradually recovers to the basal level (Fig. 3.111a). In order to eliminate the possibility that the increase of the EFCBP expression level was caused by demands of wound healing or the immune response, we conducted the LPS infection assay and tissue damage and regeneration assay. Lipopolysaccharides (LPS) could cause acute immune response, but Fig. 3.111b shows that EFCBP did not respond to the LPS infection. What's more, neither the mantle damage nor the gill damage could change the expression level of EFCBP (Fig. 3.111c–d). Thus, we supposed that EFCBP might function as an important regulator of shell growth.

*3.8.2 A Novel Ferritin Subunit Involved in Shell Formation from the Pearl Oyster (*Pinctada fucata*)*

Iron is one of the most important minor elements in the shell of bivalves. This study was designed to investigate the involvement of ferritin, the principal protein for iron storage, in shell formation. A novel ferritin cDNA from the pearl oyster (*Pinctada fucata*) was isolated and characterized. The ferritin cDNA encodes a 206 amino acid polypeptide, which shares high similarity with snail soma ferritin and the H-chains

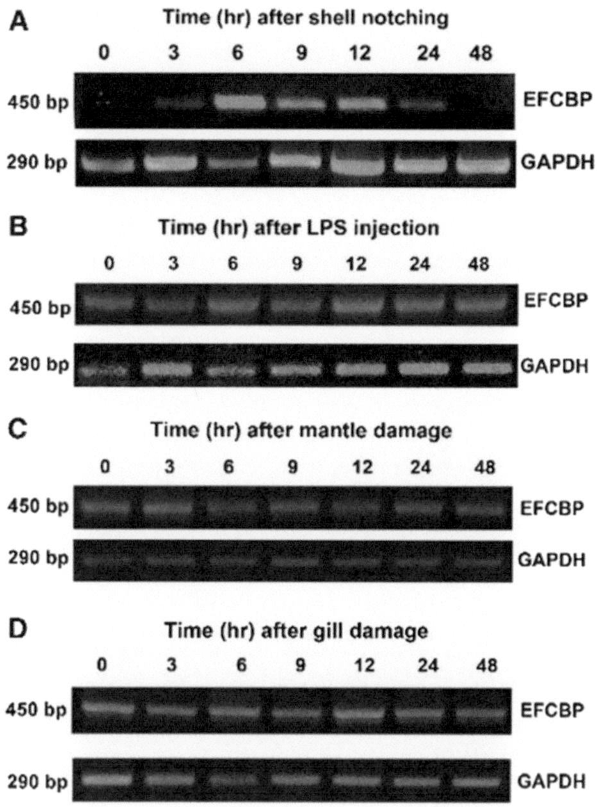

Fig. 3.111 RT-PCR analyses of the relative EFCBP gene expression after (**a**) the shell notching, (**b**) the LPS injection, (**c**) the mantle damage, and (**d**) the gill damage. The housekeeping gene GAPDH is used as a positive control. The PCR segment of EFCBP is 450 bp in length, and the full length of GAPDH is 290 bp

of mammalian ferritins [103]. Oyster ferritin mRNA shows the highest level of expression in the mantle and the organ for shell formation. In situ hybridization analysis revealed that oyster ferritin mRNA is expressed at the highest level at the mantle fold, a region essential for metal accumulation, and contributes to metal incorporation into the shell. Taken together, these results suggest that ferritin is involved in shell formation by iron storage. The identification and characterization of oyster ferritin also helps to further understand the structural and functional properties of molluscan ferritins.

3.8.2.1 Isolation and Sequence Analyses of a cDNA Clone and Phylogenetic Analyses

A 354 bp cDNA product was obtained via RT-PCR with two degenerate oligonucleotides (D1 and D2) with total mantle RNA as the template. The DNA sequence of the PCR product had high similarity with the H-chains of mammalian ferritins. According to the sequence, a pair of specific primers (G1 and G2) was designed and the mantle cDNA library was searched with PCR by these primers. With four

```
 GCACGAGGCTTGCTGCGTCAGTGAACGTACGGGCAGATTAGCCTAACTCCGTCTTACCA   -55
AATAATTCAACTTTATATTTTACACAACAACCTACTTTAAACAACTGAATCAAAATGGCA    6
                                                         M  A    2
CTGAGTCAACCTCGCCAGAATTTCCACGTTGAAAGTGAAGCAGGGATCAACAGGCAGATA   66
 L  S  Q  P  R  Q  N  F  H  V  E  S  E  A  G  I  N  R  Q  I    22
AACATGGAATTGTACGCCAGTTACACCTATCAGTCAATGTCCTTCTACTTTGATCGAGAT  126
 N  M  E  L  Y  A  S  Y  T  Y  Q  S  M  S  F  Y  F  D  R  D    42
GATGTCGCTCTACCAGGATTTGCCAAATTTTTCAAGCACTCGTCTGATGAGGAACGAGAA  186
 D  V  A  L  P  G  F  A  K  F  F  K  H  S  S  D  E  E  R  E    62
CATGCTGAGAAGCTGATGAAGTATCAGAACAAGAGAGGAGGGCGTATCGTTCTCCAAGAC  246
 H  A  E  K  L  M  K  Y  Q  N  K  R  G  G  R  I  V  L  Q  D    82
ATCAAGAAGCCTGATCGTGATGATTGGGGACTGGACTGGAGTCCATGCAGGTTGCCCTG   306
 I  K  K  P  D  R  D  D  W  G  T  G  L  E  S  M  Q  V  A  L   102
CAACTAGAGAAGTCTGTCAATCAGTCCCTGCTTGACCTTCATAAAGTGGCCGACACCCAT  366
 Q  L  E  K  S  V  N  Q  S  L  L  D  L  H  K  V  A  D  T  H   122
GGAGATGCCCAGATGATGGATTTCATAGAGAGTGAATACCTTGAGGAGCAAGTGAATGCT  426
 G  D  A  Q  M  M  D  F  I  E  S  E  Y  L  E  E  Q  V  N  A   142
ATAAAGGAGATCTCTGACCACATCACCCAGCTGAAGCGTGTAGGTCCAGGACTTGGAGAG  486
 I  K  E  I  S  D  H  I  T  Q  L  K  R  V  G  P  G  L  G  E   162
TACCAATATGACCTGAAACTCCAGTCATCTGAGTTAAGGGGATCTGGCCCCTCACTCCCC  546
 Y  Q  Y  D  L  K  L  Q  S  S  E  L  R  G  S  G  P  S  L  P   182
ACAGGCCCTCATTGTTATGGGAAATGTTATCCCATTGTTATTGGACTCCTTTTGGTGAAA  606
 T  G  P  H  C  Y  G  K  C  Y  P  I  V  I  G  L  L  L  V  K   202
ATTCATTATTGTTAATTCTGGGACTCCTAACTTTCCCTTAATTTTTAAGAAACCCATCAA  666
 I  H  Y  C  *                                                206
CCCAAAAAAAAAAAAA                                              681
```

Fig. 3.112 The oyster ferritin cDNA sequence and putative amino acid sequence. The predicted iron responsive element (IRE) in the 59 UTR is underlined. The conserved amino acid residues, according to which degenerate primers were hypothesized, are in bold. The nucleotides designed as the specific primers used for expression analysis are shaded. The stop codon is shown with an asterisk [103]

rounds of screening, six cDNA clones were separated. Sequence analyses showed that the longest cDNA sequence (794 bp) contained a 618 bp open reading frame, with a poly(A) tail at its 3′ end and a 113 base 5′ untranslated region (UTR) (Fig. 3.112). The predicted ferritin protein included 206 amino acid residues with a calculated molecular weight of 23.6 kDa, which is bigger than that of most mammalian ferritins [176] and snail soma ferritin [177].

A supposed iron responsive element (IRE) sequence (nt y106 to y80) is found in the 5′ UTR of the cDNA, which includes a characteristic CAGUGN loop and a bulged Cys and contains the overall structure of the ferritin IREs (Fig. 3.113a). IREs have been observed in almost all identified ferritin subunits mRNAs but ferritins of

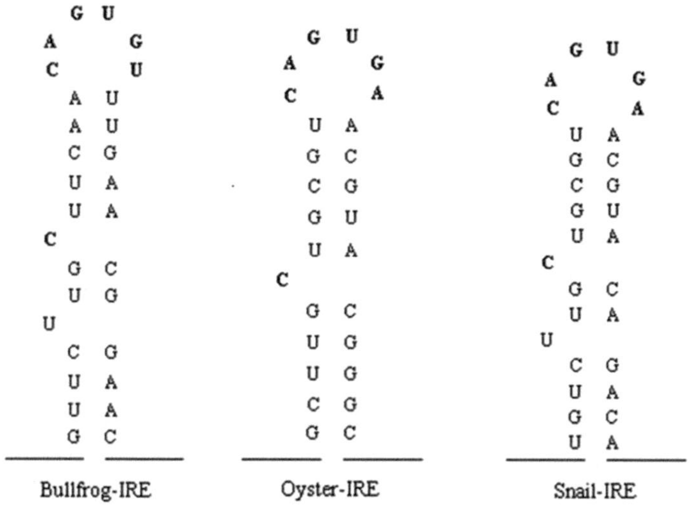


```
Pfa     GCTTGCTGCGTCAGTGAACGTACGGGC
LstS    T----------CAGTG-----A-A-A-A
RnoH    TCT----T-AACAGTGTTT-AAC-GAACC
HsaH    TCC----T-AACAGTGTTT-GAC-GAACC
Afo     TTG---GTTCGCAGTGTCG-AACCAAGCA
Ple      TCC-GGT--CCAGTGTGT-AAC-A-CTC
```

Fig. 3.113 IRE alignments and the predicted stem-loop structure. (**a**). Stem-loop structure of the predicted oyster IRE blasted with IREs of snail soma ferritin mRNA and bullfrog ferritin mRNA, indicating the classical CAGUGN loop and the bulged Cys. The amino acid residues of the loop and the bulged Cys are bolded. (**b**). Comparison of the oyster putative IRE with IREs of 59 UTRs from several ferritins. Conserved residues are shaded and residues accordant with residues in oyster IRE are indicated by hyphens. Abbreviation and sources are according to Fig. 3.114

Schistosoma mansoni and snail yolk ferritin. Sequence alignment showed that the putative IRE of oyster ferritin shared high homology with the IRE of snail soma ferritin and its mammalian homologues (Fig. 3.113b). Since IRE could be folded into a stem-loop structure which could bind iron regulatory protein (IRP), the existence of IRE in the 5′ UTR of oyster ferritin mRNA shows the expression of oyster ferritin is regulated by iron at the translation level.

The sequence of oyster ferritin has high similarity with those of ferritins from animals and plants (Fig. 3.114). The primary structure of oyster ferritin more resembled mammalian H-chains than L-chains. Seven amino acids identified as metal ligands at the H-specific ferroxidase center of mammalian ferritins are completely the same with oyster ferritin, which are Glu25, Tyr32, Glu59, Glu60, His63, Glu105, and Gln139 [176]. Alignment analyses indicate that residue Tyr27 is conserved in all oyster ferritin, snail soma ferritin, starfish ferritin, crayfish ferritin,

```
Pfa     (1)  --------------------------------------------------
LstS    (1)  --------------------------------------------------
HsaH    (1)  --------------------------------------------------
RnoH    (1)  --------------------------------------------------
Afo     (1)  --------------------------------------------------
Ple     (1)  --------------------------------------------------
Aae     (1)  ------------------------------------------------MM
Dmel    (1)  --------------------------------------------------
Gma     (1)  MALAPSKVSSFSGFSPKPSVGDALKNPTCSVSLSFANVKLGSRN-----L
Bna     (1)  MASKALSSFTAKPAVSLLPHGVSSSASPSVMSLSFS-RHTGGRGVVAASS
Sma2    (1)  --------------------------------------------------

                                                       A helix
Pfa     (1)  ------------------------------MALSQPRQNFHVESEAGINR
LstS    (1)  ------------------------------MSVSQARQNYHAESEAGINR
HsaH    (1)  ---------------------------MTTASTSQVRQNYHQDSEAAINR
RnoH    (1)  ---------------------------MTTASPSQVRQNYHQDSEAAINR
Afo     (1)  --------------------------------MATIRQNYNETSEAGVNK
Ple     (1)  -------------------------------MASQIRHNYHEDCEP-INK
Aae     (3)  KSVFFGVVAITVAILSIYQETAQAQEQTVGATDNYQWDSVDDQCLAALHR
Dmel    (1)  ---MVKLIASLLLLAVVAQAYGDFKCSLAVPEITKDWVDMKDACIKGMRN
Gma     (46) RVCASTVPLSGVIFEPFEEVKKGELAVPTAPQVSLARQNYADECESAINE
Bna     (50) TVDTNNMPMTGVVFQPFEEVKKADLAIPITSNASLARQRYADSSEAAINE
Sma2    (1)  ------------------------------MSSSRARQSFATECENAINK

                                                 B helix
Pfa     (21) QINMELYASYTYQSMSFYFDRDDVALPGFAKFFKHSSDEEREHAEKLMKY
LstS    (21) QINMELYASYSYQSMAYYFDRDDVALPGFHKFFKHQSEEEREHAEKLMKY
HsaH    (24) QINLELYASYVYLSMSYYFDRDDVALKNFAKYFLHQSHEEREHAEKLMKL
RnoH    (24) QINLELYASYVYLSMSCYFDRDDVALKNFAKYFLHQSHEEREHAEKLMKL
Afo     (19) QINLELYASYTYLSMAFYFDNTTVALPGAHKYFKKASDEEREHAMLLMKF
Ple     (19) QINLEFYASYVYMSMGHYFDRDDISLPGASKFFKDSSDEEREHGQKLMKY
Aae     (53) QINKEFDASIIYLKYAAYFAQEKINLPGFEKFFFHAAAEEREHGIKLIEY
Dmel    (48) QIQEEINASYQYLAMGAYFSRDTVNRPGFAEHFFKAAKEEREHGSKLVEY
Gma     (96) QIKVEYNASYAYHSLFAYFDRDNVALKGFAKFFKESSEEEREHAEKLMKY
Bna    (100) QINVEYNVSYVYHSMYAYFDRDNVALKGLAKFFKESSDEEREHAEKFMEY
Sma2    (21) QINVELQAAYDYMAFFTYFDRDDVSFPKAAEFFRKASHEEREHAEKLAKY
```

Fig. 3.114 Blast of the sequences of oyster ferritin with the other ferritins of invertebrates, vertebrates, and plants. The consensus residues involved in the iron-binding sites are enclosed by the box. The conserved Tyr27 and predicted glycosylation sequence are shaded. Residues Met24 and Gln33, which are consensus and unique in oyster ferritin and snail soma ferritin, are bolded and shaded. Pfasoyster (*Pinctada fucata*) ferritin, AF547223; LstSssnail (*Lymnaea stagnalis*) soma

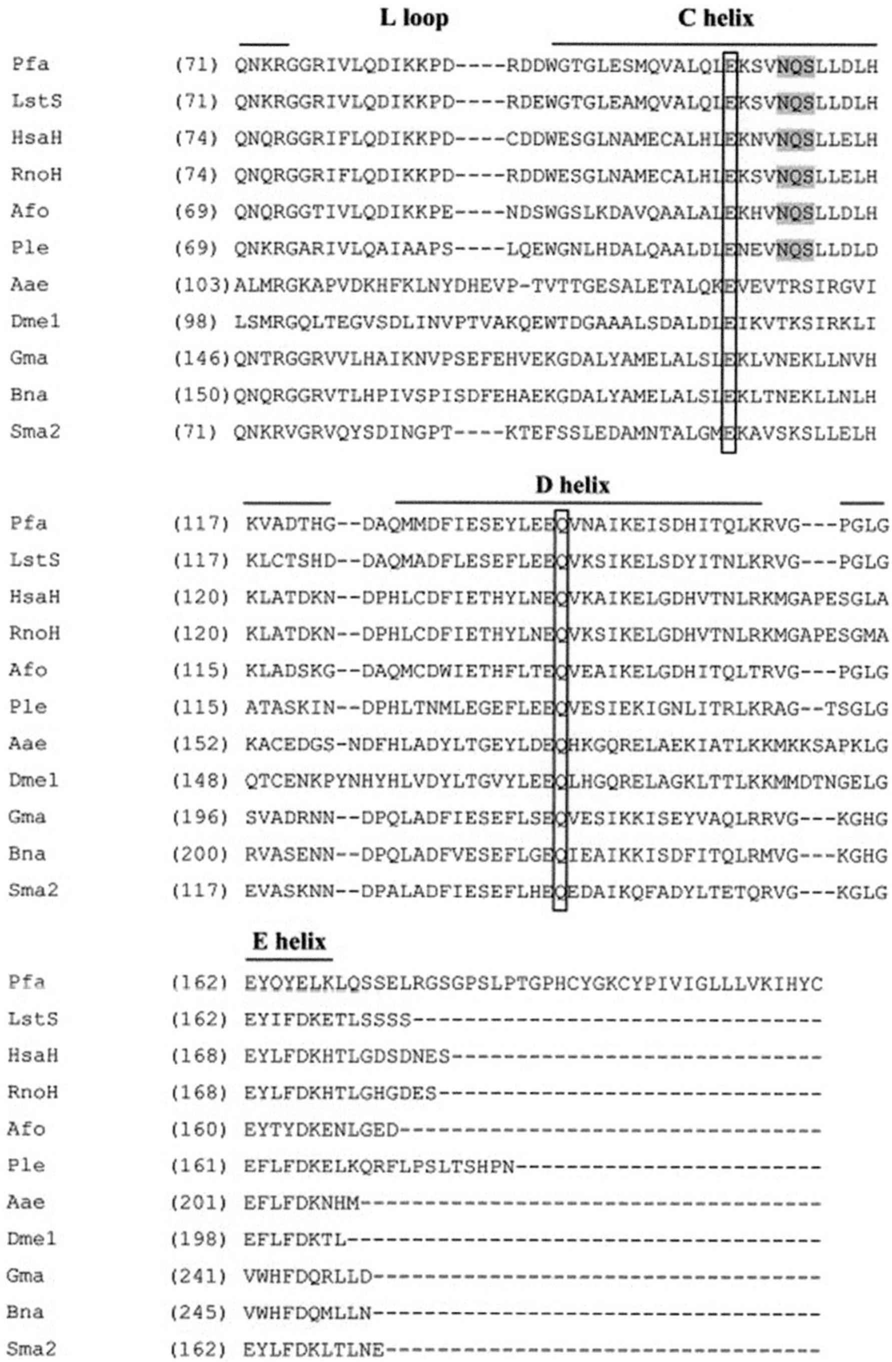

Fig. 3.114 (continued) ferritin, CAA40096; HsaHshuman (*Homo sapiens*) ferritin H-chain, AAA52437; RnoHsrat (*Rattus norvegicus*) ferritin H-chain, AAB39890; Afosstarfish (*Asterias forbesii*) ferritin, AAB60883; Plescrayfish (*Pacifastacus leniusculus*) ferritin, CAA62186; Aaesmosquito (*Aedes aegypti*) ferritin, AAA99996; Dme1sfruitfly (*Drosophila melanogaster*) ferritin chain 1, AAB70121; Gmassoybean (*Glycine max*) ferritin, AAB18928; Bnasrape (*Brassica napus*) ferritin, AAB53099; Sma2sSchistosome (*Schistosoma mansoni*) ferritin chain 2, AAA29881

and mammalian H-ferritins [176–179]. As Tyr27 was supposed to be necessary for the rapid biomineralization of iron in vertebrate H-chains [180], these suggest that these ferritins have the ability of a higher rate of iron biomineralization than other ferritins. A proposed glycosylation sequence Asn–Gln–Ser (aa 109–111) is also conserved in these ferritins. What's more, compared to amino acid sequences of pearl oyster, ferritin with its plants and animals homologues indicates several conserved regions, which include five α-helix regions (A–E), and the L-loop linking helices B and C [176] (Fig. 3.114). This evidence suggests that oyster ferritin also shares homology of two-dimensional structure with other ferritins [103].

The sequence of oyster ferritin is different from other sequences with its longer C-terminus, which leads to its larger protein weight. What's more, residues Met24 and Gln33, which are adjacent to residues Glu25 and Tyr32, respectively, are conserved only in the sequences of oyster ferritin and snail soma ferritin. As Glu25 and Tyr32 are residues related with the iron-binding site at the ferroxidase center, this shows that the ferroxidase centers of oyster ferritin and snail soma ferritin may own unique functional properties and are three-dimensional.

The similarity of ferritin is quantified as Fig. 3.115, which possess both a phylogenetic tree calculated by ClustalX program with the NJ algorithm and a matrix of sequence identities. As Fig. 3.115 had shown, oyster ferritin indicates the highest similarity of sequence with soma ferritin, snail *(Lymnaea stagnalis*) and starfish (*Asterias forbesi*) ferritin [177, 179]. Whereas, oyster ferritin and its mammalian homologues have over 50% amino acid sequence similarity. The sequence alignments of oyster ferritin with plant, vertebrate L-ferritins, and insect are relatively low. Phylogenetic analyses indicate that the oyster ferritin is related closely to snail soma ferritin and starfish (*A. forbesii*) ferritin and mammalian H-chains. Even though insect is a member of invertebrate animals, insect ferritins are clearly different from oyster ferritin. The evidence shows that the studied oyster ferritin is originated from the same ancestor of starfish ferritin, snail soma ferritin, and mammalian H-chains.

3.8.2.2 Expression Analyses and Ability of Iron Incorporation

To explore the function of oyster ferritin in vivo, the tissue-specific expression level of oyster ferritin mRNA of different tissues was detected by employing semiquantitative PCR. RT-PCR performed with the total RNA of various tissues as template and specific primers (G1 and G2) produced a 325 bp product, whereas the negative control had no product (data not shown). The PCR product was inserted into pGEM-T vector and confirmed by sequencing. RT-PCR reactions were conducted with RNA samples from mantle, gill, digestive gland, and muscle. As Fig. 3.116 had shown, oyster ferritin mRNA expressed in all tissues was tested, while the expression level of oyster ferritin mRNA was higher in the digestive gland and the mantle than in the muscle and the gill despite equal levels of 28S ribosomal RNA in all tissues. Data from three independent experiments were conducted to avoid error.

(a)

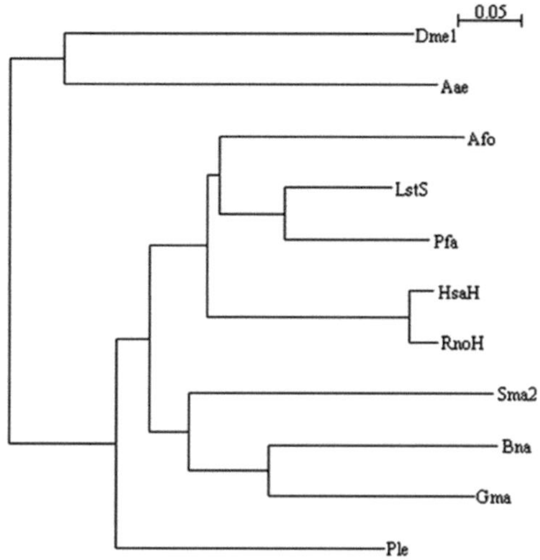

(b)

		Sequence identities (%)										
		Pfa	LstS	RnoH	HsaH	Afo	Ple	Aae	Dme1	Gma	Bna	Sma2
		1	2	3	4	5	6	7	8	9	10	11
Mollusc	Pfa	100	67.5	51.9	52.4	53.9	47.8	23.1	22.2	33.7	32.5	41.3
	LstS		100	64.8	64.5	64.4	54.1	30.5	28.1	39.5	36.6	52.3
Vertebrate	RnoH			100	94.0	57.1	47.9	27.4	28.8	35.3	34.4	46.7
	HsaH				100	57.9	47.3	27.8	29.1	34.4	34.6	47.5
Echinoderm	Afo					100	50.5	24.1	24.9	31.7	31.6	45.1
Arthropod	Ple						100	28.5	28.9	32.1	31.2	43.7
	Aae							100	39.3	23.2	20.3	23.7
	Dme1								100	20.8	20.8	39.9
Plant	Gma									100	60.6	35.7
	Bna										100	33.7
Fungus	Sma2											100

Fig. 3.115 Phylogenetic analysis and amino acid sequences alignment. (**a**) Phylogenetic tree indicating the evolutionary relationship of oyster ferritin with other proteins. The bootstrap N–J tree was built as described in section 2. (**b**) Matrix representing the percentage identities of the aligned ferritins. Abbreviations and sources are accordant with Fig. 3.114

The mantle of mollusks is included in the metal ions metabolism and takes participation actively in the addition of Ca, P, Mn, and Fe into the shell [181]. Iron in the oyster is supposed to accumulate more rapidly in the mantle compared with other tissues in other shellfishes. The possible cause is the requirement to form iron minerals continuously for structural composition of the shell. We

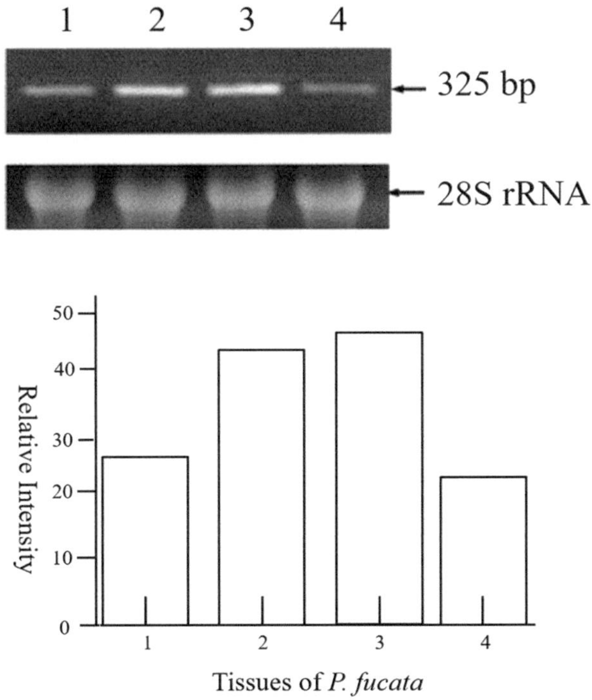

Fig. 3.116 Relative expression of ferritin mRNA in different tissues of the pearl oyster. RT-PCR was executed with RNA samples from different tissues of *P. fucata*. The PCR product (325 bp) is pointed by an arrow and the relative signals of bands were evaluated via densitometry. The 28S rRNA was used as a control of equal content of total RNA used by RT-PCR. 1, muscle; 2, digestive gland; 3, mantle; 4, gill

propose that met partially with the high expression level of oyster ferritin in the mantle. To explore the involvement of ferritin in shell formation with more detail, in situ hybridization was conducted. Strong hybridization signals were observed in the outer fold of the mantle tissue (Fig. 3.117a). Positive signals are mainly located in the epithelial cells of both sides and cells below the basement membrane in the outer fold (Fig. 3.117b), indicating ferritin involved in the formation of both prismatic layer and periostracum of the shell. Hybridization signals were also observed in the outer epithelial cells of the dorsal region of the mantle which was related with nacre formation [6], while hybridization with sense probe as the control shows no hybridization signal (Fig. 3.117c, d). This is accordant with previous reports that the mantle epithelial cells of bivalves collect Fe in lysosomes and calcified concretions including Fe present in mussel mantles [181–183]. All in all, ferritin proteins are abundant in the mantle (the outer fold especially); this may promote the course of iron storage and satisfy the requirement of iron for lasting shell formation [103].

The digestive gland is also a major tissue for metal accumulation. The high expression level of oyster ferritin mRNA in the digestive gland is accordant with its important roles in iron storage and metabolism. This is supplemented by histochemical results and microanalysis on the subcellular and cellular distribution of iron in mollusks. Even though the gill is a crucial organ for the concentration of dissolved metal irons from the water, the relative expression level of oyster ferritin mRNA in this organ is lower than that in the mantle (Fig. 3.116). This might be caused by the

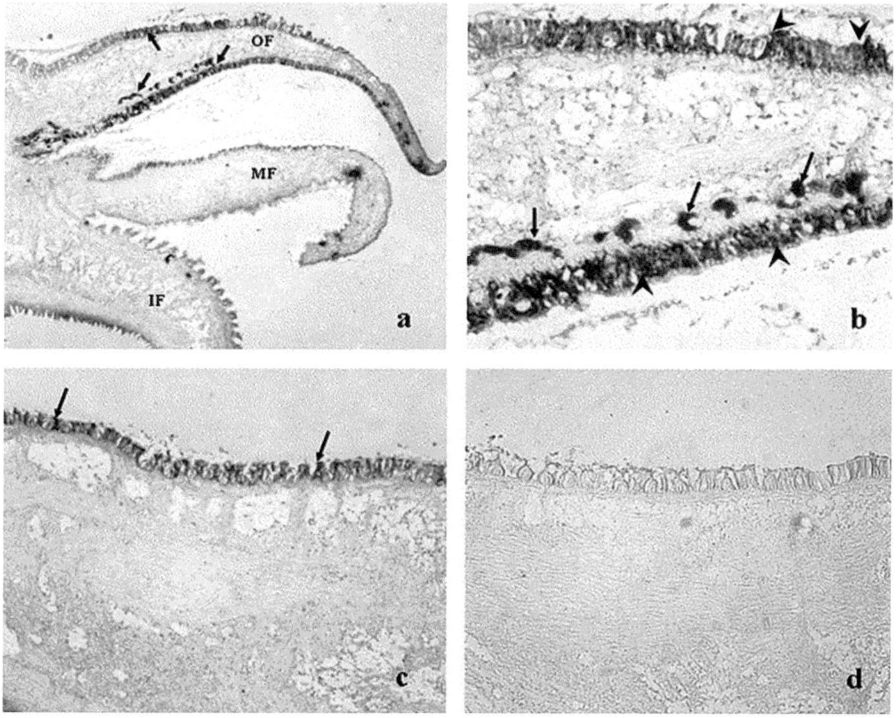

Fig. 3.117 In situ hybridization of oyster ferritin mRNA of the mantle in pearl oyster. In the three folds of the mantle, hybridization signals (arrows indicate) were found only in the outer fold in (**a**). Strong hybridization signals appeared in the epithelial cells of both sides of the outer fold (arrows indicate) and cells under the basement membrane (arrow heads indicate) in (**b**). Hybridization signals were also represented in the outer epithelial cells of the dorsal part of the mantle (arrows) in (**c**). No hybridization signal appeared in the control section incubated with the sense probe in (**d**). *OF* outer fold, *MF* middle fold, *IF* inner fold. Original amplifications are ×100 (**a**, **c** and **d**) and ×200 (**b**)

reason that the gill is not an organ of iron storage. What's more, oyster ferritin mRNA is also expressed in the muscle with a level higher than the gill, indicating that the muscle might participate in iron storage.

The characterization and purification of ferritin proteins in a few molluscan species showed that they are different from mammalian ferritins in their structure and functional properties [80, 184, 185]. Until now, molluscan ferritin sequences have been cloned and characterized only in the freshwater snail (*L. stagnalis*) [177].

Compared with the heteropolymer ferritins of mammals, bacteria, ferritin molecules isolated from plants, and the freshwater snail (*L. stagnalis*) were shown to be homological [176, 177]. It is still unclear whether oyster ferritin only includes the H-chain homologous subunit. However, with the incorporation of iron in H-chains being quicker than in L-chains [180], and the fact that in molluscan, chiton (*Clavarizona hirtosa*), ferritin has been suggested to concentrate iron about ten times more rapidly than horse spleen ferritin [186], it is possible that the oyster ferritin constitutes H-chain subunits.

3.9 Ubiquitylation Functions in the Calcium Carbonate Biomineralization in the Extracellular Matrix

Mollusks shell formation is regulated by matrix proteins and many of these proteins have been distinguished and characterized. Whereas, the mechanisms of protein regulation remain unknown. Herein, we report the ubiquitylation of matrix proteins of the prismatic layer from the pearl oyster, *Pinctada fucata*. The appearance of ubiquitylated proteins in the prismatic layer of the shells was identified with both Western blot and immunogold experiments. The coupled ubiquitins were isolated and distinguished by Edman degradation and liquid chromatography/mass spectrometry (LC/MS). Antibody injection in vivo led to large amounts of calcium carbonate randomly assembling on the surface of the nacreous layer. These ubiquitylated proteins could attach to specific faces of calcite and aragonite, which are the main mineral compositions of the shell. In the in vitro calcium carbonate crystallization experiment, they could decrease the rate of calcium carbonate precipitation and stimulate the calcite formation. What's more, when the linked ubiquitins were removed, the functions of the EDTA-soluble matrix of the prismatic layer were different. Their ability to inhibit precipitation of calcium carbonate was reduced and their effect on the morphology of calcium carbonate crystals was different. In conclusion, ubiquitylation participated in shell formation. Even though the ubiquitylation is thought to take part in every aspect of biophysical processes, our work combined the biomineralization-related proteins and the ubiquitylation mechanism of the extracellular matrix for the first time [41]. This would stimulate our understanding of the shells biomineralization and the ubiquitylation procedure.

3.9.1 Ubiquitylated Proteins in the Prismatic Layer

Reports have shown that ubiquitin may participate in the processes of biomineralization. Biochemical analyses were conducted to investigate whether ubiquitylated proteins participate in the nacreous layer and/or the prismatic layer. Monoclonal antibodies synthesized against poly-ubiquitin and mono-ubiquitin was used for immunolocalization in EDTA extracts of nacre and prisms, respectively. Western blot analyses found ubiquitylated proteins exist in the EDTA-soluble matrix (ESM) of the prismatic layer exclusively. There is no ubiquitin-specific signal in the matrix of the nacreous layer and the EDTA-insoluble matrix (EISM) of the prismatic layer (Fig. 3.118a).

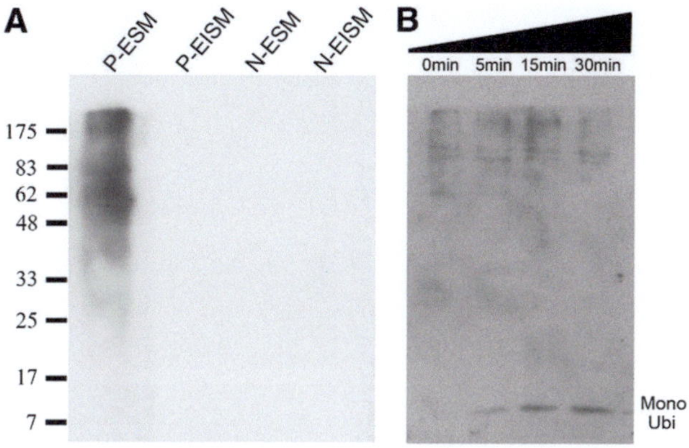

C

```
1    MQIFVKTLTG KTITLEVEPS DTIENVKAKI
31   QDEEGIPPDQ QRLIFAGKQL EDGRTLSDYN
61   IQKESTLHLV LSLRGG
```

Fig. 3.118 Ubiquitylation of matrix proteins from *P. fucata*. (**a**) The ubiquitylated proteins were identified by Western blotting of EDTA extracts of nacre and prisms extracted from the shell. The ubiquitylated proteins were mainly existed in the EDTA-soluble matrix protein of prismatic layer. P-ESM, EDTA-soluble matrix protein of the prism; P-EISM, the EDTA-insoluble matrix protein of the prism; N-ESM, EDTA-soluble matrix protein of the nacre; N-EISM, the denatured part of the EDTA-insoluble matrix protein of the nacre. (**b**) Time-course reaction of isopeptidase with the EDTA-soluble matrix segment of the prismatic layer. Reaction products were identified by Western blotting. The reaction was executed at 37 uC with a volume of 15 ml possessing 0.1 mM of isopeptidase, 2 mg of substrate, for the immediate times. Mono Ubi, mono-ubiquitin. (**c**) Amino acid sequences of ubiquitin indicating the residues identified by Edman degradation (underlined) and the peptide sequences distinguished by LC-MS analysis (red highlights)

To further affirm the presence of ubiquitylated proteins, isopeptidase was utilized to catalyze the cleavage of the isopeptide bond binding the terminal diglycine to ubiquitin [187]. A time course reaction was conducted. The anti-ubiquitin signal was stronger in the 8.5 kDa line along time, i.e., the molecular mass of ubiquitin in most animals (Fig. 3.118b). The 8.5 kDa proteins were degraded by Edman degradation and 17 amino acid residues of the N terminus (MQI FVK TLT GKT ITL EV) had 100% homology with the ubiquitin studied in various mollusks. According to the N-terminal residues, we obtained the cDNA sequence of ubiquitin by 3′-RACE and 5′-RACE to further investigate whether the 8.5 kDa protein was ubiquitin. The 8.5 kDa protein was further tested using liquid chromatography/mass spectrometry (LC/MS) (Fig. 3.118c). The LC/MS data affirmed that the 8.5 kDa protein was *P. fucata* ubiquitin.

The EDTA-etched nacreous and prismatic layers were immunogold-labeled with anti-ubiquitin antibodies as the first antibody and 5 nm gold-labeled antibodies as the

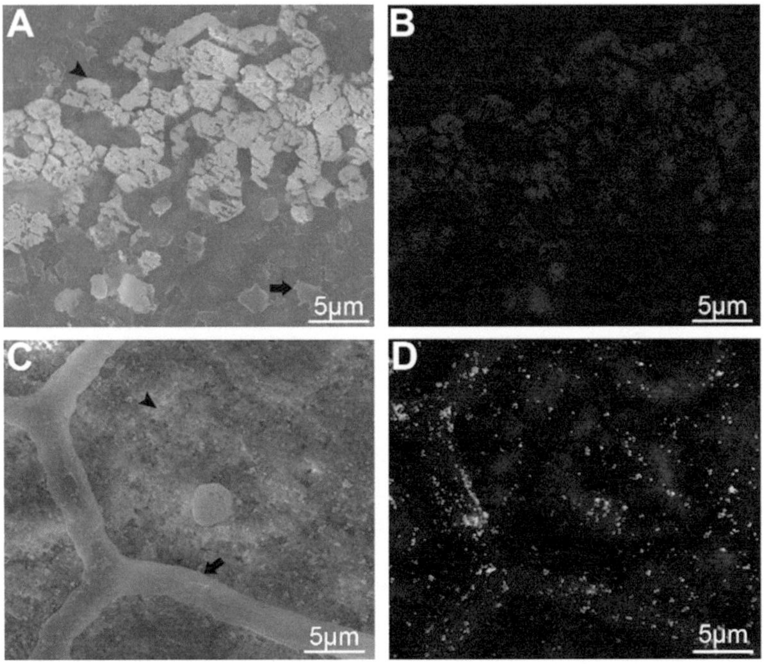

Fig. 3.119 Immunogold labeling of ubiquitylated proteins in the nacreous layer (**a** and **b**) and the prismatic layer (**c** and **d**) from *Pinctada fucata*. The complexes formed by the antigen first antibody, second gold-labeled antibody were observed as tiny bright spots via SEM-BSE. (**a**) SEM image of immunogold dying of the nacreous layer surface. The black arrow represents the EDTA-insoluble framework, while the black arrowhead shows the etched aragonitic tablets. (**b**) SEM-BSE picture of the same area indicated in (**a**). (**c**) SEM image of the immunogold-dyed prismatic layer surface. The black arrow means the EDTA-insoluble framework, while the black arrowhead points the etched prisms. (**d**) SEM-BSE image of the same area represented in (**c**). Ubiquitylated proteins were observed in the prisms and on the superficies of the EDTA-insoluble framework. Scale bars, 5 mm in (**a**)–(**d**)

secondary antibody to clarify the microstructural location of native ubiquitylated proteins of the shell. The EDTA exposure allowed the calcium carbonate in the shell to be slightly damaged away to expose proteins of the shell's structure. The aragonitic lamellae in the nacre were etched away (Fig. 3.119a, black arrowhead), but the matrix between lamellae was not influenced as it is EDTA-insoluble framework (Fig. 3.119a, black arrow). The nacreous layer was not observed of gold as there were no bright, tiny spots to show high atomic number gold elements with SEM in backscattered electron mode (SEMBSE) (Fig. 3.96b). The prism layers were etched away (Fig. 3.119c, black arrowhead) and the insoluble framework was recognized by the antibody (Fig. 3.119c, black arrow) in the prismatic layer. The tiny spots suggested that ubiquitylated proteins were present in the prisms and on the surface of the framework (Fig. 3.119d). Shell sections incubated without the anti-ubiquitin antibodies have negative control and no staining was found in these sections (data not shown). The microstructure distribution of the ubiquitylated proteins in the shell affirmed the results of the Western blot analyses [41].

3.9.2 In Vivo Investigation of the Role of Ubiquitylated Proteins During Shell Mineralization [41]

An in vivo antibody injection experiment was conducted to further identify the role of ubiquitylated proteins in shell formation. Anti-ubiquitin antibodies were injected into the extrapallial fluid of *P. fucata* at concentrations of 0.5 mg or 1 mg of protein per day per gram wet weight of oysters. The inner surfaces of the nacre were surveyed by SEM at 3 days after the antibodies were injected. The surfaces in the preimmune rabbit serum-injected group was full of small flat tablets of aragonite with a stair-like growth pattern which was resembled to untreated samples (Fig. 3.120e, f). Compared with the preimmune rabbit serum injected experiment group, small tablets with abnormal shapes were randomly assembled on the surface of the nacre in 11 of the 15 individuals in the low concentration (0.5 mg/g/d) antibodies-injected experiment group (Fig. 3.120a, b). These abnormal tablets damaged the stair-like growth pattern of the inner shell surface. This abnormal phenomenon was more obvious in the high concentration (1 mg/g/d) injected experiment group (Fig. 3.120c, d). More crystals were assembled on the surface and they bound together to form a new mineralized layer in 12 of the 15 samples. Chemical constitute analyses using energy-dispersive X-ray spectroscopy suggested that the abnormal precipitates were constituted of carbon, calcium, oxygen, and a small amount of magnesium (Fig. 3.120g). The physiological effects of ubiquitylated proteins were depressed, which caused the anomalous deposition of calcium carbonate on the surface of the nacre. What's more, the surfaces of the prism of the antibody injected experiment groups were not influenced. These findings suggested that these ubiquitylated proteins may work as a negative regulator of calcium carbonate deposition.

3.9.3 Immunoaffinity Chromatography of the Ubiquitylated Proteins

This research found that the ubiquitylated proteins of the ESM of the prism played an important role in shell formation. The effects of these proteins were further explored by isolating the ubiquitylated proteins using immunoaffinity chromatography (IAC). Anti-ubiquitin antibodies were bound on the column to purify the ubiquitylated proteins. The purified proteins were then confirmed using a dot blot assay. As Fig. 3.121 had shown, the ubiquitylated proteins of the purified sample were

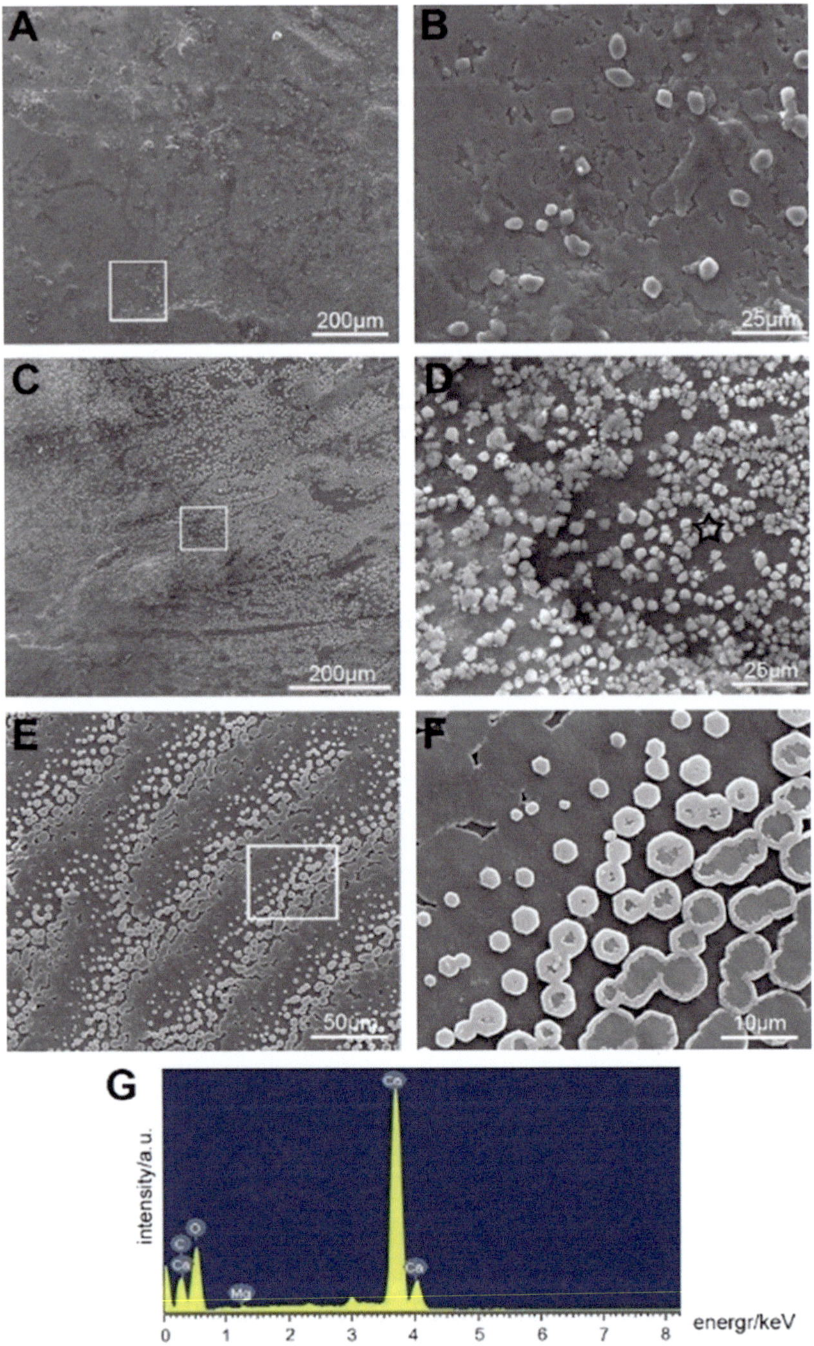

Fig. 3.120 The physiological effects of the ubiquitylated proteins were repressed by antibody injection. (**a**) SEM image of the surface of shells from the low dosage antibody-injected group. The stair-like growth pattern was destroyed. (**b**) Amplification of the box indicated in (**a**), illustrating the

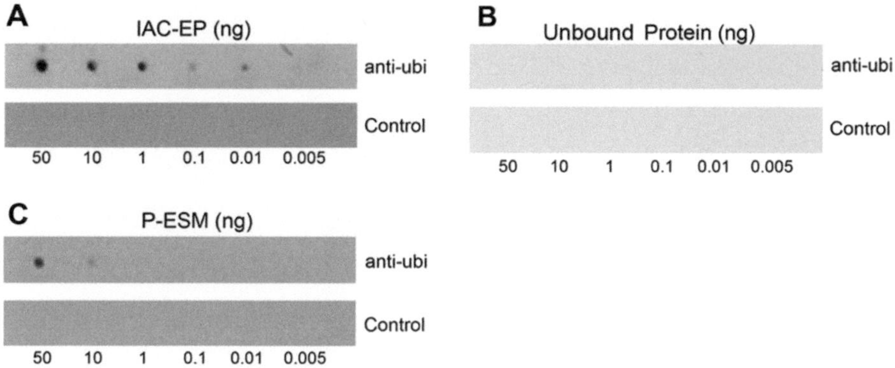

Fig. 3.121 Finding of ubiquitylated proteins in the proteins rich in IAC. The IAC-enriched proteins (**a**) and IAC unbound proteins (**b**) and EDTA-soluble matrix protein of the prismatic layer (**c**) were analyzed by a dot blot. The extracts were aligned with known amounts of IAC-enriched proteins, which were diluted serially and used as positive controls. The negative controls were executed without the anti-ubiquitin antibodies

enriched by over 1000-fold, whereas a dot of 0.01 ng enriched proteins was found in the analyses (Fig. 3.121a). The sensitivity of the substrate in the dot blot experiment was about 1~3 picograms following the manufacturer's instruction (Millipore), and this shows that the enriched proteins were ubiquitylated proteins of the ESM from the prism. The immune-affinity column unbound material was also explored by dot blot and the data exhibited no ubiquitylated proteins in the unbound material (Fig. 3.121b).

3.9.4 Interactions Between the Matrices of the Different Shell Layers

The main matrix of the *P. fucata* shell is constituted of calcium carbonate, which forms aragonite in the nacreous layer and calcite in the prismatic layer, while chitin composes the major insoluble part of the organic framework. The binding assays

Fig. 3.120 (continued) crystals deposited on the inner surface of nacreous layer. (**c**) SEM image of the inner surface of shells from the high dosage antibody-injected group. More crystals were randomly assembled. (**d**) Amplification of the box shown in (**c**), indicating that the crystals were attached together to form a new layer. (**e**) SEM image of the inner surface of shells from preimmune rabbit serum injected *P. fucata* shell showing the stair-like growth model. (**f**) Amplification of the box shown in (**e**), suggesting the flat tablets. (**g**) Energy-dispersive X-ray spectroscopy analyses of the deposition (black asterisk indicates) shown in (**d**). Scale bars, 200 mm in (**a**) and (**c**); 50 mm in (E); 25 mm in (**b**) and (**d**); 10 mm in (**f**)

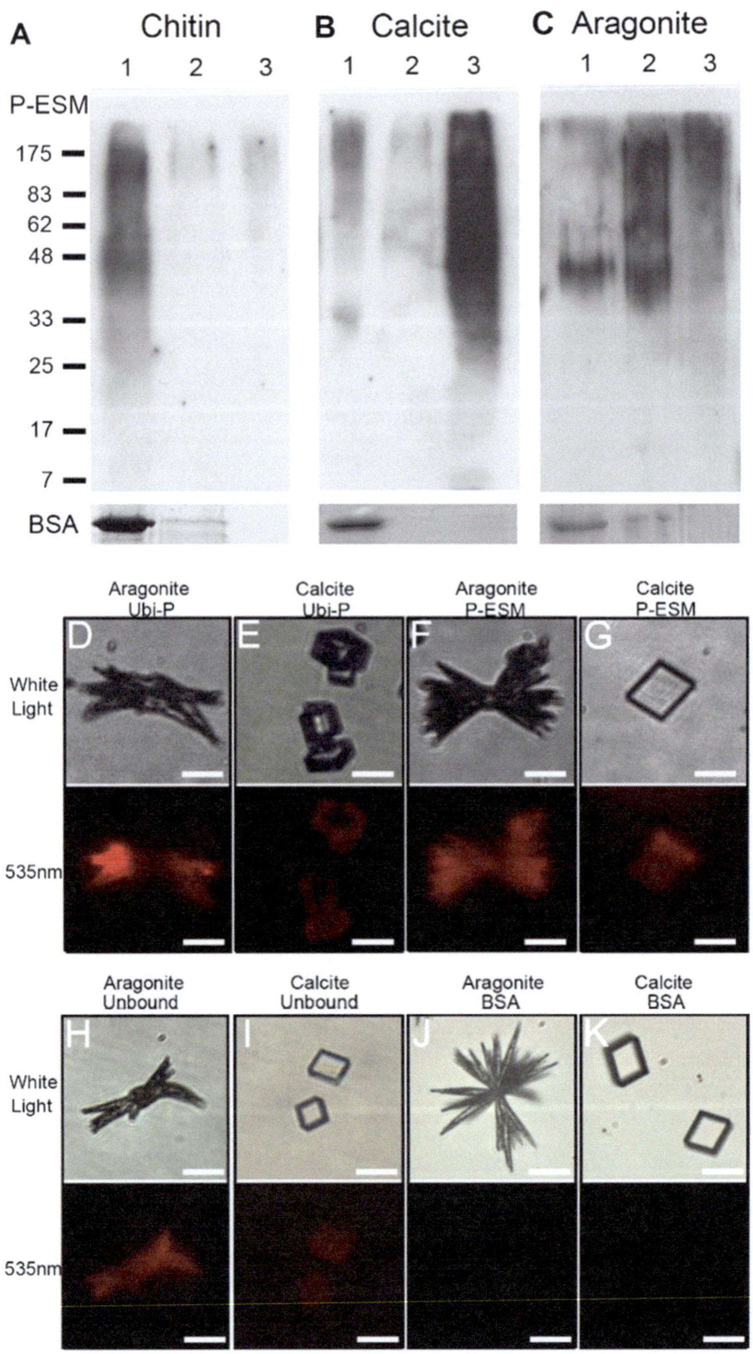

Fig. 3.122 Binding ability of ubiquitylated proteins. The binding abilities of ubiquitylated proteins with chitin (**a**), calcite (**b**), and aragonite (**c**) were detected by Western blot. Lane 1, proteins after

were performed to analyze the binding ability of the ubiquitylated proteins with the major elements of the shell. The proteins were mixed with chitin and washed with distilled water, saline, and a hot denaturing solution successively in the chitin-binding assay. Ubiquitylated proteins were completely washed by water indicating that the proteins had a very weak binding ability with chitin (Fig. 3.122a). While in the calcite-binding assay, the ubiquitylated proteins were not all washed with water and saline, but they were eluted by the denaturing solution, which showed that these proteins could bind stronger to calcite (Fig. 3.122b). While with aragonite, some proteins were washed with water, most proteins were eluted with saline, and some proteins of high molecular weight were washed with Laemmli buffer, which indicates that these proteins could also bind to aragonite tightly (Fig. 3.122c).

These results suggest that the ubiquitylated proteins have binding ability to calcite and aragonite. We confirmed this hypothesis by mixing the rhodamine-labeled ubiquitylated proteins with calcite and aragonite. The binding distribution of ubiquitylated proteins was located to the needlelike extensions of aragonite and the margins of calcite (Fig. 3.122d and e). On the contrary, the fluorescence was observed on all the aragonite and calcite surfaces with the immuno-affinity column unbound material or ESM of the prismatic layer (Fig. 3.99f–i). BSA was added in the control experiment and no fluorescent signal was observed on the surface of calcium carbonate crystals (Fig. 3.99j, k).

3.9.5 *In Vitro Effect of Ubiquitylated Proteins on Calcium Carbonate Crystallization [41]*

A series of calcium carbonate precipitation assays was conducted to explore the effect of ubiquitylated proteins. The precipitation rate of calcium carbonate was detected according to the absorbance at 570 nm of saturated calcium carbonate solution. With the addition of ESM of the prismatic layer, the rate of precipitation decreased significantly (Fig. 3.123). The same dosage of ubiquitylated proteins inhibited the rate of precipitation weakly compared with the ESM of the prismatic layer. When the immuno-affinity column unbound matrix is added, the precipitation rate decreased stronger when the ubiquitylated proteins were used in the system.

Fig. 3.122 (continued) water washing; lane 2, proteins after 0.2 M NaCl washing; lane 3, proteins extracted with SDS/b-mercaptoethanol at 100 °C for 10 min . Ubiquitylated proteins were completely washed with water in the chitin-binding experiment (**a**) and with the denaturing solution in both the calcite- and aragonite-binding experiments (**b** and **c**). BSA was added as a negative control. Ubiquitylated proteins represented different binding patterns with aragonite (**d**) and calcite (**e**) in the in vitro crystal-protein binding assay. The EDTA-soluble matrix of the prismatic layer attached to all the surfaces of aragonite (**f**) and calcite (**g**). The IAC-unbound proteins attached to all the surfaces of aragonite (**h**) and calcite (**i**). BSA was applied as a negative control indicating no signals with aragonite (**j**) or calcite (**k**). Scale bars: 20 mm in (**d**)–(**k**)

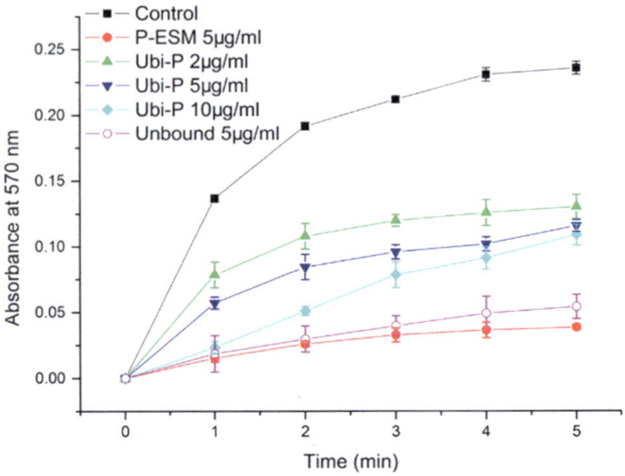

Fig. 3.123 Inhibitory ability of ubiquitylated proteins during calcium carbonate precipitation. Changes in the turbidity of the assayed solutions are recorded as figure. ■, BSA (10 μg/ml) was used as a negative control. ●, The EDTA-soluble matrix protein from the prismatic layer (5 μg/ml) was used as a positive control. ▲, 2 μg/ml ubiquitylated proteins. ⍰, 5 mg/ml ubiquitylated proteins. ◆, 10 mg/ml ubiquitylated proteins. ○, 5 mg/ml IAC unbound proteins

To identify the function of these ubiquitylated proteins on the calcium carbonate growth, two in vitro crystal growth experiments were performed to explore the formation of aragonite and calcite. The functions of ubiquitylated proteins on calcite growth were performed in a magnesium-free solution. In the control experiments added with 30 mg/ml BSA (Fig. 3.124a, b), the precipitated crystals showed typical rhombohedra of calcite with smooth surfaces. With the addition of ESM of the prismatic layer at a concentration of 30 mg/ml, the formed crystals didn't look uniform, but they were deficient on the edges and corners (Fig. 3.124e, f). Raman analyses were used to compare the same specific peaks as those obtained in the control group, which confirmed the calcitic character of these crystals (Fig. 3.124q, r). On the contrary, the crystals had spherical shapes with 30 mg/ml ubiquitylated proteins (Fig. 3.124i, j). These crystals were affirmed to be vaterite by Raman analyses (Fig. 3.124s). The immuno-affinity column unbound matrix could change the shape of the deposited crystals to form flexible morphologies of calcite (Fig. 3.124m, n and t).

Magnesium (50 mM) was added into the aragonite precipitation solution to duplicate the solution of the extrapallial fluids found in *P. fucata*. In the control groups, typical large needle-shaped crystals were found together with many small rodlike polyhedral crystals (Fig. 3.124c, d). None of these typical calcium carbonate crystals were found with ESM of the prismatic layer (30 μg/ml). Aragonite growth was depressed in this system and some gel-like aggregates appeared (Fig. 3.124g, h). The Raman spectra indicated that these aggregates were not typical crystals in the absence of characteristic peaks (Fig. 3.124r). Unlike the ESM of the prism, the

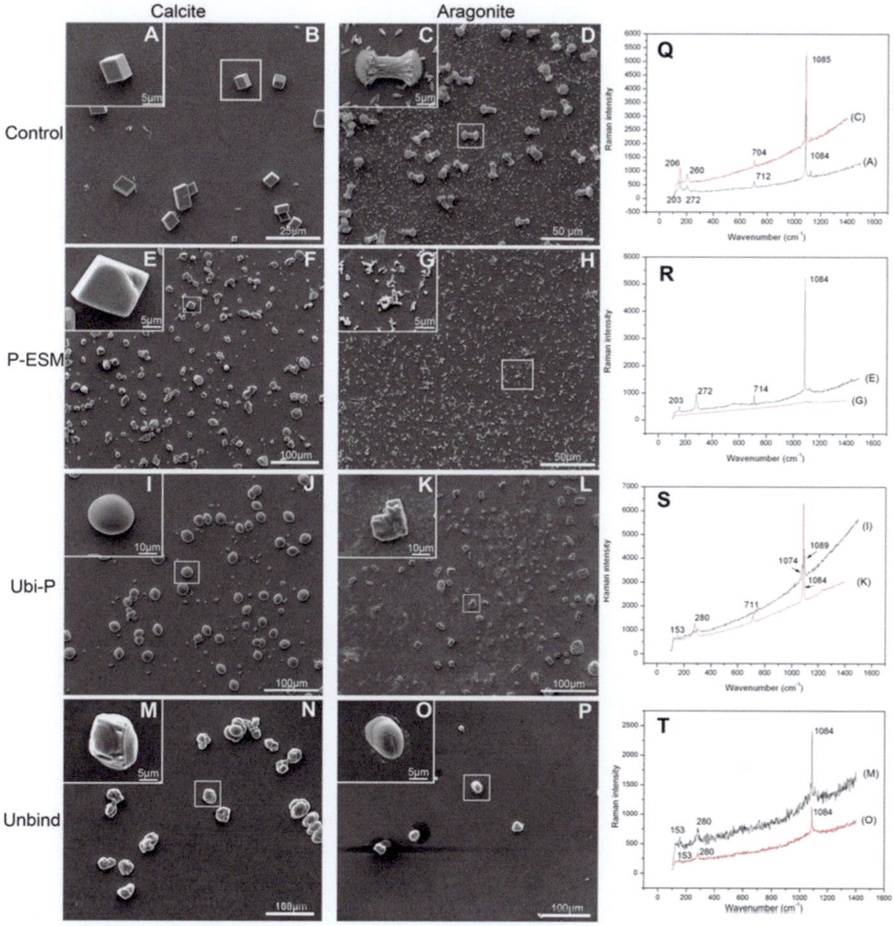

Fig. 3.124 In vitro crystallization assays with the addition of ubiquitylated proteins. The function of ubiquitylated proteins on the calcium carbonate crystals growth was detected in two crystallization systems, that is, aragonite growth in the middle column and calcite growth in the left column. (**b**) 30 mg/ml BSA was used as the negative control. (**f**) 30 μg/ml EDTA-soluble matrix of the prismatic layer was used as a positive control. The crystals were observed. (**i**) 30 μg/ml ubiquitylated proteins were used to the system to stimulate the formation of vaterite. (**n**) 30 μg/ml IAC unattached fraction was used. (**d**), (**h**), (**l**), and (**p**) are the same as those indicated in the left column, except that 50 mM Mg^{2+} was used to these systems. (**h**) No crystals were found in the aragonite growth systems, when adding EDTA-soluble matrix of the prismatic layer. (**l**) The ubiquitylated proteins stimulated calcite formation with a 30 μg/ml dosage. (**p**) The IAC unbound fraction could stimulate the formation of round crystals. Right column, Raman spectra of the crystals formed in calcite and aragonite systems. (**a**), (**e**), (**i**), (**m**), (**c**), (**g**), (**k**), and (**o**) show amplificated images of the boxed regions in (**b**), (**f**), (**j**), (**n**), (**d**), (**h**), (**l**), and (**p**), respectively. Scale bars: 100 μm in (**f**), (**j**), (**l**), (**n**), and (**p**); 50 μm in (**d**) and (**h**); 25 μm in (**b**); 10 μm in (**i**) and (**k**); 5 μm in (**a**), (**c**), (**e**), (**g**), (**m**), and (**n**)

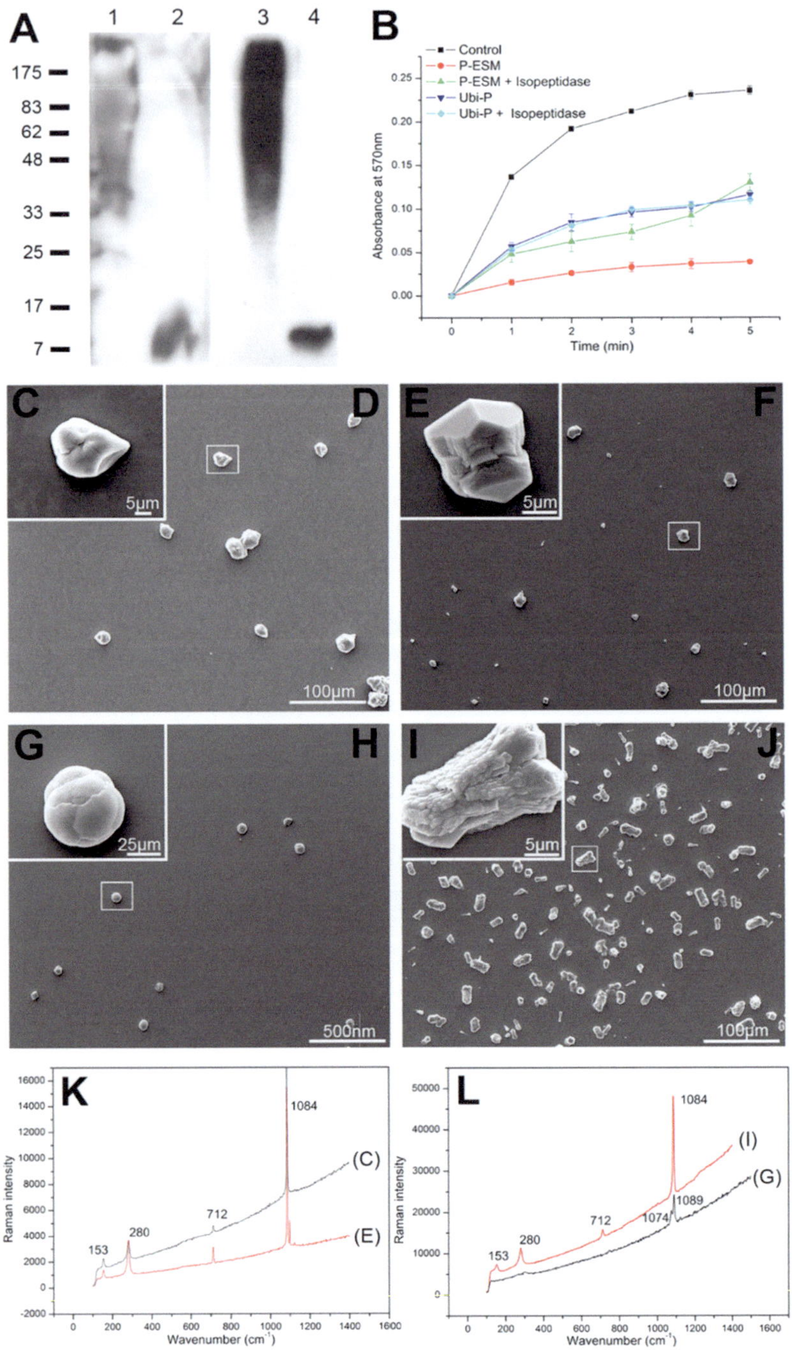

Fig. 3.125 Effect of ubiquitins in the ubiquitylated proteins. (**a**) Catalytic stimulation of isopeptidase with ESM from the prismatic layer or ubiquitylated proteins. Different segments were analyzed using Western blotting with anti-ubiquitin antibodies. Lane 1, the ESM proteins

crystals formed with ubiquitylated proteins were not needle-shaped but cuboid-like in shape with some etching at the margins and corners (Fig. 3.124k, l). These crystals were affirmed as calcite according to their Raman spectra (Fig. 3.124s). The immunoaffinity column unbound matrix could lead the formation of spherical-shaped calcite (Fig. 3.124o, p and t). These immunoaffinity column unbound matrix could induce the deposition of calcite in the aragonite precipitation solution, as Aspein was observed in the ESM of the prism and could induce the formation of calcite. The trend of calcite formation may be the function of the Aspein-like proteins in the immuno-affinity column unbound matrix.

3.9.6 The Function of Ubiquitin

To further explore the functions of ubiquitin in these proteins, the ESM of the IAC and the prismatic layer bound ubiquitylated proteins were catalyzed by isopeptidase to cleave the ubiquitin. After catalyzation for 1 h, the ubiquitin was completely cleaved off the proteins (Fig. 3.125a). After the ubiquitin was completely removed off the proteins, SDS-PAGE analyses exhibited a multiband result suggesting that more than one protein was ubiquitylated (data not shown). The ubiquitin-removal fraction was detected in the following protein function analyses. As the calcium carbonate precipitation experiments had shown, the removal of ubiquitins in the ESM of the prismatic layer could induce inhibitory effect in calcium carbonate precipitation experiment. Whereas, the ubiquitylated proteins could suppress the rate of precipitation to a resemble degree with or without ubiquitin (Fig. 3.125b).

In the calcium carbonate growth experiment, the dislodge of ubiquitins from ESM of the prismatic layer could lead to the deposition of abnormal shaped calcite with no magnesium (Fig. 3.125c, d). With the addition of magnesium, the removal of ubiquitin could induce the formation of etched calcite (Fig. 3.125e, f), while the ESM of the prismatic layer could lead to the formation of some gel-like aggregates

Fig. 3.125 (continued) from the prismatic layer. Lane 2, the ESM proteins of the prismatic layer after incubation for 1 h with isopeptidase. Lane 3, the ubiquitylated proteins. Lane 4, the ubiquitylated proteins after incubation with isopeptidase for 1 h. (**b**) Inhibitory ability of the ubiquitin-removed fraction in calcium carbonate precipitation assay. ■, BSA (10 μg/ml) was added as the negative control. ●, 5 μg/ml ESM proteins of the prismatic layer. ▲, 5 μg/ml isopeptidase reacted with ESM of the prismatic layer. ⍰, 5 μg/ml the untreated ubiquitylated proteins. ◆, 5 μg/ml isopeptidase reacted with ubiquitylated protein. The functions of the isopeptidase processed ESM of the prismatic layer (30 μg/ml) on crystal growth were detected in a calcite growth experiment (**d**) and an aragonite growth assay (**f**). The functions of the isopeptidase incubated ubiquitylated protein (30 μg/ml) on the growth of crystal were detected in a calcite growth experiment (**h**) and an aragonite growth assay (**j**). (**c**), (**e**), (**g**), and (**i**) show amplifications of the boxes in (**d**), (**f**), (**h**), and (**j**), respectively. (**k**) and (**l**) indicate the Raman spectra of the crystals grown in (**d**), (**f**), (**h**), and (**j**). Scale bars, 25 mm in (**g**); 500 nm in (**h**); 5 mm in (**c**), (**e**), and (**i**); 100 mm in (**d**), (**f**), and (**j**)

(Fig. 3.124g, h). The ubiquitin-removed IAC could bind ubiquitylated proteins leading to the formation of vaterite in the absence of magnesium and calcite with magnesium (Fig. 3.125g–j). With or without magnesium, the addition of ubiquitin or poly-ubiquitin chains (Ub2–7, K48-, or K63-linked) in the calcium carbonate crystallization solution did not influence crystal deposition or morphology (data not shown). Therefore, the removal of ubiquitin influenced the functions of the ESM of the prism but not the ubiquitylated proteins only.

This work was modified from the paper published by our group in scientific reports (Liu, C. et al. 2015, 5; Yan, Y. et al. 2017, 7), marine biotechnology (Zhang, C. et al. 2006, 8, 624–633), biomacromolecules (Yan, Z. et al. 2007, 8, 3597–3601,), journal of biological chemistry (Fang, D. et al. 2012, 287, 15776-15785; Kong, Y. et al. 2009, 284, 10841–10854; Pan, C. et al. 2014, 289, 2776-2787; Ma, Z. et al. 2007, 282, 23253–23263), biochemical and biophysical research communications (Zhang, C. et al. 2006, 344, 735–740), Plos One (Liang, J. et al. 2015, 10; Fang, D. et al. 2012, 7), comp. biochem., physiol., b-biochem., mol. biol. (Zhang, Y. et al. 2003, 135, 565–573; Zhang, Y. et al. 2003, 135, 43-54), Acta Biochimica Et Biophysica Sinica (Feng, Q. et al. 2009, 41, 955–962), biochem. J. (Su, J. et al. 2013, 454, 167–167), Crystengcomm (Su, J. T. et al. 2016, 18, 2125–2134,), crystal growth and design (Xie, J. et al. 2016, 16, 672–680), Biochimica Et Biophysica Acta general subjects (Huang, J. et al. 2007, 1770, 1037-1044), and Tsinghua Science and Technology (Ma, C. et al. 2005, 4, 499–503). The related contents are reused with the permission.

Supplementary Table 3.1

Chapter	Primer name	Primer sequence
3.5.1	YD5P	ARNCCNCCRTANCCNCC
	FSP	ACTTGACTCCACCGATATTTTTG
	kmpF	GAATGAAGTTCGCCGCTGTT
	kmpR	TTCCAATCCCARGGRTGACA
	MSI31F	GGTTCTACCTACGGAAATCTTG
	MSI31R	TCCTGAAGCCTGACCTGACA
	MSI60F	GAACAATGACTGGAATGACA
	MSI60R	GGAAAGGTATCCAATAACAAC
	MSI7F	CTATGGCAAATACGGCGGTG
	MSI7R	CCAGGTGAAAAGGTGAGTCC
	actF	CACCCCAGCCTTTTATGTAG
	actR	CGTCCG GTA ATT CGT AGT TC
3.5.2	K5	CCGGAATTCTACTGGCATAAACCTAATC
	K3	CCGCTCGAGCGGTTAGTATTTGTATTTACGATGAC
	BR3	CCGCTCGAGCGGTTAAAATTTATGAAGGCACCAT
	GY5	CCGGAATTCGGTGGGCATTATCCGTAT

(continued)

Chapter	Primer name	Primer sequence
3.5.3	YGS-F1	GGNTAYGGNGGNTAYNG
	YGS-R1	AACTATACCCTGAACGCATTCCACC
	YGS-F2	TTCACTGCAGTTTCGAACTAC
	YGS-T1	ATGCGTTCAGGGTATAGTTATTACAGC
	YGS-T2	TACTACCAGAACTGTAATATGATGG
	gapdhF	GCCGAGTATGTGGTAGAATC
	gapdhR	CACTGTTTTCTGGGTAGCTG
3.6.1	G3	GCGGATCCGGCGGAGATGGTGAC
	G4	GCCTCGAGTTATTGCAAAAAAATGAATATC
3.6.2	DGDF	GCGGATCCGGCGGAGATGGTGACTAT
	DGDR	GCCTCGAGTTATTGCAAAAAAATDAATATC
	DDF	GCGGATCCGGCGGAGATGACTATGGCAAA
	DDDF	GCGGATCGCGGATCCGGCGGAGATGATGACTATGGCAAA
	DGYDF	GCGGATCCGGCGGAGATGGTTACGACTATGGCAAA
3.6.3	PfN44-e-5	GATCCATGGGCTTTTGTTACAAAACATTTGG
	PfN44-e-3	CATCTCGAGAAATCCCGGGAAATCAGCAAATTTC
	RACE3	TTCCCTCTACTATGAAGACAAG
	RACE5	GCCGTTGATTGGCTGTCC
	LONG	CTAATACGACTCACTATAGGGCAAGCAGTGGTATCAACGCAGAGT
	SHORT	CTAATACGACTCACTATAGGGC
	NUP	AAGCAGTGGTATCAACGCAGAGT
	FULL1	AATCTCATACGGAGAAATTGGTGT
	FULL2	GAATGACACACATTTTTTATTGTA
	RNAi-P1	CTAATACGACTCACTATAGGGAGAAGGAATTCTATTTTTGATGC
	RNAi-P2	CTAATACGACTCACTATAGGGAGAAGGGTTTATTGTAATGTTCG
	RT-PfN44-F	GAGGTCGGAGAAGTAAG
	RT-PfN44-R	GAACACGATGGTAAAAA
	RT-GAPDH-F	GCCGAGTATGTGGTAGAATC
	RT-GAPDH-R	CACTGTTTTCTGGGTAGCTG
	qβ-actin-f	CTCCTCACTGAAGCCCCCCTC
	qβ-actin-R	ATGGCTGGAATAGGGATTCTGG
	qKRMP-F	AAGAAATGTCACCCTTGGGATTGG
	qKRMP-R	AATCATCGCCACCATATCCATCG
	qNacrein-F	GGCTTTGGCGACGAACCGGA
	qNacrein-R	ACACGGGGGAGTGGTCAGGG
	qPfN44-F	TTGCTGATGGATGTGAAG
	qPfN44-R	CTGTATATTGCCGTTGATTG
3.6.4	Long UPM	CTAATACGACTCACTATAGGGCAAGCAGTG GTATCAACGCAGAGT
	Short UPM	CTAATACGACTCACTATAGGGC
	NUP	AAGCAGTGGTATCAACGCAGAGT
	Y2-R1	ACTTTTACTTTCAGAAGCGCAAGTCAATTC
	Y2-R2	TGAGTAACACGGCAAATAGGACTGCTAC
	Y2-F1	ATGAAGTCAGCTACGGTAGCAGTC
	Y2-F2	CGCTTCTGAAAGTAAAAGTGTTCATCGCC
	Y2-confirm F	TACGATTCCCTGTGAAAGACAATCAAACG
	Y2-confirm R	TATTCTCATCACAACAAAGGCACACAGGC
	eY2-F1	GGAATTCCATATGAATCCTTGGAATTATCC AGG

(continued)

Chapter	Primer name	Primer sequence
	eY2-R1	CCGCTCGAGCTAGTTGTTCCAGTTATTCC
	eY2-F2	GGAATTCCATATGAATCCTTGGAATTATCC
		AGG
	eY2-R2	GGAATTCCTAGTGATGATGATGATGATGGT
		TGTTCCAGTTATTCC
	RT-Y2-F1	ATGAACATGGGTAGAGAGTTC
	RT-Y2-R1	CACCAGTTGTTCCTGTTCCA
	RT-GAPDH-F	GCCGAGTATGTGGTAGAATC
	RT-GAPDH-R	CACTGTTTCTGGGTAGCTG
	RT-KRMPI-F	AGAATGAAGTTCGCCGC
	RT-KRMPI-R	GCATTTCCAATCCCAGG
	RT-Nacrein-F	GGCGACGAACCGGATGACG
	RT-Nacrein-R	GACTCTGTACACGGGGGAGTG
	RT-P14-F	TAAGACCATTCTACGGC
	RT-P14-R	ACCATATCCTCCATATAATCCT
	RT-Aspein-F	TGATAGTGAAGACGATGA
	RT-Aspein-R	TGTCATCATCATCATCATC
	dsY2-F	GCGTAATACGACTCACTATAGGGAGATTGC
		CGTGTTACTCACAGCGT
	dsY2-R	GCGTAATACGACTCACTATAGGGAGACTAC
		CCATGTTCATGTTCCTGC
	dsGFP-F	GCGTAATACGACTCACTATAGGGAGAATG
		GTGAGCAAGGGCGAGGAG
	dsGFP-R	GCGTAATACGACTCACTATAGGGAGATTAC
		TTGTACAGCTCGTCCATG
	ISTY2-F	ATGAACATGGGTAGAGAGTTCTT
	ISTY2-R	CACCAGTTGTTCCTGTTCCAAGG
3.7.1	Sense primer	AARTGYGAYTAYCCNGARGC
	Antisense primer	CTTGCCTTTTGAATGGGGAC
	gapdhF	GCCGAGTATGTGGTAGAATC
	gapdhR	CACTGTTTCTGGGTAGCTG
	accbpF	AAATTCCTGCTGGATGACTATG
	accbpR	AACGGTCCACCATCTTTGTTCC
3.7.4	qGAPDH-F	GCCGAGTATGTGGTAGAATC
	qGAPDH-R	CACTGTTTCTGGGTAGCTG
	qACCBP-F	GACATGGAACAAAGATGGTGGA
	qACCBP-R	CTGTGGCTGGAATGGTTGG
	qPfN23-F	AGGGTCAAGGTCATCCA
	qPfN23-R	TAAGCGAGCTTTTCCAA
	qPif-F	TGCTGCCATCACGTGAGTATG
	qPif-R	GACTTCCCTTTCTCACACTTCCA
	qNacrein-F	GAGCCAGAGGATGGGGAAA
	qNacrein-R	GCCTCCATAGGTGTGAAACGA
	qKrmp-F	AGAATGAAGTTCGCCGC
	qKrmp-R	GCATTTCCAATCCCAGG
	qP39-F	CTGGAATGAGAGGATATG
	qP39-R	TGCTGCTGTAATAACTATA

(continued)

Chapter	Primer name	Primer sequence
3.8.1	EFP5	GAYGCNGAYGGNRAYGG
	EFP5′	GAYAARGAYGGNRAYGG
	EFP3	TTCTGACAGGCGTCTATCAACG
	EFgsp5	CAAAGGAAAAGCTCAATTAGGAG
	EFgsp3	ATACAGAGATAAGATCTGCTTCC
	GAPDH5	GCCGAGTATGTGGTAGAATC
	GAPDH3	CACTGTTTTCTGGGTAGCTG
3.8.2	D1	GCCWSCTACRBCTAYCWGTCIATG
	D2	CTCCTTRATIGCYTYCACYTGYTC
	G1	ACCTATCAGTCAATGTCCTTCT
	G2	GCTCCTCAAGGTATTCACTCT
3.9	3′-RACE	TGCARATHTTYGTNAARACNYTNAC
	RACE	CAATCTCTGTTGGTCTGGGGGAA

References

1. K. Simkiss, K.M. Wilbur, *Biomineralization* (Elsevier, San Diego, 2012)
2. L. Li, C. Ortiz, Pervasive nanoscale deformation twinning as a catalyst for eficient energy dissipation in a bioceramic armour. Nat. Mater. **13**, 1–7 (2014). https://doi.org/10.1038/nmat3920
3. H. Kim et al., The role of nacreous factors in preventing osteoporotic bone loss through both osteoblast activation and osteoclast inactivation. Biomaterials **33**, 7489–7496 (2012). https://doi.org/10.1016/j.biomaterials.2012.06.098
4. G. Falini, S. Albeck, S. Weiner, L. Addadi, Control of aragonite or calcite polymorphism by mollusk shell macromolecules. Science **271**, 67–69 (1996). https://doi.org/10.1126/science.271.5245.67
5. H. Miyamoto et al., A carbonic anhydrase from the nacreous layer in oyster pearls. Proc. Natl. Acad. Sci. U. S. A. **93**, 9657–9660 (1996). https://doi.org/10.1073/pnas.93.18.9657
6. S. Sudo et al., Structures of mollusc shell framework proteins. Nature **387**, 563–564 (1997). https://doi.org/10.1038/42391
7. T. Samata et al., A new matrix protein family related to the nacreous layer formation of Pinctada fucata. FEBS Lett. **462**, 225–229 (1999). https://doi.org/10.1016/s0014-5793(99)01387-3
8. M. Suzuki et al., Characterization of Prismalin-14, a novel matrix protein from the prismatic layer of the Japanese pearl oyster (Pinctada fucata). Biochem. J. **382**, 205–213 (2004). https://doi.org/10.1042/bj20040319
9. C. Zhang, L.P. Xie, J. Huang, X.L. Liu, R.Q. Zhang, A novel matrix protein family participating in the prismatic layer framework formation of pearl oyster, Pinctada fucata. Biochem. Bioph. Res. Commun. **344**, 735–740 (2006). https://doi.org/10.1016/j.bbrc.2006.03.179
10. D. Tsukamoto, I. Sarashina, K. Endo, Structure and expression of an unusually acidic matrix protein of pearl oyster shells. Biochem. Bioph. Res. Commun. **320**, 1175–1180 (2004). https://doi.org/10.1016/j.bbrc.2004.06.072
11. C. Zhang, L. Xie, J. Huang, L. Chen, R. Zhang, A novel putative tyrosinase involved in periostracum formation from the pearl oyster *Pinctada fucata*. Biochem. Biophys. Res. Commun. **342**, 632–639 (2006)

12. Z. Yan et al., N40, a novel nonacidic matrix protein from pearl oyster nacre, facilitates nucleation of aragonite in vitro. Biomacromolecules **8**, 3597–3601 (2007). https://doi.org/10.1021/bm0701494
13. Y. Kong et al., Cloning and characterization of Prisilkin-39, a novel matrix protein serving a dual role in the prismatic layer formation from the oyster Pinctada fucata. J. Biol. Chem. **284**, 10841–10854 (2009). https://doi.org/10.1074/jbc.M808357200
14. D. Fang et al., Novel basic protein, PfN23, functions as key macromolecule during nacre formation. J. Biol. Chem. **287**, 15776–15785 (2012). https://doi.org/10.1074/jbc.M112.341594
15. C. Pan et al., A novel acidic matrix protein, PfN44, stabilizes magnesium calcite to inhibit the crystallization of aragonite. J. Biol. Chem. **289**, 2776–2787 (2014). https://doi.org/10.1074/jbc.M113.504027
16. T. Takeuchi, I. Sarashina, M. Iijima, K. Endo, In vitro regulation of CaCO(3) crystal polymorphism by the highly acidic molluscan shell protein Aspein. FEBS Lett. **582**, 591–596 (2008). https://doi.org/10.1016/j.febslet.2008.01.026
17. M. Yano, K. Nagai, K. Morimoto, H. Miyamoto, Shematrin: a family of glycine-rich structural proteins in the shell of the pearl oyster Pinctada fucata. Comp. Biochem. Physiol. B **144**, 254–262 (2006). https://doi.org/10.1016/j.cpbp.2006.03.004
18. J. Seto et al., Nacre protein sequence compartmentalizes mineral polymorphs in solution. Cryst. Growth Des. **14**, 1501–1505 (2014). https://doi.org/10.1021/cg401421h
19. H. Tohse, H. Ando, Y. Mugiya, Biochemical properties and immunohistochemical localization of carbonic anhydrase in the sacculus of the inner ear in the salmon Oncorhynchus masou. Comp. Biochem. Physiol. A. Mol. Integr. Physiol. **137**, 87–94 (2004). https://doi.org/10.1016/s1095-6433(03)00272-1
20. F. Marin, B. Pokroy, G. Luquet, P. Layrolle, K. De Groot, Protein mapping of calcium carbonate biominerals by immunogold. Biomaterials **28**, 2368–2377 (2007). https://doi.org/10.1016/j.biomaterials.2007.01.029
21. M. Suzuki et al., An acidic matrix protein, Pif, is a key macromolecule for nacre formation. Science **325**, 1388–1390 (2009)
22. H. Inoue, N. Ozaki, H. Nagasawa, Purification and structural determination of a phosphorylated peptide with anti-calcification and chitin-binding activities in the exoskeleton of the crayfish, Procambarus clarkii. Biosci. Biotechnol. Biochem. **65**, 1840–1848 (2001). https://doi.org/10.1271/bbb.65.1840
23. G.F. Xu, N. Yao, I.A. Aksay, J.T. Groves, Biomimetic synthesis of macroscopic-scale calcium carbonate thin films. Evidence for a multistep assembly process. J. Am. Chem. Soc. **120**, 11977–11985 (1998). https://doi.org/10.1021/ja9819108
24. C. Liu et al., In-depth proteomic analysis of shell matrix proteins of Pinctada fucata. Sci. Rep. **5**, 17269 (2015). https://doi.org/10.1038/srep17269
25. T. Takeuchi et al., Draft genome of the pearl oyster Pinctada fucata: a platform for understanding bivalve biology. DNA Res. **19**, 117–130 (2012)
26. S. Berland et al., Coupling proteomics and transcriptomics for the identification of novel and variant forms of mollusk shell proteins: a study with P. margaritifera. ChemBioChem **12**, 950–961 (2011)
27. J.L. Drake et al., Proteomic analysis of skeletal organic matrix from the stony coral Stylophora pistillata. Proc. Natl. Acad. Sci. U. S. A. **110**, 3788–3793 (2013)
28. J.S. Evans, Aragonite-associated biomineralization proteins are disordered and contain interactive motifs. Bioinformatics **28**, 3182–3185 (2012). https://doi.org/10.1093/bioinformatics/bts604
29. B. Marie et al., Different secretory repertoires control the biomineralization processes of prism and nacre deposition of the pearl oyster shell. Proc. Natl. Acad. Sci. U. S. A. **109**, 20986–20991 (2012)
30. S.Y. Bahn, B.H. Jo, B.H. Hwang, Y.S. Choi, H.J. Cha, Role of Pif97 in nacre biomineralization: in vitro characterization of recombinant Pif97 as a framework protein for the association

of organic–inorganic layers in nacre. Cryst. Growth Des. **15**, 3666–3673 (2015). https://doi.org/10.1021/acs.cgd.5b00275

31. J. Liu et al., Microarray: a global analysis of biomineralization-related gene expression profiles during larval development in the pearl oyster, Pinctada fucata. BMC Genomics **16**, 325 (2015). https://doi.org/10.1186/s12864-015-1524-2
32. A.S. Mount, A. Wheeler, R.P. Paradkar, D. Snider, Hemocyte-mediated shell mineralization in the eastern oyster. Science **304**, 297–300 (2004)
33. X. Wang et al., Oyster shell proteins originate from multiple organs and their probable transport pathway to the shell formation front. PLoS One **8**, e66522 (2013). https://doi.org/10.1371/journal.pone.0066522
34. S. Kinoshita et al., Deep sequencing of ESTs from nacreous and prismatic layer producing tissues and a screen for novel shell formation-related genes in the pearl oyster. PLoS One **6**, e21238 (2011)
35. D. Funabara et al., Novel genes participating in the formation of prismatic and nacreous layers in the pearl oyster as revealed by their tissue distribution and RNA interference knockdown. PLoS One **9**, e84706 (2014). https://doi.org/10.1371/journal.pone.0084706
36. S. Nakayama et al., Identification and characterization of a matrix protein (PPP-10) in the periostracum of the pearl oyster, Pinctada fucata. FEBS Open Bio **3**, 421–427 (2013)
37. H.-L. Liu et al., Identification and characterization of a biomineralization related gene PFMG1 highly expressed in the mantle of Pinctada fucata. Biochemistry **46**, 844–851 (2007). https://doi.org/10.1021/bi061881a
38. F. Yan et al., Tissue inhibitor of metalloproteinase gene from pearl oyster Pinctada martensii participates in nacre formation. Biochem. Bioph. Res.Co. **450**, 300–305 (2014). https://doi.org/10.1016/j.bbrc.2014.05.118
39. M. Yano, K. Nagai, K. Morimoto, H. Miyamoto, A novel nacre protein N19 in the pearl oyster Pinctada fucata. Biochem. Bioph. Res.Co. **362**, 158–163 (2007)
40. B. Marie et al., Characterization of MRNP34, a novel methionine-rich nacre protein from the pearl oysters. Amino Acids **42**, 2009–2017 (2012). https://doi.org/10.1007/s00726-011-0932-0
41. D. Fang et al., Ubiquitylation functions in the calcium carbonate biomineralization in the extracellular matrix. PLoS One **7**, e35715 (2012). https://doi.org/10.1371/journal.pone.0035715
42. L. Addadi, D. Joester, F. Nudelman, S. Weiner, Mollusk shell formation: a source of new concepts for understanding biomineralization processes. Chem-Eur. J. **12**, 980–987 (2006). https://doi.org/10.1002/chem.200500980
43. I. Tency et al., Imbalances between matrix metalloproteinases (MMPs) and tissue inhibitor of metalloproteinases (TIMPs) in maternal serum during preterm labor. PLoS One **7**, e49042 (2012)
44. L. Bédouet et al., Heterogeneity of proteinase inhibitors in the water-soluble organic matrix from the oyster nacre. Mar. Biotechnol. **9**, 437–449 (2007)
45. G. Zhang et al., The oyster genome reveals stress adaptation and complexity of shell formation. Nature **490**, 49–54 (2012). https://doi.org/10.1038/nature11413
46. Z. Liao et al., In-depth proteomic analysis of nacre, prism, and myostracum of Mytilus shell. J. Proteome **122**, 26–40 (2015)
47. B. Marie et al., Proteomic analysis of the organic matrix of the abalone Haliotis asinina calcified shell. Proteome Sci. **8**, 54 (2010). https://doi.org/10.1186/1477-5956-8-54
48. V.P. Hytonen, B. Wehrle-Haller, Protein conformation as a regulator of cell-matrix adhesion. Phys. Chem. Chem. Phys. **16**, 6342–6357 (2014). https://doi.org/10.1039/c3cp54884h
49. P. Ramos-Silva et al., The skeletal proteome of the coral Acropora millepora: the evolution of calcification by co-option and domain shuffling. Mol. Biol. Evol. **30**, 2099–2112 (2013). https://doi.org/10.1093/molbev/mst109
50. R. Timpl et al., Structure and function of laminin LG modules. Matrix Biol. **19**, 309–317 (2000). https://doi.org/10.1016/s0945-053x(00)00072-x

51. B. Marie et al., The shell-forming proteome of Lottia gigantea reveals both deep conservations and lineage-specific novelties. FEBS J. **280**, 214–232 (2013). https://doi.org/10.1111/febs.12062
52. C. Serra-Pages et al., The LAR transmembrane protein tyrosine phosphatase and a coiled-coil LAR-interacting protein co-localize at focal adhesions. EMBO J. **14**, 2827 (1995)
53. M. Humphries, Integrin structure. Biochem. Soc.Trans. **28**, 311–339 (2000)
54. L. Mohazab et al., Critical role for alphavbeta6 integrin in enamel biomineralization. J. Cell Sci. **126**, 732–744 (2013). https://doi.org/10.1242/jcs.112599
55. D. Norman et al., Three-dimensional structure of a complement control protein module in solution. J. Mol. Biol. **219**, 717–725 (1991)
56. H. Miyamoto et al., The diversity of shell matrix proteins: genome-wide investigation of the pearl oyster, Pinctada fucata. Zool. Sci. **30**, 801–816 (2013)
57. P. Huan, G. Liu, H. Wang, B. Liu, Identification of a tyrosinase gene potentially involved in early larval shell biogenesis of the Pacific oyster Crassostrea gigas. Dev. Genes Evol. **223**, 389–394 (2013). https://doi.org/10.1007/s00427-013-0450-z
58. J. Liang et al., Dual roles of the lysine-rich matrix protein (KRMP)-3 in shell formation of pearl oyster, Pinctada fucata. PloS one **10**, e0131868 (2015). https://doi.org/10.1371/journal.pone.0131868
59. N. Nassif et al., Amorphous layer around aragonite platelets in nacre. Proc. Natl. Acad. Sci. U. S. A. **102**, 12653–12655 (2005)
60. B. Bayerlein et al., Self-similar mesostructure evolution of the growing mollusc shell reminiscent of thermodynamically driven grain growth. Nat. Mater. **13**, 1102–1107 (2014)
61. M.B. Johnstone et al., Cellular orchestrated biomineralization of crystalline composites on implant surfaces by the eastern oyster, Crassostrea virginica (Gmelin, 1791). J. Exp. Mar.Biol. Ecol. **463**, 8–16 (2015). https://doi.org/10.1016/j.jembe.2014.10.014
62. L. Xiang et al., Amorphous calcium carbonate precipitation by cellular biomineralization in mantle cell cultures of Pinctada fucata. PLoS One **9**, e113150 (2014). https://doi.org/10.1371/journal.pone.0113150
63. C.M. Giachelli, S. Steitz, Osteopontin: a versatile regulator of inflammation and biomineralization. Matrix Biol. **19**, 615–622 (2000)
64. C. Zhang, S. Li, Z. Ma, L. Xie, R. Zhang, A novel matrix protein p10 from the nacre of pearl oyster (Pinctada fucata) and its effects on both $CaCO_3$ crystal formation and mineralogenic cells. Mar. Biotechnol. **8**, 624–633 (2006). https://doi.org/10.1007/s10126-006-6037-1
65. A.P. Wheeler, J.W. George, C.A. Evans, Control of calcium-carbonate nucleation and crystal-growth by soluble matrix of oyster shell. Science **212**, 1397–1398 (1981). https://doi.org/10.1126/science.212.4501.1397
66. G. Fu, S. Valiyaveettil, B. Wopenka, D.E. Morse, $CaCO_3$ biomineralization: acidic 8-kDa proteins isolated from aragonitic abalone shell nacre can specifically modify calcite crystal morphology. Biomacromolecules **6**, 1289–1298 (2005). https://doi.org/10.1021/bm049314v
67. L.D. Quarles, D.A. Yohay, L.W. Lever, R. Caton, R.J. Wenstrup, Distinct proliferative and differentiated stages of murine mc3t3-e1 cells in culture – an invitro model of osteoblast development. J. Bone Miner. Res. **7**, 683–692 (1992)
68. J.Y. Choi et al., Expression patterns of bone-related proteins during osteoblastic differentiation in MC3T3-E1 cells. J. Cell. Biochem. **61**, 609–618 (1996). https://doi.org/10.1002/(sici)1097-4644(19960616)61:4<609::aid-jcb15>3.0.co;2-a
69. T. Mosmann, Rapid colorimetric assay for cellular growth and survival – application to proliferation and cyto-toxicity assays. J. Immunol. Methods **65**, 55–63 (1983). https://doi.org/10.1016/0022-1759(83)90303-4
70. D. Fang et al., Identification of genes directly involved in shell formation and their functions in pearl oyster, Pinctada fucata. PLoS One **6**, e21860 (2011). https://doi.org/10.1371/journal.pone.0021860

71. Y. Zhang et al., A novel matrix protein participating in the nacre framework formation of pearl oyster, Pinctada fucata. Comp. Biochem. Physiol. B-Biochem. Mol. Biol. **135**, 565–573 (2003). https://doi.org/10.1016/s1096-4959(03)00138-6
72. B.A. Gotliv et al., Asprich: a novel aspartic acid-rich protein family from the prismatic shell matrix of the bivalve Atrina rigida. Chembiochem **6**, 304–314 (2005). https://doi.org/10.1002/cbic.200400221
73. T. Miyashita, R. Takagi, H. Miyamoto, A. Matsushiro, Identical carbonic anhydrase contributes to nacreous or prismatic layer formation in Pinctada fucata (Mollusca : Bivalvia). Veliger **45**, 250–255 (2002)
74. F. Marin et al., Caspartin and calprismin, two proteins of the shell calcitic prisms of the Mediterranean fan mussel Pinna nobilis. J. Biol. Chem. **280**, 33895–33908 (2005). https://doi.org/10.1074/jbc.M506526200
75. X.Y. Shen, A.M. Belcher, P.K. Hansma, G.D. Stucky, D.E. Morse, Molecular cloning and characterization of lustrin A, a matrix protein from shell and pearl nacre of Haliotis rufescens. J. Biol. Chem. **272**, 32472–32481 (1997). https://doi.org/10.1074/jbc.272.51.32472
76. I. Sarashina, K. Endo, Primary structure of a soluble matrix protein of scallop shell: Implications for calcium carbonate biomineralization. Am. Mineral. **83**, 1510–1515 (1998)
77. M. Kono, N. Hayashi, T. Samata, Molecular mechanism of the nacreous layer formation in Pinctada maxima. Biochem. Biophys. Res. Commun. **269**, 213–218 (2000). https://doi.org/10.1006/bbrc.2000.2274
78. F. Marin, P. Corstjens, B. de Gaulejac, E.D. Vrind-De Jong, P. Westbroek, Mucins and molluscan calcification – molecular characterization of mucoperlin, a novel mucin-like protein from the nacreous shell layer of the fan mussel Pinna nobilis (Bivalvia, Pteriomorphia). J. Biol. Chem. **275**, 20667–20675 (2000). https://doi.org/10.1074/jbc.M003006200
79. T. Miyashita et al., Complementary DNA cloning and characterization of pearlin, a new class of matrix protein in the nacreous layer of oyster pearls. Mar. Biotechnol. **2**, 409–418 (2000)
80. K. Mann, I.M. Weiss, S. Andre, H.J. Gabius, M. Fritz, The amino-acid sequence of the abalone (Haliotis laevigata) nacre protein perlucin – detection of a functional C-type lectin domain with galactose/mannose specificity. Eur. J. Biochem. **267**, 5257–5264 (2000). https://doi.org/10.1046/j.1432-1327.2000.01602.x
81. I.M. Weiss, W. Gohring, M. Fritz, K. Mann, Perlustrin, a Haliotis laevigata (abalone) nacre protein, is homologous to the insulin-like growth factor binding protein N-terminal module of vertebrates. Biochem. Biophys. Res. Commun. **285**, 244–249 (2001). https://doi.org/10.1006/bbrc.2001.5170
82. J.C. Marxen, M. Nimtz, W. Becker, K. Mann, The major soluble 19.6 kDa protein of the organic shell matrix of the freshwater snail Biomphalaria glabrata is an N-glycosylated dermatopontin. Biochim. Biophys. Acta. Proteins Proteomics **1650**, 92–98 (2003). https://doi.org/10.1016/s1570-9639(03)00203-6
83. M. Michenfelder et al., Characterization of two molluscan crystal-modulating biomineralization proteins and identification of putative mineral binding domains. Biopolymers **70**, 522–533 (2003). https://doi.org/10.1002/bip.10536
84. Y. Suzuki, T. Matsuoka, Y. Iimura, H. Fujiwara, Ecdysteroid-dependent expression of a novel cuticle protein gene BMCPG1 in the silkworm, Bombyx mori. Insect Biochem. Mol. Biol. **32**, 599–607 (2002). https://doi.org/10.1016/s0965-1748(01)00136-9
85. H. Zhao, C.J. Sun, R.J. Stewart, J.H. Waite, Cement proteins of the tube-building polychaete Phragmatopoma californica. J. Biol. Chem. **280**, 42938–42944 (2005). https://doi.org/10.1074/jbc.M508457200
86. M. Suzuki, H. Nagasawa, Mollusk shell structures and their formation mechanism. Can. J. Zool. Revue Canadienne De Zoologie **91**, 349–366 (2013). https://doi.org/10.1139/cjz-2012-0333
87. F. Marin, G. Luquet, Molluscan biomineralization: the proteinaceous shell constituents of Pinna nobilis L. Mater. Sci. Eng. C Biomim. Supramol. Syst. **25**, 105–111 (2005). https://doi.org/10.1016/j.msec.2005.01.003

88. M. Suzuki, S. Sakuda, H. Nagasawa, Identification of chitin in the prismatic layer of the shell and a chitin synthase gene from the Japanese pearl oyster, Pinctada fucata. Biosci. Biotechnol. Biochem. **71**, 1735–1744 (2007). https://doi.org/10.1271/bbb.70140
89. A.S. Ghatak, M. Koch, C. Guth, I.M. Weiss, Peptide induced crystallization of calcium carbonate on wrinkle patterned substrate: implications for chitin formation in molluscs. Int. J. Mol. Sci. **14**, 11842–11860 (2013). https://doi.org/10.3390/ijms140611842
90. J.E. Rebers, L.M. Riddiford, Structure and expression of a manduca-sexta larval cuticle gene homologous to drosophila cuticle genes. J. Mol. Biol. **203**, 411–423 (1988). https://doi.org/10.1016/0022-2836(88)90009-5
91. J.E. Rebers, J.H. Willis, A conserved domain in arthropod cuticular proteins binds chitin. Insect Biochem. Mol. Biol. **31**, 1083–1093 (2001). https://doi.org/10.1016/s0965-1748(01)00056-x
92. A. Shechter et al., A gastrolith protein serving a dual role in the formation of an amorphous mineral containing extracellular matrix. Proc. Natl. Acad. Sci. U. S. A. **105**, 7129–7134 (2008). https://doi.org/10.1073/pnas.0800193105
93. C. McDougall, F. Aguilera, B.M. Degnan, Rapid evolution of pearl oyster shell matrix proteins with repetitive, low-complexity domains. J. R. Soc. Interface **10**, 20130041 (2013). https://doi.org/10.1098/rsif.2013.0041
94. J.D. Currey, The design of mineralised hard tissues for their mechanical functions. J. Exp. Biol. **202**, 3285–3294 (1999)
95. M. Suzuki, H. Nagasawa, The structure-function relationship analysis of Prismalin-14 from the prismatic layer of the Japanese pearl oyster, Pinctada fucata. FEBS J. **274**, 5158–5166 (2007). https://doi.org/10.1111/j.1742-4658.2007.06036.x
96. T. Takeuchi, I. Sarashina, M. Iijima, K. Endo, In vitro regulation of $CaCO_3$ crystal polymorphism by the highly acidic molluscan shell protein Aspein. FEBS Lett. **582**, 591–596 (2008). https://doi.org/10.1016/j.febslet.2008.01.026
97. S.V. Patwardhan, S.J. Clarson, C.C. Perry, On the role(s) of additives in bioinspired silicification. Chem. Commun. **2005**, 1113–1121 (2005). https://doi.org/10.1039/b416926c
98. M. Kozak, Interpreting cDNA sequences: some insights from studies on translation. Mamm. Genome **7**, 563–574 (1996). https://doi.org/10.1007/s003359900171
99. M. Kozak, The scanning model for translation – an update. J. Cell Biol. **108**, 229–241 (1989). https://doi.org/10.1083/jcb.108.2.229
100. A. Kuboyama, K. Wada, Pi-electronic excitation energies of anthraquinone. B. Chem. Soc. Jpn. **39**, 1874 (1966). https://doi.org/10.1246/bcsj.39.1874
101. I. Sarashina, K. Endo, The complete primary structure of molluscan shell protein 1 (MSP-1), an acidic glycoprotein in the shell matrix of the scallop Patinopecten yessoensis. Mar. Biotechnol. **3**, 362–369 (2001). https://doi.org/10.1007/s10126-001-0013-6
102. Y. Hasegawa, K. Uchiyama, cDNA clonings of shell matrix proteins from scallop shell. Fish. Sci. **71**, 1174–1178 (2005). https://doi.org/10.1111/j.1444-2906.2005.01078.x
103. Y. Zhang et al., A novel ferritin subunit involved in shell formation from the pearl oyster (Pinctada fucata). Comp. Biochem. Physiol. B-Biochem. Mol. Biol. **135**, 43–54 (2003). https://doi.org/10.1016/s1096-4959(03)00050-2
104. J.E. Hansen et al., NetOglyc: prediction of mucin type O-glycosylation sites based on sequence context and surface accessibility. Glycoconj. J. **15**, 115–130 (1998). https://doi.org/10.1023/A:1006960004440
105. N. Blom, S. Gammeltoft, S. Brunak, Sequence and structure-based prediction of eukaryotic protein phosphorylation sites. J. Mol. Biol. **294**, 1351–1362 (1999). https://doi.org/10.1006/jmbi.1999.3310
106. S.F. Altschul et al., Gapped BLAST and PSI-BLAST: a new generation of protein database search programs. Nucleic Acids Res. **25**, 3389–3402 (1997). https://doi.org/10.1093/nar/25.17.3389

107. N. Suzuki et al., Trophinin expression in the mouse uterus coincides with implantation and is hormonally regulated but not induced by implanting blastocysts. Endocrinology **141**, 4247–4254 (2000). https://doi.org/10.1210/en.141.11.4247
108. S. Weiner, W. Traub, Macromolecules in mollusk shells and their functions in biomineralization. Philos. Trans. R. Soc. Lond. Ser. B Biol. Sci. **304**, 425 (1984). https://doi.org/10.1098/rstb.1984.0036
109. Y. Levi-Kalisman, G. Falini, L. Addadi, S. Weiner, Structure of the nacreous organic matrix of a bivalve mollusk shell examined in the hydrated state using Cryo-TEM. J. Struct. Biol. **135**, 8–17 (2001). https://doi.org/10.1006/jsbi.2001.4372
110. Y. Dauphin, Soluble organic matrices of the calcitic prismatic shell layers of two pteriomorphid bivalves – Pinna nobilis and Pinctada margaritifera. J. Biol. Chem. **278**, 15168–15177 (2003). https://doi.org/10.1074/jbc.M204375200
111. R. Lakshminarayanan, S. Valiyaveettil, V.S. Rao, R.M. Kini, Purification, characterization, and in vitro mineralization studies of a novel goose eggshell matrix protein, ansocalcin. J. Biol. Chem. **278**, 2928–2936 (2003). https://doi.org/10.1074/jbc.M201518200
112. F. Marin, G. Luquet, Molluscan shell proteins. Comptes Rendus Palevol **3**, 469–492 (2004). https://doi.org/10.1016/j.crpv.2004.07.009
113. S. Li et al., Cloning and expression of a pivotal calcium metabolism regulator: calmodulin involved in shell formation from pearl oyster (Pinctada fucata). Comp. Biochem. Physiol. B **138**, 235–243 (2004). https://doi.org/10.1016/j.cbpc.2004.03.012
114. K. Nagai, M. Yano, K. Morimoto, H. Miyamoto, Tyrosinase localization in mollusc shells. Comp. Biochem. Physiol. B-Biochem. Mol. Biol. **146**, 207–214 (2007). https://doi.org/10.1016/j.cbpb.2006.10.105
115. L.M. Faircloth, T.H. Shafer, Differential expression of eight transcripts and their roles in the cuticle of the blue crab, Callinectes sapidus. Comp. Biochem. Physiol. B **146**, 370–383 (2007). https://doi.org/10.1016/j.cbpb.2006.11.008
116. S.O. Andersen, Studies on proteins in post-ecdysial nymphal cuticle of locust, Locusta migratoria, and cockroach, Blaberus craniifer. Insect Biochem. Mol. Biol. **30**, 569–577 (2000). https://doi.org/10.1016/s0965-1748(00)00029-1
117. S.O. Andersen, Characterization of proteins from arthrodial membranes of the lobster, Homarus americanus. Comp. Biochem. Physiol. A Mol. Integr. Physiol. **121**, 375–383 (1998). https://doi.org/10.1016/s1095-6433(98)10146-0
118. L. Addadi, S. Weiner, Biomineralization – crystals, asymmetry and life. Nature **411**, 753–755 (2001). https://doi.org/10.1038/35081227
119. F. Marin, G. Luquet, B. Marie, D. Medakovic, Molluscan shell proteins. primary structure, origin, and evolution. Curr. Top. Dev. Biol. **80**, 209–276 (2008). https://doi.org/10.1016/S0070-2153(07)80006-8
120. P.M. Steinert et al., Amino-acid sequences of mouse and human epidermal type-II keratins of Mr 67,000 provide a systematic basis for the structural and functional diversity of the end domains of keratin intermediate filament subunits. J. Biol. Chem. **260**, 7142–7149 (1985)
121. C.M. Condit, R.B. Meagher, A gene encoding a novel glycine-rich structural protein of petunia. Nature **323**, 178–181 (1986). https://doi.org/10.1038/323178a0
122. D. Hohl et al., Characterization of human loricrin – structure and function of a new class of epidermal-cell envelope proteins. J. Biol. Chem. **266**, 6626–6636 (1991)
123. I.M. Weiss, S. Kaufmann, K. Mann, M. Fritz, Purification and characterization of perlucin and perlustrin, two new proteins from the shell of the mollusc Haliotis laevigata. Biochem. Biophys. Res. Commun. **267**, 17–21 (2000). https://doi.org/10.1006/bbrc.1999.1907
124. Marin, F., Luquet, G., Marie, B., Medakovic, D.: Molluscan shell proteins: primary structure, origin, and evolution. In: Schatten, G. P. (ed.) Current Topics in Developmental Biology, vol. 80, pp 209-276. Elsevier Academic Press Inc Cambridge, MA (2008)
125. K.M. Wilbur, A.M. Bernhardt, Effects of amino-acids, magnesium, and molluscan extrapallial fluid on crystallization of calcium-carbonate – invitro experiments. Biol. Bull. **166**, 251–259 (1984). https://doi.org/10.2307/1541446

126. S. Weiner, I. Sagi, L. Addadi, Choosing the crystallization path less traveled. Science **309**, 1027–1028 (2005). https://doi.org/10.1126/science.1114920
127. Y. Yan et al., A novel matrix protein, PfY2, functions as a crucial macromolecule during shell formation. Sci. Rep. **7**, 6021 (2017). https://doi.org/10.1038/s41598-017-06375-w
128. D.J. Jackson et al., A rapidly evolving secretome builds and patterns a sea shell. BMC Biol. **4**, 40 (2006). https://doi.org/10.1186/1741-7007-4-40
129. L. Xiang et al., Patterns of expression in the matrix proteins responsible for nucleation and growth of aragonite crystals in flat pearls of Pinctada fucata. PLoS One **8**, e66564 (2013). https://doi.org/10.1371/journal.pone.0066564
130. B. Farre, Y. Dauphin, Lipids from the nacreous and prismatic layers of two Pteriomorpha mollusc shells. Comp. Biochem. Physiol. B Biochem. Mol. Biol. **152**, 103–109 (2009). https://doi.org/10.1016/j.cbpb.2008.10.003
131. Y. Politi, T. Arad, E. Klein, S. Weiner, L. Addadi, Sea urchin spine calcite forms via a transient amorphous calcium carbonate phase. Science **306**, 1161–1164 (2004). https://doi.org/10.1126/science.1102289
132. F. Meldrum, Calcium carbonate in biomineralisation and biomimetic chemistry. Int. Mater. Rev. **48**, 187–224 (2003)
133. J. Aizenberg, G. Lambert, L. Addadi, S. Weiner, Stabilization of amorphous calcium carbonate by specialized macromolecules in biological and synthetic precipitates. Adv. Mater. **8**, 222 (1996). https://doi.org/10.1002/adma.19960080307
134. B.H. Wilkinson, Biomineralization, paleoceanography, and the evolution of calcareous marine organisms. Geology **7**, 524–527 (1979)
135. S. Weiner, Biomineralization: a structural perspective. J. Struct. Biol. **163**, 229–234 (2008). https://doi.org/10.1016/j.jsb.2008.02.001
136. J.T. Su et al., Transformation of amorphous calcium carbonate nanoparticles into aragonite controlled by ACCBP. CrystEngComm **18**, 2125–2134 (2016). https://doi.org/10.1039/c5ce02288f
137. Q. Feng et al., Crystallographic alignment of calcite prisms in the oblique prismatic layer of Mytilus edulis shell. J. Mater. Sci. **35**, 3337–3340 (2000)
138. J. Xie et al., Influence of the extrapallial fluid of Pinctada fucata on the crystallization of calcium carbonate and shell biomineralization. Cryst. Growth Des. **16**, 672–680 (2016). https://doi.org/10.1021/acs.cgd.5b01203
139. L. Qiao, Q.-L. Feng, Z. Li, Special vaterite found in freshwater lackluster pearls. Cryst. Growth Des. **7**, 275–279 (2007)
140. N. Spann, E.M. Harper, D.C. Aldridge, The unusual mineral vaterite in shells of the freshwater bivalve Corbicula fluminea from the UK. Naturwissenschaften **97**, 743–751 (2010)
141. B.A. Wustman, D.E. Morse, J.S. Evans, Structural characterization of the N-terminal mineral modification domains from the molluscan crystal-modulating biomineralization proteins, AP7 and AP24. Biopolymers **74**, 363–376 (2004). https://doi.org/10.1002/bip.20086
142. H. Tong et al., Control over the crystal phase, shape, size and aggregation of calcium carbonate via a L-aspartic acid inducing process. Biomaterials **25**, 3923–3929 (2004). https://doi.org/10.1016/j.biomaterials.2003.10.038
143. V. Gerbaud et al., Mechanism of calcite crystal growth inhibition by the N-terminal undecapeptide of lithostathine. J. Biol. Chem. **275**, 1057–1064 (2000)
144. Z. Ma et al., A novel extrapallial fluid protein controls the morphology of nacre lamellae in the pearl oyster, Pinctada fucata. J. Biol. Chem. **282**, 23253–23263 (2007). https://doi.org/10.1074/jbc.M700001200
145. T. Ogino, T. Suzuki, K. Sawada, The formation and transformation mechanism of calcium-carbonate in water. Geochim. Cosmochim. Acta **51**, 2757–2767 (1987). https://doi.org/10.1016/0016-7037(87)90155-4
146. J.R. Clarkson, T.J. Price, C.J. Adams, Role of metastable phases in the spontaneous precipitation of calcium-carbonate. J. Chem. Soc. Faraday Trans. **88**, 243–249 (1992). https://doi.org/10.1039/ft9928800243

147. P.J. Corringer, N. Le Novere, J.P. Changeux, Nicotinic receptors at the amino acid level. Annu. Rev. Pharmacol. Toxicol. **40**, 431–458 (2000). https://doi.org/10.1146/annurev.pharmtox.40.1.431
148. A.B. Smit et al., A glia-derived acetylcholine-binding protein that modulates synaptic transmission. Nature **411**, 261–268 (2001). https://doi.org/10.1038/35077000
149. S.B. Hansen, T.T. Talley, Z. Radic, P. Taylor, Structural and ligand recognition characteristics of an acetylcholine-binding protein from Aplysia californica. J. Biol. Chem. **279**, 24197–24202 (2004). https://doi.org/10.1074/jbc.M402452200
150. N. Koga, Y.Z. Nakagoe, H. Tanaka, Crystallization of amorphous calcium carbonate. Thermochim. Acta **318**, 239–244 (1998). https://doi.org/10.1016/s0040-6031(98)00348-7
151. J. Rieger, J. Thieme, C. Schmidt, Study of precipitation reactions by X-ray microscopy: $CaCO_3$ precipitation and the effect of polycarboxylates. Langmuir **16**, 8300–8305 (2000). https://doi.org/10.1021/la0004193
152. A. OrrUrtreger et al., Mice deficient in the alpha 7 neuronal nicotinic acetylcholine receptor lack alpha-bungarotoxin binding sites and hippocampal fast nicotinic currents. J. Neurosci. **17**, 9165–9171 (1997)
153. J. Su et al., Structural characterization of amorphous calcium carbonate-binding protein: an insight into the mechanism of amorphous calcium carbonate formation (vol 453, pg 179, 2013). Biochem. J. **454**, 167–167 (2013)
154. E.F. Pettersen et al., UCSF chimera – a visualization system for exploratory research and analysis. J. Comput. Chem. **25**, 1605–1612 (2004). https://doi.org/10.1002/jcc.20084
155. F.F. Amos, M. Ndao, J.S. Evans, Evidence of mineralization activity and supramolecular assembly by the N-terminal sequence of ACCBP, a biomineralization protein that is homologous to the acetylcholine binding protein family. Biomacromolecules **10**, 3298–3305 (2009). https://doi.org/10.1021/bm900893f
156. L.J. McGuffin, K. Bryson, D.T. Jones, The PSIPRED protein structure prediction server. Bioinformatics (Oxford) **16**, 404–405 (2000)
157. Y.K. Reshetnyak, E.A. Burstein, Decomposition of protein tryptophan fluorescence spectra into log-normal components. II. The statistical proof of discreteness of tryptophan classes in proteins. Biophys. J. **81**, 1710–1734 (2001)
158. J.T. Han, X.R. Xu, D.H. Kim, K.W. Cho, Biomimetic fabrication of vaterite film from amorphous calcium carbonate on polymer melt: effect of polymer chain mobility and functionality. Chem. Mater. **17**, 136–141 (2005). https://doi.org/10.1021/048892f
159. J. Xiao, Z. Wang, Y. Tang, S. Yang, Biomimetic mineralization of $CaCO_3$ on a phospholipid monolayer: from an amorphous calcium carbonate precursor to calcite via vaterite. Langmuir **26**, 4977–4983 (2010). https://doi.org/10.1021/la903641k
160. L. Addadi, J. Moradian, E. Shay, N.G. Maroudas, S.A. Weiner, Chemical-model for the cooperation of sulfates and carboxylates in calcite crystal nucleation – relevance to biomineralization. Proc. Natl. Acad. Sci. U. S. A. **84**, 2732–2736 (1987). https://doi.org/10.1073/pnas.84.9.2732
161. L. Addadi, S. Weiner, M. Geva, On how proteins interact with crystals and their effect on crystal formation. Zeitschrift fur Kardiologie **90**(Suppl 3), 92–98 (2001)
162. Y. Levi-Kalisman, S. Raz, S. Weiner, L. Addadi, I. Sagi, Structural differences between biogenic amorphous calcium carbonate phases using X-ray absorption spectroscopy. Adv. Funct. Mater. **12**, 43–48 (2002). https://doi.org/10.1002/1616-3028(20020101)12:1<43::aid-adfm43>3.0.co;2-c
163. Y. Politi et al., Structural characterization of the transient amorphous calcium carbonate precursor phase in sea urchin embryos. Adv. Funct. Mater. **16**, 1289–1298 (2006). https://doi.org/10.1002/adfm.200600134
164. J.H.E. Cartwright, A.G. Checa, J.D. Gale, D. Gebauer, C.I. Sainz-Diaz, Calcium carbonate polyamorphism and its role in biomineralization: how many amorphous calcium carbonates are there? Angew. Chem. Int. Ed. **51**, 11960–11970 (2012). https://doi.org/10.1002/anie.201203125

165. L.B. Gower, Biomimetic model systems for investigating the amorphous precursor pathway and its role in biomineralization. Chem. Rev. **108**, 4551–4627 (2008). https://doi.org/10.1021/cr800443h
166. M.A. Bewernitz, D. Gebauer, J. Long, H. Coelfen, L.B. Gower, A metastable liquid precursor phase of calcium carbonate and its interactions with polyaspartate. Faraday Discuss. **159**, 291–312 (2012). https://doi.org/10.1039/c2fd20080e
167. S.E. Wolf et al., Strong stabilization of amorphous calcium carbonate emulsion by ovalbumin: gaining insight into the mechanism of 'polymer-induced liquid precursor' processes. J. Am. Chem. Soc. **133**, 12642–12649 (2011). https://doi.org/10.1021/ja202622g
168. V. Pipich, M. Balz, S.E. Wolf, W. Tremel, D. Schwahn, Nucleation and growth of CaCO (3) mediated by the egg-white protein ovalbumin: a time-resolved in situ study using small-angle neutron scattering. J. Am. Chem. Soc. **130**, 6879–6892 (2008). https://doi.org/10.1021/ja801798h
169. X. Wang et al., Influence of ovalbumin on $CaCO_3$ precipitation during in vitro biomineralization. J. Phys. Chem. B **114**, 5301–5308 (2010). https://doi.org/10.1021/jp1008237
170. Y. Politi, J. Mahamid, H. Goldberg, S. Weiner, L. Addadi, Asprich mollusk shell protein: in vitro experiments aimed at elucidating function in $CaCO_3$ crystallization. CrystEngComm **9**, 1171–1177 (2007). https://doi.org/10.1039/b709749b
171. E.M. Pouget et al., The initial stages of template-controlled $CaCO_3$ formation revealed by Cryo-TEM. Science **323**, 1455–1458 (2009). https://doi.org/10.1126/science.1169434
172. X. Liu et al., The role of matrix proteins in the control of nacreous layer deposition during pearl formation. Proc. R. Soc. Lond. B Biol. Sci. **279**, 1000–1007 (2012)
173. S. Weiner, P.M. Dove, An overview of biomineralization processes and the problem of the vital effect. Rev. Mineral. Geochem. **54**, 1–29 (2003)
174. J. Huang, C. Zhang, Z. Ma, L. Xie, R. Zhang, A novel extracellular EF-hand protein involved in the shell fonnation of pearl oyster. Biochim. Biophys. Acta. Gen. Subj. **1770**, 1037–1044 (2007). https://doi.org/10.1016/j.bbagen.2007.03.006
175. A. Lewit-Bentley, S. Rety, EF-hand calcium-binding proteins. Curr. Opin. Struct. Biol. **10**, 637–643 (2000). https://doi.org/10.1016/s0959-440x(00)00142-1
176. P.M. Harrison, P. Arosio, Ferritins: molecular properties, iron storage function and cellular regulation. BBA-Bioenergetics **1275**, 161–203 (1996). https://doi.org/10.1016/0005-2728(96)00022-9
177. M. Vondarl, P.M. Harrison, W. Bottke, cDNA cloning and deduced amino-acid-sequence of 2 ferritins – soma ferritin and yolk ferritin, from the snail Lymnaea-stagnalis L. Eur. J. Biochem. **222**, 353–366 (1994)
178. T.S. Huang, J.H. Law, K. Soderhall, Purification and cDNA cloning of ferritin from the hepatopancreas of the freshwater crayfish Pacifastacus leniusculus. Eur. J. Biochem. **236**, 450–456 (1996). https://doi.org/10.1111/j.1432-1033.1996.00450.x
179. G. Beck, T.W. Ellis, G.S. Habicht, S.F. Schluter, J.J. Marchalonis, Evolution of the acute phase response: iron release by echinoderm (Asterias forbesi) coelomocytes, and cloning of an echinoderm ferritin molecule. Dev. Comp. Immunol. **26**, 11–26 (2002). https://doi.org/10.1016/s0145-305x(01)00051-9
180. G.S. Waldo, J.S. Ling, J. Sanderslоehr, E.C. Theil, Formation of an Fe(III)-tyrosinate complex during biomineralization of H-subunit ferritin. Science **259**, 796–798 (1993). https://doi.org/10.1126/science.8430332
181. M. Pekkarinen, I. Valovirta, Histochemical and X-ray studies on tissue concretions and shells of Margaritifera margaritifera (Linnaeus). J. Shellfish Res. **16**, 169–177 (1997)
182. B.A. Fowler, D.A. Wolfe, W.F. Hettler, Mercury and iron uptake by cytosomes in mantle epithelial-cells of quahog clams (Mercenaria-mercenaria) exposed to mercury. J. Fish. Res. Board Can. **32**, 1767–1775 (1975)
183. J.D. Thomson, B.J.S. Pirie, S.G. George, Cellular metal distribution in the pacific oyster, crassostrea-gigas (thun) determined by quantitative x-ray microprobe analysis. J. Exp. Mar. Biol. Ecol. **85**, 37–45 (1985). https://doi.org/10.1016/0022-0981(85)90012-7

184. T.G. Stpierre, K.S. Kim, J. Webb, S. Mann, D.P.E. Dickson, Biomineralization of iron – Mossbauer-spectroscopy and electron-microscopy of ferritin cores from the chiton acanthopleura-hirtosa and the limpet Patella-laticostata. Inorg. Chem. **29**, 1870–1874 (1990). https://doi.org/10.1021/ic00335a022
185. N.D. Chasteen, P.M. Harrison, Mineralization in ferritin: an efficient means of iron storage. J. Struct. Biol. **126**, 182–194 (1999)
186. K.S. Kim, J. Webb, D.J. Macey, Properties and role of ferritin in the hemolymph of the chiton clavarizona-hirtosa. Biochim. Biophys. Acta **884**, 387–394 (1986)
187. T. Sumitomo, H. Kakisawa, Y. Owaki, Y. Kagawa, In situ transmission electron microscopy observation of reversible deformation in nacre organic matrix. J. Mater. Res. **23**, 1466–1471 (2008). https://doi.org/10.1557/jmr.2008.0184

Chapter 4
The Study on Enzymes Related to Biomineralization of *Pinctada fucata*

Abstract As an efficient, economical, and clean catalyst, enzymes are expected to have novel features. Efforts have been made to screen new enzymes in marine organisms. In our recent studies, we explored the basic characteristics and functions of alkaline phosphatases, acid phosphatases, carbonic anhydrases, tyrosinase, and a novel astacin-like metalloproteinase in *Pinctada fucata*.

An alkaline phosphatase (ALP), an acid phosphatase (ACP) and two isoenzymes (AcPase I and II), a novel carbonic anhydrase (CA), tyrosinase, and a novel astacin-like metalloproteinase were purified or isolated from *Pinctada fucata*. The specific activity of the enzymes and the optimum pH and temperature were detected. The effects of HPO_4^{2-} production and the product-analog WO_4^{3-}, MoO_4^{3-}, and AsO_4^{3-} as well as some metal ions on the related enzymes were also determined. The effect of inhibitors on the enzyme activity was also examined. Besides, the tissue distribution of CA and tyrosinase was analyzed by in situ hybridization and RT-PCR. The effects of AcPase I on $CaCO_3$ crystal formation were studied in vitro. Taken together, these results revealed the important functions and features of enzymes in *Pinctada fucata*, which would have important roles to further understand the enzymes in peal oyster.

Keywords Alkaline phosphatases · Acid phosphatases · Carbonic anhydrases · Tyrosinase · Metalloproteinase

4.1 Introduction

4.1.1 Enzymes

As an efficient, economical, and clean catalysts, enzymes are increasing used from laundry detergents and paper processing to fine chemical synthesis and diagnostic/research reagents [1]. Efforts have been made to screen new enzymes from various kinds of organisms to improve existing enzymes, optimize existing processes, or obtain marketable intellectual property. Marine life has long been neglected though it is a rich reservoir for enzymes. Since marine organisms live

R. Zhang et al., *Biomineralization Mechanism of the Pearl Oyster, Pinctada fucata*,
https://doi.org/10.1007/978-981-13-1459-9_4

in a quite different environment from land organisms, enzymes in these organisms are expected to have novel features.

4.1.2 Alkaline Phosphatase

As a substrate nonspecific phosphomonoesterase, alkaline phosphatase (ALP, EC 3.1.3.1) can catalyze hydrolysis of a wide variety of phosphomonoesters under alkaline conditions. In the presence of a phosphate acceptor, ALP can also act as a transphosphorylase [2]. Over the past decade, ALPs from mammalian sources and microorganisms have been widely investigated and used as diagnostic and research reagents [3–6].

Based on kinetic and structural studies of the native and some site-directed mutant *E. coli* ALP, a detailed catalytic mechanism of the enzyme has been proposed. The reaction follows a ping-pong replacement kinetics and proceeds through a phosphoseryl intermediate to yield inorganic phosphate or to transfer the phosphoryl group to another alcohol [7, 8]. Although the detailed physiological functions have not been fully elucidated, the widespread distribution of ALP in almost all organisms indicates its important role in cell metabolism. In marine organisms, ALPs play important roles in cell phosphate metabolism, which is related to the absorption of phosphate and calcium in seawater as well as the biomineralization process in marine organisms. ALP activity has been used as a stress marker to assess the impact of environmental exotics, such as heavy metals and organic chlorides, on marine organisms [9, 10]. As a first step in screening for new ALPs from marine organisms, we have thoroughly studied ALP from a green crab *Scylla serrata* in previous studies [11]. Here, we report the purification of the ALP from *Pinctada fucata* and the enzymological properties of this enzyme.

The crystal structures of ALP from *E. coli*, *Pandalus borealis*, and human placenta have been solved [2, 12, 13]. They are all homodimers with very similar overall topologies, and their active site architectures all had a metal triad (usually two zinc ions and one magnesium ion). Comparative studies of the primary sequence of ALPs from *E. coli*, *P. borealis*, *S. cerevisiae*, chicken, and humans showed 25–45% sequence identity among ALPs from different species [14]. Most of the important functional residues, including the serine nucleophile (Ser-112 in E. coli ALP) and an arginine residue (Arg-166 in *E. coli* ALP) involved in substrate binding, are highly conserved, which means that similar catalytic mechanisms are shared in ALPs of different species.

Although ALPs from bacteria and mammals have been widely studied, there are few reports about ALPs from Mollusca, which is the second largest phylum of the animal kingdom. A better understanding of oyster ALP would help determine the mechanism of shell and pearl formation. We illustrated the enzymatic properties of the tissue-nonspecific ALP from *P. fucata* as well as the OPA modification of this ALP. And we reported the chemical modification of three positively charged residues of this alkaline phosphatase, lysine, arginine, and tryptophan.

4.1.3 Acid Phosphatase

Acid phosphatase (AcPase) (orthophosphoric monoester phosphohydrolase, EC 3.1.3.2) is widely distributed in plant and animal cells [15]. It is primarily localized in lysosomes, but the microsomal and soluble enzyme forms have also been documented [16, 17]. As a marker enzyme of lysosomes, acid phosphatase can be altered by the presence of xenobiotics and cadmium ions [18, 19]. Enzymes can potentially be used as biomarkers of marine pollution. AcPase is a multifunctional enzyme that is suggested to be involved in food degradation in digestive diverticula cells [20], protein and organelle turnover [21], cellular defense [22], contaminant detoxication [23], macrophage inflammatory responses [24], and endochondral ossification [25].

Acid phosphatase forms were mainly studied in vertebrate tissues. In invertebrates, previous studies reported only a few acid phosphatase activities. Various enzyme forms have been purified from sea urchin eggs and embryos [26], mosquito [27], tapeworm [28], drosophila [29], and planarians previously. Recently, other acid phosphatases have also been well studied in desert locust [30], earthworm [15], and freshwater snail [19]. No acid phosphatase isolated from marine bivalves was reported before.

Pinctada fucata is a marine pearl oyster that is widely cultured in China. However, in recent years, the oyster has been affected by a variety of pollutants and diseases. Although some drugs that enhance the immunity of the pearl oyster have been developed, the immunity mechanisms of mollusks have not been thoroughly studied. In mollusks such as *Pinctada fucata*, acid phosphatase plays an important role in defense against pathogens. In our study, purification and partial characterization of two acid phosphatases from pearl oyster (*P. fucata*) were studied [20]. AcPase activity in *P. fucata* exposed to copper was also studied [31], and the results indicated that they were involved in cellular defense. In addition, the exact nature of how the AcPase works in vivo was also explored.

4.1.4 Carbonic Anhydrase

Carbonic anhydrase (CA; EC 4.2.1.1) catalyzes the reversible hydration of CO_2, accelerates the formation of bicarbonate, and plays an important role in a variety of physiological functions in mollusks, including acid–base regulation, calcification, respiration, and mineralization. The activity of CA and its effect on physiological functions of mollusks have been extensively studied [32–36]. CA was shown to play a role in mineralization of all invertebrates studied, and it was found to be essential for rapid shell development.

Miyamoto et al. purified a matrix protein from nacreous layers of the pearl oyster (*Pinctada fucata*) and named it nacrein which contains a CA domain and many characteristic Gly–Xaa–Asn repeats [35]. Their results suggested that CA may play important roles in calcium carbonate crystal formation of the nacreous layer. The nacreous layer formation model controlled by nacrein was also proposed. In another pearl oyster *Pinctada margaritifera*, which is regulated by calcitonin gene-related peptide [37], CA activity has been considered as one of the biomineralization markers during a phase of active growth. There seems to be little disagreement about the involvement of CA in the biomineralization of shells, but the mechanism involved is not fully understood.

P. fucata produce pearls in the shells. Therefore, a pearl is equivalent to the nacreous layer, which consists of the aragonite crystal and organic matrices. In order to study the role of CA in the biomineralization of pearls, we isolated and identified a novel CA from the mantle of the pearl oyster. At the same time, we investigated its distribution in mantle tissue. The study provides further evidence that CA plays a role in the process of pearl formation and biomineralization.

4.1.5 Tyrosinase

Tyrosinase (monophenol; L-DOPA, oxygen oxidoreductase; EC1.14.18.1) is one of the copper-containing phenoloxidases widely distributed in nature. It catalyzes the orthohydroxylation of monophenols (e.g., tyrosine) to o-diphenols (e.g., 3, 4-dihydroxyphenylalanine, DOPA) and the oxidation of the resultant o-diphenols to their respective o-quinones [38, 39]. Quinones formed by these reactions are able to bind free amino groups in proteins to form protein cross-linkages, eventually leading to the formation of insoluble protein polymers [40, 41]. In this way, tyrosinases or other phenoloxidases are thought to be involved in many biological processes in invertebrates, such as pigment formation [42, 43], wound healing [44], native immune response [43–47], and the sclerotization or quinone tanned of many insoluble, byssus [48, 49], chemical-resistant extracellular structure, including the egg capsules [50, 51] and the periostracum of the mollusk, which we pay attention to now [48, 52].

Periostracum is an uncalcified cuticle layer covering the surface of mollusk shell. It is one of the most powerful but most inert chemical structures in the animal kingdom. In addition to its direct protective contribution to organisms, it is also thought to play an important role in shell biomineralization. Periostracum enclosed the extrapallial space on the ventral side and separated it from the external environment, enabling supersaturation conditions inside, which is necessary for the formation of calcified layers [53]. Some studies also showed that it might act as the substrate for the initial deposition of calcium and even affect the formation of prismatic layer [54, 55]. The periostracum is secreted as a soluble precursor, and the periostracin is then cross-linked by o-diphenols and tyrosinase (or phenoloxidases) to form the insoluble periostracum. Therefore, isolation and sequence of these tyrosinases or phenoloxidases help to understand the molecular details of periostracum formation.

Although histological studies have demonstrated that tyrosinases (or phenoloxidases) are involved in the formation of periostracum [48, 52], we do not know much about the molluskan tyrosinases. So far, only one tyrosinase has been partially purified from the mantle and the periostracum of the marine bivalve *Modiolus demissus*, and several other tyrosinases have been isolated from the other tissues of mollusk [43, 51, 56]. Among these molluscan tyrosinases, only two cephalopod tyrosinase amino acid sequences have been determined recently (Accession Nos. AB107880 and AJ297474, respectively) [43]. In this study, a novel putative tyrosinase was cloned by RT-PCR, and rapid amplification of cDNA ends (RACE) from the mantle of pearl oyster *Pinctada fucata*. To the best of our knowledge, this is the first tyrosinase sequence identified from the bivalves. This tyrosinase was named OT47 because of its predicted molecular weight of 47 kDa, exhibiting a specific expression pattern, and may partially change our view on the secretion of periostracum in *P. fucata*.

4.1.6 Glycosylphosphatidylinositol–Anchored Alkaline Phosphatase

Astacin metalloproteinases are a family of zinc-dependent endopeptidases. Structurally, astacins contain several domain motifs: N-terminal signal and proenzyme sequences; catalytic domain that contains a zinc-binding motif HEXXHXXGXXH and a conserved methionine, which forms a unique "Met-turn" structure; and many contain multiple C-terminal domains [57].

Astacins function in the extracellular space and occur either as membrane-bound or secreted enzymes [58, 59]. They exhibit many different physiological functions, such as bone growth, embryonic development, digestion, morphogenesis, and cell migration [60–64]. The astacins can be roughly divided into three categories by their functions. The membrane-bound meprins found in the small intestine and kidney of mammals are thought to be involved in the processing of biologically active peptides and extracellular matrix proteins [59, 65]. Bone morphogenetic protein-1 (BMP-1) [60] and tolloid [62] form the second class and contribute to morphogenesis and embryonic pattern formation. Astacin-like enzymes also involve the hatching process of embryos. Previous studies have shown that the fish enzymes choriolysin H (HCE1 and HCE2) and choriolysin L (LCE) act synergistically to degrade the eggshell/chorion of embryos [66–68].

As an important skeletal extracellular matrix protein, collagen plays a crucial role in changing the morphology of calcite crystals [69]. Previous studies have demonstrated that the collagen can be synthesized in hemocytes and mantle of *Haliotis tuberculata* and it is also a significant matrix protein in nacre of the marine snail [70–72]. Previous work by Mount et al. [73] showed that a type of granulocytic hemocytes was directly involved in the production of oyster shell crystal. However, the function of the collagen secreted by hemocytes is not clear yet, which may be required for nacre formation. In addition, the regulation of embryo development that may contribute to the assembly progress for constructing the shells of biomineralized structures is also largely unknown [74]. On account of the potential importance of astacins in embryo development, morphogenesis, and extracellular matrix disposition, identification and characterization of members of this family in mollusks may help resolve the relative functions of these enzymes.

In our lab, we reported the cloning of pf-ALMP, a novel astacin with cysteine arrays from bivalve pearl oyster, *P. fucata*. The expression pattern was investigated and revealed that pf-ALMP was abundant in hemocytes. We expressed, purified the functional catalytic domain of pf-ALMP in *E. coli*, and analyzed its proteolytic activity. The function of the cysteine arrays was also investigated [57].

4.2 Materials and Methods

4.2.1 Materials

Adult *Pinctada fucata* were collected from Beihai Oyster Culture Centre, Guangxi Province, China. After collection, the oysters were transported to the laboratory and stored in a filtered flow-through system. The aerated seawater is used at 20 °C until the enzyme is purified.

p-Nitrophenyl phosphate (p-NPP) and phenylglyoxal (PG) were purchased from Merck. o-Phthalaldehyde (OPA) came from Amresco. 2,4,6-Trinitrobenzenesulfonic acid (TNBS), diethyl pyrocarbonate (DEPC), 5,5′-dithiobis (2-nitrobenzoic acid) (DTNB), p-aminomethylbenzene sulfonamide–agarosea, and Z-histidyldiazobenzylphosphonic acid agarose were Sigma products. DEAE-32 cellulose was a Whatman product. Diethylmalonic acid and 5-dimethylamino-1-naphthalenesulfonamide (DNSA) were purchased from Aldrich and Fluka, respectively. Sephadex G-150 and Con A Sepharose 4B were purchased from Amersham Pharmacia Biotech (Uppsala, Sweden). DEAE-Sepharose Fast Flow, Sephadex G-200 superfine, and Con A Sepharose 4B were Pharmacia products. 4-Nitrophenyl acetate (pNPA), acetazolamide (AZ), and carbonic anhydrase I (CA I) are from human erythrocytes. N-Bromosuccinimide (NBS) and other locally sourced chemicals were all analytical grade.

4.2.2 Gene Sequence Acquisition

4.2.2.1 RNA Extraction

P. fucata adults were obtained from the Guofa Pearl Farm in Beihai, Guangxi Province, China. According to the manufacturer's instructions, total RNA was extracted from the mantle tissue with the RNAzol RNA Isolation Kit (Biotecx,

Houston, TX, USA). RNA integrity was determined by separation on a 1.2% formaldehyde-denatured agarose gel and staining with ethidium bromide. The quantity of RNA was confirmed by measuring OD260nm with an Ultrospec 3000 UV/Visible Spectrophotometer (Amersham, Piscataway, NJ, USA).

4.2.2.2 Reverse Transcription–Polymerase Chain Reaction (RT-PCR)

A pair of degenerate oligonucleotide primers was synthesized based on the conserved regions within each enzyme. These are listed as follows; the details can be found in the tables of the end. A total of 2.5 μg RNA from the mantle was used as template for the RT reaction with superscript II and oligo(dT) primers (Invitrogen, CA, USA) according to the manufacturer's instructions on a Tgradient Thermocycler (Biometra, Gottingen, Germany). PCR cycles were generally conducted as follows: denaturation at 95 °C for 5 min, followed by 30 cycles at 95 °C for 45 s, 43 °C for 30 s, and 72 °C for 1 min. A final extension step was performed at 72 °C for 10 min. There are some setting modifications according to different genes. PCR products were then electrophoresed, and the desired size bands were excised and purified using Wizard PCR Preps DNA Purification System (Promega, Madison, WI, USA). The purified PCR products were then subcloned into the pMD 18-T Vector (Takara, Dalian, China) and sequenced.

4.2.2.3 Isolation and Sequencing of the *P. fucata* Enzyme cDNA

Single-stranded cDNA for all rapid amplification of cDNA ends (RACE) reactions were prepared from the mantle total RNA by using the PowerScript™ (Clontech, Palo Alto, CA, USA). 5′-RACE and 3′-RACE were performed using the SMART RACE cDNA Amplification Kit (Clontech) and Advantage 2 cDNA Polymerase Mix (Clontech) according to the manufacturer's instructions. The gene-specific primer 5RP for 5′-RACE and 3RP for 3′-RACE was prepared based on the nucleotide sequence of the cDNA fragment amplified by RT-PCR; the detailed information can be seen in the following list. All amplified products were cloned into pMD 18-T Vector (Takara, Dalian, China) for sequencing. To confirm cloning and sequencing accuracy, a pair of gene-specific primers was used to re-amplify the entire cDNA from the start codon to stop codon with high-fidelity polymerase (Takara). The purified PCR products were subcloned into pMD 18-T Vector (Takara, Dalian, China) and then resequenced.

Single-stranded cDNA for all rapid amplification of cDNA ends (RACE) reactions was prepared from the hemocytes total RNA using the PowerScript (Clontech, Palo Alto, CA, USA). 5′-RACE and 3′-RACE were performed using the SMART RACE cDNA Amplification Kit (Clontech) and Advantage 2 cDNA Polymerase Mix (Clontech) according to the manufacturer's instructions. The 5′- and 3′- regions were amplified by using specific primers based on the sequence obtained by RT-PCR and primers in the SMART RACE cDNA Amplification Kit (Clontech,

Palo Alto, CA, USA). The 5′-RACE was carried out under the conditions recommended by the kit above. To amplify the 3′-end of the cDNA, the PCR was conducted under the following conditions with some modifications: the annealing temperature decreased 2 °C every 3 cycles from 72 to 64 °C; 30 cycles were carried out at 64 °C; denaturation and extension were conducted at 94 °C and 72 °C, respectively.

4.2.2.4 Construction of Expression Vectors

The pcDNA3.1 expression vector was constructed by amplifying the full-length sequence using two primers, and then the PCR product was then inserted into the specific sites of pcDNA3.1, such as BamHI and EcoRI. The cysteine array deletions of pf-ALMP were generated by high-fidelity polymerase chain reaction with two primers and then subcloned into pcDNA3.1 as described above. To express a specific protein or an important domain, the pET-28(b) expression vector was constructed by amplifying the specific domains with two primers, and then the PCR product was inserted into the selected sites of pET-28(b), such as NcoI and XhoI. All recombination plasmids were ensured by DNA sequencing.

4.2.3 Gene Expression Analysis

4.2.3.1 DNA Sequencing and Sequence Analysis

All recombinant plasmids were sequenced using an automated DNA sequencer (Applied Biosystems 377). A list of known sequences of the number of astacins was obtained from GenBank databases using the BLAST family of programs (http://www.ncbi.nlm.nih.gov/BLAST/). The presence and location of signal peptide cleavage site were analyzed using the SignalP v3.0 program [75] available at the Center for Biological Sequence Analysis, BioCentrum-DTU, Technical University of Denmark (CBS prediction servers; http://www.cbs.dtu.dk/). Multiple alignments were created using the ClustalX program [76].

4.2.3.2 Gene Expression Analysis by RT-PCR

Expression levels of related mRNA were examined using semiquantitative RT-PCR analysis. The primers were listed in the following tables. Tissue expression of the oyster-related mRNA was investigated by RT-PCR. As described above, total RNA was isolated from each of the mantle edge, mantle pallial, foot, gill, gonad, digestive gland, adductor muscle tissues, as well as the hemocytes of the adult individual of *P. fucata*. Equal amounts (2 μg) of total RNA from different tissues were reverse transcribed into cDNA with Superscript II and oligo(dT) primers in 20 μL of reaction mixtures (Invitrogen, CA, USA). A PCR reaction was performed using Takara Taq DNA polymerase (Takara, Dalian, China), typically 95 °C for 5 min, 30 cycles of 95 °C for 30 s, 52 °C for 30 s, and 72 °C for 1 min, followed by 1 min incubation at 72 °C. The primer pairs used for amplification of related gene fragments were also listed as the following table. The PCR products were then subcloned into pMD 18-T Vector (Takara, Dalian, China) and verified by sequencing. Equal volumes of the PCR products were separated on a 1.5% agarose gel and stained with ethidium bromide. Negative controls were performed without the cDNA template to check for the cross contamination of the samples.

4.2.3.3 In Situ Hybridization

The tissues were removed from *P. fucata* and then immediately fixed in 4% paraformaldehyde overnight. In situ hybridization of oyster enzyme mRNAs was performed on frozen sections. Digoxigenin-labeled RNA probes were generated from the cDNA clone encoding the recombinant plasmid using a digoxigenin (DIG) RNA Labeling Kit (Roche, Hong Kong, China), with SP6 and T7 RNA polymerase for the sense and the antisense probe, respectively. RNA in situ hybridization was performed as described in the above chapters with modification [77]. To avoid false-positive signals, the hybridization temperature increased to 58 °C.

4.2.4 *Enzyme Purification and Activities*

4.2.4.1 Enzyme Purification

The alkaline phosphatase was prepared as previously described [78] in Sephadex G-150 gel filtration step. The active fractions were collected and assessed using affinity chromatography as described below. The active peaks were then combined and dialyzed for 72 h to remove the inorganic phosphate. All the purification procedures were performed at 4 °C. The final preparations with specific activity of 1215 unit mg^{-1} showed homogeneity in the absence and presence of SDS on polyacrylamide gel electrophoresis.

4.2.4.2 Enzyme Purification of CA

After careful separation from pearl oyster, the mantle tissue was homogenized in 25 mM Tris–sulfate buffer (pH 7.5), containing 25 mM sodium sulfate using a blade homogenizer (Moulinex, Germany), and then centrifuged (10,000 × g,40 min). The supernatant was precipitated with 70% saturated ammonium sulfate overnight and then centrifuged at 10,000 × g for 40 min again. The precipitated proteins were dissolved in the Tris–sulfate buffer (pH 7.5) and then dialyzed against distilled deionized water for about 48 h until no ammonium sulfate could be found. The solution was dialyzed against the same buffer as above (adjusted to pH 8.7) and then subjected to affinity chromatography.

The affinity chromatography with p-aminomethylbenzene sulfonamide–agarose was conducted as described by Lucas and Knapp [79]. The dialyzed solution was loaded onto the affinity chromatography column pre-equilibrated with the same Tris–sulfate buffer (pH 8.7) and then eluted by 0.1 M KSCN. The effluent fraction containing KSCN was dialyzed against buffer (25 mM Tris–sulfate pH 7.5 containing 25 mM sodium sulfate) until SCN− could not be detected. The protein was concentrated using a Nanosep Centrifugal Device (Pall, MWCO = 10 kDa).

4.2.4.3 Protein Expression and Purification In Vitro

Recombinant pET-28(b) plasmid was transformed into *E. coli* strain BL21(DE3) expressing T7 polymerase by induction with isopropyl-D-thiogalactoside (IPTG). Appropriate transformants were identified by restriction analysis and sequencing. Protein expression was induced with nearly 0.8 mM IPTG at 37 °C. When the optical density at 600 nm of the culture reached 0.6, IPTG was added. After the induction for 3 h, bacterial cells were harvested by centrifugation at 10000 rpm for 5 min. Harvested cell paste from 11 bacterial cultures was suspended in 50 mM lysis buffer (containing 8 M urea, 0.02 M NaH_2PO_4, 0.5 M NaCl, pH 7.2) and then sonicated using a Branson Sonifier. After centrifugation (10,000 × g, 30 min, 4 °C), the supernatant was filtrated through a 0.45 μm filter and loaded onto a Ni–NTA affinity column (Amersham). The column was subsequently washed with binding buffer (including 8 M urea, 0.02 M NaH_2PO_4, 0.01 M imidazole, 0.5 M NaCl, pH 7.2). The column was eluted with the elution buffer (containing 8 M urea, 0.5 M NaCl, 0.02 M NaH_2PO_4, 0.5 M imidazole, pH 7.2). For protein folding, the procedure of Sabine Reyda [80] was followed. Inclusion bodies were dissolved in 8 M urea, 1 mM EDTA (ethylenediaminetetraacetic acid), 0.1 M β-mercaptoethanol, and 50 mM Tris–HCl, pH 8.0, at a protein concentration of 10 mg/ml. This solution was diluted into 50 mM Tris–HCl buffer (pH 8.0), containing 1 mM GSH (reduced glutathione), 0.1 mM GSSG (oxidized glutathionine), and 0.8 M L-arginine, so that a final concentration was 10 μg/ml, incubated for 24 h at 4 °C and dialyzed against 50 mM Tris–HCl buffer, pH 8.0, containing 0.1 μM $ZnSO_4$, and finally against zinc-free 50 mM Tris–HCl buffer, pH 8.0.

4.2.4.4 Substrate Gel Zymography on SDS-PAGE

Substrate gel zymography was performed using a modified method of Rockey. Briefly, 8% acrylamide running gels at pH 8.8 were polymerized with gelatin at a final concentration of 0.5% (w/v). A 3% stacking gel at pH 6.8 dose not contained a substrate. The purified protein was incubated for 30 min at room temperature in SDS sample buffer containing 5% w/v SDS at pH 6.8. Samples were not boiled or treated with reducing agents to reanneal the proteins. Electrophoresis was performed on a vertical gel system (Amersham) at room temperature with constant current of 20 mA for 2 h. After electrophoresis, the gels were incubated for 1 h at 37 °C in a solution containing 2.5% (v/v) Triton X-100 and 50 mM Tris–HCl, pH 7.4. The gels were then transferred to a solution which contains 50 mM Tris–HCl, pH 7.4, and 10 mM $CaCl_2$ and incubated at 37 °C for 24 h. After incubations, gels were fixed and then stained with Coomassie Brilliant Blue G and destained to reveal clear bands from proteolytic activity. Clear proteolytic active bands were observed on a blue background of stained intact gelatin.

4.2.4.5 Western Blot

SDS-PAGE was performed using 3.75% stacking and 7.5% separating gels, according to the method of Laemmli [81]. After electrophoresis, separated proteins were electrotransferred to nitrocellulose membranes by the Multiphor II Western blotting system (Amersham, Sweden). Purified polyclonal antibody was diluted 1:50, and peroxidase-labeled anti-mouse IgG antibody (Amersham, Sweden) was used at 1:2000. Other procedures were based on Sambrook et al. [82]. For controls, membranes were incubated in normal mouse serum but not in antibody.

4.2.4.6 Detection of Enzyme Activities

The enzyme activity was measured in 0.05 M Na_2CO_3–$NaHCO_3$ buffer (pH 9.7) at 40 °C by the absorbance changes at 405 nm accompanied by substrate hydrolysis (p-NPP). One unit of enzyme activity represents the amount of enzyme that catalyzes the formation of 1 μmol *p*-nitrophenol per minute from *p*-NPP.

The alkaline phosphatase activity was determined spectrophotometrically by using *p*-NPP as a substrate [92]. The standard assay system contained 1 mM *p*-NPP in 50 mM sodium bicarbonate buffer (pH 9.00), and the reaction was performed at 25 °C. The released *p*-nitrophenol was measured by testing the absorbance increase at 405 nm using a molar extinction coefficient of 17.3 mM^{-1} cm^{-1} determined under the assay conditions. One activity unit represents the amount of enzyme that required to catalyze 1 μmol reaction per minute under the assay conditions. The K_M and k_{cat} parameters were measured by the Lineweaver–Burk analysis.

AcPase enzyme activity was measured spectrophotometrically from the rate of hydrolysis of model substrates. The amount of enzyme that increased the amount of the enzyme reaction product by 1 μmol/min was taken as 1 unit of the enzyme activity (U). The specific activity was expressed in units per 1 mg protein and was determined by Bradford [83] method with BSA as a standard.

Upon most occasions, AcPase activity was determined using *p*-NPP as a substrate after a modification of the method described by Barett for the first time. The enzyme fractions were incubated with 200 μL 2 mM p-NPP in 20 mM Na-acetate buffer (pH 5.0) for 20 min at 37 °C. The assay was stopped by the addition of 100 μL 1 M NaOH, and the amount of free p-nitrophenol was measured in a microtiter plate reader at 405 nm. The substrate specificity of AcPase was characterized by the liberation rate of inorganic phosphate using a method which was similar to that of ChandraRajan and Klein [84]. The concentration of the substrates in the assay mixture was 10 mM.

The CA activity was determined spectrophotometrically using pNPA as a substrate. The standard assay system contained 1 mM pNPA in 0.01 M diethylmalonic acid (pH 8.0), and the reaction was performed at 35 °C. The released p-nitrophenol was assessed by measuring the increase in absorbance at 400 nm using a molar extinction coefficient of $16.3 \times 10^3\ M^{-1} \cdot cm^{-1}$). One active unit represents the amount of enzyme that catalyzes 1 μmol reaction per min under assay conditions.

4.2.5 *Enzyme Modification and Functional Analysis*

4.2.5.1 Protein Folding

The protein folding follow the procedure for Sabine Reyda [80]. Inclusion bodies were dissolved in 8 M urea, 50 mM Tris–HCl, 0.1 M β-mercaptoethanol, and 1 mM EDTA (ethylenediaminetetraacetic acid), pH 8.0, at a 10 mg/ml protein concentration. This solution was diluted into 50 mM Tris–HCl buffer (pH 8.0), containing 1 mM GSH (reduced glutathione), 0.1 mM GSSG (oxidized glutathionine), and 0.8 M l-arginine, and the final concentration was 10 μg/ml, incubated for 24 h at 4 °C and dialyzed against 50 mM Tris–HCl buffer (pH 8.0), containing 0.1 μM $ZnSO_4$, and finally to zinc-free 50 mM Tris–HCl buffer, pH 8.0.

4.2.5.2 Chemical Modification

PALP (ALP from a pearl oyster, *Pinctada fucata*) reaction with different concentrations of PG was performed in 50 mM bicarbonate buffer at 25 °C (pH 9.00). The reagent was prepared in DMSO and had no effect on the activity of enzyme. The final enzyme concentration was 2.3 μM. A control tube without any PG was kept at the same time. After 30 min of modification, aliquot of HCl was added to end the reaction. The absorbance at 250 nm was determined to indicate the modification of arginine residues. The reaction mixture was then ultracentrifuged and diluted to determine the residual enzyme activity.

4.2.5.3 Modification of PALP by TNBS

All the reactions were conducted at 25 °C. The interfering Tris was removed by first passing the enzyme through a Sephadex G-50 column pre-equilibrated with 20 mM

Hepes buffer (pH 7.50). The modification process was performed by incubating PALP with diverse concentrations of TNBS in 50 mM bicarbonate buffer (pH 9.00). The final enzyme concentration was 2.3 μM. A control tube was kept with the same amount of enzyme but without any TNBS. After 30 min, the reaction was stopped by adding an equal volume of 0.2 M HCl to adjust pH to about 5.0. The reaction mixture was then ultracentrifuged using a Centricon (Millipore) apparatus to remove excess TNBS and then diluted ten times with 20 mM Tris–HCl buffer to determine the residual enzyme activity.

4.2.5.4 Modification of PALP by PG

Reactions of PALP with various concentrations of PG were performed in 50 mM bicarbonate buffer at 25 °C (pH 9.00). The reagent was prepared in DMSO and had no influence on the enzyme activity. The final enzyme concentration was 2.3 μM. A control tube without any PG was kept at the same time. After 30 min of modification, aliquot of HCl was added to end the reaction. The absorbance at 250 nm was measured to indicate the modification of arginine residues. The reaction mixture was then ultracentrifuged and diluted to determine the residual enzyme.

4.2.5.5 Modification of PALP by NBS

PALP (18 nM) was incubated with various concentrations of NBS (25–200 μM) in 50 mM acetate buffer for 30 min at 25 °C (pH 6.0), followed by determination of the residual enzyme activity and recording of the absorbance spectrum at 280 nm with an Ultrospec 3000 (Pharmacia, Sweden) spectrophotometer with a reference cell containing NBS buffer. A control contained equivalent enzyme, but no NBS. Prior to the determination of the residual enzyme activity, the reaction mixture was subjected to ultrafiltered using a Centricon (Millipore, USA) apparatus to remove excess NBS and then diluted ten times with 20 mM Tris–HCl buffer.

4.2.5.6 Influences of Temperature and pH on Enzyme Activities

The optimum temperature of the enzyme was assessed by measuring the activity at different temperatures in the buffer described above. The reaction temperatures ranged from 22 to 78 °C. Thermal stability was researched by incubating the enzyme solution at different temperatures for different periods. The activities at different pHs were measured by using 0.1 M Gly–HCl buffer (pH 2.2–3.6), 0.1 M acetate buffer (pH 3.0–6.5), and 0.1 M Tris–HCl buffer (pH 7.0–9.0). The Km values were determined by repeated experiments using a Lineweaver–Burk plot at 37 °C with p-NPP as substrate.

4.2.5.7 Effects of the Product (HPO_4^{2-}) and the Product Analog (WO_4^{3-}) on the Enzyme

The roles of the product HPO_4^{2-} and its analog WO_4^{3-} on the enzyme were measured using assays in 0.05 M Na_2CO^3–$NaHCO_3$ (pH 9.7), 2 mM Mg^{2+} 4 mM *p*-NPP at 40 °C with different concentrations of HPO_4^{2-} or WO_4^{3-}. The concentrations ranged from 1.0 to 10 mM. The HPO_4^{2-} and $WO4^{3-}$ inhibition types on the enzyme were analyzed by changing the substrate *p*-NPP concentrations with HPO_4 $^{2-}$ or WO_4^{3-}.

4.2.5.8 Influences of Metal Ions on Enzyme Activities

The influence of chemical reagents on the enzyme was analyzed. These chemical reagents included product analogs (WO_4^{3-}, MoO_4^{3-}, and ASO_4^{3-}), haloids (CI^-, Br^-, I^-, and F^-), heavy metal ions (Ag^+, Pb^{2+}, and Cu^{2+}), the high relative molecular mass ACP inhibitor tartrate, protein denaturant ethanol, chelating reagent EDTA, as well as the chemical modification reagents bromoacetic acid (BrAc), formaldehyde, and dithiothreitol (DTT). Twenty pL of enzyme was incubated at 37 °C in 0.1 $mol{\cdot}L^{-1}$ acetate buffer (pH 5.0) containing the substrate and different concentrations of chemical reagents, and the reaction was stopped by 1 mL 0.2 $mol{\cdot}L^{-1}$ NaOH. The type of inhibition was determined by varying the concentration of p-NPP with the chemical reagents.

4.2.5.9 Protection Experiments

The enzyme was preincubated with different concentrations of AMP, 2-glycerol phosphate, phosphate, or tungstate, in 50 mM sodium bicarbonate buffer (pH 9.00) at 25 °C for 5 min. After 30 min of addition of the modifier, the reaction was stopped, and the residual activity was measured after ultracentrifugation and dilution. Control experiments without modifiers were carried out simultaneously.

4.2.5.10 Stoichiometry of Inactivation

Various partially modified enzymes were obtained by mixing enzyme with modifiers (TNBS, PG, OPA, or NBS) in specific molar ratio of 1:20 to 1:1000. After some time, reactions were terminated, and samples were taken and ultracentrifuged to remove unreacted reagents. The residual activities were determined after dilution, and the number of modified residues was measured concurrently.

4.2.5.11 AZ Inhibition and DNSA Binding of CA

CA (8 μM) was incubated with different concentrations of AZ (from 0.1 to 1.0 μM) for 1 h at 35 °C. An aliquot was then used to determine CA activity as described in Sect. 4.2.4.6 above.

A solution containing 1 μM of CA was incubated with DNSA (50 μM) for 10 min at room temperature. The fluorescence spectrum was recorded at 400–600 nm using a Hitachi F-2500 spectrofluorimeter with an excitation wavelength at 326 nm. Excitation and emission bandwidths were 5 and 10 nm, respectively. And the cuvette path length was 10 mm.

4.2.5.12 Histochemical Detection of CA by Fluorescence

Mantle tissue was dissected and washed with 0.9% NaCl. 10-μm sections were prepared in a − 20 °C cryostat. The sections were mounted on slides. Twenty μL of 50 μM DNSA was added to the sections and incubated at 35 °C for 30 min. Sections used as controls (without DNSA) were incubated with phosphate buffer (50 mM, pH 7.5) alone. Sections were observed and photographed in a Leica Dialux microscope equipped with a high-pressure Hg lamp as the ultraviolet source. The sample was observed through a 460 nm barrier filter.

4.2.5.13 Calcium–Binding Assay

The calcium-binding ability of AcPase I was analyzed according to the stains-all method [85]. Gels were run, thoroughly washed with 25% isopropyl alcohol, and stained with 0.0025% stains-all, 7.5% formamide, 25% isopropyl alcohol, as well as 30 mM Tris base(pH 8.8) in the dark for 4 days. Calmodulin, which was a calcium-binding protein [86], was used as a positive control.

4.2.6 Cell Culture and MTT Assay

NIH3T3 cells were kept in DMEM (Gibco BRL) containing 10% fetal bovine serum (FBS) (hyclone) at 37 °C and 5.0% CO_2. The cells were seeded in 96-well plates at a density of 2.0×10^4 cells/well 24 h prior to transfection. Transfection was performed using pcDNA3.1 recombination plasmid using Sofast™ Transfection Reagent (Sofast, Xiamen, China). Forty-eight hours after transfection, 0.5 mg/ml 3-[4,5]-2,5-diphenyltetrazolium bromide (MTT) in fresh medium was added and incubated for an additional 4 h. And then the blue formazan crystals were dissolved in 200 μl dimethyl sulfoxide (DMSO) and measured spectrophotometrically at 490 nm.

4.2.7 Statistical Analysis

The data from the experiment involving MTT assays were analyzed using one-way determination of variance (ANOVA) in SPSS11.5 software, and $p < 0.05$ was considered as significant.

4.3 Alkaline Phosphatase

4.3.1 Purification and Enzymatic Characterization of Alkaline Phosphatase from Pinctada fucata

4.3.1.1 Purification of ALP

Purified enzyme with a specific activity of 2040 U mg^{-1} was acquired using chromatography on DEAE-32 cellulose, Sephadex G-150, and DEAEA-25. SDS-PAGE analysis of the purified ALP demonstrated a single band with a molecular mass of ~40 kDa (Fig. 4.1). The purification steps are summarized in Table 4.1.

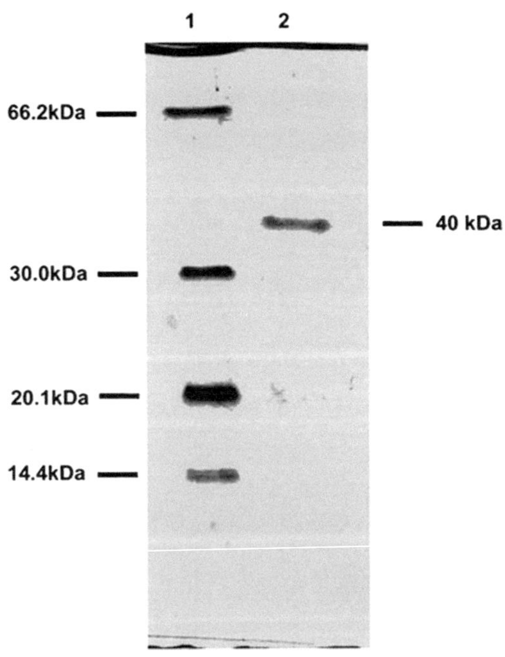

Fig. 4.1 Purification of ALP from *Pinctada fucata*. SDS-PAGE analysis was performed on 12% gels. The gel was stained with Coomassie Brilliant Blue R-250. Lane 1 was the standard molecular markers (kDa), and lane 2 was the purified ALP

Table 4.1 Purification of ALP from *Pinctada fucata*

Steps	Volume (ml)	Total protein (mg)	Total activity (U)	Specific activity (U mg − 1)	Purification (fold)	Yield (%)
Crude extract	665	3100	57,700	18.6	1	100
(NH4)SO4 fraction	139	630	34,800	55.3	3.0	60.4
DEAE-32 fraction	92	63.4	13,200	208	11.2	22.9
G-150 fraction	101	17.3	10,800	623	33.5	18.7
A-25 fraction	15	4.5	9200	2040	101	15.9

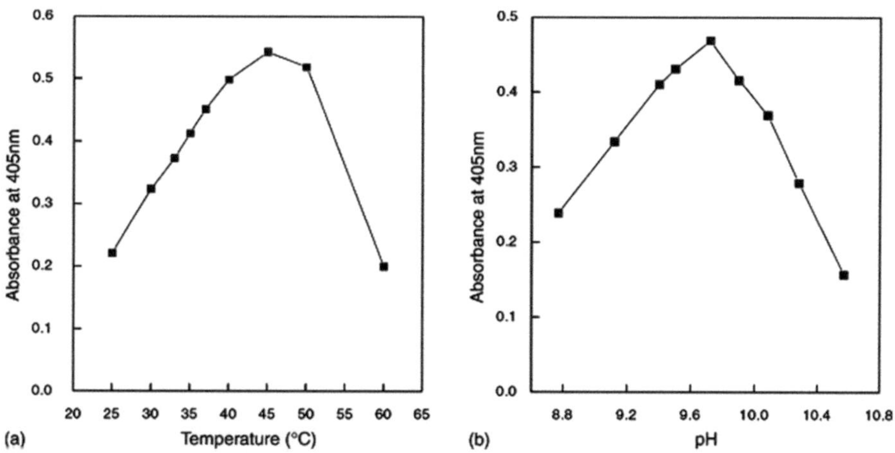

Fig. 4.2 Effect of temperature (**a**) and pH (**b**) on ALP activity [78]

4.3.1.2 Effect of Temperature and pH on ALP Activity

The measured enzyme activities at different temperatures and pH values are shown in Fig. 4.2. The optimal catalysis reaction temperature was 45 °C,and the activity rapidly decreases at or above the optimum.

The optimal pH of the enzyme in Fig. 4.2b was 9.7. The optimum pH is lower than that 10.0 for ALP.

4.3.1.3 Effects of HPO_4^{2-} and WO_4^{3-} on ALP

As shown in Fig. 4.3, increased concentrations of Na_2HPO_4 or Na_3WO_4 in the assay system decreased the enzyme activity relative to the control. The effector

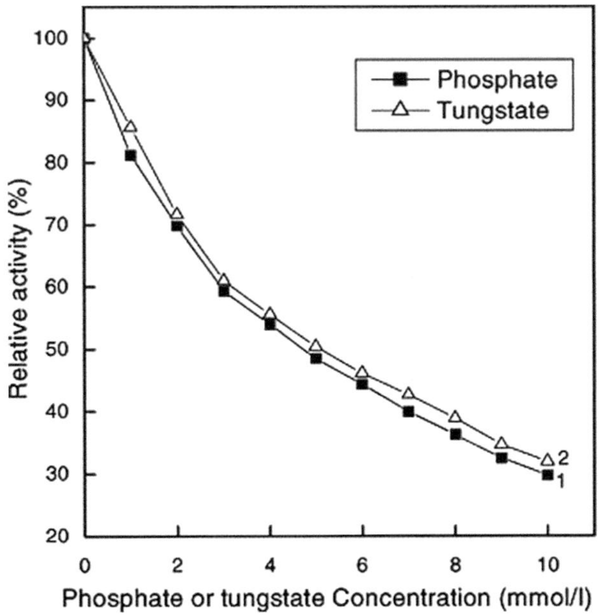

Fig. 4.3 Effect of the product HPO_4^{2-} and the product analog WO_4^{3-} on the enzyme activity. Curve 1 shows the effect of HPO_4^{2-}, and curve 2 shows the effect of WO_4^{3-}

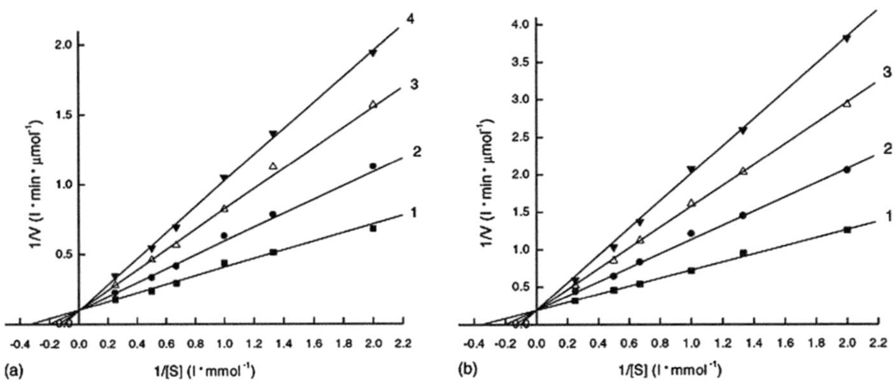

Fig. 4.4 Lineweaver–Burk plots inhibition of HPO_4^{2-} and WO_4^{3-}. The experimental conditions were shown in this figure. The HPO_4^{2-} (**a**) or WO_4^{3-} (**b**) concentrations for curves 1–4 were 0, 2, 4, and 6 mM, respectively

concentrations increased from 1 to 10 mM in the assay system, leaving only about 33% activity at 10 mM concentrations. The inhibitory effect of HPO_4^{2-} was slightly stronger than that of WO_4^{3-}.

The Lineweaver–Burk plots (Fig. 4.4) showed that both HPO_4^{2-} and WO_4^{3-} could competitively inhibit the enzyme with inhibition constants of 0.89 and 1.20 mM, respectively, which also indicates that HPO_4^{2-} inhibits the enzyme activity more strongly than WO_4^{3-}.

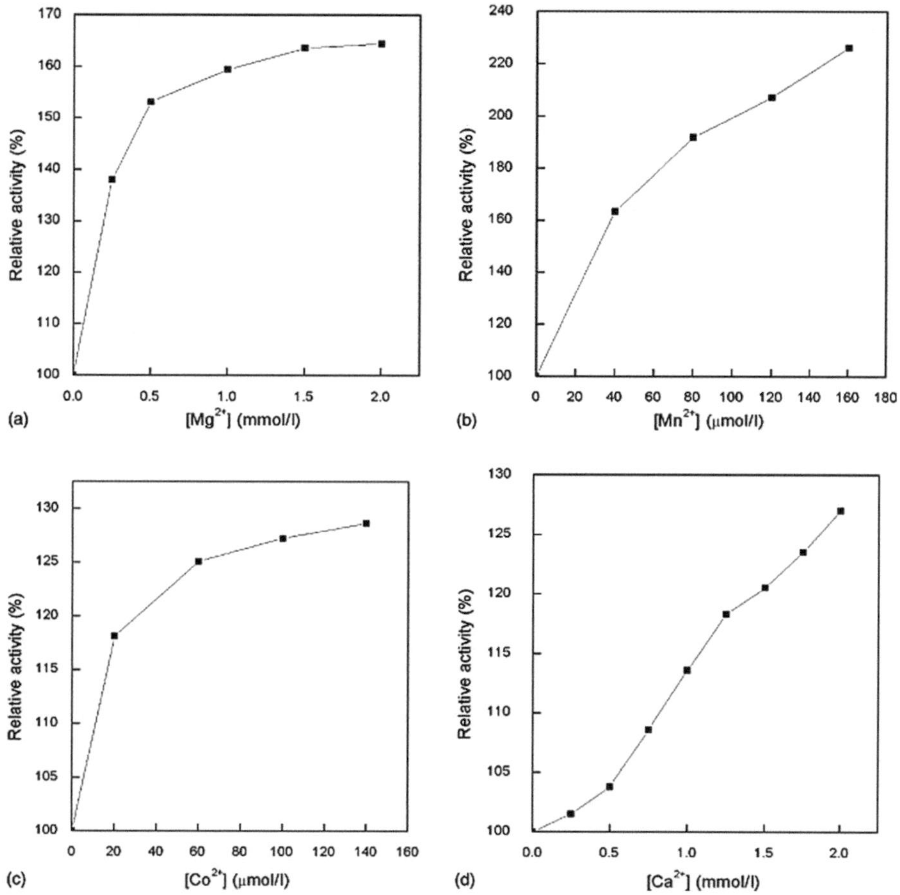

Fig. 4.5 Activation effects of ALP activity by metal ions. The experimental conditions were as follows: (**a**) Mg^{2+}, (**b**) Mn^{2+}, (**c**) Co^{2+}, and (**d**) Ca^{2+}

4.3.1.4 Effects of Metal Ions on ALP

At concentrations ranging from 5 to 50 mM, positive monovalent alkali metal ions such as Li^+, Na^+, and K^+ had no significant effect on the ALP activity of *Pinctada fucata* (data not shown). Nevertheless, treatment with positive divalent alkali metal ions such as Mg^{2+}, Mn^{2+}, Co^{2+}, and Ca^{2+} increased the enzyme activity, as shown in Fig. 4.5.

With 2 mM Mg^{2+}, the enzyme activity increased by 60% (Fig. 4.5a). Nevertheless, compared to Mg^{2+}, Ca^{2+} had a relatively low impact on *Pinctada fucata* ALP activity (Fig. 4.5d), and the enzyme activity of 2 mM Ca^{2+} was increased by 27% compared with the control. Mn^{2+} and Co^{2+} significantly increased the enzyme activity at micromole concentrations (Fig. 4.5b, c). However, the Lineweaver–Burk analysis demonstrated that the mechanisms of these two ions were different.

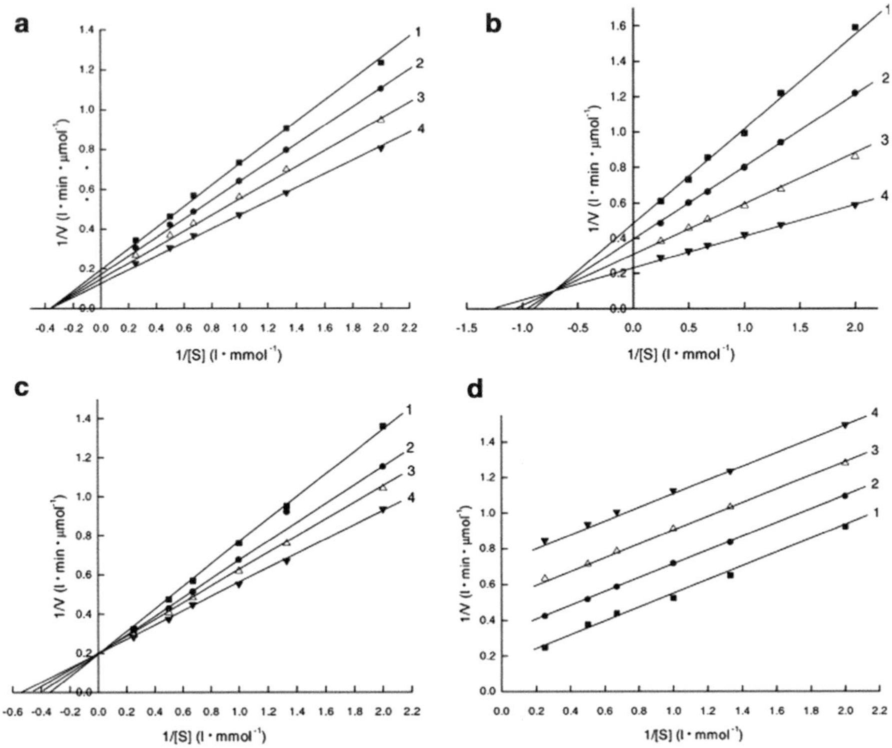

Fig. 4.6 Lineweaver–Burk plots for the effect of metal ions on p-NPP hydrolysis. (**a**) The Mg^{2+} concentrations in curves 1–4 were 0, 0.1, 0.2, and 0.3 mM, respectively. (**b**) The Mn^{2+} concentrations in curves 1–4 were 0, 40, 80, and 120 μM, respectively. (**c**) The Co^{2+} concentrations in curves 1–4 were 0, 15, 30, and 45 μM, respectively. (**d**) The Zn^{2+} concentrations in curves 1–4 were 0, 10, 20, and 30 μM, respectively

The activity of Mg^{2+} on ALP (Fig. 4.6a) is not competitive, while the activity of Co^{2+} is competitive (Fig. 4.6c). The activity of Mn^{2+} on ALP activity showed a mixed uncompetitive and competitive role (Fig. 4.6b).

Some heavy metal ions such as Zn^{2+}, Pb^{2+}, Cd^{2+}, and Cu^{2+} were demonstrated to inhibit the ALP activity (Fig. 4.7). As shown in Fig. 4.7a, treatment with 40 μM Zn^{2+} significantly reduced the ALP activity to only 30.8% of the control. Further increases of the Zn^{2+} concentration had a small additional inhibitory effect on the enzyme. When the concentration of Zn^{2+} was 200 μM, the activity was 24%, and the Lineweaver–Burk analysis using Zn^{2+} showed that the inhibition effect of Zn^{2+} on the enzyme was anti-competitive (Fig. 4.6d).

Pb^{2+} and Cd^{2+} are two major elements of marine pollution. As shown in Fig. 4.7b, c, 11 μM Pb^{2+} and 55 μM Cd^{2+} inhibited the activity of enzyme to 82.3% and 73.2%, respectively. The results also demonstrated that Cu^{2+} could reduce the enzyme activity to 30.9% at a concentration of 200 μM (Fig. 4.7d).

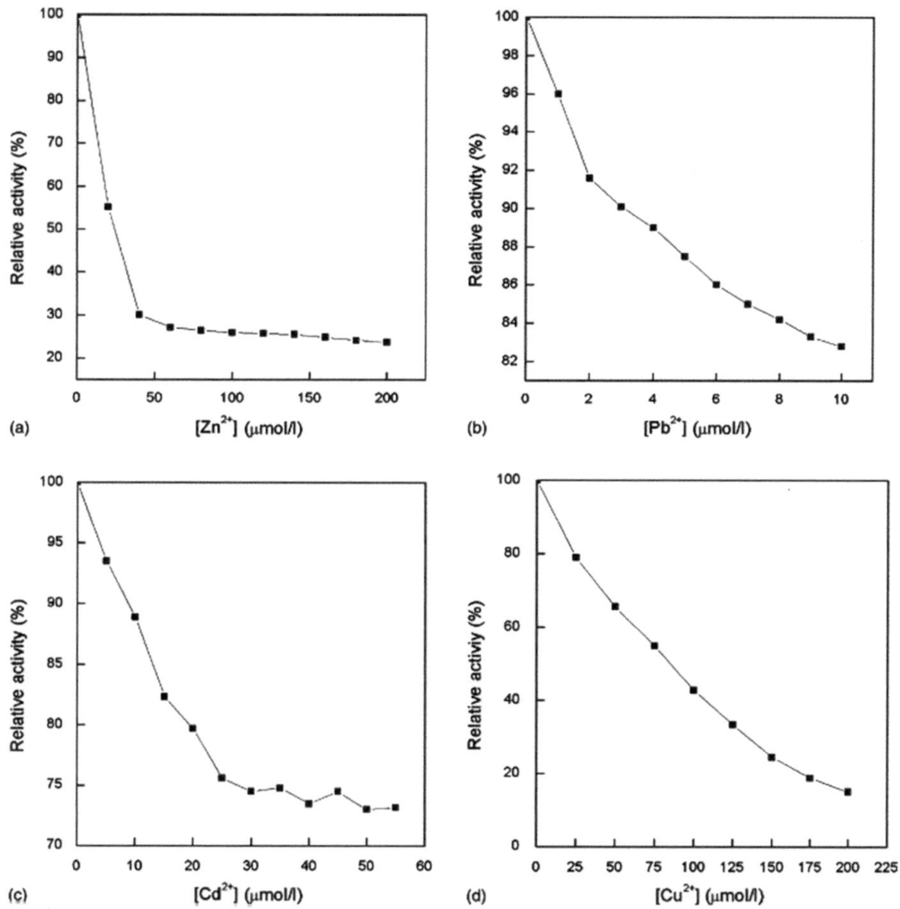

Fig. 4.7 The inhibition effects of metal ions on the enzyme activity. The experimental conditions were as for Fig. 4.4, (**a**) Zn^{2+}, (**b**) Pb^{2+}, (**c**) Cd^{2+}, and (**d**) Cu^{2+}

4.3.2 Chemical Modification Studies on Alkaline Phosphatase from **Pinctada fucata**

4.3.2.1 Inactivation of ALP by TNBS

Treatment of PALP with increasing concentrations of TNBS resulted in a gradual decrease of the enzyme activity (Fig. 4.8a). In the bicarbonate buffer at pH 9.00, the initial activity of the enzyme was reduced by 80% after incubation with 1.0 mM TNBS at 25 °C for 30 min. However, incubation of the enzyme with higher concentrations of TNBS could not result in more loss of activity, and the absorbance

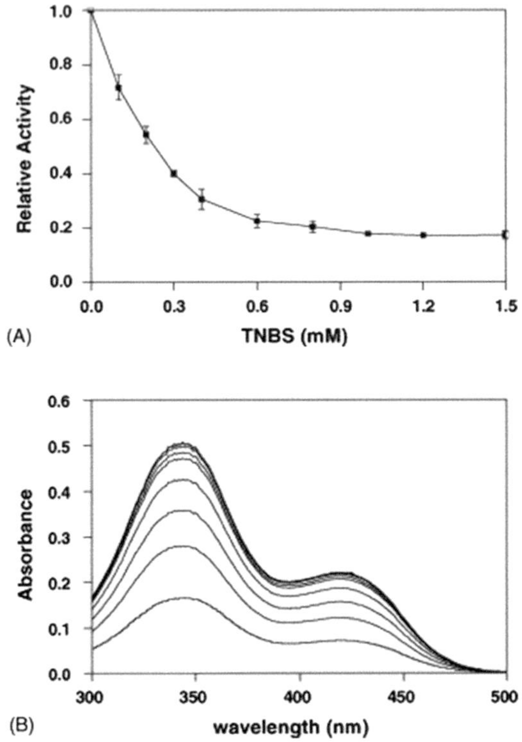

Fig. 4.8 Inactivation of PALP in different concentrations of TNBS. 2.3 μM PALP was incubated with various concentrations of TNBS (0.1–1.5 mM) in 0.5 M bicarbonate buffer (pH 9.00), at 25 °C for 30 min, and then the residual enzyme activity (**a**) was determined, and the absorption spectrum was recorded (**b**). A sample without any TNBS was used as a control for activity assay. For each concentration, three measures were performed. The spectra were recorded on an Ultraspec 3000 (Pharmacia) spectrophotometer with a reference cell containing PALP without TNBS treatment. The absorption spectra shown in B, from bottom to top, correspond to TNBS concentrations of 0.1, 0.2, 0.3, 0.4, 0.6, 0.8, 1.0, 1.2, and 1.5 mM, respectively

spectrum of the modified enzyme became unchanged (Fig. 4.8b), indicating that after complete modification the enzyme retains about 20% of its initial activity. The loss of activities could not be restored by dialysis or ultracentrifugation, which indicates an irreversible reaction. Figure 4.8b shows the absorbance spectrum of the modified enzymes. The spectra reach a peak with the maximum at 345 nm and a shoulder around 420 nm, thus indicating the formation of the trinitrophenylamino derivatives [87, 88]. The increase of absorbance at 345 nm correlates well with the inactivation of PALP. These results suggested that the inactivation of PALP by TNBS was due to modification of lysine residues [89].

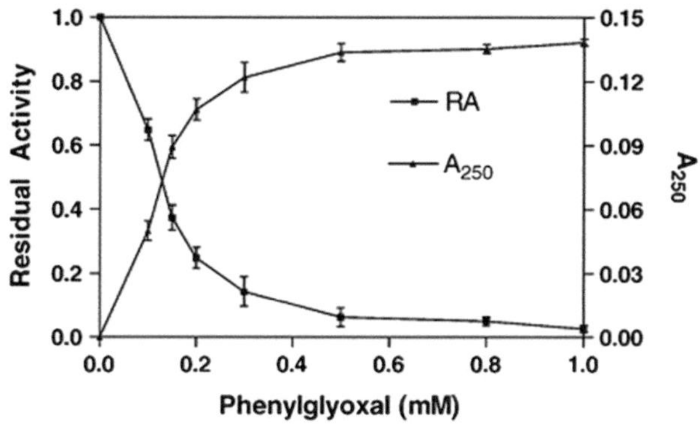

Fig. 4.9 Inactivation of PALP under different concentrations of PG. 2.3 μM PALP was incubated with various concentrations of PG (0.1–1.0 mM) in 0.5 M bicarbonate buffer (pH 9.00) for 30 min at 25 °C, followed by the measurement of A250 and determination of the residual enzyme activity. A sample without any PG was simultaneously used as control for activity assay. For each concentration, the activity assay was performed in triplicates

4.3.2.2 Inactivation of ALP by PG

PG is an arginine-specific modification reagent and has been used to characterize the arginine residues of many enzymes [90, 91]. Reaction with increasing concentrations of PG resulted in a remarkable loss of the PALP activity and an increase in absorbance at 250 nm, which indicates the modification of arginine residues (Fig. 4.9). The PALP was completely inactivated by incubation with 1.0 mM PG for 30 min in bicarbonate buffer (pH 9.00) at 25 °C. Dialysis or ultracentrifugation could not restore lost activity, which indicates an irreversible reaction. Though PG was supposed to react with amino group to some extent [90, 92], it specifically modified the arginine residues [91] in the bicarbonate buffer system. The TNBS modified and the PG-TNBS double-modified PALP shared the similar A420 (data not shown), confirming that PG did not react with lysine residues. These results suggested that the inactivation of PALP by PG was due to modification of key arginine residues [89].

4.3.2.3 Correlation Between the Number of ALP–Modified Residues and Its Residual Activity

If we assume that the reactivity of all the n modifiable residues including essential one(s) toward the residue-specific reagents is approximately the same and modification of any essential residues would inactivate the enzyme from the

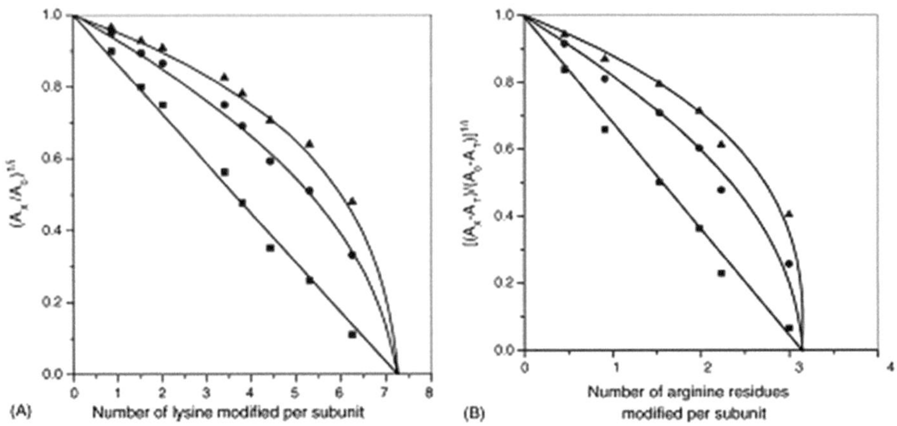

Fig. 4.10 Correlation between the number of PALP-modified residues and its residual activity of 2.3 μM PALP was incubated with a specific molar proportions ranging from 1:20 to 1:1000 of modifiers (TNBS or phenylglyoxal). After a period of incubation, the activity of residue and the number of modified residues were measured as described above. Graphs (**a**) and (**b**) are for TNBS and PG inactivation, respectively. The data are presented as the Tsou's plot for i = 1 (■), i = 2 (●), and i = 3 (▲)

initial activity of A0 to the loss of AT, the relationship between the residual activity Ax against residue modification will be as follows:

$$[(\mathrm{Ax} - \mathrm{AT})/(\mathrm{A0} - \mathrm{AT})]1/i = (n - m)/n \quad (4.1)$$

where m is the number of modified residues and i is the tryout value for the number of essential residues.

For PG modification, the enzyme was completely inactivated, so AT is equal to 0. For TNBS inactivation, the residual activity AT could be obtained from the progress curves. The number of important lysine and arginine residues is the values of i which gives a straight line when plotting (Ax − AT)/(A0 − AT), i.e., the fraction of fully active enzyme, against the number of residue(s) modified (m). In Fig. 4.10, a key lysine residue and a key arginine residue were involved in the activity of PALP.

4.3.3 An Essential Tryptophan Residue in Alkaline Phosphatase from Pearl Oyster

4.3.3.1 Inactivation of ALP by NBS

NBS is a frequently used tryptophan-specific modification reagent that oxidizes the indole residue to an oxindole derivative. Treatment of PALP with increasing concentrations of NBS led to a gradual decrease in the activity and absorbance at 280 nm

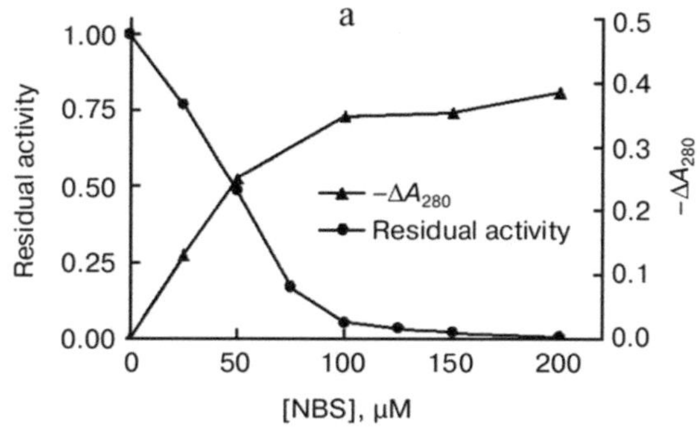

Fig. 4.11 Residual activity and absorbance change at 280 nm of PALP modified by various concentrations of NBS. For each concentration, three measurements were performed

(Fig. 4.11). The enzyme was completely inactivated with 200 μM NBS. Dialysis or ultrafiltration could not restore the loss of activity, indicating that there was an irreversible reaction.

4.3.3.2 Kinetics of Inactivation of PALP by NBS

After incubating PALP with different concentrations of NBS, the partially modified enzyme was sampled at the same time intervals, and the residual activity was assayed. By plotting the logarithm of residual activity of the modified enzyme versus the time, a series of straight lines with different slopes was obtained. This result clearly showed that the inactivation of NBS followed pseudo first-order kinetics (Fig. 4.12). A straight line was drawn from the apparent first-order inhibitory constant (kobs) against NBS concentration. The slope represented second-order inactivation constant (Fig. 4.12, insert), $k = 0.001\ \mu M^{-1}{\cdot}min^{-1}$. An analysis was performed using Levy's method [93] by plotting lnkobs against the logarithm of NBS concentration to obtain a straight line with a slope of 0.806, which was close to 1, and indicated that at least 1 mole of NBS per mole of active site was required to inactivate the enzyme [57].

4.3.3.3 Correlation Between the Fraction of Residual Trp Residues and the Fraction of Remaining Activity of ALP Modified by NBS

Tsou's plot analysis was further used to calculate the number of inactivated essential tryptophan residues. If we assume that the reactivities of all n reaction sites including essential one(s) toward the residue-specific reagents are approximately equal and modification of any essential residues would inactivate the enzyme, the relationship between the fraction of residual activity of the enzyme (a) and the fraction of residual tryptophan residues (x) will be $a^{1/i} = x$, where $x = (n - m)/n$, m is the number of modified residues and i is the tryout value for the number of essential residues.

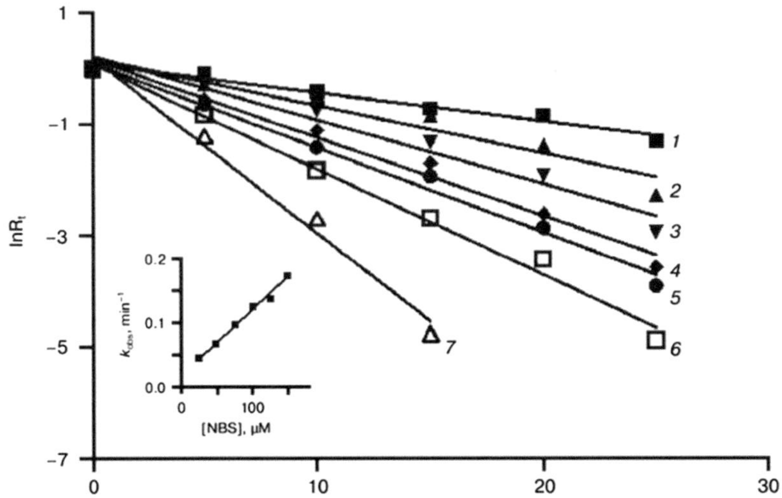

Fig. 4.12 Kinetics of inactivation of PALP by NBS. By plotting the logarithm of residual activity of modified enzyme versus the time, the apparent rate constants for inactivation by different concentrations of NBS (lines 1–7 correspond to 25, 50, 75, 100, 125, 150, and 200 μM NBS, respectively) were obtained from the slopes of lines. For each concentration, three measurements were performed. The deactivated second-order rate constant was determined from the slope of line yielded by plotting kobs against various concentrations of NBS (insert)

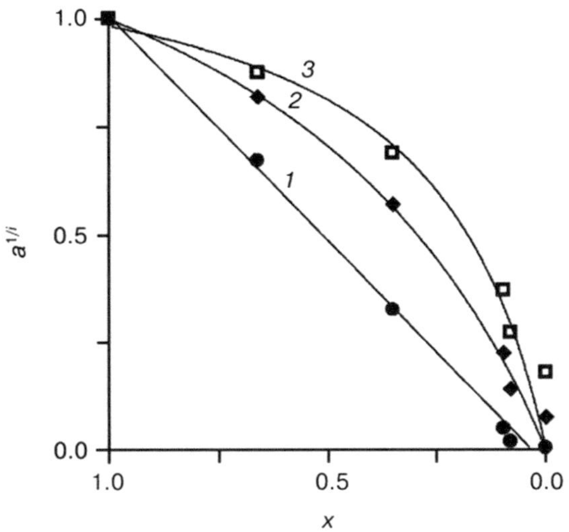

Fig. 4.13 Correlation between the fraction of residual Trp residues and the fraction of remaining activity of PALP modified by NBS. The data are presented as a Tsou's plot for i = 1, 2, and 3 (curves 1–3, respectively) [57]

Figure 4.13 shows that when i = 1, only a straight line could be drawn when plotting a1/i against x, indicating that only one of the five tryptophan residues is essential for PALP activity. Similar results for alkaline phosphatase from the green crab (*Scylla serrata*) have been reported [94].

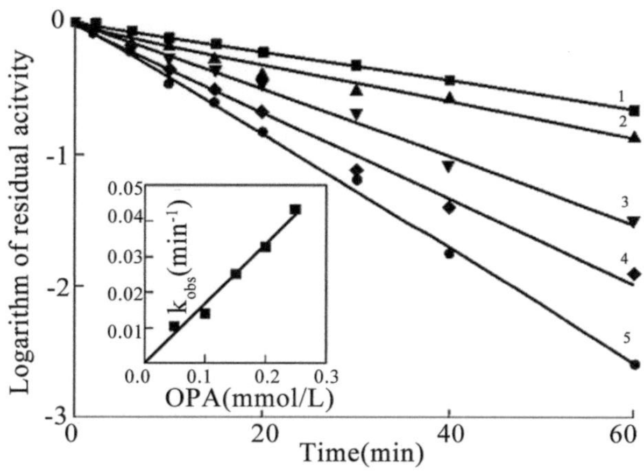

Fig. 4.14 Kinetics of inactivation of ALP from *P. fucata* by OPA. The OPA concentration for lines 1–5 was 50, 100, 150, 200, and 250 μmol/L, respectively. The inset shows the second-order plot of the pseudo-first rate constants as a function of log OPA concentration

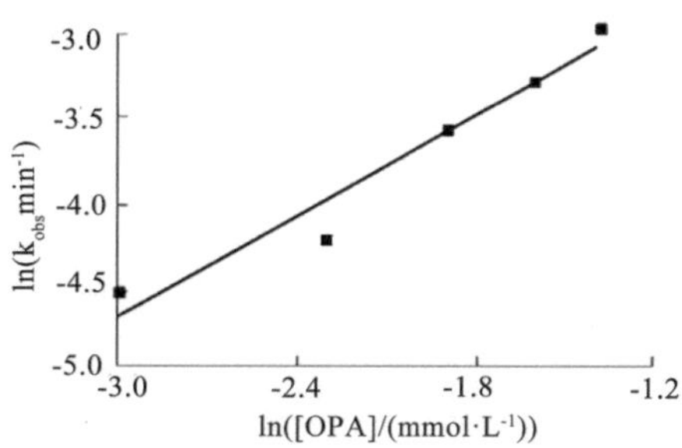

Fig. 4.15 Lery's plot of OPA inactivation. The slope gives a reaction order of 0.89

4.3.4 *Purification and Enzymatic Characterization of Inhibition of Alkaline Phosphatase from Pearl Oyster* Pinctada fucata *by o-Phthalaldehyde*

4.3.4.1 Inhibition of ALP from *P. fucata* by OPA

Incubation of ALP from *P. fucata* with increasing concentrations of OPA resulted in a gradual decrease in enzyme activity. Dialysis or ultracentrifugation could not restore the loss of activity, indicating an irreversible reaction. Figure 4.14 shows that deactivation follows pseudo first-order kinetics. The second-order rate constant was calculated by replotting the pseudo first-order rate constants (k_{Obs}) against OPA concentration. A linear relationship was observed, and the slope determined the second-order rate constant of 0.167 $(mmol/L)^{-1}\cdot min^{-1}$ (Fig. 3.14 inset). An analysis using Lery's method of plotting log kohs against the logarithm of the OPA concentration yielded a slope of 0.89 (Fig. 4.15), indicating that at least one mole of OPA per mole of active site was required to inactivate the enzyme [93].

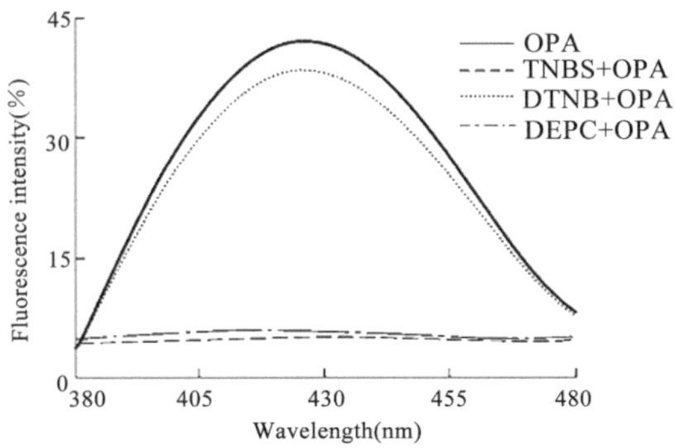

Fig. 4.16 Effect of lysine, histidine, and cysteine modifications on isoindole fluorescent of ALP from *P. fucata*. The enzyme (2.3 μηol/L) was incubated with 2.0 mmol/L DEPC, 0.2 mmol/L TNBS, or 1.0 mmol/L DTNB at 25 °C for complete modification, respectively. The modified enzyme samples were then treated with 1.0 mmol/L OPA for 1 h. The change in the isoindole fluorescence was monitored at an excitation wavelength of 337 nm

4.3.4.2 Effect of Lysine, Histidine, and Cysteine Modifiers on the Isoindole Fluorescent of ALP from *P. fucata*

Figure 4.16 demonstrates the effects of these residual specific modifiers on the isoindole fluorescence of the OPA-modified enzyme. Complete premodification of DEPC or TNBS eliminated the characteristic isoindole derivative peak at 415 nm. Nevertheless, DTNB pretreatment did not prevent the formation of isoindole fluorescence. These results strongly suggest that no cysteine residue was involved in the isoindole formation during the OPA inactivation. Therefore, the inactivation was due to the formation of an isoindole derivative which involves lysine and histidine residues.

4.4 Acid Phosphatase

4.4.1 Partial Purification and Properties of an Acid Phosphatase from Pinctada fucata

4.4.1.1 Purification of ACP

Figure 4.17 shows the elution profile of the enzyme present in the pearl oyster *Pinctada fucata* extract from gel filtration chromatography on Sephadex G-150.

The protein peak and enzyme activity peak are located in different positions, which indicate that the extract from $(NH_4)_2SO_4$ precipitation was impure. Figure 4.18 shows the profile of purified enzyme by affinity chromatography on a Con A Sepharose 4B column.

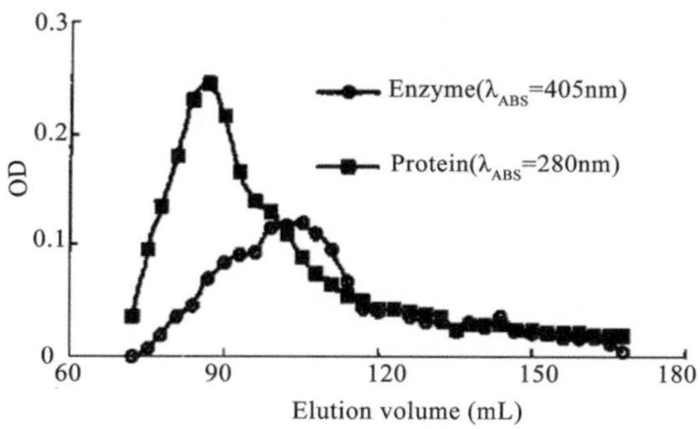

Fig. 4.17 Elution profile of pearl oyster (*Pinctada fucata*) extract on a Sephadex G-150 column. The enzyme activity is measured on p-NPP

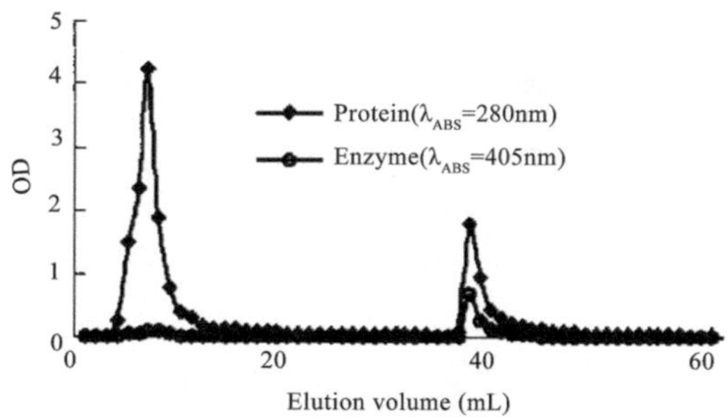

Fig. 4.18 Con A Sepharose 4B affinity chromatography of enzyme eluted from Sephadex G-150. The enzyme activity is measured on p -NPP

Table 4.2 Summary of the purification procedure of ACP

Steps	Volume (mL)	Total protein (mg)	Total activity (U)	Specific activity (U/mg)	Purification fold	Yield (%)
Extracts	765	4300	63,400	14.7	1	100
$(NH_4)_2SO_4$	147	617	36,100	58.5	4	56.9
G-150	54	54.5	11,500	211	14	18.1
Con A	15	5.7	9800	1719	117	15.4

The first protein peak was very high with little ACP, while the elution protein peak (the second protein peak) contained a higher concentration of ACP, which indicates that ACP is a glucoprotein. The purification results are shown in Table 4.2.

4.4.1.2 Effects of Pb^{2+}, Cu^{2+}, and Formaldehyde on ACP

The product analogs WO_4^{3-}, MoO_4^{3}, and AsO_4^{3-} all showed competitive inhibition of the enzyme (Fig. 4.19), and WO_4^{3-} and MoO_4^{3} were stronger inhibition reagents compared to AsO_4^{3-}.

Ag^+ and Cu^{2+} showed a strong enzyme inhibition. Only 1×10^{-7} mol·L^{-1}Ag^+ reduced the relative activity of the enzyme to 32%, while 5×10^{-4} mol·L^{-1} Cu^{2+} suppressed the relative enzyme activity to 50% (Fig. 4.20). The Lineweaver–Burk

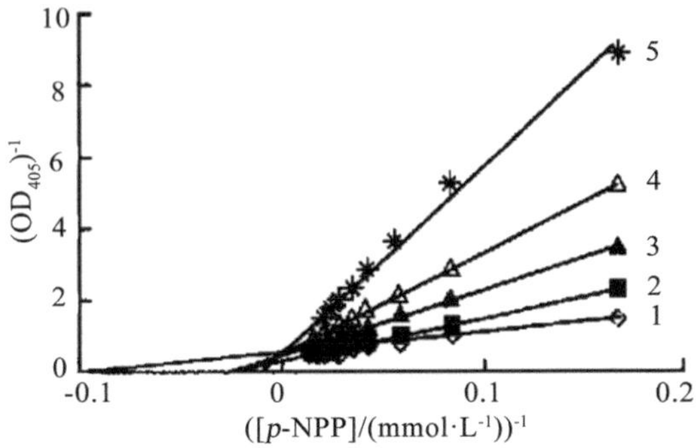

Fig. 4.19 Lineweaver–Burk plot for inhibition of ACP by WO_4^{3-}. The concentration of WO_4^{3-} for curves 1–5 were 0, 1×10^{-5}, 5×10^{-5}, 1×10^{-4}, and 5×10^{-4} $mol \cdot L^{-1}$, respectively

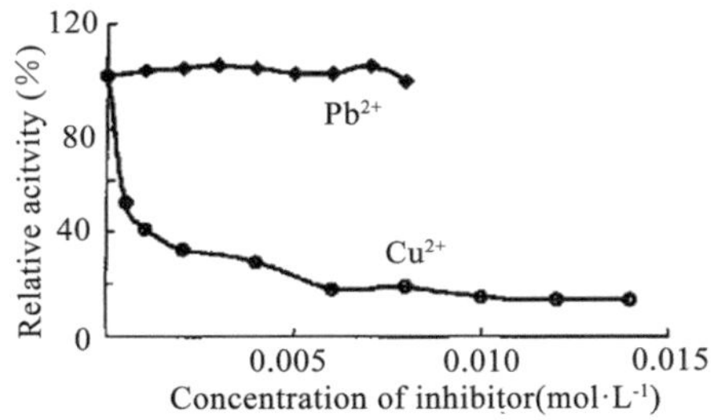

Fig. 4.20 Effect of Pb^{2+} and Cu^{2+} on ACP

plot (not shown) showed that when the Ag^+ concentration increased, K m and V m were changed, which indicates a mixed type of inhibition. In contrast, Pb^{2+} had little effect on the enzyme (Fig. 4.20).

As we all know, ethanol is a strong protein denaturant. When the ethanol concentration was close to 32%, the activity of ACP began to decrease, while at an ethanol concentration of 64%, only 28% of the relative activity remained unchanged. Haloids including CI^-, Br^-, and I^- did not inhibit or activate the enzyme, whereas F^- slightly activated the enzyme at low concentrations (10^{-8}–10^{-4} $mol \cdot L^{-1}$) and inhibited it in higher concentrations (>10^{-3} $mol \cdot L^{-1}$).

Tartrate also found to inhibit the activity of the enzyme. A Lineweaver–Burk plot (not shown) showed that when the concentration of tartrate changed, both Km and Vm were modified, which indicated a mixed-type inhibition.

However, the strong chelating reagent EDTA showed little effect on ACP, suggesting that the active site of the enzyme was free of metal atoms.

For acid conditions, bromoacetic acid (BrAc) can selectively modify histidine residues in the enzyme. In the experiment, when the BrAc concentration was 0.3 $mol \cdot L^{-1}$, the relative activity was only 20%, and when the concentration reached 0.5 $mol \cdot L^{-1}$, the relative activity was completely inhibited. These results suggest that histidine residues of the enzyme are essential for its catalytic activity and are located at the active site of the enzyme (Fig. 4.21).

Dithiothreitol (DTT) can break the bisulfide bonds. When the DTT concentration reached 0.44 $mol \cdot L^{-1}$, 50% activity was remained, suggesting that DTT can affect the activity, but the bisulfide bonds are not essential (Fig. 4.21).

Fig. 4.21 Effect of BrAc and DTT on ACP

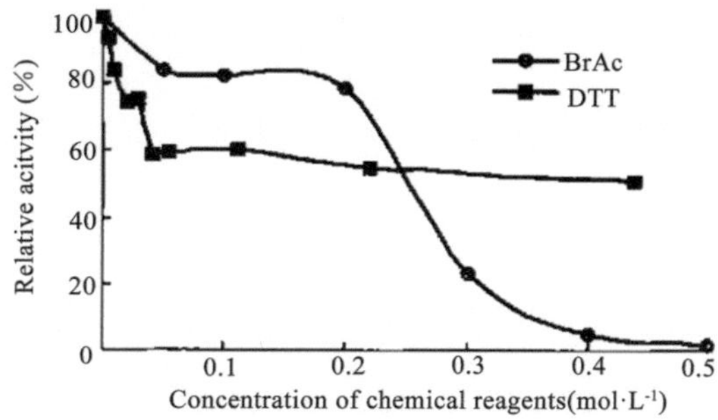

Fig. 4.22 Effect of formaldehyde on ACP

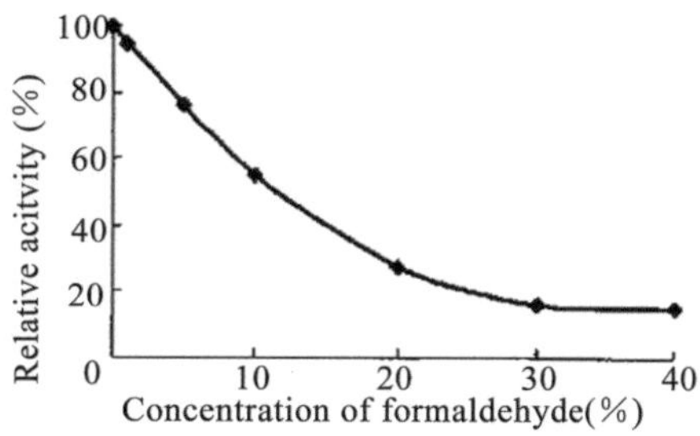

Formaldehyde has been used to modify the enzyme amido. When the formaldehyde concentration reached 40%, 20% of the activity was still present, suggesting that -NH_2 can affect the enzyme activity, but the amido is not essential (Fig. 4.22).

4.4.2 *Purification and Partial Characterization of Two Acid Phosphatase Forms from Pearl Oyster*

4.4.2.1 Purification of AcPase

Figure 4.23a showed the elution profile from Sephadex G-200 superfine of acid phosphatases in the viscera of *P. fucata*. The enzyme activity was divided into two distinct peaks called AcPases I and II. AcPase II was combined, concentrated, and further purified by affinity chromatography on a Con A Sepharose 4B column (Fig. 4.23b). The purification steps are summarized in Table 4.3.

SDS-PAGE analysis of purified AcPases I and II demonstrated a single band. The molecular masses of the subunit were ~208.8 and 29.8 kDa, respectively (Fig. 4.24). The molecular weights of the native enzymes were measured on a calibrated Sephadex G-200 superfine column. The experiments were carried out in triplicate with average molecular weights of ~197.1 and 64.3 kDa, respectively. These results indicated that AcPase I is a single polypeptide chain and AcPase II is a dimeric enzyme consisting of two equivalent subunits.

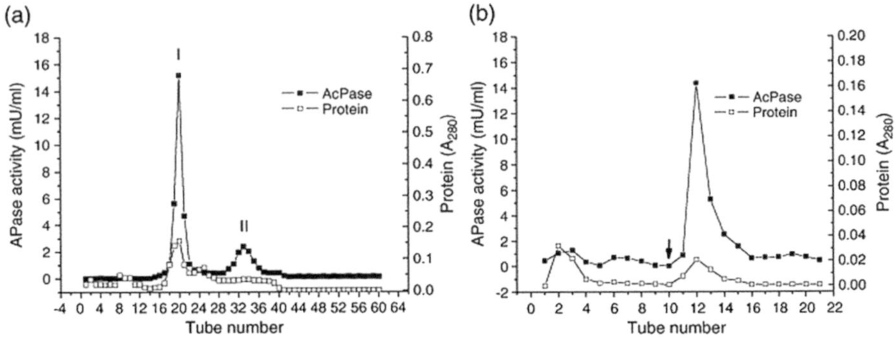

Fig. 4.23 (**a**) Sephadex G-200 superfine chromatography of pearl oyster extract. The column (1.0 × 80 cm) was equilibrated and eluted with 20 mM Tris–HCl buffer (pH 7.1, 2 mM ME). (**b**) Con A Sepharose 4B affinity chromatography of AcPase II eluted from Sephadex G-200 superfine. The column was washed with binding buffer and then eluted by the addition of 0.2 M methyl-α-d-mannopyranoside (arrow)

Table 4.3 Summary of purification procedure of AcPase from *P. fucata* [20]

Steps	Volume (mL)	Total protein (mg)	Total activity (U)	Specific activity (mU/mg)	Purification (fold)	Yield (%)
Crude extract	900	5185.67	18.772	3.62	1	100
$(NH_4)_2SO_4$ fraction	140	148.40	10.689	72.03	19.9	56.9
DEAE-Sepharose-FF	70	42.35	4.811	113.60	31.4	25.6
G200 Superfine						
AcPase I	58	1.117	0.649	581.02	160.5	3.5
AcPase II	57	1.140	0.829	727.19	200.9	4.4
Con A Sepharose 4B						
AcPase II	16	0.285	0.633	2221.05	613.6	3.4

4.4.2.2 Effect of Temperature and pH on AcPase I and II Activities

Figure 4.25 showed the enzyme activities that were measured at different temperatures and pH values. The optimal catalytic reaction temperature for AcPase I was 47 °C and for AcPase II was 57 °C (Fig. 4.25a, c). Using p-NPP as substrate, the optimal pH of AcPase I was 4.6, whereas the optimal pH of AcPase II was 3.2 (Fig. 4.25b, d). AcPases I and II exhibited more than 80% activity in the range of pH 4.0–5.2 and 2.8–4.4, respectively.

Figure 4.26 demonstrated the effect of preincubation of AcPases I and II at various temperatures for different times. After preincubation at 50 °C for 40 min for the two enzymes, the p-NPP hydrolysis only slightly decreased, but a higher temperature (60 °C) rapidly inactivated AcPase I (Fig. 4.26a). When incubated at

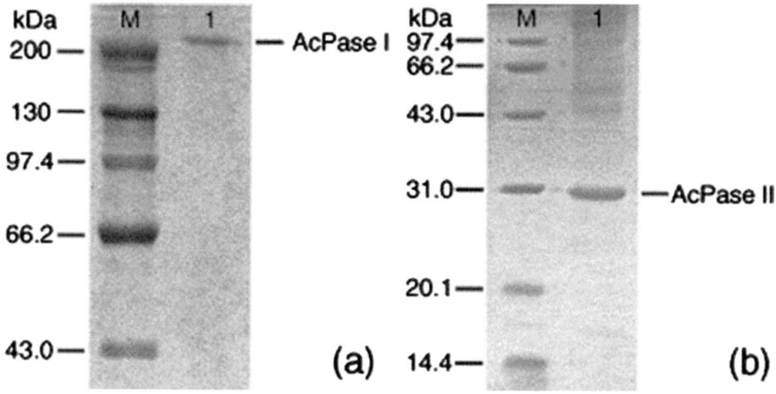

Fig. 4.24 SDS-PAGE of AcPases I (**a**) and II (**b**) from *P. fucata*. SDS-PAGE analysis was carried out on 7.5% gels for AcPase I and 15% gels for AcPase II. The gels were stained with Coomassie Brilliant Blue R-250. M, molecular mass standards; Lane 1, AcPase I (**a**) and AcPase II (**b**), respectively

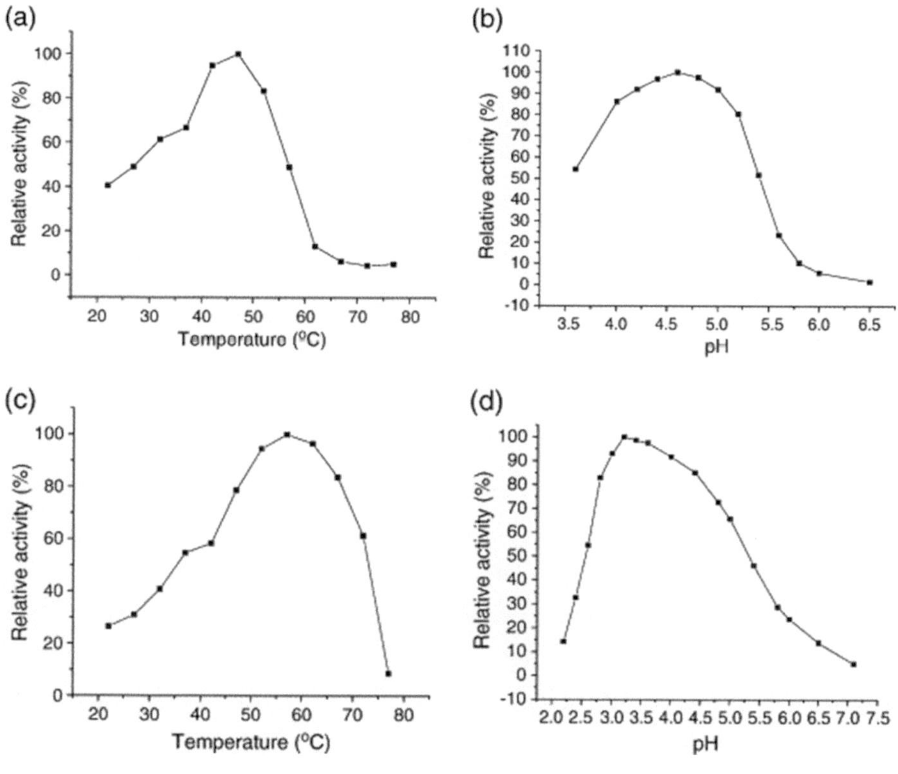

Fig. 4.25 Effects of temperature and pH on activities of AcPases I (**a**, **b**) and II (**c**, **d**) from *P. fucata*

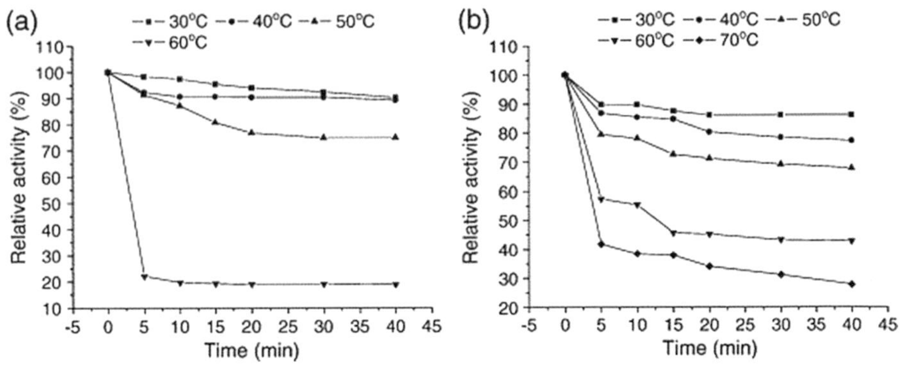

Fig. 4.26 Thermal stability of AcPases I (**a**) and II (**b**) from *P. fucata*. The values shown represent one typical experiment carried out in triplicate, and the control was not preincubated before enzyme activity was detected

Table 4.4 Effects of metal ions and inhibitors on activities of AcPases I and AcPase II from *P. fucata*

Added compound	Concentration (mM)	Relative activity (%)	
		AcPase I	AcPase II
Co^{2+}	5	101	311
Fe^{2+}	5	70	152
Fe^{3+}	5	11	7
Zn^{2+}	5	44	73
Cu^{2+}	5	8	55
Cd^{2+}	5	95	92
Cr^{3+}	5	130	98
Mn^{2+}	5	107	113
Pb^{2+}	5	46	59
HPO_4^{2-}	5	41	88
WO_4^{3-}	5	1	7
F^-	5	0	0
Tartrate	20	5	27
EDTA	5	79	93
EGTA	5	93	112

60 °C, AcPase I activity was decreased by about 81% after only 5 min, while AcPase II activity remained at 43%, even after 40 min (Fig. 4.26a, b).

4.4.2.3 Effects of Metal Ions on AcPase I and II Activities

Table 4.4 showed the effects of metal ions and inhibitors at various concentrations. Fe^{3+}, Zn^{2+}, Cu^{2+}, and Pb^{2+} inhibit the activity of AcPases I and II but to varying

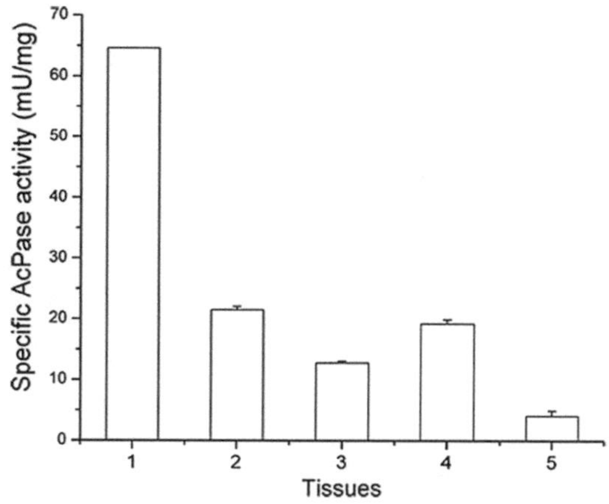

Fig. 4.27 Distribution of AcPase activity in different tissues of *P. fucata*. 1, The digestive gland; 2, gills; 3, mantle; 4, gonad; 5, foot

degrees. AcPase II was activated greatly by Co^{2+} and Fe^{2+}, while AcPase I was inhibited 30% by Fe^{2+}. Five mM Cr^{3+} had no effect on the activity of AcPase II but increased AcPase I enzyme activity to about 130%.

The product (HPO_4^{2-}) and the product analog (WO_4^{3-}) are strong inhibitors of the enzymes, and WO_4^{3-} inhibits the enzyme activity more strongly than HPO_4^{2-} (Table 4.4). Tartrate and F^- were the most effective inhibitors of enzyme activity. The effects of measuring EDTA and EGTA at 5 mM indicate that AcPases I and II are not metal-dependent enzymes [20].

4.4.3 *Immunolocalization of an Acid Phosphatase from* Pinctada fucata *and Its In Vitro Effects on Calcium Carbonate Crystal Formation*

4.4.3.1 Distribution of AcPase Activity in Different Tissues of *P. fucata*

Figure 4.27 showed the AcPase activity observed in all tested tissues, but the distribution was not uniform. The highest level of AcPase activity was detected in the digestive gland. In the gills, mantle, and gonad, the level of AcPase activity was also relatively high, whereas the AcPase activity in the foot was lower than in other tissues (Fig. 4.27).

4.4.3.2 Immunohistochemical Localization of AcPase I in Different Tissues of *P. fucata*

The cross-reactivity between the purified polyclonal antibody and AcPase I in total oyster homogenates was tested by immunoblotting. The results are demonstrated in

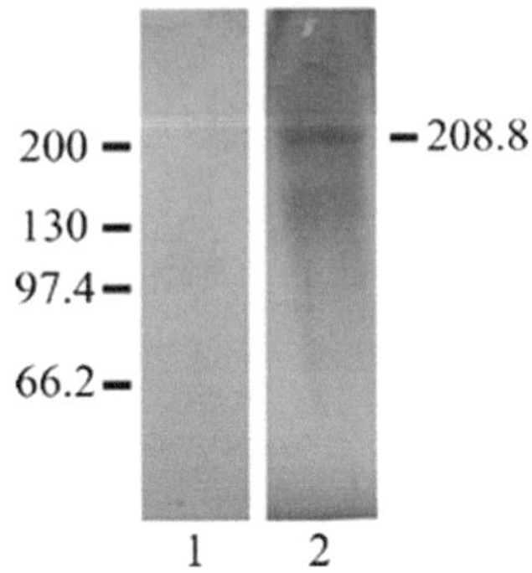

Fig. 4.28 Immunoblot analyses of AcPase I in whole homogenate of *P. fucata* with normal mouse serum (lane 1) and purified polyclonal antibodies (lane 2) [31]

Fig. 4.28. Only one visualized immunoreactive band was detected at 208.8 kDa, which was in accordance with the purified polyclonal antibody of AcPase I, whereas no band was observed in the control (Fig. 3.28).

Sections of the digestive gland showed strong positive immunostaining in the stomach, duct, and digestive tubule epithelia (Fig. 4.29a, b). In connective tissues, there was no immunostaining (Fig. 4.29a). A positive reaction was observed in the base of gill filaments (Fig. 4.29c). In the case of the mantle, strong labeling was distributed in the outer-side epithelia of the middle fold and the inner-side epithelia of the outer fold. There was also a weak reaction in the epithelia of the spermatogenic duct (Fig. 4.29e). However, no positive immunostaining was observed in the oyster foot (Fig. 4.29f). No immunostaining was observed in the controls (Fig. 4.30) [31].

4.4.3.3 Inhibitory Activity of AcPase I on Calcium Carbonate Precipitation and Ca–Binding Activity

AcPase I inhibited calcium carbonate precipitation in a dose-dependent manner and completely inhibited it at a concentration of 10 μg/ml, as shown in Fig. 4.31. Given the inhibitory activity of AcPase I on calcium carbonate precipitation, we analyzed its ability to bind calcium. We observed no positive results by stains-all method, which indicates that AcPase I probably had no affinity for calcium (data not shown).

4.4.3.4 Morphology and Polymorph Determination of Calcium Carbonate Crystals Induced by AcPase I

To study the effect of AcPase I on the morphology of calcium carbonate crystals, the in vitro precipitated crystals were examined by SEM. Crystals in the control experiment were completely hexahedral, which showed a typical morphology of calcite (Fig. 4.32e). In the presence of AcPase I, in addition to the typical morphology of calcite, crystals showed the conventional cluster needlelike morphology of aragonite

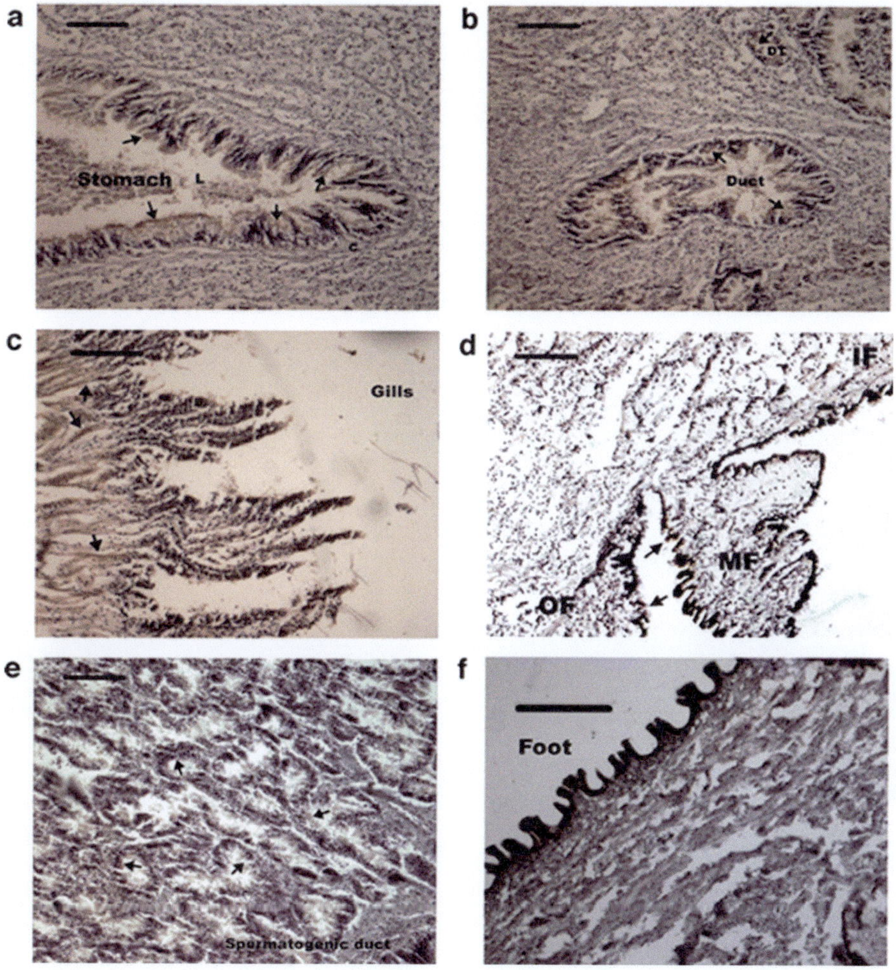

Fig. 4.29 Immunohistochemical localization of AcPase I in various tissues of *P. fucata*. (**a**) The stomach; (**b**) duct; (**c**) gills; (**d**) mantle; (**e**) spermatogenic duct; (**f**) foot bars = 100 μm in (**a**), (**c**), (**d**), and (**e**); 50 μm in (**b**), and 200 μm in (**f**). *L* lumen, *c connective tissue*, *DT* digestive tubule, *IF* inner fold, *MF* middle fold, *OF* outer fold

(Fig. 4.32a, c) [95]. In Fig. 4.32b, d, crystal morphology and size of aragonite varied when adding different concentrations of AcPase I (2.5 and 5.0 μg/ml).

The polymorphs of the above various calcium carbonate crystals were analyzed by Raman microscopy (Fig. 4.33). The needlelike crystals grown with 2.5 or 5.0 μg/ml of AcPase I generated the characteristic Raman bands of aragonite at 206, 703, and 1085 cm^{-1} (Fig. 4.33a, b), while the hexahedral crystals in all samples showed the characteristic Raman bands of calcite at 282, 712, and 1086 cm^{-1} (Fig. 4.33c) [96].

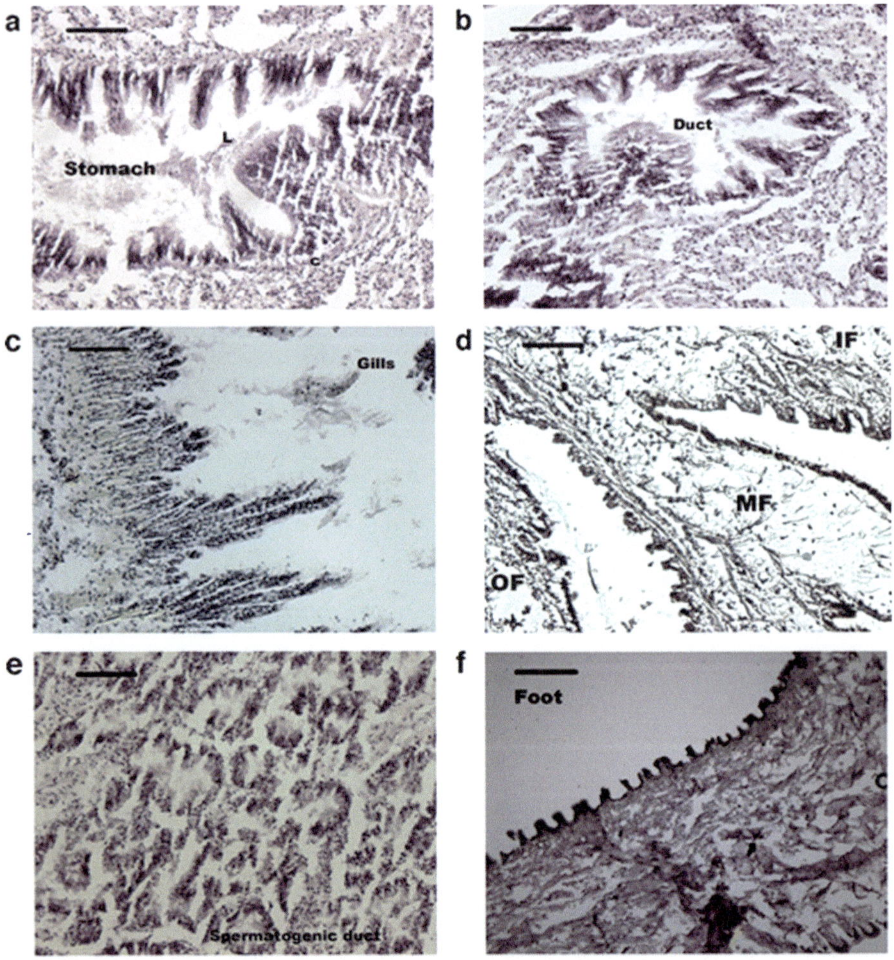

Fig. 4.30 Controls for immunohistochemical localization of AcPase I in various tissues of *P. fucata*. (**a**) The stomach; (**b**) duct; (**c**) gills; (**d**) mantle; (**e**) spermatogenic duct; (**f**) foot bars = 100 μm in (**a**), (**c**), (**d**), and (**e**); 50 μm in (**b**) and 200 μm in (**f**). *L* lumen, *c* connective tissue, *IF* inner fold, *MF* middle fold, *OF* outer fold

4.5 A Novel Carbonic Anhydrase

*4.5.1 A Novel Carbonic Anhydrase from the Mantle of the Pearl Oyster (*Pinctada fucata*)*

4.5.1.1 Purification of CA

CA was purified from the mantle following homogenization, ammonium sulfate precipitation, and affinity chromatography on p-aminomethylbenzene sulfonamide–agarose.

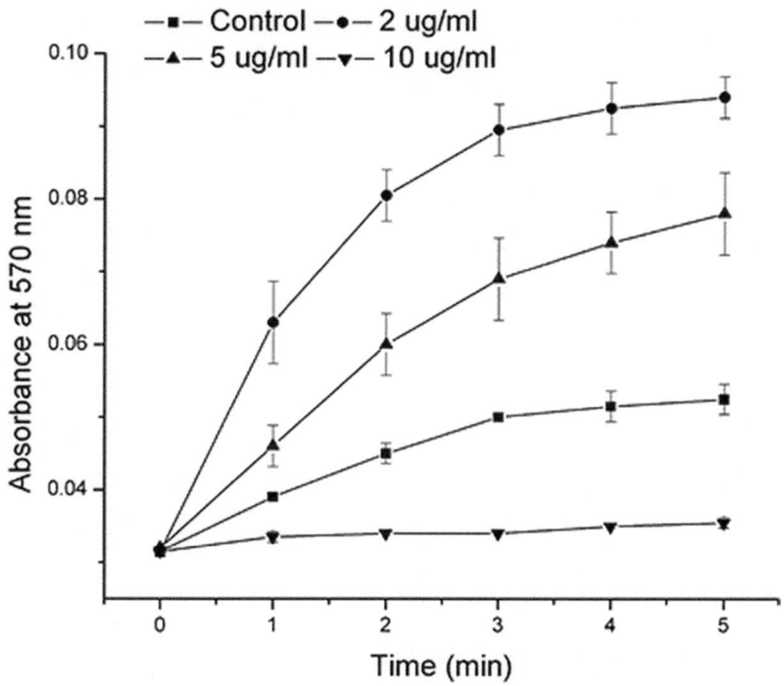

Fig. 4.31 Inhibitory activity of AcPase I on calcium carbonate precipitation changes in the turbidity of the assayed solutions are shown. ▪, control; •, 2 μg/ml of AcPase I; ▴, 5 μg/ml of L AcPase I; ▾, 10 μg/ml of AcPase I

SDS-PAGE gels (Fig. 4.34, lane b) indicated that CA is a single band with a molecular mass of 38 kDa, which is similar to previously reported mammalian, reptilian, and fish CAs [97].

DNSA (a fluorescent probe) can bind to CA, and the fluorescence of the resulting CA–DNSA complex can be used to visualize the specific CA band on polyacrylamide gels. To identify CA activity associated with the purified protein, native PAGE (as described above) was performed using the sample from mantle, which shows a single band on SDS-PAGE. A green fluorescent band was visualized under UV illumination, as shown in Fig. 4.35 (lane a), which was similar to that in the positive control CA I in Fig. 4.35 (lane b), confirming that the purified protein had the carbonic anhydrase activity. The purified CA migrated faster than CA I shown in Fig. 4.35, although the two differ by 30 kDa in molecular weight. This result demonstrates that the isoelectric point (pI) of the purified CA is lower than that of CA I (pI = 6.8), which indicates that the CA from the mantle is an acid protein.

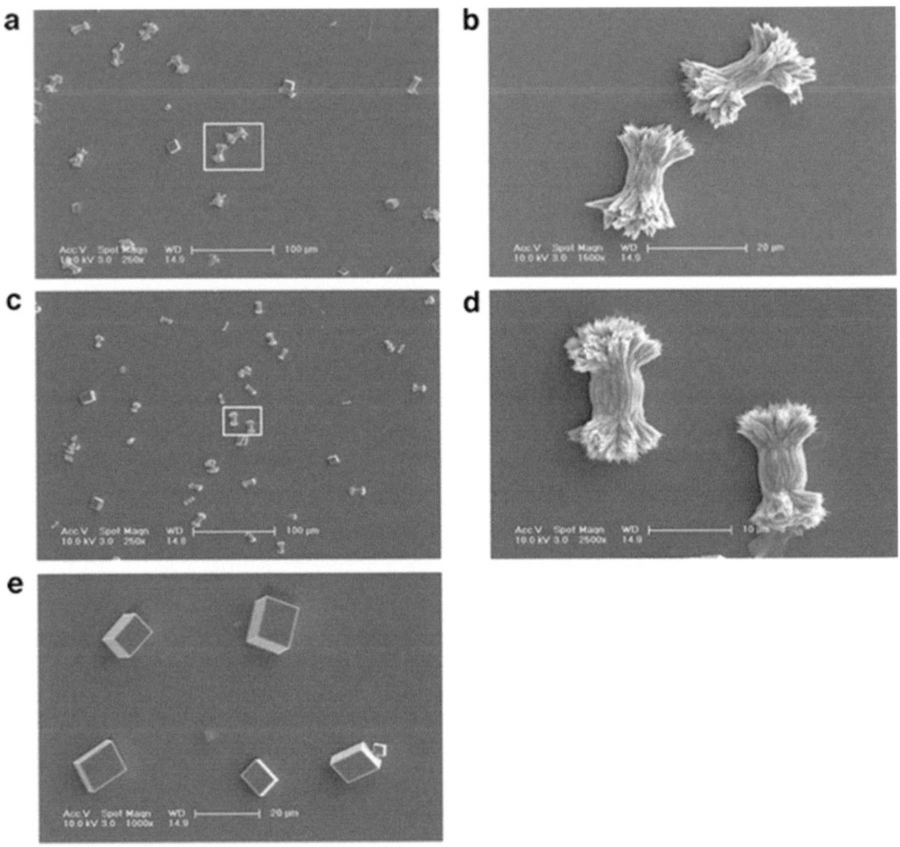

Fig. 4.32 SEM images of calcium carbonate crystals precipitated with a supersaturated calcium carbonate solution in vitro. (**a**) With 2.5 μg/ml of AcPase I. (**b**) Enlarged image of the boxed part in (**a**). (**c**) With 5.0 μg/ml of AcPase I. (**d**) Enlarged image of the boxed part in (**c**). (**e**) Without AcPase I as a control, bars = 100 μm in (**a**) and (**c**), 20 μm in (**b**) and (**e**), and 10 μm in (**d**)

4.5.1.2 AZ Inhibition and DNSA Binding of CA

AZ is a potent, reversible inhibitor of CA, which is used to distinguish CA from other esterases. With the gradual increase of AZ concentration, treatment of CA resulted in a progressive decrease of enzyme activity. After incubating with 0.5 μM AZ at 35 °C for 1 h, the initial activity of the enzyme lost by 50% (Fig. 4.36). Whereas, incubation of the enzyme with higher concentrations of AZ did not lead to further loss of activity, indicating that the enzyme was only partially inhibited by AZ. This result is in accordance with an earlier report about nacrein obtained from the prismatic layer [99].

DNSA is a fluorescent probe that specifically binds to the active site of CA. After binding to CA, the emission wavelength of DNSA shifts from 580 to 468 nm

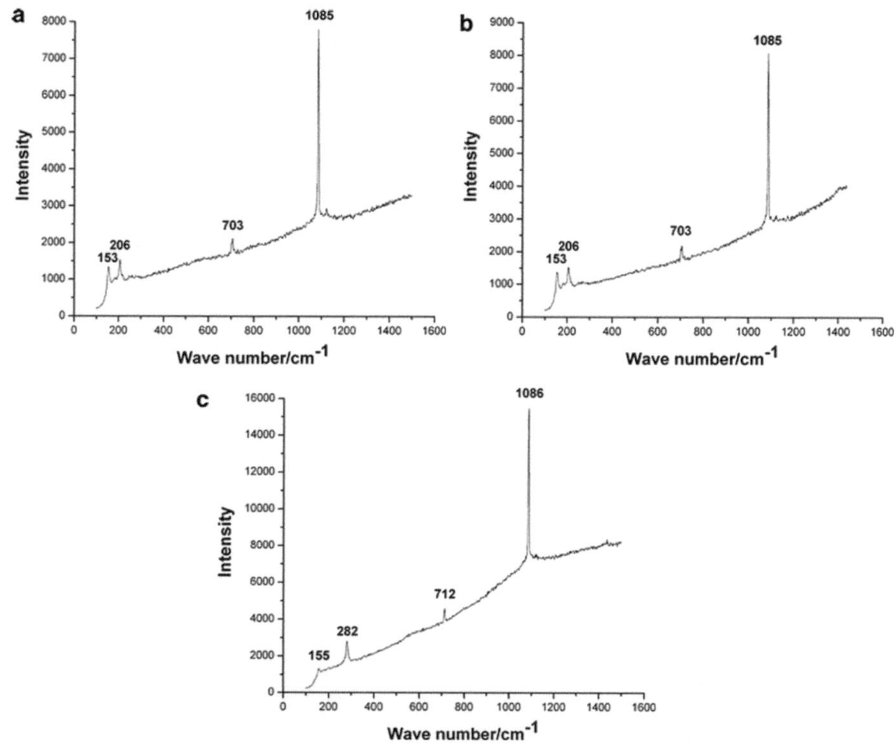

Fig. 4.33 Raman spectra of crystals with various morphologies. (**a**) Raman spectrum of single needlelike aragonite crystal grown with 2.5 μg/ml of AcPase I, (**b**) Raman spectrum of single needlelike aragonite crystal grown with 5.0 μg/ml of AcPase I. (**c**) Raman spectrum of single hexahedral calcite crystal

[100, 101]. After incubation of the purified CA with 50 μM DNSA for 10 min at room temperature, the wavelength at which the maximum intensity of blue is emitted becomes 454 nm. The relative quanta yield was 2.9 when excited at 326 nm (Fig. 4.37), while the emission of free DNSA at 454 nm was undetectable under the equal condition [98].

4.5.1.3 Distribution of CA in Mantle

The mantle of the mollusk was known from studies by Freeman and Wilbur [102] to contain CA activity. In the present paper, we detected histochemically the distribution of CA in the mantle of *P. fucata*, by the use of the fluorochrome DNSA. Strong CA–DNSA fluorescence was detected in cells along the outer edge of the mantle (Fig. 4.38) and weaker fluorescence in central zone.

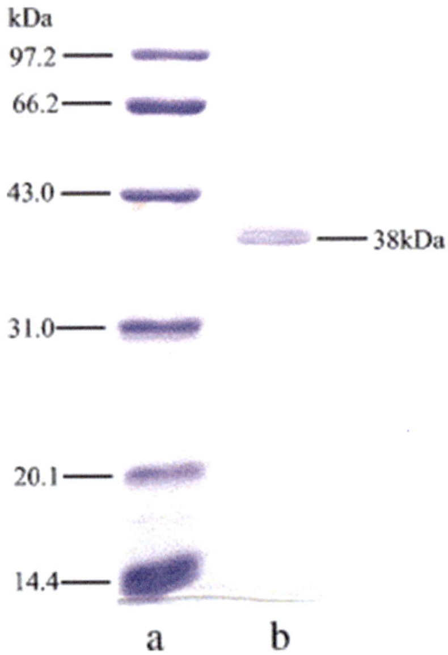

Fig. 4.34 SDS-PAGE of purified carbonic anhydrase (CA). Electrophoresis of 3.75–15% polyacrylamide gels was performed at 10 °C. Lane a, molecular mass standards; lane b, purified CA from mantle (38 kDa) [98]

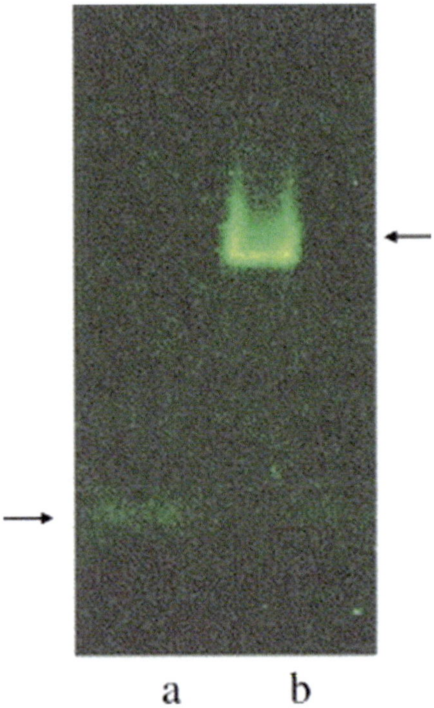

Fig. 4.35 Native PAGE of purified carbonic anhydrase (CA). Analysis of CA from mantle on polyacrylamide gels by fluorescence of carbonic anhydrase–DNSA complexes. Lane a, CA from mantle of pearl oyster; lane b, carbonic anhydrase I (CAI) from human erythrocytes

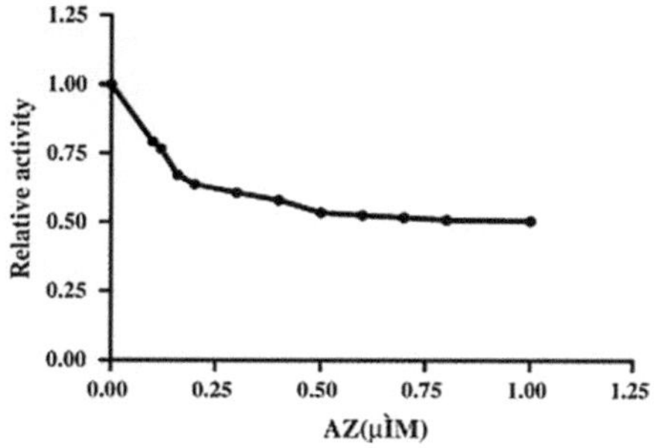

Fig. 4.36 Inhibition of CA by acetazolamide (AZ). CA (8.0 μM) was incubated with various concentrations of AZ (0.1–1.0 μM) in 0.01 M diethylmalonic acid, pH 8.0, for 1 h at 35 °C, followed by the assay for AC activity. As a control, another sample was kept for 1 h as described above, but without AZ

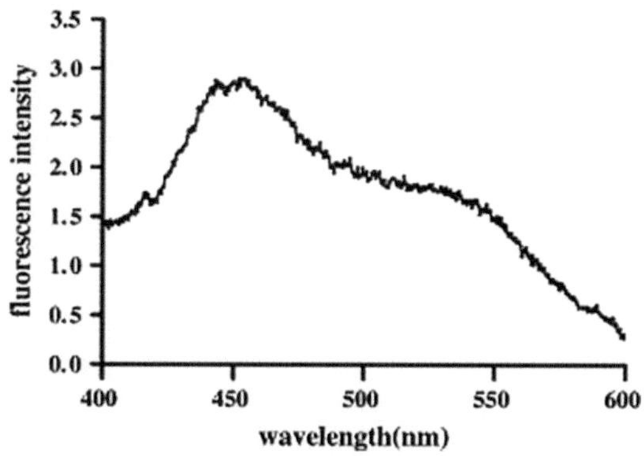

Fig. 4.37 Fluorescence spectra of carbonic anhydrase after binding to 5-dimethylamino-1-naphthalene sulfonamide (DNSA). CA (1 μM) was incubated with DNSA (50 μM) for 10 min at room temperature. The fluorescence spectrum of 400–600 nm was recorded using a Hitachi F-2500 spectrofluorimeter at an excitation wavelength at 326 nm. The emission maximum intensity wavelength of CA combined with DNSA was 454 nm, and the emission of free DNSA at 454 nm was undetectable in the same conditions

4.6 A Novel Putative Tyrosinase

*4.6.1 A Novel Putative Tyrosinase Involved in Periostracum Formation from the Pearl Oyster (*Pinctada fucata*)*

4.6.1.1 Isolation and Sequence Analysis of cDNA Sequence Encoding Oyster Tyrosinase

Using total mantle RNA as the template, a 408 bp cDNA product was obtained by RT-PCR with two degenerate oligonucleotides (DP1 and DP2). Based on the sequence, two gene-specific primers were synthesized and used to amplify the 5′ and 3′ nucleotide sequences of putative oyster tyrosinase cDNA by touchdown PCR. In order to confirm the sequence obtained by RACE, two specific primers corresponding to 5′- and 3′-UTR sequence of putative tyrosinase mRNA were

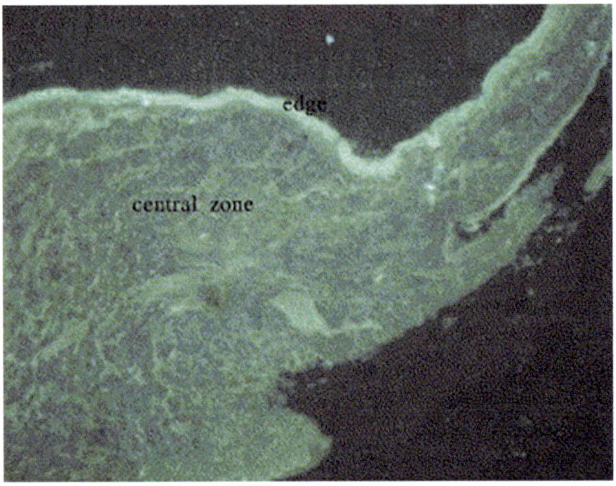

Fig. 4.38 The fluorescence of CA–DNSA in the mantle of pearl oyster. 20 μl of 50 μM DNSA was added to the mantle cryostat sections and incubated at 35 °C for 30 min. As controls, sections containing no DNSA were incubated with phosphate buffer. Photographs were taken in a Leica Dialux microscope equipped with and a high-pressure Hg lamp as ultraviolet source; the samples were observed through a 460 nm barrier filter

designed and subjected to RT-PCR. The PCR products were cloned and sequenced, which matched the sequence expected from the 5′-and 3′-RACE results.

In Fig. 4.39, the complete cDNA sequence containing the poly(A) tail is 1487 bp. It contains a 28-base 5′-untranslated sequence, an1338 bp ORF, a TAA terminator, a 106-base 3′-untranslated sequence, and a 15 nucleotides poly(A) tail. A putative polyadenylation signal (AATAAA) is recognized at position 1450, upstream of the poly(A) tail, separated by 17 nucleotides. This cDNA sequence has been submitted to GenBank under the accession number DQ112679 [103].

The cDNA-cloned ORF encodes a protein consisting of 445 amino acid residues, which has a high degree of similarity to the two putative copper-binding sites of cephalopod tyrosinases and was therefore considered to be capable of being used as tyrosinase. The first 22 amino acids of the deduced protein were predicted to be a signal peptide by the neural networks and hidden Markov models which was provided by CBS prediction servers. As the calculated molecular mass of the deduced protein is approximately 47 kDa without the signal peptide, it was named OT47. In previous studies, the phenoloxidases was found to be secreted as inactive precursors and then activated by cleavage of proteinase [43, 46, 47]. Due to a number of putative proteinase and cleavage sites, the mature form of OT47 could only be detected by purifying from the mantle and periostracum of *P. fucata*. The amino acid composition of OT47 demonstrated high proportion of Arg (9.69% mol/mol) and Gly (9.46% mol/mol). The former residue led to the basic property of OT47 with calculated pI of 9.4. The latter residues are even more abundant in periostracum (it may reach about 50% in some species [104]) than in OT47. It could also be noticed that there are 18 tyrosine residues (and 11 lysine residues) in OT47, tending to be oxidized to quinines and immobilized OT47 itself in the periostracum, which is related to the previous findings on the tyrosinase from the periostracum of *Modiolus demissus*.

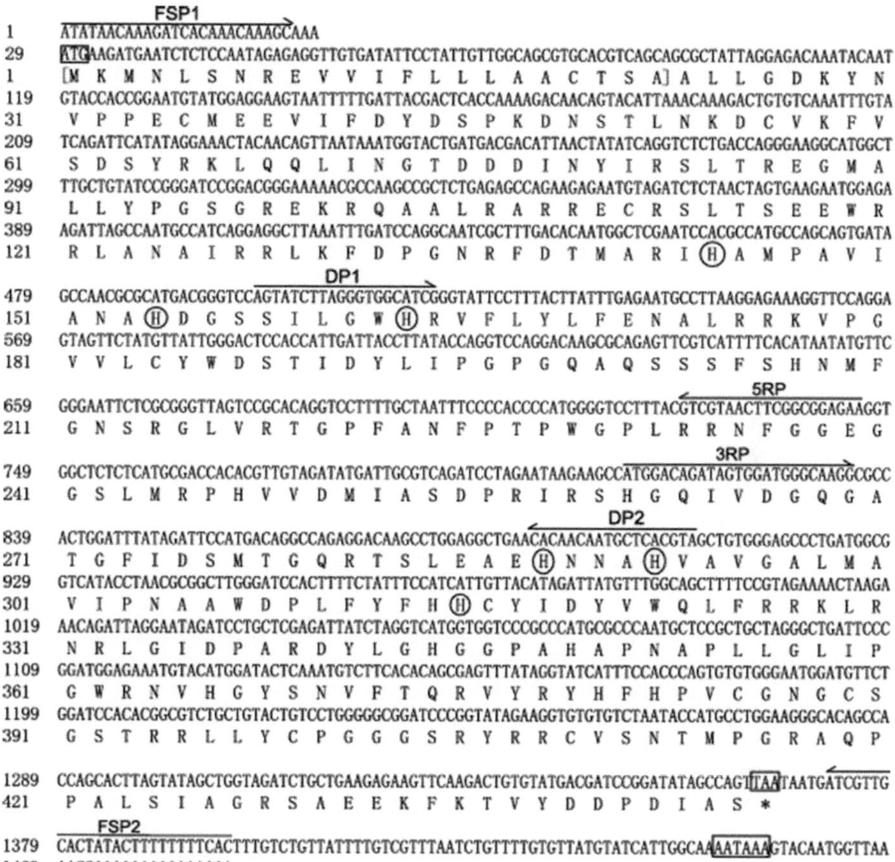

Fig. 4.39 The cDNA sequence and deduced amino acid sequence of the tyrosinase OT47of *P. fucata*. Brackets enclose the putative signal peptide. The start codon, the stop codon, and the putative polyadenylation signal (AATAAA) are boxed. The putative copper ligands are circled. Arrows indicate positions of primers used for cloning. The asterisk (∗) in the amino acid sequence also indicated the terminate codon

4.6.1.2 Structural Similarity of OT47 to Other Tyrosinase and Hemocyanins

Tyrosinases belong to the type 3 copper protein superfamily, and arthropod phenoloxidases and hemocyanins are also the superfamily members. This protein superfamily has been isolated from many organisms, and their evolutionary relationships have also been elucidated on the basis of the amino acid sequences that are conserved around two copper-binding sites that form an oxygen-binding active center [105, 106]. Figure 4.40 showed the comparison of these two conserved copper-binding sites of OT47 with hemocyanins, tyrosinases, and phenoloxidases

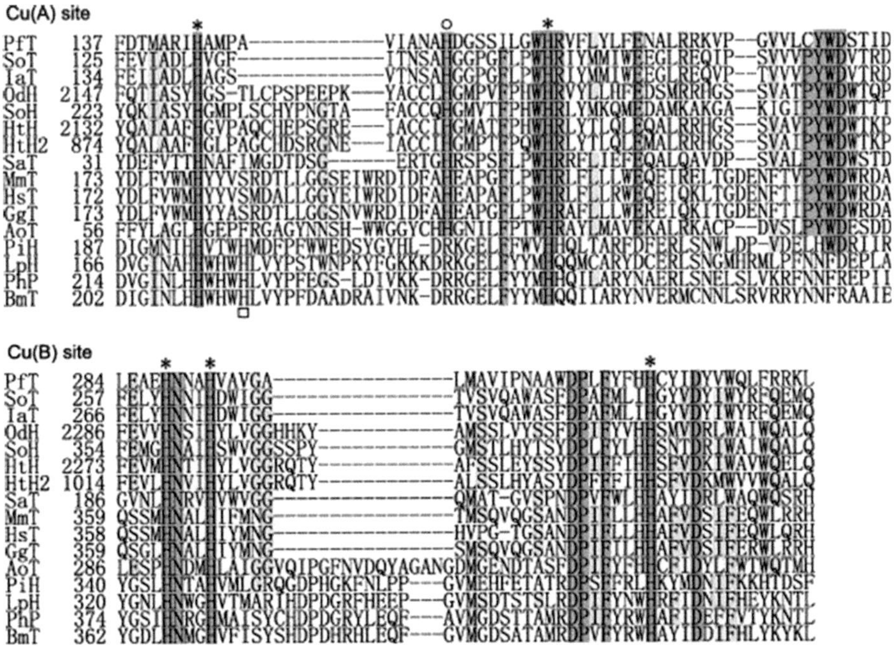

Fig. 4.40 Comparison of the amino acid sequences at the two putative copper-binding sites, Cu (A) and Cu(B), in OT47 and other type 3 copper proteins. Numbers demonstrate positions of the amino acid residues in each sequence. Gaps (−) have been introduced to optimize the alignment. The putative copper ligands of histidine residues that are conserved in all proteins are marked with asterisks (∗), those conserved in molluskan proteins and tyrosinases with ○, and those conserved in arthropod proteins with □. The same residues are shaded. PfT, *P. fucata* tyrosinase OT47; SoT, *Sepia officinalis* tyrosinase (EMBL, AJ297474); HtH, *Haliotis tuberculata* hemocyanin (EMBL, AJ252741); IaT, *Illex argentinus* tyrosinase precursor 1 (DDBJ, AB107880); HtH2, *H. tuberculata* hemocyanin type 2 protein (EMBL, AJ012048); BmT, *Bombyx mori* phenoloxidase subunit 1 precursor (SWISS-PROT, Q27451); OdH, *Octopus dofleini* hemocyanin G-type (GenBank, AF338426); SoH, *S. officinalis* hemocyanin units H (SWISS-PROT, P56826); GgT, *Gallus gallus* tyrosinase (GenBank, S81675); AoT, *Aspergillus oryzae* tyrosinase (DDBJ, D37929); HsT, *Homo sapiens* tyrosinase precursor (SWISS-PROT, P14679); SaT, *Streptomyces avermitilis* tyrosinase (DDBJ, AB070938); PhP, *Pimpla hypochondriaca* phenoloxidase I (EMBL, AJ297596); MmT, *Mus musculus* tyrosinase (SWISS-PROT, P11344); LpH, *Limulus polyphemus* hemocyanin II (SWISS-PROT, P04253); PiH, *Panulirus interruptus* hemocyanin subunit a (SWISS-PROT, P04254)

from other organisms. The two copper-binding sites of OT47 were similar to the cephalopod tyrosinase and the tyrosinases from vertebrates, microorganisms, and molluskan hemocyanins and less similar to those of phenoloxidases and hemocyanins from arthropods. Assume that six histidine residues within these sites of OT47 were copper-binding ligands that could be arranged in similar positions to molluskan tyrosinases and hemocyanins. Only the second copper ligand in the Cu(A) site of the phenoloxidases and hemocyanins from arthropods is different from other tyrosinase and molluskan hemocyanins. And several amino acid residues around the six copper ligands are also conserved in molluskan tyrosinases and hemocyanins but differ from arthropod phenoloxidases and hemocyanins. These observations confirm previous hypothesis that molluskan hemocyanins and tyrosinases genetically separated from their common ancestor and are separated from the evolutional line of hemocyanins and phenoloxidases in arthropods [107, 108]. While there also exist some differences in the amino acid sequences around the copper ligands between the hemocyanins and tyrosinases of molluscan, it is worth noting that in molluskan hemocyanins, a ligand in the Cu(A) site is involved in the formation of cysteine–histidine thioether bridge [109]. This unusual linkage in the copper-binding site has also been observed in ascomycete tyrosinases, which indicate that molluskan hemocyanin might evolve from ancient tyrosinase [110, 111], whereas no cysteine residue was detected in OT47 and other molluskan tyrosinase, despite the presence of two cysteine residues in the copper-binding regions of OT47. These phenomena indicated that the divergence of molluskan hemocyanins and tyrosinases was earlier than that of tyrosinases from bivalve and cephalopod, in accordance with the hypothesis that the divergence of molluskan hemocyanins and tyrosinases occurred at an extremely early evolutional stage of life when aerobic metabolism was established and metazoans emerged [107, 108].

Then, a phylogenetic tree was calculated based on the alignment of the copper-binding regions by ClustalX program with the NJ algorithm (Fig. 4.41). In addition to the above points, the phylogenetic tree also showed that there were great evolutional distances between the bivalves and cephalopod tyrosinases. Besides, the similarities between OT47 and cephalopod tyrosinases were also very low, for the whole entire amino acid sequence. Only 16.4% amino acid of OT47 showed an identity with the tyrosinase of *Illex argentinus* and 15.6% with that of *Sepia officinalis*. It is unclear whether this phenomenon is only due to the early differentiation of bivalves and cephalopods or these mollusk tyrosinases belong to different functional subtypes, which may diverge earlier than the divergence of bivalves and cephalopods. More structural information on tyrosinase in bivalves and other mollusks is required to further elucidate the phylogenetic relationships among type 3 copper proteins of mollusks [103].

4.6.1.3 Gene Expression Analyses of Oyster Tyrosinase OT47

Tyrosinases are thought to be involved in various biological activities of mollusks, which include native immune response [46, 56] and the formation of egg capsules [51], byssus [48, 49], shell matrix proteins [104, 112], and periostracum. In order to study the functions of oyster tyrosinase OT47 in vivo, tissue-specific expressions of OT47 mRNA in various tissues were detected by RT-PCR using specific primers. Samples of total RNA were prepared from the mantle pallial, mantle edge, foot, gill, digestive gland, gonad, adductor muscle tissues, and hemocytes of *P. fucata*. The results demonstrated that OT47 transcript is expressed specifically in the mantle edge but not in any other tissues, which contributes to the formation of the periostracum and prismatic shell layer (Fig. 4.42). The expressions of MSI31

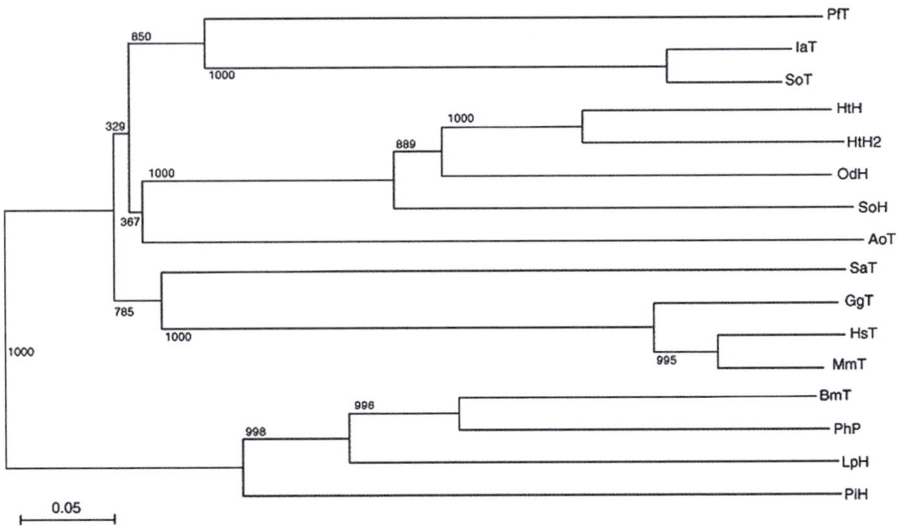

Fig. 4.41 Copper-binding regions of phylogenetic tree. The tree was deduced by an analysis of adjacent linkages based on the amino acid sequence alignment of copper-binding regions ranging from the N-terminal of Cu(A) site to the C-terminal of Cu(B) site shown in Fig .4.40, which was performed by the ClustalW program. Gaps were treated as missing data. Bootstrap values (1000 replicates) are displayed at the nodes. For abbreviated protein names, see Fig. 4.40

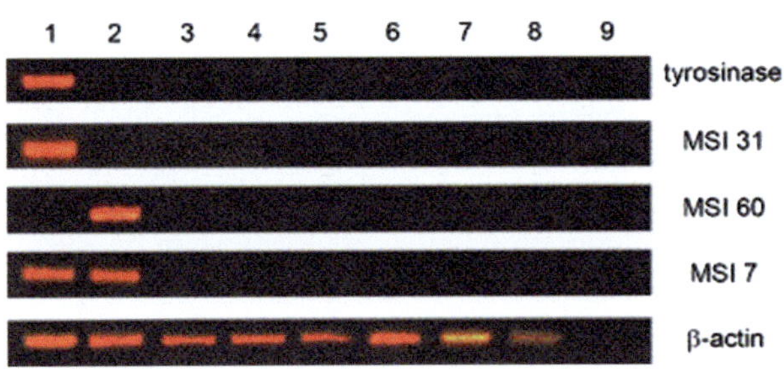

Fig. 4.42 Analysis of tyrosinase OT47 and some shell matrix protein gene expressions in *P. fucata* by RT-PCR. Lane 1, the mantle edge; lane 2, the mantle pallial; lane 3, the foot; lane 4, gill; lane 5, gonad; lane 6, digestive glade; lane 7, adductor muscle; lane 8, hemocytes; and lane 9, negative control. The housekeeping gene β-actin was regarded as a positive control

[113], MSI60 [113], and MSI7 [114] were also analyzed to compare with OT47. As expected, mRNA for MSI31, the Gly-rich framework protein of the prismatic layer, was expressed only in the mantle edge, whereas mRNA of MSI60 was expressed only in the mantle pallial; the MSI60 was a nacreous layer framework protein, and the MSI7 mRNA was expressed in both the edge and the pallial of the mantle tissue. Experiments were repeated twice, and the results were similar. It seemed that oyster tyrosinase OT47 only involves in the formation of periostracum and/or the prismatic shell layer matrix.

In order to determine the more precise expression site of OT47 mRNA at the margin of the mantle edge in *P. fucata*, in situ hybridization was carried out with frozen sections of the mantle (Fig. 4.43a). Strong hybridization signals for OT47 mRNA were found in the outer epithelial cells of the middle fold near the periostracal groove base, but no hybridization signal was found in the outer fold. This result suggested that OT47 might be involved in the formation of periostracum, but unlike previous histological studies in *Mercenaria mercenaria*, it was shown that the tyrosinase is present in the inner epithelial cells of the outer fold. To further confirm the in situ hybridization result, the tyrosinase activity was also detected in situ by incubating a portion of the tyrosinase substrate L-DOPA. The cells in the equal position produced positive signals, which also indicate that the tyrosinase is present in the outer epithelial cells of the middle fold. Similar results were obtained for experiments carried out in three batches of individuals of different seasons (in June, September, and December). Thus, the secretion of periostracum in *P. fucata* seemed to be different from that in *M. mercenaria*.

The previous cell biology model in bivalves of periostracum secretion suggested that the initial membrane of periostracum (usually called pellicle) is produced by the basal cell at the bottom of the periostracal groove; two various types of cells in the inner epidermis of the outer fold, secreting either the tyrosinases or the diphenols, can cause periostracum thickening, and the outer epithelial cells of the middle fold could function to lubricate and put the pellicle in place. In addition to the histological studies in *M. mercenaria*, this model was also supported by the observation that periostracum thickens at the surface facing the outer fold and that the outer epithelial cells of the middle fold are all lacking the secretion organelles [115, 116]. And it was believed that periostracum of *P. fucata* might also secrete by this means, whereas some later studies demonstrated that the outer epithelial cells of the middle fold also contained some member-bound vesicles [117] and in some species the periostracum within the periostracal groove is coated on both sides, suggesting that the outer epithelial cells of the middle fold also have the secretion capability. In addition, the studies of shell matrix proteins of *P. fucata* demonstrated that the insoluble matrix protein prismalin-14 purified from the prismatic layer is specifically expressed only in the inner epithelial cells of the outer fold [118] and three soluble matrix proteins involved in the prismatic layer formation, MSI31, MSI7, and nacrein, are expressed in the outer and the inner epithelial cells of the outer fold [113, 114, 119]. These results suggested that the inner epithelial cells in the outer fold of *P. fucata* participate in the formation of prismatic layer. Therefore, it is difficult to imagine that cells could involve in the secretion of periostracum at the same position, that is, another layer out of the prismatic layer simultaneously. Our results of the in situ hybridization and tyrosinase activity staining showed that the tyrosinase OT47 is derived from the outer epithelial cells of the middle fold, which would explain this paradox and might complement the model of periostracum secretion.

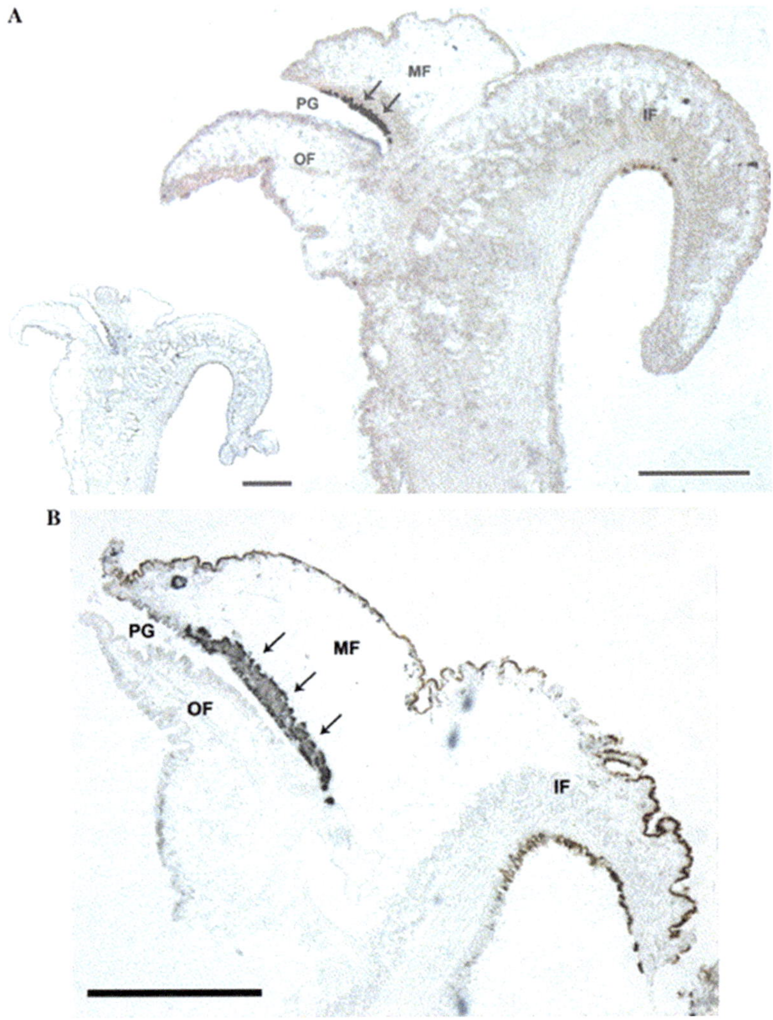

Fig. 4.43 The OT47 mRNA in the mantle of *P. fucata* detected by in situ hybridization (**a**) and tyrosinase activity staining (**b**). Hybridization signals in (**a**) and tyrosinase activity signals in (**b**) are revealed by arrow. The inset of (**a**) indicates a control section stained with a sense probe. *PG* periostracal groove, *OF* outer fold, *MF* middle fold, *IF* inner fold. Scale bar, 0.5 mm

In summary, OT47 is a novel putative tyrosinase cloned from *P. fucata* that is specifically expressed in the outer epithelial cells of the middle fold and thus appeared to participate in the periostracum formation. Further structural and functional studies might help to understand periostracum formation and quinone-tanned process in mollusk at the molecular levels.

4.7 A Novel Astacin-Like Metalloproteinase

4.7.1 Pf-ALMP, a Novel Astacin-Like Metalloproteinase with Cysteine Arrays, Is Abundant in Hemocytes of Pearl Oyster Pinctada fucata

4.7.1.1 Isolation and Sequence Analysis of cDNA Sequence Encoding Oyster Astacin–Like Metalloproteinase

A cDNA fragment of the putative pf-ALMP gene which consists 3189 nucleotides (nt) was cloned from pearl oyster hemocytes mRNA using a primer-directed, RT-PCR-based strategy and sequencing, and the GenBank accession number was AY818356. This fragment contained a 2040 nt open reading frame encoding a predicted protein that contains 679 amino acids with a molecular weight of 78 kDa (Fig. 4.44a). Besides, 28 nt of the 5′-end untranslated region (UTR) and 1121 nt of the 3′-end UTR were also found. In Fig. 4.44b, a conceptual translation of the open reading frame reveals that this novel putative pf-ALMP gene encodes [1] a hydrophobic segment of the signal peptide; [2] RTKR, basic amino acid residues which were found in the different members of astacins, suggesting furin or a trypsin-like protease serve as potential proprotease activators; [3] a conserved ZnMc catalytic domain; and [4] five cysteine array repeats [120].

4.7.1.2 Distribution and Expression of Pf–ALMP in Various Tissues

In order to study the function of oyster pf-ALMP in vivo, semiquantitative PCR was used to detect the tissue-specific expression of oyster pf-ALMP mRNA among different tissues RT PCR performed with specific primers, and the total RNA of different tissues as template yielded a 454 bp product, whereas the negative control showed no product (data not shown). The PCR product was subcloned into pGEM-T easy vector and verified by sequencing. RT-PCR reactions were carried out with RNA samples from the mantle, gill, digestive gland, muscle, and hemocytes. In Fig. 3.43, Pf-ALMP mRNA was expressed in all tissues tested. However, although the 18S rRNA levels were the same in all samples, the expression level of Pf-ALMP mRNA was the highest in the hemocytes. Data from three independent experiments were similar (Fig. 4.45).

We also explored the expression pattern of pf-ALMP transcript in hemocytes of challenged and unchallenged oysters by RT-PCR as described above. Total RNAs were extracted from hemocytes in different conditions (0, 6, 12, 24, and 48 h) respectively. The β-actin was used as a control of equal quantities of total RNA used in RT-PCR. Within 6 h, 1 μg LPS challenge significant effect on the expression level of pf-ALMP mRNA in hemocytes. The maximal expression was about three times higher than the value detected in the control, and the effect lasted more than 48 h (Fig. 4.46). Three independent experiments were conducted with similar results.

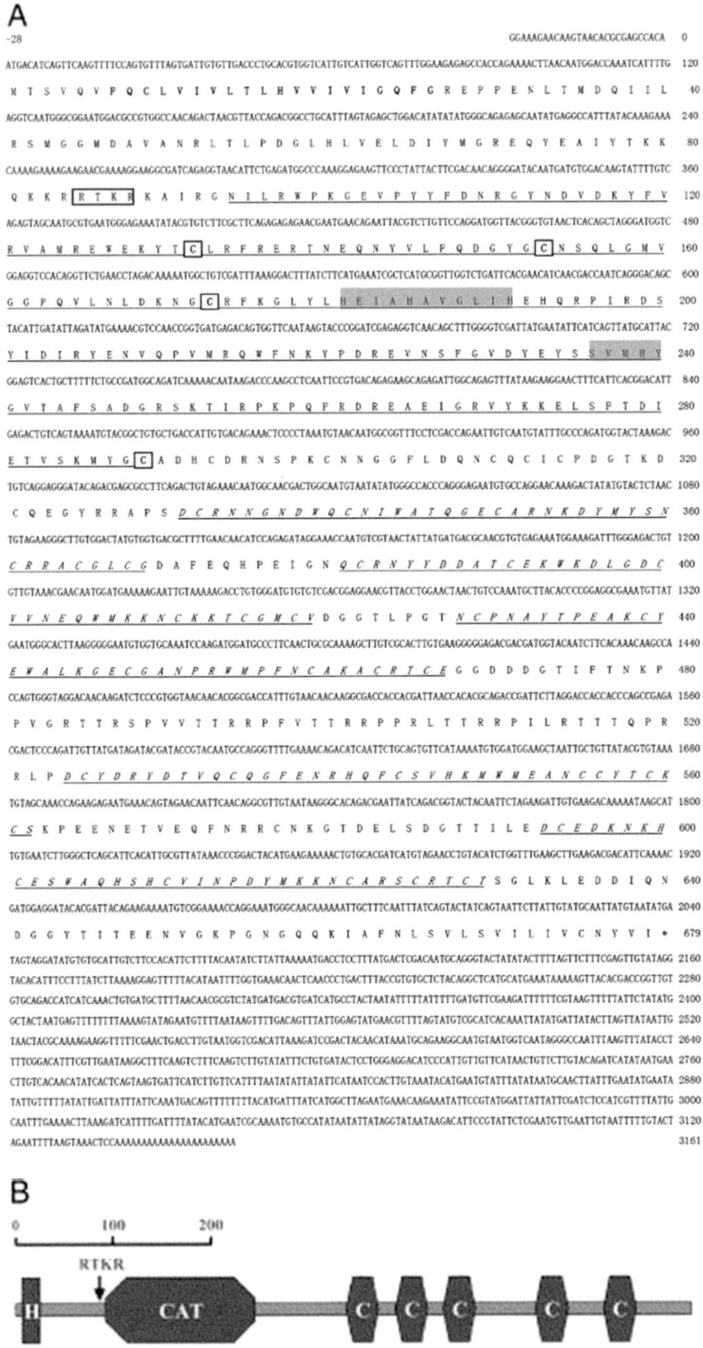

Fig. 4.44 Primary structure of pearl oyster Pf-ALMP. A: Nucleotide and amino acid sequences of Pf-ALMP. The number on the right is the nucleotide and amino acid sequence. The N-terminal

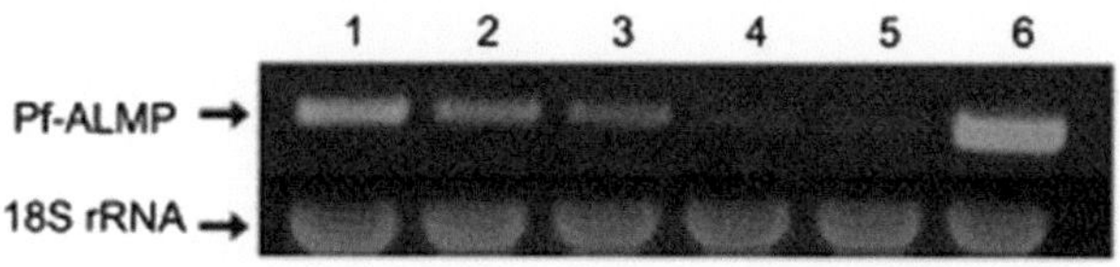

Fig. 4.45 Expression of Pf-ALMP mRNA in tissues of the pearl oyster. RT-PCR was carried out with RNA samples from the adult tissues of *Pinctada fucata*. The 18 S rRNA was used as a control of the same quantities of total RNA used in RT-PCR. Lane 1, the gill; lane 2, gonad; lane 3, digestive gland; lane 4, mantle; lane 5, muscle; lane 6, hemocyte

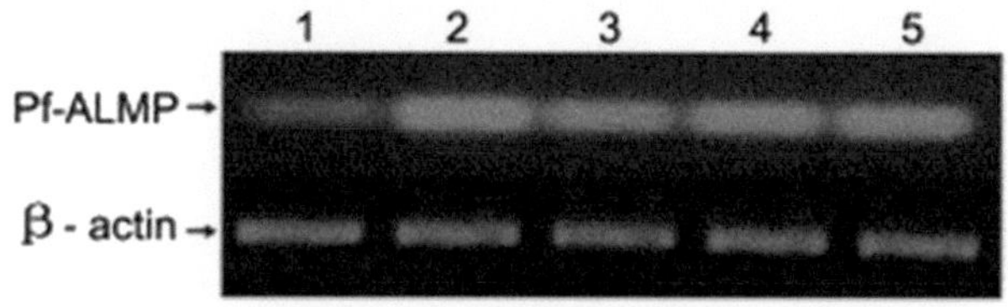

Fig. 4.46 Pf-ALMP mRNAs expression from oyster hemocytes. RT-PCR was carried out with RNA samples from unchallenged oysters or LPS-challenged oysters for several times. The expression of oyster β-actin was as a control of equal quantities of total RNA used in RT-PCR. Lane 1, unchallenged; lane 2, challenged for 6 h; lane 3, 12 h; lane 4, 24 h; lane 5, 48 h

In bivalve mollusks, the hemocytes are responsible for many physiopathologic events, such as digestion, inflammation, nutrient transport, respiratory burst, internal defenses, phagocytosis, and encapsulation of foreign bodies [121, 122]. The putative pf-ALMP mRNA expressed abundantly in hemocytes, and the level of expression affected by LPS challenge significantly showed that the pf-ALMP participated in defense mechanisms.

Pf-ALMP distribution was studied by in situ hybridization among three different tissues which includes the gill, gonad, and digestive gland. The pf-ALMP mRNA hybridization signals were dark purple. In Fig 4.47a–c, strong positive signals were found in the gonad and epithelial cells of the gill and digestion gland.

Astacins have been regarded as having roles in mature and development systems. Many of them contribute to bone/cartilage formation, pattern formation, egg hatching, and early embryo development [60–62, 66]. They are expressed in a tissue-specific manner in mature organisms and are temporally and spatially expressed in developmental systems. High levels of expression in the gonad of

Fig. 4.44 (continued) hydrophobic segment is presented in bold. The putative furin recognition motif RTKR is boxed. The catalytic domain is underlined. The zinc-binding site is shaped, followed by a Met-turn (shaped). The four cysteine residues, known to form intradomain sulfhydryl bridges in astacin, are boxed and bolded. The five cysteine array repeats are italicized and underlined. B, motifs and domain structure of Pf-ALMP. H, hydrophobic segment; RRKR, the potential furin processing signal; CAT, catalytic domain; C, cysteine array

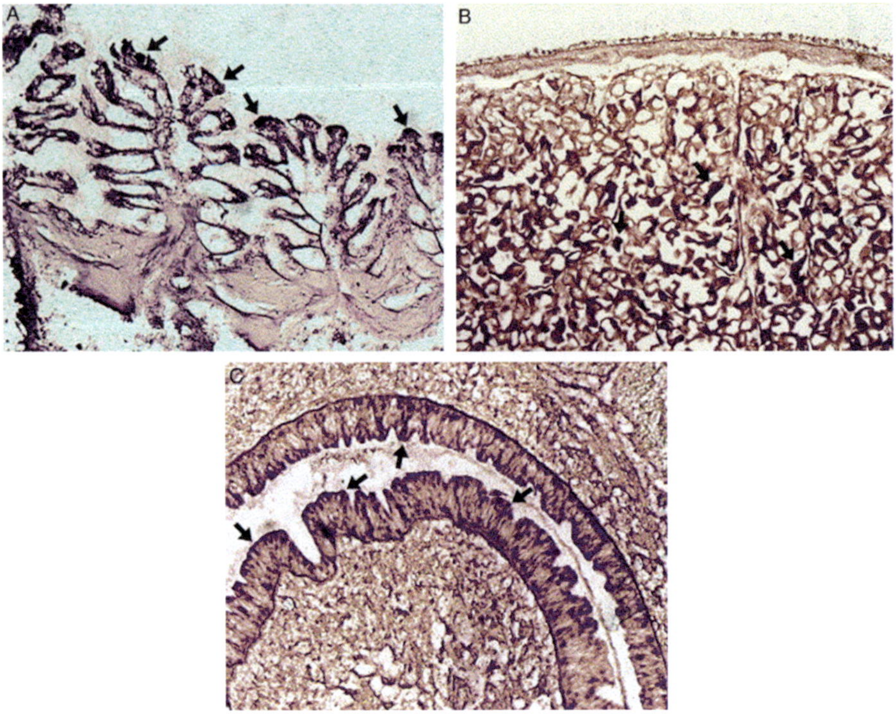

Fig. 4.47 In situ hybridization of Pf-ALMP mRNA in three different tissues of *P. fucata*. In these tissues, hybridization signals (dark purple) were observed in epithelial cells of the gill (**a**), gonad (**b**), and digestive gland (**c**)

pf-ALMP indicated that the pf-ALMP might be involved in hatching and/or embryo development, which remains to be confirmed.

4.7.1.3 Expression and Characterization of Pf-ALMP Active Proteinase

To study the proteolytic activity of pf-ALMP, the pf-ALMP catalytic domain was expressed as a His-tag fusion protein at carboxyl terminus in *E. coli* and purified using Ni–NTA affinity columns. Purified protein as judged by SDS-PAGE shows expected molecular weight and confirmed by Western blotting (Fig. 4.48a, b). The purified fusion protein was tested for its proteinase activity of gelatin. High proteolytic activity was observed with gelatin (Fig. 4.48c), whereas substrate SDS-PAGE performed with the same technique and substrate showed that serum samples had no bands of activity (data not shown).

Studies of Mannello et al. [123] and Ziegler et al. [124] have revealed that proteinases with MMP-like activity are highly expressed in hemocytes of *C. virginica* and *M. galloprovincialis*. Cg-TIMP, which was an inducible tissue

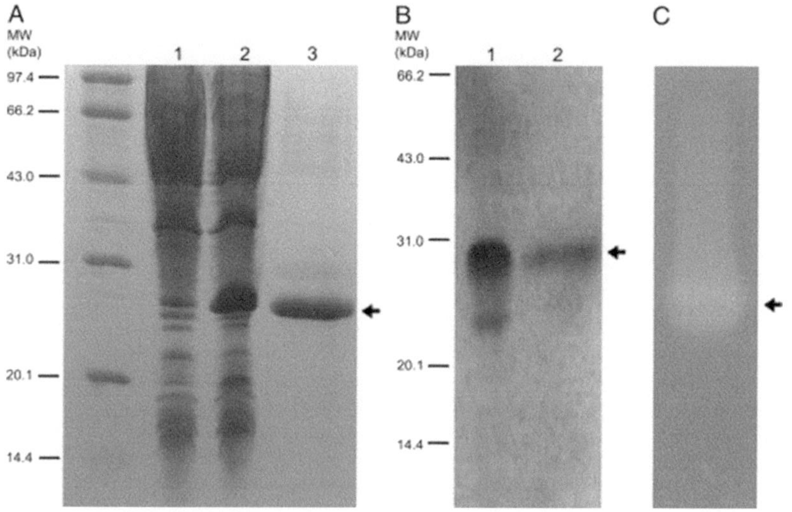

Fig. 4.48 Expression and characterization of active Pf-ALMP catalytic domain. (**a**) Expression and purification of recombinant Pf-ALMP catalytic domain. Lane 1, proteins of BL21 (DE3); lane 2, pET-28b/Pf-ALMP-CAT proteins induced by 0.5 mM IPTG for 3 h; lane 3, recombinant Pf-ALMP catalytic domain purified by Ni–NTA affinity column. (**b**) Western blot analysis of the expression of Pf-ALMP catalytic domain. Lane 1, proteins of BL21 (DE3); lane 2, purified recombinant Pf-ALMP catalytic domain. (**c**) Proteolytic activity of the recombinant Pf-ALMP catalytic domain is visualized by digestion of immobilized gelatin (0.1%) in a 10% SDS-PAGE gel, stained with Coomassie Brilliant Blue

inhibitor of metalloproteinase from the Pacific oyster *C. virginica*, was expressed only in hemocytes of mature and accumulated during shell damage and bacterial attack [125]. Based on these results, we propose that these proteases with MMP-like activity, including pf-ALMP, may be involved in the extracellular matrix deposition, which was a main function in shell repair and development.

4.7.1.4 Pf-ALMP-Enhanced NIH3T3 Cell Proliferation

Recent observations provide compelling evidence that MMP substrates include many non-ECM molecules ranging from binding proteins and growth factor precursor to cell surface adhesion receptors. TGF-α and IGFs, which are the two classes of growth factor precursors that promote cell proliferation, are released and activated by MMPs [126, 127]. To investigate the potential pf-ALMP function in cell proliferation, the plasmid of pcDNA3.1/pf-ALMP was transfected into NIH3T3 cell line. The transient expression of pf-ALMP significantly increased the proliferation of NIH3T3 cell (0.81831 ± 0.05499 vs 0.49794 ± 0.04554, $n = 16$, $p < 0.01$) (Fig. 4.49). In contrast, pcDNA3.1/Pf-ALMPΔCA, expressing a CAs-negative form of the pf-ALMP, inhibited the proliferation of NIH3T3 cell at 48 h post-transfection

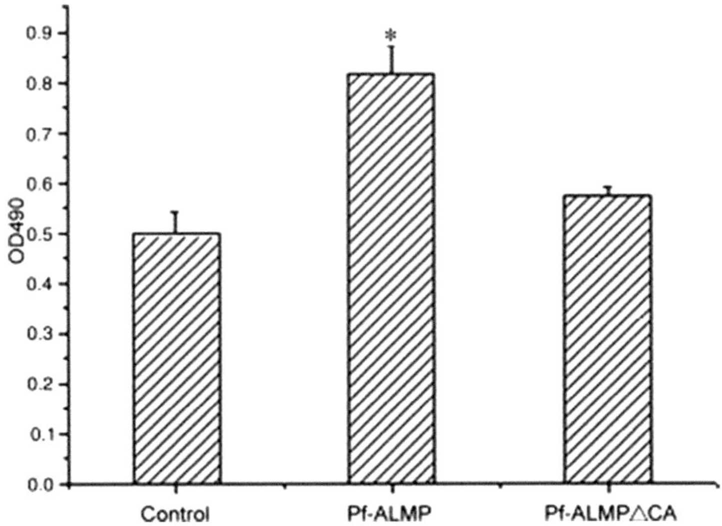

Fig. 4.49 The effect of Pf-ALMP and the CAs-negative form on NIH3T3 cell proliferation. After transfection with control pcDNA3.1 or plasmid-encoding wild type or the deletion mutant of Pf-ALMP for 48 h, the cell viability was detected using MTT assay. The data represent means ± S.D. of 16 experiments with all of the reactions carried out in duplicate. *$p < 0.01$, compared with the control and the deletion mutant

(0.57325 ± 0.01934 vs 0.81831 ± 0.05499, $n = 16$, $p < 0.01$). Though the mechanism was unclear, these results demonstrated that the cysteine arrays were necessary for pf-ALMP to promote cell proliferation and opened up a new way to study the function of astacins.

In summary, we have cloned and characterized pf-ALMP which was the first member of the astacin family identified in *P. fucata*. The enzyme shows similar protease activities as its counterparts, although it has a different C-terminal domain organization. Expression analysis reveals interesting effects of oyster immunity and development process, which require detailed investigation. This study also provides in-depth understanding of the new physiological functions of astacins. Besides, this molecular study might stimulate searches for the corresponding biomineralization processes and morphological structures in *P. fucata* [120].

This work was modified from the papers published by our group in Mar Biotechnol (Xie et al., 2007; 9: 613–623), (Jing et al., 2007; 9: 650–659), in *Biochem Biophys Res Commun* (Zhang et al., 2006; 342: 632–639), in *Comp Biochem Physiol B Biochem Mol Biol* (Yu et al., 2006; 143: 190–194), (Jing et al., 2006; 143: 229–235), in *Biochim Biophys Acta* (Xiong et al., 2006; 1759: 526–534), in *Int J Biochem Cell Biol* (Chen et al., 2005; 37: 1446–1457), in *Journal of Molecular Catalysis B-Enzymatic* (Xiao et al., 2002; 17: 65–74), and in Tsinghua Science and Technology (Chen et al., 2003; 8: 476–480), (Chen et al., 2005; 10: 414–420). The related contents are reused with permission.

References

1. D. Wahler, J.L. Reymond, Novel methods for biocatalyst screening. Curr. Opin. Chem. Biol. **5**, 152–158 (2001)
2. E.E. Kim, H.W. Wyckoff, Reaction mechanism of alkaline phosphatase based on crystal structures. Two-metal ion catalysis. J. Mol. Biol. **218**, 449–464 (1991)
3. S.E. Halford, M.J. Schlesinger, Mutationally altered rate constants in the mechanism of alkaline phosphatase. Biochem. J. **141**, 845–852 (1974)
4. J. Ahlers, The mechanism of hydrolysis of beta-glycerophosphate by kidney alkaline phosphatase. Biochem. J. **149**, 535–546 (1975)
5. J.E. Coleman, Structure and mechanism of alkaline phosphatase. Annu. Rev. Biophys. Biomol. Struct. **21**, 441–483 (1992). https://doi.org/10.1146/annurev.bb.21.060192.002301
6. T.A. Hamilton, S.Z. Gornicki, H.H. Sussman, Alkaline phosphates from human milk. Comparison with isoenzymes from placenta and liver. Biochem. J. **177**, 197–201 (1979)
7. B. Stec, K.M. Holtz, E.R. Kantrowitz, A revised mechanism for the alkaline phosphatase reaction involving three metal ions. J. Mol. Biol. **299**, 1303–1311 (2000). https://doi.org/10.1006/jmbi.2000.3799
8. K.M. Holtz, E.R. Kantrowitz, The mechanism of the alkaline phosphatase reaction: insights from NMR, crystallography and site-specific mutagenesis. FEBS Lett. **462**, 7–11 (1999)
9. M.T. Mazorra, J.A. Rubio, J. Blasco, Acid and alkaline phosphatase activities in the clam Scrobicularia plana: kinetic characteristics and effects of heavy metals. Comp. Biochem. Physiol. B Biochem. Mol. Biol. **131**, 241–249 (2002)
10. M.S. Reddy, K.V. Rao, Methylparathion-induced alterations in the acetylcholinesterase and phosphatases in a penaeid prawn, Metapenaeus monoceros. Bull. Environ. Contam. Toxicol. **45**, 350–357 (1990)
11. Q.X. Chen et al., Kinetics of inhibition of alkaline phosphatase from green crab (Scylla serrata) by N-bromosuccinimide. J. Protein Chem. **15**, 345–350 (1996)
12. M. de Backer et al., The 1.9 A crystal structure of heat-labile shrimp alkaline phosphatase. J. Mol. Biol. **318**, 1265–1274 (2002)
13. M.H. Le Du, T. Stigbrand, M.J. Taussig, A. Menez, E.A. Stura, Crystal structure of alkaline phosphatase from human placenta at 1.8 A resolution. Implication for a substrate specificity. J. Biol. Chem. **276**, 9158–9165 (2001). https://doi.org/10.1074/jbc.M009250200
14. I.W. Nilsen, K. Overbo, R.L. Olsen, Thermolabile alkaline phosphatase from Northern shrimp (Pandalus borealis): protein and cDNA sequence analyses. Comp. Biochem. Physiol. B Biochem. Mol. Biol. **129**, 853–861 (2001)
15. H.E. Stubberud, T.G. Honsi, J. Stenersen, Purification and partial characterisation of tentatively classified acid phosphatase from the earthworm Eisenia veneta. Comp. Biochem. Physiol. B Biochem. Mol. Biol. **126**, 487–494 (2000)
16. H. Janska, A. Kubicz, A. Szalewicz, J. Harazna, The high molecular weight and the low molecular weight acid phosphatases of the frog liver and their phosphotyrosine activity. Comp. Biochem. Physiol. B **90**, 173–178 (1988)
17. F. Panara, A. Angiolillo, R. Pascolini, Acid phosphatases from liver of Rana esculenta. Subcellular localization and partial characterization of multiple forms. Comp. Biochem. Physiol. B **93**, 877–882 (1989)
18. M.P. Cajaraville et al., The use of biomarkers to assess the impact of pollution in coastal environments of the Iberian Peninsula: a practical approach. Sci. Total Environ. **247**, 295–311 (2000)
19. I.L. Tsvetkov, A.P. Popov, A.S. Konichev, Acid phosphatase complex from the freshwater snail Viviparus viviparus L. under standard conditions and intoxication by cadmium ions. Biochemistry (Mosc) **68**, 1327–1334 (2003)
20. G. Jing, L. Li, Y. Li, L. Xie, R. Zhang, Purification and partial characterization of two acid phosphatase forms from pearl oyster (Pinctada fucata). Comp. Biochem. Physiol. B Biochem. Mol. Biol. **143**, 229–235 (2006). https://doi.org/10.1016/j.cbpb.2005.11.008

21. G. Owen, Lysosomes, peroxisomes and bivalves. Sci. Prog. **60**, 299–318 (1972)
22. C.M. Adema, W.P.W. Vanderknaap, T. Sminia, Molluscan hemocyte-mediated cytotoxicity – the role of reactive oxygen intermediates. Rev. Aquat. Sci. **4**, 201–223 (1991)
23. S. Nicholson, Cardiac and branchial physiology associated with copper accumulation and detoxication in the mytilid mussel Perna viridis (L.). J. Exp. Mar. Biol. Ecol. **295**, 157–171 (2003). https://doi.org/10.1016/s0022-0981(03)00292-2
24. A.J. Bune, A.R. Hayman, M.J. Evans, T.M. Cox, Mice lacking tartrate-resistant acid phosphatase (Acp 5) have disordered macrophage inflammatory responses and reduced clearance of the pathogen, Staphylococcus aureus. Immunology **102**, 103–113 (2001)
25. A.R. Hayman et al., Mice lacking tartrate-resistant acid phosphatase (Acp 5) have disrupted endochondral ossification and mild osteopetrosis. Development **122**, 3151–3162 (1996)
26. Y. Yokota, Purification and characterization of particulate acid-phosphatases from eggs of Mediterranean-sea urchins. Comp. Biochem. Physiol. B Biochem. Mol. Biol. **84**, 255–260 (1986). https://doi.org/10.1016/0305-0491(86)90073-8
27. E.J. Houk, J.L. Hardy, Acid phosphatases of the mosquito Culex tarsalis coquillett. Comp. Biochem. Physiol. B **87**, 773–782 (1987)
28. P.W. Pappas, Acid phosphatase activity in the isolated brush border membrane of the tapeworm, Hymenolepis diminuta: partial characterization and differentiation from the alkaline phosphatase activity. J. Cell. Biochem. **37**, 395–403 (1988). https://doi.org/10.1002/jcb.240370407
29. M.I. Feigen, M.A. Johns, J.H. Postlethwait, R.R. Sederoff, Purification and characterization of acid phosphatase-1 from Drosophila melanogaster. J. Biol. Chem. **255**, 10338–10343 (1980)
30. Y. Xia et al., Acid phosphatases in the haemolymph of the desert locust, Schistocerca gregaria, infected with the entomopathogenic fungus Metarhizium anisopliae. J. Insect Physiol. **46**, 1249–1257 (2000). https://doi.org/10.1016/S0022-1910(00)00045-7
31. G. Jing, Z. Yan, Y. Li, L. Xie, R. Zhang, Immunolocalization of an acid phosphatase from pearl oyster (Pinctada fucata) and its in vitro effects on calcium carbonate crystal formation. Mar. Biotechnol. (NY) **9**, 650–659 (2007). https://doi.org/10.1007/s10126-007-9018-0
32. K.M. Wilbur, N.G. Anderson, Carbonic anhydrase and growth in the oyster and busycon. Biol. Bull. **98**, 19–24 (1950). https://doi.org/10.2307/1538594
33. S.A. Nielsen, E. Frieden, Carbonic-anhydrase activity in molluscs. *Comp. Biochem. Physiol.* **41**, 461–468 (1972). https://doi.org/10.1016/0305-0491(72)90107-1
34. L. Duvail, J. Moal, M. Fouchereau-Peron, CGRP-like molecules and carbonic anhydrase activity during the growth of Pecten maximus. Comp. Biochem. Physiol. C **120**, 475–480 (1998). https://doi.org/10.1016/S0742-8413(98)10068-3
35. H. Miyamoto et al., A carbonic anhydrase from the nacreous layer in oyster pearls. P. Natl. Acad. Sci. USA **93**, 9657–9660 (1996). https://doi.org/10.1073/pnas.93.18.9657
36. R.P. Henry, S. Gehnrich, D. Weihrauch, D.W. Towle, Salinity-mediated carbonic anhydrase induction in the gills of the euryhaline green crab, Carcinus maenas. Comp. Biochem. Phys. A **136**, 243–258 (2003). https://doi.org/10.1016/S1095-6433(03)00113-2
37. M. Rousseau et al., Biomineralisation markers during a phase of active growth in Pinctada margaritifera. Comp. Biochem. Phys. A **135**, 271–278 (2003). https://doi.org/10.1016/S1095-6433(03)00070-9
38. H.S. Mason, Oxidases. Annu. Rev. Biochem. **34**, 595–634 (1965). https://doi.org/10.1146/annurev.bi.34.070165.003115
39. J.N. Rodriguezlopez, J. Tudela, R. Varon, F. Garciacarmona, F. Garciacanovas, Analysis of a kinetic-model for melanin biosynthesis pathway. J. Biol. Chem. **267**, 3801–3810 (1992)
40. J.H. Waite, The phylogeny and chemical diversity of quinone-tanned glues and varnishes. Comp. Biochem. Physiol. B **97**, 19–29 (1990)
41. T.L. Hopkins, K.J. Kramer, Insect cuticle sclerotization. Annu. Rev. Entomol. **37**, 273–302 (1992)

42. H. Sugumaran, Comparative biochemistry of eumelanogenesis and the protective roles of phenoloxidase and melanin in insects. Pigment Cell Res. **15**, 2–9 (2002). https://doi.org/10.1034/j.1600-0749.2002.00056.x
43. T. Naraoka et al., Purification, characterization and molecular cloning of tyrosinase from the cephalopod mollusk, Illex argentinus. Eur. J. Biochem. **270**, 4026–4038 (2003). https://doi.org/10.1046/j.1432-1033.2003.03795.x
44. N.A. Ratcliffe, A.F. Rowley, S.W. Fitzgerald, C.P. Rhodes, Invertebrate immunity – basic concepts and recent advances. Int. Rev. Cytol. **97**, 183–350 (1985)
45. E. Vass, A.J. Nappi, Y. Carton, Alterations in the activities of tyrosinase, N-acetyltransferase, and tyrosine aminotransferase in immune reactive larvae of Drosophila-melanogaster. Dev. Comp. Immunol. **17**, 109–118 (1993). https://doi.org/10.1016/0145-305x(93)90021-H
46. R. Asokan, M. Arumugam, P. Mullainadhan, Activation of prophenoloxidase in the plasma and haemocytes of the marine mussel – Perna viridis Linnaeus. Dev. Comp. Immunol. **21**, 1–12 (1997). https://doi.org/10.1016/S0145-305x(97)00004-9
47. K. Soderhall, L. Cerenius, Role of the prophenoloxidase-activating system in invertebrate immunity. Curr. Opin. Immunol. **10**, 23–28 (1998). https://doi.org/10.1016/S0952-7915(98)80026-5
48. C.H. Brown, Some structural proteins of Mytilus-edulis. Q. J. Microsc. Sci. **93**, 487–502 (1952)
49. J.H. Waite, M.L. Tanzer, The bioadhesive of Mytilus Byssus – a protein containing L-Dopa. Biochem. Bioph. Res. Co. **96**, 1554–1561 (1980). https://doi.org/10.1016/0006-291x(80)91351-0
50. F. Eshete, P.T. Loverde, Characteristics of phenol oxidase of Schistosoma-mansoni and its functional implications in eggshell synthesis. J. Parasitol. **79**, 309–317 (1993). https://doi.org/10.2307/3283563
51. G.X. Bai, J.F. Brown, C. Watson, T.P. Yoshino, Isolation and characterization of phenoloxidase from egg masses of the gastropod mollusc, Biomphalaria glabrata. Comp. Biochem. Physiol. B Biochem. Mol. Biol. **118**, 463–469 (1997). https://doi.org/10.1016/S0305-0491(97)00159-4
52. G.M. Jones, A.S.M. Saleuddin, Cellular mechanisms of periostracum formation in Physa spp (Mollusca-Pulmonata). Can. J. Zool. **56**, 2299–2311 (1978). https://doi.org/10.1139/z78-312
53. A. Veis, Biomineralization – cell biology and mineral deposition – Simkiss, K, Wilbur, Km. Science. **247**, 1129–1130 (1990). https://doi.org/10.1126/science.247.4946.1129
54. H. Petit, W.L. Davis, R.G. Jones, H.K. Hagler, Morphological-studies on the calcification process in the fresh-water mussel Amblema. Tissue Cell **12**, 13–28 (1980). https://doi.org/10.1016/0040-8166(80)90049-X
55. A. Checa, A new model for periostracum and shell formation in Unionidae (Bivalvia, Mollusca). Tissue Cell **32**, 405–416 (2000). https://doi.org/10.1054/tice.2000.0129
56. R.S. Cong et al., Purification and characterization of phenoloxidase from clam Ruditapes philippinarum. Fish Shellfish Immun. **18**, 61–70 (2005). https://doi.org/10.1016/j.fsi.2004.06.001
57. L.P. Xie, Y.T. Wu, Y.P. Dai, Q. Li, R.Q. Zhang, A novel glycosylphosphatidylinositol-anchored alkaline phosphatase dwells in the hepatic duct of the pearl oyster, Pinctada fucata. Mar. Biotechnol. (NY) **9**, 613–623 (2007). https://doi.org/10.1007/s10126-007-9015-3
58. G. Geier, R. Zwilling, Cloning and characterization of a cDNA coding for Astacus embryonic astacin, a member of the astacin family of metalloproteases from the crayfish Astacus astacus. Eur. J. Biochem. **253**, 796–803 (1998). https://doi.org/10.1046/j.1432-1327.1998.2530796.x
59. M.Z. Kounnas, R.L. Wolz, C.M. Gorbea, J.S. Bond, Meprin-a and Meprin-B – cell-surface endopeptidases of the mouse kidney. J. Biol. Chem. **266**, 17350–17357 (1991)
60. J.M. Wozney et al., Novel regulators of bone-formation – molecular clones and activities. Science **242**, 1528–1534 (1988). https://doi.org/10.1126/science.3201241
61. G.F.Z. da Silva, R.L. Reuille, L.J. Ming, B.T. Livingston, Overexpression and mechanistic characterization of blastula protease 10, a metalloprotease involved in sea urchin

embryogenesis and development. J. Biol. Chem. **281**, 10737–10744 (2006). https://doi.org/10.1074/jbc.M510707200
62. M.J. Shimell, E.L. Ferguson, S.R. Childs, M.B. Oconnor, The Drosophila dorsal-ventral patterning gene tolloid is related to human bone morphogenetic protein-1. Cell **67**, 469–481 (1991). https://doi.org/10.1016/0092-8674(91)90522-Z
63. J.L. Quinones, R. Rosa, D.L. Ruiz, J.E. Garcia-Arraras, Extracellular matrix remodeling and metalloproteinase involvement during intestine regeneration in the sea cucumber Holothuria glaberrima. Dev. Biol. **250**, 181–197 (2002). https://doi.org/10.1006/dbio.2002.0778
64. A.L. Williamson et al., Ancylostoma caninum MTP-1, an astacin-like metalloprotease secreted by infective hookworm larvae, is involved in tissue migration. Infect. Immun. **74**, 961–967 (2006). https://doi.org/10.1128/Iai.74.2.961-967.2006
65. E. Dumermuth, J.A. Eldering, J. Grunberg, W. Jiang, E.E. Sterchi, Cloning of the PABA peptide hydrolase alpha subunit (PPH alpha) from human small intestine and its expression in COS-1 cells. FEBS Lett. **335**, 367–375 (1993)
66. S. Yasumasu et al., Isolation of cDNAs for LCE and HCE, two constituent proteases of the hatching enzyme of Oryzias latipes, and concurrent expression of their mRNAs during development. Dev. Biol. **153**, 250–258 (1992)
67. S. Yasumasu et al., Different exon-intron organizations of the genes for two astacin-like proteases, high choriolytic enzyme (choriolysin H) and low choriolytic enzyme (choriolysin L), the constituents of the fish hatching enzyme. Eur. J. Biochem. **237**, 752–758 (1996)
68. S. Yasumasu, K.M. Mao, F. Sultana, H. Sakaguchi, N. Yoshizaki, Cloning of a quail homologue of hatching enzyme: its conserved function and additional function in egg envelope digestion. Dev. Genes Evol. **215**, 489–498 (2005). https://doi.org/10.1007/s00427-005-0007-x
69. F.H. Shen, Q.L. Feng, C.M. Wang, The modulation of collagen on crystal morphology of calcium carbonate. J. Cryst. Growth **242**, 239–244 (2002). https://doi.org/10.1016/S0022-0248(02)01376-3
70. A. Serpentini et al., Collagen study and regulation of the de novo synthesis by IGF-I in hemocytes from the gastropod mollusc, Haliotis tuberculata. J. Exp. Zool. **287**, 275–284 (2000). https://doi.org/10.1002/1097-010x(20000901)287:4<275::Aid-Jez2>3.0.Co;2-8
71. J.M. Poncet et al., In vitro synthesis of proteoglycans and collagen in primary cultures of mantle cells from the nacreous mollusk, Haliotis tuberculata: a new model for study of molluscan extracellular matrix. Mar. Biotechnol. **2**, 387–398 (2000)
72. S. Blank et al., The nacre protein perlucin nucleates growth of calcium carbonate crystals. J. Microsc-Oxford **212**, 280–291 (2003). https://doi.org/10.1111/j.1365-2818.2003.01263.x
73. A.S. Mount, A.P. Wheeler, R.P. Paradkar, D. Snider, Hemocyte-mediated shell mineralization in the eastern oyster. Science **304**, 297–300 (2004). https://doi.org/10.1126/science.1090506
74. F.H. Wilt, Developmental biology meets materials science: morphogenesis of biomineralized structures. Dev. Biol. **280**, 15–25 (2005). https://doi.org/10.1016/j.ydbio.2005.01.019
75. J.D. Bendtsen, H. Nielsen, G. von Heijne, S. Brunak, Improved prediction of signal peptides: signalP 3.0. J. Mol. Biol. **340**, 783–795 (2004). https://doi.org/10.1016/j.jmb.2004.05.028
76. J.D. Thompson, T.J. Gibson, F. Plewniak, F. Jeanmougin, D.G. Higgins, The CLUSTAL_X windows interface: flexible strategies for multiple sequence alignment aided by quality analysis tools. Nucleic Acids Res. **25**, 4876–4882 (1997). https://doi.org/10.1093/nar/25.24.4876
77. W.Q. Huang et al., Immunohistochemical and in situ hybridization studies of gonadotropin releasing hormone (GnRH) and its receptor in rat digestive tract. Life Sci. **68**, 1727–1734 (2001)
78. R. Xiao et al., Purification and enzymatic characterization of alkaline phosphatase from Pinctada fucata. J. Mol. Catal. B-Enzym. **17**, 65–74 (2002). https://doi.org/10.1016/S1381-1177(02)00007-3

79. J.M. Lucas, L.W. Knapp, Biochemical characterization of purified carbonic anhydrase from the octocoral Leptogorgia virgulata. Mar. Biol. **126**, 471–477 (1996). https://doi.org/10.1007/Bf00354629
80. S. Reyda, E. Jacob, R. Zwilling, W. Stocker, cDNA cloning, bacterial expression, in vitro renaturation and affinity purification of the zinc endopeptidase astacin. Biochem. J. **344**(Pt 3), 851–857 (1999). https://doi.org/10.1042/0264-6021:3440851
81. U.K. Laemmli, Cleavage of structural proteins during the assembly of the head of bacteriophage T4. Nature **227**, 680–685 (1970)
82. J. Sambrook, F. Fritsch, T. Maniatis, *Molecular Cloning: A Laboratory Manual* (Science Press, Beijing, 1989, in Chinese)
83. M.M. Bradford, A rapid and sensitive method for the quantitation of microgram quantities of protein utilizing the principle of protein-dye binding. Anal. Biochem. **72**, 248–254 (1976)
84. J. ChandraRajan, L. Klein, Determination of inorganic phosphorus in the presence of organic phosphorus and high concentrations of proteins. Anal. Biochem. **72**, 407–412 (1976)
85. F. Marin et al., Caspartin and calprismin, two proteins of the shell calcitic prisms of the Mediterranean fan mussel Pinna nobilis. J. Biol. Chem. **280**, 33895–33908 (2005). https://doi.org/10.1074/jbc.M506526200
86. S. Li, L. Xie, Z. Ma, R. Zhang, cDNA cloning and characterization of a novel calmodulin-like protein from pearl oyster Pinctada fucata. FEBS J. **272**, 4899–4910 (2005). https://doi.org/10.1111/j.1742-4658.2005.04899.x
87. C.J. Coffee, R.A. Bradshaw, B.R. Goldin, C. Frieden, Identification of the sites of modification of bovine liver glutamate dehydrogenase reacted with trinitrobenzenesulfonate. Biochemistry **10**, 3516–3526 (1971)
88. A. Kotaki, M. Harada, K. Yagi, Reaction between Sulfhydryl Compounds and 2,4,6-Trinitrobenzene-1-Sulfonic Acid. J. Biochem. **55**, 553–561 (1964)
89. H.T. Chen, L.P. Xie, Z.Y. Yu, G.R. Xu, R.Q. Zhang, Chemical modification studies on alkaline phosphatase from pearl oyster (Pinctada fucata): a substrate reaction course analysis and involvement of essential arginine and lysine residues at the active site. Int. J. Biochem. Cell Biol. **37**, 1446–1457 (2005). https://doi.org/10.1016/j.biocel.2005.02.002
90. U. Bhattacharyya, G. Dhar, A. Bhaduri, An arginine residue is essential for stretching and binding of the substrate on UDP-glucose-4-epimerase from Escherichia coli. Use of a stacked and quenched uridine nucleotide fluorophore as probe. J. Biol. Chem. **274**, 14573–14578 (1999)
91. J. Kaminska, A. Wisniewska, J. Koscielak, Chemical modifications of alpha 1,6-fucosyltransferase define amino acid residues of catalytic importance. Biochimie **85**, 303–310 (2003). https://doi.org/10.1016/s0300-9084(03)00074-9
92. A. Garen, C. Levinthal, A fine-structure genetic and chemical study of the enzyme alkaline phosphatase of e-coli .1. Purification and characterization of alkaline phosphatase. Biochim. Biophys. Acta **38**, 470–483 (1960). https://doi.org/10.1016/0006-3002(60)91282-8
93. H.M. Levy, P.D. Leber, E.M. Ryan, Inactivation of myosin by 2,4-dinitrophenol and protection by adenosine triphosphate and other phosphate compounds. J. Biol. Chem. **238**, 3654–3659 (1963)
94. W.Z. Zheng et al., An essential tryptophan residue of green crab (syclla serrata) alkaline phosphatase. Biochem. Mol. Biol. Int. **41**, 951–959 (1997)
95. C. Zhang, S. Li, Z.J. Ma, L.P. Xie, R.Q. Zhang, A novel matrix protein p10 from the nacre of pearl oyster (Pinctada fucata) and its effects on both $CaCO_3$ crystal formation and mineralogenic cells. Mar. Biotechnol. **8**, 624–633 (2006). https://doi.org/10.1007/s10126-006-6037-1
96. G. Fu, S. Valiyaveettil, B. Wopenka, D.E. Morse, CaCO3 biomineralization: acidic 8-kDa proteins isolated from aragonitic abalone shell nacre can specifically modify calcite crystal morphology. Biomacromolecules **6**, 1289–1298 (2005). https://doi.org/10.1021/bm049314v
97. R.E. Tashian, The carbonic anhydrases – widening perspectives on their evolution, expression and function. Bioessays **10**, 186–192 (1989). https://doi.org/10.1002/bies.950100603

98. Z. Yu, L. Xie, S. Lee, R. Zhang, A novel carbonic anhydrase from the mantle of the pearl oyster (Pinctada fucata). Comp. Biochem. Physiol. B Biochem. Mol. Biol. **143**, 190–194 (2006). https://doi.org/10.1016/j.cbpb.2005.11.006
99. T. Miyashita, R. Takagi, H. Miyamoto, A. Matsushiro, Identical carbonic anhydrase contributes to nacreous or prismatic layer formation in Pinctada fucata (Mollusca : Bivalvia). Veliger **45**, 250–255 (2002)
100. R.F. Chen, J.C. Kernohan, Combination of bovine carbonic anhydrase with a fluorescent sulfonamide. J. Biol. Chem. **242**, 5813–5823 (1967)
101. R. Dermietzel, A. Leibstein, W. Siffert, N. Zamboglou, G. Gros, A fast screening method for histochemical-localization of carbonic-anhydrase – application to kidney, skeletal-muscle, and thrombocytes. J. Histochem. Cytochem. **33**, 93–98 (1985)
102. J.A. Freeman, K.M. Wilbur, Carbonic anhydrase in molluscs. Biol. Bull. **94**, 55–59 (1948). https://doi.org/10.2307/1538209
103. C. Zhang, L. Xie, J. Huang, L. Chen, R. Zhang, A novel putative tyrosinase involved in periostracum formation from the pearl oyster (Pinctada fucata). Biochem. Biophys. Res. Commun. **342**, 632–639 (2006). https://doi.org/10.1016/j.bbrc.2006.01.182
104. E.T. Degens, D.W. Spencer, R.H. Parker, Paleobiochemistry of molluscan shell proteins. Comp. Biochem. Physiol. **20**, 553–579 (1967). https://doi.org/10.1016/0010-406x(67)90269-1
105. M.E. Cuff, K.I. Miller, K.E. van Holde, W.A. Hendrickson, Crystal structure of a functional unit from Octopus hemocyanin. J. Mol. Biol. **278**, 855–870 (1998). https://doi.org/10.1006/jmbi.1998.1647
106. H. Decker, F. Tuczek, Tyrosinase/catecholoxidase activity of hemocyanins: structural basis and molecular mechanism. Trends Biochem. Sci. **25**, 392–397 (2000). https://doi.org/10.1016/S0968-0004(00)01602-9
107. H. Decker, N. Terwilliger, Cops and robbers: putative evolution of copper oxygen-binding proteins. J. Exp. Biol. **203**, 1777–1782 (2000)
108. K.E. van Holde, K.I. Miller, H. Decker, Hemocyanins and invertebrate evolution. J. Biol. Chem. **276**, 15563–15566 (2001). https://doi.org/10.1074/jbc.R100010200
109. C. Gielens et al., Evidence for a cysteine-histidine thioether bridge in functional units of molluscan haemocyanins and location of the disulfide bridges in functional units d and g of the beta(c)-haemocyanin of Helix pomatia. Eur. J. Biochem. **248**, 879–888 (1997). https://doi.org/10.1111/j.1432-1033.1997.00879.x
110. U. Kupper, D.M. Niedermann, G. Travaglini, K. Lerch, Isolation and characterization of the tyrosinase gene from Neurospora-crassa. J. Biol. Chem. **264**, 17250–17258 (1989)
111. Y. Fujita, Y. Uraga, E. Ichisima, Molecular-cloning and nucleotide-sequence of the protyrosinase gene, melo, from Aspergillus-oryzae and expression of the gene in yeast-cells. Bba-Gene Struct. Expr. **1261**, 151–154 (1995). https://doi.org/10.1016/0167-4781(95)00011-5
112. J. Gordon, M.R. Carriker, Sclerotized protein in the shell matrix of a bivalve mollusk. Mar. Biol. **57**, 251–260 (1980). https://doi.org/10.1007/Bf00387568
113. S. Sudo et al., Structures of mollusc shell framework proteins. Nature **387**, 563–564 (1997). https://doi.org/10.1038/42391
114. Y. Zhang et al., A novel matrix protein participating in the nacre framework formation of pearl oyster, Pinctada fucata. Comp. Biochem. Physiol. B Biochem. Mol. Biol. **135**, 565–573 (2003)
115. G. Bevelander, H. Nakahara, An electron microscope study of the formation of the periostracum of Macrocallista maculata. Calcif. Tissue Res. **1**, 55–67 (1967)
116. J.M. Neff, Ultrastructural studies of periostracum formation in the hard shelled clam Mercenaria mercenaria (L). Tissue Cell **4**, 311–326 (1972)
117. C.A. Richardson, N.W. Runham, D.J. Crisp, A histological and ultrastructural study of the cells of the mantle edge of a marine bivalve, Cerastoderma edule. Tissue Cell **13**, 715–730 (1981)

118. M. Suzuki et al., Characterization of Prismalin-14, a novel matrix protein from the prismatic layer of the Japanese pearl oyster (Pinctada fucata). Biochem. J. **382**, 205–213 (2004). https://doi.org/10.1042/BJ20040319
119. H. Miyamoto, F. Miyoshi, J. Kohno, The carbonic anhydrase domain protein nacrein is expressed in the epithelial cells of the mantle and acts as a negative regulator in calcification in the mollusc Pinctada fucata. Zool. Sci. **22**, 311–315 (2005). https://doi.org/10.2108/zsj.22.311
120. X. Xiong, L. Chen, Y. Li, L. Xie, R. Zhang, Pf-ALMP, a novel astacin-like metalloproteinase with cysteine arrays, is abundant in hemocytes of pearl oyster Pinctada fucata. Biochim. Biophys. Acta **1759**, 526–534 (2006). https://doi.org/10.1016/j.bbaexp.2006.09.006
121. F.J. Genthner, A.K. Volety, L.M. Oliver, W.S. Fisher, Factors influencing in vitro killing of bacteria by hemocytes of the eastern oyster (Crassostrea virginica). Appl. Environ. Microbiol. **65**, 3015–3020 (1999)
122. A.K. Volety, W.S. Fisher, In vitro killing of Perkinsus marinus by hemocytes of oysters Crassostrea virginica. J. Shellfish Res. **19**, 827–834 (2000)
123. F. Mannello, L. Canesi, G. Gazzanelli, G. Gallo, Biochemical properties of metalloproteinases from the hemolymph of the mussel Mytilus galloprovincialis Lam. Comp. Biochem. Physiol. B Biochem. Mol. Biol. **128**, 507–515 (2001). https://doi.org/10.1016/S1096-4959(00)00352-3
124. G. Ziegler, K. Paynter, D. Fisher, Matrix metalloproteinase-like activity from hemocytes of the eastern oyster, Crassostrea virginica. Comp. Biochem. Physiol. B Biochem. Mol. Biol. **131**, 361–370 (2002). https://doi.org/10.1016/S1095-6433(01)00518-9
125. C. Montagnani, F. Le Roux, F. Berthe, J.M. Escoubas, Cg-TIMP, an inducible tissue inhibitor of metalloproteinase from the Pacific oyster Crassostrea gigas with a potential role in wound healing and defense mechanisms. FEBS Lett. **500**, 64–70 (2001). https://doi.org/10.1016/S0014-5793(01)02559-5
126. S. Manes et al., Identification of insulin-like growth factor-binding protein-1 as a potential physiological substrate for human stromelysin-3. J. Biol. Chem. **272**, 25706–25712 (1997). https://doi.org/10.1074/jbc.272.41.25706
127. S. Manes et al., The matrix metalloproteinase-9 regulates the insulin-like growth factor-triggered autocrine response in DU-145 carcinoma cells. J. Biol. Chem. **274**, 6935–6945 (1999). https://doi.org/10.1074/jbc.274.11.6935

Chapter 5
Cellular Regulation of Biomineralization in *Pinctada fucata*

Abstract Because it is widely recognized that the mantle tissue controls nacre precipitation, intensive studies have been conducted to unravel the regulatory mechanism underlying nacre formation. In this chapter, we will focus on the cellular regulation of shell formation in the pearl oyster *Pinctada fucata*. First, the morphology and proliferation rate in different parts of the mantle tissue have been investigated, and a proliferation hot spot located in the center of the thinnest mantle region was found. However, mantle tissues mainly composed of the mantle edge were used for primary cell culture, due to their high yield of migrated mantle cells. The primary mantle cell culture is used to study how the mantle cells regulate calcium carbonate precipitation. It was found that the physiological functions of the mantle cells were maintained in vitro. High expression of many shell matrix proteins, including ACCBP, Pif80, and nacrein, and high activities of carbonic anhydrase and alkaline phosphatase were detected. Numerous crystals were found inside the cultured cells by polar light microscopy and scanning electron microscopy, and FTIR and XRD analysis demonstrated that these particles were amorphous calcium carbonate (ACC). What's more, the cultured mantle cells promoted and regulated calcium carbonate precipitation in the culture dishes. These results showed that mantle cells may directly participate in shell formation. In addition, we have studied the function of hemocytes in shell formation. Hemocytes were identified from *P. fucata* and found to be present in the extrapallial space (EPS). Many components involved in immunity and calcification were identified by proteomics analysis. Poststimulation of lipopolysaccharide and shell damage, most of the tested immune genes and calcification, was upregulated. Moreover, polar light microscopy, scanning electron microscopy, and energy spectrum showed that some hemocytes carried crystals of calcium carbonate, indicating they participated in both immunity and biomineralization. Our studies elucidated the vital roles of mantle cells and hemocytes in regulating shell growth and nacre formation, which would shed light on the improvement of the cultured pearl quality.

Keywords Biomineralization · Mantle cell · Hemocyte · Nacre · Cell culture

R. Zhang et al., *Biomineralization Mechanism of the Pearl Oyster, Pinctada fucata*,
https://doi.org/10.1007/978-981-13-1459-9_5

5.1 Introduction

According to the classical theory, mantle tissue dominates shell formation process in *P. fucata*. It secretes organic matrix into the extrapallial space (EPS) and controls the ion composition in the EPS for shell growth [1]. Histological and cytological studies have revealed that the mantle tissue controls the secretion of organic matrix with remarkable precision and regularity [2, 3]. Many shell matrix protein genes are specifically expressed in the outer mantle epithelium facing the shell [4]. Generally, the pallial zone of the outer mantle epithelium secretes matrix proteins that are involved in nacreous layers formation, and the edge of the outer mantle epithelium is always responsible for the formation of prismatic layers [4, 5]. In the contrary, the central part of mantle expresses low amounts of matrix proteins [4] and has a high capability to proliferate. However, the mechanism underlying the cellular control of the outer epithelium on shell formation (i.e., the secretion of matrix proteins) is not well understood. As suggested by the progress made by using in vitro models of bone and coral mineralization, long-term primary cell cultures are innovative tools to explore mineralization at the cellular level [6–8]. Accordingly, the nature of mantle cells involved in the regulation of biomineralization and the mechanisms of their cooperation can be investigated, so that insight into shell formation may be obtained from mantle cell cultures [9].

However, one obstacle hindering the in vitro investigation of cellular regulation in biomineralization is the lack of established cell line for marine invertebrates, and it's difficult to subculture primary dissociated single cells [10, 11]. Nevertheless, some progresses in in vitro study of biomineralization have been made in certain species by using primary mantle cell culture [12] and mantle explant culture [13–15]. Explant cultures have been proved to facilitate long-term cell survival. These studies focused on long-term observation and measurement of the natures of mixed cell populations from the mantle explants. Yet the features of various cell types in vitro and their cellular metabolism in mineralization are not well studied. To further explore the cellular mechanisms of biomineralization, it needs successful cultures of single cell type rather than of a mixed cell population.

Deposition of CaCO3 crystals has been reported in mantle tissue cultures from *Pinctada fucata* [16], suggesting that in vitro cultured cells maintain their abilities with regard to biomineralization. Moreover, EDS analysis confirmed the $CaCO_3$ nature of the deposits and their biogenic origin, and the expression and secretion of some SMPs have been detected in mantle explant cultures [9]. However, the polymorph of calcium carbonate deposits in mantle tissue culture and the cellular regulation of crystal formation remain unknown. Considering the remarkable nacre structure and the central role of the mantle cells in shell formation, the application of cellular biological tools is critical for further in-depth investigation of shell formation mechanism.

In addition to the classical mantle-dominating shell formation theory, the hemocytes-mediating mineralization pathway emerged. Mount et al. [17] found that in eastern oysters, some granulocytes of the hemocytes carried and transported calcium to the mineralization front in crystal form. And these $CaCO_3$ crystals were reconstructed to form the shell layers. Similarly, in *Pinctada fucata*, hemocytes were proposed to be involved in calcium transportation, and some cellular proteins identified by proteomic analysis were related to calcification [18]. Crystal-bearing hemocytes were reported in several bivalves, though the composition of the crystals varied. The role of hemocytes serving as an additional calcium pool is plausible. However, whether or not could hemocytes mediate shell formation independently remains unclear. And the mechanism underlying the precipitation of intracellular CaCO3 in hemocytes is far from understood.

In this part, we will summarize our works on developing long-term primary cell culture of mantle tissue and its application in exploring the cellular regulation of shell formation in *P. fucata*. The role of hemocytes in shell formation will also be discussed.

5.2 Materials and Methods

5.2.1 Chemicals

Ethylene diamine tetra acetic acid (EDTA), glutaraldehyde, and paraformaldehyde (PFA, electron microscope grade) were Sigma products. Adenosine triphosphate (ATP), BrdU, Percol, HEPES, and BSA were supplied by Pharmacia. Other chemicals, such as alcohol, methanol, H_2O_2, NaCl, $MgCl_2$, $CaCl_2$, $SrCl_2$, NaOH, HCl, Na_2CO_3, and Na_2SO_4 were local products of analytical grade.

5.2.2 Primers

Refer to Supplementary Table 5.1 for details.

5.2.3 Histochemistry of Mantle Tissue

Alcian blue/PAS cytochemistry method was applied to stain the mantle tissue paraffin sections and study the different secretory activities of four regions in the outer mantle epithelium. The mantle tissues were dissected and fixed in Carnoy's solution. After embedded in paraffin, the fixed mantles were subjected to routine paraffin sectioning. Alcian blue/PAS staining was performed to detect the secretory activities. Different outer epithelia could be identified by the colors: red color indicated neutral mucopolysaccharide secreting outer epithelia; blue color indicated acid mucopolysaccharide secreting outer epithelia. Those cell groups including both neutral and acid mucopolysaccharide secreting outer epithelia would be amethyst-stained.

5.2.4 TEM and SEM Observation of Mantle Tissue

Small mantle fragments of four different regions were immersed in 0.1 M phosphate buffer saline (PBS, pH 7.4) at 20 °C for 12 h, containing 1.5% glutaraldehyde and 3.4% paraformaldehyde. After thorough washing with 0.1 M PBS, the mantle pieces were subjected to a graded series dehydration of alcohol. These tissues were embedded with Epon812 at 60 °C for 48 h and cut using Leica Ultracut microtome. Then the 65 nm ultrathin sections were counter-stained with lead citrate and uranyl acetate. Finally, the samples were observed using transmission electron microscope (Philips CM120 Biotwin) at 100 kV.

To characterize the mantle surface, *Pinctada fucata* was sacrificed using sterilized surgical blade, and the mantle margin and center were immediately taken and fixed in the mixture of 1% glutaraldehyde and 4% paraformaldehyde for 24 h at 4 °C. Then dehydration was performed in a graded ethanol series (10%, 25%, 50%, 75%, 90%, and 100%), and CO_2 was used as intermediate to dry the samples in a critical point dryer (CPD 030, Leica, Microsystems, Buffalo Grove, IL, USA). The dried samples were coated with gold for 60 s in a sputter coater (SCD 050, Bal-Tec AG, Balzers, Liechtenstein), and observed with SEM (Quanta 200, FEI company, Eindhoven, Netherland) with an accelerating voltage of 15 kV. Mantle samples from 30 individuals were observed in SEM.

5.2.5 BrdU Labeling of Mantle Epithelia

The pearl oyster *P. fucata* with an average shell length of 55–60 mm were kept in artificial sea water as described above. In this experiment, the seawater was carefully sterilized using UV light and filtered with a 0.22 μm membrane. The seawater was not changed throughout the experiment. Then BrdU exposure was conducted for 12 h. We used the concentration of BrdU according to Zaldibar et al. [19] Initially, a pulse in 4 mg BrdU/L seawater was performed, and then additional 2 mg BrdU/L seawater was added every 4 h so that the continuous effectivity of BrdU was maintained but not reaching cytotoxic levels. Slices of the four mantle zones were cut off separately and fixed in 4% PFA at 4 °C for 8 h and then dehydrated in 22.5% sucrose solution for 10 h. Frozen sections (8 μm thick) were cut in a Leica cryo-ultramicrotome. The sections were placed on polylysine-coated slides. Later, the sections were washed with PBS.

In order to eliminate endogenous peroxidase activity, the sections were immersed in a 30% H_2O_2–methanol solution (1:50) bath at room temperature for 30 min and then rinsed in PBS. DNA was denatured using 2 N HCl at 37 °C for 30 min. The sections were treated with pepsin in 1 ml 3% citric acid solution for 30 s. After rinsed with PBS, the sections were incubated in 5% BSA blocking serum supplied in an SABC Kit (Boster, Wuhan, P. R. China) at RT for 20 min. After washing off the excessive BSA, the specific antibody for BrdU (Roche; 1:50 in PBS with 0.1% BSA and 0.1% Triton-X 100) was added, and the sections were incubated at 37 °C for 1 h. Later, the solution was replaced with the biotinylated goat anti-rat IgG secondary antibody, and the slides were kept at room temperature.

To avoid false-positive signals of the pigment in the mantle margin, the negative control experiment was performed using the developed DAB method. The experiment procedure of the control was following the abovementioned steps, except the incubation of primary antibody.

In order to quantify the BrdU-positive cells, the proportion of BrdU-positive cells was counted in 10 vision fields containing 100 cells each, based on the previous studies of Okudela et al. [20] and Zaldibar et al. [19]

5.2.6 Changes of Mantle Cell Proliferation During the Tidal Cycle

BrdU was applied to pearl oysters for 2 days with the same pulse addition described above. Sample sections were achieved at different intervals and treated using the same method as applied previously in the outer epithelia proliferation rate analysis and quantification. The proportion of BrdU-positive outer epithelia of different regions at different times was calculated by counting the outer epithelia from 10 vision fields with 100 cells in each vision field.

5.2.7 Primary Cell Culture of Mantle Tissue

5.2.7.1 Culture Medium

Pf-CM2.5 medium for *P. fucata* primary cell culture was prepared as follows: an equal volume of 2× Medium 199 (Gibco, Invitrogen, USA) and 2× Leibovitz-15 medium was mixed and complemented with 128.9 μg/ml taurine, 170.6 mM NaCl, 40 μg/ml ascorbic acid, 3.57 mg/ml HEPES with antibiotics, 10 μg/ml ATP, 400 μg/ml lactalbumin hydrolysate (Hyclone, Utah, USA), 50 μg/ml kanamycin, 1 μg/ml nystatin, 100 μg/ml streptomycin, and 100 IU/ml penicillin. The mixture was filtered through 0.22 μm Millipore membrane. This resulted medium contains 2.5 mM Ca^{2+} and 2.5 mM Mg^{2+} with pH 7.6.

5.2.7.2 Tissue Operation

The oyster was sacrificed to cut off the mantle tissue under aseptic conditions. Mantle folds were removed, and the mantle pallial epithelia were used for cell culture. The tissues were sterilized for 20 min in an antibiotic solution containing 100 IU/ml gentamicin, 2 μg/ml nystatin, 0.5 mg/ml streptomycin, and 500 IU/ml penicillin in molluskan balanced salt solution (MBSS, Machii and Wada [21]). After washed eight times with D-MBSS (MBSS without Ca^{2+} and Mg^{2+}), the fragments were minced into pieces (2 mm × 2 mm). Ten pieces of tissue fragments were placed in a 25 cm^2 flask (Greiner Bio-One, Germany) coated with rat-tail collagen. Culture medium (Pf-CM2.5 with 10% fetal bovine serum) was added to the explants. The tissues were maintained in a 25 °C incubator without CO_2, and an additional 200 μL aliquot of culture medium was added every 3 days. Cells migrating from the mantle explants were collected after 7 days.

Gonadal cells were collected from gonads using a sterile syringe and immersed in disinfecting solution for 30 min. After rinsed in D-MBSS for several times, the gonadal cells were cultured in the same medium as mentioned above for mantle tissues.

The cell viability was measured by using the trypan blue dye exclusion test (Phillips 1973). Cellular metabolic activity was determined using the MTT reduction assay [22]. The cell cycle was detected using flow cytometry (BD FACSCalibur, USA).

5.2.8 Density Gradient Centrifugation of Mantle Cells

A discontinuous Percoll (Amersham Bioscience) gradient was applied to separate the migrated mantle cells collected from 4-day-old mantle explant cultures according to their densities [23]. A 15 ml centrifuge tube was prepared by treating with 1% bovine serum albumin (BSA; in water, w/v) for 15 min. Percoll was diluted into 60%, 36%, 30%, and 15% (vol/vol) with D-MBSS (respective densities: 1.055, 1.042, 1.039, and 1.029 g/ml), and layered in the centrifuge tube in the order of density. One milliliter of dissociated mantle cells (approximately 2×10^7 cells) was loaded carefully on top of the gradient. Centrifugation was performed first at 1000 rpm for 5 min and subsequent continued at 1500 rpm for 10 min. A fine pipette was used to collect the separated cells at different gradient interfaces between 15% to 30% (15%F), 30% to 36% (30%F), and 36% to 60% (36%F). After being rinsed in basic medium, 2×10^6 cells from the various fractions were transferred to12-well plates (Greiner Bio-One, Germany) and cultured for 3 days after adding 0.5 ml growth medium. The culture media were subjected to ELISA assays. To explore the effect of calcium on nacrein secretion, the cells of 36%F were cultured with the same techniques but grew in media containing calcium of 2.5 mM, 6 mM, 11.5 mM, or 14 mM. Each data point contained three independent cultures for assay.

The cells in 36% F were sampled for observation using a scanning electron microscope (FEI Quanta 200 FEG) and transmission electron microscope (Philips CM120 Biotwin) according to routine protocols described in Sect. 5.2.2.

5.2.9 Immunolabeling Analysis of Epithelial Cells

Cells from 15%F, 30%F, and 36%F were subjected to immunolabeling analysis based on the method previously described by Puverel et al. [24] Because pancytokeratin was regarded as a cellular marker of epithelial cells, we used mixed monoclonal antibodies to pancytokeratin (anti-pancytokeratin, 1:100; Boster, China) as primary antibodies to characterize the epithelial cells. Fluorescein isothiocyanate (FITC)-conjugated goat anti-mouse antibodies (Santa Cruz, USA) were applied as secondary antibodies. Finally, 4′,6-diamidino-2-phenylindole (DAPI; Sigma- Aldrich) were used to visualize the nuclear of the cells, and the cells were observed under a fluorescence microscope (Leica DIMR).

5.2.10 RNA Extraction and RT-PCR

RNA extraction and RT-PCR from mantle tissue and hemocytes were operated as described above.

Three genes encoding shell matrix proteins (nacrein, Pif80, and ACCBP) were detected, and the housekeeping gene actin was chosen as the reference. The primer sequences for each gene are listed in Supplementary Table 5.1.

5.2.11 Alkaline Phosphatase and Carbonic Anhydrase Activities

The cultured mantle cells were rinsed twice in D-MBSS and sonicated for 30 s to detect the activity of alkaline phosphatase (ALP) and carbonic anhydrase (CA). The cellular protein concentration was determined using the bicinchoninic acid assay (BCA, Pierce, Sigma), with bovine serum albumine (BSA) as a standard. The ALP activity was detected according to Sud et al. [12] The cell lysates were incubated at 37 °C for 1 h in 0.2 M bicarbonate buffer (pH 10.0) containing 4 mM MgCl2 and 2 mM p-nitrophenolphosphate. NaOH (1 M) was added to stop the reaction, and the absorbance at 405 nm was detected. We assayed the CA activity according to Antonino et al. [25] with slight modifications. The cell lysate (10 mg/ml) diluted in a solution of Tris–HCl pH 7.4 (134 ml, 46 mM) was mixed with the substrate p-nitrophenyl acetate (66 ml, 30 mM) (Sigma, USA). The reaction was carried out at 25 °C for 30 min, and the absorbance at 405 nm was detected. For comparison, the ALP and CA activities were normalized to the cell protein content.

5.2.12 Preparation of Recombinant and Natural Proteins of Nacrein

Two recombinant proteins N1 and N2 were expressed in *Escherichia coli* BL21 (Novagen), representing amino acid residues 79–217 and residues 300–360 of nacrein (GenBank accession no. BAA11940), respectively. For the detailed operation of protein expression and purification, please see in Sect. 2.1.12.

5.2.13 Antibody Preparation

See Sect. 2.1.13.

5.2.14 Immunoaffinity Reaction Analysis

Purified McAb was coupled to CNBr-Sepharose (Amersham Bioscience) and packed into an immunoaffinity column. The extrapallial fluid was extracted from the EPS and adjusted to pH 7.5 subsequently loaded onto the prepared immunoaffinity column. Similarly, the media from mantle explant cultures and fresh medium were also subjected to the immunoaffinity column separately. The column was thoroughly washed with PBS containing 0.5 M NaCl. The elution process was performed with 100 mM glycine-HCl (pH 3.0) and immediately neutralized with 1 M TRIS–HCl (pH 8.0). The eluates were subjected to 12% SDS-PAGE and subsequent Western blot analysis.

5.2.15 Sandwich ELISA

Anti-N2 was applied as the capture antibody (6 μg/ml in 20 mM $NaHCO_3$, pH 9.6) and for pretreatment of the 96-well microtiter plates (Corning, USA) at 4 °C overnight. After blocked with 200 μl blocking reagent (5% bovine serum and 5% goat serum in PBS) for 1.5 h, the wells were added with 100 μl standard protein and samples for 2 h at 37 °C. Later, the wells were incubated for 2 h at 37 °C with 100 μl biotin-labeled McAb (5 μg/ml in 1% BSA, PBS) as the detection antibody and followed by 1 h incubation of 100 μl horseradish peroxidase streptavidin (1 μg/ml in 1% BSA, PBS; Santa Cruz, USA). The wells were washed four times with PBST (PBS plus 0.05% Tween 20) after each ELISA step. Reaction mixture of 100 μl 0.2% 1,2-phenylenediamine (Amresco, USA) in potassium citrate buffer (0.1 M, pH 5.8) containing H_2O_2 was added to each well for coloration, and 30 μl of 2 M H_2SO_4 was added to stop the reaction. Absorbance of the reactions at 490 nm was measured using an ELISA plate reader (Bio-Rad).

5.2.16 *Western Blot Analysis*

See Sect. 2.1.28.

5.2.17 *Calcium Ion Detection*

To determine intracellular and extracellular calcium ions, mantle cells from a 25-day-old culture were incubated in Pf-CM2.5 containing 8 mM Fluo-3 AM (Sigma) or 50 mM Calcein (Sigma) at 24 °C for 1 h. After rinsed with Pf-CM2.5 three times, the mantle cells were examined using a fluorescence microscope (Leica, DMIRB). Polarized microscope (Leica, DMIRB) was used to observe the intracellular calcium carbonate crystals in the 25-day-old cells. For SEM with back-scattered electron (SEM-BSE) and energy-dispersive spectrometry (EDS), the mantle cells were grown on a cover glass in a Petri culture dish with Pf-CM2.5, Pf-C10M50, or Pf-S10 M50 adding 10% fetal bovine serum. The Pf-S10 M50 and Pf-C10M50 medium were prepared similar to Pf-CM2.5 but contained 50 mM Mg^{2+} with 10 mM Sr^{2+} plus 2.5 mM Ca^{2+} or 10 mM Ca^{2+}, respectively. These cultured cells were fixed in 4% paraformaldehyde for 15 min. The samples were gently washed with sterile water, subsequently subjected to gradient ethanol dehydration (50–100%) and liquid CO_2-mediating critical point drying. The samples were coated with carbon and analyzed with a scanning electron microscope (FEI Quanta 200 FEG).

5.2.18 *In Vitro Calcium Carbonate Crystallization Assay*

See Sect. 2.1.29.

5.2.19 *Cell–Mediated Calcium Carbonate Crystallization*

Aseptic and siliconized cover glasses were placed in 6-cell plates. Primary cultured mantle cells or gonadal cells in medium containing 25 mM Ca^{2+} were transferred to the 6-well plate (n = 6/group). The cell concentration was 1 × 10^5/m containing 500 μL culture medium in each well. The plates were placed in an incubator (25 °C without CO_2), and an additional 200 μL aliquot of culture medium was added to each well every 3 days.

5.2.20 SEM and Energy–Dispersive Spectrometry (EDS)

Two cover glasses in each group were taken out from the wells and rinsed two times with D-MBSS to analyze the structural features and elemental composition of the deposits. The samples were fixed with 4% paraformaldehyde for 15 min and washed two times with Milli-Q water followed by gradient ethanol dehydration. After freeze drying for 1 h, all samples were coated with gold and analyzed under SEM-EDS (JSM-7001F field emission SEM, Tokyo, Japan) at an acceleration voltage of 15 kV.

To validate $CaCO_3$ crystal formation in hemocytes, hemolymph was extracted from *Pinctada fucata* according to Li et al. (2015) [26]. The hemolymph from ten individuals was merged together as one biological replicate. A total volume of 200 μL hemolymph was dropped onto a slide glass and air-dried at room temperature for 1 h. The sample was then coated with gold, and the element compositions were measured as described above. The dry artificial $CaCO_3$ crystal was added onto the slide glass as a positive control. Observations and determinations represented three biological replicates, and each replicate had five microscope fields.

5.2.21 Examination of Calcium Deposition

The media containing different concentrations of Ca^{2+} and Mg^{2+} were used to study inorganic effects on calcium deposition. Media C10M5, C10M10, and C10M23.5 were prepared similar to Pf-CM2.5 but containing 10 mM Ca^{2+} with Mg^{2+} of 5 mM, 10 mM, and 23.5 mM, respectively. Medium C5M5 was prepared by adding 5 mM Ca^{2+} and 5 mM Mg^{2+} to the medium Pf-CM2.5. Mantle and heart tissues were cultured in these media for 25 days. In an additional experiment, mantle tissues were kept in Pf- CM2.5 for 15, 25, and 45 days.

To analyze the calcium deposits, the cultures were washed two times with D-MBSS and fixed in 4% paraformaldehyde for 15 min, thoroughly washed with sterile water, and air-dried. These cultures were examined under an SEM (FEI Quanta 200 FEG, Netherlands) without gold sputtering to avoid heat damage. Calcium deposits in the cultures were determined by back-scattered electron (BSE) imaging at an accelerating voltage of 20 kV. The chemical composition of the deposits was measured by EDS as previously described.

After cultured in Pf-C10M50 for 25 days, cell culture from ten dishes were collected and centrifuged. The pellet was washed with Milli-Q deionized water and stored in ethanol before used. For Fourier transform infrared spectroscopy (FTIR) analysis, the ethanol was removed and cell powder was prepared using critical point drying with liquid CO_2, and then subjected to a Nicolet 560 Fourier transform infrared spectrometer. The cell powder was also detected by X-ray diffraction (XRD) using a D/max 2500 X-ray diffractometer. Synthetic ACC was prepared according to the method of Koga et al. [27] and used as a reference in XRD analysis.

5.2.22 Hemolymph Collection

Hemolymph was gently intermittently sucked from the pericardial cavity or the adductor muscle sinuses with using a 1 mL syringe. The hemolymph from ten oysters was transferred to a 1.5 mL sterile centrifuge tube, mixed, and placed on ice as a pooled replicate. Hemolymph from 30 individuals was collected to obtain three biological replicates. For flow cytometry analysis, the sample was mixed with one volume of precooled Marine Alsever's solution (MAS, 145 mM NaCl, 27 mM sodium citrate, 9 mM EDTA, 115 mM glucose in sterile water, pH 8.0) once extracted and placed on ice to prevent cell clumping. The dispersity of hemocytes was checked under an optical microscope (Nikon Ti-U, Tokyo, Japan).

5.2.23 Hemocyte Morphology and Classification

A sample of 500 μL pooled hemolymph was centrifuged, and the pellets were dropped on a glass slide, air-dried, and then subjected to hematoxylin-eosin staining (Beyotime Institute of Biotechnology, Haimen, China). To observe the granules in hemocytes, 30 μL fresh hemolymph sample was dropped on slide, covered with cover slip, and examined with an optical microscopy (Nikon Ti-U, Tokyo, Japan) equipped with a Nomarski DIC prism. Next, 500 μL pooled hemolymph was used for SEM and TEM as described above in Sects. 5.2.4 and 5.2.20. Diluted hemolymph sample (100 μL) was characterized using a flow cytometer (BD FACSCalibur, San Jose, USA) following Castellanos-Martínez et al. [28] The hemocytes were grouped based on side scatter (SSC) and forward scatter (FSC), and the percentages of each population were calculated.

5.2.24 Shell Notching and Staining

To obtain direct evidence that the hemocytes participate in shell formation, a shell notching experiment for *Pinctada fucata* was performed.

One month after notching, the regenerated together with mature shells near the notching site were collected from thirty notched individuals and washed with sterile seawater. The shells were fixed in 4% paraformaldehyde for 2 h at 4 °C and rinsed three times with sterile seawater before H&E staining as described above. The stained shell samples were examined immediately using optical microscopy (LM, TieU, Nikon, Tokyo, Japan).

Calcein AM (Dojindo Laboratories, Kumamoto, Japan) was also used to stain the shell samples, according to Mount et al. (2004) [17], with slight modifications. Calcein AM was dissolved in dimethyl sulfoxide (DMSO) and diluted to final concentration of 20 mM with phosphate buffered saline (PBS, 20 mM sodium phosphate, 200 mM NaCl, and pH 7.4). The shell samples were placed in a petri 9 cm dish and completely immersed in 2 mL diluted calcein AM. After being incubated for 3 h at room temperature, the shell samples were washed two times with PBS, transferred to glass slides, and then examined under a fluorescent microscope (TieU, Nikon, Tokyo, Japan).

5.2.25 Fluorescence–Tracking of EPS Hemocytes

Hemocytes extracted from the EPF were mixed to an equal volume of precooled MAS immediately. The pellets were centrifuged and resuspended in MAS and were used to label the hemocytes. The collected cells were labeled with a green living cell fluorescent dye carboxyfluorescein succinimidyl ester (CFSE, BD-Pharmingen) for 10 min, with the final concentration of 5 μM diluted in MAS and rinsed twice with MMBSS. The labeled cell suspension was adjusted to 10^6/ml.

For labeled cell injection, 12 recipient pearl oysters were anesthetized in seawater containing 0.2% 1-phenoxy-2-propanol (Sigma-Aldrich). When the two valves opened, 50 μL of labeled hemocytes suspension was injected into the EPS or adductor muscle, and each treatment contained six individuals. The pearl oysters were left for 10 min and then returned to the tank. After 24 h, body fluids from both the EPS and adductor muscle were extracted, and the labeled hemocytes were counted under a Nikon fluorescence microscope.

5.2.26 Determination of Immune Parameters in the Hemolymph

Hemolymph of 300 μL, 500 μL, and 300 μL was used to determine the neutral red uptake (NR), total protein content (TP), and total hemocyte count (THC), respectively. Th TP e and NR were measured at 595 nm and 550 nm using a spectrophotometer (Ultrospec 7000, GE, USA) as previous described [29]. The results were denoted as mg protein per mL hemolymph and OD per mL hemolymph, respectively. The THC in the hemolymph was determined using Absolute Counting Tubes (BD Trucount™, San Jose, USA) following the manufacturer's instructions. The THC was estimated with the following formula and presented as cells per mL hemolymph:

$$THC = \frac{Numbers\ of\ acquired\ haemycotes}{Numbers\ of\ acquired\ haemycotes} \times \frac{Numbers\ of\ beads\ in\ count\ tube}{Volume\ of\ hemolymph}$$

5.2.27 Determination of the pH Value in Extrapallial Fluid and Hemolymph

The extrapallial fluid from ten individuals exposed to various stressors was sucked by a disposable syringe and transferred into a sterile centrifuge tube to make one pooled replicate. Each treatment contains three replicates. The pH values of the extrapallial fluid and hemolymph samples were measured directly by the pH controller described above.

5.2.28 Calcium Contents in Hemocyte and Hemolymph

The cytoplasmic free calcium (Ca^{2+}) contents of hemocytes were measured by Fluo-3/acetoxymethyl ester (Fluo-3 AM) and subsequently detected by the flow cytometry method as described above. The calcium concentration of the hemolymph was determined by an inductively coupled plasma mass spectrophotometer (ICP-MS 7900, Agilent Technologies, CA, USA) according to Fernandez-Turiel et al. [30]

5.2.29 Extrapallial Fluid (EPF) Observation

To characterize the hemocytes in EPF, 0.5 mL EPF from one pearl oyster was sucked with a 5 mL syringe. The EPF from ten individuals were mixed together to obtain one biological replicate. The merged EPF for each replicate was observed by LM (TieU, Nikon, Tokyo, Japan) and transmission electron microscopy (TEM, Hiachi H-7650B, Japan) according to the method mentioned above.

5.2.30 Proteome Analysis

To analyze the molecular basis of mineralization microenvironment in hemocytes, liquid chromatography tandem mass chromatography (LC-MS/MS) method was applied to determine the total cellular proteins. Pooled hemolymph was obtained as described above and centrifuged at 800 g and 4 °C for 5 min to collect to pellets. After being washed with 0.1 M of PBS three times, the precipitant was resuspended in 200 mL of RIPA lysis buffer (150 mM NaCl, 50 mM Tris–HCl (pH 7.4), 0.1%

SDS, 1% NP-40) and 1.7 mg/ml of a protease inhibitor (Sangon Biotech, Shanghai, China). The hemocytes lysate was placed on ice for 10 min and then vortexed for 30 s every 5 min. The mixture was centrifuged at 12000 g for 5 min. The supernatant component was transferred to a new 1.5 mL tube and kept in boiling bath for 10 min with protein loading buffer. The resulting sample was loaded for 12% SDS-PAGE and then stained with Coomassie Brilliant Blue R-250. The selected bands were excised for further investigation by NanoLC-MS/MS with Q Exactive MS (Thermo Fisher Scientific, Waltham, MA, USA), according to Liu et al. (2015) [31].

The output data were blasted against the *P. fucata* draft genome database (Predicted Proteins for Genome Assembly version 1.0, http://marinegenomics.oist.jp/genomes/downloads) and oyster peptides database (Version 1.0, http://www.oysterdb.com/) by Proteome Discoverer software version 1.4.1.14 (Thermo Fisher Scientific, San Jose, CA). The tolerance was set to 0.5 Da. The sequences with at least two matched peptide fragments and with Mascot Scores >5.0 were considered to be valid.

5.2.31 Statistical Analysis

Statistical analysis was performed with SPSS statistical package. One-way analyses of variance (ANOVA) were employed to determine the effect of sampling time. Nonlinear regression analyses were carried out with the SigmaPlot statistical package.

5.3 Morphology of the Mantle Tissue

5.3.1 Histology and Ultrastructure

5.3.1.1 Histochemistry

The mantle tissue in *P. fucata* is the thin membrane organ beneath the covering shell. Generally, it is divided into three regions based on their functions: marginal zone, submarginal zone, and central zone (Fig. 5.1). No morphological boundary exists between the submarginal zone and marginal zone. The marginal zone has outer fold (*of*), middle fold (*mf*), and inner fold (*if*) in structure. The thickness of mantle declines from distal axis to proximal axis. Some nodes can be found in mantle pallial separating the central zone and marginal zone. These nodes are formed by the extending connect tissue and mediate the connection of the shell and the mantle.

The epithelial cells lining the outer surface of mantle are believed to be responsible for shell formation. We applied histochemistry to characterize the cell types and corresponding functions. Alcian blue/PAS staining revealed that a layer of columnar epithelia was surrounding the outer fold of mantle margin (Fig. 5.2a). Both Alcian

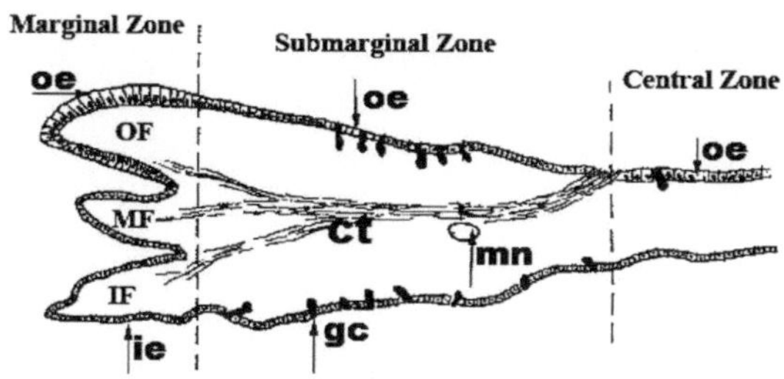

Fig. 5.1 Diagram of transverse section of the mantle of *P. fucata*. *if* inner fold, *mf* middle fold, *of* outer fold, *oe* outer epithelium, *ie* inner epithelium, *gc* globlet cells secreting mucopolysaccharides, *ct* connective tissue, *mn* mantle nerves

blue- and PAS-positive outer epithelia coexist in this part. In submarginal zone, the outer epithelia did not array so compactly with each other as those in the marginal zone. The density and amount of secretory granules significantly deepen the Alcian blue/PAS staining. Outer epithelia of Alcian blue- and PAS-positive were not clearly separated any more (Fig. 5.2b). The outer epithelia in the central zone containing almost one cell type and the shape were relatively cubic. The region was bloom and succulent in Alcian blue/PAS-positive cells (Fig. 5.2c). The outer epithelia in the visual field were dominated by amethyst-stained cells. The cells in the outer epithelia of mantle isthmus were tall and columnar. The secretory cells were comparative absence in this zone, but the connective tissue was well developed (Fig. 5.2d).

Our results of cytochemistry investigation shared some common features consistent with some earlier studies in other mollusks. For example, the outer epithelia contained mainly columnar epithelial cells in the outer fold of the marginal zone; then, the height gradually decreased in the outer epithelia toward the submarginal zone; the outer epithelia were constant in size and displayed in low columnar form in the central zone; and at last in the mantle isthmus, the secretory cells were absent [32–34]. Using TEM observation, it was found that in the mantle outer epithelia, the protein synthetic apparatus undergoes a progressive reduction, while the quantities of glycogen and mucopolysaccharides increased from the marginal toward the central zone. It should be noted that this interesting progressiveness was also found in *Mytilus edulis*, *Lymnaea stagnalis* [35], *Nucula sulcata* [33], *Cardium edule*, and *Pinctada maxima* [36].

5.3.1.2 The Pellet Was Washed with Milli-Q 3.1.2 Electron Microscopy

TEM observations were used to further confirm the configuration of the epithelium established by the optical microscope [16]. The outer epithelia lining the outer fold were rather compact and lay parallel to each other in the marginal zone (Fig. 5.3a). These cells had round nuclei and contained well-developed granular endoplasmic reticulum with numerous mitochondria. What's more, numerous membrane-bounded electron-dense vesicles fused with each other and dominated the whole

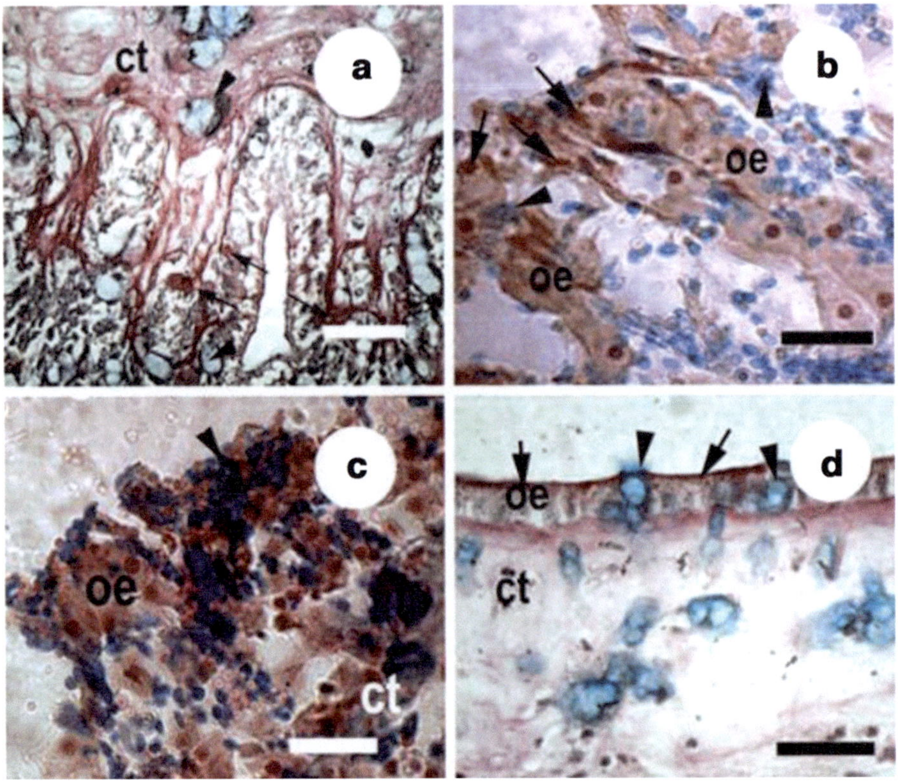

Fig. 5.2 Histology of outer epithelia in marginal, submarginal, central, and isthmus mantle of *Pinctada fucata* based on paraffin serial sections (cross section). LM. (**a**) Alcian blue/PAS staining, showing outer epithelial cells in outer fold of marginal zone composed of a layer of columnar epithelia. Both Alcian blue-positive (*black arrowheads*) and PAS-positive (*black arrows*) outer epithelia were detected. *ct* connective tissue, *bars* 10 μm. (**b**) Secretory granules in outer epithelia of submarginal zone. The amount and density of secretory granules increased greatly in this zone. *oe* Outer epithelia, *bars* 20 μm. (**c**) Amethyst-stained granules in outer epithelia of the central zone dominated this region (*black arrows*). *oe* Outer epithelia, *ct* connective tissue, b*ars* 20 μm. (**d**) Isthmus Alcian blue/PAS staining showing the columnar outer epithelia and much lower secretory activity compared with central zone (*arrows*). *oe* Outer epithelia, *ct* connective tissue, *bars* 30 μm

apical half in these outer epithelia (Fig. 5.3b). Measuring about 0.5 μm in diameter, these membrane-bounded vesicles were found only in the outer epithelia of outer fold and directly responsible for shell formation. In Fig. 5.3c, which showed another outer epithelium in the outer fold, small granules were nipped off from the marginal tips of the outermost cisterna of Golgi complex, and several such granules were found around the Golgi complexes. In the submarginal zone, numerous secretory vesicles were found and supposed to be responsible for shell formation. The extensive granular endoplasmic reticulum (ER) suggested that active metabolism and secretory activities was also present in the outer epithelia of this zone (Fig. 5.3d). But the organelles in the outer epithelia of mantle isthmus, such as the Golgi

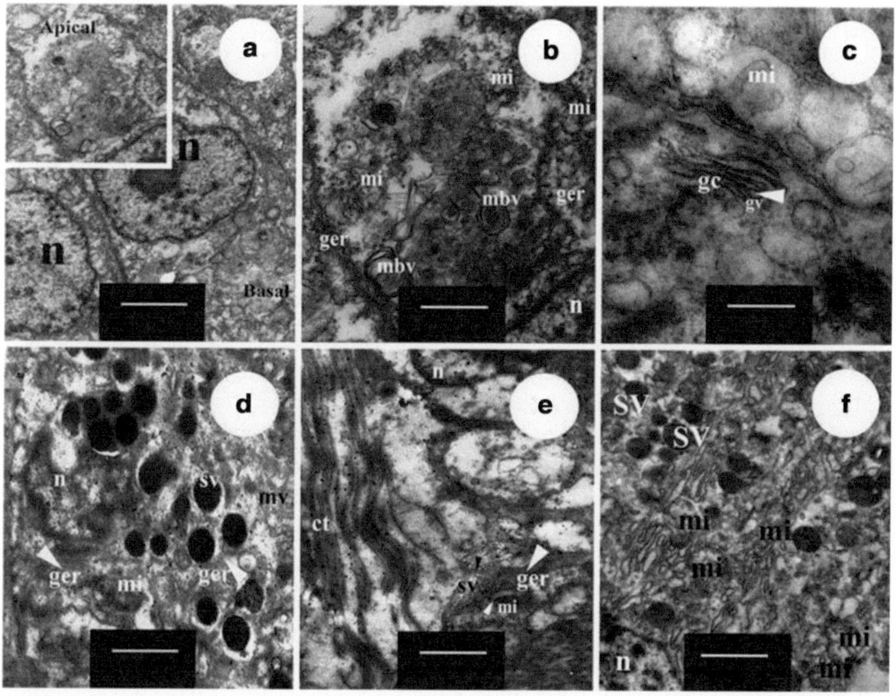

Fig. 5.3 TEM observations of the outer epithelia in four zones. An overview of outer epithelia in mantle outer fold. The cytoplasm especially the whole apical half of these epithelial cells was occupied by double membrane-bounded vesicles. *n* Nucleus, *bars* = 1 μm. (**b**) Magnification of inset of (**a**), details of apical half of the outer epithelia in outer fold mantle margin: its well-developed granular endoplasmic reticulum (*ger*), the mitochondria (*mi*) with well-developed cristae, and double membrane-bounded vesicles (*mbv*). *Bars* = 0.5 μm. (**c**) Outer epithelia in mantle margin outer fold containing Golgi complex (*gc*), Golgi secretory granules (*gv*), and the mitochondria (*mi*), *Bars* = 0.5 μm. (**d**) Observations of the outer epithelia in mantle submarginal zone: prominent secretory vesicles(*sv*), well-developed granular endoplasmic reticulum (*ger*), and mitochondria (*mi*) indicating active metabolism and active secretory activities responsible for shell formation. *n* Nucleus, *mv* microvilli, *bars* = 1 μm. (**e**) Sparse organelles such as granular endoplasmic reticulum (*ger*) and the mitochondria (*mi*) of outer epithelia in mantle isthmus indicating inactive secretory activities. *ct* connective tissue, *n* nucleus, *bars* = 1 μm. (**f**) Outer epithelia in mantle central zone entirely filled with secretory vesicles (*sv*) and numerous mitochondria (*mi*). *n* nucleus, *bars* = 2 μm

complex, endoplasmic reticulum, and especially the secretory vesicles, were small (about 0.05 μm in diameter) and translucent (Fig. 5.3e) when compared with submarginal and marginal zones. In the central zone showed in Fig. 5.3f, the entire cytoplasm of the outer epithelial cells was filled with numerous granules of varied electron densities and diameters (about 0.5–2 μm). Furthermore, other organelles in the outer epithelia of the central zone, such as ER and the Golgi complex, were drastically reduced in number, except that of mitochondria.

5.3.2 Cell Types in the Outer Surface of Mantle

5.3.2.1 Epidermal Cells

The epidermal cells dominate in the outer and inner surface of the mantle, which was revealed by transmission electron microscope observation of primary cell culture and the ultrathin section. It was supposed that the epidermal cells play a critical role in SMPs and ions secretion during shell formation. Numerous vesicles accompanied by plenty of mitochondria were found to occupy near the epical and basal membranes of the epidermal cell. Moreover, fraction of epidermal cells was obtained from the cultured mantle cells using Percoll density gradient centrifugation and further confirmed by the detection of the cellular marker pancytokeratin. When cultured in dishes, these cells attached to the substrate and several pseudopodia stretched out from the cell surface (Fig. 5.4a). The cell size was about 10–15 μm in diameter. Western blot results indicated that these epidermal cells maintained their ability to synthesize and secrete shell matrix proteins in vitro.

5.3.2.2 Mucous Cells

Mucous cells represent another important cell type respected to shell mineralization, lining the outer epithelia of the mantle among the epidermal cells. They distribute heterogeneously on the outer surface of mantle, with higher proportion in the pallial and central zone and declined in number toward the mantle edge. Large vesicles containing homogenous fiber-like mucus occupy the whole cells and bulge in to the extrapallial space. These mucous cells might be responsible for the polysaccharide secretion during shell formation.

5.3.2.3 Other Cell Types

Other cell types involved in biomineralization include hemocytes and special secreting cells in the periostracum groove, whose features and functions were not fully understood. Muscle cells were characterized in the mantle tissue and cultured cells, as well as some unknown cell types.

5.3.3 Cell Proliferation in Mantle Tissue

5.3.3.1 Identification and Quantification of Proliferating Mantle Outer Epithelia

After exposure of BrdU for 12 h and immunohistochemistry on the tissue, the four mantle regions displayed different proliferation activities in all the specimen. In the

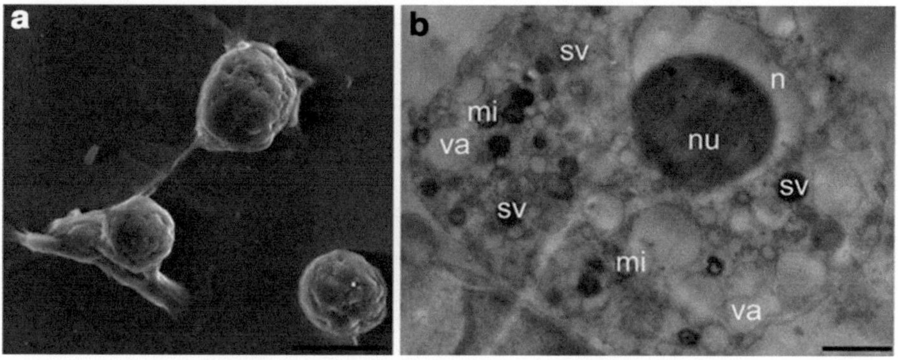

Fig. 5.4 Electron microscopic observation of epithelial-like cells [23]. (**a**) Scanning electron micrograph of epithelial-like cells, which are half-adherent to the substrate and spherical in vitro. *Bar* 10 μm. (**b**) Transmission electron micrograph of an epithelial-like cell showing prominent secretory vesicles (*sv*), vacuoles (*va*), mitochondria (*mi*), and a large nucleus (*n*) with a prominent nucleolus (*nu*), indicating active metabolism and secretory activities. *Bar* 1 μm

marginal zone, no positive signal was detected in the outer epithelium of the outer fold (Fig. 5.5a). Whereas in the submarginal zone (Fig. 5.5b) and mantle isthmus (Fig. 5.5c), immunopositive signals were found were under the visual field. At last, the outer epithelium nuclei of mantle central zone exhibited very strong immunopositive signals (Fig. 5.5d). As seen with the confocal microscope (Olympus FV500), the nucleus to cytoplasm ratio was clearly much higher in outer epithelia of central zone than those of the isthmus, submarginal and the marginal zones (inset of Fig. 5.5d).

According to multiple comparisons between the mean values by the one-way ANOVA and F-test, BrdU-labeling values of the four mantle regions differed significantly from each other. Specifically, the outer epithelia of central zone had a significantly higher proliferation rate ($P < 0.01$) than the other three regions (Fig. 5.6). So we propose that in the mantle of *P. fucata,* there is a proliferation hot spot, located in the center of the thinnest mantle region. The proliferation rate declines from the center zone to the margin zone, and the mantle margin almost loses the ability of proliferation. The mantle continuously differentiates as the oyster grows, and the direction is from the central part to the marginal zone.

The progressiveness of mantle's proliferation and secretion features was an interesting finding. Though there was no very solid evidence to prove that the epithelia in the central zone were definitely stem cells, our observation did find some characteristic features of the stem cells, the high proliferation rate and high content of saccharides. So it deserves more research to search for stem cell markers such as CD9, CD34, ABCG2/Bcrp1, and AC133 in this region. When the human bone marrow mesenchymal stem cells (hMSCs) differentiate into endothelial cells, a very important feature is that the cells form a glycogen pool in the ectoplasm and are positive for glycogen stain. And the glycogen decreased or disappeared after differentiation [37]. Much research has been looking into the proliferation of molluskan

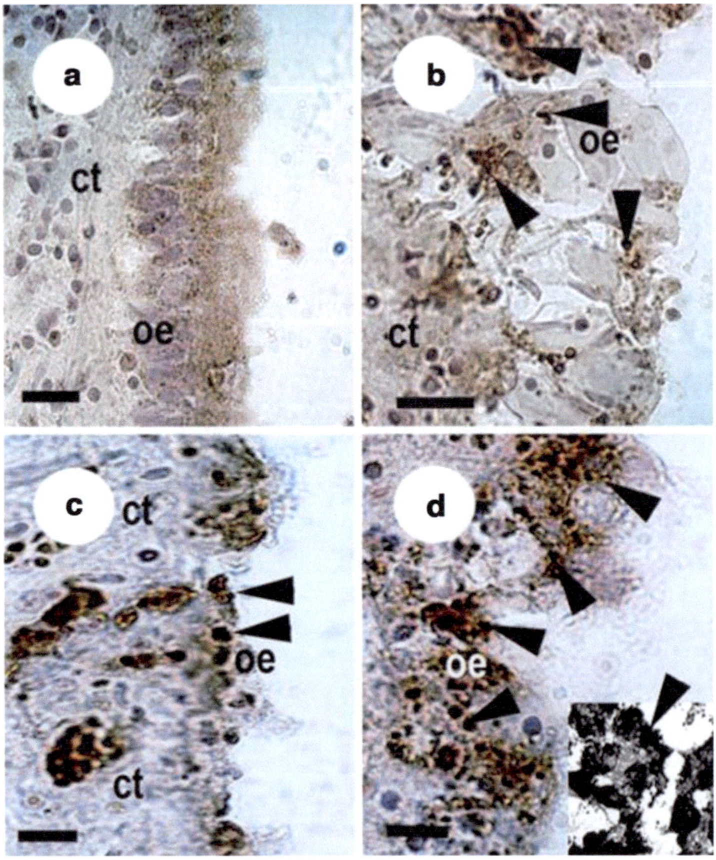

Fig. 5.5 BrdU labeling of mantle epithelia in different regions of *Pinctada fucata*. LM. (**a**) Negative BrdU signal in nuclei of marginal zone outer epithelia (*oe* outer epithelia, *ct* connective tissue). (**b**) BrdU-positive nuclei (*black arrowheads*) in submarginal zone outer epithelia (*oe* outer epithelia, *ct* connective tissue). (**c**) BrdU-positive nuclei (*black arrowheads*) in mantle isthmus outer epithelia (*oe* outer epithelia, *ct* connective tissue). (**d**) Very strong BrdU-positive signals (*black arrowheads*) in the outer epithelia of central zone (*oe* outer epithelia). *Inset* Part of the central zone outer epithelia exhibiting a high ratio of nucleus to cytoplasm under confocal microscope. *Bars,* 20 μm (**a**–**d**)

cells, including stomach cells in *Crenomytilus grayanus* [38], hemocytes in *Mercenaria mercenaria* [39], and digestive gland cells of mussels [40–43]. Yonge supposed that certain kinds of basophilic cells may be stem cells and later reports in some crustacean hepatopancreas [44, 45] indicated a similar proliferation model. Marigómez et al. [46] found that both basophilic and digestive cells were able to dedifferentiate, similar to that some mature cell types in vertebrates undergo autologous division [47]. Usually, epithelial cell renewal relies on the autologous division of mature cells [19] but Hanselmann and Smolowitz [39] showed direct proliferation

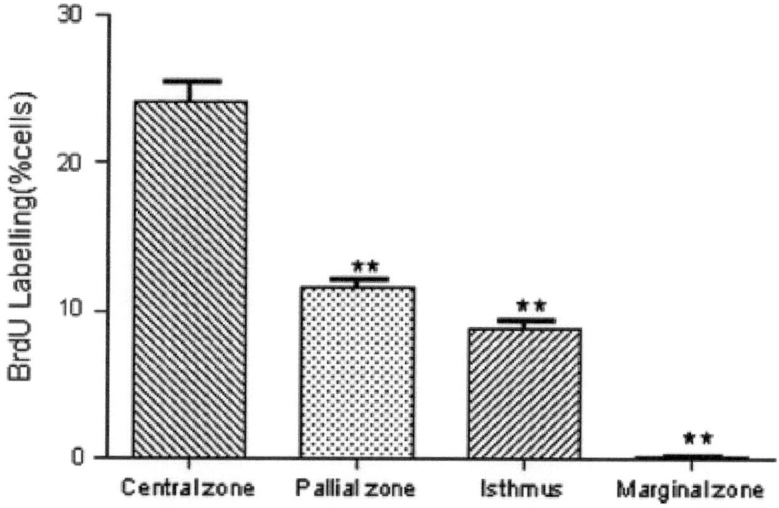

Fig. 5.6 Comparison of the cell proliferation rate in different mantle regions. The central zone exhibited a higher proliferation rate than the other regions. ** $P < 0.01$, significant difference compared with that in the central zone. Each *bar* represents the mean ± SE of different regions

of hemocytes even after parasite damage. So the model of proliferation may vary in different mollusk cells and can't be summarized simply.

There has been no research about stem cells in invertebrates, especially in mollusks. And this has become the hindering factors for cell culture of invertebrates. Checa [48] proposed that the mantle margin is the main morphogenetic shell-secreting unit. However, the submarginal zone is usually selected as the origin of mantle outer epithelia in primary mantle cell culture, based on the knowledge from the aquaculture of pearl. The quality and color of the cultured pearl is closely related to the choice of the donor oyster for mantle slices. The reason for the choice of such slices for pearl aquacultures may be that the submarginal region not only secretes actively but also has the ability to generate a pearl sac and display well-differentiated biological functions. However, based on the results of our investigation, the submarginal zone may not be the ideal region for developing a cell line because it has low proliferation rate compared with that of the outer epithelia in the central zone.

5.3.3.2 Changes in Cell Proliferation During the Tidal Cycle

In the light of the multiple comparisons between mean values ($P < 0.01$) by one-way ANOVAs and subsequent Duncan's tests, BrdU-labeling values were significantly different between the varied sampling times. It was found that both the central and submarginal zones exhibited pattern of rhythmic variation with the low and high tides over a 48-h time period (Fig. 5.7a, b). Based on nonlinear regression analysis,

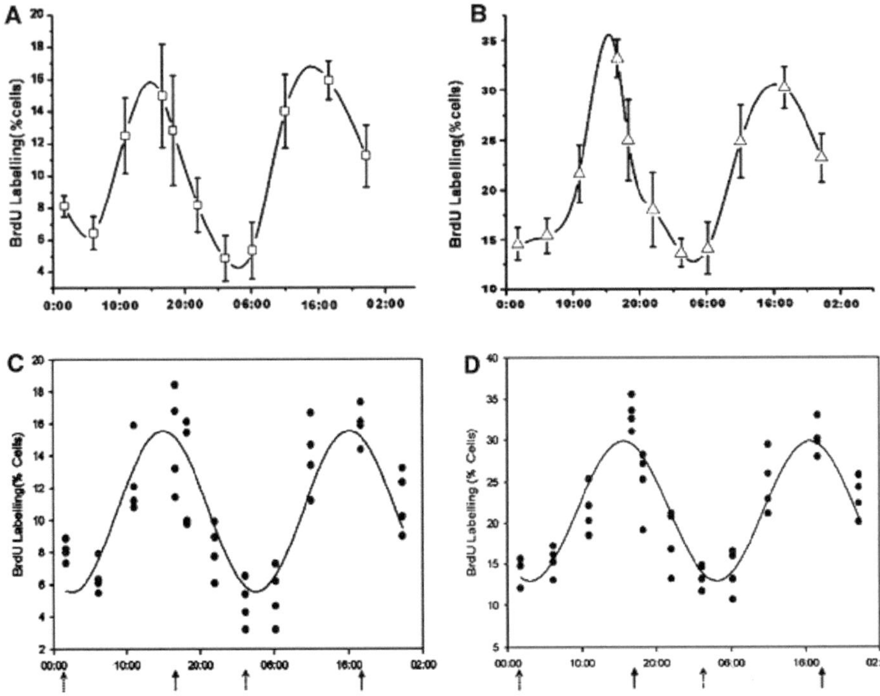

Fig. 5.7 BrdU labeling (% ± standard deviation) of the outer epithelium of *P. fucata* in (**a**) submarginal zone and (**b**) central zone relative to 48 h. Nonlinear regression of BrdU-labeling values (% cells) recorded against experimental time (in 48 h) relative to the time; (**c**) the submarginal zone's best fitness corresponded to a waveform function; $Y = 10.5515 + 4.9923\sin[(2\pi X/25.1082) + 4.0982]$; $R^2 = 0.7041$ where all regression and correlation coefficients were significant ($P < 0.01$). The distance between the peaks and between the troughs is approximately 24 h. (**d**) The central zone's best fitness corresponded to a waveform function: $Y = 21.4276 + 8.4160\sin[(2\pi X/25.0405) + 3.9506]$; $R^2 = 0.7736$ where, all regression and correlation coefficients were significant ($P < 0.01$). The distance between the peaks and between the troughs is approximately 24 h (*dotted arrows* low tide, *arrows* high tide)

the data apparently adjust to a waveform function of an approximately 24 h periodicity (Fig. 5.7c, d).

It is also quite interesting to compare our results with the previous work of Zaldibar et al. [19], which reported the proliferating cells in the stomach and digestive gland in *Mytilus galloprovincialis*. They found that cell renewal in digestive diverticula was synchronized with the tidal pattern. So it is obvious that the mantle, stomach, and digestive gland epithelia showed a similar fluctuating pattern that perfectly resembles a tidal rhythm. The was consistent with the study of Pannella [49], who found that the rapid shell deposition occurred in 2–3 h during rising tides. In addition to the shell mineralization rate, many physiological processes exhibit similar fluctuation of tidal rhythms, which have been reported in other marine invertebrates [50, 51]. To avoid any interference between the photoperiod and tidal cycles, we closely investigated the subtidal pearl oyster according to Zaldibar et al. [19] We supposed that the proliferation fluctuation of the mantle cell may be a direct result of tidal rhythms.

There are quite a lot of physiological and environmental factors including temperature, tides, day/night cycles, salinity, and food supply that could affect shell calcification process, either in combination or independently [52, 53]. Many studies have focused on the formation of growth lines, and different methods have been applied to understand this rhythmicity [52, 54, 55]. Thompson supposed that even though the bivalve shell ultimately mirrors the cycles of these rhythms, this signal transfer was neither immediate nor passive, and there might be a "rhythmic" tissue in the organism involved. Thompson also showed that this organism maintains its growth rhythm in the absence of these environmental cycles [56]. The excellent consistency of cell proliferation fluctuation with the tidal rhythm is a very interesting phenomenon, indicating that the mantle may be a "rhythmic" tissue, as Thompson suggested.

5.4 Primary Cell Culture of the Mantle Tissue

5.4.1 Maintenance of Primary Cell Culture

After 24 h of culture initiation, mantle cells actively migrated from explants, and formed cell sheets surrounding the explants after several days of cultivation (Fig. 5.8b). Three types of cells dominated the cell culture, namely, epithelial-like cells, hyalinocytes, and hemocyte-like cells (Fig. 5.9a). Generally, $1–2 \times 10^6$ of migrated cells were obtained from the explant cultures of per oyster after cultivation for 4 days; the total viability of the cells ranged from 86% to 90%. Flow cytometry analysis indicated that 83.80% $\pm$ 1.95% of the migrated cells were distributed in the G0/ G1 phase, 1.59% $\pm$ 1.15% in the G2/M phase, and 1 4.61% $\pm$ 0.93% in the S phase of the cell cycle. These results suggested that the mantle cells maintained their viability but did not actively proliferate.

To obtain single cells, enzyme-dissociation method was also adopted to dissociate the tissue. Approximately 2×10^6 cells of one oyster were enzymatically obtained from mantle tissue fragments; the total viability of the cells ranged from 90% to 95%. However, after a 3-day cultivation, the MTT activity of the dissociated cells declined to 55–70% of that in the first day, whereas the MTT activity of the migrated cells showed no significant change (Fig. 5.9b). Therefore, mantle explant culture was applied for further experiments. Moreover, although wide-spectrum antibiotics–antimycotics were added in the media, some of the primary explant cultures had to be discarded due to serious contamination of amoebae, fungus, or protozoan originating from the explants. So we cultured the explants from several oysters separately to select non-contaminated cultures in order to obtain single cells and reduce the contamination risk in subsequent cultivation.

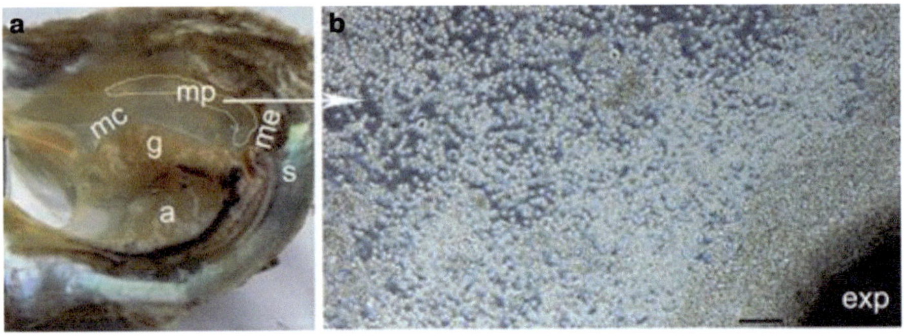

Fig. 5.8 Mantle explant culture. (**a**) The valve of *Pinctada fucata*. The mantle pallial (mp) was used for explant culture (mc mantle center, me mantle edge, s shell, a adductor muscle, g gill). (**b**) Cell sheets formed around the explant after incubation for 4 days (exp explant). Bar = 100 μm

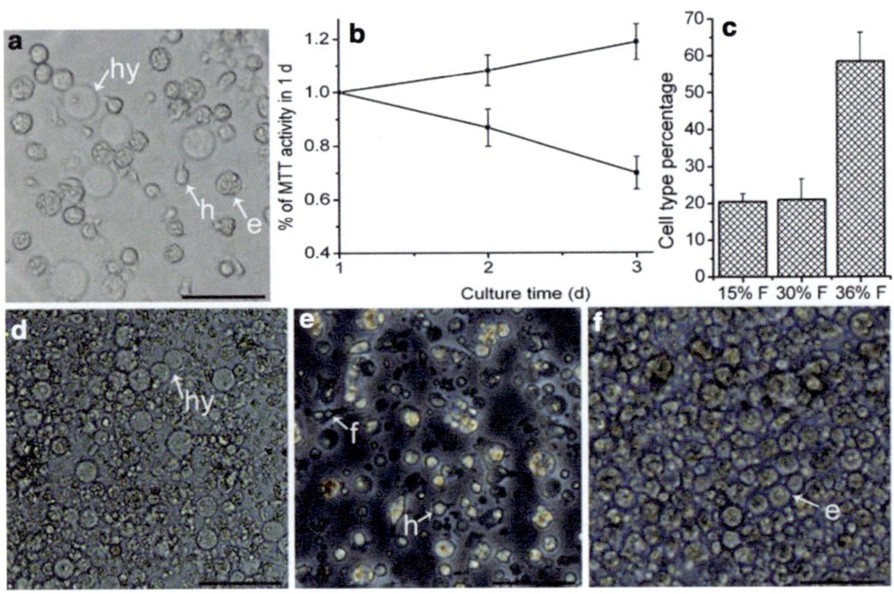

Fig. 5.9 Cell culture and separation. (**a**) Phase-contrast light micrograph of a mixed cell population in a mantle explant culture from *P. fucata* (*e* epithelial-like cell, *h* hemocyte-like cell, *hy* hyalinocyte). (**b**) Relative MTT activity of migrated cells from explant culture (*circles*) and of enzyme-dissociated cells (*squares*). Data are the mean ± SE of three independent experiments. (**c**) Percentage of cells in Percoll fractions of 15%F, 30%F, and 36%F in the total (*vertical lines* standard deviations). (**d**) Hyalinocytes (*hy*) distributed in Percoll fraction of 15%F. (**e**) Hemocyte-like cells (*h*) distributed in Percoll fraction of 30%F; sometimes, fibroblast-like cells (*f*) were observed. (**f**) Epithelial-like cells (*e*) distributed in Percoll fraction of 36%F. *Bars* 50 μm

In mantle tissue culture, cells began to migrate from the explants 6–8 h after culture initiation and the migration became active after 24 h. Numerous cells, mainly the epithelial-like cells (8–10 μm), round hyalinocytes (10–14 μm), and hemocyte-like cells (6–8 μm) migrated out and began to grow around the mantle explants after 2–3 days. The adherence of mantle explants to the substrate was critical for cell migration. To promote their adherence, the explants were cultured in reduced culture medium so that they were just covered during the first 2 days. The non-adhered tissue explants were discarded after a 2-day incubation, and the adhered explants continued to be cultured with enough media. Frequently, fibroblast-like cells were observed to be derived from the explants after 4–5 days. After 7–10 days, the migrated mantle cells containing dominant epithelial-like cells would shape confluent cell sheets covering most area of the culture dish, and the cell viability counted 88–90% (Fig. 5.10a). The epithelial-like cell confluent would last for several weeks unless serious contamination occurred and maintain a viability of 60–70% within 45 days accompanied with spindle-like muscle cells when the explants were disrupted (Fig. 5.10b). Generally, the epithelial-like cells would remain alive for up to 2 months. Modifications of the Ca^{2+} and Mg^{2+} concentrations in the medium did not have adversely effect on cell migration.

During the first 2–3 days, some fragments in heart tissue cultures would keep beating in vitro. Numerous granular and hyalinocytes cells surrounded near the fragments at day 9 after the tissue fragments adhered to the substrate (Fig. 5.10c). As the culture continued, epithelial-like cells would dominate the migrated cell surrounding the tissue fragments (Fig. 5.10d) and survive for about 25 days.

5.4.2 Cell Separation and Identification

The single mantle cells were collected from explant cultures and mixed together for separation by density. Most of the cells were divided into three Percoll fractions, namely, 15%F, 30%F, and 36%F, lied at the interfaces between gradient 15% to 30%, 30% to 36%, and 36% to 60%, respectively. The relative proportion of cells in the 15%F, 30%F, and 36%F fractions were 20% ± 2%, 21% ± 5%, and 58% ± 8% of the total cells, respectively (Fig. 5.9c). In the 15%F, the main proportion was round hyalinocytes with diameter of 10–14 μm, (Fig. 5.9d). Most cell debris and dead cells also fell in this fraction. The hemocyte-like cells with 6–8 μm in diameter mainly settled in 30%F, together with some fibroblast-like cells (Fig. 5.9e). The epithelial-like cells (8–10 μm of diameter) were mainly separated into 36%F (Fig. 5.9f).

In order to characterize the epithelial cells in the separated fractions, immunolabeling of anti-pancytokeratin was performed. DAPI staining of cell nuclei was used to indicate the presence of cells (Fig. 5.11a, d, g). As a result, the cells in 15%F and 30%F were seldom stained by FITC (Fig. 5.11c, f), while positively staining was seen in most of the cells in 36%F (Fig. 5.11i). Because pancytokeratin

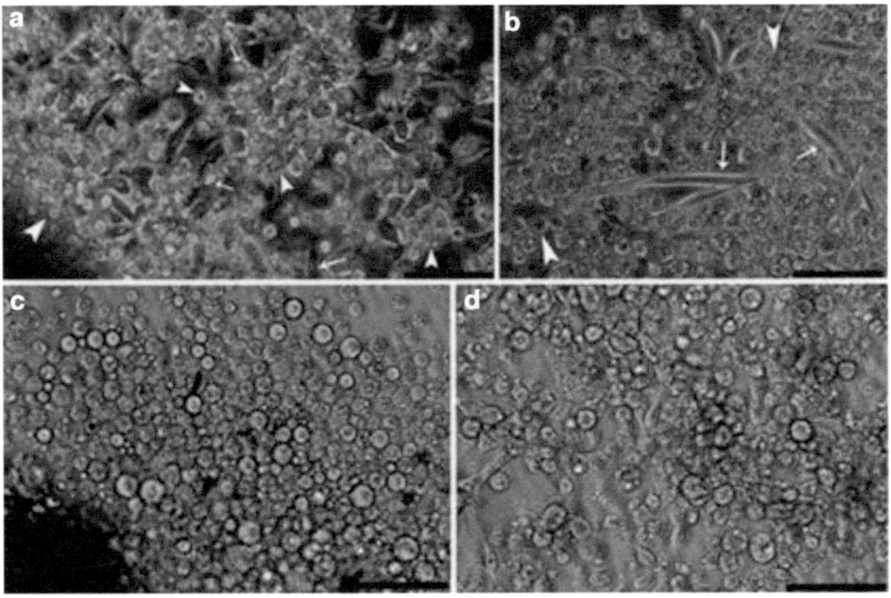

Fig. 5.10 Primary mantle and heart tissue culture. (**a**, **b**) Confluent cell sheet of migrated cells from mantle explants after 10 days and 45 days, respectively; the arrowhead indicates the epithelial-like cells. (**a**) White and black arrows indicate hemocyte-like cells and fibroblast-like cells, respectively. (**b**) White arrows indicate spindle-like muscle cells. (**c**) Migrated cells around a heart tissue fragment in 9 days; the shadow at the bottom left corner indicates the tissue fragment. (**d**) Migrated cell sheet in heart tissue cultures after 25 days. Scale bars, 50 μm

was regarded as a cellular marker of epithelial cells, we assumed that the FITC positive cells in 36%F were epithelial cells origin from the mantle epithelium.

Previous studies have showed that cells lining the outer surface of the mantle pallial actively synthesize and secrete shell matrix proteins [4, 16, 57]. Furthermore, mantle explant cultures using the pallial zone are often applied for the in vitro study of biomineralization and displayed active secretion of organic matrix in vitro [9, 15, 21]. Therefore, we tried to develop long-term primary cell culture from the mantle pallial to study shell matrix protein secretion in *P. fucata*.

Enzyme digestion is a common used technique for single cells isolation from tissue. Awaji [58] treated mantle tissue fragments from *Pinctada fucata* with dispase II to obtain separated mantle epidermal cells. Most dissociated cells survived at the beginning of the cell culture but suffered from decline of cells viability about 60% in 3 days [59]. Chen and Wen [11] have found that collagenase treatment resulted in harvest of dissociated cells from oyster heart maintaining good growth in vitro and the damage was less than trypsin treatment. In our study, we employed collagenase and hyaluronidase to dissociate viable single cells from mantle tissue. However, MTT activity assays showed that the dissociated mantle cells exhibited poor growth in vitro, only 55%–70% of MTT activity was retained after 3 days. On the opposite, the cells migrated from the mantle explant retained high MTT activity after 3 days.

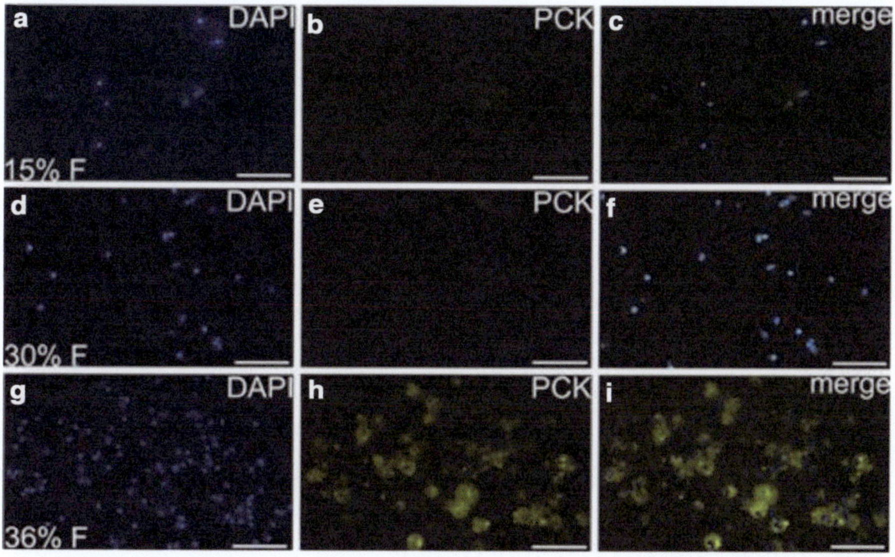

Fig. 5.11 Immunolabeling of the cells from 15%F, 30%F, and 36%F with mixed monoclonal antibodies to pancytokeratin as primary antibodies and FITC anti-mouse antibodies as secondary antibodies. The presence of cells from the 15%F (**a**), 30% F (**d**), and 36%F (**g**) was indicated by DAPI staining (DAPI; blue) of cell nuclei. The presence of pancytokeratin-labeled cells (PCK) was indicated by FITC labeling (green/yellow). The cell types in 15%F (**b**) and 30%F (**e**) were poorly stained by FITC, whereas most of epithelial-like cells in 36%F were positively stained (**h**). (**c**, **f**, **i**) Merged images of DAPI and PCK staining. Bars 25 μm

Therefore, we chose the explant cultures, but not enzyme treatment, to achieve single cells for cellular metabolism study.

Active cell migration has been observed all over mantle explant culture period [8, 9]. Four days since culture initiation, the migrated cell population mainly consist of epithelial-like cells, hemocyte-like cells, and hyalinocytes. Immunolabeling for anti-pancytokeratin indicated that the epithelial-like cells might originate from the mantle epithelium. The ultrastructure of the epithelial-like cells revealed by TEM observation exhibited numerous secretory vacuoles and vesicles, similar to the epithelial cells in mantle pallia [3, 16]. The particular cellular organelles in this type of cells had a higher cell density, leading to their isolation from the mixed cell population by using discontinuous Percoll gradient centrifugation. Taken the immunolabeling and ultrastructure results together, we deduced that the isolated epithelial-like cells should be the outer epithelial cells in mantle tissue, which were further confirmed by ELISA assay. In addition, the epithelial-like cells in 36%F have the activity of secreting nacrein, a matrix protein highly expressed in the outer mantle epithelium [60]. In contrast, trace amounts of this protein was detected in the cells from 30% F or 15% F. Characterized by tissue specificity, the expression of SMPs is always confined to specific areas of the mantle tissue according to previous studies [1, 4]. Therefore, the ability to synthesize and secrete SMPs can be regarded as a functional feature in identifying outer epithelial cell of the mantle in vitro.

Most of the migrated mantle cells were detained at the G0/G1 phase and did not actively proliferate in vitro, as revealed by the cell cycle analysis. Efforts to obtain a proliferating cell culture from mantle tissue have never succeeded [8, 14]. This might be due to the lack of vital information about the physiology and the nutrient requirements of the cells in vitro [30]. Even though the cultured cells we obtained did not actively proliferate in vitro, they maintained their viability, physiological capabilities, and intrinsic ultrastructure, thus having potential applications for characterizing cellular physiology and regulatory mechanism of biomineralization in vitro, which are poorly understood and difficult to accomplish solely in in vivo models.

5.5 Cellular Control of Biomineralization by Mantle Tissue

5.5.1 Shell Matrix Protein Secretion

5.5.1.1 Expression Level of Matrix Proteins During Cell Culture

The expression of SMPs in the cultured mantle cells in vitro was measured by RT-PCR. The fragments of *nacrein* (484 bp), *MSI7* (156 bp), and *MSI60* (354 bp) were detected in the mantle cells from 5-day-old mantle cultures (Fig. 5.12a), 12-day-old mantle cultures (Fig. 5.12b), and fresh mantle tissue (Fig. 5.12c). Results from heart tissue showed negative signal of the above genes but the positive signal of GAPDH (Fig. 5.12d). These results suggested that the SMPs nacrein, MSI7, and MSI60 were expressed in the cultured mantle epithelia in vitro, but not in heart tissue.

We further examined the expression of MSI7, MSI60, N19, and N16 in the in vitro cultured mantle cells and found that they were retained after 25 days since the cell culture initiated. The expression levels increased significantly with a maximum at 7.5 days (Fig. 5.13). Further, nacrein, ACCBP, and Pif80 in the mantle culture medium were also detected by Western blot analysis (Fig. 5.14). These results indicated that the mantle cells synthesized matrix proteins intracellularly and transported to the mineralization front to control crystal nucleation and growth. The function of the matrix proteins detected in the culture medium was further investigated by in vitro crystallization experiments, which will be described later in the text.

5.5.1.2 Specific Secretion of Nacrein by In Vitro Cells

A purified recombinant protein N61/217 (20 kDa; Fig. 5.15, lane 1) was used to immunize rabbit to obtain polyclonal antibodies against nacrein. The prepared

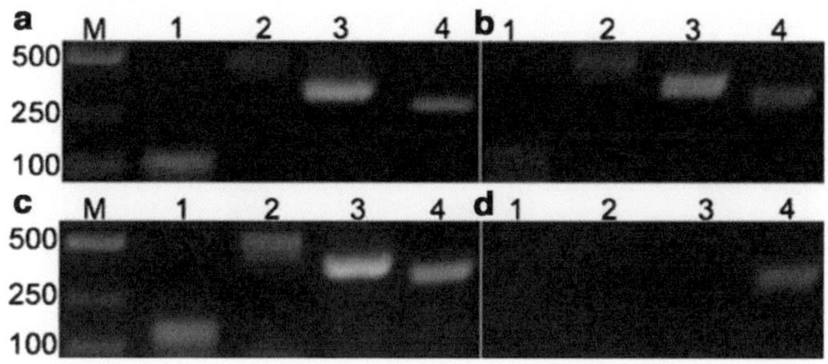

Fig. 5.12 RT-PCR analysis of gene expression in 5-day-old mantle cultures (**a**), 12-day-old mantle cultures (**b**), mantle tissue (**c**), and heart tissue (**d**). Lane 1, *MSI7* (156 bp); lane 2, *nacrein* (484 bp); lane 3, *MSI60* (354 bp); lane 4, GAPDH (290 bp); negative amplification of *MSI7*, *nacrein*, and *MSI60* from heart tissue (d)

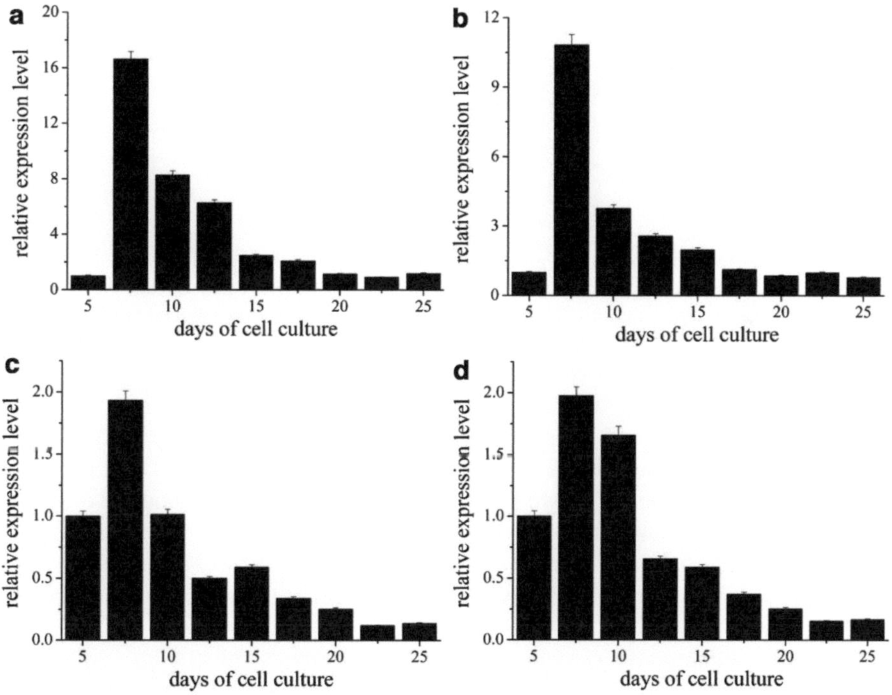

Fig. 5.13 mRNA expression levels of SMPs during primary cell culture of mantle. (**a**) MSI7; (**b**) MSI60; (**c**) N19; (**d**) N16

antibodies were employed for Western blot analysis. A protein band near molecular mass of 60 kDa (Fig. 5.15, lane NM) was seen in the immunoblotting of water-soluble matrix proteins, consistent with the molecular mass of nacrein [61], indicating that the antibodies are specific and can be used for the detection of nacrein.

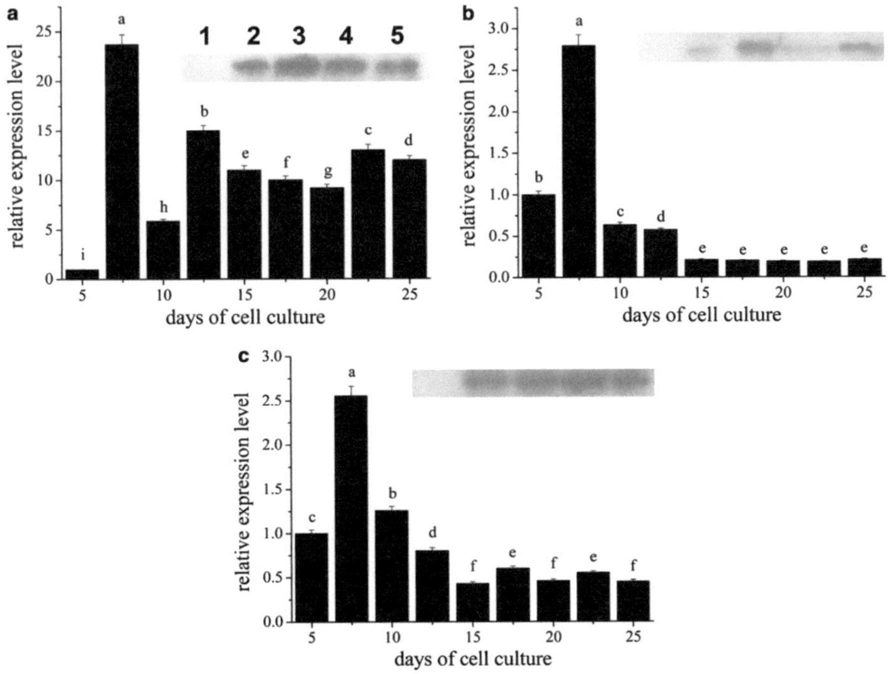

Fig. 5.14 mRNA expression levels and western blot of SMPs during primary cell culture of mantle. (**a**) ACCBP; (**b**) nacrein; (**c**) Pif80. Inset in each histogram represents the Western blot respectively. Lane 1, control group of uncultured medium; lanes 2–5, cultured medium for 10 days, 15 days, 20 days, and 25 days, respectively

Similar bands around 60 kDa were observed in the immunoblotting of the extrapallial fluid (Fig. 5.15, lane EPF) and mantle culture medium (Fig. 5.15, lane MC) but not in the uncultured medium (Fig. 5.15, lane UC) and heart culture medium (Fig. 5.15, lane HC). Due to the high specificity of the antibody, we deduce that the detected protein corresponds to nacrein, which can be secreted by the mantle epithelial cells in vitro and in vivo. A similar band was also seen in Western blot analysis of the medium collected from a 1-month-old mantle explant culture (data not shown).

For further quantitative detection of nacrein, a sandwich ELISA was developed for measurement of nacrein. Two recombinant proteins of His-tagged N1 and GST-tagged N2 were expressed and purified corresponding for two fragments of the nacrein protein, namely, N1 and N2, respectively (Fig. 5.16a, lanes 1, 2). The corresponding antibodies (McAb and anti-N2) were produced and checked for quality by Western blotting (Fig. 5.16b, lanes 1, 2). The antibodies showed high affinity for the natural nacrein of approximately 60 kD [62] and applied for subsequent sandwich ELISA method. Because two extra faint bands were also determined in lane 2 of Fig. 5.16b, suggesting that anti-N2 might have immunoreactivity with other proteins except nacrein, we carried out immunoaffinity analysis to validate the

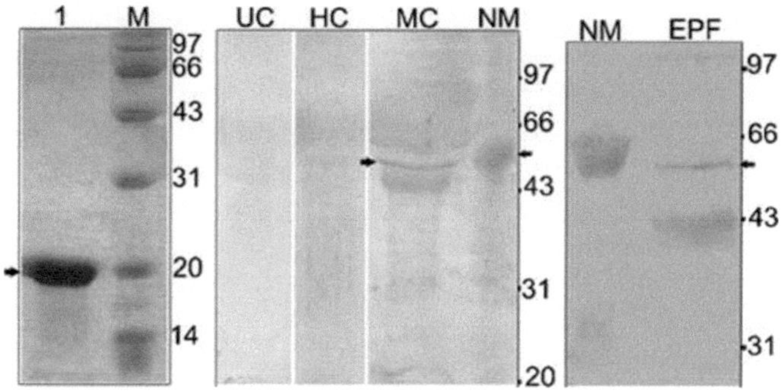

Fig. 5.15 SDS-PAGE analysis of purified N61/217 (lane 1, arrow indicating 20 kDa) and Western blot analysis of uncultured medium (lane UC), medium from heart tissue culture (lane HC), medium from mantle tissue culture (lane MC), water-soluble matrix proteins (lane NM), and extrapallial fluid (lane EPF); M indicates standard proteins; the arrow indicates the band of approximately 60 kDa

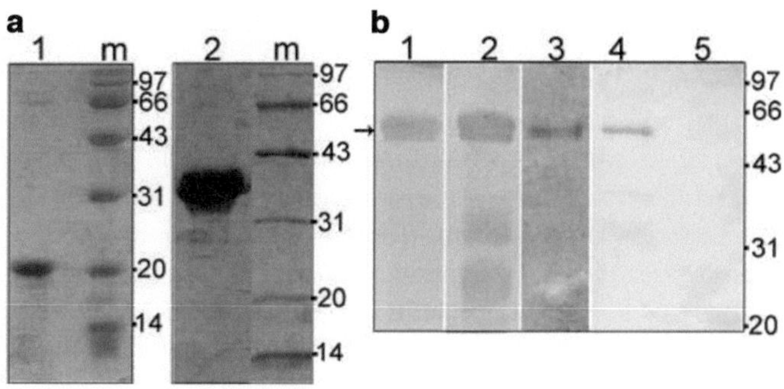

Fig. 5.16 SDS-PAGE and Western blot analysis. (**a**) SDS-PAGE analysis of the purified recombinant proteins N1 (*lane 1*) and N2 (*lane 2*) with protein markers (*lane m*) in kDa. b Western blot analysis of natural nacrein by using McAb (*lane 1*) or anti-N2 (*lane 2*) as primary antibody (*arrow* band of around 60 kDa) and analysis with anti-N2 as primary antibody to determine the eluate of extrapallial fluid (*lane 3*), mantle explant culture medium (*lane 4*), and uncultured medium (*lane 5*) from the immunoaffinity column with the ligand of McAb

specificity of the sandwich ELISA. Western blot results (Fig. 5.16b, lanes 3, 4) showed that a single band at around 60 kDa was observed using anti-N2 as a primary antibody to identify nacrein in the explant culture medium and extrapallial fluid eluted from the immunoaffinity column containing the ligand of McAb. No band was observed in the eluate of uncultured medium (Fig. 5.16b, lane 5). These results suggested that biotinylated McAb as the detection antibody and anti-N2 as the capture antibody were specific for nacrein detection in the developed sandwich ELISA, both in vitro and in vivo. The detection limit was 120 pg/ml of natural nacrein; the inter- and intra-assay variability of our ELISA were 21% and 6.8%, respectively.

Viable cells (2×10^6) from 36%F, 30%F, and 15%F were cultured for 3 days in 12-well plates containing 0.5 ml culture medium in each well. The cultured medium was transferred to the microtiter plate and subjected for ELISA assay. In the cultured medium of 36%F cells, the average concentration of nacrein was 320 pg/ml. The nacrein concentrations in the cultured media of the cells from 30%F to 15%F were far below the limit of detection in our ELISA method. These results validated that cells in 36%F were able to secrete nacrein.

5.5.1.3 Effect of Ca^{2+} on Nacrein Secretion

The effect of calcium on nacrein secretion was tested using the 36%F cells containing dominant epithelial-like cells. The cells (2×10^6) were placed in a 12-well plate adding 0.5 ml culture medium. After a 3-day cultivation, the cell viability maintains 90%–95%. Increased concentration of calcium in the medium resulted in increased nacrein production: 11.5 mM and 14 mM Ca^{2+} promoted 1.5-fold and 1.65-fold higher nacrein production than 2.5 mM Ca^{2+}, respectively.

The cell viability and functionality for biomineralization [9, 15, 21] have been successfully maintained in mantle explant culture, although in vitro cellular metabolism of the mantle cell remains to be elucidated to date. The shortage of techniques for investigation of cellular metabolism in biomineralization was one of the major problems in previous studies. As the critical regulating effectors of shell growth, matrix protein secretion by the mantle should be an important marker of in vitro cellular metabolism in biomineralization. We first used Western blot analysis to verify the presence of nacrein in the culture medium of outer mantle epithelia and then developed an ELISA method to measure the amount of nacrein. We used an immunochemical method to quantify nacrein secretion rather than by using its carbonic anhydrase activity [61]. As shown in our ELISA results, the concentration of nacrein in the cultured medium was too low to measure enzyme activity. Therefore, the ELISA method should be a sensitive assay in quantitative analysis of nacrein and other shell matrix proteins as well.

During biomineralization process, calcium counts an indispensable element. Rousseau et al. [63, 64] have found that the concentration of circulating calcium in a 48-month-old *P. margaritifera* is 12.3 mM, 0.8 mM higher than that in 24-month-old animals with 11.5 mM circulating calcium, indicating that the calcium requirement in shell formation increases as the animal grows. Hence, to meet the increasing requirement for shell growth, more matrix protein must be produced. Our ELISA results show that the increased calcium concentration in the medium stimulates the increased nacrein production in a concentration-dependent manner. The stimulation of nacrein production by calcium indicates that the oyster might regulate matrix protein expression by adjusting the circulating calcium concentration. Furthermore, these results suggest that the purified mantle epithelial cells could display detectable responses to special regulators in vitro.

Isolated mantle cells have been showed to be able to secrete some organic matrix, such as collagen, proteoglycans, and glycosaminoglycans in vitro [6]. Based on our experiments, we infer that the expression of shell matrix proteins, nacrein, MSI60, and MSI7 was retained in in vitro cultured mantle epithelia within at least 12 days after culture began. Nacrein in mantle culture media could be detected by Western blot analysis and ELISA assay within a month. Results of Western blotting and ELISA assay of the mantle culture medium were in agreement with that in the extrapallial fluid, which demonstrated that in vitro cultured mantle epithelia could synthesize some SMPs similar to the mantle epithelia in vivo. Therefore, this cell culture system could be employed for elucidating the regulative mechanism in secretion of nacrein and other matrix proteins.

5.5.2 *Control of the $CaCO_3$ Precipitation In Vivo and In Vitro*

5.5.2.1 Intracellular Precipitation of ACC in Cultured Mantle Cell

No deposits were present in the mantle cells cultured in Pf-CM2.5 for 10 days (Fig. 5.17a). On the opposite, mineral deposits in irregular or spherical shape were observed inside the mantle cells from 15 to 25 days, with the number of deposits significant increased during the cell culture period (Fig. 5.17b, c, d). When the mantle cells were grown in Pf-C10M50 containing Ca^{2+} and Mg^{2+} similar to the EPF in concentrations [65], the mineral deposits became larger and abundant (Fig. 5.17e). The spot EDS results revealed that the elemental compositions were similar in these deposits, containing dominant calcium, oxygen, carbon, and phosphorus, with magnesium in small amount (Fig. 5.17h, i); the composition of the cells was mainly oxygen and carbon (Fig. 5.17g). The detection of phosphorus in the calcium deposits is consistent with the presence ACC phase, as Jacob et al. recently proposed [66]. To verify that the calcium deposits were formed de novo, we tested the incorporation of Sr into the deposits. When the mantle cells were cultured with Pf-S10 M50 for 25 days, numerous deposits were seen in the mantle cell (Fig. 5.17f). The spot EDS data showed that the deposits contained mainly carbon, phosphorus, strontium, and oxygen, with small amounts of calcium and magnesium (Fig. 5.17j).

To determine the polymorph of intracellular calcium deposits, the powdered cell components were analyzed using FTIR and XRD [67]. The FTIR spectrum indicated that ACC could be present in the mineral deposits come from the mantle cells (Fig. 5.18a). Furthermore, a broad hump spanning between 20u and 40u of 2θ was found in the XRD profile, which was similar to that of synthetic ACC (Fig. 5.18b). This result was consistent with the previous report for the early veliger stages of *Pinctada fucata* [68] but different from our previous study of calcium carbonate formed in the mantle explants culture [13]. Usually, amorphous materials do not show birefringence under polarized light, surprisingly enough, ACC in mantle was birefringent in our study. We supposed that birefringent areas might act as centers of mineralization containing amorphous and crystalline material of different ratios [69], but the trace of crystalline materials might be omitted due to the detection limit; another possibility is that some highly organized structure or special organic

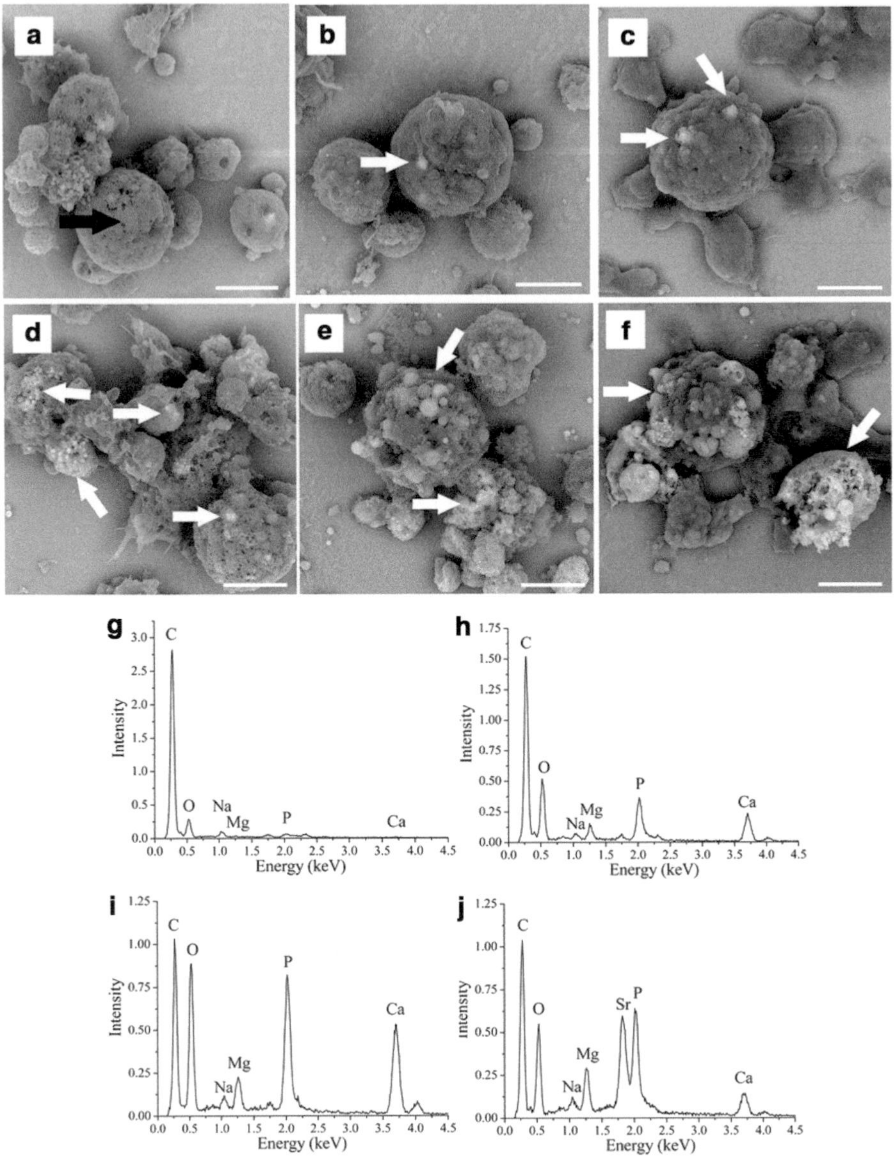

Fig. 5.17 Backscatter electron imaging combined with energy-dispersive X-ray spectral analysis of the mantle cells [67]. In the backscatter electron (BSE) mode, elements with a higher atomic number scatter more electrons (such as calcium and strontium), thus appearing lighter, whereas carbon appears black. (**a**), (**b**), (**c**), and (**d**) SEM-BSE images of cells at 10 days, 15 days, 20 days, and 25 days, respectively, in Pf-CM2.5. (**e**) and (**f**) SEM- BSE images of cells at 25 days in Pf-C10M50 and Pf-S10 M50, respectively; the intracellular crystals are indicated with white arrows. (**g**) Energy-dispersive X-ray spectral (EDS) of the cells in (**a**), the elemental composition of the cells consisted of C and O. The black arrow on the SEM image indicates the sample point of the EDS. (**h**) EDS of the crystals in (**b**), (**c**), and (**d**). The elemental composition of the crystals is C, O, Mg, P, and Ca. (**i**) and EDS of the crystals in (**e**) and (**f**), respectively. The elemental composition of the crystals is C, O, Mg, P, and Ca in (**e**) and C, O, Mg, P, Ca, and Sr in (**f**). The white arrows on the SEM image indicate the sample point of the EDS. Scale bars 10 mm

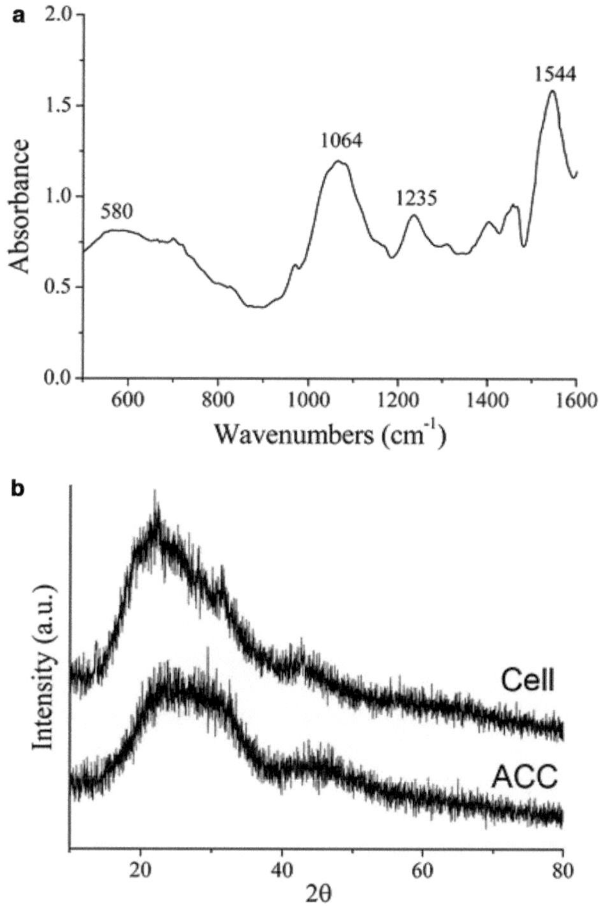

Fig. 5.18 FTIR and XRD of intracellular crystals showing characteristic bands indicating the presence of ACC. (**a**) FTIR spectra of powdered intracellular crystals; the broad bands at 580, 1064, 1235, and 1544 cm21 are not related to a distinct calcium carbonate polymorph (calcite or aragonite). (**b**) X-ray powder diffraction (XRD) pattern of intracellular crystals (cell) and synthetic ACC (ACC); the intracellular crystals and synthetic ACC exibit the same characteristic bands for ACC

compounds in the ACC-containing vesicles may exhibit birefringence. [70] Similar results have been reported by Auzoux-Bordenave et al. [71] Moreover, the gene expression of ACCBP was relatively high from day 10 to day 25 during cell culture, compared with nacrein and Pif80, indicating that ACCBP might be involved in the intracellular ACC synthesis because ACC could be stabilized by ACCBP [72]. Further studies are needed to elucidate how ACC transforms into aragonite or calcite in shell formation.

ACC synthesis is an important event not only in shell formation of pearl oyster but also other mineralization processes, such as spicule formation. Beniash et al. [73] found that in sea urchin embryos, ACC-containing vesicles were present in the mineralizing cells and speculated that ACC is transported into the spiculogenic compartment when the vesicle membrane fused with the cell membrane of the syncytium so that each spicule grows at the expense of ACC. Based on our study, we propose that shell formation may begin with ACC formed in the intracellular vesicles of the mantle cell, and then ACC is exported to the EPS and subsequently transforms into aragonite or calcite. In other words, shell formation may proceed in a cell-mediated manner that crystal nucleation occurs inside the mantle cells and the crystallogenic cells continue transporting crystals to the mineralization front. Such hypothesis challenges the classical matrix-mediated shell formation mechanism, which claim that crystal nucleation and growth take place only in the extrapallial fluid with the guide of the self-assembly shell matrix. An ultrastructural study of the mantle in the gastropod *Haliotis asinina* was also compatible with the cell-mediated hypothesis [74]. In our previous study, some "imprints" of aragonite were observed, together with "shaped" aragonite, and the "imprint" might be the trace that the mantle cells had abutted to the inner shell surface [75]. Addadi et al. [76] proposed that mantle cells must be juxtaposed to the mineralizing matrix in illustrating how shell is being produced. What's more, oyster shell formation is involved in complex assembly of diverse proteins and variant cellular processes, biomineralization, suggested by Zhang et al. [77, 78]

5.5.2.2 $CaCO_3$ Deposition in the Mantle Cell Cultured

Mantle and heart tissue explants were cultured in C10M23.5 medium with Ca^{2+} in similar concentration to the body fluid of oyster [63]. After 25 days, mineral deposits in irregular or spherical shape were observed around the mantle explants (Fig. 5.19a, b) and in the cell outgrowth area (Fig. 5.19c) [13]. Spot EDS analysis revealed that the elemental compositions of these deposits were quite similar, including large amounts of calcium, oxygen, and carbon, together with small amounts of magnesium, phosphorus, nitrogen, and sulfur (Fig. 5.19d). Surprisingly, mineral deposition was also present in heart tissue cultures, as showed in the light area (Fig. 5.19e), which was clustered by numerous fine grains (magnification in Fig. 5.19f). EDS analysis indicated that calcium composition in this light area was higher than that in the dark area. In the heart tissue culture, when the concentration of Mg^{2+} was reduced (media C10M10 and C10M5), many spherical granules with diameter of approximately 2–10 μm were observed in the cell outgrowth area after a 25-day cultivation (Fig. 5.19g, h). Spot EDS analysis revealed that the elemental composition of the granules was similar to the calcium deposits grown in medium C10M23.5 (Fig. 5.19i).

When the Ca^{2+} and Mg^{2+} concentrations in the medium were both decreased to 5 mM, calcium deposits observed in the heart cultures after 25 days (Fig. 5.20a, arrow) were significantly less than those generated in the media containing 10 mM Ca^{2+}. Few deposits were found in heart tissue cultures after a 25-day incubation in Pf-CM2.5 medium (Fig. 5.20b). In contrast, many deposits scattered throughout the mantle explants (Fig. 5.20c, d) and cell outgrowth region (Fig. 5.20e) after a 15-day cultivation with sizes generally no more than 10 μm in length; these deposits contained carbon, calcium, and oxygen in major and some minor elements, phosphor, sulfur, magnesium, nitrogen, silicon, and sodium (Fig. 5.20c–s). In the 15-day-old mantle cultures, calcium deposition was seen at the tip of some cell membrane (Fig. 5.20f). After 45 days, more mineral deposits (5–10 μm in length)

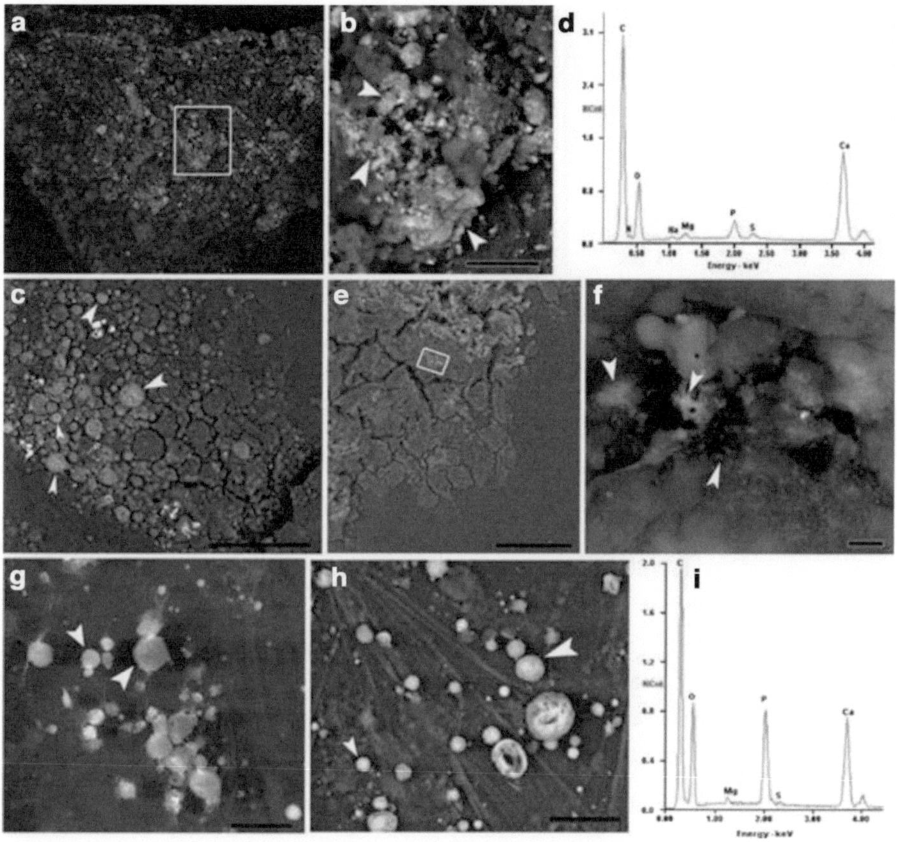

Fig. 5.19 SEM-BSE imaging calcium deposits in mantle and heart tissue cultures incubated in the media with 10 mM Ca^{2+}. (**a**) Calcium deposits in mantle tissue explants incubated in medium C10M23.5 for 25 days. (**b**) Magnified view of the frame shown in (**a**); the arrowhead indicates deposits with irregular morphology. (**c**) Calcium deposits in the outgrowth area of mantle tissue culture; the arrowhead indicates granules with spherical morphology. d X-ray spectrometry of the deposits in (**b**). (**e**) Calcium deposits in heart explants, shown in the light area. (**f**) Magnified view of the frame in (**e**); the arrowhead indicates fine grains. (**g**) and (**h**) Spherical granules in 25-day-old heart tissue cultures incubated in C10M10 and C10M5 media, respectively; the arrowhead indicates the spherical granules. (**i**) X-ray spectrometry of the deposits in g and h. Scale bars in a–c and e–h: 50, 10, 50, 75, 2, 10, and 20 μm, respectively

were generated in the explants (Fig. 5.20g, h) and the cell outgrowth area (Fig. 5.20i), most of which were larger than those deposits in 15-day-old cultures. It should be noted that these calcium deposits consisted of many nanograins and exhibited nanogranular structure (Fig. 5.20h, i). The elemental composition of these deposits was similar to those found in 15-day-old mantle cultures (Fig. 5.20h–s, i–s).

Significant calcium deposits were observed in 25-day-old mantle cultures with C10M23.5 medium containing similar concentrations of Ca^{2+} and Mg^{2+} to the oyster body fluid. [63] The deposits were in irregular and spherical morphology, like those

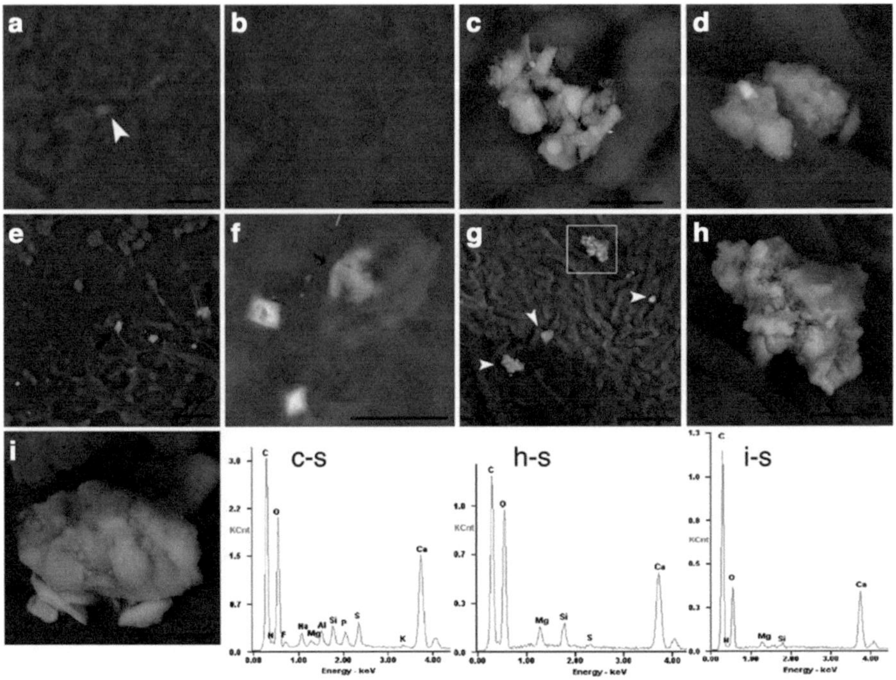

Fig. 5.20 SEM-BSE imaging of calcium deposits in heart and mantle cultures incubated in media containing low concentration of Ca^{2+}. (**a**) Small calcium deposit (indicated by arrowhead) found in heart explants incubated in C5M5 medium. (**b**) No deposit found in heart explants incubated in Pf-CM2.5 medium. (**c–i**) Calcium deposits forming in Pf-CM2.5 medium. (**c**), (**d**) Calcium deposits in 15-day-old mantle explants. (**e**) Calcium deposits in the cell outgrowth area of 15 day-old mantle tissue culture (indicated by arrows). (**f**) Calcium deposits at the tip of mantle cell membranes (indicated by arrow). (**g**) Calcium deposits in 45-day-old mantle explants (indicated by arrow). (**h**) Magnified view of the frame in (**g**). (**i**) Calcium deposit in the cell outgrowth area of a 45-day-old mantle tissue culture. (**c-s**), (**h-s**) and (**i-s**) X-ray spectrometry of the deposits in (**c**), (**h**), and (**i**), respectively. Scale bars in (**a–i**): 10, 5, 5, 2, 20, 10, 10, 5, and 5 μm, respectively

described by Samata et al. [15] However, remarkable calcium deposits were also observed in heart tissue cultures, clustered by fine grains. When the concentration of Mg^{2+} reduced, large granules were deposited instead of fine grains, indicating that the inhibition of Mg^{2+} on calcium carbonate crystallization was weakened [79, 80]. It has been proved that amorphous calcium carbonate-binding protein (ACCBP) was able to stabilize supersaturated $CaCO_3$ solution in vivo by inhibiting ACC crystallization thus preventing undesired crystal growth [72]. We deduced that safeguards in in vitro tissue cultures were not sufficient as in in vivo situation. Therefore, calcium would precipitate when bicarbonate was continued supplying by alive cells to increase the saturation of Ca $(HCO_3)_2$. However, this process would be remarkably weakened when the concentration of calcium was reduced. Therefore, in heart cultures, calcium deposits were significantly reduced in C5M5 medium compared with mantle tissue culture, and few deposits were observed in Pf-CM2.5 medium.

In mantle cultures for Pf-CM2.5 medium, many calcium deposits were observed after15 days and became larger after 45 days, suggesting that the deposits gradually grew in the culture. The chemical composition was confirmed to be mainly calcium carbonate based on EDS results. Composed of smaller granule buildups, the nanogranular structure of these deposits displayed the nature of biogenic minerals (Fig. 5.20h, i), indicating their biogenic origin. Because nanograined texture is a typical feature in calcium carbonate biominerals but not in abiogenic crystals [81].

Calcium deposits generally located near the adherent mantle explants as well as the cell outgrowth area. In some cases, calcium deposits were present between cells or at the tip of cell membranes (Fig. 5.20e, f), consistent with the calcium nodules grown on the extracellular fibrils network extending from coral calicoblast cells [82]. Therefore, we deduced that organic matrixes secreted from mantle epithelia might initiate $CaCO_3$ nucleation by enriching Ca^{2+} on their surfaces, thus promoting calcium deposition in a low-calcium environment. Based on the results of RT-PCR and Western blot analysis, we proposed that some matrix proteins present in mantle tissue culture, such as nacrein, MSI7, and MSI60, might be involved in calcium deposits nucleation and growth. Alternatively, hemocytes located in mantle explants might induce crystal nucleation, because hemocytes contain intracellular $CaCO_3$ granules [17].

Our experiments show that mantle tissue cultures could, at least, partially resemble mantle-mediated shell formation process in vivo. So mantle tissue culture would be a useful tool for understanding biomineralization at the cellular level. Our results also reveal that the cell culture system containing low concentration of calcium and magnesium is more applicable for investigating biological mechanisms in biomineralization than media with high calcium/magnesium concentrations, because inorganic effects on mineralization were remarkably weakened in the former case.

5.5.2.3 Cultured Mantle Cells Regulate $CaCO_3$ Precipitation

$CaCO_3$ crystals grew in neither the gonadal cell culture nor the uncultured cell medium, indicating that calcium carbonate does not nucleate in the culture medium spontaneously and that normal cellular metabolism does not induce calcium carbonate crystallization solely. However, numerous crystals were found in the primary mantle cell culture and deposited throughout the entire cover slip. We noticed that crystals nearest the cell aggregates formed in priority both in size and number. Some adjacent crystals fused with each other.

The primary mantle cells mediating growth of $CaCO_3$ crystals is demonstrated in Fig. 5.21. Both the size and morphology of the crystal changed over time. Crystals were not observed by light microscopy until day 7, and then they grew over time. The $CaCO_3$ crystals were in polyhedron shape at first with one sharp end (Fig. 5.21a). Bright ridges were found on the surface of these crystals and were

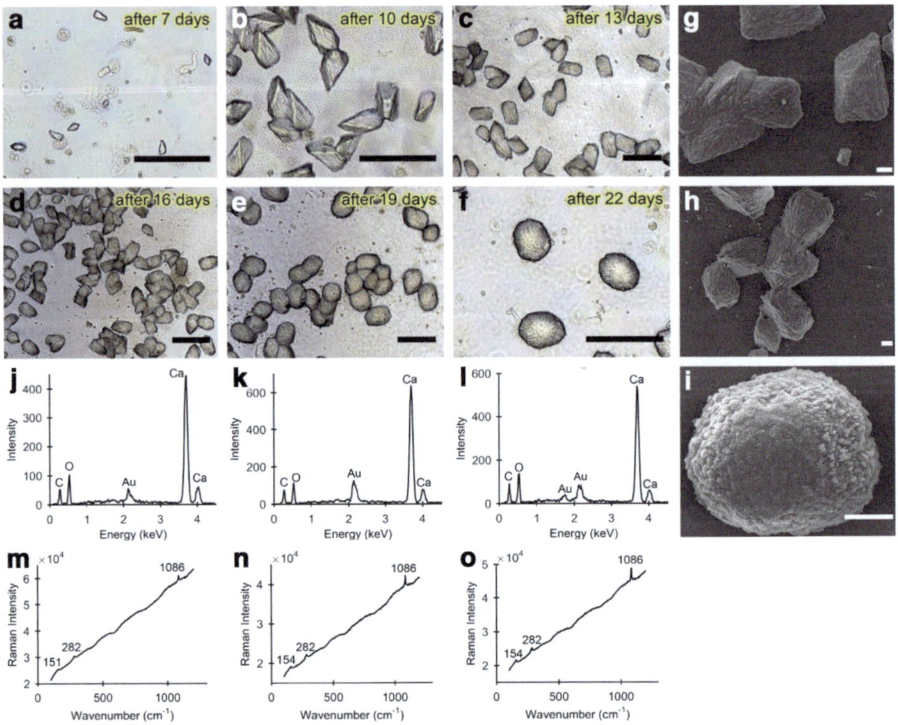

Fig. 5.21 Crystal growth process in the mantle cell culture [83]. (**a**) Crystals began to appear after 7 days. (**b**) The number of polyhedron-shaped crystals increased after 10 days. (**c**) The polyhedron-shaped crystals became flat after 13 days. (**d**) Adjacent crystals aggregated after 16 days. (**e**) Round rectangular-shaped crystals appeared after 19 days. (**f**) Round crystals appeared after 22 days. Scale bar (AeF) 100 mm. (GeI) Scanning electron microscopic images of polyhedron-shaped crystals, round rectangular-shaped crystals, and round crystals, respectively. Scale bar ¼ 20 mm. (JeL) Energy-dispersive X-ray spectra of the crystals in (GeI); these crystals were composed of Ca, C, and O. (MeO) Raman spectrum of the crystals in (GeI); these crystals were all composed of calcite

only a few microns in length at first but grew to more than 50 μm ultimately (Fig. 5.21b). The EDS analysis indicated that the crystals were mainly composed of calcium, carbon, and oxygen (Fig. 5.21j) and their polymorphs were calcite, as Raman spectroscopy showed (Fig. 5.21m). After 13 days, the polyhedron-shaped crystals gradually grew into larger crystals with a flatter appearance (Fig. 5.21c) and length of 50–100 μm. Adjacent crystals began to aggregate together as the time passed by (Fig. 5.21d), and the edges in some crystals became round (Fig. 5.21e). These crystals were calcite and composed of calcium, carbon, and oxygen (Fig. 5.21k, n). Round crystals were found under optical microscope after 22 days (Fig. 4.21f). Raman spectroscopy and EDS analysis suggested that the round crystals were composed of calcite calcium carbonate (Fig. 5.21l, o).

The SEM images showed more details about the calcium carbonate crystals (Fig. 5.21). Each crystal was piled up by numerous highly ordered tiny pieces,

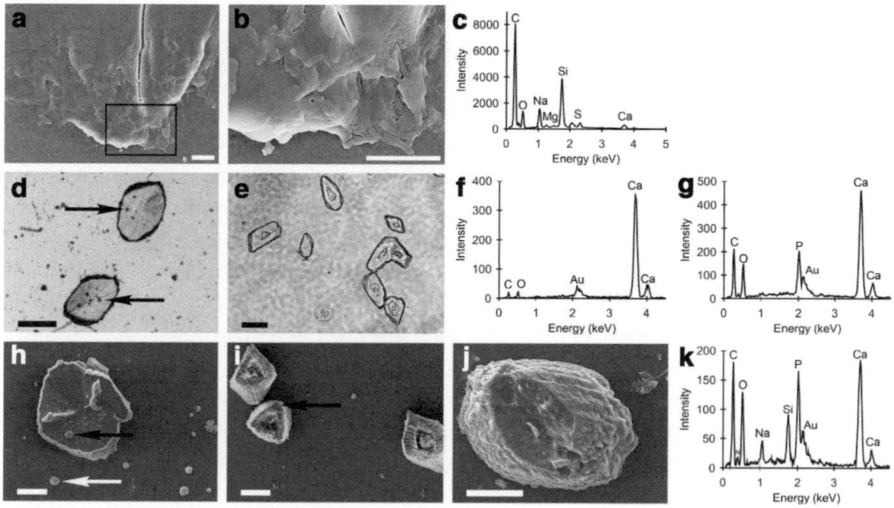

Fig. 5.22 Mantle cell remodeled crystals. (**a**) The organic frame of the crystals was fragile and was cracked. (**b**) Magnification of the box in (**a**); little holes were observed on the organic frame. (**c**) The energy-dispersive X-ray spectral (EDS) analysis of the organic frame showed that it contained C, O, S, Na, Si, Mg, and Ca. Scale bar (**a, b**) 4 mm. (**d**) Crystals with flat and smooth surfaces. Cell-like particles attached on to the top of the crystals (arrows). (**e**) Crystals with a hole inside. (**h**), (**i**) Scanning electron microscopic (SEM) images of the crystals in (**d**) and (**e**). (**j**) SEM image of crystal with a partly etched surface. Scale bar (DeJ) 40 mm. (**f**) EDS analysis of crystals in (**h**); the crystals were composed of Ca, C, and O. (**g**) EDS analysis of the cell-like particles in (**h**) (black arrow); the particles were composed of Ca, C, O, and P. (**k**) EDS analysis of the cell-like particles in (**h**) (white arrow); the cell-like particles were composed of Ca, C, O, P, N, Na, and Si

which is the characteristics of typical biomineralization product. When treated with 10% EDTA (pH 8.0) solution, a semi-transparent fragile film containing cracks and holes was observed (Fig. 5.22a, b), whose outlines resembled the shapes of the corresponding crystals. The EDS results showed that the film consisted of 73% C (Fig. 5.22c). Therefore, this film was likely organic framework for mineralization.

As culture time passed by, the surfaces of some crystals appeared to have been reformed by the mantle cells. Fig. 5.22d showed one crystal of such type. These crystals had smooth and flat surfaces, very similar to transparent glass disc. The size of these crystals was 50–100 μm in diameter, larger than the average size of the total crystals. Round particles were found on the surfaces of some crystals (Fig. 5.22d, arrows) and supposed to be mantle cells. SEM showed that the glass disc-shaped crystals also consisted of tiny pieces, and the edges were rough, just like other crystals (Fig. 5.22h). Some round particles, adhered on the smooth surfaces of the crystals, were 10 μm in diameter, consistent with the mantle cells as previously reported [84]. The EDS analysis further confirmed the idea that the two particles were mantle cells (Fig. 5.22f, g, k), due to the significant composition ratio of phosphorus, which is an important component of cell membrane.

The crystals shown in Fig. 5.22e were similar to those polyhedron-shaped crystals in Fig. 5.21b, except that they had a small hole inside. These crystals had holes about 20 μm in diameter on the upper surface and were also piled up with thin sheets. The crystal surfaces were partially smooth (Fig. 5.22i, arrow). There was another interesting type of crystal (Fig. 5.22j), which was quite similar to that shown in Fig. 5.21i, except that it was largely concave. The inner surface of this crystal was relatively smooth compared with the outer surface, indicating that it might be remodeled into glass disc-shaped crystal.

Molluskan shell formation is generally considered to be an extracellular process mediated by mantle-secreted matrix proteins. However, whether mantle cells directly participate in the biomineralization remains unknown. In vitro cultured mantle cells have been showed to retain their capacity of secreting matrix proteins and mineralization-related enzymes [84]. However, it was showed that no calcium carbonate crystals grew in cultured medium without mantle cells. So the substances secreted from mantle cells, such as shell matrix proteins and enzymes as well as the inorganic ions cannot mediate crystal nucleation and growth solely.

Crystal nucleation can process intracellularly in biologically controlled mineralization, such as spicule formation in sea urchin [73]. Mount et al. demonstrated that the oyster hemocytes are involved in intracellular crystal formation [17, 85]. In our study, $CaCO_3$ only precipitated when mantle cells were present, strongly indicating that mantle cells directly participate in crystal nucleation process. This finding is in accordance with the existence of ACC in mantle cells [84]. ACC is a highly amorphous phase and a precursor of aragonite and calcite [69, 86]. We deduced that mantle cells intracellularly synthesize and transport ACC to the mineralization front, where ACC was transformed into aragonite or calcite later.

EDTA (10%) is widely employed for mineral dissolving in shell research [87]. The semi-transparent film that found after EDTA treatment was probably organic matrix providing frames for crystal growth. The mantle is responsible for synthesizing matrix proteins which would self-assemble into shell framework [85, 88]. So it is reasonable to assume that the mantle cells are the source of the framework-forming matrix, leading to more crystals deposited in regions closer to the cell aggregates. In addition, the crystals near the cell aggregates were always larger than in other area and aggregated gradually as time passed by. This phenomenon may due to the relatively high concentration of matrix protein near the cell aggregates.

All the calcium carbonate deposits in the mantle cell culture were calcite, which might ascribe to the region of mantle tissue chosen for cell culture. The mantle tissue is divided into the edge, pallial, and center zone [4]. The mantle edge is responsible for depositing the prismatic layer, which consists of calcite, whereas the pallial and center zone are responsible for forming the nacreous layer, which is composed of aragonite [89]. We have previously found that tissues from the mantle edge generated more cells; thus, we cultured the tissue from this region to obtain isolated cells [83]. These mantle cells mainly secrete matrix proteins that are responsible for the formation of calcite, which explains why all the crystals were calcite.

The $CaCO_3$ crystals that grew in the mantle cell culture were typical products of biomineralization [81]. They were all composed of superimposed units, which is quite different from pure calcium carbonate formed under inorganic conditions. Crystals in such morphology may be the result of cooperation between organic matrix and cells. Mantle cells may be involved in the complex assembly and modification processes of diverse proteins, leading to the formation of the oyster shell [78]. Organic matrix could help control the growth pattern of biominerals [90]. The morphology of crystals changed over time in our cell culture. Polyhedron-shaped crystals appeared first and gradually transformed into round rectangular-shaped crystals. At last, round crystals were present after a 20-day culture. The changes in crystal morphology may due to accumulation of different matrix proteins or some unknown crystal reconstruction mechanism.

At the early stage of cell culture, most crystals had rough surfaces but later some crystals with smooth facets appeared. The crystals in Fig. 5.22h had a special flat and smooth appearance. Mantle cells were found to adhere onto the smooth surfaces of these crystals, indicating the participation of mantle cells in the crystal remodeling process. The crystals showed in Fig. 5.22i, J had partially smooth surfaces, and they seemed to be etched. If the etch proceeds, they would likely become the glass disc-shaped crystals in Fig. 5.22h. Therefore, these crystals are regarded as "semi-finished products." Crystals remodeled by cells have also been reported by Mount et al. [17], who showed that hemocytes in eastern oyster reformed the crystals into platelike structures.

Mount et al. [17] and Xiang et al. [84] both proposed that mantle cells export minerals to the mineralization front and remain attaching to the front. We propose that mantle cells are directly involved throughout shell formation process, including $CaCO_3$ nucleation, growth, and remodeling. Our results demonstrate that studying shell formation at cellular level using a mantle cell culture is practical and could shed light on the cellular regulation of calcium carbonate mineralization.

5.6 Roles of Hemocytes in Shell Formation

Shell formation is dominated by mantle tissue in *Pinctada fucata*. The cells lining on the outer epithelium of mantle tissue secrete organic components such as shell matrix proteins (SMP) and inorganic ions such as calcium and magnesium into the extrapallial spaces (EPS) for CaCO3 precipitation. In addition to this widely recognized mantle-mediated shell formation theory, biomineralization mediated by hemocytes was proposed very recently. Mount et al. [17] showed that in the eastern oyster *Crassostrea virginica*, special refractive hemocytes could mediate the mineralization process and are responsible for shell damage. These hemocytes carried considerable calcium in crystal form and were competent to fulfill the vacancy of the large nonexchangeable calcium pool left by mantle tissue. Furthermore, hemocytes bearing $CaCO_3$ crystals or Ca-positive granular contents were also found to participate in shell repair in oysters [18] and mussels [91], indicating that hemocytes may play a role in shell formation. Besides, hemocytes in some oysters were responsible for part of the SMP secretion [77] and appeal to more studies about the exact role hemocytes play in shell formation.

5.6.1 Identification and Classification of Hemocytes

5.6.1.1 Morphology of the Hemocytes in *P. fucata*

After H&E staining, all the hemocytes looked blue in the cell nucleus and pink in the cytoplasm; visual observation characterized differences in the N/C, size, and shape. Small, middle, and large-sized hemocytes were observed according to the cell N/C and size (Fig. 5.23a, b). Moreover, hemocytes could be grouped into granulocyte, small hyalinocyte, and large hyalinocyte, by DIC microscope (Fig. 5.23c, d). With no granules in the cytoplasm, hyalinocytes in smaller diameter were round, and those in larger diameter looked like nerve cells with many pseudopodia at the cell edges. With the cytoplasm packed with granules, granulocytes exhibited an irregular or round shape, larger cell size, and smaller N/C and had few pseudopodia extending from the cytoplasm edges.

SEM images of hemocytes revealed three cell types with diameters of 1–3 μm, 3–5 μm, and 5–8 μm, respectively (Fig. 5.23e, f). These hemocytes exhibited differences on the shape of pseudopodia and the cell surface. Those elliptical hemocytes with four protrusions on the head and a tail at the bottom were most likely to be the small hyalinocytes. These small hyalinocytes were the most abundant cell group of the hemocytes. Round hemocytes with small pseudopodia and a smooth surface were most likely to be the middle-size hemocytes or large hyalinocytes. Other hemocytes, with small pseudopodia and similar diameters to large hyalinocytes, were most likely to be granulocytes and had irregular surfaces with shallow protrusions. TEM observation was used to further confirm the identified three types of hemocytes, large hyalinocytes, small hyalinocytes, and granulocytes (Fig. 5.24), based on the cellular contents and organelles. The predominant hemocytes were small hyalinocytes with large nuclei and scanty cytoplasm. The chromatin was condensed in electron-dense granules in the cell nucleus and in small clumps near the periphery. Abundant mitochondria, Golgi apparatus, endoplasmic reticulum, and vesicles were found in the cytoplasm (Fig. 5.24a). It should be noted that blast-like cells have been observed among the small hyalinocytes, having few, if any, membranous organelles and a large nucleus (Fig. 5.24b). The large hyalinocytes contained smaller nuclei and denser cytoplasm and were larger in diameter. Abundant vesicles and mitochondria could be seen in the cytoplasm, and the chromatin in the cell nucleus was highly condensed (Fig. 5.24c). The granulocytes generally had small nucleus and contained electron-lucent or electron-dense granules and a few vesicles in the cytoplasm. This type of hemocytes exhibited an irregular shape and rough cell membrane with various protrusions (Fig. 5.24d, e, f).

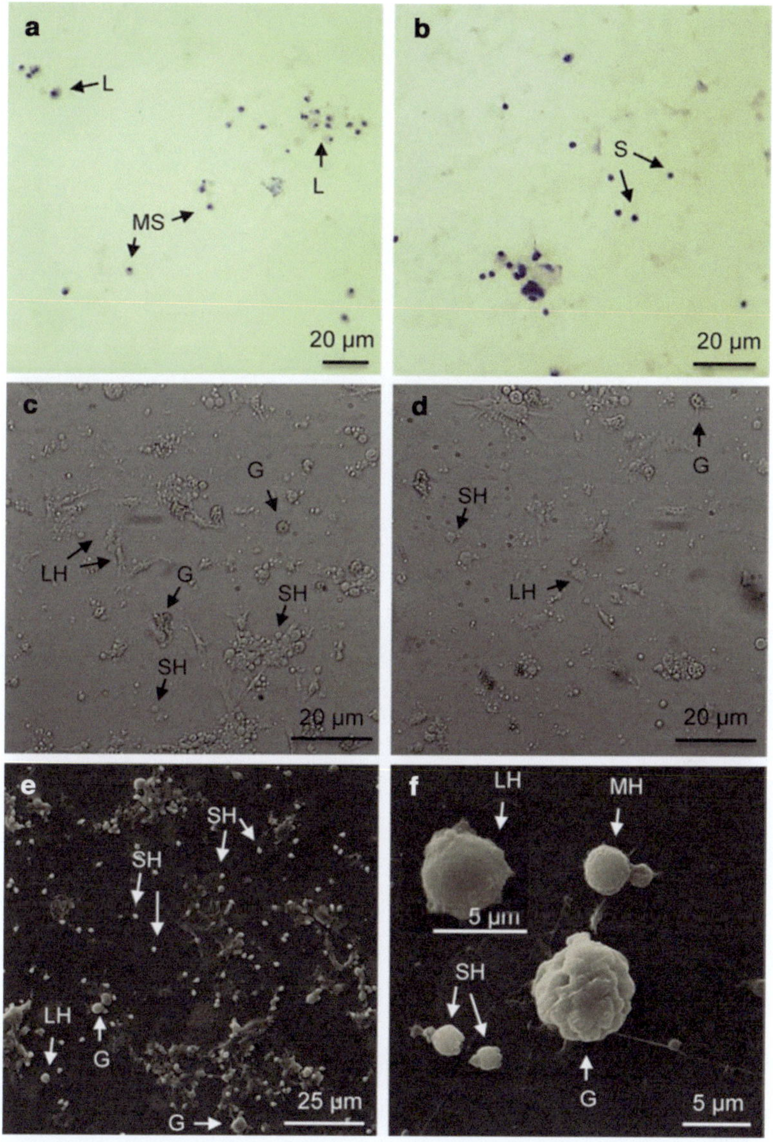

Fig. 5.23 Light micrographs and scanning electron micrographs of hemocytes in *P. fucata*. (**a**) and (**b**) light micrographs of hemocytes stained by hematoxylin-eosin; (**c**) and (**d**) differential interference contrast light micrographs of hemocytes; (**e**) and (**f**) scanning electron micrographs of hemocytes. L, large hemocytes; MS, middle-size hemocytes; S, small hemocytes; LH, large hyalinocyte; SH, small hyalinocyte; G, granulocyte

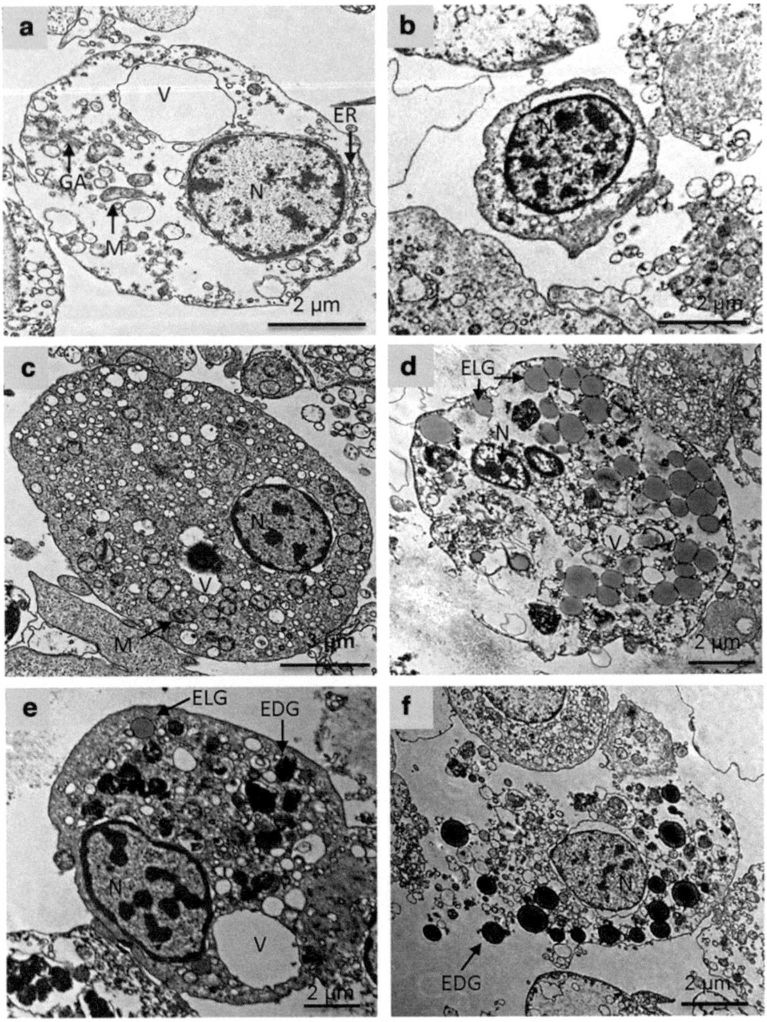

Fig. 5.24 Transmission electron micrographs of hemocytes in *P. fucata*. (**a**) Small hyalinocyte; (**b**) blast-like cell; (**c**) large hyalinocyte; (DeF) granulocyte. N, nucleus; M, mitochondria; GA, Golgi apparatus; V, vesicle; ER, endoplasmic reticulum; EDG, electron-dense granule; ELG, electron-lucent granule

5.6.1.2 Flow Cytometry Analysis

In flow cytometry, FSC and SSC were commonly employed to determine cell size and internal complexity. As shown in Fig. 5.25, three subpopulations were identified in hemocytes of *Pinctada fucata*: small hyalinocytes (R1), large hyalinocytes (R2), and granulocytes (R3). R1 counted for 78.23% of the total hemocytes population and exhibited the least internal complexity and smallest cell size. R2 (9.32%) and R3

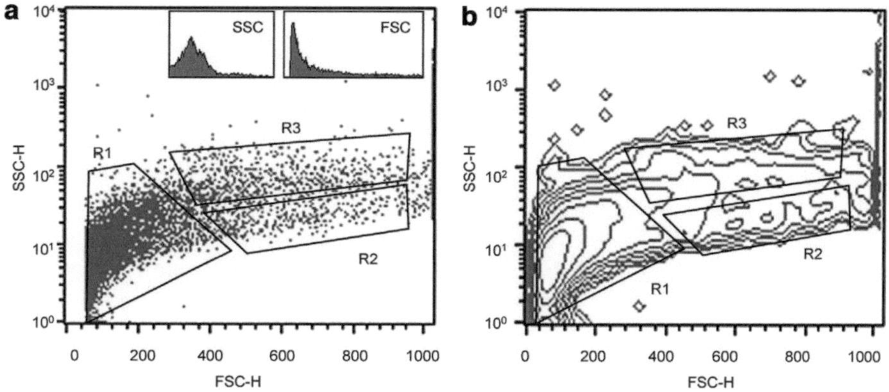

Fig. 5.25 Scatter plot (**a**) and contour map (**b**) showing the size and internal complexity of hemocytes in *Pinctada fucata* obtained by flow cytometry using forward scatter (FSC) and side scatter (SSC). Three subpopulations of hemocytes can be differentiated: small hyalinocytes (R1), large hyalinocytes (R2), and granulocytes (R3)

(12.22%) were similar in cell size but with different particle complexity, and the boundary between these two subpopulations was clear.

There is an ongoing contention about hemocyte classification in bivalve mollusks, because widely accepted nomenclature criteria are lacking [92–94]. In previous studies, two basic types of hemocytes, namely, granulocytes and hyalinocytes, have been defined in the Mollusca phylum based on the morphological features [95, 96]. More hemocyte types have been defined in some bivalves, such as *Cristaria plicata* [97],, *Perna viridis* [98], and *Chlamys farreri* [99], using flow cytometry, light microscopy, and electron microscopy. For the hemocytes in marine pearl oysters, granulocytes, hyalinocytes, and serous cells were found in *Pinctada imbricate* [93] and granulocytes, hyalinocytes, and small agranulocytes were identified in *Pinctada fucata* [100]. The previous research on hemocytes of *P. fucata* [100] only focused on the classification and had no further descriptions of the feature details about the hemocytes. In our study, the features of the hemocytes in *P. fucata* were successfully characterized by multiple techniques, and the hemocytes were divided into three groups: small hyalinocytes, large hyalinocytes, and granulocytes. Our classification results using flow cytometry were consistent with the previous study on this species [100]. The differences between the large hyalinocytes and small hyalinocytes were N/C ratio, the cell size and the types of organelles. In the small hyalinocytes subpopulation, blast-like cells were figured out as a special type of small hyalinocytes and similar to the characteristics of undifferentiated cells. The blast-like cell has been identified in *Ruditapes decussatus*, *Ruditapes philippinarum*, [101] and *Saccostrea glomerata* [102] but was reported in pearl oysters for the first time.

In our study, SEM and TEM not only confirmed the observations from light microscopy but also provided additional ultrastructural characteristics of hemocytes in *P. fucata*. Two types of granules (electron-lucent granules and electron-dense) were found in the granulocytes; these granules are the primary immunomodulation sites in hemocytes, containing defensive enzymes and antimicrobial compounds [93]. A previous study on hemocyte development hypothesized that hyalinocytes in bivalves might be the precursor of granulocytes and these two types of hemocytes may both originate from undifferentiated cells [103]. In addition, the abundant distribution of Golgi apparatuses was related to granule formation and proposed as a feature of cell differentiation [104]. However, Golgi apparatuses were widely distributed in various types of hemocytes in *P. imbricate*, suggesting that the Golgi apparatuses may not faithfully reflect the development state of hemocytes [93]. In our study, Golgi apparatuses were observed in both large and small hyalinocytes, supporting the conclusion from *P. imbricate*.

The total hemocyte count in *Pinctada fucata* (4.9×10^6 cells mL^{-1}) was higher than in some bivalves, such as *Crassostrea gigas* (up to 1.6×10^6 cells mL^{-1}) [105] and *Callista chione* (1.2×10^6 cells mL^{-1}) [92], but lower than that in *Scylla olivacea* (16.2×10^6 cells mL^{-1}) [106], *Mytilus edulis* (5.68×10^6 cells mL^{-1}) [107], *Argopecten irradians* (37.5×10^6 cells mL^{-1}) [108], and *P. viridis* (5.54×10^6 cells mL^{-1}) [98]. In addition, the abundance of granulocytes and hyalinocytes varied in different bivalves, as reported by Matozzo et al. [92] In contrast to the pearl oyster *P. imbricate* [93], the number of hyalinocytes in *P. fucata* was more than granulocytes. The differences of THC between these species may be ascribed to different environmental conditions. *P. fucata* are naturally growth in subtidal zone, whereas mussels and oysters inhabit in intertidal zone.

5.6.2 *Migration of the Hemocytes in Between the EPS and the Circulation System*

5.6.2.1 Comparison of the Hemocytes from the EPS and the Circulation System

Hemocytes in *Pinctada fucata* have been classified into three cell types: small hyalinocytes, large hyalinocytes, and granulocytes [26]. These three types of hemocytes were also found by light microscope in the EPF. Granulocytes and hyalinocytes were prominent in the EPF, accompanied by other minority groups such as blast-like cells and huge hyalinocytes in bulliform shape, which was slightly different from the hemocytes from hemolymph. The granular contents were Ca-positive in the hemocytes of the deep-sea mussel *B. azoricus* [91] and the green ormer *H. tuberculata* [109], indicating that the granulocytes in *Pinctada fucata* may serve as calcium pool.

5.6.2.2 The Hemocytes from the EPS Originate from the Circulation System

The mantle edge of *P. fucata* had threefolds: inner fold, middle fold, and outer fold (Fig. 5.26a). The inner fold faces the mantle cave and the visceral mass, and the outer fold faces the inner surface of the shell. Abundant secretory cavities and vesicles were observed on the outer surface of the mantle outer fold based on the SEM images (Fig. 5.26b–e). However, the secretory cavities were found alone on the outer mantle surface of the center zone (Fig. 5.26f), while hemocytes and crystal-like structures were also observed in this area (Fig. 5.26g, h). The morphological features of the hemocytes observed on the surface of the center zone were consistent with the granulocytes in the hemolymph of *Pinctada fucata* [26]. These hemocytes closely contacted with the secretory cavities, indicating that the granulocytes may migrate out from the cavities on the surface (Fig. 5.26h). The ultrastructure of the mantle inner surface differed from that of the outer surface by the presence of short microvilli and hyalinocytes but lacks granulocytes. In the inner fold margin, only secretory cavities were found on the inner surface (Fig. 5.26j). On the inner surface of the mantle center, there were full of bulges (Fig. 5.26k), where numerous short microvilli and small hyalinocytes were found (Fig. 5.26l). Granulocytes were not found on the inner surface in this area.

In several bivalves, the secretory cavities are responsible for secreting mucous, vesicles, or organic matrices onto the mantle surface, which are supposed to contribute to water flow, mantle lubrication, and $CaCO_3$ crystal deposition [110]. $CaCO_3$ deposits can be constantly precipitated in these cavities under the regulation of organic matrix and after that, the crystals are transported onto the surface of the mantle outer fold. This was also the case in the previous study of the pearl oyster *P. margaritifera*, where the granular contents were found in the secretory cells of the mantle's outer fold [111]. Our finding was in agreement with this conclusion, and crystal-like deposition was detected on the outer mantle surface of the central zone in *Pinctada fucata*. In our study, granulocytes were observed near the secretory cavities in *P. fucata*, suggesting that the hemocytes may be released from these cavities. If that's the case, the origin of the granulocytes in the EPF will be clear. In contrast, on the surface of the mantle inner fold, short microvilli and small hyalinocytes were present, and the short microvilli are propitious to and breathe air filter food by promoting the passage of seawater [112]. More and more evidences indicate that it is the granulocytes, not the hyalinocytes, that are responsible for $CaCO_3$ precipitation in oysters [17, 113–115]. Therefore, the ultrastructure of *Pinctada fucata* mantle reflected the participation of the mantle outer surface in shell biomineralization, as well as the involvement of granulocytes.

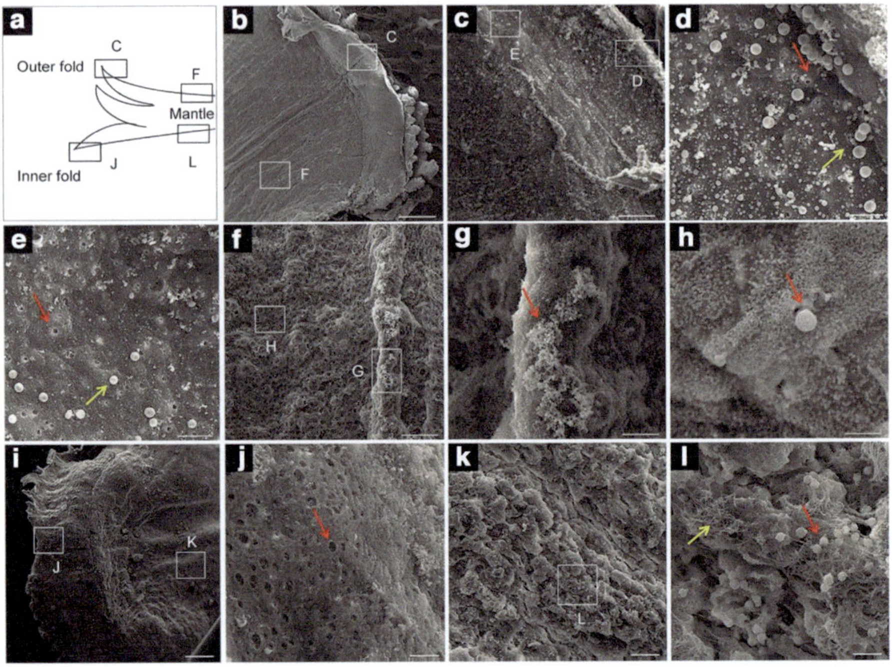

Fig. 5.26 Scanning electron micrographs of mantle in *P. fucata*. (**a**): Schematic presentation of the mantle and the sampling areas. The black boxes indicate the sampling areas: outer surface of outer fold margin (C), outer surface of mantle center (F), inner surface of inner fold margin (J), and inner surface of mantle center (L). (**b**): Outer fold margin and mantle center. The white boxes indicate the sampling areas for (C) and (F). Bar one-fourth 650 mm. (**c**): Magnification of (B). Bar one-fourth 30 mm. (**d**) and (**e**): Magnification of (C). The red and yellow arrows indicate secretory cavity and vesicle, respectively. Bar one-fourth 6 mm. (**f**): Magnification of (B). Bar one-fourth 30 mm. (**g**): Magnification of (F). The red arrow indicates crystals produced by the outer fold. Bar one-fourth 6 mm. (**h**): Magnification of (F). The red arrow indicates a granulocyte released from a cavity. Bar one-fourth 3 mm. (**i**): Inner surface of the inner fold margin and mantle center. The white boxes indicate the sampling areas for (J) and (K). Bar one-fourth 1 mm. (**j**): Magnification of (I). The red arrow indicates the secretory cavity. Bar one-fourth 6 mm. (**k**): Magnification of (I). Bar one-fourth 30 mm. (**l**): Magnification of (K) showing abundant small hyalinocytes (red arrow) and short microvilli (yellow arrow). Bar one-fourth 6 mm (For interpretation of the references to color in this figure legend, the reader is referred to the web version of this article)

5.6.3 Hemocytes Participate in Shell Regeneration

5.6.3.1 Hemocytes Maintain the Homeostasis in EPF

Phagocytosis by hemocytes is an important way to clear invaded pathogens in mollusks. As shown in Fig. 5.27a–c, many hemocytes containing at least one yeast cell in the cytoplasm were seen when the EPS was injected with yeasts. Granulocytes were the predominant cell type undergoing phagocytosis, while

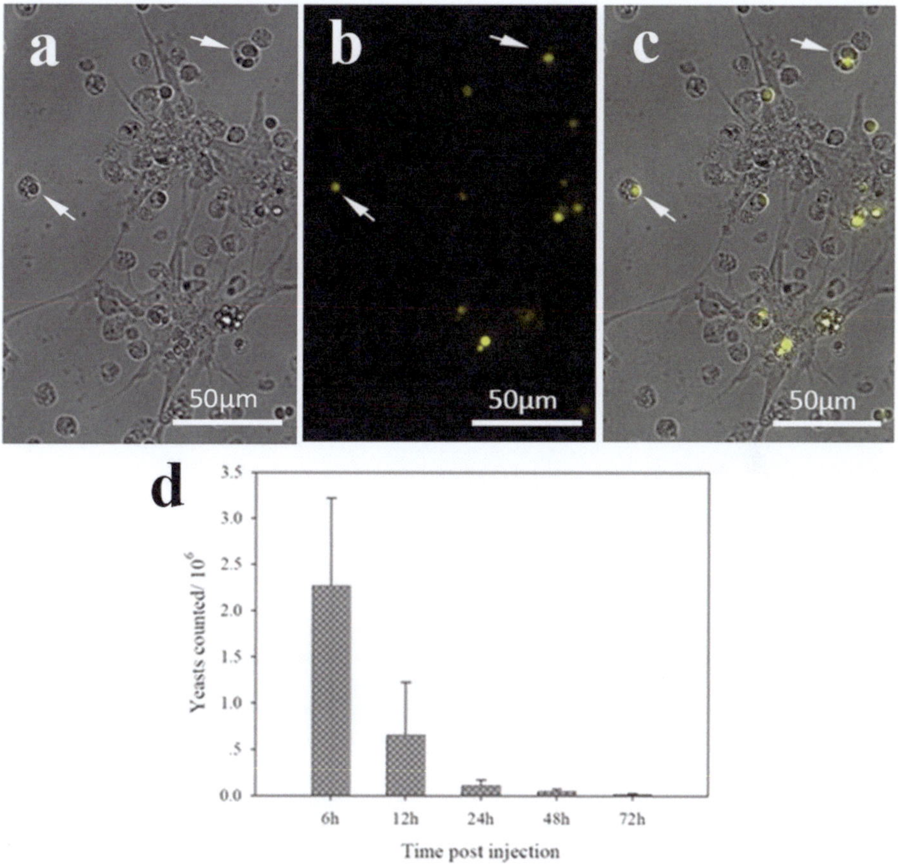

Fig. 5.27 Phagocytosis of yeast by hemocytes in EPS. (**a**) bright field; (**b**) YFP channel (542 nm); (**c**) merged picture of a and b. Some yeast cells with YFP signal were present inside the hemocytes (white arrows). (**d**) time series of the residual yeast in the EPS. The data represent an average of three to five replicates

hyalinocytes were seldom found to engulf any labeled yeasts. The hemocytes underwent active phagocytosis and cleared most of the yeast within 24 h (Fig. 5.27d).

The rich nutrients (proteins and polysaccharides) in the EPF may favor the exogenous intruders, such as microbes and fungi to burst. Uncontrolled rapid proliferation of microbes in the EPS has led to shell disease or even death in clams and oysters [116, 117]. Therefore, hemocytes are critical for maintaining homeostasis in the EPS to ensure the shell formation process by clean up the intruding pathogens.

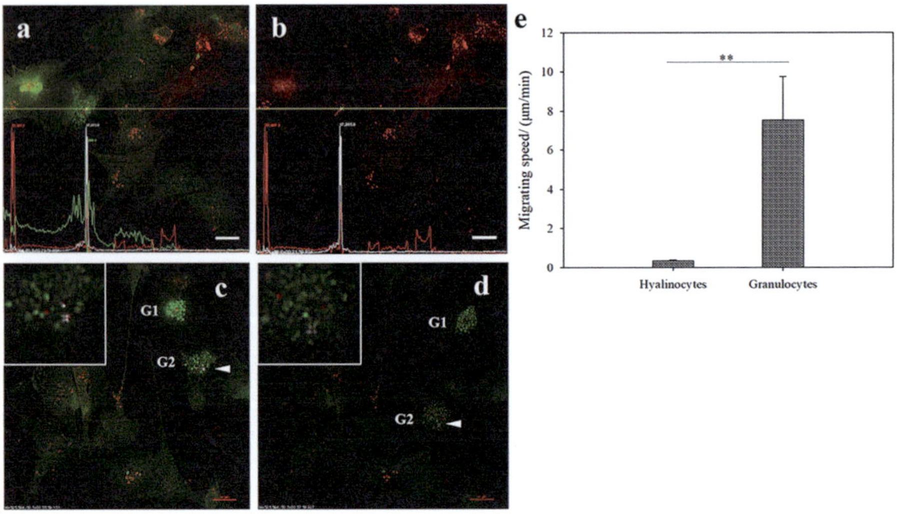

Fig. 5.28 Crystal-bearing granulocytes in the EPS of *P. fucata*. (**a**) Micrograph of the transparent hyalinocytes and the crystal-bearing granulocytes in different channels, 488 nm for calcein-AM, 542 nm for DiI, and polar light for the crystal. The signal intensity distribution across the four-leaf clover spot was shown on the bottom. (**b**) merged micrograph of 542 nm and polar light channels. (**c**) and (**d**), timing diagrams of two migrating granulocytes (G1 and G2), showing the birefringent particle within the granulocyte (white arrow head). (**c**) corresponds to 00: 56 min, and d corresponds to 00:58 min in the DeltaVision raw data. Scale bar = 10 μm. (**e**) migrating speed of hyalinocytes and granulocytes in the EPS. The results represent the average of six duplicates

5.6.3.2 Hemocytes Are Calcium Pool for Shell Formation

A special type of granulocytes was found to carry crystals both in the hemolymph and the EPS. In these granulocytes there were numerous vesicles containing high concentrations of calcium inside the cytoplasm, compared to the transparent hyalinocytes (Fig. 5.28a). Interestingly, some birefringent particles were observed in these granulocytes. The polar light signals and peaks of DiI co-localized in the same granulocyte (Fig. 5.28b), indicating the birefringent particles located in the vesicles. Another characteristic of these granulocytes was the rapid mobility. As shown in Fig. 5.28c, d, two granulocytes migrated in between the hyalinocytes at a high speed, one of which carried birefringent particles. Their migration speed was 7.54 μm/min on average, while the hyalinocytes barely moved (Fig. 5.28e).

Crystal-bearing hemocytes were further studied by SEM observations. Crystal-like structures in various shapes were found in some hemocytes, and one cell contained no more than one crystal (Fig. 5.29a). EDS results showed that the crystal-like structures were composed of carbon, calcium, oxygen, silicon, and phosphorus (Fig. 5.29b), similar to those of the $CaCO_3$ crystal in the surrounding (Fig. 5.29c). Phosphorus was detected in all living hemocytes (Fig. 5.29b, d). The characteristic peaks of $CaCO_3$ were not found in the hemocytes without crystal-like structure (Fig. 5.29d). In *Pinctada fucata*, cultured mantle cells promoted the

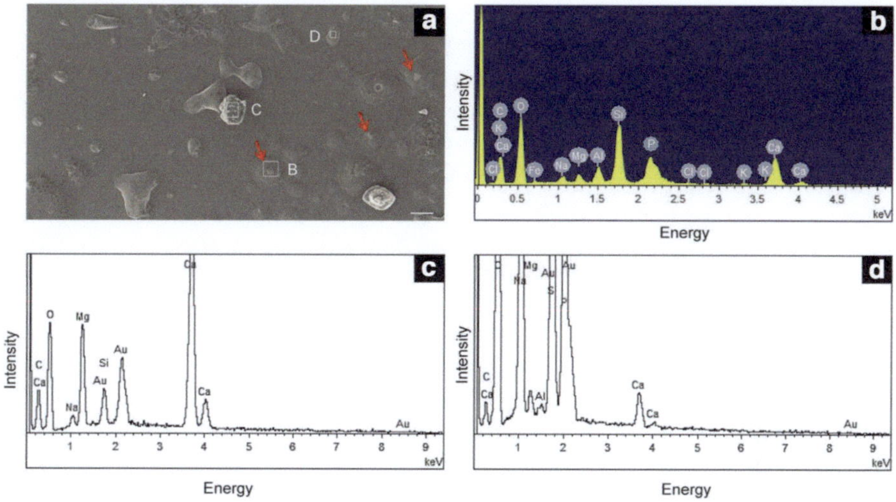

Fig. 5.29 Scanning electron microscope-energy-dispersive spectroscopy (SEM-EDS) analyses of hemocytes in *Pinctada fucata*. (**a**): SEM micrograph of hemocytes. The red arrows indicate the hemocytes with crystal-like structures. The white boxes indicate the detection areas for EDS in (B), (C), and (D). Bar = 20 μm. (**b**): EDS analysis of the crystal-bearing hemocyte. (**c**): EDS analysis of the $CaCO_3$ crystal. (**d**): EDS analysis of the hemocyte (For interpretation of the references to color in this figure legend, the reader is referred to the web version of this article)

formation of $CaCO_3$ crystals in vitro [67, 83] by secreting SMPs into the culture medium. In other marine calcifiers, such as the sea urchin *Paracentrotus lividus*, the calcified tissue formation was involved in intracellular mineralized vesicles during embryonic development [118]. These results clearly proved the existence of cellular regulation in biomineralization in these species. Similarly, there were also suitable environments in the hemocytes for $CaCO_3$ crystallization in shelled mollusks, such as the hemocytes of the *Crassostrea virginica*. In this species, the crystals in hemocytes ranged from 0.5 to 1 μm in length, which were smaller than what we observed in the hemocytes of *P. fucata*. The crystal shapes in *P. fucata* were diverse, while only hexahedron crystals were observed in *C. virginica*, which may due to species specificity.

The mechanism of calcium transportation from mantle tissue to the shell mineralization front in shelled mollusks has not been elucidated. Previous studies proposed two possible transfer pathways. One is vesicle transportation, which speculates that the amorphous phase of $CaCO_3$ is synthesized in specialized vesicles in mantle cells and transported to the mineralization front [119]. The alternative hypothesis is matrix transportation, which suggests that organic matrix- or mucus-enriching calcium is secreted from mantle cells and transported to the mineralization front [72, 91, 114]. What's more, recent studies highlighted the important roles of hemocytes in the calcium transportation process. Mount et al. [17] reported that granulocytes transferred and released calcium carbonate crystals to the repairing site for shell regeneration [17]. The finding of Ca-positive deposits in the EPS of the deep-sea mussel *Bathymodiolus azoricus* indicated the mobilization of calcium-bearing hemocytes during shell regeneration process [91]. These results demonstrated that there might be a hemocyte-mediated pathway for calcium transportation in the EPF, although the type of hemocytes that is involved was unclear. Our results validated this hypothesis and provided some direct evidences that granulocytes play an important role in the calcium transportation.

5.6.3.3 Gene Expressions in Hemocytes During Shell Regeneration

To understand the molecular regulation of hemocytes in calcium metabolism during shell regeneration, we tested gene expressions of some proteins related to calcification as well as some immune genes after notching stimulation. Generally, oysters began to regenerate the shell in several hours [120], and a newly born prism was observed as early as 96 h after shell notching in our study. As a result, significant upregulation of gene expressions was detected for carbonic anhydrase (62fold), calponin-2 (17fold), and calponin-3 (tenfold) after 48 h of notching, and the expression levels remained to be very high after 96 h (Fig. 5.30a). Mantle gene 11 and calreticulin showed relatively small fluctuations. Besides, some immune genes, including ALMP, macroglobulin, and AIF, also revealed significant upregulation (Fig. 5.30b).

Calcium supply counts a critical constraint in $CaCO_3$ precipitation. We speculated that the hemocytes in *P. fucata* provide a replenishing pathway to the mantle tissue for the calcium supply during shell regeneration. Indeed, we found that some granulocytes in *P. fucata* carried vesicles containing concentrated calcium and crystals, as showed above. Moreover, gene expressions of some components involved in calcium metabolism were upregulated dramatically after shell notch, and the response patterns were similar to that of SMPs, such as nacrein, PfN44, and KRMP in mantle tissues during shell regeneration [120], These results demonstrated that hemocytes might accelerate calcium transportation to assist the mantle tissue in shell regeneration when stressors were detected. Interestingly, the injurious operation also activated some immune genes, such as AIF, galectin, and macroglobulin, which play crucial roles during wound repair process in mollusks [121–123].

To ascertain the capability of hemocytes in secreting SMP, we determined the gene expressions of some important SMPs in mantle and hemocytes, and some immune genes were also tested. As shown in Fig. 5.31, the expression levels of most tested genes differed remarkably between the two samples. SMPs including MSI31 and KRMP3 from prism, MSI60 from nacre, and Pif80 and nacrein from both shell layers were mainly expressed in mantle tissue, in agreement with previous studies [60, 88, 124]. Meanwhile, macroglobulin, mantle gene 11, and SOD were enriched in hemocytes.

Then the question comes up: can hemocytes mediate shell formation independently without the mantle tissues? Because SMPs are indispensable in shell formation, hemocytes will not be able to accomplish this job without secreting essential SMPs. As a result, most of the SMPs were not present in hemocytes of *P. fucata*

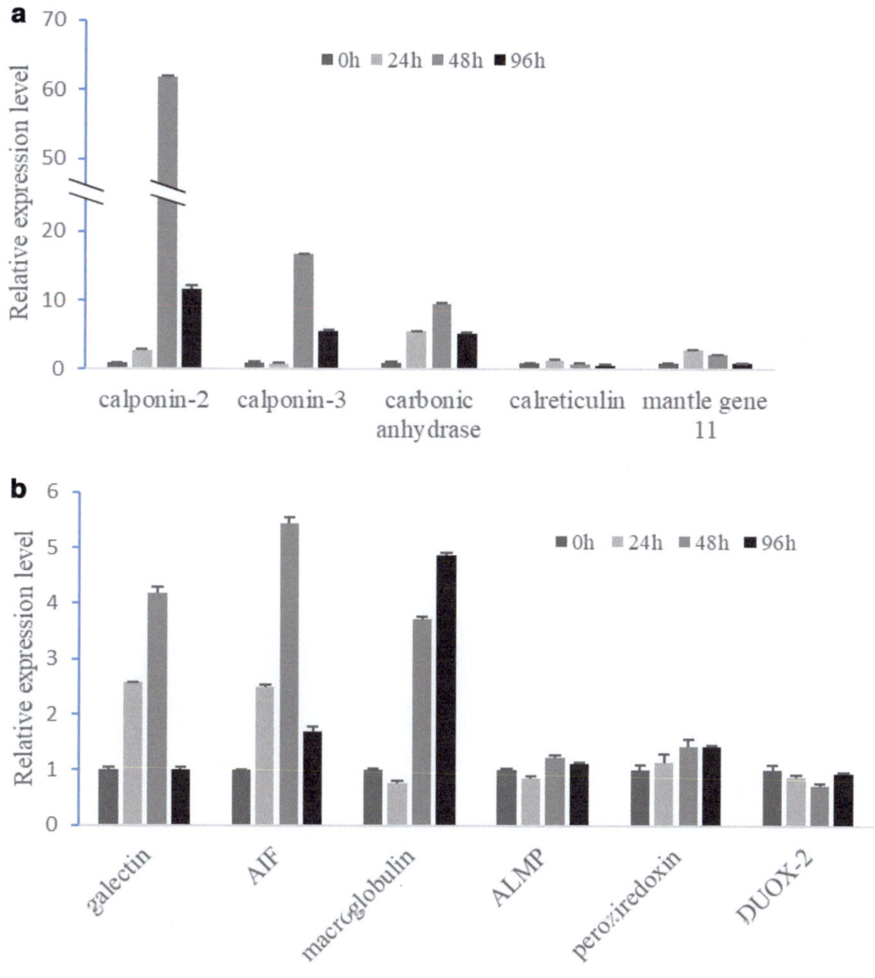

Fig. 5.30 Gene expression of hemocytes in the EPS during shell regeneration. Genes involved in immunity (**a**) and the calcification process (**b**) were determined by RT-PCR. ALMP, astacin-like metalloproteinase; DUOX-2, dual oxidase 2; AIF, allograft inflammatory factor

based on our proteomics analysis. Although we have found some calcification-related proteins in hemocytes [18], these are barely SMPs buried in the oyster shells. In addition, the expression levels of all the tested SMPs were extremely low in hemocytes (Fig. 5.31). Though mantle gene 11, identified first from mantle, was highly expressed in hemocytes, its exact role in shell formation was not known, nor is it an SMP present in the shells [125]. Two possibilities might lead to detection of individual SMP in the hemocyte proteome. Firstly, trace amount of SMPs from the EPF was included in the aggregated hemocytes samples. Indeed, the identified SMPs in hemocytes had relatively low scores and abundances. Secondly, hemocytes might encapsulate a handful of SMPs when phagocytosing microbes in the EPF, leading to

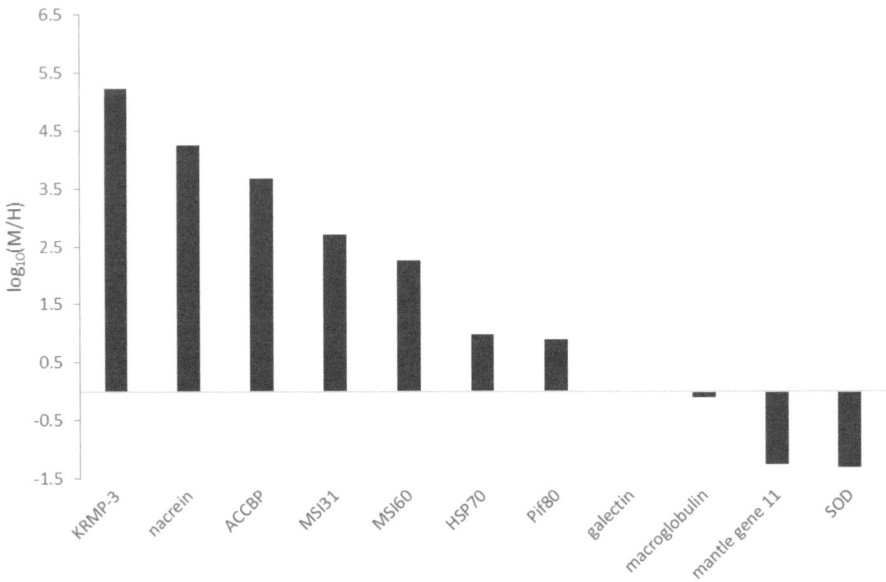

Fig. 5.31 Comparison of some gene expressions between hemocyte in EPS and mantle. Eleven genes were selected, including seven genes participating in the calcification process and four participating in immune response. The upper and lower panels indicate gene expression levels in the mantle were higher or lower than those in the hemocytes, respectively. M, mantle; H, hemocytes. KRMP-3, lysine-rich matrix protein 3; ACCBP, amorphous calcium carbonate-binding protein; HSP70, heat-shock protein 70; SOD, superoxide dismutase

be detected by LC-MS/MS. Taking previous studies into account [72], we concluded that hemocytes are not the major source of SMPs in *P. fucata*. So we argued that, confined to supply calcium when necessary, hemocytes in oysters were unlikely to mediate the shell formation solely, which was slightly different from the previous study [17]. Nevertheless, hemocytes could synthesize intracellular $CaCO_3$ crystals, the molecular mechanism of which may completely differ from that controls the extracellular shell deposition. Further studies will be needed to elucidate this interesting phenomenon.

5.7 Conclusion

Shell matrix proteins secreted by mantle tissue in *Pinctada fucata* self-assembly into framework and promote CaCO3 precipitation. Despite the intensive studies on the discovery of hundreds of SMPs buried in the calcified shells, the mechanism underlying shell and pearls formation did not emerge as expected. Apparently, the SMPS do not hold the entire truth. Increasing evidences show that the mantle directly and precisely regulates the shell growth throughout the process.

One of the main problems in elucidating the cellular regulation is the lack of efficient and appropriate tools and techniques of cell biology. To overcome the absence of immortal cell lines in bivalves, we established and improved the long-term cell culture of mantle tissue to investigate biomineralization. The primary cultured cells persisted high viability and capacity to synthesize and secrete SMPs as long as 2 months. Calcium carbonates were deposited in the culture of variant conditions, partially reproducing in vivo shell mineralization, as showed by our experiments. So mantle tissue culture could provide a useful tool for studying shell formation at the cellular level. Interestingly, the secreted SMPs in the culture medium by the mantle explants could not induce $CaCO_3$ precipitation solely, indicating the directly involvement of the mantle cells. Furthermore, de novo intracellular synthesis of ACC was detected in the cultured mantle cells, suggesting that at least part of the $CaCO_3$ was nucleated in the cytoplasm and then secreted to the mineralization front. However, since the primary cell culture contains a cocktail of many cell types, it's difficult to figure out the exact mineralizing cells. To make matters more complex, hemocytes were constitutive cell types in the cell culture, because hemocytes run over the whole body. And they were supposed to participate in calcium nucleation and transportation. So it's impossible to clarify the roles of mantle cells and hemocytes in the cell culture.

Since the primary cell culture contains heterogeneous cell types, our results about cellular metabolism and regulation in calcium carbonate mineralization represent a mixture output of the mineralized tissue. Subculture of proliferating mantle cells may provide a solution to obtain relative pure cell type. We have found a proliferation hot spot in the mantle tissue of *Pinctada fucata*. This hot spot locates in the center of the mantle pallial. The proliferation rate in mantle tissue decreases from the center to the margin. So cell culture of mantle fragments from the hot spot may facilitate the proliferation of the cultured cells.

Hemocytes are the main bodyguards in bivalves and supposed to be involved in shell formation. Our results show that hemocytes are crucial for maintaining homeostasis in the EPS by cleaning up the intruding pathogens, so that the shell formation process is ensured. Indeed, incontrollable rapid proliferation of microbes in the EPS led to serious shell disease, even death in the clams and oysters [116, 117]. In addition, some granulocytes are enriched in calcium and carry $CaCO_3$ crystals. Such granulocytes were also found in eastern oyster and responded to shell notching treatment. It's reasonable to speculate that, during shell regeneration process, more calcium is required to accelerate $CaCO_3$ mineralization than in normal situation, which may be fulfilled by hemocytes. However, as for SMPs secretion, hemocytes are incompetent for this primary mission of mineralizing cells. So the role of hemocytes in controlling shell formation is second to mantle cells.

We conclude that mantle cells are likely to participate in the entire shell formation process, including nucleation, growth, and reconstruction. Our results demonstrate the practicability of studying biomineralization at the cellular level using a primary cell culture of mantle tissue and provide evidences that mantle cells participate in the calcium carbonate mineralization process. The classical theory on shell formation is therefore reaffirmed: the mantle cells directing the shell growth by regulating the ions and SMPs secretion, assisted by hemocytes in maintaining homeostasis in the EPS and supplying calcium.

This work was modified from the papers published by our group in *PLoS One* (Xiang, L., Kong, W., Su, J., Liang, J. & Zhang, G. Amorphous Calcium Carbonate Precipitation by Cellular Biomineralization in Mantle Cell Cultures of Pinctada fucata (vol 9, e113150, 2014). *PLoS One* **9**, doi: ARTN e116034https://doi.org/10.1371/journal.pone.0116034 (2014).), *Biochemical and Biophysical Research Communications* (Kong, W., Li, S. G., Xiang, L., Xie, L. P. & Zhang, R. Q. Calcium carbonate mineralization mediated by in vitro cultured mantle cells from Pinctada fucata. *Biochem Bioph Res Co* **463**, 1053–1058, doi: https://doi.org/10.1016/j.bbrc.2015.06.057 (2015).), *Marine Biotechnology* (Gong, N. et al. Characterization of calcium deposition and shell matrix protein secretion in primary mantle tissue culture from the marine pearl oyster Pinctada fucata. *Mar Biotechnol (NY)* **10**, 457–465, doi: https://doi.org/10.1007/s10126-008-9081-1 (2008).), *Cell and Tissue Research* (Gong, N. P. et al. Culture of outer epithelial cells from mantle tissue to study shell matrix protein secretion for biomineralization. *Cell and Tissue Research* **333**, 493–501, doi: https://doi.org/10.1007/s00441-008-0609-5 (2008).), *Fish and Shellfish Immunology* (Li, S. et al. Hemocytes participate in calcium carbonate crystal formation, transportation and shell regeneration in the pearl oyster Pinctada fucata. *Fish Shellfish Immunol* **51**, 263–270, doi: https://doi.org/10.1016/j.fsi.2016.02.027 (2016).), and *Marine Biology* (Fang, Z., Feng, Q. L., Chi, Y. Z., Xie, L. P. & Zhang, R. Q. Investigation of cell proliferation and differentiation in the mantle of Pinctada fucata (Bivalve, Mollusca). *Marine Biology* **153**, 745–754, doi: https://doi.org/10.1007/s00227-007-0851-5 (2008).). The related contents are reused with the permission.

Supplementary Table 5.1

Chapter	Primer name	Primer sequence
5.2.2	MSI7F	CTATGGCAAATACGGCGGTG
	MSI7R	GTCCATATGGTAATTCCACC
	nacreinF	CACTTTGAACCTACACGC
	nacreinR	TATAAGCACATTCCAGGATCC
	MSI60F	GAACAATGACTGGAAT GACA
	MSI60R	GGAAAGGTATCCAATAACAAC
	GADPH-F	GCCGAGTATGTGGTAGAATC
	GADPH-R	CACTGTTTT CTGGGTAGCTG

References

1. C. Zhang, R.Q. Zhang, Matrix proteins in the outer shells of molluscs. Mar. Biotechnol. **8**, 572–586 (2006). https://doi.org/10.1007/s10126-005-6029-6
2. Z. Fang, Q. Feng, Y. Chi, et al., Investigation of cell proliferation and differentiation in the mantle of Pinctada fucata (Bivalve, Mollusca). Mar. Biol. **153**, 745–754 (2008)
3. B. Westermann, H. Schmidtberg, K. Beuerlein, Functional morphology of the mantle of Nautilus pompilius (Mollusca, Cephalopoda). J. Morphol. **264**, 277–285 (2005). https://doi.org/10.1002/jmor.10321
4. T. Takeuchi, K. Endo, Biphasic and dually coordinated expression of the genes encoding major shell matrix proteins in the pearl oyster Pinctada fucata. Mar. Biotechnol. (N.Y.) **8**, 52–61 (2006). https://doi.org/10.1007/s10126-005-5037-x
5. S. Sudo et al., Structures of mollusc shell framework proteins. Nature **387**, 563–564 (1997). https://doi.org/10.1038/42391
6. J.M. Poncet et al., In vitro synthesis of proteoglycans and collagen in primary cultures of mantle cells from the nacreous mollusk, Haliotis tuberculata: a new model for study of molluscan extracellular matrix. Mar. Biotechnol. **2**, 387–398 (2000)
7. I.J. Domart-Coulon, D.C. Elbert, E.P. Scully, P.S. Calimlim, G.K. Ostrander, Aragonite crystallization in primary cell cultures of multicellular isolates from a hard coral, Pocillopora damicornis. Proc. Natl. Acad. Sci. U. S. A. **98**, 11885–11890 (2001). https://doi.org/10.1073/pnas.211439698
8. C.P. Suja, N. Sukumaran, S. Dharmaraj, Effect of culture media and tissue extracts in the mantle explant culture of abalone, Haliotis varia Linnaeus. *Aquaculture* **271**, 516–522 (2007). https://doi.org/10.1016/j.aquaculture.2007.04.086
9. N. Gong et al., Characterization of calcium deposition and shell matrix protein secretion in primary mantle tissue culture from the marine pearl oyster Pinctada fucata. Mar. Biotechnol. (N.Y.) **10**, 457–465 (2008). https://doi.org/10.1007/s10126-008-9081-1
10. B. Rinkevich, Marine invertebrate cell cultures: New millennium trends. Mar. Biotechnol. **7**, 429–439 (2005). https://doi.org/10.1007/s10126-004-0108-y
11. C.M. Wen, G.H. Kou, S.N. Chen, Establishment of Cell-Lines from the Pacific Oyster. In Vitro Cell Dev-An. **29a**, 901–903 (1993)
12. D. Sud, D. Doumenc, E. Lopez, C. Milet, Role of water-soluble matrix fraction, extracted from the nacre of Pinctada maxima, in the regulation of cell activity in abalone mantle cell culture (Haliotis tuberculata). Tissue Cell **33**, 154–160 (2001). https://doi.org/10.1054/tice.2000.0166
13. N.P. Gong et al., Characterization of calcium deposition and shell matrix protein secretion in primary mantle tissue culture from the marine pearl oyster Pinctada fucata. Mar. Biotechnol. **10**, 457–465 (2008). https://doi.org/10.1007/s10126-008-9081-1
14. S.K. Barik, J.K. Jena, K.J. Ram, CaCO3 crystallization in primary culture of mantle epithelial cells of freshwater pearl mussel. Curr. Sci. India **86**, 730–734 (2004)
15. T. Samata, H. Somiya, C. Horita, S. Akera, SEM observation of microcrystals developed over black secretion on the cultured tissue of the pearl oyster Pinctada fucata. Fish. Sci. **60**, 343–344 (1994)
16. Z. Fang, Q.L. Feng, Y.Z. Chi, L.P. Xie, R.Q. Zhang, Investigation of cell proliferation and differentiation in the mantle of Pinctada fucata (Bivalve, Mollusca). Mar. Biol. **153**, 745–754 (2008). https://doi.org/10.1007/s00227-007-0851-5
17. A.S. Mount, A.P. Wheeler, R.P. Paradkar, D. Snider, Hemocyte-mediated shell mineralization in the eastern oyster. Science **304**, 297–300 (2004). https://doi.org/10.1126/science.1090506
18. S. Li et al., Hemocytes participate in calcium carbonate crystal formation, transportation and shell regeneration in the pearl oyster Pinctada fucata. Fish Shellfish Immunol. **51**, 263–270 (2016). https://doi.org/10.1016/j.fsi.2016.02.027
19. B. Zaldibar, I. Cancio, I. Marigomez, Circatidal variation in epithelial cell proliferation in the mussel digestive gland and stomach. Cell Tissue Res. **318**, 395–402 (2004). https://doi.org/10.1007/s00441-004-0960-0

20. K. Okudela, T. Ito, Y. Kameda, N. Nakamura, H. Kitamura, Immunohistochemical analysis for cell proliferation-related protein expression in small cell carcinoma of the esophagus. Histol. Histopathol. **14**, 479–485 (1999)
21. A. Machii, K.T. Wada, Some marine invertebrates tissue culture. Invertebrate cell system applications **2**, 225–233 (1989)
22. I. Domart-Coulon, S. Auzoux-Bordenave, D. Doumenc, M. Khalanski, Cytotoxicity assessment of antibiofouling compounds and by-products in marine bivalve cell cultures. Toxicol Vitro **14**, 245–251 (2000). https://doi.org/10.1016/S0887-2333(00)00011-4
23. N.P. Gong et al., Culture of outer epithelial cells from mantle tissue to study shell matrix protein secretion for biomineralization. Cell Tissue Res. **333**, 493–501 (2008). https://doi.org/10.1007/s00441-008-0609-5
24. S. Puverel et al., Antibodies against the organic matrix in scleractinians: a new tool to study coral biomineralization. Coral Reefs **24**, 149–156 (2005). https://doi.org/10.1007/s00338-004-0456-0
25. A. Natoli et al., Bio-vaterite formation by glycoproteins from freshwater pearls. Micron **41**, 359–366 (2010). https://doi.org/10.1016/j.micron.2010.01.002
26. S. Li et al., Morphology and classification of hemocytes in Pinctada fucata and their responses to ocean acidification and warming. Fish Shellfish Immunol. **45**, 194–202 (2015). https://doi.org/10.1016/j.fsi.2015.04.006
27. N. Koga, Y.Z. Nakagoe, H. Tanaka, Crystallization of amorphous calcium carbonate. Thermochim. Acta **318**, 239–244 (1998). https://doi.org/10.1016/S0040-6031(98)00348-7
28. S. Castellanos-Martinez, M. Prado-Alvarez, A. Lobo-da-Cunha, C. Azevedo, C. Gestal, Morphologic, cytometric and functional characterization of the common octopus (Octopus vulgaris) hemocytes. Dev. Comp. Immunol. **44**, 50–58 (2014). https://doi.org/10.1016/j.dci.2013.11.013
29. V. Matozzo et al., First evidence of immunomodulation in bivalves under seawater acidification and increased temperature. PLoS One **7** (2012). ARTN e33820301371/journal.pone.0033820
30. J.L. Fernandez-Turiel et al., Strategy for water analysis using ICP-MS. Fresen. J. Anal. Chem. **368**, 601–606 (2000). https://doi.org/10.1007/s002160000552
31. C. Liu et al., In-depth proteomic analysis of shell matrix proteins of Pinctada fucata. Sci. Rep. **5**, 17269 (2015). https://doi.org/10.1038/srep17269
32. Tsujii, T. Studies on the mechanism of shell-and pearl-formation in Mollusca. 三重県立大学水産学部紀要 **5.1** (1960)
33. A. Bubel, An electron-microscope investigation of the cells lining the outer surface of the mantle in some marine molluscs. Mar. Biol. **21**, 245–255 (1973)
34. A. Garciagasca, R.I. Ochoabaez, M. Betancourt, Microscopic anatomy of the mantle of the pearl oyster Pinctada-Mazatlanica (Hanley, 1856). J. Shellfish Res. **13**, 85–91 (1994)
35. E. Kniprath, Growth of the Shell-field in Mytilus (Bivalvia). Zool. Scr. **7**(1), 119–120 (1978)
36. T.G. Dix, Histochemistry of mantle and pearl sac secretory cells in Pinctada maxima (Lamellibranchia). Aust. J. Zool. **20**, 359–368 (1972)
37. M. Morigi et al., Human bone marrow mesenchymal stem cells accelerate recovery of acute renal injury and prolong survival in mice. Stem Cells **26**, 2075–2082 (2008). https://doi.org/10.1634/stemcells.2007-0795
38. N.L. Leibson, L.T. Frolova, Winter-spring essential reorganization of cell proliferation in the digestive tract epithelia in the mussel Crenomytilus grayanus. Mar. Biol. **118**, 471–477 (1994)
39. R. Hanselmann, R. Smolowitz, Identification of proliferating cells in hard clams. Biol. Bull. **199**, 199–200 (2000). https://doi.org/10.2307/1542896
40. C.M. Yonge, XV—The digestive diverticula in the Lamellibranchs. Earth Environ. Sci. Trans. R. Soc. Edinb. **54**, 703–718 (1926)
41. R.W. McQuiston, Cyclic activity in the digestive diverticula of Lasaea rubra (Montagu) (Bivalvia: Eulamellibranchia). J. Molluscan Stud. **38**, 483–492 (1969)

42. M.P. Cajaraville, G. Diez, I.A. Marigomez, E. Angulo, Responses of basophilic cells of the digestive gland of mussels to petroleum hydrocarbon exposure. Dis. Aquat. Org. **9**, 221–228 (1990). https://doi.org/10.3354/dao009221
43. L.V. Salvini-Plawen, The structure and function of molluscan digestive systems. The Mollusca **11**, 301–379 (1988)
44. J.D. Icely, J.A. Nott, Digestion and absorption: digestive system and associated organs. Microscopic anatomy of invertebrates **10**, 147–201 (1992)
45. G. Vogt, Life-cycle and functional cytology of the Hepatopancreatic cells of Astacus-Astacus (Crustacea, Decapoda). Zoomorphology **114**, 83–101 (1994). https://doi.org/10.1007/Bf00396642
46. I. Marigomez, X. Lekube, I. Cancio, Immunochemical localisation of proliferating cells in mussel digestive gland tissue. Histochem. J. **31**, 781–788 (1999). https://doi.org/10.1023/A:1003950003381
47. P.C. Denny, C. Yang, D.K. Klauser, P.A. Denny, Parenchymal-cell proliferation and mechanisms for maintenance of granular duct and acinar cell-populations in adult male-mouse submandibular-gland. Anat. Rec. **235**, 475–485 (1993). https://doi.org/10.1002/ar.1092350316
48. A.G. Checa, Fabricational morphology of oblique ribs in bivalves. J. Morphol. **254**, 195–209 (2002). https://doi.org/10.1002/jmor.10028
49. G. Pannella, *Paleontological Clocks and the History of the Earth's Rotation. Growth Rhythms and the History of the Earth's Rotation* (Wiley, London, 1975), pp. 253–284
50. E. Naylor, Crab clockwork: the case for interactive circatidal and circadian oscillators controlling rhythmic locomotor activity of Carcinus maenas. Chronobiol. Int. **13**, 153–161 (1996). https://doi.org/10.3109/07420529609012649
51. P. Abello, C.G. Warman, E. Naylor, Circatidal moulting rhythm in the shore crab Carcinus maenas. J. Mar. Biol. Assoc. U. K. **77**, 277–279 (1997)
52. G.R. Clark, *Periodic Growth and Biological Rhythms in Experimentally Grown Bivalves. Growth Rhythms and The History of the Earth's Rotation* (Wiley, London, 1975), pp. 103–117
53. B. Schöne, K. Tanabe, D.L. Dettman, S. Sato, Environmental controls on shell growth rates and d 18 O of the shallow-marine bivalve mollusk Phacosoma japonicum in Japan. Mar. Biol. **142**, 473–485 (2003)
54. L.M. Joll, Daily growth rings in juvenile saucer scallops, Amusium balloti (Bernardi). J. Shellfish Res. **7**, 73–76 (1988)
55. G.J. Parsons, S.M.C. Robinson, J.C. Roff, M.J. Dadswell, Daily growth-rates as indicated by valve ridges in Postlarval Giant scallop (Placopecten-Magellanicus) (Bivalvia, Pectinidae). Can. J. Fish. Aquat. Sci. **50**, 456–469 (1993). https://doi.org/10.1139/f93-053
56. I. Thompson, *Biological Clocks and Shell Growth in Bivalves. Growth Rhythms and the History of the Earth's Rotation* (Wiley, London, 1975), pp. 149–161
57. C. Jolly et al., Zonal localization of shell matrix proteins in mantle of Haliotis tuberculata (Mollusca, Gastropoda). Mar. Biotechnol. **6**, 541–551 (2004). https://doi.org/10.1007/s10126-004-3129-7
58. M. Awaji, A. Machii, Fundamental studies on in vivo and in vitro pearl formation—Contribution of outer epithelial cells of pearl oyster mantle and pearl sacs. Aqua BioSci. Monogr. **4**, 1–39 (2011). https://doi.org/10.5047/absm.2011.00401.0001
59. J. Mitsuhashi, *Invertebrate Tissue Culture Methods* (Springer, New York, 2002)
60. H. Miyamoto, F. Miyoshi, J. Kohno, The carbonic anhydrase domain protein nacrein is expressed in the epithelial cells of the mantle and acts as a negative regulator in calcification in the mollusc Pinctada fucata. Zool. Sci. **22**, 311–315 (2005). https://doi.org/10.2108/zsj.22.311
61. H. Miyamoto et al., A carbonic anhydrase from the nacreous layer in oyster pearls. Proc. Natl. Acad. Sci. U. S. A. **93**, 9657–9660 (1996). https://doi.org/10.1073/pnas.93.18.9657
62. H. Miyamoto et al., A carbonic anhydrase from the nacreous layer in oyster pearls. Proc. Natl. Acad. Sci. U. S. A. **93**, 9657–9660 (1996)

63. M. Rousseau, C. Milet, E. Plouguerne, E. Lopez, M. Fouchereau-Peron, Biomineralisation markers during a phase of active growth in pinctada margaritifera. J. Bone Miner. Res. **18**, S107–S107 (2003)
64. M. Rousseau et al., Multiscale structure of sheet nacre. Biomaterials **26**, 6254–6262 (2005). https://doi.org/10.1016/j.biomaterials.2005.03.028
65. M.A. Crenshaw, The inorganic composition of molluscan extrapallial fluid. Biol. Bull. **143**, 506–512 (1972). https://doi.org/10.2307/1540180
66. D.E. Jacob et al., Nanostructure, composition and mechanisms of bivalve shell growth. Geochim. Cosmochim. Acta **72**, 5401–5415 (2008). https://doi.org/10.1016/j.gca.2008.08.019
67. L. Xiang et al., Amorphous calcium carbonate precipitation by cellular biomineralization in mantle cell cultures of Pinctada fucata. PLoS One **9** (2014). UNSP e11315069137l/journal.pone.0113150
68. Y. Miyazaki, T. Nishida, H. Aoki, T. Samata, Expression of genes responsible for biomineralization of Pinctada fucata during development. Comp. Biochem. Phys. B. **155**, 241–248 (2010). https://doi.org/10.1016/j.cbpb.2009.11.009
69. I.M. Weiss, N. Tuross, L. Addadi, S. Weiner, Mollusc larval shell formation: amorphous calcium carbonate is a precursor phase for aragonite. J. Exp. Zool. **293**, 478–491 (2002). https://doi.org/10.1002/jez.90004
70. R. Collin, J. Voltzow, Initiation, calcification, and form of larval "archaeogastropod" shells. J. Morphol. **235**, 77–89 (1998)
71. S. Auzoux-Bordenave et al., Ultrastructure, chemistry and mineralogy of the growing shell of the European abalone Haliotis tuberculata. J. Struct. Biol. **171**, 277–290 (2010). https://doi.org/10.1016/j.jsb.2010.05.012
72. Z. Ma et al., A novel extrapallial fluid protein controls the morphology of nacre lamellae in the pearl oyster, Pinctada fucata. J. Biol. Chem. **282**, 23253–23263 (2007). https://doi.org/10.1074/jbc.M700001200
73. E. Beniash, L. Addadi, S. Weiner, Cellular control over spicule formation in sea urchin embryos: A structural approach. J. Struct. Biol. **125**, 50–62 (1999). https://doi.org/10.1006/jsbi.1998.4081
74. C. McDougall, K. Green, D.J. Jackson, B.M. Degnan, Ultrastructure of the mantle of the gastropod Haliotis asinina and mechanisms of shell regionalization. Cells Tissues Organs **194**, 103–107 (2011). https://doi.org/10.1159/000324213
75. Z.G. Yan et al., In vivo and in vitro biomineralization in the presence of the inner-shell film of pearl oyster. Acta Oceanol. Sin. **30**, 87–93 (2011). https://doi.org/10.1007/s13131-011-0094-7
76. L. Addadi, D. Joester, F. Nudelman, S. Weiner, Mollusk shell formation: a source of new concepts for understanding biomineralization processes. Chem. Eur. J. **12**, 981–987 (2006). https://doi.org/10.1002/chem.200500980
77. X. Wang et al., Oyster shell proteins originate from multiple organs and their probable transport pathway to the shell formation front. PLoS One **8**, e66522 (2013). https://doi.org/10.1371/journal.pone.0066522
78. G. Zhang et al., The oyster genome reveals stress adaptation and complexity of shell formation. Nature **490**, 49–54 (2012). https://doi.org/10.1038/nature11413
79. S. Raz, S. Weiner, L. Addadi, Formation of high-magnesian calcites via an amorphous precursor phase: Possible biological implications. Adv. Mater. **12**, 38–42 (2000). https://doi.org/10.1002/(Sici)1521-4095(200001)12:1<38::Aid-Adma38>3.0.Co;2-I
80. D. Kralj, J. Kontrec, L. Brecevic, G. Falini, V. Nothig-Laslo, Effect of inorganic anions on the morphology and structure of magnesium calcite. Chem. Eur. J. **10**, 1647–1656 (2004). https://doi.org/10.1002/chem.200305313
81. J. Stolarski, M. Mazur, Nanostructure of biogenic versus abiogenic calcium carbonate crystals. Acta Palaeontol. Pol. **50**, 847–865 (2005)

82. P.L. Clode, A.T. Marshall, Calcium associated with a fibrillar organic matrix in the scleractinian coral Galaxea fascicularis. Protoplasma **220**, 153–161 (2003). https://doi.org/10.1007/s00709-002-0046-3
83. W. Kong, S.G. Li, L. Xiang, L.P. Xie, R.Q. Zhang, Calcium carbonate mineralization mediated by in vitro cultured mantle cells from Pinctada fucata. Biochem. Bioph. Res. Commun. **463**, 1053–1058 (2015). https://doi.org/10.1016/j.bbrc.2015.06.057
84. L. Xiang, W. Kong, J. Su, J. Liang, G. Zhang, Amorphous calcium carbonate precipitation by cellular biomineralization in mantle cell cultures of Pinctada fucata (vol 9, e113150, 2014). PLoS One **9** (2014). ARTN e116034871371/journal.pone.0116034
85. M.B. Johnstone et al., Cellular orchestrated biomineralization of crystalline composites on implant surfaces by the eastern oyster, Crassostrea Virginica. J. Shellfish Res. **34**, 644–644 (2015)
86. L. Addadi, S. Raz, S. Weiner, Taking advantage of disorder: amorphous calcium carbonate and its roles in biomineralization. Adv. Mater. **15**, 959–970 (2003)
87. A.P. Wheeler, K.W. Rusenko, D.M. Swift, C.S. Sikes, Regulation of in vitro and in vivo CaCO 3 crystallization by fractions of oyster shell organic matrix. Mar. Biol. **98**, 71–80 (1988)
88. C. Zhang, L. Xie, J. Huang, X. Liu, R. Zhang, A novel matrix protein family participating in the prismatic layer framework formation of pearl oyster, Pinctada fucata. Biochem. Biophys. Res. Commun. **344**, 735–740 (2006). https://doi.org/10.1016/j.bbrc.2006.03.179
89. S. Weiner, L. Addadi, Design strategies in mineralized biological materials. J. Mater. Chem. **7**, 689–702 (1997). https://doi.org/10.1039/a604512j
90. A.G. Fincham, J. Moradian-Oldak, J.P. Simmer, The structural biology of the developing dental enamel matrix. J. Struct. Biol. **126**, 270–299 (1999). https://doi.org/10.1006/jsbi.1999.4130
91. E. Kadar, A. Lobo-da-Cunha, C. Azevedo, Mantle-to-shell CaCO3 transfer during shell repair at different hydrostatic pressures in the deep-sea vent mussel Bathymodiolus azoricus (Bivalvia: Mytilidae). Mar. Biol. **156**, 959–967 (2009). https://doi.org/10.1007/s00227-009-1140-2
92. V. Matozzo, L. Bailo, A first insight into haemocytes of the smooth venus clam Callista chione. Fish Shellfish Immunol. **42**, 494–502 (2015). https://doi.org/10.1016/j.fsi.2014.11.034
93. R.P. Kuchel, D.A. Raftos, D. Birch, N. Vella, Haemocyte morphology and function in the Akoya pearl oyster, Pinctada imbricata. J. Invertebr. Pathol. **105**, 36–48 (2010). https://doi.org/10.1016/j.jip.2010.04.011
94. C. Lopez, M.J. Carballal, C. Azevedo, A. Villalba, Morphological characterization of the hemocytes of the clam, Ruditapes decussatus (Mollusca: Bivalvia). J. Invertebr. Pathol. **69**, 51–57 (1997). https://doi.org/10.1006/jipa.1996.4639
95. W.P.W. Vanderknaap, C.M. Adema, T. Sminia, Invertebrate blood-cells - morphological and functional-aspects of the Hemocytes in the pond snail Lymnaea-Stagnalis. Comp. Haematol. Int. **3**, 20–26 (1993)
96. P.M. Hine, The inter-relationships of bivalve haemocytes. Fish Shellfish Immunol. **9**, 367–385 (1999). https://doi.org/10.1006/fsim.1998.0205
97. Y.H. Xie, B.C. Hu, C.G. Wen, S.N. Mu, Morphology and phagocytic ability of hemocytes from Cristaria plicata. Aquaculture **310**, 245–251 (2011). https://doi.org/10.1016/j.aquaculture.2010.09.034
98. Y.J. Wang, M.H. Hu, M.W.L. Chiang, P.K.S. Shin, S.G. Cheung, Characterization of subpopulations and immune-related parameters of hemocytes in the green-lipped mussel Perna viridis. Fish Shellfish Immunol. **32**, 381–390 (2012). https://doi.org/10.1016/j.fsi.2011.08.024
99. W. Zhang, X. Wu, J. Sun, D. Li, Morphological and functional characterization of the hemocytes of the scallop, Chlamys farreri. J. Shellfish Res. **24**, 931–936 (2005)
100. C. Dang, T. Tan, D. Moffit, J.D. Deboutteville, A.C. Barnes, Gender differences in hemocyte immune parameters of bivalves: The Sydney rock oyster Saccostrea glomerata and the pearl oyster Pinctada fucata. Fish Shellfish Immunol. **33**, 138–142 (2012). https://doi.org/10.1016/j.fsi.2012.04.007

101. L. Donaghy, C. Lambert, K.S. Choi, P. Soudant, Hemocytes of the carpet shell clam (Ruditapes decussatus) and the manila clam (Ruditapes philippinarum): Current knowledge and future prospects. Aquaculture **297**, 10–24 (2009). https://doi.org/10.1016/j.aquaculture.2009.09.003
102. M.D. Rebelo et al., New Insights from the Oyster Crassostrea rhizophorae on Bivalve Circulating Hemocytes. PLoS One **8** (2013). ARTN e573841061371/journal.pone.0057384
103. M.C. Mix, A general model for leukocyte cell renewal in bivalve mollusks. Mar. Fish. Rev. **38** (1), 37–41 (1976)
104. S.J. Chang, S.M. Tseng, H.Y. Chou, Morphological characterization via light and electron microscopy of the hemocytes of two cultured bivalves: a comparison study between the hard clam (Meretrix lusoria) and pacific oyster (Crassostrea gigas). Zool. Stud. **44**, 144–152 (2005)
105. M.B. Duchemin, M. Fournier, M. Auffret, Seasonal variations of immune parameters in diploid and triploid Pacific oysters, Crassostrea gigas (Thunberg). Aquaculture **264**, 73–81 (2007). https://doi.org/10.1016/j.aquaculture.2006.12.030
106. S. Mangkalanan, P. Sanguanrat, T. Utairangsri, K. Sritunyalucksana, C. Krittanai, Characterization of the circulating hemocytes in mud crab (Scylla olivacea) revealed phenoloxidase activity. Dev. Comp. Immunol. **44**, 116–123 (2014). https://doi.org/10.1016/j.dci.2013.11.018
107. E.C. Wootton, E.A. Dyrynda, N.A. Ratcliffe, Bivalve immunity: Comparisons between the marine mussel (Mytilus edulis), the edible cockle (Cerastoderma edule) and the razor-shell (Ensis siliqua). Fish Shellfish Immunol. **15**, 195–210 (2003)
108. W.Z. Zhang, X.Z. Wu, M. Wang, Morphological, structural, and functional characterization of the haemocytes of the scallop, Argopecten irradians. Aquaculture **251**, 19–32 (2006). https://doi.org/10.1016/j.aquaculture.2005.05.020
109. C. Fleury et al., Shell repair process in the green ormer Haliotis tuberculata: a histological and microstructural study. Tissue Cell **40**, 207–218 (2008). https://doi.org/10.1016/j.tice.2007.12.002
110. J.A. Audino, J.E.A.R. Marian, A. Wanninger, S.G.B.C. Lopes, Mantle margin morphogenesis in Nodipecten nodosus (Mollusca: Bivalvia): new insights into the development and the roles of bivalve pallial folds. Bmc. Dev. Biol. **15** (2015). ARTN 221151186/s12861-015-0074-9
111. R. Jabbour-Zahab, D. Chagot, F. Blanc, H. Grizel, Mantle histology, histochemistry and ultrastructure of the pearl oyster Pinctada margaritifera (L.). Aquat. Living Resour. **5**, 287–298 (1992)
112. P. Southgate, J. Lucas (eds.), *The pearl oyster* (Elsevier, Oxford, 2011)
113. M.B. Johnstone et al., Cellular orchestrated biomineralization of crystalline composites on implant surfaces by the eastern oyster, Crassostrea virginica (Gmelin, 1791). J. Exp. Mar. Biol. Ecol. **463**, 8–16 (2015). https://doi.org/10.1016/j.jembe.2014.10.014
114. J.M. Myers, M.B. Johnstone, A.S. Mount, H. Silverman, A.P. Wheeler, TEM immunocytochemistry of a 48kDa MW organic matrix phosphoprotein produced in the mantle epithelial cells of the Eastern oyster (Crassostrea virginica). Tissue Cell **39**, 247–256 (2007)
115. S.-M. Cho, You-me lee, and woo-Geon Jeong. Effect of Polycyclic Aromatic Hydrocarbon (PAH) on shell repair in the Pacific oyster, Crassostrea gigas. Korean J. Malacol. **27**(1), 35–42 (2011)
116. B. Allam, C. Paillard, S.E. Ford, Pathogenicity of Vibrio tapetis, the etiological agent of brown ring disease in clams. Dis. Aquat. Org. **48**, 221–231 (2002). https://doi.org/10.3354/dao048221
117. A. Ralph, D.C. Elston, P. Frelier, D. Lynn, Invasive orchitophryid ciliate infections in juvenile Pacific and Kumomoto oysters, Crassostrea gigas and Crassostrea sikamea. Aquaculture **174**, 1–14 (1999)
118. N. Vidavsky et al., Initial stages of calcium uptake and mineral deposition in sea urchin embryos. Proc. Natl. Acad. Sci. U. S. A. **111**, 39–44 (2014). https://doi.org/10.1073/pnas.1312833110
119. S. Weiner, L. Addadi, Crystallization pathways in biomineralization. Annu. Rev. Mater. Res. **41**, 21–40 (2011). https://doi.org/10.1146/annurev-matsci-062910-095803

120. C. Pan et al., A novel acidic matrix protein, PfN44, stabilizes magnesium calcite to inhibit the crystallization of aragonite. J. Biol. Chem. **289**, 2776–2787 (2014). https://doi.org/10.1074/jbc.M113.504027
121. Y. Li et al., DNA methylation is associated with expression level changes of galectin gene in mantle wound healing process of pearl oyster, Pinctada fucata. Fish Shellfish Immunol. **45**, 912–918 (2015). https://doi.org/10.1016/j.fsi.2015.06.016
122. M. De Zoysa et al., Allograft inflammatory factor-1 in disk abalone (Haliotis discus discus): molecular cloning, transcriptional regulation against immune challenge and tissue injury. Fish Shellfish Immunol. **29**, 319–326 (2010). https://doi.org/10.1016/j.fsi.2010.04.006
123. B. Peter, J. Armstrong, P. Q. Humoral immunity: a2 macroglobulin activity in the plasma of mollusks. Veliger S **35**, 161–164 (1992)
124. M. Suzuki et al., An acidic matrix protein, Pif, is a key macromolecule for nacre formation. Science **325**, 1388–1390 (2009). https://doi.org/10.1126/science.1173793
125. H.L. Liu et al., Identification and characterization of a biomineralization related gene PFMG1 highly expressed in the mantle of Pinctada fucata. Biochemistry **46**, 844–851 (2007). https://doi.org/10.1021/bi061881a

Chapter 6
Molecular Regulation Mechanism of Biomineralization of *Pinctada fucata*

Abstract To elucidate the mechanism of biomineralization in *Pinctada fucata*, most of the researchers put their attention on the roles of matrix proteins in shell formation. Our group has identified and characterized many essential matrix proteins in the regulation of deposition of calcium carbonate crystals in both prism and nacre layers, as described in the former chapter. Meanwhile, the regulation of the transcription, translation and expression of matrix proteins and how these regulatory factors mediate the biomineralization have become a hot spot in recent years. In this chapter, we mainly assayed RACE to obtain the sequence of the members of signaling pathways and the transcriptional factors; real time-qPCR to analyze the expression pattern of these factors in different tissues and/or distinct time during shell repair and pearl sac; in situ hybridization to find out the expression location of specific genes in mantle tissue; luciferase assay and electrophoretic mobility assay (EMSA) to clarify the recruitment of transcriptional factors on promoters of matrix protein; yeast two hybridization to explore the interactions between different pathways. We demonstrate the function of NF-κB, TGFβ, Wnt signal pathway, G protein-mediated pathway and several transcriptional factors in mediating the biomineralization in *Pinctada fucata*. The mechanism of transcriptional regulation can give deep sight to the matrix protein expression pattern which enrich the theory of bionimeralization in *Pinctada fucata*.

Keywords Biomineralization · Signal pathway · Transcriptional factors · Transcription mechanism

6.1 Introduction

At present, the research on the regulation of biomineralization is mainly focused on mammals (humans and mice) and some model animals (*Xenopus*, zebrafish), with the most detailed research on disease-related bone or tooth formation.

In mammals, multiple signaling pathways have been found to act on the formation of bones. TGF-β, BMP, IGF, FGF, Hedgehog, Notch, IL, and Wnt signaling pathways and their inhibitory factors have been found to be associated with

R. Zhang et al., *Biomineralization Mechanism of the Pearl Oyster, Pinctada fucata*,
https://doi.org/10.1007/978-981-13-1459-9_6

osteogenesis, and mutations in these genes have been shown to be responsible for various bone diseases. The Wnt/β-catenin signal can control cell differentiation, proliferation, and survival by influencing the balance between osteoblast and osteoclast differentiation. In the early stage of skeletal development, Wnt/β-catenin signal is the key to the transformation of mesenchymal stem cells into osteoblasts. Wnt/β-catenin promotes the survival, differentiation, and cartilage ossification of chondrocytes in bone growth. Mouse and human body Wnt signal element function acquisition and deletion mutation can lead to increased or decreased bone mass. Wnt/β-catenin signal inactivation accelerates osteoclast osteoprotegerin (OPG) levels and leads to imbalance between bone formation and bone resorption [1]. The important function and multifunctional regulation of TGFβ/BMP in mammalian development have been widely recognized. TGFβ/BMP can specifically pass through classical Smad-dependent signal transduction pathways (TGFβ/BMP ligands, receptors, and Smads) and nonclassical Smad-free signal transduction pathways such as p38 mitogen-activated protein kinase (MAPK) signaling pathway. Since Smad signaling pathway and p38 MAPK signaling pathway act on Runx2 gene to induce the differentiation of mesenchymal cells under TGFβ/BMP induction, Runx2's synergistic activity and TGFβ-/BMP-activated Smads are critical to bone formation [2].

In invertebrate, such as sea urchins, corals, and bivalve mollusks, the mechanism of biomineralization is mainly focused on the clonal identification and functional study of matrix proteins. In sea urchins, it is known that SM50 protein can stabilize amorphous calcium carbonate (ACC), which can guide the growth of bone needles by raising calcium ions while self-assembling the bone matrix to guide the growth of bone needles [3], while the matrix proteins SM37 and PM27 also have similar mineralization regulation; coral matrix protein CARPs 1–4 can bind calcium ions and can be in the pH value of 8.2 and 7.6 in seawater by hydrogenated anion on calcium sedimentation interaction in vitro deposition of aragonite [4]. In the bivalve mollusks, dozens of matrix proteins have been cloned and identified. Our lab has found that in nacreous layer, MSI17, rich of glycine, serves as the framework protein, N40 induces aragonite nucleation, and PfN23 promotes the formation of aragonite crystals. While in prismatic layer, KRMP family controls the crystal form, inhibiting the formation of aragonite, and Prisilkin-39 exercises silk protein-like function. But little is known about the regulation of these matrix proteins, and there is a lack of understanding of the signal regulation network.

We tried to detect the gene expression of fertilized eggs, trochophore, and D-shaped and umbonal stage larvae, as well as juvenile stage through microarray. Kyoto Encyclopedia of Genes and Genomes (KEGG) pathway terms are represented significantly (q-value < 0.05) between the two consecutive stages. The results of KEGG pathway analysis are shown below (Table 6.1): between consecutive stages, there are many signaling pathways having fluctuating gene expression. In particular, we put our attention on D-shaped stage, in which the first mineralized organism, prodissoconch I, appears. Some typical signaling pathways related to osteogenesis in

Table 6.1 The KEGG pathway enrichment of several signaling pathways

Names	Stages	Gene number	*P* value	Q value
ko03320://PPAR signaling pathway	Jstage vs Ustage	40	0.26	0.63
	Ustage vs Dstage	41	0.37	0.95
	Dstage vs Tstage	40	0.40	0.82
	Tstage vs Fstage	65	0.15	0.52
ko04010://MAPK signaling pathway	Jstage vs Ustage	79	0.99	1.00
	Ustage vs Dstage	87	0.98	1.00
	Dstage vs Tstage	130	0.00	0.01
	Tstage vs Fstage	167	0.16	0.52
ko04012://ErbB signaling pathway	Jstage vs Ustage	39	0.70	1.00
	Ustage vs Dstage	41	0.75	1.00
	Dstage vs Tstage	37	0.90	1.00
	Tstage vs Fstage	75	0.11	0.45
ko04013://MAPK signaling pathway—Fly	Jstage vs Ustage	16	0.48	0.93
	Ustage vs Dstage	15	0.71	1.00
	Dstage vs Tstage	14	0.78	1.00
	Tstage vs Fstage	23	0.71	1.00
ko04020://calcium signaling pathway	Jstage vs Ustage	103	0.14	0.47
	Ustage vs Dstage	91	0.87	1.00
	Dstage vs Tstage	147	0.00	0.00
	Tstage vs Fstage	150	0.53	0.98
ko04062://chemokine signaling pathway	Jstage vs Ustage	54	0.96	1.00
	Ustage vs Dstage	68	0.63	1.00
	Dstage vs Tstage	74	0.25	0.66
	Tstage vs Fstage	89	0.98	1.00

(continued)

Table 6.1 (continued)

Names	Stages	Gene number	*P* value	Q value
ko04115://p53 signaling pathway	Jstage vs Ustage	48	0.00	0.03
	Ustage vs Dstage	47	0.02	0.18
	Dstage vs Tstage	56	0.00	0.00
	Tstage vs Fstage	65	0.03	0.18
ko04150://mTOR signaling pathway	Jstage vs Ustage	27	0.36	0.78
	Ustage vs Dstage	12	1.00	1.00
	Dstage vs Tstage	28	0.39	0.82
	Tstage vs Fstage	43	0.32	0.77
ko04310://Wnt signaling pathway	Jstage vs Ustage	61	0.87	1.00
	Ustage vs Dstage	72	0.59	1.00
	Dstage vs Tstage	99	0.00	0.00
	Tstage vs Fstage	108	0.62	1.00
ko04330://Notch signaling pathway	Jstage vs Ustage	76	0.04	0.21
	Ustage vs Dstage	57	0.95	1.00
	Dstage vs Tstage	55	0.96	1.00
	Tstage vs Fstage	79	1.00	1.00
ko04340://Hedgehog signaling pathway	Jstage vs Ustage	33	0.30	0.70
	Ustage vs Dstage	36	0.24	0.79
	Dstage vs Tstage	37	0.15	0.53
	Tstage vs Fstage	41	0.93	1.00
ko04350://TGF-beta signaling pathway	Jstage vs Ustage	39	0.47	0.93
	Ustage vs Dstage	30	0.98	1.00
	Dstage vs Tstage	43	0.33	0.79
	Tstage vs Fstage	70	0.09	0.39

(continued)

Table 6.1 (continued)

Names	Stages	Gene number	*P* value	Q value
ko04370://VEGF signaling pathway	Jstage vs Ustage	26	0.98	1.00
	Ustage vs Dstage	44	0.14	0.60
	Dstage vs Tstage	48	0.03	0.14
	Tstage vs Fstage	58	0.45	0.93
ko04620://toll-like receptor signaling pathway	Jstage vs Ustage	33	0.73	1.00
	Ustage vs Dstage	48	0.04	0.36
	Dstage vs Tstage	42	0.23	0.66
	Tstage vs Fstage	65	0.12	0.46
ko04621://NOD-like receptor signaling pathway	Jstage vs Ustage	28	0.46	0.93
	Ustage vs Dstage	29	0.54	1.00
	Dstage vs Tstage	30	0.41	0.82
	Tstage vs Fstage	49	0.17	0.54
ko04622://RIG-I-like receptor signaling pathway	Jstage vs Ustage	18	0.89	1.00
	Ustage vs Dstage	22	0.73	1.00
	Dstage vs Tstage	17	0.96	1.00
	Tstage vs Fstage	33	0.79	1.00
ko04630://Jak-STAT signaling pathway	Jstage vs Ustage	28	0.48	0.93
	Ustage vs Dstage	22	0.96	1.00
	Dstage vs Tstage	26	0.76	1.00
	Tstage vs Fstage	51	0.11	0.43
ko04660://T cell receptor signaling pathway	Jstage vs Ustage	25	1.00	1.00
	Ustage vs Dstage	38	0.84	1.00
	Dstage vs Tstage	40	0.70	1.00
	Tstage vs Fstage	64	0.59	0.98

(continued)

Table 6.1 (continued)

Names	Stages	Gene number	*P* value	Q value
ko04662://B cell receptor signaling pathway	Jstage vs Ustage	27	0.97	1.00
	Ustage vs Dstage	38	0.57	1.00
	Dstage vs Tstage	36	0.67	1.00
	Tstage vs Fstage	53	0.82	1.00
ko04664://fc epsilon RI signaling pathway	Jstage vs Ustage	18	1.00	1.00
	Ustage vs Dstage	30	0.74	1.00
	Dstage vs Tstage	34	0.38	0.82
	Tstage vs Fstage	51	0.39	0.86
ko04722://Neurotrophin signaling pathway	Jstage vs Ustage	55	0.92	1.00
	Ustage vs Dstage	68	0.56	1.00
	Dstage vs Tstage	80	0.04	0.22
	Tstage vs Fstage	119	0.03	0.18
ko04910://insulin signaling pathway	Jstage vs Ustage	82	0.25	0.62
	Ustage vs Dstage	67	0.98	1.00
	Dstage vs Tstage	85	0.29	0.73
	Tstage vs Fstage	119	0.69	1.00
ko04912://GnRH signaling pathway	Jstage vs Ustage	60	0.25	0.62
	Ustage vs Dstage	49	0.95	1.00
	Dstage vs Tstage	82	0.00	0.00
	Tstage vs Fstage	100	0.07	0.35
ko04920://Adipocytokine signaling pathway	Jstage vs Ustage	21	0.99	1.00
	Ustage vs Dstage	29	0.81	1.00
	Dstage vs Tstage	29	0.78	1.00
	Tstage vs Fstage	49	0.57	0.98

mammals also function in *P. fucata*, giving evidence that signaling pathways like Wnt pathway, TGFβ pathway, etc. may regulate the formation of the shell prism or nacre layers.

Besides the signaling pathway, Japanese researchers have predicted the presence of multiple transcription factors, including homeobox gene, Tbx, Fox, Ets, HMG, NF-κB, bZIP, and C2H2 zinc finger proteins, by gene annotation of transcriptome [5, 6] in *P. fucata*. These transcription factors may play an important regulatory role in the expression of matrix proteins in biomineralization.

Therefore, this chapter sums up the signal pathways and transcription factors serving as regulatory elements related to biomineralization of *P. fucata*, setting up a regulatory network toward the nacre formation.

6.2 Experimental Materials and Methods

6.2.1 Molecular Cloning of the Promoter from Oyster

To obtain the promoter regions of the matrix protein genes, genome-walking PCR was performed as described by the Genome Walking Kit (Takara, Tokyo, Japan). Genomic DNA was isolated from the adductor muscle of pearl oyster using a TIANamp Marine Animals DNA Kit (Tiangen, Beijing, China), and the quality of the DNA was determined by agarose gel electrophoresis and use of a Nanodrop spectrophotometer. Before performing genome-walking PCR, the sequence of the partial intron region was verified by semi-nested PCR using KOD polymerase. Primers for genome-walking PCR were designed based on the results of the semi-nested PCR. The final sequences of the gene promoters were confirmed using KOD polymerase. All primers used are listed in Supplementary Table 6.1. The transcription start sites of the promoters were predicted using BDGP software (http://www.fruitfly.org/seq_tools/promoter.html).

6.2.2 Cell Culture and Transient Transfection

MC3T3-E1 cells and HEK293T cells were maintained in Dulbecco's modified Eagle's medium (DMEM, Gibco BRL) supplemented with 10% fetal bovine serum (FBS, HyClone), at 37 °C under humidified 5.0% CO_2.

For transient transfection, the cells were plated in 96-well plates at a density of 2.5×10^5 cells 24 h before transfection, or 60 mm plates at $2–6 \times 10^4$ cells/plate, or in 6-well at a density of 8.5×10^5 cells. The transfections were mediated by Lipofectamine 2000 Transfection Reagent (Invitrogen), Sofastk (Sofast, Xiamen, China), or VigoFect (Vigorous Biotechnology, Beijing, China) according to the manufacturer's instructions. To normalize the transfection efficiency, pRL-TK (Promega) or renilla was cotransfected as an internal control.

For simple transfection, 500 ng plasmid and 5 ng pRL-TK were used for each well of a 48-well plate, 0.3 μg plasmid for 96-well plate, and 2 μg plasmids DNA for 6-well plate. In the meantime, the pEGFP plasmid, as control, was also transiently transfected into the cells to evaluate the transfection efficiency. Correct expression of each tagged protein was verified by Western blot using the appropriate antibody.

For cotransfection, 500 ng of recombinant/empty pcDNA3.1(+)-myc expression vector, 100 ng of recombinant/empty pGL4.10 luciferase vector, and 2 ng of pRL-TK were used for each well of a 48-well plate. Empty pcDNA3.1(+)-myc expression vector was used to adjust the total DNA input in dose–effect experiments.

6.2.3 Luciferase Assay

Luciferase reporter analysis was performed 36 h after transfection. Cells were washed with PBS quickly at room temperature and then lysed using passive lysis buffer (Promega). The relative activity of the promoter was determined using the Dual-Luciferase Assay System (Promega) according to the manufacturer's instructions and a Varioskan™ Flash multimode reader (Thermo Scientific).

6.2.4 Electrophoretic Mobility Assay (EMSA)

HEK-293 T cells were plated on 100 mm dishes and transiently transfected with 10 μg recombinant pcDNA3.1-myc expression vector. Untransfected cells were prepared as a control. After incubation for 36 h, the cells were washed with PBS and scraped into microcentrifuge tubes. Nuclear extracts of transfected HEK-293T cells were prepared using a total nuclear protein extraction kit (Xinghan) or NE-PER Nuclear and Cytoplasmic Extraction Reagents (Thermo Scientific) according to the manufacturer's instructions and stored at −80 °C until use. The protein concentrations of the nuclear extracts were determined using the bicinchoninic acid method or a Pierce™ BCA Protein Assay Kit (Thermo Scientific).

For probe preparation, complementary oligonucleotides were all 50 end-labeled using biotin or using DIG Gel Shift Kit, second Generation (Roche), and then annealed to obtain double-stranded oligonucleotides. Unlabeled competitors or mutant competitors were also prepared. Details of the oligonucleotides used are provided in Supplementary Table 6.1. EMSA was performed using a LightShift™ Chemiluminescent EMSA Kit (Thermo Scientific) according to the manufacturer's instructions. Briefly, 5 μg nuclear extract and 20 fmol biotin end-labeled target DNA were used per 20 μL binding reaction. For the supershift assay, 1 μL anti-myc serum or 1 μL polyclonal antibody against transcriptional factors was added to the reaction. Final detection of biotin-labeled DNA was performed using a chemiluminescent imaging system (Thermo Scientific) and exposure to X-ray film.

6.2.5 Yeast Two Hybridization (Y2H)

For investigating the interaction between two proteins, yeast two-hybrid analysis was performed using the Matchmaker GAL4 Two-Hybrid System 3 and *Yeast Protocols Handbook* according to the manufacturer's instructions (Clontech, Japan). Protein A cDNA was amplified using a pair of primers and then subcloned into pGADT7 plasmid to construct A-AD fusion genes. Protein B cDNA was amplified using a pair of primers and then subcloned into pGBKT7 plasmid to construct B-BD fusion genes. Primers used are listed in Supplementary Table 6.1. After construction, constructs were independently transformed into the yeast strain AH109 and plated on SD/−Leu/X-α-Gal and SD/−His/−Leu and SD/−Trp/X-α-Gal and SD/−His/−Trp, respectively, to verify that they do not activate reporter genes. Then, a simultaneous cotransformation of constructs was performed into strain AH109 and plated on SD/−Leu/−Trp, SD/−Leu/−Trp/X-α-Gal, and SD/−Ade/−His/−Leu/−Trp/X-α-Gal to test their interaction with low and high stringency. Cotransformations of pGADT7-T and pGBKT7–53 were used as a positive control, and cotransformations of pGADT7-T and pGBKT7-Lam were used as a negative control.

6.2.6 Pearl Sac Selection

Artificial nuclei were implanted in oysters near the gonad. After implanting the nuclei, the pearl sacs were isolated carefully from the gonad tissue every 5 days from days 0 to 30, and the samples were stored in liquid nitrogen. Ten oysters were used in each treatment group.

6.2.7 Subcellular Localization in HEK-293T Cells

HEK-293 T cells plated on cover slips were transfected with gene expression vectors. After incubation for 36 h, the cells were washed with PBS and fixed in 4% paraformaldehyde in PBS for 20 min, washed with PBS again, and permeabilized using 0.5% Triton X-100 in PBS for 25 min. For transfection with myc epitope expression vectors, the cells were blocked in 2% BSA for 1 h. The cells were subsequently incubated with mouse anti-myc serum (dilution 1:60) for 1 h or with preimmune mouse serum (dilution 1:60) for 1 h as a control. The cells were then washed with PBS and incubated with fluorescein isothiocyanate (FITC)-conjugated goat anti-mouse IgG for 1 h. Finally, cells were washed again with PBS, stained with 40,6-diamidino-2-phenylindole (DAPI) for 25 min in the dark, washed with PBS, and mounted and observed using a TCS SP5 confocal microscope (Leica, Heidelberg, Germany). For transfection with GFP fusion expression vectors, cells were stained with DAPI after permeabilization, washed, mounted, and observed as described above. All these were performed at room temperature.

6.2.8 Measurement of Alkaline Phosphatase (ALP) Activity

After 48 h transfection with the recombination expression plasmids, the NIH3T3 cells were harvested followed by washing with physiological saline three times. The ALP activity was measured by incubation in 0.1 M $NaHCO_3$–Na_2CO_3 buffer, pH 10, containing 0.1% Triton X-100, 2 mm $MgSO_4$, and 6 mm pNPP at 37 °C for 30 min. The reaction was terminated by adding 1 M NaOH, and the absorbance was assayed at 405 nm. The percent changes of the ALP activity with respect to the value found in the control were calculated according to the formula: M = value of absorbance at 405 nm/value of absorbance at 550 nm. Percent change = [(M of the test-M of the control) M of the control] $\times$ 100.

6.2.9 cAMP Assay

cAMP accumulation in intact cells was assayed as described [7]. The cells were plated at a density of 2.5×10^5 per well in 24-well plates and transiently transfected with pcDNA4c encoding either wild-type or mutant protein. After 48 h, the cells were incubated with the phosphodiesterase inhibitor 3-isobutyl-1-methylxanthine (IBMX; 10 μm) for 30 min and stimulated with 1 μg/ml of cholera toxin (CTX, Sigma) that is known to be an activator for Gs, a GTP-binding protein that activates adenylyl cyclase (AC), for 2 h. Then the cells were quickly lysed in 6% perchloric acid and scraped from the plates. The cAMP content in each sample was determined by the 3H-cAMP assay system (Beijing Isotopes, Beijing, China), according to the manufacturer's instructions.

6.3 NF-κB Pathway

6.3.1 Introduction

Nuclear factor-kappa B (NF-κB) was first identified as a binding protein that interacted with the specific DNA sequence within the immunoglobulin heavy chain and kappa light chain enhancers in B cells. It was also shown that NF-κB could be induced by bacterial lipopolysaccharide and active phorbol ester, and the process of induction did not require protein synthesis [8]. During the two decades since its discovery, Rel/NF-κB has been identified in most cell types, and much progress on this subject has been made. It is now known that different kinds of biological factors can induce NF-κB, which then regulates the expression of a wide variety of genes that are involved in many cellular processes: ontogeny of the immune system, immune and inflammatory responses to pathogens, cell survival and apoptosis, oncogenesis, skin physiology and pathology, hemopoiesis, and neuronal survival [9–11].

All the Rel/NF-κB family proteins share the highly conserved region called Rel homology domain (RHD), which is essential in dimerization, nuclear location, and DNA binding [12]. As shown in Fig. 6.1, Rel/NF-κB proteins are sequestrated in the cytoplasm as homodimers or heterodimers, binding to the inhibitory proteins IκB in the majority of cell types. The interactions between the ankyrin repeats of IκB and the RHD in Rel/NF-κB dimers inhibit the translocation of Rel/NF-κB proteins to the nucleus. However, with the effects of many extracellular stimuli, the serine kinase IκB kinase (IKK) is activated [13]. It then catalyzes the phosphorylation, ubiquitination, and degradation of IκB proteins, which leads to the translocation of the released Rel/NF-κB dimers to the nucleus. The activated Rel/NF-κB proteins then bind to DNA and activate gene transcription.

The focus of most studies has been on the involvement of the NF-κB signaling pathway in both vertebrate and invertebrate immunity. However, the NF-κB signaling pathway was found to not only regulate genes involved in the inflammatory and immune responses but also play an important role in bone homeostasis, osteoclast differentiation, and vertebrate bone formation [14]. In the process of vertebrate bone formation, NF-κB signaling pathway can participate in the formation of osteoclasts and intramuscular homeostasis regulation [15]. In mammalian chondrocytes, NF-κB binds to the BMP-2 gene promoter κB site in the TGFβ pathway and activates its expression. In vivo studies have found that P50/P52 knockout mice in the level of BMP-2 expression decreased by 80%, indicating that NF-κB signaling pathway in the bone growth process plays a critical role in regulation [16]. Therefore, we put our focus on the NF-κB pathway to find the mechanism of biomineralization of mollusk.

6.3.2 *Pf-Rel: A Rel/Nuclear Factor-κB Homolog*

We have reported the identification and characterization of a Rel/NF-κB homolog from pearl oyster, *P. fucata.* A cDNA clone was isolated based on RT-PCR, and amino acid sequence analysis was conducted. The complete cDNA sequence of *Pf-rel* including the poly(A) tail is 2737 bp. It contains a 100 bp 5′-untranslated sequence, an open reading frame consisting of 1938 bp, a TGA stop, a 688 bp 3′-untranslated sequence, and a poly(A) tail of 11 nucleotides. The cDNA sequence has been submitted to GenBank with the accession number EF121959. The putative Pf-Rel sequence consists of the Rel homology domain (RHD), IPT domain of the transcription factor NF-κB (IPT-NFκB) domain, nuclear localization signal (NLS), and C-terminal transactivation domain (TD). The RHD and IPT-NFκB domains, which are two characteristic domains in Rel/NF-κB family proteins, function in

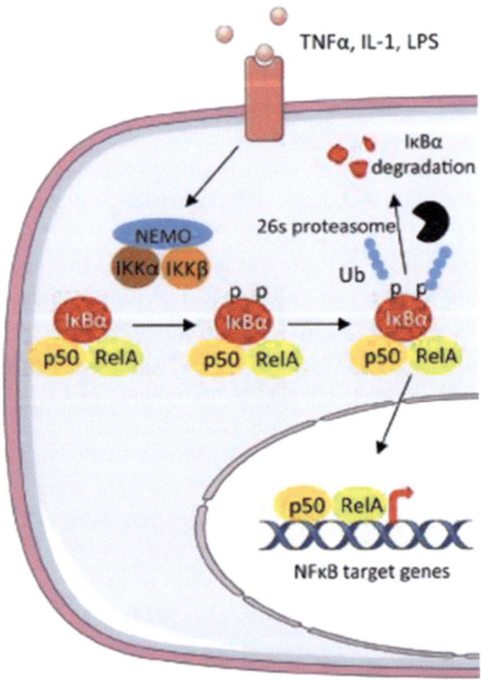

Fig. 6.1 The model of NF-κB signal transduction [12]

DNA binding, dimerization, and ankyrin protein binding. The putative Pf-Rel sequence shows high similarity in the two conserved domains with other Rel/NF-κB proteins. Nevertheless, the TD, which is essential in the transactivating ability of Rel/NF-κB, shows no similarity with known proteins. The TD of Pf-Rel is also rich in serine, one characteristic of this domain (Fig. 6.2).

The gene expression analyses help to investigate the pattern and location of *Pf-rel* expression. As expected, mRNA of *Pf-rel* was expressed in all the tissues studied (also including hemocytes, Fig. 6.3), which is the case for most of the Rel/NF-κB family proteins. This expression pattern suggests the involvement of Pf-Rel in diverse cellular processes. To investigate the precise expression site of the *Pf-rel* gene, in situ hybridization was carried out (Fig. 6.4). Strong hybridization signals were detected in the outer epithelial cells of the middle fold near the base of the periostracal groove and the inner epithelial cells of the outer fold. In the previous studies, the two sorts of cells were considered to be involved in the formation of the prismatic layer and periostracum. Pf-Rel might function in these processes.

In conclusion, a putative NF-κB homolog Pf-Rel was identified from pearl oyster, *P. fucata*, and it was expressed ubiquitously. Further structural and functional investigations could help to understand the involvement of the NF-κB signaling pathway in mollusk immunity and biomineralization.

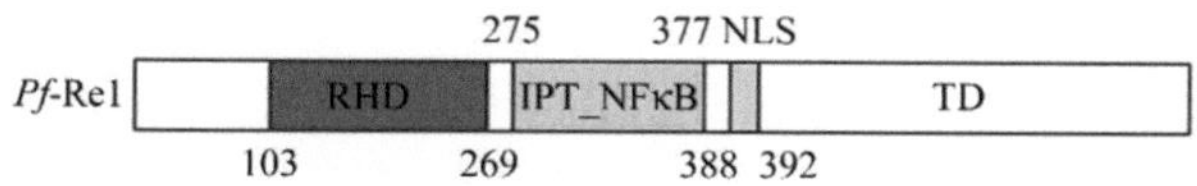

Fig. 6.2 Amino acid sequence analysis of *Pinctada fucata* Rel (Pf-Rel), a cDNA clone encoding transcription factor Rel/nuclear factor-κB homolog

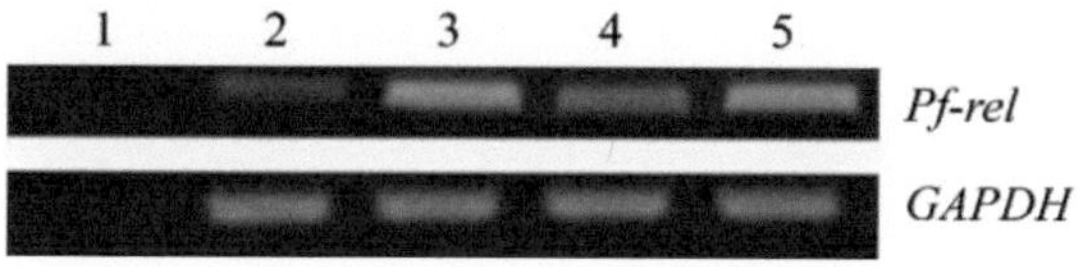

Fig. 6.3 Reverse transcription–polymerase chain reaction for analysis of *Pf-rel* gene expression in four tissues from *P. fucata*. 1, Negative control; 2, mantle; 3, gill; 4, adductor muscle; 5, gonad. The housekeeping gene encoding glyceraldehyde 3-phosphate dehydrogenase (GAPDH) was included as a positive control

6.3.3 Pf-IKK: The IκB Kinase Homolog in P. fucata

We present the isolation of a pearl oyster homolog of mammalian IKK-α. As shown in Fig. 6.5, the complete cDNA sequence is 2546 bp long. It contains a 173-base 50-untranslated sequence, an ORF consisting of 2214 bp, a TGA stop, and a 159-base 30-untranslated sequence. This cDNA sequence has been submitted to GenBank with the accession number AY818355. The ORF of this cDNA clone encodes a protein consisting of 737 amino acid residues, composed of an amino-terminal kinase domain (KD), a leucine zipper amphipathic a-helix (LZ), and a helix–loop–helix (HLH) motif, which is similar to IKK proteins from other organisms. We also demonstrated that the ability of Pf-IKK activated NF-κB-dependent reporter gene expression in a dose-dependent manner, and the activation was induced through IKK/IκB/NF-κB cascade (Fig. 6.5). Transfected by 4 mg Pf IKK recombinant plasmid, the luciferase activity increases more than 20-fold compared to the control vector (Fig. 6.6a) and was decreased sharply by 100 mm pyrrolidine dithiocarbamate (PDTC). A specific NF-κB DNA-binding activity was determined as shown in Fig. 6.6b, which appears to be increased in Pf-IKK transient expression cells. Pf-IKK-induced nuclear accumulation of NF-κB was also inhibited by PDTC. The fact that the NF-κB signaling system is present in both vertebrates and invertebrates (such as arthropods and mollusks) indicates that it is an ancient signaling system that originated prior to the deuterostome–protostome divergence.

The involvement of NF-κB in the expression of numerous cytokines and adhesion molecules has supported its role as an evolutionarily conserved coordinating element in organism's response to situations of infection, stress, and injury. Being stimulated by LPS, IκBα degradation was detected in hemocytes, suggesting that a similar cascade may exist in mollusks. In order to better understand the biological roles of Pf-IKK, we analyzed the gene expression pattern (Fig. 6.7). Expression analysis demonstrated that Pf-IKK mRNAs were present in all pearl oyster tissues studied,

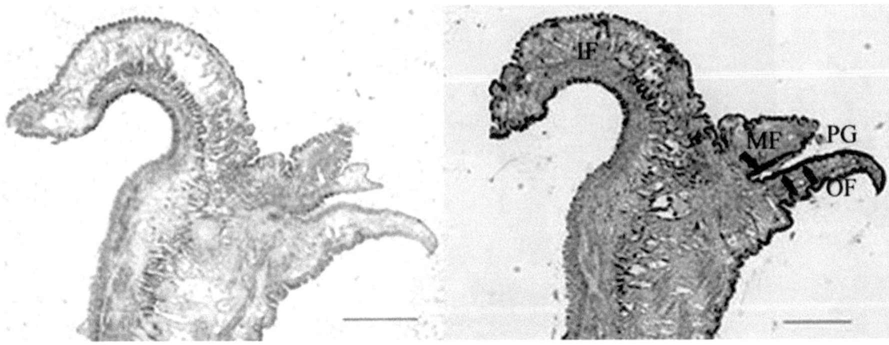

Fig. 6.4 Detection of *Pf-rel* mRNA distribution in the mantle of *Pinctada fucata* by in situ hybridization. A frozen section stained with a sense probe was used as the control (left frame). Strong hybridization signals are indicated by arrows (right frame). *IF* Inner fold, *MF* Middle fold, *OF* Outer fold, *PG* Periostracal groove. Bar = 0.5 mm

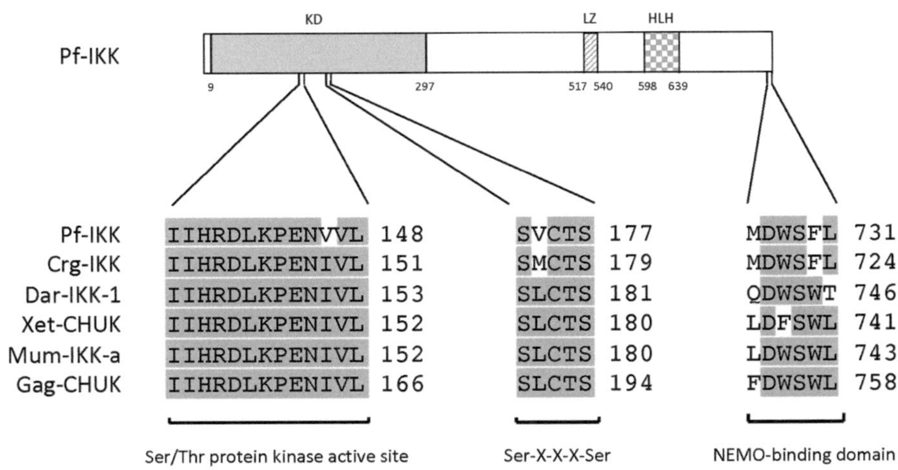

Fig. 6.5 Primary structure of pearl oyster Pf-IKK. (**a**) Motifs and domain structure of Pf-IKK. (**b**) Comparison of the amino acid sequences of the conserved motifs. Numbers indicate positions of the amino acid residues of each sequence. Pf-IKK, *P. fucata* I-kappa-B kinase (GenBank, AY818355); Crg-IKK, *Crassostrea gigas* IkB kinase (GenBank, AF051320); Dar-IKK-1, *Danio rerio* IkB kinase subunit I (GenBank, XM_686926); Xet-CHUK, *Xenopus tropicalis* conserved helix–loop–helix ubiquitous kinase (GenBank, genbank:NM_001016900); Mum-IKK-a, *Mus musculus* IkB kinase subunit alpha (GenBank, genbank:MMU12473); Gag-CHUK, *Gallus gallus* conserved helix–loop–helix ubiquitous kinase (GenBank, genbank:NM_001012904)

suggesting that Pf-IKK was ubiquitously expressed, which strongly implicates their role in numerous physiological processes. Being like that for oyster and abalone NF-κB pathway elements, Pf-IKK was equally expressed in challenged and unchallenged pearl oysters suggesting that transcription control plays minor role in regulating the expression of Pf-IKK.

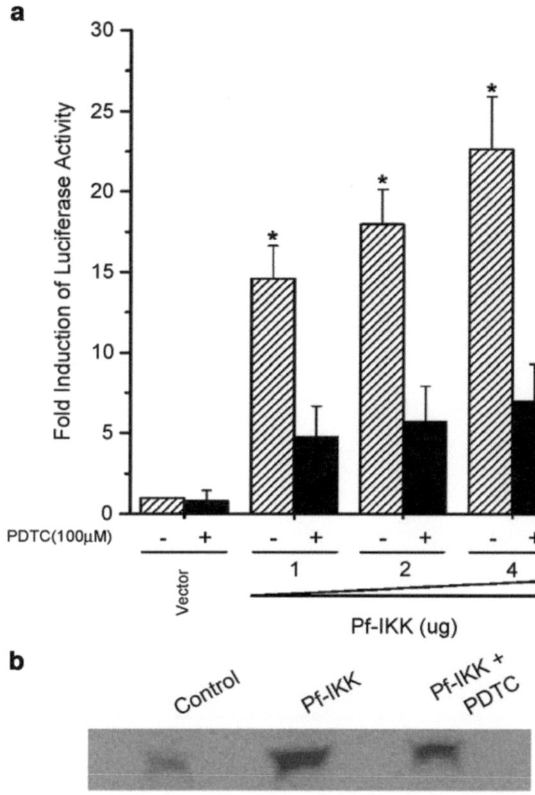

Fig. 6.6 Pf-IKK stimulated the expression of NF-kB-controlled luciferase reporter gene. (**a**) NIH3T3 cells were transiently cotransfected with pNFkB-TA-Luc reporter gene plasmid and pcDNA4/HisMax/lacZ, together with the variable amounts of pcDNA4/HisMax A/Pf-IKK vector. PcDNA4/HisMax A was used as controls. After 24 h, the cells were treated by 100 mm PDTC (or untreated) and incubated for another 24 h. Luciferase activity was normalized according to the level of β-galactosidase expression. Data are means and standard errors of five independent experiments. Asterisks denote significant (*for $P < 0.05$) differences from the control samples. (**b**) Electrophoretic mobility shift assays were carried out with nuclear extracts prepared from NIH3T3 cells. The nuclear extracts from untransfected cells used as control

In mammals, the inducible transcription factor NF-κB not only regulates gene expression during inflammatory and immune responses [17] but also is important for bone homeostasis and osteoclast differentiation [18]. The shells of mollusks are favored objects for the study of biomineralization of bones. Mollusk hemocytes are phagocytic cells involved in many functions, including wound healing, shell repair, as well as internal defense, which implied their potential role in biomineralization. Since NF-κB signal pathway is evolutionarily conserved in vertebrates and invertebrates, it gets twice the result with half the effort to study the counterpart function of

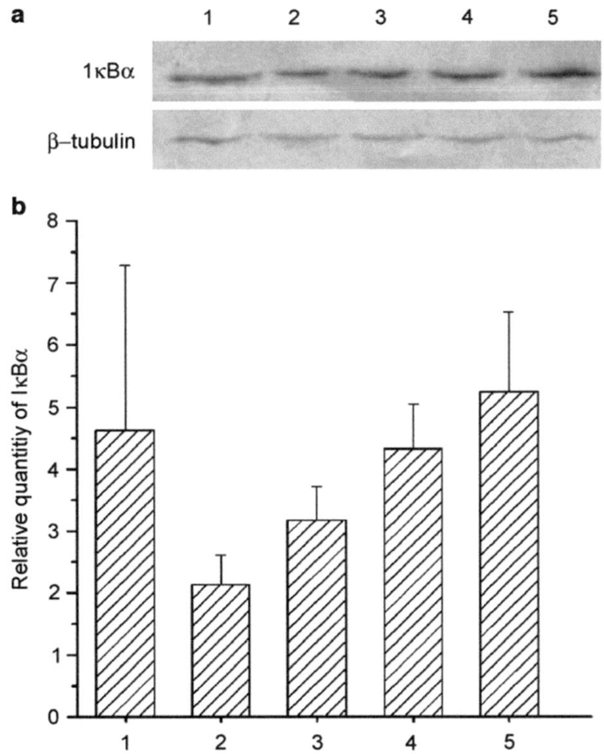

Fig. 6.7 LPS induces IκBα degradation in hemocytes. (**a**) Western blot analysis of hemocyte proteins probed with anti-IκBα antibody showing IκBα degradation levels with time, following LPS treatment. Equal loading (10 mg protein per lane) of membranes was confirmed by re-probing with anti-b-tubulin antibody. (**b**) Quantification of IκBα degradation levels with time following LPS treatment. Immunoblots (n ¼ 3) were scanned and band intensities measured to give mean IκBα levels (7 ± SDM.). Lane 1, unchallenged; lane 2, challenged for 10 min; lane 3, 30 min; lane 4, 60 min; lane 5, 120 min

NF-κB pathway in mollusks. To investigate whether the Pf-IKK functions as its mammalian counterparts, we analyzed the affection of Pf-IKK on the expression of genes involved in immune response and bone homeostasis. ALP is considered to be one of the first markers of the osteoblastic phenotype (Fig. 6.8). In this study, NIH3T3 cells transfected with Pf-IKK showed significant increase in ALP activity, suggesting that some NF-κB target genes may be regulated by activation of NF-κB pathway, which leads to the differentiation of fibroblast cells into "osteoblast-like" cells.

In conclusion, we have characterized the first NF-κB pathway element from pearl oyster *P. fucata*, the expression of which is ubiquitous. We also studied the function of Pf-IKK in NIH3T3 cells by a transient transfection strategy. Current investigations should focus on the cloning and characterization of other members of NF-κB signaling cascade and genes controlled by this signaling pathway. To better understand the involvement of NF-κB in pearl oyster shell formation processes and defense mechanism, it would be interesting to follow Pf-IKK and other NF-κB members' expression during *P. fucata* development processes as well as the shell damage in adults.

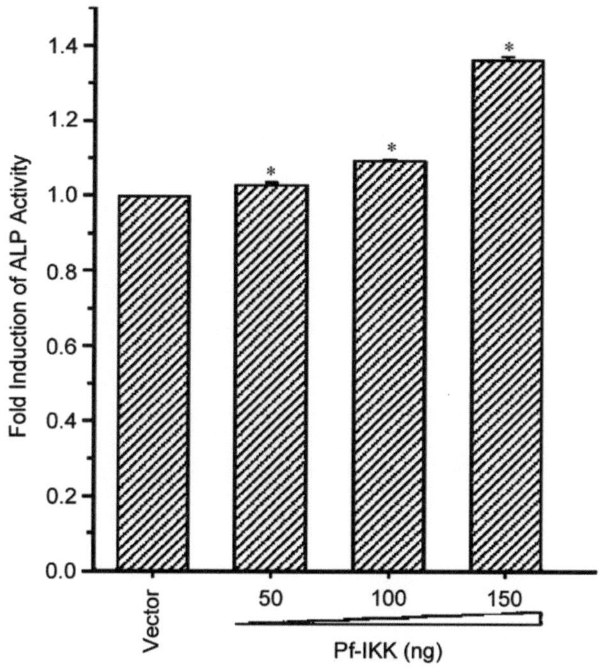

Fig. 6.8 The effect of Pf-IKK on ALP activity in NIH3T3 cells. Cells were seeded in 96-well plate and transfected with control pcDNA4 vector or various amounts of plasmid encoding Pf-IKK for 48 h. The data represent means ±SEM of eight independent experiments. *, $P < 0.01$, compared with the control

6.3.4 NF-κB Signaling Regulates the Transcription of Nacrein

As mentioned above, we found that Pf-IKK activated the expression of NF-κB-controlled reporter genes and induced NF-κB translocation. Sequence analysis of Pf-Rel shows that it shares high similarity with other Rel/NF-κB family proteins, especially within conserved domains. A conserved degradation motif and six ankyrin repeats were identified in the poIκB, which shares significant homology with other IκB proteins [19]. Here, we have shown that the NF-κB signaling pathway exists and functions to regulate nacrein, an important matrix protein capable of regulating the formation of oyster shells [20], transcription in *P. fucata*.

An in silico analysis of the nacrein promoter sequence identified two possible NF-κB binding sites between nucleotides 429–438 and 882–891. We were able to confirm that the NF-κB signaling pathway could regulate the activity of the nacrein promoter in *P. fucata* by cotransfection PGL3-Nacrein and pcDNA4.0A/Pf-IKK or pcDNA4.0A/Pf-Rel (Fig. 6.9). Expression of both of these proteins promoted nacrein transcription, while the NF-κB inhibitor PDTC inhibited nacrein transcription (Fig. 6.10).

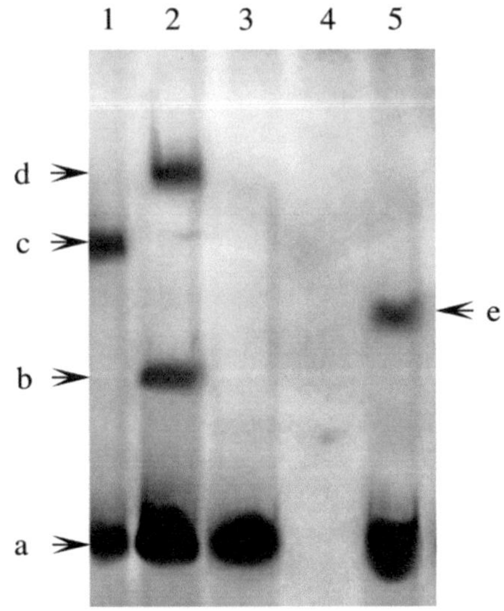

Fig. 6.9 Pf-Rel could bind to the promoter of nacrein (EMSA). Nuclear proteins extracted from the gills of *P. fucata* were incubated with oligonucleotide probes labeled with DIG-ddNTP to perform EMSAs. Lane 1: labeled DNA probes, nuclear protein, and antibody of Pf-Rel. Lane 2: labeled DNA probes and nuclear protein. Lane 3: labeled DNA probes. Lane 4: unlabeled DNA probes and nuclear protein. Lane 5: positive control. Band a: free DNA. Band b: DNA and Pf-Rel. Band c: DNA, Pf-Rel, and antibody of Pf-Rel. Band d: DNA, Pf-Rel, and other interactive nuclear proteins. Band e: positive control band. Compared to lane 2, there is a super shift after Pf-Rel antibody was added in lane 1, suggesting that Pf-Rel present in the total nuclear protein extracts is capable of binding to the nacrein promoter probes

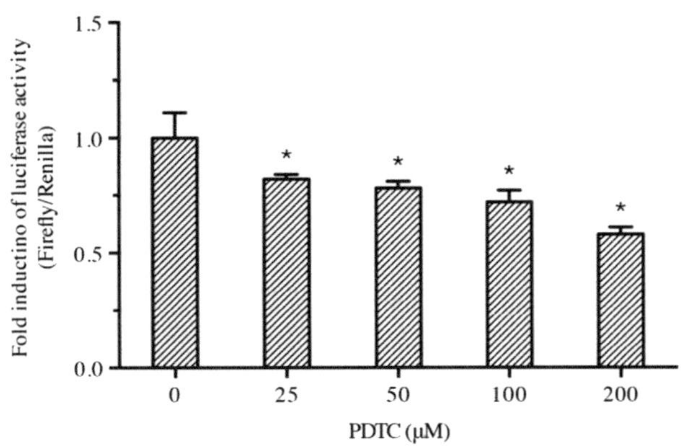

Fig. 6.10 PDTC inhibits the nacrein promoter activity. Equal amounts of pGL3-Nacrein luciferase and pcDNA4.0A/Pf-IKK plasmids were cotransfected into HeLa cells that were 24 h previously treated with increasing concentrations of PDTC (0, 25 μm, 50 μm, 100 μm, and 200 μm). 0.5 ng of *Renilla* was used as an internal reference. Each reaction was repeated in triplicate. Increasing amounts of PDTC lead to a decrease in luciferase activity, suggesting a decrease in nacrein promoter activity. Significant differences were identified by one-way ANOVA. The symbol "*" indicates a significant reduction ($P < 0.05$), compared to the control

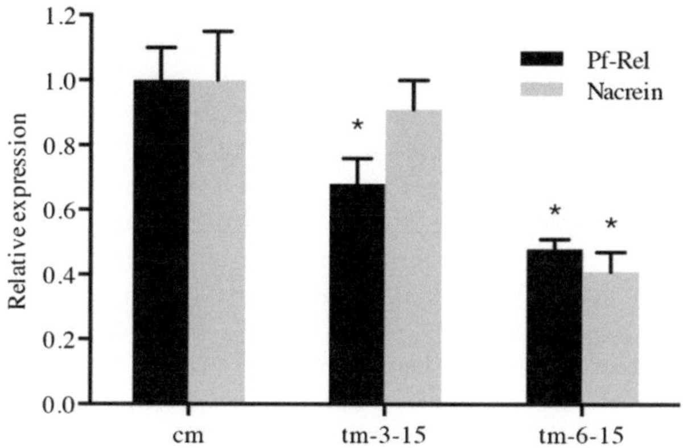

Fig. 6.11 Pf-Rel knockdown decreased the nacrein gene expression level. The expression levels of nacrein (gray columns) and Pf-Rel mRNA (black columns) in oyster mantle were measured by real-time PCR, 3 or 6 days after injection of Pf-Rel dsRNA. cm: NaCl control solution. Both nacrein and Pf-Rel mRNA expression levels in controls are attributed a relative value of 1.0. tm-3-15 and tm-6-15: samples injected with 15 ng Pif-Rel dsRNA 3 or 6 days, respectively. The expression levels of nacrein slightly decreased after dsRNA injection, compared to the control group. Six days after injection, the expression level of both Pf-Rel and nacrein was suppressed by nearly 50%. Significant difference was identified by one-way ANOVA. The symbol"*"indicates a significant reduction ($P < 0.05$), compared to control oysters

Although NF-κB signaling is able to regulate the nacrein promoter, we were not sure whether this pathway could regulate the expression level of nacrein. Real-time quantitative PCR results (Fig. 6.11) showed that mRNA transcription of nacrein is depressed when Pf-Rel mRNA is partially silenced, suggesting that Pf-Rel indeed is involved in regulating nacrein transcription. In conclusion, we have shown how the important NF-κB signaling pathway protein Pf-Rel regulates the transcription of nacrein.

The NF-κB signaling pathway has been extensively studied in mammals. Here, we focused on the effect of one component of NF-κB signaling in regulating *nacrein* transcription in pearl oysters. In the future, it will be interesting to investigate if an interaction exists between the different NF-κB signaling components Pf-IKK, poIκB, and Pf-Rel. Because of a lack of mollusk cell lines, transfection experiments have to be performed on mammalian cells instead. It would be interesting to confirm these results experimentally within primary cell cultures of oyster tissue.

6.4 TGFβ Signaling Pathway

6.4.1 Introduction

The transforming growth factor beta (TGFβ) superfamily has a wide range of frequently different effects in different cells and tissues, of which the members include TGFβs, bone morphogenetic proteins (BMPs), activins and inhibins, growth differentiation factor, nodal, decapentaplegic (dpp), and glial cell line-derived neurotrophic factor. TGFβ superfamily members are involved in controlling a variety of physiological and disease processes of vertebrates and invertebrates including cell proliferation, differentiation, fibrosis, and apoptosis. They also play important roles in the specification of developmental fate during embryogenesis and in mature tissues, extracellular matrix synthesis, bone formation and rebuilding, angiogenesis, immunoregulation, and hormone function [21].

The ligands, receptors, and the regulators in TGFβ signaling pathway interact with each other to set a regulation network as follows. After the soluble ligand (like BMP2) binding to the type II receptor, the ligand–type II receptor (TRII) complex interacts with GS box of type I receptors (TRI), inducing the formation of the heterologous tetramer. Phosphorylated to activate, the tetramer complex in turn phosphorylates the downward receptor-regulated Smads (R-Smads, Smad1, 5, 8). And then R-Smad combines with the common Smads (Co-Smads) to form smad protein complex, which moves into the cell nucleus to regulate the transcription of target genes directly or cooperate with transcriptional factors [22].

In contrast to abundant findings in vertebrates, we have limited knowledge on the roles of members of TGFβ family in mollusk biomineralization. Although related studies have successfully cloned and characterized BMPR1, TGFR1, ACVRI, ACVRII, and ALR1 in the Pacific oyster *C. gigas* [23–26] and TGFR1 in the scallop *Chlamys farreri* [27], there was no direct evidence to support their roles in the biomineralization process. Therefore, it is largely needed to perform detailed investigation in shelled mollusks to supply reliable data for the involvement of TGFβ signaling pathway in shell biomineralization (Fig. 6.12).

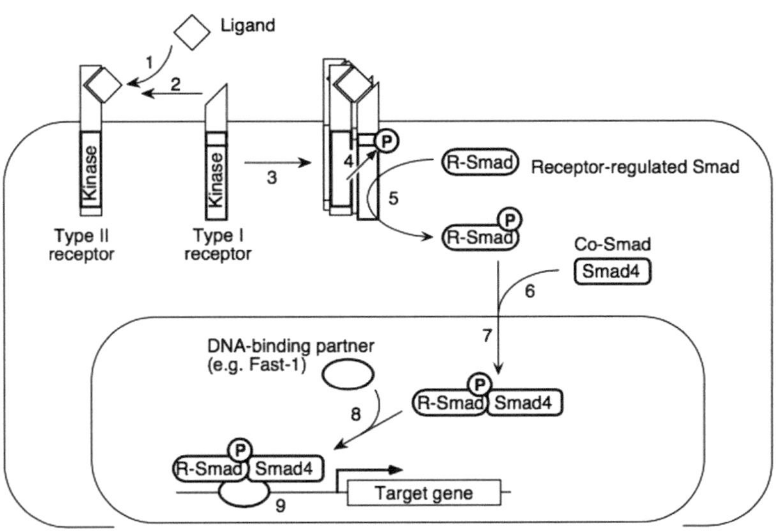

Fig. 6.12 The transforming growth factor β (TGF-β)/SMAD pathway [28]

6.4.2 The Receptors of the Pathway

6.4.2.1 PfALR1: An Activin-Like Receptor 1 Homolog

In a classical TGF-β signal pathway, the ligand first binds to its type II cell surface transmembrane receptor, leading to the recruitment, phosphorylation, and activation of its type I activin receptor. Then, the activated receptor complex transiently interacts with downstream factors. Therefore, it is of great scientific value to identify as many other TGFβ signaling receptors in oysters as possible.

Here we report a cDNA encoding an activin-like receptor 1 homolog (designated as Pf-ALR1) cloned from the oyster, *P. fucata*. A 2289 bp sequence contains an 1641 bp open reading frame that encodes a 547-amino acid protein, a 628 bp 3′-untranslated sequence, and a poly(A) tail of 9 nucleotides. The cDNA sequence has been submitted to GenBank with the accession number GQ280384. Its deduced amino acid sequence has over 60% similarity with vertebrate activin A type I receptors in their conserved regions. A putative signal peptide cleavage site is predicted between amino acids 17 and 18. The cleavage would release a 529 residue mature peptide. Pf-ALR1 comprises an extracellular domain (residues 18–147) that contains 9 cysteine residues that align in all serine–threonine kinases. In addition, a potential extracellular glycosylation site is predicted at residues 61. Hydrophobicity plots reveal a potential transmembrane segment between amino acids 148 and 181. The intracellular region is characterized by a serine/threonine kinase domain in which phosphorable serines [29] may be present at positions 207 and 216. An ATP-binding site (GKGRYG) between residues 244 and 249 is predicted [30]. Immediately preceding the protein kinase domain is the SGSGSG signature (residues 219–224), a unique feature of TGF-b superfamily type I receptors [31]. Following this is a GS box, a leucine–proline motif [32]. The cytoplasmic domain also comprises several other motifs such as EIV/F, RIKKTL, and GRNNK boxes that have been previously reported as hallmarks of all serine/threonine type I receptors [33]. More importantly, the Pf-ALR1 also consists of many characteristic motifs that are considered as hallmarks of activin type I receptors [30]. Usually, the L45 loop in a type I receptor kinase domain specifies the receptor-regulated Smads (R-Smad) interaction [34]. The L45 loop kinase domain of Pf-ALR1 is very similar to the L45 loops that are specific to the receptors such as ALK (activin receptor-like kinase) 1, ALK2, and Sax but different from the two kinds of L45 loops for TGFβ and BMP receptor groups. Therefore, we hypothesize that Pf-ALR1 may play similar functions as ALK1 and ALK2. It was reported that an activin receptor fusion protein showed promise as an anabolic agent in early human trials on new drug targets for osteoporosis [35]. It indicates that activin receptor could be developed as a reagent to adjust the biomineralisation process in oysters (Fig. 6.13).

To figure out the distribution of the Pf-ALR1 mRNA expression in the oyster, semiquantitative reverse transcription–polymerase chain reactions with the cDNA templates of the different tissues of the oyster were conducted. The mRNA for Pf-ALR1 is expressed in all the seven tissues analyzed (Fig. 6.14), which suggest

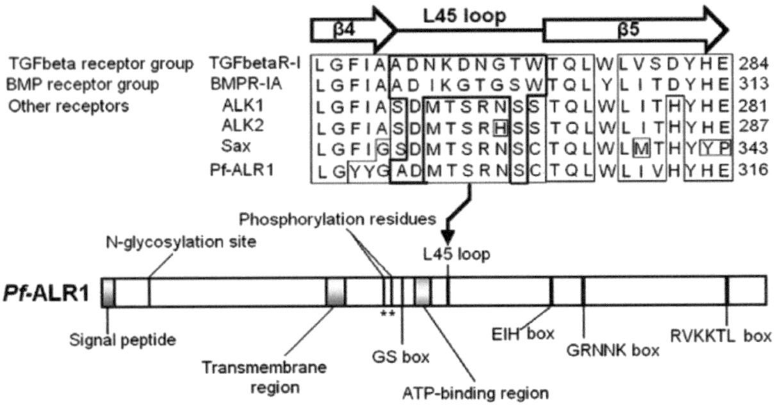

Fig. 6.13 Schematic representation of Pf-ALR1

that the Pf-ALR1 gene is ubiquitously expressed in adult *P. fucata*. Moreover, the mRNA expression levels of Pf-ALR1 in the gonad, the mantle, the viscera, and the hemocytes of the oyster are much higher, which indicates Pf-ALR1 may play important roles mainly in those tissues.

In situ hybridization was carried out to locate the precise site of Pf-Smad3 mRNA expression in the mantle (as shown in Fig. 6.15). Slight signals were detected in the outer epithelial cells of the middle fold as well as the inner epithelial cells of the outer fold of the mantle. These expression locations had overlapping areas with Pf-Smad3 and some matrix proteins in the oyster. It suggests Pf-ALR1 may also play important roles in primastic layer and periostracum formation.

To detect the response of Pf-ALR1 in the shell regeneration process, the mRNA expression level of Pf-ALR1 was assayed at each time point after shell repair was induced. As shown in Fig. 6.16, the mRNA expression level of Pf-ALR1 is remarkably increased near the moment when the shell is notched. Then, the mRNA level of Pf-ALR1 decreases gradually, while the Pf-Smad3 mRNA level increases gradually. At 12 h after shell notching, the Pf-ALR1 mRNA decreases to a very low level, and the Pf-Smad3 mRNA reaches a high peak. After that, the Pf-ALR1 and Pf-Smad3 mRNA levels increase to their high expression levels at 24 h and decrease again to their lowest levels at 36 h after the shell notching. The mRNA levels of Pf-ALR1 and Pf-Smad3 return to their high level again at 48 h after the shell notching. These results indicate that Pf-ALR1 and Pf-Smad3 mRNA levels changed quickly after shell injury. The change tendency of the Pf-ALR1 is opposite to Pf-Smad3 during the first 12 h after shell notching. According to the report by Finnson that ALK1 opposes ALK5/Smad3 signaling and expression of extracellular matrix components in human chondrocytes [36], we suggest that Pf-ALR1 might repress the Pf-Smad3 expression during the initial shell repair. The relationship between the expressions of these two genes in the oyster needs to be deeply investigated.

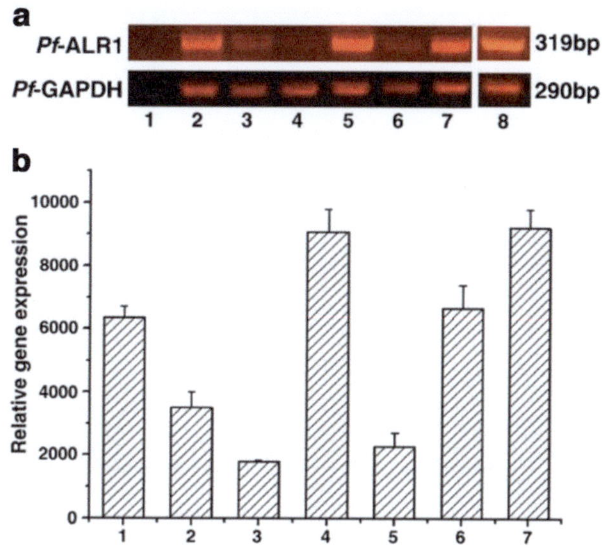

Fig. 6.14 (**a**) RT-PCR for analysis of Pf-ALR1 gene expression in seven tissues from *Pinctada fucata*. 1, negative control; 2, gonad; 3, gill; 4, foot; 5, mantle; 6, adductor muscle; 7, viscera; 8, hemocytes (done separately). The housekeeping gene encoding glyceraldehyde 3-phosphate dehydrogenase (GAPDH) was included as a positive control. (**b**) Relative expression of Pf-ALR1 mRNA in tissues of the pearl oyster. 1, Gonad; 2, gill; 3, foot; 4, mantle; 5, adductor muscle; 6, viscera; 7, hemocytes

It has been reported that dexamethasone (DXM) and hydrogen peroxide (H_2O_2) can be employed to adjust the Smad3 expression levels in mammal cells [37, 38]. Thereby, DXM or H_2O_2 was added to the culture medium of the in vitro cultured oyster mantle tissue cells. Then, real-time quantitative PCR was conducted. The results show that the mRNA levels of Pf-Smad3, Pf-ALR1, and the two oyster NF-κB members, Pf-Rel and Pf-IKK, are all changed and correlated. Taken together, Pf-ALR1 is a pivotal TGFβ member in the oyster. It correlates with the Pf-Smad3. They would play important roles in many physiological processes.

All our observations suggest that there should be similar TGFβ signal pathways in the oyster and vertebrate. However, the potential functions Pf-ALR1 has and the relations of TGFβ and NF-κB members in the oyster all need to be thoroughly investigated.

6.4.2.2 PfBMPR1B: One Type I Receptor in BMP Pathway

The BMPs, one of the largest subfamilies of TGFβ superfamily, play the indelible roles in the formation, differentiation, and development of cartilage and bones [39]. BMPR1B, bone morphogenetic protein receptor type 1b, is the type I serine/threonine receptor of the BMP signal pathway which cannot function without the help of the type II receptors. The Bmpr1b receptor abides by the typical route of the TGFβ signal pathway. After the soluble ligand binding to the type II receptor, the ligand–type II receptor (TRII) complex interact with GS box of type I receptors (TRI), inducing the formation of the heterologous tetramer. Phosphorylated to activate, the tetramer complex in turn phosphorylates the downward receptor-

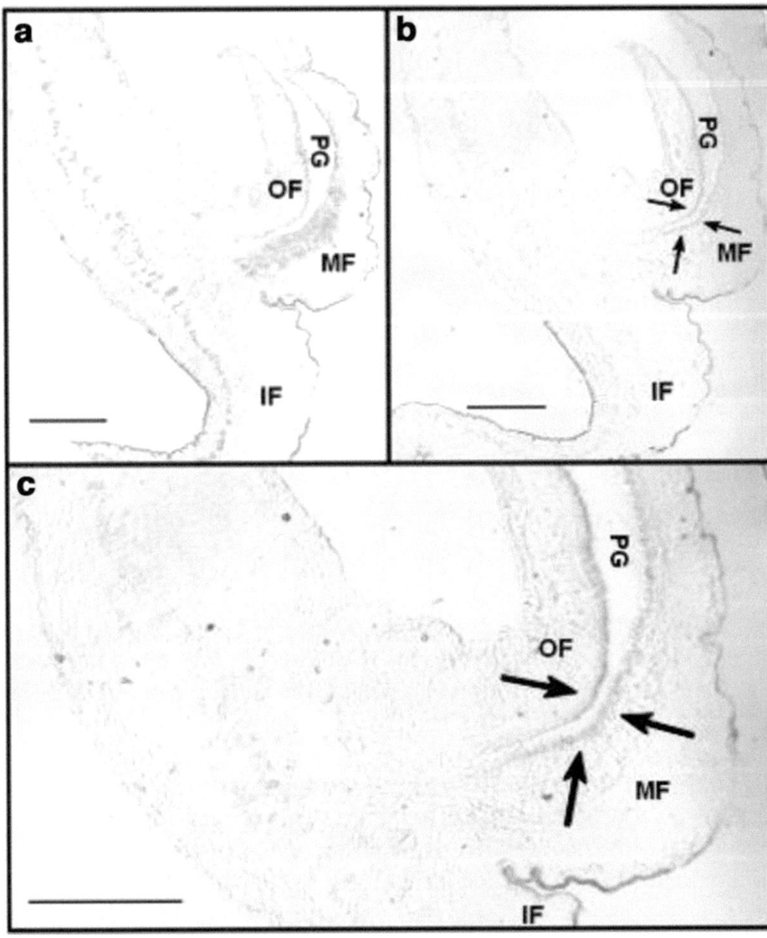

Fig. 6.15 Detection of Pf-ALR1 mRNA distribution in the mantle by in situ hybridization. A frozen section stained with a sense probe was used as the control (frame a). Slight hybridization signals are indicated by arrows (frames b and c). *IF* Inner fold, *MF* Middle fold, *OF* Outer fold, *PG* Periostracal groove. Bar = 500 μm

regulated Smads. And then R-Smad combines with the common Smads to form smad protein complex, which moves into the cell nucleus to regulate the transcription of target genes directly or cooperate with transcriptional factors [22]. The BMPR1B gene has been widely studied that transforms of BMPR1B prove the ovulation rate with the precocious development of a large number of small antral follicles in the Booroola animals [40]. However, researchers have become increasingly interested in the BMP signal pathway roles in chondrogenesis and osteogenesis. The findings suggest that the disruption of BMPR1B reduced the prechondrogenic cells and chondrocyte differentiation in the phalangeal region of

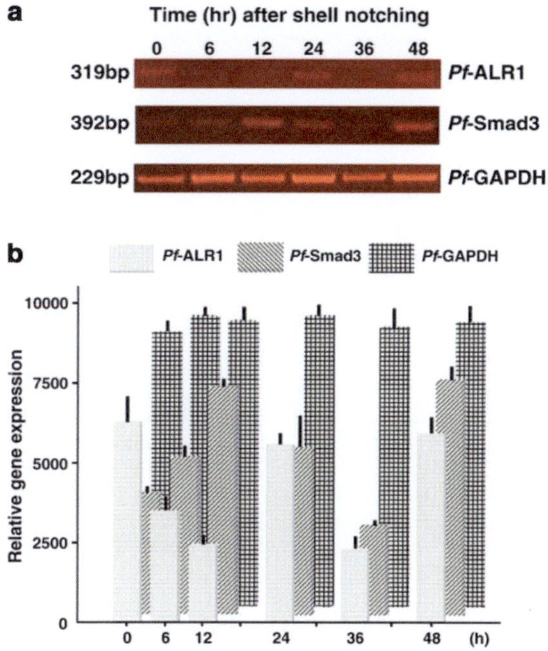

Fig. 6.16 (**a**) RT-PCR analysis of the oyster ALR1, Smad3 expression after the shell notching. The housekeeping gene Pf-GAPDH is adopted as a positive control. (**b**) Relative expression of the oyster ALR1, Smad3 expression after the shell notching. The housekeeping gene Pf-GAPDH is adopted as a positive control

mice limb [41]. Overall, the BMPR1B may mainly account for the development of gonad and the formation of the bones.

Pf-BMPR1B gene was identified with a full-length 1925 bp including a single long ORF of 1602 bp encoding a 533 aa protein, with a 22 bp 5′ UTR and a 301 bp 3′ UTR. The Bmpr1b protein has a 28aa long signal peptide, which also demonstrates that this certain protein is a kind of secretary protein. As a receptor loading on the cell membrane, Bmpr1b protein has a transmembrane region of site 138–160. Meanwhile the intracellular region ranging from 161 to 533 has much more phosphorylation sites than BAMBI, which is the specific characteristics of the BMP signal pathway type I receptors. The serine and threonine predictions are circled in Fig. 1, and the protein kinase domain exists in 213–507. GS domain, the feature of the type I receptors, is characterize in the range of 181–212.

Besides the understanding of the sequence information, the functions of certain gene wait to be explored, from which we investigated the specificity expression in different tissues (adductor, mantle edge, mantle pillial, gonad, gill, foot, viscus) by RT-PCR and the distribution by in situ hybridization. Setting the gene expression in adductor as the control, the results in Fig. 6.17 show that the Pf-BMPR1B gene expression level in gonad was far more than the other tissues, although there were also high levels in gill, foot, and viscus. It has proved that the members of TGFβ family play the corresponding roles in different organisms, like affecting the development of the heart, digestive gland, kidney, liver, and lung and functioning in the fibroblast growth wound healing, inducing angiogenesis, and regulating the

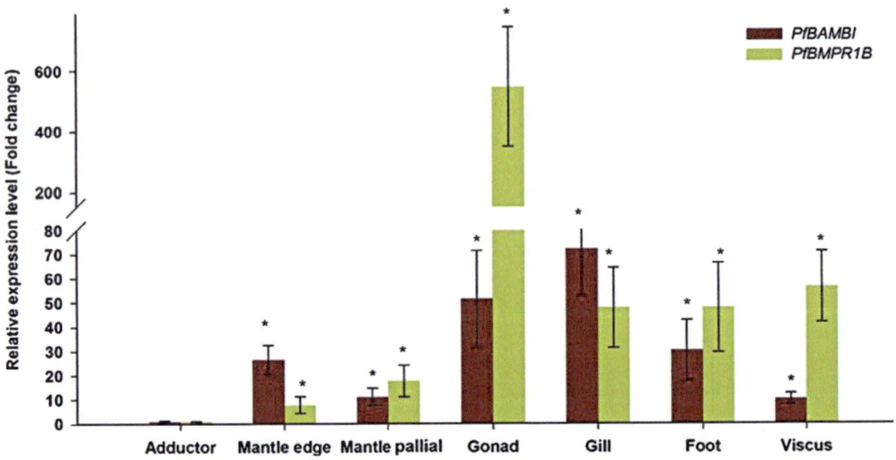

Fig. 6.17 The relative expression levels of PfBMPR1B and PfBAMBI in different tissues of *Pinctada fucata*. Mantle edge, mantle pallial, gonad, gill, foot, and viscus were selected as study tissues, and the expression levels of PfBMPR1B and PfBAMBI in the adductor were used as controls. $*P < 0.05$

hematopoiesis. The varied expressions in distinct tissues prove that BMPR1B gene may be the multifunctional gene to maintain the normal physiological function of diverse organisms.

To further investigate the gene expression pattern in the mantle tissue, we analyze the special expression of Pf-BMPR1B mRNA through in situ hybridization. In contrast with the undetected signals in controls (Fig. 6.18), the positive signal of Pf-BMPR1B gene appeared at the epithelial cells of the outer fold and the mantle pallial, suggesting that it is related to the formation of both the prismatic and nacreous layer.

6.4.2.3 PfBAMBI: A Pseudoreceptor to Compete with Type I Receptors

BAMBI, BMP, and activin membrane-bound inhibitor are the pseudoreceptor, resembling the homodimerization interface of TGFβ family type I receptors but lacking an intracellular serine/threonine kinase domain required for signaling transduction. In the process of the combination of TGFβ family ligand and receptors, Bambi competitively binds with the ligand and type II receptors replacing Bmpr1b position in the heterologous tetramer complex and then preventing the downward phosphorylation because of the absence of the kinase domain. It is definitely a sort of negative feedback regulation mechanism. BAMBI is an essential regulator of cell proliferation and differentiation that represses TGFβ and enhances Wnt/beta-catenin signaling in various cell types [42]. The BAMBI gene of vertebrate has been widely

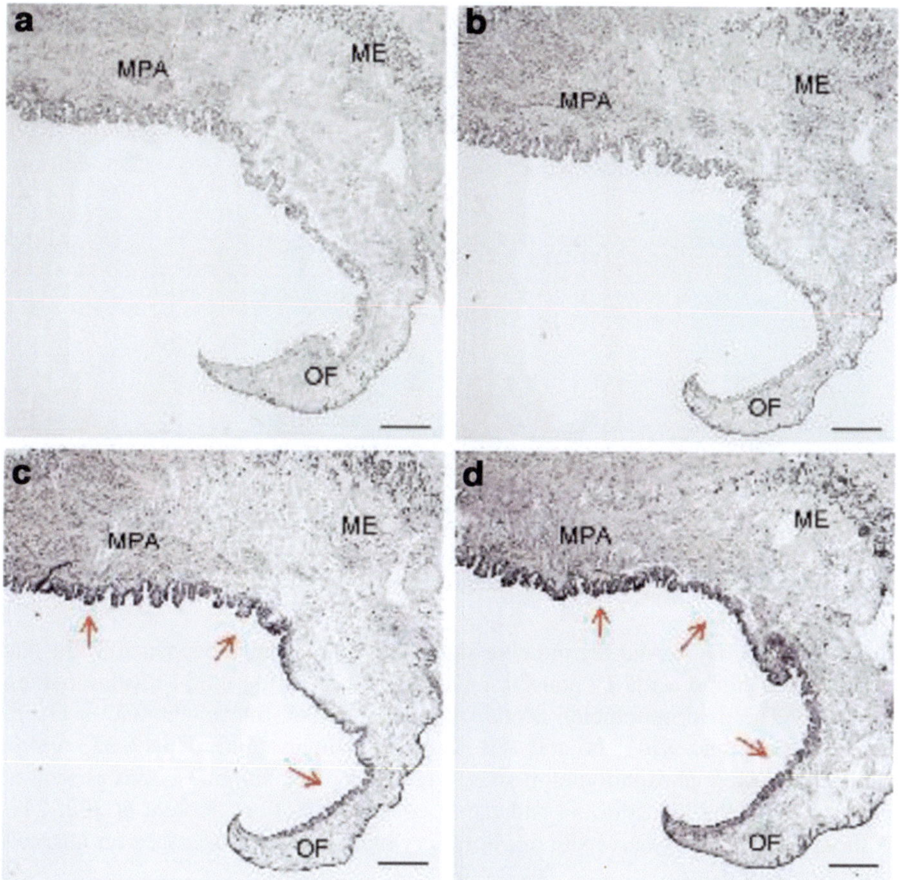

Fig. 6.18 The localizations of PfBMPR1B and PfBAMBI in the mantle of *Pinctada fucata* detected by in situ hybridization. The sense probes for PfBMPR1B (**a**) and PfBAMBI (**b**) are used as controls. *MPA* Mantle pallial, *ME* Mantle edge, *OF* Outer fold. The red arrows indicate the positive signals (dark blue) detected in the outer epithelia of the MPA and OF using antisense probes of PfBMPR1B (**c**) and PfBAMBI (**d**). Bar = 200 μm

researched; most exciting, the previous research indicates the interactions between BAMBI and biomineralization. It has been found that the Bambi functions as a dominant-negative protein that regulates reciprocal epithelial–mesenchymal interactions during dentinogenesis [43]. However, it still remains blank and has no relevant scientific reports on BAMBI of the mollusks.

After the cloning and sequencing, we obtained Pf-BAMBI gene which presents a full-length 1939 base pair DNA with the region of 822 bp to encode an open reading frame of a 273 amino acid long protein, except for a 331 bp long 5'UTR and a 786 bp 3′UTR. The predicted Bambi amino acid sequence contained the specific features of

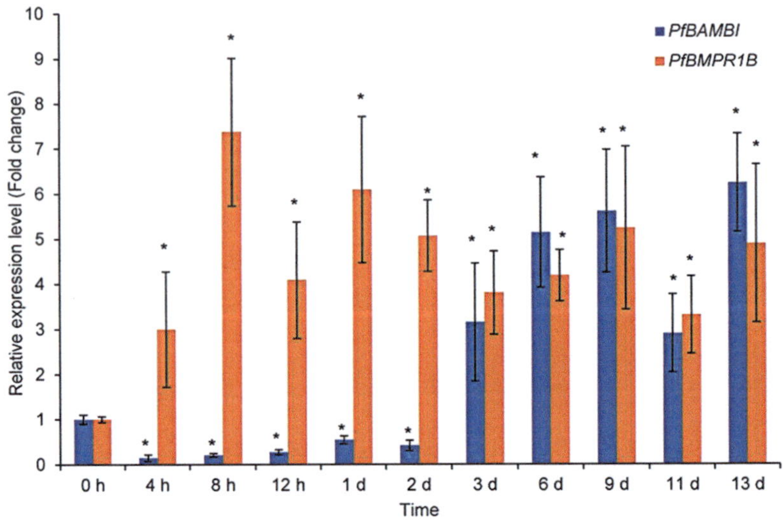

Fig. 6.19 The relative expression levels of PfBMPR1B and PfBAMBI in the mantle of *Pinctada fucata* during shell notching. $*P < 0.05$

the BAMBI domain. And the putative signal peptide cleavage happened in the site between the amino acids 19 and 20, as a result of releasing a 254 residue mature peptides. The hydrophobicity plots suggest potential transmembrane segment between the amino acids 166 and 188. The intracellular region of 189–273 amino acids contain few phosphorylation sites, where the phosphorable serines present at position of 192, 207, 248, 254 and the threonine predictions present at 206, 241. Although it has the putative site of phosphorylation, BAMBI does lack an intracellular kinase domain.

When it occurs to the Pf-BAMBI expression level in different tissues, the result showed in Fig. 6.17 is highly expressed in gill, gonad, foot, and mantle edge, which may demonstrate the various functions of the gene. To investigate the specific location in mantle tissue, we chose to take out in situ hybridization experiment. As shown in Fig. 6.18, the Pf-BAMBI gene is located in the outer epithelial cells of both the outer fold and the mantle pallial, implying that the function participates in the analogous process with Pf-BMBR1B. The similar location also inferred that the function of the two receptors was associated with each other.

To further confirm of the role Pf-BAMBI and Pf-BMPR1B genes play in the process of the mineralization, we performed the shell notching experiments that excise V-shaped notch on both sides of the shell, which mimic the rapid shell growth. According to Fig. 6.19, the expression pattern of Pf-BAMBI gene displays a downward and then upward trend. Simultaneously, Pf-BMPR1B gene expression was upregulated at first, reached a peak in 8 h, and then declined but still higher than the basal level after the shell notching. Pf-BAMBI immediately responded to the notching that from the beginning to 2 days, the Pf-BAMBI gene was repressed. The high level of Pf-BMPR1B gene expression at the same time infers that as a competitive inhibitor, decreased Pf-BAMBI expression promoted to express more Bmpr1b protein to function with less competition. Consecutively, from 3 days to 13 days, the increased Bambi protein might compete with the type I receptor causing the slight decline of the Pf-BMPR1B expression. Overall, the Pf-BMPR1B expression does respond to this rapid shell growth, and the converse pattern of Pf-BAMBI expression speculates a competitive relation between Pf-BAMBI and Pf-BMPR1B. It brings the evidence that both the Pf-BAMBI and Pf-BMPR1B are involve in biomineralization.

In conclusion, we cloned and identified two homologous BMP receptor genes, PfBMPR1B and PfBAMBI, from the pearl oyster *Pinctada fucata*. The pearl oyster homologs Pf-BAMBI and Pf-BMPR1B distribute ubiquitously in varied tissues, participate the rapid shell growth, have the effect on the matrix proteins, and play essential roles in the shell formation at embryonic and larva development stage. Collectively, these results clearly show that PfBMPR1B and PfBAMBI are involved in regulating shell biomineralization in *P. fucata*. Our study therefore provides the direct evidence that BMP receptors participate in mollusk biomineralization.

6.4.3 Downstream Response Protein

6.4.3.1 PfSmad3: Downstream Signal of BMP Pathway

Smad3 has two highly conserved domains, DNA-binding mothers against dpp (MAD) homology 1 (MH1) domain and Runx2/Cbfa1-binding MH2 domain. The MH1 domain can bind specific DNA sequences, called Smad3-binding sequences, which always contain the CAGA sequence in an unusual way, including a β-hairpin structure binding to the major groove. The MH2 domain can bind other transcriptional factors to provide specificity and selectivity to Smad function as well as the interactions in Smad oligomers that phosphorylate the last two Ser of a conserved SSXS motif located at the C-terminus of MH2.

In our study, we have reported the identification and characterization of a Smad3 homolog from pearl oyster, *P. fucata*. A cDNA clone was isolated based on RT-PCR, and amino acid sequence analysis was conducted. The complete cDNA sequence of *Pf*-Smad3, including the poly(A) tail, is 1515 bp. It contains a 109 bp 5′-untranslated sequence, an open reading frame consisting of 1242 bp, a TAA stop, and a 164 bp 3′-untranslated sequence which contains a poly(A) tail of 12 nucleotides. A putative polyadenylation signal (AATAAA) was recognized at position 1295–1300, located upstream of the poly(A) tail and separated by 94 nucleotides. The cDNA sequence was submitted to GenBank with the accession no. EU137731. The open reading frame of the cDNA clone encodes a protein consisting of 413 amino acid residues. Amino acid sequence analysis (using BLAST) showed that Pf-Smad3 possesses the characteristic primary structure of Smad3 proteins. Two putative conserved domains have been detected, MH1 (MAD homology 1: Pfam03165.13) of 123 amino acid residues (aa 10–131) and

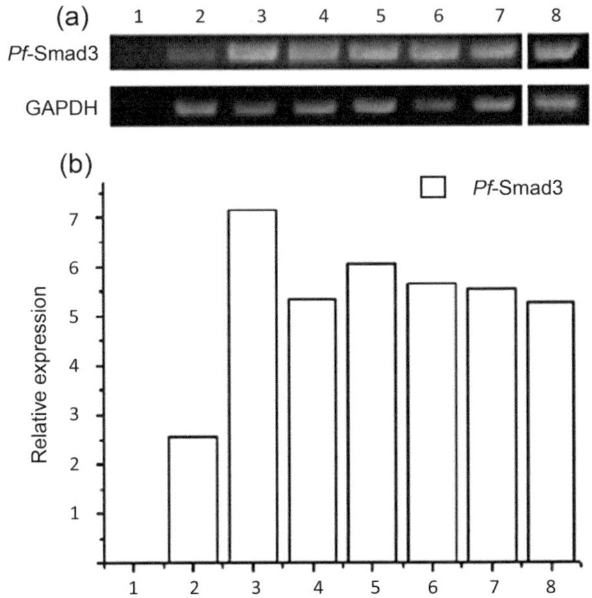

Fig. 6.20 Relative expression of the Pf-Smad3 gene in seven tissues from *P. fucata*. (**a**) Reverse transcription–polymerase chain reaction analysis was carried out with mRNA samples from the adult tissues of *P. fucata*. The pearl oyster housekeeping gene encoding glyceraldehyde-3-phosphate dehydrogenase (GAPDH) was included as a positive control. (**b**) Relative expression of Pf-Smad3 mRNA in tissues of *P. fucata*. The pearl oyster GAPDH was used as a control of equal quantities of total RNA. 1, Negative control; 2, gonad; 3, gill; 4, foot; 5, mantle; 6, adductor muscle; 7, viscera; 8, hemocytes

MH2 (MAD homology 2: Pfam03166.13) of 176 amino acid residues (aa 214–389), both of which are characteristic of Smad proteins.

Compared with other Smad3 proteins, Pf-Smad3 has extremely high similarity with the Smad proteins in vertebrates. The identity of the conserved regions (including the MH1 and MH2 domains and the C-terminal) is conserved up to 83.9%; Smad proteins in *Drosophila* and *C. elegans* are very dissimilar to the vertebrate Smad3 proteins. Based on the theory of evolution, the remarkable similarity of Pf-Smad3 to vertebrate Smad3 proteins indicates that Pf-Smad3 might play an important role in the oyster through pathways conserved from, or at least similar to, the vertebrate.

The RT-PCR showed that mRNA of Pf-Smad3 was expressed in all the tissues analyzed (Fig. 6.20), consistent with the expression pattern of the mouse Smad3. This expression pattern indicates the involvement of Pf-Smad3 in diverse cellular processes. The changes of the mRNA expression of Pf-Smad3 during the D-sharp larva stage and the attachment stage of the oyster embryo were detected by semiquantitive PCR (Fig. 6.21). The relationship between the change in the level of Pf-Smad3 and the process of shell formation, particularly the D-sharp larva stage when oysters start to secrete nacre, should be further studied.

To investigate the precise expression site of the Pf-Smad3 gene in the mantle, in situ hybridization was carried out. Obvious hybridization signals were detected in the outer epithelial cells of the middle fold near the base of the periostracal groove and the inner epithelial cells of the outer fold, as Fig. 6.22 shows. Furthermore, the signals around the gutter in the bottom of the periostracal groove were even stronger.

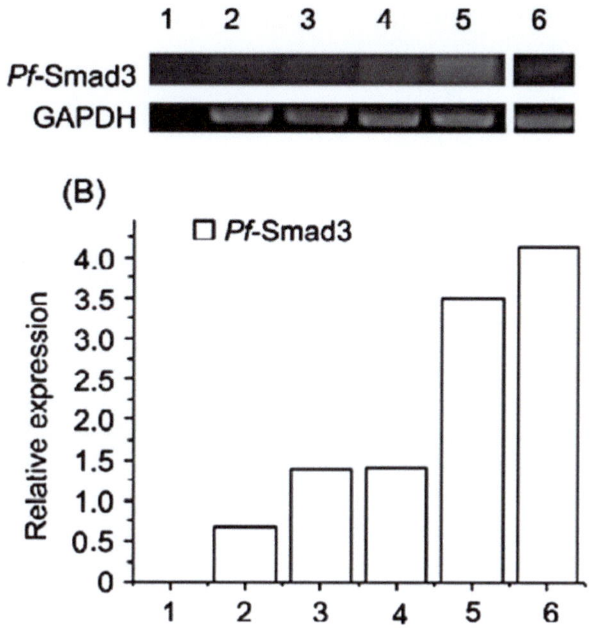

Fig. 6.21 Relative expression of the Pf-Smad3 gene in six different developmental stages of *P. fucata*. (**a**) Reverse transcription–polymerase chain reaction analysis for the Pf-Smad3 gene in different developmental stages. The pearl oyster housekeeping gene encoding glyceraldehyde-3-phosphate dehydrogenase (GAPDH) was included as a positive control. (**b**) Relative expression of Pf-Smad3 mRNA in different developmental stages of *P. fucata*. 1, Ovum; 2, cleavage 1 h; 3, gastrula stage; 4, D-sharp larva stage; 5, attachment stage; 6, adult

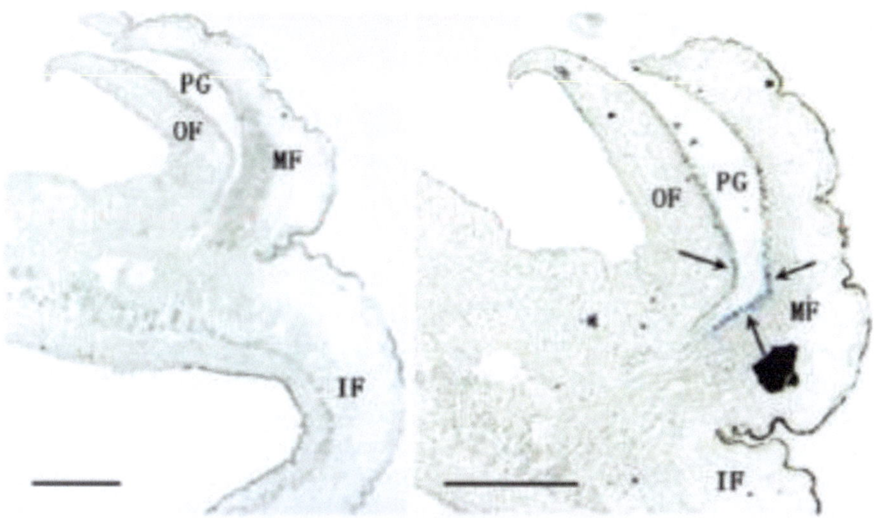

Fig. 6.22 Detection of Pf-Smad3 mRNA distributions in the mantle of *P. fucata* by in situ hybridization. A frozen section stained with a sense probe was used as the control (left frame). Strong hybridization signals are indicated by arrows (right frame). *IF* Inner fold, *MF* Middle fold, *OF* Outer fold, *PG* Periostracal groove. Bar = 500 μm

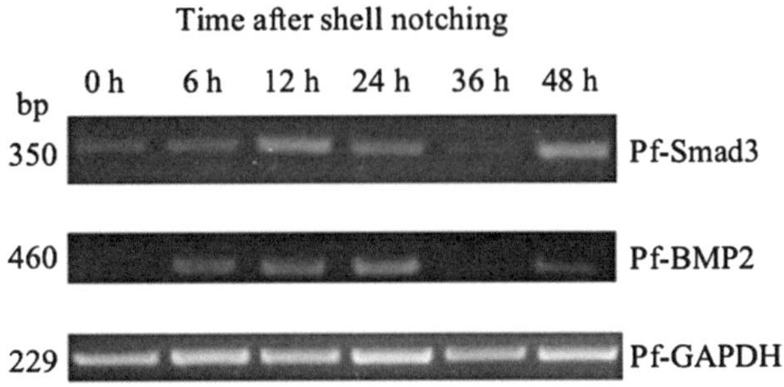

Fig. 6.23 The oyster Smad3 and BMP2 mRNA expressions after shell notching. The pearl oyster housekeeping gene encoding GAPDH was included as a positive control

We compared these results with the results of in situ hybridization of several matrix proteins of the oyster in the mantle. There are overlapping signal areas in the mantle where the cells were considered to be involved in the formation of the prismatic layer [44–46] and periostracum [47].

As shown in Fig. 6.23, the mRNA level of Pf-Smad3 increases gradually and reaches a peak at 12 h after the shell notching. Then, the Pf-Smad3 mRNA level decreases gradually and reaches the lowest level at 36 h after the shell notching. However, the mRNA expression of Pf-Smad3 increases to the highest level again at 48 h after the shell notching. These results suggest that the oyster Smad3 participates in shell repair. The change pattern of the mRNA levels of Pf-BMP2 changed in a different pattern. It increases to its highest expression level at 6 h after the shell notching and stays at that level until 36 h after the shell notching. The shell notching experiment results indicate that BMP2 also participates in shell repair, and the response of BMP2 is faster.

As it has been reported that dexamethasone (DXM) and hydrogen peroxide (H_2O_2) can be added into the culture medium, respectively, to adjust the Smad3 expression level in the mammal cells [37, 38], we have investigated the effects of DXM and H_2O_2 on the Smad3 and the BMP2 mRNA expression levels in primary cultured oyster mantle tissue cells. Real-time quantitative PCR results show that the oyster Smad3 and BMP2 expression levels are all changed in the drug-treated groups: DXM represses the mRNA expressions of the Smad3 and the BMP2 of the oyster mantle cells, while H_2O_2 both increases the two genes of mRNA expressions (Fig. 6.24). These results demonstrate there are significant correlations between the Smad3 and the BMP2 mRNA expressions in the oyster mantle tissue, and the drugs may be used to adjust pearl sac formation and nacre secretion.

In conclusion, our study showed that Pf-Smad3 is highly similar to the Smad3 proteins in vertebrates and revealed the ubiquitous expression of Pf-Smad3 in many tissues and at several developmental stages. Therefore, Pf-Smad3 might participate in many physiological processes in the oyster. DXM and H_2O_2 experiments indicate that the mRNA expression of the Smad3 and the BMP2 is correlated, and there may be similar signal transduction mechanism on these gene expressions. These genes are most likely to play important roles in the oyster biomineralization although they are also important to other physiological processes. This study is the initial step to

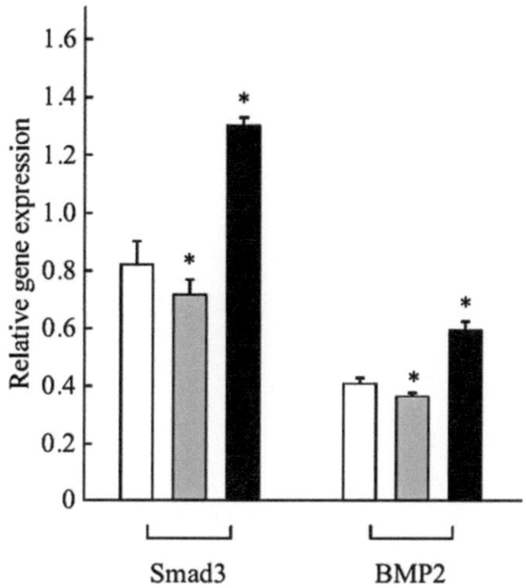

Fig. 6.24 Relative expression levels of the oyster Smad3 and BMP2 mRNAs. Data are means and standard errors of four independent experiments. Asterisks indicate that significant difference exists between the experimental samples and the control samples (*$P < 0.01$). White: NaCl; gray: DXM; black: H_2O_2

give light on the signal transduction and developmental mechanisms on molluskan shell formation. Further identification of the DNA-binding sequences of the Smads in the promoter regions of the matrix proteins will provide more clues to explore their function in biomineralization.

6.4.3.2 PfSox9: An Important Transcription Factor

PfSox9, the homolog of Sox (SRY-related high-mobility-group box gene)9, plays a vital role in maintaining the characteristics of chondrocytes and regulating cartilage development [48]. SOX9 is an important member of the SOX family and the first to discover the intron containing SOX gene. SOX9 can bind to a variety of proteins, in which the interaction with the matrix protein is associated with the mineralization process. In the previous biomineralization studies, it has been found that SMAD3 will induce cartilage differentiation by activating SOX9-dependent transcription in the TGF [beta] signaling pathway [49]; SOX9 also activates KRMP1 to participate in the biomineralization process; in the process of cartilage formation, SOX9 binds to SSP1 to inhibit the mathematic mineralization of the matrix [50].

We cloned and characterized the DNA and protein sequence. The length of open reading frame was 1389 bp, the length of 5′UTR was 196 bp, the length of 3′UTR was 761 bp, and the length of poly (A) was 13 bp. The cDNA sequence has been submitted to GenBank under accession number KC462556.

As a general transcription factor, PfSox9 does have universal expression in different tissues. The organism specificity of SOX9 gene was analyzed by real-time quantitative PCR. The results in Fig. 6.25 showed that the expression levels of SOX9 gene were the highest in the foot, followed by the gill, adductor muscle, gonad, mantle edge, mantle pallial, visceral, and heart. It indicated that SOX9 gene might be involved in the process of exercise, respiration, reproduction, and

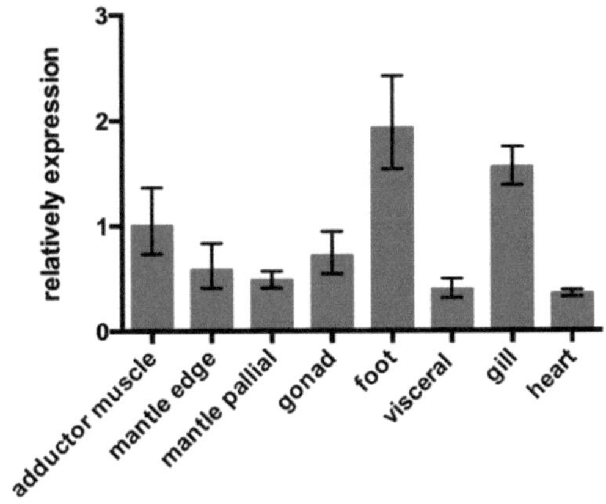

Fig. 6.25 Tissue expression distribution of PfSox9

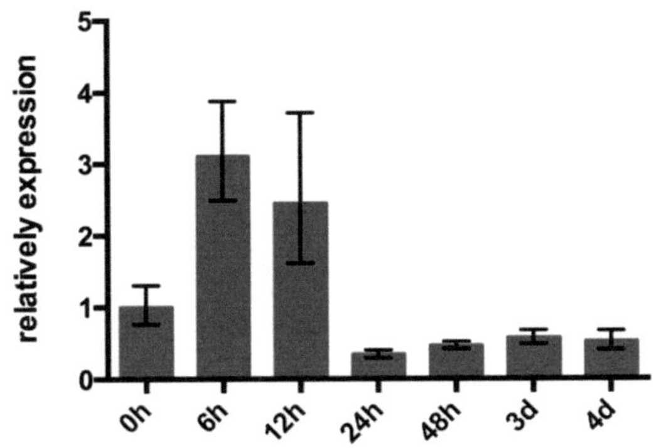

Fig. 6.26 Expression pattern of PfSox9 in damage repair process

mineralization in *Pinctada fucata*. When it comes to the response of shell regeneration, we also detect the expression pattern of SOX9 gene during the repair of shell damage. The expression level of SOX9 gene was increased first and then decreased during the repair of shell damage (shown in Fig. 6.26). The results indicated that SOX9 gene responded as a transcription factor during shell regeneration.

The target of the PfSox9 is important to find out its vital function. So we use the online prediction software PROMO to analyze the transcription factors and its binding sites on the potential gene promoter. It was found that SOX9 was one of the possible transcription factors on the Prismalin-14 gene promoter. We designed a series of experiments to confirm. First, the recombinant vector of Prismalin-14 and SOX9 was cotransfected into HEK-293T cells, and the activity of Prismalin-14 promoter was detected by Dual-Luciferase Reporter Assay System. The results in Figs. 6.27 and 6.28 showed that SOX9 gene could activate the activity of Prismalin-14 gene promoter and with a dose effect. Subsequently, two luciferase reporter vectors of the Prismalin-14 promoter truncation were constructed and transfected

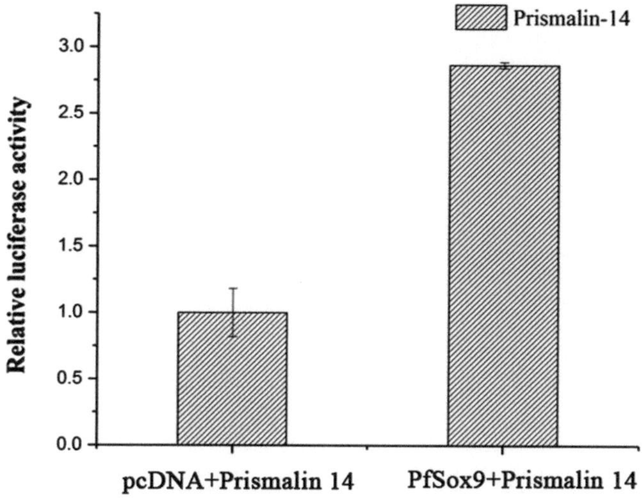

Fig. 6.27 SOX9 active Prismalin-14 promoter. PfSox9 was constructed on pcDNA 3.1 (+) – myc vector and cotransfected into HEK-293 T cells. We used pcDNA3.1 (+) – myc vector as control group. The dual-luciferase assay results show PfSox9 significantly promotes the promoter activity of Prismalin-14

Fig. 6.28 PfSox9 gene activates Prismalin-14 gene promoter with a dose effect

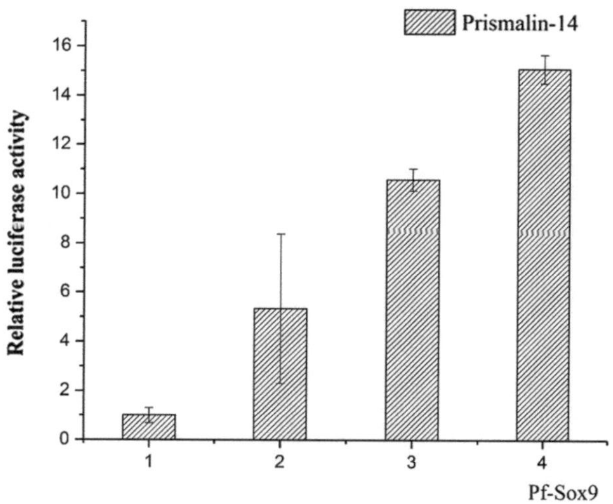

into HEK-293T cells with the eukaryotic expression vector of SOX9. The transcriptional activity of Prismalin-14 and its truncation were analyzed by Dual-Luciferase Assay System. The −415 bp to −405 bp regions on Prismalin-14 gene promoter are the key region of SOX9 activation (Fig. 6.29). Finally, the eukaryotic expression vector of two SOX9 truncations was constructed, and the effect of the integrity of the SOX9 on the Prismalin-14 promoter was analyzed by Dual-Luciferase Assay

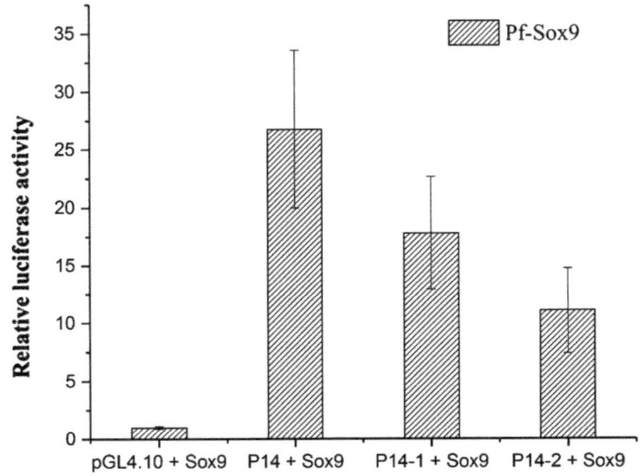

Fig. 6.29 Identification of the binding site of PfSox9 and Prismalin-14 promoter. P14 promoter activity is the highest. The activity of P14-1 promoter is a little lower than P14 promoter. However, the activity of P14-2 promoter that without binding site is the lowest; its promoter activity was 49% of the P14 promoter. Among them, the P value of P14 and pcDNA was 0.02, the P value of P14-1 and P14 was 0.5, and the *P* value of P14-2 and P14-1 was 0.01

System. The results in Fig. 6.30 showed that the HMG domain in SOX9 is necessary for the transcriptional regulation of the Prismalin-14.

Based on the above experimental results, it is suggested that SOX9 gene may involve in the process of biomineralization by regulating the transcription of the matrix protein Prismalin-14 gene. This evidence provides some basic data for the upstream regulation mechanism of biomineralization in *Pinctada fucata*, which will help to enrich the theory of biomineralization of pearl oyster.

Different doses of PfSox9 were cotransfected into HEK-293 T cells with 100 ng Prismalin-14 per well. 1 represents 0 ng of PfSox9 eukaryotic expression vector; 2 represents 125 ng of PfSox9 eukaryotic expression vector; 3 represents 250 ng of PfSox9 eukaryotic expression vector; 4 represents 500 ng of PfSox9 eukaryotic expression vector.

6.4.4 The Relationship of the TGFβ Signaling Pathway and the Biomineralization

6.4.4.1 From Cell Level

To study the effects of the BMP type I receptor on the expression of the MP genes in vitro, the primary cells of mantle tissue were cultured and treated with the receptor inhibitor LDN193189. The mantle cells in all groups grew healthy throughout the duration of exposure. Western blot analysis indicates that the phosphorylation levels of Smad1/5/8 significantly decreased in mantle cells exposed to 2 μm LDN193189 for 12 h and 24 h (Fig. 6.31a). When exposed to seawater with 2 μm LDN193189 for 12 h, the relative expression levels of PfKRMP, PfPRISMALIN-14, PfPIF, and PfMSI60 exhibit a decrease of 67.59%, 58.29%, 90.37%, and 76.66%, respectively,

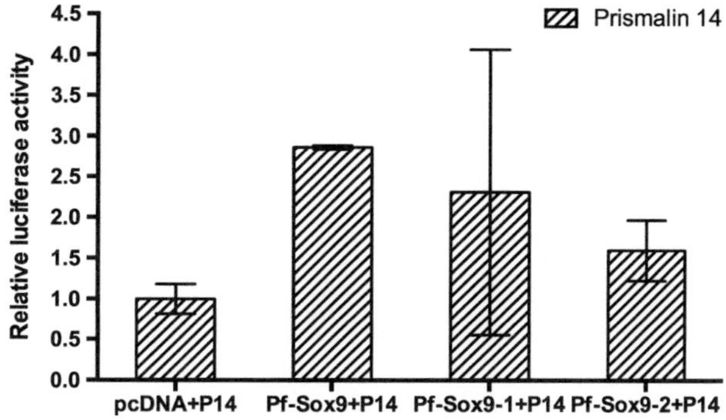

Fig. 6.30 Activation of Prismalin-14 promoter by PfSox9 and its truncation. The decrease in the length of PfSox9 resulted in a decrease in the activity of the Prismalin-14 promoter. When the HMG domain of PfSox9 was completely removed, Prismalin-14 promoter activity was almost not activated. These results showed the structural integrity of PfSox9 was essential for the promoter of the Prismalin-14 gene activation function

compared with the control (Fig. 6.31b, $P < 0.05$). The results confirm that Bmpr1b and the pathway it may participate are involved in the regulation of MP genes in mantle cells of *P. fucata*.

6.4.4.2 From Individual Level

To obtain the direct evidence of the function of Pf-BAMBI and Pf-BMPR1B in vivo, the knockdown of the two gene expressions by methods of RNA interference was performed. As the results that Fig. 6.32 shows, if we attribute the Pf-BAMBI and Pf-BMPR1B expression of the PBS group a relatively value of 1, respectively, the Pf-BAMBI and Pf-BMPR1B gene expression levels of each the experimental groups were both inhibited. And the Pf-BAMBI and Pf-BMPR1B gene expression level decreased with increasing injection doses of 80, 160 μg dsRNA. Owing to the interference that the dsRNA induces the degradation of the certain mRNA affecting the gene translation level, the Bambi or Bmpr1b cannot function normally. To confirm the phenotype of the gene disfunction, we choose several matrix proteins related to the formation of prismatic layer (KRMP, Prismalin-14) and nacreous layer (PIF, MSI60), respectively. Responding to BMPR1B gene repression, the KRMP, Prismalin-14 (P14), PIF, and MSI60 expression levels were all reduced suggesting that the knockdown of type I receptors in BMP signal pathway affects the integral pathway function. The depressed pathway decreased the matrix protein in turn indicating that the Pf-BMPR1B was related to both the formation of prism and nacre. But actually, the prism and the nacre formation actually have the complex network of multiple matrix protein, and the single matrix protein regulation does have time and polygenic effect. So more detailed Pf-BAMBI and Pf-BMPR1B function remains to be excavated in the complicated phenomenon.

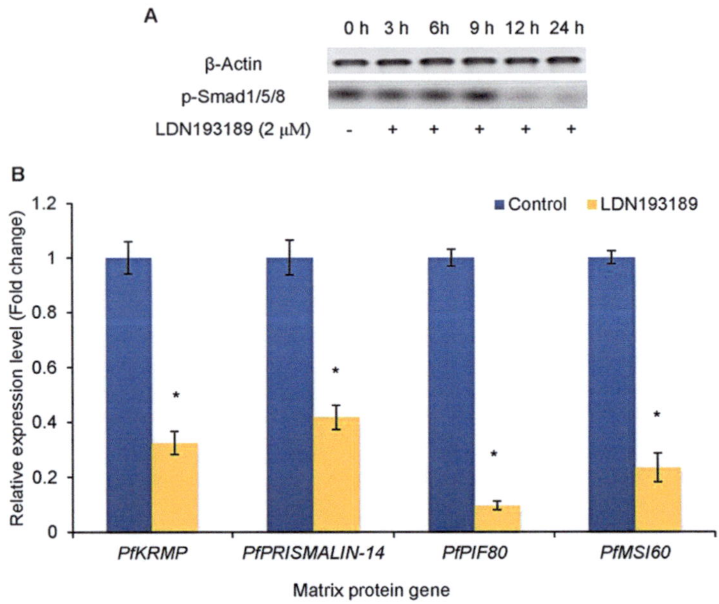

Fig. 6.31 Effects of LDN193189 on the phosphorylation level of Smad1/5/8 and the expression level of matrix protein (MP) genes in mantle primary cells. (**a**) Changes in phosphorylated Smad1/5/8 in mantle cells detected by Western blot. P-Smad1/5/8, phosphorylated Smad1/5/8. β-Actin is used as control. (**b**) The relative expression levels of MP genes in the mantle cells exposed to LDN193189. The expression levels of MP genes in unexposed mantle cells were used as corresponding controls. $^*P < 0.05$

To verify the role of the BMP type I receptor in shell germination, LDN193189 was used to disturb Smad phosphorylation at the early developmental stage of *P. fucata*. Artificial insemination was employed to obtain oosperm from sexually mature pearl oysters. Many investigations showed that inhibition of the BMP signaling pathway influenced bone and tooth biomineralization in vertebrates. Similar to these effects, a deep dent was observed on the shell margin at the prodissoconch I stage in the larvae exposed to LDN193189 (Fig. 6.33d–f). After exposure for 48 h, more than 95% of the larvae developed to the D-shaped stage, and 73.15%, 69.45%, and 66.35% of the D-shaped larvae were alive in the control, 2 μm LDN193189 and 10 μm LDN193189 groups, respectively (Fig. 6.33a). There were no significant differences in the survival rates among these three groups ($P < 0.05$). For the surviving larvae, both 2 μm and 10 μm LDN193189 significantly decreased the phosphorylation level of Smad1/5/8 (Fig. 6.33b). When exposed to seawater with LDN193189 for 48 h, the relative expression levels of PfKRMP, PfPRISMALIN-14, PfPIF, and PfMSI60 exhibit a significant decrease of 64.36%, 93.05%, 77.87%, and 87.12% for the concentration of 2 μm and 89.53%, 88.29%, 87.37%, and 94.65% for the concentration of 10 μm compared with the control (Fig. 6.33c, $P < 0.05$). Although our result cannot ensure which receptor was

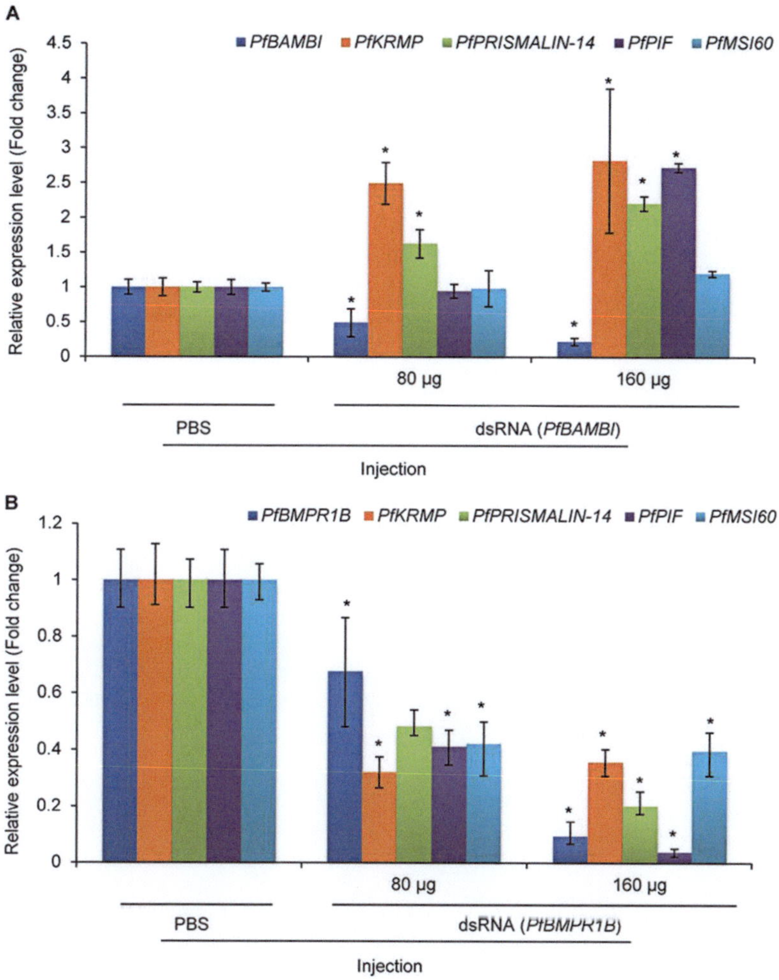

Fig. 6.32 The relative expression levels of PfBMPR1B, PfBAMBI, and shell matrix protein (MP) genes inhibited by RNA interference. (**a**) The expression of PfBAMBI and the MP genes PfKRMP, PfPRISMALIN-14, PfPIF, and PfMSI60 after injecting 80 and 160 μg of double-stranded RNA (dsRNA) of PfBAMBI for 6 d. (**b**) The expression of PfBMPR1B and the MP genes PfKRMP, PfPRISMALIN-14, PfPIF, and PfMSI60 after injecting 80 and 160 μg of double-strandeded RNA (dsRNA) of PfBMPR1B for 6 d. PBS-injected groups were used as controls. $*P < 0.05$

influenced by LDN193189, it shows that the TGFβ/BMP signaling pathway acts on shell germination of *P. fucata*.

The involvement of the BMP signaling pathway in larval shell biomineralization was also observed in *L. anatine*, a brachiopod with phosphate biomineralization products [51]. The larval shell of shelled mollusks is formed by a shell gland, which is composed of a special kind of ectodermal cell in early embryos [52]. The inner

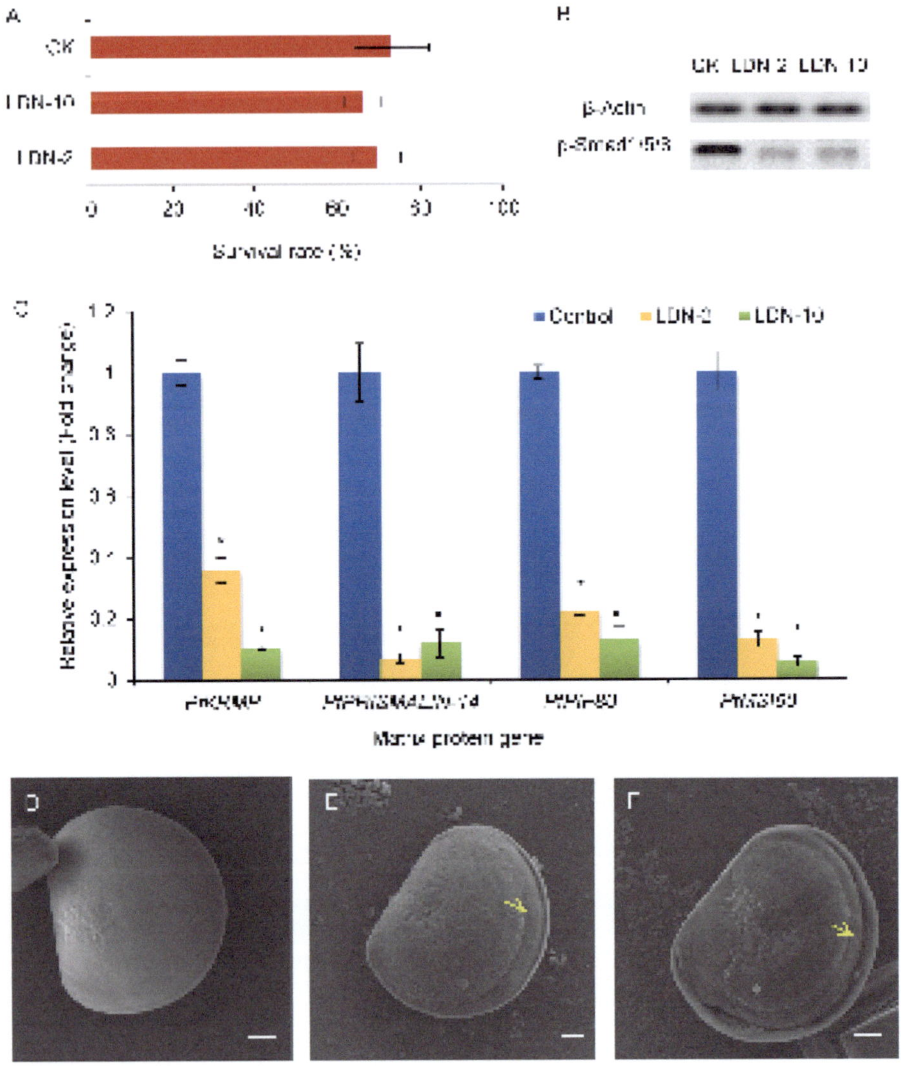

Fig. 6.33 Effects of LDN193189 on shell germination during larval development in *Pinctada fucata*. (**a**) Survival rates of larvae exposed to different concentrations of LDN193189. CK, the control larvae incubated in normal seawater. LDN-2, the larvae exposed to seawater with 2 μm of LDN193189 for 48 h. LDN-10, the larvae exposed to seawater with 10 μm of LDN193189 for 48 h. (**b**) Western blot analysis of the phosphorylation level of Smad1/5/8 in the larvae exposed to 2 μm and 10 μm LDN193189. p-Smad1/5/8, phosphorylated Smad1/5/8. (**c**) The relative expression levels of MP genes in the larvae exposed to LDN193189. The expression levels of MP genes in unexposed larvae were used as corresponding controls. $^*P < 0.05$. Differences were assessed by one-way ANOVA. (**d–f**) Scanning electron microscope (SEM) images of the shell surface of D-shaped larvae (prodissoconch I) incubated in control (**d**) and 2 μm (**e**) and 10 μm (**f**) LDN193189. The yellow arrows indicate the wrinkled shell. Bar = 20 μm

part of these ectodermal cells is subsequently transformed into mantle epithelium of the larva. Studies on the mussel *M. galloprovincialis* and the European oyster *Ostrea edulis* [53] have shown that the organic matrix of the shell begins to be secreted by the shell gland during the late trochophore stage. Meanwhile, biomineralization products were detected at the inner side of the organic matrix at the early veliger stage. As a component of the organic matrix, MPs play important roles in shell germination 48. The morphological changes in the prodissoconch I in the larvae of *P. fucata* may be caused by alterations of MPs, because in vitro study in mantle primary cells and in vivo study in larvae illustrate that LDN193189 decreases the expression levels of these genes. Interestingly, LDN193189 did not induced significant changes in the survival rate of *P. fucata* larvae. The reason for this phenomenon is largely unknown and needs further investigation. Although not conclusive, one piece of evidence implies that TGFβ signal pathway type I receptors, particularly PfBMPR1B, are critical for biomineralization during the shell germination stage.

6.4.5 The Network of the TGFβ Pathway

It is necessary to investigate the interaction between the transmembrane receptors (Pf-Bambi, Pf-Bmpr1b) and the upstream ligand (BMP2) or downstream signal (Smad5, Smad6) to verify the essential roles Pf-Bambi and Pf-Bmpr1b played in the whole TGFβ signal pathway. When we transform the single plasmid into the AH109, after shaking for 15 h, we detected the OD 600 to ensure that they did not have the toxicity. All these results indicated that the yeast two-hybrid (Y2H) system was efficient. As shown in Fig. 6.34, the white colonies grown on the synthetic dropout (SD) medium SD-Trp-Leu represent successful transformations of the plasmids pGBKT7 (BD)-PfBAMBI, BD-PfBMPR1B, pGADT7 (AD)-PfBMP2, AD-PfSMAD1/5/8, AD-PfSMAD4, and AD-PfBMPR1B. The blue clones are observed for the yeast AH109 cotransformed with AD-PfBMP2 and BD-PfBAMBI, AD-PfSMAD1/5/8 and BD-PfBAMBI, AD-PfBMP2 and BD-PfBMPR1B, and AD-PfSMAD1/5/8 and BD-PfBMPR1B on the SD-Trp-Leu-His-Ade/X-gal medium. However, no clones grew for the yeast cotransformed with AD-PfSMAD4 and BD-PfBAMBI or AD-PfSMAD4 and BD-PfBMPR1B on the SD-Trp-Leu-His-Ade/X-gal medium. The results indicate that Bmpr1b and Bambi interact with members of the TGFβ/BMP pathway in *P. fucata*.

Based on the findings from the investigations on ligands, receptors, and Smads in the TGFβ/BMP signaling pathway and the mechanism of MP-mediated biomineralization in shelled mollusks, we propose a hypothesis about the regulatory mechanism of the TGFβ/BMP signaling pathway in shell biomineralization (Fig. 6.35). In the mantle, the ligand Bmp2 or Bmp7 binds to the complex of type I receptor Bmpr1b and type II receptor Bmpr2. Bmpr1b is phosphorylated by the constitutively

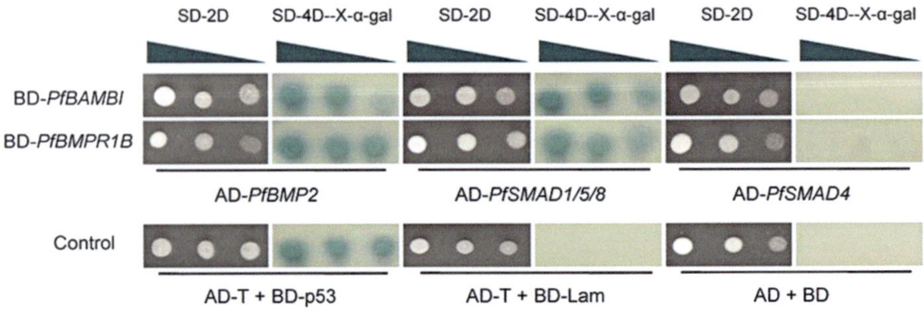

Fig. 6.34 Interactions among Bambi, Bmpr1b, and members of the TGFβ/BMP signaling pathway in *Pinctada fucata* detected by a yeast two-hybrid system. The plasmids are cotransformed into the yeast AH109 and spread on the dropout synthetically defined (SD) medium. The white colonies on SD-Leu-Trp medium (SD-2D) indicate that the plasmids are successfully transformed. The blue colonies on SD-Ade-His-Leu-Trp/X-α-Gal medium (SD-4D- X-α-Gal) indicate the presence of an interaction between two proteins, comparable with that of the positive (AD-T + BD-p53), negative (AD-T + BD-Lam), and empty plasmid (AD + BD) controls. The triangle indicates that the yeast is gradiently diluted five times with sterilized water before spreading on the medium

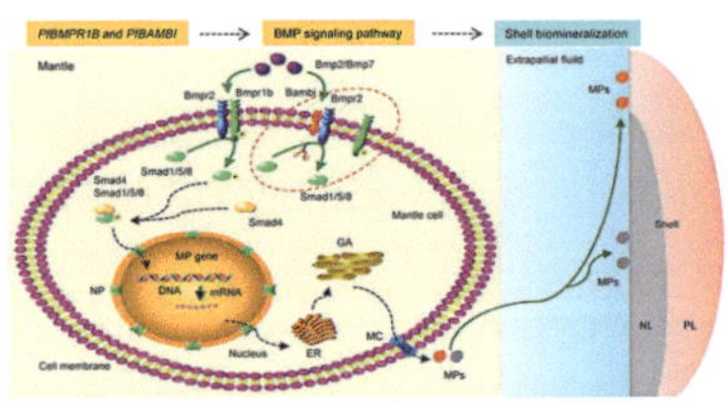

Fig. 6.35 Schematic presentation of the mechanism for shell biomineralization mediated by the TGFβ/BMP signaling pathway in mollusks. The green arrows indicate known pathways, and the gray dotted arrows indicate potential pathways. The red dotted circle indicates the interactions of Bambi, Bmpr2, and Bmpr1b. *NP* nuclear pore, *MC* membrane channel

active Bmpr2 kinase and thereby recruits a cytoplasmic Smad1/5/8 (R-Smad), which is then phosphorylated by activated Bmpr1b. Smad1/5/8 may dissociate from the receptors to form a heteromeric complex with Smad4 (Co-Smad). The Smad complex is transported into the nucleus, where it regulates the expression of specific matrix protein (MP) genes alone or by binding certain transcription factors (either corepressors or coactivators). The MPs may be synthesized through conventional ways throughout the endoplasmic reticulum (ER) and Golgi apparatus (GA) and secreted out of the mantle cells. Finally, the MPs arrive at the mineralization site on the surface of the shell, and the MPs with different functions regulate the formation of the nacreous layer (NL) and prismatic layer (PL). Bambi is a negative regulator and competitively binds to the ligands and Bmpr2, which cannot trigger the phosphorylation of Smad1/5/8, inhibiting the signal transduction in this pathway. The green arrows indicate known pathways, and the gray dotted arrows indicate potential pathways.

6.5 Wnt Signaling Pathway

6.5.1 Introduction

WNTs are secreted cysteine-rich glycoproteins that act as short-range ligands to locally activate receptor-mediated signaling pathways. Their functions have been the subject of investigation for more than 30 years [54]. Wnt signaling, a critical intercellular pathway in all multicellular animals, is involved in the regulation of various processes including embryonic development, stem cell proliferation, organogenesis, and maintenance of mineralized tissues, such as tooth development and regulation of bone mass [55, 56], though there is little research on the Wnt signaling pathway in pearl oyster.

The canonical Wnt signaling pathway is characterized by the stability of β-catenin in the cytoplasm and its nuclear entry [54]. The regulatory mechanism of this pathway is as follows (Fig. 6.36). In the absence of a Wnt signal, the "β-catenin destruction complex" is formed in the cytoplasm and contains β-catenin, casein kinase I (CKI), glycogen synthase kinase 3β (GSK3β), adenomatous polyposis coli (APC), and Axin. In this complex, β-catenin is phosphorylated and ubiquitylated, then degraded by the proteasome. In contrast, in the presence of a Wnt signal, cytoplasmic Dishevelled (Dvl) is activated, leading to inhibition of GSK3β phosphorylation and dissociation of β-catenin from the destruction complex. β-catenin is then transferred into the nucleus where it binds to T-cell factor (TCF)/lymphoid enhancer factor (LEF), which results in the activation of Wnt target gene transcription.

6.5.2 *Pf-β-Catenin: A Key Factor Wnt Pathway*

β-Catenin is the key factor of canonical Wnt signaling pathway; before that, it was originally isolated as a protein associated with E-cadherin that functioned in mediating cell–cell adhesion [57]. A considerable amount of accumulated evidence indicates that β-catenin exists widely in various species from human to *Porifera*, the lowest metazoan phylum [58, 59, 60]. Structurally, the central armadillo (ARM) repeat domain within β-catenin mediates its binding to cadherins, APC, Axin, and TCF-family transcription factors [61], which provide the structural basis for its function.

The 3993 bp full-length cDNA sequence of Pf-β-catenin was cloned from RNA isolated from the mantle tissue (GenBank accession no. KM374536). It comprised a 153 bp 5′-untranslated region (UTR), a 1359 bp 3′ UTR, and a 2481 bp open reading frame (ORF) encoding a protein of 826 amino acids with a predicted molecular mass of 91.04 kDa and an estimated isoelectric point of 5.79. Sequence analysis showed that genomic DNA of Pf-β-catenin was composed of 15 exons and 14 introns (Fig. 6.37a).

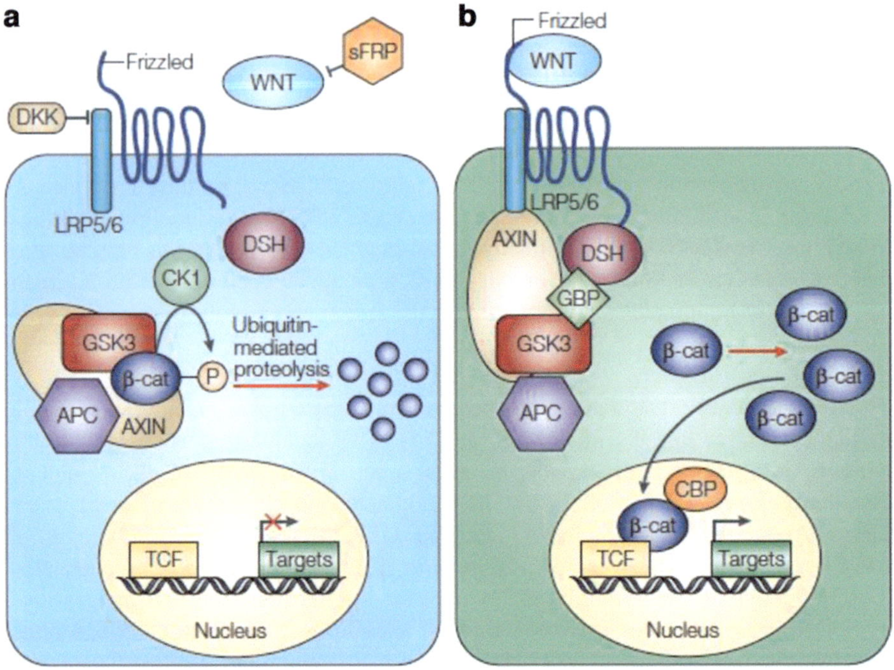

Fig. 6.36 The pathway of WNT/β-catenin signaling [54]. In the absence of active WNT (**a**), β-catenin is degraded, and prospective target genes are in a repressed state. If WNT signaling (**b**) is active, β-catenin degradation is reduced. As β-catenin accumulates, it enters the nucleus, interacts with DNA bound T-cell factor (TCF) and lymphoid enhancer-binding protein (LEF) family members to activate the transcription of target genes. The components shown are described in more detail in the text; additional pathway components are described on websites that are linked to the main text. *APC* Adenomatous polyposis coli, *β-cat* β-catenin, *CBP* CREB-binding protein, *CK* casein kinase, *DKK* Dickkopf, *DSH* Dishevelled, *GBP* GSK3-binding protein, *GSK* Glycogen synthase kinase, *LRP* LDL receptor-related protein, *P* Phosphorylation, *sFRP* Secreted frizzled-related protein, *TCF* T-cell factor

Domain architecture analysis indicated that Pf-β-catenin contained 12 ARM repeats, which are the most important feature of β-catenin. In addition, two conserved ubiquitination/acetylation residues (Lys-34 and Lys-66) and one conserved GSK3β phosphorylation motif (DSGXXS) were found in the Pf-β-catenin N-terminal sequence by alignment of β-catenin amino acid sequences from six other species (Fig. 6.37b). Four amino acid residues, defined in human as GSK3β phosphorylation sites [62], were highly conserved in Pf-β-catenin: three serine residues (S33, S37, and S45 in human) and one threonine (T41 in human).

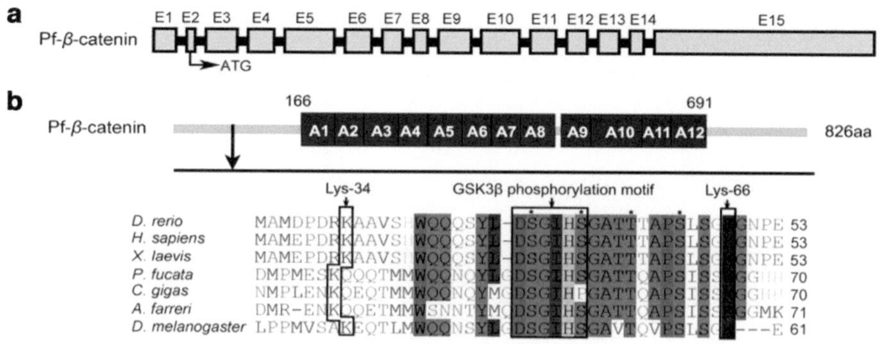

Fig. 6.37 Sequence analysis of Pf-β-catenin (**a**) genomic structures. Exons are indicated with gray boxes, while introns are indicated with straight black lines. (**b**) Characteristics of the deduced amino acid sequence of Pf-β-catenin. Twelve ARM repeats are denoted in the black box and are labeled A1 to A12. Alignment of the N-terminal sequence of Pf-β-catenin with the amino acid sequence of six other species revealed two conserved ubiquitination/acetylation residues (Lys-34 and Lys-66) and one conserved GSK3β phosphorylation motif (DSGXXS), denoted in the black pane. Four consensus GSK3β phosphorylation sites, three serine (S) residues, and one threonine (T) residue are denoted by an asterisk. Amino acids critical for binding β-catenin are denoted by an asterisk. *H. sapiens*: *Homo sapiens*; *X. laevis*: *Xenopus laevis*; *D. rerio*: *Danio rerio*; *C. gigas*: *Crassostrea gigas*; *A. farreri*: *Azumapecten farreri*; *L. variegatus*: *Lytechinus variegatus*; *D. melanogaster*: *Drosophila melanogaster*

6.5.3 Pf-Dvl: A Member of Wnt Pathway

Dvl is another component of the Wnt signaling pathway and has also been thoroughly studied in recent years. Functionally, on the one hand, based on its capacity to propagate canonical Wnt signals, Dvl participates in bone development and early embryogenesis [63, 64], similar to β-catenin and TCF/LEF, while on the other hand, it can mediate other signaling pathways, such as the planar cell polarity (PCP) pathway, via interaction with various proteins due to its multiple domains [65].

The 2698 bp full-length cDNA sequence for pearl oyster Pf-Dvl was also cloned from the mantle tissue (GenBank accession no. KM374538). It contained a 76 bp 5′ UTR, a 390 bp 3′ UTR, and a 2232 bp ORF encoding a protein of 743 amino acids with a predicted molecular mass of 81.61 kDa and an estimated isoelectric point of 5.70. Sequence analysis showed that genomic DNA of Pf-Dvl was composed of 16 exons and 15 introns (Fig. 6.38a).

As in other species, domain architecture analysis indicated that Pf-Dvl has three conserved domains, an N-terminal DIX (Dvl and Axin) domain, a central PDZ (postsynaptic density 95, discs large, and zonula occludens-1) domain, and a C-terminal DEP (Dvl, Egl-10, and pleckstrin) domain [66]. What is more, alignment of Pf-Dvl with six other species revealed one nuclear import signal (NLS) near the PDZ domain and one leucine-rich nuclear export signal (NES) (M\LXXLXL) near the DEP domain (Fig. 6.39b).

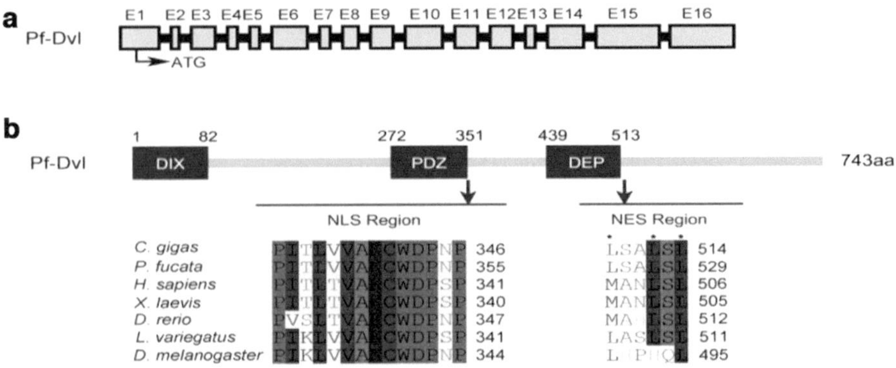

Fig. 6.38 Sequence analysis of Pf-Dvl. (**a**) Genomic structures. (**b**) Characteristics of the deduced amino acid sequence of Pf-Dvl. The N-terminal DIX domain, the central PDZ domain, and C-terminal DEP domain are predicted. After aligning the amino acid sequence of Pf-Dvl with that of six other species, one conserved NES (M\LXXLXL) and one NLS were identified

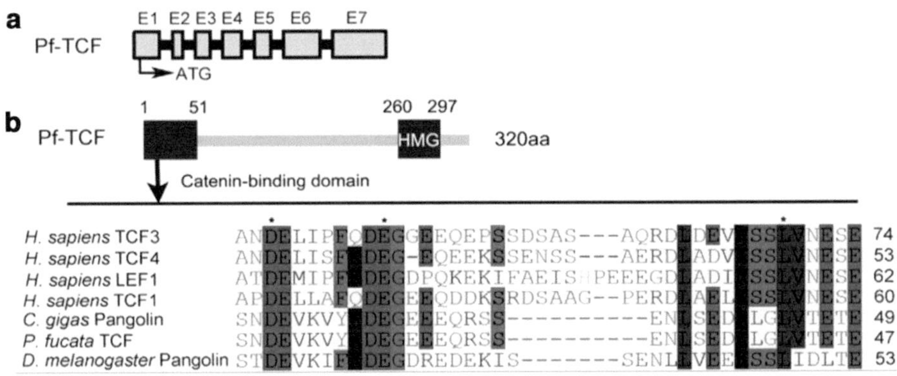

Fig. 6.39 Sequence analysis of Pf-TCF. (**a**) Genomic structures. Exons are (**b**) characteristics of the deduced amino acid sequence of Pf-TCF. The β-catenin-binding domain and HMG box were predicted

6.5.4 *Pf-TCF: Component of Wnt Pathway*

TCF/LEF is a downstream effector of Wnt signals, and its involvement in Wnt signaling was first addressed in *Xenopus laevis* [67]. This transcription factor family has four members in chordates: LEF1, TCF1, TCF3, and TCF4. The direct interaction of TCF/LEF with β-catenin is essential for the multiple roles they played, which have been extensively studied [68].

The full Pf-TCF cDNA was 1052 bp in length and comprised of a 5′ UTR of 21 bp, an ORF of 963 bp, and a 3′ UTR of 68 bp (GenBank accession no. KP875380). The deduced Pf-TCF protein is a single polypeptide of 320 amino acids with a predicted molecular mass of 34.57 kDa and an estimated isoelectric point of 6. Sequence analysis showed that genomic DNA of Pf-TCF was composed of 7 exons and 6 introns (Fig. 6.39a).

Analysis of the functional domain showed that Pf-TCF has the conserved β-catenin-binding domain and DNA-binding high mobility group (HMG) box domain (Fig. 6.39b). The three residues playing critical roles in interaction with β-catenin defined in human [69] were also conserved in Pf-TCF (Fig. 6.39b).

6.5.5 The Three Wnt Pathway Members Interact with Each Other

The direct interaction of β-catenin with TCF/LEF is essential for the multiple roles they play. Thus, whether Pf-β-catenin interacts with Pf-TCF was investigated using a GAL4-based yeast two-hybrid system. First, we verified that the constructs of pGADT7-Pf-β-catenin and pGBKT7-Pf-TCF did not autonomously activate the reporter gene (Fig. 6.40a). Then, the α-galactosidase assay of cotransformants of pGADT7-Pf-β-catenin and pGBKT7-Pf-TCF showed that there were blue colonies on SD/−Leu/−Trp/X-α-Gal and SD/−Ade/−His/− Leu/−Trp/X-α-Gal, comparable to that of the positive control (pGADT7-T/pGBKT7–53), suggesting that Pf-β-catenin interacted with Pf-TCF (Fig. 6.41b). As negative controls, the results of cotransformants of pGADT7-T and pGBKT7-Lam, pGADT7-Pf-β-catenin and pGBKT7, pGADT7 and pGBKT7-Pf-TCF, and pGADT7 and pGBKT7 showed white colonies on SD/−Leu/−Trp/X-α-Gal and no colonies on SD/−Ade/−His/−Leu/−Trp/X-α-Gal (Fig. 6.40b).

As these three genes were interacted together, their expression pattern was similar too. As shown in Fig. 6.41, these three genes were ubiquitously expressed in the seven tissues assayed. The mRNA expression levels were all highest in the gill and lowest in the adductor muscle. In addition, in the mantle pallial and mantle edge, which are considered to be biomineralization-related tissues, there were certain amounts of expression of these three genes.

To detect the response of Pf-β-catenin, Pf-Dvl, and Pf-TCF to the short-term shell injury and in the regeneration process, these gene expressions were determined in different time after shell notching. As shown in Fig. 6.42, in general, the dynamic changes of Pf-β-catenin, Pf-Dvl, and Pf-TCF gene expression were relatively similar to each other during the tested period. Although they behaved differently within 48 h, they were all upregulated significantly and reached the maximum expression at 4 days. These results suggested that Pf-β-catenin, Pf-Dvl, and Pf-TCF might participate in the process of shell regeneration and function in pearl oyster biomineralization.

Relative gene expression of Pf-β-catenin, Pf-Dvl, and Pf-TCF during embryonic and larval development at five different stages, including oosperm, trochophore stage, D-shaped stage, umbonal stage, and juvenile stage, was also detected. As

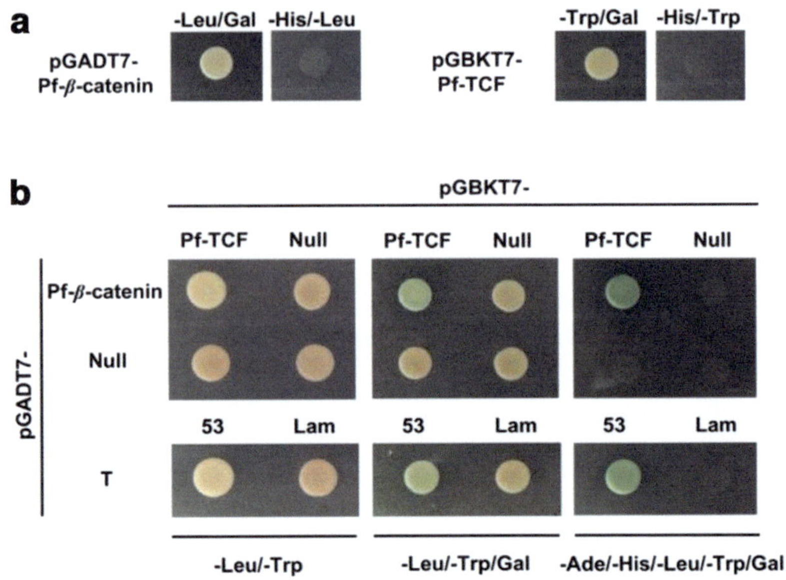

Fig. 6.40 Interaction of Pf-β-catenin with Pf-TCF by yeast two-hybrid analysis. (**a**) Independently transformed the constructs of pGADT7-Pf-β-catenin and pGBKT7-Pf-TCF into the yeast strain AH109 to test self-activating effect. White colonies on SD/−Leu/X-α-Gal or SD/−Trp/X-α-Gal and no colonies on SD/−His/−Leu or SD/−His/−Trp illustrated that they cannot activate reporter genes individually. (**b**) Cotransformation of pGADT7-Pf-β-catenin and pGBKT7-Pf-TCF into strain AH109. The interaction is shown by blue colonies on SD/−Leu/−Trp/X-α-Gal and SD/−Ade/−His/−Leu/−Trp/X-α-Gal, comparable with that of the positive control (pGADT7-T/pGBKT7–53)

shown in Fig. 6.43, they were expressed during all five developmental stages, with the highest level at the first oosperm stage. These results indicated that these genes might be essential factors in the early embryonic development of the pearl oyster.

In summary, the cDNA of three Wnt signaling pathway genes from the pearl oyster, encoding Pf-β-catenin, Pf-Dvl, and Pf-TCF, was cloned and sequenced. The interaction of Pf-β-catenin with Pf-TCF was verified. They were ubiquitously expressed in the seven tissues examined with certain amounts in the mantle tissue and were capable of responding to shell regeneration. Their expression during embryonic and larval developmental stages was also studied. Additional research will be required to determine the function of the Wnt signaling pathway and of Pf-β-catenin, Pf-Dvl, and Pf-TCF in shell formation and embryonic development in the pearl oyster.

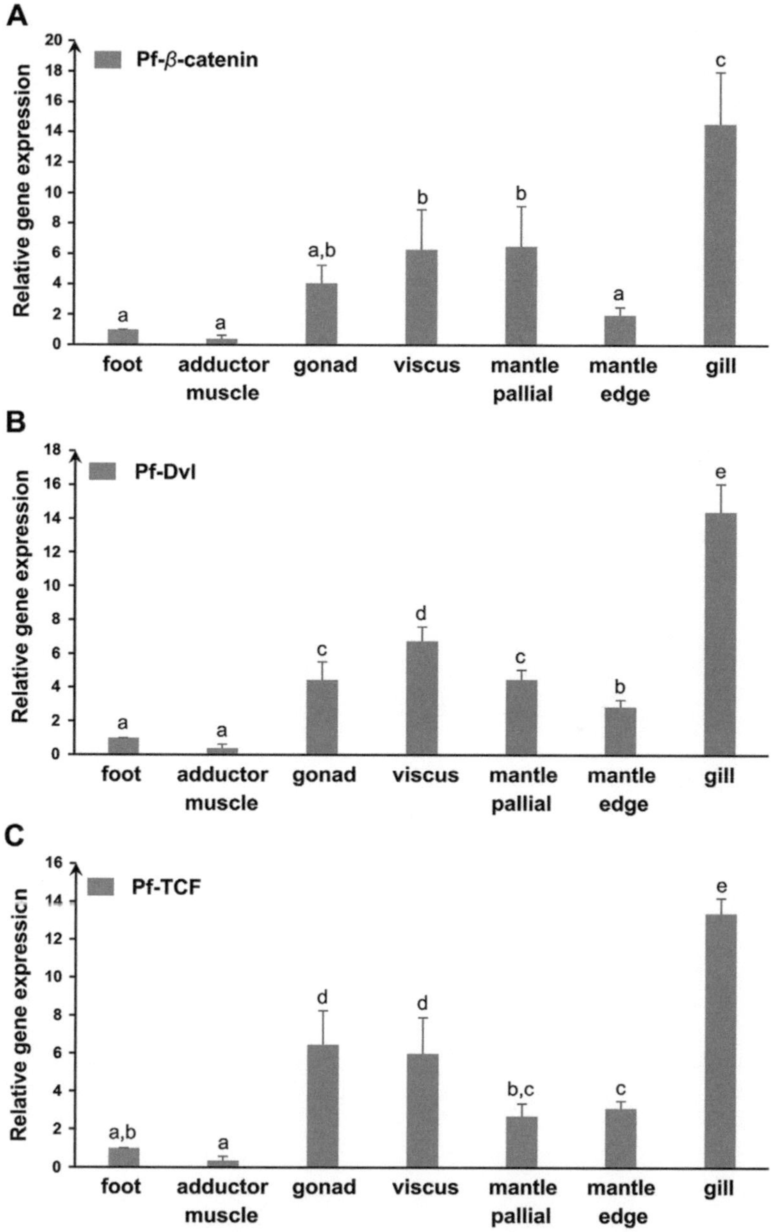

Fig. 6.41 Relative mRNA expression of Pf-β-catenin (**a**), Pf-Dvl (**b**) and Pf-TCF (**c**) in seven pearl oyster tissues measured

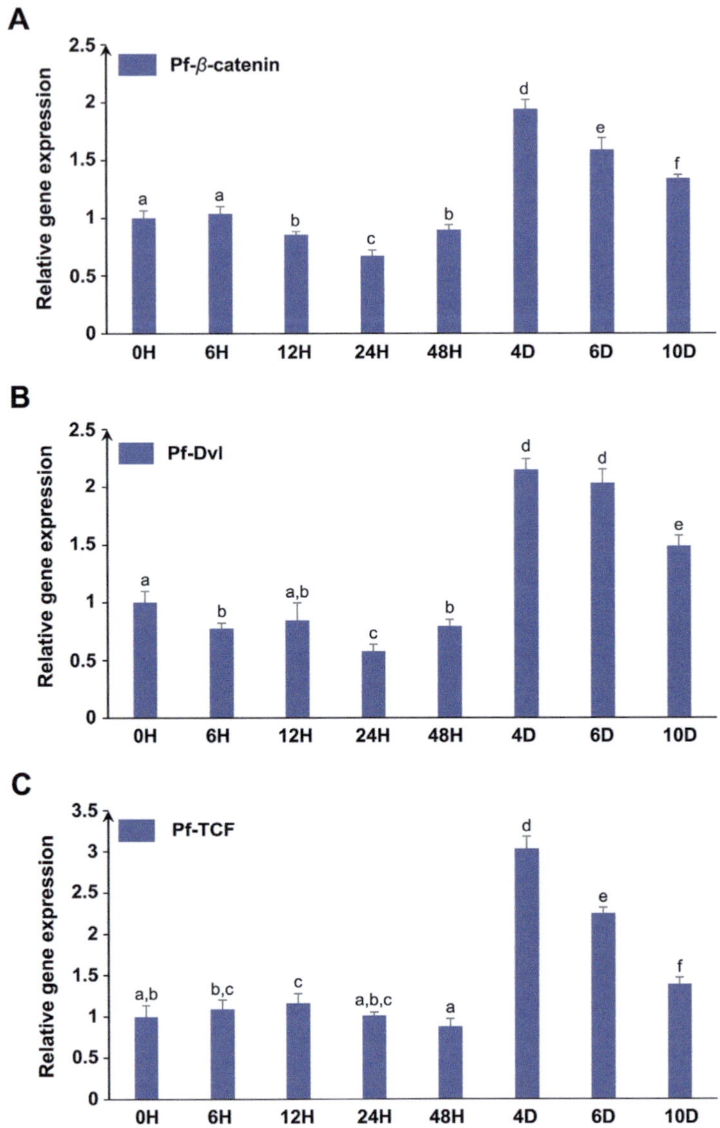

Fig. 6.42 Relative mRNA expression of Pf-β-catenin (**a**), Pf-Dvl (**b**), and Pf-TCF (**c**) in the mantle tissue of pearl oyster after shell notching

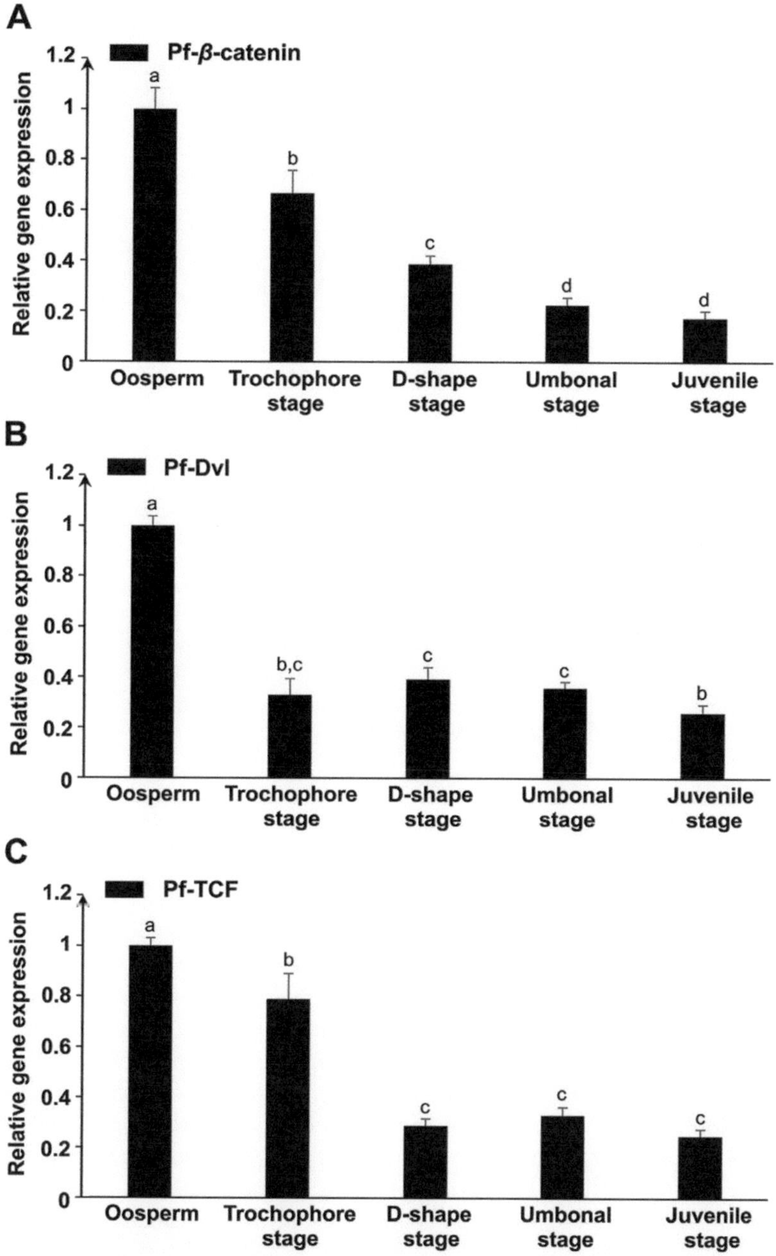

Fig. 6.43 Relative mRNA expression of Pf-β-catenin (**a**), Pf-Dvl (**b**) and Pf-TCF (**c**) during embryonic and larval development at five different stages

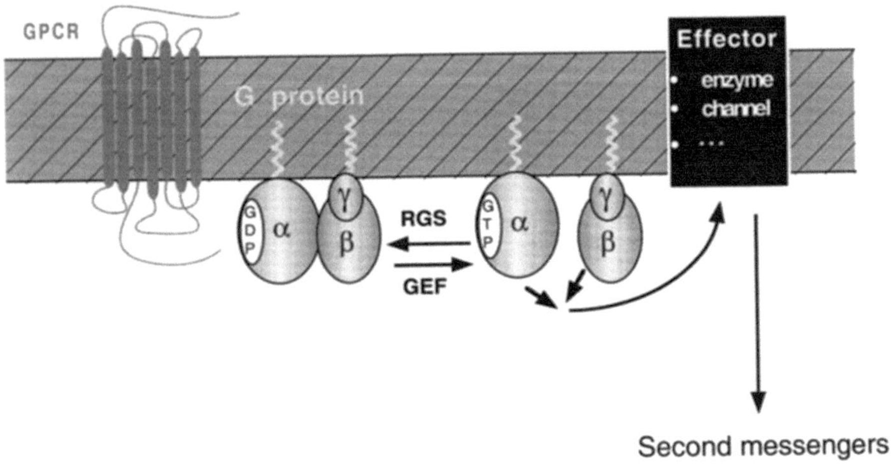

Fig. 6.44 Effect of regulator of G protein signaling (RGS) proteins on the classical G protein cycle at the plasma membrane. A G protein-coupled receptor (GPCR) serves as a guanine nucleotide exchange factor (GEF) that activates the G protein by enhancing GDP dissociation from the Gα subunit. Gα and Gβγ dissociate and stimulate their respective effectors. RGS proteins serve as GTPase activating proteins that accelerate GTP hydrolysis and thereby return the Gα subunit to its inactivated GDP-bound form, followed by reassembly of the heterotrimer [73]

6.6 G Protein-Mediated Pathway

6.6.1 Introduction

G proteins composed of α-, β-, and γ-subunits form a superfamily of signaling molecules that modulate a huge repertoire of cell surface receptors for diverse extracellular and developmental cues [70] (Fig. 6.44). Sensory stimuli, hormones, and neurotransmitters drive the receptors into a conformational change in α-subunit and catalyze the exchange of GDP or GTP, generating two potential signaling molecules, a GTP-bound α-subunit and a free βγ dimer. Each released component can activate a variety of downstream effectors. The signal termination is through the hydrolysis of GTP to GDP by GTPase activity intrinsic to α-subunit, allowing its reassociation with βγ dimer. The best characterized are the G protein α-subunits which are divided into four classes (G_s, G_i, G_q, G_{12}) according to their amino acid sequence identity, and each class is composed of several subclasses. When activated, $G_s\alpha$ can stimulate membrane-bound adenylyl cyclases (AC) and calcium channels, resulting in the elevation of intracellular levels of key second messengers cAMP and Ca^{2+} [71]. These second messengers can mediate diverse physiological processes such as metabolism, development, cell proliferation, and gene transcription. In fact, previous studies have implicated that $G_s\alpha$ play a role in biomineralization. In mammals, mutations of $G_s\alpha$ can cause many kinds of diseases that are related to biomineralization. β-Subunit has been identified in five isoforms up to now, and they showed 50%–90% identity in primary structure. β-Subunit depends upon γ-subunit to fold correctly [72]. The top surface of the propeller structure interacts with G protein α subunit, and the bottom surface of the torus is the major site for interaction with γ-subunit, which set the fundamental elements for its function. However, in mollusk, the relationship between Gα, Gβ, and biomineralization has not been elucidated.

6.6.2 G Protein α-Subunit Involved in Regulation of Biomineralization

We have cloned and characterized a novel G protein α-subunit, designated as PfG$_s$α which the full-length cDNA is 1391 bp in size, with the translation start site at nt 260 (ATG) and stop codon at nt 1391 (TGA), encoding a polypeptide of 377 aa in length. The cDNA sequence and its predicted translation have been deposited in GenBank under accession number AY391226.

The deduced amino acid sequence of PfGsα showed higher similarity to snail Gsα (86%) and octopus G$_s$α (84%), in particular, the well-conserved amino acid motifs characteristic of G protein α-subunits: A, C, and G and switch II functional sites are completely identical to other G protein α-subunits. In addition, other conserved amino acids include Arg^{185}, that is, the site of ADP-ribosylation by cholera toxin and carboxyl terminus an important site for interaction with receptors is also conserved [74].

By using in situ hybridization, we studied the distribution of the hybridization signal of PfG$_s$α mRNA and found that the G$_s$α mRNA is widely expressed in the mantle, gill, spermatogenic epithelia, digestion gland epithelia, and muscle cells in *P. fucata* (Fig. 6.45). Northern blotting analysis (Fig. 6.46) indicated different expression levels among the five kinds of tissues. The signals in mantle and gill are higher than that in the other tissues with the digestive gland being the lowest. The mantle of mollusks is a center of periostracum and shell formation [75]. Strong hybridization signals were detected in the outer fold of the mantle (Fig. 6.45d) in which the hybridization signals were seen mainly in the epithelia cells, indicating that the PfG$_s$α may be involved in biomineralization in *P. fucata*.

G$_s$α-mediated signaling is important in numerous biological processes, by generating the second messenger cAMP upon stimulation. We found that in the presence of CTX, PfG$_s$α-transfected MC3T3-E1 had a massive increase of cAMP (Fig. 6.47). CTX-induced increase in cAMP occurs as a result of ADP-ribosylation of the α-subunit of stimulatory G protein (G$_s$α) and constitutive activation of AC [76]. These findings indicate that PfG$_s$α has the same biochemical properties as mammalian G proteins [7].

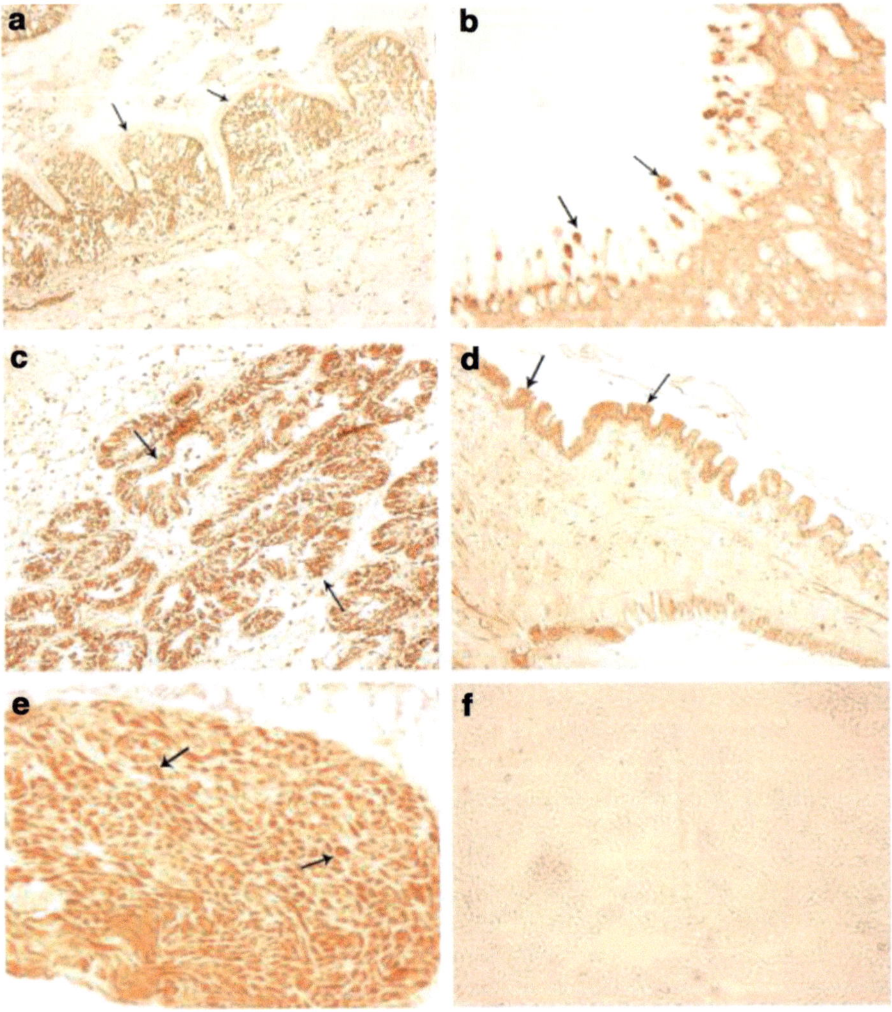

Fig. 6.45 In situ hybridization of oyster PfG$_s\alpha$ mRNA in five different tissues of *P. fucata*. Among these tissues, hybridization signals (arrows) were observed in epithelial cells of digestive gland (**a**), gill (**b**), spermatogenic duct (**c**), the outer fold in the mantle (**d**), and muscle cells (**e**). No hybridization signal is apparent in the control section stained with the same probe in (**f**). Original magnifications are ×200 (**a**, **c**, **d**, and **f**) and ×400 (**b** and **e**)

Here, we also found that PfG$_s\alpha$ can stimulate ALP activity in MC3T3-E1 (Fig. 6.48). The previous data also showed that ALP's primary function in mineralizing is to fine-tune inorganic pyrophosphate (PPi) concentrations and maintain steady-state levels of PPi adequate for controlled mineralization. Additionally, through hydrolysis of PPi and nucleoside triphosphates, ALP also may contribute to the pool of Pi available for deposition as hydroxyapatite [77]. The increase of ALP activity was also correlated to the function of osteoblast-like cells in proliferating

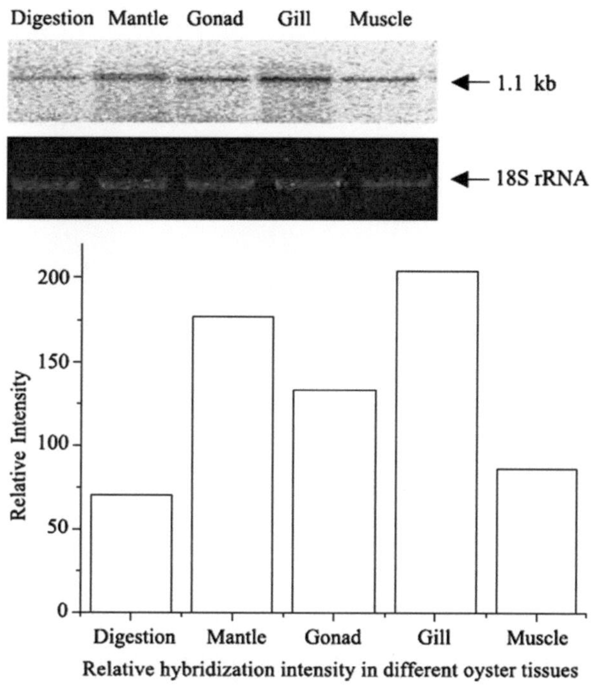

Fig. 6.46 Northern blot and its relative intensity of PfG$_s\alpha$ in various tissues in *P. fucata*

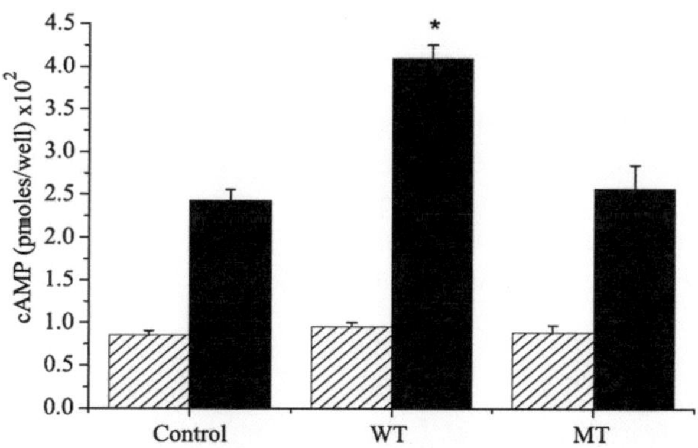

Fig. 6.47 Effect of CTX on cAMP accumulation in transfected MC3T3-E1. After transfection with control PcDNA4c or plasmids encoding WT or MT PfG$_s\alpha$ for 36 h, the cells were incubated with 10 μm IBMX for 30 min and then stimulated with 1 μg/ml CTX for 2 h (shaded bar) or no drug (hatched bars), and cAMP accumulation was measured. All values represent means ±S.D., and each sets of results is representative of at least three additional experiments. Key: *$P < 0.05$, compared with the control and mutant

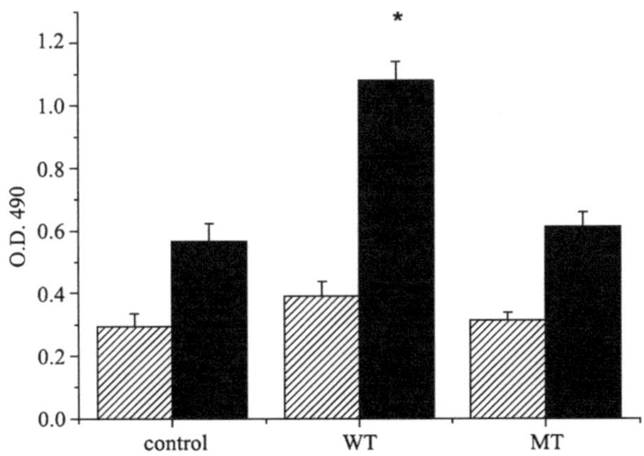

Fig. 6.48 Effects of CTX on ALP activity in transfected MC3T3-E1 cells. After transfection with control PcDNA4c or plasmids encoding wild-type or mutant PfG$_s\alpha$ for 24 h, the cells were incubated with 1 μg/ml CTX for additional 36 h (shaded bar) or no drug (hatched bars), and ALP activity was measured. All values represent means ±S.D., and each set of results is representative of at least three additional experiments. There are no differences among them in the absence of CTX in MC3T3-E1. Key: *$P < 0.05$, compared with the control and MT

and differentiation, but their signal transduction is not well clear. Recent information revealed that signal molecules through several mitogen-activated protein kinases (MAPK), such as ERK, p38 MAPK, and JNK, regulate proliferation and differentiation of osteoblast-like cells, including in controlling ALP activity [78]. Our results give evidence that G$_s\alpha$-mediated activation may in turn activate downstream signal molecules, such as MAPK family, in process of mineralization.

In summary, PfG$_s\alpha$ is widely expressed in tissues. Detailed comparison of the deduced amino acid sequence among different G$_s\alpha$s suggests that PfG$_s\alpha$ is a novel subunit within the G$_s\alpha$ class. Furthermore, PfG$_s\alpha$ may have a stimulatory effect on ALP activity, which is a vital marker in the process of biomineralization.

6.6.3 G Protein β-Subunit Interacted with CaM and Modulating Cellular Functions

We first cloned and characterized the G protein β-subunit in bivalve pearl oyster, *Pinctada fucata* (PfGβ). The nucleotide sequence was submitted to GenBank with the accession no. AY524667. The sequence of oyster Gβ subunit shared high similarity with those of Gβ subunits from mollusk, amphibian, and mammalian (94% identity to squid, 90% to tiger salamander, and 86% to human). The well-conserved amino acid domains in G protein β-subunit, seven WD repeats, were found in the deduced amino acid sequence. The 3D structure of PfGβ was predicted by the following Internet address: http://www.expasy.org/swissmod/SWISS-

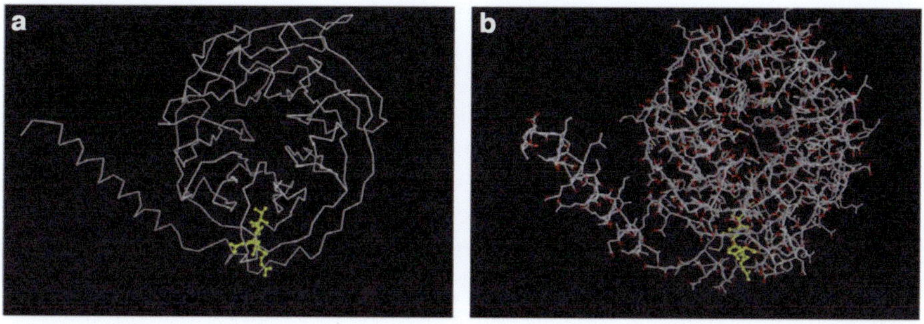

Fig. 6.49 The 3D structure prediction from the deduced amino acids: (**a**). Stick-and-stick model of the seven WD repeats in PfGβ. Each WD repeat forms one β-propeller, (**b**) Ball-and-stick model of the seven WD repeats in PfGβ

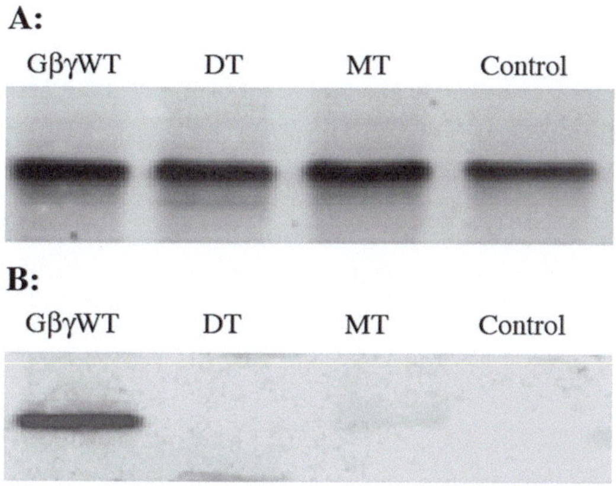

Fig. 6.50 Mutation analysis of putative CaM-binding domain of PfGβ subunit: (**a**) similar amounts of wild-type (WT), mutants (deletion 49–53 amino acids, DT; R49, 50A, MT), and positive control proteins were obtained by in vitro translation in 293T cell line. (**b**) CaM binding of in vitro translated pfGh protein, its mutants, and negative control

MODEL.html. The results showed that PfGβ also belongs to β-propeller family of proteins whose members contain 4–8 antiparallel β-sheets resembling the blades of a propeller [79]. The Gβ consists of seven β-sheets, each containing four antiparallel strands radiating outward which form a central core with approximate sevenfold symmetry (Fig. 6.49). The inner strands of each sheet run roughly parallel to central axis and form the walls of a water-fill tunnel, whereas the outer strands tilt as they radiate away from the center. Each WD repeat forms one β-propeller. However, a single WD repeat does not directly correspond to a single stable fold domain; rather, the most variable segment of each WD repeat, corresponding to its N-terminal portion, forms the outermost, C-terminal strand of one blade, whereas the conserved core of the same repeat forms the innermost strand of the nest blade. The overall propeller structure of β is closed by juxtaposition of strands from the first to the seventh WD repeat. This structure may contribute to the resistance to cleavage by trypsin despite many tryptic cleavage sites. As shown in Fig. 6.50, yellow highlights the different structures, the variable region of the seventh WD repeat which is different from any other Gβ subunits that may be involved in the closing β-propeller and formation.

G protein h subunits must depend upon γ-subunits to fold correctly, and its signal transduction in vitro also bound to γ-subunits [72]. We also cloned mouse γ2 subunit and ligated expression vector. Figure 6.50a shows the in vitro translated Gβ protein and mutant Gβ proteins separated by Western blot using anti-His-tag monoclonal antibody. The in vitro translated protein mixtures were incubated with anti-HA-tag monoclonal antibody, and the agarose beads were also added. The precipitated proteins were detected by Western blot, using Gβ rabbit anti-h subunit antibody. As shown in Fig. 6.50b, unlike the wild-type Gβ protein, the binding affinities of DT and MT for CaM were dramatically reduced. The data suggested that the mutant residues (the N-terminal 49–53 amino acids) are critical to the interaction with CaM.

We also assessed the effects of ALP activity in WT, WTC, DTC, MTC, and CaM. The data showed that the ALP activity in WTC is significantly higher than that of the other groups ($P < 0.05$). No prominent difference among the groups of DTC, MTC, and CaM was observed ($P < 0.05$), while they were all higher than that of the control group (only transfected vector without cDNA fragment) (Fig. 6.51).

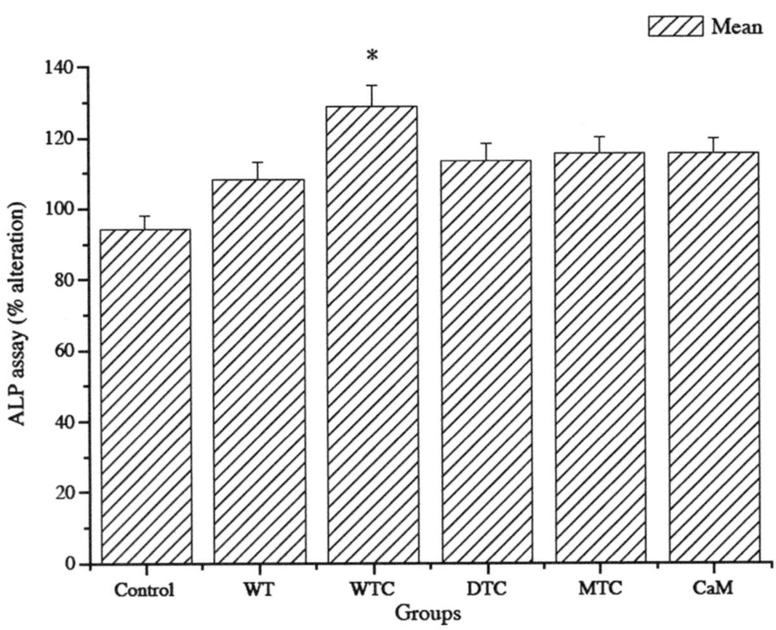

Fig. 6.51 The interaction of Gβγ subunits with CaM by measuring the ALP activity in cotransfected MC3T3-E1. ALP activity in WTC is significantly higher than that of the other groups ($P < 0.05$). There are no differences among the groups of WT, DTC, MTC, and CaM ($P < 0.05$), while they are all higher than that of the control group (only transfected vector without cDNA fragment). (*) $P < 0.05$, compared with any other group

In WT group, overexpression of Gβγ may cause general changes in the metabolic state of the cell and activate various signal transduction pathways, including stimulating MAPK kinase pathway through Ras and Raf [80] and causing permanent changes in Ca^{2+} channels state [81], which all improved the proliferation and further improved the ALP activity. In WTC group, the affinity of Gβγ subunits and CaM is sufficiently high so that much of the free βγ in the cells should be in the CaM-bound form. Therefore, an increase in intracellular Ca2+ concentration could selectively regulate the interactions of Gβγ with serials of other proteins through binding with CaM. CaM not only increases ALP activity but also participates in the biomineralization [82]. In DTC and MTC groups, mutants Gβγ failed to associate with CaM and, as a result, did not activate the downstream signal molecules. So there are no differences among the DTC, MTC, and CaM groups, but all of their ALP activities are higher than the control because of action with CaM.

In conclusion, we have cloned oyster Gβ subunit and found it widely expressed in many tissues. By comparing the deduced amino acid sequence among Gβ family members, we considered that PfGβ should be classified in a novel subunit within Gβ1 class. Further research indicates that PfGβ might interact with CaM and point to the important physiological function in modulating cellular functions.

6.7 Transcription Factors

6.7.1 Introduction

Biomineralization is precisely regulated by the matrix proteins, which account for less than 5% of the weight of shells. Since 1996, researchers have investigated the identification, separation, purification, and functions of matrix proteins in *P. fucata*. However, the transcriptional regulation of matrix proteins has seldom been reported, and few of the transcription factors involved in this regulation have been identified. Furthermore, it is extremely difficult to investigate transcription regulation without well-established cell lines of *P. fucata*. So it is lack of studies on the transcriptional regulation in biomineralization process. PfMSX is the first transcription factor in *P. fucata* which had been studied and demonstrated having the ability to activate transcription of certain matrix protein [83]. As more and more matrix proteins had been cloned and characterized, the research in transcriptional regulatory network of these matrix proteins become apparently requisite.

6.7.2 Pf-POU3F4 Regulating the Expression of the Matrix Proteins

The transcription factors POU3F4 are members of the POU domain family, which was initially defined by three mammalian transcription factors, the pituitary-specific

Pit-1 protein, the octamer-binding proteins Oct-1 and Oct-2, and the neural protein Unc-86 from *Caenorhabditis elegans* [84]. This family of transcription factors shares a common DNA-binding domain, the POU domain, which consists of an N-terminal POU-specific domain (POUS), a C-terminal homeodomain (POUH), and a linker region of varying lengths connecting the two domains. Since their initial discovery, more than 1000 POU sequences have been identified. Based on the amino acid sequence of the POU domain and conservation of the variable linker region, this POU domain family is classified into six major classes with various important functions in neurogenesis, pluripotency, and cell-type specification [85]. POU3F4, also known as brn-4, belongs to POU class III and is expressed mostly in the brain, inner ear, and pancreas in mammals [86]. POU3F4 mainly participates in the regulation of neural stem cells, hypothalamus differentiation, and inner ear development [87]. Targeted mutagenesis of POU3F4 causes abnormal temporal bone development in the mouse's inner ear [88]. In contrast to the large number of POU transcription factors with various functions in vertebrates, POU transcription factors are limited in mollusks and have received less attention; only four POU family classes have been identified and mainly in development of the nervous system [85]. However, given that transcription factors are usually involved in various biological processes, further study is required to identify which other biological processes POU transcription factors participate in in mollusks.

To better understand the mechanism that regulates shell formation, we investigated whether POU family members are involved in biomineralization of the pearl oyster. We isolated and cloned a member of the POU family, Pf-POU3F4. The full-length Pf-POU3F4 cDNA is 1961 bp long (GenBank accession number KM519606) and comprises a 5′ UTR of 137 bp, an ORF of 1233 bp that encodes a protein of 410 amino acids with a predicted molecular mass of 45.7 kDa and an estimated isoelectric point of 7.19 and a 3′UTR of 591 bp. Functional domain analysis revealed that Pf-POU3F4 has a POUS domain and a POUH domain; these domains play important roles in the function of POU family members as transcription factors.

POU3F4 has been reported that it mainly participates in the regulation of neural stem cell and hypothalamus differentiation and inner ear development [87]. Unexpectedly, our results indicated that Pf-POU3F4 directly binds the promoters of the Aspein and Prismalin-14 genes to enhance their transcriptional activities (Fig. 6.52) and that Pf-POU3F4 knockdown leads to a decrease in Aspein and Prismalin-14 gene expression (Fig. 6.53). These results provide a new perspective and also suggest that the mechanism of shell formation in the invertebrate pearl oyster has some unique features in addition to its similarities with bone and teeth formation in mammals.

According to its structure, initially, the ATGCAAAT motif was considered to be a conserved DNA-binding site for this POU domain family, with the POU_S and POU_H domains associating with the ATGC and AAAT regions, respectively [89]. However, this family of POU domain proteins was subsequently shown to recognize diverse DNA sequences due to the structural flexibility of the bipartite POU domain, which permits flexible interactions with DNA [87, 90]. In addition, there has been very little research on the proteins of the POU domain family in

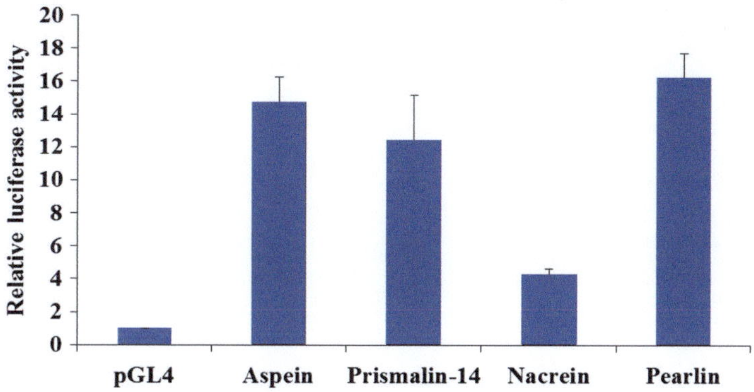

Fig. 6.52 Transcriptional activities of the Aspein and Prismalin-14 gene promoters in HEK-293 T cells

mollusks. Thus, it is difficult to predict the potential binding sites for Pf-POU3F4 in the Aspein and Prismalin-14 gene promoters. In our study, two binding sites for Pf-POU3F4 in the Prismalin-14 gene promoter and one region including three sites in the Aspein gene promoter were identified experimentally (Figs. 6.54 and 6.55). Experiments using truncated mutants of the Prismalin-14 promoter suggested that A/T-rich sites might be high-affinity binding sites for Pf-POU3F4, roughly coinciding with the binding sites for the POU_H domain in mammals. The DNA-binding sites of Pf-POU3F4 do not contain strictly conserved ATGCAAAT motifs. The multitude of roles played by POU domain family members in various processes related to biomineralization in invertebrate mollusks may be related to the diversity of their binding sites due to their flexible structures.

In sum, a POU family member, POU3F4 homolog, and the promoters of two matrix protein genes, Aspein and Prismalin-14, were cloned and identified in the pearl oyster. Interestingly, we observed that Pf-POU3F4 significantly enhanced the promoter activities of the Aspein and Prismalin-14 genes. Next, we identified the binding sites of Pf-POU3F4 in the promoters of the Aspein and Prismalin-14 genes. These results indicate that the transcription factor Pf-POU3F4 regulates the expression of the matrix protein genes Aspein and Prismalin-14 in the pearl oyster.

6.7.3 Pf-AP-1 Activates Transcription of Multiple Biomineral Proteins

Activator protein-1 (AP-1) is a ubiquitous and important transcription factor that participates in many physiological processes. AP-1 is a homodimeric or heterodimeric protein consisting of Jun family members (Jun, JunB, and JunD) and Fos family members (c-Fos, FosB, Fra1, and Fra2), and all of these members have a conserved bZIP domain [91]. AP-1 interacts with other bZIP proteins or

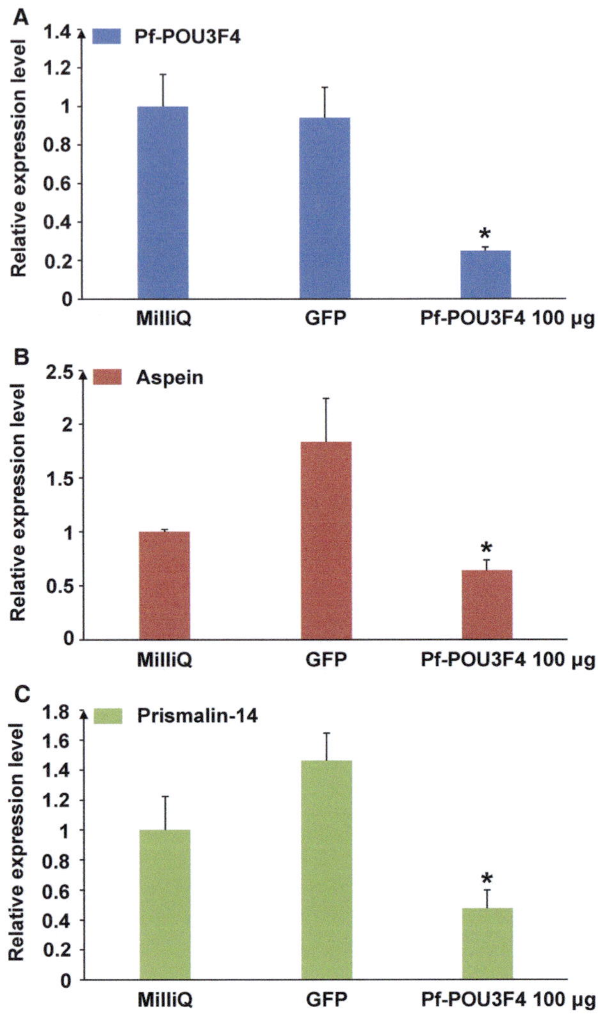

Fig. 6.53 RNAi knockdown of the Pf-POU3F4 gene. The relative expression levels of Pf-POU3F4 (**a**), Aspein (**b**), and Prismalin-14 (**c**) mRNAs in the mantle of the pearl oyster were determined 2 days after injection with Milli-Q water and dsRNAs against GFP and the Pf-POU3F4 gene

cofactors and regulates cell differentiation, cell cycle progression, apoptosis, tumorigenesis, and bone mineralization [92]. The regulatory ability of AP-1 is cell- or tissue-specific.

Most studies on AP-1 have focused on mammals and model organisms. They demonstrated that Jun/AP-1 is a key transcription factor during bone formation and reconstruction. Jun/AP-1 and its phosphorylated form are essential for efficient osteoclastogenesis [93]. Jun/AP-1 is recruited and activated by the receptor activator of nuclear factor kappa-B ligand (RANKL) in osteoclast precursors [94]. In addition, the signaling pathways mediated by RANK/AP-1 are critical for osteoclastogenesis [95]. In mollusks, most studies on molluskan AP-1 have focused on immune function. Though human AP-1 has been reported with regulatory function in

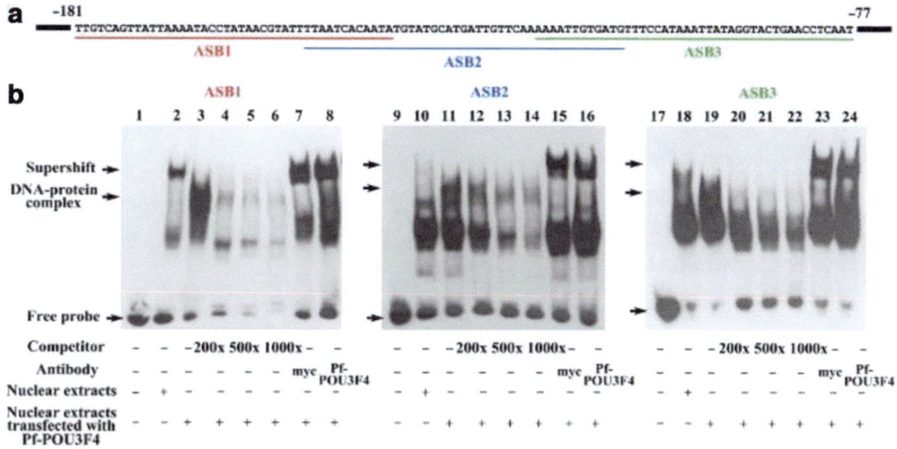

Fig. 6.54 Pf-POU3F4 directly binds to the −181 to −77 bp region of the Aspein gene promoter, as determined by EMSA. (**a**) Sequences of three probes, ASB1, ASB2, and ASB3, for the −181 to −77 bp region of the Aspein gene promoter. (**b**) EMSA results for the three probes

P. fucata matrix protein transcription in vitro [96], the existence and sequence of AP-1 homologs in *P. fucata* remain unknown, and more importantly, the function of *P. fucata* AP-1 protein in vivo had little been illustrated yet.

We isolated the cDNA of the c-Jun homolog by PCR and named it Pf-AP-1. The cDNA was 1361 bp long, with a 51 bp 5′-untranslated region (UTR) and a 380 bp 3′ UTR with a poly-A tail (GenBank accession No: KP347629). The open reading frame (ORF) was 930 bp and encoded a 309-amino acid (aa) bZIP family protein. As a hallmark, a base domain and a leucine zipper were identified in the C-terminal region (aa 230–293), which is highly conserved among mammals. Many of the modified Pf-AP-1 residues were highly conserved compared with those of human AP-1, particularly residues phosphorylated by p21 protein-activated kinase 2 (PAK2; aa 7, 76, and 80), MAPK8 (aa 51 and 61), polo-like kinase 3 (PLK3; aa 51 and 61), dual-specificity tyrosine-(Y)- phosphorylation-regulated kinase 2 (DYRK2; aa 221), and glycogen synthase kinase 3-beta (GSK3-β; aa 217 and 221), whereas aa 44 may be a potential acetylation site. Thus, Pf-AP-1 was very similar to homologs of other species.

As shown in Fig. 6.56, the bivalve mantle, which is the most important biomineral tissue, expressed the highest levels of Pf-AP-1. Similar results have been reported for *Crassostrea hongkongensis* and *Haliotis discus discus*, suggesting a potential relationship between Pf-AP-1 and biomineralization. The significant positive correlation we found between Pf-AP-1 and N19, Prisilkin-39, Pearlin, and KRMP suggests a biomineral protein transcription-regulating function for Pf-AP-1. The results of in situ hybridization showed that Pf-AP-1 had a strong expression in inner epithelial cells of outer fold and the outer epithelial cells in the mantle, suggesting that Pf-AP-1 had similar location with matrix proteins and might have important function in regulating matrix proteins.

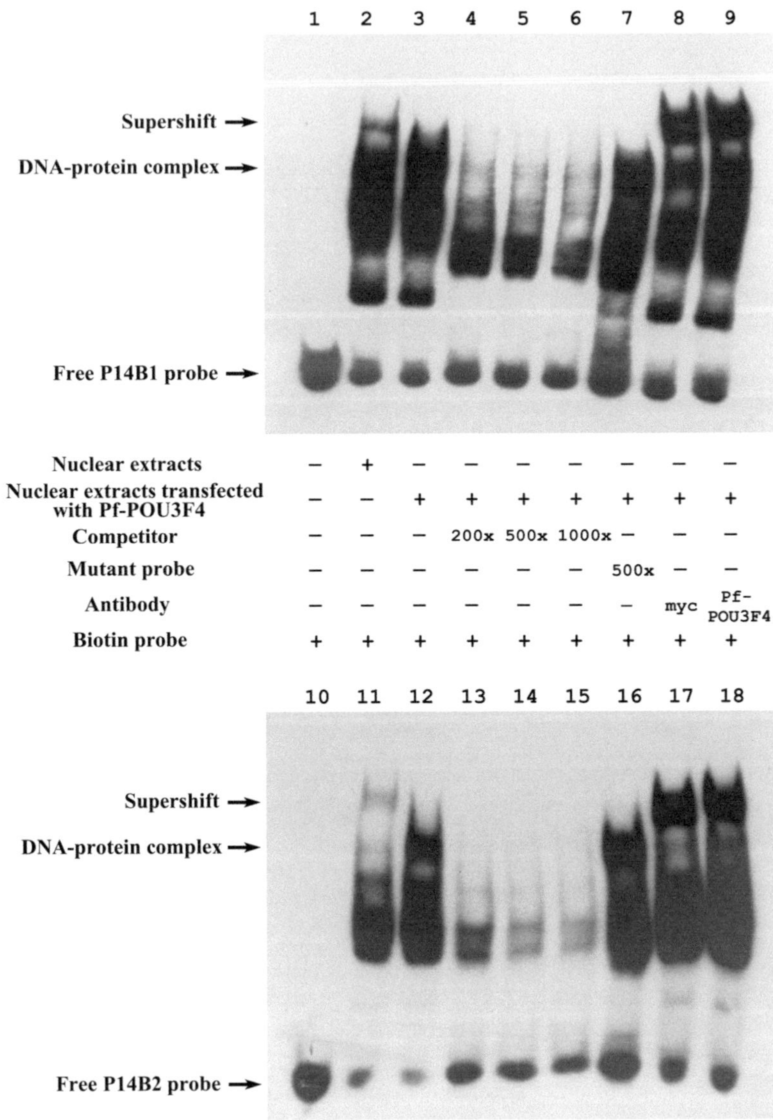

Fig. 6.55 Pf-POU3F4 directly binds to two sites of the Prismalin-14 gene promoter, as determined by EMSA. The EMSA results for the first site (−359 to −337 bp) of the Prismalin-14 gene promoter are shown in the upper panel, and those for the second site (−100 to −73 bp) are shown in the lower panel

The luciferase reporter assay results (Fig. 6.57) showed transactication ability of Pf-AP-1 to matrix proteins in vitro. And the luciferase reporter assay results of Pearlin promoter 5′ deletion confirmed the Pf-AP-1 regulation ability to matrix protein and suggested potential functional site in regions −191 to −81 bp in Pearlin

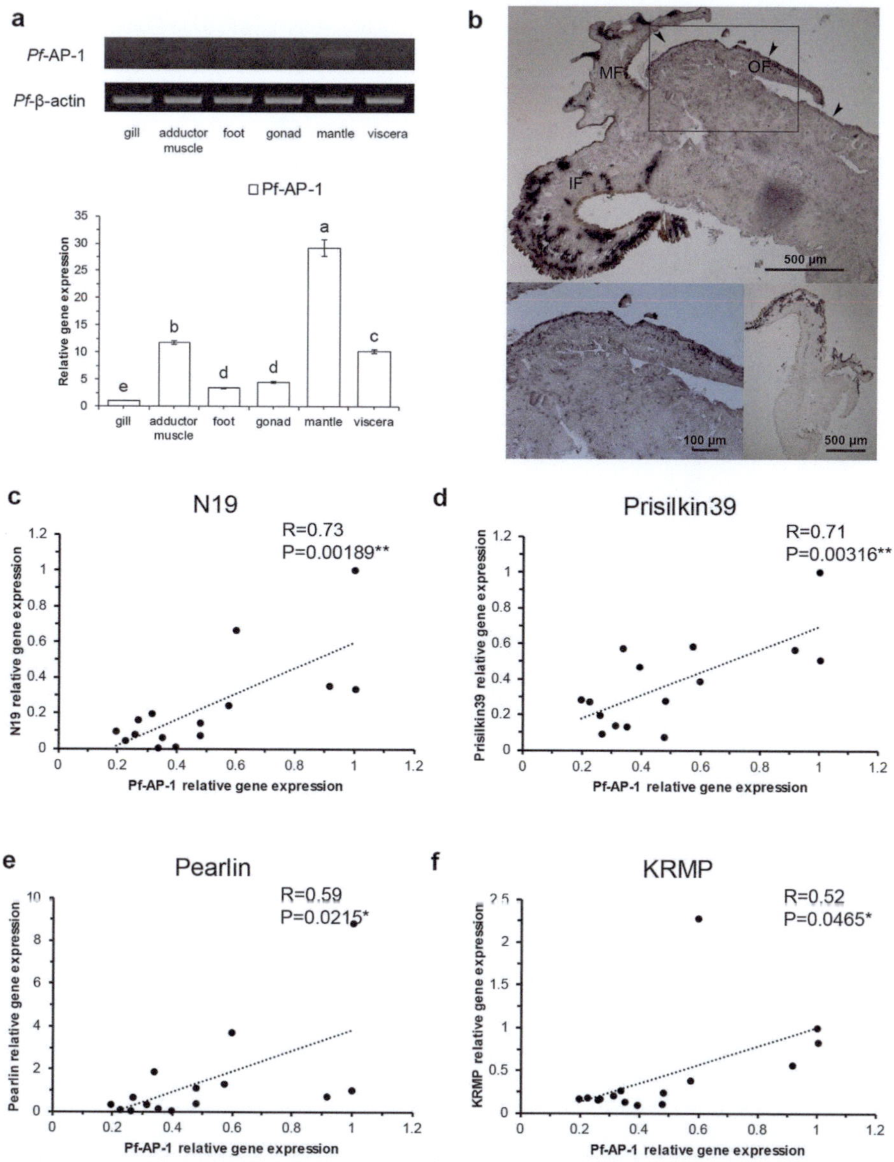

Fig. 6.56 (**a**) Tissue distribution of Pf-AP-1 gene expression in *P. fucata*. Relative Pf-AP-1 mRNA expression in tissues was assessed by semiquantitative and qRT-PCR analyses. (**b**) Detection of Pf-AP-1 mRNA in the mantle of *P. fucata* by in situ hybridization. *OF* Outer fold, *MF* Middle fold, *IF* Inner fold. The lower left showed the enlargement of the box in the upper. The lower right showed the control section stained with the sense probe. (**c–f**) Relevance of Pf-AP-1 and N19, Prisilkin-39, Pearlin, and KRMP mRNA expression in normal oyster ($n = 15$), respectively, as analyzed by t-test

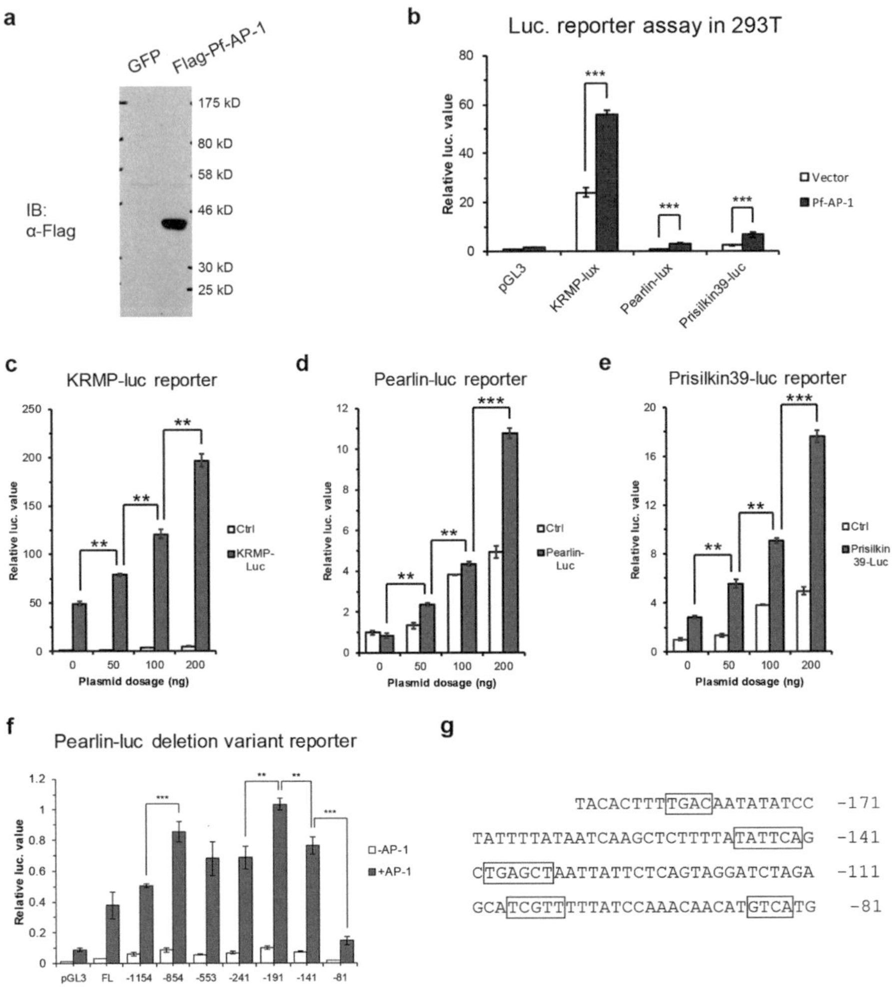

Fig. 6.57 (**a**) Western blot assay of Pf-AP-1. (**b**) Luciferase reporter assay to assess the ability of Pf-AP-1 to activate transcription of the KRMP, Pearlin, and Prisilkin-39 promoter. (**c–e**) Luciferase reporter assay to assess the ability of different concentrations of Pf-AP-1 to activate the transcription of KRMP, Pearlin, and Prisilkin-39. (**f**) Luciferase reporter assay of Pearlin promoter deletion variant in the presence or absence of Pf-AP-1. (**g**) Putative AP-1 binding sites in Pearlin promoter predicted by Transcription Element Search System (TESS). Putative binding sites were boxed

promoter. TESS analysis showed that there were 5 putative AP-1 binding sites in this region. However, this analysis was made based on the AP-1 binding sites of human but not on the mollusks. Whether the promoter of mollusk had conserved transcription regulation binding motif with those in mammals had not been reported. So further studies were needed and were in process. The AP-1 inhibition experiment supported the transactivation ability of Pf-AP-1 to KRMP, N19, Pearlin, and Prisilkin-39 in vivo, according to Fig. 6.58. And the nacrein had no response to

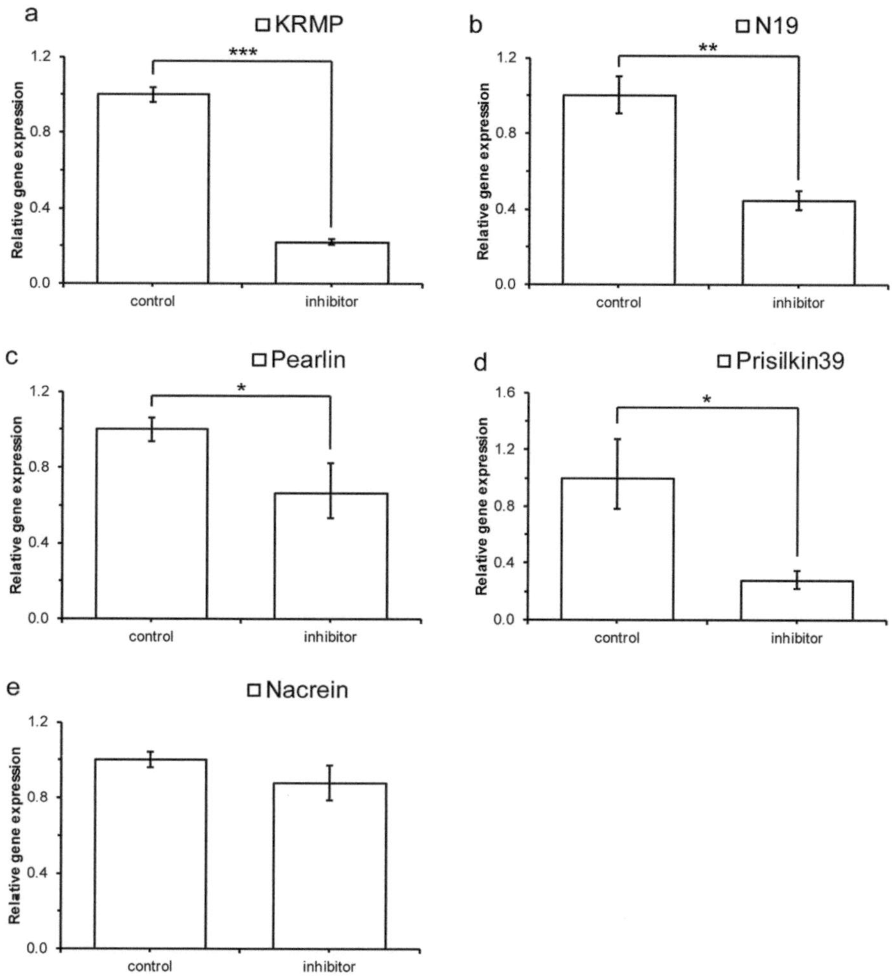

Fig. 6.58 Relative mRNA expression of matrix proteins under AP-1 inhibitior. qRT-PCR analysis was performed to evaluate the expression of each gene with or without the AP-1 inhibitor SR11302. (**a**–**e**) Relative expression of KRMP, N19, Pearlin, Prisilkin-39, and nacrein mRNAs

AP-1 inhibitor. This result supported that nacrein had no correlation with Pf-AP-1 and that the transcativartion ability of Pf-AP-1 was gene specific.

The shell regeneration experiment is a routine method to study the physiological activities during shell repair and formation. We found increased Pf-AP-1 expression from 4 h to 1 day after shell notching in the shell regeneration experiment (Fig. 6.59). This finding may have been due to stimulation from injury that caused an immune reaction. The shell resists attack from external parasites or pathogens. Previous studies have reported that the mantle probably participates in immunity. As a response to bacterial infection, the expression of immune-related factors in the

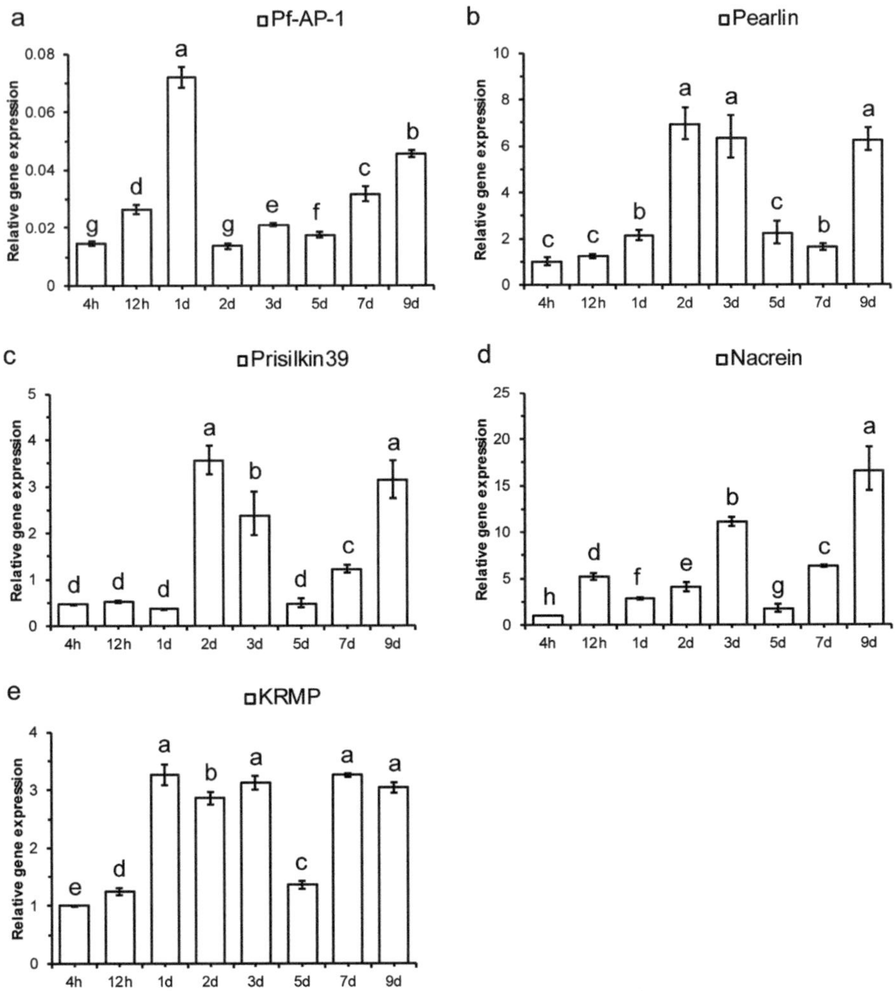

Fig. 6.59 Relative mRNA expression of Pf-AP-1 and biomineral proteins during shell regeneration

mantle had an increase in 48 h after bacterial challenge [97]. These results suggest that increased Pf-AP-1 expression in the mantle is due to the immune response and shell repair following activation of biomineral protein transcription. We found a linear increase in Pf-AP-1 expression on days 5–9 after notching. A similar pattern was observed for Prisilkin-39 and nacrein expression. This period is a rapid shell regenerating stage, in which Pf-AP-1 may specifically activate several biomineral proteins. Pearlin expression increased from day 7 rather than from day 5, which may have been due to a transcription regulation delay. The KRMP expression pattern was less similar to that of Pf-AP-1 than the other abovementioned proteins. Thus, other more specific or unknown but important transcription factors may regulate the expression of these proteins. Nevertheless, the transactivation ability of AP-1 had a strong time–space specification and was cell- and tissue-specific, so it may show that a gene regulated by Pf-AP-1 in one physiological process has no response in another process. Thus, Pf-AP-1 may regulate different biomineral proteins by forming different AP-1 dimeric proteins and interacting with diverse coactivators or cofactors during the different biomineralization processes.

In conclusion, Pf-AP-1 had a highly conserved structure with known AP-1 family members and was highly expressed in tissues that were important during biomineralization. Pf-AP-1 expression was positively correlated with that of several biomineral proteins, which suggest that Pf-AP-1 is an important transcription factor that regulates several biomineral proteins and participates in several biomineralization processes in the bivalve *P. fucata*.

6.7.4 **Pf-*YY-1's Role in Biomineralization***

YY-1 (Yin Yang-1) belongs to the zinc finger protein family, with four C2H2-type zinc fingers at the C-terminus, possessing a sequence-specific DNA-binding and repression abilities: both the C- terminus and N-terminus have transcriptional activation domains [98]. YY-1 is a well-studied transcription factor in mammals, especially in humans and mice. YY-1 is universally expressed in most human tissues (GEO: GSE2361), and it has multiple functions, including activation, repression, or initiation transcription in different tissues and under different stimulations, depending on the cofactors it recruits [99]. Interestingly, several studies showed that YY-1 could regulate gene transcription during bone formation in mammals. YY-1 binds to and activates the Col1a1 proximal promoter, integrating an upstream enhancer function with the core promoter [100]. YY-1 induces histone deacetylation by recruiting histone deacetylase, resulting in decreased expression of the cartilage-specific gene chondromodulin-1 in mesenchymal stem cells [101]. YY 1 also interacts with Runt and the C-terminal domain of Runx2 to inhibit Runx2's transcriptional activity, causing repression of osteoblast differentiation mediated by BMP-2 [102]. Meanwhile, YY-1 also affects cancer and nervous system development by activating and repressing related gene expression [103, 104]. YY-1 is conserved in many species, and we hypothesized that if this were the case in *P. fucata*, it would suggest that YY-1 was also involved in biomineralization.

We cloned and reported a cDNA encoding an YY-1 homolog from *P. fucata*, termed Pf-YY-1 (GenBank accession no: KM502551). Meanwhile, we also identified another isoform (GenBank accession no: KM502552), which is slightly longer and contains two insertion fragments. The putative Pf-YY-1 protein has a classical zinc finger domain comprising four C2H2-type zinc fingers.

Compared with YY-1 s from model animals, Pf-YY-1 has some differences; however, its C-terminal is very similar to those from other species. The identity of the whole YY-1 from *P. fucata* vs human or mouse is 54%; *P. fucata* vs *Xenopus laevis* is 62%. By contrast, the identities of the conserved region (C-terminus zinc fingers) from *P. fucata* vs human, mouse, or *X. laevis* are all 99%. Pf-YY-1, like the

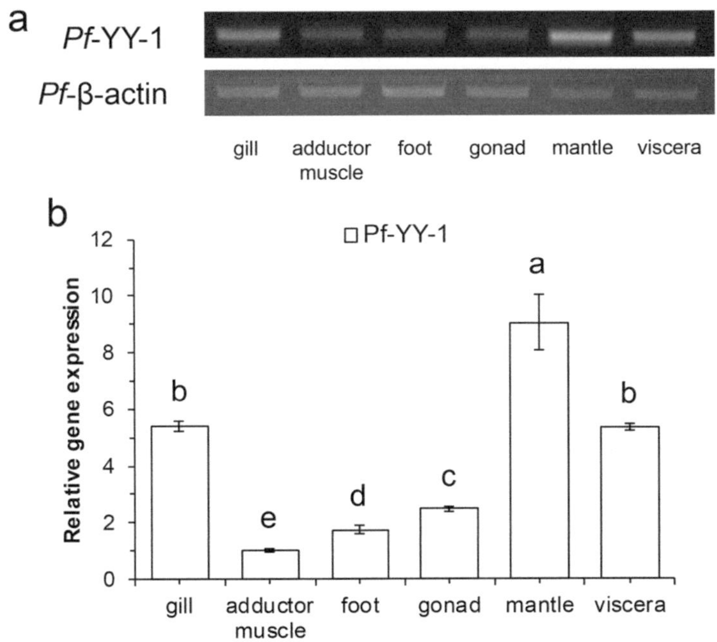

Fig. 6.60 Tissue distribution of the YY-1 gene transcript in *Pinctada fucata*. (**a**) Relative expression of Pf-YY-1 mRNA in tissues from *P. fucata* as assessed by semiquantitative RT-PCR analysis. (**b**) Relative expression of Pf-YY-1 mRNA in tissues from *P. fucata* as assessed by qRT- PCR analysis

YY-1 s from nonmammalian species, does not have the 11 consecutive acidic amino acid repression domains. In addition, Pf-YY-1 and Cg-YY-1 do not have the interesting GGKKSGK or GAGAAG sequence motif that exists in most vertebrate YY-1 s. This suggested that Pf-YY-1 has relatively weak repression ability or it may possess another motif with that function. The whole sequence of Pf-YY-1 is more similar to vertebrate YY-1 s than in- vertebrate YY-1 s, which suggested that the function of Pf-YY-1 resembles more that of the vertebrate YY-1 s.

Semiquantitative RT-PCR and RT-PCR showed that Pf-YY-1 was ubiquitously expressed, and its expression pattern was similar to that in humans (GEO: GSE2361) according to Fig. 6.60. Considering the importance of the gill to the mollusk and its complex digestion activity, we hypothesized that the high expression in the gill and viscera was related to the multiple physiological activities in these organs. The highest expression level in the mantle suggested that YY-1 might be involved in matrix protein expression and contribute to the regulation of biomineralization.

We also found that the matrix protein mRNAs showed a similar transcription pattern to Pf-YY-1 (Fig. 6.61). Previous studies in higher organisms described the activation of transcription by YY-1 using three different models: through direct interaction with general transcription factors, by inhibiting repressor activity and/or unmasking activation domains, and by recruiting co-activators. Thus, we

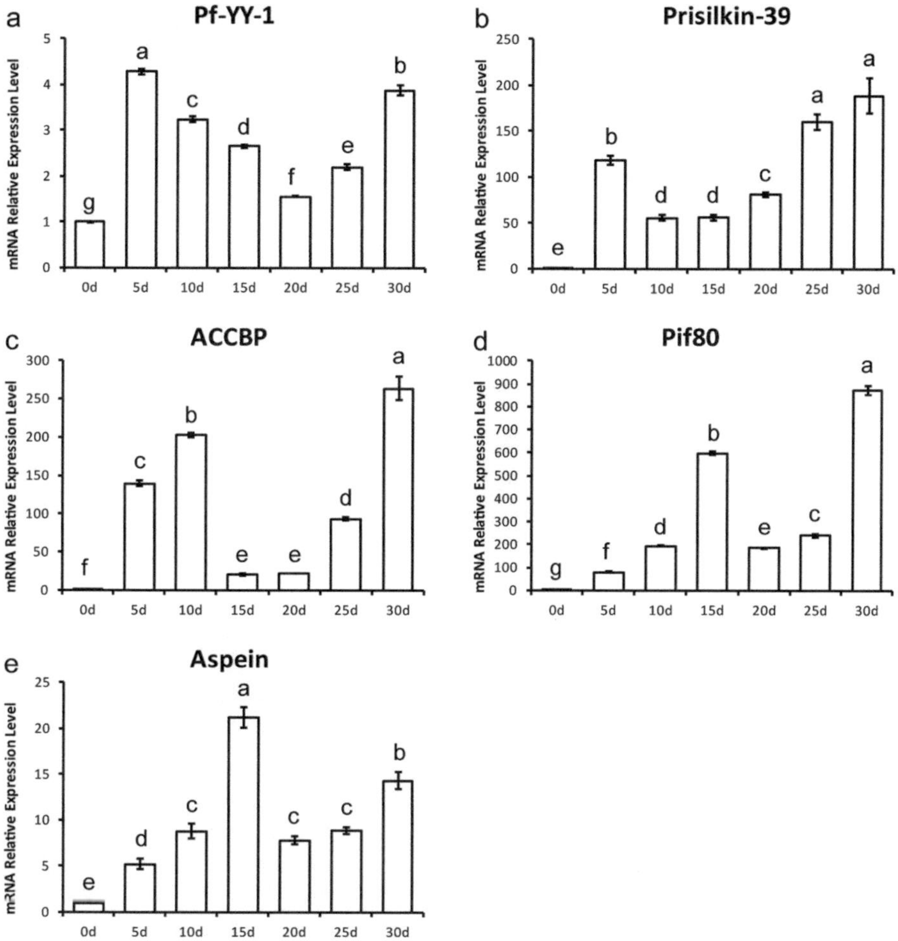

Fig. 6.61 Relative mRNA expression of *Pinctada fucata* YY-1 and matrix proteins during pearl sac development. qRT-PCR was performed to analyze the expression of each gene at different developmental stages. The abscissa shows the days after implantation. (**a**, **b**, **c**, **d**, and **e**): the relative expression levels of Pf-YY-1, Prisilkin-39, ACCBP, Pif80, and Aspein mRNAs in the pearl sac, respectively

hypothesized that the delay of matrix protein expression change in the first stage of pearl formation may reflect the different ways in which Pf-YY-1 regulates their transcription. In the second stage of pearl formation, the formation of the organic framework has been completed, and Pf-YY-1 may activate matrix protein expression to adapt the fast aragonite deposition.

In summary, we report a gene encoding an YY-1 homolog (designated Pf-YY-1) from the *P. fucata* cDNA library. Sequence alignment and bioinformatics analysis indicated that it has a conserved DNA-binding domain with a typical zinc finger motif. Pf-YY-1 has high similarity to that of *C. gigas* and that the YY-1 protein homologs are similar and conserved from invertebrates to vertebrates. The tissue distribution of the gene transcription showed that Pf-YY-1 is ubiquitously expressed in adult *P. fucata*. In the different stages of pearl sac development, the expression of Pf-YY-1 displayed a distinct pattern that correlated with multiple matrix proteins. These results indicated that Pf-YY-1 might play a role in many physiological processes especially biomineralization in *P. fucata*. The transcriptional activity of Pf-YY-1 in biomineralization requires further investigations.

6.7.5 Pf-Sp8/9 Involved in Mantle-Related Physiological Activities

Specificity proteins (Sp) are members of the classical zinc finger transcription factor family. All Sp proteins have a highly conserved DNA-binding domain with three continuous zinc finger domains in their C-terminus. Sp1 was the first transcription factor to be identified and cloned in mammals [105]. Other mammalian Sp proteins have been identified and cloned according to their conserved zinc finger domains, and further studies have described their essential functions. Sp8 affected the growth of limbs and tail, as well as closure of the neuropore in mice [106]. Sp8/9 homolog identified in beetle had a function similar to the mouse Sp8 [107]; mouse Sp9 also affected the growth of limbs and was shown to be regulated by the fibroblast growth factor Fgf10. However, studies on Sp family members in invertebrates are limited.

We cloned and identified a Sp gene from *P. fucata*, in which the complete mRNA (GenBank: KR057959) was 2343 bp long, with a 749 bp 5′ UTR, a 385 bp 3′ UTR with a poly-A tail, and a 1209 bp ORF encoding a 402 aa protein. Analysis of the translated protein sequence showed that it was a Sp8/9 homolog. The predicted Pf-Sp8/9 sequence shared high similarity with Sp8–9 proteins from six model organisms, and the BTD-box and three continuous zinc fingers were highly conserved. Studies on Sp protein zinc fingers have revealed that the conserved amino acids KHA, RER, and RHK in the three zinc finger motifs are responsible for ecognizing DNA sequences and interacting with DNA fragments. These amino acids are highly conserved across species and were also conserved in the Pf-Sp8/9 sequence. In addition, the core cysteine and histidine residues that interact with zinc ions and the residues that formed the a-helixes and b-sheets were identical in all the tested species. Interestingly, the Sp-box, which is conserved only in mammals, was absent in Pf-Sp8/9, as well as in *G. gallus*, *X. laevis*, and *D. rerio*. We have shown that Pf-Sp8/9 contains the highly conserved DNA-binding and transcription regulatory motifs found in other known Sp8/9 proteins, implying that Pf-Sp8/9 may have similar functions to other Sp8/9 proteins. The luciferase reporter assays supported the transactivation ability of Pf-Sp8/9 and the possible interaction between Pf-Sp8/9 and the promoters of the genes encoding matrix proteins (Fig. 6.62).

The high Pf-Sp8/9 expression levels in the mantle tissue suggest the potential function of the encoded protein in physiological processes in the mantle (Fig. 6.63).

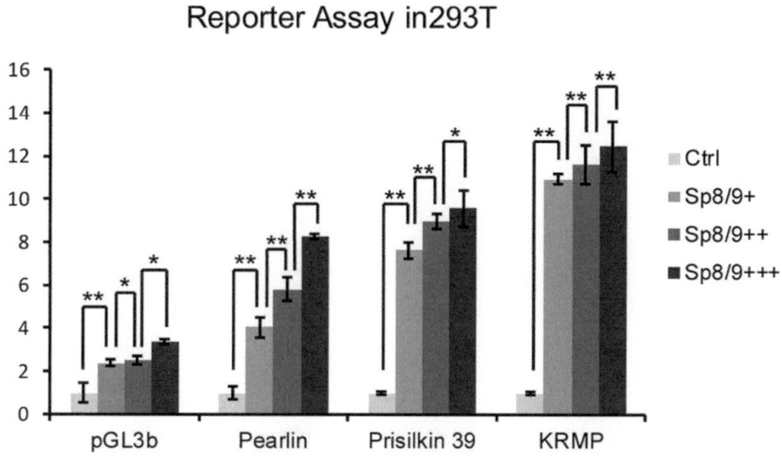

Fig. 6.62 Luciferase reporter assay to assess the ability of Pf-Sp8/9 to activate transcription of the genes encoding Pearlin, Prisilkin-39, and KRMP. The Flag-pcDNA3.1 vector was used as a blank control. HEK293T cells were transfected with Flag-Sp8/9-pcDNA3.1 with gradients of 50 ng, 100 ng, or 150 ng. The results were analyzed by the unpaired two-tail student's t-test. $*P < 0.05$, $**P < 0.01$

Fig. 6.63 Tissue distribution of Pf-Sp8 gene expression in *Pinctada fucata*

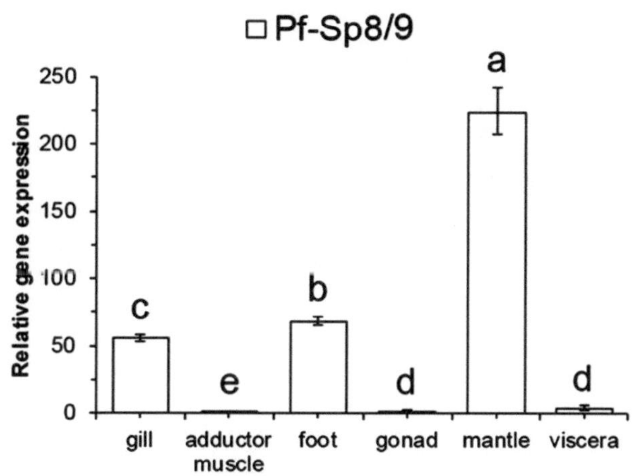

Pf-Sp8/9 expression was significantly induced after implanting artificial nuclei to produce pearls (Fig. 6.64). Because the pearl sac proliferates from mantle fragments, the high Pf-Sp8/9 expression level may be related to proliferation of the mantle tissue. However, because the artificial nucleus is foreign to the oyster and may induce an immune response, that may activate Pf-Sp8/9 expression during early pearl sac development. Thus, the high Pf-Sp8/9 expression levels during the early stage of pearl development may be a result of complex interactions among several factors. However, we found that Pf-Sp8/9 expression increased rapidly during the late stage after day 25 when the immune response would have been completed, the

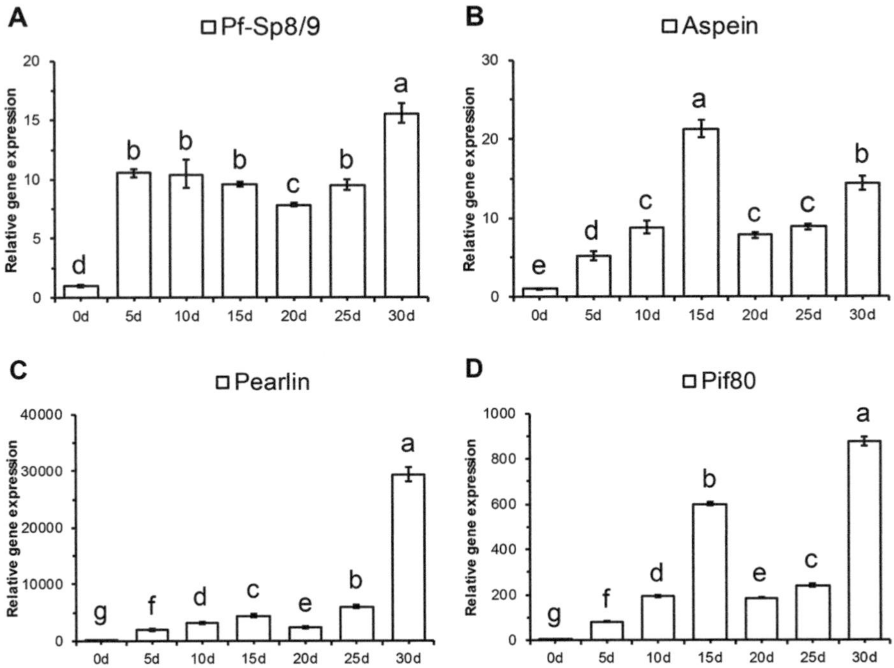

Fig. 6.64 Relative mRNA expression levels of Pf-Sp8/9 and genes encoding biomineral proteins during pearl sac development. (**a–d**) Relative Pf-Sp8/9, Aspein, Pearlin, and Pif80 mRNA expression levels in the pearl sac

pearl layer appeared on the surface of the artificial nucleus, and aragonite was deposited with accurate regulation by the matrix proteins. At this stage, high expression of matrix proteins is required, and the pearl sac is actively involved in the secretion of these proteins. The increase in Pf-Sp8/9 expression at this late stage was synchronized with the reported increased abundance of the matrix proteins Aspein [108], Pearlin [109], and Pif 80 [110], suggesting that Pf-Sp8/9 may be adapted to the physiological activity in the mantle with the potential to regulate the transcription of genes encoding the matrix proteins.

In conclusion, we identified the first Sp protein family member in *P. fucata*, and its alignment with Sp proteins from other organisms showed that it was a Sp8/9 homolog; therefore, we named it Pf-Sp8/9. Pf-Sp8/9 expression was high in the mantle compared with other tissues and was upregulated rapidly during the late stage of pearl sac development, which seemed to be synchronized with the secretion of some matrix proteins. Pf-Sp8/9 also showed the potential to regulate transcription of a number of matrix proteins. So it is suggested that Pf-Sp8/9 is associated with physiological processes that occur in the mantle and may participate in biomineralization in *P. fucata*.

This work was modified from the papers published by our group in *Acta Biochimica et Biophysica Sinica* (Wu Xi et al., 2007, 39 (7): 533–539; Zhou Y et al., 2008, 40 (3): 244–252), *Developmental & Comparative Immunology* (Xiong Xunhao et al., 2008, 32 (1): 15–25), *PLOS One* (Sun Juan et al., 2015, 10 (7): e0131711), *Progress in Biochemistry and Biophysics* (Zhou Y et al., 2010, 37 (7): 737–746), *Comparative Biochemistry and Physiology Part B: Biochemistry and Molecular Biology* (Chen L et al., 2004, 139 (4): 669–679; Chen L et al., 2005, 142 (2): 142–152; Zhou Y et al., 2010, 156 (3): 158–167; Gao J et al., 2016, 196: 92–101), *Gene* (Zheng X et al., 2015, 572 (1): 108–115), *Scientific Reports* (Zheng X et al., 2015 (5): 14408), *The FEBS Journal* (Gao J et al., 2016, 283 (10): 1962–1978), and *Journal of Structural Biology* (Zheng X et al., 2016, 196 (2): 119–126). The related contents are reused with the permission.

Supplementary Table 6.1

Chapter	Primer name	Primer sequence
6.3.2	5′-GSP1	TTATGGGGGTGGGGGTCGTATGGTGTGT
	5′-GSP2	TCTTCTCATTTGTGCTGCTCTCC
	3′-GSP	GCTAAAGGTGGAGATGAGGTGTT
	sense	AACAACACTGGAAAGCCCAGACT
	antisense	ATGCCATTGAGATGAGGGGGAAT
	RT-FP	AGCAGCACAAATGAGAAGAA
	RT-RP	TCCCAAATAGTATCACCAGAGC
6.3.3	5RP	CCGACAAAGGTGCGTAGGTGATGTGATT
	3RP	CCTGGAGAAGAAGATTGGGTGG
	FLGSPF	GGCACAATTTAGGTGGAACGATCA
	FLGSPR	GCTTTCTCTCTCTCTCTCTCTCATC
	EPF	CCGGAATTCGGATGGCAGTCAATAGTAAAG
	EPR	CCGCTCGAGCTATCAGGTATTGTCTAACGGTG
	RTGSPF	AAGCAGTAGTAATCCTGGTAG
	RTGSPR	GTGTATTCGTCTTGTTCCCT
6.3.4	Nacrein-F	GGGGTACCTGAGTATCGACGAGAAACGCTTA
	Nacrein-R	CCCAAGCTTCAGTCAAATGAAGATACATCACC
	Rel-exp-F	GCGAGCGAGCGGTGACTTA
	Rel-exp-R	GCACAGGCGCACACATACG
	Sense-Rel	CATCCCCGGGGAGAGCAGCAC
	Antisense-Rel	GTGCTGCTCTCCCCGGGATG
	Nacrein-RT-F	GGCTTTGGCGACGAACCGGA
	Nacrein-RT-R	ACACGGGGGAGTGGTCAGGG
	Rel-RT-F	CTCGAGTGAAAGCTTCAACA
	Rel-RT-R	CGAGCTATACGAGCACGCAGC
	actin-RT-F	CTCCTCACTGAAGCCCCCCTCA
	actin-RT-R	ATGGCTGGAATAGGGATTCTGG
	KBWTL-5	GATCACTGCAACAGGGTCTGGTTGGAGATCCCCTCCCTTCTTGTAAGA

(continued)

Chapter	Primer name	Primer sequence
		AGTTGAGGGGACTTTCCCAGGC
6.4.2.1	AR1	GGSACDGGVTCCTGGACNCAG
	AR2	GAGCCATGTAACGTTTGGTTCC
	AR5R2	CTTTGGAGATAATCATATAATGACCCG
	AR3R1	TTGTTACGTCACGACTCGATCCTGGG
	AR3R2	CGGAAAGTGGAAAAATAGATTACGG
	ALR-cf1	TATCAACGCAGAGTACGCGGGGAG
	ALR-cf2	CTGAGCTTCAAATTCAGTACCGTGG
	AR3	CGAGGTGTAGGGTGTTTGTGTTC
	AR4	TAGGTTTTCCTGGCTCTCTTGTCG
	SB1	TATTGAAGCACCAGTGTTGCCTCC
	SB2	TCTAAAGGTGGGCTTGGTGAAACG
	RTREL1	AGCAGCACAAATGAGAAGAA
	RTREL2	TCCCAAATAGTATCACCAGAGC
	RTIPF	AAGCAGTAGTAATCCTGGTAG
	RTIPR	GTGTATTCGTCTTGTTCCCT
6.4.2.2	BMPR1B-GSP1	TACAGTAATCCTCGTCATCACAGCAC
	BMPR1B-NGSP1	GAAGTTGTTGTATTGCTGATCTGGGATTAGC
	BAMBI-GSP1	GCAATAGGGCCTCCATAAACTGTCAACC
	BAMBI-NGSP1	TCCATAAACTGTCAACCCCCTCGTCCC
	RT-PfACTIN-F	TACCGCCGCGTCATCATCAT
	RT-PfACTIN-R	TGCCTCGGGACATCTGAACC
	RT-PfBAMBI-F	ACTGTACAAGGAAGGCGTGTCAC
	RT-PfBAMBI-R	GAGCTGGCATTTGTTTGGACGTG
	RT-PfBMPR1B-F	GCACATAACACACGGCAAGGAAC
	RT-PfBMPR1B-R	TCTGCTTGTCGGTAGGCTTCAAA
	RT-KRMP-F	GAATGAAGTTCGCCGCTGTT
	RT-KRMP-R	TTCCAATCCCARGGRTGACA
	RT-PRISMALIN14-F	AAAGAAATACTTAACTGGTGCTA
	RT-PRISMALIN14-R	CATGAGCAGCCCGGGTC
	RT-PIF-F	TGCTGCCATCACGTGAGTATG
	RT-PIF-R	GACTTCCCTTTCTCACACTTCCA
	RT-MSI60-F	GAACAATGACTGGAATGACA
	RT-MSI60-R	GGAAAGGTATCCAATAACAAC
	ISH-BAMBI-F	GAGAAGAGGTGCTACTGC
	ISH-BAMBI-R	GCTGGATTCTGCTTCTCG
	ISH-BMPR1B-F	GAGTGCTGTGATGACGAGGATTAC
	ISH-BMPR1B-R	TTGATTCACTTATATATCGCACTGC
	RNAi-BAMBI-F	GCGTAATACGACTCACTATAGGGAGATGCGATGTTGTAAAGAGGA
	RNAi-BAMBI-R	GCGTAATACGACTCACTATAGGGAGATTTCACGACGGATTTGTAT
	RNAi-BMPR1B-F	GCGTAATACGACTCACTATAGGGAGACTGGCGACTCAATGTCTCAA
	RNAi-BMPR1B-R	GCGTAATACGACTCACTATAGGGAGACTGGCATTGGTGTTGTTGAC

(continued)

Chapter	Primer name	Primer sequence
	BAMBI-EcoRI	CGGAATTCATGGAGGCCCTATTGCTTC
	BAMBI-PstI	AACTGCAGCTATACAGAAGCCACCAAGTC
	BMPR1B-EcoRI	CGGAATTCATGGCAGACCTCTGCTGG
	BMPR1B-PstI	AACTGCAGCTAGCTTTCTCCGGGTTTTATGAC
	BMPR1B-BamHI	CGGGATCCCTAGCTTTCTCCGGGTTTTATGAC
	SMAD1/5/8-ClaI	CCATCGATATGAGTTCACCCATCTCC
	SMAD1/5/8-BamHI	CGGGATCCTCATGATACAGATGAAATTGGG
	BMP2-ClaI	CCATCGATATGATTTACGGATTTGGACATTACC
	BMP2-BamHI	CGGGATCCCTACCGACATCCGCATCC
	SMAD4-EcoRI	CGGAATTCATGTTTCGGTCTAAAAGATCTACCCTC
	SMAD4-BamHI	CGGGATCCTCACCTGTGGACGTTCAATAAAATTTC
6.4.4.1	SD5-1	GCAAACTCCTGATTGTTAAATATC
	SD5-2	GCATGAAATGGTTCTCCAACACG
	SD3-1	GGTGCTCGATAGCTTACTATGAGC
	SD3-2	CAGTCTGCAAAATACCCCCAGG
	Smad-cf1	AATCGTCGTTTTGTTCTCTTCTAG
	Smad-cf2	CGGTATCAGTCAATATCACACAG
	SB1	TATTGAAGCACCAGTGTTGCCTCC
	SB2	TCTAAAGGTGGGCTTGGTGAAACG
	SI1	TTCCCCATTTACTCCACCCATC
	SI2	GCAACACTGGTGCTTCAATACGA
	Pf-engrailed-ES1	CGGCATTCTCTAACGATCAGC
	Pf-engrailed-ES2	CCATGAGATGTAACGCTAACAAG
	Pf-Smad3-SB1	TTCCCCATTTACTCCACCCATC
	Pf-Smad3-SB2	TCGTATTGAAGCACCAGTGTTGC
	Pf-BMP2-BMP2F	GCGGTCGAAGAACTAAAA
	Pf-BMP2-BMP2R	ATCCGCATCCTTCAACAA
6.5	BCAT-3F1	CGAGGAATGAAGGGGTTGC
	BCAT-3F2	TCACGGTAGTCAGGGAAGCAT
	BCAT-5R1	GGGTGGAGGGGATGGGTAC
	BCAT-5R2	CGGCGTGTATGCCTGTTCT
	BCAT-5	ACAAATAAGGAACCACCAAGATG
	BCAT-3	CATGGTTACAGAAGTGGAATATCC
	DVL-3F1	CCCCTCAGCCTGGGTAGCA
	DVL-3F2	GAAAGACTCCGCTGAAACTCC
	DVL-5R1	CCAGCACTTGGCAACGACT
	DVL-5R2	TGATGTGCGGTTTGTGGTGT
	DVL-5	AATCGGACGCCATTTTACATCA
	DVL-3	TTGCGGTACTGCGTTGAGGT
	TCF-3F1	GGGAATCCACCAGAAGACAAA
	TCF-3F2	AAAGAAGTGCGAGCCCAAGT
	TCF-5R1	TGGGCTCGCACTTCTTTCAT
	TCF-5R2	ACATCAAGCCTGGGTGGG

(continued)

Chapter	Primer name	Primer sequence
	TCF-5	TTGTTTTTAACTTTGACAATGCCG
	TCF-3	AATGTCTACTACCGTTGCCATGTTC
	BCAT-F	AACCATTCTACAAGCAGGCGATA
	BCAT-R	TCCCTCAGTGGTGCGTGGTTAG
	DVL-F	TATTAGCAGGGCTTCGTCATTCA
	DVL-R	CCCGTCTCCTCCTTTGTTACTCT
	TCF-F	TGAAGGCGAGGAAGAGCAGA
	TCF-R	CCCAAGGGAGAAGATCCATTAGG
	actin-F	CTCCTCACTGAAGCCCCCCTCA
	actin-R	ATGGCTGGAATAGGGATTCTGG
	18S-F	AGGACCTCGGTTCTATTTTGTTGG
	18S-R	TTTCACCTCTAACACCGCAATACC
6.6.2	5AGSP1	GTACA GGGGGATTCAGTGTGCTCATGG
	3AGSP2	CATGAGCACACTGAATCCCCCTGTACC
	AeF	CGCGGATCCATGGGATGTTTTAAGCCTAAGGGAG
	AeR	CCGCTCGAGCTATCACAGAAGCTCGTATTGTCGCAG
6.6.3	GSP3	CCTGTAAATGCTGTCGTCTGCTGACCT
6.7.2	POU3-3F1	GCCATCCGCCCAAGAAAT
	POU3-3F2	ACCAAATGGAGAACTGCTGATG
	POU3-5R1	CCTGTGGTCGAGTCCGCTTCTTCAA
	POU3-5R2	CCAGGAATCACCGCCCGTAATCG
	POU3F4-5	AATAAGCCAGGACCAAGTAAAGG
	POU3F4-3	CATTCCCAACACCTTACTTTCTG
	AS-1F	GGGATAGCCATTCTTATGTGTCT
	AS-1R1	AGCATCACTGGGCTCCGATA
	AS-1R2	ACTACAGCACCAGCGGCAG
	P14-1F	GGGACACAGTCCCAAAGAAATAC
	P14-1R1	AACAAACATGAGCAGCCCG
	P14-1R2	GATTCGCTGGATTCCGCTA
	AS-GWR1	TGACATCAGAAACGAATAAACAGGA
	AS-GWR2	TATCAGCATTTCCTTGAGCAGCCAT
	AS-GWR3	ACAATACCAAACAAACGAAATCACT
	AS-GWR4	GTCCGTTACCGAAAGCGAAAGA
	AS-GWR5	TAACTGACAAGGATCGGGCTAG
	AS-GWR6	CTAGAGCGTGCTTTGTGGAACT
	ASP-F	TCGGATGCCCTGTAGAGT
	ASP-R	AATCACTTACAAGTAACGCTTA
	P14-GWR1	CACAAGCAGCTAAGGCAAGGA
	P14-GWR2	GAATGGAAATGTAGCACCTGTTGAT
	P14-GWR3	GCTTGCCGAAAATCTTTTACGA
	P14P-F	AAGGCTTCTGCGGGGTTT
	P14P-R	TCGAACTCCGGTGCCAGAG
	POU2F1-F	CTCCGATTTCCGCCCCTC

(continued)

Chapter	Primer name	Primer sequence
	POU2F1-R	GCTAGAATGATTTGTCCGTTAGTCG
	POU3F4-F	AGATGTGGGACTTGCTTTGGG
	POU3F4-R	TTCTTTCGCTTTCTACCTTGTGC
	AS-F	GAAGGGGATAGCCATTCTTATGTG
	AS-R	GCATCACTGGGCTCCGATACTA
	P14-F	CTATTTCCCGCGTTTCTCCTATC
	P14-R	TCCTCCGTAACCACCGTTAAATC
	actin-F	CTCCTCACTGAAGCCCCCCTCA
	actin-R	ATGGCTGGAATAGGGATTCTGG
	18S-F	AGGACCTCGGTTCTATTTTGTTGG
	18S-R	TTTCACCTCTAACACCGCAATACC
	dsPOU3F4- 1F	GGATCCTAATACGACTCACTATAGGTTGGCGGA CCCTACTCA
	dsPOU3F4- 1R	CTTCTTTCGCTTTCTACCTT
	dsPOU3F4- 2F	TTGGCGGACCCTACTCA
	dsPOU3F4- 2R	GGATCCTAATACGACTCACTATAGGCTTCTTTCG CTTTCTACCTT
	dsGFP-1F	GGATCCTAATACGACTCACTATAGGATGGTGAG CAAGGGCGA
	dsGFP-1R	ACTTGTACAGCTCGTCCATG
	dsGFP-2F	ATGGTGAGCAAGGGCGA
	dsGFP-2R	GGATCCTAATACGACTCACTATAGGACTTGTAC AGCTCGTCCATG
	P14B1F	GTCAAAATTGCAAATTTAATTAGATTCAAAAAC GCACGG
	P14B1R	CCGTGCGTTTTTGAATCTAATTAAATTTGCAATT TTGAC
	P14B1WTF	GTCAAAATTGCAAATGGAAGGAGAGGCAAAA ACGCACGG
	P14B1WTR	CCGTGCGTTTTTGCCTCTCCTTCCATTTGCAATT TTGAC
	P14B2F	GATTAGACTATAATTAATTTTAATTATTCTGTGCA ATAT
	P14B2R	ATATTGCACAGAATAATTAAAATTAATTATAGTC TAATC
	P14B2WTF	GATTAGACTATGGTTGGTTTTAATGGTTCTGTGC AATAT
	P14B2WTR	ATATTGCACAGAACCATTAAAACCAACCATAGT CTAATC
	ASB1F	TTGTCAGTTATTAAAATACCTATAACGTATTTTA ATCACAATA
	ASB1R	TATTGTGATTAAAATACGTTATAGGTATTTTAATA ACTGACAA
	ASB2F	TTAATCACAATATGTATGCATGATTGTTCAAAAA ATTGTGATG
	ASB2R	CATCACAATTTTTTGAACAATCATGCATACATAT TGTGATTAA
	ASB3F	AAAATTGTGATGTTTCCATAAATTATAGGTACTG AACCTCAAT
	ASB3R	ATTGAGGTTCAGTACCTATAATTTATGGAAACAT CACAATTTT
	P14B1F	GTCAAAATTGCAAATTTAATTAGATTCAAAAAC GCACGG
	P14B1R	CCGTGCGTTTTTGAATCTAATTAAATTTGCAATT TTGAC
	P14B1WTF	GTCAAAATTGCAAATGGAAGGAGAGGCAAAA ACGCACGG
	P14B1WTR	CCGTGCGTTTTTGCCTCTCCTTCCATTTGCAATT TTGAC
	P14B2F	GATTAGACTATAATTAATTTTAATTATTCTGTGCA ATAT
	P14B2R	ATATTGCACAGAATAATTAAAATTAATTATAGTC TAATC
	P14B2WTF	GATTAGACTATGGTTGGTTTTAATGGTTCTGTGC AATAT
	P14B2WTR	ATATTGCACAGAACCATTAAAACCAACCATAGT CTAATC
	ASB1F	TTGTCAGTTATTAAAATACCTATAACGTATTTTA ATCACAATA
	ASB1R	TATTGTGATTAAAATACGTTATAGGTATTTTAATA ACTGACAA

(continued)

Chapter	Primer name	Primer sequence
	ASB2F	TTAATCACAATATGTATGCATGATTGTTCAAAAA ATTGTGATG
	ASB2R	CATCACAATTTTTTGAACAATCATGCATACATAT TGTGATTAA
	ASB3F	AAAATTGTGATGTTTCCATAAATTATAGGTACTG AACCTCAAT
	ASB3R	ATTGAGGTTCAGTACCTATAATTTATGGAAACAT CACAATTTT
6.7.3	c-Jun186-f	TCTCCKCACGTKGGKCTKCTCAARC
	c-Jun873-f	CAAGTAGCWGAGCTCAARCARAAAG
	c-Jun35-r	TCATCGTAGAAAGTAGTTTCCAT
	c-Jun489-r	GAGTGATGAAATGACGAGTTAGAATCATC
	c-Jun1152-f	AATGGCAATATTATGTCTCTTGTTAGCAC
	cjun-qf	CGGTGGCAAAACAAAACAGC
	cjun-qr	TGGGGCACGGGTAAAACTT
6.7.4	yy10-f	GGCGACACCCTCTACATTGCC
	yy95-f	CGACAGAAATCAAACATGGCGTC
	yy1391-r	CCTCGTATGTGTATCCTATTAACATGGC
	yy833-3f	TAGCAGAATTTGCTAGTCTCCAGCC
	yy1098-3f	TTTTCAGTGTACGTTTGAGGGATGC
	yy1251-3f	TATACTGACACATGCAAAAGCAAATAGC
	yy776-5r	AATTCCCCCTGGGGGAAGTTTTTTCC
	yy545-5r	CTCCCAAAGTTCCAGCCCCGAGAAG
	yy238-5r	TGGAAGCGGTTGTAACGCTATCATAGGTTG
	yy92-5r	CGACGCCATGTTTGATTTCTGTCGAAG
	yy-confirmf	AGCAGTGGTATCAACGCAGAG
	yy-confirmr	CCGATTCCCAAAACAGAGTAA
	yy-ef	CCATCGATATGGCGTCGGTTATCGGTTC
	yy-er	AACCTCGAGCTATGACTCCTCGTATGTGTATCCTATTAAC
	yy-sf	ATCCCAAACAGTTAGCAGA
	yy-sr	CTTACCGCATCCCTCAA
	β-actin-qf	CTCCTCACTGAAGCCCCCCTCA
	β-actin-qr	ATGGCTGGAATAGGGATTCTGG
	yyqf	AGAGGAAGATGCGTCAGTTTCAG
	yyqr	GTTTGGGATCACTAAGGTCAAGG
	P39-qf	CTGGAATGAGAGGATATG
	P39-qr	TGCTGCTGTAATAACTATA
	ACCBP-qf	GACATGGAACAAAGATGGTGGA
	ACCBP-qr	CTGTGGCTGGAATGGTTGG
	Pif-qf	TGCTGCCATCACGTGAGTATG
	Pif-qr	GACTTCCCTTTCTCACACTTCCA
	Aspein-qf	TGATAGTGAAGACGATGA
	Aspein-qr	TGTCATCATCATCATCATC
6.7.5	sp157-3f	CTAGCCGCTCAATGTAACAAGATC
	sp1311-3f	TACSTGGCAGACCATCCAGCT
	sp1324-3r	TGGATGGTCTGCCASGTAACTTG
	sp1318-3r	CASGTAACTTGGCCATTWGGRCC

(continued)

Chapter	Primer name	Primer sequence
	sp180-5r	GATCTTGTTACATTGAGCGGCTAGCATTG
	sp1705-5r	CATTGAGCGGCTAGCATTGCTAATGG
	sp-85r	AAGGATGGAACCCCTTCCCTACA
	sp-43r	GCGGGCTTTTGTTTGTGATCTTG
	spqf	CTCACGGCATCAAGTCTGAAAT
	spqr	CGGACTGCTGAACGGGTAAT
	Pif-qf	TGCTGCCATCACGTGAGTATG
	Pif-qr	GACTTCCCTTTCTCACACTTCCA
	aspein-qf	TGATAGTGAAGACGATGA
	aspein-qr	TGTCATCATCATCATCATC
	Pearlin-qf	TACTCATACTGCTGGATA
	Pearlin-qr	TATCATCATCGGTGTAAC
	KRMP-ef1	AAAGGTACCGTACATTTTTGACATATTTTTCACACTCTTTG
	KRMP-er1	AACCTCGAGTCTGAACCTGTAGAAAAATATTATTCATTAGATTC
	pearlin-ef	AAAGGTACCAAAATTACTGTGTCCAAAGG
	pearlin-er	AAACTCGAGAGACAAAAGATATTTCTCTTTAACC
	p39-ef	AAAGGTACCCCGTACTTCATCATTGTAGCTCCG
	p39-er	AACCTCGAGGACAATTTTGTAACCTGAAACAAAG
	spef	AAAGGTACCATGCTAGCCGCTCAATGTAACAAGA
	sper	GGATCGATTCACTTCATAGCACGGTTATCCCTG

References

1. G. Colaianni, G. Brunetti, M.F. Faienza, S. Colucci, M. Grano, Osteoporosis and obesity: Role of Wnt pathway in human and murine models. World J. Orthop **5**, 242 (2014)
2. G. Chen, C. Deng, Y.-P. Li, TGF-β and BMP signaling in osteoblast differentiation and bone formation. Int. J. Biol. Sci. **8**, 272 (2012)
3. A. Rao et al., Roles of larval sea urchin spicule SM50 domains in organic matrix self-assembly and calcium carbonate mineralization. J. Struct. Biol. **183**, 205–215 (2013)
4. T. Mass et al., Cloning and characterization of four novel coral acid-rich proteins that precipitate carbonates in vitro. Curr. Biol. **23**, 1126–1131 (2013)
5. Y. Morino et al., A genome-wide survey of genes encoding transcription factors in the Japanese pearl oyster, Pinctada fucata: I. Homeobox genes. Zool. Sci. **30**, 851–857 (2013)
6. H. Koga et al., A genome-wide survey of genes encoding transcription factors in Japanese pearl oyster Pinctada fucata: II. Tbx, Fox, Ets, HMG, NFκB, bZIP, and C2H2 zinc fingers. Zool. Sci. **30**, 858–867 (2013)
7. K. Kalavantavanich, C.M. Schramm, Dexamethasone potentiates high-affinity β-agonist binding and G s α protein expression in airway smooth muscle. Am. J. Phys. Lung Cell. Mol. Phys. **278**, L1101–L1106 (2000)
8. R. Sen, D. Baltimore, Inducibility of κ immunoglobulin enhancer-binding protein NF-κB by a posttranslational mechanism. Cell **47**, 921–928 (1986)
9. M. Grossmann, Y. Nakamura, R. Grumont, S. Gerondakis, New insights into the roles of ReL/NF-κB transcription factors in immune function, hemopoiesis and human disease. Int. J. Biochem. Cell Biol. **31**, 1209–1219 (1999)

10. A. Denk, T. Wirth, B. Baumann, NF-κB transcription factors: critical regulators of hematopoiesis and neuronal survival. Cytokine Growth Factor Rev. **11**, 303–320 (2000)
11. S. Bell et al., Involvement of NF-κB signalling in skin physiology and disease. Cell. Signal. **15**, 1–7 (2003)
12. M.S. Hayden, S. Ghosh, Signaling to NF-κB. Genes. Dev. **18**, 2195–2224 (2004)
13. S.T. Whiteside, A. Israël, in *Seminars in cancer biology* (Elsevier), pp. 75–82
14. B. Boyce, L. Xing, G. Franzoso, U. Siebenlist, Required and nonessential functions of nuclear factor-kappa B in bone cells. Bone **25**, 137–139 (1999)
15. A. Robaszkiewicz et al., ARTD1 regulates osteoclastogenesis and bone homeostasis by dampening NF-κB-dependent transcription of IL-1β. Sci. Rep. **6** (2016)
16. J.Q. Feng et al., NF-κB specifically activates BMP-2 gene expression in growth plate chondrocytes in vivo and in a chondrocyte cell line in vitro. J. Biol. Chem. **278**, 29130–29135 (2003)
17. S. Ghosh, M.J. May, E.B. Kopp, NF-κB and Rel proteins: Evolutionarily conserved mediators of immune responses. Annu. Rev. Immunol. **16**, 225–260 (1998)
18. T. Nakashima, J.M. Penninger, RANKL and RANK as novel therapeutic targets for arthritis. Curr. Opin. Rheumatol. **15**, 280–287 (2003)
19. D. Zhang et al., Molecular characterization and expression analysis of the IκB gene from pearl oyster Pinctada fucata. Fish Shellfish Immunol. **26**, 84–90 (2009)
20. H. Miyamoto et al., A carbonic anhydrase from the nacreous layer in oyster pearls. Proc. Natl. Acad. Sci. **93**, 9657–9660 (1996)
21. D. Chin, G.M. Boyle, P.G. Parsons, W.B. Coman, What is transforming growth factor-beta (TGF-β)? Br. J. Plast. Surg. **57**, 215–221 (2004)
22. K. Miyazono, S. Maeda, T. Imamura, BMP receptor signaling: Transcriptional targets, regulation of signals, and signaling cross-talk. Cytokine Growth Factor Rev. **16**, 251–263 (2005)
23. A. Herpin et al., Structural and functional evidence for a singular repertoire of BMP receptor signal transducing proteins in the lophotrochozoan Crassostrea gigas suggests a shared ancestral BMP/activin pathway. FEBS J. **272**, 3424–3440 (2005)
24. H. Le Quéré, A. Herpin, A. Huvet, C. Lelong, P. Favrel, Structural and functional characterizations of an Activin type II receptor orthologue from the pacific oyster Crassostrea gigas. Gene **436**, 101–107 (2009)
25. A. Herpin et al., Structural and functional evidences for a type 1 TGF-β sensu stricto receptor in the lophotrochozoan Crassostrea gigas suggest conserved molecular mechanisms controlling mesodermal patterning across bilateria. Mech. Dev. **122**, 695–705 (2005)
26. A. Herpin, P. Favrel, C. Cunningham, Gene structure and expression of cg-ALR1, a type I activin-like receptor from the bivalve mollusc Crassostrea gigas. Gene **301**, 21–30 (2002)
27. H. Guo et al., Molecular characterization of TGF-β type I receptor gene (Tgfbr1) in Chlamys farreri, and the association of allelic variants with growth traits. PLoS One **7**, e51005 (2012)
28. J. Massagué, TGF-β signal transduction. Annu. Rev. Biochem. **67**, 753–791 (1998)
29. S. Souchelnytskyi, P. Ten Dijke, K. Miyazono, C. Heldin, Phosphorylation of Ser165 in TGF-beta type I receptor modulates TGF-beta1-induced cellular responses. EMBO J. **15**, 6231 (1996)
30. T. Xie, A.L. Finelli, R.W. Padgett, The Drosophila saxophone gene: A serine-threonine kinase receptor of the TGF-beta superfamily. Science **263**, 1756–1760 (1994)
31. J.L. Wrana et al., Two distinct transmembrane serine/threonine kinases from Drosophila melanogaster form an activin receptor complex. Mol. Cell. Biol. **14**, 944–950 (1994)
32. Y.G. Chen, F. Liu, J. Massague, Mechanism of TGFβ receptor inhibition by FKBP12. EMBO J. **16**, 3866–3876 (1997)
33. C. Lelong, M. Mathieu, P. Favrel, Structure and expression of mGDF, a new member of the transforming growth factor-β superfamily in the bivalve mollusc Crassostrea gigas. FEBS J. **267**, 3986–3993 (2000)

34. Y.-G. Chen, J. Massagué, Smad1 recognition and activation by the ALK1 group of transforming growth factor-β family receptors. J. Biol. Chem. **274**, 3672–3677 (1999)
35. C. Deal, Future therapeutic targets in osteoporosis. Curr. Opin. Rheumatol. **21**, 380–385 (2009)
36. K.W. Finnson, W.L. Parker, P. ten Dijke, M. Thorikay, A. Philip, ALK1 opposes ALK5/Smad3 signaling and expression of extracellular matrix components in human chondrocytes. J. Bone Miner. Res. **23**, 896–906 (2008)
37. M.-F. Iu et al., Dexamethasone suppresses Smad3 pathway in osteoblastic cells. J. Endocrinol. **185**, 131–138 (2005)
38. D. Yang, Y. Ding, B. Ying, Effects of hydrogen peroxide on the expression of transforming growth factor-苗 1 and Smad3 in A549 cells. Chin. J. Respir. Crit. Care Med. **7**, 43–46 (2008)
39. T. Liu et al., BMP-2 promotes differentiation of osteoblasts and chondroblasts in Runx2-deficient cell lines. J. Cell. Physiol. **211**, 728–735 (2007)
40. C.J. Souza, C. MacDougall, B.K. Campbell, A.S. McNeilly, D.T. Baird, The Booroola (FecB) phenotype is associated with a mutation in the bone morphogenetic receptor type 1 B (BMPR1B) gene. J. Endocrinol. **169**, R1–R6 (2001)
41. S.E. Yi, A. Daluiski, R. Pederson, V. Rosen, K.M. Lyons, The type I BMP receptor BMPRIB is required for chondrogenesis in the mouse limb. Development **127**, 621–630 (2000)
42. A. Liu, L.A. Niswander, Bone morphogenetic protein signalling and vertebrate nervous system development. Nat. Rev. Neurosci. **6**, 945–954 (2005)
43. C.B. Gonzales, D. Simmons, M. MacDougall, Competing roles of tGFβ and nma/bAMbI in Odontoblasts. J. Dent. Res. **89**, 597–602 (2010)
44. Y. Zhang et al., A novel matrix protein participating in the nacre framework formation of pearl oyster, Pinctada fucata. Comp. Biochem. Physiol. B: Biochem. Mol. Biol. **135**, 565–573 (2003)
45. H. Miyamoto, F. Miyoshi, J. Kohno, The carbonic anhydrase domain protein Nacrein is expressed in the epithelial cells of the mantle and acts as a negative regulator in calcification in the Mollusc Pinctada fucata. Zool. Sci. **22**, 311–315 (2005). https://doi.org/10.2108/zsj.22.311
46. M. Suzuki et al., Characterization of Prismalin-14, a novel matrix protein from the prismatic layer of the Japanese pearl oyster (Pinctada fucata). Biochem. J. **382**, 205–213 (2004)
47. C. Richardson, N. Runham, D. Crisp, A histological and ultrastructural study of the cells of the mantle edge of a marine bivalve, Cerastoderma edule. Tissue Cell **13**, 715–730 (1981)
48. T. Ikeda et al., Distinct roles of Sox5, Sox6, and Sox9 in different stages of chondrogenic differentiation. J. Bone Miner. Metab. **23**, 337–340 (2005)
49. T. Furumatsu, T. Ozaki, H. Asahara, Smad3 activates the Sox9-dependent transcription on chromatin. Int. J. Biochem. Cell Biol. **41**, 1198–1204 (2009)
50. J.D. Peacock, D.J. Huk, H.N. Ediriweera, J. Lincoln, Sox9 transcriptionally represses Spp1 to prevent matrix mineralization in maturing heart valves and chondrocytes. PLoS One **6**, e26769 (2011)
51. Y.-J. Luo et al., The Lingula genome provides insights into brachiopod evolution and the origin of phosphate biomineralization. Nat. Commun. **6**, 8301 (2015)
52. F.H. Wilt, Developmental biology meets materials science: morphogenesis of biomineralized structures. Dev. Biol. **280**, 15–25 (2005)
53. T.R. Waller, Functional morphology and development of veliger larvae of European oyster, Ostrea edulis Linne. Smithsonian Contrib. Zool. **328**, 1–70 (1981)
54. R.T. Moon, A.D. Kohn, G.V. De Ferrari, A. Kaykas, WNT and beta-catenin signalling: diseases and therapies. Nat. Rev. Genet. **5**, 691–701 (2004). https://doi.org/10.1038/nrg1427
55. T.P. Rao, M. Kuhl, An updated overview on Wnt signaling pathways: a prelude for more. Circ. Res. **106**, 1798–1806 (2010). https://doi.org/10.1161/CIRCRESAHA.110.219840
56. K.A. Tompkins, Wnt proteins in mineralized tissue development and homeostasis. Connect. Tissue Res. **52**, 448–458 (2011). https://doi.org/10.3109/03008207.2011.616610

57. P.D. McCrea, C.W. Turck, B. Gumbiner, A homolog of the armadillo protein in drosophila (plakoglobin) associated with e-cadherin. *Science* **254**, 1359–1361 (1991). https://doi.org/10.1126/science.1962194
58. M. Adamska et al., Structure and expression of conserved Wnt pathway components in the demosponge Amphimedon queenslandica. Evol. Dev. **12**, 494–518 (2010). https://doi.org/10.1111/j.1525-142X.2010.00435.x
59. C.D. Galetto, M.F. Izaguirre, V. Bessone, V.H. Casco, Isolation and nucleotide sequence analysis of the of Rhinella arenarum beta-catenin: an mRNA and protein expression study during the larval stages of the digestive tract development. Gene **511**, 256–264 (2012). https://doi.org/10.1016/j.gene.2012.09.030
60. F. Nollet, G. Berx, F. Molemans, R.F. Van, Genomic organization of the human beta-catenin gene (CTNNB1). Genomics **32**, 413 (1996)
61. A.H. Huber, W.I. Weis, The structure of the beta-catenin/E-cadherin complex and the molecular basis of diverse ligand recognition by beta-catenin. Cell **105**, 391–402 (2001). https://doi.org/10.1016/s0092-8674(01)00330-0
62. C.M. Liu et al., Control of beta-catenin phosphorylation/degradation by a dual-kinase mechanism. Cell **108**, 837–847 (2002). https://doi.org/10.1016/s0092-8674(02)00685-2
63. H.E. Weitzel et al., Differential stability of beta-catenin along the animal-vegetal axis of the sea urchin embryo mediated by dishevelled. Development **131**, 2947–2956 (2004). https://doi.org/10.1242/dev.01152
64. N. Zhong, R.P. Gersch, M. Hadjiargyrou, Wnt signaling activation during bone regeneration and the role of Dishevelled in chondrocyte proliferation and differentiation. Bone **39**, 5–16 (2006). https://doi.org/10.1016/j.bone.2005.12.008
65. C.C. Malbon, H.Y. Wang, Dishevelled: a mobile scaffold catalyzing development. Curr. Top. Dev. Biol. **72**(72), 153 (2006). https://doi.org/10.1016/s0070-2153(05)72002-0
66. W. Kim, M. Kim, E.H. Jho, Wnt/beta-catenin signalling: from plasma membrane to nucleus. Biochem. J. **450**, 9–21 (2013). https://doi.org/10.1042/BJ20121284
67. M. Molenaar et al., XTcf-3 transcription factor mediates beta-catenin-induced axis formation in Xenopus embryos. Cell **86**, 391–399 (1996). https://doi.org/10.1016/s0092-8674(00)80112-9
68. T. Kubota, T. Michigami, K. Ozono, Wnt signaling in bone metabolism. J. Bone Miner. Metab. **27**, 265–271 (2009). https://doi.org/10.1007/s00774-009-0064-8
69. T.A. Graham, C. Weaver, F. Mao, D. Kimelman, W.Q. Xu, Crystal structure of a beta-catenin/Tcf complex. Cell **103**, 885–896 (2000). https://doi.org/10.1016/s0092-8674(00)00192-6
70. H.E. Hamm, A. Gilchrist, Heterotrimeric G proteins. Curr. Opin. Cell Biol. **8**, 189–196 (1996)
71. J.H. Hurley, Structure, mechanism, and regulation of mammalian adenylyl cyclase. J. Biol. Chem. **274**, 7599–7602 (1999)
72. S.R. Sprang, G Protein mechanisms: insights from structural analysis. Annu. Rev. Biochem. **66**, 639–678 (1997)
73. L. De Vries, B. Zheng, T. Fischer, E. Elenko, M.G. Farquhar, The regulator of G protein signaling family. Annu. Rev. Pharmacol. Toxicol. **40**, 235–271 (2000)
74. D.G. Lambright, J.P. Noel, H.E. Hamm, P.B. Sigler, Structural determinants for activation of the α-subunit of a heterotrimeric G protein. Nature **369**, 621–628 (1994)
75. A. Checa, A new model for periostracum and shell formation in Unionidae (Bivalvia, Mollusca). Tissue Cell **32**, 405–416 (2000). https://doi.org/10.1054/tice.2000.0129
76. F.-H. Chang, H. Bourne, Cholera toxin induces cAMP-independent degradation of Gs. J. Biol. Chem. **264**, 5352–5357 (1989)
77. L. Hessle et al., Tissue-nonspecific alkaline phosphatase and plasma cell membrane glycoprotein-1 are central antagonistic regulators of bone mineralization. Proc. Natl. Acad. Sci. **99**, 9445–9449 (2002)
78. J. Guicheux et al., Activation of p38 mitogen-activated protein kinase and c-Jun-NH2-terminal kinase by BMP-2 and their implication in the stimulation of osteoblastic cell differentiation. J. Bone Miner. Res. **18**, 2060–2068 (2003)

79. J. Sondek, A. Bohm, D. Lambright, H. Hamm, P. Sigler, Crystal structure of a G-protein bg dimer at 2.1 AĘ resolution. Nature **379** (1996)
80. A. Ito, T. Satoh, Y. Kaziro, H. Itoh, G protein βγ subunit activates Ras, Raf, and MAP kinase in HEK 293 cells. FEBS Lett. **368**, 183–187 (1995)
81. T. Ivanina, Y. Blumenstein, E. Shistik, R. Barzilai, N. Dascal, Modulation of L-type Ca2+ channels by Gβγ and calmodulin via interactions with N and C termini of α1C. J. Biol. Chem. **275**, 39846–39854 (2000)
82. M. Kondo, In vitro effect of chlorpromazine on the mineralization of tooth germ in mice--comparison with that of retinoic acid and HEBP. Nihon Yakurigaku Zasshi. **97**, 85–95 (1991)
83. M. Zhao, M. He, X. Huang, Q. Wang, A homeodomain transcription factor gene, PfMSX, activates expression of Pif gene in the pearl oyster Pinctada fucata. PLoS One **9**, e103830 (2014)
84. W. Herr et al., The POU domain: a large conserved region in the mammalian pit-1, oct-1, oct-2 and Caenorhabditis elegans unc-86 gene products. Genes Dev. **2**, 1513–1516 (1988)
85. D.A. Gold, R.D. Gates, D.K. Jacobs, The early expansion and evolutionary dynamics of POU class genes. Mol. Biol. Evol. **31**, 3136–3147 (2014)
86. D. Tantin, Oct transcription factors in development and stem cells: insights and mechanisms. Development **140**, 2857–2866 (2013)
87. B. Andersen, M.G. Rosenfeld, POU domain factors in the neuroendocrine system: lessons from developmental biology provide insights into human disease. Endocr. Rev. **22**, 2–35 (2001)
88. S. Raft, T.M. Coate, M.W. Kelley, E.B. Crenshaw III, D.K. Wu, Pou3f4-mediated regulation of ephrin-B2 controls temporal bone development in the mouse. PLoS One **9**, e109043 (2014)
89. J.D. Klemm, M.A. Rould, R. Aurora, W. Herr, C.O. Pabo, Crystal structure of the Oct-1 POU domain bound to an octamer site: DNA recognition with tethered DNA-binding modules. Cell **77**, 21–32 (1994)
90. W. Herr, M.A. Cleary, The POU domain: versatility in transcriptional regulation by a flexible two-in-one DNA-binding domain. Genes Dev. **9**, 1679–1693 (1995)
91. T. Curran, B.R. Franza, Fos and Jun: the AP-1 connection. Cell **55**, 395–397 (1988)
92. J. Hess, P. Angel, M. Schorpp-Kistner, AP-1 subunits: quarrel and harmony among siblings. J. Cell Sci. **117**, 5965–5973 (2004)
93. J.-P. David, K. Sabapathy, O. Hoffmann, M.H. Idarraga, E.F. Wagner, JNK1 modulates osteoclastogenesis through both c-Jun phosphorylation-dependent and-independent mechanisms. J. Cell Sci. **115**, 4317–4325 (2002)
94. Y.-H. Zhang, A. Heulsmann, M.M. Tondravi, A. Mukherjee, Y. Abu-Amer, Tumor necrosis factor-α (TNF) stimulates RANKL-induced osteoclastogenesis via coupling of TNF type 1 receptor and RANK signaling pathways. J. Biol. Chem. **276**, 563–568 (2001)
95. L. Bakiri et al., Role of heterodimerization of c-Fos and Fra1 proteins in osteoclast differentiation. Bone **40**, 867–875 (2007)
96. T. Miyashita, A. Takami, R. Takagi, Molecular cloning and characterization of the 5′-flanking regulatory region of the carbonic anhydrase nacrein gene of the pearl oyster Pinctada fucata and its expression. Biochem. Genet. **50**, 673–683 (2012)
97. X. Zhao et al., Identification of genes potentially related to biomineralization and immunity by transcriptome analysis of pearl sac in pearl oyster Pinctada martensii. Mar. Biotechnol. **14**, 730–739 (2012)
98. S. Gordon, G. Akopyan, H. Garban, B. Bonavida, Transcription factor YY1: structure, function, and therapeutic implications in cancer biology. Oncogene **25**, 1125 (2006)
99. Z. Deng, P. Cao, M.M. Wan, G. Sui, Yin Yang 1: a multifaceted protein beyond a transcription factor. Transcription **1**, 81–84 (2010)
100. F.B. Riquet et al., YY1 is a positive regulator of transcription of theCol1a1 gene. J. Biol. Chem. **276**, 38665–38672 (2001)

101. T. Aoyama et al., Histone modifiers, YY1 and p300, regulate the expression of cartilage-specific gene, chondromodulin-I, in mesenchymal stem cells. J. Biol. Chem. **285**, 29842–29850 (2010)
102. H.M. Jeong, Y.H. Choi, S.H. Lee, K.Y. Lee, YY1 represses the transcriptional activity of Runx2 in C2C12 cells. Mol. Cell. Endocrinol. **383**, 103–110 (2014)
103. M. Yao, G. Niu, Z. Sheng, Z. Wang, J. Fei, Identification of a smad4/yy1-recognizedand bmp2-responsive transcriptional regulatorymodule in the promoter of mouse gaba transporter subtype i (gat1) gene. J. Neurosci. **30**, 4062–4071 (2010)
104. F. Romeo et al., Negative transcriptional regulation of the human periostin gene by YingYang-1 transcription factor. Gene **487**, 129–134 (2011)
105. J.T. Kadonaga, K.R. Carner, F.R. Masiarz, R. Tjian, Isolation of cDNA encoding transcription factor Sp1 and functional analysis of the DNA binding domain. Cell **51**, 1079–1090 (1987)
106. S.M. Bell et al., Sp8 is crucial for limb outgrowth and neuropore closure. Proc. Natl. Acad. Sci. **100**, 12195–12200 (2003)
107. N.D. Schaeper, N.-M. Prpic, E.A. Wimmer, A conserved function of the zinc finger transcription factor Sp8/9 in allometric appendage growth in the milkweed bug Oncopeltus fasciatus. Dev. Genes Evol. **219**, 427 (2009)
108. T. Takeuchi, I. Sarashina, M. Iijima, K. Endo, In vitro regulation of CaCO 3 crystal polymorphism by the highly acidic molluscan shell protein Aspein. FEBS Lett. **582**, 591–596 (2008)
109. T. Samata et al., A new matrix protein family related to the nacreous layer formation of Pinctada fucata. FEBS Lett. **462**, 225–229 (1999)
110. M. Suzuki et al., An acidic matrix protein, Pif, is a key macromolecule for nacre formation. Science **325**, 1388–1390 (2009). https://doi.org/10.1126/science.1173793

Chapter 7
Ecological Study on Biomineralization in *Pinctada fucata*

Abstract *Pinctada fucata* inhabits in tropical and subtropical seas, so the sea water conditions have great impacts on its growth and shell formation. To determine the potential effects of global climate change and the water pollution on the oyster growth in the near future, we have conducted a series of studies about the metabolism and shell formation of *P. fucata* under varied stressors. Various factors controlling biomineralization, such as genes, environmental factors, and ions that ocean acidification (OA), ocean warming (OW), and metal pollution have great effects on the metabolism, biomineralization process, as well as the immune defense, were pointed by our study. It was found that, during short-term exposure, the impact of CO_2 and temperature stresses was not manifested in the shell ultrastructure, although it affected the process of biomineralization. If the stressors are present in the long-term, they will have adverse results on the biomineralization of pearls, most likely affecting pearl quality, which results in substantial economic losses for the aquaculture industry. In addition, OA and OW have a great impact on the physiological conditions of hemocytes, including altering pH value of hemolymph and increasing the total hemocyte count, total protein content, and percentage of large hyalinocytes and granulocytes, while decreasing phagocytosis ability. We query the metal accumulations and the corresponding enzymatic responses in the pearl oyster *Pinctada fucata* after copper exposure. We found that the gills and the digestive gland played different roles associated with distinct copper concentrations and observed the adaptation and recovery of the oysters. Our studies have indicated potential implications for predictions of the effects of environmental changes on pearl aquaculture. Moreover, it also provides theoretical basis for precautions to avoid adverse impact of environmental challenges on the marine pearl aquaculture in the near future.

Keywords Ocean acidification · Ocean warming · Copper · Mantle · Hemocyte

R. Zhang et al., *Biomineralization Mechanism of the Pearl Oyster, Pinctada fucata*,
https://doi.org/10.1007/978-981-13-1459-9_7

7.1 Introduction

Because *Pinctada fucata* inhabits in tropical and subtropical seas, the water conditions have great impacts on its growth and shell formation. Therefore, the success of the pearl aquaculture depends largely on the finely monitored environmental parameters, such as pH, temperature, and salinity. In the South China Sea (SCS), the pH and temperature of the surface seawater fluctuate from a maximum of 7.6 and 30 °C in summer to a minimum of 8.1 and 19 °C in winter [1, 2], and the large fluctuations in seawater pH and temperature impose demands on the ability of pearl oysters to adapt to these changes [3]. Beside the natural periodic changes in temperature and pH, *P. fucata* is also encountered by the detrimental effects of global warming. The increasing carbon dioxide (CO_2) levels, mostly anthropogenically derived, has great impacts on the seawater conditions of the whole world [4]. The increased partial pressure of CO_2 (pCO_2) has changed the dissolved inorganic carbon (DIC, the sum of inorganic carbon species in seawater), and then the pH value of surface seawater decreased, which would lead to ocean acidification (OA) [5]. What's more, ocean warming emerges along with the OA. Since year 1750, a 30% increase in pCO_2 has reduced the pH of surface seawater to 8.08 on average (the preindustrial value was 8.18). Further pH decreases of 0.3–0.5 unit concomitant with seawater warming of 2–6 °C will occur by the end of this century [6, 7]. OA and ocean warming have the potential to affect the physiology and behavior of a range of organisms in marine ecosystems, causing by the shift in seawater chemistry and carbonate saturation [8–10]. Importantly, the effects of concurrent OA and ocean warming on marine calcifiers differ among species (adverse and/or beneficial) [11–13]. Given the interdependence among climate change stressors, their combined effects on marine calcifiers are complex and remain poorly understood. Thus, there is a critical need for a comprehensive understanding of the strategies used by pearl oysters to adapt to these changes. This is particularly important in the SCS, which is the center for pearl production in China and worldwide.

Increasing evidence has indicated that OA and ocean warming does not act in isolation; rather, they usually interact with marine pollutants [14], salinity [13], and light [15]. Environmental poisoning by metals has increased in the last decades due to extensive use of metals in agricultural, chemical, and industrial processes that are becoming threats to living organisms [16, 17]. Copper is an essential element for bivalve development, but at elevated concentrations it is particularly toxic [18, 19]. As a major marine pollutant, copper induces different responses at different levels in the bivalves [19–22]. To determine the effect of these pollutants on bivalves, several strategies have been used [23]. In recent years, there has been a rapid development of enzyme biomarkers [24], and the changes of the key enzymatic activities have been employed in many bivalves [25–27]. The pearls produced by *P. fucata* contain trace metal ions, and the color of the pearls is due to trace metal elements in the oyster's environment. However, the safety of the pearls produced by oysters in metal polluted environment was still one problem. Also, some parts of soft tissue such as adductor muscle have been processed for human consumption

[28]. The metals accumulated in them are potential poisons to the human health [28, 29]. Also, oysters as well as other bivalve mollusks have been regarded as ideal indicator organisms for the assessment of environmental pollution [30–32]. Hence the correlations between metals and pearl oysters need to be well studied.

The access to the mantle transcriptomes and the draft genome facilitates the study on the response mechanisms of these species to environmental stressors [33]. In this chapter, we will focus on the impact of environmental parameters (pH, temperature, and copper exposure) on the biomineralization in *P. fucata* and provide some suggestions to improve the pearl aquaculture in the future.

7.2 Materials and Methods

7.2.1 Chemicals

Ethylenediaminetetraacetic acid (EDTA), p-nitrophenylphosphate (p-NPP), dithiothreitol (DTT), and NaN_3 were purchased from Merck. Hypoxanthine, cytochrome c, reduced glutathione (GSH), β-nicotinamide adenine dinucleotide phosphate (NADPH), L-3,4-dihydroxyphenylalanine (L-DOPA), xanthine oxidase, and glutathione reductase were Sigma products. Sephadex G-25 was supplied by Pharmacia. All other reagents were local products of analytical grade.

7.2.2 Primers

Refer to Supplementary Table 7.1 for details.

7.2.3 Experimental Design for OA and OW Treatments

Two individual experiments designs were applied to investigate acute short-term and mild long-term stressors on the oysters, respectively. Prior to the experiment, *P. fucata* was cultured at 19.0 ± 0.5 °C, pH 8.1 ± 0.05, and salinity 33.0 ± 0.5 psu in the artificial seawater (Formula Grade A Reef Sea Salt, Formula, Japan) for acclimation. The experiments were conducted in 50 L seawater circulation tanks with maintained seawater temperature and pH levels by automatically heating the seawater and continuously pumping a CO_2-gas mixture into the tanks. The oysters were daily fed with the microalgae *Platymonas subcordiformis*.

For the short-term experiments, oysters of similar size (6–7 cm length) were subjected to varied CO_2 and temperature stressors. The stresses of mimicked CO_2

and temperature for the seasonal changes in the SCS and the predicted levels were set for the years 2100 (pH 7.8) and 2300 (pH 7.5) and 2100 (medium level at 25 °C) and 2300 (high level at 31 °C). The oysters were cultured for 72 h under the following conditions: 19.0 °C, pH 7.8, and salinity 33.0 psu (group denoted P78); 19.0 °C, pH 7.5, and salinity 33.0 psu (P75); 25.0 °C, pH 8.1, and salinity 33.0 psu (T25); and 31.0 °C, pH 8.1, and salinity 33.0 psu (T31). The control group was kept in 50 L tanks at 19.0 °C, pH 8.1, and salinity 33.0 psu. Each treatment was carried out with three replicates in three independent tanks, and each tank contained 30 individuals.

Mantle tissues from three randomly selected individuals in each tank were merged to obtain one biological replicate. For physiological and molecular analysis, mantles in the control and treatment groups were sampled at 0, 0.5, 2, 6, 12, 24, 48, and 72 h, washed with sterilized seawater, and tested immediately. For shell analyses, the shells in the control and treatment groups were collected at 0, 24, 48, and 72 h, washed with deionized water. Then the organic components were removed by immersing in 5% sodium hydroxide for 12 h to be attached to the inner surface. The shells were then washed thoroughly with deionized water, air-dried, and stored in a desiccator before used.

For the long-term experiments, two pH levels (pH 8.1 and pH 7.8) and two temperatures (23.0 and 28.0 °C) were designed to mimic the predicted levels for the ocean scenarios by the year 2100 (IPCC, 2007). These values also fall within the natural seawater pH and temperature fluctuations of the sampling sea. Pearl oysters exposed to these conditions were divided into four groups: 28.0 °C, pH 7.8, and salinity 33.0 psu (28.0 °C × pH 7.8); 28.0 °C, pH 8.1, and salinity 33.0 psu (28.0 °C × pH 8.1); 23.0 °C, pH 7.8, and salinity 33.0 psu (23.0 °C × pH 7.8); and 23.0 °C, pH 8.1, and salinity 33.0 psu (23.0 °C × pH 8.1). The ambient condition (23.0 °C × pH 8.1) served as the control. These treatments represented the individual and interactive effects of temperature and pH on pearl oysters. Prior to the experiment, the pearl oysters were acclimatized to the experimental conditions by gradually increasing the seawater temperature from 23 to 28 °C (1 $°C{\cdot}d^{-1}$) and gradually decreasing the pH from 8.1 to 7.8 (0.05 $units{\cdot}d^{-1}$). All treatments (including the control group) were performed for 60 days. Each treatment comprised three biological replicates in three independent tanks, and each tank contained 30 individuals. For the details of the experimental treatments, please see *Shiguo Li* et al.'s primitive work [34].

The temperature and pH value (NBS scale) were continuously monitored in all tanks and recorded twice a day. For total alkalinity (TA) determination, 25 mL seawater samples were filtered through a 0.45 μm Millipore filter and tested with a Gran titration on with a Kloehn digital syringe pump (Kloehn, Las Vegas, Nevada, with a precision of 0.1%). Total dissolved inorganic carbon (DIC) was measured using a DIC analyzer (AS-C3, Apollo SciTech Inc., GA) on the filtered seawater samples. Temperature, pH, salinity, DIC, and TA with the dissociation constants K_1, K_2, and K_{SO4} were applied to calculate the partial pressure of carbon dioxide (pCO_2) and other carbonate system parameters using the software CO2SYS [35].

7.2.4 *Transcriptome Sequencing*

Ten pearl oysters were randomly sampled from each tank. Equal mantles (including edge, pallial, and center zone) were cut from each individual, and one sample mixture contained mantles collected from ten oysters. The mantle mixtures from four experiments were washed with RNase-free water and cryopreserved in liquid nitrogen immediately. Total RNA was extracted using the TRIZOL Reagent (Life Technologies, Carlsbad, CA) according to the manufacturer's directions. The concentration and quality of all the RNA samples were measured using a NanoDrop spectrophotometer (ND-1000, Wilmington, DE) and 1.0% formaldehyde-denatured agarose gel electrophoresis, respectively. Transcriptome sequencing, sequence assembly, and cDNA library construction were accomplished by BGI-Shenzhen (Shenzhen, China) using Illumina HiSeq 2000 (Illumina, Inc.) applying routine protocols that are suitable for pearl oyster [36]. The transcriptome data could be accessed in the Sequence Read Archive (SRA) at the National Center for Biotechnology Information (NCBI), and the accession number is SRP064812. The sequences were aligned to the nucleotide (nt), non-redundant (nr), and Swiss-Prot databases ($e < 1.00 \times 10^{-5}$) in NCBI and annotated with known nucleotide and protein sequences with the highest homology to mollusks and mammals by Blastn and Blastx.

To identify differentially expressed unigenes (DEGs) between the treatments and the control, the RPKM (reads per kilobase of transcript per million uniquely mapped reads) values were determined using MA plot-based methods using a random sampling model (MARS) in the DEGseq R package [37]. The representative DEGs (Benjamini–Hochberg correction) were estimated by measuring the false discovery rate (FDR)-adjusted p value. Those DEGs meet the criteria of a fold change >2 and $p < 0.001$. The DEGs were submitted to the Kyoto Encyclopedia of Genes and Genomes (KEGG) database for analyzing metabolic pathway enrichment ($e < 1.00 \times 10^{-5}$). The Bidirectional Best-Hit (BBH) method was applied to query the set of organism representative for eukaryotes with default settings. Fisher's exact test was used to determine DEG enrichment. The most representative pathways were identified with an FDR-adjusted p value <0.001. To determine the enrichment of functional categories, gene ontology (GO) annotations were performed with the Blast2GO program with default parameters ($e < 1.00 \times 10^{-5}$) [38]. In addition, specific DEG categories (including ion and acid–base regulation, thermal stress responses, and biomineralization) were also artificially added to study the responses of pearl oysters to seawater environment stressors based on previous studies [39–41].

To confirm the differentially expressed genes obtained from transcriptome sequencing, ten genes (five upregulated and five downregulated) in each treatment were tested by real-time quantitative PCR (RT-qPCR). The degree of consistency between the RT-qPCR and transcriptome sequencing was calculated by least-squares linear regression [42]. For the full details regarding gene selection, primer design, and RT-qPCR, please see *Shiguo Li* et al.'s primitive work [43].

7.2.5 Net Calcification Rate (NCR)

To measure the changes in calcification in *P. fucata* exposed to the temperature and pH stressors, the alkalinity anomaly technique [44] was used to determine the NCR. The NCR was estimated using the variations in TA values during the 60-day experiment and compared between each treatment and control group at each time point. The NCR was denoted as μmol $CaCO_3 \cdot g^{-1}\ d^{-1}$.

7.2.6 ALP Activity Measurement

Mantle tissue of 0.1 g was sampled to measure ALP activity using the colorimetric ALP assay kit (Beyotime Institute of Biotechnology, Haimen, China) according to the manufacturer's instructions. After adding p-nitrophenol phosphate to each sample and 10-min incubation at 37 °C, OD_{405} was measured using an Ultrospec 7000 (GE Healthcare Life Sciences, Uppsala, Sweden). The p-nitrophenol concentrations of 0.02, 0.04, 0.08, 0.12, and 0.16 mM were set to draw the standard curve. To validate the stability of the control, ALP activities of the control group at 0.5, 2, 6, 12, 24, 48, and 72 h were compared to that at 0 h. To eliminate the interference of other factors, ALP activities in T25, T31, P78, and P75 were compared with that of the control at corresponding time point of 0.5, 2, 6, 12, 24, 48, and 72 h.

7.2.7 X-ray Photoelectron Spectroscopy (XPS)

To measure shell-surface $CaCO_3$ deposition in *Pinctada fucata* exposed to the varied treatments, shells from three oysters in each tank were randomly selected and pooled to obtain one mixture. The shells were washed with deionized water and treated with 5% sodium hydroxide for 12 h to remove organic components on the shell surface. The shells were then washed thoroughly with deionized water and air-dried. Three shell samples (around 7 mm × 7 mm) were cut from pearl oysters in each treatment and control group. To avoid individual variation across oysters, three regions covering the nacre–prism transition region were observed for each shell sample. To measure the calcium and carbon contents of the nacreous and prismatic layers on the inner shell surface near the nacre–prism transition zone, XPS (ESCALAB 250Xi, Thermo Scientific, USA) with an incident angle of 45°and monochromatic A1 Ka radiation (1486.7 eV) was applied. The adventitious C (1 s) peak at 285 eV was used to charge-correct to all the binding energies, and the measurement accuracy of the binding energy was 0.2 eV. High-resolution C (1 s) and Ca (2p) spectra were obtained at a pass energy of 50 eV. The signal intensities of the elements and sensitivity factors were used to estimate the relative atomic concentration ratio of the elements on the inner shell surface by Advantage V4 (Thermo-VG Scientific, Sussex, England). The element content was revealed as the peak area.

7.2.8 Scanning Electron Microscopy (SEM)

See in Sect. 5.2.20.

7.2.9 Liquid Chromatography–Tandem Mass Spectrometry (LC-MS/MS)

The amino acid composition of the mantle in *P. fucata* was measured by LC-MS/MS. The selected amino acids were based on the enriched pathway results: methionine (Met), cysteine (Cys), alanine (Ala), phenylalanine (Phe), and tyrosine (Tyr). Moreover, arginine (Arg), asparagine (Asn), glycine (Gly), glutamic acid (Glu), and glutamine (Gln) were also chosen as reference amino acids.

7.2.10 Microarray Experiment

A 58,583-feature microarray representing 58,583 unigenes was conducted based on the mantle transcriptome of *Pinctada fucata*. The resulted data have been submitted to Gene Expression Omnibus (http://www.ncbi.nlm.nih.gov/geo/; GSE57171) and Dryad (doi:https://doi.org/10.5061/dryad.958 sr). For detailed methods of RT-qPCR validation and the microarray, please see the primitive work of Li, Shiguo et al. [43].

To identify differentially expressed unigenes, the normalized signal values of four treatments (P78 P75, T25, and T31) with the control at the time point 72 h were compared using unpaired student's t-tests. The false discovery rate (FDR)-adjusted p value was applied to determine the representative DEGs using the Benjamini–Hochberg method. Unigenes with fold change >2 and $p < 0.05$ were regarded as DEGs. All DEGs in the four treatments were collected together, and then Cluster 3.0 and Java TreeView software were used to analyze the hierarchical clustering of the corresponding normalized signal values in T25, T31, P78, and P75, as well as the control. The changes of unigenes in T25, T31, P78, and P75 were measured by comparing the normalized signal values with that in the control group. To identify the underlying pathways related to elevated temperature and CO_2 concentration, Kyoto Encyclopedia of Genes and Genomes (KEGG) pathway enrichment analyses of the DEGs in the four treatment groups (T25, T31, P78 and P75) were conducted in the pathway database (http://www.genome.jp/kegg/). Fisher's exact tests were used to analyze DEG enrichment. The FDR-adjusted p value was applied to determine the most representative pathways using the Benjamini–Hochberg method with $p < 0.05$.

Besides KEGG pathways, specific categories were classified artificially based on previous studies [41, 45], in which gene enrichment analyses for microarray or transcriptome data were successfully employed to study the responses of marine animals to the seawater environment changes. The genes in these reported functional categories were searched against our microarray results in the four treatments (T25, T31, P78, and P75), and the fold changes of the matched-gene expression were listed individually. Genes with fold change >2 and $p < 0.05$ were considered to be differentially expressed. The category enriched in these genes was regarded as the principal category.

7.2.11 Hemolymph Collection

See in Sect. 5.2.22.

7.2.12 Determination of the pH Value in Extrapallial Fluid and Hemolymph

The extrapallial fluid from ten individuals in each tank was extracted with a 1 mL syringe and merged into a sterile centrifuge tube as one replicate. Each treatment has three replicates. The pH values of the hemolymph and extrapallial fluid samples were determined by the pH controller as previously described.

7.2.13 Calcium Contents in Hemocyte and Hemolymph

Fluo-3/acetoxymethyl ester (Fluo-3 AM) was used to stain the cytoplasmic free calcium (Ca^{2+}) of the hemocyte samples, and the quantitative analysis was accomplished by flow cytometry. The pooled hemolymph samples (300 μL) were centrifuged at 3000 g for 10 min to collect the cell pellets. The pellets were then washed twice for 5 min in phosphate-buffered saline (PBS) and resuspended in 200 μL PBS containing 2 μM fluo-3 AM, and the mixtures were kept in the dark for 15 min at room temperature. Then, the cells were collected at 3000 g for 10 min. The pellets were washed twice for 5 min in PBS and resuspended in 200 μL PBS. Then a flow cytometer was applied to analyze the stained hemocytes using the FL1 (green fluorescence) parameter following Aton et al. [46] The cytoplasmic free calcium was represented as the fluorescence intensity. The calcium concentration of the hemolymph samples was determined using an inductively coupled plasma-mass spectrophotometer (ICP-MS 7900, Agilent Technologies, CA, USA) according to Fernandez-Turiel et al. [47].

7.2.14 Flow Cytometry Analysis

See in Sect. 5.2.23.

7.2.15 Copper Exposure

Pearl oysters were randomly divided into three groups of 80 individuals each. Two groups were treated with copper ($CuSO_4$) of 0.05 μM and 0.5 μM, respectively. The third group was used as a control. The oysters were kept in 50 L tanks at constant temperature of 20–22 °C and salinity of 30%. Seawater was renewed daily [20] and the oysters were not fed during the experiments [19].

7.2.16 Sample Preparations, Metal Accumulation, and Biochemical Assays

At different exposure time point (12, 24, 48, and 72 h), ten individuals from each group were sacrificed, and digestive glands and gills were collected immediately. After being washed with cold sterile water, tissues were accurately weighed and homogenized in two volumes of cold 20 mM Tris–HCl buffer (pH 7.1), containing 0.5 mM phenylmethylsulfonyl fluoride (PMSF). Then the homogenate was centrifuged at 12000 × g for 30 min, and the supernatant was transferred and used for subsequent biochemical analysis. To measure superoxide dismutase (SOD) activity, the supernatant was purified with Sephadex G-25 to remove small molecules that could interfere the assay. All the steps were performed at 4 °C.

After exposure for 12, 24, 48, and 72 h, six individuals from each group were sacrificed to collect digestive glands and gills. After washed with cold sterile water, tissues were dehydrated to constant mass at 80 °C in a drying oven. Samples were dissolved in concentrated nitric acid at 100 °C until mineralized [48]. Copper concentrations in the samples were measured by inductively coupled plasma atomic emission spectrometry (Varian, Vista-MPX) [48].

To assay acid phosphatase (AcPase, EC 3.1.3.2) activity, the method [49] was modified using *p*-NPP as substrate. The samples were incubated at 37 °C for 20 min, after adding 200 μL 20 mM Na-acetate buffer (pH 5.0) containing 2 mM *p*-NPP. The reaction was terminated by adding 100 μL 1 M NaOH, and the amount of generated *p*-nitrophenol was calculated by OD_{405}. One enzyme activity unit is defined as the enzyme amount generating the reaction product by 1 μM/min.

Alkaline phosphatase (ALP, EC 3.1.3.1) activity was measured according to the method of AcPase activity described above in Sect. 5.2.11, except that the buffer was replaced with 20 mM Na_2CO_3–$NaHCO_3$ (pH 10.0). One enzyme activity unit represents the same as that of AcPase.

7.2.17 Statistical Analysis

All assays were conducted in three replicates. Data analysis was accomplished by one-way analysis of variance (ANOVA) using Tukey's test. Statistically significant differences between groups represented by p values lower than 0.05.

7.3 The Impact of Ocean Warming and Oceanic Acidification on Biomineralization

The aims of this part were to study the physiological responses of *Pinctada fucata* to pH and temperature stress, thus proposing the potential effects of these stressors on biomineralization. Usually, we use the microarray-based approach. A microarray-based approach was employed to investigate the transcriptome responses of *Pinctada fucata* to elevated CO_2 and temperature. To determine the effects of these stressors on biomineralization, the alkaline phosphatase activity, net calcification rate, and calcium content were detected, and the shell ultrastructure was measured after treatments. We tested the mantle tissue, because it is responsible for accessory respiration, sensory functions, and biomineralization in pearl oysters [50]. Moreover, hemocytes were studied to clarify the immune response to these stressors.

7.3.1 Responses of Mantle Tissue to OA and OW

7.3.1.1 Transcriptomic Analysis of the Mantle Tissue

A total of 5862, 6555, 3181, and 3730 differentially expressed unigenes (DEGs, change fold >2 and $p < 0.05$) for T25, T31 P78, and P75 are identified from a microarray, accounting for 5.43%, 6.49%, 10.01%, and 11.19% of the total genes, respectively (Fig. 7.1a, b). Clustering analysis for the DEGs reveals excellent repeatability and characteristic gene expression profiles for the different treatment groups (Fig. 7.1c). The expression data of microarray and real-time quantitative PCR (RT-qPCR) showed the same trend, with similar expression levels. Highly

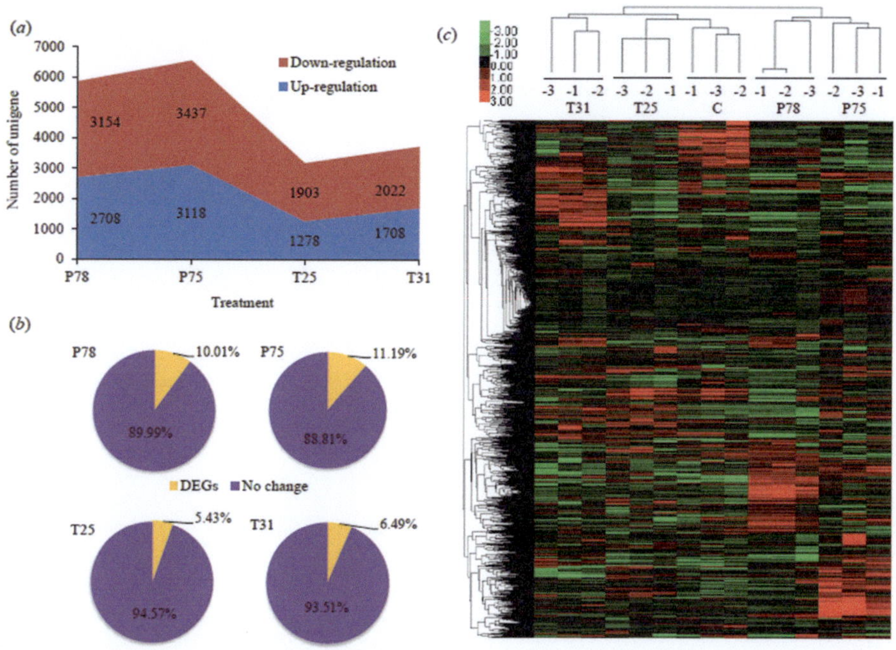

Fig. 7.1 The numbers of up- and downregulation (**a**), ratios (**b**), and heat map (**c**) of differentially expressed unigenes (DEGs) in *Pinctada fucata* exposed to elevated CO2 and temperature. In the heat map, red indicates upregulation; green indicates downregulation; and black indicates insignificant changes. The three biological replicates in each cluster are referred to as −1, −2, and −3

significant correlations (R2 = 0.9449 in T25, 0.8399 in T31, 0.9429 in P78, and 0.8885 in P75) between the RT-qPCR and microarray results are procured, confirming the reliability of the microarray results.

Enrichment analyses indicate 29, 115, 56, and 67 DEGs are assigned to 1, 5, 3, and 2 overrepresented pathways ($p < 0.05$) in P78, P75, T25, and T31, respectively (Fig. 7.2a). It is obvious that a total of 58.62% and 72.62% of the unigenes involved in the category "amino acid metabolism" are downregulated in P78 and P75, indicating a significant metabolic depression. The most representative pathway in P78 and P75 is "phenylalanine metabolism." The "Toll-like receptor signaling pathway" in T25 and the "lysosome" and "other glycan degradation" pathways in T31 are activated, as assessed by unigene upregulation; nevertheless the other pathways are suppressed in the four treatments.

The unigenes involved in the categories "ion and acid–base regulation," "apoptosis," "cell stress responses," and "cell division" are highlighted in this study (Fig. 7.2b). As for the ion and acid–base regulation, few gene expression alterations were observed after the temperature treatments, but the expression levels of these genes were significantly upregulated after the CO_2 treatments, specifically, for vacuolar-type H^+-ATPase (V-ATPase), sodium/potassium-transporting ATPase (NKA), chloride channel (CLC), sodium/hydrogen exchanger (NHE3/5), sodium/

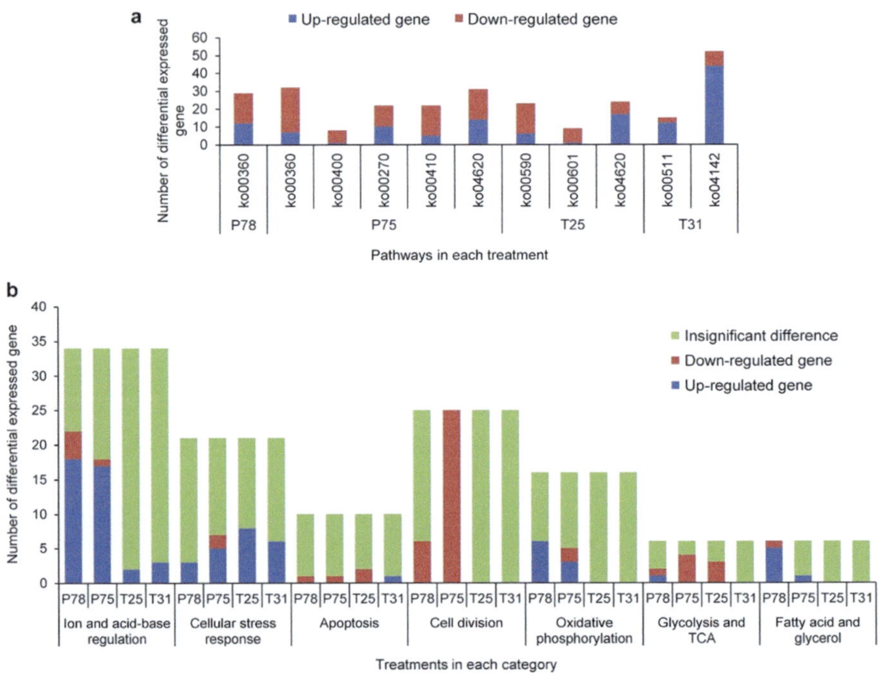

Fig. 7.2 The pathway (**a**) and category (**b**) enrichments of differentially expressed genes (DEGs) in *Pinctada fucata* exposed to CO_2 and temperature stress. (**a**) ko00360, phenylalanine metabolism; ko00400, phenylalanine, tyrosine, and tryptophan biosynthesis; ko00270, cysteine and methionine metabolism; ko00410, beta-alanine metabolism; ko00590, arachidonic acid metabolism; ko00601, glycosphingolipid biosynthesis; ko00511, other glycan degradation; ko04142, lysosome; ko04620, Toll-like receptor signaling pathway

potassium/calcium exchanger (NCKX), voltage-dependent calcium channel (VDCC), anion/bicarbonate transporter family members (SLC26A3/4/6), and sarco-endoplasmic reticulum calcium transport ATPase (SERCA). The expression levels of anion exchanger (AE2) and a sodium bicarbonate cotransporter (NBC1/3) do not reveal significant differences. For the category of "cellular stress response," the expression levels of heat shock protein 70 (HSP70), which related to the heat shock response, were upregulated in response to the four treatments. Besides, the unigenes involved in "antioxidation and oxidation" were also upregulated in response to increased temperature: these included cytochrome P450 (CYP) in T25 and T31 and glutathione S-transferase (GST) and glutathione peroxidase (GPO) in T25. The gene expression patterns indicated that cell apoptosis might occur in T31. The "cell division"-related unigenes were unchanged in T25 and T31, although downregulation was showed in P78 and P75.

To understand biomineralization responses, changes in 51 biomineralization-related unigenes from the microarray are analyzed separately (Fig. 7.3): 80.39% are differentially expressed in response to at least one treatment. Downregulation of

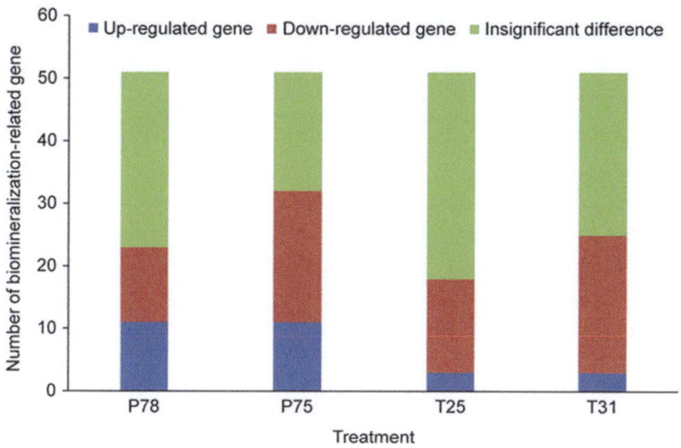

Fig. 7.3 Expression changes in biomineralization-related genes in *Pinctada fucata* exposed to CO_2 and temperature stress

the unigenes encoding tyrosinase (TYR) in P75, T25, and T31 and upregulation of tyrosinase-like proteins (TYRL1 and TYRL2) in P78 and P75 are observed. Moreover, the expression levels of unigenes encoding chitin synthase (CHS), nacrein, and carbonic anhydrase precursor (CA) are downregulated, although CA12 isoform is not affected.

7.3.1.2 Combined Effects of OA and OW on Pearl Oysters

In Fig. 7.4a, b, the alkaline phosphatase (ALP) activity in the control shows no significant difference between the each experimental time point ($p > 0.05$) and 0 time point. At the 72 h time point, the ALP activities decrease by 71.94% in P78 and 82.53% in P75 (from 17.14 nmol pNP•min^{-1}•mg^{-1} to 4.8 nmol pNP•min^{-1}•mg^{-1} and from 17.14 nmol pNP•min^{-1}•mg^{-1} to 2.91 nmol pNP•min^{-1}•mg^{-1}, respectively) compared with that of the control ($p < 0.05$). Although the temperature stress in T25 at 48 h and 72 h does not affect the ALP activities compared with the control (from 16.58 nmol pNP•min^{-1}•mg^{-1} to 15.63 nmol pNP•min^{-1}•mg^{-1} and from 17.24 nmol pNP•min^{-1}•mg^{-1} to 16.04 nmol pNP• min^{-1}•mg^{-1}, respectively), the ALP activity decreases by 42.66% at 48 h and 66.31% at 72 h in T31 (from 16.58 nmol pNP•min^{-1}•mg^{-1} to 9.45 nmol pNP• min^{-1}•mg^{-1} and from 17.24 nmol pNP•min^{-1}•mg^{-1} to 5.86 nmol pNP• min^{-1}•mg^{-1}, in several, $p < 0.05$). The net calcification rate (NCR) in P78 reduces by 83.12% at 24 h, 95.26% at 48 h, and 102.28% at 72 h compared with the control at the corresponding time points. For P75, the NCR decreases by 85.07% at 24 h, 126.67% at 48 h, and 129.71% at 72 h ($p < 0.05$, Fig. 7.4c). As shown in Fig. 7.4d, temperature stress in T25 causes the NCR to increase by 0.64% at 24 h and 8.28% at 48 h and decrease by 3.13% at 72 h compared with the control at the corresponding time points ($p < 0.05$).

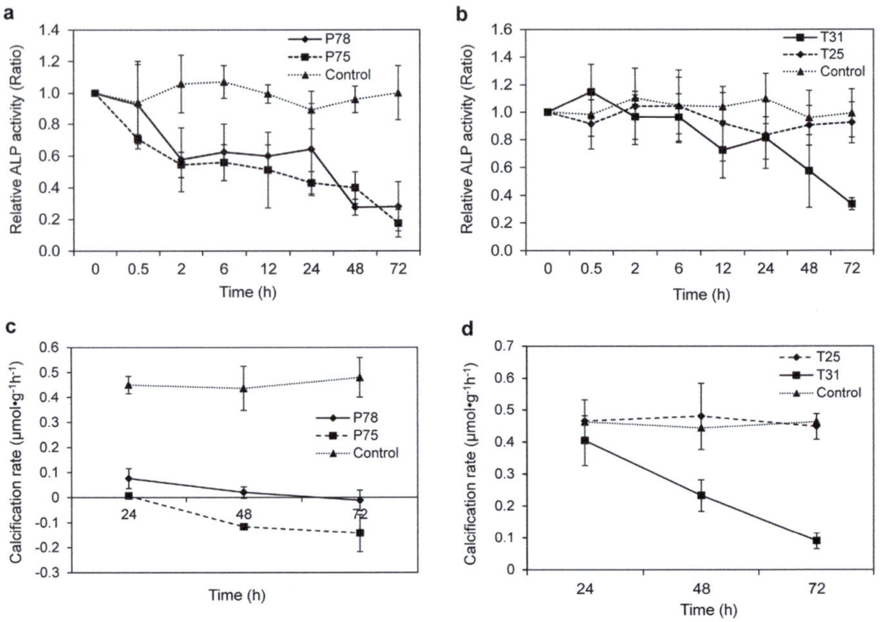

Fig. 7.4 The relative activity of mantle alkaline phosphatase (ALP, (**a** and **b**)) and net calcification rate (NCR, (**c** and **d**)) of *Pinctada fucata* under CO_2 and temperature stress. The ALP and NCR data are presented as the ratios between the treatment group and the control at the corresponding time point

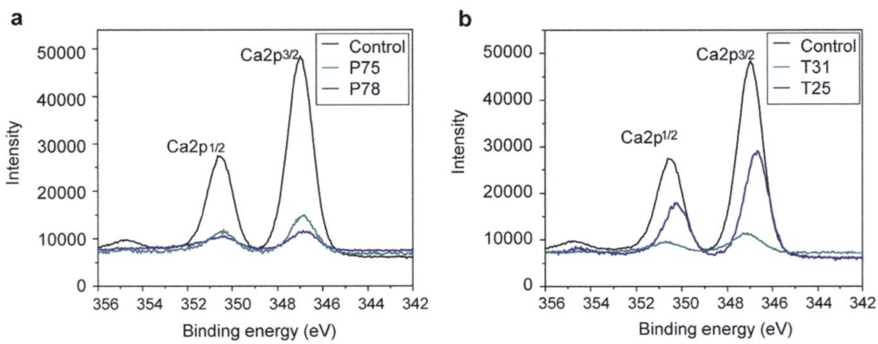

Fig. 7.5 The relative calcium content on the inner surface of the shell near the nacre–prism transition region in *Pinctada fucata* under CO_2 (**a**) and temperature (**b**) stress

Besides, the NCR reduces by 12.54% at 24 h, 47.54% at 48 h, and 80.54% at 72 h, in T31 compared with the control at the corresponding time points ($p < 0.05$).

Reduced pH (pH 7.8 in P78 and pH 7.5 in P75) has an important negative effect on the relative calcium content on the surface of the shell (Fig. 7.5a). Likewise, the relative calcium content decreases with an increase in temperature from 19 °C to

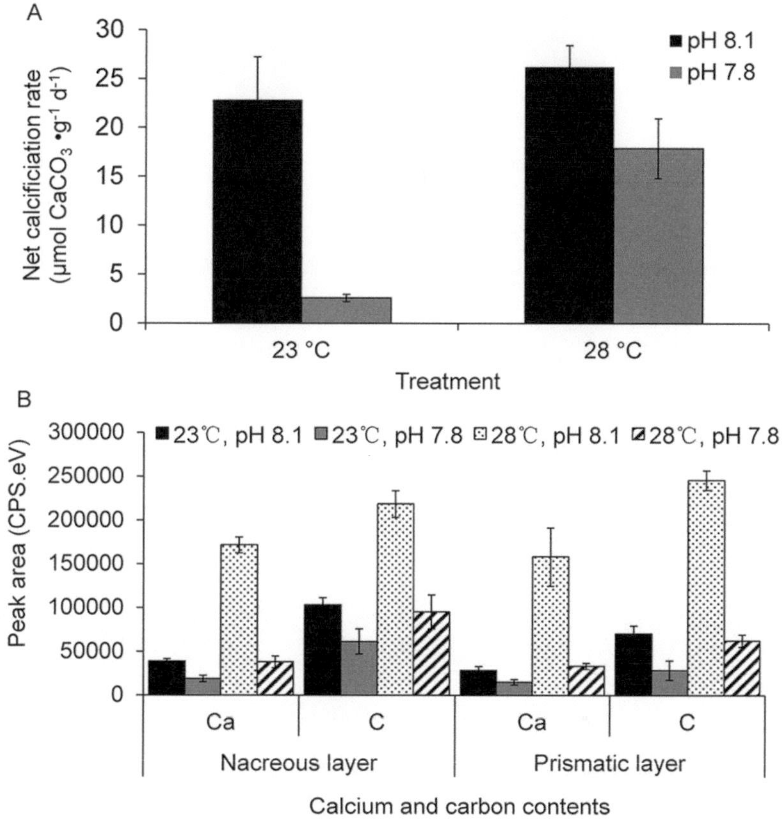

Fig. 7.6 Effects of seawater acidification and elevated temperatures on (**a**) the net calcification rate and (**b**) the calcium and carbon contents on the *Pinctada fucata* shell surface

31 °C in T31 and 19 °C to 25 °C in T25 (Fig. 7.5b). Normal prism-like growth patterns of calcite on the prismatic layer and normal stair-like growth patterns of aragonite on the nacreous layer are observed in the shells in the control group. With growth patterns similar to the control, temperature and CO_2 do not result in significant changes in shell ultrastructure.

The morphological and physiological changes support the conclusion predicted from the gene expression analysis. In the 23 °C × pH 7.8 group, there was an 88.70% reduction in NCR compared with the control (from 22.87 to 2.58 μmol CaCO3$\cdot g^{-1}\cdot d^{-1}$, Fig. 7.6a). On the nacreous layer, the calcium content decreased by 52.01%, and the carbon content decreased by 40.49% compared with the control. On the prismatic layer, the calcium content decreased by 47.19%, and the carbon content decreased by 31.02% compared with the control (Fig. 7.6b). The electron micrographs showed disorganization of the aragonite tablets on the nacreous layer and incomplete calcite crystal with obvious holes on the prismatic layer in *Pinctada fucata* (Fig. 7.7c, g) compared with the control (Fig. 7.7b, f). The changes in the

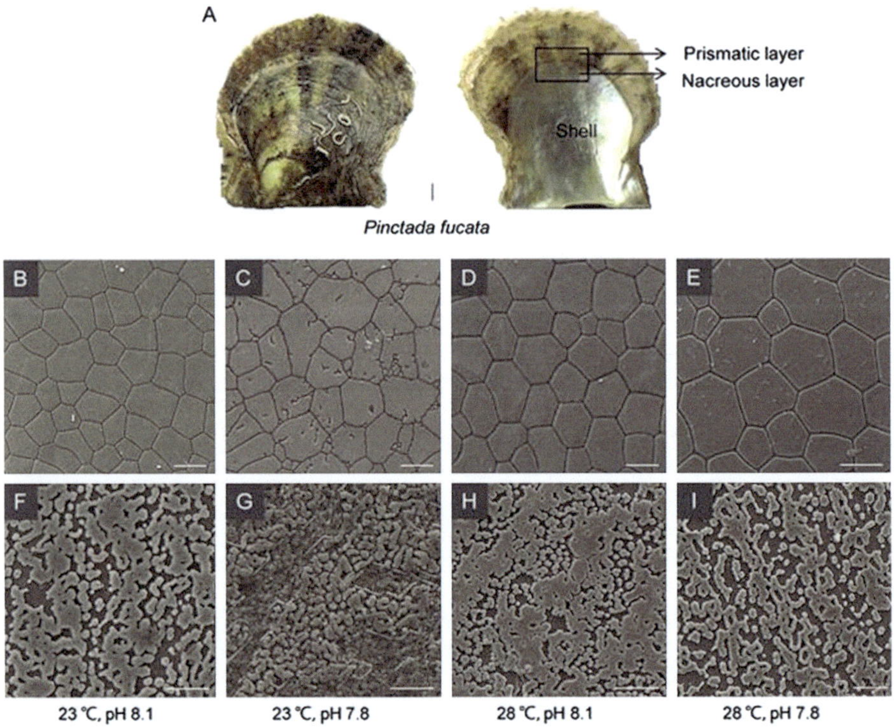

Fig. 7.7 Scanning electron micrographs of *Pinctada fucata* shells exposed to seawater acidification and elevated temperatures [34]. (**a**) Scanning region of the shell: nacre–prism transition region. (**b**)–(**e**) Prismatic layer. (**f**)–(**i**) Nacreous layer. Bars: (**a**) 0.5 cm; (**b**)–(**i**) 25 μm

nacreous layer differed from those seen in a previous study in this species under the same pH condition after a 28-day treatment but were similar to the changes at pH 7.6 [51], meaning time-dependent effects of seawater acidification on shell ultrastructure. The detrimental effects of OA on biomineralization are a common response of marine organisms to environmental changes [10, 11, 51]. OA has been shown to be involved in shell decalcification in marine mollusks [52], and the disordered crystal growth in our study is probably consistent with this observation.

7.3.1.3 The Mechanisms Underlying the Response of Mantle Cells to CO_2 and Temperature Stress

By comparing changes in the expression levels of genes involved in the categories "ion and acid–base regulation," "cell stress responses," "apoptosis," and "cell division," the regulation of genes involved in "ion and acid–base regulation" emerges as the principal transcriptome response of adult *Pinctada fucata* to CO_2 stress. This discovery is consistent with findings in sea urchins but not in corals

[53, 54]. Therefore, we assume that the specific mechanism is involved in this response (Fig. 7.8). Based on the results of functional studies in marine organisms and vertebrates, we suggest that the homologs of genes involved in the response perform similar functions in the *Pinctada fucata* mantle.

Carbonic anhydrases, which catalyze CO_2 and HCO_3^- interconversion, appear to be sensitive to OA in marine calcifiers [39]. Indeed, the downregulation of cytoplasmic CA and normal expression of membrane-associated CA in *P. fucata* are proposed to prevent an excessive increase of intracellular HCO_3^- and H^+. Such trends are also found for CA and calcium binding-related genes. The significant consequence of these changes is the downregulation of NCR, which is strongly responsive to increased temperature and CO_2 temperature in marine calcifiers [33].

Studies on the responses of biomineralization-related genes to OA and ocean warming have found diverse expression patterns [54–56], and these observations are corroborated in the current study. Although most of these genes are differentially expressed in response to increases in CO_2 and temperature, individuals respond differently, indicating a complex relationship between gene expression and stressors. The genes encoding tyrosinase and chitin synthase are highlighted here, as they show larger fold changes in the microarray experiments than did other responsive genes. Tyrosinase plays an important role in the formation of the prismatic layer and the periostracum, which is a protective layer in the shells of pearl oysters. The function-specific response hypothesis about *M. edulis* [57] might aid in understanding our results: TYR is responsible for melanogenesis, which is negatively regulated by CO_2 and temperature stresses; contrary, increased expression of TYRL1 and TYRL2 enhances the synthesis of tyrosinase, strengthening the periostracum and prismatic layer to resist stress. In mollusk shells, chitin is proposed to form the matrix structure on which matrix proteins can control $CaCO_3$ crystal growth [58]. In contrast to *M. edulis* and *Laternula elliptica* [57, 59], the expression of CHS in *Pinctada fucata* exposed to acidified seawater and temperature stress (high levels) was downregulated, which might result in serious effects on frame organization during biomineralization. In *Pinctada fucata*, phenylalanine metabolism was enriched during exposure to increased CO_2 concentrations.

Phenylalanine, a biologically essential amino acid, generates metabolic products comprising a range of different substances, including phenethylamine. Interestingly, phenethylamine degradation has been considered to depress the expression of calcification-related genes in *S. purpuratus* exposed to OA [39]. Therefore, the depression of phenylalanine metabolism in *Pinctada fucata* might restrain its biomineralization ability by influencing the expression of related genes that might be regulated by phenethylamine.

Combined with changes in biomineralization ability and ion status, changes in the temperature and CO_2 levels will have a negative influence on crystal growth. Although the calcium content on the surface of the shell and the NCR reduced, SEM showed that the ultrastructure of the nacreous and prismatic layers was not affected. Previous investigation has indicated that biomineralization is a process, and the effects of CO_2 levels and temperature depend upon the duration of exposure [60].

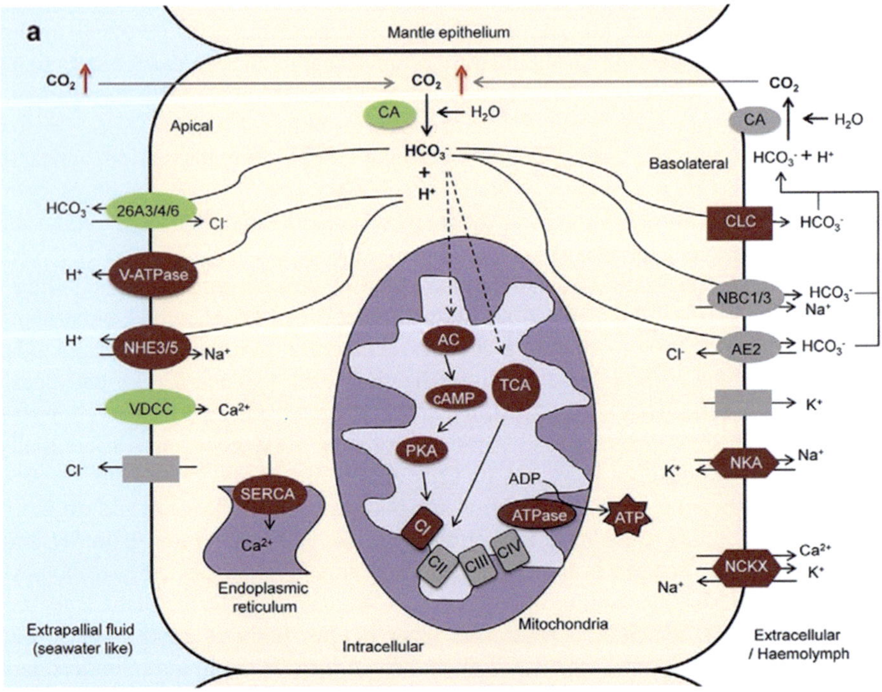

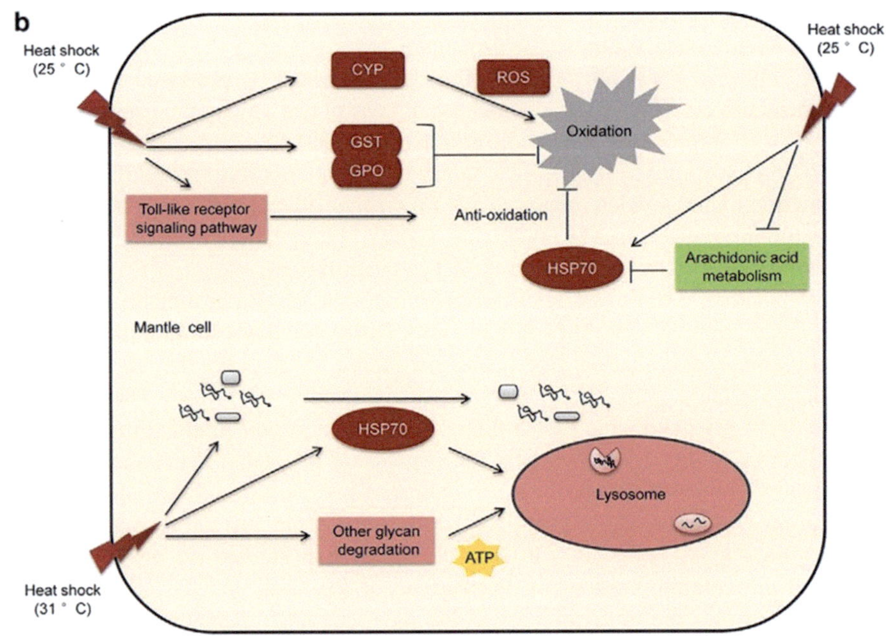

Fig. 7.8 Schematic representations of the mechanisms underlying the response of mantle cells in *Pinctada fucata* to increased temperature and CO_2 stress. (**a**) Acid–base regulatory strategy in *P. fucata* is involved in the response to elevated CO_2. The homeostasis of intracellular H^+ and HCO_3^- is regulated by membrane-associated CA12, cytoplasmic carbonic anhydrase (CA), and

7.3.2 *Responses of Hemocytes to OA and OW*

7.3.2.1 Physiological Changes

In the EPF (Fig. 7.9a), the pH values in the control groups (8.14 at 24 h, 8.13 at 48 h, and 8.14 at 72 h) were not significantly different from that in ambient seawater (8.18). Comparing with the control, the pH values of the EPF had no differences in T25, but a significant reduction was observed in P78, P75, and T31 ($p < 0.05$). In the hemolymph (Fig. 7.9b), the pH values in the control were lower than in seawater (7.98 at 24 h, 8.00 at 48 h, and 7.99 at 72 h). In comparison with the control, the pH values of hemolymph were not significantly different from T25, while the pH values in P78 and P75 at 24 h were higher than the control ($p < 0.05$), suggesting a stress response of the pearl oysters to the pH decrease.

Decreased pH and increased temperature have positive effects on the THC, TP, and percentages of large hyalinocytes and granulocytes but a negative effect on NR. THC significantly increased from 0.49×10^{10} cells/L in the control to 0.97×10^{10} cells/L, 0.84×10^{10} cells/L, and 0.82×10^{10} cells/L in T31, P78, and P75 at 72 h, respectively ($p < 0.05$, Fig. 7.10a). The extracted TP of the hemocytes was 306.39 mg/mL, 345.07 mg/mL, 420.78 mg/mL, 767.41 mg/mL, and 888.10 mg/mL in the control, T25, T31, P78, and P75 at 72 h, respectively ($p < 0.05$). These significant increases of TP were also detected at 24 h and 48 h (Fig. 7.10b). The large hyalinocytes and granulocytes (R2 and R3) increased with the decreasing pH and

Fig. 7.8 (continued) apical and basolateral proton pumps, exchangers and transporters, and ion channels. Among them, the expressions of vacuolar-type H^+-ATPases (V-ATPase), chloride channels (CLC), sodium/hydrogen exchangers (NHE3/5), sodium/potassium-transporting ATPases (NKA), and sodium/potassium/calcium exchangers (NCKX) were up-regulated, while the expressions of anion/bicarbonate transporter family members (SLC26A3/4/6) were down-regulated. And the expressions of sodium bicarbonate cotransporters (NBC1/3), K^+ channels, and anion exchangers (AE2) showed no significant changes. In these processes, increased energy demand is supplemented by enhancement of the tricarboxylic acid cycle (TCA), and oxidative phosphorylation (OP) is stimulated by HCO_3^- via adenylate cyclase (AC)-induced signaling and membrane transport, including increased expression of complex I (CI), AC, F0F1-type ATPases (F-ATPase), cAMP, and protein kinase A (PKA). In the diagram, gray indicates insignificant changes, crimson indicates upregulation, and green indicates downregulation. The solid arrows show the direction of CO_2 and ions diffusion. The dotted arrows show the hypothetical signaling pathways. (**b**) Antioxidative and lysosome pathways in *P. fucata* are involved in the response to elevated temperature. Elevated temperature induces the production of reactive oxygen species (ROS) via increased expression of cytochrome P450 (CYP), leading to oxidative stress. Under the medium-temperature stress (25 °C), the increased expression of "antioxidation"-related genes [e.g., glutathione S-transferase (GST), glutathione peroxidase (GPO), and heat shock protein 70 (HSP70)], activation of the "Toll-like receptor signaling pathway," and depressed "arachidonic acid metabolism" promote antioxidative regulation in mantle cells. Under high temperature stress (31 °C), the "lysosome" pathway is the main defense system for clearing damaged proteins and organelles, with assistance of HSP70. The energy supplement might come from the enhanced degradation of other glycans. In this diagram, green indicates downregulated genes, crimson indicates upregulated genes, blue indicates inhibited pathways, pink indicates enhanced pathways, and gray indicates oxidative stress

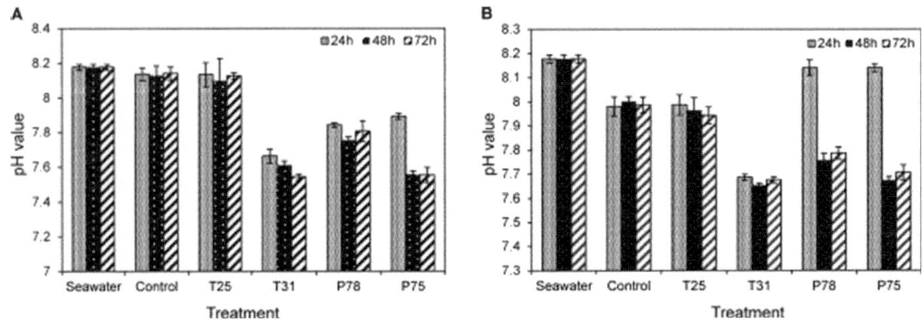

Fig. 7.9 Effects of decreased pH and elevated temperature on the pH values of extrapallial fluid (**a**) and hemolymph (**b**) in *Pinctada fucata*. The symbols T25, T31, P78, and P75 are the abbreviations of the different treatment groups, which represent the exposed seawater conditions of increased temperature and decreased pH: 25 °C, 31 °C, pH 7.8, and pH 7.5, respectively

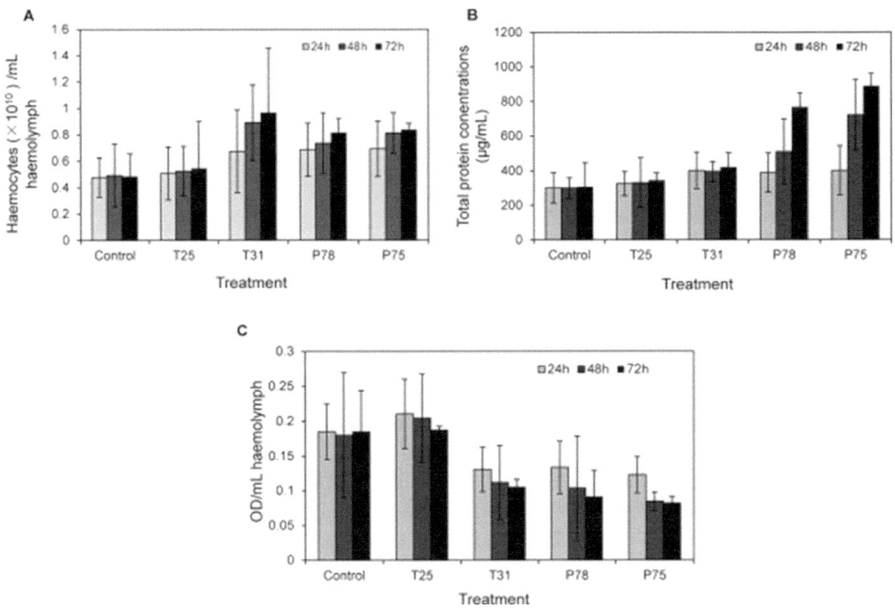

Fig. 7.10 Effects of decreased pH and elevated temperature on the total hemocyte count, total protein concentration, and neutral red uptake of hemocytes in *Pinctada fucata*

increasing temperature, while they could not be clearly differentiated from each other in T31, P78, and P75, with the exception of T25. No significant differences were found for the NR of the hemocytes in T25. In contrast, the decreases of NR were significant at all time points in T31, P78, and P75 ($p < 0.05$, Fig. 7.10c).

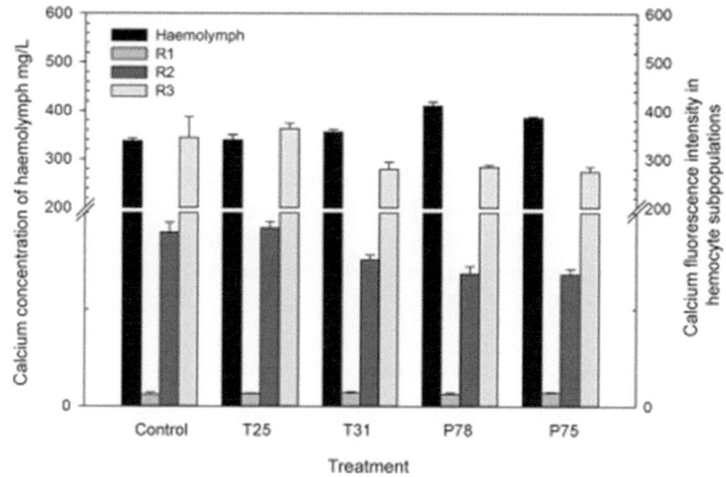

Fig. 7.11 Effects of decreased pH and elevated temperature on the contents of calcium ions in hemocytes and hemolymph of *Pinctada fucata* [61]. R1, R2, and R3 are the subpopulations of small hyalinocytes, large hyalinocytes, and granulocytes classified by flow cytometry

7.3.2.2 Effects of OA and OW on Calcium Content

As shown in Fig. 7.11, the fluorescence intensities of the small hyalinocytes for different treatments reveal no statistical significance ($p < 0.05$) after 72 h of exposure. Comparing with the control, there were decreases of 18.98%, 17.57%, and 20.51% for the granulocytes in T31, P78, and P75, respectively, and 15.80%, 23.84%, and 24.20% for the large hyalinocytes in T31, P78, and P75, respectively ($p < 0.05$). Significant increases of the calcium contents in the hemolymph were detected in T31, P78, and P75 ($p < 0.05$).

The impacts of environmental changes on marine organisms are generally revealed by immunological parameters in hemocytes. Therefore, the morphological characterizations of hemocytes are helpful to understand the immune response of pearl oysters to environmental stressors. The modification of the immune parameters in the hemocytes of *Pinctada fucata* implied that short-term OA and OW (31 °C) have positive effects on THC, TP, and the percentage of granulocytes and large hyalinocytes but have negative effects on NR. However, exposure to OW (25 °C) shows no differences for these parameters, suggesting an excellent regulation of the pearl oysters to heat stress at this level [61]. The increase of circulating hemocytes in bivalves can reflect the movement of cells from tissue into hemolymph, suggesting an enhanced immune reaction when detecting environmental stresses [62].

In our study, the increased hemocytes are granulocytes and large hyalinocytes that have been shown to have an immunological capacity (e.g., phagocytosis) in pearl oysters [63]. Accordingly, as reported in gastropod abalones [64], this increase of hemocytes number is considered as active stress responses of *Pinctada fucata* to short-term exposures. However, the phagocytic activity of hemocytes decreased in

Pinctada fucata according to NR results. In contrast, it increased in green mussel *P. viridis* [65] and had no significant changes in the Pacific oyster *C. gigas* [66], showing that the accumulation of hemocyte counts is the basic characteristic for *Pinctada fucata* to respond to short-term OA and OW, rather than the enhancement of phagocytic capacity. The TP in hemolymph is also a useful parameter to evaluate the health status of bivalves exposed to environmental stresses [67]. In bivalves hemolymph, TP contains numerous proteins, such as lysosome enzymes and anti-microbial peptides [65, 68, 69], responsible for protein degradation and antibacterial function in immunomodulation. The release of various enzymes from the hemocytes might cause the increase of the TP shown in our data, which were conductive to resisting OA and OW.

Immunoregulation, acid–base regulation, and biomineralization are the principal responses of marine calcifiers to OA and OW [70, 71]. Due to its open circulatory system, *Pinctada fucata* cannot accumulate enough bicarbonate to neutralize the hydrion in hemolymph. Therefore, the pH value in hemolymph is sensitive to CO_2 concentration in the seawater. A linear relationship between the hemolymph pH and seawater pH has been revealed in the mussel *M. edulis* [72], and similar results were observed in the EPF and hemolymph under the stress of OA and OW (31 °C) in our study. Importantly, the pearl oysters seem to have an active stress response when exposed to elevated temperature and CO_2, especially in the duration of the first 24-h exposure to elevated CO_2, showing that *Pinctada fucata* may implement compensation for acid–base disturbance. However, the compensation is an energy consuming process [71], resulting in energy redistribution, which might affect biomineralization in the pearl oysters. Because the hemocytes were involved in the biomineralization and shell repair processes of oysters [60, 73, 74], the calcium content in hemocytes, therefore, is critical for the formation of $CaCO_3$.

In our study, the calcium content decreased in the hemocytes and increased in the hemolymph of *Pinctada fucata* exposed to short-term OA and OW (31 °C), indicating the outflow of calcium from hemocytes to hemolymph, which will eventually affect the biomineralization process. In addition, calcium is a signaling molecule involved in intracellular signal transductions. The decrease of calcium in granulocytes and large hyalinocytes might alter the immune reactions of hemocytes by disturbing the calcium-dependent signaling pathway in key physiological processes, such as phagocytosis.

7.4 The Impact of Pollution on Biomineralization

In this part, the enzymatic responses in the digestive gland and gills of pearl oysters *Pinctada fucata* exposed to copper at different concentrations were investigated by us. We assayed enzyme activities at the metal accumulations, and different exposure times in different tissues were also determined at the same time. We researched five enzymes belonging to three different classes. The proper use of these enzymes as biomarkers and pave the way for investigation of the defense mechanism of the pearl oyster may be supported by the present result.

7.4.1 The Impact of Copper on Immune Defense

In gills, 0.5 μM copper exposure indicated a linear relationship with time, and after exposing in 0.5 μM copper for 72 h, the metal concentration reached 65.93 μg/g dry wt. (Fig. 7.12a). Copper concentrations in gills of oysters exposed to 0.05 μM were not significantly different from controls throughout the period exposure. Rate and levels of metal accumulation in gills were higher than in digestive gland exposing to 0.5 μM copper, and we only detected 16.46 μg/g dry wt. copper after 72 h (Fig. 7.12b). However, in low exposure oysters, metal accumulated more show in gills than in the digestive gland.

7.4.1.1 Acid Phosphatase

It showed different pattern of changes in acid phosphatase activity between the gills and digestive gland. In gills, significant high acid phosphatase activity has been detected in the two experimental groups at most times (Fig. 7.13a). We obtained the highest activity of 0.05 μM exposure at 72 h and found the enzyme activity kept increasing after 24 h. It indicated the highest activity of acid phosphatase in oysters exposed to 0.5 μM copper was at 12 h. The enzyme activity decreased thereafter but remained higher than in controls even after 72-h exposure (Fig. 7.13a). In the digestive gland, in low exposure oysters, acid phosphatase activity was significantly high at 12, 24, and 72 h (Fig. 7.13b). After exposing in 0.5 μM copper for 72 h, the

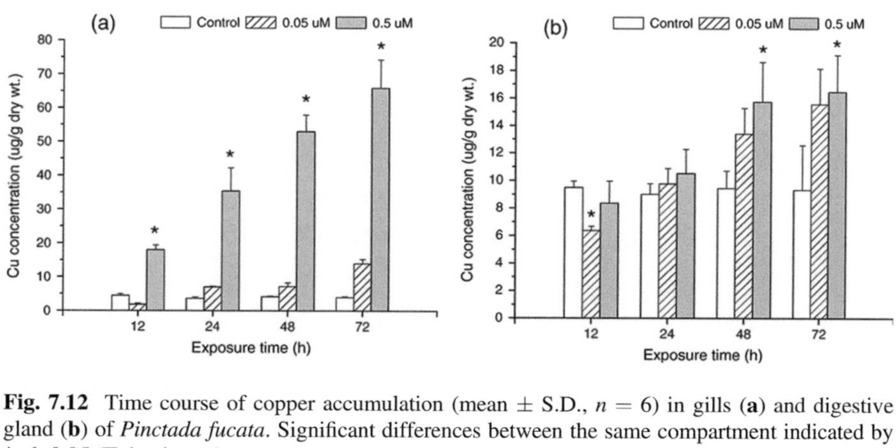

Fig. 7.12 Time course of copper accumulation (mean ± S.D., $n = 6$) in gills (**a**) and digestive gland (**b**) of *Pinctada fucata*. Significant differences between the same compartment indicated by **p* b 0.05 (Tukey's test)

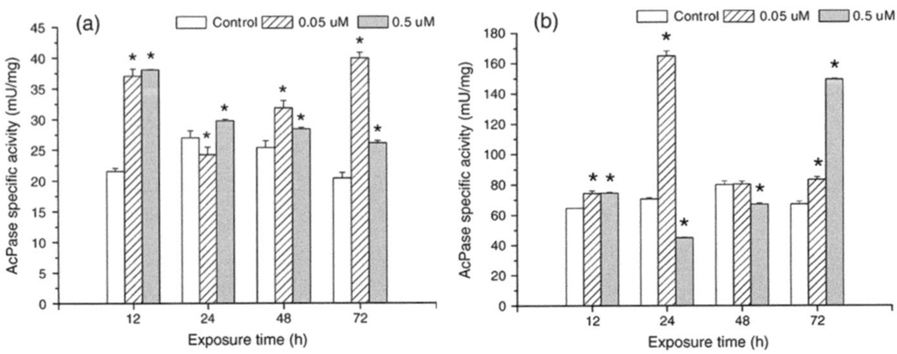

Fig. 7.13 Time course of acid phosphatase-specific activity (mean ± S.D., $n = 10$) in gills (**a**) and digestive gland (**b**) of *Pinctada fucata*. Significant differences between the same compartment indicated by *p b 0.05 (Tukey's test)

enzyme activity was low but increased from 24 h and became higher than the control at 72 h (Fig. 7.13b).

7.4.1.2 Alkaline Phosphatase

In gills, alkaline phosphatase activity became higher than the control after 24 h exposure to 0.05 μM and after 48 h exposure to 0.5 μM, respectively (Fig. 7.14a). The gill alkaline phosphatase activity only showed a little bit difference between 48 and 72 h exposing to the varied concentration of copper in the two experimental groups. In the digestive gland exposed to 0.05 μM copper, the highest value of the alkaline phosphatase activity was at 24 h, and then the value decreased until they were lower than the control at 72-h exposure (Fig. 7.14b).

As recognized in the last years, any indication of deleterious effects of contaminants on the biota cannot be provided by risk assessment cannot solely base on chemical analysis of environmental samples because this approach does not provide. Therefore, measurements of the biological effects of pollutants have become of major importance for the assessment of the quality of the environment [75]. Bivalves are useful monitoring organisms because they are ubiquitous, sedentary, filter-feeders inhabiting coastal and estuarine areas [32]. But there are only a few studied biomarkers of pearl oysters [76]. Thus, an increasing need to develop a set of applied biomarkers of pearl oysters to assess levels of environmental pollution appeared.

In the present work, the copper concentrations were acute and sublethal levels for bivalves (equals 12.5 and 125 μg/l) [48]. The sample times were 12, 24, 48, and 72 h because the responses of short-term exposure to copper were focused on [77]. We were investigating the gills and digestive gland as the important target organs for metal accumulation [20]. We selected five enzymes of three classes to be investigated for a detailed picture of the general metabolic condition of the organisms [20, 78]. Acid phosphatase and phenoloxidase are involved in the immune defense

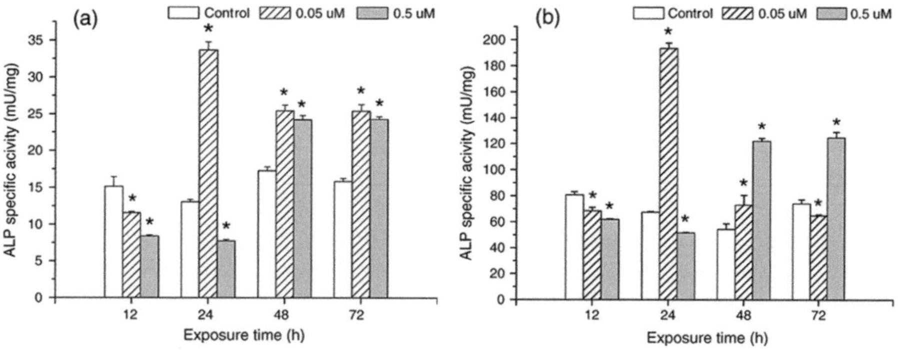

Fig. 7.14 Time course of ALP-specific activity (mean ± S.D., $n = 10$) in gills (**a**) and digestive gland (**b**) of *Pinctada fucata*. Significant differences between the same compartment indicated by $*p$ b 0.05 (Tukey's test)

of the oysters [79–81]. Superoxide dismutase and Se-dependent glutathione peroxidase are two main antioxidant enzymes to protect organisms against damages induced by oxyradical production [82–84]. Alkaline phosphatase, which is sensitive to metals, was also examined to obtain a better picture of the general metabolic condition of the organisms [20, 78].

The bivalves immediately close their valve to response the addition of copper to seawater as a protective strategy, but the timing of this varies with the copper concentration [85]. Generally, quicker valve closure was caused by higher concentrations [86]. Bivalves can remain closed in seawater polluted by copper during the initial period, but because they need feeding and respiration, they tend to open again. These facts might be the reasons why metal accumulations and enzymatic responses at 12 h always differed from the results at other sample times.

The digestive gland represents the major site of metal accumulation in bivalves [77]. Gills are also frequent targets of environmental pollutants because they are the main interfaces between the environment and the organisms [81]. The results of the copper accumulation in *Pinctada fucata* imply that the two tissues may play different roles when exposed to different copper concentrations. The digestive gland was the main copper accumulation organ when oysters were exposed to low concentrations, whereas gills became the target organ in oysters exposed to high concentrations (Fig. 7.12) [87]. To our knowledge, previously there is no report to describe the different copper accumulation patterns in gills and digestive gland. One possible reason was that feeding and thus the rate of copper accumulation in the digestive gland may have been affected by higher concentration [77].

Acid phosphatase is a lysosomal marker enzyme and can be taken as a reliable tool for the biological assessment of metal pollution [81]. The higher gill acid phosphatase activity in mussels was always associated with the damage of lysosomal membranes [20, 81]. Hence, in the gills of *Pinctada fucata*, the stimulation of cell defense mechanisms likely causes the higher acid phosphatase activity of 0.05 μM exposure [87]. That the copper concentrations were too high (Fig. 7.12a) and interfered with the defense mechanisms was the reason for the depression of the enzyme activity when exposed to 0.5 μM. In the digestive gland, the results of enzyme activity exposed to 0.5 μM copper indicate that the severe presence of copper may inhibit the activity at the beginning, but as oysters adapted to the environment and cell defense recover, the acid phosphatase activity would increase slowly. The acid phosphatase activity in the digestive gland exposed to 0.05 μM did not increase as in gills. Because under this condition metal was accumulated more quickly in the digestive gland than in gills (Fig. 7.12b) and normal defense mechanisms are disturbed by the high copper. Inhibition of acid phosphatase activity in digestive cells by copper has been reported in mussels [81], but enzyme activity recovery was not observed. The adaptation and recovery of the oysters were reported for the first time in our studies.

Alkaline phosphatase is an intrinsic plasma membrane enzyme of almost all animal cells. Alkaline phosphatase is sensitive to metals, and the effects of metals on alkaline phosphatase have been reported [76]. Our previous study has purified and characterized alkaline phosphatase of *Pinctada fucata* [78], and the results also implied that it was sensitive to metals including copper. Changes of alkaline phosphatase activity in the tissues were difficult to explain. For example, the gill alkaline phosphatase activity varied little between 48 and 72 h of exposure in the two experimental groups; however, at those times, the copper concentrations differed greatly (Figs. 7.12a and 7.13a). One possible reason was that alkaline phosphatase activity was influenced by complex factors in *Pinctada fucata* and copper was only one of them. Alkaline phosphatase activity was decreased in the digestive gland of mussels exposed to copper [20], which may be the normal response of alkaline phosphatase. The enzymatic mechanisms of alkaline phosphatase in *Pinctada fucata* need further studies.

7.5 Conclusion

The increasing anthropogenic carbon emission has great effects on the chemical and physical conditions of the oceans, leading to global warming and ocean acidification. The influence of global warming and ocean acidification on the pearl oysters are complex. Potential implications for predictions of the effects of global climate change on pearl aquaculture might be indicated in our study. During short-term exposure, the impact of CO_2 and temperature stresses were not manifested in the shell ultrastructure, although we affected the process of biomineralization. Importantly, the nacreous layer of pearl oyster shells is structurally similar to pearls, and the shells can actually reflect the effect of environmental stressors on pearls. If the stressors are long-term events, they will have adverse results on the biomineralization of pearls, most likely affecting pearl quality, which result in substantial economic losses for the aquaculture industry. We need further studies to directly test the biochemical and physiological changes that are demonstrated by our gene expression data (e.g., the physiology of ion transporters), as the abundance of mRNAs may not absolutely coincide with the changes in the corresponding cellar proteins [88].

As for the copper exposure, we inquire the metal accumulations and the corresponding enzymatic responses in the pearl oyster *Pinctada fucata*. That the gills and the digestive gland played different roles associated with different copper concentrations was found. We observed the adaptation and recovery of the oysters. Levels of the copper accumulation and the sensitivity to copper were the main, if not, part of the reasons for the various responses of the selected enzymes. Potentially, we may use many of those enzymes as biomarkers in testing of marine metal pollutions. The results supported the proper use of these enzymes as biomarkers and paved the way for the investigation of the defense mechanism of the pearl oyster.

Various factors control biomineralization, such as genes, environmental factors and ions. The idea that ocean acidification, ocean warming and metal pollution have great effects on the metabolism, biomineralization process, and immune defense in *P. fucata* was pointed by our study. Theoretical basis for precautions to avoid adverse impact of environmental challenges on the marine pearl aquaculture in the near future is provided.

This work was modified from the papers published by our group in *Environmental Science & Technology* (Li, S. G. et al. Interactive Effects of Seawater Acidification and Elevated Temperature on the Transcriptome and Biomineralization in the Pearl Oyster Pinctada fucata. Environmental Science & Technology 50, 1157–1165, doi:https://doi.org/10.1021/acs.est.5b05107 (2016)); *Scientific Reports* (Li, S. G. et al. Transcriptome and biomineralization responses of the pearl oyster Pinctada fucata to elevated CO2 and temperature. Sci Rep-Uk 6, doi:ARTN 18943https://doi.org/10.1038/srep18943 (2016)); *Fish and Shellfish Immunology* (Li, S. et al. Hemocytes participate in calcium carbonate crystal formation, transportation and shell regeneration in the pearl oyster Pinctada fucata. Fish Shellfish Immunol 51, 263–270, doi:https://doi.org/10.1016/j.fsi.2016.02.027 (2016)); *Comparative Biochemistry and Physiology*, Part C (Jing, G., Li, Y., Xie, L. P. & Zhang, R. Q. Metal accumulation and enzyme activities in gills and digestive gland of pearl oyster (Pinctada fucata) exposed to copper. Comp Biochem Phys C 144, 184–190, doi: https://doi.org/10.1016/j.cbpc.2006.08.005 (2006)); and *Fish and Shellfish Immunology* (Li, S. et al. Morphology and classification of hemocytes in Pinctada fucata and their responses to ocean acidification and warming. Fish Shellfish Immunol 45, 194–202, doi:https://doi.org/10.1016/j.fsi.2015.04.006 (2015)). The related contents are reused with the permission.

Supplementary Table 7.1

Chapter	Primer name	Primer sequence
	Prismalin-14-F	TGGGTATGGCGGATTTAACGGTG
	Prismalin-14-R	AATCCGCCATCATCGTCACCAAA
	Prisilkin-39-F	TAGCTAGTGGTGGTTTAGGTGCC
	Prisilkin-39-R	TACCCGCTATATCCGCCGTAACC
	Shematrin-7-F	TGGAGCCGGAAGTTTGTCATCT
	Shematrin-7-R	CCGAGGTTTACTGATGGTCCGT
	Shematrin-6-F	ACACCTTCGGAACATTGGCAAC
	Shematrin-6-R	CACCTCCAAGACCGAGACTGAG
	Lysine-rich matrix protein 5-F	TCACCCTTGGGATTGGAAATGCA
	Lysine-rich matrix protein 5-R	GCCAAAGTTGTAATCATCGCCACC
	Nacrein-F	GAGCCAGAGGATGGGGAAA
	Nacrein-R	GCCTCCATAGGTGAAACGA
	Pif80-F	TGCTGCCATCACGTGAGTATG
	Pif80-R	GACTTCCCTTTCTCACACTTCCA
	Mantle gene 1-F	GTCCTTGGCGTCGTAGTTTGTG
	Mantle gene 1-R	TCCTCCTCTCCCTATTTCCGGT
	Lustrin A-F	AGCCACTAGTAGCAGCTTCTCG
	Lustrin A-R	TAGGGGAAGGTTGCTGTTGAGG
	Pearlin-F	CCTGGAATCATGACGTGCACAC
	Pearlin-R	AGCAGTATGAGTAACGTCCGCA
	Calponin-F	CAGGCAGGCATGAGAGGTTTTG
	Calponin-R	TTTGCTGCATTCTGAGCTGCTC
	Calmodulin-F	ATCCAACCGAGGCTGAACTTCA
	Calmodulin-R	GTGTCTTTCATCTTTCGCGCCA
	Calreticulin-F	ACAAGCCAGAACACATCCCTGA
	Calreticulin-R	CCATTCTCCCTTGTACTCGGGG
	Actin-F	TACCGCCGCGTCATCATCAT
	Actin-R	TGCCTCGGGACATCTGAACC
	18srRNA (AB214463.1) -F	GCTCGTAGTTGGATCTCGGG
	18srRNA (AB214463.1) -R	AGTAAACGTTCCGGTCACCC
	CL15630.Contig2-F	GCCCCTGTTTGTGGATCAGA
	CL15630.Contig2-R	ACATGGACCTTCAGAGCGTAC
	CL4407.Contig2-F	ACACGGACTTTGAGGAGCTG
	CL4407.Contig2-R	ACTGAGGAGTCGATGTGTGTG
	CL7659.Contig1-F	TCGCGACAAAGGTGAGGATG
	CL7659.Contig1-R	TCGGTGTTGTCCATTTCCGA
	Unigene508-F	AATGTCGCCAAGGAAGTCGT
	Unigene508-R	TTGTCATTCCCGCCAGATCC
	Unigene52885-F	GAACCCCGTATGAAGCTGCT
	Unigene52885-R	ACATCACAGAAGCCGCATGT
	CL754.Contig2-F	AGGGTCCGCTGATCACAATG

(continued)

Chapter	Primer name	Primer sequence
	CL754.Contig2-R	GCATGGCAAGGATCCACATC
	Unigene37660-F	CCTCAGAGATGGTAGCGAACA
	Unigene37660-R	TTTCCCGTCGTCTACTGCAC
	Unigene58966-F	TGCATGCATTCAAAGCCCAG
	Unigene58966-R	ACACTCAGCGTCACACCATT
	Unigene58161-F	GGCTCTCCGTGATGACAACT
	Unigene58161-R	TGAGTACGTGAAAACCGGGG
	CL1162.Contig3-F	AGTATGGCACGGGGAGTTTG
	CL1162.Contig3-R	GTGGGTAGAAGGTTGGGACG
	Unigene44471-F	AGCTCTTAGTAATCCCTGAGGC
	Unigene44471-R	ACCTTGGTGACCTGATGCTG
	Unigene44607-F	AGGGAAGTAAAACGCCTCCA
	Unigene44607-R	TGCCACTAGAAGCTCACCAG
	Unigene24893-F	GAAGTCGACCCATAGCGTCT
	Unigene24893-R	TGTCCCAGCAATGTCACCTC
	Unigene16645-F	GGATCATGTGGCGTCATTGC
	Unigene16645-R	CATTTCGGACACCTGCTCCT
	Unigene19167-F	TTTGTTGCGATCGGTCCTCA
	Unigene19167-R	AGATGTCTGTTCCTCGTGCG
	Unigene37643-F	TGTGGCGAAGGATTCTCTGG
	Unigene37643-R	AGGTGCTCGTACTTTGGTCG
	Unigene33618-F	GGAGGAGTTGCTGAGACACG
	Unigene33618-R	TGCCAGTATTTCCTCGTGCA
	CL1779.Contig2	GGACGTCTCCCTCTGTTGTC
	Treatment Gene ID Primers	ACTGGCATTGCTGGTGTGTA
	Unigene31731-F	CGTGGAGCAATGCGATTACC
	Unigene31731-R	TCGACTTTGGTGCGTGAAGA
	Unigene23402-F	GGGACAAGGAGCAGGTCTTC
	Unigene23402-R	CCCGTCACCATTCCACTTGA
	Unigene27741-F	ACTGACAAACGAGGCGACAA
	Unigene27741-R	TTGGCTTGCCACCAATGTTC
	Unigene32821-F	CCTTGCGGTGAGTTTGTGTG
	Unigene32821-R	CCGGAACCTATTGTGGCTGT
	Unigene11478-F	TGCTTCTTTGTTTTGCGCCT
	Unigene11478-R	CGCTGATTTCCTCGCACAAA
	Unigene21946-F	GGGGAGGGAGAGAAGAGTGA
	Unigene21946-R	TGCTGGCAAACACACTCTCT
	Unigene766-F	GTTCCCAGCGACTTACACCA
	Unigene766-R	AGTTCCAGGCAAGTGTTGGT
	CL3909.Contig2-F	GGAGGAGCTACAATGGCCAG
	CL3909.Contig2-R	TGTGAGATGGGTTGTGACGT
	CL8302.Contig3-F	ATCCCGTACCTTGCCATGAC
	CL8302.Contig3-R	CTGCAAAGAACCACCAAGCC

(continued)

Chapter	Primer name	Primer sequence
	CL3078.Contig3-F	CATCGATACGTCCACCAGCA
	CL3078.Contig3-R	TGCAGCTCTTGGTCAGTCAG
	Unigene33622-F	TTATTTCAGAGAATGGTAA
	Unigene33622-R	CAAATATGGCCATTCCTT
	CL1722.Contig1-F	AGGGCTTTAACACCAGAGCC
	CL1722.Contig1-R	CCTCAGTAAATGCGAGGCCA

References

1. H. Yan, W. Soon, Y. Wang, A composite sea surface temperature record of the northern South China Sea for the past 2500 years: a unique look into seasonality and seasonal climate changes during warm and cold periods. Earth Sci. Rev. **141**, 122–135 (2015)
2. Y. Liu et al., Acceleration of modern acidification in the South China Sea driven by anthropogenic CO(2). Sci. Rep. **4**, 5148 (2014). https://doi.org/10.1038/srep05148
3. Z.Y. Jing, Y.Q. Qi, Z.L. Hua, H. Zhang, Numerical study on the summer upwelling system in the northern continental shelf of the South China Sea. Cont. Shelf Res. **29**, 467–478 (2009)
4. S.C. Doney, V.J. Fabry, R.A. Feely, J.A. Kleypas, Ocean acidification: the other CO2 problem. Annu. Rev. Mar. Sci. **1**, 169–192 (2009). https://doi.org/10.1146/annurev.marine.010908.163834
5. C.L. Sabine et al., The oceanic sink for anthropogenic CO2. Science **305**, 367–371 (2004). https://doi.org/10.1126/science.1097403
6. K. Caldeira, M.E. Wickett, Oceanography: anthropogenic carbon and ocean pH. Nature **425**, 365 (2003). https://doi.org/10.1038/425365a
7. B. Bates, Climate change and water: IPCC technical paper VI world health organization, 2009
8. K.M. Mostofa, C.Q. Liu, M. Minella, D. Vione, Balancing of ocean acidification by superoxide redox chemistry? Environ. Sci. Technol. **47**, 11380–11381 (2013). https://doi.org/10.1021/es4039177
9. S.A. Pedersen et al., Multigenerational exposure to ocean acidification during food limitation reveals consequences for copepod scope for growth and vital rates. Environ. Sci. Technol. **48**, 12275–12284 (2014). https://doi.org/10.1021/es501581j
10. L.B. Azevedo, A.M. De Schryver, A.J. Hendriks, M.A. Huijbregts, Calcifying species sensitivity distributions for ocean acidification. Environ. Sci. Technol. **49**, 1495–1500 (2015). https://doi.org/10.1021/es505485m
11. K.J. Kroeker et al., Impacts of ocean acidification on marine organisms: quantifying sensitivities and interaction with warming. Glob. Chang. Biol. **19**, 1884–1896 (2013). https://doi.org/10.1111/gcb.12179
12. C. Hiebenthal, E.E. Philipp, A. Eisenhauer, M. Wahl, Effects of seawater pCO2 and temperature on shell growth, shell stability, condition and cellular stress of Western Baltic Sea Mytilus edulis (L.) and Arctica islandica (L.). Mar. Biol. **160**, 2073–2087 (2013)
13. G.W. Ko et al., Interactive effects of ocean acidification, elevated temperature, and reduced salinity on early-life stages of the pacific oyster. Environ. Sci. Technol. **48**, 10079–10088 (2014). https://doi.org/10.1021/es501611u
14. A.L. Campbell, S. Mangan, R.P. Ellis, C. Lewis, Ocean acidification increases copper toxicity to the early life history stages of the polychaete Arenicola marina in artificial seawater. Environ. Sci. Technol. **48**, 9745–9753 (2014). https://doi.org/10.1021/es502739m
15. D. Xu et al., Effects of CO_2 and seawater acidification on the early stages of Saccharina japonica development. Environ. Sci. Technol. **49**, 3548–3556 (2015). https://doi.org/10.1021/es5058924

16. R. Prego, A. Cobelo-Garcia, Twentieth century overview of heavy metals in the Galician Rias (NW Iberian Peninsula). Environ. Pollut. **121**, 425–452 (2003). https://doi.org/10.1016/S0269-7491(02)00231-2
17. K.C. Cheung, B.H.T. Poon, C.Y. Lan, M.H. Wong, Assessment of metal and nutrient concentrations in river water and sediment collected from the cities in the Pearl River Delta, South China. Chemosphere **52**, 1431–1440 (2003). https://doi.org/10.1016/S0045-6535(03)00479-X
18. P. Chavez-Crooker, N. Garrido, P. Pozo, G.A. Ahearn, Copper transport by lobster (Homarus americanus) hepatopancreatic lysosomes. Comp. Biochem. Phys. C **135**, 107–118 (2003). https://doi.org/10.1016/S1532-0456(03)00103-0
19. S. Nicholson, Cardiac and branchial physiology associated with copper accumulation and detoxication in the mytilid mussel Perna viridis (L.). J. Exp. Mar. Biol. Ecol. **295**, 157–171 (2003). https://doi.org/10.1016/S0022-0981(03)00292-2
20. F. Regoli, M. Nigro, E. Orlando, Lysosomal and antioxidant responses to metals in the Antarctic scallop Adamussium colbecki. Aquat. Toxicol. **40**, 375–392 (1998). https://doi.org/10.1016/S0166-445x(97)00059-3
21. A. Anandraj, D.J. Marshall, M.A. Gregory, T.P. McClurg, Metal accumulation, filtration and O-2 uptake rates in the mussel Perna perna (Mollusca : Bivalvia) exposed to Hg2+, Cu2+ and Zn2+. Comp. Biochem. Phys. C **132**, 355–363 (2002). https://doi.org/10.1016/S1532-0456(02)00081-9
22. H.E. Parry, R.K. Pipe, Interactive effects of temperature and copper on immunocompetence and disease susceptibility in mussels (Mytilus edulis). Aquat. Toxicol. **69**, 311–325 (2004). https://doi.org/10.1016/j.aquatox.2004.06.003
23. M.P. Cajaraville et al., The use of biomarkers to assess the impact of pollution in coastal environments of the Iberian Peninsula: a practical approach. Sci. Total Environ. **247**, 295–311 (2000). https://doi.org/10.1016/S0048-9697(99)00499-4
24. S. Nicholson, P.K.S. Lam, Pollution monitoring in Southeast Asia using biomarkers in the mytilid mussel Perna viridis (Mytilidae : Bivalvia). Environ. Int. **31**, 121–132 (2005). https://doi.org/10.1016/j.envint.2004.05.007
25. A. Viarengo, L. Canesi, Mussels as biological indicators of pollution. Aquaculture **94**, 225–243 (1991). https://doi.org/10.1016/0044-8486(91)90120-V
26. A. Orbea, M. Ortiz-Zarragoitia, M. Sole, C. Porte, M.P. Cajaraville, Antioxidant enzymes and peroxisome proliferation in relation to contaminant body burdens of PAHs and PCBs in bivalve molluscs, crabs and fish from the Urdaibai and Plentzia estuaries (Bay of Biscay). Aquat. Toxicol. **58**, 75–98 (2002). https://doi.org/10.1016/S0166-445x(01)00226-0
27. R. Company et al., Effect of cadmium, copper and mercury on antioxidant enzyme activities and lipid peroxidation in the gills of the hydrothermal vent mussel Bathymodiolus azoricus. Mar. Environ. Res. **58**, 377–381 (2004). https://doi.org/10.1016/j.marenvres.2004.03.083
28. S. Katano, Y. Matsuo, K. Hanaoka, Arsenic compounds accumulated in pearl oyster Pinctada fucata. Chemosphere **53**, 245–251 (2003). https://doi.org/10.1016/S0045-6535(03)00447-8
29. J. Gailer, K.A. Francesconi, J.S. Edmonds, K.J. Irgolic, Metabolism of arsenic compounds by the blue mussel mytilus-edulis after accumulation from seawater spiked with arsenic compounds. Appl. Organomet. Chem. **9**, 341–355 (1995). https://doi.org/10.1002/aoc.590090408
30. D.E. Conners, A.H. Ringwood, Effects of glutathione depletion on copper cytotoxicity in oysters (Crassostrea virginica). Aquat. Toxicol. **50**, 341–349 (2000). https://doi.org/10.1016/S0166-445x(00)00092-8
31. T. Elfwing, M. Tedengren, Effects of copper on the metabolism of three species of tropical oysters, Saccostrea cucullata, Crassostrea lugubris and C-belcheri. Aquaculture **204**, 157–166 (2002). https://doi.org/10.1016/S0044-8486(01)00638-X
32. B. Gagnaire, H. Thomas-Guyon, T. Renault, In vitro effects of cadmium and mercury on Pacific oyster, Crassostrea gigas (Thunberg), haemocytes. Fish Shellfish Immunol. **16**, 501–512 (2004). https://doi.org/10.1016/j.fsi.2003.08.007
33. M.W. Kelly, G.E. Hofmann, Adaptation and the physiology of ocean acidification. Funct. Ecol. **27**, 980–990 (2013)

34. S.G. Li et al., Interactive effects of seawater acidification and elevated temperature on the transcriptome and biomineralization in the pearl oyster Pinctada fucata. Environ. Sci. Technol. **50**, 1157–1165 (2016). https://doi.org/10.1021/acs.est.5b05107
35. E. Lewis, D. Wallace, L.J. Allison, Program developed for CO2 system calculations. Tennessee: carbon dioxide information analysis center, managed by Lockheed Martin Energy Research Corporation for the US Department of Energy, 1998
36. X. Zhao et al., Identification of genes potentially related to biomineralization and immunity by transcriptome analysis of pearl sac in pearl oyster Pinctada martensii. Mar. Biotechnol. (NY) **14**, 730–739 (2012). https://doi.org/10.1007/s10126-012-9438-3
37. L. Wang, Z. Feng, X. Wang, X. Wang, X. Zhang, DEGseq: an R package for identifying differentially expressed genes from RNA-seq data. Bioinformatics **26**, 136–138 (2010). https://doi.org/10.1093/bioinformatics/btp612
38. A. Conesa et al., Blast2GO: a universal tool for annotation, visualization and analysis in functional genomics research. Bioinformatics **21**, 3674–3676 (2005). https://doi.org/10.1093/bioinformatics/bti610
39. T.G. Evans, F. Chan, B.A. Menge, G.E. Hofmann, Transcriptomic responses to ocean acidification in larval sea urchins from a naturally variable pH environment. Mol. Ecol. **22**, 1609–1625 (2013). https://doi.org/10.1111/mec.12188
40. A.E. Todgham, G.E. Hofmann, Transcriptomic response of sea urchin larvae Strongylocentrotus purpuratus to CO2-driven seawater acidification. J. Exp. Biol. **212**, 2579–2594 (2009). https://doi.org/10.1242/jeb.032540
41. L. Harms et al., Gene expression profiling in gills of the great spider crab Hyas araneus in response to ocean acidification and warming. BMC Genomics **15**, 789 (2014). https://doi.org/10.1186/1471-2164-15-789
42. J. Hou, X. Liu, J. Wang, S. Zhao, B. Cui, Microarray-based analysis of gene expression in lycopersicon esculentum seedling roots in response to cadmium, chromium, mercury, and lead. Environ. Sci. Technol. **49**, 1834–1841 (2015). https://doi.org/10.1021/es504154y
43. S.G. Li et al., Transcriptome and biomineralization responses of the pearl oyster Pinctada fucata to elevated CO2 and temperature. Sci Rep-Uk **6**, (2016). doi:ARTN 18943; https://doi.org/10.1038/srep18943
44. F. Gazeau, C. Quiblier, J.M. Jansen, J.P. Gattuso, J.J. Middelburg, C.H. Heip, Impact of elevated CO2 on shellfish calcification. Geophys. Res. Lett. **34** (2007)
45. P. De Wit, S.R. Palumbi, Transcriptome-wide polymorphisms of red abalone (Haliotis rufescens) reveal patterns of gene flow and local adaptation. Mol. Ecol. **22**, 2884–2897 (2013). https://doi.org/10.1111/mec.12081
46. E. Aton et al., A flow cytometric approach to study intracellular-free Ca(2+) in Crassostrea gigas haemocytes. Fish Shellfish Immunol. **20**, 493–502 (2006). https://doi.org/10.1016/j.fsi.2005.06.008
47. J.L. Fernandez-Turiel et al., trategy for water analysis using ICP-MS. Fresenius J. Anal. Chem. **368**, 601–606 (2000). https://doi.org/10.1007/s002160000552
48. S. Nicholson, Lysosomal membrane stability, phagocytosis and tolerance to emersion in the mussel Perna viridis (Bivalvia : Mytilidae) following exposure to acute, sublethal, copper. Chemosphere **52**, 1147–1151 (2003). https://doi.org/10.1016/S0045-6535(03)00328-X
49. J.T. Dingle, *Lysosomes: A Laboratory Handbook* (North-Holland Publishing Company, Amsterdam, 1972)
50. L.D. Gardner, D. Mills, A. Wiegand, D. Leavesley, A. Elizur, Spatial analysis of biomineralization associated gene expression from the mantle organ of the pearl oyster Pinctada maxima. BMC Genomics **12**, 455 (2011). https://doi.org/10.1186/1471-2164-12-455
51. H.M. Welladsen, P.C. Southgate, K. Heimann, The effects of exposure to near-future levels of ocean acidification on shell characteristics of Pinctada fucata (Bivalvia: Pteriidae). Molluscan Res. **30**(3), 125 (2010)
52. S. Goffredo et al., Biomineralization control related to population density under ocean acidification. Nat. Clim. Chang. **4**, 593–597 (2014). https://doi.org/10.1038/Nclimate2241

53. T.G. Evans, P. Watson-Wynn, Effects of seawater acidification on gene expression: resolving broader-scale trends in sea urchins. Biol. Bull. **226**, 237–254 (2014). https://doi.org/10.1086/BBLv226n3p237
54. A. Moya et al., Whole transcriptome analysis of the coral Acropora millepora reveals complex responses to CO(2)-driven acidification during the initiation of calcification. Mol. Ecol. **21**, 2440–2454 (2012). https://doi.org/10.1111/j.1365-294X.2012.05554.x
55. C. Joubert et al., Temperature and food influence shell growth and mantle gene expression of shell matrix proteins in the pearl oyster Pinctada margaritifera. PLoS One **9**, e103944 (2014). https://doi.org/10.1371/journal.pone.0103944
56. W. Liu, X. Huang, J. Lin, M. He, Seawater acidification and elevated temperature affect gene expression patterns of the pearl oyster Pinctada fucata. PLoS One **7**, e33679 (2012). https://doi.org/10.1371/journal.pone.0033679
57. A.K. Hüning, F. Melzner, J. Thomsen, et al., Impacts of seawater acidification on mantle gene expression patterns of the Baltic Sea blue mussel: implications for shell formation and energy metabolism. Mar. Biol. **160**, 1845–1861 (2013)
58. H. Ehrlich, Chitin and collagen as universal and alternative templates in biomineralization. Int. Geol. Rev. **52**, 661–699 (2010). https://doi.org/10.1080/00206811003679521
59. V. Cummings et al., Ocean acidification at high latitudes: potential effects on functioning of the Antarctic Bivalve Laternula elliptica. PLoS One **6**, (2011). doi:ARTN e16069; https://doi.org/10.1371/journal.pone.0016069
60. G. Zhang et al., The oyster genome reveals stress adaptation and complexity of shell formation. Nature **490**, 49–54 (2012). https://doi.org/10.1038/nature11413
61. S. Li et al., Morphology and classification of hemocytes in Pinctada fucata and their responses to ocean acidification and warming. Fish Shellfish Immunol. **45**, 194–202 (2015). https://doi.org/10.1016/j.fsi.2015.04.006
62. R.K. Pipe, J.A. Coles, Environmental contaminants influencing immunefunction in marine bivalve molluscs. Fish Shellfish Immunol. **5**(**8**), 581–595 (1995)
63. R.P. Kuchel, D.A. Raftos, D. Birch, N. Vella, Haemocyte morphology and function in the Akoya pearl oyster, Pinctada imbricata. J. Invertebr. Pathol. **105**, 36–48 (2010). https://doi.org/10.1016/j.jip.2010.04.011
64. S. Liang, X. Luo, W. You, L. Luo, C. Ke, The role of hybridization in improving the immune response and thermal tolerance of abalone. Fish Shellfish Immunol. **39**, 69–77 (2014). https://doi.org/10.1016/j.fsi.2014.04.014
65. L. Donaghy, A.K. Volety, Functional and metabolic characterization of hemocytes of the green mussel, Perna viridis: in vitro impacts of temperature. Fish Shellfish Immunol. **31**, 808–814 (2011). https://doi.org/10.1016/j.fsi.2011.07.018
66. B. Gagnaire, H. Frouin, K. Moreau, H. Thomas-Guyon, T. Renault, Effects of temperature and salinity on haemocyte activities of the Pacific oyster, Crassostrea gigas (Thunberg). Fish Shellfish Immunol. **20**, 536–547 (2006). https://doi.org/10.1016/j.fsi.2005.07.003
67. V. Matozzo et al., First evidence of immunomodulation in Bivalves under seawater acidification and increased temperature, PLoS One **7** (2012). doi:ARTN e33820; https://doi.org/10.1371/journal.pone.0033820
68. G. Mitta, F. Vandenbulcke, P. Roch, Original involvement of antimicrobial peptides in mussel innate immunity. FEBS Lett. **486**, 185–190 (2000)
69. E. Bachere et al., Insights into the anti-microbial defense of marine invertebrates: the penaeid shrimps and the oyster Crassostrea gigas. Immunol. Rev. **198**, 149–168 (2004). https://doi.org/10.1111/j.0105-2896.2004.00115.x
70. C.L. Yao, G.N. Somero, The impact of ocean warming on marine organisms. Chin. Sci. Bull. **59**, 468–479 (2014). https://doi.org/10.1007/s11434-014-0113-0
71. F. Gazeau et al., Impacts of ocean acidification on marine shelled molluscs. Mar. Biol. **160**, 2207–2245 (2013). https://doi.org/10.1007/s00227-013-2219-3

72. J. Thomsen, I. Casties, C. Pansch, A. Kortzinger, F. Melzner, Food availability outweighs ocean acidification effects in juvenile Mytilus edulis: laboratory and field experiments. Glob. Chang. Biol. **19**, 1017–1027 (2013). https://doi.org/10.1111/gcb.12109
73. A.S. Mount, A.P. Wheeler, R.P. Paradkar, D. Snider, Hemocyte-mediated shell mineralization in the eastern oyster. Science **304**, 297–300 (2004). https://doi.org/10.1126/science.1090506
74. M.B. Johnstone et al., Cellular orchestrated biomineralization of crystalline composites on implant surfaces by the eastern oyster, Crassostrea virginica (Gmelin, 1791). J. Exp. Mar. Biol. Ecol. **463**, 8–16 (2015). https://doi.org/10.1016/j.jembe.2014.10.014
75. J.S. Gray, Biological and ecological effects of marine pollutants and their detection. Mar. Pollut. Bull. **25**, 48–50 (1992). https://doi.org/10.1016/0025-326x(92)90184-8
76. M.T. Mazorra, J.A. Rubio, J. Blasco, Acid and alkaline phosphatase activities in the clam Scrobicularia plana: kinetic characteristics and effects of heavy metals. Comp. Biochem. Physiol. B **131**, 241–249 (2002). https://doi.org/10.1016/S1096-4959(01)00502-4
77. R.K. Pipe, J.A. Coles, F.M.M. Carissan, K. Ramanathan, Copper induced immunomodulation in the marine mussel, Mytilus edulis. Aquat. Toxicol. **46**, 43–54 (1999). https://doi.org/10.1016/S0166-445x(98)00114-3
78. R. Xiao et al., Purification and enzymatic characterization of alkaline phosphatase from Pinctada fucata. J. Mol. Catal. B-Enzym. **17**, 65–74 (2002). https://doi.org/10.1016/S1381-1177(02)00007-3
79. P.C. Das, S. Ayyappan, B.K. Das, J.K. Jena, Nitrite toxicity in Indian major carps: sublethal effect on selected enzymes in fingerlings of Catla catla, Labeo rohita and Cirrhinus mrigala. Comp. Biochem. Physiol., Part C: Toxicol. Pharmacol. **138**, 3–10 (2004)
80. P. Muñoz, J. Meseguer, M. Ángeles Esteban, Phenoloxidase activity in three commercial bivalve species. Changes due to natural infestation with Perkinsus atlanticus. Fish Shellfish Immunol. **20**(1), 12–19 (2006)
81. S. Rajalakshmi, A. Mohandas, Copper-induced changes in tissue enzyme activity in a freshwater mussel. Ecotox. Environ. Safe **62**, 140–143 (2005). https://doi.org/10.1016/j.ecoenv.2005.01.003
82. A. Orbea, H. Dariush Fahimi, M.P. Cajaraville, Immunolocalization of four antioxidant enzymes in digestive glands of mollusks and crustaceans and fish liver. Histochem. Cell Biol. **114**, 393–404 (2000)
83. C. Dautremepuits, S. Paris-Palacios, S. Betoulle, G. Vernet, Modulation in hepatic and head kidney parameters of carp (Cyprinus carpio L.) induced by copper and chitosan. Comp. Biochem. Physiol. Part C: Toxicol. Pharmacol. **137**, 325–333 (2004)
84. R. Chandran, A.A. Sivakumar, S. Mohandass, M. Aruchami, Effect of cadmium and zinc on antioxidant enzyme activity in the gastropod, Achatina fulica. Comp. Biochem. Phys. C **140**, 422–426 (2005). https://doi.org/10.1016/j.cca.2005.04.007
85. H.B. Akberali, J.E. Black, Behavioural responses of the bivalve Scrobicularia plana (da Costa) subjected to short-term copper (Cu II) concentrations. Mar. Environ. Res. **4**(**2**), 97–107 (1980)
86. D. Tran, E. Fournier, G. Durrieu, J.C. Massabuau, Copper detection in the Asiatic clam Corbicula fluminea: optimum valve closure response. Aquat. Toxicol. **65**, 317–327 (2003)
87. G. Jing, Y. Li, L.P. Xie, R.Q. Zhang, Metal accumulation and enzyme activities in gills and digestive gland of pearl oyster (Pinctada fucata) exposed to copper. Comp. Biochem. Phys. C **144**, 184–190 (2006). https://doi.org/10.1016/j.cbpc.2006.08.005
88. B. Schwanhäusser, D. Busse, N. Li, G. Dittmar, J. Schuchhardt, J. Wolf, et al., Global quantification of mammalian gene expression control. Nature **473**, 337 (2011)

Chapter 8
Study of Shell Structure

Abstract In this chapter, multiple methods were explored to study protein distribution, inner films, shell growth theory, and pearl production process.

First, we will describe how the proteins are distributed inside the calcite using different techniques:

1. The immunolocalization approach, in which antibodies against biomineralization proteins are used to localize specific proteins in the shells.
2. Confocal laser scanning microscopy and super-resolution microscopy were used to investigate the distribution of proteins inside the calcite. In addition, soluble shell matrix from the shells was shown to tune magnesium content in calcite.

Second, the inner shell films located in the shell-mantle zone were found to be important in shell formation. Furthermore, by using flat pearl method, the inner film was found to facilitate the transition of calcite to aragonite as nucleation surface and predefined mold. Third, by observing the microstructure of shells, jumping development theory and nacre-prism transition layer theory were developed. Finally, to study the function of matrix proteins in nacreous layer biomineralization in vivo, the deposition on pearl nucleus and the expression of matrix protein genes in the pearl sac during the early stage of pearl formation were studied. It is found that matrix proteins controlled the crystal phase, shape, size, nucleation, and aggregation of $CaCO_3$ crystals.

Keywords Shell structure · Protein distribution · Pearl · Shell growth theory · Confocal laser scanning microscopy

8.1 Material and Methods

8.1.1 Shell Preparation and Protein Extraction

All methods were carried out in accordance with the approved guidelines. All experimental protocols were approved by the Animal Experimental Ethics Committee of Tsinghua University, Beijing, China. Pearl oysters, *Pinctada fucata* (the

R. Zhang et al., *Biomineralization Mechanism of the Pearl Oyster, Pinctada fucata*,
https://doi.org/10.1007/978-981-13-1459-9_8

length of shells is 5.5–5 cm, wet weight is 30–40 g, and around 2 years of age), were obtained from Zhanjiang Pearl Farm (Guangxi Province, China). The cleaned shells were immersed in a 5% sodium hydroxide solution for 24 h followed by rinsing with DI water to avoid possible contamination. The two layers of the shells, the outer calcitic prismatic layer and the inner aragonitic nacreous layer, were separated mechanically by abrasion after air-drying. 30 g of the fragments were pulverized and then were decalcified with 0.8 M EDTA at 4 °C for 60 h with continuous stirring. For the extraction of the EDTA-soluble matrix (ESM-P/N, P, and N are prismatic and nacreous layers, respectively), the supernatant was collected by centrifugation at 13,000 rpm at 4 °C for 30 min and then desalted by ultrafiltration (3 K, 4000 rpm, 1–2 h). The obtained extracts were run on 12% SDS-polyacrylamide gels and stained with Coomassie Brilliant Blue R-250. The concentration of proteins was measured by NanoDrop 2000 [1].

8.1.2 Preparation of FITC–Labeled Proteins

1 mL of 250 μg·mL^{-1} proteins was mixed with 75 μL of 1 mg·mL^{-1}FITC. Then, they were incubated for 1 h, followed by dialysis against deionized water for 72 h with exchange of water every 24 h. The entire process was conducted in the dark. The resultant proteins are designated as FITC-ESM-P/N [1].

8.1.3 Preparation of Cy5-Labeled Proteins

N-Hydroxysuccinimide (NHS) esters and other activated esters (sulfo-NHS; sulfotetrafluorophenyl, STP) are reactive compounds suitable for the modification of amino groups. 0.14 mg of NHS-Cy5 (LiTTLE-PA Sciences Co., China) was dissolved in 100 μL of DMSO. Proteins were dissolved in 0.1 M $NaHCO_3$ buffer (pH 8.2). A 100 μL aliquot of NHS-Cy5 ester was added into 900 μL of the 1.0 mg·mL^{-1} protein, and the mixed solution was vortexed, followed by incubation at room temperature for 4 h. To remove unreacted agents, the conjugates were dialyzed against water for 48 h at 4 °C under agitation in the dark. The resultant proteins were designated Cy5-ESM. The average ratios of Cy5 to protein were characterized to be 1.3:1 for Cy5-ESM using a NanoDrop 2000 spectrophotometer (Thermo Fisher Scientific, USA) [2].

8.1.4 Crystallization of $CaCO_3$ Using FITC Proteins as the Additives

$CaCO_3$ was formed by the slow diffusion of around 2 g $(NH_4)_2CO_3$ vapor into six cell culture dishes in a closed desiccator for 14 h at room temperature. Each well had

180 μ L of 20 mM $CaCl_2$ and 20 μL of 25 $\mu g \cdot mL^{-1}$ proteins on a silica cover glass. The proteins are ESM-P/N or FITC-ESM-P/N. For the control, the proteins were replaced with DI water or FITC. The samples for further test were rinsed twice with DI water and air-dried [1].

8.1.5 SEM

The morphologies of samples were observed by a scanning electron microscope (SEM, FEI Quanta 200, 15 kV) after sputter coating with a thin layer of gold nanoparticles.

8.1.6 Raman Spectroscopy

By using a LabRAM HR Evolution spectrometer (HORIBA Jobin Yvon), the Raman spectra were recorded at an excitation wavelength of 514 nm. The spectra were scanned three times for 20 s in the range 100–1400 cm^{-1}.

8.1.7 CLSM Imaging

By using a LSM 710 META confocal microscope (Zeiss), the CLSM images were obtained, and the excitation wavelength for FITC-labeled samples was set at 488 nm [1].

8.1.8 The X–ray Diffraction

By using a Bruker D8 Advance X-ray diffractometer, the X-ray diffraction (XRD) patterns were obtained (Cu Kα, $\lambda = 0.154$ nm, 40 kV, 200 mA). The scattered radiation was detected in the angular range 20–80° (2θ) with a scan rate of $1.2° \cdot min^{-1}$.

8.1.9 Stochastic Optical Reconstruction Microscopy Imaging

8.1.9.1 STORM Sample Preparation

Glass bottom cell culture dishes (Nest) were used for all STORM imaging. Imaging buffers supplemented with an oxygen scavenging system are critical for reliable photoswitching of the fluorophores imaging biological samples. It usually consists of 10% (w/v) glucose (Sigma), 0.1% (v/v) β-mercaptoethanol (Sigma), 500 μg/mL glucose oxidase (Sigma), and 10 μg/mL catalase (Roche). However, this imaging buffer seems to be unsuitable for Cy5-labeled proteins in the crystals, and the bleaching time of fluorophores will take more than 30 mins. Thus, the imaging buffer containing the oxygen scavenging system was replaced with phosphate-buffered saline buffer. It is believed that the single-crystalline hosts can serve as dense shells to reduce the diffusivity of oxygen to the dyes, which is a possible mechanism occurring in our system [2].

8.1.9.2 Microscope Image Acquisition

A Nikon Ti-E inverted microscope equipped with an N-STORM system was used for STORM imaging. The fluorescence emission of Cy5 was collected with a NA 1.49100 × oil immersion objective. The setup was controlled by Nikon NIS-Elements AR software with a N-STORM module. To obtain images, the field of view was selected based on the live electron-multiplying charge-coupled device (EMCCD) image under 647 nm illumination. A confocal stack was then collected using 647 nm excitation. The field of view of the confocal scan area and the EMCCD were set to be identical. Next, the samples were bleached by scanning in the z-axis using maximal illumination from the 647 laser line. Finally, the STORM image was acquired using a STORM filter cube (Nikon) and the EMCCD. The laser power used was 40–50 mW, and the exposure time was 20–30 ms. Oblique illumination was applied using the TIRF illuminator of the microscope. The total number of frames acquired was at least 10,000. The reconstructed 3D STORM image was created using algorithms for molecule identification and drift correction as previously described and employed by Nikon Instruments. The images were analyzed by ImageJ 1.6.0 [2].

8.1.10 Rabbit Polyclonal Antibodies (P–anti–NCA) and a Mouse Monoclonal

Antibodies against CA domain of nacre were developed in New Zealand rabbits and BALB/c mice [3]. The DNA encoding TGF-b domain (amino acids 346–447) of Pf-BMP-2 (GenBank accession no. AB176952) was cloned into pET-28b expression vector (Novagen) and expressed in bacterial strain of *Escherichia coli* BL21 (DE3). The recombinant protein was highly purified using a HiTrap chelating HP affinity column (Amersham Biosciences). Polyclonal antibodies against the recombinant BMP-2 (anti-P43) or total SM proteins (anti-SM) were raised in New Zealand rabbits. The antisera were then purified by 33% saturation ammonium sulfate precipitation and a protein A agarose column (Calbiochem). The purified antibodies were used for immunoblot analysis, immunolabeling, and antibody inhibition assay [3].

8.1.11 Immunoblot Analysis

The proteins were electrophoretically transferred to nitrocellulose membranes using a Hoefer semidry blotter (Pharmacia Biotech). The nitrocellulose membranes were blocked with 5% skimmed milk and incubated with 5 μg/mL primary antibody for 2 h. After washing and incubated with an appropriate second antibody-conjugated alkaline phosphatase (Santa Cruz), detection was conducted using BCIP/NBT solution (Roche) [3].

8.1.12 Shell Decalcification

The decalcified nacre was thoroughly washed with DI water. The lamellar sheets were peeled off from the fragments, mounted on glass slides, air-dried, and ready for immunolabeling.

Partial decalcification of nacre was performed as follows. The fresh nacre was mechanically fractured into small fragments and cleaned in diluted sodium hypochlorite solution for 2 min. The samples were then partially etched by 10 mM acetic acid (pH 3.5) for 1–3 min and neutralized with 10 mM $NaHCO_3$. The partially etched nacre lamellae were thoroughly washed with DI water and were ready for immunolabeling [3].

8.1.13 Immunolabeling of Nacre

Completely decalcified lamellar sheets, as well as partially etched nacre lamellae, were blocked in 10% goat serum for 2 h. Then, they were incubated with a primary antibody at concentration of 5 μg/mL or preimmune serum for 2 h. The samples were then incubated with a secondary antibody (anti-SM, P-anti-NCA, or anti-P43) or a mouse monoclonal antibody that recognized nacrein (M-anti-NCA). The secondary antibodies were goat anti-rabbit/mouse coupled to rhodamine (Santa Cruz) and goat anti-rabbit coupled to 12 nm gold particles (Jackson). In every step of immunolabeling, the samples were washed three times with PBS containing 0.05% Tween 20 for 5 min. At last, samples were washed in water. The rhodamine-labeled samples were examined by a fluorescence microscope (Leica DIMR). The immunogold-labeled samples were dried overnight, carbon coated, and examined using a scanning electron microscopy [3].

8.1.14 In Vitro Crystallization Experiment

The CO_2 diffusion method was adopted to synthesize crystals. The samples were prepared by placing two 35 × 10 mm CELLSTAR cell culture dishes, respectively, filled with 1 mL 10% H_2SO_4 and 4 mL 0.075 g/mL NH_4HCO_3, each into a 100 × 20 mm glass Petri dish. The initial solution contained 10 mM $CaCl_2$ and different concentrations of $MgCl_2$. Crystallization experiments were carried out by mixing 50 μL initial solutions and 1 μL ESM solution on silanized glass cover slips laid in the Petri dish. After 4 h precipitation, the crystallizing solution was removed. $CaCO_3$ crystals were rinsed in ethanol and stored in a desiccator for further use [1].

8.1.15 X-ray Photoelectron Spectroscopy (XPS) Analysis

XPS was used for surface analysis. $CaCO_3$ crystals collected from crystallization experiments were collected for XPS analysis. A PHI-5300 Quantera XPS spectrometer (ULVAC-PHI, Japan) with a spherical sector analyzer and unmonochromatized Al Kα (E = 1486.7 eV) as an excitation source was used. Adventitious carbon, with a binding energy of 284.8 eV on clean metal surfaces, was utilized as an internal standard for charge correction for all samples [4].

8.1.16 Mg Content of the Prismatic Layer

3.0 g of shells separated from the prismatic layer of *P. fucata* was dissolved with 500 μL 37.5% HCl for 3 h. The resulting solution was filtered with a 0.22 μm membrane filter and diluted to 50 mL. Quantifications of Ca and Mg were performed using an ICP-OES (Thermo Electron Corp., Franklin, USA). A reagent blank was prepared by diluting 500 μL 37.5% HCl to 50 mL with DI water [4].

8.1.17 Isolation of the Film Proteins

After the shells being rinsed with EDTA (0.5 M) for 30 min, the film was separated from the shells. The detached film was rinsed again with EDTA (0.5 M) until it was completely decalcified. The decalcified film was washed extensively with water and collected by centrifugation. After that, the film was ground in a mortar with continuous addition of fluid nitrogen. The powder-like products were dissolved in water and the soluble and insoluble fractions were collected by centrifugation. The insoluble fraction was stored at −20 °C, while the soluble fraction was subjected to ammonium sulfate fractionation. The precipitate was dissolved in water, and then it was extensively dialyzed against DI water and concentrated by ultrafiltration (Millipore; cutoff, 5 kDa) [5].

8.1.18 Amino Acid Analysis of the Film Proteins

For amino acid analysis, the soluble and insoluble fractions of the film were hydrolyzed under vacuum in HCL (6 N) at 110 °C for 24 h. Then, samples and amino acid standards were analyzed on the amino acid analyzer. The amino acid compositions of the samples, expressed as mole percentage, represent the average of three independent determinations [5].

8.1.19 $CaCO_3$ Precipitation Experiment

In brief, 10 mL of sample solution was mixed with $NaHCO_3$ (100 mL, 40 mm, pH 8.7). After the addition of $CaCl_2$ (100 mL, 20 mm) to the mixed solution, the formation of $CaCO_3$ precipitates was monitored by recording the turbidity. Changes in the turbidity of the solutions were measured every minute for 5 min by the absorbance at 570 nm using a spectrophotometer [5].

8.1.20 Crystallization in the Presence of the Inner–Shell Film

After being immersed in 0.5 M EDTA for 12 h, the inner-shell film was detached from the shell with a pair of forceps. The detached film was rinsed thoroughly with distilled water. Then the film was immersed in the calcitic crystallization solution with or without the water-soluble proteins of the nacre. After 24 h, the crystal precipitated on the film was characterized [5].

8.1.21 Implantation Procedure and Sample Collection

The implantation procedure was conducted as described by the UC Santa Barbara group, with some modifications. Square microscope cover glass (approximately 36 mm^2 by 0.15 mm thick) covered by inner-shell film and the cover glass used as a control were inserted into the extrapallial cavity (the region between the mantle and shell) of a total of 100 individuals. Care was taken to avoid damage to the mantle tissue. After implantation, oysters were returned to the seawater tanks for 5, 10, 15, 20, or 25 days. Implants remained in contact with the shell-facing side of the mantle organ throughout the duration of their incubation. At each sampling time, 18 oysters were randomly divided into three groups and sacrificed, and the mantle facing sides of the cover glass were gathered and kept in liquid nitrogen for total RNA extraction. Mantle tissues without facing sides of the cover glass were collected as controls. These implants were excised from the shell with a scalpel and dried for examination [6].

8.1.22 Typical Pearl and Pearl Sac Selection

P. fucata were cultured at the Guofa Pearl Farm, Beihai, Guangxi Province, China. Aragonitic nuclei and mantle grafts were implanted in the gonads of a total of 800 individuals on 10 April 2009. On the 5th, 10th, 15th, 20th, 25th, 30th, and 35th day following implantation, 60 oysters with aragonitic nuclei were harvested, and the pearl sacs were carefully isolated from the gonad tissues. The nuclei were then separated from the pearl sacs for study [6].

8.1.23 FTIR

For FTIR investigations, the crystals were collected from the cover glass, and KBr pellets were produced with a 1% sample and analyzed on a Perkin–Elmer model 1600 FTIR spectrometer. All the spectra were recorded at 4 cm^{-1} resolution with 64 scans with a strong Norton–Beer apodization [6].

8.1.24 Statistical Analysis

The cDNA of the three groups was detected at sampling time and repeated three times for each sample. After data analysis, we standardized the $2 - \Delta\Delta Ct$ of each gene by taking the fifth day as the zero point. Data were compared by one-way analysis of variance (ANOVA) and with post hoc comparisons by Duncan's Multiple Range Test using the software SPSS 20.0. All tests were performed at a significance level of 5% [6].

8.1.25 Primers

Refer to Supplementary Table 8.1 for details.

8.2 Study Biomineralization by Using In Vitro Model

8.2.1 Localization of Shell Matrix Proteins in Biominerals or $CaCO_3$

Although identification of biomineralization-related proteins is a prerequisite to better understand biomineralization, their spatial location and function are also important. Knowing the spatial location of these proteins is challenging because of their small size and diversity.

Annular dark-field scanning transmission electron microscopy (ADF-STEM) coupled with three-dimensional (3D) reconstruction has been applied for nanoscale visualization of biomacromolecules within calcite from *Atrina rigida* shells [7] and calcite single crystals grown in agarose hydrogels [8], respectively. The organic matters are found to exist as network or aggregates inside the calcite [8]. Electron tomography has also been used to show organic-containing voids in the biominerals. X-ray absorption near-edge structure (XANES) spectroscopy is used to characterize the organic contents along the axis of prisms, indicating the presence of concentrated organic regions [9]. XANES has also been used to show that Asp-rich peptides can be occluded in calcite and permanently disorder crystals at both short- and long-range distances [10]. Pulsed-laser atom-probe tomography (APT) has been used to construct 3D chemical maps of the organic fibers in nanocrystalline magnetite mineral in the tooth of the chiton *Chaetopleura apiculata* [11]. These methods enable us to visualize 3D structures of organics occluded in biominerals, but cannot show how one type of biomacromolecule is distributed. X-ray powder diffraction (XRD) is a powerful tool to show that incorporation of organic matters into biominerals can cause anisotropic lattice distortions along the c-axis compared with their non-biogenic counterparts [12]. Using this approach, Boaz Pokroy et al. screened the incorporation of 20 amino acids into calcite, and they found that amino acid charge, size, rigidity, and the relative pKa of the carboxyl and amino functional groups were important factors for the interaction with the mineral phase [13, 14].

As a complementary method, Weber et al. recently used recombinant green fluorescent protein (GFP) fusion proteins to study the effects of perlucin—a lectin protein from the nacre of *Haliotis laevigata*—on the crystallization of $CaCO_3$ and visualized them by confocal laser scanning microscopy (CLSM) [15, 16]. By XRD, they found that GFP–perlucin can be incorporated into the calcite and induced concentration-dependent anisotropic lattice distortions along the host's c-axis [16]. However, GFP is a relatively large protein with a molecular weight (MW) of 26.9 kDa compared to 17.9 kDa perlucin.

8.2.1.1 Localization of Matrix Proteins in Biominerals or $CaCO_3$ by Immunolocalization

In this chapter, two methods will be introduced. The first is the immunolocalization approach, in which antibodies against biomineralization proteins are used to localize the specific proteins in the shells. This method can indirectly observe the spatial relationship between the organic and mineral phases, but it cannot reflect how proteins are distributed within the mineral. Immunolocalization has been widely used to study the functions of matrix proteins in the growth of biomaterials, such as sea urchin, coral skeletons, and shells [17, 18]. Antibody inhibition assay is proved to be a reliable method for investigation of the functions of matrix proteins in the growth of nacre lamellae in vivo [19]. When injected into the extrapallial fluid between shell and mantle tissue, where nacre calcification occurs, the antibodies against a specific matrix protein could interrupt the action of this protein in calcification, resulting in abnormal growth of nacre.

The functions of this protein in nacre formation will be reflected from the abnormal growth pattern. Nacrein is a major soluble matrix (SM) protein in the nacre of the *P. fucata* and has carbonic anhydrase activity [20], which is essential for rapid shell development [21]. Moreover, nacrein-like proteins are conserved in bivalves and gastropods and thus are thought to play important roles in nacre formation [22]. In this study [3], we conducted immunolocalization and antibody inhibition assays to examine its function in nacre formation. We also characterize an insoluble matrix (ISM) protein by the antibodies raised against the recombinant protein of *P. fucata* bone morphogenetic protein 2 (Pf-BMP-2). A better understanding of the structural and functional roles of SM and ISM proteins in nacre biomineralization will provide knowledge to mimic nacre formation.

After complete decalcification of lamellae, only organic interlamellar sheets remained, and individual crystal imprints on the sheets were no longer detected by a microscope (Fig. 8.1a). Because high molecular weight of antibody complex for immunolabeling makes them hardly diffusing through the sheets, immunolabeling of matrix proteins in decalcified nacre lamellae mostly reflects their distributions in the surface layer of interlamellar sheets [18].

Immunolabeling experiment showed that SM proteins existed in polygonal crystal imprints in the organic matrix sheet (arrowhead) as well as the central regions of the imprints (arrow, Fig. 8.1c), suggesting that SM proteins constitute the intertabular matrix and the nucleation sites. Immunolabeling using P-anti-NCA and M-anti-NCA both discerned polygonal crystal imprints (arrowheads), but indiscernible staining was observed in the central area (Fig. 8.1d, e), indicating that nacrein may be localized to intertabular matrix. In the negative control of immunolabeling, in contrast, no staining was shown in the matrix sheet (Fig. 8.1b). Figure 8.2 showed the distributions of nacrein and P43 in the cross section of nacre lamellae (section parallel to the c-axis). The immunogold labeling confirmed that nacrein widely distributes in nacre lamellae and is embedded in the mineral phase as calcium carbonate crystals growing (Fig. 8.2c, d). P43 distributes

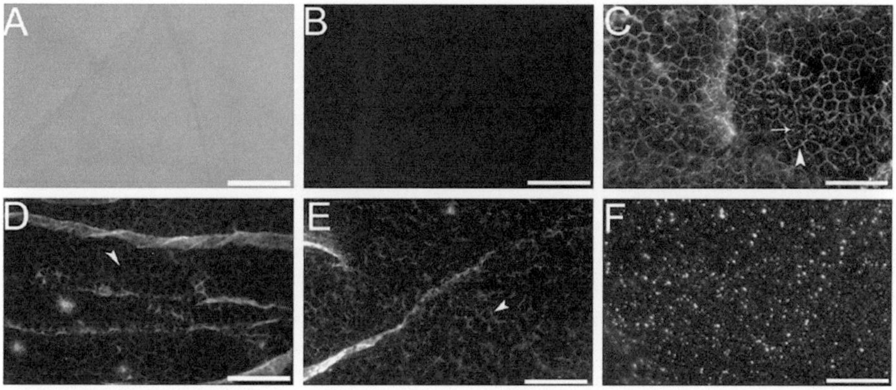

Fig. 8.1 (**a**) Light micrograph of completely decalcified interlamellar sheets, in which only organic matrix remained; the lines in the picture are due to folds in the interlamellar sheet. (**b–f**) Fluorescence micrographs of decalcified interlamellar sheets labeled by preimmune serum, anti-SM, P-anti-NCA, M-anti-NCA, and anti-P43, respectively. Arrowheads indicate the stained intertabular matrix. Arrow in (**c**) indicates the stained area in the center of the crystal imprints. Bar, 20 μm

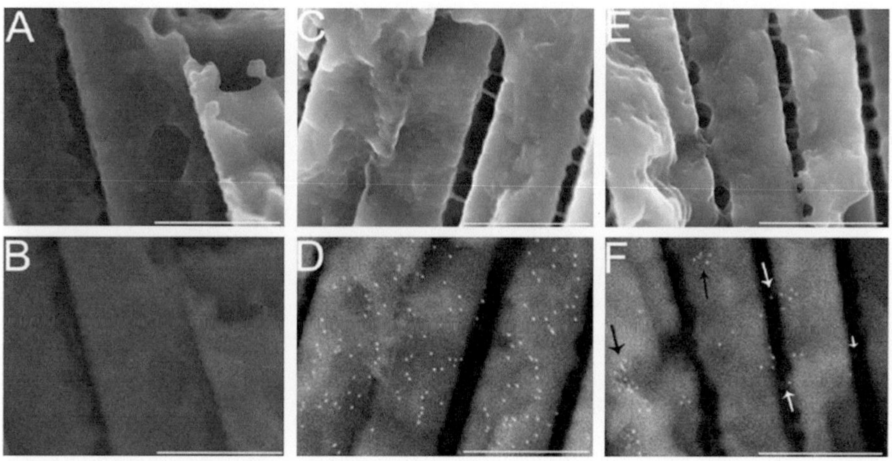

Fig. 8.2 Immunogold labeling the cross section of partially etched nacre lamellae (section parallel to the c-axis). (**a, c,** and **e**) The lamellae labeled by preimmune serum, P-anti-NCA, and anti-P43, respectively; (**b, d,** and **f**) SEM-BSE imaging the immunogold staining in the lamellae of (**a, c,** and **e**), respectively. Bars, (**a** and **b**) 500 nm; (**c** and **d**) 400 nm; (**e** and **f**) 500 nm

within the lamellae (Fig. 8.2e, f, black arrows) and is also localized to interlamellar matrix at interface with the mineral phase (Fig. 8.2e, f, white arrows). As expected, the negative controls displayed a low background (Fig. 8.2a, b).

The results show that SM and ISM proteins differentially distribute in nacre lamellae. Some regulating activities of matrix proteins in the growth of aragonitic tablets would be indicated by antibody inhibition assays. The analysis of abnormal growth pattern of aragonitic tablets along with immunolocalization would further reveal the regulatory mechanisms of matrix proteins in nacre formation.

8.2.1.2 Localization of Shell Matrix Proteins in Biominerals or $CaCO_3$ by Immunolocalization by Light Microscopy

Another possible approach to visualize proteins in crystals is to label macromolecules with small fluorescent molecules. Fluorescein isothiocyanate (FITC) is a derivative of fluorescein with a MW of 389.38 Da and has excitation and emission spectrum peak wavelengths of approximately 495 nm and 519 nm. In order to probe the distribution mode of biomineralization proteins inside crystals, FITC were used to label proteins extracted from the shells of *P. fucata*. By using CLSM in 2D mode and Z-stack imaging, we were able to determine the distribution of FITC-labeled proteins inside the artificial calcite according to their fluorescence intensity [1].

The extracted proteins were divided into two groups according to their origins from the prismatic calcite or nacreous aragonite layer: ESM-P and ESM-N (ESM, EDTA-soluble matrix; P, prismatic layer; N, nacreous layer). Before determining the location and distribution of these proteins in synthesized $CaCO_3$, their effects on the crystal growth and morphology were investigated. Compared with the rhombohedral crystals of the control sample (Fig. 8.3a), ESM-P significantly affected the morphology of $CaCO_3$, forming round crystals with diameters of 5–10 μm (Fig. 8.3b). Moreover, the surfaces of the crystals were rougher than the control. When ESM-N was used as the additive, truncated patterns were obtained at the edge of the crystals (Fig. 8.3c).

Confocal laser scanning microscopy (CLSM) is a technique for obtaining high-resolution optical images with depth selectivity, allowing imaging at a controlled and highly limited depth of focus [23]. CLSM is widely used in numerous biological science disciplines, from cell biology and genetics to microbiology and developmental biology by labeling biological objects with fluorescent markers [23]. Therefore, CLSM can be a powerful tool to visualize interaction between organic matters

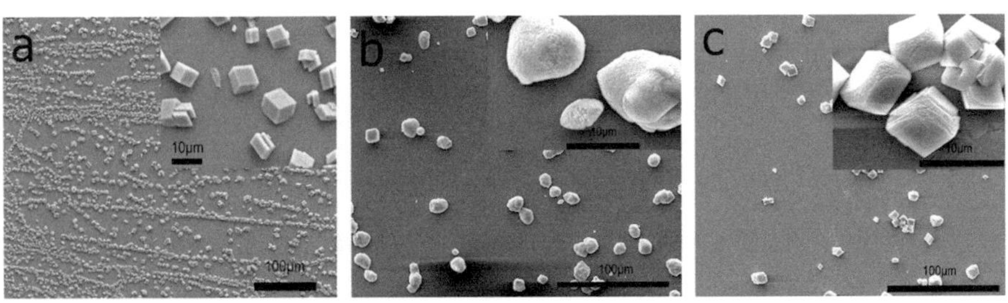

Fig. 8.3 Scanning electron microscope (SEM) images of $CaCO_3$ crystals obtained without the addition of any proteins (**a**) and in the presence of ~2.5 $\mu g \cdot mL^{-1}$ ESM-P (**b**) and ESM-N (**c**). Insets are their magnified images. ESM-P, EDTA-soluble matrix from the prismatic layers of *Pinctada fucata*; ESM-N, EDTA-soluble matrix from the nacreous layers of *Pinctada fucata*

with inorganic materials such as crystals [24, 25]. Figure 8.3 shows the confocal laser scanning micrograph of calcium carbonate precipitated with dye-labeled proteins as the additives.

FITC was used because of its availability and small molecular weights compared to GFP. FITC is negatively charged and has a deprotonated carboxyl group on the phenyl ring at pH 8.0 (the pH of reaction solution). In the control group in which no additives were added, no fluorescence was observed at 488 nm with intensity of 18% (Fig. 8.4a-a2). Under the same condition, when FITC was added in the $CaCl_2$, homogenous fluorescence was observed inside the calcite (Fig. 8.4b-b2), and the morphology was identical to the control group, indicating FITC has no preference binding to specific faces of calcite. Interestingly, when FITC-ESM-P protein was added, the rounded particles were formed, and heterogeneous fluorescence was observed. Specifically, the fluorescence was more concentrated at the edge of particles (0.6–1.7 μm in depth below the surfaces of particles) (Fig. 8.4c-c2). When FITC-ESM-N protein was added, rhombohedral crystals with blur surfaces under optical microscope were formed, and heterogeneous fluorescence was found. The fluorescence was concentrated at the center (with an area of 1–4 μm^2) of crystals (Fig. 8.4d–d2). To exclude the possible misinterpretation caused by the different orientation of $CaCO_3$ on the glass cover, we performed z-stack imaging in a region of $200 \times 200\ \mu m^2$ with 9 μm height (interval 0.75 μm).

Calcites in the ESM-P group showed a hole in the center of globular particles (Fig. 8.5a); calcites in the ESM-N group showed strong fluorescence at the center of crystals (Fig. 8.5b). Figure 8.5c shows that in the ESM-P group, fluorescence intensity at the edge of particles is 2–5 times than the other parts. By contrast, in the ESM-N group, fluorescence intensity at the center of particles is 5–10 times than the other parts (Fig. 8.5d).

In summary, by using FITC-labeled proteins extracted from the shell of *P. fucata* as the additives, a heterogeneous distribution of biomineral-associated matrix proteins inside synthetic calcite at the microscale was observed for the first time according to the fluorescence intensity. The distribution may be related to their secondary structures and functions. This study shows the use of small-molecule dye-labeled matrix proteins as the additives for the visualization of protein distributions in crystals. Moreover, it indicates that different distribution patterns of proteins may occur in natural biominerals and may inspire synthesis of biocomposite materials.

Although it is possible to visualize the distribution of proteins at the microscale, the information obtained is still limited due to the diffraction limit of traditional light microscopy (approximately 200 nm), which is much larger than the sizes of a single protein molecule or protein complexes. Super-resolution light microscopy such as stochastic optical reconstruction microscopy (STORM) has achieved three-dimensional multicolor fluorescence imaging of cells and tissues with subdiffraction limit resolution of approximately 10 nm, leading to the discoveries of previously unknown cellular structures and the structures of many molecular assemblies in cells [26, 27]. Direct STORM has been used to visualize the distribution of gelatin in vaterite and calcite as well to investigate the process by which gelatin is excluded

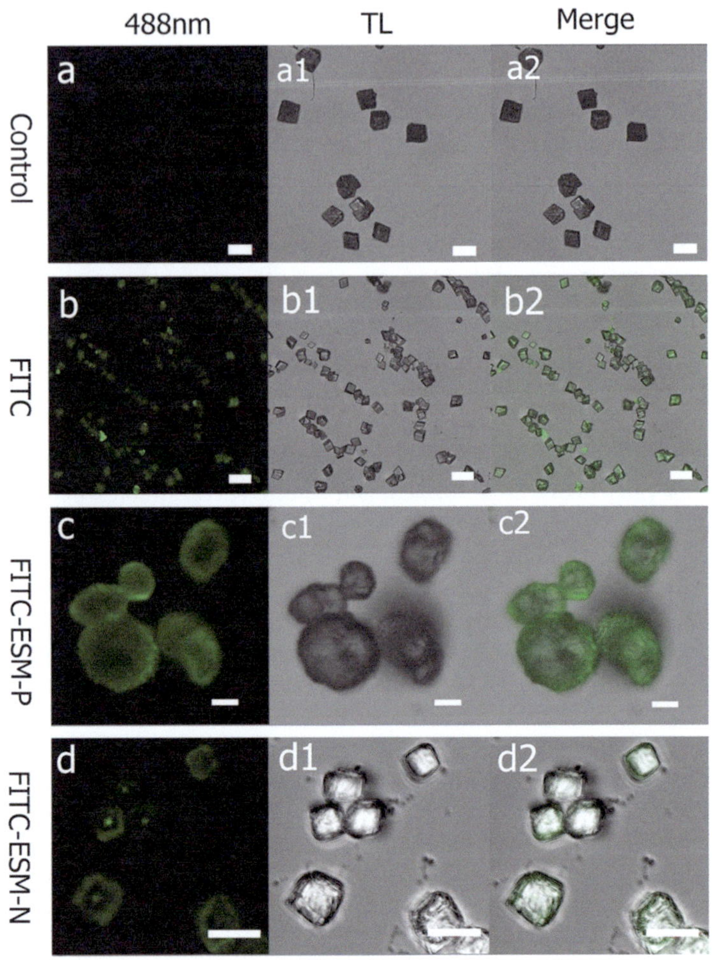

Fig. 8.4 Confocal laser scanning micrographs of $CaCO_3$ precipitates. (**a**, **a1**, **a2**) The control $CaCO_3$ formed without the addition of proteins. (**b**, **b1**, **b2**). The $CaCO_3$ formed in the presence of fluorescein isothiocyanate. (**c**, **c1**, **c2**). The $CaCO_3$ formed in the presence of FITC-labeled ESM-P proteins. (**d**, **d1**, **d2**). The $CaCO_3$ formed in the presence of FITC-labeled ESM-N proteins. (**a–c**) are images taken at 488 nm laser excitation (intensity, 18%); **a1**, **b1**, and **c1** are bright-field images; **a2**, **b2**, and **c2** are the merge images. Scar bars, **a–a2** and **b–b2**, 20 μ m; **c–c2** and **d–d2**, 5 μm

from the crystals [28]. However, direct observation of biomineralization-related protein occlusions inside calcium carbonate with high-resolution optical microscopy is lacking.

Then, we labeled soluble nacre proteins extracted from the nacreous layers of the *P. fucata* with cyanine 5 (Cy5) and then directly visualized the distribution patterns of nacre proteins inside the whole synthetic calcite with STORM imaging [2]. The samples were still the same as those used in the CLSM mode (Fig. 8.6a) but were

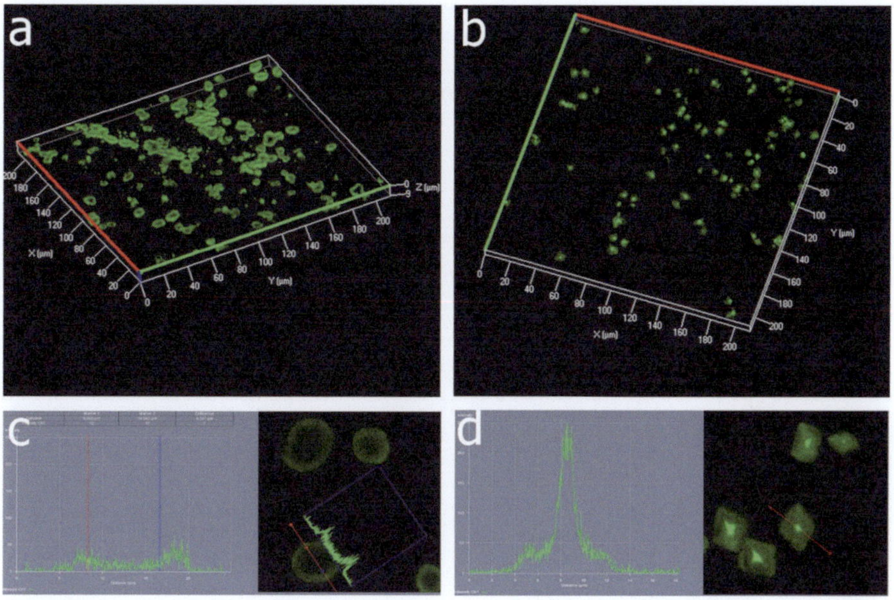

Fig. 8.5 Z-stack images of $CaCO_3$ formed in the presence of FITC-labeled ESM-P (**a**) and ESM-N proteins (**b**) (scanning depth 9 μm, interval 0.75 μm, area 200 × 200 μm^2). Fluorescence signal intensity across the lines in the crystals in the presence of FITC-labeled ESM-P (**c**) and ESM-N proteins (**d**) (The analysis was performed using ZEN 2012, Zeiss)

observed under STORM imaging, and it can be seen that detailed distribution patterns of proteins were revealed in the 2D projections of 3D imaging (Fig. 8.6b). Based on the full width at half maximum (FWHM) of gray values in a single fluorescence spot, the resolution of the images was approximately 31 nm. Clearly, three forms of fluorescent clusters, which reflect the distribution of proteins, were observed: chain-like with lengths of 0.4–1.8 μm (Fig. 8.6b, a green ellipse), islet-like with diameters of 30–400 nm (Fig. 8.6b, green arrows), and haze-like clusters (Fig. 8.6b, a green rectangle). These three forms of fluorescence clusters can be found in the crystal in the presence of 50 (Fig. 8.6c, d) and 100 $\mu g \cdot mL^{-1}$ ESM (Fig. 8.6e, f) as well. From the plot profiles of gray values in the yellow lines (Fig. 8.6a, b), it can be seen that the gentle curves of gray values in the CLSM mode became peaks in the STORM imaging (Fig. 8.6g), indicating distinct protein-rich and protein-poor regions. In addition, it is found that as the concentration of ESM increased, the average diameters of islet-like particles improved, which were 74, 157, and 212 nm for 20, 50, and 100 $\mu g \cdot mL^{-1}$, respectively (Fig. 8.6h). Notably, from STORM data, small crystals do not have homogeneous distribution of protein, rather the outer edge. It is possible that, in CLSM, due to light diffraction limitation, two spots with distance less than 200 nm cannot be distinguished and were visualized as blurred areas as shown in Fig. 8.6a, while STORM has a resolution of about 30 nm and the same areas in the CLSM can be distinguished as separated areas. From

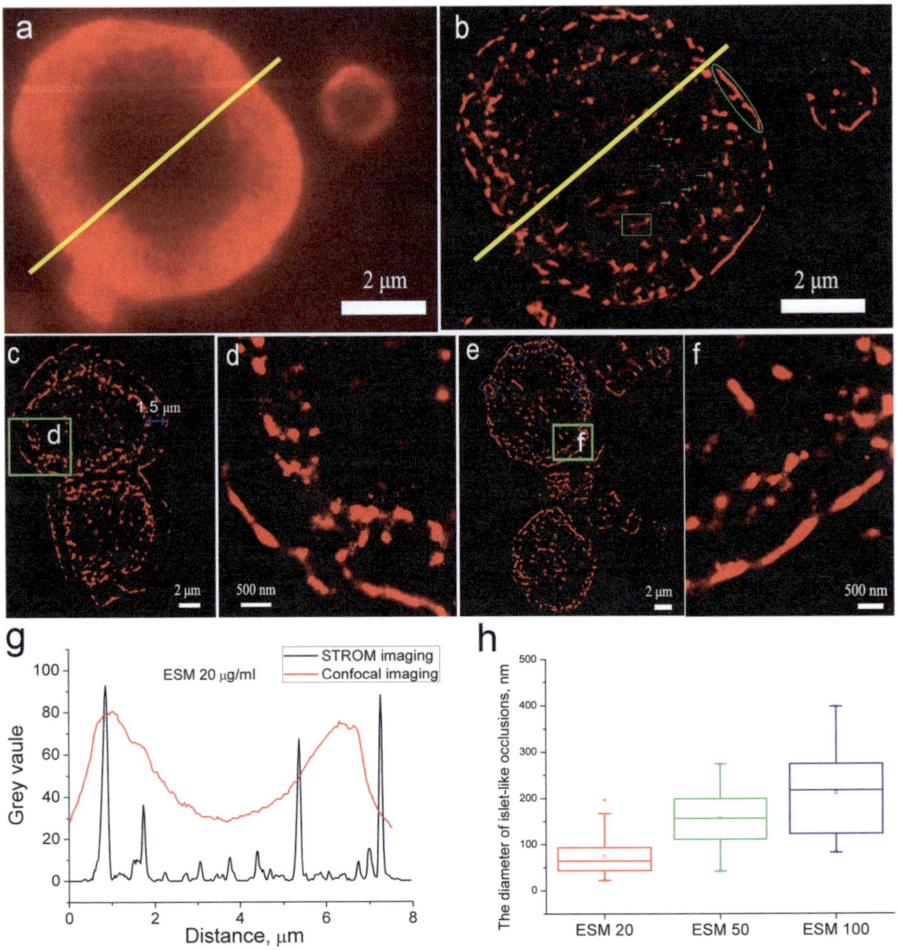

Fig. 8.6 STORM imaging of $CaCO_3$ formed in the presence of ESM. (**a**) Confocal images of $CaCO_3$ crystals formed in the presence of 20 μg·mL^{-1} Cy5-labeled ESM. STORM images of $CaCO_3$ crystals formed in the presence of 20 (**b**), 50 (**c, d**), and 100 μg·mL^{-1} (**e, f**) Cy5-labeled ESM. (**g**) Plot profile of gray values corresponding to the yellow lines in panels a and b. (**h**) Diameter of islet-like occlusions in the $CaCO_3$ crystals. Each box plot presents the measured distribution of values per group. The box encloses the range from the first to third quartile. The horizontal line and the square within the box represent the median and the mean of the distributions, respectively

the magnification of the STORM pictures (Fig. 8.6d, f), it can be seen that the fluorescence of one particle was continuous rather than being made up of small fluorescence dots, indicating that proteins possibly aggregated inside the crystals. Interestingly, as the concentration of ESM increased, a two-ring pattern with a protein-poor region flanked by protein-rich regions appeared with an average diameter of 1.5 μm, especially in the presence of 50 μg·mL^{-1} ESM (Fig. 8.6c, blue arrows). The protein-poor region in the crystals should correspond to the decreased fluorescence region between the outer edge and inner edge observed in the CLSM, showing that STORM imaging can enable us to identify and corroborate unprecedented and unique structures or patterns in the whole crystals. At the concentration of 100 $\mu g \cdot mL^{-1}$, the two-ring patterns in the crystal became less obvious. However, it is found that, at the edges of large crystals, fluorescence patterns similar to the edges of small rhombohedral calcium carbonates were observed (Fig. 8.6c, blue rectangles).

In summary, we used the super-resolution microscopy method STORM to directly observe the Cy5-labeled nacre proteins in the synthetic calcite. Due to the improved resolution, we found three forms of occlusions in the calcite and a unique and unprecedented two-ring pattern with a protein-poor region flanked by protein-rich regions in the whole calcite. This study will improve our understanding of the internal structures of biominerals or synthetic minerals and has implications for synthesizing nacre-like materials with increased mechanical properties.

8.2.2 Tuning Calcite Magnesium Content by Soluble Shell Matrix

Mg as the principal impurity ion accommodated by the calcite lattice, Stephenson et al. found that its incorporation is susceptible to additive effects. In the presence of a carboxyl-rich synthetic peptide, the calcite Mg content showed an increase of up to 3 mol% [29]. Calcite Mg content has been extensively studied because the Mg/Ca ratio of marine sediments and calcified organic structures is a widely used paleothermometer when reconstructing the past earth environment [30]. This finding provided a probable explanation for the unexplained deviations observed in predictions using biogenic calcites. However, if this mechanism of impurity control is also utilized in biomineralization, similar activities would be associated with biomineralization-related natural biomolecules, and in consequence of physiological control rigor, negative regulators must therefore exist with positive regulators. Here, we extracted the two soluble matrices from a bimineralic bivalve, *Pinctada fucata*, seeking to ascertain whether or not they may affect the incorporation of Mg into calcite differently [4].

Morphology of the precipitated crystals was characterized by SEM observation. Despite varying in size, SEM observation confirmed that the morphology characteristics were homogeneous for crystals within the same group. Rhombohedral crystals were generated by the initial solution alone (Fig. 8.7A, a), with their surfaces slightly modified, which may be ascribed to the poisoning and incorporation of Mg^{2+}. For crystals grown in the presence of either ESM (Fig. 8.7B, b, C, c), the morphologies were intensively modified. Nonetheless, they retained some polyhedral characteristics that resembled calcite traits. Raman spectra have demonstrated that all of the analyzed single crystals were calcitic (characteristic peak at about

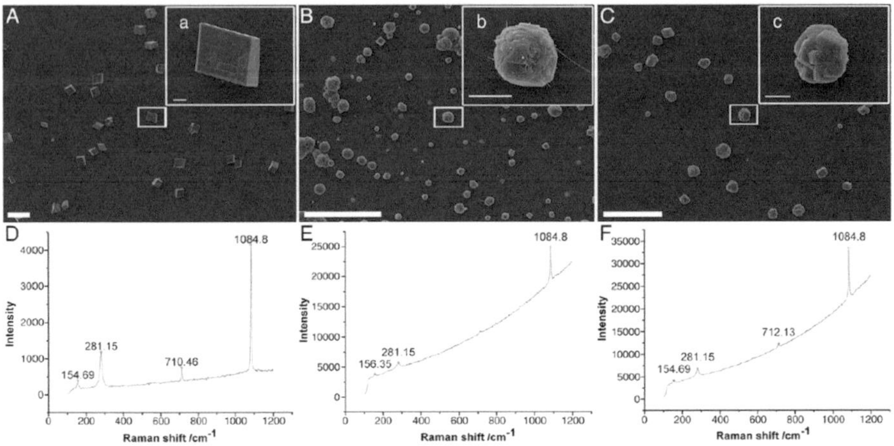

Fig. 8.7 Morphology and polymorph of $CaCO_3$ crystals. SEM images of $CaCO_3$ crystals and corresponding Raman spectra. (**A**, a) Crystals grown in the initial solution without additive. (**B**, b) Crystals grown with prismatic ESM in the initial solution. (**C**, c) Crystals grown with nacreous ESM in the initial solution. (**D, E, F**) Raman spectra corresponding to (**A**), (**B**), (**C**). Scale bars, 100 μm in (**A**), (**B**), (**C**); 10 μm in insets. Crystals shown in this figure were prepared at a Mg^{2+} concentration of 1.25 mM

281 cm^{-1} in the lattice mode region); therefore, their respective Mg contents were comparable. The high baseline of the Raman spectra in Fig. 8.7E, F indicates the adsorption of organic ESM components on the crystal surface. The crystals shown in Fig. 8.7 were grown at a Mg^{2+} concentration of 1.25 mM, with similar morphology and polymorph characteristics observed for crystals prepared with lower Mg^{2+} concentrations. The prismatic EDTA-soluble matrix (ESM) of *Pinctada fucata* was found to significantly reduce calcite Mg content when presented as an additive. Together with the Mg exclusion effect identified in the prismatic layer of *P. fucata*, a novel mechanism was proposed that organisms might use the additive effect to control the crystal impurity content in a solution of defined element composition. The precipitation inhibition assay and the direct ion-binding assay demonstrated that the ESM effect is realized neither by altering the precipitation rate nor by regulating the relative ion activity in the crystallizing solution. A plausible correlation between crystal impurity control and polymorphism determination was also discussed.

The result of the overall EDS analysis is indicated in Fig. 8.8. Long error bars are observed because a relatively large deviation is associated with EDS analysis, and only six parallel measurements were performed for each point. However, with such precision, it is clearly indicated in Fig. 8.8 that the calcite Mg contents of the prismatic ESM groups were significantly reduced. Values of nacreous ESM groups also showed a slight reduction. In order to test the significance of the reduction

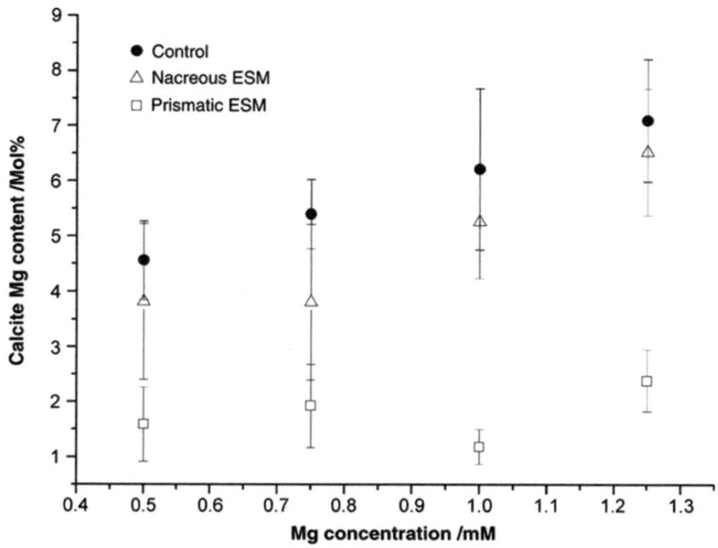

Fig. 8.8 ESM effects on calcite Mg content. Mg contents of calcites precipitated without additive, with prismatic ESM, and with nacreous ESM are marked with different symbols. Calcite Mg content is dependent on the solution Mg concentration. The values decreased significantly when prismatic ESM presented in solution. A slight reduction effect is also observed for nacreous ESM

effects, another trial was conducted with a Mg concentration of 1.25 mM (higher concentrations will result in severe crystal growth inhibition). One-way ANOVA analysis showed that the Mg-content difference between the control and the prismatic ESM groups was highly significant ($p = 0.000$) and the difference between the control and the nacreous ESM groups was significant but very close to the critical p value ($p = 0.049$, significance level: $p = 0.05$). Crystals grown with 1 mM EDTA in the crystallizing system, which would be regarded as a substantially high concentration if there were any EDTA remains, displayed no significant difference with Mg content.

Our major findings in this study are as follows: [1] Significant Mg-content reduction was observed for calcite grown with additive prismatic ESM. Light reduction of bulk phase Mg content but an increase of surface Mg content was observed with crystals grown with nacreous ESM. (2) The prismatic layer of *P. fucata* is subject to a physiological Mg exclusion effect. The additive effect of prismatic ESM may have contributed to this phenomenon [3]. Preliminary mechanism studies showed that the Mg exclusion effect of ESM is achieved neither by altering the crystal precipitation rate nor by directly altering the solution's relative ion activity.

8.3 The Function of Inner Film in Biomineralization

8.3.1 The Inner–Shell Film Is an Immediate Structure Participating in Shell Formation

In mollusks, the inner-shell films are located in the shell-mantle zone and they are important in shell formation. Previous studies showed that the films play an intermediary role in shell formation or dissolution [31]. In the pearl oyster *P. fucata*, an organic sheet is observed in pearl sac–nacre preparation [32]. However, no further characterization has been performed. In this study [5], we found that a thin organic film indeed formed the inner surface of the shell in *P. fucata*. Furthermore, sometimes, a two-layer structure of the film was definitely observed in the extrapallial zone of the shell. The analysis of the amino acid composition indicated that the film proteins may contain shell framework proteins. The in vitro $CaCO_3$ crystallization and precipitation experiments showed that the film proteins not only could inhibit the growth of calcite but also could promote the precipitation of $CaCO_3$.

8.3.1.1 SEM Observation of the Film

To detect the film in the pearl oyster, the shell section samples were observed by SEM. Comparison of the samples processed with and without EDTA revealed a film that forms the inner surface of the shell (Fig. 8.9). As shown in Fig. 8.9, the film exhibited different appearance in different zones of the shell. Mostly, the film of the central and extrapallial zone has a similar appearance, where a distinct polygonal shape can be seen (Fig. 8.9a, c). This suggested that the underlying substance was aragonite crystal. However, in a few cases, the film in these two zones seemed too thick to easily distinguish the shape of the underlying substance (Fig. 8.9b, d). In these cases, a vague polygonal shape of the underlying aragonite tablets can be seen in the central zone (Fig. 8.9b, indicated by arrows), whereas a vague round-like shape can be observed in the extrapallial zone (Fig. 8.9d). When the film was reversed with forceps, some distinct aragonitic tablets were observed. From the sample of a partially reversed film (Fig. 8.10), it is seen that the aragonitic tablets not only form the surface of the nacre (Fig. 8.10, left) but also adhere to the shell-facing surface of the film (Fig. 8.10, right).

8.3.1.2 Amino Acid Compositions of the Film Proteins

The amino acid analysis showed that both the soluble and insoluble proteins of the film were rich in Gly and Ala, and they were composed of 16 kinds of amino acids (Table 8.1). The total percentage of Gly and Ala in soluble and insoluble proteins of the film is 55.92 mol% and 59.88 mol%, respectively. No Cys was detected in the film proteins.

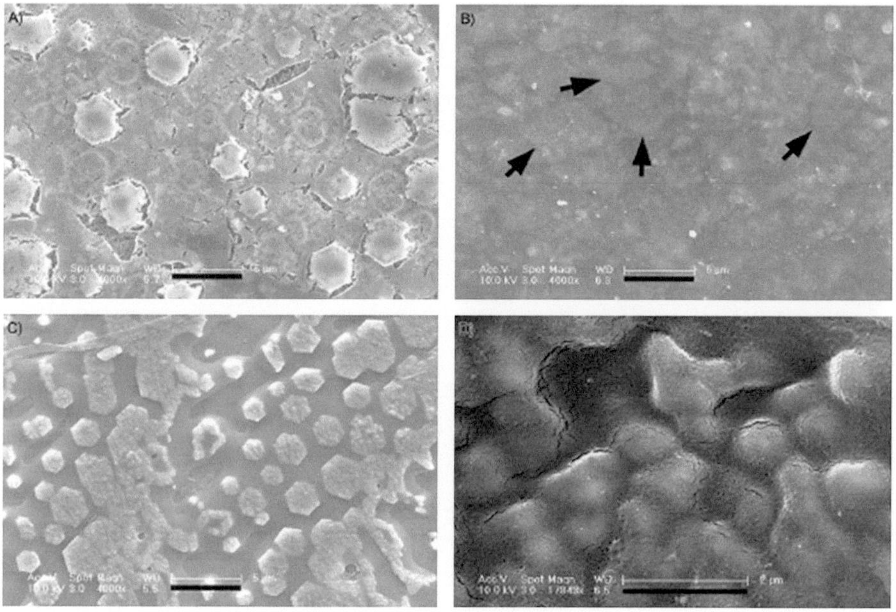

Fig. 8.9 SEM images of inner-shell film (in situ). The samples were treated without the use of EDTA. (**a**) and (**b**), the appearance of the film of the central zone. Some distinct and vague shapes (indicated by arrows) of aragonite are shown. (**c**) and (**d**), the appearance of the film of the extrapallial zone. Some distinct shapes of aragonite are shown in C, whereas some underlying round-like shapes are shown in (**d**). The bars are 5 μm in (**a**), (**b**), and (**c**) and 2 μm in (**d**)

Fig. 8.10 SEM images of the nacre (left) and the shell-facing surface of the film (right). The film was reversed with forceps. Bar, 2 μm

8.3.1.3 Soluble Film Proteins' Promotion of $CaCO_3$ Precipitation

A series of precipitation experiments with or without the soluble film proteins were performed (Fig. 8.11). The effect of the soluble film proteins on the rate of precipitation of $CaCO_3$ was determined by recording the increase of absorbance at 570 nm

Table 8.1 Amino acid analysis of the soluble and insoluble film proteins

Amino acid	Soluble film proteins (mol%)	Insoluble film proteins (mol%)
Asp	10.45	9.40
Glu	3.00	2.52
Gly	37.67	43.76
Ala	18.25	16.12
Thr	0.98	0.89
Ser	6.23	5.89
Val	2.28	1.93
Met	1.19	0.95
Lle	1.06	1.02
Leu	6.57	6.35
Tyr	1.97	1.78
Phe	2.34	1.92
Lys	2.12	1.83
Arg	4.12	3.90
His	0.19	0.19
Pro	1.58	1.55

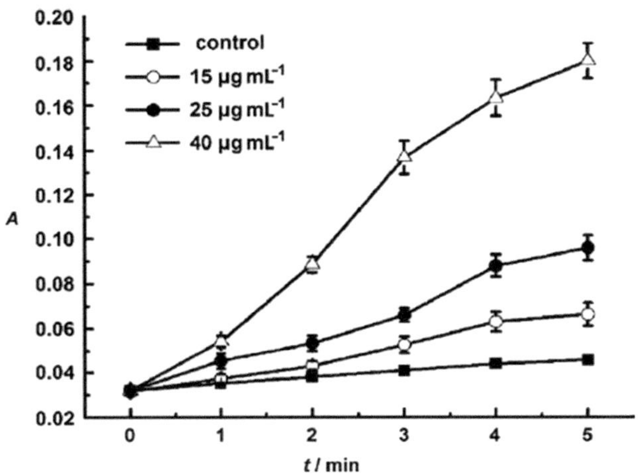

Fig. 8.11 Promotion activity of the film proteins on $CaCO_3$ precipitation. Changes in the turbidity of the assayed solutions are shown. The vertical axis: absorbance at 570 nm

in a saturated $CaCO_3$ solution. The initial absorbance, which depended on the $CaCO_3$ concentration, was about 0.032, while $CaCO_3$ was formed at room temperature in less than 5 mins. When the soluble film proteins were added to the $CaCO_3$ solution, the absorbance of solution at 570 nm significantly increased in a concentration-dependent manner compared with that of the control (shown in Fig. 8.11). Water or BSA did not affect the absorbance intensity.

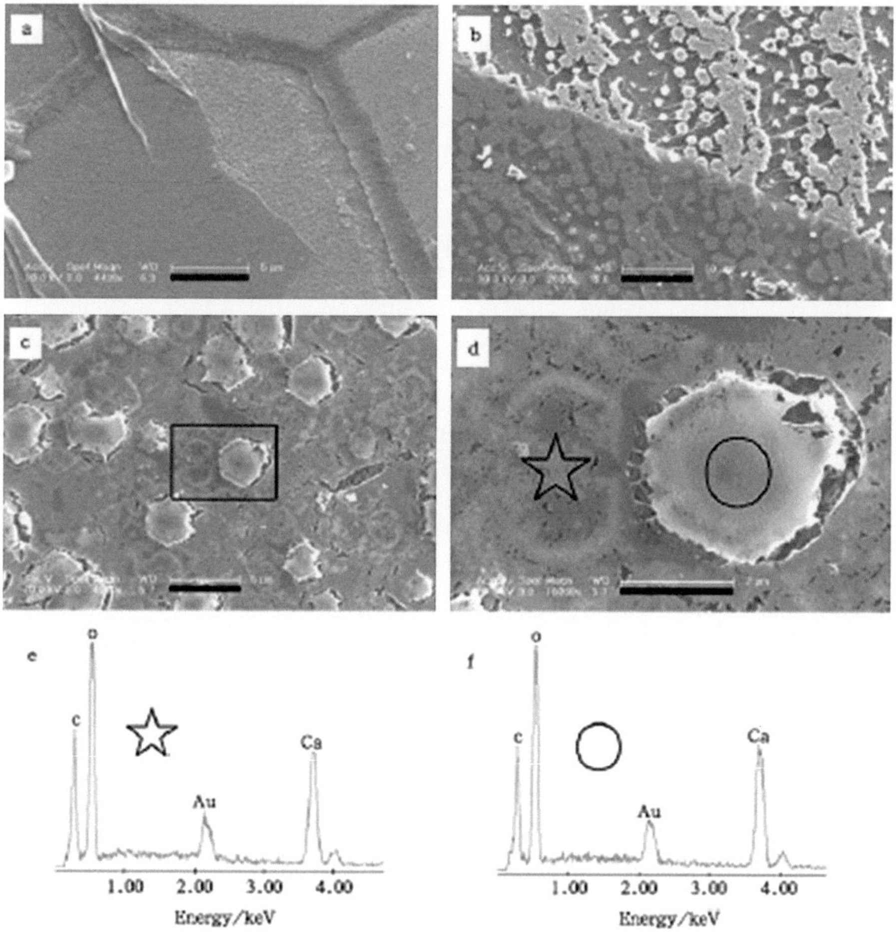

Fig. 8.12 Partial characterization of the inner shell surface. (**a**) The surface of calcite; b: the surface of aragonite; **a** and **b** are both covered by the inner-shell film. (**c**) The surface of the central region of the shell; (**d**) enlarged images of the boxed part in **c**. The star indicates the "imprint" of aragonite and the circle indicates "shaped" aragonite. Bars: **a** and **c**, 5 μm; **b**, 10 μm; **d**, 2 μm. Bottom: EDS analysis of the region of **d**

8.3.2 In Vivo and In Vitro Biomineralization in the Presence of the Inner-Shell Film

8.3.2.1 Morphology and Characteristic of the Inner–Shell Surface

In Fig. 8.12, it can be seen that the surfaces of both calcite (Fig. 8.12a) and aragonite (Fig. 8.12b) covered the inner-shell film. In addition, besides the spiral aragonite, some dispersed aragonites were also found in the central region. Furthermore, some imprints of aragonite were found (Fig. 8.12d, indicated by a star), accompanied with shaped aragonite (Fig. 8.12d, indicated by a circle). EDS analysis showed that the imprint and shaped aragonite were composed of $CaCO_3$ (Fig. 8.12, bottom) [33].

8.3.2.2 Induction of Aragonite by Combination of the Inner–Shell Film and Nacre Proteins

The inner-shell film was detached from the shell (Fig. 8.13a) to determine its functions in aragonite formation. When the inner-shell film was immersed in the calcite crystallization solution, some crystals grew on it (Fig. 8.13c). No significant influence on calcite formation was observed. On the other hand, the water-soluble proteins of nacre could inhibit calcite growth, and the crystal was rounded (Fig. 8.13d) compared with the control, which produced typical rhombic calcite (Fig. 8.13b). When the inner-shell film and the nacre proteins were both presented in the solution, some spherical crystals were formed on the film (Fig. 8.13e). Raman spectra showed that these spherical crystals were aragonite (Fig. 8.13f).

In summary, we found the following: [1] By the observation of inner-shell surface of the pearl oyster, amalgamated calcite crystals and aragonitic imprints were found in the calcite regions and the central part of the aragonitic regions, respectively. The imprints are probably due to the trace of mantle cells that adhered to the inner shell surface when the shells were produced [2]. To build a new in vitro biomineralization system, the inner-shell film was detached and was introduced to the calcite crystallization system. Crystallization experiments showed that nacre proteins induced aragonite crystals but inhibited calcite growth in the absence of the inner-shell film. These data suggested that the inner-shell film may induce aragonite growth in vivo with nacre proteins.

8.3.3 Matrix Proteins Regulate Aragonite Nucleation and Growth with the Inner–Shell Film In Vivo

The initial growth of the nacreous layer is critical for understanding the formation of nacreous aragonite. A flat pearl method in the presence of the inner-shell film was performed to evaluate the role of matrix proteins in the initial stages of nacre biomineralization in vivo [6].

8.3.3.1 Microstructure of Growth Surfaces in the Flat Pearl

To investigate the process of nacre growth in the flat pearl, implantation experiments in vivo were performed. The transitory morphologies leading to steady-state crystal growth after nacre growth interruption were observed. The sequential morphologies of this growth sequence are presented in Fig. 8.14. In the inner-shell film system, the

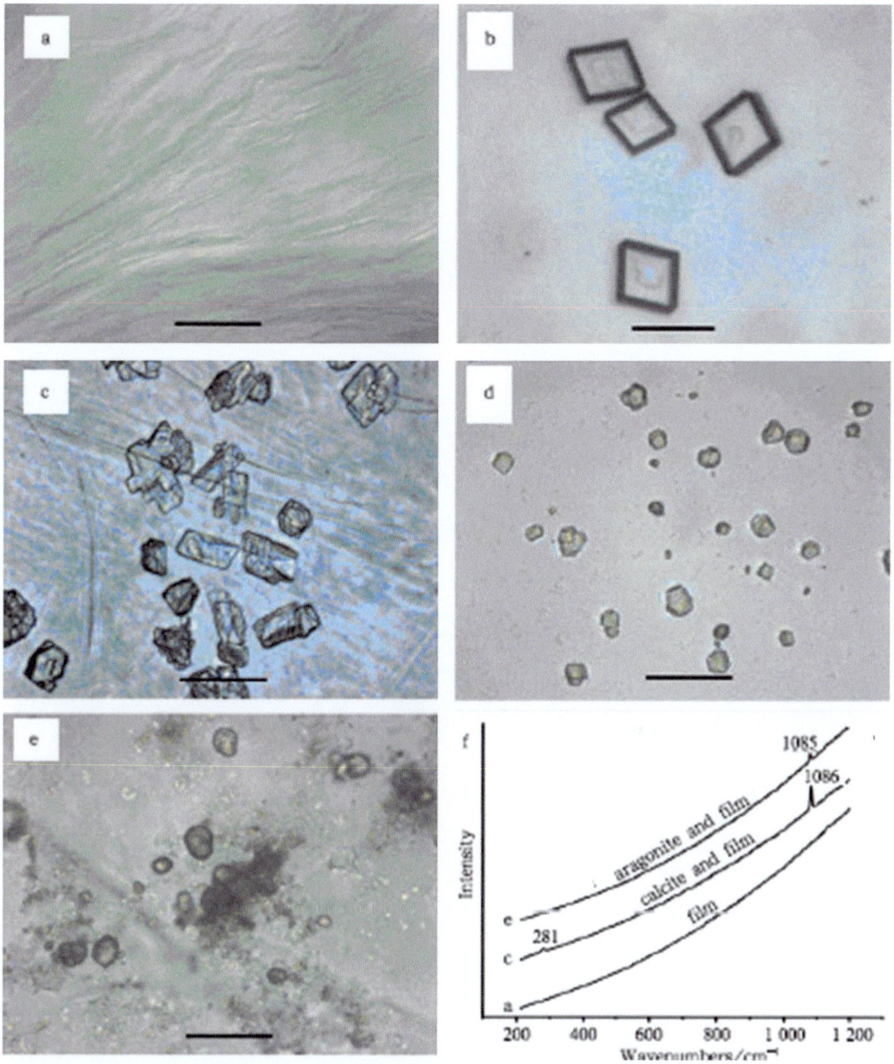

Fig. 8.13 Crystallization experiments in the presence of the inner-shell film and nacre proteins. (**a**) The inner-shell film in water; (**b**) calcite induced in the crystallization solution without the film or proteins; (**c**) calcite induced in the presence of the film; (**d**) the modified calcite in the presence of nacre proteins; (**e**) aragonite induced by the combination of the film and nacre proteins. Bars: **a**, **c**, **d**, and **e**, 50 μm; **b**, 30 μm. (**f**) Raman spectra of the crystals precipitated on the film

crystals randomly nucleated on the implanted inner-shell film, and they coalesced together to form the initial layer after 5 days of growth (Fig. 8.14a), when the inner-shell film around the crystals can be observed (Fig. 8.14a, arrow). On the tenth day after implantation, the crystals had spread across the entire substrate, and the deposited layer had coarse surfaces (Fig. 8.14b). Figure 8.14c showed that at 15th

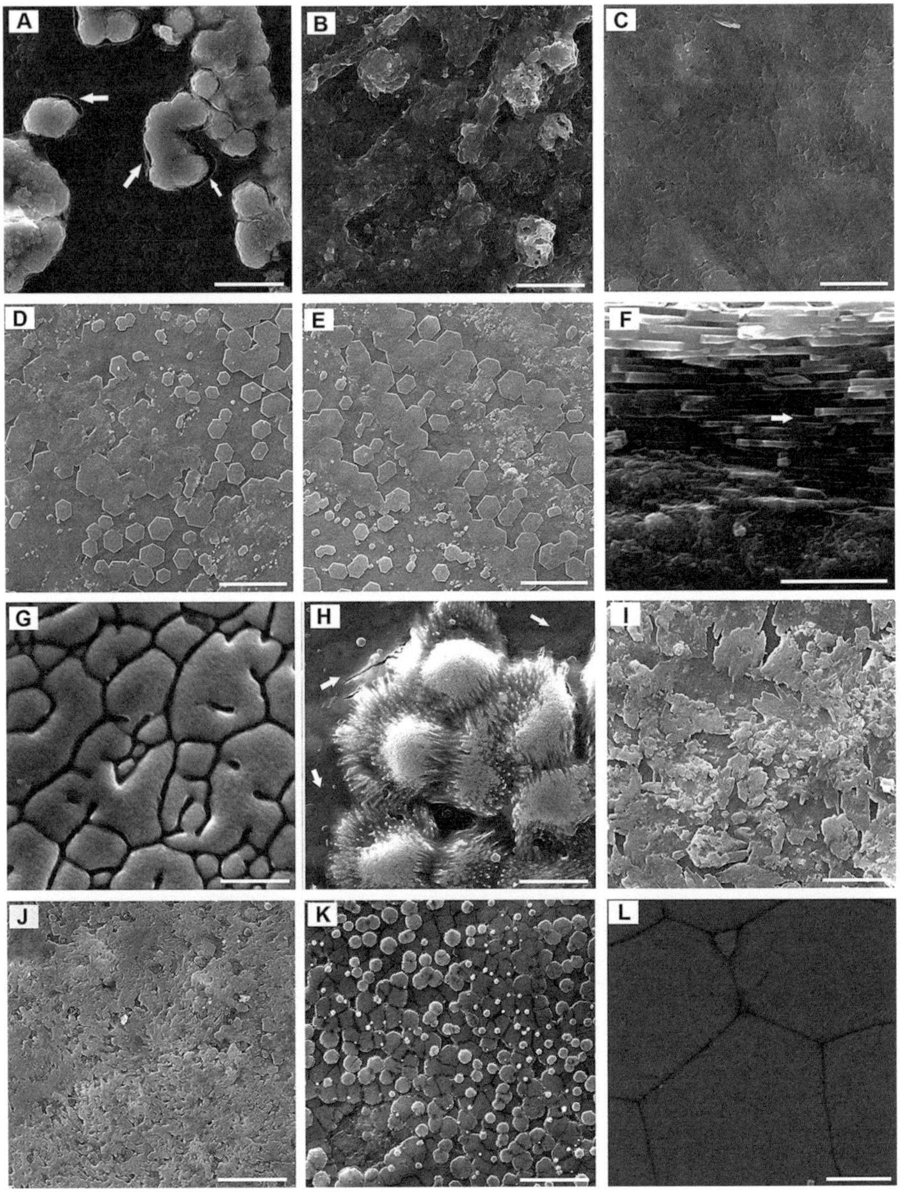

Fig. 8.14 SEM images of the deposition on the surface. (**a**) 5d (the inner-shell film indicated by arrows), (**b**) 10d, (**c**) 15d, (**d**) 20d, (**e**) 25d, and (**f**) cross section of 25d (brick and mortar structure indicated by arrows) in the inner-shell film system. Growth direction is toward the top in (**f**). (**g**) 5d, (**h**) 10d (the new forming organic film indicated by arrows), (**i**) 15d, (**j**) 20d, and (**k**) 25d in the film-free system. (**l**) Prismatic layer of adult shell. Scale bars, 10 μm in (**a**), (**b**), (**c**), (**d**), (**e**), (**g**), (**h**), (**i**), (**j**), (**k**), and (**l**), 4 μm in (**f**)

day, the morphology of deposited minerals became smoother than that at 10th day. The crystals were packed into a compact aggregation and the surface appeared flattened. After 20 days, the hexagonal flat tablets appeared on the surface (Fig. 8.14d), and some bigger tablets were formed by the fusion of crystals with disordered patterns. After 25 days, the surface showed a normal stair-like growth pattern, and the deposited crystals changed into the typical hexagonal flat tablets (Fig. 8.14e) that were structurally similar to the nacre of *P. fucata*. In addition, a brick–mortar arrangement was also observed in the cross section of the crystals after 25 days (Fig. 8.14f arrow).

In the film-free system, compared with the inner-shell film system, the surface of the cover glass was completely covered by prism-like crystals at fifth day (Fig. 8.14g). This prismatic structure was different from the calcitic prisms in the outer shell of adult *P. fucata* (Fig. 8.14l). After 10 days, the spherical particles with acicular margins nucleated randomly on the surface, similar to the process at 5 days in the inner-shell film system (Fig. 8.14h). The newly formed organic layer around the spherical particles was also visible (Fig. 8.14h, arrow). After 15 days, the surface was completely covered by deposited minerals which were composed of irregularly shaped crystals (Fig. 8.14i). The crystals became smaller and packed into compact aggregation, which resulted in a smooth surface after 20 days (Fig. 8.14j). After 25 days, the crystals were amalgamated to form round tablets (Fig. 8.14k), and the surface did not appear to have a normal stair-like growth pattern.

8.3.3.2 Polymorph of Deposited Crystals on the Surface

To confirm the polymorph of crystals deposited on the surface of the substrate, the same stages were analyzed by Raman and FTIR spectroscopy. The Raman spectra are presented in Fig. 8.15. In the inner-shell film system, all of the crystals from the growth periods were aragonite (Fig. 8.15a), which exhibited characteristic bands at 154, 206, 706, and 1085 cm^{-1}. The baseline of Raman spectra increased significantly from 5 to 10 days, decreased at day 15, and then showed a relatively steady level between days 15 and 25 (Fig. 8.15a). In the film-free system, Raman analysis confirmed the crystals to be aragonite (Fig. 8.15b) except for the deposition on day 5. The Raman spectra revealed the crystals on the cover glass as calcite, with three bands at 282, 712, and 1086 cm^{-1} (Fig. 8.15b). This was different from the polymorph of calcium carbonate in the inner-shell film system at 5 days. The baseline of Raman spectra also showed a different pattern from that in the inner-shell film system (Fig. 8.15b). The baseline increased gradually from days 5 to 15, suddenly decreased at day 20, and then increased again at day 25 (Fig. 8.15b). Because Raman spectra can only identify small areas of crystals on the surface of the inner-shell film, it is necessary to characterize the entire crystal. Thus, we collected crystals grown on the inner-shell film and characterized them by FTIR. The crystals on the surface of the film at each period demonstrated bands at 700, 712, 858, and 1082 cm^{-1} (Fig. 8.16a), indicating that the crystals deposited in this system were all aragonite. In the film-free system, the spectrum of the crystals on the surface of the

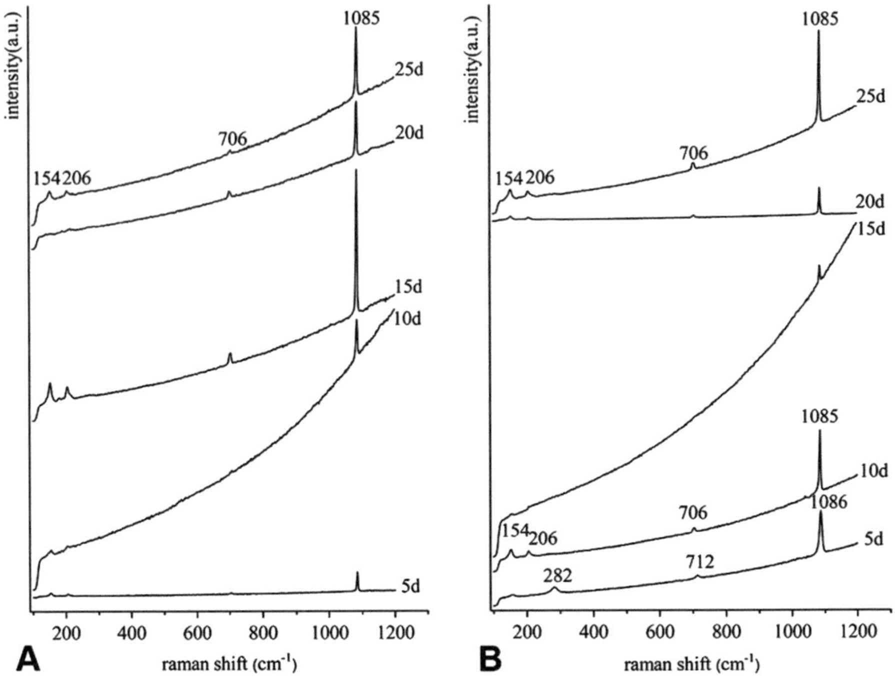

Fig. 8.15 Raman spectra of the crystals deposited on the surface. (**a**) The inner-shell film system, (**b**) the film-free system. The characteristic peaks of calcite are at 282, 712, and 1086 cm^{-1} (5d in B), and the characteristic bands of aragonite are at 154, 206, 706, and 1085 cm^{-1}

cover glass at day 5 contained the typical bands of calcite at 712 and 876 cm^{-1} (Fig. 8.16b). The crystals over the other periods were confirmed to be aragonite by FTIR. The characteristic bands of calcite and aragonite were observed to occur simultaneously at day 10 because of the mixture of calcite at day 5 and aragonite at day 10, and the characteristic bands of calcite disappeared with increasing aragonite content after day 15 (Fig. 8.16b).

8.3.3.3 Expression Patterns of Genes during the Initial Stages of Nacre Growth

In the inner-shell film system, the expression levels of Pif80, MSI7, nacrein, MSI60, N19, and N16 showed similar patterns (Fig. 8.17a, b, d, e, f, g), which increased significantly and reached a maximum at 10 days, and suddenly decreased at 15 days to their minimum value at 20 days, and then increased at 25 days again. The expression level of ACCBP increased significantly and reached a maximum at 10 days and then suddenly decreased at 15 days and remained at low levels from 15 to 25 days (Fig. 8.17c). In the film-free system, the expression level of MSI7 decreased gradually to a minimum value at day 25 (Fig. 8.17h). The expression

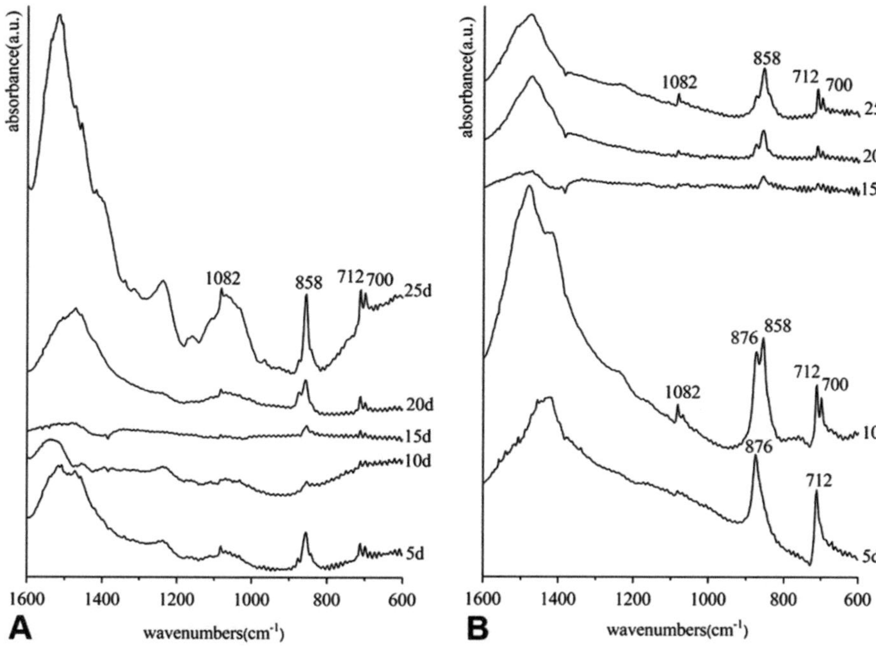

Fig. 8.16 FTIR spectra of the crystals collected from the surface. (**a**) The inner-shell film system, (**b**) the film-free system. The characteristic bands of calcite are at 712 and 876 cm^{-1} (5d in **b**), and the characteristic bands of aragonite are at 700, 712, 858, and 1082 cm^{-1}

patterns of nacrein, MSI60, N19, N16, and Pif80 were similar and increased gradually from 5 to 15 days and then decreased sharply at 20 days to their maximum value at 25 days (Fig. 8.17i, k, l, m, n). The expression of ACCBP increased gradually from day 5 to day 15 and remained high from 15 to 25 days (Fig. 8.17j). The aragonite crystals nucleated on the surface at 5 days in the inner-shell film system. In the film-free system, the calcite crystals nucleated at 5 days, a new organic film covered the calcite, and the aragonite nucleated at 10 days. Matrix proteins (nacrein, MSI60, N19, N16, and Pif80) had similar expression patterns in controlling the sequential morphologies of the nacre growth in the inner-film system, while these proteins in the film-free system also had similar patterns of expression.

In summary, we showed that the switch in phase from calcite to aragonite requires the deposition of a new nucleating organic sheet in vivo. This finding may be explained by the strong interaction between the growing crystal and the surface of the inner-shell film, which may directly participate in calcite-to-aragonite transition as nucleation surface and predefined mold. The formation mechanism of the composite nacre is still under debate including the polymorph switching between calcite prism and aragonite nacres. One explanation is based on a hetero-epitaxial nucleation model of aragonite tablets on the organic chitin sheet proposed by Weiner et al [34]. In this study the aragonite crystals are formed on the surface of the inner-shell film at 5 days, and the inner-shell film is an assembly of β-chitin frameworks, silk-

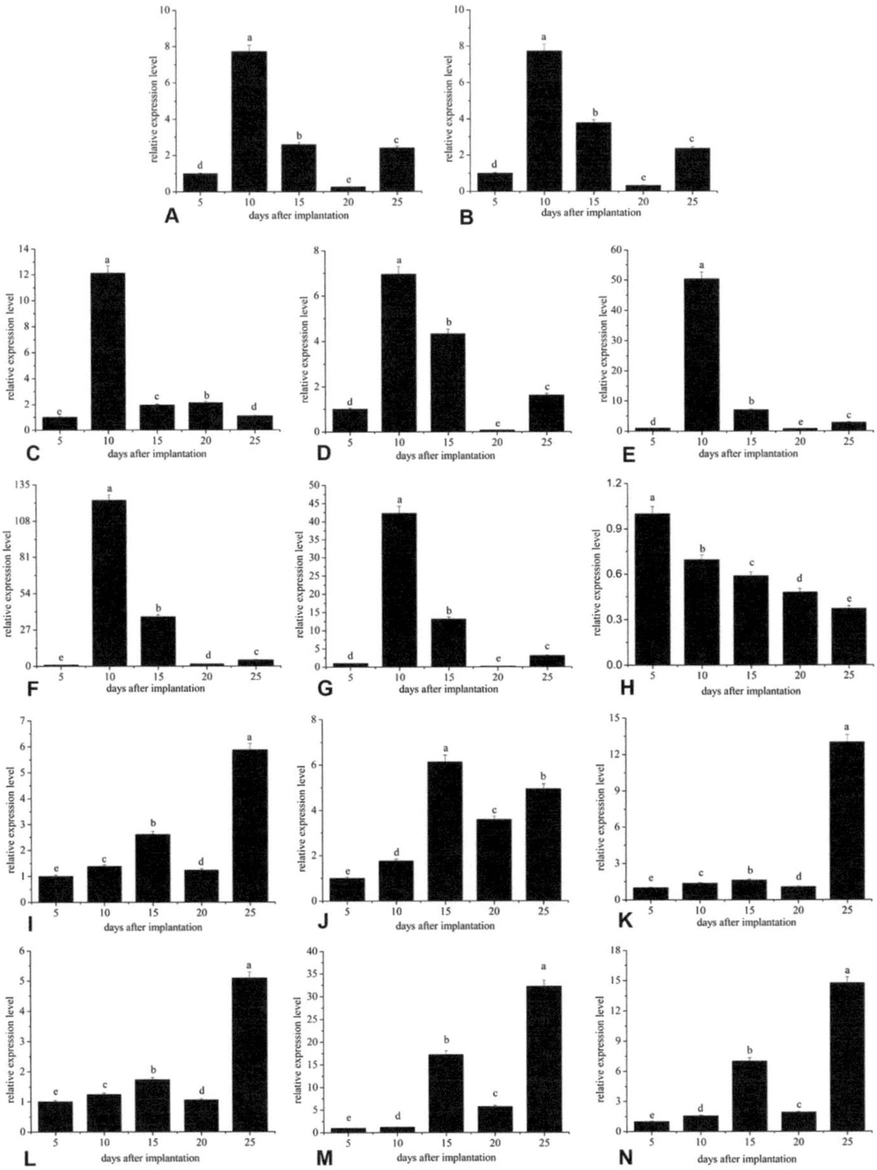

Fig. 8.17 The relative expression of genes encoding matrix proteins at different stages of nacre growth. (**a**), (**b**), (**c**), (**d**), (**e**), (**f**), and (**g**) The relative expression levels of MSI7, nacrein, ACCBP, MSI60, N19, N16, and Pif80 in the inner-shell film system, respectively. (**h**), (**i**), (**j**), (**k**), (**l**), (**m**), and (**n**) The relative expression levels of MSI7, nacrein, ACCBP, MSI60, N19, N16, and Pif80 in the film-free system, respectively. Values in the same figure with a different superscript are significantly different ($p < 0.05$)

like protein, and acidic macromolecules. These results imply that nucleation of aragonite crystals also can be explained by organic template or hetero-epitaxy at planar organic–inorganic interface.

8.4 Theories for Shell Growth and Pearl Production

8.4.1 Jumping Development Theory

Although shell growth rate is affected by environment factors, a one-to-one relationship between a given factor and a given rate has not been established. This is partly due to that the mechanism of the formation of annual growth lines is poorly understood. Based on our analysis of a unique growth band discovered in the shell margin of *P. fucata*, we proposed a concept of "jumping development" in bivalve shells to explain the formation of annual growth lines [35]. During jumping development, the center of growth in the bivalve shell jumps forward over a short time interval when the position of the mantle changes. Jumping development explains the unusual structure of the anomalous shell and the development of annual growth lines in typical shells. Annual growth lines are the result of a discontinuity in the shell microstructure induced by jumping development.

The surface of the growing prism was formed and exhibited a concentric pattern, which is distinctly different from the flat surface of a mature prism (Fig. 8.18a). The unique margin contained two organic layers separated by aragonite crystals. The first organic layer covered the aragonites. The organic layer between the exposed growth band and the mature prismatic layer (Fig. 8.18a) constituted the inner surface of the shell. The transition between the organic layer and mature prismatic layer was visible 8 mm from the shell margin (Fig. 8.18b) illustrating the formation of the organic layer. The organic envelope surrounding the prism is formed by the addition of an organic matrix to its surface (Fig. 8.18c). The organic matrix is continuously added and grows toward the center of the surface of the prisms (Fig. 8.18d) until the surface is completely covered by the organic matrix. The trail of the raised organic envelope is ultimately invisible (Fig. 8.18e).

The most unusual pattern of the margin in the anomalous shell is that the nacreous layer occurs in the shell margin. Furthermore, the pattern of growth of the nacre in the growth band was unusual (Fig. 8.19a). The growth did not resemble simple ladder-like growth (Fig. 8.19b), the typical pattern of a generic nacreous layer, and some layers appeared to shrink regularly. We named this unique pattern "complex ladder-like growth." When the height of the two kinds of nacreous layer is equal ($h1 = h2$), the breadth of the growth surface is far narrower than that seen in simple ladder-like growth ($b1 < b2$, Fig. 8.19a, b). Complex ladder-like growth resulted in a more stable and harder nacreous layer. We proposed that it can also lead to a step-growth pattern of the nacre at the inner surface of the shell (Fig. 8.19c).

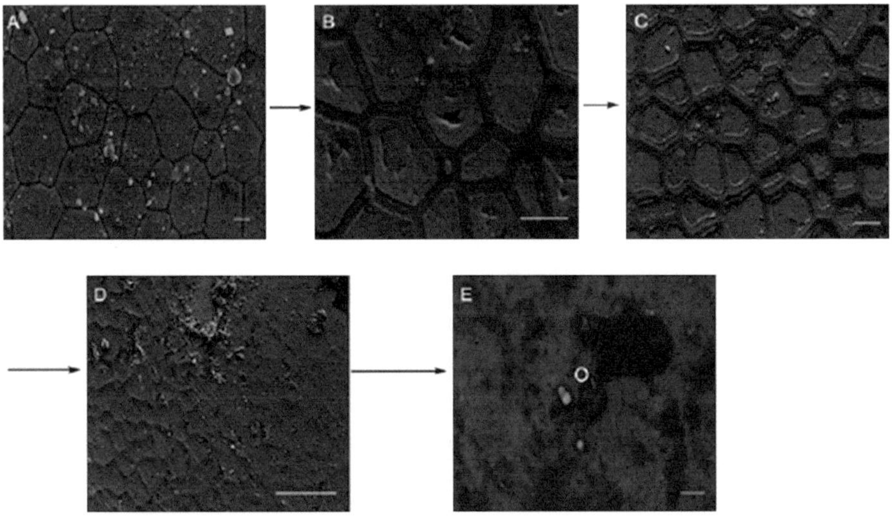

Fig. 8.18 SEM images of the formation of the organic layer covering the nacreous layer in the unique growing shell margin, the organic layer covering the newly formed nacreous layer. (**a**) The mature prismatic layer about 8 mm from the special growing shell margin. (**b–d**) The transition of the mature prismatic layer and the organic layer covering the nacreous layer. (**e**) The organic layer covering the nacreous layer. Scale bar, 20 μm in **a–c**; scale bar, 100 μm in **d**; scale bar, 10 μm in **e**. The direction of growth is from the left to the right

8.4.2 Nacre–Prism Transition Layer Theory

The nacre–prism transition layer is important because it provides an initial nucleating surface for aragonite and it alters the mode of shell growth. There are great differences in mineralogy, microstructure, and organic matrix between the prisms and the nacre. For example, the major matrix proteins found in the prismatic layer have lower pI than those found only in the nacreous layer [36, 37]. Thus, a series of changes must occur on the surfaces and within the prismatic layer as the nacre–prism transition occurs. However, little is known concerning these changes, and the mechanism by which these two different layers become closely joined remains known.

We investigated this process by observing the changes that occur in the organic matrix during the formation of the nacre–prism transition layer in the shells of *P. fucata* [38]. Three variants of the sequence of formation of the nacre–prism transition layer were observed. In each case, the layer was formed by the organic matrix secreted by the mantle, together with the interprismatic organic envelope. The continuity of the organic phase throughout the shell was maintained as the new nacreous layer was formed on the nacre–prism transition layer. Changes in the interprismatic organic envelopes on either side of the nacre–prism transition zone indicated that the organic matrix of the nacre–prism transition layer becomes

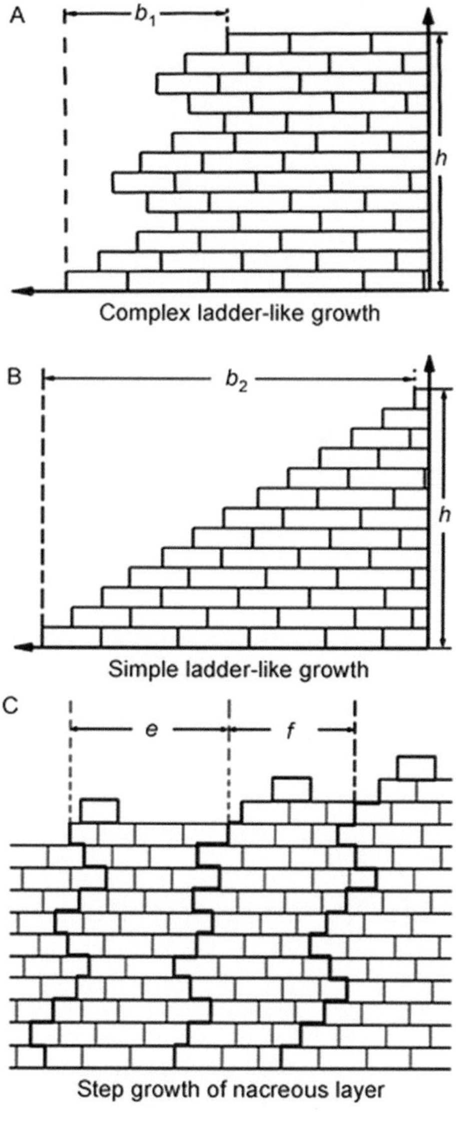

Fig. 8.19 Models of growth in the nacreous layer of the growth band in the anomalous shell and typical shells. (**a**) The complex ladder-like growth of the nacreous layer in the unique band of the anomalous shell. (**b**) The simple ladder-like growth of the nacreous layer in a typical shell. (**c**) The step growth of the nacreous layer in the unique band of the anomalous shell

embedded into the organic phase of the prismatic layer. It is concluded that penetration and erosion of the prisms by the organic matrix generate a strong bond between the prismatic and nacreous layers.

8.4.2.1 Three Variants of the Nacre–Prism Transition Sequence

The nacre–prism transition layer is an organic layer that covers the internal surface of the prismatic layer (Fig. 8.20a). The transition layer provides a surface for aragonite

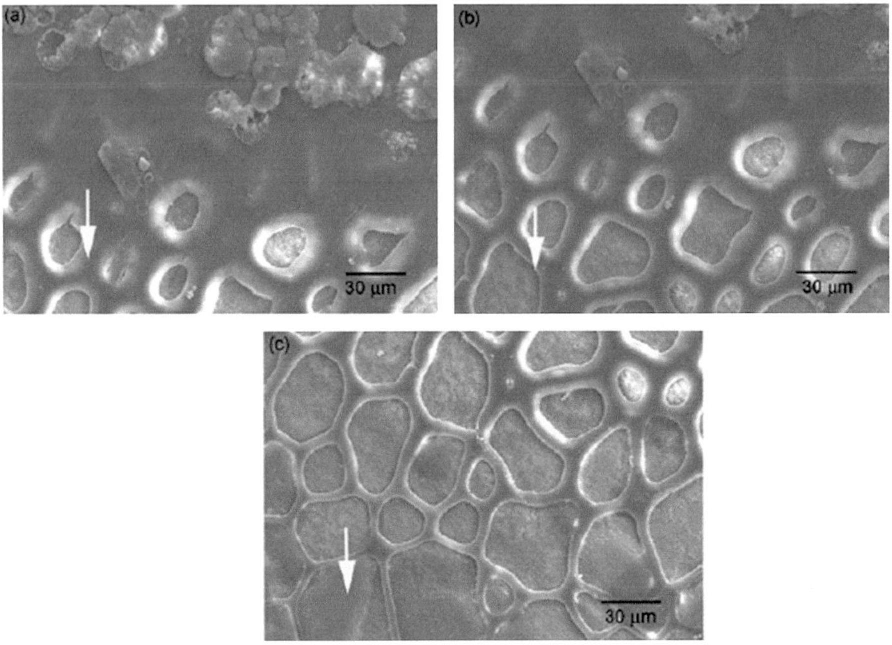

Fig. 8.20 The first type of nacre–prism transition layer formation. (**a**) An organic layer representing the nacre–prism transition layer forms on the prismatic layer; (**b**) the diameter of the visible surface of the prisms decreases before the formation of the nacre–prism transition; (**c**) the addition of the organic matrix to the interprismatic organic envelope of the prismatic layer near to the transition zone. The shell was slightly decalcified in 5% EDTA for 10 min. The white arrows indicated the growth direction of the shell

growth. We identified three different configurations of the nacre–prism transition zones: [1] Increased thickness of the organic matrix forming the interprismatic organic envelopes. This form of the nacre–prism transition layer has been reported in previous studies [39] and was observed here in slightly decalcified shells. The surfaces of the prisms adjacent to the nacre–prism transition zone were not completely visible, and regions close to the interprismatic organic envelope were covered by organic matrix (Fig. 8.20b). Because the mantle secreted organic matrix onto the internal shell surfaces during the process, the surface of the interprismatic organic envelope was elevated (Fig. 8.20c). Organic matrix was continuously added and appeared to converge toward the center of the prisms (Fig. 8.20b) until the surface was entirely covered by the organic matrix. Thus, the nacre–prism transition layer was formed at the end of this process [2]. Transformation of the prismatic layer into the nacre–prism transition layer. This second type of transition layer formation was observed in untreated shells. In surface view, the shape of the prisms located outside the nacre–prism transition zone appeared to change (Fig. 8.21a, 8.21b). Extending further into the transition zone, the visible surface area of the prisms was reduced or completely disappeared; i.e., they were visible when still partially

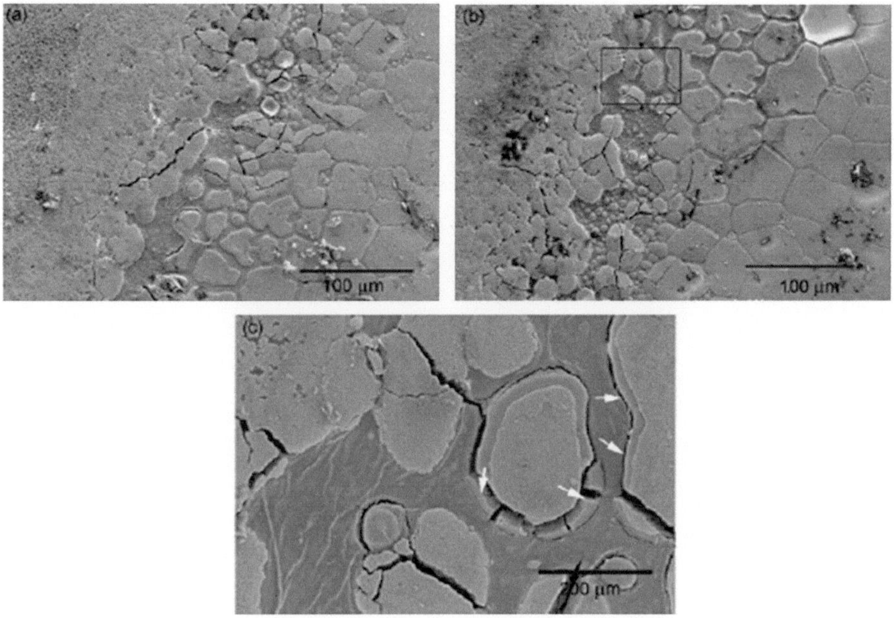

Fig. 8.21 The second type of nacre–prism transition layer formation. (**a**) and (**b**) The transformed prisms before the formation of the nacre–prism transition. (**c**) Enlargement of the area indicated in (**b**); dissolution of the interprismatic organic envelope as the nacre–prism transition layer forms. The growth direction of the shell is from the left to the right

invested by the organic matrix but became invisible as they were covered by the matrix. The interprismatic organic envelope surrounding each prism seemed to be incorporated into the organic matrix (Fig. 8.21c). Before this stage, the thickness of the interprismatic organic envelopes remained unchanged although their shape was distorted. Thus, it seemed that new organic matrix was secreted into the inner surface of the prismatic layer [3]. Decreased organic matrix in the interprismatic organic envelopes in the formation of the nacre–prism transition layer. This variation of transition layer formation was observed in slightly decalcified shells. On the inner surface close to the nacre–prism transition zone, the thickness of the organic envelope of prisms decreased (Fig. 8.22a). The double-layered interprismatic organic envelopes were elevated above the surface in areas distant from the transition zone (Fig. 8.22b, c), while in areas near to the transition zone, the surface was rather smooth (Fig. 8.22d). This latter phenomenon reflected a decrease in the thickness of the organic matrix of the interprismatic organic envelopes and the retreat of surface of the interprismatic organic envelopes below the inner surface of the shell (Fig. 8.22d). In areas close to the nacre–prism transition zone, the prisms seemed to grow over both layers of organic matrix of the interprismatic envelopes, leaving only the central regions exposed (Fig. 8.22e). In these parts, the prisms were

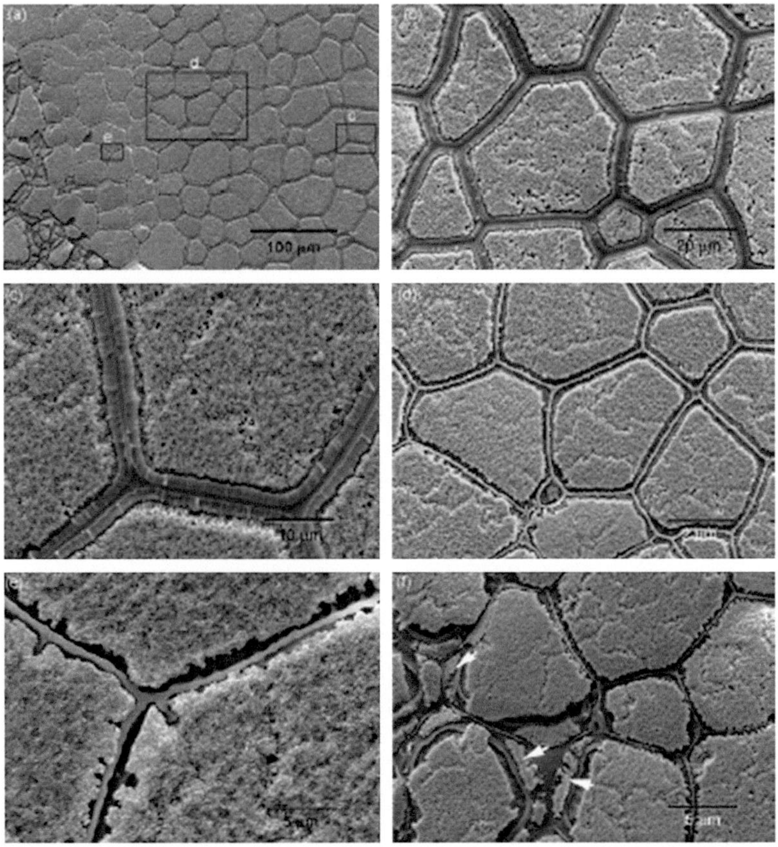

Fig. 8.22 The third type of nacre–prism transition layer formation. (**a**) The prismatic layer in the transition zone. The shell was slightly decalcified in 5% EDTA for 10 min. (**b**) The prismatic layer is distant from the Fig. (**a**). (**c**) The interprismatic organic envelopes of the prisms far from the transition. (**d**) The prismatic layer near the transition zone. (**e**) The interprismatic organic envelopes of the prisms near the transition zone. (**f**) The disappearance of the interprismatic organic envelopes on the surface of the organic matrix composing the transition layer. The growth direction of the shell is from the right to the left

not different in shape, but, adjacent to the transition zone, the visible surface area of the prisms decreased suddenly, and the space between neighboring prisms was filled with organic matrix (Fig. 8.22f). This newly organic matrix is below the inner surface of the shell.

During the formation of the nacre–prism transition layer in the shells of *P. fucata*, some of the newly secreted organic macromolecules cover the surface of the prisms to form the nacre–prism transition layer, and some are embedded into the interprismatic organic envelopes to provide an adhesive. The prismatic and nacreous layers are closely bonded together by this interaction.

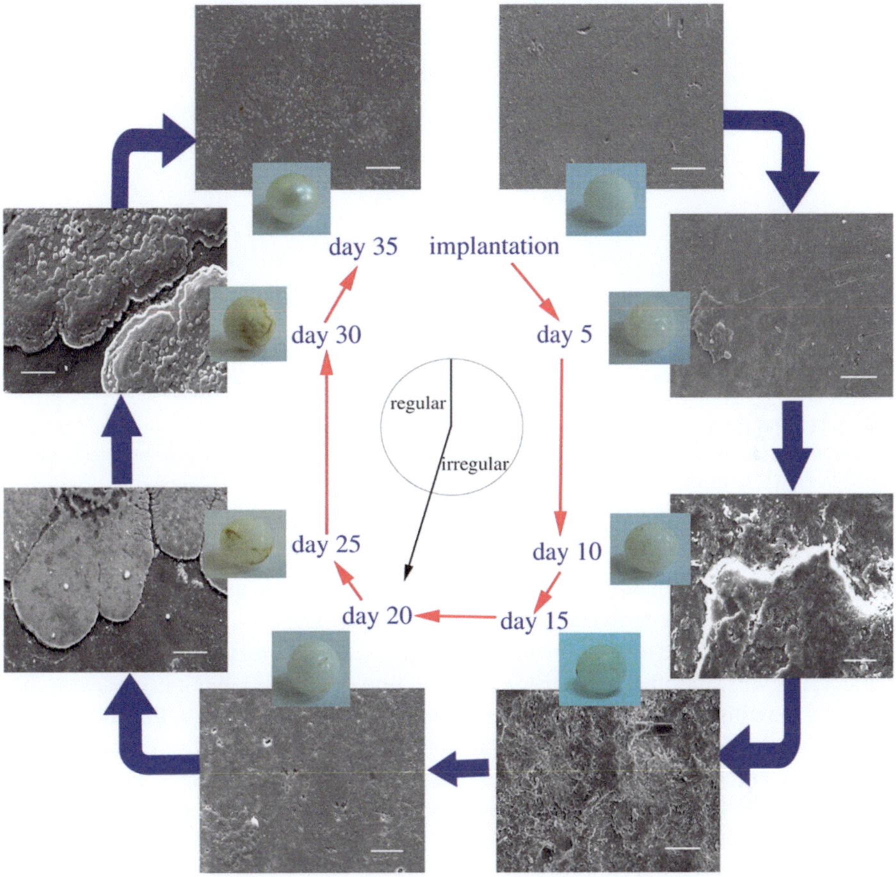

Fig. 8.23 SEM images of the deposition on the nucleus at different stages of the pearl development. Smaller photos showed the typical nuclei in different developmental periods. Scale bars, 20 μm

8.4.3 The Role of Matrix Proteins in the Control of Nacreous Layer Deposition During Pearl Formation

To study the function of matrix proteins in nacreous layer biomineralization in vivo, the deposition on pearl nucleus and the expression of matrix protein genes in the pearl sac during the early stage of pearl formation were studied [40].

On the fifth day after implantation, the mantle and the gonad formed the completed pearl sac. Over the course of pearl maturation, it is revealed that the early nucleus had a bare surface which was gradually enveloped by the mature nacreous layer (Fig. 8.23). In the early stages, the $CaCO_3$ deposition was irregular, but it became smooth on the 20th day (Fig. 8.23). At this point, the deposition pattern started to change. On the 25th day, large crystals with shapes similar to the aragonite

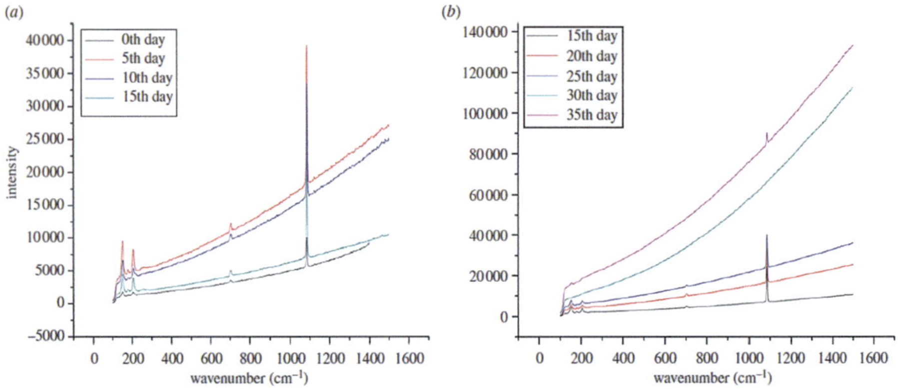

Fig. 8.24 The Raman spectra of the deposition of calcium carbonate at different stages of pearl development. (**a**) The baseline changed suddenly on the fifth day after implantation and decreased at the 10th and 15th days. (**b**) The baseline continuously increased after the 15th day

formed on the nacre–prism transition in bivalve shells were observed [39]. On the 30th day, divided nucleation occurred on these large crystals as they grew larger and formed aggregates. A new layer covered these large crystals. On the 35th day, the nacreous layer formed around the nucleus (Fig. 8.23). On the fifth day, the surface of the nucleus was smoother than that at implantation (Fig. 8.23), but it became rougher by day 10 owing to the deposition of aragonite crystals. By day 20, the surface was smooth again. From days 25 to 30, the surface was covered with yellow deposits (Fig. 8.23), and by day 35 these deposits were covered by the nacreous layer. On day 35, photographs showed a smoother surface and a mature pearl luster. Raman analysis confirmed the aragonitic nature of the crystals over the seven time periods (Fig. 8.24), with the exception on day 30 (Fig. 8.24b). On day 30, Raman spectra showed no aragonitic or calcitic peaks. We speculated that the deposition onto the nuclear surface at day 30 was ACC. The baseline of Raman spectrum on day 5 was higher than that on day 0. From days 5 to 15, the baseline of the Raman spectra decreased (Fig. 8.24a). However, from days 15 to 35, the baseline of Raman spectra increased continually (Fig. 8.24b) [40].

Significant expression of matrix proteins was detected by the 30th day after implantation (Figs. 8.25 and 8.26). During the earlier stages, the expression levels of N19, N16, and nacrein were very low, but they increased to a maximum by the 30th day, and the value remained high until the end of the experiment (i.e., the 35th day; Fig. 8.25a–c). The expression of MSI7 also increased during development and was highest on day 30, but it decreased by day 35 (Fig. 8.26a). Similarly, expression of ACCBP was the highest at day 30 and decreased dramatically by day 35 (Fig. 8.26b). In summary, we found that the process of pearl formation involves two consecutive stages: [1] irregular $CaCO_3$ deposition on the bare nucleus and [2] $CaCO_3$ deposition that becomes more and more regular until the mature nacreous layer was formed on the nucleus. The low expression level of matrix proteins in the pearl sac during the periods of irregular $CaCO_3$ deposition suggested that the

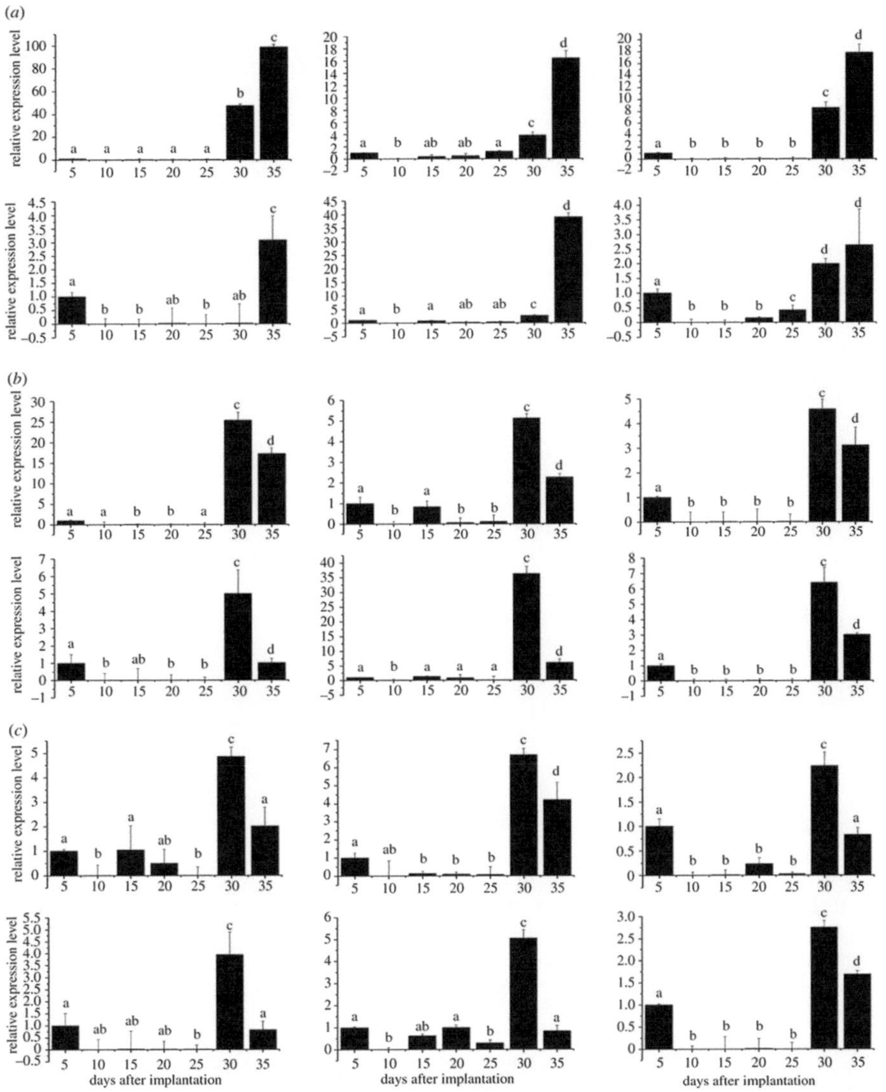

Fig. 8.25 The relative expression level of matrix proteins in the pearl sac at different stages of pearl development. (**a**) The relative expression level of nacrein. The highest expression level appeared on the 35th day after implantation. (**b**) The relative expression level of N19. (**c**) The relative expression level of N16. The highest expression level of N19 and N16 both appeared on the 30th day after implantation. Values in the same figure with a different superscript are significantly different ($p < 0.01$)

deposition may not be controlled by the organic matrix during this stage of the process. However, significant expression of matrix proteins in the pearl sac was found by days 30–35 after implantation. On day 30, a thin layer of $CaCO_3$, which we

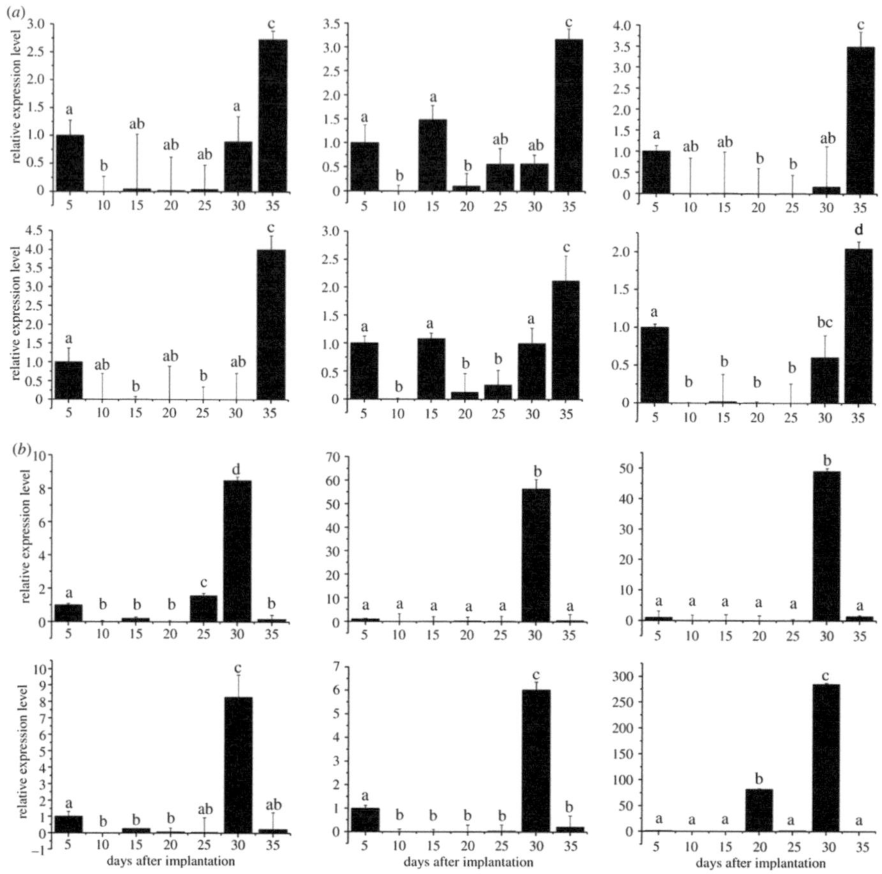

Fig. 8.26 The relative expression level of matrix proteins in the pearl sac at different stages of pearl development. (**a**) The relative expression level of MSI7. The highest expression level appeared on the 35th day after implantation. (**b**) The relative expression level of ACCBP. The highest expression level appeared on the 30th day after implantation. Values in the same figure with a different superscript are significantly different ($p < 0.01$)

believed was ACC, covered large aragonites. By day 35, the nacreous layer formed. The whole process is similar to that observed in shells, and the temporal expression of matrix protein genes indicated that their activities were important for the pearl development. Therefore, matrix proteins controlled the crystal phase, shape, size, nucleation, and aggregation of $CaCO_3$ crystals [40].

This work was modified from the papers published by our group in *Scientific Reports* (Liu, C. et al. 2015, 5, 18338), *Cryst. Growth Des.* (Liu, C. et al., 2017, 17 (4), 1966–1976), *Journal of Structural Biology* (Gong, N. et al., 2008, 164 (1), 33–40;), *Materials Science and Engineering: C* (Zhu, F. et al., 2010, 30 (7), 963–969), *ChemBioChem* (Yan, Z. et al., 2008, 9 (7), 1093–1099), *PLOS One* (Xiang, L. et al., 2013, 8 (6), e66564), *J. Biol. Chem.* (Ma, Z. et al., 2007,

282 (32), 23253–23263), *Acta Oceanologica Sinica* (Yan, Z. et al., 2011, 30 (1), 87–93), *Science China Life Sciences* (Liu, X. et al., 2011, 54 (2), 175–180), *Chinese Science Bulletin* (Liu, X. et al., 2011, 56 (9), 869–876), and *P. Roy. Soc. B-Biol. Sci.* (Liu, X. et al., 2012, 279 (1730), 1000–1007). The related contents are reused with the permission.

Supplementary Table 8.1

Chapter	Primer name	Primer sequence
	nacrein-F	GAGCCAGAGGATGGGGAAA
	nacrein-R	GCCTCCATAGGTGTGAAACGA
	N19-F	CCAGATTTCAACTCGATCTAAGGA
	N19-R	CGCCATACCCATCAAAAGTG
	N16-F	TGCGGACGTTACTCATACTGCT
	N16-R	CCACATCTAAGCCACTCATACCTCT
	Pif80-F	TGCTGCCATCACGTGAGTATG
	Pif80-R	GACTTCCCTTTCTCACACTTCCA
	MSI60-F	AGGTGGACGAAGAGGTAGAGGTAG
	MSI60-R	AGCGGCAGCGGCAACATC
	MSI7-F	GATCGTCTTAATTGCCTCTGTTTCT
	MSI7-R	CTCCAAGCCCGATAACACCT
	ACCBP-F	GACATGGAACAAAGATGGTGGA
	ACCBP-R	CTGTGGCTGGAATGGTTGG
	actin-F	CTCCTCACTGAAGCCCCCCTCA
	actin-R	ATGGCTGGAATAGGGATTCTGG

References

1. C. Liu, L. Xie, R. Zhang, Heterogeneous distribution of dye-labelled biomineralizaiton proteins in calcite crystals. Sci. Rep. **5**, 18338 (2015). https://doi.org/10.1038/srep18338. http://www.nature.com/articles/srep18338#supplementary-information
2. C. Liu, J. Du, L. Xie, R. Zhang, Direct observation of nacre proteins in the whole calcite by super-resolution microscopy reveals diverse occlusion patterns. Cryst. Growth Des. **17**, 1966–1976 (2017). https://doi.org/10.1021/acs.cgd.6b01897
3. N. Gong et al., Immunolocalization of matrix proteins in nacre lamellae and their in vivo effects on aragonitic tablet growth. J. Struct. Biol. **164**, 33–40 (2008). https://doi.org/10.1016/j.jsb.2008.05.009
4. F. Zhu et al., Tuning calcite magnesium content by soluble shell matrices: Insights into biomineral impurity control. Mater. Sci. Eng. C **30**, 963–969 (2010). https://doi.org/10.1016/j.msec.2010.04.013
5. Z. Yan et al., The inner-shell film: An immediate structure participating in pearl oyster shell formation. Chembiochem **9**, 1093–1099 (2008). https://doi.org/10.1002/cbic.200700553

6. L. Xiang et al., Patterns of expression in the matrix proteins responsible for nucleation and growth of aragonite crystals in flat pearls of Pinctada fucata. PLoS One **8**, e66564 (2013). https://doi.org/10.1371/journal.pone.0066564
7. H. Li et al., Calcite prisms from mollusk shells (Atrina Rigida): Swiss-cheese-like organic–inorganic single-crystal composites. Adv. Funct. Mater. **21**, 2028–2034 (2011)
8. H. Li, H.L. Xin, D.A. Muller, L.A. Estroff, Visualizing the 3D internal structure of calcite single crystals grown in agarose hydrogels. Science **326**, 1244–1247 (2009). https://doi.org/10.1126/science.1178583
9. S. Younis, Y. Kauffmann, L. Bloch, E. Zolotoyabko, Inhomogeneity of nacre lamellae on the nanometer length scale. Cryst. Growth Des. **12**, 4574–4579 (2012). https://doi.org/10.1021/cg3007734
10. K. Gries, R. Kröger, C. Kübel, M. Fritz, A. Rosenauer, Investigations of voids in the aragonite platelets of nacre. Acta Biomater. **5**, 3038–3044 (2009)
11. L.M. Gordon, D. Joester, Nanoscale chemical tomography of buried organic-inorganic interfaces in the chiton tooth. Nature **469**, 194–197 (2011). https://doi.org/10.1038/nature09686
12. B. Pokroy et al., Anisotropic lattice distortions in biogenic calcite induced by intra-crystalline organic molecules. J. Struct. Biol. **155**, 96–103 (2006). https://doi.org/10.1016/j.jsb.2006.03.008
13. B. Pokroy, J.P. Quintana, E.N. Caspi, A. Berner, E. Zolotoyabko, Anisotropic lattice distortions in biogenic aragonite. Nat. Mater. **3**, 900–902 (2004). https://doi.org/10.1038/nmat1263
14. S. Borukhin et al., Screening the incorporation of amino acids into an inorganic crystalline host: The case of calcite. Adv. Funct. Mater. **22**, 4216–4224 (2012). https://doi.org/10.1002/adfm.201201079
15. E. Weber, C. Guth, I.M. Weiss, GFP facilitates native purification of recombinant perlucin derivatives and delays the precipitation of calcium carbonate. PLoS One **7**, e46653 (2012)
16. E. Weber et al., Incorporation of a recombinant biomineralization fusion protein into the crystalline lattice of calcite. Chem. Mater. **26**, 4925–4932 (2014). https://doi.org/10.1021/cm500450s
17. F. Marin, B. Pokroy, G. Luquet, P. Layrolle, K. De Groot, Protein mapping of calcium carbonate biominerals by immunogold. Biomaterials **28**, 2368–2377 (2007). https://doi.org/10.1016/j.biomaterials.2007.01.029
18. F. Nudelman, B.A. Gotliv, L. Addadi, S. Weiner, Mollusk shell formation: Mapping the distribution of organic matrix components underlying a single aragonitic tablet in nacre. J. Struct. Biol. **153**, 176–187 (2006). https://doi.org/10.1016/j.jsb.2005.09.009
19. Z. Ma et al., A novel extrapallial fluid protein controls the morphology of nacre lamellae in the pearl oyster, Pinctada fucata. J. Biol. Chem. **282**, 23253–23263 (2007). https://doi.org/10.1074/jbc.M700001200
20. H. Miyamoto et al., A carbonic anhydrase from the nacreous layer in oyster pearls. Proc. Natl. Acad. Sci. U. S. A. **93**, 9657–9660 (1996). https://doi.org/10.1073/pnas.93.18.9657
21. J.A. Freeman, Influence of carbonic anhydrase inhibitors on shell growth of a fresh-water snail, Physa heterostropha. Biol. Bull. **118**, 412–418 (1960). https://doi.org/10.2307/1538819
22. H. Miyamoto, F. Miyoshi, J. Kohno, The carbonic anhydrase domain protein Nacrein is expressed in the epithelial cells of the mantle and acts as a negative regulator in calcification in the Mollusc Pinctada fucata. Zool. Sci. **22**, 311–315 (2005). https://doi.org/10.2108/zsj.22.311
23. S. Inoué, *Handbook of Biological Confocal Microscopy* (Springer, New York, 2006), pp. 1–19
24. Z. Deng et al., Fluorescein functionalized random amino acid copolymers in the biomimetic synthesis of $CaCO_3$. Soft Matter **7**, 9685 (2011). https://doi.org/10.1039/c1sm05644a
25. Y. Celik et al., Microfluidic experiments reveal that antifreeze proteins bound to ice crystals suffice to prevent their growth. Proc. Natl. Acad. Sci. U. S. A. **110**, 1309–1314 (2013)
26. M.J. Rust, M. Bates, X.W. Zhuang, Sub-diffraction-limit imaging by stochastic optical reconstruction microscopy (STORM). Nat. Method. **3**, 793–795 (2006). https://doi.org/10.1038/nmeth929

27. B. Huang, S.A. Jones, B. Brandenburg, X. Zhuang, Whole-cell 3D STORM reveals interactions between cellular structures with nanometer-scale resolution. Nat. Method. **5**, 1047–1052 (2008)
28. M. Fu, A. Wang, X. Zhang, L. Dai, J. Li, Direct observation of the distribution of gelatin in calcium carbonate crystals by super-resolution fluorescence microscopy. Angew. Chem. Int. Ed. **55**, 908–911 (2016). https://doi.org/10.1002/anie.201508010
29. A.E. Stephenson et al., Peptides enhance magnesium signature in calcite: Insights into origins of vital effects. Science **322**, 724 (2008)
30. J.A.D. Dickson, Fossil echinoderms as monitor of the mg/ca ratio of Phanerozoic oceans. Science **298**, 1222 (2002)
31. M.E. Marsh, R.L. Sass, Calcium-binding phosphoprotein particles in the extrapallial fluid and innermost shell lamella of clams. J. Exp. Zool. **226**, 193–203 (1983). https://doi.org/10.1002/jez.1402260204
32. F. Marin, G. Luquet, Molluscan shell proteins. Comptes Rendus Palevol **3**, 469–492 (2004). https://doi.org/10.1016/j.crpv.2004.07.009
33. Z. Yan et al., In vivo and in vitro biomineralization in the presence of the inner-shell film of pearl oyster. Acta Oceanol. Sin. **30**, 87–93 (2011). https://doi.org/10.1007/s13131-011-0094-7
34. S.W.T. Weiner, Mineral phases in biology – macromolecules in mollusc shells and their functions in biomineralization. Philos. Trans. R. Soc. Lond.. B. Biol. Sci. **304**, 425–434 (1984). https://doi.org/10.1098/rstb.1984.0036
35. X. Liu et al., A possible mechanism for the formation of annual growth lines in bivalve shells. Sci. China Life Sci. **54**, 175–180 (2011). https://doi.org/10.1007/s11427-010-4132-z
36. T. Samata et al., A new matrix protein family related to the nacreous layer formation of Pinctada fucata. FEBS Lett. **462**, 225–229 (1999). https://doi.org/10.1016/s0014-5793(99)01387-3
37. M. Suzuki et al., Characterization of Prismalin-14, a novel matrix protein from the prismatic layer of the Japanese pearl oyster (Pinctada fucata). Biochem. J. **382**, 205–213 (2004). https://doi.org/10.1042/bj20040319
38. X. Liu et al., Erosion of the prismatic layer by the organic matrix during the formation of the nacre-prism transition layer in the shell of Pinctada fucata (Bivalvia, Mollusca). Chin. Sci. Bull. **56**, 869–876 (2011). https://doi.org/10.1007/s11434-011-4348-8
39. A.G. Checa, A.B. Rodríguez-Navarro, Self-organisation of nacre in the shells of Pterioida (Bivalvia: Mollusca). Biomater. **26**, 1071–1079 (2005). https://doi.org/10.1016/j.biomaterials.2004.04.007
40. X. Liu et al., The role of matrix proteins in the control of nacreous layer deposition during pearl formation. Proc. R. Soc. B Biol. Sci. **279**, 1000–1007 (2012). https://doi.org/10.1098/rspb.2011.1661

MIX
Papier aus verantwortungsvollen Quellen
Paper from responsible sources
FSC® C105338

If you have any concerns about our products,
you can contact us on
ProductSafety@springernature.com

In case Publisher is established outside the EU,
the EU authorized representative is:
Springer Nature Customer Service Center GmbH
Europaplatz 3, 69115 Heidelberg, Germany

Printed by Libri Plureos GmbH
in Hamburg, Germany